Coastal and Estuarine Studies

Coastal and Estuarine Studies

formerly Lecture Notes on Coastal and Estuarine Studies

34

S.J. Neshyba Ch.N.K. Mooers
R.L. Smith R.T. Barber (Eds.)

Poleward Flows Along Eastern Ocean Boundaries

Springer-Verlag
New York Berlin Heidelberg London Paris Tokyo Hong Kong

Editors

S.J. Neshyba
College of Oceanography, Oregon State University
Corvallis, OR 97331-5503, USA

Ch.N.K. Mooers
Institute for Naval Oceanography
Stennis Space Center, Mississippi 39529-5005, USA

R.L. Smith
College of Oceanography, Oregon State University
Corvallis, OR 97331-5503, USA

R.T. Barber
Monterey Bay Aquarium Research Institute, 160 Central Avenue
Pacific Grove, CA 93950, USA

ISBN-13:978-1-4613-8965-1 e-ISBN-13:978-1-4613-8963-7
DOI: 10.1007/978-1-4613-8963-7

Softcover reprint of the hardcover 1st edition 1989

2837/3140-543210 – Printed on acid-free paper

PROLOGUE

This is the first volume to deal exclusively with an ubiquitous feature of all the major ocean basins--the poleward-flowing surface currents and undercurrents along the continental shelf and slope of their eastern boundaries. The volume emanates from the First International Workshop held on this topic, and which was hosted by the Naval Postgraduate School, Monterey, California, in December 1986.

Organized initially by Steve Neshyba, Robert Smith, Dick Barber, and Chris Mooers, the list of those interested in the workshop topic grew to more than 90 investigators, worldwide, who were then directly involved in or had previously carried out studies of eastern ocean boundary flows. We had come to realize, ca. 1984, that such poleward flows were being observed and described, and that hypotheses were being generated to account for their cause, by many investigators within the context of descriptive theoretical and physical oceanography of regional phenomena.

It was timely to convene the scientific community involved for the broad purpose of comparing the "anatomy" of eastern boundary flow concepts, of deciding how to integrate theoretical and modeling tools more effectively into future eastern boundary studies, and of recommending future sampling recipes from which more wholesome descriptions could come.

And it was also high time to elevate the study of eastern boundary flow processes to its rightful importance alongside that of intense western boundary currents and of their concomitant rings and eddy dynamics. Indeed, it was the discovery of features along eastern boundaries that we now call squirts and jets (made possible by the technology of satellite ocean surveillance) that led us to conclude how the complexity of eastern flow regimes might be organized into an analytical framework, just as the study of the dynamics of western boundary regimes has been given over to its broad-scale description and theory. Eastern ocean zones formerly labeled simply as "transition zones" for lack of adequate tools of analysis, now become recognized as dynamic ocean process regimes worthy of study in their own analytical domain.

And it was the development of the scientific understanding of Pacific basin-wide equatorial waves, as processes integral to the large-scale ocean-atmosphere coupling, that gave birth to explanations of events like El Niño, and that also elevated eastern ocean boundary processes to their proper place in the dynamics of total ocean circulation.

Just as eastern boundary flow processes are important to the physical understanding of ocean circulation, they are equally important to the study of aphysical phenomena. It happens that eastern ocean boundary regimes of both the Atlantic and the Pacific also contain the four major zones of coastal upwelling driven by gyre-dominated wind systems: Baja California-California-Oregon-Washington in the NE Pacific; Peru to the central Chilean coast in the SE Pacific; the Senegal-Mauritania region in the NE Atlantic; and the Angola-Namibia coastal states in the SE Atlantic. Major upwelling zones of the Indian Ocean are driven by monsoon-dominated wind systems.

These are zones of major fisheries, particularly of the clupeoid species from which the major part of world fishmeal production derives: herring, anchovy and menhaden. One of the central problems, posed to appropriate management of these fisheries, is to define the physical constraints to the spawning-egg-larval-juvenile-adult life cycles of members of their biotic communities. Accompanying the physical-biological aspect of these biomes is the chemistry of abiotic cycling on which the biota depend. We recognized early in our design of the workshop agenda how the several disciplines might contribute important insights to one another, to the common goal of understanding how eastern ocean boundary regimes function.

On a larger scale, there is the question of how eastern ocean boundary flows participate in the basin-wide distribution of heat, salt, nutrients, and water, i.e., the "budget" oceanography studies that are also key aspects of global circulation models. An integral part of this same question is whether or not the poleward flow is continuous in transport from low to high latitudes, whether or not it is continuous in time, how it is linked to the intensity and time scale of gyre circulation, and how it is linked to the coastal upwelling process.

All such questions came to the fore during this workshop. We did not find answers to them all. What we did accomplish, however, was a first global scale recommendation for the methodology of proceeding with both the field work of observations in situ and for the refining of theory and models which will eventually need to assimilate these data. This volume presents both of these achievements.

ACKNOWLEDGEMENTS

The authors express appreciation to all those persons whose efforts have made this volume a reality.

Steve Neshyba deserves special recognition among the editors for "staying the course" from inception of the Workshop on Poleward Flowing Undercurrents on Eastern Boundaries held in Monterey December 1986, to its completion in this text.

The Naval Postgraduate School (NPS) provided excellent facilities for the workshop, and this within ten kilometers of an Eastern Boundary Poleward Undercurrent, the (presumed) California Undercurrent off Monterey Bay. Special thanks to Chris Mooers, then the Oceanography Chairman at NPS, for arranging workshop details at the NPS.

Special thanks also to Bob Smith for assembling the initial list of participants, from whose replies to inquiry we eventually compiled a mailing group of over 70 scientists the world over, and to Florence Beyer for her patient mailing effort. More than a third of us gathered in Monterey.

Chris Mooers mobilized support for assembly of the camera-ready copy through the Institute for Naval Oceanography (INO). A very special thanks goes to Ms. Lydia Harper, INO, who performed exceptionally in coordinating the final version with the authors and editors, as well as with Ms. Sabine Kessler, technical editor of Lecture Notes on Coastal and Estuarine Studies.

We thank the special efforts of Richard Barber in providing a biological and chemical continuity to the workshop deliberations of otherwise physics-oriented oceanographers.

We express appreciation also to the College of Oceanography of Oregon State University for sustaining the costs of paper work and mailings and telecommunication needed to bring this international effort to success.

The National Science Foundation supported travel for one international participant to the workshop, and has supported the research of several of the participants which resulted in their contributions to this volume.

This volume is a contribution of the Institute for Naval Oceanography, which is sponsored by the Navy and administered by the Office of the Chief of Naval Research. Any opinion, findings and conclusions or recommendations expressed in this publication are those of the authors and do not necessarily reflect the views of the Chief of Naval Research.

Lastly, we give our best thanks to all colleagues who participated in the reviews of papers.

Table of Contents

PART I:

WORKSHOP SUMMARIES

WORKSHOP SUMMARY: POLEWARD FLOW-OBSERVATIONAL AND THEORETICAL ISSUES

Christopher N.K. Mooers
Institute for Naval Oceanography
Stennis Space Center, MS 39529-5005

I will summarize from my own perspective what I think the essence of the workshop has been, to help set the stage for discussion of scientific futures. With that, let me fill in some of the historical background leading to this workshop.

Some of us, who over the past 10 to 30 years of association with the coastal upwelling studies which "bumped" into poleward undercurrents, have long wondered about their greater importance, but not a great deal has been done about the topic in general. I always thought the topic deserved more attention on its own. Beginning a half-dozen years ago, through the Eastern Pacific Oceanic Conference (EPOC) where western North American physical and other oceanographers gather annually for informal discussions, the undercurrent continually "popped up" as a topic for discussion. And we kept hoping that someone, like Bob Smith or Dudley Chelton or Andrew Willmott, who were "designated hitters" for awhile, would produce an organized research effort on poleward undercurrents.

The opportunity for this workshop arose about two years ago when Steve Neshyba proposed holding a topical session on poleward flows in the context of a Fall AGU meeting. Both Joe Reid and I were then involved as officers of the Ocean Sciences Section of the AGU, and we supported the idea. As it was, the date was a bit late for scheduling it that year, and the event was rescheduled as a poster session for the following year, the 1986 Fall AGU. In the sequence of events that followed, I offered the facilities of the Naval Postgraduate School for an expanded workshop on the topic just prior to the AGU meeting, the better to organize a more comprehensive discussion than is possible within the AGU context, but still designed to interact with the AGU. Bob Smith joined the organizers, drawing up a first list of invited participants who were then encouraged to suggest others with demonstrated interest in the poleward undercurrent topic. The list grew and here we are.

The 1986 AGU meeting next week has special interest to our group because of its sessions celebrating the centennial of scientific investigations of the Equatorial Undercurrent—this first cousin of the poleward undercurrents of interest to us is to be given special treatment in recognition of 100 years of its somewhat systematic study.

Along the way to this workshop, we discussed how the results might be published. It is always good to hold a forum like this, to examine old ideas and old results, and put forth new ideas and new results. But it is my opinion that we have at times lost the benefits of the progress made in the workshop mode by failing to establish a permanent record. In addition to the papers we all will submit to refereed journals, we plan also to compile our workshop papers, notes, and discussions as a volume of the Springer-Verlag series of Lecture Notes on Coastal and Estuarine Studies. One of the most important ingredients of this volume will be our collective thoughts on scientific futures.

I have, therefore, organized my talk accordingly. By taking a systematic approach, I want to bemuse you a bit with some fairly fresh results that will remind you that my research group here at the Naval Postgraduate School does not understand everything about poleward flows in eastern boundary current regimes! Just at the time that we thought we had begun to understand the regime off Northern California, through the NPS/Harvard OPTOMA Program's studies conducted over a period of four-and-a-half years, (cf. Rienecker and Mooers, 1989a) we encountered a big surprise. I want to go back in time, to show some highlights, though at the risk of being very superficial. You will hear more about these later—at the moment I wish to paint a picture for you.

During 1986 we acquired a series of quasi-synoptic maps off Northern California, as a "swan song" burst of activity in the last phase of the OPTOMA Program. Over about a 270 kilometer square domain off Northern California, the surface dynamic topography relative to 450 db in late March/early April indicated the development of a cyclonic coastal upwelling center, just south of Point Arena, Figure 1a, at the beginning of the upwelling season. By late April/end of May, it had separated from the coast, Figure 1b. In mid-July, instead of sampling a regular grid of some 200 stations, as done for the other maps shown here, a cool anomaly moving offshore in association with a jet system was tracked over a more limited domain, Figure 1c. The cool anomaly and jet system flow between an anticyclone and a cyclonic belt extending offshore from near Point Arena; it extends offshore, until it turns sharply to the south in a convergence with an offshore-to-onshore-oriented front and flow.

The shocking result appears in the surface dynamic height field from the next cruise acquired two weeks later, Figure 1d, which covers a

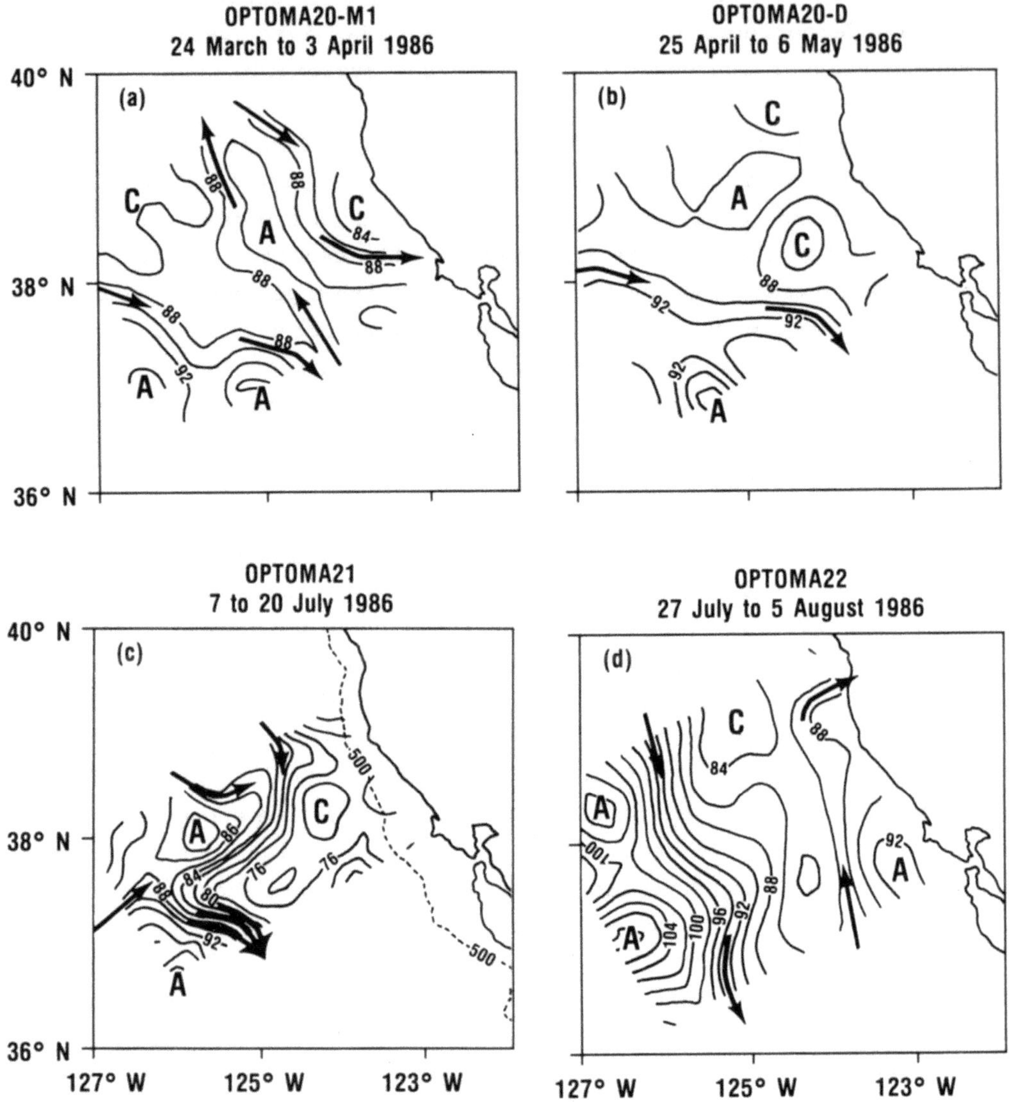

Figure 1. Surface Dynamic Height Relative to 450 m; C.I. = 2 dyn. cm.

somewhat larger domain of 320 kilometers square. There is no offshore flowing jet remaining--basically, all flow is coast-parallel within 200 kilometers of the coast. There has been a massive change in the flow regime, and we do not know why--it seems to be associated with an enhanced poleward flow over the continental slope. While the wind field fluctuated during this period, the fluctuations were not of such magnitude as to provide an obvious explanation for the reversed coastal circulation and onset of strong poleward flow in late July, the latter disrupting the upwelling center off Point arena as well as the offshore-flowing jet. We had never before encountered such abrupt changes. (N.B. Subsequent analyses suggested that relaxation of the local equatorward winds and an increase in the large scale cyclonic wind stress and may have played a role, Rienecker and Mooers (1989b)). The moral of this story is that we ought to keep "looking," that there are probably more surprises linked to poleward flows to be found in these eastern boundary regimes.

WHAT I THINK I SAW AND HEARD DURING THESE PAST TWO DAYS

With the above preface, I turn now to summarize what I have gathered from our workshop. My notes are organized with these topic headings:

1. Occurrence of Poleward Flow

2. Scientific Issues

3. Observational Implications

4. Theoretical Implications

5. Overall Implications

OCCURRENCE--THE GEOGRAPHY OF POLEWARD (PUC) FLOW REGIMES

Spatial. The set of coarsely-defined pigeon-holes into which we have characterized the spatial domains of poleward flows has consisted of these elements:

- There is evidence for flow, or the dynamical equivalent of poleward undercurrents, along all eastern continental margins, plus the Mid-Atlantic Ridge, and perhaps extending into polar regions.

- They clearly occur from tropical to polar latitudes, but we do not imply continuity across this broad latitudinal range.

- They can occur nearshore, typically as near-surface, buoyancy-driven poleward flows. The most common experience and interest relates to flows on the outer shelves and upper slope. Examples are given in this workshop by Jane Huyer et al., [Poleward Flow in the California Current System], Howard Freeland [Observations of the Low-Frequency Circulation Off the West Coast of British Columbia], and others. One must wonder if there are not also flows over the deep slope; we know little of this because no one has yet made the definitive measurements off the eastern boundaries. Some measurements have recently been made on the deep slope off Iberia by Des Barton [The Poleward Undercurrent on the Eastern Boundary of the Subtropical North Atlantic], and Greville Nelson reports some data taken on the deep slope in the region of the Benguela Current. The need to extend our studies of poleward flows to the boundary regions of the deep slope, and even farther offshore, has arisen repeatedly as an issue in our discussions. It is not yet clear whether the poleward flows observed over the deep offshore margins are topographically trapped or not.

Most of the data described here relate to subsurface flows. But the poleward flow regime clearly incorporates surface flows as well; these probably include surface manifestations of coastally trapped wave modes and surfaced undercurrents. Flow that is trapped topographically to the bottom, has also been described by Greg Holloway [Topographic Stress in Coastal Circulation Dynamics].

Temporal. From the temporal view, we have compared much data on poleward flows from margins around the world ocean, mostly on the basis of seasonal variation.

- Overall, our data bases are "short" of the extended measurements needed to focus on the mean poleward flow characteristics; this remark applies across the entire shelf, near-shore and deep-slope, as well as to the extended offshore zones. Once such data are acquired, we shall then also have solid descriptions of seasonal cycles in poleward flows throughout the coastal ocean.

- The problem of variability definition extends to interannual periods as well. Wherever eastern boundary flow data cover two years or more, we see evidence of interannual variability.

Surprisingly, not a great deal has been said in this workshop of variation on the time scale of El Niño events.

- At the other end of the temporal scale, there is much evidence of variability in poleward flow on the storm scales associated with atmospheric forcing. The corresponding coastal ocean response occurs at periods from several-days-to-a-few-weeks.

- We have heard a few reports describing oscillations in coastal flows with periods of the order of one to two months. This particular range of time scales in ocean motion is currently receiving much attention in studies of equatorial flows in connection with tropical meteorology. This point highlights another rationale for more extensive monitoring plans in eastern boundary regimes; i.e., the potential for "connecting" poleward flows on eastern boundaries with flow patterns in equatorial current systems at these time scales. Here we find also the phenomena of coastal trapped waves.

Quantitative Characteristics. The following are quantitative descriptions derived from our discussions:

- Typical along-shore speeds of poleward flows range between 10 to 30 cm/sec; I estimate that the range includes more than 50% of the measured values.

- In spatial terms the flows are found in the depth range 100 to 300 m, over a zone extending 10 to 30 km (or more) offshore.

- If we translate these data into transports, these range from 1 to 2 Sverdrups; again, such characterization applies to the central half of the transport distribution, with values easily extended to 0.5 Sv greater or less. What our workshop has defined clearly is that we do not know transport numbers at all well, and this deficiency could be one of the more important results of our discussions.

- Des Barton has pointed out that the core of the undercurrent off Northwest Africa appears to sink as it moves poleward; this apparent behavior is based on rather short-term observations. I do not know if anyone else has so organized their data as to draw analogous views of other poleward flow regimes, from which one might extract such behavior as a general characteristic. Jane Huyer reports no obvious trend of this type in her data; Steve

Neshyba states that the Gunther Undercurrent data does not show such a trend either. Nevertheless, the depth-latitude behavior of the core of poleward undercurrents, and its variation in time, is a factor which we ought to watch for in the future, as a possible clue to the governing dynamics. Core-depth variation on a local scale can yield clues to topography-connected dynamic terms as well.

SCIENTIFIC ISSUES IN POLEWARD FLOW STUDIES

I break this topic into four categories: kinematics, dynamics, energetics, and roles.

Kinematics. Here there is a rich lode of features to observe:

- The continuity of the poleward flow features, both in space and time, is receiving more and more attention, and rightfully so. This characteristic is a rich one to study, because even though there may be a continuous flow at any one time and over a very large domain, this does not mean that a water parcel entering at one end of the domain necessarily moves continuously to the other end. The motion pattern *per se* can be "moving" through the coastal circulation system, and we need to know more of how such "motion translation" can occur. Cross-flow processes need further study to understand fully the along-shore continuity topic.

- Based on our discussions, we know little about the sources and sinks of water masses involved in the poleward transport process. Some mention was made of a possible tie-in to the equatorial undercurrents, and, while this "connection" is a most interesting one to explore, its existence is not yet well documented, let alone understood. More work is needed because the source-sink question spills over into several of the topics previously mentioned; e.g.,continuity, transports, core depth behavior, and more.

- Of course, a long-standing issue in kinematic descriptions of poleward flows is that of undercurrents versus counter-currents. There is also an intermediate point of view, i.e., that countercurrents are due to "surfacing" of an undercurrent for short times or long times, over short or long distances. Here we encounter problems of communication, largely due to a lack of adequate data from which to derive a more precise description.

- Concomitant with continuity and transport topics is that of water mass anomalies associated with poleward flows. In many upwelling regimes the poleward currents transport anomalously warm, salty water, high in nutrients and low in dissolved oxygen. This is not a universal rule, for as Des Barton pointed out for the Northwest African case, the water mass carried in the poleward flow is relatively fresh, producing a salinity minimum in vertical profiles. In either case, we use the salinity and nutrient characteristics as tracers for the poleward flow.

- Then there seem to be some real puzzles associated with the along-shore descriptions, in the sense that the poleward flow must traverse transition zones which are themselves described in terms of water-mass transitions. Examples are (1) the fronts off the Gulf of California [where the California Current converges with the mainstream of the Equatorial Countercurrent (the latter diverted around the Costa Rican Dome to become itself a "poleward" stream) and both veer westward into the North Equatorial Current] and (2) the zone of extrusion of Mediterranean Intermediate Water westward from the Strait of Gibraltar.

Dynamics. There should be no surprises in the following list of controlling factors:

- The forcing functions include heat and mass fluxes, as discussed above. Equally important is the atmospheric wind stress and curl. A good deal of information has been developed on the space and time structure of the coastal wind stress fields. We are accumulating information on the along-shore position and seasonal variation in wind stress at size scales from local and regional to basin, and on time scales from the mesoscale and synoptic scale to the seasonal and interannual. But it is not yet clear that we have successfully incorporated these observations into our models. For one thing, we do not know how accurate the wind curl charts are or need to be, i.e., from the point of view of the sensitivity of the ocean's response to such forcing.

  River runoff has to be included for its key role in controlling local buoyancy flow. Several papers demonstrated the importance of the strong internal tides, particularly due to their breaking in the vicinity of the shelf break, which could lead to setup of radiation stresses to play a role in driving poleward flows. The subject provokes much discussion; one can immediately talk of these effects in terms of Sverdrup transports and poleward flows to the (unknown) extent that those dynamics are involved.

- We have heard much talk on various aspects of topographic control of poleward flows, particularly of capes and canyons. In the case of the Benguela regime, there is the influence of the Walvis Ridge, a submarine ridge that runs seaward from the slope. There seems to be other, less well appreciated, transverse ridges extending outward from continental margins, e.g., the Mendocino Escarpment. We ought to remain alert to the possible influence such submarine ridges may have on flows higher in the water column.

- Greg Holloway, Ken Brink, and others have made us well aware of the influence of topographic drag on poleward flows.

- Repeatedly, we have heard of the importance of the along-shore pressure gradient in the dynamics of poleward flow, both in observations and in models. The impression is "... if we just knew THE along-shore pressure gradient ...!" everything would come out fine; what we soon appreciate, however, is that there is a frequency dependence, and a horizontal wavenumber spectrum, to this gradient, so that, until we know much more about its behavior, many of our questions will remain unsettled. This subject is not well-treated in our theoretical models either.

- Coastal trapped waves offer a mechanism for transferring momentum and other properties along coastal boundaries. These occur together with along-shore pressure gradients, but one can encounter the latter without associated trapped waves. Similarly, coastal trapped waves can be associated with topographic drag, but can exist without the latter in the case of a smooth bottom.

- There is of course the dynamic factor of dissipation of turbulent energy. As yet there is not much of an empirical or theoretical basis for understanding the turbulent mixing as Julian McCreary parameterizes it in his models. (It is unfortunate that Rolf Lueck and Hidekatsu Yamazaki could not be here because they have actually made measurements of turbulent dissipation in poleward undercurrents, and would undoubtedly have added to the discussion of this aspect; however, they provided a manuscript to this volume in absentia.)

- The strong, meso-scale eddies that we see in eastern boundary current regimes entrain coastal waters. They may also entrain undercurrent waters. We do not know this for fact, but the possibility remains and this process ought to be kept in mind.

- As usual, issues of linear versus non-linear dynamics and of viscid versus inviscid fluid dynamics crop up. Even though the world is non-linear and viscid, there is always the hope and belief that the processes associated with the undercurrent may be somewhat linear and inviscid. On the other hand, we may just be victims of seduction to that which we want to see, as Greg Holloway, Ken Brink, and others have argued forcefully that non-linear processes are fundamental to the existence of poleward flows on eastern boundaries.

Energetics in Poleward Flows. In the event that the undercurrent is unstable, in the sense of baroclinic and/or barotropic instability, it probably plays a role in the genesis of the cross-shore mesoscale jets and eddies that are rather ubiquitous on eastern boundaries. If this transfer mechanism is substantial, we then face the greater task of evaluating the impact of poleward flows on the general ocean circulation of eastern boundary regions, with all their associated cyclic biological impacts or its role in offshore-directed fluxes of materials introduced at coastal margins. We saw a glimpse of this process in the paper by Batteen (Model Simulations of a Coastal Jet and Undercurrent in the Presence of Eddies and Jets in the California Current System), in watching the evolution of contorted streamfunctions at the level of the undercurrent during the genesis of an eddy. This was probably a glimpse into the real world.

Roles of Poleward Undercurrents in Ocean Science. I will try to identify some of the known, or probable, important roles and effects of eastern boundary poleward undercurrents in ocean behavior:

- In many instances, the poleward undercurrents provide "feedwater" for coastal upwelling regimes. Thus, we ought to advance research into the sources from which these undercurrents draw their water masses. It is in this sense also, that our studies gain strong inter-disciplinary harmonics and overtones. Our chemical and biological colleagues are dependent on us for even more detailed descriptions of the physics of poleward undercurrents than we have provided to date.

- In the same vein of interdependency, we ought to be aware that other transport processes are relevant; for example, biogeochemical transports associated with nutrient, oxygen, and other budgets. Clearly, the upcoming GOFS Program that plans to define global ocean fluxes of all major chemical constituents must depend heavily on the description of sources and sinks of poleward undercurrents, and how these feed into and extract from zones of rich production-like coastal upwelling centers. As Antoine Badan notes, the entry of warm waters into the Gulf of California [via poleward-flowing coastal waters from the south] have a curious effect of greatly increasing productivity.

- Of paramount importance is that we focus on learning how the observed "upwelling centers," fed by poleward undercurrents, are coupled into the mesoscale circulation features of eastern boundary regions; e.g., the squirts, jets, and eddies, both cyclonic and anticyclonic, that are ubiquitous in eastern oceans.

- The oceanographic community is gaining appreciation of the ocean's complex response to atmospheric forcing, as in the case of the El Niño-Southern Oscillation system. This response involves poleward-propagating disturbances from the equatorial region and sub-tropical, zonal, surface Ekman mass transports, as well as responses to anomalous atmospheric forcing in the mid-latitude and sub-polar zones. In other words, there are regional scales of motion through which the poleward undercurrent may pass.

- At least part of the mid-latitude El Niño response of poleward undercurrents seems to be connected with poleward-propagating disturbances. As another part, does the poleward undercurrent transport of warm water make a significant contribution to basin-scale heat budgets, for example?

- There are seasonal "transitions" that occur on eastern boundaries that are interconnected with poleward flows, but about which we have not developed substantive explanations. An example is the occurrence of the Davidson Current off California and Oregon-Washington in the fall-early winter months. Perhaps these phenomena will be better understood once we have answered a related puzzle aired many times in this workshop ... "does the poleward undercurrent regime consist of two distinct streams—the break-slope submerged flow stream and the shelf-shore surface flow, or a single stream that exhibits seasonal variation in depth and position?"

- There remains the uncertain role that the high primary productivity in eastern boundary zones may have on total ocean production. Lou Codispoti (Do Nitrogen Transformations in the Poleward Undercurrent off Peru and Chile Have a Globally Significant Influence?) has posed for us the problem that, whereas high production in undercurrent-fed upwelling centers is well-known, the denitrification that accompanies this phenomenon can lead eventually, over time scales of 100,000 years, to a depression in available nitrogen the ocean over, and therefore limit total oceanic production.

OBSERVATIONAL IMPLICATIONS IN PUC STUDIES

- A clear message was developed in this workshop: we absolutely require long time series [many years in length] of current measurements at several along-shore locations, so as to acquire a regional [ca. 1,000 km] view of flow regimes on the outer shelf and upper slope.

- Further, there is much value gained in short-term, areal "case studies," of the type outlined by Alan Bratkowich for the Central California regime, in which a variety of observing systems are deployed in a specific area for a specific time and experiment; these sensor packages produce shorter time series coupled with spatial images, current meter and drifter data, and so on.

- Perhaps as important as any other aspect of our future studies is the need to focus more attention on the "evolution" of vertical and horizontal structure in poleward undercurrents in response to well-documented forcing. We are still dealing with many fragments of information; we are limited by under-resolved descriptions of the structure, both in the vertical and horizontal. In the few cases where a good oceanic array has been deployed and the *in situ* measurements are adequate, the forcing fields are not adequately and correspondingly defined. It is the familiar story of questions asked after the fact.

- We ought to devise ways, and include them in our research programs, to test better the utility of so-called "easy indicators" as "proxy" monitors of the undercurrent behavior. I include among these such data series as (1) the upwelling index, (2) coastal sea level, and (3) coastal sea surface temperature.

- Lastly, we ought to explore the utility of "biotracers" in undercurrent research. Howard Freeland (Observations of the Low-Frequency Correlation off the West Coast of British Columbia, Canada) gave us one example in his use of a "zooplankton stream" in evaluating observations of physical variables off British Columbia.

THEORETICAL IMPLICATIONS IN PUC STUDIES

- Clearly, as McCreary has shown us, three-dimensional linear models are needed to produce poleward undercurrents. These lead to interesting implications for the influence of the wind stress curl, particularly in accounting for the surface poleward flow.

- We are aware that mixing parameterization plays a crucial role in these models, as does also thermohaline forcing [river outflow, rainfall, evaporation, etc.]. We are also aware of the need to incorporate coastal trapped wave phenomena into more modeling studies.

- Numerical simulation provides guidance in other ways, too, in that it helps both in planning and interpretation of real observations.

- Not well-developed at this workshop, but of equal value in guidance of future work, is the use of diagnostic analysis of analytical and numerical model output. Much more needs to be developed here. Such diagnostic analyses could help pinpoint dynamical balances, mixing zones, and more; hence, they would foster more significant interaction between the theoretical and observational efforts.

- A suite of topics that enter into modeling was dicussed: vortex-stretching, vorticity waves, bottom friction, and coastal topography. Greg Holloway and others (Topographic Stress in Coastal Circulation Dynamics) led us in a discussion of how coastal topography yields canyon upwelling and cape downwelling, and how poleward flow results from the interaction among these factors. Topographic drag issues were raised. Bob Haney described his exploration of the impact on slope circulation of such features as the Mendocino Escarpment.

OVERALL IMPLICATIONS

- Clear from our discussions is that both observational and theoretical modeling studies of poleward flows have reached useful, interesting levels of competence. It is now time to work even harder than before to bring the models and observations together. While the complexity of nature forces us to rely heavily on numerical simulations, this ought not to delay further theoretical studies on how to incorporate basic processes into these models. This consideration only underscores a need on the part of observationalists to enhance our ability to describe the pressure fields which form such important elements of models. Jack Whitehead reminded us of the potential that laboratory modeling has for studies of deep, poleward flow along the Mid-Atlantic Ridge—can we also look toward laboratory models for application to specific problems in coastal flows?

- The need for more diagnostic analyses, to obtain the structure and evolution of various fields, and dominant balances in the eastern boundary regimes, is re-emphasized.

- None of us at this workshop has addressed the task of "data assimilation" into models. We work with somewhat inadequate models and incomplete observations--this is the real world. What we must do next is to meld these techniques, piece-by-piece if necessary, to achieve the highest efficiency of the capacity of both employed together against the limitations of using each alone. Data assimilation is a process of inserting observations into fields computed by models, to serve as "constraints" upon the modeled processes themselves. In the process of assimilation, one faces the weaknesses that exist in each technique separately; subgrid seale processes are not simulated in the models, yet much of our observed data are point values influenced by subgrid scale processes--the problems of data assimilation are indeed not trivial ones.

- We might give serious thought to activating more fully the communications among our group, particularly insofar as long-term planning and effort are concerned. The rationale is that the resources required to advance the topic significantly, at least on the observational side, are formidable. We face great distances along the several coastlines where observations must be made, and these cross national boundaries. We face the problem that the time scales over which observations ought to be sustained can exceed the productive life span of an investigator. To progress boldly, we ought to collectivize bravely.

We have heard a great deal of geographical explorations, with reports from every eastern boundary regime of the world ocean, including the Arctic Ocean. The excitement we find here is that poleward flows occur in all these situations. But we are coming to the end of pure geographical exploration. We ought now to change to some frame of comparison other than geographical. We require a system of comparative anatomy, so to speak. We need to devise a set of transforms through which the separate geographical cases might be folded into a "mother model space." This new parameter space will have to conform to physical laws of course, but it also ought to accommodate the diversity of cases that characterize our individual "study" areas. There are many of these individual characterizations: narrow shelf, wide shelf, shallow shelf with shallow break, prevailing wind directions, shoreline orientation and length, coastal topography broken with canyons and capes, and so forth.

POLEWARD FLOWS ALONG EASTERN OCEAN BOUNDARIES: An Introduction and Historical Review

Robert L. Smith
College of Oceanography
Oregon State University
Corvallis. OR 97331

As the first speaker at the workshop on "Poleward Flows Along Eastern Ocean Boundaries," I was asked to "get things started" by giving an historically based overview. I felt that exposing my naivete was not inconsistent with the task and stated that two questions hovered in my mind: What is meant by "poleward flows" in the context of this workshop?; and why does the topic warrant a workshop? My geographical focus is the eastern ocean boundaries at low- and mid-latitudes; other authors reddress this parochialism.

One reason for singling out poleward flow along the ocean's eastern boundaries at low- and mid-latitudes is that it is, in some sense, unexpected, i.e., not easily or fully explained by our present knowledge. We were brought up in the wind-driven school of ocean circulation in which the received view was that the global-scale wind system drove the large scale subtropical gyres and the local winds drove the coastal processes. Tomczak (1981) nicely summarizes it all: "The subtropical eastern boundary currents extend from the eastern boundaries of the oceans into their central areas. They are driven by the wind stress curl and modified by frictional boundary layers; compared to the basic circulation, these modifications are of only little influence on the overall dynamics. Along the eastern boundaries where the direction of the wind stress is meridional – while on the oceanic scale it is mainly zonal – coastal upwelling currents are found superimposed on the eastern boundary currents, driven by the meridional wind stress. These currents are elements of an independent dynamic process and not just modifications of the eastern boundary currents." Is poleward flow along the eastern boundary part of the large scale dynamics?, or part of the boundary effects (perhaps related to the coastal upwelling process)?, or due to yet another independent process?

All five of the ocean's eastern boundaries at the latitudes of the subtropical gyres, or the semi-permanent anti-cyclones of the atmospheric circulation, (Iberian Peninsula/Northwest Africa, Southwest Africa, South and North America, and the west coast of Australia) show evidence of poleward flow manifested in water properties, in the ecology, or in the movement of passive drifters. These occur in the region where the

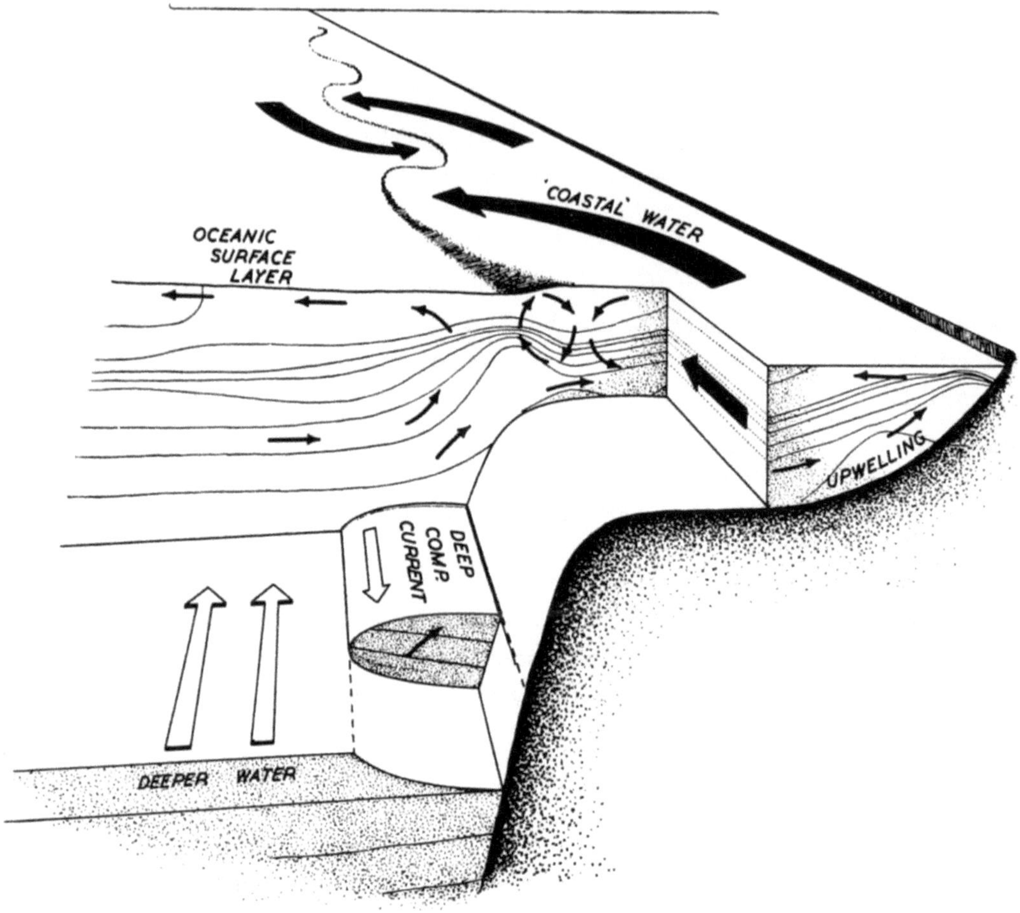

Figure 1: Schematic of circulation during upwelling in Benguela Current (southwest Africa) region; features shown appear to be common to all major coastal upwelling regions (mirror images in Northern Hemisphere). Note the sinuous frontal boundary indicative of eddies between upwelled coastal water and oceanic surface water, the circulation in vertical section, and subsurface counter-current or 'compensation' current. From T.J. Hart and R.I. Currie. The Benguela Current. Discovery Reports 31, 123-298, 1960. (Cambridge Univ. Press) Figure 97.

winds are equatorward and thus the poleward flow raises a dynamical question. But the poleward flow is more than simply a "dynamical curiosity" since it affects the distribution of heat, salt, chemical properties, the sediment transport and the life cycles of things that live in the sea. Indeed, until recently, the evidence for the existence of poleward flow along the ocean's eastern subtropical boundaries was just these effects. Perhaps, the existence of discernable effects is the warrant for having the workshop, and the criterion for considering a particular poleward flow feature in the workshop. The poleward flow does not have to be a continuous ever-flowing river, but clearly must have some temporal and/or spatial persistence to cause a discernible effect...or to even interest modelers. The poleward flow may be a counter-current (counter to the wind and expected surface flow) or a subsurface undercurrent at modest depths; the abyssal circulation is beyond our focus.

During the workshop several theoretical papers were given, revealing that several possible answers to the dynamical questions exist and that modelers are, finally, giving the topic serious consideration. Field experiments during the past decade or so, while rarely setting out to study poleward flow per se, have added to our descriptive knowledge...and added new questions for both observationalists and theoreticians. Present field studies over the continental margins off northern California (the Coastal Transition Zone and MMS studies) and the west coast of Australia (LUCIE: Leeuwin Current Interdisciplinary Experiment) are indirectly or directly, respectively, confronting poleward flow. Since these recent studies will be discussed in the papers of this volume, I shall return to the historical background.

<u>The California Current</u> region can perhaps be taken as paradigmatic. It gives evidence of both a countercurrent and an undercurrent, and has been widely, even if not adequately, observed. Jones (1918) wrote a monograph, "The Neglected Waters of the Pacific Coast," in which he recounts that George Davidson spent many years studying "the subject of coastwise currents. As a result of such study, Professor Davidson concluded that there existed, from 50 to 100 miles offshore, a south setting current of unknown width and velocity, and that inside of this, closely following the general trend of the coast, was a north setting current which he named the 'Davidson Inshore Eddy Current.' The data...were necessarily meager. He had observed such north setting currents at various anchorages while engaged on survey work along the coast, and had also ascertained that logs of the redwood (which do not grow north of California) were frequently found on the shores of Washington, British Columbia, and even Alaska, the wood being well know

to the natives of these regions...Davidson was too good a scientist to attempt to promulgate a theory more definite than was justified by the available data..." Both Jones and Davidson were aware of the insufficiency of observations and "had in mind nothing more than a general tendency to a current flowing in a north direction, which tendency might at any time disappear or be reversed by the prevailing weather conditions." Tibby (1941), using water mass analysis, traced 'Equatorial Pacific' water up the coast to at least 45°N, and stated: "The transport of warm, saline Equatorial water to the north appears to be largely accounted for by the inshore northward flow of the California Coastal Counter Current. Present data indicate that this current exists throughout the year at about 200 meters and below, and that above 200 meters it is encountered only during the winter months when it appears at the surface as a narrow northward coastal current. It is within the region immediately adjacent to the coast and at intermediate depths that the highest percentages (of equatorial water) are found."

As hydrographic observations increased, it became clear that there was poleward geostrophic flow off central and southern California even when the winds were strongly southward. The CalCOFI Atlas No. 4 (Wyllie, 1966) shows northward flow at 200 m (relative to 500 m) throughout the year, except for April, from the tip of Baja, California to Cape Mendocino. Munk (1950) pointed out that the curl of the wind stress was positive (cyclonic) within 200 km of the central California coast and thus "the average wind distribution off Monterey is consistent with the existence of a south-flowing current offshore and a countercurrent inshore...a consequence of the local stress curl."

Beginning in the 1960s, physical oceanographical studies in the California Current region, from British Columbia to Baja, California, increased in number and employed the modern methods available to oceanography. The work from then to the present is reviewed by Huyer et al. (this volume). None of the results of recent studies contradict the above general picture. A strong tendency for poleward flow all along the coast is observed (Strub et al., 1987a). The strongest poleward flow seems to be at about 100 to 200 m over the slope (Huyer et al., this volume) with the flow further seaward being highly variable without a significant mean (Stabeno and Smith, 1987). The cyclonic wind stress curl is manifested in the long-term mean and seems to be a robust feature of the eastern boundary regions (Bakun, 1988). It is in the observations of the variability and the details of the structure, and in the details of the theories, that the richness and importance of modern work appears.

The Leeuwin Current region off western Australia is geographically and topographically the analogue of the other eastern boundary current regions and, as in those regions, the winds off western Australia are predominantly equatorward. However, the behavior of the ocean off western Australia seems to be quite unlike the other eastern boundary regions. There is no evidence of a regular continuous equatorward flow within 1000 km of the coast (Andrews, 1977) nor of coastal upwelling, despite the strong equatorward wind stress (Wooster and Reid, 1963). However, for us the region may be an exception that proves a rule. The observations are consistent in showing a poleward (southward) flow against the equatorward wind, over the continental slope, near the shelf edge, from at least 22°S to the southwest tip of Australia. An experiment (LUCIE: Leeuwin Current Interdisciplinary Experiment) had just begun at the time of the workshop and will provide an order of magnitude more observations (Church et al., this volume).

The earliest suggestion of southward flow of tropical water past western Australia came from the observations of tropical marine fauna (including coral) at subtropical latitudes (Saville-Kent, 1897). More recently, satellite tracked buoys (e.g., Cresswell and Golding, 1980) and the evidence for poleward flow in satellite VHRIR photos (Legeckis and Cresswell, 1981) emboldened Cresswell to name the poleward surface flow "The Leeuwin Current." Prior to LUCIE, the only detailed sections across the Leeuwin Current and direct subsurface current measurements were made by Thompson (1984). The axis of the poleward current (speeds up to about 50 cm/sec) seemed to coincide with a maximum in the surface mixed layer depth, which was near the shelf edge and sometimes exceeded 90 m. Below about 150 m, Thompson observed an equatorward (northward) undercurrent. Thompson (1984) argued that the poleward (southward) surface current was driven by the large along-shore pressure gradient, which is larger than in any other eastern boundary region, probably because the eastern tropical Indian Ocean has an open connection with the western equatorial Pacific (Godfrey and Ridgway, 1985). As Godfrey and Ridgway (1985; their Figure 12) show, the along-shore steric sea level slope is negative poleward in all eastern boundary regions, but greatest by far in the Leeuwin Current. It seems clear that it is the along-shore pressure gradient, and not the wind stress curl, that is the driving mechanism for the Leeuwin Current. The recent theories of McCreary et al. (1986) and Thompson (1987) model the Leeuwin Current and consider the effect of the along-shore pressure gradient. Much will be heard about the Leeuwin Current in the immediate future as a result of LUCIE.

The Peru Current seems the closest analogue to the California Current, with the historical literature referring to coastal currents, countercurrents, and undercurrents. The first "modern" study of the Peru Current was made by Gunther (1936) in his monograph entitled "A Report on the Oceanographical Investigations in the Peru Coastal Current," based on the cruise of the WILLIAM SCORESBY in 1931. The report is rich in history and scientific observations. Gunther observed a high salinity warm-water wedge which "had a breadth of some 50 miles and seemed to extend along the greater part of the Peruvian coast...(with) south flow...recorded off San Juan (15°S) and Northern Peru. Upon these facts, the conception of a counter-current is based, though its acceptance as a continuous counter-current leaves much to be explained." The text and the figures indicate that Gunther wasn't sure whether the counter-current(s) might not be part of 'swirls' (eddies). Perhaps Davidson had the same thoughts in mind when he named his current an 'Eddy Current.' Gunther also describes an undercurrent off Chile, based on the observations of high temperature and salinity and low oxygen.

The STEP-I Expedition on the HORIZON in 1960 was a latitudinally extensive study with necessarily wide station spacing. An analysis of the hydrography by Wyrtki (1963) and the drogue measurements by Wooster and Gilmartin (1961) provided evidence for both a Peru Undercurrent and a Peru Countercurrent. Wooster and Gilmartin state that an undercurrent, underlying the Peru Current at depths of several hundred meters, extends from northern Peru (about 6°S to 41°S) with its coastal edge paralleling the shelf edge. They establish its presence by direct measurements with parachute drogues, geostropic computations and analysis of water properties. Like Gunther, they find high salinity and temperature and low oxygen south of 15°S to be evidence of the undercurrent. Wyrtki, using the ship's wind data to obtain the Ekman transport, interprets the dynamic height field to obtain an absolute geostrophic current. He states that the undercurrent exists, but is weak compared to the Peru Countercurrent, which is distinct from, and extends farther seaward than the undercurrent. Wyrtki seems to assign no important role to the undercurrent stating that "south of 15°S upwelling is supplied by Subantarctic water flowing north with the Peru Coastal Current, and north of 15°S by Equatorial subsurface water flowing south with the Peru Countercurrent. It is shown that the Peru Countercurrent and the Peru Undercurrent, although both flow south, are different currents." A recent thorough analysis of a large collection of hydrographic data by Lukas (1986) suggests that water from the Equatorial Undercurrent reaches the coastal upwelling area off Peru by two paths: "One is along a line from the Galapagos Islands to the coast at 5°S (the Peru Countercurrent), and the other is along the equator to the coast, and then along the coast as the Peru-Chile Undercurrent. The Peru-Chile Undercurrent begins near 5°S; north of this latitude, the maximum southward current is at the surface." Silva and Neshyba (1979), using hydrographic data from STEP-I and other cruises, concluded that the "Peru-Chile Undercurrent can now be traced from its origin near 10°S off Peru to 48°S off Chile. The criterion for continuity is the association of a subsurface maximum in the southward geostrophic flow profile with the salinity maximum and oxygen minimum properties..." Thus, all the authors who have considered hydrographic data conclude that poleward flow is a major feature of the circulation (and perhaps a major factor in the ecology) of the Peru-Chile Current region.

The time-series current measurements from moorings off the Peru coast during the past decade (e.g., Brockmann et al., 1981; Brink et al., 1983) suggest that poleward flow is a nearly ubiquitous feature at depths of about 50 to 400 m on the shelf and over the continental slope from 5°S to 15°S. It is this poleward flow that supplies the upwelling water along the Peru coast (Brink et al., 1980). During El Niño the poleward flow initially intensifies and transports anomalously warm water poleward (Smith, 1983). The normal tendency for poleward flow over the continental shelf off Peru, in spite of remarkably steady and relatively strong equatorward winds, appears similar to the tendency for poleward flow on the central California shelf (Strub et al., 1987).

The Canary Current and the Benguela Current regions are reviewed in papers by Barton (this volume) and Shannon (1985), respectively. As in the other eastern boundary regions, early evidence for subsurface poleward flow in the Canary Current region came from the distribution of properties (see Wooster and Reid, 1962). There is now considerable evidence, including direct current measurements, for a poleward undercurrent off Northwest Africa and the Iberian Peninsula. A complete review of the recent work is found in Barton (this volume). Whether the subsurface poleward flow is continuous over this extent is an open question posed by Mittelstaedt (this volume).

Hart and Currie (1960) presented the first evidence for poleward flow along the coast of Southwest Africa, based on a cruise by the recommissioned WILLIAM SCORESBY in 1950. They calculated the geostrophic flow at 200 m relative to 600 m between about 22°S and 28°S and noted poleward flow along the edge of the continental shelf. From water mass analyses they inferred that the upwelling water on the shelf came from depths of 200 to 300 m, and referred to the poleward flow as a "compensation current" which provided the replacement source for the water that was upwelled. This is analogous to what is observed, at shallower depths, off Peru. While a number of papers present evidence of poleward flow off the west coast of South Africa (see Shannon, 1985), the region has not yet had the benefit of a large-scale coordinated field experiment that would help synthesize the previous observations. One can only hope that the political and social problems will be resolved in that troubled part of the world, and that Southwest African physical oceanography can again become an international field.

Now the reader should consider the papers in this volume. They show the advances made since the earlier work, but in general do not contradict the necessarily "broad-brush" and speculative results of the earlier investigators. Indeed, the wonderful schematic that Hart and Currie (1960, their Figure 97 and used as Figure 1 in this paper) present includes almost every mesoscale feature of research interest today. Today, as the papers in this volume demonstrate, we have a better knowledge of the spatial and temporal location and extent of the features, but we are still asking to what extent the features are dynamically inter-related. For example, the development of the 'squirts and jets' (cold filaments) seen off many eastern boundaries may depend on the existence of a poleward undercurrent (Ikeda and Emery, 1984). What is also demonstrated in this volume is that poleward flow is a common and important feature of the ocean's eastern boundary and that this topic is now ready for, and deserving of, a concentrated research effort. It is of a scale, geographically, and in both oceanographic and social-economic importance, that justifies the study of the poleward flow along the ocean's eastern boundaries being included in WOCE (see the What Next? summary in this volume).

From: Joseph L. Reid, Scripps Institution of Oceanography

On: Review and commentary to paper **POLEWARD FLOWS ALONG EASTERN OCEAN BOUNDARIES** by Robert L. Smith

"I wish to point out that the author in his opening paragraphs discusses only those poleward flows that lie in mid-latitude, alongside but opposed to the great anticyclonic gyres. The remarks made therein are meant to apply to those latitudes, but not necessarily to the eastern boundary currents of the higher latitudes, the Alaska Current, which is a part of a cyclonic gyre, and the poleward flow south of 40°S along the coast of Chile. These, of course, are not always opposed to an equatorward wind, and although they are currents at the eastern boundary, they seem to be part of the great west-wind drifts."

THEORETICAL UNDERSTANDING OF EASTERN OCEAN BOUNDARY POLEWARD UNDERCURRENTS

Allan J. Clarke
Department of Oceanography and
Geophysical Fluid Dynamics Institute
Florida State University
Tallahassee, Florida 32306-3048

INTRODUCTION

Mean flows along eastern ocean boundaries may be generated by several mechanisms, including tidal rectification [e.g., Georges Bank, Loder (1980)]; melting ice and rivers emptying into the sea [e.g., the northern Gulf of Alaska, Royer (1981)]; forcing by offshore mean density fields [e.g., the Leeuwin Current, Huthnance (1984), and McCreary et al. (1986)]; the nonlinear interaction of forced barotropic shelf waves over a continental margin [Denbo and Allen (1983)]; oscillatory flow over small scale bumps in the shelf bottom topography [Brink (1986), Haidvogel and Brink (1986), and Holloway (1987)]; and mean wind-stress and wind-stress curl [McCreary (1981), Wang (1982), Suginohara (1982), Philander and Yoon (1982), Suginohara and Kitamura (1984), McCreary and Chao (1985), and McCreary et al. (1986)]. We shall concentrate our discussion primarily on the dynamics of poleward-flowing undercurrents on eastern ocean boundaries. Consequently, since mean flows produced by melting ice and rivers are generally surface flows near the coast, this mechanism is not relevant. Tidally rectified flows only exist in special areas where tidal currents and bottom topography slope are large (e.g., Georges Bank), so this mechanism is also not important in the generation of the major poleward-flowing undercurrents. Generation of poleward flow by the nonlinear interaction of forced barotropic shelf waves would also appear not to be a relevant mechanism as the mean flows generated have been estimated to be very small, only ~0.1 cm s^{-1} . Mean flows produced by offshore mean density fields do produce flows of the right strength, but on eastern boundaries these offshore fields drive equatorward rather than poleward undercurrents (McCreary et al., 1986).

Of the two remaining mechanisms, let us first consider mean flow generation by oscillatory flow over small amplitude, small scale bumps in the shelf bottom topography. Coastally trapped waves always propagate poleward on eastern ocean boundaries, so when the oscillatory current is flowing equatorward, a coastally trapped lee wave disturbance can form behind a bump. Energy propagates off to infinity and, thus, creates a drag on the flow. When the oscillatory current is flowing poleward, no coastally trapped waves can remain stationary with respect to the obstacle, so there is no drag on the current. The asymmetry in the wave drag results in a poleward mean flow. Holloway (1987) discusses this generation in a more general context. For realistic parameters, barotropic theory gives currents of about 2.5 cm s^{-1} (Haidvogel and Brink, 1986), considerably smaller than observed undercurrent speeds (~10 cm s^{-1}). Transports are on the order of 10 percent or less than observed coastal undercurrent transports. Calculations still need to be performed for the more realistic stratified ocean case.

Undercurrents can also be driven by the wind. More work has been done on this mechanism than any other, and it will be discussed in the next section.

WIND-FORCED EASTERN BOUNDARY POLEWARD UNDERCURRENTS

McCreary (1981), Wang (1982), Suginohara (1982), Philander and Yoon (1982), Suginohara and Kitamura (1984), and McCreary and Chao (1985) have all considered wind-driven eastern boundary flows. Many fundamental aspects of the dynamics are illustrated by McCreary (1981), and this model is discussed below.

MCCREARY'S CONSTANT DEPTH, LINEAR MODEL

The equations of motion, linearized about a state of rest and horizontally uniform stratification, are

$$u_t - fv = -p_x + (\nu u_z)_z, \tag{2.1}$$

$$v_t + fu = -p_y + (\nu v_z)_z, \tag{2.2}$$

$$u_x + v_y + w_z = 0, \tag{2.3}$$

$$\rho'_t + w\bar{\rho}_z = (\kappa\rho')_{zz}, \tag{2.4}$$

$$p_z = -g\rho', \tag{2.5}$$

where the vertical eddy viscosity and diffusivity coefficients are given by

$$\nu = A/N^2, \ \kappa = \sigma^{-1}A/N^2. \tag{2.6}$$

In these equations x, y, and z refer to distances eastward, northward, and vertically upward from the ocean surface at z = 0; u, v, and w are velocities in the x, y, and z directions; and t, f, p, $\bar{\rho}$, ρ', g, N, and σ refer, respectively, to time, Coriolis parameter, pressure divided by the mean (constant) water density, the water density in the absence of motion, the perturbation density due to the motion, the acceleration due to gravity, the buoyancy (Brunt-Väisälä) frequency and the Prandtl number. Note that in Eq. (2.4) McCreary uses an unconventional form for the diffusion of mass so that the problem can be separated into vertical modes.

The surface boundary conditions are

$$\nu u_z = \tau^x, \quad \nu v_z = \tau^y, \quad w = \kappa\rho' = 0 \quad \text{at } z = 0, \tag{2.7}$$

where τ^x and τ^y are the eastward and northward components of the wind stress at the ocean surface. The ocean is of constant depth H and at the bottom z = -H

$$\nu u_z = 0 = \nu v_z = w = \kappa\rho'. \tag{2.8}$$

Solutions of (2.1)-(2.6) can be written as sums of vertical modes $F_n(z)$ as

$$u = \sum_{n=1}^{\infty} u_n F_n, \quad v = \sum_{n=1}^{\infty} v_n F_n, \quad p = \sum_{n=1}^{\infty} p_n F_n,$$
$$w = \sum_{n=1}^{\infty} w_n \int_{-H}^{z} F_n \, dz, \quad \rho' = \sum_{n=1}^{\infty} \rho_n F_{nz}, \tag{2.9}$$

where the expansion coefficient functions u_n, v_n, p_n, w_n, and ρ_n are functions only of x, y, and t. The eigenfunctions $F_n(z)$ satisfy

$$(F_z/N^2)_z + F/c^2 = 0, \tag{2.10a}$$

$$F_z = 0 \quad \text{on } z = 0, -H, \tag{2.10b}$$

and are normalized so that

$$F_n(0) = 1. \tag{2.11}$$

The barotropic (n = 0) mode is missing from the sums in (2.9) as it typically makes a negligible contribution to the solution. Substitution of (2.9) into (2.1)-(2.5) and use of the orthogonality relation for the eigenfunctions leads to the following equations for the expansion coefficient functions:

$$(\partial/\partial t + \lambda_n)\, u_n - f v_n = -p_{nx} + X_n, \tag{2.12a}$$

$$(\partial/\partial t + \lambda_n)\, v_n + f u_n = -p_{ny} + Y_n, \tag{2.12b}$$

$$(\partial/\partial t + \lambda_n/\sigma)\, p_n/c_n^2 + u_{nx} + v_{ny} = 0, \tag{2.12c}$$

$$w_n = (\partial/\partial t + \lambda_n/\sigma)\, p_n/c_n^2, \quad \rho_n = -p_n/g, \tag{2.12d}$$

where

$$\lambda_n = A/c_n^2, \tag{2.13}$$

$$(X_n, Y_n) = (\tau^x, \tau^y)/D_n, \tag{2.14}$$

and

$$D_n = \int_{-H}^{0} F_n^2\, dz. \tag{2.15}$$

We are interested in steady solutions $\partial/\partial t = 0$) driven by an along-shore windstress $\left(Y_n \neq 0, X_n = 0\right)$ Under these simplifications, u_n and v_n can be expressed in terms of p_n:

$$\left(\lambda_n^2 + f^2\right) u_n = -f p_{ny} - \lambda_n p_{nx} + f Y_n$$

$$\left(\lambda_n^2 + f^2\right) v_n = f p_{nx} - \lambda_n p_{ny} + \lambda_n Y_n .$$

Since the along-shore (y) scale is much greater than the offshore (x) scale and all modes contributing significantly to the response have

$$\lambda_n << f,$$

the above equations can be simplified to

$$u_n = -p_{ny}/f - \lambda_n p_{nx}/f^2 + Y_n/f, \tag{2.16a}$$

$$v_n = p_{nx}/f. \tag{2.16b}$$

Substitution of these expressions into (2.12c) gives the field equation for the pressure as

$$p_{nxx} + \beta\lambda_n^{-1} p_{nx} - \sigma^{-1} f^2 p_n/c_n^2 = 0 \tag{2.17}$$

where it has been assumed that the wind stress Y_n is independent of x, and we have taken $\beta = f_y$. This field equation is subject to the boundary conditions that the motion remains finite at large distances from the coast and that u_n vanish at the coast x = 0. In terms of p_n, these conditions are

$$p_n \text{ remains finite as } |x| \to \infty, \tag{2.18a}$$

$$\lambda_n p_{nx} + f p_{ny} = fY_n \text{ at } x = 0. \tag{2.18b}$$

The solution of (2.17) subject to (2.18) is

$$p_n = \exp\left(k_n x\right) \varphi_n\left(y\right), \tag{2.19}$$

where

$$k_n = \beta\left(2\lambda_n\right)^{-1}\left[\left(1 + 4\sigma^{-1} f^2 \lambda_n^2 c_n^{-2} \beta^{-2}\right)^{\frac{1}{2}} - 1\right], \tag{2.20}$$

and $\varphi_n(y)$, which satisfies

$$\varphi_{ny} + \lambda_n k_n \varphi_n/f = Y_n \quad , \tag{2.21}$$

is given by

$$\varphi_n = \exp\left(-\Lambda_n(y)\right)\int_{y_0} \exp\left(\Lambda_n(y_*)\right) Y_n(y_*)\, dy_* \quad , \tag{2.22}$$

with

$$\Lambda_n(y) = \int_{y_0}^{y} \lambda_n k_n / f \, dy_* \ . \tag{2.23}$$

From (2.19) and (2.16)

$$v_n = k_n p_n / f \tag{2.24}$$

and

$$u_n = \left(Y_n/f\right)\left(1 - \exp\left(k_n x\right)\right) \ . \tag{2.25}$$

The solutions for p, v, and u are now simply obtained by summing the modes as in (2.9).

DYNAMICS

As in the time dependent no dissipation case (Gill and Clarke, 1974), the velocity perpendicular to the coast is determined entirely by the local wind-stress (see (2.25)), while the along-shore velocity and pressure depend on wind-stress integrated along the coast (see (2.19), (2.22), and (2.24)). Thus, the along-shore velocity and pressure may be nonzero even in regions where the wind-stress is negligible.

To further understand the dynamics, first let us consider some idealized cases. For very strong dissipation (2.18b) reduces to

$$p_{nx} = fY_n/\lambda_n \quad ,$$

i.e.,

$$v_n = Y_n/\lambda_n \quad . \tag{2.26}$$

In this case, the pressure and along-shore velocity are locally determined. At the other extreme, as $\lambda_n \to 0$ and dissipation is negligible, by (2.20), $k_n \to 0$,and, hence, in the limit of no dissipation

$$u_n = v_n = 0 \tag{2.27}$$

and

$$p_{ny} = Y_n \ , \tag{2.28}$$

i.e., there are no currents and the wind-stress is balanced by a pressure gradient. The initial value problem (Anderson and Gill, 1975) shows that this is the balance left after long Rossby waves have propagated westward from the initial wind-generated disturbance on the eastern boundary. When dissipation is included, the Rossby waves cannot completely escape from the eastern boundary, and nonzero currents are present.

Since $c_n \sim 1/n$ for large n, from (2.13)

$$\lambda_n \sim n^2 \quad . \tag{2.29}$$

Consequently, we expect the dominant balance in (2.18b) to be (2.26) for the higher order modes. We would not expect these modes to contribute much to the alongshore velocity field because (2.26) and (2.29) show that for these modes

$$v_n \sim Y_n n^{-2} \sim n^{-2} \tag{2.30}$$

and n is large. [The result $Y_n \sim n^0$ in (2.30) follows from (2.15) and results for the asymptotic eigenfunctions $F_n(z)$ in Clarke and Van Gorder

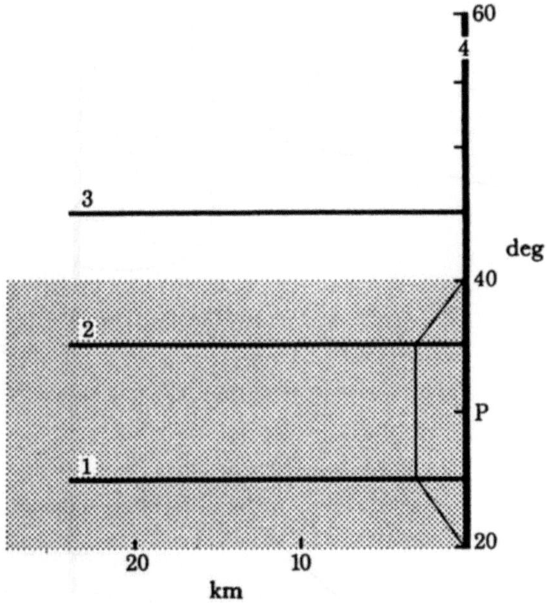

Figure 1: A schematic diagram showing the location of sections 1-3, the position of the coast, and the structure of the wind forcing. The shaded region indicates the latitudinal extent of the wind field, and the thin line its meridional profile. The wind field does not weaken offshore, and so it is without curl. Wind stress is entirely equatorward and reaches a maximum amplitude of 0.5 dyn/cm^2. (After McCreary, 1981).

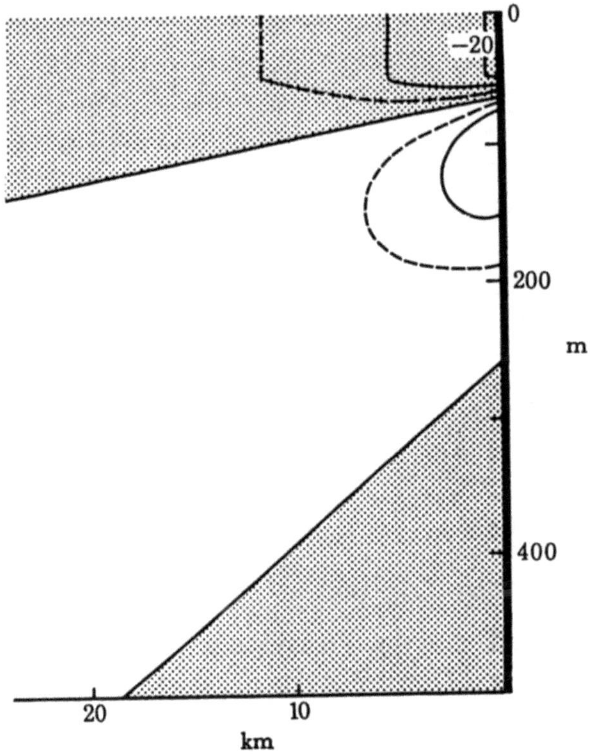

Figure 2a: Vertical sections of along-shore current along section 1 of Figure 1. The velocity contour interval is 10 cm s^{-1} and the dashed contours are ±5 cm s^{-1}. The shaded region indicates equatorward flow. (After McCreary, 1981).

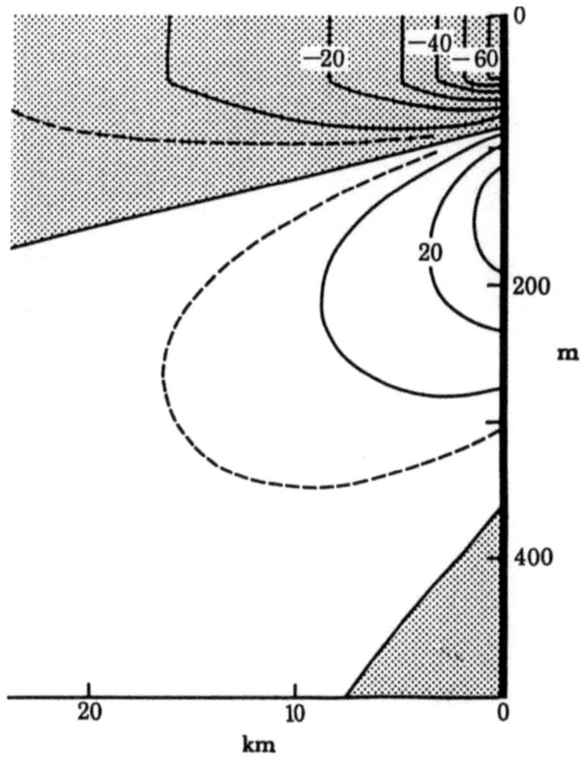

Figure 2b: As for Figure 2a, but along section 2.

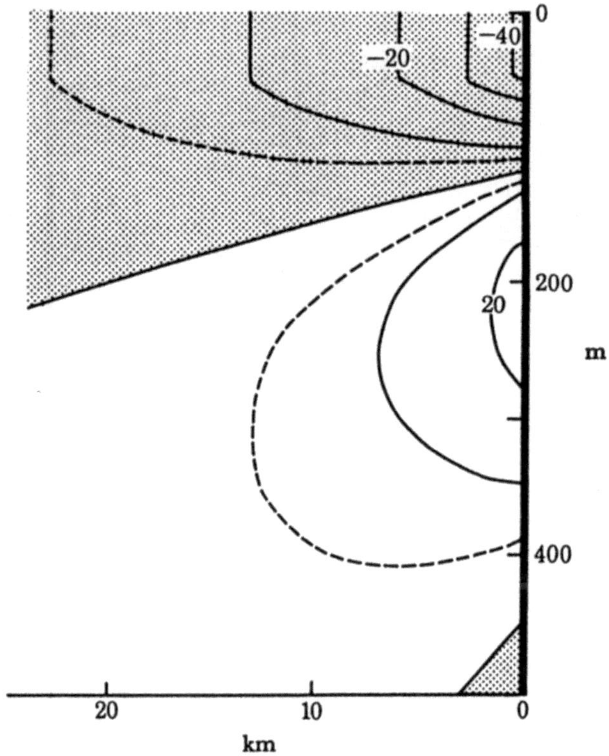

Figure 2c: As for Figure 2a, but along section 3.

(1986)]. For realistic parameters, the lower order modes satisfy (2.27) and (2.28), so again the contribution to the along-shore velocity is small. Therefore, the along-shore velocity field is principally due to the "intermediate modes" and these consequently set the horizontal and vertical scales for the motion.

Figure 2 shows McCreary's calculations of the along-shore velocity field for realistic parameters and a wind field resembling the steady component of the winds off the western coast of North America (see Figure 1). The undercurrent is clearly visible and has a strength and horizontal and vertical structure comparable with the observed undercurrent. The horizontal and vertical scales correspond to scales for the "intermediate modes" ($n \approx 6$).

The undercurrent's existence depends strongly on the existence of an alongshore pressure gradient; when calculations are carried out with $\partial/\partial y = 0$, no undercurrent forms. The undercurrent's existence is not, however, crucially dependent on ß-plane dynamics; the f-plane solution and the ß-plane solution are very similar.

Some observations (e.g., Mittelstaedt, 1978) suggest that the coastal undercurrent may deepen and weaken toward the poles. Figures 1 and 2 show that the model undercurrent deepened and weakened north of the wind patch. What is the physical reason for this? North of the wind patch the solution of (2.21) is

$$\varphi_n(y) = \varphi_n(y_0) \exp\left(-\Lambda_n(y)\right) . \tag{2.31}$$

Since $\Lambda_n \sim n^3$, the higher order modes are much more rapidly damped than the lower order ones, so north of the wind patch the motion is increasingly dominated by the lower order intermediate modes. These modes are less tightly trapped to the coast and have deeper z structures; so north of the wind patch, the undercurrent is deeper, less tightly trapped to the coast and, because there are fewer modes, it is weaker.

CRITIQUE

Suginohara and Kitamura (1984) and McCreary and Chao (1985) showed that, due to an equatorward barotropic shelf current induced by the wind, the continental shelf acts to strengthen the equatorward jet and to weaken or eliminate the undercurrent of the constant depth model. However, McCreary and Chao (1985) showed that the positive wind-stress curl which exists near the western coast of North America drives a coastal poleward shelf flow (as would be expected from Sverdrup balance). This poleward shelf flow counteracts the equatorward shelf barotropic flow and restores the undercurrent.

Other problems still exist, however. Figure 6 of McCreary and Chao (1985) suggests that when the bottom stress is realistic, the undercurrent is much weaker than observed. This is cause for concern because the wind-driven undercurrent must be strong enough to overcome likely equatorward undercurrents induced by offshore mean density fields (McCreary et al., 1986). In their study of the California Current System, McCreary et al. (1987) imposed a no-slip condition at the "coast" which strengthened the undercurrent, presumably because the equatorward shelf barotropic flow was lessened by the no-slip condition. But both the slip and no-slip conditions are not really justified, since they are not applied at the real coast, but rather at about 100 m depth near the edge of the true shelf topography. Another problem is that the models discussed are linear and nonlinear terms and are not negligible in parts of the flow. Finally, many of the important results depend on quite large values for the eddy viscosity and diffusivity, and we do not know whether it is reasonable to use such values.

RECOMMENDATIONS FOR FUTURE WORK

I believe that the theory of wind-driven mean eastern boundary flows is sufficiently developed, the theoretical results sufficiently encouraging, and observations sufficiently scarce that much would be gained from a large scale Eastern BOundary Circulation Experiment (EBOCE). Because of its proximity and the measurements we already have, the western coast of the United States should be the site of the experiment. Models already exist for a prior estimate of the flow to provide design criteria and focus for the experiment (e.g., McCreary et al., 1987). Some important unanswered questions about eastern boundary flows are listed below:

(i) The linear dissipation wind-driven models mainly discussed in this paper suggest that mean along-shore flows are not driven locally, but rather they are at least partially forced by an integral along the coast of the along-shore component of the wind-stress [see Eqs. (2.19) and (2.22)]. This suggests that eastern boundary currents are of large along-shore scale. What do observations say? Can present theory explain (c.f. the earlier discussion in section 2) observed changes in the width, depth, and strength of eastern boundary currents?

(ii) Current structures and strengths in the linear wind-driven models depend crucially on the use of large values for eddy viscosity and diffusivity coefficients. Would observations of turbulent fluctuations in the flow warrant the use of such large values?

(iii) Are the calculated mean eastern boundary current fields stable? The eddies, jets, and squirts observed along the U.S. west coast suggest that they are not. Perhaps the instabilities will result in the required large values for the eddy coefficients.

(iv) How important is the generation of mean flow by the mean large-scale offshore density field and by oscillatory flow over small bumps in the shelf and slope topography?

(v) Alongshore changes in the shelf and slope bottom topography will affect the structure and strength of mean or very low frequency flows. Can theory like that of Gill and Schumann (1979) predict such changes?

(vi) Pacific Eastern boundary flow is strongly affected by interannual variations associated with El Niño. At these frequencies, coastal sea level has been observed to propagate poleward at about 40 cm s^{-1} (Enfield and Allen (1980) and Chelton and Davis (1982)). Associated El Niño currents strongly affect eastern ocean boundary biology (e.g., Chelton et al., 1982). What is the horizontal and vertical structure of these El Niño currents? Does theory suggest that these currents can be estimated by using convenient and cheap sea level measurements at the coast?

ACKNOWLEDGEMENTS

This work was supported by NSF Grants OCE-8500669 and OCE-8515979. Paula Tamaddoni-Jahromi ably typed the manuscript.

The editors of this volume express appreciation to Dr. Ken Brink, Woods Hole Oceanographic Inst., for his helpful review and suggestions to the manuscript.

POLEWARD FLOW ALONG EASTERN BOUNDARIES: WHAT NEXT?

Adriana Huyer

TIME VARIABILITY OF POLEWARD UNDERCURRENTS

We need long-term (> 1 year) monitoring of poleward flows in each of the five subtropical eastern boundary regions, and very long records (> 5 years) in at least one region. Observations should include moored current measurements across the continental margin, repeated hydrographic and ADCP sections, and remote and in situ measurements of the wind and pressure fields. These observations might be situated at the eastern end of each of the WOCE principal zonal long-lines; i.e., at 28°S and 24°N in the Pacific, 30°S and 24°N in the Atlantic, and 30°S in the Indian Ocean.

ROLE OF POLEWARD UNDERCURRENTS IN GLOBAL AND GYRE-SCALE PROCESSES

We need experiments to determine the along-shore structure and continuity of an eastern boundary undercurrent, and to determine the associated transports of mass, heat, nutrients, and tracers. Observations should include current meter moorings at several latitudes, repeated hydrographic and ADCP surveys, and remote and in situ measurements of the wind and pressure fields. Designing the optimum experiment and interpreting results will require both theoretical and numerical modeling. These experiments will address the question of how the major gyres communicate with their eastern boundaries, and therefore they should be included in the WOCE Gyre Dynamics Experiment.

We also need to determine whether and how processes associated with poleward flow on eastern boundaries affect the global nutrient budgets: The high productivity of these regions gives them the potential for affecting chemical and biological cycles in the open ocean; e.g., the southeast tropical Pacific may play a major role in the global nitrogen cycle.

FORCING AND DYNAMICS OF POLEWARD UNDERCURRENTS

We need to continue to develop analytical, numerical, and laboratory models to increase our understanding of the roles of mixing, along-shore pressure gradients, wind-stress curl, form-drag, and wave-mean-flow interactions in driving undercurrents and countercurrents. We need to develop paradigms for choosing among different models, and to design field experiments that can discriminate among different forcing/response hypotheses.

A more detailed outline of the discussion under each heading follows:

I. Time Variability of Poleward Undercurrents

What To Do

(a) Long-term (> 1 year) monitoring of poleward flows in each of the five subtropical eastern boundary regions (viz., California, Canary, Benguela, Peru, and Leeuwin Current regions); very long-term (> 5 years) monitoring in one region.

(b) Comparative anatomy of poleward undercurrents in different "regions" with different physics; viz., latitude, climate (wind, runoff, and heat budget), and bathymetry. This might be accomplished by studies at several latitudes within one eastern boundary current region.

How

(a) Long-term current-meter moorings across the continental margin.

(b) Repeated hydrographic and ADCP sections across the margin and beyond.

(c) Measure wind and wind-stress curl fields (scatterometers and buoys).

(d) Measure along-shore pressure gradients (altimeters, hydrographic surveys, pressure, and tide gauges).

(e) Use support from local agencies and countries (including LDC's) where possible.

Where

(a) At the eastern end of each of the WOCE CORE-I zonal long-lines (i.e., at about 28°S and 24°N in the Pacific, 30°S and 24°N in the Atlantic, and 30°S in the Indian Ocean). At each of these locations, moored current meters should be in place for at least one year.

In at least one of these locations, current meters should be in place for five years or more.

(b) On the eastern margin of the gyre chosen as the object of the WOCE CORE III: Gyre Dynamics Experiment. (For the North Atlantic subtropic gyre, we would recommend an intensive study off both Portugal and North Africa.)

(c) Continue monitoring at sites presently (or recently) occupied;

(i) In the California Current:
- La Perousse Bank at 49°N (1983-1993?, by IOS, Canada).
- Off northern and central California (MMS studies, NPS).
- Off Oregon 43°N (OSU).

II a. Role of Poleward Undercurrents in Gyre Dynamics

What To Do

(a) Conduct process-oriented studies of poleward undercurrent kinematics and dynamics as part of the WOCE CORE-III Gyre Dynamics Experiment. (How do gyres communicate with their eastern boundaries?)

(b) Determine the along-shore structure and continuity of poleward undercurrent; explore evidence for flow instabilities.

(c) Determine transports of mass, heat, and tracers by poleward undercurrent/countercurrent.

Where

Off North Africa and Portugal assuming the North Atlantic subtropical gyre is chosen for WOCE CORE-III. (Ideally, similar studies would also be conducted off North America and South America).

How

- Subsurface drifters (RAFOS or Pop-Up).
- Moorings (conventional and ADCP).
- Altimeter and scatterometer (to measure surface pressure gradients and wind fields).
- Hydrography and tracers (dye?).
- Models.

II b. Role of Undercurrents in Global Chemical/Biological Budgets/ Transports

What To Do

Determine the transports and uptake/removal rates of nitrogen and carbon in the southeastern tropical Pacific.

Why

There is evidence that materials advected from the continental margin strongly affect global budgets of nitrogen, carbon, etc. (Chemical processes in the ocean interior may be influenced by what happens on continental margins.)

How

- Measure uptake/removal rates directly.
- Measure distributions and currents to estimate transports.
- Hydrographic and ADCP surveys.
- Surface and subsurface drifters.
- Conventional and ADCP moorings.
- Modeling.
- Take advantage of WOCE and GOFS plans.

III. Forcing and Dynamics of Poleward Undercurrents

What To Do

(a) Continue to develop the different theoretical and numerical models to understand undercurrent dynamics (e.g., McCreary, Brink, Haidvogel, Clarke, and Holloway).

(b) Continue to work on understanding the important physics contained in the results of numerical models.

(c) Develop diagnostics from theoretical and numerical models which can serve to focus observational efforts.

(d) Design laboratory experiments to test particular hypotheses.

(e) Develop paradigms for choosing among different models. (What feature(s) of each model can we expect to observe?)

(f) Design field experiments that can discriminate between different models. (Different models might give similar results for entirely different reasons.)

PART II:

GEOGRAPHICAL REPORTS:

THE ATLANTIC OCEAN
THE PACIFIC OCEAN
THE INDIAN OCEAN

THE ATLANTIC OCEAN

THE BAROCLINIC CIRCULATION OF THE WEST SPITSBERGEN CURRENT

Robert H. Bourke and Alan M. Weigel
Naval Postgraduate School
Monterey, CA 93943

ABSTRACT

The baroclinic circulation of the West Spitsbergen Current (WSC) is described based on a dense network of CTD stations acquired in September 1985. Near 78°N the baroclinic flow extends to 200 m with a transport of 1.3 Sv northward. Westward turning of the WSC to join the southward flowing East Greenland Current occurs from 76° to about 81° N, being interrupted between 79° and 80°N by the Molloy Fracture Zone. Near 80°N the WSC bifurcates to form an eastern branch which transports about 20% of its initial flow into the Arctic basin, observed as the warm, salty Atlantic Layer. No northward flow of WSC water is observed along the western flank of the Yermak Plateau. A cyclonic recirculation of the eastern branch of the WSC is noted east of the Yermak Plateau.

INTRODUCTION

During September 1985 the ice breaker NORTHWIND conducted an oceanographic measurement program in the northern Greenland Sea. A relatively dense network of CTD stations (Figure 1), approximately 10 to 20 km spacing, provided extensive information on the circulation and water mass structure of the polar waters of this region. It is the purpose of this paper to describe the characteristics of the relatively warm, saline West Spitsbergen Current, a poleward flowing eastern boundary current. An investigation of the baroclinic nature of the flow and a detailed assessment of the bifurcation of the current into a westward branch connecting with the East Greenland Current and a northeastward branch flowing into the Arctic basin were the two principal objectives of the cruise, and these features are described herein.

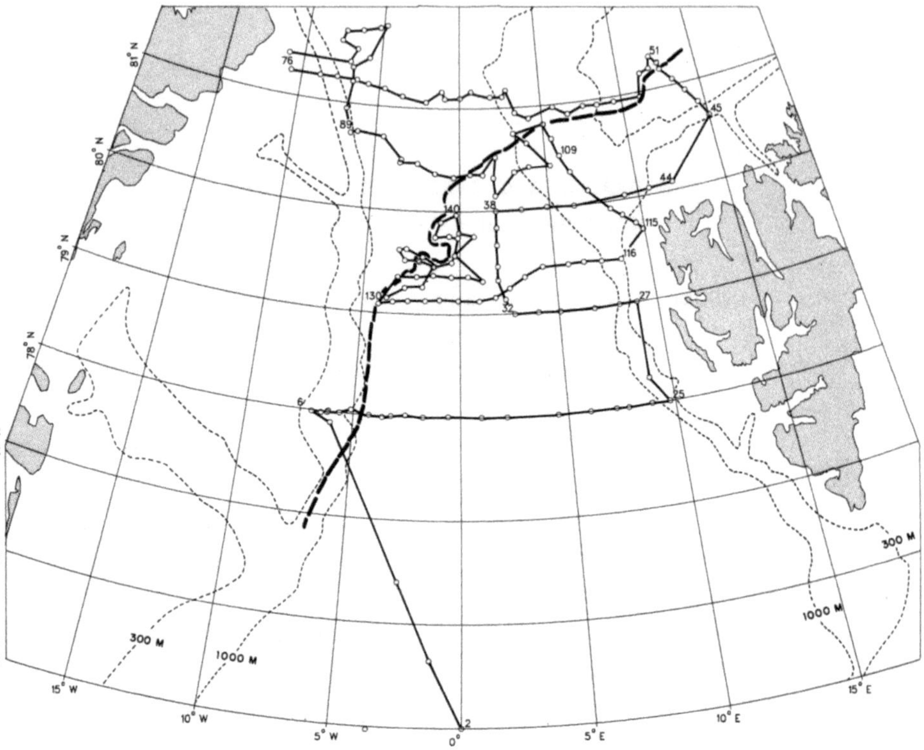

Figure 1: The NORTHWIND September 1985 cruise track and location of CTD stations. The heavy dashed line shows the position of the ice edge. The Yermak Plateau is the broad, shallow area to the northwest of the Svalbard archipelago near 80°N.

HISTORIC OVERVIEW

The circulation within the Greenland Sea is dominated by a cyclonic gyre centered on the prime meridian between 74°-76°N (Figure 2). The northward flowing West Spitsbergen Current (WSC) forms the eastern boundary of this gyre. Aagaard and Griesman (1975) have shown that the WSC provides the major inflow of mass, salt, and thermal energy into the Arctic basin. This flow enters the Arctic Ocean through Fram Strait, the narrow, moderately deep passage between Greenland and Svalbard Archipelago to the east. Along the Greenland coast the southward flowing East Greenland Current (EGC) forms the western boundary. Its characteristic features, e.g., temperature, salinity, speed, baroclinicity, depth distribution,

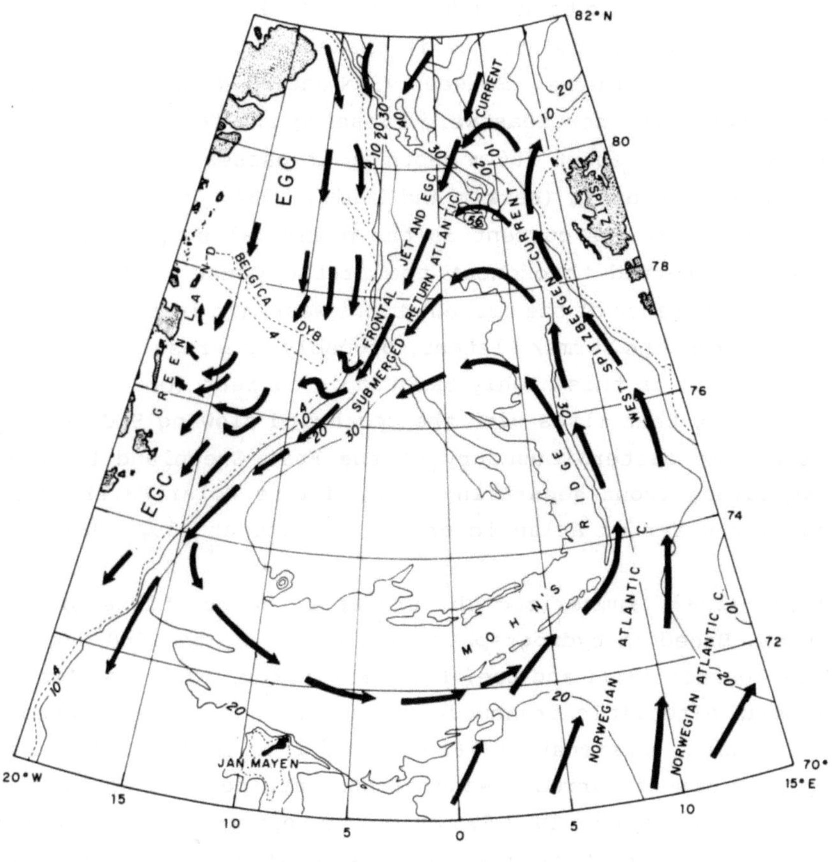

Figure 2: The general circulation and bathymetry of the Greenland Sea (from Paquette et al., 1985). The complex bathymetry of the Molloy Fracture Zone is seen near 79°N, 3°E.

etc., are almost completely in opposition to those of the WSC, a contrast which dramatically illustrates the differences between eastern and western boundary currents. The eastward zonal flow known as the Jan Mayen Polar Current closes the circulation to the south while the westward turning of the WSC into the EGC closes the circulation to the north in a flow known as the Return Atlantic Current (RAC) (Gladfelter, 1964; Paquette et al., 1985).

The WSC is formed when warm, saline water from the North Atlantic enters the Norwegian Sea and flows northward along the continental slope as the Norwegian Atlantic Current. Off the southern tip of Spitsbergen near 74°N it mixes with colder, fresher water from the Jan Mayen Polar Current and the East Spitsbergen Current and continues flowing north along the west coast of Spitsbergen. Over the continental slope the flow is relatively deep, extending to about 800 m (Hanzlick, 1983). North of about 75°N it loses heat and is diluted as it spreads out in a thin layer over the top of the Greenland Gyre. Between 79°N and 80°N the WSC bifurcates. An inshore branch turns northeastward crossing the Yermak Plateau where it eventually sinks to about 200 m depth as it passes under the ice edge and enters the Arctic basin (Perkin and Lewis, 1984; Johannessen et al., 1983). This branch of the current is the principal supply for the warm Atlantic layer observed throughout the Arctic Ocean (Aagaard et al., 1987). The second branch of the current turns westward initially following the western flank of the Yermak Plateau, probably as a result of conserving potential vorticity in this highly barotropic current (Quadfasel et al., 1987). This branch then mixes with the southward flowing EGC as the RAC. The RAC forms the eastern boundary of the East Greenland Polar Front (EGPF), the strong front separating cold, dilute Polar Water (PW) from warmer, saltier water of Atlantic origin (Bourke et al., 1987).

Hanzlick (1983) summarized the transport calculations of earlier investigators. Based on hydrographic data from 1933 to 1960, as gleaned from various sources (Hill and Lee, 1957; Kislyakov, 1960; and Timofeyev, 1962), the mean baroclinic transport across the WSC was estimated to be 3 to 4 Sv northward. Greisman (1976) estimated the mean total transport, using long-term direct current measurements obtained during 1971-72, as about 7 Sv northward, with a 10.7 Sv maximum in September and a 3 Sv minimum in February. Based upon hydrographic data obtained during this period, Greisman estimated the mean baroclinic transport to be 3.5 Sv, or about one-half of the total. During 1976-77 Hanzlick (1983) estimated a mean transport of 5.6 Sv from current meter data, with a maximum in the fall and a minimum in the spring. The mean baroclinic transport for the same period was only 1 Sv, significantly less than in 1971-72.

The WSC is rich in temporal and spatial variability, perhaps its principal characteristic feature. Greisman (1976) and Hanzlick (1983) have shown that both volume flow and core temperatures of the WSC are variable over time scales that range from days to years. For example, Hanzlick showed evidence, based largely on hydrographic data, that the interannual variability in the volume and heat transport was of the order of 50%. Shorter-term variability is noted in the direct current measurements made in Fram Strait during 1984-85 (Aagaard et al., 1985a) which showed a northward flow in the core of the WSC in excess of 0.4 m/s in the latter half of August 1984. By the first week in September 1984, the flow had reversed, with southward speeds of approximately 0.05 m/s being observed. An important conclusion which arises from these various data sets is that the changes between successive yearly surveys may in fact be due to short term variability and are not true interannual changes. Various investigators have also shown evidence of spatial variation in the WSC flow and temperature. Dickson (1972) and Hanzlick (1983) indicate this variation takes the form of banded structures or northward flowing filaments with the scale of the lateral structures being about 15 km.

The mean wind stress over the Greenland-Norwegian sea region is apparently the major factor affecting the interannual and shorter time scale variability of the WSC (Greisman, 1976; Hanzlick, 1983). Greisman and Aagaard (1979) suggested that the WSC is driven by large scale variations in the atmospheric forcing over the Greenland and Norwegian Seas on time scales of less than one month and that there was considerable month-to-month variation in the forcing. Hanzlick suggested that baroclinic instability and coastally trapped waves were also possible contributors to shorter time scale variability.

Water Mass Characteristics

Four water masses are found in the WSC. Their basic properties have been defined by Coachman and Aagaard (1974), but modifications to the temperature and salinity limits have frequently been imposed. Atlantic Water (AW) extends from the surface to about 800 m and is warmer than 0°C. Surface temperatures in summer may be higher than 6°C, while during the winter the surface temperatures are typically warmer than 2°C keeping the west coast of Spitsbergen ice free. The salinity of the AW increases in the upper 100 m so that in the core of the WSC the salinity may be greater than 35.2. Below 100 m the salinity decreases to about 34.95. Warm water, properly called AW, can also be found with salinities as low as 34.7. This low salinity fraction of AW, termed Atlantic Surface Water by Swift (1986), is primarily a surface phenomenon and is the result of dilution by ice melt. The area of low salinity is generally found near the ice edge or in more southerly areas where ice melt water has been advected.

Beneath the AW is found Greenland Sea Deep Water (GSDW). This water mass forms a large dome at the center of the Greenland Gyre with temperatures colder than 0°C and salinity between 34.87 and 34.95. Carmack (1972) proposed that GSDW was formed by convective subsurface modification of AW. Carmack and Aagaard (1973) and Aagaard et al. (1985b) suggest that chimney formation and double-diffusive mixing between PW and AW may also be involved with the production of GSDW.

From the CTD data acquired between 6 and 26 September 1985, a series of vertical cross sections, or transects, have been constructed to characterize the properties and flow of the WSC. The AW core of the WSC will be traced with the 3°C isotherm in order to distinguish AW from colder (< 3°C) Atlantic Intermediate Water (AIW), a cooled and slightly diluted variant of AW associated principally with the EGC (Paquette et al., 1985; Bourke et al., 1987). The lower salinity limit for AW was earlier defined as 34.9, which is also the upper limit for AIW. To further distinguish it from AIW in the following transects, AW will be traced using water more saline than 35.0.

From a transect along 78°N, the structure of the northward flow of the WSC along the west coast of Spitsbergen can be observed (Figure 3). In this figure AW can be seen as a wedge-shaped core along the coast, extending seaward to a distance of 60 km (near Station 22) and to a depth of almost 275 m near the coast. Associated with the warm wedge is a core of saline (> 35.1) water between 50 and 100 m. Seaward of the wedge the warm water spreads out, for almost 200 km, into a surface layer 50-75 m deep. The saline water defined by the 35 isohaline is also found as a tongue spreading out almost halfway across Fram Strait, but generally deeper than the 3°C isotherm, at 100 to 400 m. At the western end of the wedge, the surface salinity decreases to less than 34.9 as the AW mixes with PW of the EGC and becomes diluted. The temperature-salinity features of the WSC observed in Figure 3 are quite similar to those observed by Bourke et al. (1987) in August 1984, suggesting minimal interannual variability was experienced during these two summers. As the WSC flows farther northward, the warm wedge remains closer to the Spitsbergen coast becoming thicker, extending in depth to 300 m. The doming of the deep and intermediate waters in the center of the Greenland Gyre is seen in the downward sloping of the isotherms towards the coast. Figure 3 and other transects off the west Spitsbergen coast illustrate that in August the warm core of the WSC is at the surface. However, observations made later in the year show that surface cooling forces the warm core subsurface between a 100 and 400 m depth (Johannessen et al., 1983).

The distribution of AW in transects which cross the WSC between 78° and 80°N appears to be controlled, at least in part, by bathymetry. In Figure 3 the isotherms and isohalines, although slightly offset to the right, appear to be pushed up by the 2000 m isobath defining the Knipovich Ridge. The core of the WSC is located to the west of the shelf break, while the largest part of the 35.0 isohaline tongue is located over the western slope of the ridge. In this and other transects, the western slope of the ridge appears to control the western limit of AW becoming more predominant with increasing latitude.

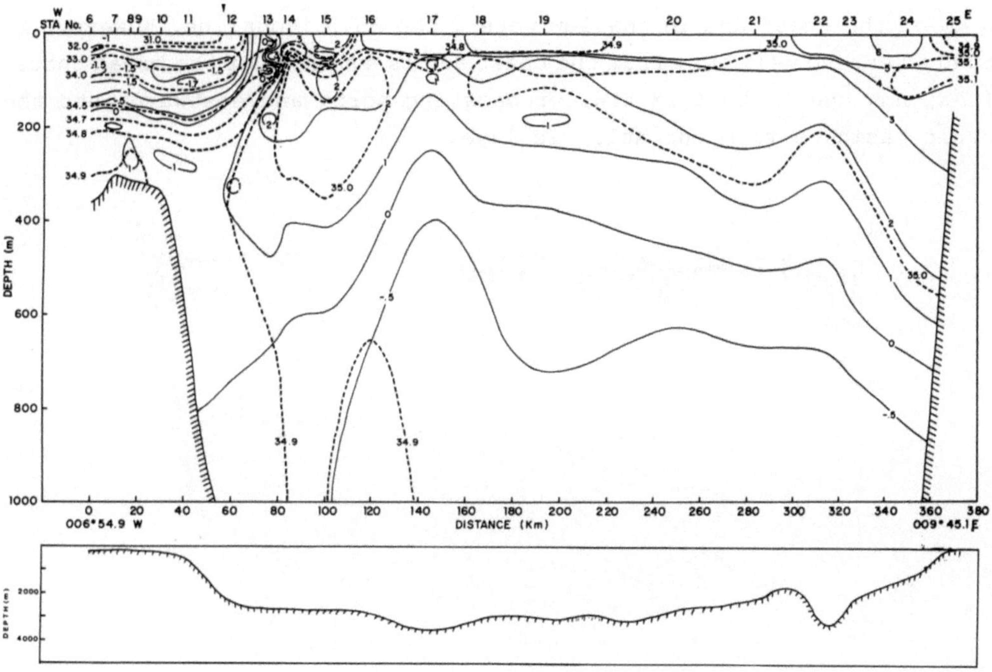

Figure 3: A temperature (solid line) and salinity (dashed line) transect along 78°N. The 2000 m feature between Stations 21 and 22 is the Knipovich Ridge. The arrow at the top indicates the location of the ice edge.

Branching of the WSC

Just south of 80°N, the WSC branches into two streams. The left branch turns westward and appears to sink below the cold surface water. The right branch turns eastward crossing the Yermak Plateau and flows along the north coast of Spitsbergen and into the Arctic basin. The branching appears to be controlled in part by bathymetry. A transect (Figure 4) which extends northwestward along the ridge of the Yermak Plateau shows that the warm saline core, defined by the 3°C isotherm and the 35.1 isohaline, is positioned along the northwest coast of Spitsbergen with the 35.0 isohaline extending beyond the 1000 m isobath into deep water. The properties of the eastern branch of the WSC can be seen in a transect which extends northwestward from the continental shelf north of Spitsbergen into the Arctic basin (Figure 5). AW is found here as a narrow stream, within 40 km of the coast and in shallow water. The 35.0 isohaline, however, is found to extend well away from the coast and is evidence of the broad thrust of AW, now cooled to form AIW, propagating north and eastward into the Arctic basin to form the Atlantic layer.

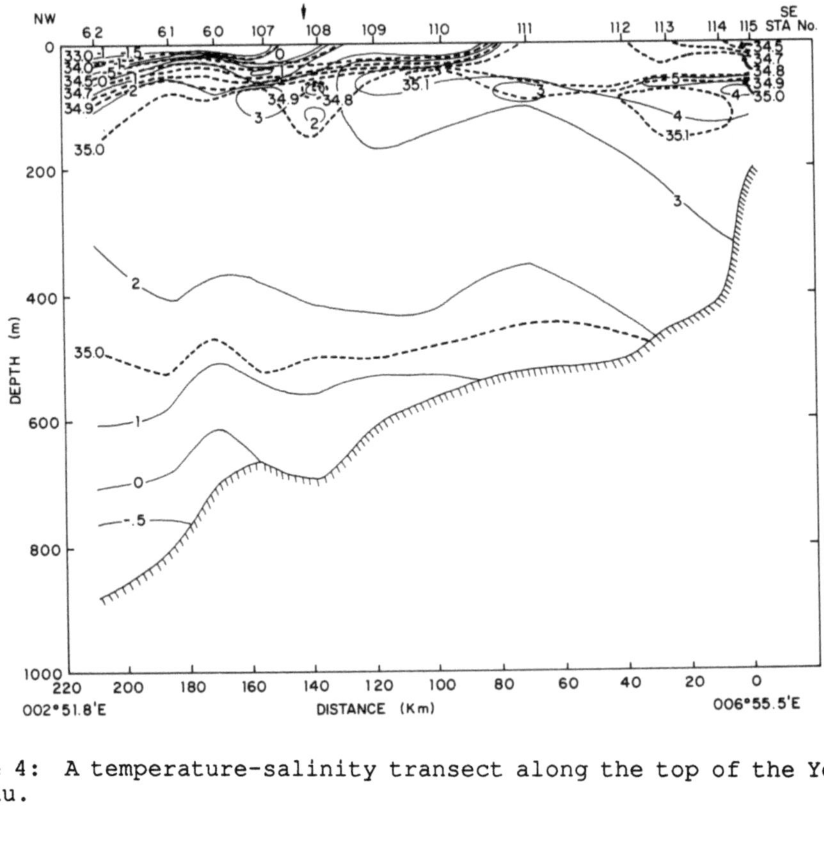

Figure 4: A temperature-salinity transect along the top of the Yermak Plateau.

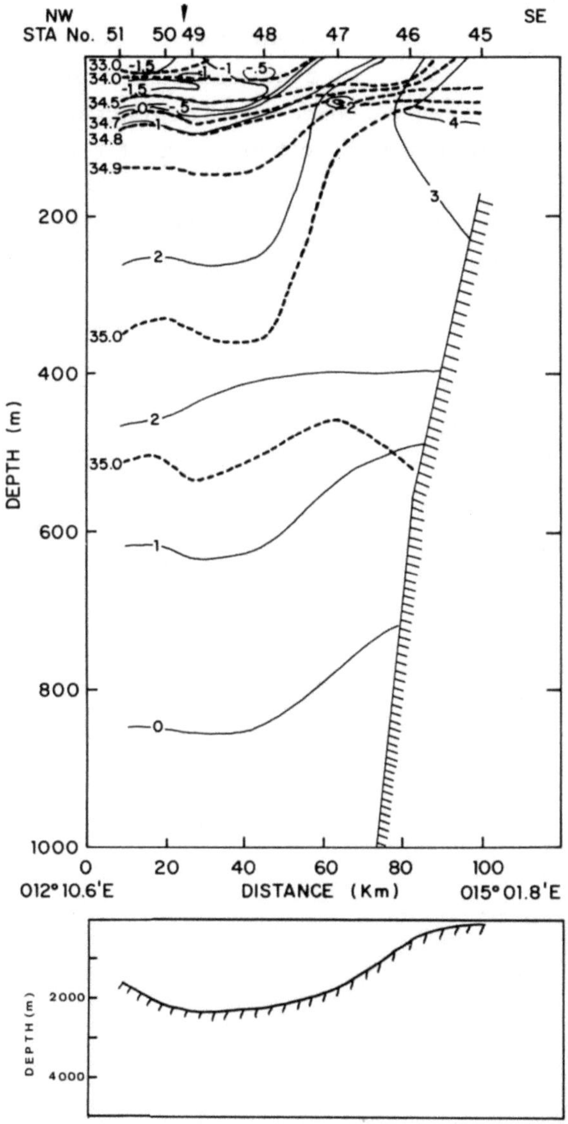

Figure 5: A temperature-salinity transect north of Spitsbergen. The WSC is seen to hug the north coast and shelf of Svalbard.

As previously discussed, south of 80°N, AW is found along the west Spitsbergen coast close to the 1000 m isobath. North of this latitude the left branch of the WSC leaves the shallow water along the west slope of the Yermak Plateau, turns westward into deep water, and joins the southward flowing EGC. The westward turning of the WSC can be traced by following the AW where most of the baroclinicity occurs. The northward extent of the westward turning is found just north of 81°N where only a small trace of AW at 100 m near the prime meridian is seen. This is illustrated in a transect which crosses through the center of the westward turning along the 2°E meridian (Figure 6). In this transect the core of the warm westward-turning WSC is seen centered at 100 m between Stations 101 and 99, just at the ice edge, and well to the north of the seamounts between Stations 34 and 36. This is the farthest north that warm AW is found. Westward turning is also indicated south of the complex bathymetry of the Molloy Fracture Zone south of Station 33.

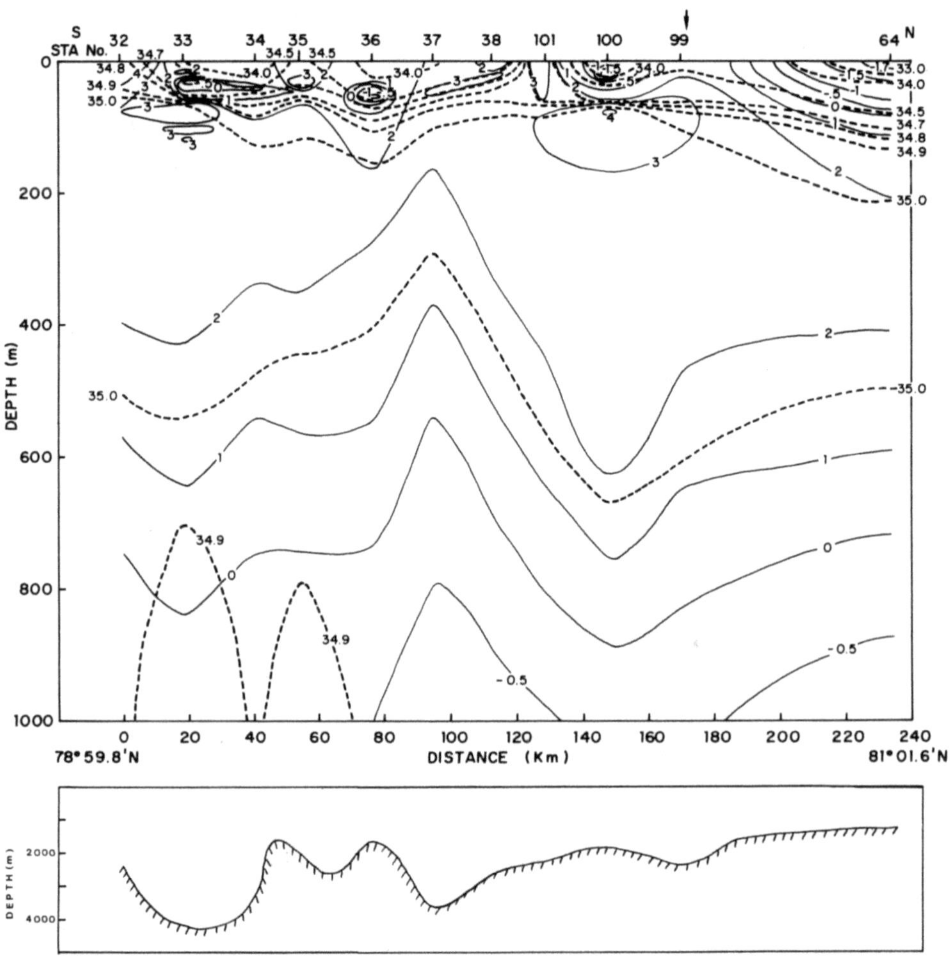

Figure 6: A temperature-salinity transect along 2°E. Westward flow of the WSC is indicated by the lens of 3°C water between Stations 37 and 99 and also south of Station 33.

The core of the warm westward turning is found below the surface and can be accounted for in two possible ways. The warm saline AW may have submerged under the cold, fresh PW. This could occur if the salinity of the AW were high enough to offset the density of the cold PW. The AW would then sink as it turned westward into the EGC. Another possibility is that the AW is cooled and freshened by the presence of the ice edge as it turns westward. This would then result in finding a temperature maximum below the cold surface layer.

Horizontal plots of water mass properties may also be used to describe the branching and westward turning of warm AW. Although plots of surface properties are of limited usefulness because of the strong influence of the ice edge, Aagaard and Coachman (1968) have shown that plotting isotherms on a sigma-t surface can be effective to illustrate the westward turning. An analysis of the property profiles at each station revealed that in the WSC the maximum temperature in the water column (the core of AW) was nearly coincident with the maximum salinity. The density at the depth of maximum temperature, expressed as sigma-t, has a mean between 27.9 and 28.0. For this reason, and to compare with historical data presented by Aagaard and Coachman (1968), the temperature of the 28.0 sigma-t surface and the depth of the 28.0 sigma-t surface were plotted (Figures 7 and 8). The westward spread of the 2°C isotherm can be seen between 80° and 81°N, with 81°N as the maximum extent of the westward flow. Aagaard and Coachman demonstrated virtually identical features based on data from 1965, although the maximum temperatures were warmer in 1965. The depth of the 28.0 sigma-t surface perhaps shows the best evidence of westward turning based on horizontal distributions of temperature and salinity. This figure indicates that a zone of westward turning has taken place north of 78°N with little westward flow north of 81°N. To the north and west, behind the ice edge, the depth of this surface was greater than 500 m.

Dynamic Topography

Dynamic topographies derived from the September 1985 cruise were computed for both the surface and 150 decibar levels referenced to 500 decibars, the average deepest depth of the sensor and a depth which included most of the baroclinicity. The surface dynamic topographies show the weak baroclinicity of the WSC in comparison to the EGC (Figure 9). The northern limb of the cyclonic Greenland Sea Gyre can be seen in the southern part of Fram Strait near 79°N. The dynamic height gradient across the EGPF would suggest a baroclinic flow of between 0.20 and 0.30 m/s, while the sluggish WSC baroclinic flow is 0.06 to 0.08 m/s. This figure

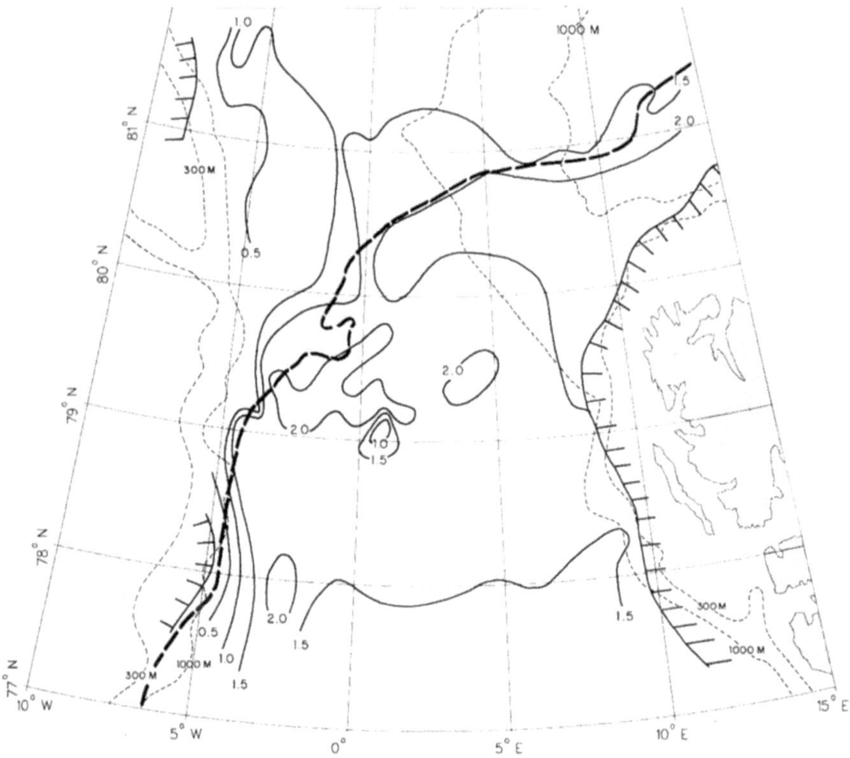

Figure 7: The temperature of the 28.0 sigma-t surface.

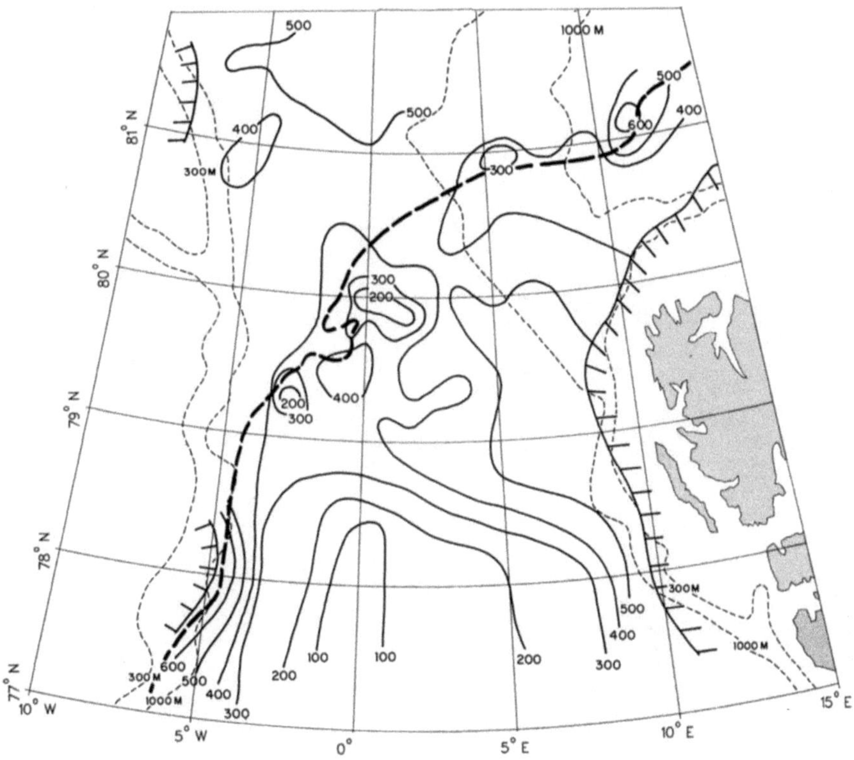

Figure 8: The depth of the 28.0 sigma-t surface illustrating the westward turning of the WSC.

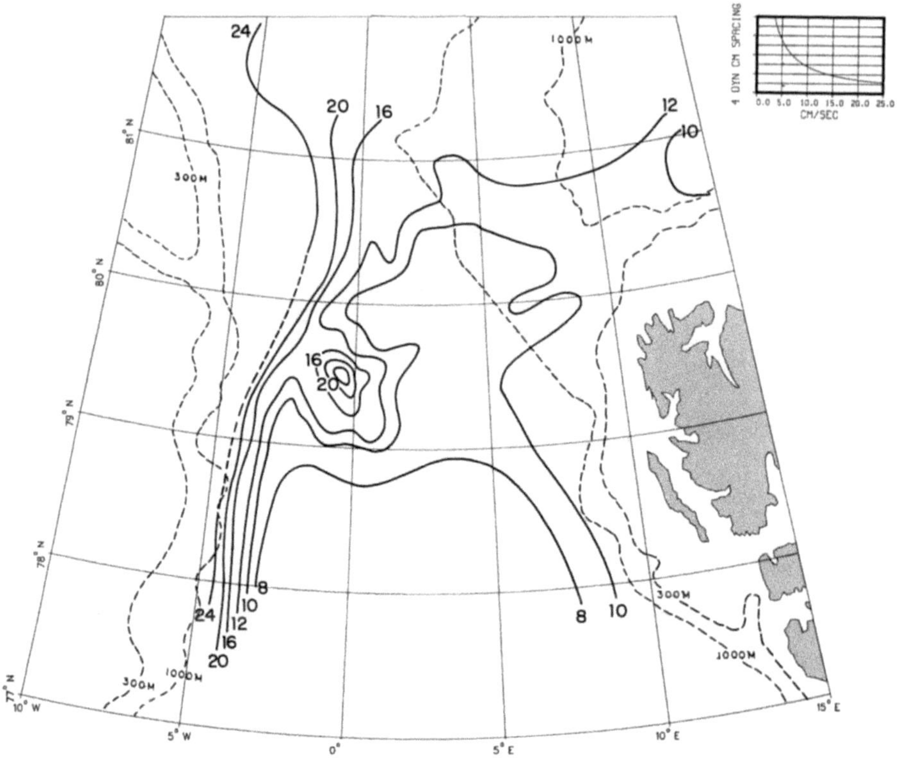

Figure 9: The surface dynamic topography referenced to 500 decibars (in dynamic centimeters).

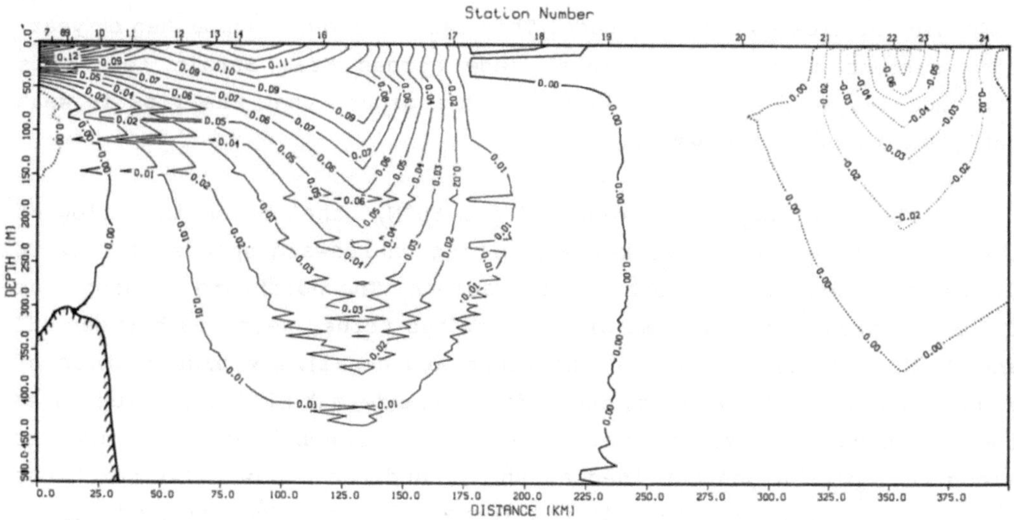

Figure 10: Baroclinic velocity section along 78°N. Southward flow is indicated by positive isotachs spaced at 0.01 m/s intervals, but at 0.02 m/s intervals in the high speed jets. The northward flowing WSC is observed near Stations 21 and 24. The RAC is centered near Station 14, while the EGC is found over the Greenland Shelf between Stations 7 and 10.

illustrates the branching of the WSC in the vicinity of the Yermak Plateau just north of 80°N. Westward flow is indicated between 79° and 81°N with the core of the westward turning crossing 5°E just north of 80°30′N and ultimately merging with the southward flowing EGC. This figure and dynamic topographies at 150 decibars show that the surface and subsurface components of the WSC turn westward at or south of 81°N.

To further investigate the geostrophic velocity and volume transport in Fram Strait, several vertical baroclinic velocity sections were constructed. A section along 78°N, based on the properties of Figure 3, shows the weakly baroclinic WSC with a core speed of 0.07 m/s located over the center of the Knipovich Ridge (Figure 10). Little baroclinic flow is noted at depths below 200 m. The northward transport resulting from this flow is 1.3 Sv. In contrast, the EGC in the extreme left part of the figure shows strong baroclinic flow over the Greenland continental shelf with a speed of almost 0.2 m/s. The core of the RAC is located at a depth of 50 m at Station 15 with a maximum speed of 0.13 m/s. The combined transport for this section for the EGC and RAC is 3.1 Sv southward. Figure 10 also shows that the circulation about the center of the Greenland Gyre, as observed between Stations 18 and 21, has virtually no north-south component of flow at this latitude. This lack of flow is supported by year-long current meter measurements, made across Fram Strait near this latitude (Aagaard et al., 1985a), which indicate that the flow is weak but predominantly westward.

Farther northward, between 79° and 80°N, the baroclinic flow of the WSC becomes even weaker exhibiting northward speeds of 0.02 to 0.03 m/s, based on geostrophic calculations. The northward transport is 0.5 Sv, indicating that some of the flow has turned westward between 78° and 79°N. A section based on the temperature-salinity distribution of Figure 4, which extends along the ridge of the Yermak Plateau, illustrates the branching of the WSC into two streams. One branch flows north and east along the Spitsbergen continental shelf, and the other flows northwest following the western slope of the Yermak Plateau. The northward speed in both branches is 0.02 to 0.03 m/s, while the net transport for this section is 0.5 Sv northward.

Currents computed from the section in Figure 5, which extends northwestward from the north coast of Spitsbergen, show the WSC positioned over the continental slope with a speed of 0.02 m/s. The transport through this section is 0.1 Sv toward the northeast, indicating that approximately 20% of the baroclinic flow of the WSC has turned into the Arctic around the northern coast of Spitsbergen. Southerly flow is observed farther offshore between Stations 46 and 49 suggesting the existence of a cyclonic filament of this branch. Such a recirculation is plausible due to vortex stretching as the current moves off the shelf into deeper water.

Current speeds calculated from a section along 81°N (Figure 11), all of which is behind the ice edge, indicate the maximum speed of the WSC over the Yermak Plateau is less than 0.01 m/s, i.e., the northward baroclinic component of the WSC has virtually disappeared. The existence of a northwestward component of flow over the plateau, as inferred to maintain continuity with the flow crossing 80°N, cannot be demonstrated due to the orientation of the hydrographic stations. However, Aagaard et al. (1987) report a similar absence of baroclinic flow over the plateau based on late November 1977 data.

The above section also illustrates the rapid (0.16 m/s) southward flow of the EGC, centered between Stations 67 and 69 just to the west of the prime meridian. The southward flow observed just to the east of the EGPF, between Stations 60 and 63, is of some interest as this flow incorporates a small amount of AW and AIW and may be considered the first trace of the RAC. This section and other nearby stations indicate that there is no substantial baroclinic component of the western branch of the

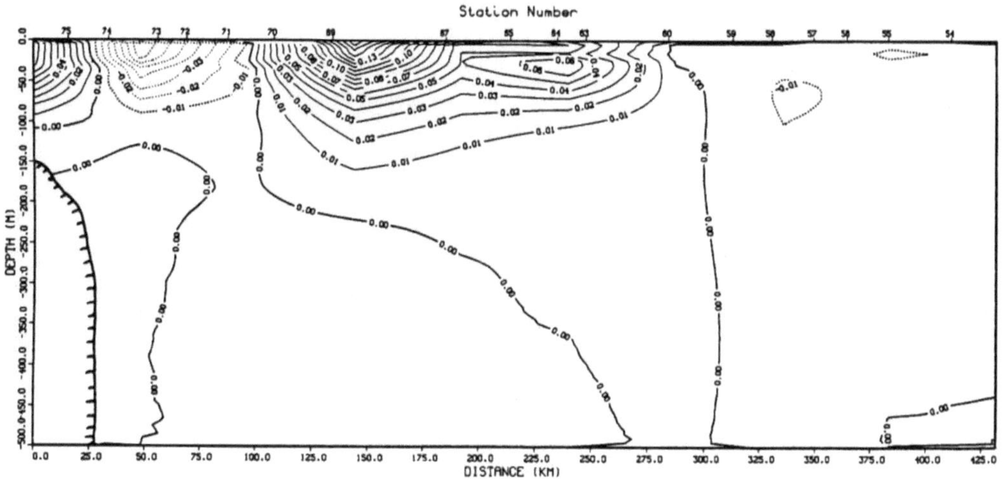

Figure 11: Baroclinic velocity section along 81°N. The southward flowing EGC is centered near Station 69. The southward flow near Station 64 is the first trace of the RAC.

WSC which continues northward past 81°N following the contours of the western flank of the Yermak Plateau to continue into the Arctic basin as suggested by Perkin and Lewis (1984) and Johannessen et al. (1983). The existence of such a northward component is questionable in light of reinterpretation of the Perkin and Lewis data which indicate a northward component cannot be substantiated.

A meridional section along 2°E between 79° and 81°N (Figure 12) shows that the core of the westward turning is predominantly north of 80°N, between Stations 38 and 99, with a speed of 0.03 m/s. Westward turning is also observed to take place south of 79°N. The transport for the northern core of westward turning is 0.4 Sv, while that south of Station 34 is about 0.1 Sv westward. The separation of the westward turning into two filaments may also be seen in the dynamic topography of Figure 9 and in the depth of the 28.0 sigma-t surface (Figure 8). The separation is attributed to the presence of the Molloy Fracture Zone (Perry, 1986), a region of complex bathymetry which includes the Molloy Deep, with depths exceeding 5000 m, and several seamounts rising to within 1500 m of the sea surface. The waters in between the two westward excursions of the WSC are cooler and less saline and may be associated with a cyclonic vortex rotating in the vicinity of the Molloy topographic structure (Quadfasel et al., 1987; Johannessen et al., 1987).

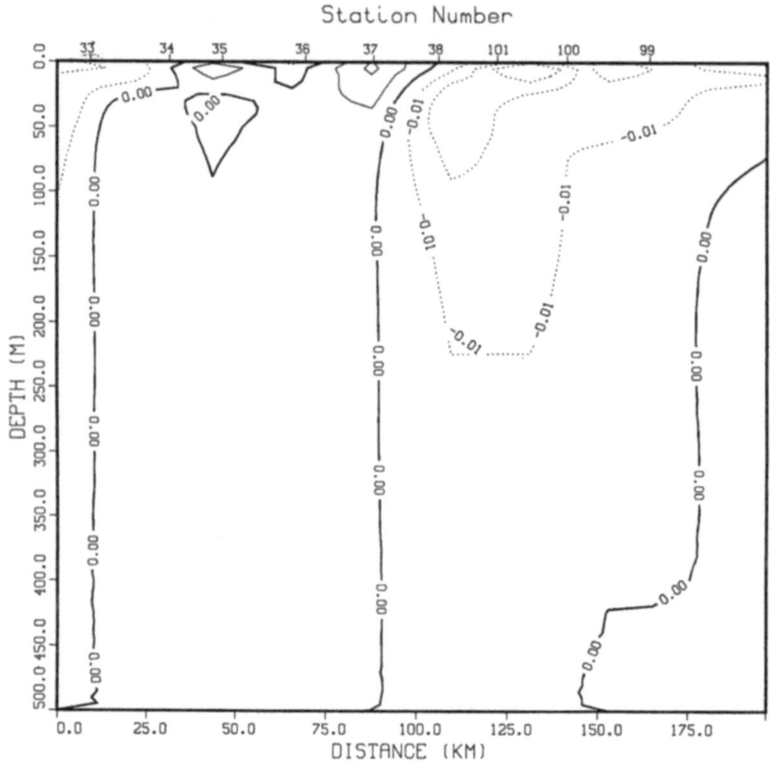

Figure 12: Velocity cross-section along 2°E showing westward turning of the WSC north of 80°N (Station 38) and south of 79°N (Station 33). The flow is weak and variable in between possibly due to the presence of the Molloy Fracture Zone.

CONCLUSIONS

The water properties and circulation of the West Spitsbergen Current (WSC) have been examined based on a dense network of CTD stations occupied by NORTHWIND in September 1985. Our principal findings are summarized below.

1. The WSC is found as a weakly baroclinic flow in several streams along the Spitsbergen continental slope. Northward geostrophic velocity at the surface along 78°N is found to be 0.07 m/s. Little baroclinic flow is observed at depths below 200 m.

2. Bathymetric steering appears to play an important role in the location of the separate streams of the WSC.

The Knipovich Ridge, a 2000 m deep feature to the west of the Spitsbergen continental shelf, exerts an influence on the distribution of water mass properties, especially north of 78°N.

3. North of 80°N the WSC splits into two branches, with approximately 20% of the baroclinic flow following the Spitsbergen continental shelf into the Arctic Basin and 80% of the flow turning westward over the western slope of the Yermak Plateau.

4. Some westward turning of the WSC takes place south of 79°N. A core of westward turning was located between 80° and 81°N along the prime meridian with baroclinic speeds of 0.02 m/s. Several filaments of AW were found turning westward between 79° and 80°N, although these contributed little to the net westward transport.

ACKNOWLEDGEMENTS

The authors are pleased to acknowledge the support of the Arctic Submarine Laboratory, Naval Ocean Systems Center, San Diego, CA, for their continued support of our marginal sea-ice zone studies, the latest under contract number N66001-86-WR-00131. We are grateful for the help provided by the officers and men, especially the marine science technicians, of the USCGC NORTHWIND.

From: Mary L. Batteen, Naval Postgraduate School, Monterey, CA

On: Review and Commentary to paper **THE BAROCLINIC CIRCULATION OF THE WEST SPITSBERGEN CURRENT**, by R.H. Bourke and A.M. Weigel

This is an interesting study and review of observations of the West Spitsbergen Current (WSC), which forms the eastern boundary of the cyclonic (subpolar) gyre circulation of the Greenland Sea. Unlike typical eastern boundary currents, e.g., the California or Peru Currents, which flow along the eastern boundaries of anticyclonic (subtropical) gyres, this surface current is relatively warm and saline and flows poleward (rather than being cold and fresh and flowing equatorward). Because of these characteristics, it would be of scientific interest to compare and contrast the characteristics of the WSC with those of typical eastern boundary coastal undercurrents, since these flows are also typically warm and saline and flow poleward. One could also make comparisons with other poleward surface currents, such as the Alaska, Davidson, or Leeuwin Currents. Insofar as the currents are wind-driven, as, for example, the Alaska Current, one would expect that the WSC and these currents would be dynamically similar. In the case of the Leeuwin Current, however, since recent results (McCreary et al., 1986) indicate that the Leeuwin Current is driven by thermohaline rather than by wind forcing, the WSC and the Leeuwin Current would not be expected to be dynamically similar.

Bourke and Weigel describe the water properties and baroclinic circulation of the WSC, based on a "relatively dense" network of CTD stations. Even though the station spacing is about 20 km, it is not dense enough to resolve eddies, which have horizontal space scales as small as 5 to 10 km in the Arctic region. However, the station spacing is dense enough to resolve the larger scale features such as the WSC.

In Bourke and Weigel's comprehensive historical overview of the WSC, they mention that the mean baroclinic transport can be about one-half of the total, but about 20% of the total is more typical. This suggests that the barotropic transport is quite significant. Bourke and Weigel also call the WSC a "highly barotropic" current, when reviewing the results of Quadfasel et al. (1987) which pertain to a branch of the WSC that follows the western flank of the Yermak Plateau. These results indicate that the barotropic component of the WSC can be quite substantial. Since Bourke and Weigel's findings also show that bathymetric steering and topographic effects can influence the WSC, and it is well known that barotropic effects are well felt by bottom topography, it is essential that future observations of the WSC include both the baroclinic and barotropic components of the velocity field.

Overall, as summarized in the conclusions, this paper brought out some important points, which may also be applicable to other poleward flows on eastern boundaries. I thought the point that changes in the WSC flow field can be construed as interannual rather than as short-term variability was especially worthy of note.

A BRIEF SKETCH OF POLEWARD FLOWS AT THE EASTERN BOUNDARY OF THE NORTH ATLANTIC

T.A. McClimans
Norwegian Hydrotechnical Laboratory
and
Division of Port and Ocean Engineering
Norwegian Institute of Technology
Trondheim, Norway

The circulation of the surface waters of the North Atlantic is shown in Figure 1 (Sverdrup, et al., 1942). The eastward drift current at mid-latitudes shows a secondary flow as shown in Figure 2, with a poleward component at depth (Dietrich, et al., 1980). Impingement of the drift current upon the European Continental Shelf results in both an equatorward flow toward the Mediterranean and a poleward flow past the British Isles. Three branches of this current system join to form the Norwegian Current (Figure 3). The early work of Helland-Hansen and Nansen (1909) implied a transport of 3 Sv, but newer results (Gould et al., 1985) imply a flow of ~7-1/2 Sv with maxima in winter. The large poleward flow of warm Atlantic Water provides a favorable climate for Norwegians. A branch of the Norwegian Current flows to the east in the Barents Sea while the major part drives north through the Fram Strait past Svalbard, diving under the brackish, ice-covered surface layer in the Arctic.

Most of the fresh water runoff from northern Europe flows to the Baltic Sea. This is the major source of the poleward flow of the Norwegian Coastal Current of Figure 4 (Braathen and Saetre, 1973). This current is colder in the spring and warmer in the fall. Satellite images reveal how large mesoscale eddies cause an interaction with the Norwegian Current, especially north of Stad (Figure 5) (NOAA-6). Some of this flow is engulfed in the Norwegian Current while much of it follows the coast eastward to the Siberian Shelf.

The third poleward flow of interest and perhaps a major driving force for the poleward circulation is the dense bottom flow generated by salt rejection during ice formation in the shelf seas of the Arctic Basin (Figure 6) (Midtun, 1985). This process extracts fresh water (in the form of ice flows) which joins the equatorward flow mainly within the East Greenland Current. All of these flows follow the continental slopes to the right due to the earth's rotation.

The hydrographic sections in Figure 7 (Dietrich, et al., 1980) show the currents described in this brief sketch.

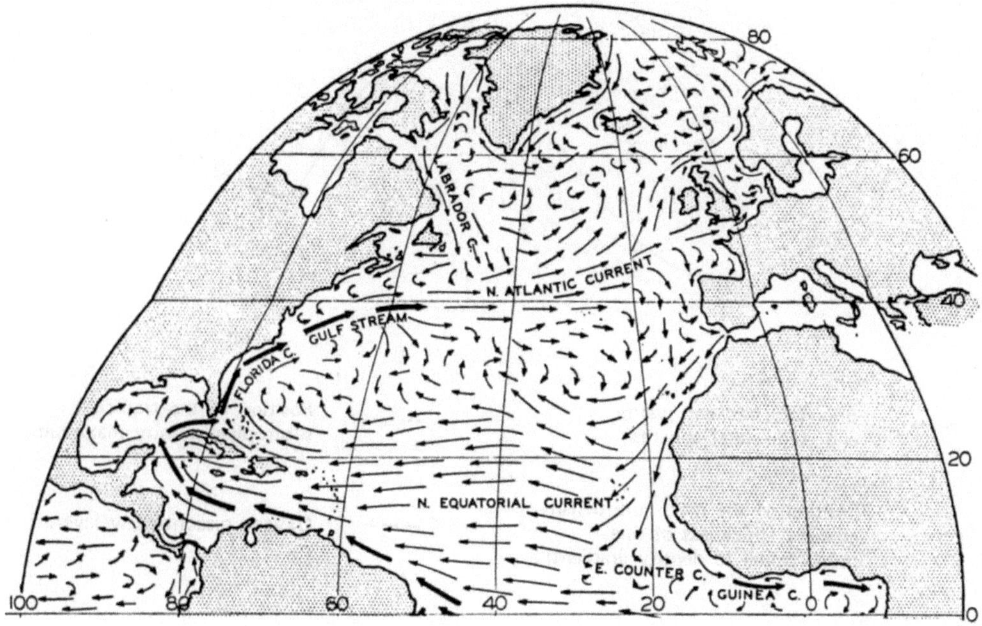

Figure 1: Surface currents of the North Atlantic in February-March. (Sverdrup, et al., 1942)

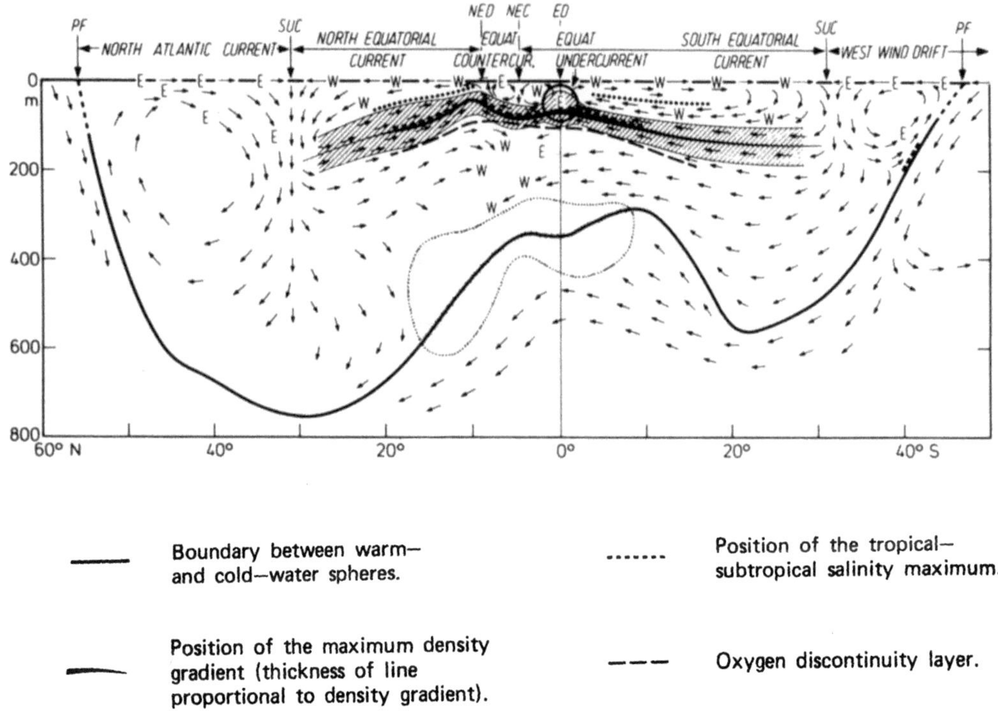

Figure 2: Longitudinal section of the poleward circulation of the Atlantic. (Dietrich, et al., 1980)

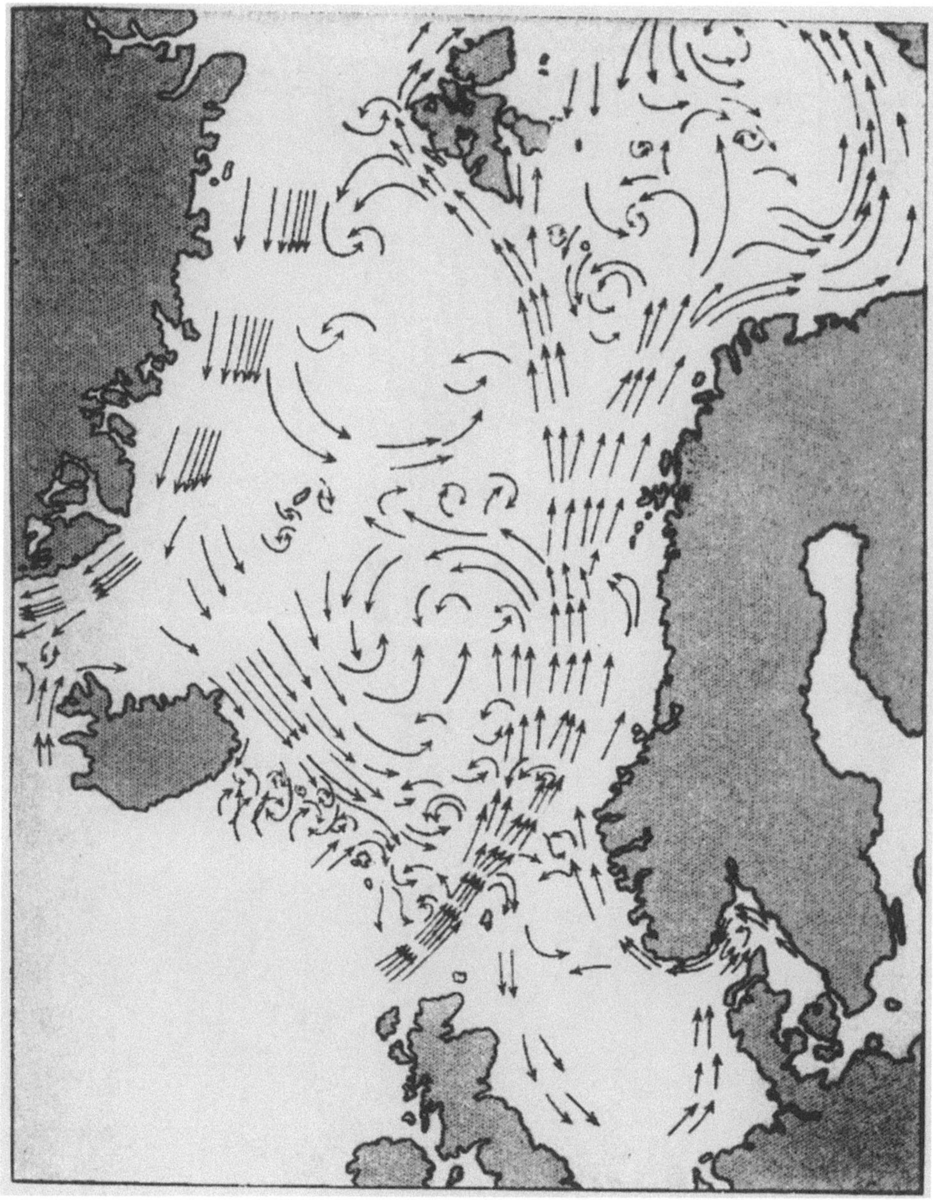

Figure 3: Surface currents of the Norwegian Sea. (Helland-Hansen & Nansen, 1909)

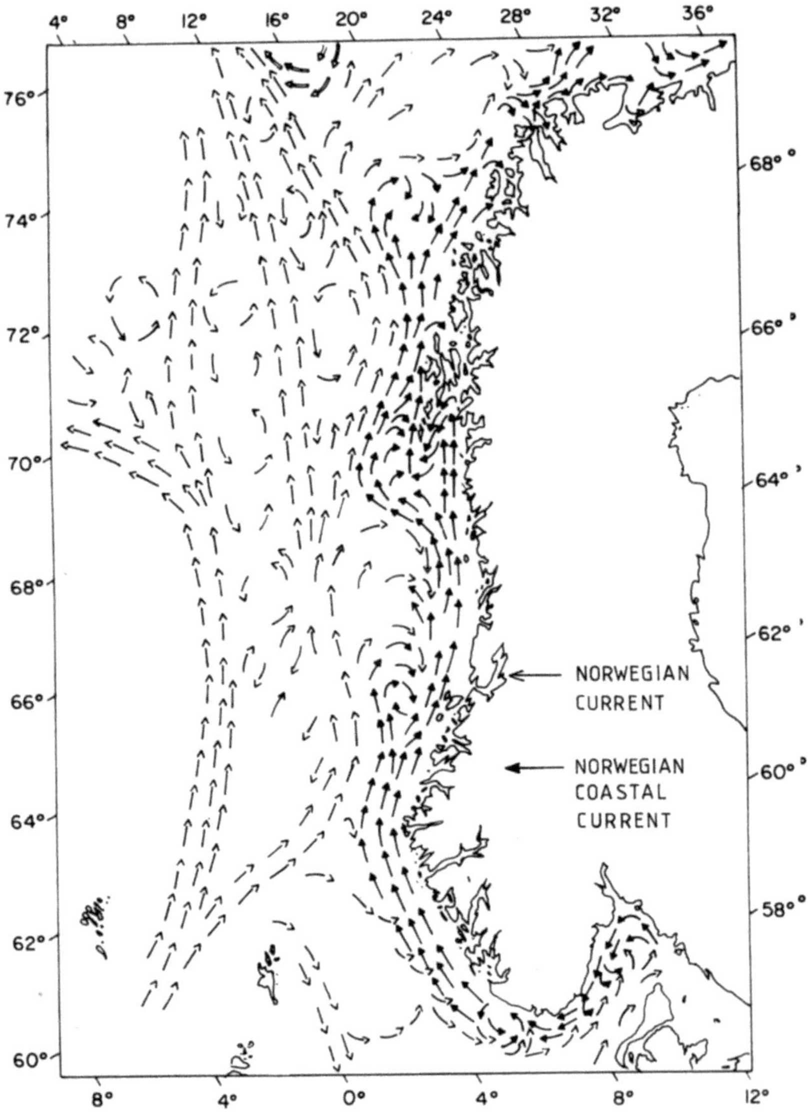

Figure 4: Surface currents west of Norway. (Braathen & Saetre, 1973)

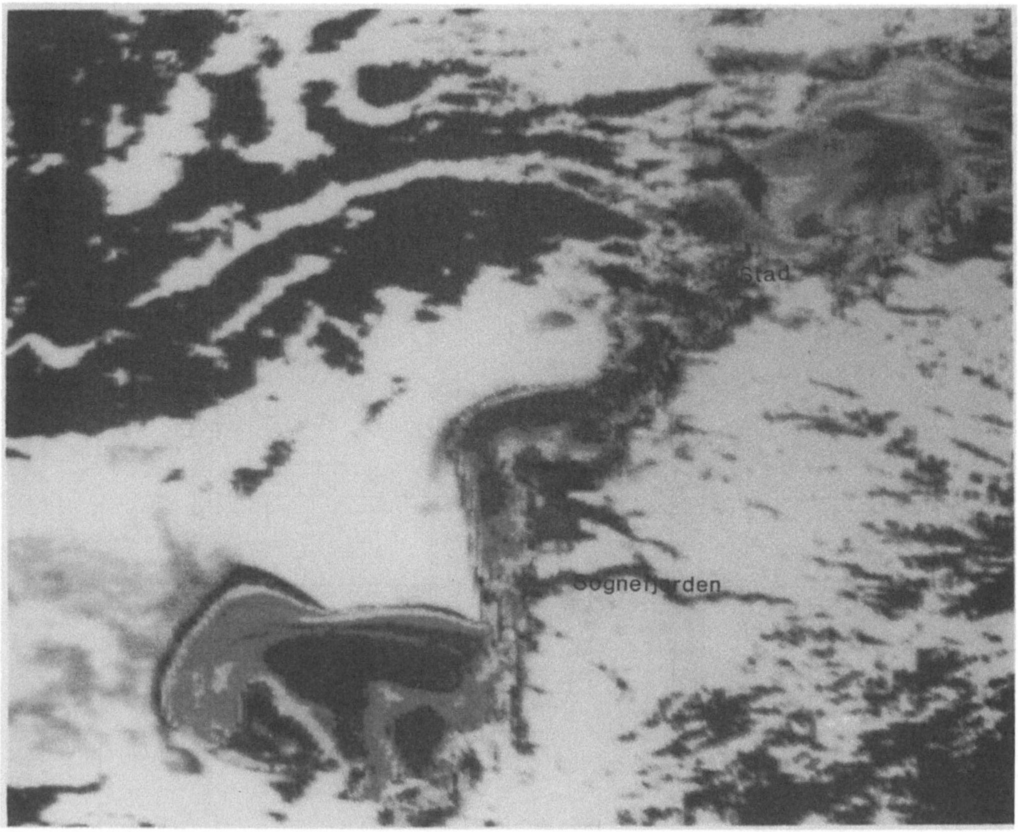

Figure 5: NOAA thermal image of the Norwegian Coastal Current off the west coast of Norway. (Data from Tromsø Telemetry Station)

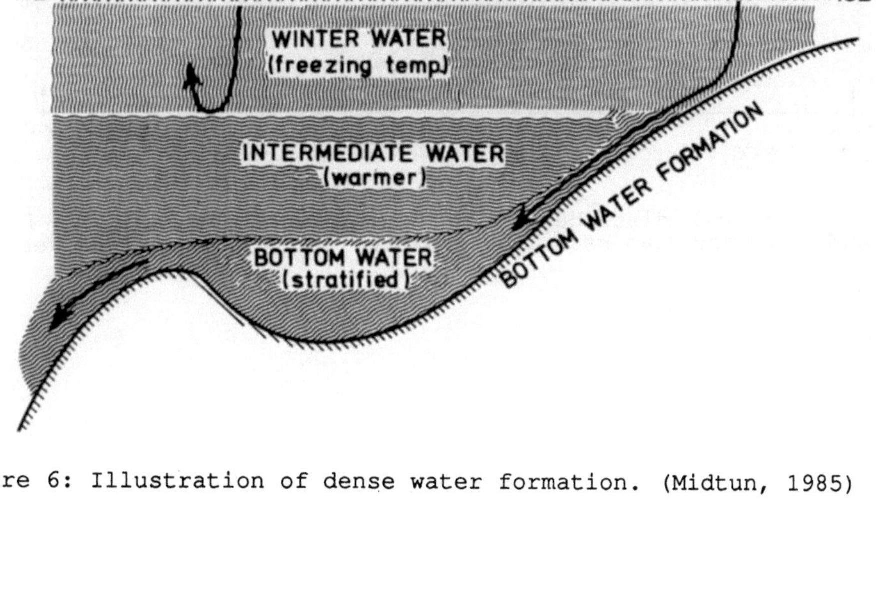

Figure 6: Illustration of dense water formation. (Midtun, 1985)

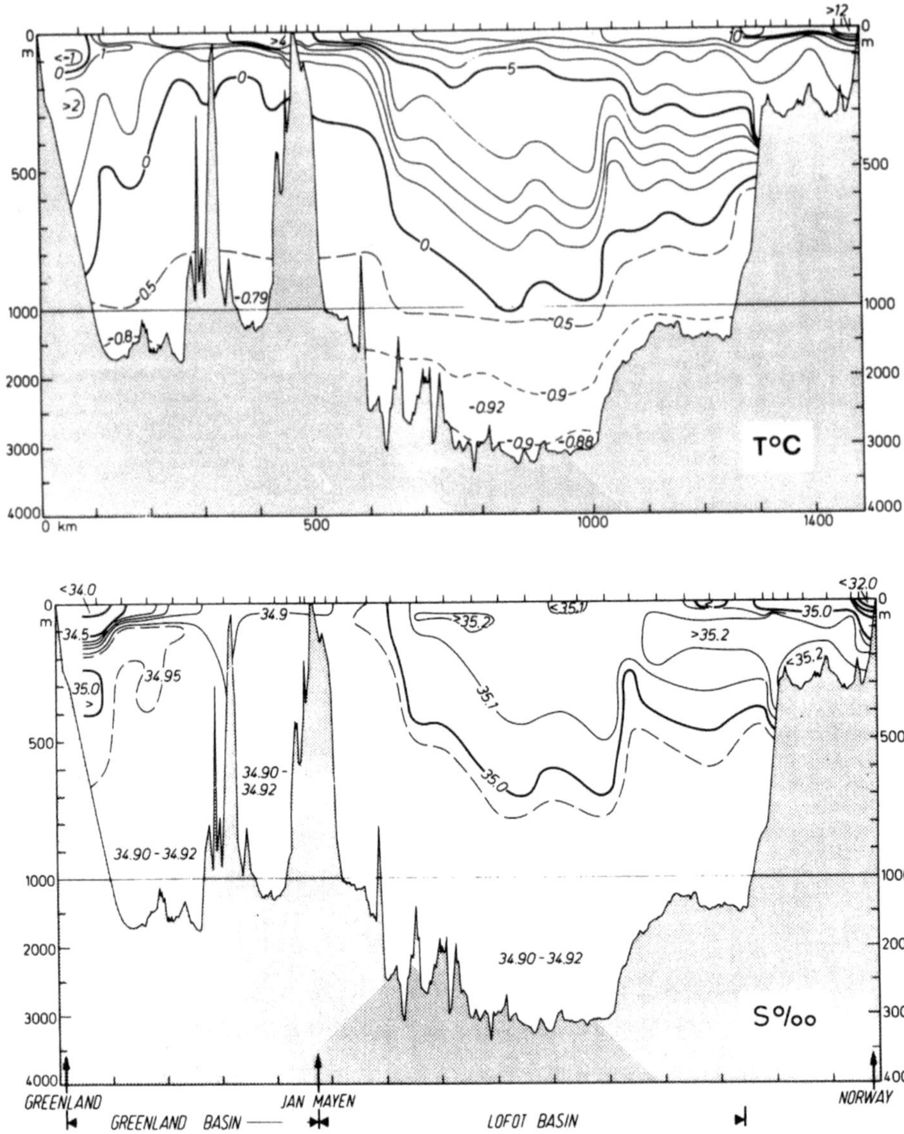

Figure 7: Vertical distributions of temperature and salinity in the Greenland and Norwegian seas in the winter of 1958. (Dietrich, et al., 1980)

From: J.M. Huthnance, Proudmand Oceanographic Laboratory, UK

On: Review and commentary to paper **A BRIEF SKETCH OF POLEWARD FLOWS AT THE EASTERN BOUNDARY OF THE NORTH ATLANTIC,** by T. A. McClimans

This "brief sketch" outlines the main contributions to poleward flow in the eastern North Atlantic and the Norwegian Sea. More detail regarding the poleward flow past the British Isles, and specifically along the continental slope, is given in Huthnance and Gould (this volume). Some of this 7 1/2 Sv (referred to) flows along the upper continental slope from west of Shetland north-eastwards and then south-eastwards into the Norwegian Trench along its western slope, before eventually turning across the Trench to join the poleward flow adjacent to the Norwegian Coast. Such flow appears to be considerably guided along the depth contours.

The fresh water runoff from northern Europe includes a contribution from the southern North Sea and German Bight (notably from the Rhine and Elbe outflows) joining the Baltic outflow at the Skagerrak.

It should perhaps be emphasized that none of the flows are steady; transports may vary seasonally by 50 percent or more and cease altogether on a few occasions during a year's current measurements; moreover, the large eddies alluded to may severely distort the path of flow from any long-term mean of simple form.

ON THE NORTHEAST ATLANTIC SLOPE CURRENT

J.M. Huthnance
Proudman Oceanographic Laboratory
Bidston Observatory
Birkenhead, Merseyside L43 7RA, United Kingdom
and
W.J. Gould
IOS Deacon Laboratory
Wormley, Godalming, Surrey, GU8 5UB, United Kingdom

ABSTRACT

A year-round current flows northwards along the steep continental slope of the NE Atlantic. Transport estimates average $1.5 \times 10^6\ m^3s^{-1}$ west of the UK, but increase greatly in the Faeroe-Shetland Channel. There is evidence of larger transports in winter. Density and associated longshore pressure gradients are thought to force the flow.

INTRODUCTION

In the context of poleward boundary currents in the eastern Atlantic, we are here concerned with flows around the west and north of the British Isles. Currents further south are considered elsewhere, the nearest observations being off Portugal (Ambar and Fiuza, 1987). By considering a slope current specifically, a distinction is made excluding a broader northwards drift of the upper waters over the Rockall Trough and Plateau (e.g., Ellett et al, 1986; also see "Discussion" below). To the north, we also discuss a slope current in the Faeroe-Shetland Channel. However, additional flows and recirculation there, and the Channel's distinctive deep thermocline and division from the Rockall Trough by the Wyville-Thomson Ridge, warrant separate treatment.

SOUTH OF THE WYVILLE-THOMSON RIDGE

Observations from 47-1/2° to nearly 60°N at the Wyville-Thomson Ridge are listed and results plotted in Huthnance (1986, Table 1 and Figure 1). In addition, drogues have been tracked along the Rockall Slope and further north (Booth and Meldrum, 1987a and b) and southwest of Britain (R.D. Pingree, personal communication) where currents were again recorded in 1984 in 500 m and 1500 m depth near 47-1/2°N, 6-1/2°W.

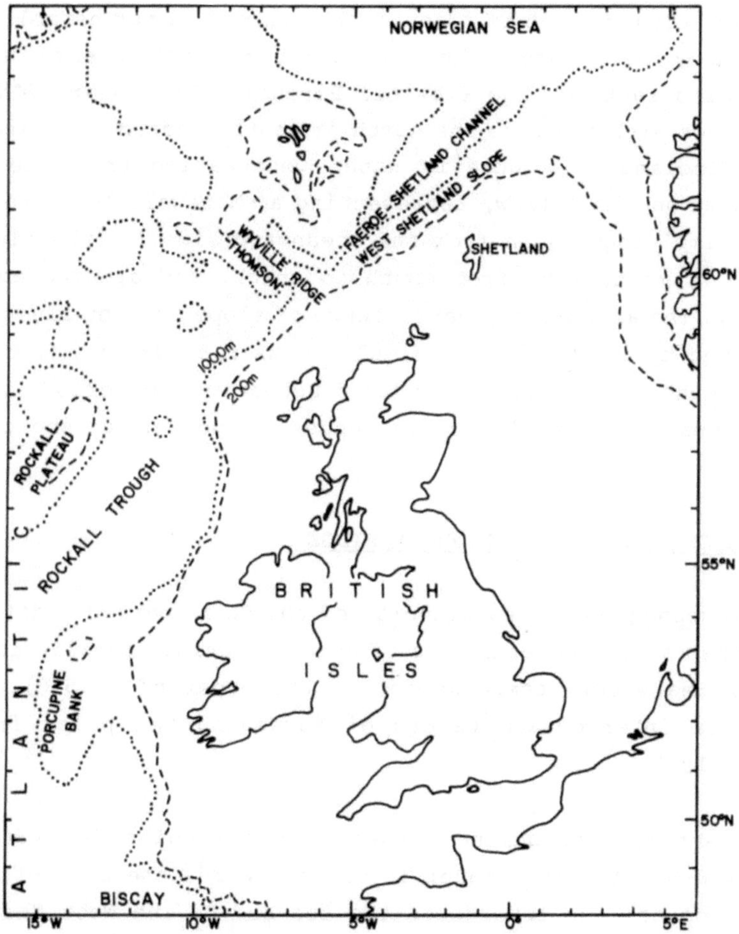

Figure 1: Locations referred to in text.

Mean currents are found closely following the depth contours from 47-1/2°N to 60°N along the continental slope. Speeds range from 3 to 30 cm s^{-1}(values exceeding 10 cm s^{-1} are mostly north of 57°N), but transport estimates are more consistent, between 1.2 and 2.2 × $10^6 m^3 s^{-1}$ inshore of the 2000 m depth contour. The mean currents are essentially barotropic (albeit decreasing with depth) and are confined to the continental slope width 20 to 50 km, being strongest usually in 500 to 1000 m water depth. Below tidal frequencies, the currents over the slope are relatively steady in that they rarely reverse, unlike the local meteorological forcing and associated currents over the shelf. However, there is some evidence of a winter maximum in transports.

Transports appear to persist over 1000 km or more. However, true continuity (the movement of identifiable water) is not established to the same extent by the few drogue tracks available. In this sense, with current moorings also lacking, the flow off Porcupine Bank near 52°N might not be continuous with flows to the north or south. Indeed, salinity patterns north of Ireland show oceanic water penetrating onto the shelf and entrained in on-shelf flow, representing a shoreward loss from the slope currents, implying a gain from the oceanward side (if the slope current transport is maintained from south to north) and arguing against true continuity. The along-slope persistence of slope current fluctuations can be estimated—up to 500 km—from simultaneous records only in the northern Rockall Trough. This estimate is for fluctuations of period exceeding two days or actually on the shelf; coherences are poorer for shorter periods.

NORTHEAST OF THE WYVILLE-THOMSON RIDGE

The oceanographic regime north of the UK is markedly different from that further south. Water masses of Atlantic origin overlie the deep cold Norwegian Sea water (temperatures on the order of -0.5°C). Sandwiched between are intermediate waters of Norwegian Sea polar front origin (Meincke, 1978).

The strong thermocline between the Norwegian Sea water and the typically 8-9°C Atlantic water is domed towards the axis of the Faeroe-Shetland Channel and intersects the sea bed on the West Shetland Slope at around 500 m. The main core of the warm saline Atlantic water is confined between the thermocline and the continental slope. This picture is confirmed by classical hydrographic and CTD stations and by satellite infrared imagery of sea surface temperatures.

There have been many estimates of the volume transport of water masses in the Faeroe-Shetland Channel, but almost all suffer from the usual uncertainties of the assumption of a reference level of no motion. Where direct measurements are available, they are for the most part of short duration (< one month).

Estimates of the inflow of Atlantic water to the Norwegian Sea derived from geostrophy alone and with a variety of assumed reference levels give mean values on the order of $2 \times 10^6\ m^3s^{-1}$ and from these figures Aas (1977) suggests a seasonality related to the strength of the wind stress having maximum transports in winter. The analysis of Aas suffers from the scarcity of winter observations (27 estimates from September to April compared with 126 for the remaining four summer months). Estimates by

Dooley and Meincke (1981) based on the combination of geostrophic calculations with brief current measurements give an estimate for the transport of Atlantic water in the slope current of 4.1×10^6 $m^3 s^{-1}$ in September 1973.

More recent year-long direct measurements of currents over the continental slope northwest of Shetland in 1983-4 showed an annual mean northeastward (along-slope) transport of 7.8×10^6 m^3s^{-1} with highest values in January/February (~9.5×10^6) and minima in June/July/August (~5.0×10^6) (Gould, Loynes, and Backhaus, 1985). These figures probably represent an (perhaps 10%) overestimate of the Atlantic water transport since they include some water which is clearly colder than 8°C, but they are a good indication of the transport of the slope current. The mean transports from these direct measurements are close to the values estimated for the Atlantic inflow to the Norwegian Sea calculated on the basis of heat budget by Worthington (1976) and McCartney and Talley (1984).

The large increase in volume transport from $1\text{-}2 \times 10^6$ $m^3 s^{-1}$ to 7×10^6 m^3s^{-1} on crossing the Wyville-Thomson Ridge suggests that there is an effective mechanism for entraining the additional Atlantic water from the west into the boundary flow. The mechanism is not clear, but there is circumstantial evidence of a persistent northward flux of Atlantic water across the entire length of the Wyville-Thomson Ridge in the permanence of warm water over the ridge axis. The added inflow implies a mean current on the order of 6 cm s^{-1} across the ridge.

Besides the seasonal signal in the slope current transport in the Faeroe-Shetland Channel, there are shorter-period fluctuations with periods in the range from 2 to 20 days but with the most pronounced spectral peaks in the current records being near 5 and 14 days.

DISCUSSION

Various possibilities for currents along the northeast Atlantic boundary include a diffusive boundary layer as described by McCreary (1981) and the results of forcing by rectified tides, winds, freshwater runoff, or longshore density gradients. Discussion by Huthnance (1984, 1986) suggests that any slope current forced by tidal rectification is concentrated close to the shelf break and is only a small fraction of the total observed; freshwater runoff is also thought to be weak in this context. McCreary's (1981) boundary layer width of several hundred kilometers (the distance offshore from the eastern boundary for baroclinic Rossby wave decay by thermal diffusion) corresponds to the breadth of the

Rockall Trough and perhaps to the overall northwards drift there (Ellett et al, 1986) rather than to the slope current specifically described here. Winter wind-forcing is suggested by mean northerly currents found from November to April over several years just east of the shelf edge at 57°N (Ellett and Booth, 1983). Otherwise, the current is confined to the slope rather than the shelf, and its relatively steady character (with only rare reversals) suggests density forcing. The mechanism (northwards driving by a pressure gradient imbalance associated with less-dense lower-latitude waters "standing high") is discussed by Huthnance (1984, 1986).

A near-geostrophic balance in cross-slope momentum is expected for the nearly steady flow; friction is relatively weak. Hence the vertical structure corresponds to cross-slope density gradients, which are set by a balance between northward advection and cross-slope diffusion. Northward advection of warmer water is apparent in hydrographic sections and satellite infrared images (Booth and Ellett, 1983; Huthnance, 1986); the resulting baroclinicity tends to reinforce the northward flow at upper levels offshore, without apparently altering the flow's essentially barotropic character. Near-barotropic geostrophic flow also corresponds with the observed close following of depth contours, and hence, with approximate conservation of along-slope transport.

Along-slope persistence of identifiable water is a separate question, however. For example, barotropic and almost frictionless flow would conserve transport well, but would be consistent with large cross-slope diffusion, with water properties advected only short distances along the slope (Huthnance, 1986). Thus, the observations of a poleward current off Portugal (Ambar and Fiuza, 1987) and west of Britain do not establish continuity of a northeast Atlantic slope current in this sense.

Along-slope persistence of fluctuations in the current is another question again, but in theory (Huthnance, 1987) corresponds to the frictional decay distance of coastal-trapped waves, being also the along-shelf evolution scale for establishing the slope current's transport and form after a barrier (say). The eventual transport, balancing forcing and friction, evolves after the decay of coastal-trapped waves accommodating the different (barrier) geometry. Observed scales of 500 km or less for along-slope coherence suggest that second- and higher-mode coastal-trapped waves are significant in the slope current's evolution; the first mode decay distance is O(1000 km) (Huthnance, 1987) and is better related to coherence over the shelf (rather than the slope).

Stability of the slope current is suggested in Huthnance (1986).

The Wyville-Thomson Ridge apparently marks the beginning of a much larger slope current transport to the northeast. Evidence discussed by Huthnance (1986) suggests that the slope current from the south is partly deflected at depth along the Wyville-Thomson Ridge, but gains overall as a broad upper-level flow from Rockall Trough is deflected anticyclonically over the Ridge (conserving potential vorticity) to follow the West Shetland Slope. Recirculation in the Faeroe-Shetland Channel is also thought to contribute.

Plans for the next year or two include current moorings and drogue tracking, to investigate continuity of the slope current, from the UK south-west approaches (48°N) northwards to Porcupine Bank (52°N) and to the south in the Bay of Biscay. A more intensive cross-slope mooring array and acoustic doppler profiling is also planned, northwest of Scotland, to better define the slope current structure there for comparison with models.

ACKNOWLEDGEMENTS

The co-authors express appreciation to Dr. D.J. Ellett, Dunstaffnage Marine Research Laboratory, Scotland, for review and suggestions to this manuscript.

THE POLEWARD UNDERCURRENT ON THE EASTERN BOUNDARY OF THE SUBTROPICAL NORTH ATLANTIC

E.D. Barton
Marine Science Laboratories
University College of North Wales
Menai Bridge, Gwynedd LL59 SEY
United Kingdom

ABSTRACT

Evidence of a persistent poleward undercurrent has been reported frequently in studies off the African and Iberian coasts. Most direct observations originate from investigations into upwelling between 17 and 26°N. The undercurrent appears to be strongest at about 300 m depth, but extends down through the Central Water Mass layer to about 1000 m. Off the Iberian Peninsula, the Mediterranean Water layer is also moving poleward, while off Africa it appears to be spreading equatorward. The undercurrent has a width of less than 100 km and a mean speed of about 10 cm s^{-1} at the level of strongest flow. South of Cap Blanc (20°N) and off northern Portugal and Spain, poleward flow extends for at least part of the year to the sea surface to form a countercurrent. Many reports are from single moorings on the continental slope, and there have been few observations of the offshore and depth structure of the flow. Moreover, most observations have been over periods of only a few weeks, and so variability of the flow is poorly defined.

INTRODUCTION

A poleward undercurrent may be defined as a persistent, poleward flow of restricted width and thickness, which is bound to the continental slope, runs counter to the dominant regional circulation, and may at certain times and locations extend upwards to the sea surface. It is generally accepted that such a flow exists along the eastern boundary of the North Atlantic, but knowledge of the flow is sketchy. The arguments for its existence as a continuous entity are based upon relatively few direct current observations, some interpretations of temperature and salinity data, and a degree of speculation.

The locations of direct observations of an undercurrent (Figure 1) are concentrated in the area around 20°N off northwest Africa. Despite some twenty years of interest in the region, investigations have been limited, in general, to short term CTD surveys and current meter deployments of a few weeks duration. After the period of intensified observational effort in the northeast central Atlantic during the seventies, interest has waned partly as a result of the territorial disputes in the area.

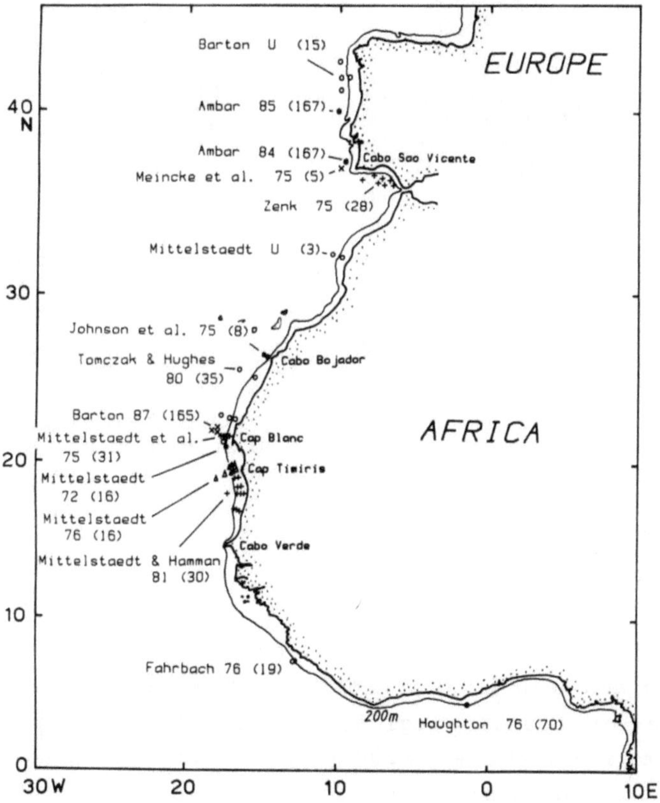

Figure 1: Locations of direct current observations referred to in the text. Author, year, and length of observation period in days are indicated 'U' signifies unpublished data.

The lack of long time series observations of the undercurrent is a major shortcoming because the eastern margin of the North Atlantic exhibits strong seasonality in response to the annual north-south shift of the Azores High. Coastal upwelling, driven by the Trade Winds, varies strongly with time of year throughout the region in question (Wooster et al., 1976). Estimates of mean monthly surface drift indicate northward flow at the surface during the absence of upwelling, both off Portugal and south of Cap Blanc (Figure 2). Moreover, significant interannual variability is apparent in observations of nearshore temperature off Cap Blanc between 1955 and 1982 (Arfi, 1987).

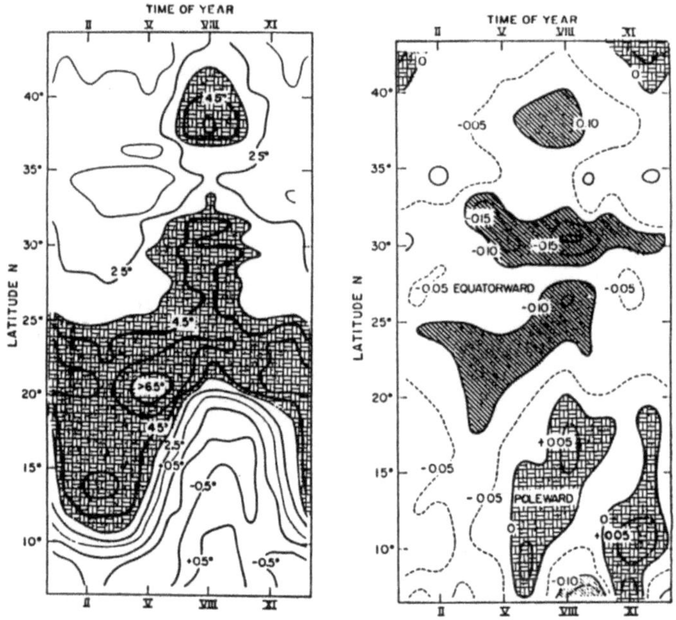

Figure 2: Monthly variation of surface zonal temperature anomaly and along-shore current calculated from wind stress on the eastern boundary of the subtropical North Atlantic (Wooster et al., 1976).

INDIRECT EVIDENCE

The first mention of an undercurrent per se off northwest Africa is by Wooster and Reid (1963), who stated that it appeared to be a common feature of eastern boundary regions. They noted that the frequent downward inclination of isopleths below 200 m depth towards the coast was indicative of a geostrophic northward current around that level. Montgomery (1938) produced charts of the salinity distribution on various sigma-t surfaces, which indicated northward flow between about 200 m and the sea surface along the slope south of Cap Blanc. Defant (1941) also deduced poleward flow at shallow levels along the African coast up to 20°N from geostrophic calculations. In neither case was there any indication, however, of a continuous flow along the eastern margin to the north of Cap Blanc.

As interest in upwelling off northwest Africa was burgeoning in the early seventies, a number of workers searched for evidence of an undercurrent in temperature-salinity data from extensive regional surveys. Tomczak (1973) and Hughes and Barton (1974) traced an anomaly of low salinity at about 300 m depth from Cap Vert, 15°N, to Cabo Bojador, 26°N (Figure 3). To the north of Cap Blanc, the dominant water mass in the layers above 700 m is the North Atlantic Central Water, but to the south is found the lower salinity and slightly cooler South Atlantic

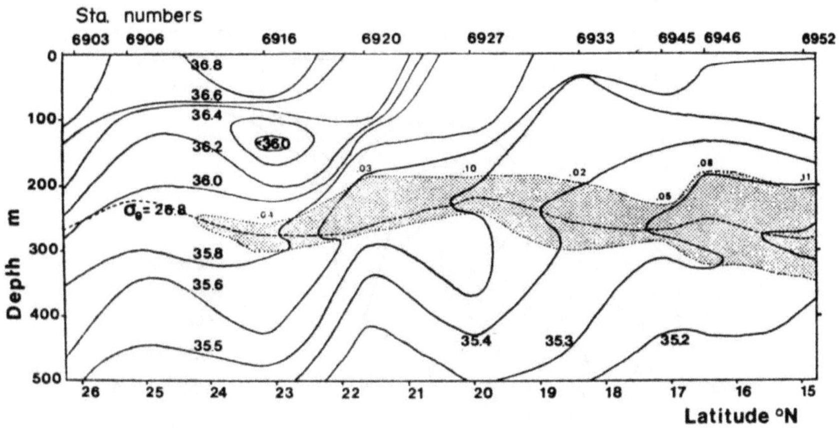

Figure 3: Along-slope section of salinity. The potential density surface $\sigma_\theta = 26.8$ is dashed, the salinity anomaly is shaded, and its magnitude is indicated above (Hughes and Barton, 1974).

Central Water. Lower salinity waters (and nutrient tracers, Gardner,1977) are apparently advected northward along the slope in the undercurrent. From the distribution of the salinity anomaly, it was concluded that the undercurrent was restricted to within less than 100 km of the coast and centered around 300 m depth. Beneath the core of the undercurrent, it seems that a slow northward drift exists to depths as great as 1000 m, where traces of Antarctic Intermediate Water have been identified (e.g., Fiuza and Halpern, 1982). Deeper still lies the presumed southward spreading Mediterranean Water layer centered at 1200 m.

Subsequent work by Tomczak and Hughes (1980) on mixing between the North and South Atlantic Central Water masses, clearly demonstrated a core of fresher water of southern origin being advected poleward along the slope at several latitudes. North of Cap Blanc, traces of southern water were found at depths down to at least 600 m (Figure 4). However, the same authors concluded that horizontal exchanges often resulted in the advection of isolated bodies of differing mixtures of the Central Waters by the undercurrent. In sections to the south of Cap Blanc, Hamann et al. (1981) showed that the southern water mass extended completely up to the sea surface.

This and other accumulated evidence, largely from water mass analysis and geostrophic calculations by Mittelstaedt (1972), Hughes and Barton (1974), Fraga (1974), Allain (1970), and other workers gave rise to an idealized circulation scheme shown in Figure 5 (Mittelstaedt, 1983). A permanent cyclonic eddy south of Cap Blanc produces a slope countercurrent

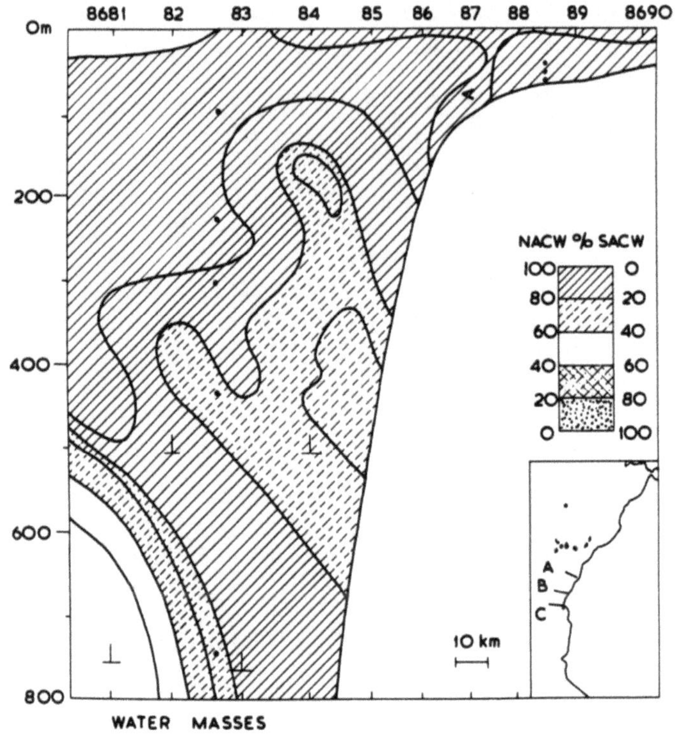

Figure 4: Percentage of North and South Atlantic Central Waters near 25°N (line A inset). The water masses are defined by conditions at 21°N and 25°N, respectively (Tomczak and Hughes, 1980).

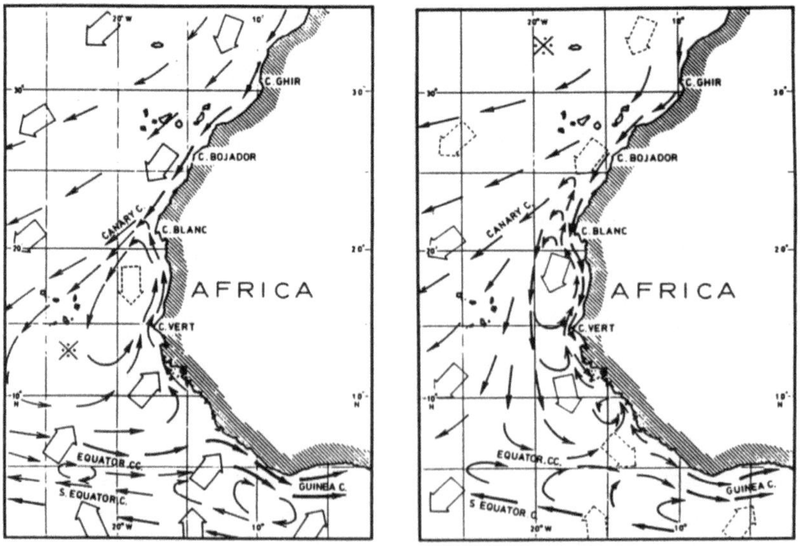

Figure 5: Summer (left) and winter (right) surface circulation. Dominant winds are shown as open arrows. Dashed arrows represent weaker, intermittent winds (Mittelstaedt, 1983).

throughout the upper layers. During summer, when the trade wind belt recedes northward and upwelling ceases, the counterflow extends inward to the coast; but during winter upwelling it is separated from shore by an equatorward coastal current. The main flow of the Canary Current separates from the coast near Cap Blanc. Evidence for this general scheme of surface flow can be found in Stramma's (1984) objective maps of annual mean surface geostrophic transport, although detail over the slope is not resolved by the analysis. At depths around 300 m, it is supposed that the slope undercurrent continues northward despite the recirculation southward of the surface waters.

DIRECT OBSERVATIONS

Near and south of Cap Blanc, the above idealization indicated that during the upwelling season (winter) three bands of meridional current could be differentiated. A section of 24 current meters on six moorings slightly south of Cap Blanc revealed the depth and offshore structure of the flow during three weeks in February-March 1972 (Mittelstaedt, 1976). Equatorward flow was evident both nearshore over the continental shelf and offshore in the southward drift of the Canary Current (Figure 6). The maximum of poleward flow appeared at about 200 m, but might have been shallower in the unsampled surface layers. Poleward flow was confined within 100 km of the shelf break and extended down through the Central Water layer to about 700 m. Similar conditions were seen in February 1975

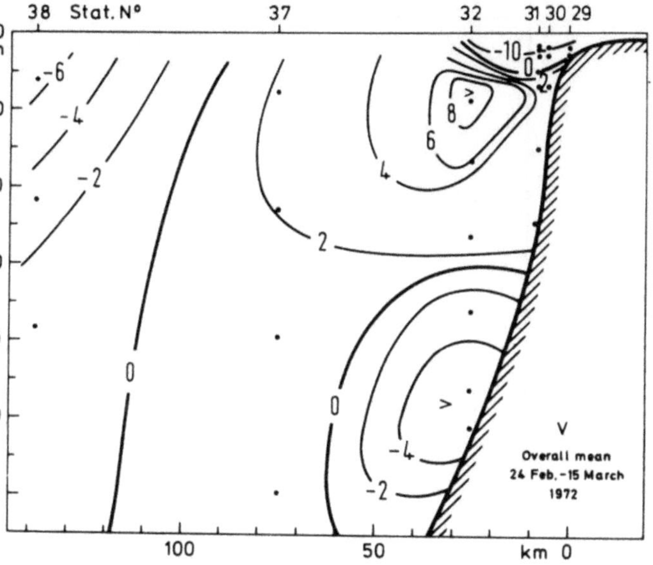

Figure 6: Mean along-shore flow at 19°30'N, February 24 to March 15, 1972. Solid dots mark current meters (Mittelstaedt, 1976).

(Barton and Hughes, 1982) within 100 km of the shelf in the area of 23°N just north of Cap Blanc.

Direct evidence of the along-shore confluence in the upper layers near Cap Blanc was reported by Barton (1987). Currents were observed at 22°N in the upper 350 m layer between November 1981 and April 1982 at three moorings less than 30 km apart and over 100 km offshore. The mean velocities at all sampling depths indicated an offshore diversion of weak flows converging from the north and south (Figure 7). The absence of observations at deeper levels and closer to the slope left unresolved the relationship between the supposed northward continuation of the slope undercurrent and the observed near-surface recirculation.

The JOINT-I current section, also near 22°N, was observed between February and April 1974 within 100 km of shore. A poleward flow restricted to the layers below 100 m depth within 30 km of the slope was observed (Barton et al., 1976). The maximum velocity reached a value of about $10\,\mathrm{cm\,s^{-1}}$ in the core of the undercurrent centered near 200-300 m (Figure 8). Further offshore, the mean along-shore flow was weak and actually poleward in a shallow eddy-like feature. On the continental shelf, strong equatorward flow representative of the coastal upwelling jet was seen. A similar picture was reported for eight days observations off Cabo Bojador, 26°N in August 1972 (Johnson et al., 1975); but strong

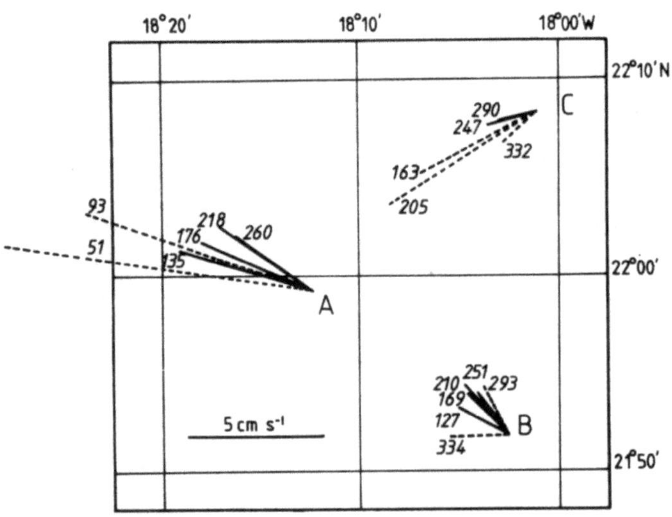

Figure 7: Mean current vectors November 1981-April 1982. Incomplete records are indicated by broken lines. Observation depths in meters are shown (Barton, 1987).

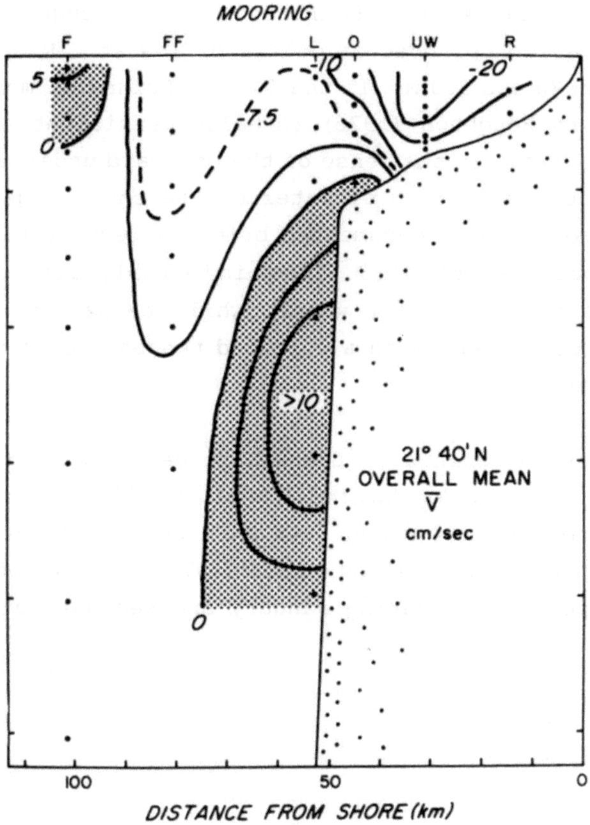

Figure 8: Mean along-shore flow at 21°40'N, February to April 1974. Solid dots mark current meter depths (Barton et al., 1976).

equatorward winds had enhanced the equatorward flow, resulting in a weak undercurrent centered at 400 m within 10 km of the slope.

In general, it has been found off Africa that poleward flow does not extend onto the continental shelf during the upwelling season, unlike off Oregon (Huyer, 1976). At times, however, this does occur when the equatorward winds weaken sufficiently. Mittelstaedt and Hamann (1981) reported that an array of moorings on the slope and shelf south of Cap Blanc between 17 and 19°N revealed an interlude of poleward flow at all but one of twenty-six current meters during a wind relaxation (but not reversal). This was also briefly evident during the JOINT-I experiment at times of weak equatorward winds (Mittelstaedt et al., 1975).

Outside the region of 17 to 26°N, there have been even fewer reports of the undercurrent off Africa. Fahrbach's (1976) roughly three weeks of observations revealed episodes of poleward flow at 7°N off Sierra Leone at depths between 80 and 900 m. In the Gulf of Guinea, almost three months of observations by Houghton (1976) revealed persistent subsurface flow towards the west, i.e., in the sense of the poleward undercurrent, beneath a shallow thermocline in 87 m of water on the shelf edge. More to the north, Mittelstaedt (1987) documented brief current observations during 60 hours off Morocco in April 1983. These indicated poleward flow at depths below about 500 m in 730 m of water, while temperature and salinity properties were consistent with a poleward tendency down to the depth of the Mediterranean Water layer.

Off the Iberian Peninsula, there have been a number of reports of poleward flow at various depths. Meincke et al. (1975) observed northward flow at levels between 234 m and 2046 m in 2450 m of water 100 km west of Cabo Sao Vicente, while Swallow et al. (1977) found a general northward flow in the upper 1500 m during January and February at a distance of

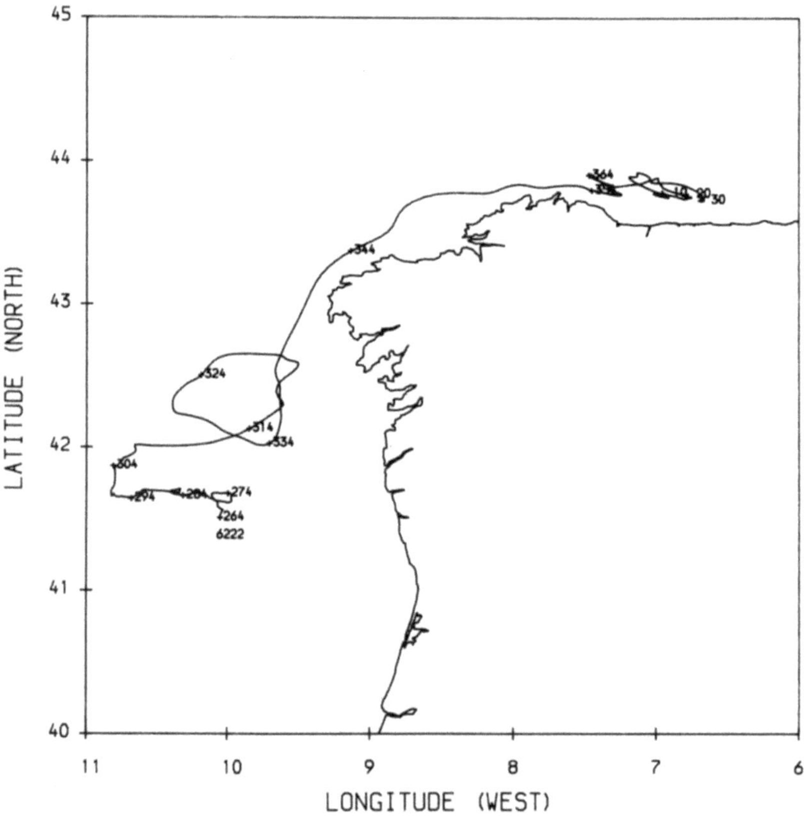

Figure 9: Trajectory of a drifter drogued at 5 m, September 1986 to February 1987. Positions every ten calendar days are marked.

95 km offshore. Closer to shore, two six-month-long series of observations were executed off Portugal in 1100 m of water by Ambar (1984 and 1985). At latitude 37°20′N, a mean flow of 10 cm s^{-1} towards the north was found over the period December to May at 750 m. The speed of the along-shore northward component decreased with decreasing depth to almost zero at 200 m. At 40°N, a more uniform mean northward flow in excess of 3 cm s^{-1} was found between 200 and 750 m between July and January. Recent current meter observations off northern Portugal and Spain showed weak poleward flow between 50 and 600 m over the slope in September 1986. Argos drifters, drogued at 5 m and released at the moorings, moved poleward along the slope around Cape Finisterre into the Bay of Biscay (Figure 9).

The eastern boundary of the North Atlantic is complicated by the physical discontinuity where the Mediterranean Outflow occurs. The warm, saline outflow from the Strait of Gibraltar itself constitutes an undercurrent (e.g., Thorpe, 1976), which is constrained to hug the northern margin of the Gulf of Cadiz. The path of the outflow through the Gulf has been investigated by several workers (e.g., Ambar and Howe (1979); Zenk (1975); Madelain (1970)). These workers have documented the splitting of the outflow into several cores which flow west at different depths to turn northward around Cabo Sao Vicente. Ambar (1983) has shown that another shallow vein of Mediterranean Water was identifiable at depths as shallow as 400 m off southern Portugal, while Fiuza (1982) has detected the same water mass on the outer shelf. The impact of the Mediterranean Outflow on the presumed northward continuation of the poleward undercurrent along the African coast has not been considered.

DISCUSSION

There are a number of unanswered questions about the nature of the undercurrent and its role. For example, it is unclear whether one can meaningfully distinguish between an undercurrent and a countercurrent. In certain locations, a counterflow adjacent to the coast may appear to be topographically generated, while in others it may appear as the temporary surfacing of a permanent undercurrent during a period of poleward wind stress. One might also ask whether the enhanced poleward flow observed bound to the continental slope further north in the subpolar gyre represents the same feature. It seems likely that a combination of different forcings is responsible for what appears as a single phenomenon.

The observations of an undercurrent reviewed here are spread quite widely along almost the entire eastern boundary of the subtropical gyre, but there are large gaps between sightings. Tentative conclusions are that the speed of the undercurrent core is about 10 cm s^{-1}, that it has a width of 30 to 100 km, and that it is concentrated at about 300 m, but extends downwards through the Central Water and Antarctic Intermediate Water layers. Off Africa, the lower limit of northward motion is the Mediterranean Water layer, but off Portugal northward flow may extend much deeper. However, many reports are based on single moorings located in various water depths; and the offshore and deep structure of the flow over the slope has been determined in only a few cases.

With few exceptions, observations have been limited to several weeks duration. Consequently, little is known of the temporal variability of the undercurrent. In particular, the seasonal and interannual fluctuations remain essentially unsampled, and even the several day time scale is poorly sampled. This could well result in misunderstanding of basic features of the flow regime. One example might be the apparent northward deepening of the undercurrent core, which is reported near the surface at Cap Blanc at 200 m near 22°N, at 400 m near 26°N, and below 500 m near 32°N. These observations were made in different years, in different months, and over short periods, and therefore may be unrepresentative.

Finally, the continuity of the flow is unknown. The effectiveness of the undercurrent as a transporter of, say nutrients or fish larvae, is undetermined. Although the tracing of water mass anomalies has been used as evidence for the existence of the undercurrent, it has been shown that these do not uniquely identify the flow at any particular time and place. Horizontal exchange with surrounding waters is evident from the presence of isolated bodies of anomalous water within and without the undercurrent. Moreover, wherever the poleward flow is shallow enough, vertical exchange by upwelling onto the shelf may occur.

While further study of the existing data base may well advance understanding of the undercurrent, there is a clear need for systematic, long term observations of the eastern boundary of the North Atlantic.

ACKNOWLEDGEMENTS

This research was supported by Natural Environment Research Council Grant GR3/5872. The drifters were provided courtesy of the United States Office of Naval Research.

From: M. Tomczak, The University of Sydney, Australia

On: Review and commentary to paper **THE POLEWARD UNDERCURRENT ON THE EASTERN BOUNDARY OF THE SUBTROPICAL NORTH ATLANTIC,** by E.D. Barton

This review of the poleward undercurrent along the continental slope of northwest Africa and southern Europe shows rather convincingly how little is known about this prominent feature of the Canary Current upwelling system, despite the high level of field activity in the area during the seventies and early eighties. I pointed out some years ago that most of the field projects were only loosely connected with the systematic investigation proposed by the Intergovernment Oceanographic Commission (Tomczak, 1979) and did not result in a comprehensive data set. The review by E.D. Barton does its best to extract what can be learned about the undercurrent from rather disparate data. Hopefully, it will stimulate oceanographers to go back to the area and produce some long time series of currents and temperature. The poleward undercurrent of the eastern North Atlantic Ocean remains one of the permanent but highly variable features of the oceanic circulation which have not yet been covered by long-term current measurements.

The R.V. "Alexander von Humboldt" was used between 1970 and 1984 by oceanographers from the Institute for Marine Research at Rostock-Warnemünde to collect data in the Canary Current upwelling region in a very systematic way. The early cruises, from 1970 to 1974, all followed a standard station pattern and formed the basis for an investigation into seasonal upwelling variability (Wolf and Kaiser, 1978). The data set consists exclusively of hydrographic casts, taken along seven sections roughly perpendicular to the coastline, with stations spaced about 35 km apart (Figure 1). This does not make it particularly suitable for the study of a feature believed to be between 30 km and 100 km wide, which may explain why Barton's review ignores the data set completely. However, following his suggestion that "further study of the existing data base may well advance understanding of the undercurrent" I extracted from the R.V. "Alexander von Humboldt" data the density difference between adjoining stations next to the continental shelf at 100 m, 200 m, and 300 m depth (at deeper levels differences were too small to give reliable results). The station pairs used are shown in Figure 1. Linear interpolation between the nearest samples was used to determine σ_t at the selected depth levels. The sign of the density difference was graphed as a function of latitude and month, with no regard to the year when the data were obtained. This should give a reasonable pattern if interannual variations are small

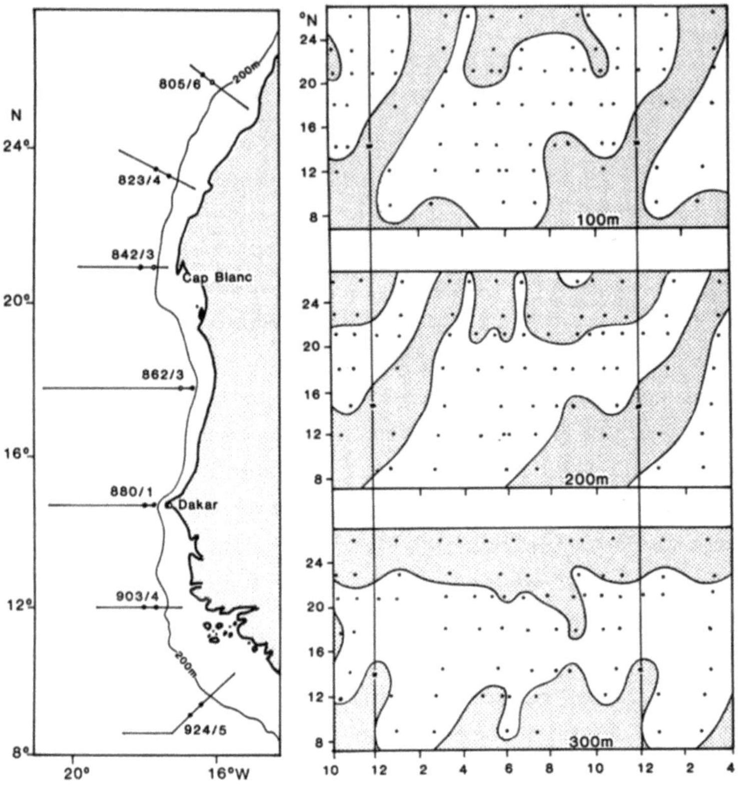

Figure 1: Standard sections of R.V. "Alexander von Humboldt" cruises 1970-1974 and station pairs used for the calculation of the zonal density gradient (left); sign of the zonal density gradient at three depth levels as a function of latitude and season (right). The area with positive gradients (isopycnals slope downwards towards the coast) is indicative of the presence of an undercurrent and is shaded. The areas outside the vertical lines are repeats of the data inside.

compared to seasonal variations. Support for this assumption was presented by Schemainda et al. (1975).

Several points should be kept in mind when looking at the graph. Interannual fluctuations are neglected. The distance of the station pairs from the continental slope is not uniform. The distance between stations in some pairs may not be sufficient to resolve the undercurrent. Consequently, while the shaded area in Figure 1 is strong evidence for the presence of the undercurrent, the current may also be present at times where the analysis does not show it. The points that can be made from the graph are:

(a) The undercurrent is generally deeper north of Cap Blanc (21°N) where it was found during all months at 300 m depth and in most cruises at 200 m depth. This is in agreement with Barton's conclusions.

(b) South of 14°S, the undercurrent is distributed more uniformly throughout the 100-300 m layer but appears to be varying in strength, being strongest in late summer when upwelling is confined to the region north of 18°N (Schemainda et al., 1975).

Note however that data in the southern part are sparse and other interpretations are possible.

(c) Coinciding with the southward progression of the upwelling regime during autumn, the strongest undercurrent signal is observed to move northward, being confined to the upper 200 meters. The same northward movement can be seen in the analysis of alongshore surface flow by Wooster and Reid (1976; figure 2 in Barton's review), but it appears there about four months earlier than in this analysis for the 100-200 m depth layer and is confined to the region south of 18°N. This may indicate a vertical migration of the poleward current in response to the southward progression of the upwelling.

These results are as tentative as Barton's conclusions, but the combined information will hopefully stimulate new oceanographic activity in the region and lead to a more cohesive data base.

I thank Eberhard Hagen for helpful input to these comments and Neil Trenaman for data processing and analysis.

THE SUBSURFACE CIRCULATION ALONG THE MOROCCAN SLOPE

Ekkehard Mittelstaedt
Deutsches Hydrographisches Institut
Postfach 220
2000 Hamburg 4 (FRG)

ABSTRACT

CTD data and short-term current data collected on METEOR Cruise No. 64 in January and April 1983 off Morocco are used to discuss the wind-induced upwelling situation at 32°N during the time of observation and the subsurface circulation off the Moroccan coast.

During the station work at 32°N, the vertical cross-sections of temperature and salinity clearly suggest upwelling during January and April. The short current records over about 60 hours in April indicate offshore flow in the surface layer. At depths around 250 m, the water moves onshore toward the shelf compensating the upwelling. The maximum depth from where the water ascends up the slope onto the shelf is deeper than 300 m.

Below 500 m depth the observed currents have a north component of a few centimeters per second extending down to the upper boundary of the Mediterranean water at about 1000 m depth. This poleward-going undercurrent is the deep part of the same subsurface flow which occurs along the continental slope off Mauritania. But in the south, the undercurrent extends from depths around 1000 m upwards to the lower boundary of the wind-driven surface layer or even to the surface. The deeper part of the undercurrent (approximately between 700 m and 1000 m) carries Antarctic Intermediate Water (AIW) toward the north. The upper part has a subsurface velocity maximum between 100 and 400 m depth and is characterized by substantial admixtures of South Atlantic Central Water (SACW), off Senegal and Mauritania from about 14°N to 21°N. This flow represents the major subsurface source for the waters upwelling along the shelfbreak and the coast.

Further north, between 21°N and 24°N, the upper undercurrent hits the Canary Current flowing toward the southwest at these latitudes. The convergence produces the frontal zone between NACW and SACW and forces the poleward-going under/counter-current within the upper 400 m to flow parallel to the Canary Current and the frontal zone toward the southwest.

The deeper part of the undercurrent merges below the Canary Current and continues its way toward the north. The transport of AIW along the continental slope can be identified at depths between 800 and 900 m as far north as the Canary Islands. At these latitudes the identity of AIW vanishes due to mixing, but the deep northward-going flow still occurs north of this latitude at 32°N.

INTRODUCTION

During her Cruise No. 64, from January to April 1983, RV METEOR investigated, among others, the waters off Morocco. The station work comprised oceanographical, chemical, and biological observations. The investigations concentrated along a section across the Moroccan continental slope at about 32°N (Figure 1). This section was surveyed twice: the first time in January, the second time in April 1983. STD-data and short-term current measurements at 32°N from this cruise, as well as other information, are used here:

- to describe the actual upwelling situation in January and April 1983, and
- to discuss the indications of a poleward-going undercurrent off Morocco.

Weikert (1986) pursues the same objective using certain mesopelagic species of zooplankton sampled on METEOR Cruise No. 64 as a tracer for the undercurrent.

OBSERVATIONS

Methods

Temperature and salinity distributions are measured by means of a CTD-probe (bathysonde). The data are adjusted to reversing thermometer-temperature readings and bottle data (for salinity) taken with the CTD-probe at every station. Two simultaneous moorings (5 to 8 April) provide short-term current information, recorded by means of Aanderaa instruments.

Hydrographic Conditions

The vertical sections of temperature and salinity at 32°N suggest upwelling over the upper slope and the shelf in January, as well as in

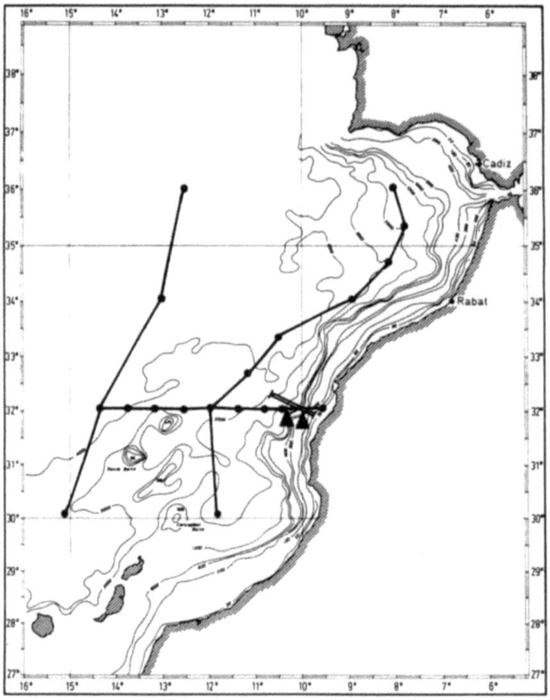

Figure 1: Stations and sections carried out by RV METEOR during winter 1983 and 1973. The double bar marks the section surveyed during January and April 1983. The two triangles give the positions of the short-term current meter moorings from 5 to 8 April 1983. The black dots and the thick lines represent a cross-section along 32°N and two meridional sections surveyed during February 1973 in the framework of CINECA.

April (Figure 2). The lower figure, showing April's near-bottom temperature at the shelf break, indicates that the upwelling has been stronger at this time because of stronger northeast winds before and during the observations. During the measurements in April, the winds blew from the NNE with a strength of 6 Beaufort (22 to 25 knots). The water temperature of the near-surface layer over the shelf corresponds to the temperatures at a depth of about 150 m, 20 to 30 nautical miles offshore of the shelf break. The near-bottom water temperature and salinity over the outer shelf compares with that one at a depth of about 200 m (in January) to 300 m (in April). Taking in account diffusion while the water ascends onto the shelf, the maximum upwelling depth is somewhat deeper than 300 m.

Due to horizontal advection, the salinity of the upper layer is higher over the shelf in April than in January. The temperature/salinity characteristics of the upwelled water agree fairly well with North Atlantic Central Water (NACW).

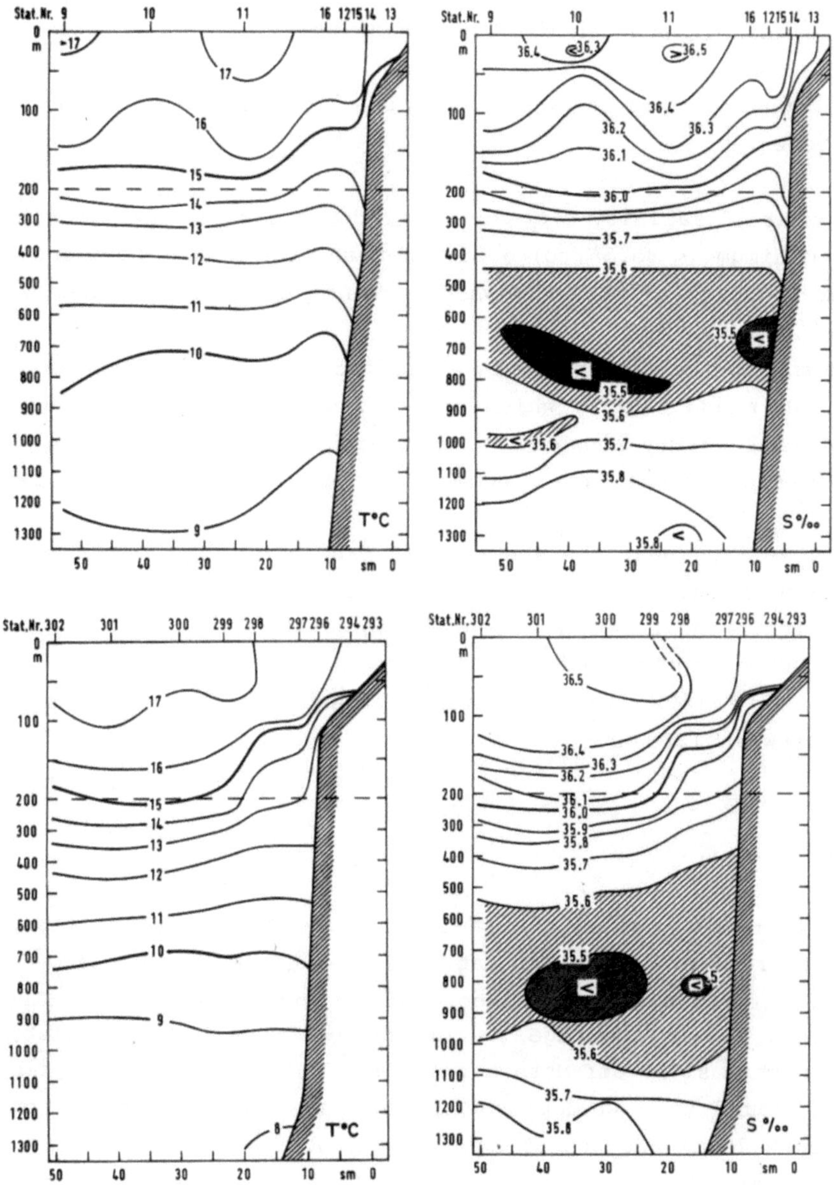

Figure 2: Sections of temperature and salinity at about 32°N from 15 to 16 January (above) and 5 to 6 April (below) 1983. The salinity minimum is hatched.

Below the shelf break the isotherms tend to incline downwards toward the slope instead of upwards, as in the layer above, due to upwelling. The down-warping within a narrow boundary zone along the slope is especially pronounced during January, when upwelling is weaker than it is in April. In January the salinity distribution also shows downtilted isohalines at the slope between 200 m and 600 m depth, but not in April when the signal is weak in temperature as well.

The downwarping of the isolines is the signature of a poleward-going undercurrent.

In general, salinity decreases with depth down to about 800 m. The salinity minimum (< 35.5 $^{o}/oo$) occurs between 600 m and 900 m and forms the lower boundary of the NACW which extends up to the surface layer.

Underneath the NACW there is a distinct salinity maximum at depths between 1200 m and 1300 m, which signalizes the Mediterranean outflow. A hydrographic survey along 32°N ten years earlier, February 13, 1973, exhibits basically the same phenomena close to the slope (Figure 3):

- upwelling near the shelf break from a depth of about 200 m,
- a vertical "divergence" of the isolines close to the slope within the layer between 400 m and 600 m (in salinity down to 1000 m depth), and
- a core layer with maximum influence of Mediterranean Water between 1200 and 1300 m depth.

Currents

Two moorings, one on the shelf and the other at the slope, were supposed to record the currents at 32°N over three months, from January to April. Unfortunately, we discovered in April that the moorings had disappeared. To our knowledge, no current measurements are available from the continental slope off Morocco, so far. In order to get at least some data from around 32°N, we deployed another two moorings, one on the shelf (water depth 70 m) and one at the slope (water depth 730 m) for short-term measurements. The moorings stayed out from April 5 to 8, over 56 to 60 hours, while METEOR did station work. They consisted of an instrument leg with subsurface buoyancy, ground line, buoy line, and surface buoy. The instruments were Aanderaa current meters; the sampling interval was five minutes.

RESULTS

At the slope the records are dominated by the semi-diurnal tidal currents (Figure 4). Their amplitudes are highest near the bottom and in the near-surface layer. In both layers, maximum speeds in connection with

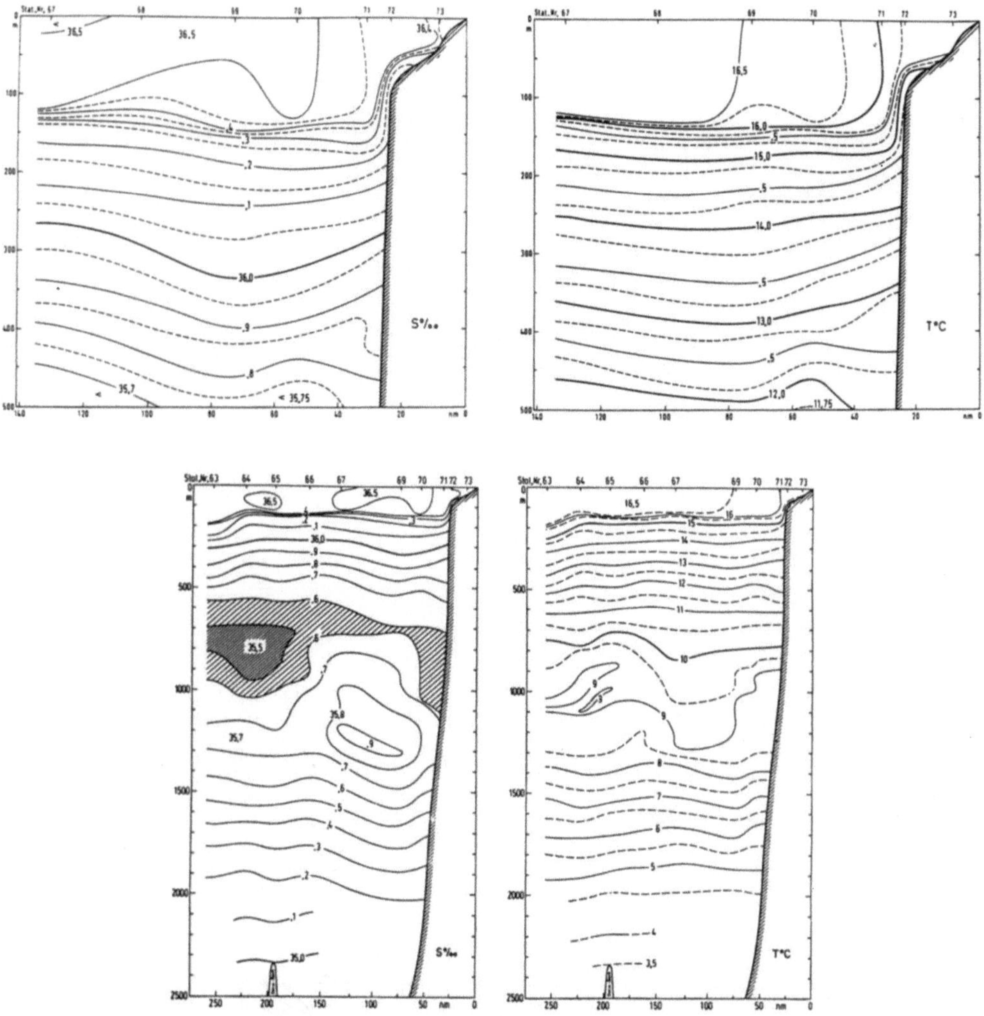

Figure 3: Sections of salinity and temperature at 32°N from 12 to 14 February 1973 (Huber, Mittelstaedt, Weichart, 1977). Distributions within the upper 500 m near the shelf break (above). Total sections (below). The salinity minimum is hatched.

the tidal currents exceed 40 cm s^{-1} (hourly mean). The tidal speed variations of the two layers are opposite to each other. While in the surface layer, the semi-diurnal variations of current directions are covered by a fairly strong mean flow, the records 20 m above the bottom exhibit strong periodical changes of current directions. At intermediate layers, the tidal current speeds are distinctly weaker than within both boundary layers near the surface and the bottom.

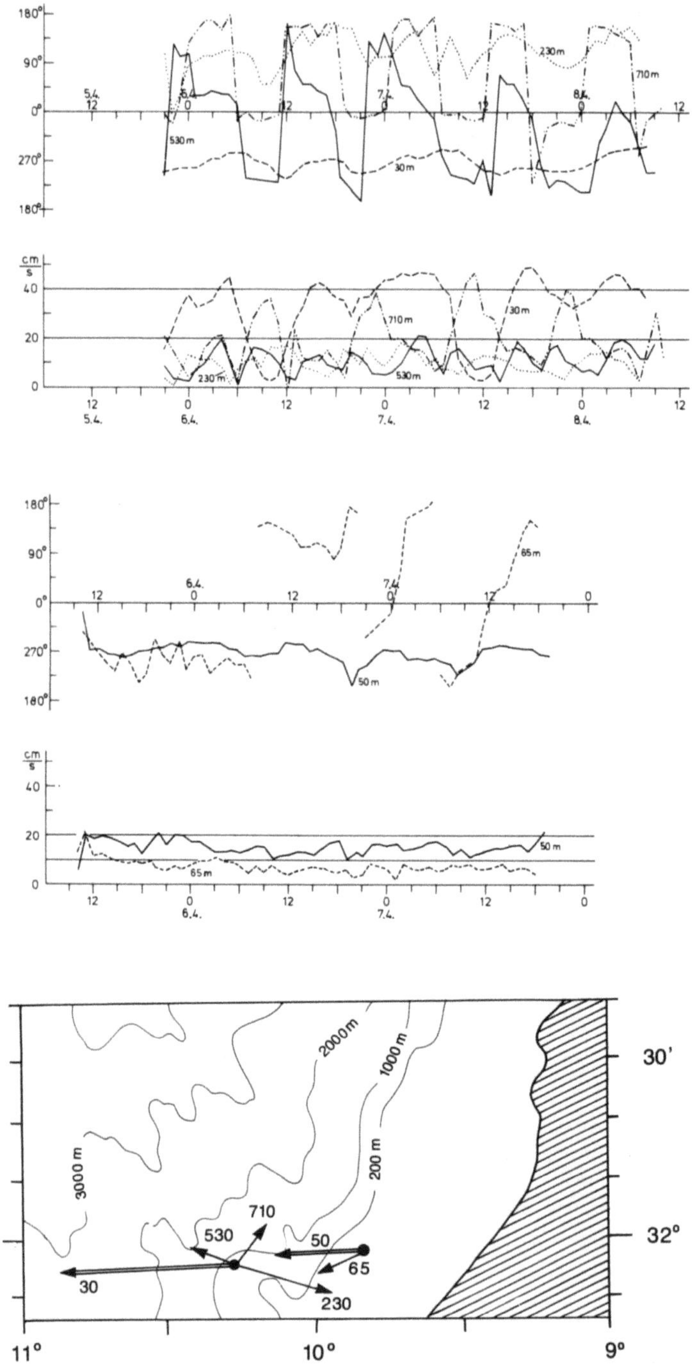

Figure 4: Current data from 5 to 8 April 1983: a.) Time series of direction and velocity (hourly means) over the slope at the observation depths of 30 m, 230 m, 530 m, and 710 m. b.) Time series of direction and velocity (hourly means) over the shelf at the observation depths of 50 m and 65 m. c.) Mean flow averaged over the observation period. Double arrows mark velocities above 10 cm s^{-1} (for numbers see Table 1).

Table 1
Mean Flow from April 5 to 8, 1983

	Observation Depth (m)	Flow Direction	Velocity (cm s^{-1})
Shelf Mooring	50	267°	15
(water depth: 70 m)	65	239°	4
Slope Mooring	30	268°	31
(water depth: 730 m)	230	115°	8
	530	290°	3
	710	34°	4

Within the lower layer over the shelf, tidal fluctuations are relatively small.

This qualitative description agrees with a number of tidal current analyses off Northwest Africa further south (Fahrbach, 1976; Horn et al., 1976; Gordon, 1978; Mittelstaedt et al., 1980).

In this report our interest concentrates on the mean water motions. The flow averaged over the observation period is given in Table 1 and Figure 4C.

The strong offshore flow near the surface indicates a pronounced upwelling situation during the period of observation, which is supported by the hydrographic section (Figure 2) from the same time.

The eastward-going and onshore current, as observed at the depth of 230 m, is most likely the subsurface compensation flow feeding the upwelling at the shelf break from below. At 530 m the flow is westward with a slight north component, and at 710 m the water movements are directed toward the NE with velocities around 4 cms^{-1}. We assume a northward deep flow within the layer, say, between 500 m and 1000 m along the continental slope off Morocco, to be a characteristic feature and not just a random event during our observations (see discussion below).

With this assumption, the vertical structure of the subsurface flow along the continental slope at 32°N can be described as follows:

1. A flow toward southerly directions between 1000 m and 1400 m depth (Mediterranean Water). These deep water movements are basin-wide and not restricted to the slope region.
2. A flow toward northerly directions between 500 m and 1000 m depth. This flow is assumed to be a boundary current with an intensification the closer one comes to the slope from offshore.
3. A flow toward southerly directions with the general Canary Current offshore and the wind driven shelf circulation.

DISCUSSION

From the METEOR cruise in the framework of CINECA during February 1973, two north-south sections of temperature and salinity are shown in Figure 5. One section runs about 250 nautical miles offshore the continental slope; the other is further inshore and follows approximately the water depths between 2000 m and 3000 m.

The salinity distributions from winter 1973 (Figure 5) and 1983 resemble each other very much at depths. The comparatively weak vertical gradient within the layer from 600 m to 1000 m suggests intense mixing between the southward flowing Mediterranean Water (MW) and the poleward moving water masses thereabove. The tongue shaped advective/diffusive spreading of the MW produces a salinity minimum (S_{min}) at its upper boundary.

From south to north the minimum salinity increases. The thickness of the S_{min}-layer and its mean depth decreases as a consequence of the MW merging underneath the NACW. At the lower latitude the S_{min} (between 700 m and 900 m depth) results, according to Wüst (1935), from admixtures of Antarctic Intermediate Water (AIW). Farther north the S_{min} weakens, and it becomes more and more difficult to associate it with influences of the AIW. A conspicuous northward orientation of the isolines within the S_{min}-layer offshore the slope, however, suggests a northward transport of AIW concentrating as boundary phenomenon along the slope (Le Floch, 1973, 1974; Fiuza and Halpern, 1982). Great parts of this water mass come with the tropical eastward flows to the eastern margin where it contributes to the deep poleward flow at depths between 700 m and 900 m.

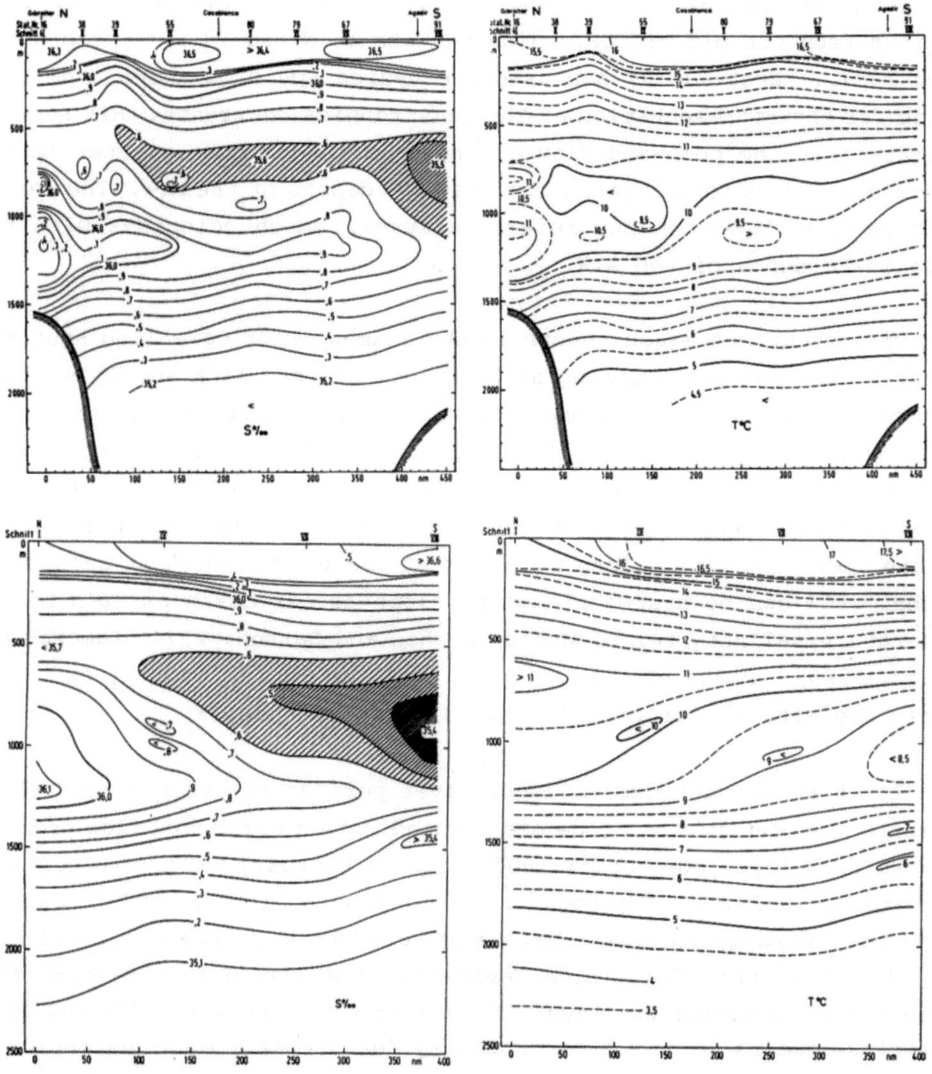

Figure 5: Meridional sections of salinity and temperature between 36°N and 30°N from February 1973 along the continental slope (above) and about 250 nautical miles offshore (Huber, Mittelstaedt, and Weichart, 1977). The salinity minimum is hatched.

Le Floch (1973, 1974) takes the view that weak admixtures of AIW are still perceivable between the Canary Islands and the African continent at about 29°N. Further north he finds that the influences of AIW disappear due to mixing with adjacent water masses. Off Mauritania, between 19°N and 22°N, Weichart (1974) finds at depths between 600 m and 1000 m maximum values of phosphate and nitrate in spring 1968. The values he gives are indicative for AIW as can be seen, for example, from the GEOSECS section (Bainbridge, 1983).

The poleward-going flow along the slope also emerges in the vertical sections from the old METEOR cruise between 14°N and 19°N (in February/ March 1927). Wüst (1957) has used the hydrographic data set from this cruise for the representation of the geostrophical currents. His results indicate that the poleward flow at these latitudes concentrates in the subsurface and surface layer, but not at the depths of the AIW. The reason might be his choice of the reference level, which coincides with the approximate depth of the AIW.

Direct current measurements in the framework of upwelling studies, during winter and early spring at latitudes between 16°N and 22°N, show a concentration of a narrow poleward undercurrent within the upper 500 m along the slope (e.g., Mittelstaedt, 1976, 1983) which may also surface. The undercurrent carries substantial admixtures of South Atlantic Central Water (SACW) toward the north, and represents the source for the upwelling water along the coast at these latitudes. At subsurface depths, SACW can be traced along the slope as far north as 26°N by its temperature/salinity relationship (Hughes and Barton, 1974; Tomczak and Hughes, 1980). Parts of the SACW come from the Gulf of Guinea, and other parts come with the tropical eastward flows as the AIW.

At 18°N, current data measured over periods of about 3 to 4 weeks offshore the continental slope clearly indicate a northward flow from the near surface layer down below 1000 m during February 1977 (Mittelstaedt and Hamann, 1981). Off Cap Bojador (26°N) Johnson et al. (1975) observed a weak northward-going flow below a depth of 400 m in summer 1972. Roemmich and Wunsch (1985) found clear evidence of a weak poleward-going flow along the slope within the layer of 400 m to 1000 m between 24°N and 28°N. This flow appears to be a broad boundary feature of a zonal extension of 1000 km. Within the layer of northward flow, below 700 m depth, oxygen is minimum and the phosphate content is maximum. In the south, the nutrient-rich water masses occur at about the same depths and are associated to AIW, while the oxygen minimum is to be found in the domain of SACW at subsurface depths within the upper 500 m (Wattenberg, 1939; Weichart, 1974; Kremling et al., 1984; Lenz et al., 1985).

Although the temperature/salinity characteristics at 32°N are not suitable to explain a deep water transport to the north, it might well be that there is still a persistent northward-going deep flow. The short-term measurements at 32°N described above seem to support this idea.

From the distribution of certain species of mesopelagic zooplankton samples during the same METEOR cruise, Weikert (1986) concludes that these species come with the poleward-going undercurrent all along the northwest

African continental slope from low latitudes up to the southwest European slope off Portugal.

The upper boundary of the flow component toward the north seems to be at depths around 500 m at 32°N. Assuming an upwelling depth of 300 to 400 m, this flow is probably no significant source for the water masses ascending up the continental slope with the wind-induced upwelling circulation.

But there might be processes such as strong internal tides or edge waves along the slope contributing to a turbulent vertical transport of deep water into shallow upwelling layers along the continental slope.

To our knowledge, there is no oceanographic evidence of the deep flow north of 32°N. Whether there is a continuous transition between the deep poleward-going flow along the African continental slope and the poleward-going flow, which occurs along the Iberian slope (see, for example, Ambar, 1982, 1985) is uncertain. The zooplankton samples by Weikert (1986) could be an indication of such a transition.

CONCLUSION

The data suggest a deep poleward-going flow within the layer of 500 m to 1000 m along the northwest African continental slope from the tropical North Atlantic to higher latitudes off Morocco. At its lower boundary this flow carries AIW toward the north along the slope. North of the Canary Islands, around 29°N, the identity of AIW seems to disappear due to mixing. But the undercurrent continues to flow toward the north at these depths.

South of 22°N the poleward-going undercurrent extends from below 1000 m depth up into the near-surface or even into the surface layer throughout the year. The upper part of the flow represents a narrow current band along the continental slope with a subsurface mean velocity maximum of 5 to 15 cm s^{-1} during the upwelling season in winter and spring. This "upwelling"-undercurrent transports substantial admixtures of SACW toward the north, which represents the major source for the upwelling water masses at lower latitudes.

On its way toward the north, the upper "floors" of the poleward-going undercurrent hit the Canary Current flowing toward the southwest at latitudes between 21°N and 24°N. This convergence produces the frontal zone between NACW and SACW. The Canary Current forces the upper poleward-

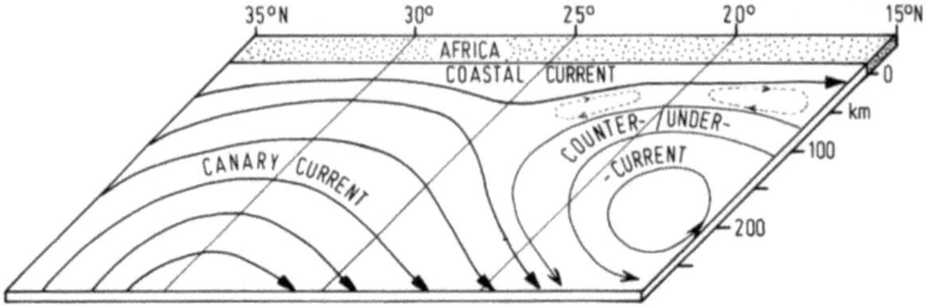

Figure 6: Schematic representation of the dominant near-surface circulation within the upper 200 m during winter and spring.

going undercurrent to turn into its own direction parallel to the frontal zone toward the southwest.

This situation is schematically represented for the winter and spring seasons in Figure 6. The circulation pattern in this figure is to describe the predominating flows within the upper 200 m. Apart from the southward flowing coastal current, this circulation scheme agrees with the representation of the general geostrophical flow (Stramma, 1984 a, b) in the eastern subtropical North Atlantic. The coarse data grid of 3° x 3° used by Stramma does not allow resolution of the narrow coastal current.

The deeper part of the undercurrent continues its way toward the north underneath the Canary Current within the layer of 400 to 1000 m depth. The small proportions of SACW which pass underneath the frontal zone at depth with the undercurrent disappear at higher latitudes around 26°N because of mixing with NACW. Due to its depth the poleward-going flow is not a major source for the upwelling waters within the surface layer off Morocco.

The question whether this flow is linked to the poleward-going subsurface flow along the Iberian Peninsula is still open and to be investigated.

ACKNOWLEDGEMENTS

The data collection off Morocco during METEOR Cruise No. 64 has been supported by the Deutsche Forschungsgemeinschaft.

From: Lothar Stromma, Inst. fur Meereskunde, Kiel University

On: Review and Commentary to paper **THE SUBSURFACE CIRCULATION ALONG THE MOROCCAN SLOPE**, by E. Mittelstaedt

Today there exists only a little knowledge regarding the oceanographic situation of the North Atlantic near Africa. This paper adds some nice additional information on the currents near the African coast. Of course we have to keep in mind that the main information is taken from some short-term observations and can't be regarded as the typical permanent flow field.

In section 2.2 the wind observations for January should also be presented for comparison with the April data. These data should be available from the ship measurements.

The schematic representation of the dominant near-surface circulation (Figure 6) is o.k., but I wonder whether it is not possible to give a schematic presentation of the subsurface circulation, which is of more interest here.

Although there is not much new information on upwelling in the ocean, this paper gives an example of how to interpret upwelling features from single sections. I also like the additional information on biological hints for the currents, which are often forgotten in oceanographic descriptions. Finally, the main point of this paper, the subsurface circulation near Africa, gives some new information on the currents in the area investigated.

POLEWARD MOTION IN THE BENGUELA AREA

G. Nelson
Sea Fisheries Research Institute,
Private Bag X2, Roggebaai 8012
South Africa

ABSTRACT

Available evidence of poleward and equatorward flow in the Benguela area is summarized and new evidence of poleward flow from direct current measurements between Cape Point (35.30°S) and Chamais Bay (27.80°S on the west coast of southern Africa is presented. Records were obtained across the shelf outwards to a depth of 3000 m. Except in the upper layers and at the shelf edge where equatorward baroclinic jets occur, these records invariably show time-averaged poleward flow along and at the base of the shelf and are indicative of cyclonic flow in the Cape Basin.

Closer to the coast, low frequency motions with periods of roughly three days display the characteristics of coastal-trapped waves. When averaged over several cycles, an ambient poleward flow of approximately 5 km d^{-1} is found for this inner shelf zone.

INTRODUCTION

In this article, poleward flow in the southern Benguela region is investigated in the broad sense of motion over the whole shelf and slope zones of southwestern Africa and over the Cape Basin. In a more restricted sense, various authors, for example Hart and Currie (1960) proposed a poleward undercurrent along the shelf edge which would compensate water removed from the 200 to 300 m level by perennial upwelling. Evidence of such a current was provided by the advection of low oxygen water as far south as the Orange River from Angola.

On a wider scale, dynamic topography (e.g., Defant, 1941) showed a substantial poleward movement of water below 2000 m in the southeastern Cape Basin. Further evidence for cyclonic motion over the Cape Basin has been obtained from marine geological surveys by Connary (1972), Embley and Morley (1980), Tucholke and Embley (1984), and Rogers (1987).

Since 1980, deployments of current meters by the Sea Fisheries Research Institute in deeper water over the shelf and continental rise of the southern Benguela region have revealed direct evidence of poleward flow. The zone of investigation is shown in Figure 1 together with geographical details, bottom topography and some of the deployment sites (see also Figure 6).

The existing data set does not allow seasonal features to be distinguished. The aim of current meter deployments was rather to identify permanent large-scale features such as bottom circulation and flow along the base of the shelf, or at the other extreme, to examine transient but repetitive phenomena such as coastal-trapped waves on time scales of a few days. The seasonal behavior of wind stress in the Benguela area has been discussed by Nelson and Hutchings (1983), Parrish et. al. (1983) and Kamstra (1985). The most significant feature is the presence of perennial southerly winds which are modulated by a three-day synoptic pressure wave, strongly so at Cape Point in the extreme south (34°S) and more weakly so as one progresses north to Möwe Point (approx 20°S). Seasonal changes in the ambient southerly wind occur as the permanent South Atlantic High pressure cell shifts or strengthens and weakens. Thus a seasonal change in Ekman transport occurs and consequently a seasonal change in a poleward compensation current would occur.

The use of current meters has revealed poleward motion near the bottom, extending to an as yet unknown height above the sea floor, but the interface between the deeper poleward moving water and the upper equatorward moving water is known to broach the surface near the coast. This has been shown both in drogue motions noted by Lamberth and Nelson (1987) and by current profiling studies reported by Nelson (1985). At one site west of Cape Columbine, the only one probed with a multiple flight of current meters, the interface showed fluctuations in height, as evidenced by low frequency switching from poleward to equatorward flow in the upper current meters, with periods of weeks.

Some details from current meter records are presented below, but first it is useful to distinguish the general physical characteristics which could influence flow over the Cape Basin and shelf zone of the Benguela area of southern Africa.

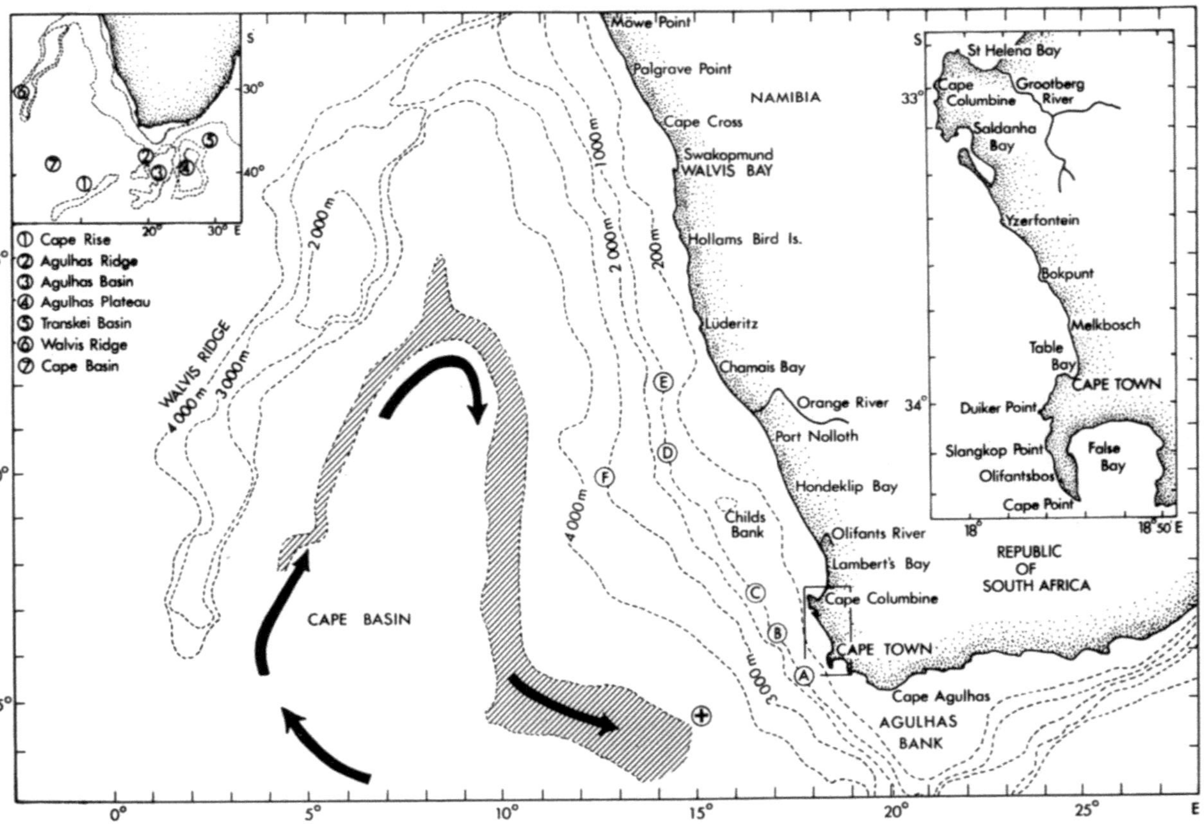

Figure 1: Geographical details, bathymetry and sites of current meter moorings (labeled A to F). The circled cross marks the site of the spot current measurement by Connary (1972) and the hatched area is the erosion zone described by Tucholke and Embley (1984). Arrows indicate the direction of bottom flow measured or inferred from various geological surveys described in the text.

The gross features of circulation based on recent current meter deployments appear to be as shown schematically in Figure 2. Although the bottom layers seem to move polewards as a uniform body of water, the physical causes of this motion are likely to be quite separate in separate domains and must be distinguished. There are numerous factors which could influence currents, some of which are as follows:

1. Density driven bottom circulation in the Cape Basin which tends to carry with it water above and water along the continental rise to the shelf-edge. The flow is cyclonic.

2. Baroclinicity over the shelf break and along the edges of terraces. An uplift in isopycnals creates equatorward baroclinic jets in these zones.

3. Net poleward motion of water over the inner shelf-zone, modulated by coastal-trapped waves. Current meter records show clear signs of wave-like currents with periodicities of three to five days and net poleward transport over periods of weeks.

4. A surface wind-driven equatorward current over the open ocean and shelf region with, additionally, a deeper geostrophic equatorward flow in the deep ocean.

5. Budgetary compensation for wind-driven Ekman divergence, presumably in the form of a poleward undercurrent along the shelf as in the classic model of Hart and Currie (1960).

6. Momentum advected onto the shelf from the deep ocean arising through changes in wind stress which generate local currents along the shelf edge, or through eddies moving against the shelf edge from the deep ocean.

Each of these will now be discussed with emphasis on items 1 and 3 and brief commentary on the others.

THE BOTTOM CIRCULATION IN THE CAPE BASIN

Of particular significance is the cyclonic motion indicated by the 2000 m relative topography of Defant (1941), based on the METEOR expedition of 1925-27 in the South Atlantic Ocean. A southeastward trend is apparent at the eastern boundary of the Cape Basin along the shelf edge and continental rise. Connary (1972) reported a southeastward set of

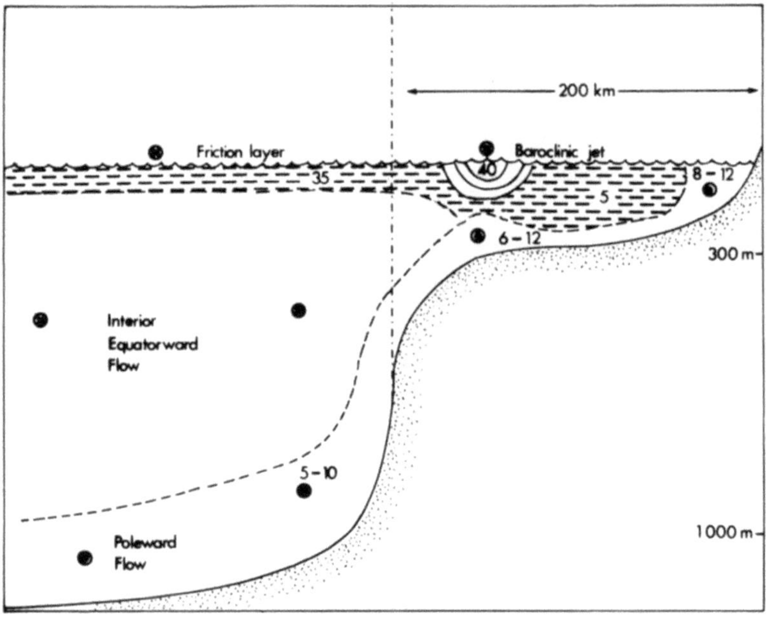

Figure 2: Schematic of poleward and equatorward flow over the shelf and eastern Cape Basin with typical current speeds (cm s^{-1}).

15 cm s^{-1} on the sea floor in this region from a spot reading instrument (see Figure 1). One should bear in mind, however, that inertial signals are strong near the bottom in this region and spot measurements of this nature could be quite misleading.

More recently, Embley and Morely (1980) have inferred cyclonic motion from bottom potential temperature. Tucholke and Embley (1984) show a band of eroded basement rock (Figure 1) claimed to be the result of scouring by strong poleward bottom currents along the eastern rise of the Cape Basin, while Rogers (1987) reported extensive erosion in a moat at about 4800 m along the northern edge of the Agulhas Ridge. Photographs of the moat show well-developed ripples. The asymmetry of the crests on these ripples indicates a westward current in excess of 20 cm s^{-1}. A camera-grab station in 4749 m of water at about 38°20'S, 17°E revealed southeastward contour currents, indicated by bending of seapens (Penatulaciae).

There seems to be little doubt that cyclonic motion occurs in the Cape Basin. Bottom currents would be deflected to the left by the Coriolis force, but contained by the topography of the Walvis Ridge to the west and north, and the continental rise to the east (Figure 1). In

Figure 3, the barrier effect of the Walvis Ridge on potential temperature is seen. Antarctic Bottom Water flows northwards from the Weddell Sea area into the basins around South Africa through fracture zone conduits (Tucholke and Embley, 1984). The Walvis Ridge, which runs obliquely from 35°S, 10°W to the African coast at about 18°S, 12°E forms an effective barrier to equatorward flow of deep water, which is consequently diverted along the continental rise, exiting the Cape Basin in the scour zone east of the Cape Rise.

Several water types are found in the Cape Basin and along the continental margin. (See Shannon, 1985, for a detailed review). South Atlantic Central Water (SACW) occurs to a depth of 600 to 800 m. This body of water is characterized by a line in TS space connecting the points (6°C; 34.5 ppt) and (16°C; 35.5 ppt) and is found along the entire continental shelf from Cape Point to Möwe Point. It is formed in the Subtropical Convergence zone by sinking and equatorward spreading into the Atlantic Ocean.

Below the SACW, Antarctic Intermediate Water (AIW) is found with a salinity minimum of 33.8 ppt at 2.2°C marking its core. With distance from the source, the salinity minimum rises to between 34.3 ppt and 34.5 ppt at temperatures of 4 to 5°C, so that a meridional change in the salinity minimum occurs.

The concept of a northward spreading of AIW along the base of the African continent which arises from TS analysis, is at variance with observed direct current measurements reported here which clearly show poleward movement. This is discussed further in Section 5.

Below 2000 m North Atlantic Deep Water and Antarctic Bottom Water are found. This is characterized by salinities lying between 34.7 ppt and 34.9 ppt at temperatures less than 3°C.

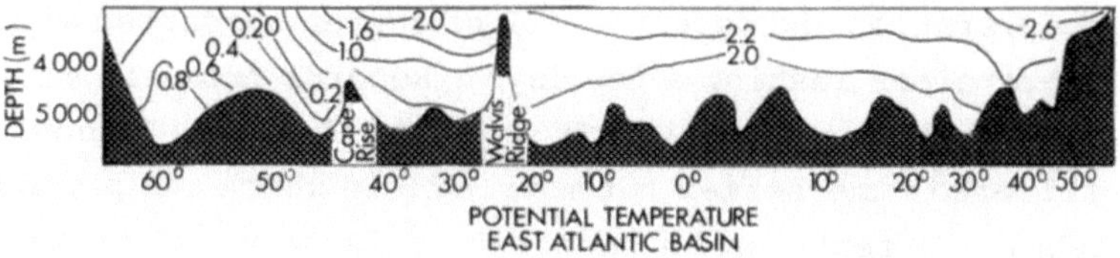

Figure 3: The barrier effect of the Walvis Ridge on potential temperature.

The recent direct current measurements in the eastern Cape Basin demonstrate that poleward flow occurs along the base of the continental shelf. Figure 1 shows the positions of five mooring sites, A to E along the upper slope in roughly 1000 m of water. Site F in the Cape Basin was at roughly 3000 m. Except at C and D which carried flights of current meters, these moorings consisted of a single current meter some 20 m off the sea floor with sufficient floatation to ensure that a tilt angle of 20° would not be exceeded in currents up to 50 cm s^{-1} .

Two types of current meter were used, Aanderaa RCM4 units, sampling over 30 or 60 minutes and Neil Brown ACM2 units vector averaging over 1, 2, or 6 minutes. The data was further vector averaged to produce hourly values.

The poleward motion along the eastern boundary is clearly seen in the form of progressive vector diagrams, some of which are shown in Figure 4. Because these records show a strong energy component at the inertial period, the data as presented here have been low-pass filtered and are displayed with supra-tidal frequencies removed. For this purpose, a Cosine-Lanczos filter with 121 weights and a half-power point at 46 hours was used. The parameters correspond to those of the Oregon State University filter as described in Woods Hole Technical Report 85-35 (1985), subsequently referred to as the OSU filter in this text.

The current at site E near Chamais Bay (Figure 4f) is seen to flow parallel to the topography over most of the period between 10 April - 5 November 1986 with two notable eastward excursions, labelled a and b in Figure 4f, that are suggestive of momentum advected onto the shelf by deep ocean disturbances. Such excursions are observed in most records, but are infrequent and not simultaneous from one site to another (See below).

Likewise, the current ran parallel to the topography at site A near Cape Point, (Figure 4a). What is clearly apparent in this record is that a periodic acceleration and deceleration of current occurred with reversal of direction, at times lasting a few days. Records from all sites exhibit this feature with periods ranging from 2.5 to 4 days. Figure 5 shows a segment of filtered time-series for mooring B in which the periodic nature of the longshelf current can be seen. In this case, 29 peaks or points of inflection can be identified over the 74 days of the record, giving a periodicity of 2.5 days. Spectra of longshelf velocity components show significant peaks between 2.5 and 3.5 days.

In Figures 4c and 4d, decreasing spatial coherence can be seen as the surface is approached. At 943 m the flow is marked simply by the

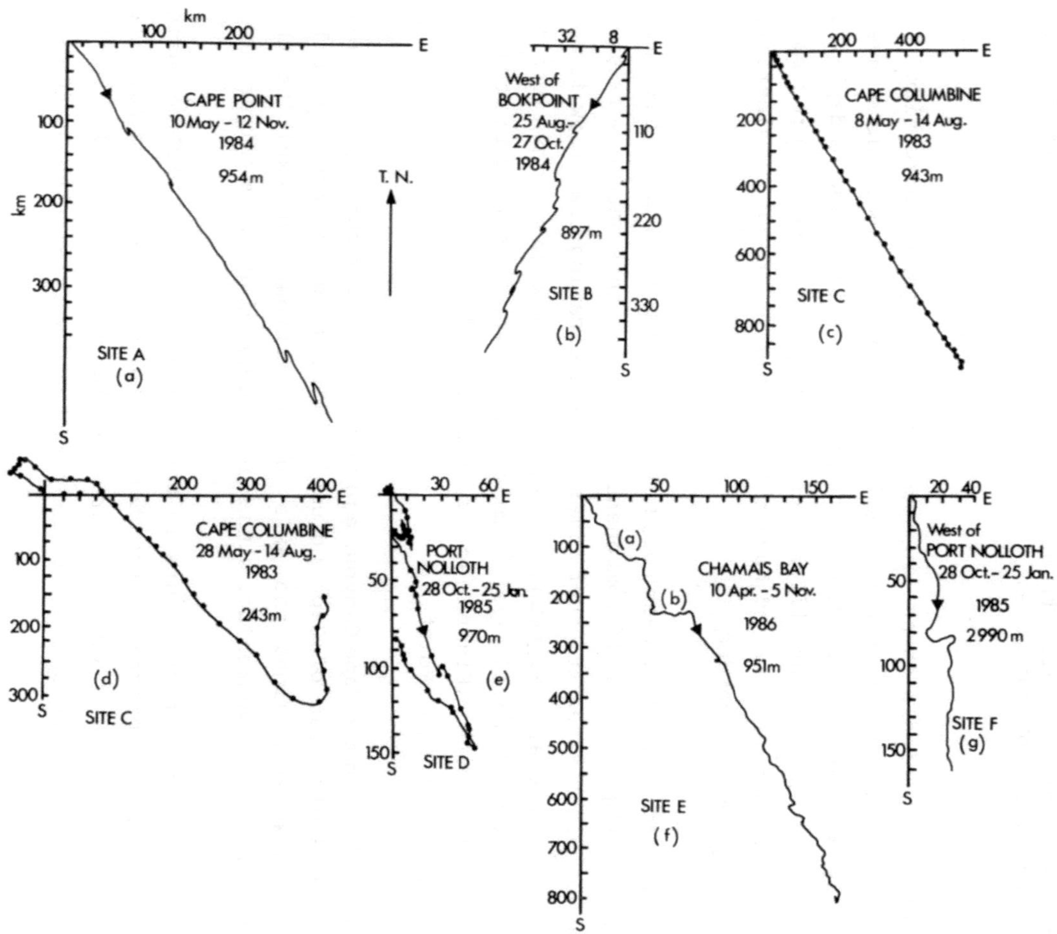

Figure 4: A selection of progressive vector displacement diagrams for sites A to F (Figure 1). Dots on (c) and (e) mark 3-day intervals and on (d) 2-day intervals. Time zero is at the origin.

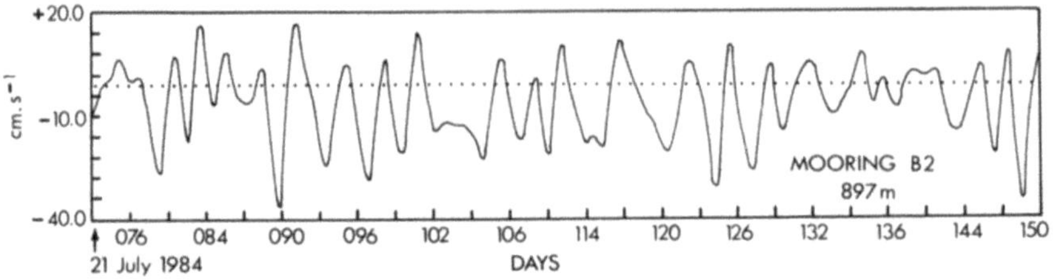

Figure 5: Segment of filtered current velocity component along the principal axis of flow (220°T) at site B (see Figure 1), showing the oscillatory nature of the flow.

characteristic periodic change in speed in the longshelf direction. At 243 m on the same mooring, two events in which reverse flow occurred over several days can be identified. Current meters at this site were deployed four times, and significant differences were noted in successive records. In particular, between May and August 1983, current speeds considerably above average were recorded.

The record from Site D near Port Nolloth shows several days of reverse flow with a slight off-shelf component. Events of this nature were seen to occur at other sites at other times, thus indicating that flow is not consistently poleward, although poleward when averaged over periods of the order of a few hundred days. Similar events were observed in records from Cape Columbine (Figure 4c is exceptional in its directional stability) and at Cape Point.

Site B lies in the Cape Canyon, a prominent incision cutting obliquely across the shelf west of Cape Columbine. The flow is observed to be uniformly downwards along the axis of the canyon, but exhibits the characteristic 3-day weakening, with slight reversal at times (Figure 4b).

That cyclonic motion occurs in the deeper water of the Cape Basin is supported by the progressive vector diagram from the record for Site F at 2990 m in 3020 m of water (Figure 4g). The flow ran parallel to the local bathymetry, with one significant eastward excursion about 40 days into the record.

Some statistical details giving vector-averaged velocities and the periodicity near 3 days for various sets of data are given in Table 1. The periodicities were obtained by counting turning points and points of inflection in OSU-filtered data and dividing by the length of record. The inclusion of points of inflection resulted in smaller variance between

Site and deployment period		Depth and sounding m		Poleward vector averaged speed		Periodicity days per peak d
				$cm\ s^{-1}$	$km\ d^{-1}$	
A	Cape Point May-Nov 84	954	983	4.3	3.7	2.5
B1	Bok Point Dec-Mar 84	1224	1253	1.6	1.4	2.1
B2	May-Oct 84	897	923	6.6	5.7	2.6
C1	Cape Columbine May-Aug 83	943	1023	16.1	13.9	3.0
C2	Sep-Nov 83	1061	1162	4.7	4.1	3.7
C3	Dec-May 83-84	889	980	5.1	4.4	3.1
C4	May-Aug 84	897	997	2.3	2.0	3.4
D1	Port Nolloth Dec-Mar 83-84	898	997	5.2	5.0	3.0
D2	Oct-Jan 84-85	970	1001	1.0	0.9	2.8
E	Chamis Bay Mar-Nov 86	951	1016	4.5	3.8	2.1
F	West of Port Nolloth Oct-Jan 84-85	2990	3001	2.1	1.8	3.4

Table 1: Statistics for poleward flow at base of the shelf edge, showing current speed vector averaged over several cycles of modulating waves and average period of waves (day).

records than was obtained from spectra in which the inflection points contributed little. Quite obviously, the acceleration of the fluid rather than its velocity is an indicator of external forcing.

BAROCLINICITY OVER THE SHELF BREAK AND TERRACE EDGES

Cross-shelf sections reveal an uplift of isopycnals at the shelf-edge or along the edge of terraces such as those west of the Cape Peninsula. Correspondingly, equatorward baroclinic jets form in these regions. Two examples of this have been discussed by Bang (1973) and Nelson (1985). The jets appear to be strongest where the topography is steepest, in particular near Cape Columbine where the bottom falls away sharply on the eastern wall of the Cape Canyon (Figure 6). Since the velocity in these jets along the shelf-edge is not uniform, local horizontal divergence must occur. This may in part account for the large offshore gyres sometimes seen on satellite images of the region west of the Cape Peninsula and particularly off Cape Columbine. Negative divergence here would result in an offshore movement of water.

The baroclinic jets off the Cape Peninsula and Cape Columbine have been studied using drogues. Notably, Harris and Shannon (1979) and Nelson and Hutchings (1983) reported accelerated motion past Cape Columbine, while Shelton and Hutchings (1982), Brown and Hutchings (1985) and Nelson (1985) have described various drogue trajectories in the vicinity of the shelf edge jets.

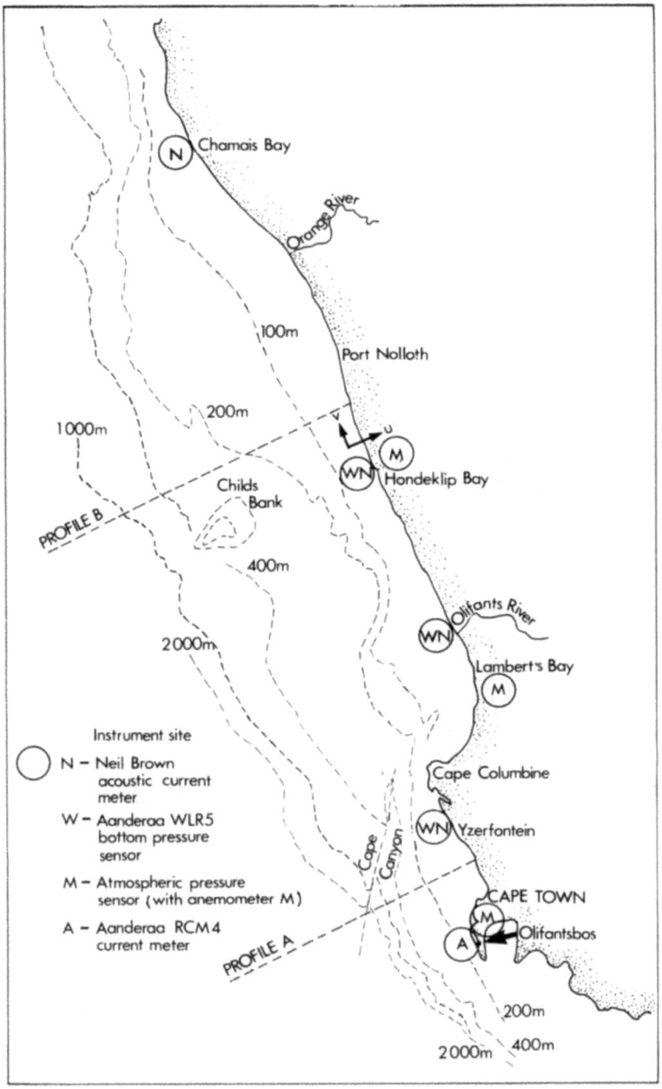

Figure 6: Geographical details for the coastal current and coastal-trapped wave experiments.

POLEWARD MOTION OF WATER OVER THE INNER SHELF

Several current meter moorings have been maintained on the shelf zone between Cape Point and Hondeklip Bay (Figure 6). Net poleward flow is observed at depths below 40 m, but the longshelf current exhibits wavelike motions with periodicities of approximately three days, similar to those of the deep shelf-edge flow. Close to the coast, poleward flow is detected in the near surface layers, and current profiling (Nelson, 1985) indicated the barotropic nature of these currents. Elsewhere, the surface layers are wind driven and show a general equatorward motion, but with poor spatial stability. A single record from Chamais Bay, north of the Orange River, showed net equatorward flow close to the coast.

A possible explanation for the observed flow over the inner shelf is that poleward-propagating coastal-trapped waves modulate an ambient poleward flow. Based on data from current meters near Cape Columbine, Holden (1987) asserted that forced modes with a periodicity of ten days occur in that area, but analysis of data from further afield, which is presented here now, indicates that free wave modes with shorter periodicities are more general for the west coast.

In an attempt to resolve such waves, three Neil Brown ACM2 current meters were deployed, together with Aanderaa WLR4, bottom pressure, and Aanderaa atmospheric pressure recorders in the vicinity. The water depth was 50 m and the current meters were approximately 25 m below the surface. The location of these moorings is shown in Figure 6. Data was obtained over the period December 1983 to January 1984.

As anticipated, the inverse barometer condition was not satisfied, and fairly large changes in bottom pressure were observed; the changes were equivalent to a variation of 20 cm in water level for a constant density water column. Figure 7a shows the OSU-filtered variation in bottom pressure in roughly 50 m of water at Hondeklip Bay and Yzerfontein. Figure 7b shows the Hondeklip Bay record with atmospheric pressure removed, and gives an indication of the magnitude of sea level variation. Large changes in sea-level such as this have been documented by Brundrit (1984) and Brundrit et. al. (1987). In these works, evidence of counter-clockwise movement of sea-level change is presented. Similar disturbances can be observed in the CODE data (Coastal Ocean Dynamics Experiment, Brown et. al., 1983) for bottom pressure. Interestingly, the periodicity for the CODE data and the Benguela data are quite different, the former being on the order of 7 days, and the latter 4 days.

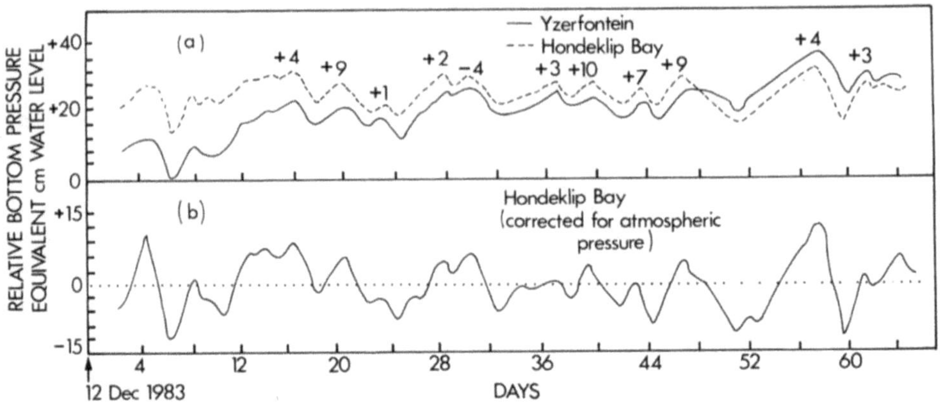

Figure 7: (a) Bottom pressure in equivalent relative cm water level variation at Yzerfontein and Hondeklip Bay showing peak-to-peak time difference in hours, Hondeklip Bay leading when positive. (b) The Hondeklip Bay record with atmospheric pressure removed showing the magnitude of sea level variation.

Changes in bottom pressure were nearly simultaneous between the northern Hondeklip Bay site and the southern Yzerfontein sites, a distance of 330 km. The peak-to-peak time difference in hours as the disturbance propagated, is shown in the annotation of Figure 7a with Hondeklip Bay leading when positive. These figures were obtained by comparing maxima in the OSU-filtered time series, as for the shelf edge data.

The range in sea level, seen here to be some 20 cm (Figure 7b), is larger than might be expected of barotropic coastal-trapped waves, as predicted by the Brink and Chapman (1985) model. Variations of a few centimeters associated with such waves would be masked by these large changes in sea level. In the classical manner, longshore wind stress would induce cross-shelf transport resulting in the generation of coastal waves (e.g., Gill and Schumann, 1974; Clarke, 1977).

Figure 8 shows low-pass OSU-filtered time-series of long-shelf currents at four nearshore sites. Three of these correspond to the synoptic shelf wave experiment done in December-January 1983-1984; net poleward transport is apparent in each of these cases, while at Chamais Bay equatorward flow was observed (Table 2). On this is superposed a fluctuating signal with an approximate three-day period, its amplitude varying from site to site.

The figure of three days is particularly significant as it is now known that three and six-day peaks occur in atmospheric pressure spectra

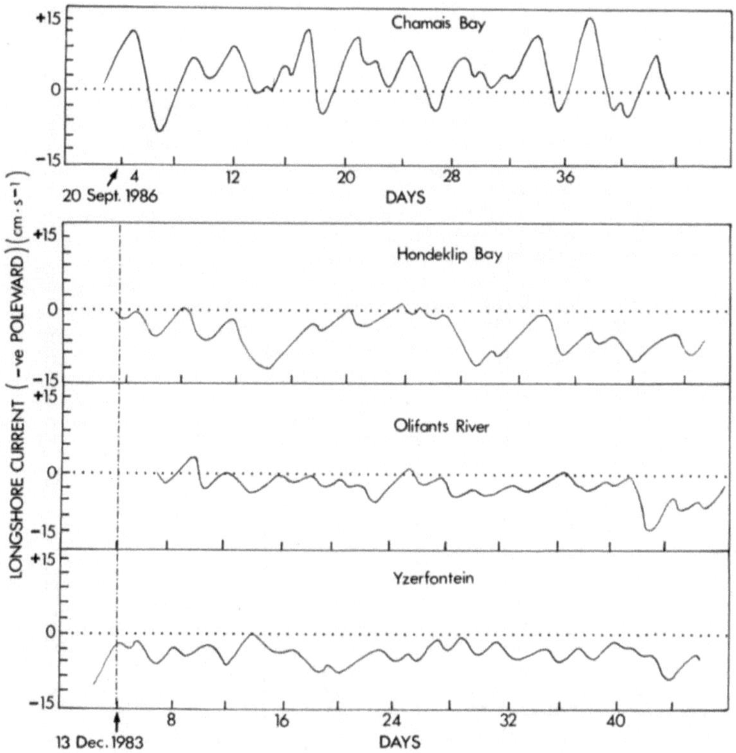

Figure 8: Filtered longshore current components along the principal axes of flow for Chamais Bay, Hondeklip Bay, Olifants River and Yzerfontein.

in the region around South Africa (Preston-Whyte and Tyson, 1974; Kamstra, 1987; see also Tyson, 1986 for more detailed references on this matter). Higher order baroclinic travelling Rossby waves in the westerlies periodically weaken the permanent high pressure cell adjacent to the west coast with a consequent periodic attenuation of ambient southeasterly winds. A short discussion of these waves is given by Tyson (1986).

Figure 9 shows segments of longshore wind time-series at Hondeklip Bay and Olifantsbos in which an approximate three-day periodicity is apparent, but with large variation in maximum velocity. The corresponding longshore currents are also shown. At times, there is a reasonable correlation between equatorward wind and poleward current, allowing for a lag of one inertial period. At other times the correlation is absent. This suggests that a mixture of free and locally forced longshore waves occurs, with local forcing and stronger poleward flow being more apparent in the narrow shelf region between Cape Columbine and Cape Point. However, one would expect the forced mode to correspond to poleward winds (Gill and Schumann, 1974) rather than equatorward winds as in the case here.

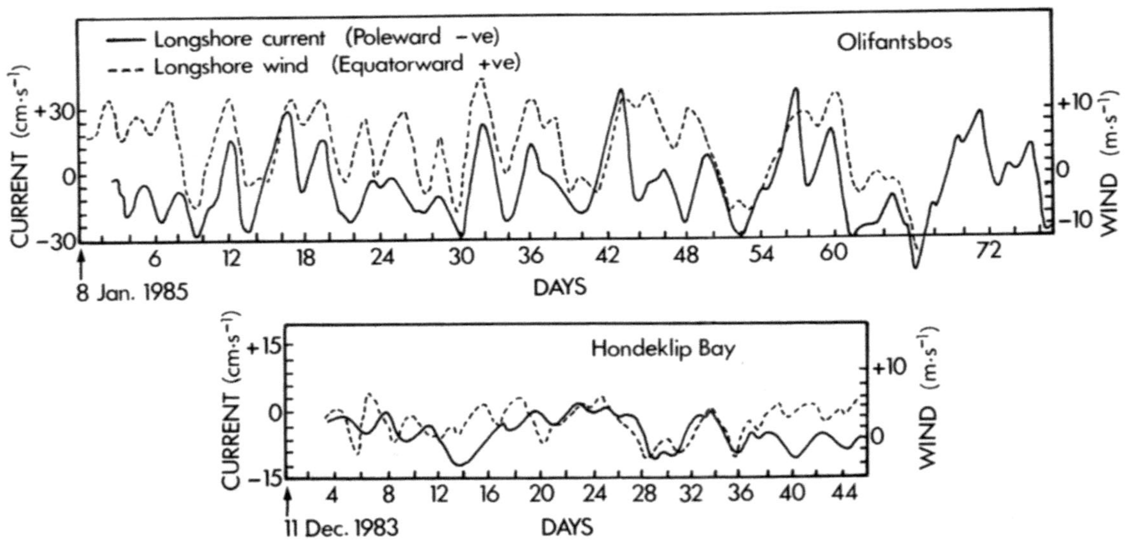

Figure 9: Filtered longshore current and wind components at (a) Olifantsbos and (b) Hondeklip Bay.

Site and deployment period (days)	Depth and sounding m		Vector averaged Poleward speed ($km\ d^{-1}$)	Periodicity days per peak d
Chamais Bay Apr-Nov 86 (208)	44	66	-1.5	2.7
Hondeklip Bay Dec-Jan 83-84 (47)	27	43	3.8	2.8
Olifants River Dec-Jan 83-84 (45)	21	37	2.6	2.8
Yzerfontein Dec-Jan 83-84 (48)	25	40	4.2	2.5
Olifantsbos Jan-Feb 85 (140)	58	73	2.6	3.0

* Note weak equatorward flow at Chamais Bay.

Table 2: Statistic for inner-shelf flow showing poleward current speed ($km\ day^{-1}$) averaged over several periods of modulation by coastal waves and average period of waves (day). Note that the data is only partly synoptic.

Further, the response time of a few days is rather short for the generation of shelf waves.

Not only is the shelf geometry distinctly different in this region, but the wind forcing is more intense than it is further north. Steep atmospheric pressure gradients occur along the coast in the summer months, particularly along the Cape Peninsula (Nelson and Hutchings, 1983) giving rise to strongly modulated longshore wind stress with a three-day periodicity.

Using programs developed by Brink and Chapman (1985), dispersion relationships were calculated for free barotropic waves over two topographic sections labelled A and B in Figure 6. The results and sections are shown in Figures 10a and 10b. Notably, periods of three to five days are possible over a wide range of wave numbers corresponding to wavelengths of 200 to 500 km for the first mode. This lends support to the hypothesis of a poleward inner-shelf current modulated by coastal-trapped waves.

The different characteristics for the shelf zones north and south of Cape Columbine, in particular the stronger wind forcing in the south, may decouple these regions, producing intermittent horizontal divergence at Cape Columbine. This would provide a second possible explanation for the large gyral tongue just south of Cape Columbine, so characteristic

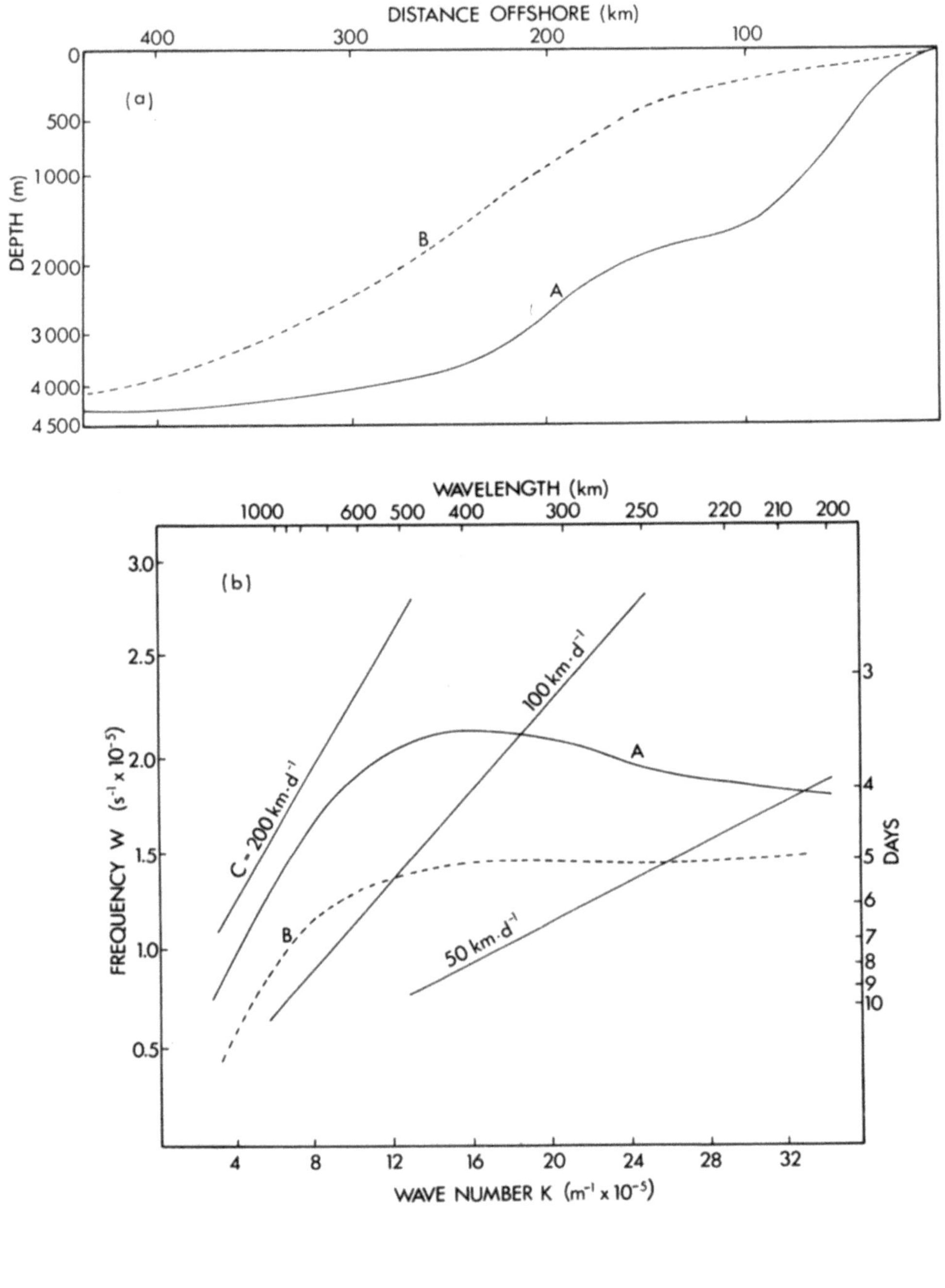

Figure 10: (a) Depth profiles for the sections A and B of Figure 6. (b) Dispersion relationships for profiles A and B for first mode free barotropic coastal-trapped waves as calculated using the Brink and Chapman (1985) routines.

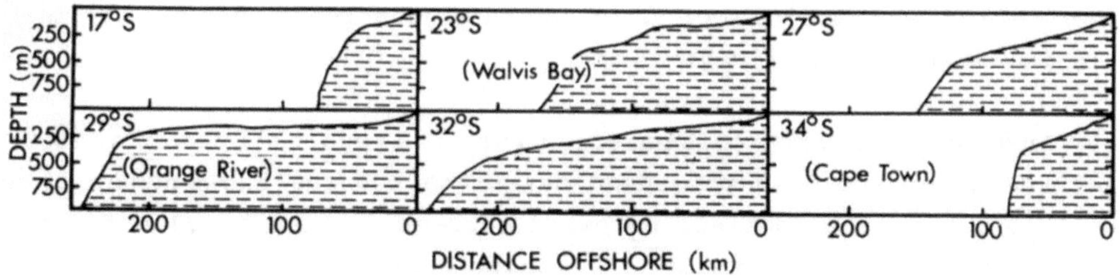

Figure 11: Zonal shelf profiles at different latitudes. (After Shannon, 1985)

of satellite images in both the visible and infra-red bands (See for example Shannon et. al., 1985).

Estimates of poleward flux are given by progressive vectors of low-pass OSU-filtered data over several wave periods. Some details are given in Table 2. Poleward flow was strong in the vicinity of the Yzerfontein mooring position, which lies on the narrow shelf above the eastern wall of the Cape Canyon. Comparing the synoptic data for the mooring near the Olifants River, where the shelf is wide, with the Yzerfontein data, one notes that the flow is weaker there. However, the width of the shelf does not appear to be a factor influencing the strength of flow, as the synoptic data for Hondeklip Bay, where the shelf is again wide, gives a figure comparable to that for Yzerfontein. The data in Table 2 is not all synoptic and further, the records are comparatively short so that considerable variation in seasonal and interannual estimates of this nature should be anticipated. A mooring placed at the same Yzerfontein site in December-January 1984 (Nelson and Polito, 1987) gave a figure of 6.2 km d^{-1} for poleward flow.

Thus, it appears that on the strength of available data, a figure of 4 to 5 km d^{-1} (4.6 to 5.7 cm s^{-1}) is a reasonable estimate for poleward flow in water 50 to 80 m deep along the coast. This is to be understood as averaged over several approximate three-day cycles of the wave-like motion observed in flow over the inner shelf zone.

The causes of the poleward flow along the inner shelf are unknown. The baroclinic structure with isopycnals lifting towards the coast from the mid-shelf region, favours equatorward flow, but both the surface Ekman and the bottom friction layers comprise a substantial fraction of the total water column, even in 100 m of water, so that geostrophy must play a small role in the inner shelf regions.

In view of the barotropic nature of currents reported by Nelson (1985) over the inner shelf, a permanent sea slope towards the equator is a possible cause of an ambient poleward current. Taking a value for the Coriolis parameter of $7.3 \times 10^{-5}\ s^{-1}$ which corresponds to latitude 30°, and an ambient flow of 5 cm s^{-1}, the gradient equation yields a slope of 3.7 cm per 100 km. There are two possible mechanisms by which such a sea slope may be maintained. One is the perennial frictional drag of southeasterly winds. Because of the orientation of winds and coastline, the Ekman transport has a small equatorward component. The other is rectification of tidal flow onto the shelf by bottom friction which would be more pronounced where the shelf is widest between a point south of Luderitz and Cape Columbine.

EQUATORWARD CURRENT

Current meters moored less than 40 m below the surface in the mid-shelf zone show equatorward motion, but with weak directional stability. Average speeds of less than 8 cm s^{-1} are characteristic of most of the mid-shelf zone north of Cape Columbine. These records are indicative of a high eddy kinetic energy density and are consistent with the convoluted patterns observed in satellite images of the west coast.

In the deeper oceanic water, equatorward geostrophic flow occurs to a depth of a few hundred meters (Defant, 1941).

The surface water over the whole southeastern Cape Basin is influenced by the permanent South Atlantic High (SAH) pressure cell and its associated southeast trade winds and thus, seasonal variations in flow can be expected.

EKMAN DIVERGENCE

The concept of poleward undercurrent along the shelf compensating and induced by Ekman divergence needs attention. In the winter months, the southeasterly wind field is disturbed by eastward propagating cyclones, but as the summer progresses, a more stable longshore wind field develops, with upwelling occurring over the whole shelf zone at discrete sites. De Decker (1970) described the poleward progression of a low oxygen tongue as the summer progressed. However, oxygen is non-conservative, particularly in the biologically and chemically active waters of the mid and inshore shelf zones in the Benguela region, and care must be exercised in using oxygen tongues to infer currents. This matter has been discussed in more detail by Chapman and Shannon (1985).

Furthermore, direct current measurements in the winter months (e.g., Nelson and Polito, 1987, their mooring H) indicate that poleward flow over the mid and inner shelf is perennial, so that a compensation current alone would not account for the observed poleward motion. One would expect the compensation current to be non-existent or considerably weaker in the winter when the wind is less favourable for continuous upwelling.

Because of the discrete zones of upwelling within the Benguela system, demands on the compensation current would be latitudinally variable. Kamstra (1985) gives figures for averaged summer wind speeds along the west coast and demarcates zones of cyclonic wind curl. The essential feature of this work is a lobe of lower wind speed north of Cape Columbine to the Olifants River, the implication being that Ekman divergence would be greater north and south of this lobe. This is suggested also in the work of Parrish et. al. (1983).

The region near Cape Columbine should thus be a source for upwelling water, with local demands on compensation current which need not necessarily be continuous along the shelf. The possibility of cross-shelf compensation from the narrow ledge at the head of the Cape Canyon should be considered.

The appearance of Antarctic Intermediate Water (AIW) below 600 m west of Cape Columbine is curious. Since the current meter records demonstrate a net poleward flow in this vicinity, the AIW cannot move in from the south and must therefore move slowly eastwards from mid-depths in the Cape Basin. A detailed TS analysis of waters in the area by Shannon and Hunter (1988), does indeed show a tongue of relatively fresh AIW touching on the continental shelf at just this point. The eastward movement could be a response to the strong upwelling in the Cape Columbine area.

ADVECTED MOMENTUM

Nothing is yet known about the scales of eddies in the Cape Basin or what effect they might have when impinging on the shelf edge. Chapman and Brink (1987), in considering the influence of the deep ocean on shelf circulation have shown that forcing in the longshore direction, which is periodic in space and time, can result in resonance with free coastal-trapped waves when the period is less than ten days.

Reference has been made above to the two eastward excursions in the progressive vector diagram obtained from data at Site E (Figures 1 and 4f). The nature of these events is unknown. They may be associated with low frequency changes in wind stress over the Cape Basin. Such excursions are common in many current meter records over the whole area explored thus far. Their frequency is on the order of 50 days and there appears to be no correlation between sites so that their spatial scales, taking into consideration their duration, would be on the order of 40 to 100 km.

SUMMARY

In conclusion, the salient features of the work on currents in the Southern Benguela zone as revealed to date are as follows:

1. Deep cyclonic motion occurs in the Cape Basin with a typical speed of 5 to 10 cm s^{-1}.

2. Wave-like currents with roughly three-day periodicities occur on the inner shelf with net poleward transport resulting from an ambient poleward current. Speeds as high as 30 cm s^{-1} are attained, with net mean flow values of typically 5 km d^{-1} (5.7 cm s^{-1}) averaged over several cycles.

3. Baroclinic jets occur along the shelf break and terrace edges with typical equatorward velocities of 30 to 60 cm s^{-1}.

4. Equatorward flow occurs in the surface layers to a depth of some 40 m over the shelf away from the coast and to depths of several hundred to thousands of meters over the deep ocean. In the surface friction layer over the open ocean, drift is typically 20 to 35 cm s^{-1}.

ACKNOWLEDGEMENTS

I am grateful to Dr. Eckart Schumann of the University of Port Elizabeth for his constructive comments in the preparation of this manuscript, to Dr. Ken Brink for suggestions and review, and to the technical staff of the Sea Fisheries Research Institute for assistance in the deployment of instrumentation and routine data processing.

THE PACIFIC OCEAN

OBSERVATIONS OF THE LOW-FREQUENCY CIRCULATION OFF THE WEST COAST OF BRITISH COLUMBIA, CANADA

Howard J. Freeland
Institute of Ocean Sciences
P.O. Box 6000
Sidney, B.C. V8L 4B2
Canada

INTRODUCTION

A variety of experiments conducted off the coast of British Columbia (BC) since 1979 have produced a general description of the circulation patterns. In this note I will present current meter data from several sources merged to form a coherent picture of the annual cycle of the current field. Figure 1 shows a diagram of the BC coast with the locations of current meter moorings indicated. Many more sites have been occupied; but to show the major features of the flow regime, I will limit discussion to two cross-shelf lines of moorings and an array deployed along the shelf edge. The two cross-shelf lines are near La Pérouse Bank, off southern Vancouver Island, and Estevan Point, halfway up the coast of the island.

Some aspects of these data bases have been discussed elsewhere, such as Freeland and Denman (1982) and Freeland et al. (1984). Data reports have been published on these and other data by Freeland (1983) and Thomson et al. (1985).

DATA REDUCTION

The stations indicated by squares on Figure 1 belonged to an experiment called CODE, Coastal-Ocean Dynamics Experiment, that extended for almost two years from 1979. The triangle locations belonged to SUPERCODE, which lasted for one and one-half years overlapping somewhat with CODE. One location was in common between the two experiments (the triangle on the shelf edge along the La Pérouse Line, we will refer to that mooring as LPB) and lasted for about three and one-half years. Initially, the time series were reduced to monthly averages for a cursory examination of the seasonal cycle; however, it immediately became obvious that the time series were very heavily dominated by the mean and annual cycle. Performing a least squares fit of a mean and 12-month cycle to the LPB data yields a simple description of that site. Subtracting the mean and annual cycle fit from the original data gives a measure of the goodness of fit. For LPB a mean and annual cycle reduces the variance of the time series by 77 percent, 69 percent, and 80 percent for time series at depths 50 m, 100 m and 150 m, respectively. (In terms of standard deviations that is 88 percent, 83 percent, and 89 percent, respectively.) This is an impressive data compression; so in the remainder of this note, the data will be represented in this form. Furthermore, for the 3 meters discussed, LPB at 50, 100, and 150 m, we find 96 percent, 98 percent, and 96 percent, respectively, of the total variance in the along-shore component. Hence, we will display only that part of the velocity field.

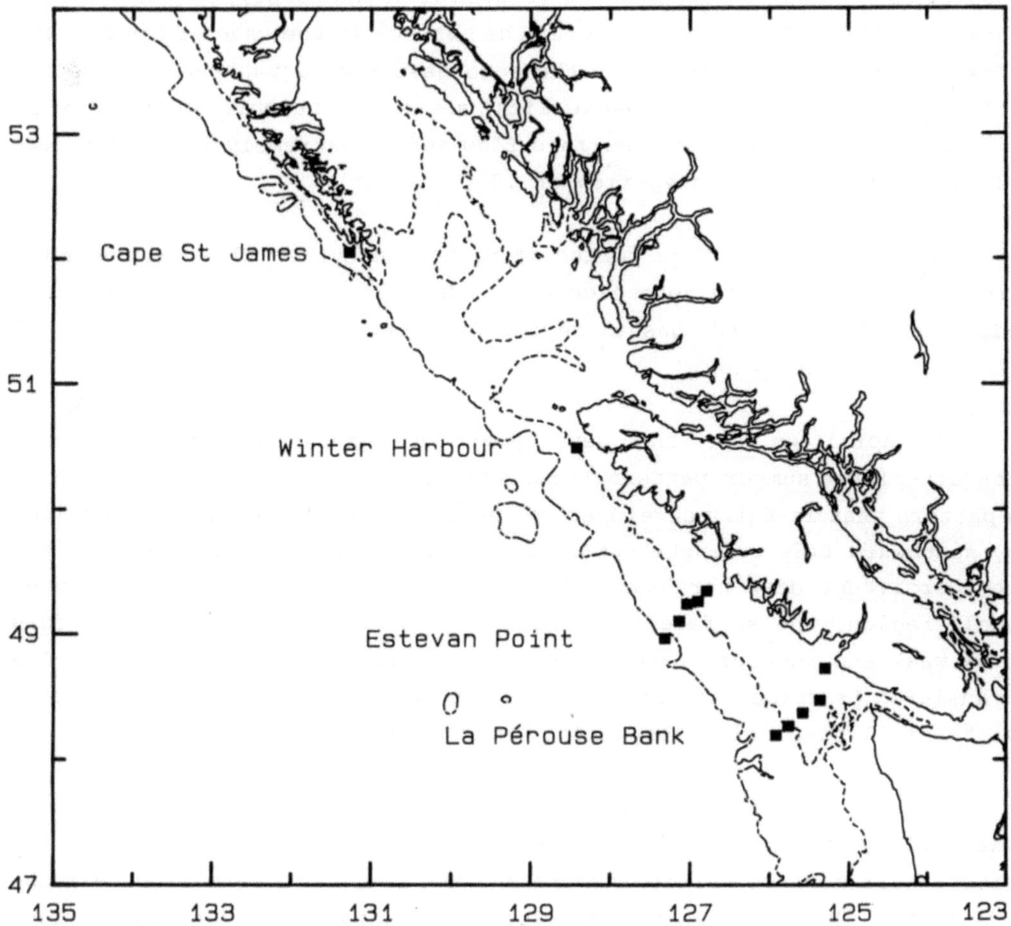

Figure 1: Map of the coast of British Columbia showing the locations of the current meter moorings used in this paper. The dotted and dashed lines indicate depths of 100 f and 1000 f, respectively.

LA PÉROUSE BANK CURRENTS

Figure 2 shows the annual cycle of currents along the La Pérouse line of current meters in the perspective plot. Each of the panels shows the distribution of currents for one month across the continental shelf and part way down the continental slope. For the January panel a scale of speed is indicated. The direction sense is such that flow to the northwest is represented by arrows to the left on each panel.

In winter, October through February, surface currents show an intense northward flow over the shelf edge and slope and intense northward flow very close in to shore. Generally current speeds decrease with depth. However, at the shelf edge mooring (fourth from the coast), we see a strong northward flow, at 150 m depth, near the bottom. In the case of the January flow regime, this appears to be the strongest flow anywhere in the water column. At the coastal site the vertical shear is particularly strong with a very large northward near-surface flow and an extremely weak (reversed but essentially zero) deep flow at 100 m.

March appears to be a period of rather unsettled flow; all vectors are small and close to zero except for the northward flows at the shelf edge, near-bottom, and coastal regime, near-surface. This is the time of the "spring transition."

In April we see the development of what appears to be the characteristic summer pattern of currents on the Vancouver Island shelf, a pattern that is fully developed in June. This pattern is characterized by southward flow over the outer continental shelf and slope except for the persistent deep northward flow at LPB at 150 m depth. In the inner shelf region, the surface flow remains strongly northward and the deep flow weak and indeterminate. Through the rest of the summer and fall a transition is gradually effected back to the winter conditions, the time of the fall transition being late September or early October.

In a region that at first sight seems to be dominated by seasonal variability, the most striking feature is the lack of variability at two sites, near surface, close inshore and the deep northward flow near the shelf edge. The former is a fresh water current named the Vancouver Island Coastal Current. This originates at the mouth of the Juan de Fuca Strait and is maintained by buoyancy sources distributed homogeneously along the coast of Vancouver Island. The second strong northward flow is of more direct interest to this conference, and appears to be a northern part of the California Undercurrent. The northward flow has temperature and salinity of 6.7°C and 33.9 ppt giving a sigma-t of 26.6. This is close

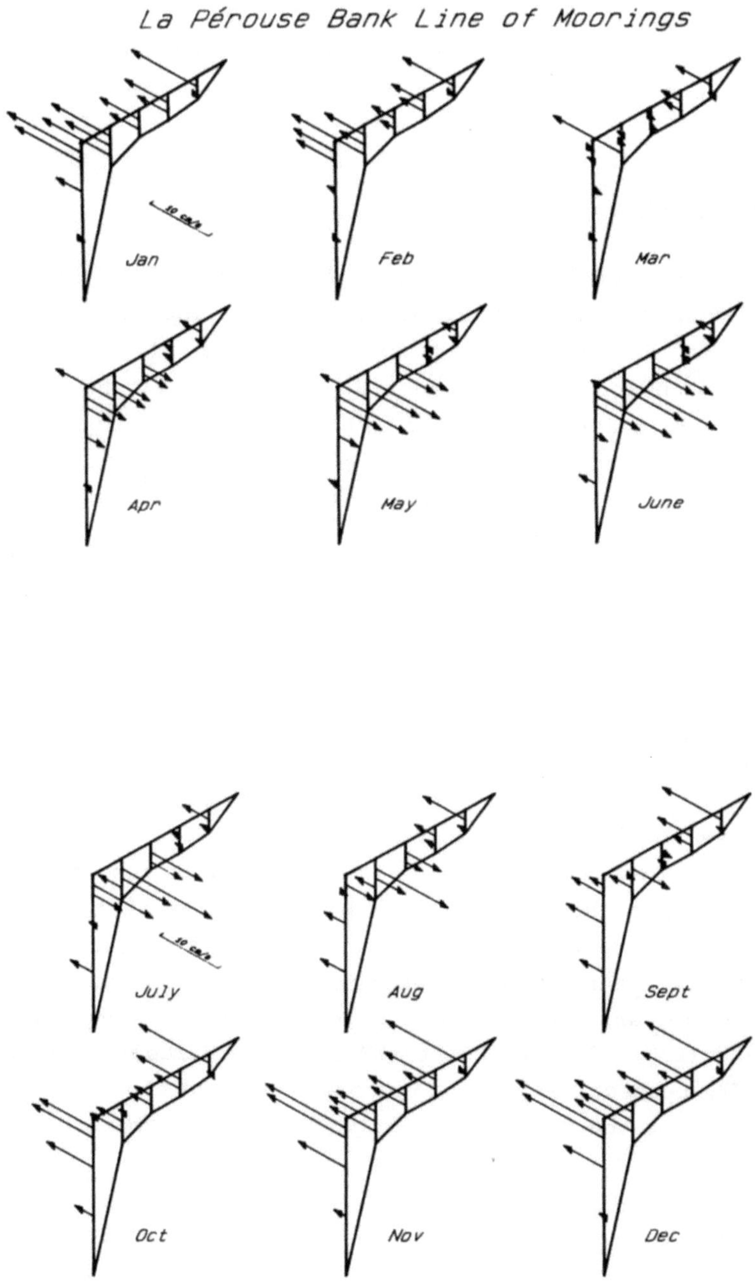

Figure 2: The annual cycle of along-shore currents off the La Pérouse Bank. The actual depths are proportional to the height of arrows on the individual panels. The top two arrows at each mooring site are at 50 m and 100 m depth; the water depth at each site from the coast is 120 m, 150 m, 140 m, 210 m, and 800 m, respectively.

to the standard values described by Hickey, 1979. The northward flow also has a high nutrient content and very low dissolved oxygen content, properties that are all standard for the California Undercurrent.

ESTEVAN POINT CURRENTS

Figure 3 shows a profile of currents for the Estevan Point line of moorings displayed in a form analogous to Figure 2. The shelf profile is a little different. The shelf is only about 40 km wide off Estevan Point, compared with 60 km near La Pérouse Bank. However, Figure 3 shows a profile of currents that is remarkably similar to the description of the previous section. Persistent northward flows are evident that can be identified as the Vancouver Island Coastal Current and the California Undercurrent. The shelf edge flows are otherwise northward in the winter and southward in the summer. During the winter all flows are northward, but a distinct mid-shelf minimum in flow speed is evident as it is on the La Pérouse section.

ALONG-SHORE STRUCTURE

Lest anyone should think that the circulation off the BC coast is simple and well determined, we present in Figure 4 a sketch of the annual cycle of currents at three depths at four sites distributed along the shelf edge. The edge of the continental shelf is, in this case, defined rather arbitrarily as the 210 meter depth contour. This contour lies about 60 km, 40 km, 20 km, and 4 km offshore from the coast at the 4 sites near La Pérouse Bank (LPB), Estevan Point (EP), Winter Harbour (WH) and Cape St. James (CSJ), respectively. In Figure 4 northward flow is denoted by leftward arrows, and the scale to the right of each month name indicates 10 cm/sec.

As indicated by earlier discussion, the currents at EP and LPB are very similar. Both show northward flow at all depths during the winter, with an indeterminate or transitionary behavior in March, followed by strong southward near-surface flow and a maintenance of a deep northward undercurrent year round. At WH, northern Vancouver Island, the pattern is similar but differences are appearing. The near-surface flow does become southward during the summer, but the period of southward flow is relatively brief, and the flow speeds relatively weak. At 150 m the differences are more striking; apparently a bias towards northward flow does exist but it is weak. The evidence for continuation of the California Undercurrent to the northern tip of Vancouver Island is marginal.

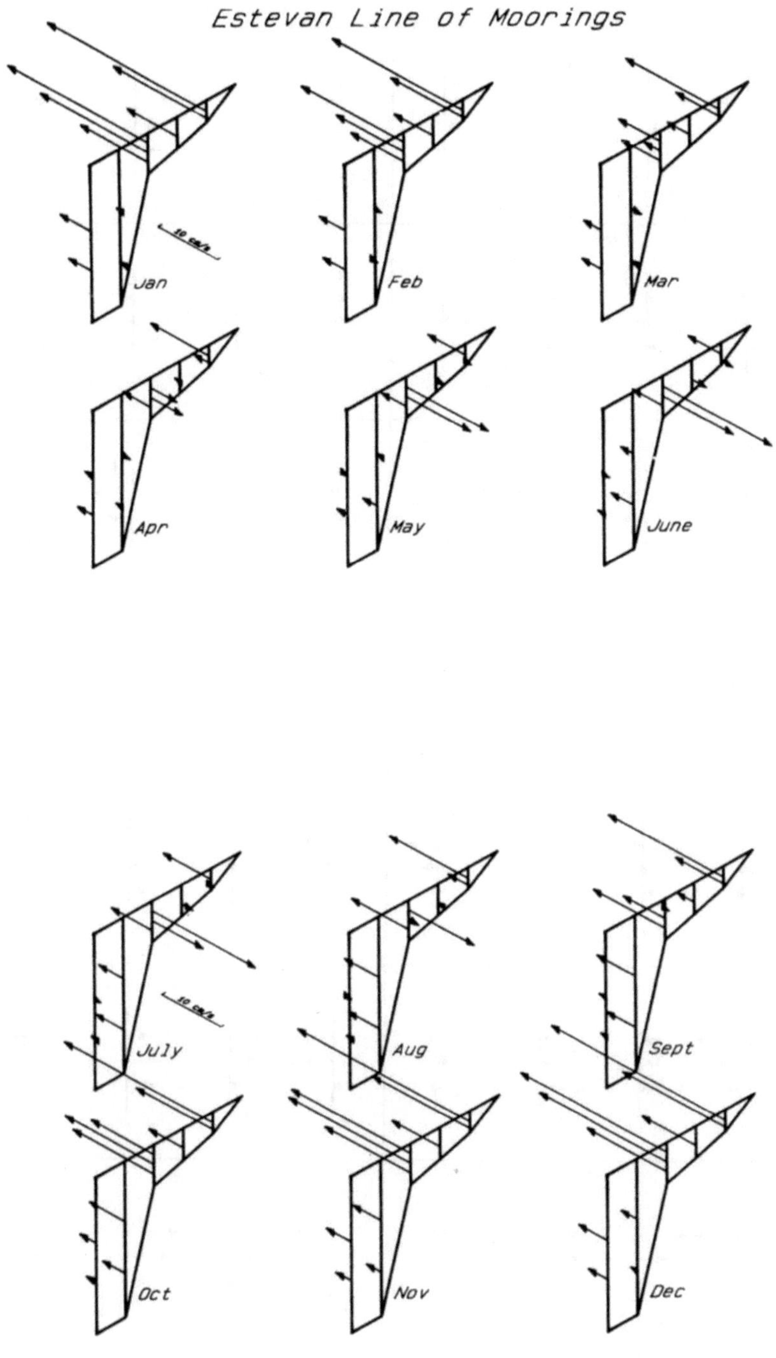

Figure 3: The annual cycle of along-shore currents off Estevan Point. The water depth at each site from the coast is 110 m, 160 m, 210 m, 800 m, and 2000 m.

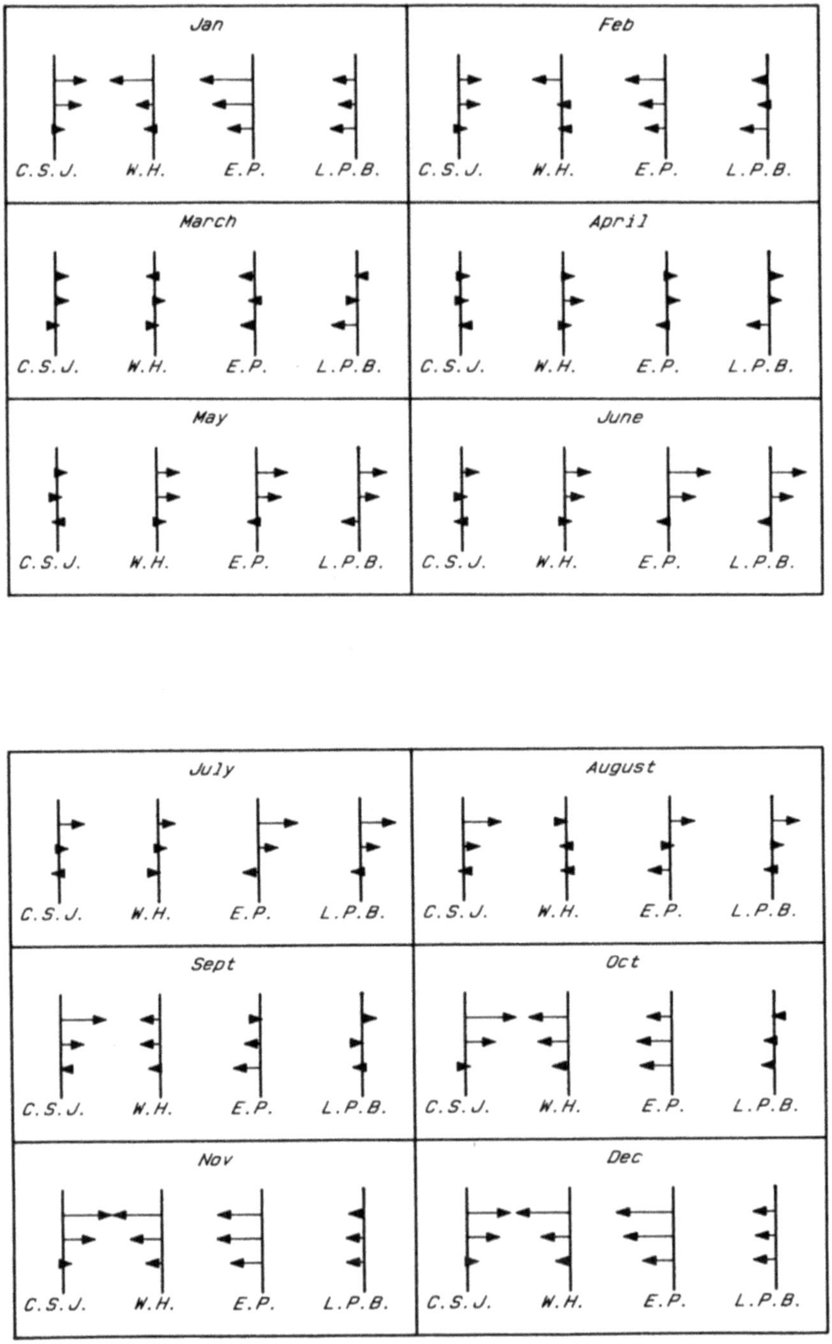

Figure 4: The annual cycle of along-shore currents at four sites distributed along the edge of the continental shelf. The sites are indicated by the solid triangles in Figure 1.

The flow patterns exhibited off the west coast of the Queen Charlotte Islands, exemplified by the mooring CSJ, have no apparent relationship to the other current regimes. The currents at 50 m and 100 m are southward year round. Furthermore, the phase of the annual cycle is different from that off Vancouver Island by about 180°, the strongest flows to the south appearing in mid-winter. At the deep meter, 150 m depth, there is no evidence of a preferred sense of flow, and so no evidence of any deep undercurrent. The general sense of flows off CSJ is baffling. The period of strongest southward flow also corresponds to the period of strongest winds to the north. A clue may exist in the detailed structure of the currents. A strong oscillation in the along-shore currents exists at the tidal period MSf. This suggests that the currents off CSJ may result from tidal rectification and so may be a very local feature not typical of the general circulation off the west coast of the Queen Charlotte Islands.

From: John Church, CSIRO Marine Laboratories, Hobart, Tasmania

On: Review and Commentary to paper **OBSERVATIONS OF THE LOW-FREQUENCY CIRCULATION OFF THE WEST COAST OF BRITISH COLUMBIA** by Howard J. Freeland

The author has assembled an impressive series of current meter observations for the continental shelf and slope adjacent to British Columbia, Canada. This data set is probably the only current meter data set for which a seasonal cycle can be determined on an eastern boundary for two cross-shelf arrays of five moorings (each with between two and four current meters) plus two additional shelf edge moorings.

In regards to the topic of this volume, the stability of the poleward undercurrent seen near the bottom at the shelf edge at latitudes of about 48°N and 49°N is quite remarkable. In fact, the poleward undercurrent is present during all seasons and appears to be the "continuous ever-flowing river" referred to by Smith in the Introduction of this volume. However, the undercurrent does not seem to be present at locations further north (at about 50.5°N and 52°N) along the coast of British Columbia. However, this may be a problem of sampling a narrow current with only a few current meter moorings.

These observations immediately raise some further questions. How continuous is the poleward flow off British Columbia with the California Undercurrent? How stable is the California Undercurrent? What happens to the poleward undercurrent between 49°N and 52°N? These are questions which could be addressed with a series of CTD/ADCP sections spaced along the eastern boundary of the North Pacific."

How the water properties varied along the core of any continuous poleward flow would be a useful diagnostic tool for examining mixing processes in this ecologically important region.

I feel the paper would be far more complete if the current meter data was supplemented by CTD data.

Author's reply:

> "[Reviewer Church was] interested in the observations at the lone current meter off the northern end of Vancouver Island and felt that if we had CTD sections suitable for geostrophic calculations, then the structures [revealed] would form a

useful extension of the report. Unfortunately, when we deployed the moorings off Winter Harbour and north, we were exploring the range of flow fields in an area that has to date had scant attention. No suitable CTD sections are available. CTD sections made in the past have extremely coarse resolution, entirely unsuitable for exploring coastal currents. Past practice has been to sample with one station on the continental shelf, one on the slope, and then a string of stations in the deep ocean. Only very recently have CTD sections had the resolution needed to examine the structure of shelf and slope flows."

POLEWARD FLOW IN THE CALIFORNIA CURRENT SYSTEM

Adriana Huyer
P. Michael Kosro
College of Oceanography
Oregon State University
Corvallis, OR 97331

Steven J. Lentz
Robert C. Beardsley
Woods Hole Oceanographic Institution
Woods Hole, MA 02543

INTRODUCTION

In the California Current System, poleward flow has been observed at different levels and in different seasons. In winter, there is broad poleward flow at the surface along most of the coast along Washington, Oregon, and northern California. During the upwelling season, in spring and summer, there is a narrow, inshore poleward surface current which appears off northern California whenever upwelling-favorable winds weaken; off southern and central California, this inshore poleward flow seems to persist through most of the upwelling season. During the upwelling season, there is also subsurface poleward flow, both near the bottom over much of the continental shelf and at depths of a few hundred meters along the upper continental slope. Whether and how these "branches" of poleward flow are related to each other is still unknown. We have some ideas of how the surface poleward flows are driven, but still very little information on the continuity and the driving of the subsurface undercurrents. In this note, we shall summarize briefly the main characteristics of each of these "branches," and present some recent observations of the California Undercurrent in the CODE region near San Francisco.

POLEWARD FLOW AT THE SURFACE: THE DAVIDSON CURRENT

Broad Northward Flow in Winter

The first evidence of a northward surface current along the west coast of North America was the use of redwood logs by North Coast Indian tribes in making dugout canoes and totem poles. This inferred northward flow was called the Davidson Current in honor of Professor George Davidson who was one of the early surveyors of the west coast. Systematic observations of the surface currents at lightships between San Francisco and the Strait of Juan de Fuca (Marmer, 1926) showed predominantly northward flow from October through March at all locations, with the strongest monthly-mean northward flow (20-30 cm/sec) occurring in January or February each year. Drift bottle releases (Wyatt et al., 1972), near-surface drogues (Reid and Schwartzlose, 1962), repeated hydrographic sections (Huyer, 1977; Chelton, 1984), and current measurements over the midshelf and shelfbreak (Hickey, 1981; Strub et al., 1987) all indicate that this winter northward surface current is considerably wider than the continental shelf, at least at latitudes north of Point Conception. The strongest northward flow (with a typical monthly mean of 15 cm/sec at the surface) occurs over the inner shelf, adjacent to the coast (Huyer et al., 1978). There seems to be considerable year-to-year variability in the strength of this current; during the El Nino winter of 1982-1983, the northward current was twice as strong as in the preceding and subsequent "normal" years (Huyer and Smith, 1985).

This winter northward surface current seems to be forced by the winter southeasterly winds which cause downwelling of the surface waters at the coast. This downwelling, combined with the increased coastal runoff caused by winter rains, results in an offshore density gradient which is in approximate geostrophic balance with the vertical shear of the northward current (e.g., Huyer et al., 1979).

Inshore Northward Flow

Poleward surface currents have also been observed during the upwelling season, with particular frequency and persistence along the coast of southern and central California. These northward flows apparently occur inshore of the coastal upwelling jet which flows equatorward along the front between dense, freshly-upwelled coastal waters and lighter oceanic surface waters. In the Southern California Bight, this inshore northward flow prevails most of the year, disappearing for only a month or so each year in March or April (Tsuchiya, 1980). The inshore poleward flow here may be as wide as 100 km or more; it seems to form the inshore limb of a large cyclonic eddy that fills most of the Bight (Reid et al., 1958). Further north, in the CODE region near San Francisco, an inshore countercurrent appears whenever upwelling-favorable winds relax (Send et al., 1987) and disappears when they become unusually intense (Winant et al., 1987). This inshore countercurrent is only 10-20 km wide and can have velocities of > 30 cm/sec (Kosro, 1987); it seems to become more persistent as the upwelling season progresses (Winant et al., 1987). Off central Oregon, a similarly narrow, inshore countercurrent is frequently observed in mid- and late summer (Stevenson et al., 1974; Kundu and Allen, 1976).

These inshore countercurrents may be driven partly by along-shore pressure gradients which are set up to balance the strong equatorward wind stress: when the winds relax, the residual pressure gradient causes a northward acceleration. This mechanism has been invoked to explain the rapid reversals of flow over the Pacific Northwest shelf that occur whenever previously strong winds relax (Hickey, 1984). However, the details of the dynamics involved are not yet clearly understood: in particular, the along-shore scales of the relevant pressure gradients are unknown, and both the sign and magnitude of estimates of the pressure gradient are sensitive to the choice of scale. There may well be other processes and dynamics involved.

UNDERCURRENTS OVER THE SHELF

Moored current measurements over the continental shelf off Washington and central Oregon have consistently shown a poleward undercurrent along the bottom over the mid- and outer-shelf during the summer upwelling season (Mooers et al., 1976; Huyer et al., 1975a, b). This undercurrent has its maximum strength within 20-30 m of the bottom (Huyer et al., 1978). Maximum velocities are relatively weak: monthly means of 2-5 cm/sec are typical. Similar undercurrents have been observed over the shelf in the CODE region just north of Point Reyes (Winant et al., 1987; their Figure 19b), and off Half Moon Bay at 37.4°N and Purisima Point at 34.8°N (Strub et al., 1987; their Figure 7). This undercurrent is absent immediately after the spring transition which marks the onset of seasonal upwelling (Huyer et al., 1979; Lentz, 1987), but it appears within a few weeks and slowly intensifies through the summer. In fall, the depth of the current maximum seems to shoal (Reid, 1987), and the shelf undercurrent becomes indistinguishable from the northward surface current which extends across the entire shelf in late autumn and winter.

Why the maximum velocities are observed so near the bottom and how this shelf undercurrent is driven remain uncertain. It seems increasingly likely that what we are observing here is merely the upper and inshore portion of the larger undercurrent that flows poleward along the upper continental slope, which is often called the California Undercurrent.

POLEWARD FLOW ALONG THE CONTINENTAL SLOPE: THE CALIFORNIA UNDERCURRENT

A poleward undercurrent over and along the continental slope has been observed at several latitudes between Baja, California and Vancouver Island. Indirect evidence for this flow is clearly visible in the large-scale temperature-salinity characteristics of coastal waters as northward-tending tongues of relatively warm, saline water (Tibby, 1941; Reid et al., 1958). More detailed studies of particular regions (e.g., Wickham, 1975; Reed and Halpern, 1976; Freitag and Halpern, 1981) also show a concentration of waters of more southerly origins along the continental margin. The undercurrent can also be clearly seen in the dynamic topography (ΔD) of isobaric surfaces of 150-300 dbar (relative to a deeper isobaric surface): in the maps of $\Delta D_{200/500}$ from repeated CalCOFI surveys off Baja, California and southern California (Wyllie, 1966); in a map of $\Delta D_{300/900}$ covering slope waters off northern California in May 1977 (Freitag and Halpern, 1981); and in maps of $\Delta D_{150/1000}$ off Washington and Oregon in September 1973 (Reed and Halpern, 1976) and July 1975 (Halpern et al., 1978). Freitag and Halpern (1981) estimate a northward transport of 1-3 Sv along the slope. Analysis of repeated CalCOFI sections shows that the undercurrent off Point Conception (Chelton, 1984; his Figure 3) persists year-round, has a width of 50-100 km, is strong (> 7 cm/sec) and shallow (core at 100 m) in winter, and weak (< 3 cm/sec) and deep (core at 250 m) in early spring; the lateral and vertical shears along the undercurrent are strongest from May through August. Similar analysis of sections off Point Sur (Chelton, 1984; his Figure 2) shows only a weak (3 cm/sec) undercurrent in late summer (July to September) that seems to shoal, intensify, and merge with the Davidson Current in winter, and to disappear in spring and early summer (March to June).

Direct measurements of this northward flow were first made by deploying parachute drogues and tracking them for several days. Reid (1962) found a 20 cm/sec northward current, about 75 km wide, at a depth of 250 m off Monterey (36°N) in November 1961. Wooster and Jones (1970) found a very narrow (20 km) undercurrent with a speed of about 30 cm/sec at about 31°N off Baja, California in August 1966. However, Stevenson et al. (1969) failed to find a northward undercurrent off Oregon in most of their 15 drogue deployments between January 1962 and September 1965.

Longer-term direct measurements of the northward flow required the installation of moored current meters. Prior to the CODE experiment conducted in 1981 and 1982, moorings of at least a few weeks duration were successfully deployed at several locations along the continental slope (Figure 1). The most complete sections, showing the most detailed vertical and offshore structure, were obtained off southern Washington in the

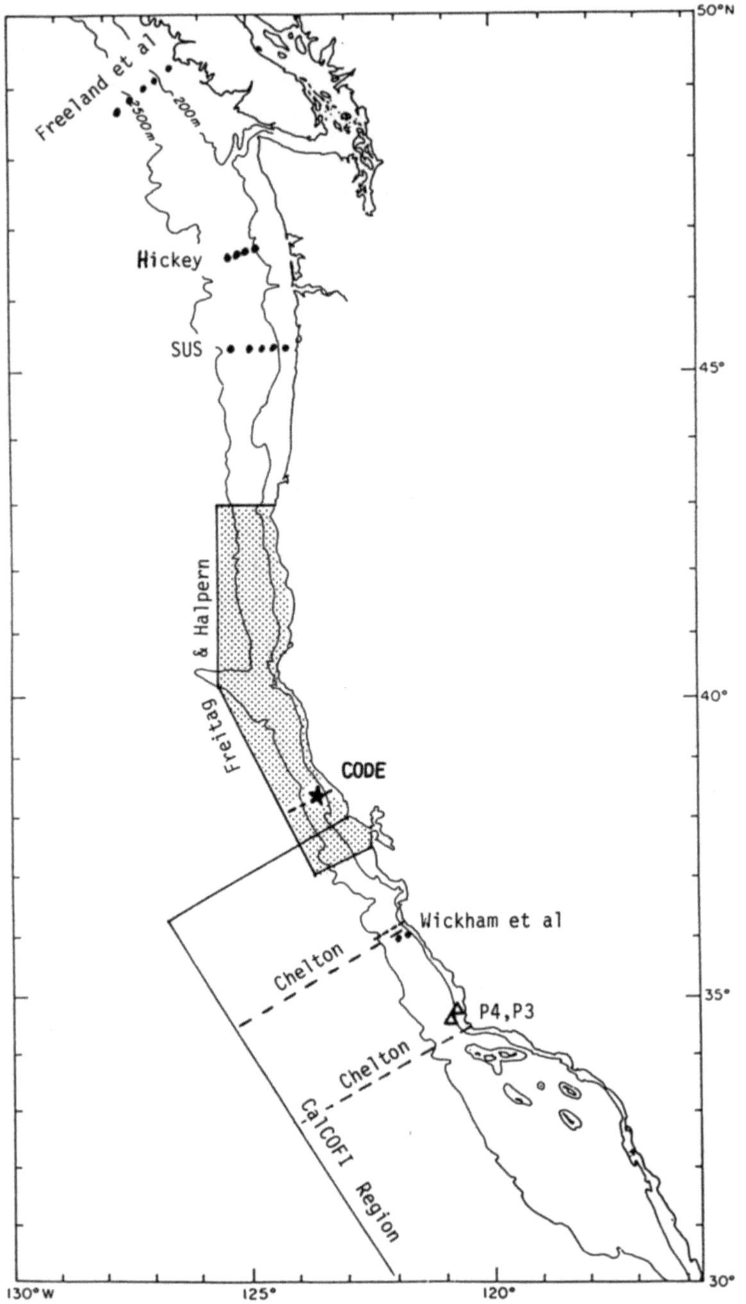

Figure 1: Map of a portion of the west coast of North America, showing the location of the principal historical studies of the poleward undercurrent along the continental slope. Repeated hydrographic sections are indicated by dashed lines, and moorings of at least a few weeks duration are indicated by dots. The CODE C-5 mooring is indicated by a star and the shelf moorings off Purisima are indicated by triangles.

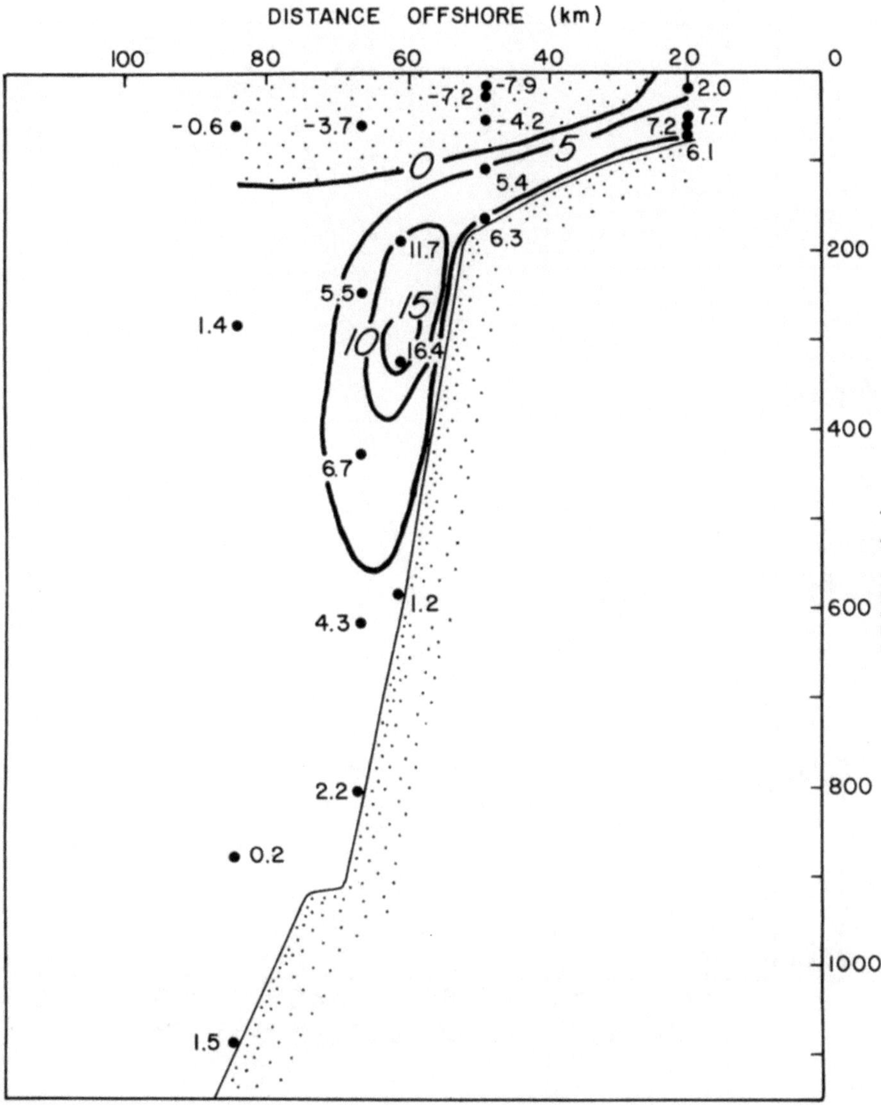

Figure 2: The distribution of the mean along-shore flow off southern Washington, 21 July to 28 August 1972. Adapted from Hickey (1979).

summer of 1972 (Hickey, 1979) and off Vancouver Island in the summer of July 1980 (Freeland et al., 1984). Average along-shore velocity sections from these arrays are shown in Figures 2 and 3; in both cases, the average was calculated over a period of about a month, in summer. Both show that the poleward velocity has a definite maximum adjacent to the continental slope, but the depth and strength of this core are rather different. Off Washington in 1972, the 15 cm/sec maximum occurs at a depth of about 300 m; off Vancouver Island in 1980, the 4 cm/sec maximum occurs at a depth

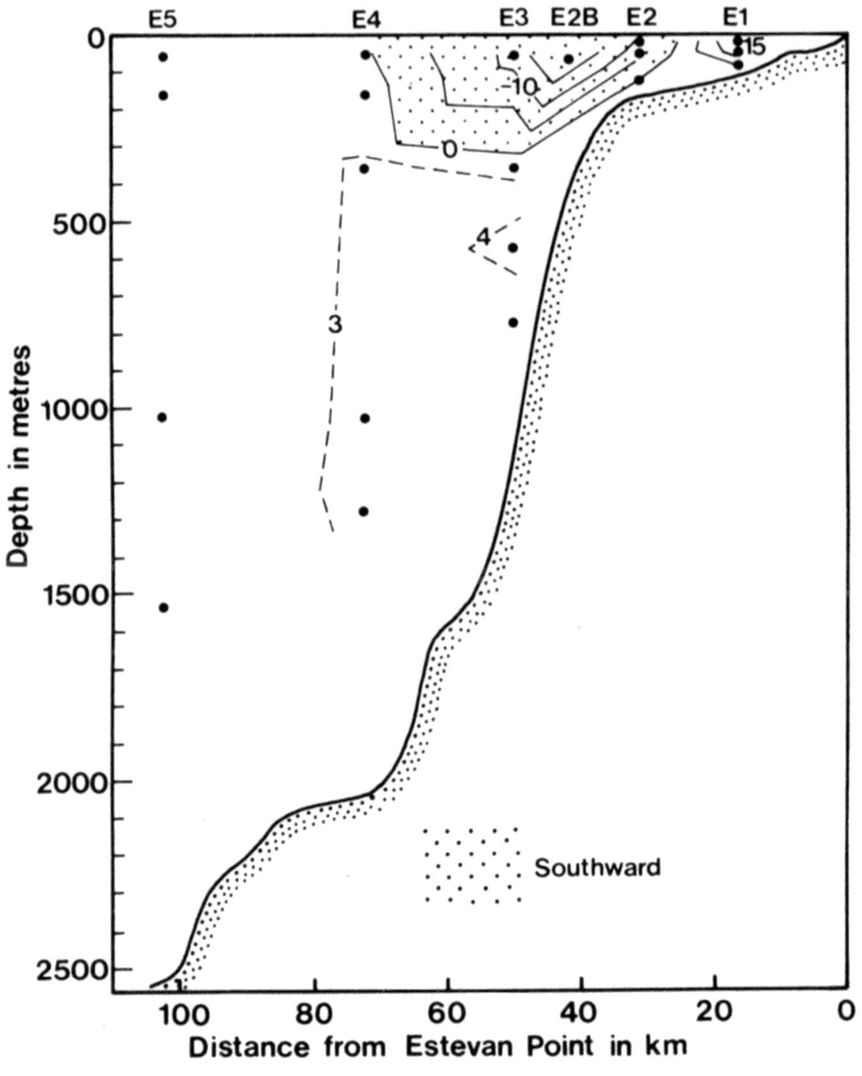

Figure 3: The distribution of the mean along-shore flow off Vancouver Island, July 1980. From Freeland et al. (1984).

of 600 m. We do not know whether these differences are associated with the difference in location or with the difference in years; i.e., whether they reflect spatial or interannual variations. The difference is probably not seasonal, since the data sets were taken at very nearly the same time of year.

A Slope Undercurrent Study (SUS) designed to observe the seasonal variation of the undercurrent off central Oregon in 1977 and 1978 met with limited success. An earlier mooring there had shown a definite undercurrent over the upper slope (500 m isobath), with the strongest northward current (5 cm/sec) near the bottom (Halpern et al., 1978). There were five moorings in this array, spanning the entire continental margin; unfortunately, there were more instrument failures and losses to fishing than usual, and a strong warm-core eddy remained over the offshore moorings for more than two months in January and February 1978 (Huyer et al., 1984). Nevertheless, the results do indicate some seasonal differences (Figure 4): in winter, there is broad northward flow (presumably the Davidson Current) over the entire margin; in spring, there is predominantly southward flow with a weak (5 cm/sec) undercurrent near the shelf-break; and in summer, there is still a weak undercurrent over the shelf-break and an apparently separate broad poleward flow over the outer margin. These sections look very different from those from the southern Washington and Vancouver Island sites (Figures 2 and 3). The data from the upper slope during summer 1978 is very similar to the summer 1975 data (Halpern et al., 1978) taken at a nearby location. We, therefore, think that much of the difference should be attributed to the shape of the continental margin, which is much wider here than farther north.

Current meter moorings were deployed for several two-month intervals at three sites over the upper slope near Point Sur between July 1978 and July 1980 (Wickham et al., 1987). Although there was little temporal overlap in the data from different moorings, the results indicate that the strongest poleward flow is adjacent to the upper slope. The strongest average flow (15 cm/sec) was obtained at the shallowest current meter, at a depth of about 100 m, only 10 km from shore; geostrophic velocity sections indicate there was southward flow at shallower depths. If these observations are assumed to be representative of the flow regime off Point Sur, we would infer that the undercurrent is much shallower there than off Oregon, Washington, and Vancouver Island. This shoaling of the undercurrent to the south is supported by the long-term shelf moorings deployed along the west coast in 1981 and 1982; the upper instruments from these moorings show an increasing tendency for northward flow at more southerly sites (Strub et al., 1987a). In particular, the moorings off Point Purisima at 34.8°N show persistently northward flow (Figure 5) which is interrupted for only a few weeks in early spring (in March 1981 and April 1982).

Data from the CODE C-5 Mooring

During the Coastal Ocean Dynamics Experiment conducted over the continental shelf and slope between Point Arena and Point Reyes, a pair of current meter moorings were maintained at 38.5°N, 123.7°W over the 400 m isobath on the upper slope (Winant et al., 1987). One of the moorings

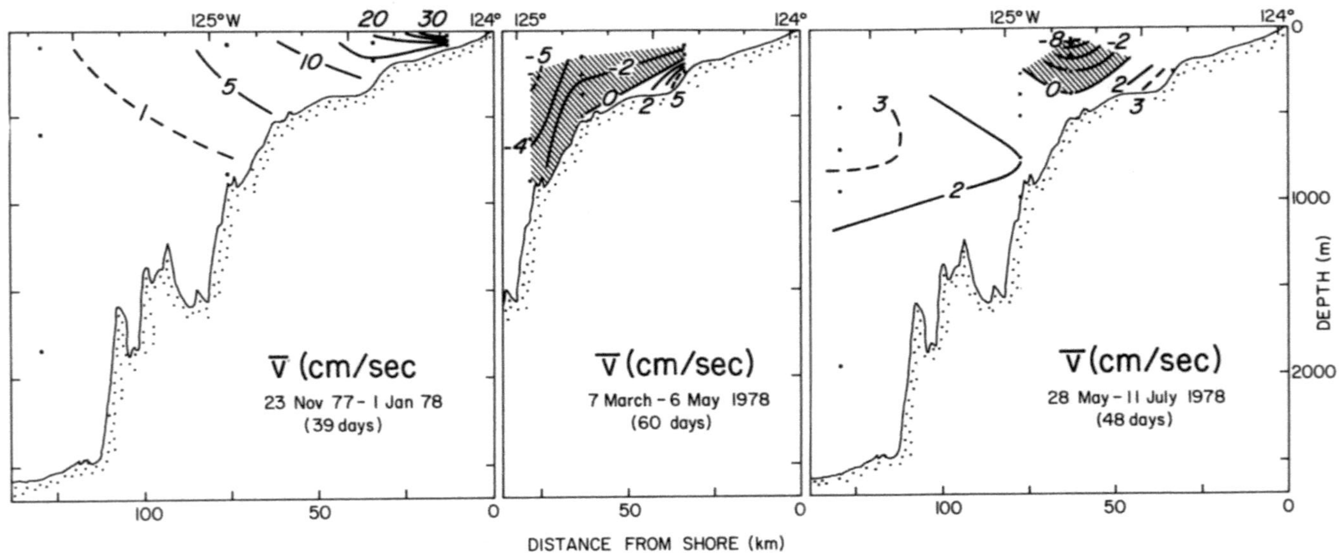

Figure 4: The distribution of average along-shore flow off central Oregon for three periods between November 1977 and July 1978. From Huyer et at. (1984).

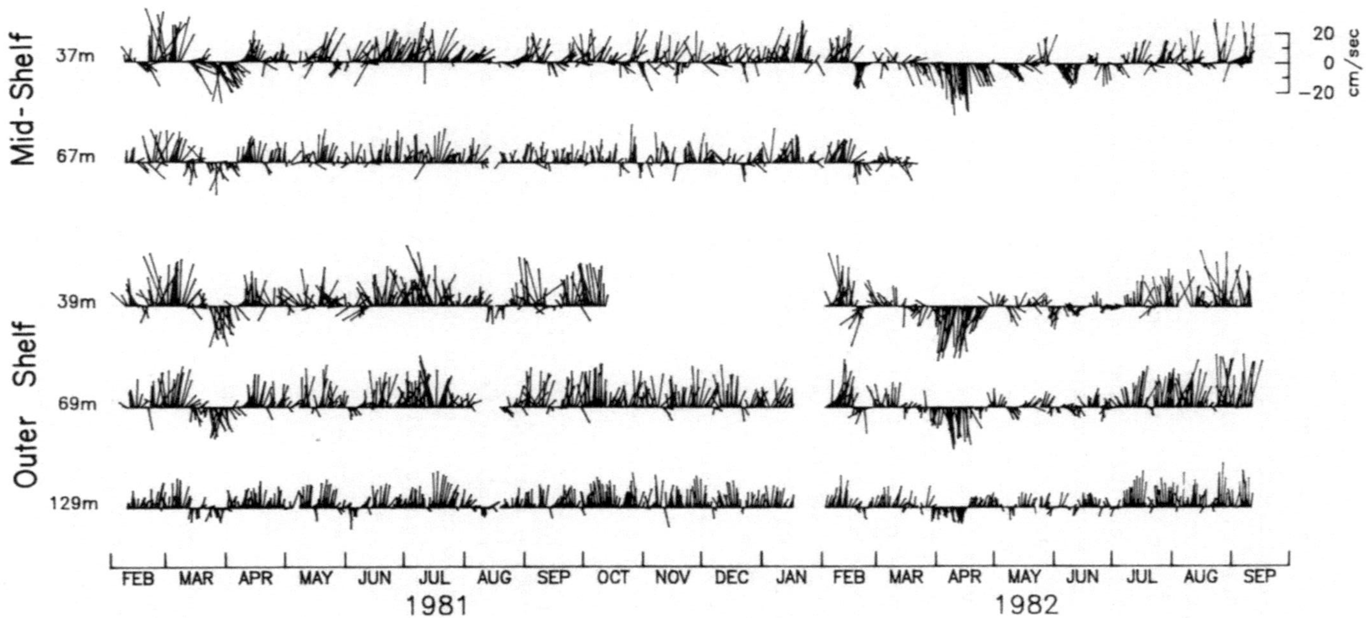

Figure 5: Time series of daily low-passed (<0.6 cpd) current vectors over the mid-shelf and outer-shelf (P3 and P4, over the 90 and 155 m isobath, respectively), off Purisima Point at 34.8°N, February 1981 to September 1982. Vectors are oriented so that true North is at the top of the page. Replotted from Denbo et al. (1984).

had surface flotation and a vector-averaging current meter (VACM) at 9 or 10 m; the other mooring was entirely subsurface, with instruments at or very near 75, 150, 250, and 350 m. The moorings were installed in April 1981, serviced in August, December, and April, and finally recovered in August 1982; these observations preceded the onset of El Nino conditions in October 1982 (Huyer and Smith, 1985). Current records at the same (or similar) depth from consecutive installations were joined by applying a predictive filter (Ulrych et al., 1973) to fill the short gaps in the hourly data. The 9 m record has a two-month gap due to an instrument failure in October 1981, but the records at the other depths are complete. The joined records were low-pass filtered (half-power at 0.6 cpd) to suppress tidal and inertial fluctuations; a coordinate rotation yielded onshore (toward 65°T) and along-shore (toward 355°T) components of the current. Time series of twelve-hourly low-passed current vectors (Figure 6) show considerable variability on time scales of days and weeks; nevertheless, it is clear that equatorward flow predominates at the surface and poleward flow predominates at depths greater than 100 m (Figure 6, Table 1). Both the equatorward surface current and the poleward undercurrent seem to be strongest in the season when the local wind stress (calculated from winds measured at nearby NDBC Buoy 13; Beardsley et al., 1987) is persistently favorable for upwelling. This is seen more clearly in the vertical profiles of three-month averages of the along-shore current (Figure 7, Table 2) which show that the undercurrent is strongest (averaging about 10 cm/sec) in spring and summer, and absent or very weak in winter. Since the spring profiles from the two years are very similar (both show strong mean vertical shear and an undercurrent core-depth of 250 m), it seems likely that they are representative for this location. However, we do not know how large a region the C-5 mooring site represents—it may be only a few kilometers across. Repeated current measurements by means of a shipborne doppler acoustic profiler along sections at 38.5°N and 39°N indicate that there are significant differences in the strength and structure of the under-current between lines separated by only 50 km (Huyer and Kosro, 1987; their Figure 27). Preliminary analysis of T-S characteristics along a section through C-5 suggests that the undercurrent core-depth is more than 100 m deeper over the mid-slope than at C-5, a separation of only a few kilometers; more work will need to be done to verify this result and to grasp its implications. On the other hand, preliminary results from recent moorings over the upper slope at other locations between Monterey and Point Conception (Chelton et al., 1987) seem to indicate vertical profiles that are similar to those observed at C-5; again, more work will be needed to verify this.

As yet, we still have a very incomplete picture of the California Undercurrent that flows poleward along the continental slope. We think

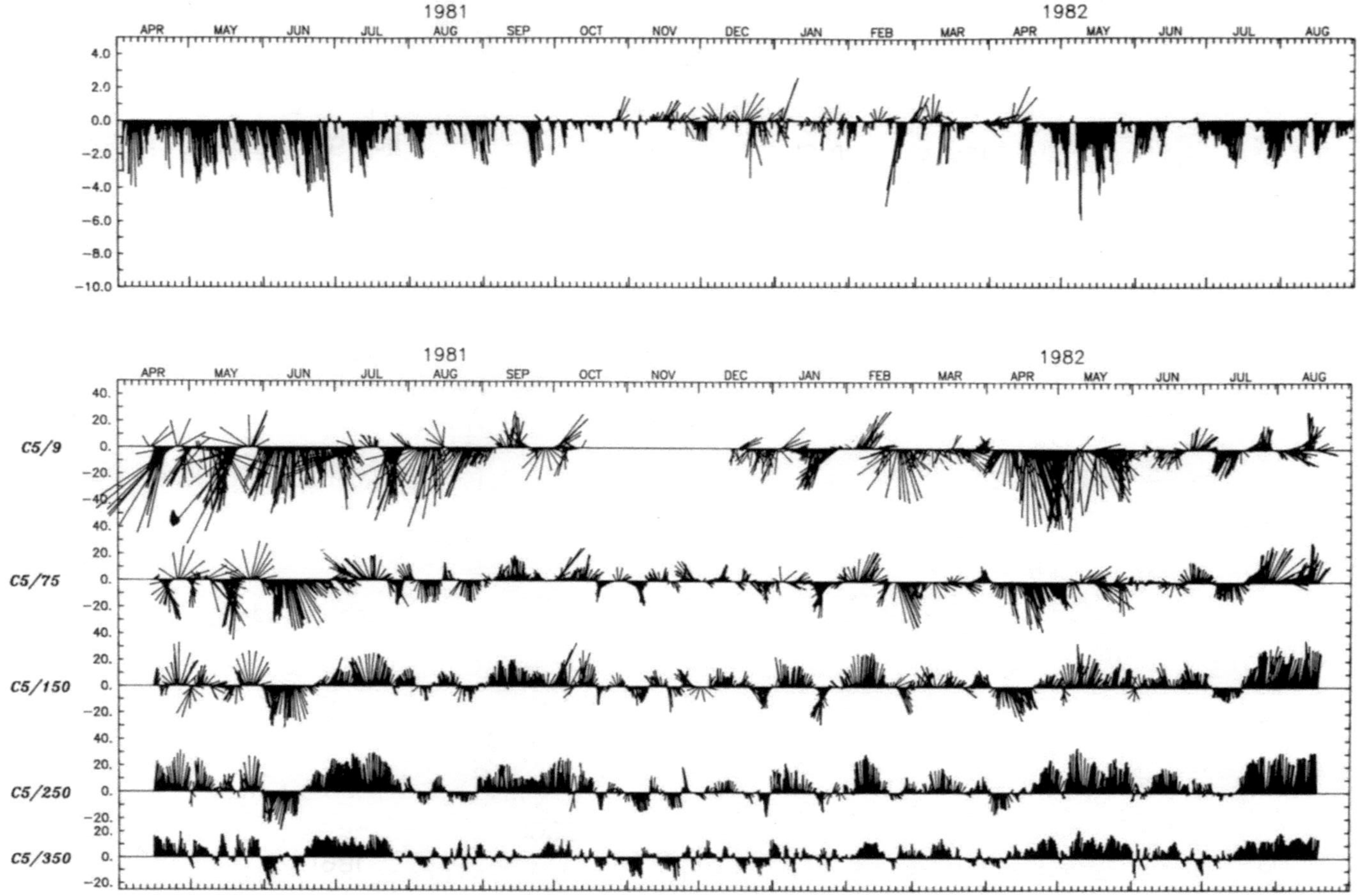

Figure 6: Time series of 12-hourly low-passed (<0.6 cpd current vectors at the CODE C-5 mooring over the 400 m isobath at 38°31'N, 123°40'W from April 1981 to August 1982. Low-passed vectors of wind stress are shown at the top. Vectors are oriented so that 335°T is at the top of the page and 65°T is to the right.

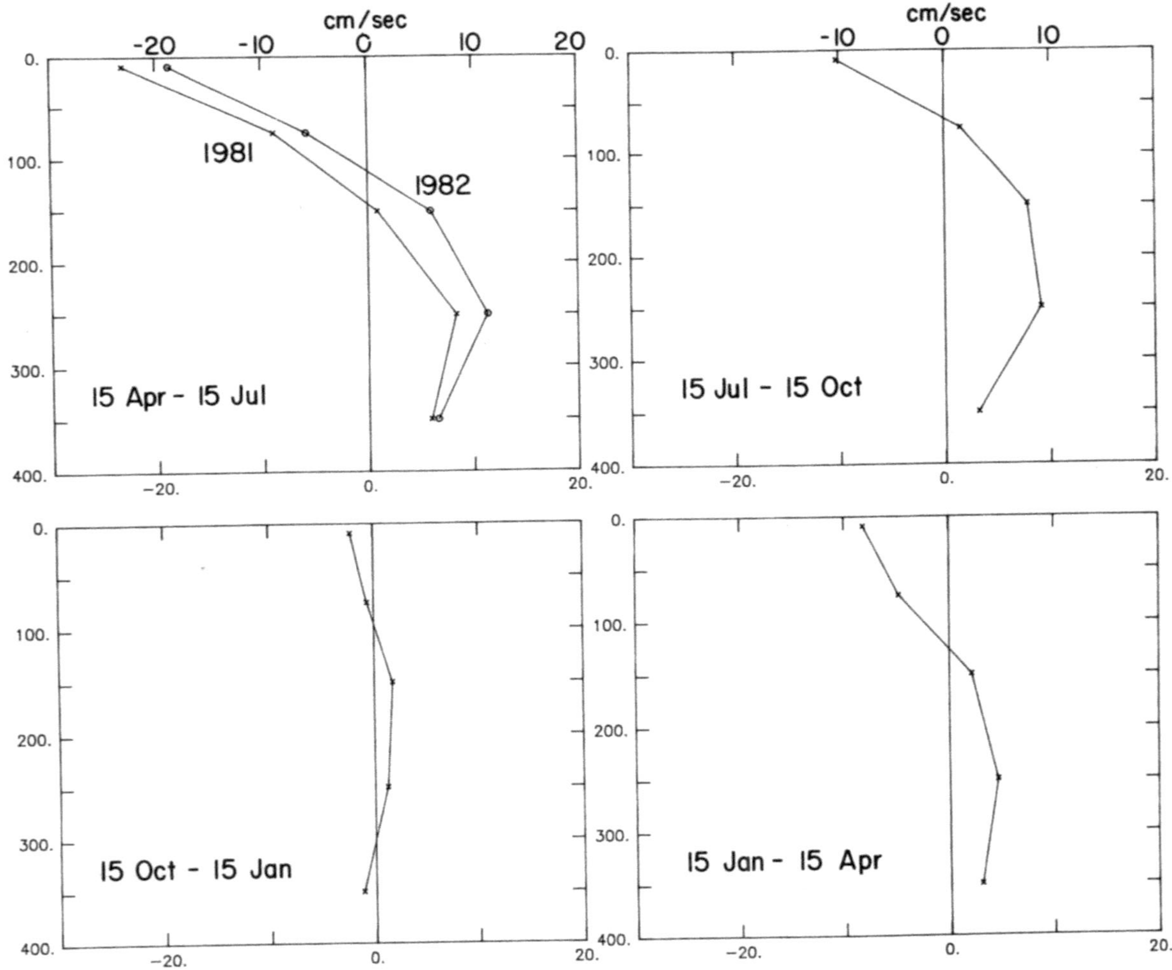

Figure 7: Vertical profiles of the three-month average along-shore current (positive toward 335°T) at C-5.

Table 1. Overall means and standard deviations of the current and temperature at the C-5 mooring.

Direction of Major Axis			Onshore Current (65°T)		Along-shore Current (335°T)		Temperature		
Depth	N	(°T)	mean	std dev	mean	std dev	mean	std dev	N
9	1717	317	-1.1	13.1	-12.8	18.8	11.30	1.18	1968
75	1960	337	2.9	7.3	-2.6	12.7	9.75	0.81	1960
150	1959	333	1.2	4.4	5.0	12.3	8.48	0.13	1959
250	1960	335	0.4	2.0	8.1	11.9	7.84	0.11	1960
350	1959	338	0.5	0.3	4.2	7.5	7.08	0.07	1959

Table 2. Three-month means and standard deviations of the onshore (65°T) and along-shore (335°T) components of the current at C-5.

	15 Apr- 15 Jul 1981		15 Jul- 15 Oct 1981		15 Oct- 15 Jan 1982		15 Jan- 15 Apr 1982		15 Apr- 15 Aug 1982	
depth	mean	std dev	mean	std dev	mean	std dev	mean	std dev	mean	std dev
u 9	-6.0	18.8	-4.9	10.0	(2.7)	(9.1)	-0.6	11.3	2.5	9.8
75	+4.2	10.5	1.2	4.8	1.3	5.4	1.5	7.5	3.6	5.9
150	1.6	5.2	0.4	3.4	0.6	4.4	1.6	5.2	1.0	3.8
250	0.3	2.8	0.0	1.4	0.7	1.8	0.7	2.0	0.3	1.7
350	0.0	1.1	0.1	0.9	0.6	0.7	0.5	0.8	0.6	0.8
v 9	-22.8	18.3	-10.2	19.3	(-6.3)	(8.3)	-8.0	13.8	-18.9	20.0
75	-9.2	15.5	1.6	10.5	-0.6	7.5	-4.6	11.5	-6.0	11.0
150	0.7	14.6	7.9	9.4	1.8	9.4	2.3	12.4	5.9	10.0
250	8.2	14.6	9.3	9.9	1.2	3.0	4.6	3.0	11.4	9.7
350	5.9	8.7	3.2	6.1	-1.2	2.1	3.0	1.6	6.5	6.6

it is strongest during the upwelling season and that it disappears or merges with the Davidson Current in winter. We also think that its core occurs at different depths (and densities) at different latitudes, but this is really only a guess since simultaneous observations from different latitudes have only recently become available. We do not know what determines the undercurrent variability on time-scales of weeks which does not seem to be directly related to variations in the wind. With so little known about the structure and variability of the undercurrent, we obviously do not know how it is forced.

CONCLUDING REMARKS

There is a great deal we still do not understand about the undercurrent that flows poleward along the continental slope off California and the Pacific Northwest. Many questions remain: Is the undercurrent continuous along the coast? What determines the width, strength, and the core-depth of the undercurrent? How and why do these vary with location and from season to season? Are the northward-tending tongues seen in water property distributions due primarily to advection, or are mixing and interleaving more important? Is the poleward flow observed near the bottom over the outer shelf merely a manifestation of the slope undercurrent, or is it an independent phenomenon with its own characteristics and forcing mechanism? We hope that these and other questions about undercurrents will be addressed by an integrated program of additional observations and improved models in the decade to come.

ACKNOWLEDGEMENTS

We wish to express our appreciation for many fruitful discussions with colleagues over the years, especially to Bob Smith, John Allen, and Barbara Hickey. Thanks also to Steve Ramp for his thorough review of this article. This paper was completed with support from the National Science Foundation through Grants OCE-8410546 and OCE-8709930 to Oregon State University.

From: Steven Ramp, Naval Postgraduate School, Monterey, CA

On: Review and commentary to paper **POLEWARD FLOWS IN THE CALIFORNIA CURRENT SYSTEM** by A.Huyer, P.M. Kosro, S.J. Lentz, and R.C. Beardsley

I feel that I must tread lightly when commenting on a review paper written by a group of authors whose combined experience in the California Current exceeds my own by many years. Indeed, the authors have done a very complete job of summarizing the available observations in the California Current System. (They found some I had not even heard of!). My primary criticism would be needless heavy emphasis on the CODE experiment, which occupies fifty percent of the figures and a good portion of the text. This is perhaps understandable in view of the very close association of all the authors with that experiment; but being primarily a shelf experiment, it was no more illuminating than many other experiments when it comes to the California Undercurrent. While recognizing the need for brevity, I would include a few more figures. There are no figures illustrating the section on "undercurrents over the shelf." This is where CODE shines, and I would include perhaps Figure 19b from Winant et al. (1987), or Figure 7 from Strub et al. (1987), although there may be better ones somewhere. I would also include Figure 2 from Chelton (1984), which shows the mean seasonal distribution of along-shore currents off Point Sur, from the CalCOFI data and represents a very nice summary of what is known about that section of the coast. This paper could also be strengthened by including some comments on the paper by Lynn and Simpson which just appeared in the most recent issue of JGR (Vol. 92, No. C12, Nov. 1987). This paper is perhaps the most complete analysis so far of the CalCOFI data, and thus, covers a spatial area far greater than the earlier work, which is useful for things like examining the continuity and seasonal variability of the coastal countercurrent and the undercurrent. Two of the more interesting results of Lynn and Simpson (1987), as far as the poleward flows are concerned, are: 1) observational evidence that the coastal countercurrent represents a surfacing of the UC, and 2) that the time of maximum flow in the UC is in fall and winter, rather than when the surface equatorward winds and currents are strongest as suggested by the CODE data.

I suspect that dynamics are being covered more completely in other papers being submitted to the volume, but on page 3, paragraph 2, I would at least list the possibilities for forcing of the poleward surface flow in the face of equatorward wind stress. McCreary (1987) lists four possibilities: remote forcing, relaxation of equatorward wind, the positive curl of equatorward wind stress, and thermohaline forcing, and comes down heavily in favor of the wind stress curl for the California Current case. Likewise, at the bottom of page 9, we do not know how the undercurrent is forced, but some possibilities could be mentioned. Two which immediately come to mind are the alongshore pressure gradient established by the equatorward wind stress at long time scales, and Kelvin waves excited by periodic forcing at periods of 20 days or less (Philander and Yoon, 1982).

To summarize, this paper makes a nice contribution to the "lecture notes" series. I feel it could be strengthened by a slightly more balanced approach that focuses less on the CODE experiment, and includes some comments on the most recent paper by Lynn and Simpson, which are very relevant to the subject at hand.

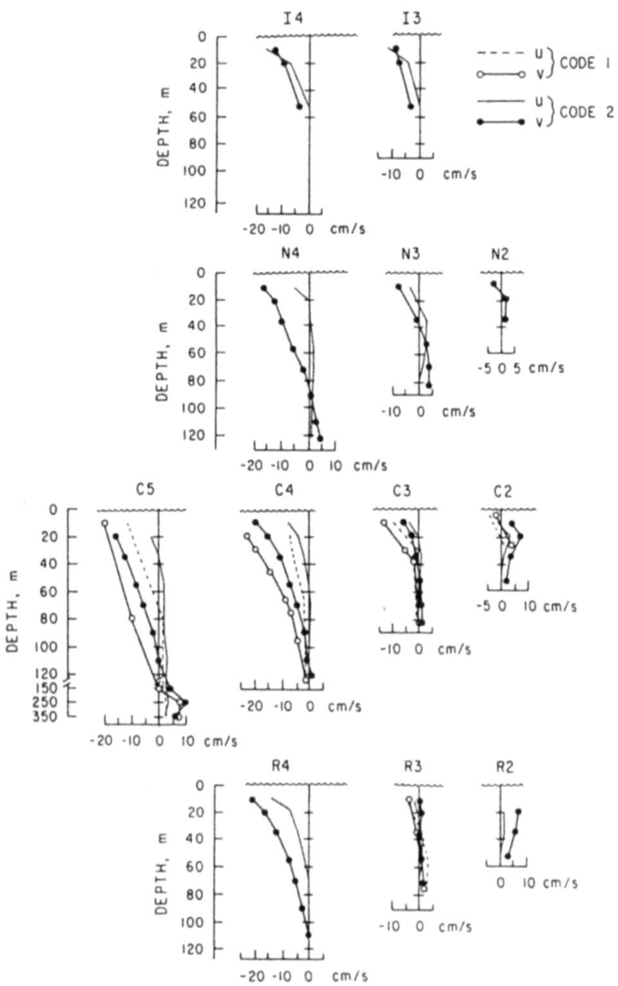

Spatial structure of time average currents during the CODE 1 and CODE 2 common analysis periods in vertical profile. (from Winant, et al., 1987; their Fig. 19b).

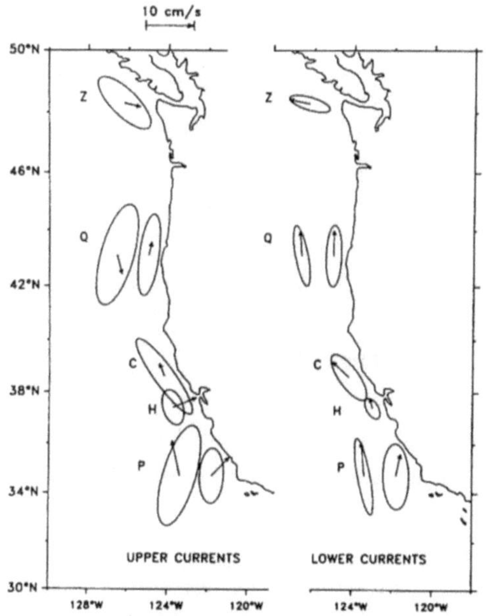

Annual vector means (arrows and principal axis ellipses, calculated from one complete year of 6-hourly current data (May 1, 1981 to May 1, 1982), for the top and bottom instruments on each mooring. The centers of the ellipses are at the latitudes of the moorings, but the longitudinal positions have been displaced for better viewing. (from Strub, et al., 1987a; their Fig. 7).

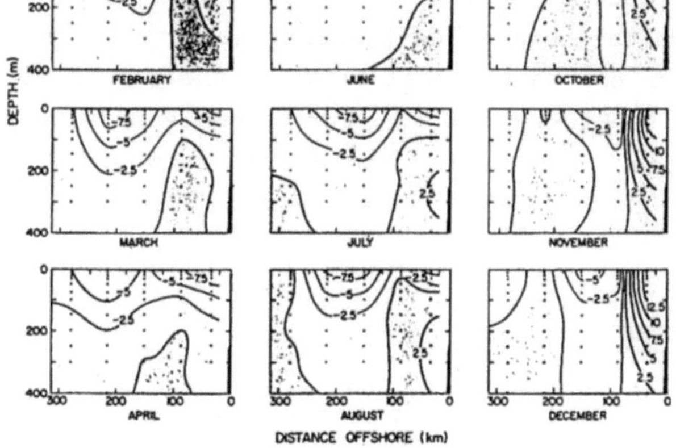

Vertical sections of seasonal alongshore geostrophic velocity relative to 500 db along CALCOFI line 70 off Point Sur. Shaded regions correspond to poleward flow (from Chelton, 1984; his Fig. 2).

POLEWARD FLOW NEAR THE NORTHERN AND SOUTHERN BOUNDARIES OF THE U.S. WEST COAST

Barbara Hickey
School of Oceanography
University of Washington
Seattle, WA 98195

ABSTRACT

Poleward flow over the shelf and slope is a common feature of west coast circulation. This paper describes the present state of knowledge of poleward flow near the end points of the U.S. West Coast, Washington State, and southern California, and compares the occurrence of poleward flow in the two regions. Attention is given to the occurrence of equatorward flow, which is now becoming the anomaly rather than the norm in west coast current observations. In particular, the occurrence of equatorward flow over the shelf during winter in relation to the Columbia River plume, and an equatorward undercurrent over the slope during winter off the Washington coast are described.

INTRODUCTION

Poleward flow over the shelf and slope is a common feature of west coast circulation. This paper uses existing data to describe the present state of knowledge of poleward flow near the end points of the U.S. West Coast, namely, Washington State and southern California. Attention is given to the occurrence of equatorward flow, which is now becoming the anomaly rather than the norm in west coast current observations. For a complete description of the seasonal flow off the Washington coast, the reader is referred to Hickey (1989a). For a complete description of the seasonal flow in the Santa Monica/San Pedro basin and shelf area off southern California, the reader is referred to Hickey (1989b).

DATA AND DATA SOURCES

Results in this paper are based primarily on current meter records. The meters were generally Aanderaa's and moorings were generally a subsurface taut wire type. Nearsurface and nearshore data were collected with EG&G Vector Measuring Current Meters suspended from surface toroids. Washington data were collected over a 15-year period from 1971 to 1984. Southern California Bight data were collected over the period 1985 to 1987. The sampling interval was generally 10 or 20 minutes. Data were filtered to obtain hourly values of current speed and direction. Because the flow generally follows the isobaths, the data were rotated into a local isobath system for most of the figures. The majority of the data are presented as monthly means. Since typical time scales are on the order of 10 days, only records whose length exceeded nine days were included in these averages.

RESULTS

Washington Shelf

Poleward flow is the rule rather than the exception on the Washington shelf for the below-pycnocline (20 m) water column. This is illustrated in Figure 1, which depicts monthly mean along-isobath velocity for selected depth intervals at inner-, mid-, and outer-shelf locations. The data available off Washington are insufficient to define the seasonal cycle in the upper 20 m of the water column. The meager data that do exist (two short summer data sets over the mid- and inner-shelf) indicate that considerable vertical shear and even flow reversals occur over the upper 20 m (Figure 2) (Hickey, 1989a). Equatorward flow occurs at all depths below the surface layer and at all locations only during spring. Equatorward flow is more persistent on the outer-shelf than at mid-shelf, occurring from spring through late summer on the outer-shelf, but only through early summer at mid-shelf. This differs from the Central Oregon shelf (~45°N), where measurements indicate that flow at mid-shelf is primarily equatorward throughout the summer (Kundu and Allen, 1976; Huyer et al., 1978). Thus, during late summer, the equatorward coastal jet is located farther offshore off Washington than off Oregon. The apparent offshore migration of the location of the equatorward coastal jet as the summer advances is illustrated in Figure 2, which gives monthly mean velocity across the Washington shelf during early (June) and late (August) summer. A second mid/late summer (July/August) example is given in Figure 5, which is presented in the next section. The shelf edge, upper slope, and location of the equatorward jet during mid-to-late summer off Washington is similar to that reported for the coast of Vancouver Island (Freeland et al., 1984), off southern Oregon (Strub et al., 1987) and off northern California (Winant et al., 1987). The occurrence of stronger monthly mean mid-shelf poleward flow off Washington more than off Oregon during mid-to-late summer may be related to the poleward decrease in

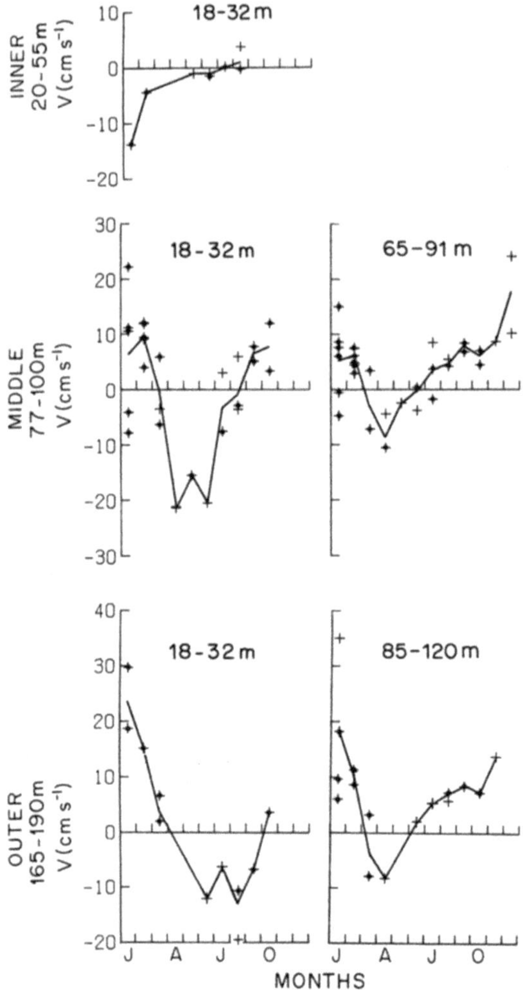

Figure 1: Monthly mean along-isobath velocity in several depth intervals at inner-, mid-, and outer-shelf locations off the Washington coast. The weighted monthly mean is shown as a solid line. Values calculated using less than 15 days of data are indicated with simple crosses.

equatorward wind stress that occurs between Oregon and Washington during the summer season (Huyer, 1983).

Poleward flow occurs at most shelf locations along the west coast during the winter season (Strub et al., 1987). An exception to this rule appears to occur off the Washington coast on the inner-shelf during late winter (Figure 3). This equatorward flow may be related to the Columbia River plume, which has a seasonal maximum in winter, as well as in spring. In support of this hypothesis, we note that geostrophic velocities

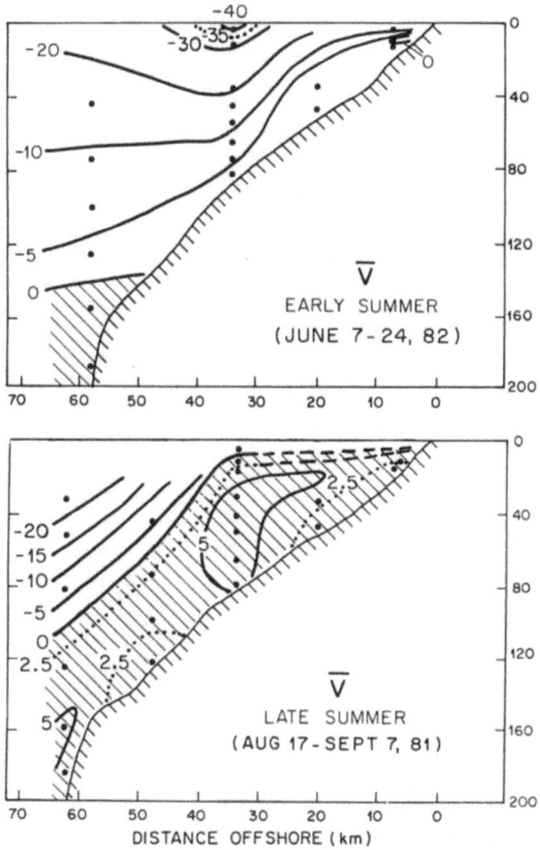

Figure 2: Cross-shelf sections of mean along-isobath velocity off the Washington coast during early and late summer. Units are cm s^{-1}. Regions of poleward flow are shaded. Dots indicate locations of current meters.

computed from seasonal mean hydrographic data also indicate equatorward flow (below a thin, poleward flowing surface layer) on the inner-shelf off Washington where the plume occurs, but not off Oregon where the plume is generally absent during this season (Figure 4).

Monthly mean poleward flow during winter often increases from the mid- to outer-shelf or upperslope off the Washington coast. This result is confirmed in mean geostrophic data (Figure 4) and in individual data sets 1975, Figure 6; 1977 and 1973, Figure 3; 1980, Figure 7), as well as in the statistical mean (Figure 1). Off Oregon, on the other hand, both mean geostrophic data (Figure 4) and direct measurements (1975, Huyer et al., 1978; 1978, Huyer et al., 1984) indicate that monthly mean poleward flow decreases from the mid- to the outer-shelf.

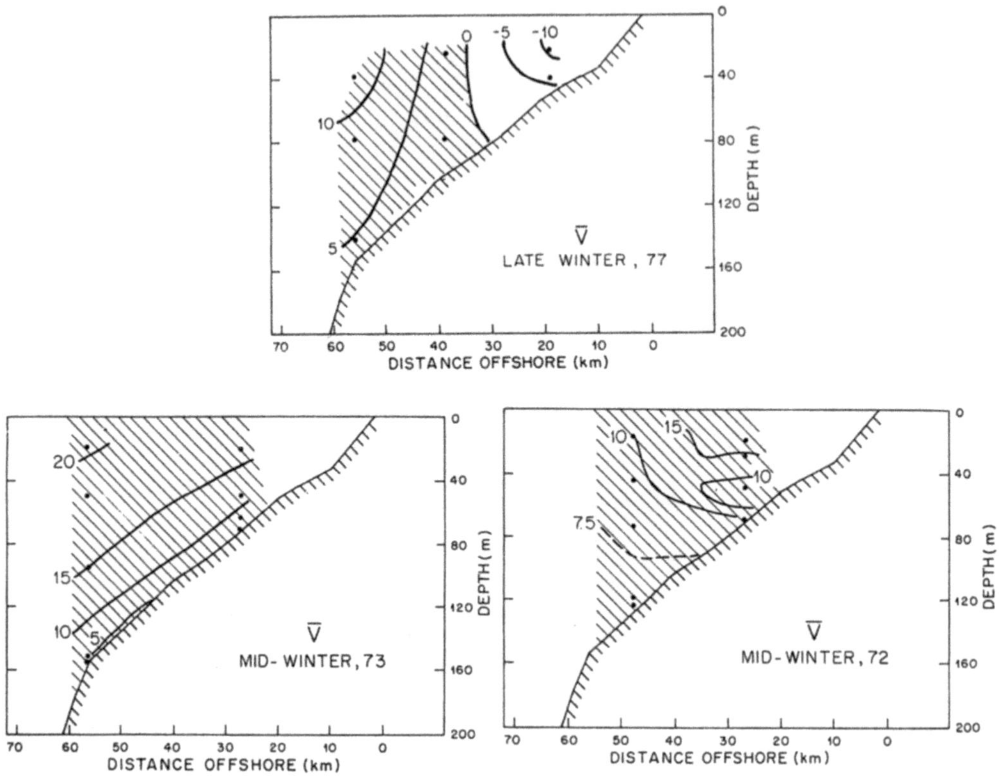

Figure 3: Cross-shelf sections of mean along-isobath velocity off the Washington coast during winter. Units are cm s^{-1}. Regions of poleward flow are shaded. Dots indicate locations of current meters.

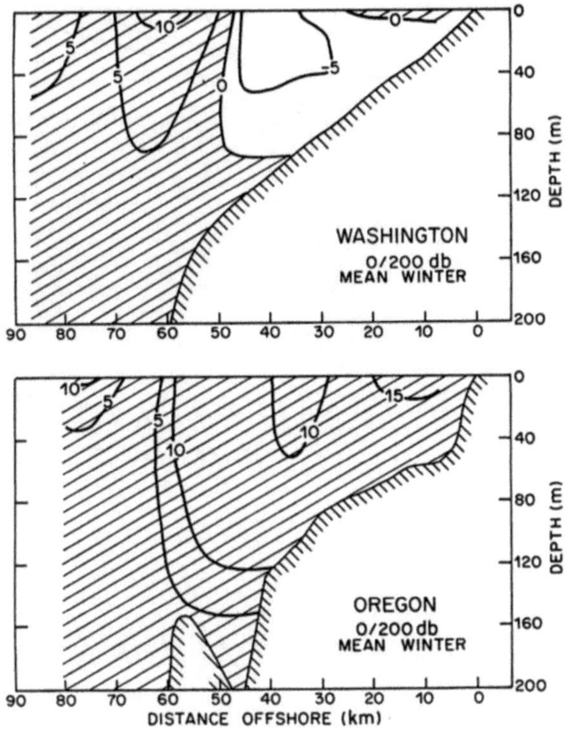

Figure 4: Geostrophic along-isobath velocity (0/200 dB) corresponding to seasonally averaged sigma-t sections for mid-Washington and mid-Oregon during winter. Units are cm s^{-1}. Data have been extrapolated into regions shallower than the reference level according to the method of Montgomery (1941). Regions of poleward flow are shaded. (Figure 2.13 in Hickey, 1989a.)

Washington Slope

Measurements have indicated that poleward flow occurs over the Washington slope below the surface layer (usually below ~100 m) throughout most of the year Figure 5, summer; Figure 6, winter). No data are available over the Washington slope during spring. However, data from the Oregon slope suggest that during spring, poleward flow is replaced with equatorward flow to depths of about 400 m (Huyer and Smith, 1976). During summer the flow over the Washington slope has the structure of a poleward jet, with maximum speeds occurring at shelf break depths (Figure 5). The jet appears to be trapped over the slope. During winter, there is no evidence to suggest that a poleward undercurrent persists; that is, that a subsurface maximum occurs (Figure 6). Rather, the limited data that are available off the Washington coast suggest that poleward flow below the

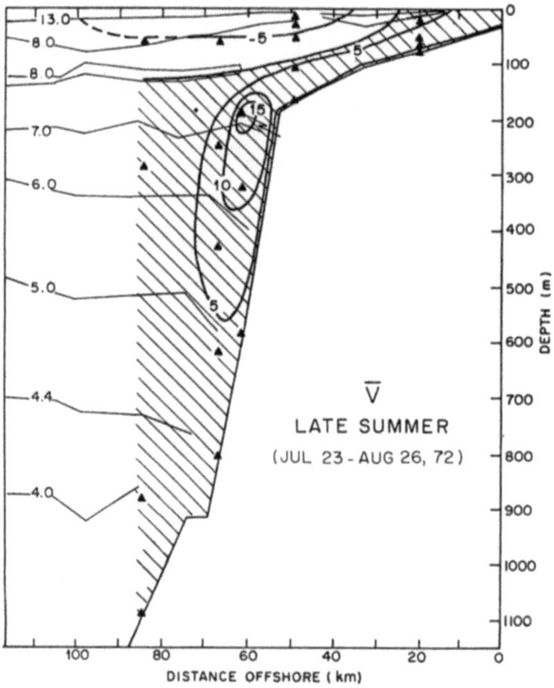

Figure 5: Cross-shelf section of seasonal mean along-isobath velocity off Washington during late summer 1972 and of temperature on 27 July 1972. Units of velocity and temperature are cm s^{-1} and °C, respectively. The data illustrate the cross-shelf structure of the poleward undercurrent. Regions of poleward flow are shaded. Locations of current meters are indicated by triangles. (Adapted from Figure 10 in Hickey, 1979.)

shelf break has a relative minimum near ~400 m (Figure 6) or that an equatorward undercurrent may even occur (Figure 7). This result seems consistent with available information on wind and along-shore pressure gradient (as measured by adjusted sea level slope); namely, in winter off Washington, along-shore wind stress and along-shore pressure gradient force are opposite in direction, and both are reversed in direction from the summer case (Figure 8). If a simple along-isobath balance of momentum is considered, it seems reasonable to expect the resulting current distributions to be opposite during winter and summer, as observed. Data from the Oregon slope during October and November 1977 clearly demonstrate the existence of persistent vertical shear in the water column below the shelf break (Figure 9). The shear is such that the tendency toward equatorward flow increases with depth. Maximum equatorward flow appears to occur at least as deep as 325 m (from the Oregon data) or 392 m (from the Washington data). Vertical shear of the same sign is also evident in mid-shelf data (Figure 9). However, the shear on the shelf is insufficient to produce persistent equatorward flow. Although one example of

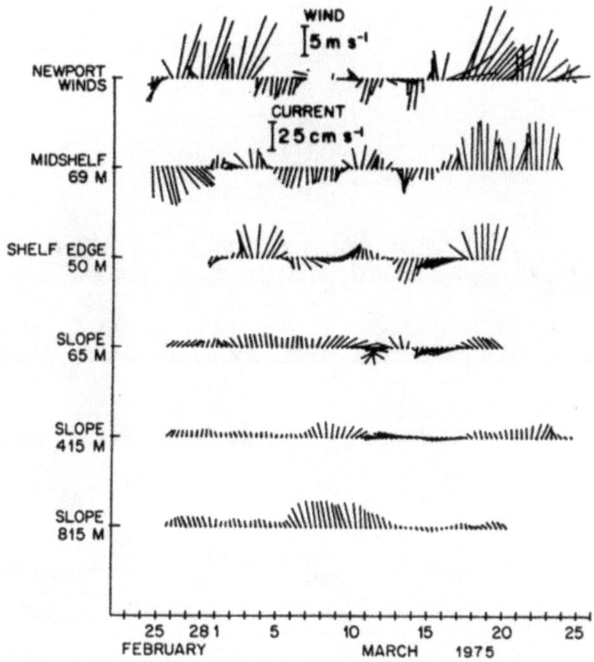

Figure 6: Selected sub-tidal (8-hourly) velocity vectors over the Washington shelf and slope during late winter 1975. Mooring depths are 94,182, and 1000 m for mid-shelf, shelf edge and slope locations, respectively. Winds at Newport, Oregon are also shown. (Adapted from Figure 12 in Hickey, 1979.)

equatorward flow over the slope during late winter has been related to the presence of an eddy (Huyer et al., 1984), we note that in this case the equatorward flow decreased rather than increased with depth. Since each of the three data sets available for the Pacific Northwest indicate the presence of persistent vertical shear over the slope below the shelf break during winter, it seems reasonable to expect that this phenomenon may be related to a seasonal reversal of the wind stress/sea level slope relationship rather than to the random occurrence of eddies.

In the above discussion concerning the origin of the poleward flow, the assumption has been made that poleward flow occurs as a result of an applied along-shore pressure gradient force in the presence of bottom stress (as in Csanady, 1978). The ultimate driving mechanism of the poleward flow is whatever is creating the along-shore sea level slope. Hickey and Pola (1983) demonstrated that both the amplitude and the phase of the seasonal and long-term mean signal in along-shore sea level slope in the Pacific Northwest could be reproduced using seasonal and mean along-shore wind stress data and the simple physics of the arrested topographic wave model (Csanady, 1978). The driving mechanism is a combination of local

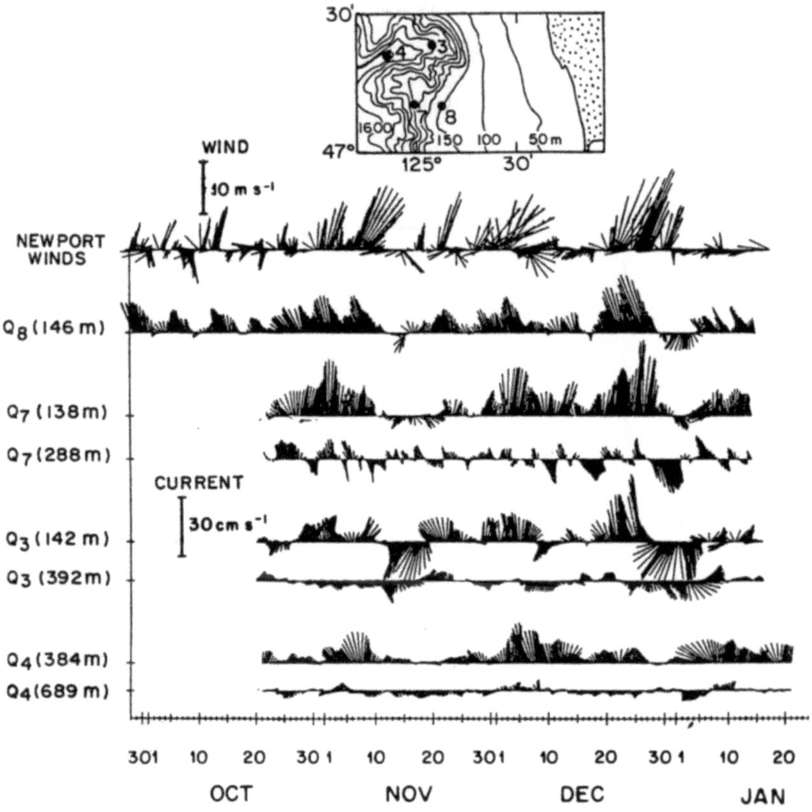

Figure 7: Selected sub-tidal velocity vectors in a north-south reference frame at locations on a section across the Washington shelf and slope during fall and winter 1980. Mooring locations are shown on the inset map. Winds at Newport, Oregon are also shown. (Adapted from Figure 9 in Werner and Hickey, 1983.)

and remote wind stress forcing. The model was not very successful off southern California where, nevertheless, a large along-shore sea level slope occurs. Recently, several other theories for poleward flow have been presented. These theories indicate that poleward flow can be generated by a variety of mechanisms; e.g., the interaction of nonlinear shelf waves or eddies with along-shore variations in topography (Holloway et al., this volume) or topographic form drag (Samelson and Allen, 1987). It is unclear at present what fraction of the signal can be attributed to each mechanism. However, intuitively, it seems that the availability of regions of distinct types of topography and distinct seasonal forcing should allow the relative importance of the proposed mechanisms to be separated.

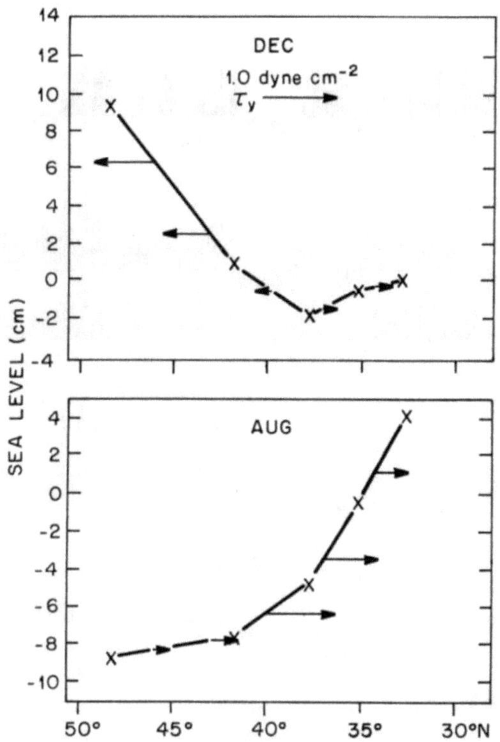

Figure 8: The north-south component of selected sub-tidal velocity time series on a section across the Oregon shelf and slope at 45°20'N during fall and winter 1977. Note scale change for current time series. The north-south component of wind on the Columbia River Lightship (46°10'N) is also displayed. Shelf (E) and slope (O) moorings are in water depths of 110 and 600 m, respectively. At the 50 m depth over the slope only direction was measured. Time series were constructed for that meter using a uniform speed of 10 cm s^{-1}.

Southern California Bight Slope

Poleward flow in the California Bight over the upper slope inshore of the Channel Islands is remarkably more persistent than that over the Washington slope. Poleward flow occurs at 40 m and 100 m depths everywhere except on the seaward side of the Santa Monica basin during winter and over the slope off San Onofre in the upper water column during late fall and spring (compare Figure 10 for southern California with Figures 5 and 6 for Washington). Off Washington, on the other hand, equatorward flow occurs from April through September in the upper 100 m over the slope. We note that flow over the open slope off San Onofre is less poleward than that off Santa Monica, but still more poleward than that off the Washington slope. Results similar to those for Santa Monica have been reported just north of the Santa Monica basin in another constricted region; namely,

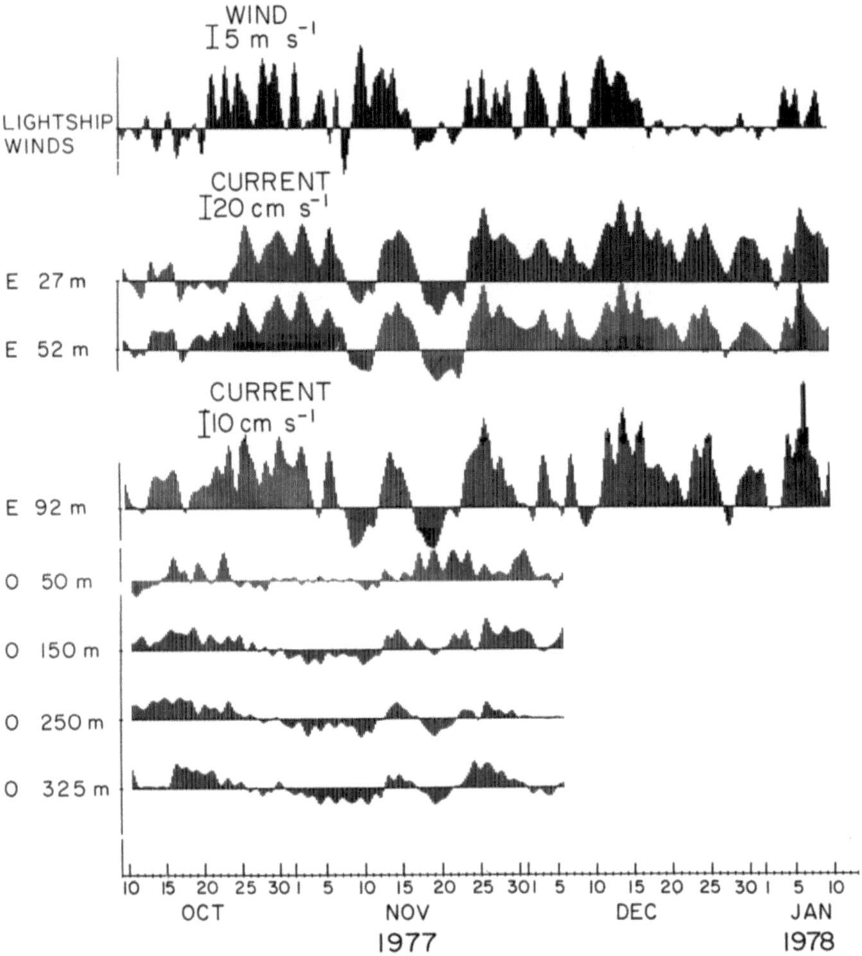

Figure 9: Monthly mean adjusted sea level height and the north-south component of wind stress along the west coast during typical winter and summer months. Wind stress data are from Nelson (1977). Sea level height data are from Hickey and Pola (1983).

the Santa Barbara Channel (Brink and Muench, 1985). Only a single ~one week reversal to equatorward flow was observed along the north side and eastern entrance to the channel at a depth ~30 m during the observation period of April to June 1983. Data from year-long (1984) current records in the channel also indicate poleward monthly means throughout the year at the ~30 m depth. Only infrequent reversals to equatorward flow occur even at this relatively shallow (30 m) depth (Gunn et al., 1986). At least at the eastern entrance to the channel, reversals are more common during spring (several occurred during that period) and almost never occur after July (only two occurred between July and February). The reversals almost never reached the deeper measurement depths of 80 m and 126 m.

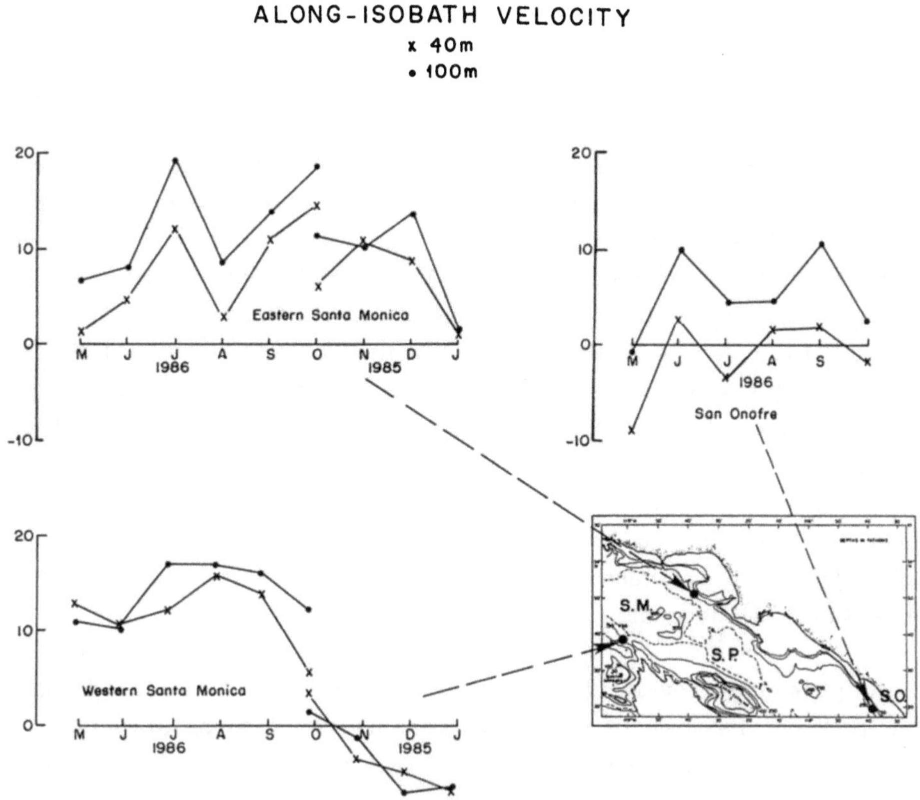

Figure 10: Monthly mean speed at selected locations over the continental slope (Z_b = 800 m) in Santa Monica/San Pedro basin and off San Onofre. Units are cm s^{-1}. Mooring locations are shown on the inset map. The data years are presented out of order to allow the semblance of a continuous record. A positive speed indicates flow directed poleward; a negative speed indicates flow directed equatorward. Since the flow is roughly paralled to local isobaths, positive and negative speeds conform almost exactly to poleward and equatorward components of along-isobath velocity.

No directly measured data are yet available over the Santa Monica slope from February to April, and no data are available above the pycnocline over the slope during any season. While it is possible that equatorward flow may occur in a thin (<20 m) surface layer during spring and summer, available satellite data usually suggest that the surface layers move poleward through the region during the spring season (not shown). Instruments presently in the water will resolve the question of the direction of the surface layer flow over the slope. Deployments during 1988 will determine whether flow reversals occur at deeper depths during the spring, as reported for <u>shelf</u> locations just outside the Bight (Strub et al., 1987). We note, however, that large-scale geostrophic data presented by Hickey (1979) suggests that no reversals occur in the seasonal averages at depths 200 m in the northern half of the Bight, even during spring.

The amplitude of the seasonal variation is similar to that reported for the shelf outside the Bight (Figure 10) (Strub et al., 1987). The seasonal variation of the amplitude is not inconsistent with the existence of a semi-annual signal in the poleward flow. A semi-annual signal in this region, with maxima in late summer/early fall and winter, has been observed in large scale geostrophic data (Hickey, 1979). These large scale results have recently been substantiated with direct measurements on the Point Conception Shelf (Strub et al., 1987). In the present data set, the seasonal maxima along the eastern boundary of the basin occur in July (in the Santa Monica-San Pedro Basin) or June (south of the basin) and between September and December. The scatter in the data makes the evidence barely convincing. The flow on the western side of the basin is directed equatorward after October. Note that the December maximum is obtained from data in 1985. During 1986 poleward flow stronger than the winter maximum in 1985 occurs as early as September. Clearly inter-annual variability is significant in the seasonal cycle in the Bight, as it is off the Washington coast.

The poleward flow in the Bight appears to have a subsurface maximum during much of the year (Figure 10); that is, it has the structure of an undercurrent. This is in contrast to the case off Washington, where the poleward undercurrent structure disappears or perhaps even reverses during fall and winter. However, as off Washington, the vertical shear is generally a maximum during summer and early fall and a minimum in late fall and winter. The difference between the two locations is qualitatively consistent with the observed differences between the wind and sea level slope distributions. Off Washington both along-shore wind stress and along-shore sea level slope reverse direction seasonally (Figure 8); off southern California they do not reverse except perhaps during January; rather they are always directed the same as occurs off Washington during summer, so that the tendency for the existence of a poleward undercurrent persists throughout the year. Note that the amplitude of the sea level slope has a seasonal cycle that is similar to that of the seasonal cycle of vertical shear: the maximum poleward decreasing sea level slope occurs in late summer and early fall in both regions, corresponding to the maximum in vertical shear in both regions; the minimum sea level slope occurs in winter (near zero off southern California, large, but decreasing equatorward off Washington) corresponding to the minimum in vertical shear off southern California and a minimum or a reversal in shear direction off Washington (Hickey and Pola, 1983).

The flow in the Bight does not appear to be trapped over the slope as is the case off Washington, except perhaps during winter (Figure 10). During the period of strongest poleward flow, the flow extends at least

40 km to the other side of Santa Monica Basin, with almost uniform strength across the basin. The strength of the seasonal mean flow is of the same order as that off Washington.

Southern California Bight Shelf

The shelf in the Southern California Bight is not continuous like the Washington shelf; rather, several bays cut into the coastline in the Northern Bight (Figure 10). The shelves in these bays are relatively wide (~20 km), although still only half the width of most of the Washington shelf. Between the San Pedro shelf and Northern Baja, the shelf width is relatively uniform, but narrow (~3-10 km). Over both types of shelves, flow appears to be equatorward, at least in the near-surface layers, during much of the year, in contrast to the flow over the Washington shelf. For example, on the wide Santa Monica shelf, measurements to date suggest that the seasonal mean flow is equatorward at 5 m and 10 m from the surface in 30 m of water from October to February, and offshore or equatorward at a depth of 25 m from May to October (Hickey, 1989b). Similar results have been reported for the narrow (~3 km) shelf just north of San Diego (Winant and Bratkovitch, 1981). Equatorward flow was observed 5 m from the surface in 15, 30, and 60 m water depths in every season, and in spring and fall as deep as 10 m from the surface at the deeper measurement sites. The strongest equatorward flow occurred during winter, when the equatorward flow extended throughout the water column at each of the three sites. The magnitude of the monthly mean equatorward flow on both the narrow and wide shelves (~5-10 cm s^{-1}) is much less than that of the poleward flow over the upper slope. The only significant poleward flow on the narrow shelf occurs during summer (August data, Winant and Bratkovitch, 1981) in the lower half of the water column at the 30 m and 60 m bottom depths: the typical undercurrent that we expect on this eastern boundary. No undercurrent was observed in the monthly means at 5 m above the bottom in a depth of 30 m on the wider Santa Monica shelf.

On the Santa Monica shelf, equatorward flow during fall and winter decreases offshore: flow even in the surface layers at the outer shelf monthly mean flow is poleward rather than equatorward (Hickey, 1989b). On the narrower San Diego shelf, equatorward flow during winter and summer increases between the mid- and outer-shelf measurement sites (Winant and Bratkovitch, 1981). The data are insufficient to determine the location of the flow maximum.

SUMMARY

In this paper differences and anomalies in west coast seasonal flow structures have been highlighted. In particular, it was emphasized that flow off Washington has significant differences from that of Oregon; namely, during summer, flow at mid-shelf is more poleward off Washington, and during winter, flow on the inner-shelf is more equatorward off Washington than off Oregon. The former result may be related to the poleward decrease in the longshelf wind stress; the latter result may be related to the presence of the Columbia River plume. Off southern California the near-surface flow over the shelf is more persistently equatorward than that off Washington. Conversely, the flow over the slope in the upper 100 m of the water column is more persistently poleward than that off Washington. Also, the undercurrent structure, that is, a subsurface maximum, is maintained at least from summer to early winter off southern California (no data are yet available from spring), but only during summer and early fall off Washington. We note that the seasonal cycle of vertical shear in the two locations is similar, although a reversal in sign sometimes occurs off Washington.

ACKNOWLEDGEMENTS

This work was supported by the Department of Energy under Grant DE-FG05-85ER60333#4 and by the National Science Foundation under Grant OCE 86-01058#1.

From: Adriana Huyer, College of Oceanography, Oregon State University, Corvallis, OR.

On: Review and Commentary to paper **POLEWARD FLOW NEAR THE NORTHERN AND SOUTHERN BOUNDARIES OF THE U.S. WEST COAST,** by Barbara Hickey.

Although this paper covers some of the same ground (or water) as my own, I think the two are not redundant. This paper gives a rather detailed view of the structure and variability of the poleward flow at the "ends" of the California Current System as it flows along the U.S. West Coast. Although the data sets are rather fragmentary, Hickey does manage to draw conclusions about local differences in structure. She clearly shows the variability of the poleward flow on time scales of days-to-weeks by including several time-series plots of current vectors.

POLEWARD FLOWS OFF MEXICO'S PACIFIC COAST

A. Badan-Dangon
J.M. Robles
J. García

CICESE. Apdo. 2732
Ensenada, B.C., Mexico

ABSTRACT

Poleward flows are predominant off the west coast of Mexico. Off the peninsula of Baja California, the northern third of the Mexican Pacific coastline, one finds the southern extension of the California Current System. Poleward flows in this region occupy a broad domain of about 200 km width, adjacent to the coast, and include an intensified countercurrent that can reach 50 cm s^{-1}, often within 20 km of the shelf and slope. From late winter to early summer, the upwelling favorable northwesterly winds oppose the poleward flow at the surface and the current proceeds as an undercurrent that hugs the continental slope. This poleward flow advects Subtropical Subsurface Water from the tip of Baja California into the Southern California Bight and farther north. The coast of mainland Mexico is exposed, on the contrary, to a broad extension of the equatorial return flow that feeds the North Equatorial Current from the Equatorial Countercurrent. This flow is known as the Costa Rica Coastal Current; it extends around the Costa Rica Dome, and northwestwards to the mouth of the Gulf of California, where the Eastern Pacific Transition Zone usually abutts the continent. This poleward flow is most prominent during the fall and may reverse directions in the spring, appearing as an intensified California Current extension. The strength of the large scale drifts, the surface expression of the counterflows, and the latitude at which the currents meet, all vary seasonally and depend on the intensity of the dominant winds. Direct observations of the poleward flows off Mexico are few and restricted to northern Baja California. They confirm a counterflow is probably present throughout the year, mostly over the shelf and slope regions, but with its surface expression obliterated by the upwelling favorable northwesterly winds that blow strongly in the late winter and spring. There are no direct measurements to provide a detailed description of the Costa Rica Coastal Current and, in particular, to know whether it possesses a counterflow.

MOTIVATION

Poleward flows are important, maybe even dominant, components of eastern boundary current systems. They are noteworthy for a number of reasons, the principal one being that their direction of motion often opposes that of the dominant winds and wind-driven currents, whence they must be propelled by rather robust sources of momentum. These flows are known as countercurrents if they flow at the surface, or undercurrents when they do not (Hickey, 1979; this volume). Counterflows are effective at distributing water properties over long distances (Tibby, 1941; Wooster and Reid, 1963; Cannon, Laird, and Ryan, 1975; Reed and Halpern, 1976) and compete against the wind-driven advection to determine the hydrographical composition of nearshore waters (Bernal, 1981; Bernal and McGowan, 1981; Chelton, 1981). Counterflows often provide the source water for the Ekman pumping offshore in eastern boundary currents (Chelton, 1982) and for upwelling at the coast (Yoshida and Tsuchiya, 1957; Smith, 1968,1981; Huyer, 1983); they thus contribute importantly to coastal ocean frontogenesis and to the nature of the jets and squirts that result from cross-shore exchanges with the open ocean (Mooers and Robinson, 1984; Davis, 1985; Huyer and Kosro, 1987). They also influence the propagation of internal tides and inertial waves (Mooers, 1975), and alter the exchanges with the atmosphere in the coastal zone (Tont, 1981). All this, of course, has substantial implications in matters such as the chemistry of nearshore waters (Codispoti, 1981) and variations in the structure of the coastal ecosystem as manifested, for example, by the alterations of the salmon migration routes (McLain and Thomas, 1983) or by the modification of plankton distribution patterns (Peláez and McGowan, 1986; Chelton, Bernal, and McGowan, 1982). The west coast of Mexico constitutes the southeastern boundary of the North Pacific Anticyclonic Gyre, and the northeastern limit of the eastern equatorial circulation, which contributes water from the Equatorial Countercurrent into the North Equatorial Current. As a result, poleward flows are prevalent in the entire region, in the first case as a classical example of a counterflow to midlatitude eastern boundary currents (Wooster and Reid, 1963), and in the second because the equatorial return flow constitutes the Costa Rica Coastal Current. The Costa Rica Coastal Current is a broad poleward flow that originates around the Costa Rica Dome, proceeds along the coast of the Mexican mainland, and merges in various proportions with the southern extension of the California Current to feed the North Equatorial Current (Wyrtki, 1965, 1966).

The present report summarizes the configuration and scales of variation of poleward oceanic motions off the Pacific seaboard of Mexico. It is organized as follows: Section 2 provides a broad geographical description, based mostly on large scale hydrographical surveys and ship drift compilations reported in the literature; Section 3 describes observations that detail the nearshore scales of the counterflows off northern Baja California, the only region off Mexico where direct measurements have been conducted; Section 4 summarizes some of the efforts that have been made to explain the origins of counterflows, and hypotheses from which future studies in the region might draw support.

THE POLEWARD FLOWS ON THE LARGE SCALE

Most of the Mexican west coast (Figure 1) is found in North America, extending from the border with the United States at 32°N, to the Gulf of Tehuantepec at 16°N; only a portion of about 300 km, the eastern side of the Gulf of Tehuantepec, is in Central America (Urrutia-Fucugauchi, 1988). The continental shelf is seldom wide, with the notable exceptions of some portions off Baja California, where it exceeds 100 km in width, and on the eastern side of the Gulf of Tehuantepec. This otherwise regular coastline, with a northwest to southeast orientation, is interrupted between Cabo San Lucas (23°N) and Cabo Corrientes (21°N) by the broad entrance to the Gulf of California, a marginal sea over 1500 km in length that lies to the east of the peninsula of Baja California. Two major wind-driven current systems compose the oceanographic makeup off the west coast of Mexico. These are the North Pacific Anticyclonic Gyre, whose southeastern limit constitutes the California Current System, and the extension of the Equatorial System into the Eastern Tropical Pacific, bounded by the coasts of Central and North America. The North and South Equatorial Currents flow to the west, separated by the eastward flowing Equatorial Countercurrent, located just north of the equator, between 2°N and 12°N. This major current penetrates seasonally into the Eastern Tropical Pacific, and a branch of it turns around the Costa Rica Dome, close to 8°N; 89°W, proceeds along the coast of Mexico, and feeds the North Equatorial Current.

The large scale configuration of the wind and current systems that encompass oceanic motions of the Eastern Tropical Pacific Ocean have been described abundantly. Early maps of these flows were offered by Puls (1895), and more detailed studies have been reported by Cromwell and Bennett (1959), Austin (1960), Reid (1961), and by Bennett (1963). Extensive data compilations and reviews of the circulation and water masses of the region have been prepared by Wyrtki (1964a, 1964b, 1965, 1966, 1967) and Tsuchiya (1974). Our description is taken largely from these latter works, and we reproduce the convenient summary made by Baumgartner and Christensen (1985) for their study of interannual

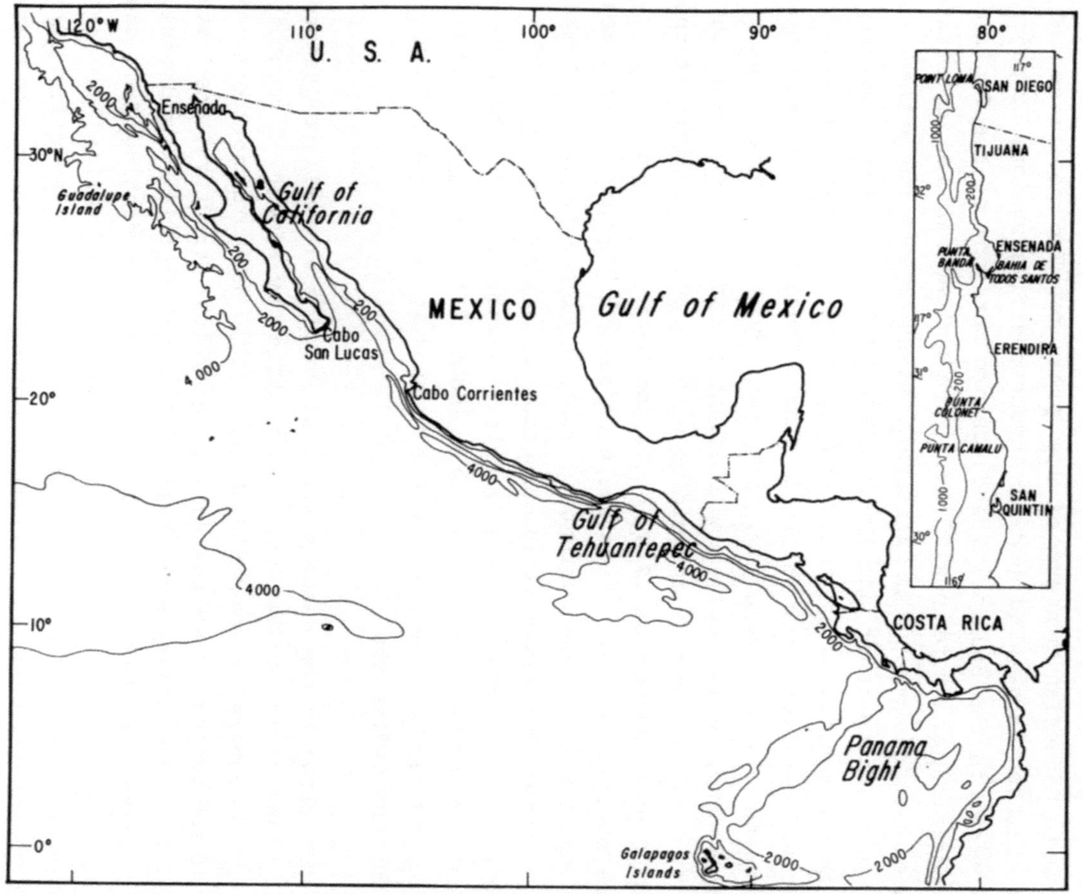

Figure 1: Configuration of the coast and of the continental shelf in the California Current and Eastern Tropical Pacific Regions off Mexico. The inset details the portions of the coast where direct measurements of poleward flows have been conducted off northern Baja California.

fluctuations in the Gulf of California (Figure 2). Other studies have included the Eastern Tropical Pacific because of the interest in the role of the Equatorial region in the generation and propagation of the El Niño signal (e.g. see Journal of Geophysical Research, Vol. 92C13).

The major pressure cells that determine the conditions of the wind field are the North and South Pacific centers of high pressure, and the Aleutian and Indonesian centers of low pressure. The positions and strengths of these large wind systems induce the fluctuations of major current systems in the equatorial and eastern portions of the Pacific Ocean. In addition, in a more local context, the Mexican Low is an important generator of fluctuations in the California Current System and in the Gulf of California (Baumgartner and Christensen, 1985). Eastern boundary currents in the subtropical gyre are characterized by equatorward winds that shift progressively into the Trade Winds. These are separated close to the equator by the doldrums, centered about the Intertropical Convergence Zone, whose position fluctuates considerably with the seasons; it is found roughly coinciding with the northern limit of the Equatorial Countercurrent at 12°N in the northern summer, and close to 2°N in the northern winter. These two extremes correspond to the times when the Equatorial Countercurrent is most developed in the northern hemisphere summer, but essentially non-existant in winter east of 130°W. Therefore, the equatorial climate is shifted to the north by about 10°; subtropical conditions reach only to about 20°N in the northern hemisphere, but can be found reaching the equator at the Galapagos Islands in the southern hemisphere.

Oceanic flows in the Eastern Tropical Pacific are well related to the local winds (Reid, 1948); they are typified by a pair of counter-rotating gyres centered at about 8°N; 89°W, and 5°N; 88°W. The northernmost is cyclonic and projects a well known hydrographical structure, the Costa Rica Dome. The circulation in the Eastern Tropical Pacific is intimately linked to these structures; an eastward current proceeding from the Equatorial Countercurrent flows between the two and splits into two branches (Roden, 1962). The southern one turns west, and eventually joins the South Equatorial Current. The northern branch flows around the Costa Rica Dome, whence it becomes the Costa Rica Coastal Current, a flow along the coast of Central America and off mainland Mexico. It reaches a latitude of confluence with the California Current, which is the eastern terminal of the North Pacific Transition Zone (Roden, 1970, 1971, 1972), and separates from the coast to feed the North Equatorial Current. The Costa Rica Coastal Current is of extreme importance in the context of the present discussion, for not only is it a swift poleward current in its own right, it is also the most likely process by which the Tropical Surface Water

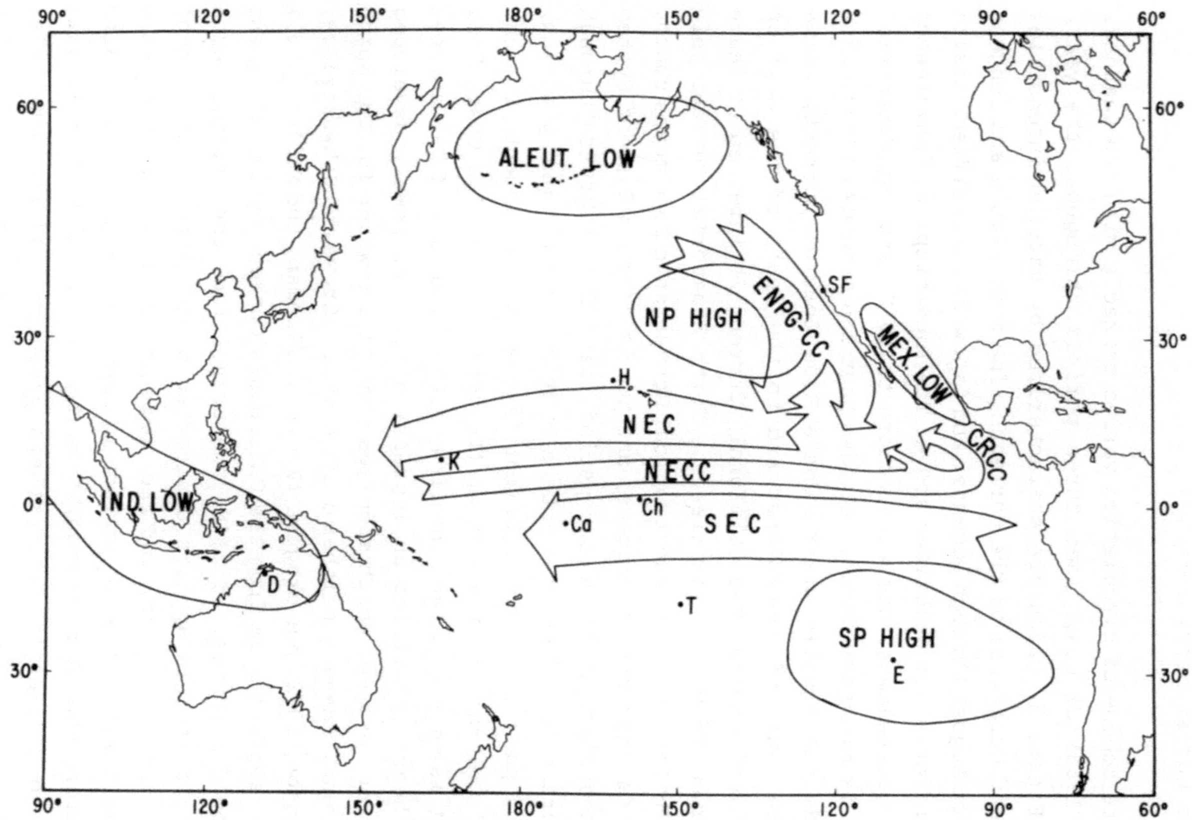

Figure 2: Locations of atmospheric pressure centers (Aleutian, Mexican, and Indonesian Lows; North Pacific and South Pacific Highs) and ocean surface current systems (Eastern North Pacific Gyre-California Current, North Equatorial Current, North Equatorial Countercurrent, South Equatorial Current, and Costa Rica Coastal Current) that contribute to the circulation off the Pacific coast of Mexico. The locations of some coastal (San Francisco) and mid-Pacific island sea level stations are also shown (H: Honolulu; K: Kwajalein; Ch: Christmas; Ca: Canton; T: Tahiti). From Baumgartner and Christensen (1985).

from the equatorial region and the Subtropical Subsurface Water, which originates in the Central Pacific, are recirculated and brought to the mouth of the Gulf of California (Wyrtki, 1967, Reid and Mantyla, 1978); there, a portion of the Subtropical Water feeds the California Countercurrent through a path that remains unknown (Simpson, 1983; Baumgartner and Christensen, 1985).

This entire scheme of circulation varies markedly with the seasons. Wyrtki (1965, 1966) distinguishes three periods throughout the year, summarized in Figure 3, which are characterized by varying intensities and changing configurations of the currents. The most stable and longest period lasts from August through December and is a well intensified version of the basic configuration described above. The Intertropical Convergence Zone is at its northernmost position, close to 10°N, and the Equatorial Countercurrent is fully developed; it flows past and around the Costa Rica Dome, and feeds the equally strong Costa Rica Coastal Current, which reaches past the Gulf of Tehuantepec and to the mouth of the Gulf of California. From the north, the California Current is inferred to leave the coast at about 25°N, and supplies the North Equatorial Current only north of 20°N. By January, the Intertropical Convergence Zone moves south, the Equatorial Countercurrent weakens, and the California Current System intensifies accordingly. The second period lasts from February through April. The Intertropical Convergence Zone is at its southernmost position, about 3°N, the Equatorial Countercurrent is essentially absent, but the two eddies off Central America remain (Tsuchiya, 1974). The Costa Rica Coastal Current is very weak, and the flow north of the Gulf of Tehuantepec is to the southeast, providing an impression of an extremely developed California Current System, which supplies most of the water for the North Equatorial Current. However, neither drift bottles (Crowe and Schwartzlose, 1972; Badan-Dangon, 1972) nor drogues (Poulain, Illeman, and Niiler, 1987) have traced the surface flow from the California Current System along the coast past Cabo Corrientes, at 21°N; the transition zone off the mouth of the Gulf of California appears unaffected by the intensification of the California Current System, and the composition of the southeastward flow off the mainland of Mexico in late winter and early spring remains unexplored. During the third period, from May to July, the Intertropical Convergence Zone returns to its northern position, and the Equatorial Countercurrent forms again, flows past the Costa Rica Dome and drives the Costa Rica Coastal Current to Cabo Corrientes. The California Current remains strong and contributes still a considerable fraction of the North Equatorial Current, although the southeastward flow does not penetrate much into the Eastern Tropical Pacific. From July to August the California Current weakens progressively.

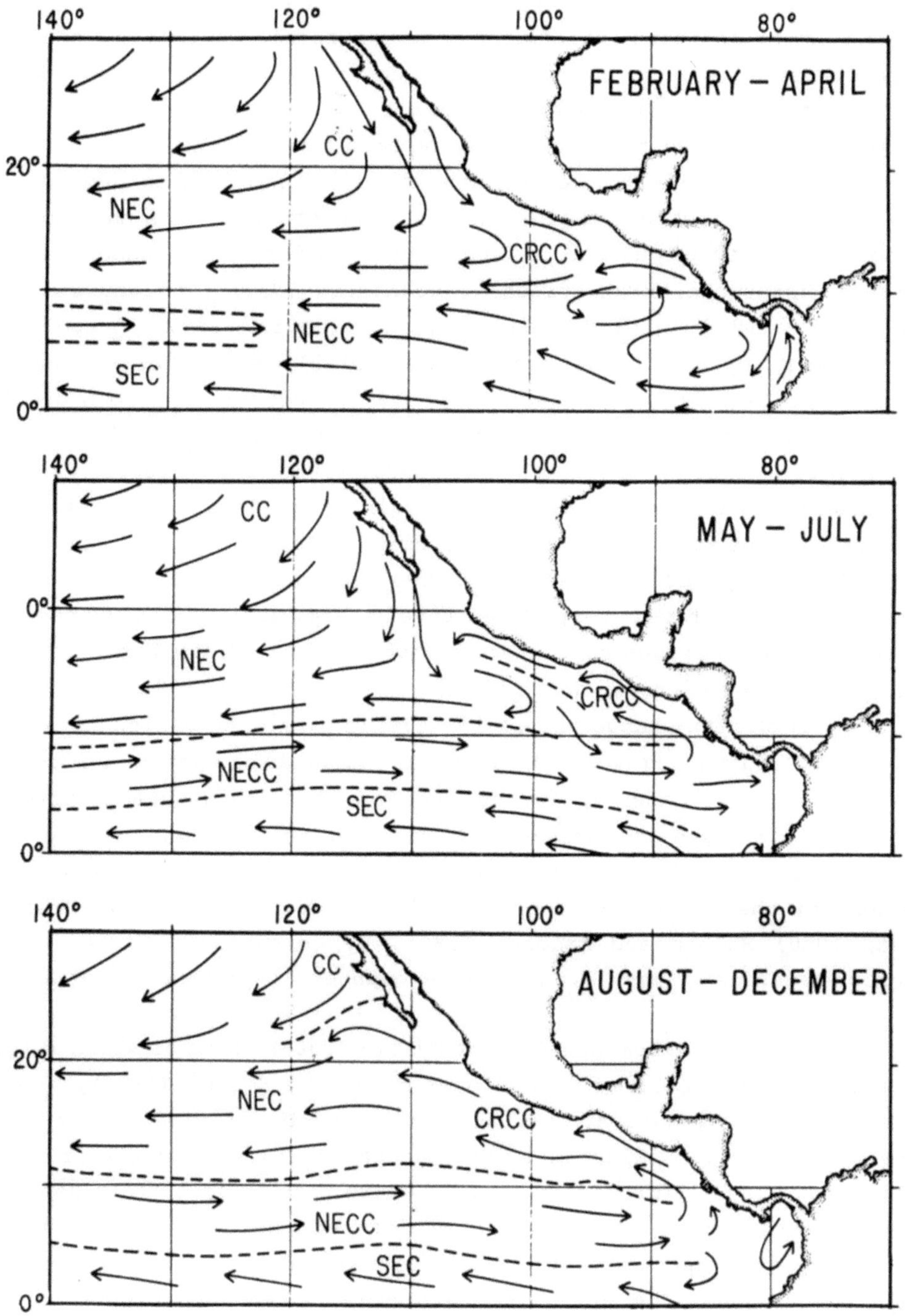

Figure 3: Principal seasonal circulation patterns of wind driven surface currents in the Eastern Tropical Pacific. Abbreviations are as in Figure 2. Redrawn from Wyrtki (1965).

In the northern portion, off Baja California, the circulation consists of the southern extension of the California Current and conforms to the classical scheme of an eastern boundary current, as described by Wooster and Reid (1963). Careful summaries of earlier knowledge on the California Current System have been prepared by Reid, Roden, and Wyllie (1958), and Hickey (1979) later integrated additional data to an extensive literature review, which remains the most complete report on this current system, although various studies have added to its recent understanding (Chelton, Bernstein, Bratkovitch, and Kosro, 1987; Lynn and Simpson, 1987; Kosro, 1987; Winant, Beardsley, and Davis, 1987). A poleward undercurrent appears to be present essentially everywhere along the peninsula of Baja California (Chelton, 1982), with its maximum usually at about 20 km from the slope, but which can reach 100 to 150 km in width (Chelton, Bratkovitch, Bernstein, and Kosro, 1988), and at depths probably not exceeding 600 m. This flow must originate at least as far south as the southernmost extension of the California Current itself, since the common hydrographical signature of the undercurrent is that of the Subtropical Subsurface Water, of relatively warm, saline, of high nutrient but low dissolved oxygen contents; the hydrographic distribution suggest this flow becomes narrower north of 30° (Figure 4). The typical T-S values in the undercurrent change remarkably little along the peninsula of Baja California, indicating that lateral mixing is probably not very important along its path and due mostly to the impingement of eddies offshore (Wickham, 1975; Lynn and Simpson, 1987); close to 9.5°C and 34.6 ppt near Cabo San Lucas (Hickey, 1979), they remain above 9.0°C and 34.4 ppt in the core of the undercurrent off northern Baja California (Wooster and Jones, 1970; Barton and Argote, 1980). By comparison, Hickey (1979) reports values of 7°C and 33.0 ppt in the counterflow off Vancouver Island.

The geostrophic computations of Wyllie (1966), some of which (Figure 5) have been recalculated with closer detail by Hickey (1979), and the compilations of historical ship drift data kindly provided to us by Mr. A. Bakun (Figure 6), illustrate the seasonal variations of the principal features of the California Current. Poleward flow is present at depth off Baja California most of the year, manifested mostly as a sequence of cyclonic gyres along the coast, of a scale comparable to the supposed width of the counterflow. The Southern California Countercurrent (Sverdrup and Fleming, 1941) or eddy (Schwartzlose, 1963) is the most important of these gyres. It occurs off northern Baja California and southern California as the flow of the California Current, partially shielded from the dominant northerly winds by the change in direction of the coast in the Southern California Bight (Figure 1), moves onshore and a portion branches north in association with the countercurrent (Reid, Schwartzlose, and Brown, 1963). Oddly, this onshore tendency of the California Current and its

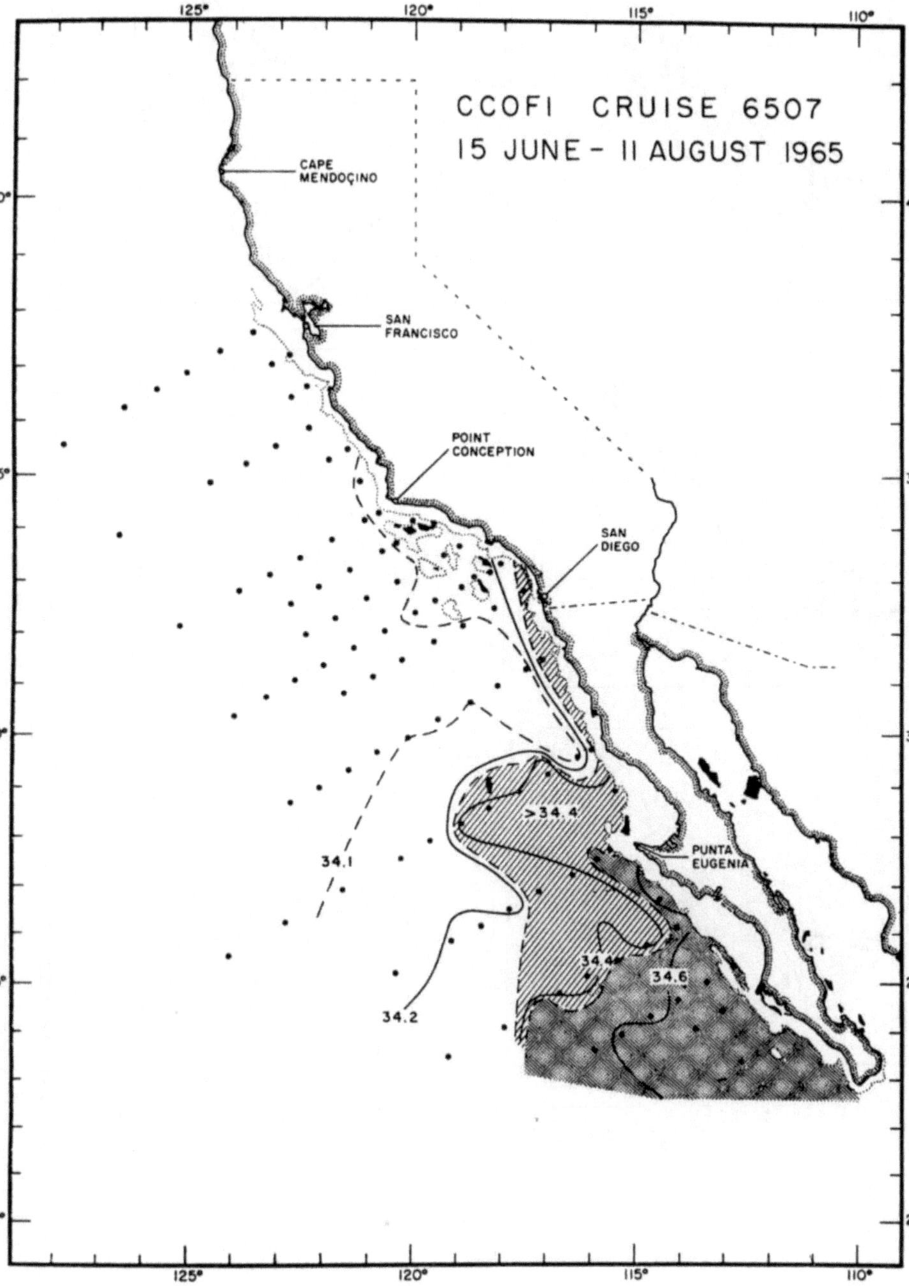

Figure 4: The signature of the California Undercurrent in the salinity field on the isanosteric surface 150 cl t-1. From Wooster and Jones (1970).

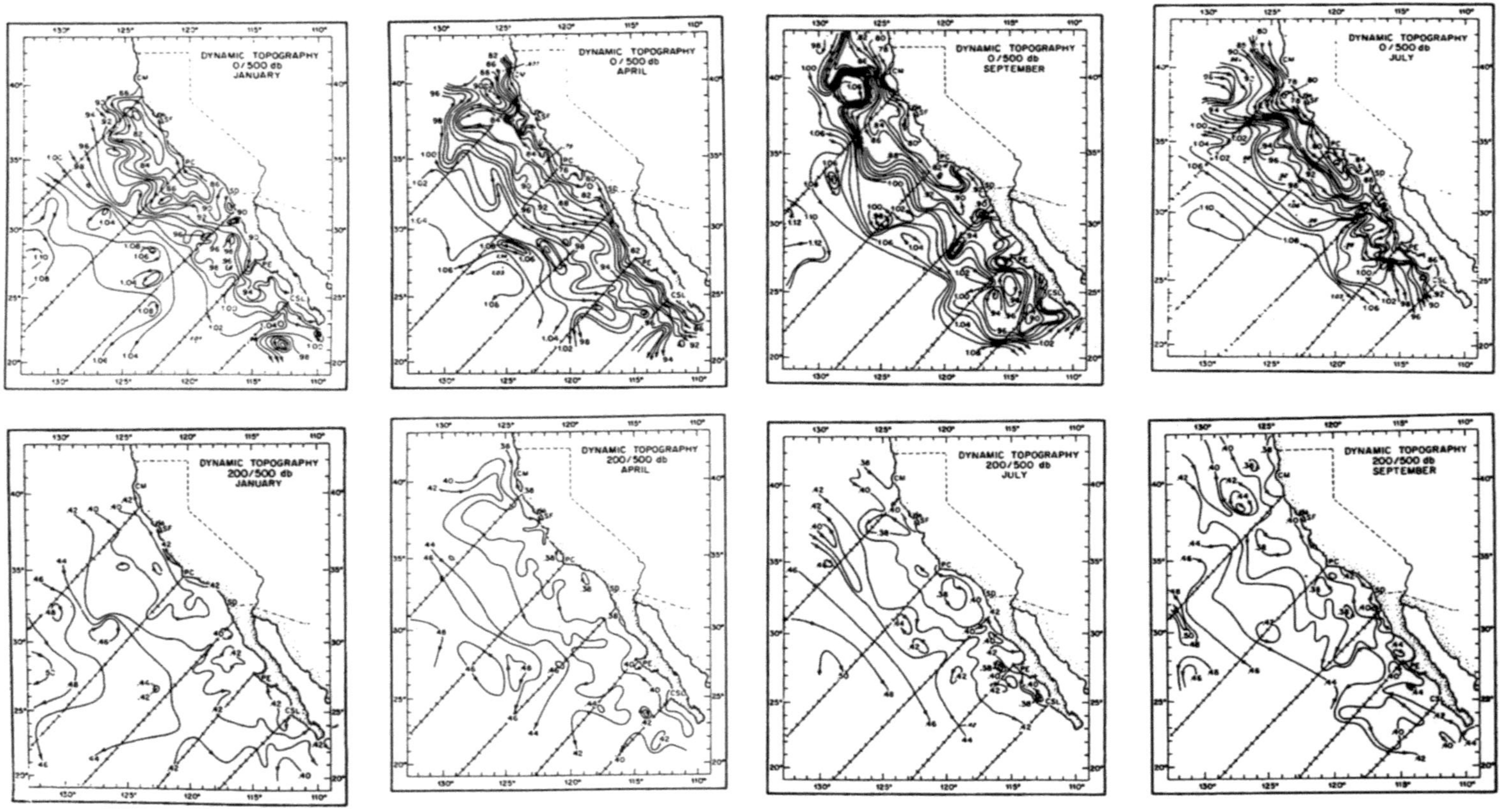

Figure 5: The seasonal variations of the California Current system at the surface and at 200 m depth. From Hickey (1979).

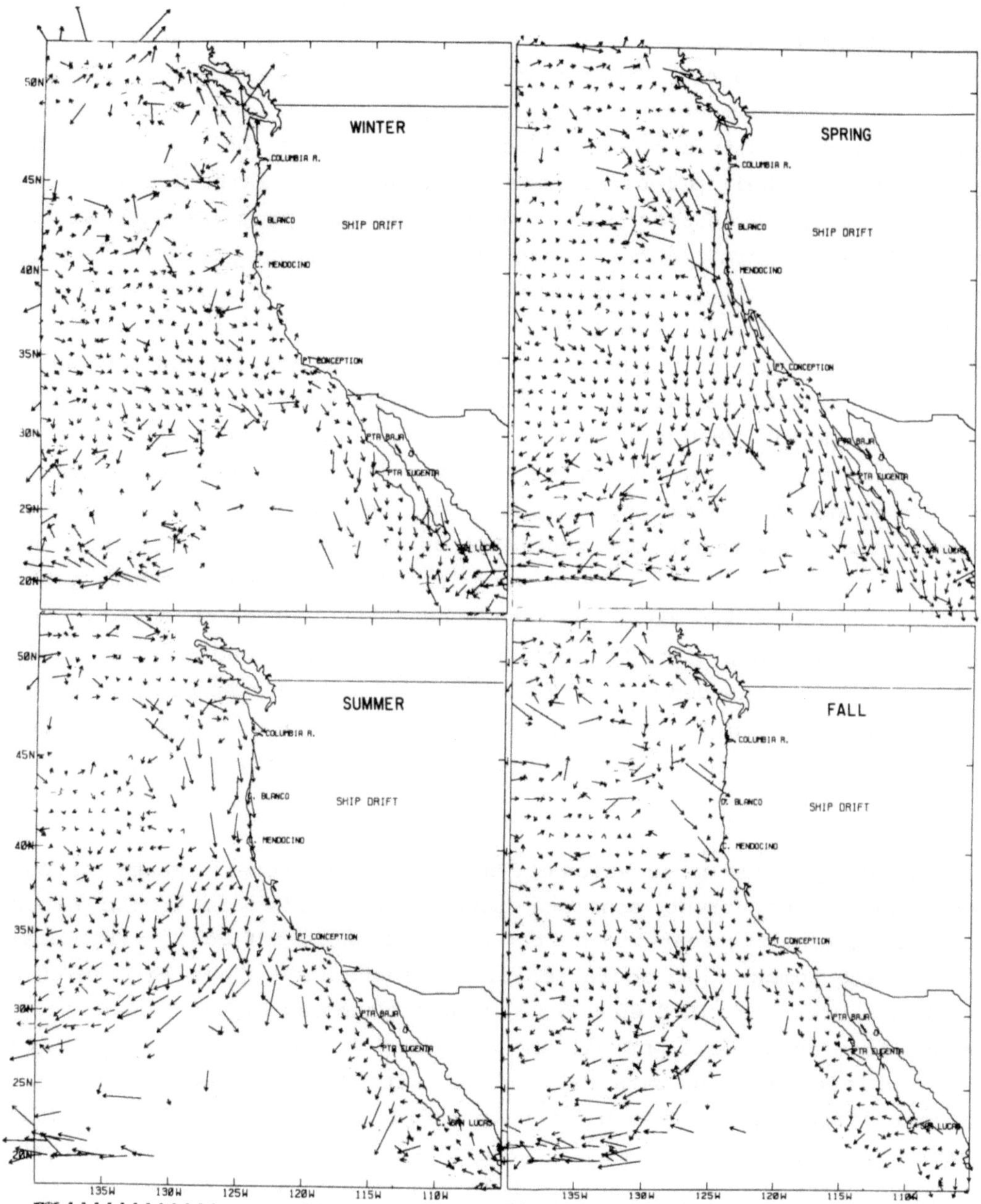

Figure 6: Historical surface ship drift in the California Current region, compiled by seasons in 1° squares. Each vector includes a minimum of five observations, the variance ellipses represents the standard error of the mean, and the entire set contains over one hundred years of data. Courtesy of A. Bakun.

resulting diffluence, which have been reported variously from geostrophic calculations (Wyllie, 1966; Reid and Mantyla, 1976, Hickey, 1979), inferred from drift bottle data (Schwartzlose, 1963; Badan-Dangon, 1972) and are supposed to cause important frontogenesis (Peláez and McGowan, 1986), are reflected neither in the ship drift data (Figure 6) nor in the mean flow computed from repeated drogue observations, but rather in the orientation of their principal axes of variance (Poulain, Illeman, and Niiler, 1987), suggesting that fluctuating flux processes should be responsible for this feature of the California Current. During the fall and early winter the shoreward side of this series of eddies appear to coalesce into a single, continuous northward flow with speeds that can attain 25 cm s^{-1}, and the Southern California Countercurrent may well flow past Point Conception and connect with the poleward flows north of that point (Reid, 1960, McLain and Thomas, 1983). Except during spring, and particularly in April, when the northwesterly winds blow persistently (Figure 6) and cause substantial upwelling, the undercurrent surfaces and constitutes a countercurrent that extends from Cabo San Lucas into the southern California Bight. The varying intensity of the winds modify the rates of advection and cause the interannual differences observed in the California Current System (Chelton, 1981). Lynn (1983) and Simpson (1983) suggest that an enhanced poleward advection results during El Niño years, in accordance with the results of McCreary (1976); Simpson (1984), however, proposes an intensified onshore advection as the principal modifying factor.

The large scale circulation off Mexico involves several water masses. The California Current consists mostly of Pacific Subarctic Water proceeding from the westwind drift; close to shore off Baja California, the surface temperature increases to the south from about 15°C to 20°C in the winter, and from 20°C to 25°C during the summer. The surface salinity ranges from about 33.5 ppt off the border with the United States, to 34.0 ppt off southern Baja California, with little seasonal changes. Farther offshore, the excess of evaporation over precipitation in the Central Pacific forms the mass of Pacific Central Water of subtropical, warm and salty, characteristics; it mixes laterally with the Subarctic Water on the edge of the California Current. The Equatorial System, and therefore the Costa Rica Coastal Current, is dominated by an excess of precipitation over evaporation; the Tropical Water carried to the north is warm and less salty. It meets the Pacific Subarctic Water of the California Current in a transition zone indicated by sharp horizontal salinity gradients (Roden, 1970, 1971). This system of fronts is usually found in the vicinity of Cabo San Lucas, at the tip of the Baja California Peninsula, where it includes some of the water that originates in the Gulf of California (Griffiths, 1965, 1968). Its position fluctuates seasonally with the strength of the currents involved and with the type of water flowing onto the North Equatorial Current. Other water masses found in the region include various proportions of Pacific Subarctic and Equatorial Waters that have combined at depth and are brought to the surface by upwelling; this explains the occasional presence nearshore of cool, high salinity water (Gómez Valdéz, 1984). No evidence has been reported of the presence of Gulf of California Water beyond a few tens of kilometers away from the mouth of the Gulf itself.

The upper structure of the California Current is now known to be dominated by intense geostrophic turbulence (Davis, 1985; Poulain, Illeman, and Niiler, 1987) and this is also probably the case with the Costa Rica Coastal Current. Gyres, whirls, and cross-shelf jets contribute most of the exchange between the nearshore coastal ocean and the offshore circulation climate (Stumpf and Legeckis, 1977). In certain locations, such as off Tehuantepec, the intense atmospheric forcing by the Tehuano winds that cross over the mountain ranges from the Gulf of Mexico (Blackburn, 1962, Alvarez, Badan-Dangon, and Valle, 1989), or from the incidence of tropical hurricanes (Enfield and Allen, 1980) provides not only peculiar localized conditions, but affect the entire coastal ocean of the Norteastern Pacific through the possible induction of a pressure gradient along the coast (Csanady, 1980) and the poleward propagation of sea-level events (Christensen, de la Paz, and Gutierrez, 1983). The mouth of the Gulf of California acts as a process of selection to the propagation of these subinertial events (Mejía, 1985); waves of certain frequencies enter the Gulf, where they dissipate and should contribute to its internal circulation. Lower frequencies appear to bypass the Gulf and proceed poleward along the west coast of the Baja California Peninsula (Enfield and Allen, 1983). The exact nature of this frequency selection has not yet been elucidated and may be accompanied by complex wave scatterings off the tip of Baja California.

DIRECT OBSERVATIONS OF POLEWARD FLOWS

The only direct observations of the coastal poleward flows off Mexico have been done on the shelf and slope off northern Baja California. They are relatively few but all provide useful measures of some of the scales involved. Amongst the earliest experiments, Reid (1963) released parachute drogues at 250 m depth, in a line extending offshore from San Quintin Bay, near 30°N (Figure 1). Only those within 50 km of the coast showed motions towards the northwest, with mean speeds of about 8 cm s^{-1}. Motions farther offshore were either to the southeast, as would be expected of the California Current drift, or close to zero. The results were compared to those made earlier near Cape Mendocino, off northern California (Reid, 1962), where drogues deployed within 70 km of the coast indicated similar net poleward motions, with speeds close to 25 cm s^{-1}. Both sets of measurements were interpreted as indicating the top of the California Counterflow, and some form of continuity between the two was inferred. However, they each lasted only a few days and could also have been a manifestation of the numerous meanders and gyres that were later described by Reid, Schwartzlose, and Brown (1963) and extensively documented by Poulain, Illeman, and Niiler (1987). The first study specific of the undercurrent off Baja California was done by Wooster and Jones (1970) off Punta Colnett, at 31°N, and included hydrographic surveys and short sets of current measurements. Their work, later complemented by Robles, Morales, García, and Flores (1981), García (1983), and Torres-Moye and Acosta (1986), all illustrate the undercurrent as an intrusion of warmer, saltier water of high nutrient and dissolved oxygen contents within 20 km of the slope, centered at about 300 m depth, and detectable to within 150 m of the surface (Figure 7). Above that level, isotherms rise to the surface near the coast, but within the undercurrent, they sink rapidly towards the slope. In those same sections, the southerly flow of Subarctic Water of the California Current is indicated by the layer of lower salinity water above 100 m. The relative currents computed from geostrophy suggested speeds close to 10 cm s^{-1} at 200 to 300 m depth. Direct measurements indicated that speeds can reach 40 cm s^{-1} in the core of the undercurrent, and at 400 m the speeds were still poleward at 20 cm s^{-1}. The hydrographic sections suggest that high levels of mixing can occur locally within the undercurrent. This configuration is remarkably like the one described later by Hickey (1979) off the northwestern United States.

Motivated by the evidence of well developed upwelling off the coast of northern Baja California, Barton and Argote (1980) developed a series of measurements near 31°N, off Punta Colonet, in summer 1976. This field effort, compiled by Morales, Argote, Amador, and Barton (1978), included detailed hydrographic measurements over the shelf and slope during times of strong upwelling, and provided direct measurements of the countercurrent. The hydrographical sections illustrate a configuration of the isopleths very similar to those reported by Wooster and Jones (1970), or those shown in Figure 7, with salinity values above 34.2 ppt typical of the Subtropical Water. The indications of this study are that at this latitude the counterflow did rarely climb onto the shelf, possibly prevented by the strong northwesterly upwelling favorable winds. Short term direct current measurements, made over the shelf with a current meter profiler, showed the water column to be flowing south most of the time, with the exceptions

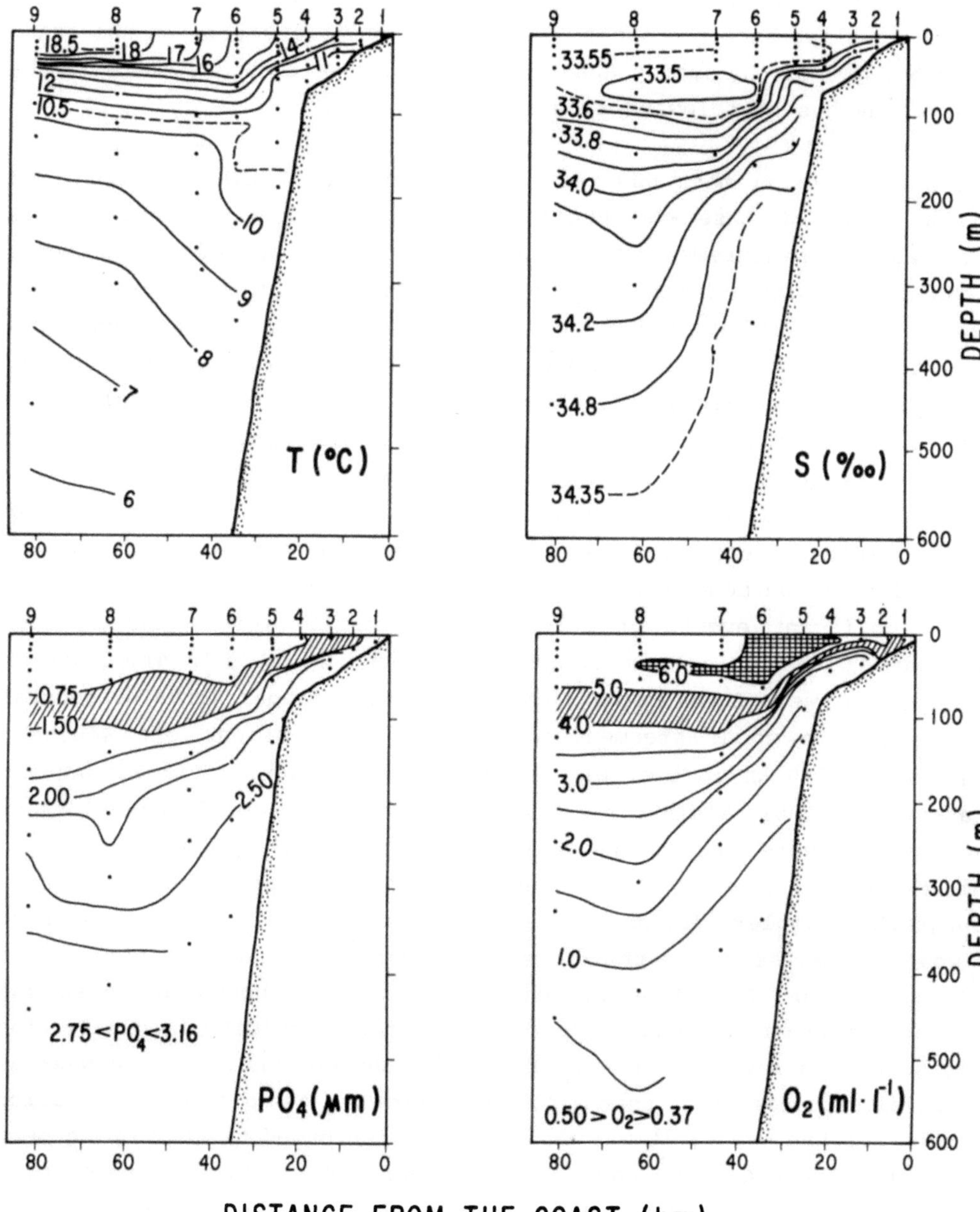

Figure 7: Hydrographic cross-sections of the California counterflow off northern Baja, California. The T and S sections are from Garcíá (1983), the PO_4 and O_2 sections were redrawn from Torres Moye and Acosta (1986).

of a few small events of northward flow that suggested the countercurrent is able to climb onto the shelf. The short term variations, on the order of a few days, of the hydrography over the slope were consistent with the observed events of poleward flow over the shelf.

A later experiment was performed by Barton, Robles, Amador, and Morales (1980) off Ensenada and off Punta Camalú, straddling the locations of the earlier measurements at Punta Colnett. Direct moored current meter measurements, with a duration of one year, were made in 75 m of water over a complicated shelf off Todos Santos Bay. From October 1978 through April 1979, the mean flow was equatorward at 25 m depth, but poleward at 60 m. From April through September, the flow was poleward at all levels, with typical values of 5 cm s^{-1}. A second set of measurements was done at about 100 km to the south, from April through July, 1979. The mean flow there was more equatorward than at the northern station, but the 60 m mean flow was still directed poleward. If the mean flow suggested that a weak protrusion of the countercurrent generally exists over the shelf, there were significant events of poleward flow at all levels and at both stations, with typical amplitudes of about 40 cm s^{-1}. The current fluctuations were found to be uncorrelated with either winds or coastal sea level, in good agreement with other observations off southern California, 140 km to the north of Ensenada (Winant, 1980; Winant and Bratkovitch, 1981; Winant, 1983). The longshore current fluctuations on event scales were found to be weakly correlated with the sea level, but this relationship broke down during the summer. It appears unlikely that topographic complexities should be the source of such large fluctuations; it would rather seem that the large scale turbulent flows offshore should be responsible. Indeed, this coastal experiment showed at least one instance of anomalously large poleward events, coinciding initially with a seasonal sea level minimum at Ensenada, and continuing for a week as sea level recovered from that minimum (Figure 8). As no clear explanation existed for the event, it was tentatively attributed to the impingement of a large scale eddy on the coast (Barton, 1985). This interpretation is supported by Christensen and Rodriguez (1979), who leveled the tide gauges at Ensenada and at Guadalupe Island, distant some 300 km offshore, by considering the effects of spatially averaged subinertial currents on the sea level differences. They showed that sea level fluctuations at the offshore station, which is located in the middle of the California Current, are of the same magnitude than those at the coast, indicating that the effects of poleward currents are not necessarily confined to a narrow coastal zone. Most of the energy was found in the 15 to 30 day band, and the relative strength of the currents varied by as much as 10 cm s^{-1}. Clearly, poleward flows can occur away from the coast, but are probably intermittent and associated with eddy structures of the sort shown by

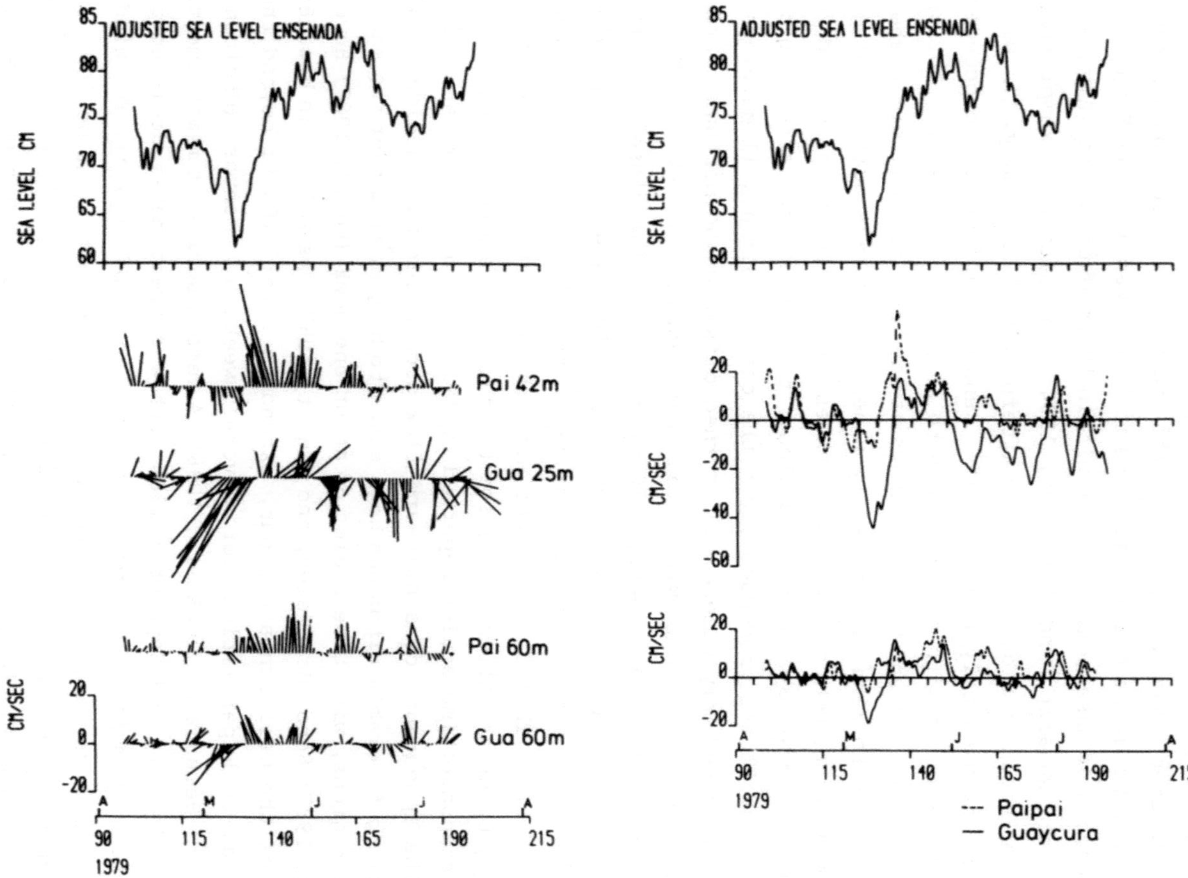

Figure 8: Low-pass filtered adjusted sea level at Ensenada and currents off Ensenada (Paipai) and off Punta Colnett (Guaycura), during the supposed passage of and eddy offshore. Left: Daily current vectors; vertically up is positive poleward along the coast. Right: Longshore component of the currents. Julian days 90 and 215 are 31 March and 3 August. From Barton (1985).

Poulain, Illeman, and Niiler (1987). Additional measurements following that period, which should confirm these results and provide more stable statistics are being analyzed.

Other studies of a more local character have been done recently off northern Baja California, consisting of short series of shallow water current measurements aimed at observing the coastal response and pollutant dispersal effects in the small scale. The first one was done off Eréndira, some 50 km south of Ensenada, in a band within 10 km of the coast (Alvarez, Durazo, Pérez, Navarro, and Hernández, 1984). The data are exemplified by the series shown in Figure 9; in the afternoon of 22 February, the wind was a steady northwesterly around 3 m s^{-1} , and the surface flow proceeded in the same direction with a speed close to 20 cm s^{-1} (Series A). During the evening of that day and the early hours of the next, the wind diminished to average close to zero; the recently redeployed floats quickly turned offshore and proceeded towards the northwest at a speed that reached 20 to 40 cm s^{-1} (Series B). Redeployed closer to shore on the afternoon of the 23rd , the floats proceeded along the coast under a well developed afternoon northwesterly breeze, but turned offshore and northwestwards during the night, as the wind died down again (Series C). These observations indicate the presence of the poleward flow close to shore, but locally obliterated at the surface by the diurnal winds, and provide a relaxation time scale from the effects of the wind in this region of only a few hours. This is congruent with the response characteristics proposed by Brink and Allen (1978) for the flow over a shallow shelf, and with the response time scales reported by Badan-Dangon, Brink, and Smith (1986). The appearance of a poleward flow soon after the subsidence of the upwelling favorable winds has been shown to happen also over the shelf and slope regions, with time scales of about one week (Kosro, 1987; Huyer and Kosro, 1987; Send, Beardsley, and Winant, 1987), as well as in broader portions of the California Current, extending over 100 km beyond the shelf break and persisting for several weeks or months (Chelton, Bratkovitch, Bernstein, and Kosro, 1988). A second series of measurements were done in July and September 1986 with a current meter deployed at 5 m off the bottom in 20 m of water, off the coast of Tijuana. The winds were light and variable northwesterlies throughout both periods of measurements. The flow, however, remained quite strong poleward at about 20 cm s^{-1} (Figure 10). The poleward flow diminished and reversed direction in late September, again without apparent relation to the wind, as had been noted earlier by Barton (1985). Accompanying Lagrangian series, not shown here, confirm the presence of poleward flow across the shallow shelf at this location (Alvarez, 1986).

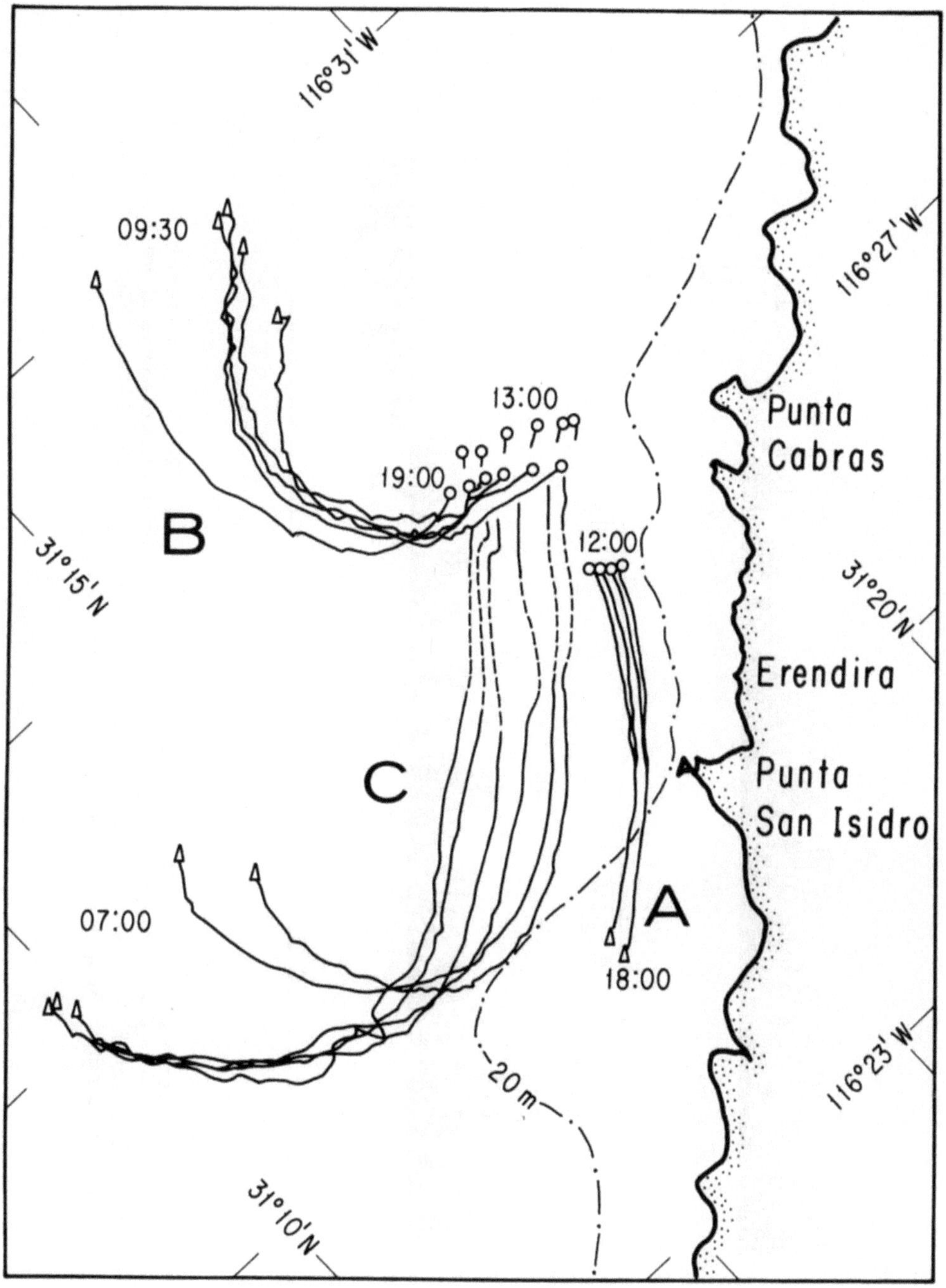

Figure 9: Lagrangian measurements of nearshore currents off Eréndira, Baja California, made in 1981. Series A begins at 1200 on 22 February 1981 and ends at 1800 on the same day. Series B starts at 1900 on 22 February and ends at 0930 on the next day. Series C begins at 1300 on 23 February and ends at 1700 on the 24 February. Redrawn from Alvarez et al (1984).

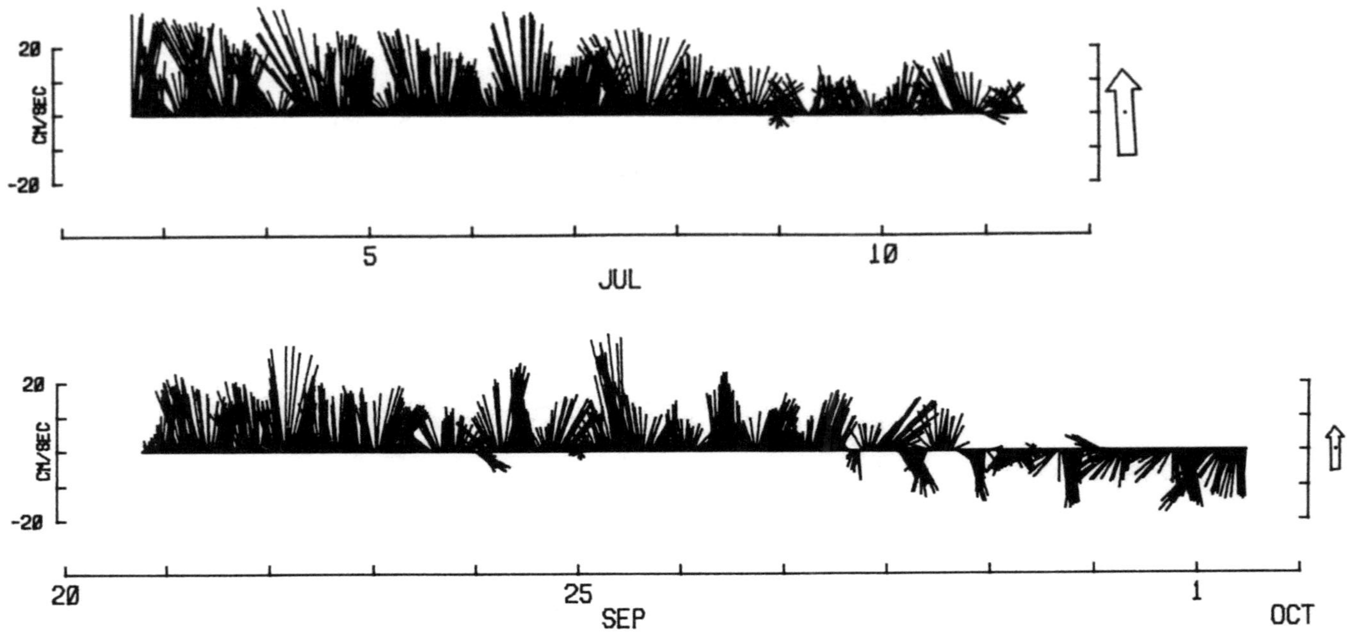

Figure 10: Vector time series of currents 5 m off the bottom, in 20 m of water off Tijuana, for the months of July and September, 1986. Vectors are drawn at 20 minute intervals and vertically on the page is positive to the north, which is also the local orientation of the coastline. The arrow on the right is the mean over the record length. Redrawn from Alvarez (1986).

In summary, direct measurements have shown that poleward flow is common over the shelf and slope off northern Baja California. The local winds obliterate the poleward flow at the surface, with a typical response time of only a few hours, but currents are otherwise weakly correlated to the wind. Barton (1985) concludes that the encroachment of the poleward undercurrent onto the shelf fluctuates seasonally, but not to the extent of being totally absent from it. Still, the distinction between undercurrent, surface countercurrent, and poleward flow over the shelf is confused, and the nature of their sources of momentum remains elusive.

THOUGHTS ON THE COUNTERFLOW

Numerous efforts have been made to understand the origins of the counterflow, undercurrent or countercurrent. Early investigations (Yoshida and Tsuchiya, 1957; Yoshida, 1967) noticed the close association of the poleward undercurrent with a rising thermocline at the coast and the presence of coastal upwelling under the influence of equatorward winds. Mass and vorticity adjustments were invoked for their nearly simultaneous development. Pedlosky (1974) indicated that equatorward winds induce equatorward currents, but that a positive wind stress curl near the coast drives a poleward undercurrent that might be coincident with upwelling, but not necessarily associated with it. Other studies have focused on the effects of the wind stress curl on scales larger than the coastal boundary, acting on the flow. Munk (1950) attempted the application of a Sverdrup balance on the California Current off Cape Mendocino, and obtained qualitative agreement with observations available then. Hickey (1979) noticed a good correspondence between the mean configuration of the poleward flow and the spatial configuration of positive wind stress curl along the coast (Nelson, 1977), but could not match their time variations on a seasonal scale. Chelton (1982), however, showed that the large scale positive wind stress curl that results from the offshore intensification of the northwesterlies is able to drive an Ekman pumping at 200 to 300 km off Baja California, and probably drives the broad poleward flow that is observed off the coast.

A longshore pressure gradient force is a well known manifestation of the inherent three-dimensionality of the coastal circulation maintained by the wind (Suginohara, 1974; Allen and Smith, 1981), and many investigators have examined it as the probable cause for the counterflow. Hickey and Pola (1983), following a formulation by Csanady (1978), provided estimates of a longshore pressure gradient off the northwestern United States, that matched the sea surface slope proposed by Sturges

(1974). Chelton (1984) indicates the poleward flow in the California Current is coherent with a large scale along-shore pressure gradient computed from differences in dynamic height, but which appears to be independent of the wind field. Models by McCreary (1981), McCreary, Shetye, and Kundu (1986), and Wang (1982), successfully developed indications of a counterflow as a result of a longshore pressure gradient force. Werner and Hickey (1983) added a longshore pressure gradient as an external force to a two-dimensional, baroclinic finite difference model of the coastal zone first developed by Hamilton and Rattray (1978) and modified by Hickey and Hamilton (1980). Their results indicate that an undercurrent can be accelerated by the longshore pressure gradient force, until bottom friction arrests its further development. In the steady state, the longshore flow is balanced by a cross-shelf pressure gradient, and the longshore pressure gradient drives a depth-dependent onshore flow, which returns offshore through a bottom frictional layer. McCreary and Chao (1985) obtained realistic coastal flows through a linear, three-dimensional, stratified model by assuming that the alongshore flow is in geostrophic balance. They show that the resulting longshore pressure gradient can drive the poleward undercurrent. However, the extension of the shelf weakens the poleward undercurrent by inducing a barotropic equatorward flow, in good agreement with the results of Suginohara (1974); it is found that a wind stress curl over the coastal region, such as results from the increase of the wind field in the offshore direction, is required to strengthen the undercurrent to realistic values. Suginohara and Kitamura (1984) speculate, further, that dissipation is required to sustain a poleward current in the steady state and that it is improbable that the California Undercurrent should be driven by the regional winds, but results rather from the more global tropical circulation interacting with the continental shelf-slope.

Early numerical formulations of the nearshore response to along-shore winds were done on an f-plane and did not produce a poleward undercurrent (O'Brien and Hurlburt, 1972). Hurlburt and Thompson (1973) and McNider and O'Brien (1973) confirmed the importance of an along-shore pressure gradient and illustrated the potential role of the variation of the Coriolis force with latitude. Indeed, McCreary (1981) and Philander and Yoon (1982) showed that for time scales long enough for ß effects to develop, the westward dispersion of the along-shore flow induces a poleward flow not unlike that observed off the coast of the northwestern United States. Chelton (1982) attributes the three to four month lag he observes between the wind stress curl and the poleward undercurrent to such a ß effect.

Other recent investigations have given much attention to nonlinear effects associated with offshore forcing processes and with coastal-trapped waves. Denbo and Allen (1983) consider an inviscid ocean adjoining an equally inviscid shelf interior bounded by frictional surface and bottom layers. The model is forced separately by a zero-mean periodic wind stress at the coast and by prescribed open ocean motions that impinge onto the shelf. The coastal winds induce an undercurrent of about the correct magnitude, but progressing in the same direction as the wind. The offshore forcing generates an undercurrent of the appropriate characteristics, except for its speed which is about an order of magnitude too small. Chapman and Brink (1987) nonetheless further examined the problem of offshore periodic forcing; typical periods larger than about ten days unlock the coastal system from near resonances that are otherwise found and promote a bottom-trapped along-shore jet that extends away from the forcing in the direction of propagation of coastally-trapped waves, with a velocity opposite to that within the forcing structure. That is, a poleward undercurrent should be propelled by offshore elements possessing anticyclonic vorticity. A numerical investigation by Suginohara (1974,1982), of a coastal ocean with cross-shelf topography and vertical stratification underlines the importance of coastal-trapped waves in the development of an upwelling circulation, including that of the poleward undercurrent. The onset of the wind promotes a rapid response over the shelf and slope; the passage of the first mode coastal-trapped wave contributes to confine the upwelling to the coastal area, as both the coastal jet and the poleward undercurrent grow. The arrival of the second mode trapped wave blocks further development of these currents. These results have been extended to an n-layer configuration (Yoon and Philander, 1982), and a steady state is obtained after the passage of the nth baroclinic mode. The undercurrent results from the disparity between the vertical configuration of the waves and that of the coastal jet initially produced by the wind. Simons (1983) confirms that the nonlinear effects of coastal waves can provide an explanation for the set-up of the longshore pressure gradient required to drive the poleward undercurrent; Thompson and Wilson (1983) have examined tidal rectification effects as a source of a poleward flow near the coast.

Another possible driving process for a poleward flow is the drag produced by an irregular topography (Brink, 1986). A steady current flowing in a direction opposite to that of the free shelf wave propagation generates a lee wave off each bump, whose growth constitutes the drag mechanism. When the steady current reverses, no lee waves can be formed, and thus no momentum is removed from the obstacle. Haidvogel and Brink (1986) confirm that this drag asymmetry, which is a form of flow rectification, is able to produce a mean flow in the direction of free shelf wave propagation, poleward for northern hemisphere eastern boundary currents. This principle is extended by Holloway, Brink, and Haidvogel (this volume) to the specific case of the undercurrent. The speeds that are obtained are of a few cm s^{-1} in the correct direction, but too small to justify it as an entire explanation by itself. Samelson and Allen (1987) indicate that, for certain parameter ranges, this response of the coastal flow is chaotic.

Still, the exact causes of counterflows are not yet known unequivocally, and further studies of the region we have examined should contribute useful insight into the physics of the poleward flows, in addition to complementing the geographical description of a very important segment of the world's ocean circulation. It appears fundamental to examine the composition and configuration of the Costa Rica Coastal Current, for which a specific study has not been reported, and in particular, to elucidate whether there exists a counterflow to that current. In that region, nonlinear processes of the coastal waves, topographic drag, and offshore forcings should contribute a poleward flow indistinguishable from the current itself, except for a possible secondary maximum in velocity. The longshore pressure gradient should generate an equatorward counterflow, which would therefore have a visible signature. Wind stress curl could contribute a flow in either direction, depending on the configuration of the wind field and on the relative intensity of the winds offshore; these are poorly known in the area. The presence or absence of a counterflow would suggest which of these generating mechanisms should be discarded.

The general path of the Tropical and Subtropical Water should be investigated, together with the configuration of the frontal structure near Cabo San Lucas, in order to understand how the California Countercurrent is provided for and how this influence is enhanced during ENSO events (McCreary, 1976). What water constitutes the flow to the southward of Cabo Corrientes should be known; Wyrtki (1965) suggests the flow is composed of an extension of the intensified California Current, but Hickey (1979) concludes that this current never reaches that far south, at least near the coast, in good agreement with the drogue observations of Poulain, Illeman, and Niiler (1987). The along-coast continuity and seasonal fluctuations of the California Undercurrent should be addressed; this, maybe more than any other observations, could direct to the causes of counterflows. Finally, the apparent low values of lateral mixing within the counterflow off Baja California pose a striking question. A possible explanation is that the influence of the Subarctic Water is much greater off the coast of the United States than farther to the south. Also, the flow appears to be extremely trapped dynamically to the slope and shelf, and mixing with offshore waters might depend on major disruptions of the counterflow, which are possibly a function of the offshore eddy climate, or may happen around large capes. Off Baja California, Punta Eugenia appears as the only major point where this might be possible, whereas off the United States several more capes are found.

ACKNOWLEDGEMENTS

Many useful comments were received from P.P. Niiler, M. Lavin, and E.D. Barton; unpublished data were kindly provided by A. Bakun and by L.G. Alvarez. The figures were prepared by Sergio Ramos and José María Dominguez, and Joan Semler assisted with the preparation of the manuscript. This research was funded by the Secretaría de Programación y Presupuesto of Mexico.

From: E.D. Barton, University College of North Wales, UK

On: Review and commentary to paper **POLEWARD FLOWS OFF MEXICO's WEST COAST,** by A. Badan-Dangon, J.M. Robles and J. Garcia

This paper reviews most thoroughly the knowledge of currents along the eastern boundary represented by the Pacific coast of Mexico. Because the area is one of few direct current observations, and, south of the peninsula of Baja California, of sparse hydrographic measurements, understanding of the region is at best sketchy. Even so, it has provided one of the most striking examples of the efficacy of a poleward subsurface flow in transporting waters from a remote source along the continental slope against the prevailing equatorward drift of the subtropical gyre (Wooster and Jones, 1970). The interaction between the tropical circulation on the eastern boundary and the subtropical gyre remains to be investigated in any detail, as is also the case in the Atlantic Ocean. There are a number of intriguing questions alluded to in the text. For example, what happens to poleward flow on encountering the mouth of the Gulf of California? How do the intense wind driven events which occur in the Gulf of Tehuantepec in winter affect the alongshore current regime? The lack of specific data on the nature of the poleward flow along the Mexican coast precludes useful comparison with the bulk of existing theoretical work. Nevertheless, it is to be hoped that progress in modelling such flow will lead to the formation of hypotheses which can be tested by systematic measurements in the region.

AN OVERVIEW OF THE POLEWARD UNDERCURRENT AND UPWELLING ALONG THE CHILEAN COAST

Tomas R. Fonseca
Universidad Catolica De Valparaiso
Casilla 1020, Valparaiso-Chile

ABSTRACT

INTRODUCTION

The Chilean coast, with its long meridional coastline and consistent equatorward wind stress, is well recognized as one of the most biologically productive areas of the global ocean. This highly productive region is part of the Southeastern Pacific upwelling system extending from 5°S off Peru to 40°S off southern Chile. Exporting more fish meal than any other country in the world, Chile has long relied on the national fishing industry as a significant factor in its economy. Yet, in spite of its obvious importance, the waters south of 15°S are remarkably little known. Research efforts, either by Chilean investigators or by foreign expeditions have been, for the most part, isolated and modest.

Present information indicates that there are two important factors for the high marine productivity along the Chilean coast: A poleward flowing undercurrent, which introduces nutrient-rich waters to the south, and wind driven upwelling, which brings it upward toward the surface. This paper reviews literature concerned with these two processes south of 18°S, i.e., Chilean waters. Most of the references were published locally, with a few published in international journals. With this background information and the present high level of interest in upwelling systems, it is expected that new research efforts can be triggered in Chilean waters, especially international cooperative work.

THE POLEWARD UNDERCURRENT ALONG THE CHILEAN COAST

Early Results

Two different names have commonly been used for the poleward flowing undercurrent along the eastern boundary of the South Pacific: the "Gunther Current" [e.g., Brandhorst (1963); Robles (1976)], and the "Peru-Chile Undercurrent" [e.g., (Wooster and Gilmartin (1961); Sievers and Silva (1975); Silva and Fonseca (1983)]. The latter name has gained most general acceptance and will be used in the remainder of this paper, but shortened for convenience to "PCU."

The first to report on the PCU and to recognize its importance was Gunther (1936). During the WILLIAMS SCORESBY expedition in September 1931, he explored the sea between Peru and the Golfo de Penas (47°S), sampling at depth with water bottles and inverting thermometers. From an alongshore section, Gunther clearly distinguished a layer of water of apparent subequatorial origin, with high salinity and very low oxygen content, at a core depth of 150 to 200 m. Downward bending of the isopycnals toward the coast below this core depth suggested, from geostrophic balance, a poleward flow within the layer. From above, the PCU is overlain at times, by a layer of low salinity water of subantarctic origin, modified by precipitation in the south of Chile (although Figure 1 shows distinct broaching of the surface by the PCU, particularly offshore). From below, the PCU is bounded by Antarctic Intermediate Water flowing equatorward.

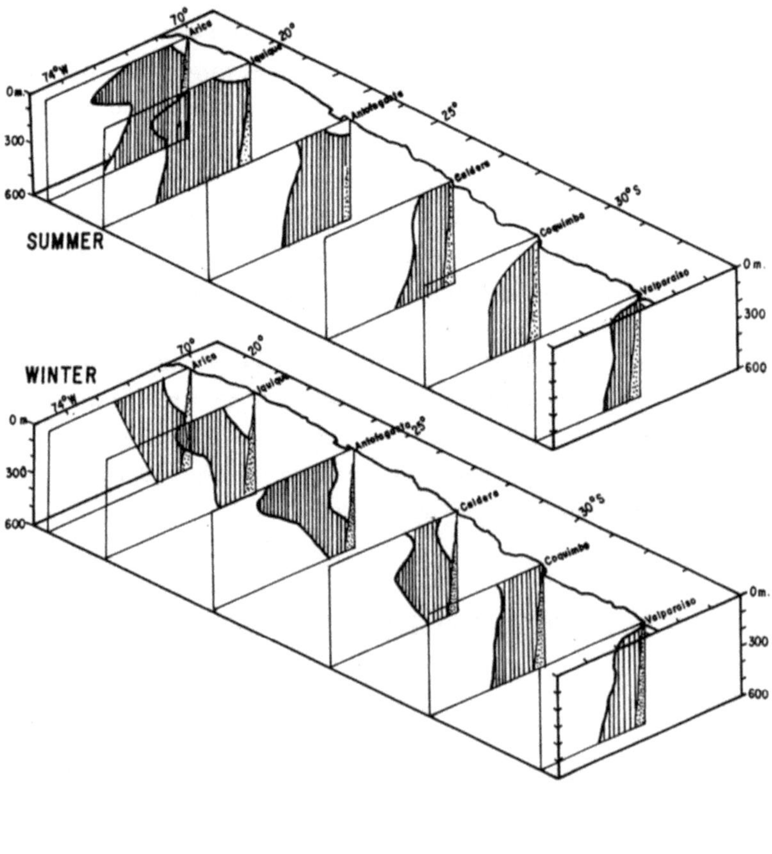

Figure 1: Simplified representation of the Peru-Chile Undercurrent (PCU) water (shaded) between the coast and 180 nm offshore in summer and winter [from Silva and Fonseca (1983)].

In the years between 1960 and 1982, twelve cruises along the Chilean coast were made with classical hydrographic stations and making indirect estimations of the flow through the dynamic method. These cruises were named MARCHILES, and were reported in Brandhorst (1971), Inostrosa (1972), Sievers and Silva (1975), Robles (1976), Silva and Fonseca (1983) and others. Silva and Fonseca (1983) processed data from several of the cruises between 1972 and 1982, looking for the most characteristic features of the currents in the region. Large variations were observed from cruise to cruise, yet there were gross features of the PCU which could be clearly distinguished: The undercurrent is present all year; it is distinguishable from near the surface to 600 m depth; its core is located around 150 to 200 m with a characteristic geostropic velocity maximum (by the dynamic method) of about 20 cm/s; and it is principally located between the coast and 200 km offshore.

A southern limit of the PCU, which can be distinguished from about 10°S off Peru, was proposed by Silva and Neshyba (1979). Based on estimations of the mass field, they suggested that 48°S (Golfo de Penas) is the southward limit, with the relative strength of the flow decreasing markedly from Valparaiso (33°S) southward (Figure 2).

Due to its characteristics and its variability, the PCU is one of the most interesting dynamic features of the Southeastern Pacific. Because of its coastal location and its association with large pelagic fish stocks along the north coast of Chile, it is a relevant and timely phenomena to study.

Direct Measurements

Direct measurements of the PCU flow are rare. Wooster and Gilmartin (1961), however, have contributed what has become a classic work. Using parachute drogues at stations off Peru and the north coast of Chile, they observed the undercurrent flowing parallel to the edge of the continental shelf at speeds of 4 to 10 cm/s. Figure 3 shows one of their stations 27 km off Antofagasta, Chile, where the poleward flow is clearly seen at drogue depths of 100 and 250 m. Coincident with the drogue measurements, a high salinity, low oxygen tongue of water was observed and reported to reach "at least to 41°S."

Further parachute drogue measurements by the Chilean oceanographer, Inostroza (1972), confirmed the presence of the PCU. Inostroza found that drogue velocities off Africa (18°S) were in agreement with computations from the dynamic method and indicated a southward flow

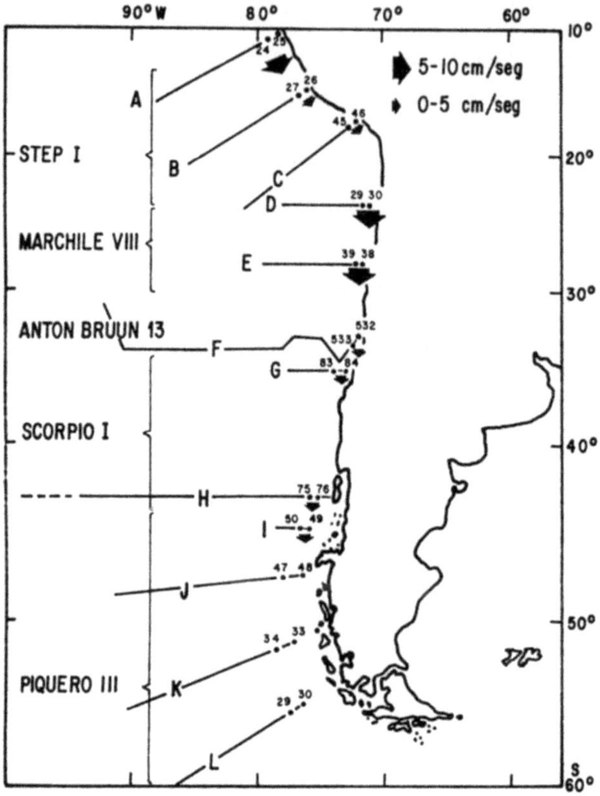

Figure 2: Composite showing the southernmost extension of the PCU [from Silva and Neshyba (1979)].

maximum of 25 cm/s at 100 m depth. In addition, a drogue at 250 m depth was recovered 13 days later 75 nm to the south, giving a minimum average drift of 12 cm/s. These measurements were taken during the winter season (July 1962).

Johnson (1980) reported direct measurements from shipboard at several stations over the shelf break off the northern coast of Chile during the MARCHILE X expedition in July 1976. Depths of poleward speed maxima were observed in clear association with the PCU water core of maximum salinity (Figure 4). Near 18°S the poleward flow was relatively weak (< 10 cm/s at 150 m), but grew stronger toward the south between 23°S and 33°S to a maximum of about 50 cm/s at 150 m.

With such sparse direct measurements, annual variations in the current are difficult to detect. However, from hydrographic data, Silva and Fonseca (1983), proposed that the undercurrent does have an annual

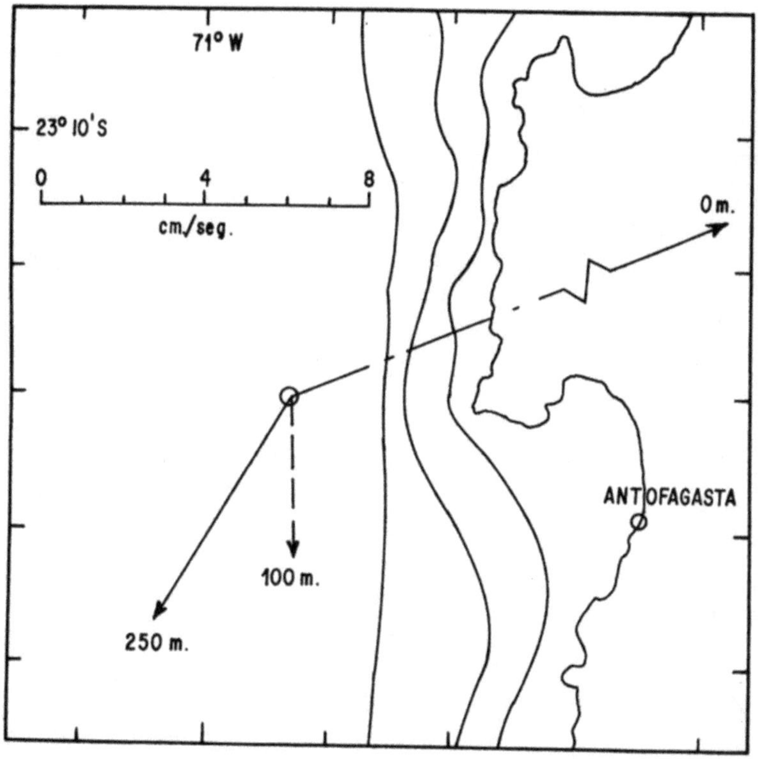

Figure 3: Mean current vectors (and bottom topography) obtained from drogues deployed at three different depths off the continental slope [from Wooster and Gilmartin (1961)].

signal. From Figure 1, it appears to be closer to the coast and broader in summer (at least off the northern coast) than in winter.

In Figure 5, salinity and temperature at 5 and 200 m depths are given for a yearly cycle over the continental shelf near Valparaiso, Chile. While the near surface signature exhibits an annual cycle, the deeper layer is relatively constant. In contrast, however, the sea level at Valparaiso (Figure 9) contains a distinctive annual signal. Furthermore, surface drifters in the nearshore region (Figure 6) appear to show seasonal reversals in direction which can be related to the variations in sea level under the geostrophic balance assumption. Sea level, then, may be a more useful indicator of flow than water mass characteristics, at least in the nearshore environment.

The predominant winds along the central and northern coast of Chile are from the SE-SW [Bakun and Parrish (1980)]. It should be expected that

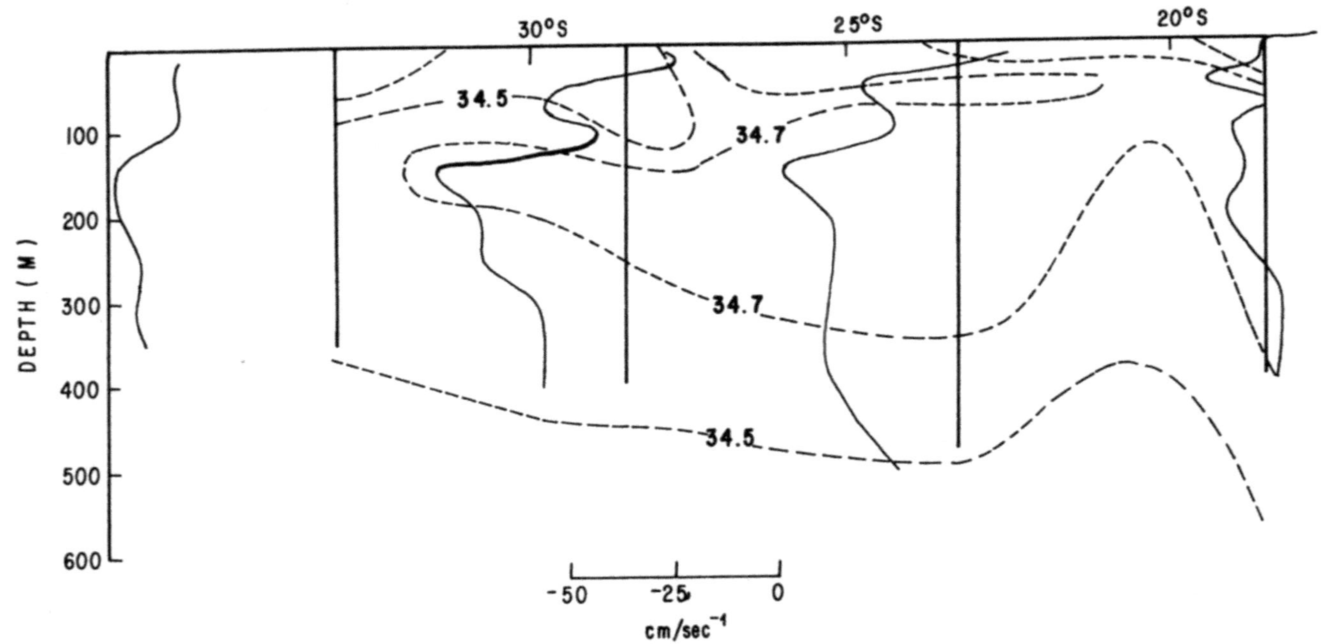

Figure 4: Along-shore salinity section (over the 1000 m isobath) with along-shore current profiles superimposed at four locations. The vertical lines representing the current axis are placed at the along-shore location of the profiles [from Johnson (1980)].

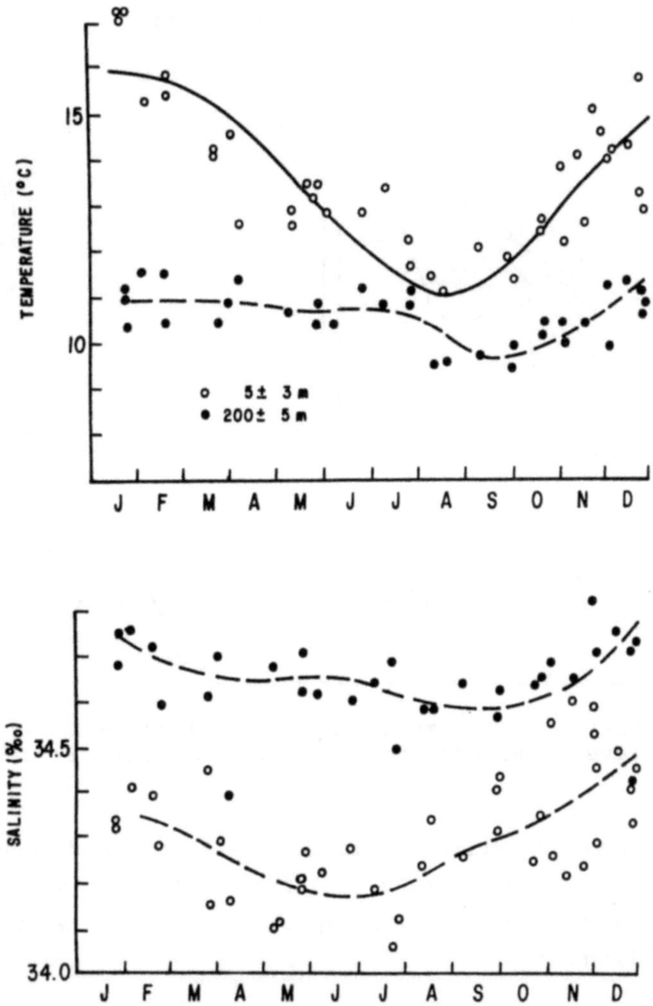

Figure 5: Annual cycle of temperature and salinity in shelf waters off Valparaiso [from Fonseca (1985)].

the northward flowing Humboldt Current would be important. However, this current has not shown a clear signal, in terms of velocity, in the first 100 miles offshore [Silva and Fonseca (1983); Espinoza, et al. (1983)]. In contrast, the PCU flowing counter to the wind stress, has been seen to be an important feature in ocean circulation off Chile, even at the surface itself (see Figure 1 and the 33°S profile in Figure 4).

In close interaction with the PCU, coastal upwelling is another feature with great influence on oceanographic conditions. Espinoza and Neshyba (1983) hypothesized that, in Chilean waters, the cross-shelf flow and local eddies associated with coastal upwelling are more important to surface conditions than the alongshore flow associated with the Humboldt Current. It is possible, and even probable, that the interaction between

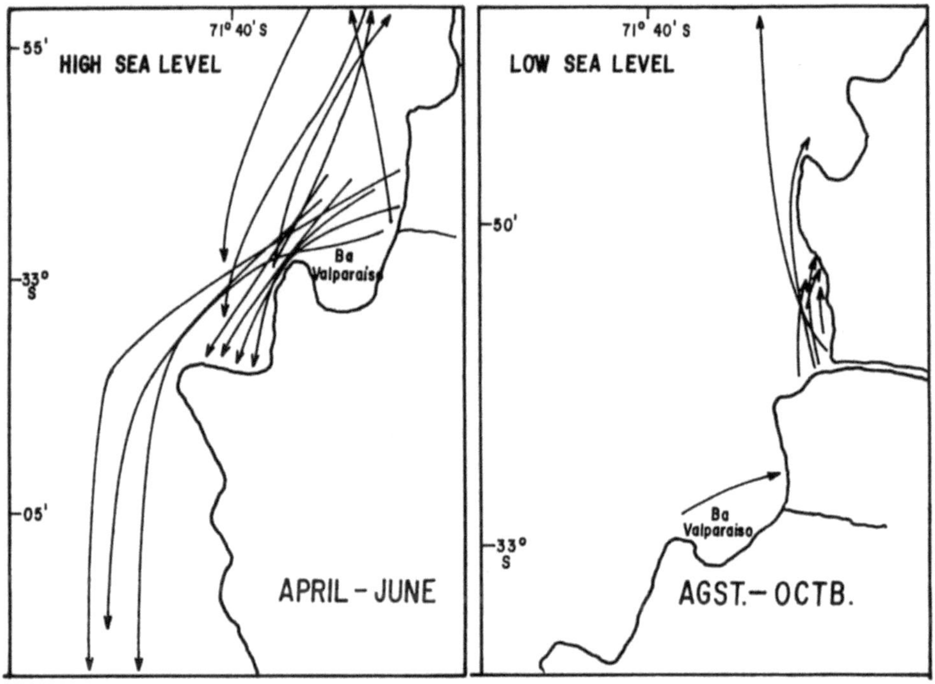

Figure 6: Coastal currents implied from drift bottles and associated with subtidal sea level measured in Valparaiso Bay [from Fonseca (1985)].

the PCU and coastal upwelling forms the basis for the intense biological growth in the area. Therefore, I feel that it is not possible to make an adequate review of the oceanography of the area without reference to coastal upwelling.

UPWELLING AND UNDERCURRENT

Centers of Upwelling

The northern and central coast of Chile, where the PCU is dominant (18°S to 36°S), is highly active in terms of fish production. Found principally in the first 100 nm off the coast, sardine, anchovy and mackerel are some of the most abundant resources. In addition, embayments close to upwelling centers support highly productive spawning areas [Rojas, et al. (1983)].

Winds are upwelling favorable at least 5 months per year. They are driven by the Southern Pacific Subtropical High, which in winter season is located off Iquique (20°S) and migrates southward to 30°S in spring-summer. South of Valparaiso (33°S), the upwelling season occurs in late spring and summer, later than on the northern coast.

In addition to its temporal displacement, upwelling is not spacially uniform, but occurs with more vigor in specific localities or centers. Satellite images are useful in locating those centers of enhanced upwelling, as can be seen in several satellite IR interpretations along the coast of Chile given in Figure 7. These centers appear to be related to the curvature of the coast, to bathymetry or simply to locally strong wind centers, and are clearly associated with areas of high productivity in the fisheries industry. In Figure 7 it can be seen that the upwelled water flows many miles offshore forming, at times, "hammer-head" or "mushroom" shaped structures.

Various authors have reported enhanced upwelling in different localities of the Chilean littoral zone. Neshyba and Mendez (1976), using surface temperature produced by the National Environmental Satellite Services (NESS), described the oceanographic conditions in relation to fisheries resources. Their maps, covering a 6-month period, show consistently colder waters at specific localities. In addition, Fonseca (1977) and Johnson, et al. (1980) refer to the upwelling center off Valparaiso (Figure 7c), and report a 2-cell type cross shelf circulation with intensive poleward undercurrent at the shelf break. Ahumada, et al. (1983) described an upwelling process off Talcahuano and showed that low oxygen waters, associated with the PCU, are raised by upwelling favorable winds toward the surface and subsequently enter Talcahuano Bay. Blanco (1984) reported on the measurements of a current meter near the surface at Talcahuano, and showed high coherence between winds and currents at 1.5 days, with an "Ekman" drift toward the west under the influence of SW winds.

The few direct measurements of the velocity field in upwelling areas [e.g., Johnson, et al. (1980) and Blanco (1984)] seem to indicate that upwelling occurs in the form of pulses or events. Those events last for five to seven days in response to intensification of SW winds. It can be expected that such events pump the nutrient rich PCU water into the surface layer and result in tongues of cold water, as seen in the satellite pictures of Figure 7, extending 30 to 40 km offshore.

In the Benguela region, Jury (1985) reported that upwelling occurs in association with capes and peninsulas similar to the Chilean experience. Also, the structure of the upwelling events observed off Valparaiso [Johnson, et al., (1980)] seems to be similar to those observed off South Africa by Nelson (1985):

Figure 7: A,B,C satellite derived sea surface temperature maps at enhanced upwelling centers on the Chilean coast. Temperature is given in degrees Centigrade.

01 NOV 85 VALPARAISO (33° S)

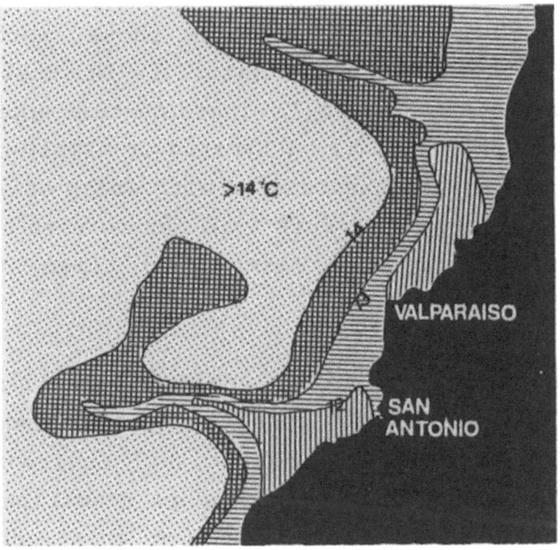

11 FEB 85 TALCAHUANO (36°40' S)

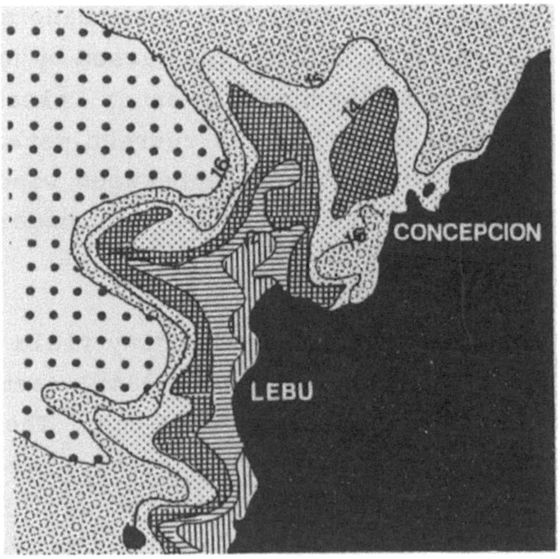

Figure 7A.

15 DIC 85 ANTOFAGASTA (23°30' S)

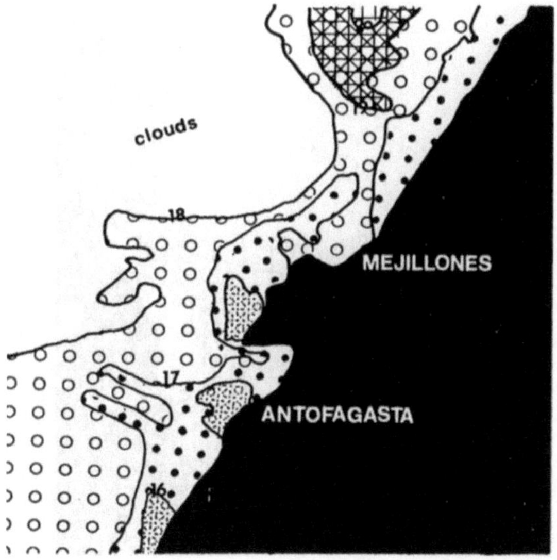

08 DIC 85 COQUIMBO (30° S)

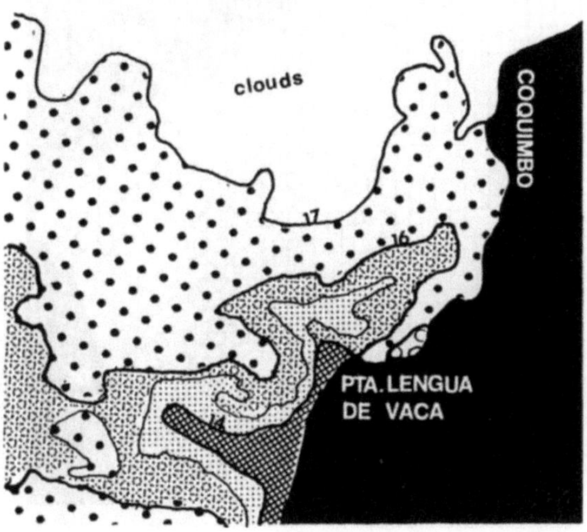

Figure 7B.

14 EN 86 IQUIQUE (20°10' S)

14 EN 86 IQUIQUE (20°10' S)

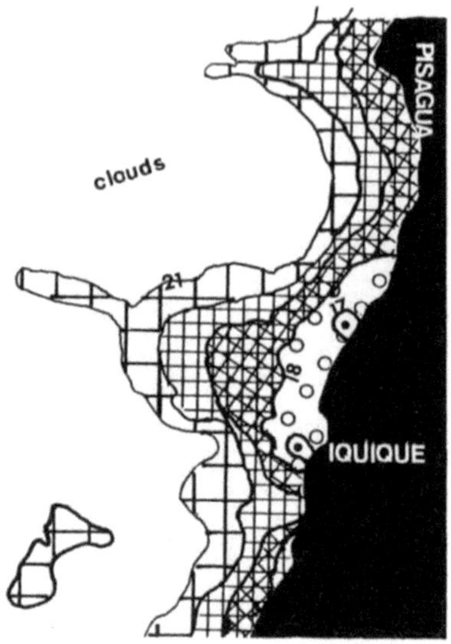

01 DIC 85 ARICA (18°10'S)

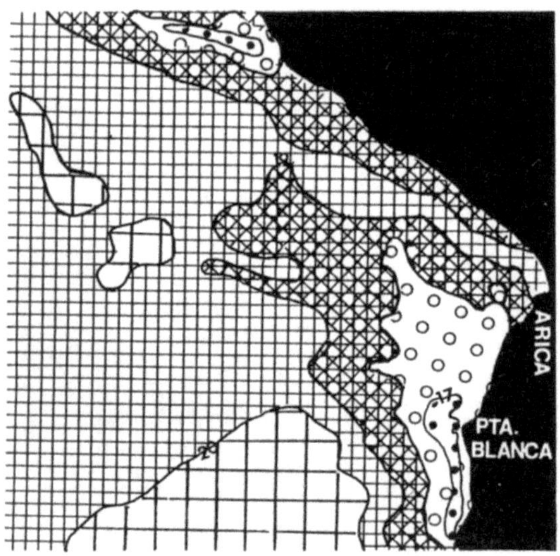

Figure 7C.

a. Curvature of the coast enhances upwelling through an adjustment of relative vorticity,

b. Zonal passage of atmospheric cyclones south of upwelling centers are associated with upwelling events of around one week in duration.

Poleward Flow, Upwelling, and Ecological Effects

Gunther (1936) and Brandhorst (1963) presented information of the characteristics of the water that is advected southward in the PCU. This water has high salinity, low oxygen, and is rich in nutrients. In Figure 8, from Brandhorst (1963), the tight association can be seen between low oxygen and high salinity of the PCU in the area between 33°S to 35°S. It would seem probable that fluctuations in the PCU must have an important effect on the demersal organisms because of the low oxygen. In fact, Brandhorst showed that Hake are concentrated near the bottom on the coast side of the 1 ml/l line. They do not occur in waters with dissolved oxygen less that 1 ml/l. Futhermore, Ahumada (1983) showed that upwelling waters with oxygen values of 0.5 ml/l penetrate inside Concepcion Bay after strong upwelling events. This penetration, clearly associated with the PCU, causes mortality and fish stranding on the beaches, apparently due to the low oxygen content.

Relationships of the physical environment to living organisms show interesting aspects of the ecological effect of the undercurrent. An interesting and provocative relationship between the depth of maximum catch of shrimp to sea level (associated with direction of surface flow) and surface temperature is presented in Figure 9. The depth of maximum catch decreases during the period of southward surface flow and concomitant decreasing sea surface temperature. The upward migration, then, appears to be associated with a rising PCU. This figure also shows the percentage of oviferous females of shrimp. The season of maximum percentage of oviferous females coincides well with low sea level and surface northward flow. It can be speculated that spawning occurs with maximum northward flow and the adults return with the PCU, thus maintaining the shrimp genetic material more or less in the same area.

McLain and Thomas (1983) presented an interesting discussion on the effect of the variability of the California Undercurrent on marine organisms inhabiting that region. Since the undercurrent systems have similarities and most of the marine organisms of each region are similar, it would be extremely valuable to compare the upwelling/undercurrent biomes in each region.

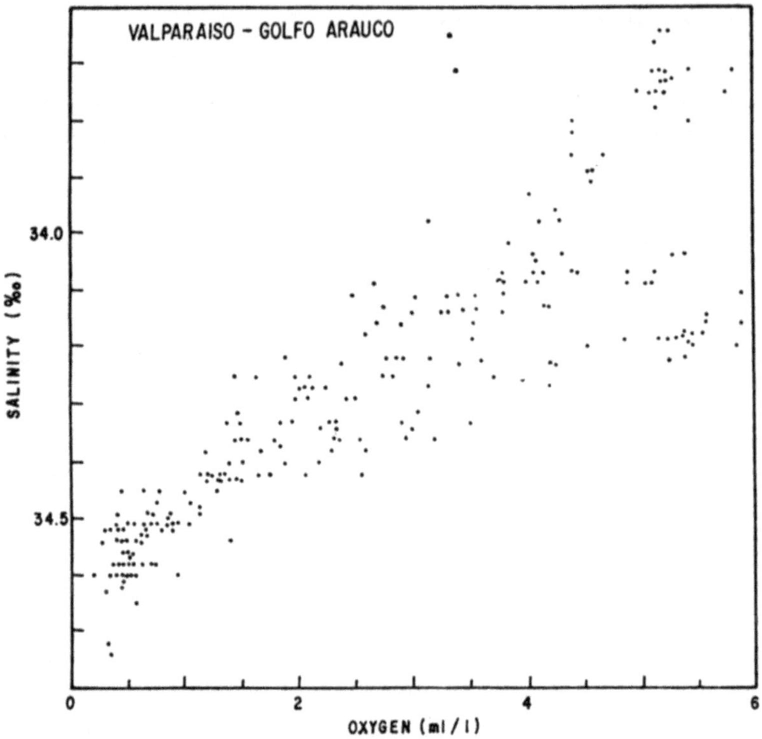

Figure 8: Oxygen/salinity diagram of shelf waters between 33°S and 35°S. The PCU has water with properties in the lower left, i.e., salinity over 34.6 parts per thousand and oxygen less than 1 ml/l [from Brandhorst, 1963].

CONCLUSIONS

The PCU or poleward undercurrent is an important feature in the current system off Chile. It dominates the flow in the first 100 nm, and it causes a great part of the oceanic variability in the area. Originating off northern Peru, it can be traced to at least 48°S. It is noticeable from the surface to at least 300 m depth. At its core, velocities range up to 50 cm/s at a depth of 100 to 150 m. There are evidences that the PCU maintains an annual cycle, with the largest transport (off the shelf) occurring during summer.

The PCU is important in the southward transport of low oxygen water, a phenomena that has serious effects on commercially important bottom organisms such as hake and shrimp. It is also an important source of upwelled water. Upwelling supplies this cold, high salinity and low oxygen water to the upper layers along the entire northern coast of Chile. However, this process is not uniform in time or space; it occurs in the

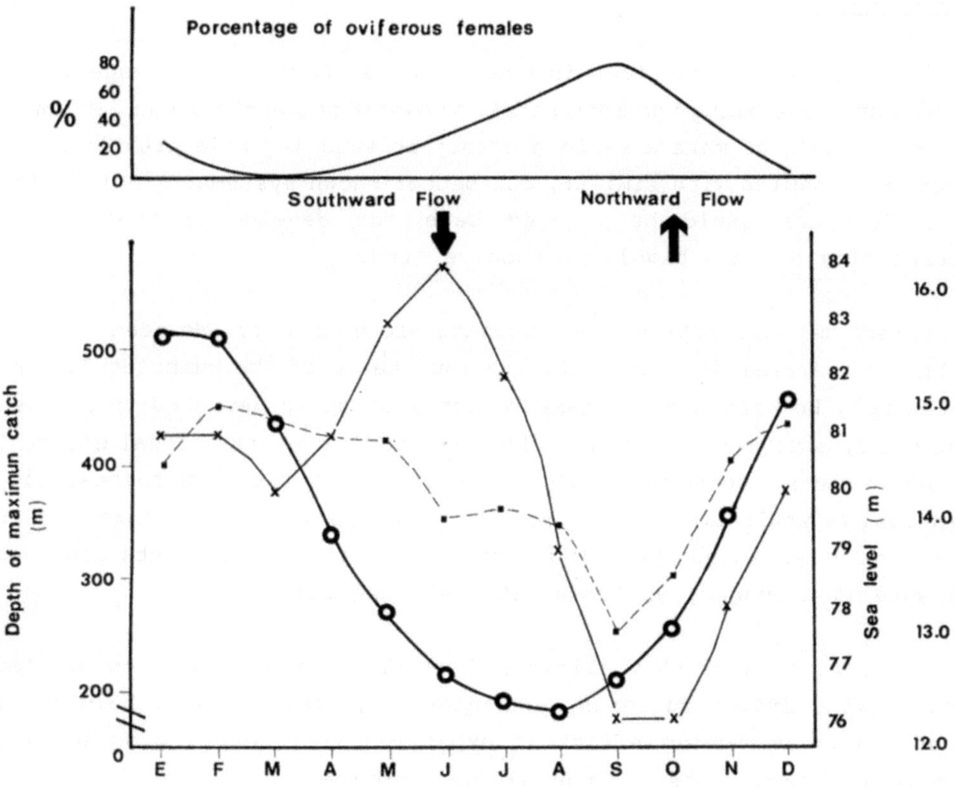

FIGURE 9 Depth of maximun catch of shrimp (■) V/s sea level (X) and surface temperature (o) at Valparaiso

Figure 9: Percentage of oviferous females and depth of maximum catch of shrimp versus surface temperature and sea level (with implied surface flow). The shrimp fishery data are from Arana and Nakanishi (1971), and the physical data from Fonseca (1985).

form of events or pulses in specific centers along the coast. From satellite images covering centers of enhanced upwellings, the surface plumes can be detected as far as 30 to 60 nm offshore.

Oceanographic conditions in this sector of the southeast Pacific coast, associated with the PCU and upwelling, seem to be analogous to those reported for the west coast of North America and the Benguela Current system, but with sufficient differences to provide important insight into the dynamics of upwelling physical/biological systems.

FINAL REMARKS

Oceanographic conditions in the Southeast Pacific, and especially the currents, are among the most poorly known of the world ocean. It would be most valuable to make a serious effort at studying this area in order to compare dynamics with similar, but better known systems. Similarities and differences could be a good base for developing theoretical understanding of this highly productive area.

Direct measurements of the currents are necessary. We need to know the limits, variability, and relative importance of the Humboldt Current and the PCU. Do branches of these currents exist as detected in studies using the dynamic method? Or, are they just eddies or other local effects? We need to have a better picture of the east-west flow and coastal eddy formation. We would like to have more information on the structure of flow conditions at enhanced upwelling centers. In-situ measurements combined with satellite imagery would be extremely helpful.

Theoretical work that allows understanding and prediction of flow changes, wind driven currents, and upwelling dynamics will help local scientists in designing efficient experimental research programs and allow more effective use of the marine resources.

ACKNOWLEDGEMENTS

I wish to express my thanks to Dr. Steve Neshyba for making this presentation possible and for his continuous cooperation with Chilean scientists. I also thank Dr. Victor Neal, Prof. Hellmuth Sievers, and Dr. Donald Johnson for their valuable comments to the manuscript. My participation in the Symposium on Poleward Undercurrents was funded by the U.S. National Science Foundation. For this support, I am grateful.

From: Donald R. Johnson Naval Ocean Research and Development Activity, Stennis Space Center, MS.

On: Review and Commentary to paper: **AN OVERVIEW OF THE POLEWARD UNDERCURRENT AND UPWELLING ALONG THE CHILEAN COAST** by T. Fonseca

INTRODUCTION

Fonseca synthesizes various pieces of broadly interdisciplinary data to form an interesting and credible picture of the Chilean coastal ocean. With a country 4270 km long, running in a nearly north/south direction across latitudinal changes from 8°S to 56°S, clearly the Chilean ocean provides an enormous challenge for that country's oceanographers. A number of Chilean universities, plus several government agencies and a small, but economically important oil industry are all engaged in a range of ocean-related science and engineering activities. If I have one negative comment about Fonseca's paper, it is that it does not include enough of the results of Chilean research activities, and consequently, does not provide an adequate view of the Chilean efforts in marine sciences.

Fonseca's paper presents three topics for consideration: the Peru-Chile Undercurrent (PCU), coastal circulation, and some concepts of ecological interactions with the PCU, upwelling and the Chilean fish stocks. My comments will take the same route.

PERU-CHILE UNDERCURRENT

Fonseca presents an overview of the geographic extent of the PCU along the Chilean coast (Figure 1) using a three-dimensional water mass representation for summer and winter seasons. (In this discussion we will conform to Fonseca and use PCU in reference both to Equatorial Sub-Surface Water and to the undercurrent.) Regardless of interpretations concerning seasonal variations or flow conditions, the most striking feature is the dominant presence of the PCU along a major portion of the Chilean coast. Second, and equally striking, is the surface broaching of the "undercurrent" (offshore in the northern sections and inshore in the southern sections). Because of the broaching, nutrient-rich water is in contact with the euphotic zone, and the stage is set for dynamic growth in the biota.

A third, and somewhat obvious, point that should be made with Figure 1, is that the strength, direction, and dynamics of the flow system cannot be easily determined from "classical" measurements. Not only does the water mass carry a complex history with it, which is difficult to unravel from its characteristic tags, but implied dynamics are least certain over the slope and shelf regions where the flow is normally most vigorous. Unfortunately, with limited resources, the Chilean experience with the PCU has mostly had to concentrate on surveys using the classical method.

In Figures 2 to 4, Fonseca provides a rough view of PCU flow characteristics from three different, but limited, methods: dynamic calculations, lagrangian drogues, and shipboard current meter profiles. Obviously missing, and much needed, are long-term moored current meter measurements. From dynamic calculations and drogue measurements, Fonseca indicates that the PCU seems to flow between 5 and 25 cm/sec. These measurements were made in deeper water just off the slope, and at discrete depths (drogues) which were poorly resolved in the vertical. The shipboard current profiles were measured over the slope itself, and are intriguing in that they suggest a vigorous current of 30 to 50 cm/sec. Although these latter measurements were "snapshots" of the current system, some rational speculations can be made after examining them in more detail.

The current profiles presented in Fonseca's Figure 4 were obtained on MARCHILE X, a hydrographic survey during July, 1976, which covered the eastern boundary regime from Valparaiso to the border with Peru, and offshore to about 200 nm. At four stations along the coast (Arica, Antofagasta, Huasco, and Valparaiso), current profiles were made over a period of 12 hours, and averaged in order to eliminate tidal variations. In addition, a similar station was made over the O'Higgins Seamount (approximately 120 nm offshore of Valparaiso). The three northern coastal stations were made near the 1000 m isobath along the continental slope, and the Valparaiso station was obtained near the 400 m isobath, very near the shelf break.

Figure A-1 shows the current profiles from Antofagasta, Huasco, and Valparaiso plotted together. In addition, a profile using the dynamic method at the central station (Huasco) is presented since this method tends to filter out non-geostrophic "noise." The similarity in shape and amplitude of the measured profiles both in the along-shore and in the cross-shelf components is striking and suggests that the snapshot-type measurements may be providing a characteristic look at the currents over the slope region. Furthermore, although the calculated profile does not reflect smaller scale structure in the upper layer, its general adherence to the measured profiles gives additional assurance that the measurements were not spurious. The Arica current profile was relatively weak in comparison and will be discussed separately.

Referring to Figure A-1, it can be seen that the maximum southward flow occurs at about 150 m in depth, in direct association with the salinity maximum and dissolved oxygen minimum. Current speeds in the core of the undercurrent are in the 30 to 50 cm/sec range. At Valparaiso, the southward flow reached the surface although high salinity water at that location did not do so (the strong southward surface flow was confirmed by measured ship drift toward the south against the windstress). It should be recalled that the Valparaiso profile was taken in shallower water and may be as much indicative of a cross-shelf variation as an along-shore variation.

In order to obtain a better feeling for areal distribution of the PCU, we can examine some of its characteristics through isentropic analysis of the MARCHILE X data. In Figure A-2, the 26.5 potential density surface has been chosen as characteristic of the level of maximum salinity and minimum oxygen indicative of the PCU. Since this surface nominally lies at approximately 175 m in deep water, the 175 m level was also chosen for comparative current analysis. Apparent Oxygen Utilization (AOU) was used as a water mass tag in Figure A-2 instead of dissolved oxygen since it provides a parameter which is primarily independent of both temperature and salinity tags. Dynamic topography at 175 m referred to 1000 m is represented in Figure A-2 as maxima (ridges) and minima (valleys), with superimposed direct current measurements interpolated to 175 m.

From Figure A-2, PCU water appears to cover a broad offshore area from Antogasta northward, but decays southward into "fingers" of water turning away from the coast. The alignment of maxima and minima in dynamic topography and, consequently, the calculated flow alignment tend to conform to the water mass tags.

A close examination of the dynamic topography in Figure A-2 together with measured currents (note the inclusion here of the current measurement over O'Higgins Seamount) reveals at least one interesting feature: the presence of the maximum just offshore in the northern section suggests a northward flow between this maximum and the continental slope. Whether it is a semi-permanent or a temporary feature, the point is that there appears to be (not too surprisingly) a high speed core of the PCU, locked to the continental slope (as indicated in the direct current measurements of 30 to 50 cm/sec), and a weaker southward drift in deeper water (as indicated in dynamic calculations and parachute drogue measurements of 5 to 10 cm/sec).

A second noteworthy feature in Figure A-2 involves the splitting of the PCU south of Chanaral. Is this an indication of gradual slowdown and decay of the PCU, or is it being dynamically wedged apart by a northward

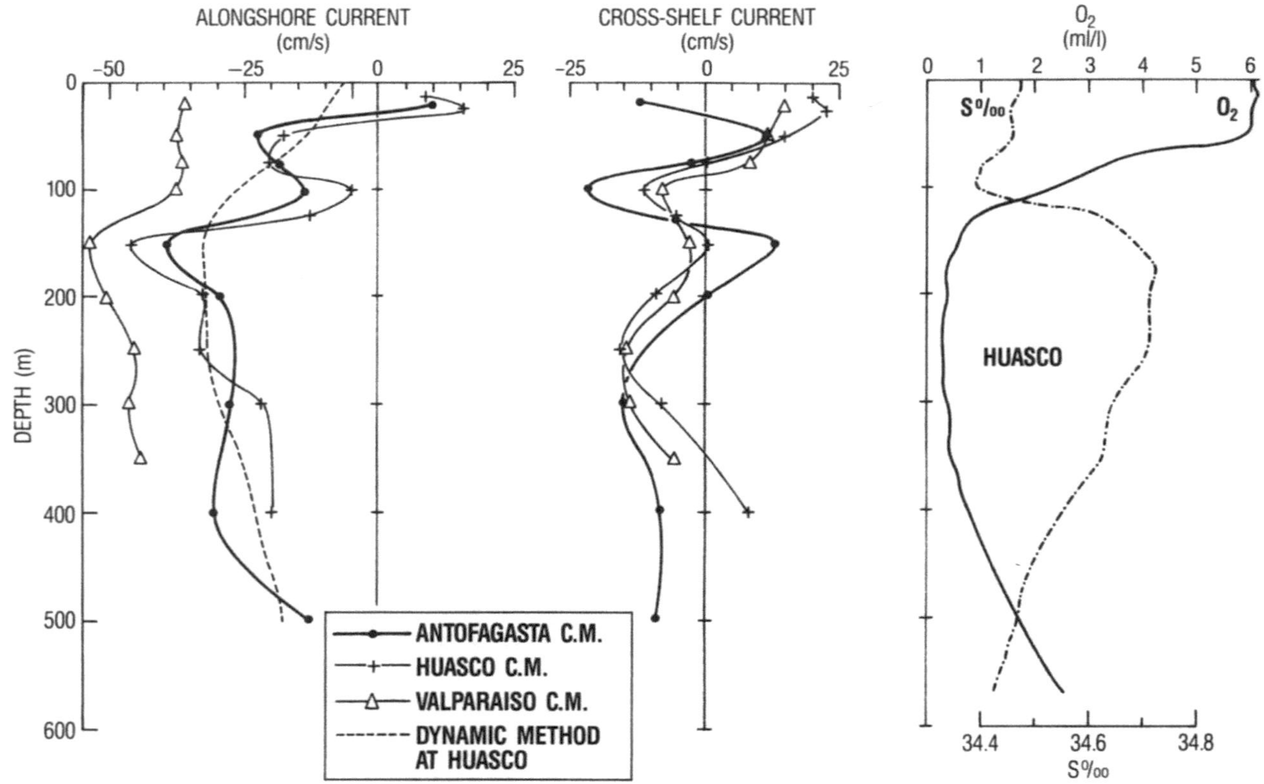

Figure A-1: Current profiles at continental slope/shelf break stations (see Figure A-2 for locations), along with a salinity/oxygen profile and a profile of currents using the dynamic method at Huasco. Equatorward and shoreward are positive directions.

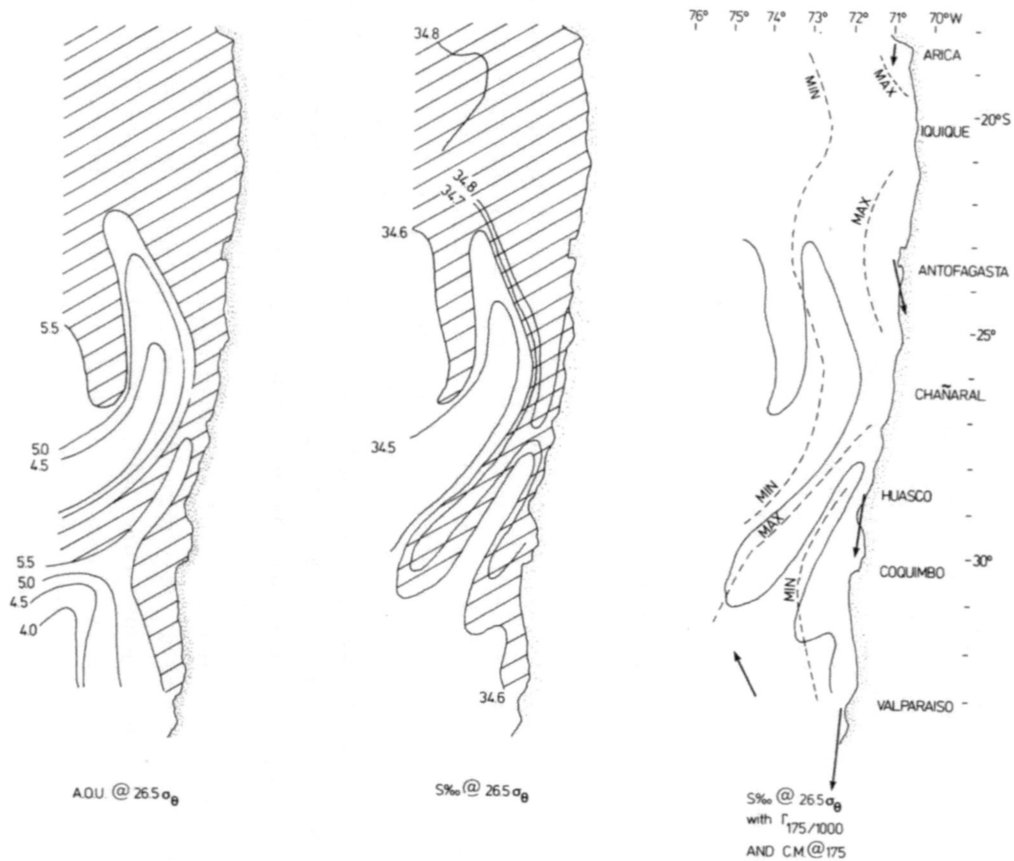

Figure A-2: Isentropic analysis on the 26.5 potential density surface. Left-Apparent oxygen utilization. Center-Salinity. Right-Salinity and dynamic topography sketch with measured current vectors at five stations.

intrusion? Note that the current measurement over O'Higgins Seamount is in alignment with dynamic topography and, at approximately 30 cm/sec toward the north, is relatively strong. Does the PCU gradually decay, or does it end in more vigorous interactions? In Figure A-3, three sections of salinity are presented along 33°S (Valparaiso), from data obtained over a 49-day period in early 1960. Changes in the PCU during this time were dramatic. Although a variety of interesting speculations can be made on the geometry and dynamics of the PCU from these sections, they do serve to illustrate the rapidity and scale of changes which occur over a relatively short time. Since hydrographic cruises, such as MARCHILE X, cover the northern and central coast area in approximately 30 days, it is apparent that such smeared snapshots may not provide more than a generalized picture of the area, and may be somewhat misleading in detailed interpretation. It should be noted that Fonseca has been quite cautious in his presentation of large scale hydrography.

COASTAL CIRCULATION AND UPWELLING

From Hastenrath and Lamb (1977), the climatological monthly wind direction along the north and central Chilean coast is clearly upwelling favorable over the entire year. In addition, a relatively strong negative windstress curl (upwelling favorable) covers the coastal area out to 200 km offshore. From the persistence of the winds, then, we should expect consistent upwelling conditions. Fonseca, however, makes the point that upwelling is both spatially and temporally varying along the Chilean coast, with a propagating seasonal signal in evidence. Fonseca also suggests that spatial/temporal changes in windstress, along with local curvature of the coastline and mid-scale along-shore changes in bathymetry tend to produce localized areas of enhanced upwelling. In spite of the first impression of a relatively straight coastline, Fonseca's satellite derived sea surface temperature maps demonstrate graphically that upwelling on the Chilean coast is indeed spatially dependent and geographically highly complex. In fact, as far back as Gunther (1936), it has been suggested that permanent "centers" of upwelling occur along the Chilean/Peruvian coast.

In a further examination of Hastenrath and Lamb's (1977) atlas, the wind, windstress curl, and sea surface temperature fields along the South American coast (to 30°S) show minima in wind and windstress curl, and a maximum in SST occurring around 18° to 22°S. This area is located at the "bend" in the coastline, separating Peru and Chile. Wooster (1970) makes an interesting speculation of this area: "...that most of the Chile Current goes off to the west before reaching 20°S, that the Peru Current

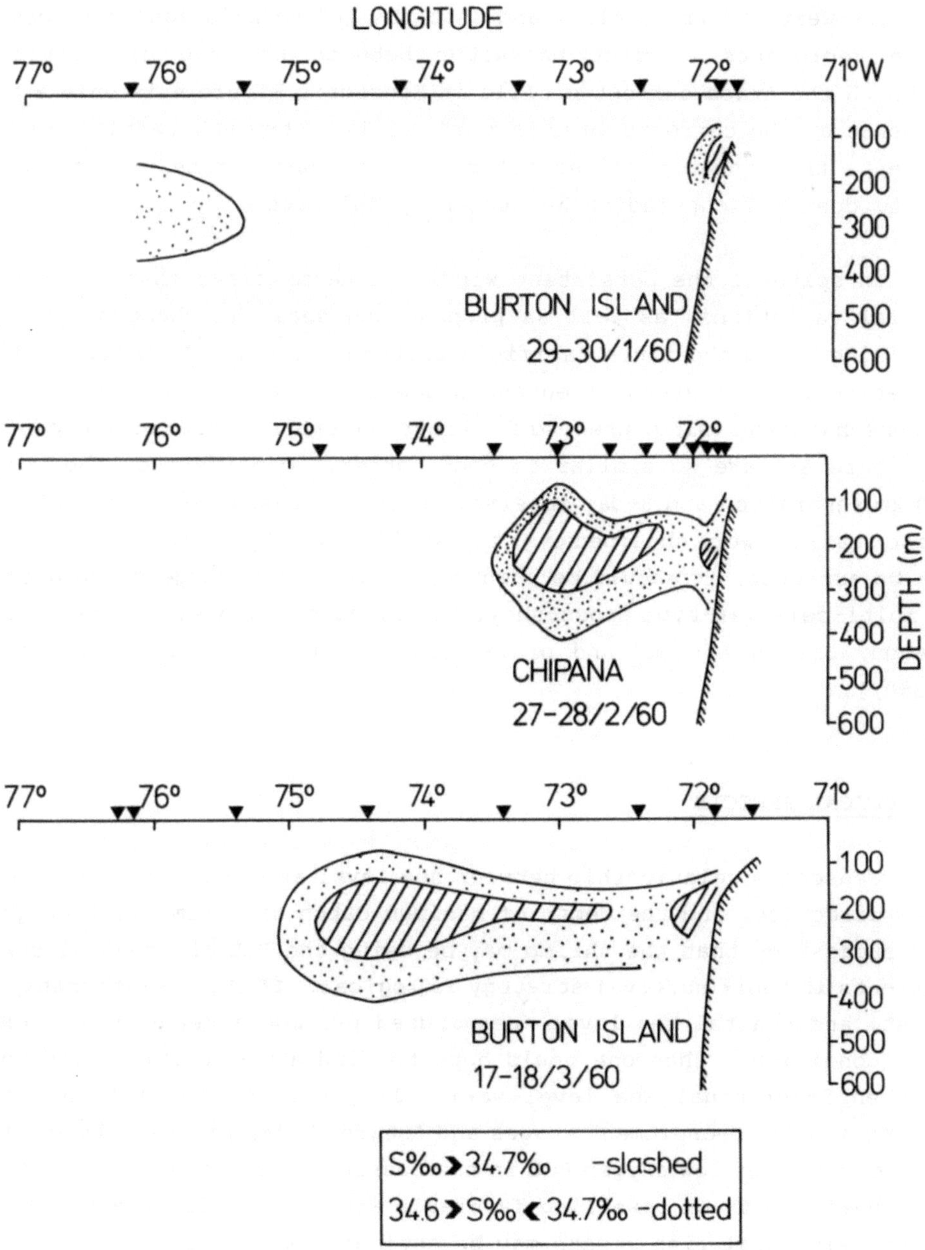

Figure A-3: Cross-sections of salinity over 49-day period off Valparaiso showing rapid change in PCU core water.

arises along the southern coast of Peru, that this current is fed largely from the west and from below, and that the 18° to 22°S zone off northern Chile represents a quiet backwater between the two circulations." Although the Arica current profile in Fonseca's Figure 4 is only a brief snapshot of the currents in this zone, it does show substantially weaker currents than those profiles taken at locations further south and, at least, does not contradict Wooster's speculation.

In spite of the persistent winds, it seems clear that large scale spatial variations, as well as propagating seasonal changes, occur in upwelling along the South Pacific's eastern boundary. Embedded in these variations, local areas of enhanced upwelling appear to be evident. It is somewhat surprising, then, to find that the cross-shelf current profiles in Figure A-1 are so similar to each other, considering the nearly 600 km separation and 8-day intervals between coastal stations. Although these stations were obtained over the shelf break where high vertical modes can be expected, it should be pointed out that they suggest the presence of multi-cellular-type cross-shelf flows, such as have been observed off Oregon, off NW Africa, and previously, off Chile [Johnson and Mooers (1980)].

ECOLOGICAL EFFECTS

Fonseca's relationship between sea level at the coast, with implied flow direction, and the depth of maximum catch of shrimp is intriguing. His suggestion that the shrimp may be using the PCU and coastal surface currents in their survival strategy is logical. If this relationship does exist, and coastal sea level does indeed provide a means of discerning flow conditions, then one would hope to find a predictive relationship between interannual sea level variability during the critical larval dispersion and recruitment stages and future shrimp harvest. In addition, if the amount of low oxygen PCU in the coastal zone can be predicted based upon variability of upwelling favorable winds, then detrimental effects on certain local fish stocks may be estimated and considered in fishery management plans.

Although it should be clearly noted that fisheries develop according to very complex biological and environmental interactions, some simple relationships as noted above can be quite useful to fishery managers in estimating the relative effects of environmental change and fishing pressure on the fish stocks. Coastal tide gauges, wind, and sometimes, coastal temperature records are relatively accessible sources of data for comparison with fishery records. Although many flaws exist in both types of data sets, statistics can be decently forgiving if applied toward long-term trends.

RECOMMENDATIONS

Chile's remote location makes collaboration with northern hemisphere colleagues a difficult problem. Although such collaboration is earnestly desired on both sides, the sharp reduction in funding of International Programs in the National Science Foundation and in NOAA Sea-Grant has considerably hampered development of collaborative investigations. International cooperative programs, such as that maintained by Oregon State University with Chilean counterparts, should be commended for persistence in spite of bleak funding, and should be encouraged. These kinds of programs are essential in order to set the stage for meaningful large scale cooperative field programs as requested by Fonseca.

But Chile's remote location has one advantage that should be exploited. During the first half of the next decade, several oceanographic satellites with all-weather sensors will be launched. The European Space Agency will launch its ERS-1 satellite with altimeter, scatterometer, and synthetic aperature radars on board. Japan will launch its J-ERS-1 SAR/optical satellite. The U.S. Navy will continue its GEOSAT series (altimeter); TOPEX/POSEIDON (altimeter) will be launched in a cooperative U.S./French effort; and Canada will launch its RADARSAT. Not only will the all-weather sensors aboard these satellites provide a means of sampling through the cloudy conditions along the Chilean coast, but satellite oceanography may be the best method of extracting information to match the spatial and temporal scales.

All of these satellites will need <u>in-situ</u> validation, particularly during wintertime extreme conditions. If the launches are made during northern hemisphere summer, the validations will either be delayed or moved to southern hemisphere locations. Participation in the validation effort can lead to access to satellite data, as well as to the entire validation data base. I recommend that Chilean oceanographers begin vigorous efforts to contact the program directors for these satellites and to offer a Chilean national response to the announcements of opportunity. The Union of Marine Sciences of Chile would be an excellent forum to begin developing such a plan.

Probably more than their northern hemisphere counterparts, Chilean scientists depend on applicability of their research to attract internal funding. With this in mind, the need for management and effective exploitation of fish stocks depend, to a large extent, on understanding the environmental effects during critical stages of biological development. As mentioned in the previous section, long term measurements of relatively accessible parameters such as coastal sea-level, wind, and temperature can be related to fisheries and, at the same time, can provide insight and direction for more focused field studies. But establishing quality long-term measurements will require a national cooperative effort.

Fonseca's paper lays a good foundation for investigation of this interesting, but relatively neglected (by northern hemisphere oceanographers) area of the world's ocean. Like Fonseca, I hope that new international cooperative work can be triggered in Chilean waters.

THE INDIAN OCEAN

THE LEEUWIN CURRENT

John A. Church
George R. Cresswell
and
J. Stuart Godfrey

CSIRO Division of Oceanography
GPO Box 1538
Hobart Tasmania 7001 Australia

ABSTRACT

The Leeuwin Current is a surface stream of warm, low-salinity tropical water that flows poleward (southward), against the climatological mean equatorward winds, from northwestern Australia to Cape Leeuwin and then eastward towards the Great Australian Bight. It flows principally, but not exclusively, in autumn and winter. In the north, it is broad and shallow (200 km by 50 m), tapering and deepening in the south to a relatively narrow current (less than 100 km wide), with a top speed of 1.8 ms^{-1} and strong vertical and horizontal shears. There is an equatorward undercurrent below. The warm, low-salinity Leeuwin Current water intrudes between the continent and the southgoing high-salinity flow described by Andrews (1977). As the current approaches Cape Leeuwin, it extends down to about 200 m and is most commonly centered at the shelf edge. After rounding Cape Leeuwin, it spreads onto the shelf and develops seaward offshoots that result in cyclone/anticyclone eddy pairs. A strong poleward geopotential gradient is thought to drive the Leeuwin Current.

INTRODUCTION

Equatorward winds characterize the ocean off Western Australia and the Benguela, Canary, Peru, and Californian current systems of the Atlantic and Pacific Oceans. However, the behavior of the ocean off western Australia seems to be different: there is no evidence of a steady, continuous equatorward flow near western Australia (Andrews, 1977) or of coastal upwelling, despite the strong equatorward wind stress. Instead, historic oceanographic observations have consistently shown a poleward (southward) surface flow in winter, against the equatorward wind stress, near the shelf edge from at least 22°S to 34.5°S (the southwest tip of Australia). The strong southward flow of warm, low-salinity water, called the Leeuwin Current by Cresswell and Golding (1980), is readily apparent in satellite infrared images (Figure 1). Further offshore, the flow in the upper 1000 m is clearly equatorward (Warren, 1981).

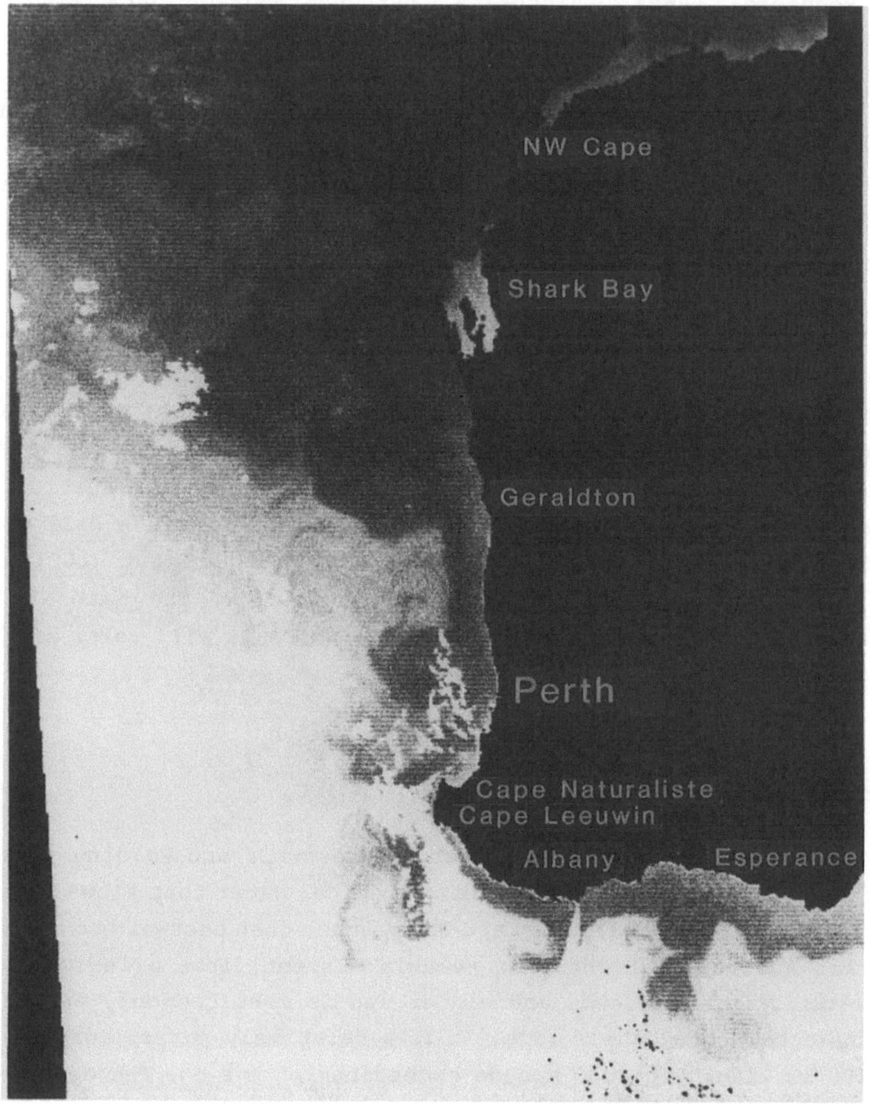

Figure 1: A satellite infrared image of western Australia showing the Leeuwin Current starting as a broad fan-shaped form off northwestern Australia and then tapering and hugging the continental shelf edge as it heads southward to Cape Leeuwin and then eastward towards the Great Australian Bight. The shallow water of Shark Bay is cool in response to mid-winter cooling; an eddy is located off Perth; and two offshoots of the Leeuwin Current can be seen off the south coast. Note that black shading indicates warm water and grey indicates cool water; there are some clouds (white) in the bottom left of the image.

The earliest evidence for the Leeuwin Current came from the observations by Saville-Kent (1897)—later confirmed by Dakin (1919)—of tropical marine fauna and warm waters around the Abrolhos Islands (28.5°S to 29.0°S). More recent evidence comes from ship drift observations (Nederlandsch Meteorologisch Institut, 1949), time series water property data (Rochford, 1969a), historical bathythermograph data (Gentilli, 1972), research vessel surveys (Kitani, 1977), and biological data sets (Wood, 1954; Colborn, 1975; Krey and Babenerd, 1976; Markina, 1976). The Leeuwin Current acts as a conduit to bring tropical marine fauna and flora to southern Australia, and appears to play an important role in the life cycles of a number of subtropical marine creatures (Maxwell and Cresswell, 1981; Wells, 1985; Cresswell, 1986). It may also influence the range and breeding seasons of seabirds (Dunlop and Wooller, 1986).

In this review, we will discuss the water masses of the Leeuwin Current and then consider various aspects of it in three regions: the region northwest of Australia, the region directly west of Australia, and its eastward extension south of Australia. We will also discuss the effect of the Leeuwin Current on the continental shelf off western Australia and the various attempts to model it. In the next section, we will briefly describe the LeeUwin Current Interdisciplinary Experiment (LUCIE) and, then present some preliminary results from LUCIE. The nature of the geography of the region and place names to which we will refer are shown in Figure 2.

THE LEEUWIN CURRENT—GENERAL FEATURES

The Leeuwin Current was defined by Cresswell and Golding (1980) as a surface stream of warm, low-salinity tropical water that flows southward from northwestern Australia to Cape Leeuwin and then eastward to and across the Great Australian Bight. The Leeuwin Current flows principally, but not exclusively, in autumn and winter and is most commonly encountered near the continental shelf edge. It is a relatively narrow current (less than 100 km wide) with top speeds exceeding 1.5 m s^{-1}. The depth of the current varies with time and position, and there is a tendency for the current to deepen as it moves southward. Despite this temporal and spatial variability, it is clear that the current is shallow and that below it there is an equatorward undercurrent. Godfrey and Ridgway (1985) indicate that the intermediate level of minimum motion is between 200 m and 300 m.

Andrews (1977) described a northeastward flow that turns at 29° to 31°S and flows poleward near the continental shelf. This flow dominates

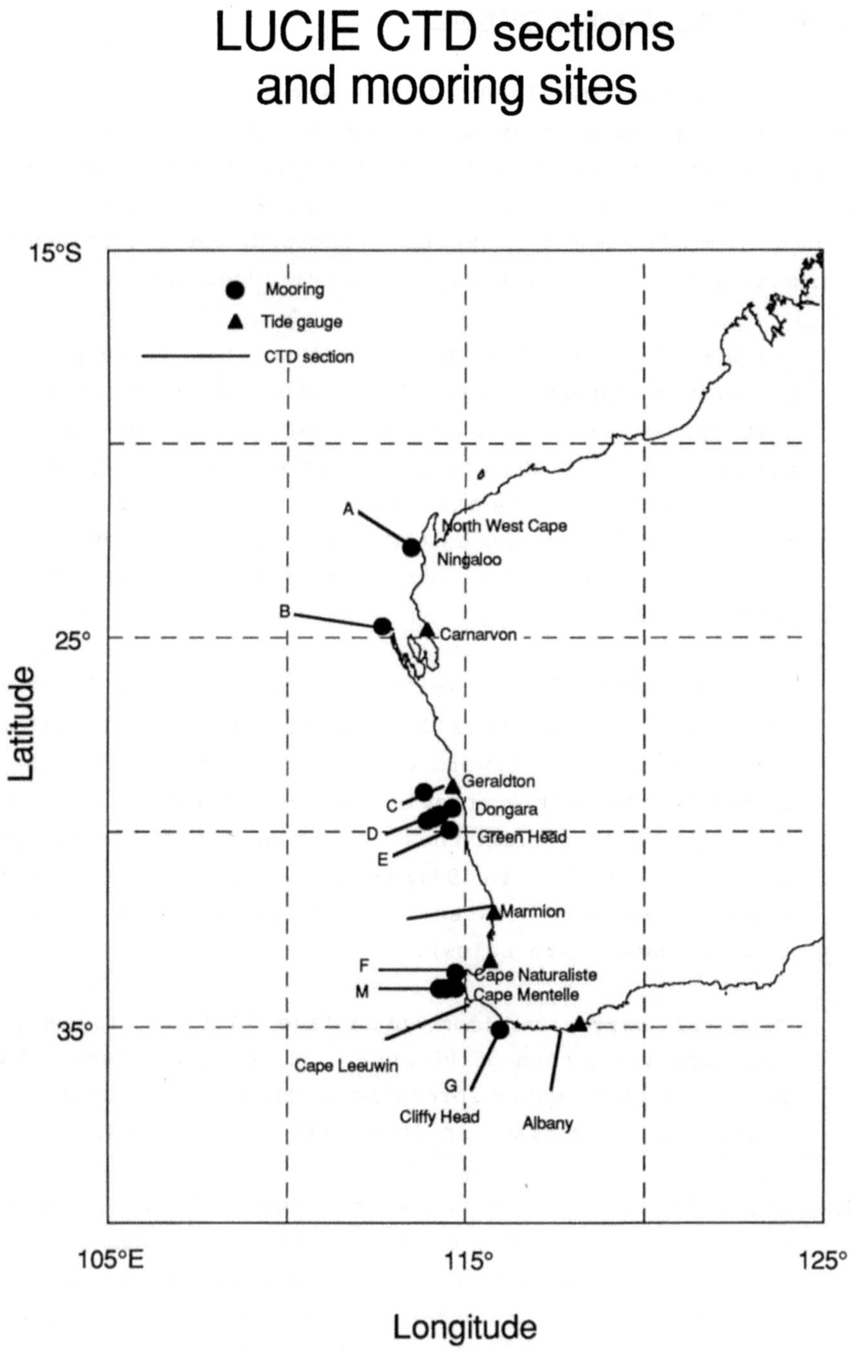

Figure 2: The location of the CTD sections and moorings used in LUCIE.

during the summer period, carries high-salinity water, and was called the West Australian Current by Andrews. It can be seen in Godfrey and Ridgway's (1985) steric sea-level maps.

WATER MASSES OF THE LEEUWIN CURRENT

The hydrological regime in which the Leeuwin Current flows was first described from a series of seasonal biological cruises conducted along the 100°E meridian between 9° and 32°S with HMAS "Diamantina" and HMAS "Gascoyne" in 1962/63. These enabled various water types to be identified in the upper 500 m off western Australia (Rochford, 1969a). Confining our attention mainly to the subtropics, the water types were:

1. Surface (0-50 m), low-salinity (less than 35.00 ppt), high temperature (greater than 25°C), low-nutrient, tropical waters that spread south of 20°S in autumn and winter—the Leeuwin Current. At 32°S the corresponding winter salinity was 35.20 to 35.40 ppt, with the increase from 20°S being due to mixing; the Leeuwin Current holds the sea surface temperature at 32°S above 22°C until after June (Rochford, 1969b; Cresswell and Golding, 1980).

2. Surface (0-50 m), high-salinity (greater than 35.90 ppt), lower-temperature (20-22°C), low-nutrient, subtropical waters at 30°S carried northward to 25°S and down to 100 m depth in summer by the West Australian Current. Webster, Golding, and Dyson (1979) identified these waters during late summer surveys between 29° and 32°S in 1974; the waters gave rise to a secondary salinity maximum between the surface and the South Indian Central Water (see below).

3. Tropical oxygen minimum—subsurface (100 to 150 m), low-salinity (less than 35.00 ppt), low-oxygen (less than 3.5 ml/l), tropical water spreading southward to about 26°S on the 25.00 sigma-t surface in late summer and autumn.

4. Subsurface high-salinity (greater than 35.80 ppt), subtropical water of 17° to 19°C of the South Indian Central region, spreading northward on about the 26.00 sigma-t surface to about 12°S in summer, and to about 16°S in winter. At 33°S this water is at less than 100 m while at its northern limit it has sunk to 300 m. Webster, Golding, and Dyson (1979) found that the Leeuwin Current spread southward over, and mixed with, the South Indian Central Water, thereby becoming progressively more saline and cooler with increasing latitude. Seasonal variations in solar heating, evaporation-precipitation effects, dynamical uplifting of the South Indian Central Water, mixing, and the supply of Leeuwin Current Water were found to be

controlling factors for the temperature and salinity of the surface layer.

5. Subtropical oxygen maximum—subsurface (400 to 500 m), low-salinity (less than 35.00 ppt) waters of the subtropical oxygen maximum (greater than 4.50 ml/l) drifting northward on about the 26.80 sigma-t surface to about 14°S in winter.

Ridgway and Loch (1987) determined mean TS relationships and confirmed a number Rochford's results. For the offshore region, they found that the high-salinity South East Indian Central Water penetrated as far north as 15°S. North of 15°S, this penetration is inhibited by the westward flow of relatively low-salinity water within the South Equatorial Current. Closer to the coast between 5°S and 20°S, the TS properties are more continuous and indicate the presence of a southward-flowing current of low-salinity water which prevents the high-salinity water from approaching the shelf.

THE NORTHWEST REGION

The waters northwest of Australia appear to contain the source region for the Leeuwin Current. Historical bathythermograph data were interpreted by Gentilli (1972) to show that the throughflow from the Pacific to the Indian Ocean in autumn and winter was isolated by a reversal of the flow in spring. This water then achieved thermal homogeneity over the summer to become a "raft" of warm water, which spread southward during the following autumn and winter. A qualitatively similar seasonal progression of the Leeuwin Current was deduced from objective maps of steric sea level through the year (Godfrey and Ridgway, 1985). A near-synoptic picture of this kind can be seen in unpublished research vessel sections ("Diamantina" cruise 4/71) taken at 2° steps of latitude from 20°S to 34°S in August 1971. It was broad and shallow (200 km by 50 m) and mainly off the continental shelf at 20°S, and then tapered, deepened, and spread onto the shelf as it moved southward as the Leeuwin Current. Satellite infrared images such as Figure 1 in this paper and also Figure 1 of Legeckis and Cresswell (1981) support this general picture.

The nature of the Leeuwin Current at the southern part of the North West Shelf was determined by Holloway and Nye (1985) from current meters moored at a number of sites from January 1982 to July 1983. The current ran parallel to the bottom topography and was strongest at the shelf break, reaching a maximum speed of 0.25 m s^{-1} . It was strongest between February and June. The southeast trades, which blow from March to August, did not generate alongshore currents that strengthened the Leeuwin Current.

Satellite drifters with 20 m tethers to their sea anchors showed weak random currents in January and February, 1983 (Cresswell, unpublished data). In March and April the situation changed rapidly with the onset of the Leeuwin Current: on the shelf it had speeds up to 0.5 m s^{-1}; after it rounded North West Cape the maximum speed reached was 1.0 m s^{-1}. At 20°S, 113°E one drifter was trapped for a month in an anticyclonic eddy 80 km in diameter and completed three revolutions.

THE LEEUWIN CURRENT OFF WESTERN AUSTRALIA

Southward flow above the continental slope from 29°S to Cape Leeuwin appears to be a common feature, although the source can vary. Surveys by Andrews (1977) in the summers of 1972 and 1973 revealed that high-salinity surface water from the West Australian Current was carried by a large-scale zonal flow (cf. Hamon, 1965) northeastwards towards the coast between 30° and 33°S, where it turned to flow southward, meandering and breaking into eddies. When it reached Cape Leeuwin, it turned to run to the east. Research vessel and satellite-tracked drifter data suggested to Cresswell and Golding (1980) that the warm, low-salinity Leeuwin Current water intruded between the continent and the south-going high-salinity flow described by Andrews (1977). The drifters showed that the waters of the Leeuwin Current and the cyclonic eddies on its seaward side were mixing. These drifters registered accelerations and water temperature changes as they interchanged between the current and the eddies. While the eddies that trapped the drifters were most frequently cyclonic, two drifters were carried into an energetic anticyclonic eddy and held in it for two months (Cresswell, 1977). The maximum speeds travelled by drifters were 1.7 m s^{-1} along the south coast (Cresswell and Golding, 1980) and 1.8 m s^{-1} near Cape Leeuwin (Godfrey, Vaudrey, and Hahn, 1986).

Eddies were a feature of Kitani's (1977) observations from "Kaiyo Maru" in November 1975. They occurred at 30°S where south-going, high-temperature (21°C), low-salinity (35.2 ppt), northern water encountered north-going low-temperature (18° to 19°C), high-salinity (35.5 to 35.6 ppt) southern water. The south going flow split so that part of it was on the continental shelf and part was farther out to sea. Kitani reasoned that if the south-going flow was a stable feature of the region, then it could be defined as a countercurrent to the north-going West Australian Current farther offshore.

Thompson (1984) took observations of the Leeuwin Current between 22°S and 28°S and suggested that the strong along-coast geopotential gradient could drive the current. He found the Leeuwin Current to be low in dissolved oxygen and high in nutrients (where it touched bottom at the shelf edge) and that only a small portion of its $4 \times 10^6\ m^3 s^{-1}$ flux came from the North West Shelf. Several hundred meters below the Leeuwin Current was an equatorward flow of salty, high-oxygen, low-nutrient water—or what we would identify as Rochford's (1969a) South Indian Central Water. The undercurrent was also identified in mean steric height patterns from historic Nansen bottle and bathythermograph data (Godfrey and Ridgway, 1985).

EFFECTS ON SHELF WATERS

The effects of the various currents on the waters of the continental shelf were first monitored at the Rottnest Island 50 m hydrology station by Rochford (1969b). From the drift cards he released over a year, he concluded there was a southward flow of low-salinity water in autumn/winter and northward flow of high-salinity water in summer. The northward flow is opposite to the southward open ocean summer flow reported by Andrews (1977) and is probably the effect of a predominantly southerly wind stress on the shallow waters of the shelf. A region of shear at the outer part of the shelf was shown by the drifters of Cresswell and Golding (1980). The arrival each autumn of the low-salinity waters of the Leeuwin Current at the Rottnest Island 50 m hydrology station produces an annual salinity variability of roughly 0.5 (Figure 3). An exception, for which we have no explanation, occurred in 1954 when there was no low-salinity winter signal.

Changes in the strength of the Leeuwin Current—or at least of the net poleward geostrophic flow—are reflected in mean sea level changes at coastal stations. These typically show an annual amplitude of 15 to 20 cm, with highest values in March (North West Shelf) to May or June (southwestern Australia); this seasonal movement of the sea-level maximum reflects the southward passage of the Leeuwin Current pulse.

THE SOUTHERN AUSTRALIAN COASTLINE

Satellite images of the Leeuwin Current along the south coast of western Australia showed that it could have five or six jet-like offshoots to seaward in various stages of development (Legeckis and Cresswell, 1981). These were separated by about 200 km and were probably associated

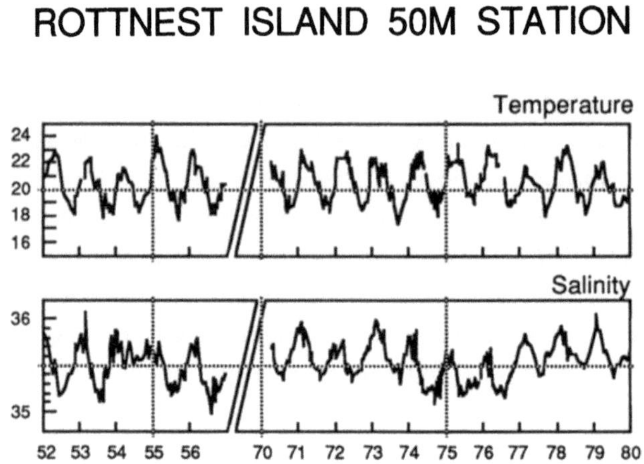

Figure 3: Surface salinities and temperatures from the Rottnest Island 50 m hydrology station.

with the eddies detected earlier by the drifters. Griffiths and Pearce (1985) found that eruptions of the Leeuwin Current to seaward along the south coast of western Australia temporarily diverted the flow into unstable baroclinic waves and cyclone-anticyclone pairs, similar to a phenomenon observed in the laboratory.

The Leeuwin Current was found from research vessel and satellite drifter data to extend along the south coast of Australia from Cape Leeuwin (115°E) to Portland (142°E) (Godfrey, Vaudrey, and Hahn, 1986). Cyclonic eddies breaking away from the shelf west of 124°E were thought to have probably removed most of the warmest water from the continental shelf. The research vessel data revealed fronts having very low Richardson numbers and, therefore, turbulent interfacial stresses.

INTERANNUAL VARIABILITY

The extent of interannual variations in the strength of the Leeuwin Current is not known. Allan and Pariwono (1987) showed that anti-ENSO (ENSO: El Niño---Southern Oscillation) events are accompanied by higher than normal sea level at Darwin and Port Hedland. Pearce and Phillips (1988) have suggested that the enhanced sea level along western Australia is associated with a stronger than normal Leeuwin Current. This suggestion finds support in the Rottnest Island data (Figure 3), which, for the 1973-76 anti-ENSO event, show low salinities suggestive of an enhanced flow of low-salinity water from the tropics.

MODELING THE LEEUWIN CURRENT

Thompson and Veronis (1983) were the first to attempt to model the Leeuwin Current. They suggested that the winter winds on the North West Shelf and slope generated the Leeuwin Current, and they also predicted a southward flow off the west coast of Tasmania. However, this model is inconsistent with Holloway's observations that the southeast trades, which blow from March to August, did not strengthen the Leeuwin Current. Thompson (1984) rejected this model and suggested instead that the strong along-coast geopotential gradient could drive the current. He felt that winter deepening of the mixed layer might allow the geopotential gradient to overcome the equatorward wind stress.

Godfrey and Ridgway (1985) show that seasonal variations of both the alongshore pressure gradient and the wind stress favor a late autumn maximum in the speed of the Leeuwin Current. In particular, they conclude that a 27 cm rise of steric sea level off North West Cape from January to May is the primary driving agency of the current. While each of the eastern boundary current regions have poleward surface pressure gradients, Godfrey and Ridgway (1985) show that this gradient is largest in the eastern Indian Ocean region.

The idea that the southward geopotential gradient drives the Leeuwin Current is a central element of the models of McCreary et al. (1986) and Thompson (1987). These models invoke this poleward pressure gradient as the principal force driving the Leeuwin Current against the equatorward wind stress. In the McCreary et al. model, vertical mixing is important, and all currents vanish as the vertical viscosity and diffusivity go to zero. This model also produces an equatorward undercurrent. In both models, the Leeuwin Current is fed by a geostrophic inflow from the west rather than an inflow from the north.

The Leeuwin Current is different from other eastern boundary currents in a number of ways, which appear to be linked in a feedback loop. The first unusual feature of the Leeuwin Current is that the current flows directly into the wind, thereby advecting warm water polewards. This is probably the reason for a second unusual feature: there is a strong heat flux out of the ocean from the waters of the Leeuwin Current, rather than into the ocean as in all other eastern boundary currents (e.g., see Figure 1 of Hsiung, 1985). This heat loss favors convection, and hence, the formation of deep mixed layers, which are a third unusual feature of the Leeuwin Current (Thompson, 1984). Thompson (1984) further argues that these deep mixed layers are essential to the maintenance of the observed large alongshore pressure gradient (a fourth unusual feature of the

Leeuwin Current). But—as already noted—all authors agree that it is the alongshore pressure gradient that drives the Leeuwin Current into the prevailing wind, while the deep mixed layers allow the alongshore pressure gradient to overcome the equatorward wind; qualitatively at least, this closes the loop.

So far, no modelers have successfully explained why all these linked features occur off western Australia and not in other eastern boundary currents. Godfrey and Ridgway (1985) suggested that it might relate to the existence of an opening to the warm waters of the equatorial Pacific, north of New Guinea; this might maintain the unusually high steric heights (warm waters) off the North West Shelf, and hence, the gradient driving the current. In Kundu and McCreary's (1987) model, an Indonesian throughflow of 7.3 Sverdrups produced a secondary flow with Leeuwin Current-like properties, but was about a factor of five too weak. This question is still very much open.

THE LEEUWIN CURRENT INTERDISCIPLINARY EXPERIMENT

The Leeuwin Current Interdisciplinary Experiment (LUCIE) was a field experiment coordinated by the CSIRO Division of Oceanography during 1986/87. LUCIE was aimed at improving the description of the Leeuwin Current and understanding the processes that drive it, along with understanding its role in the combined ocean/atmosphere system and its influence on the life cycles of marine species. The work is expected to result in realistic models of the current.

LUCIE begin in September 1986 when an array of self recording instruments were deployed. Some of the instruments were recovered and a large array was deployed in January/February 1987. The moorings included an alongshore array of pressure gauges and current meters (Figure 2) to measure the alongshore pressure gradient and to examine the continuity of the current along the shelf edge. There were also arrays of current meters at 29.5°S (Dongara) and 34°S to examine the cross-shelf structure of the current (Figure 4). Buoys recording meteorological data were also deployed at these two latitudes. The moorings were finally recovered in August 1987, and the data are presently (December 1987) being processed. Additional time series data are being gathered from coastal meteorological and tidegauge stations.

On the mooring recovery and deployment cruises, a series of CTD and acoustic Doppler current profiler (ADCP) sections were completed. CTD and ADCP data were also collected on two other cruises. Data was also

Figure 4a, 4b, 4c: Locations of current meters on (a) the Dongara section for the first deployment, and (b) the Dongara section for the second deployment, and (c) the 34°S section. A solid circle indicates useful current data was obtained and an open circle implies no useful current data was obtained. The record from the top of the D3 mooring has not yet been processed.

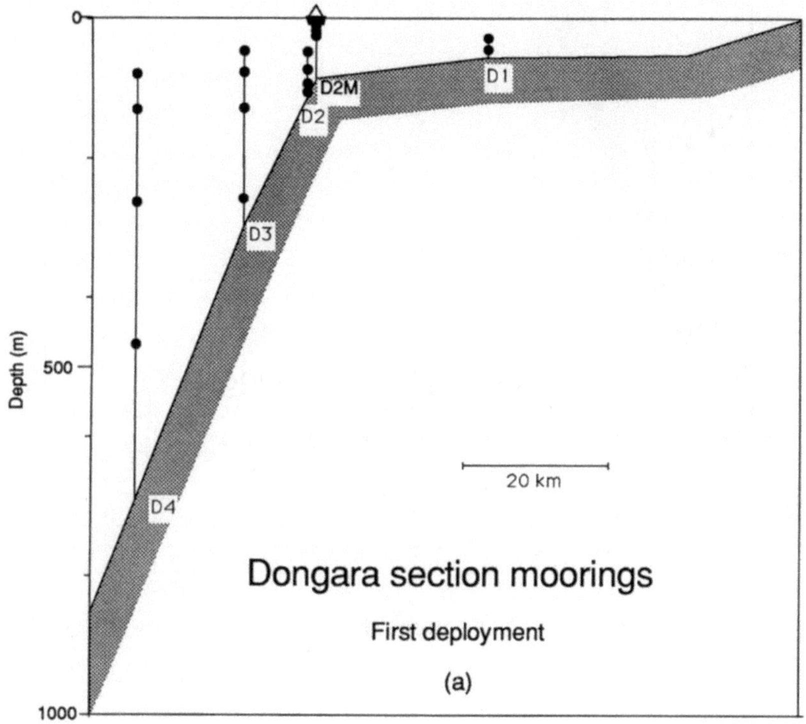

Figure 4a.

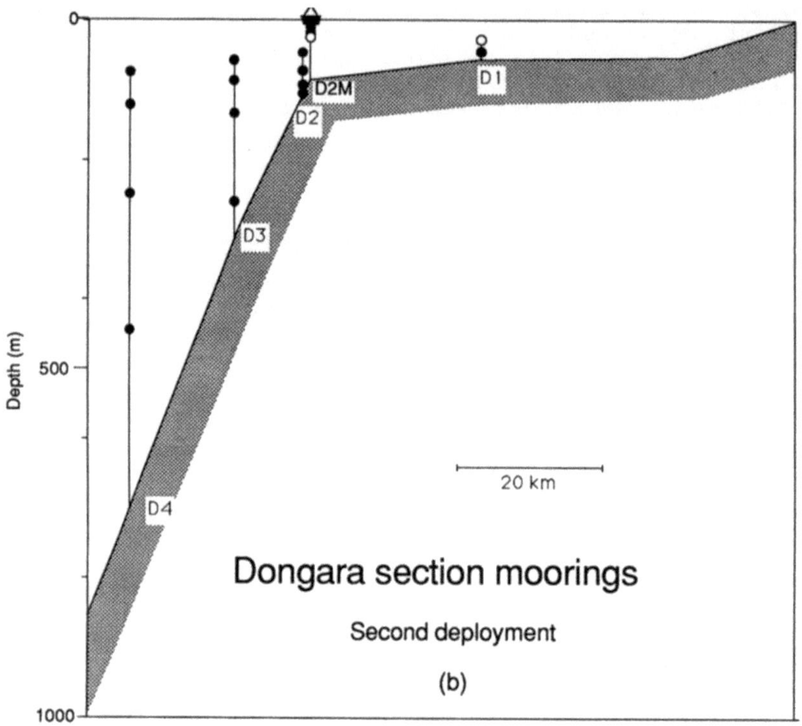

0
500
1000
Depth (m)
D2M
D1
D2
D3
D4
20 km
Dongara section moorings
Second deployment
(b)

Figure 4b.

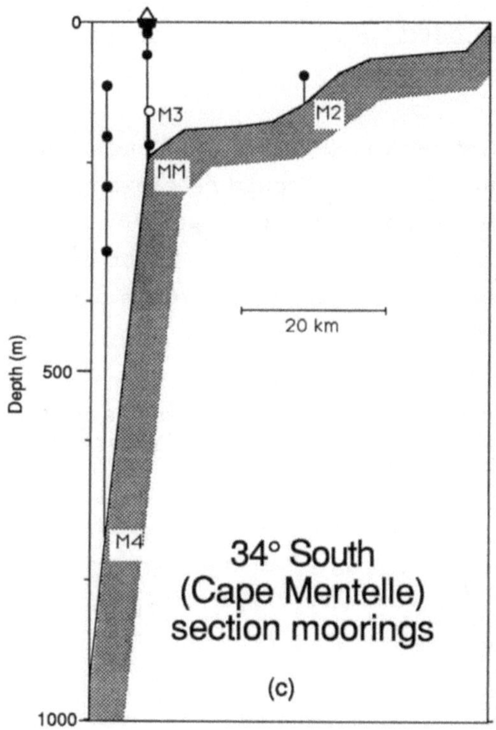

Figure 4c.

collected from satellite-tracked drifters and from BUNYIP (a towed vehicle measuring temperature, conductivity, and microstructure data as it follows an undulating path).

RESULTS

The observational phase of the experiment is now complete and the data are being processed. Some preliminary results are presented here.

CURRENT METERS OBSERVATIONS

The current meters were first deployed (September 1986 to January 1987) at a time when the Leeuwin Current was expected to be weak or non-existent. However, the mean currents at 29.5°S over this period (Figure 5a) do show a southward flow. The southward flow seems to be confined to a narrow stream at the shelf edge and over the upper slope. The maximum mean southward flow recorded was also 20 cms^{-1} at a depth of 75 m over the 300 m isobath. On the slope below this southward stream, there was a northward flow of 15 cms^{-1} and the flow of the shelf was weakly northward.

The second deployment period (February 1987 to August 1987) included the period when the Leeuwin Current should be flowing strongly. The mean currents (Figure 5b) are southward in the upper 400 m of the water column and indicate a southward jet of almost 40 cms^{-1} on the upper slope. Only the deepest current meter, at 450 m on the 700 m isobath, shows a weak northward undercurrent.

Figure 5a and 5b: The mean along-shore currents at Dongara for (a) the first period and for (b) the second period. Only records from the current meters indicated have been used.

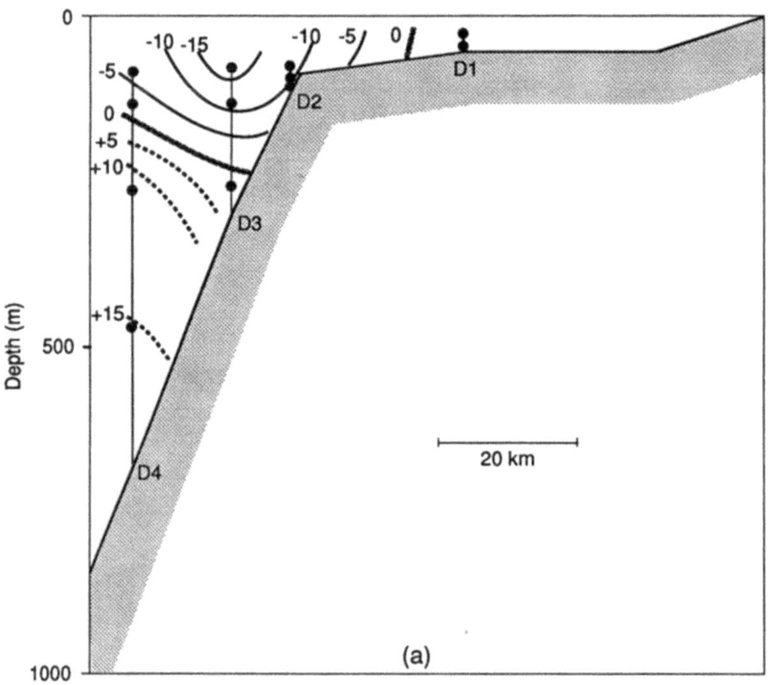

Figure 5a.

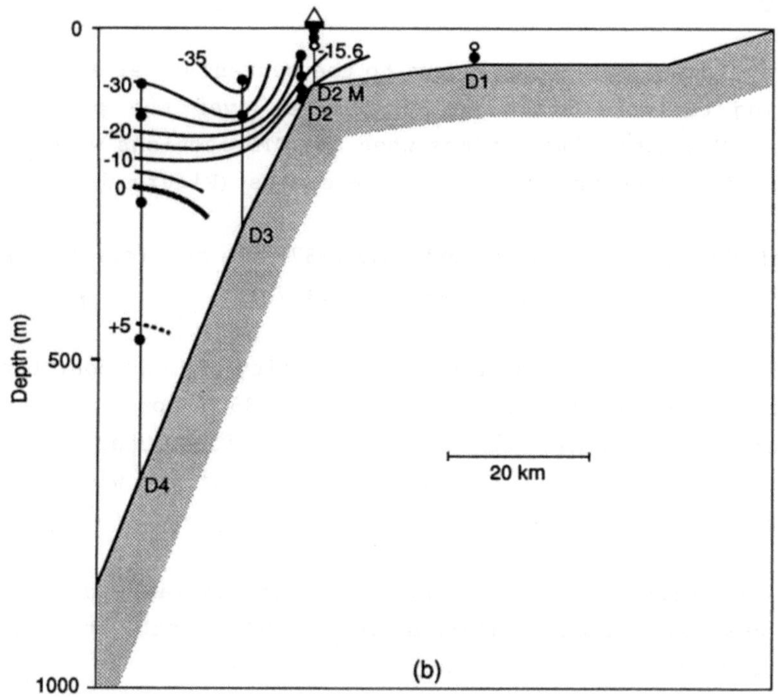

Figure 5b.

The record from the top current meter on the 300 m mooring (Figure 6) indicates a difference between spring/summer and autumn/ winter. During spring/summer, the currents were very variable with periods of southward and northward flow. However, during autumn/winter, the first four months of the record show an almost continuous southward flow, and only in the last two months of the record was there a significant weakening of the flow. The currents from the slope and shelf break mooring appear to be correlated, but this correlation has not yet been quantified.

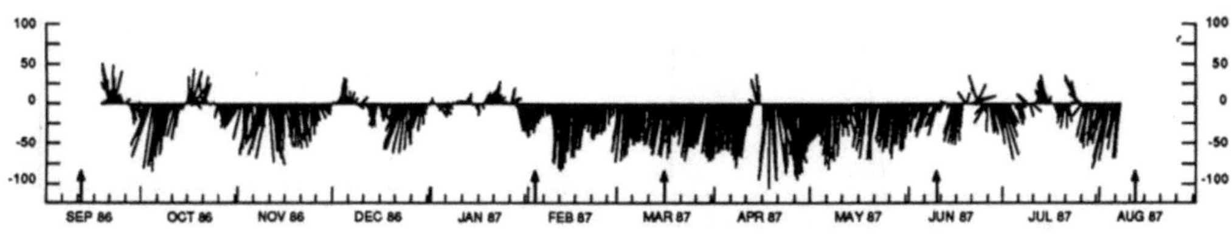

Figure 6: Currents recorded by the meter at a depth of 79 m over the 300 m isobath. A short gap between the two periods has been artificially filled. The arrows indicate the times when CTD sections were occupied.

CTD SECTIONS

CTD sections completed adjacent to the current meter array indicate the changing currents and water properties over the period of the experiment. The approximate times when the CTD sections were completed are indicated on the current meter time series (Figure 6).

Two of the sections (March and June 1987) are sufficient to show the main water types in the study area (Figure 7):

The Leeuwin Current, which has low salinity and was confined to the upper waters at the shelf-edge. Using 35.5 ppt salinity as an arbitrary upper limit for its presence, we found that it was absent in the summer months of February and March. It was widest (> 70 km) and deepest (~200 m) in June.

The South Indian Central Water, with salinities as high as 35.9 ppt. This was a subsurface feature at 100 to 200 m depth; it sank beneath the Leeuwin Current near the continental slope.

Figure 7: CTD sections of salinity and temperature from the Dongara line: (a) March and (b) June 1987, taken as part of LUCIE.

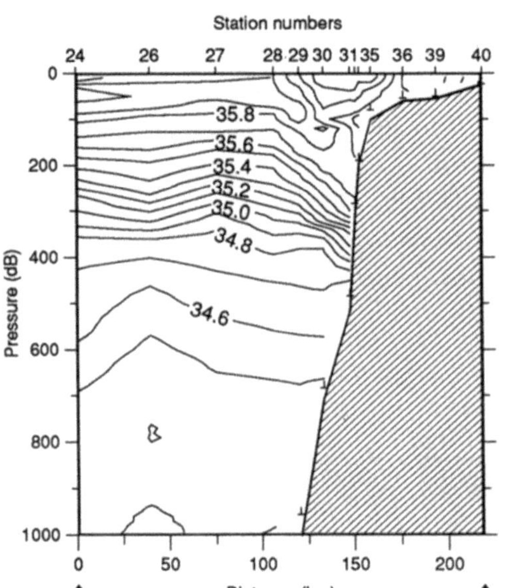

Figure 7a: Dongara section
Salintiy
14-16 March 1987.

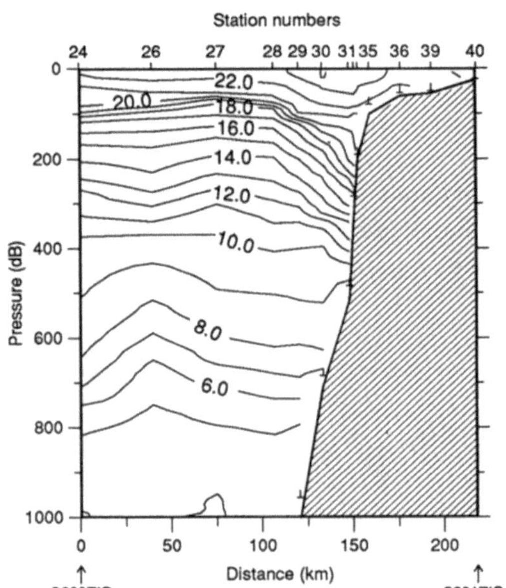

Figure 7a: Dongara section
Potential temperature
14-16 March 1987.

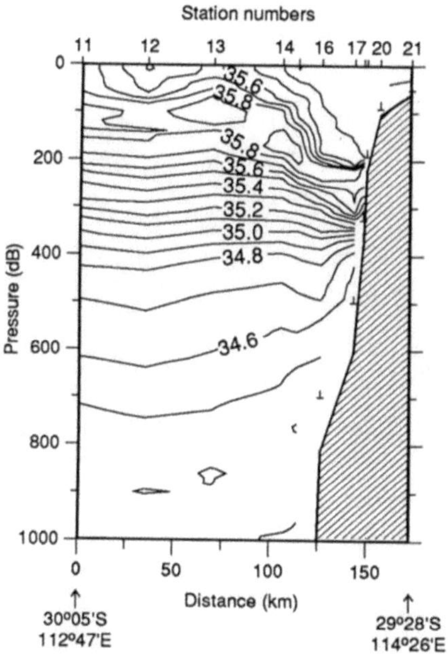

Figure 7b: Dongara section
Salinity
7-9 June 1987

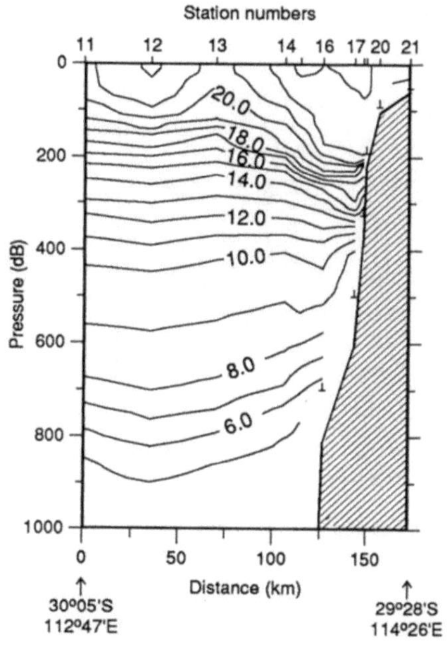

Figure 7b: Dongara section
Potential temperature
7-9 June 1987

The waters of the West Australian Current, which were found at the surface in February and March. These overrode the South Indian Central Water, as Webster et al. (1979) have described.

The σ_θ sections (not shown) are qualitatively similar to the temperature and salinity sections: the geostrophic current shears above the continental slope inferred from them are southward down to about 400 m in September 1986; southward down to about 400 m in February 1987; to about 600 m in March; to about 350 m in June; and to about 400 m in August. In February and June there are strong equatorward shears below 400 m.

The June 1987 survey included sections out from Cape Mentelle, between Capes Naturaliste and Leeuwin, and Cliffy Head on the south coast. At Cape Mentelle (Figure 8), the Leeuwin Current could be best defined as having salinity less than 35.8 and temperature greater than 20°C; in other words, cooler and more saline than farther north at Dongara. The Leeuwin Current extended from very near the coast to some 130 km offshore

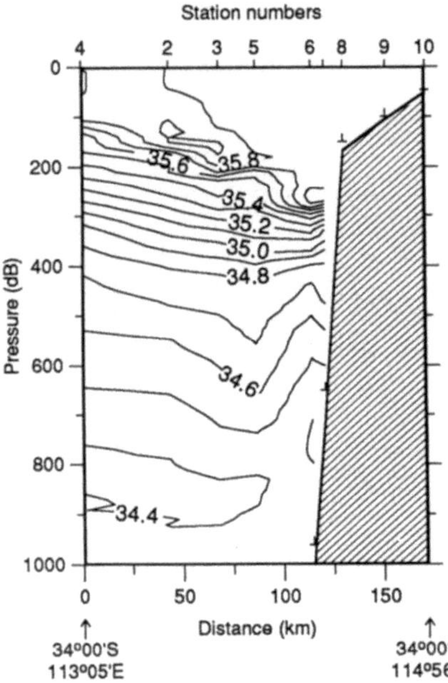

Figure 8a: Cape Mentelle section
Salinity
3-5 June 1987

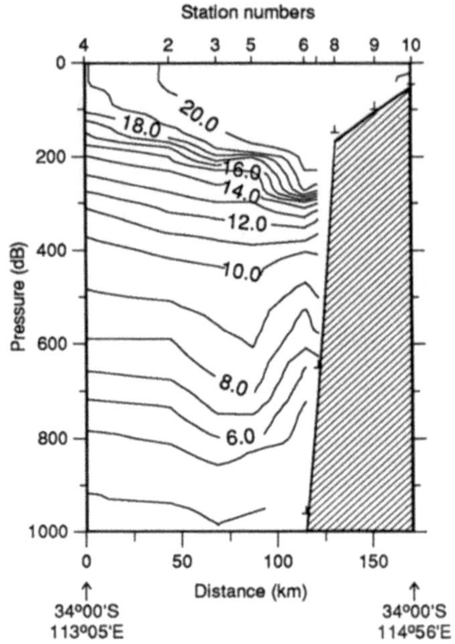

Figure 8b: Cape Mentelle section
Potential temperature
3-5 June 1987

Figure 8: CTD sections of salinity and temperature for the Cape Mentelle line in June 1987.

and was mixed down to 220 m. At Cliffy Head (Figure 9), the situation was interesting because the Leeuwin Current had flowed from a regime where the offshore waters were more saline to one where they were less saline. The Leeuwin Current was narrower and slightly cooler and the 18° to 19°C surface isotherms that marked its offshore edge descended to the shelf break. Preliminary results from an acoustic Doppler current profiler obtained on the same day and roughly along the Cliffy Head section are shown in Figure 10. The Leeuwin Current can be seen to have considerable vertical and horizontal shear.

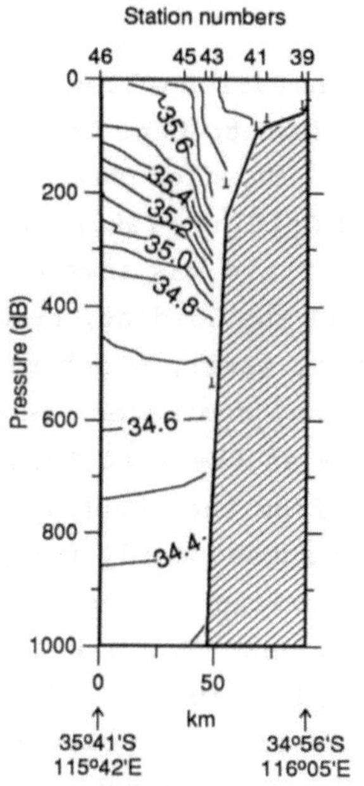

Figure 9a: Cliffy Head section
Salinity
15-17 June 1987

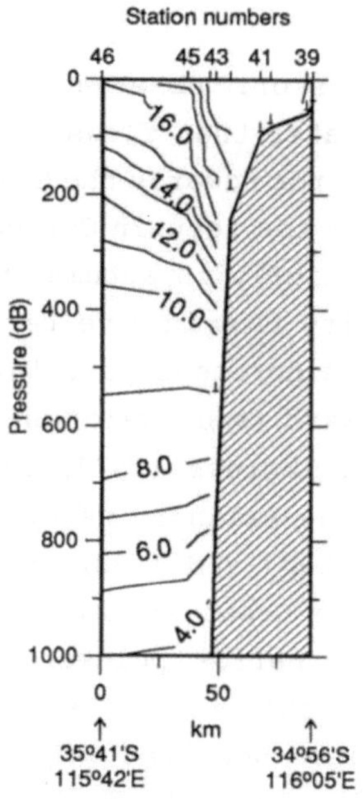

Figure 9a: Cliffy Head section
Potential temperature
15-17 June 1987

Figure 9: CTD sections of salinity and temperature for the Cliffy Head line in June 1987.

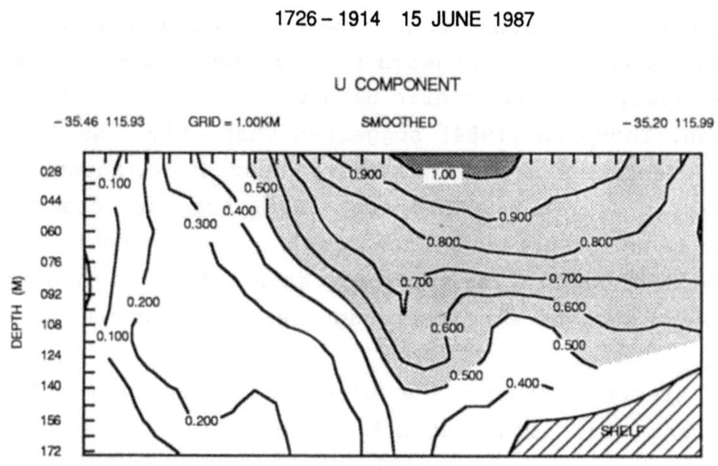

Figure 10: A section of the east component of the current measured by an acoustic Doppler current profiler roughly along the Cliffy Head CTD section on the same day as for Figure 9.

DISCUSSION

The historic observations clearly show the intrusion of low-salinity tropical water that flows poleward (southward) down to Cape Leeuwin and then eastwards towards the Great Australian Bight. This low-salinity flow (defined as the Leeuwin Current by Cresswell and Golding, 1980) was found to occur chiefly in autumn and winter, although there was some poleward low-salinity flow in November/December (Cresswell and Golding, 1980). In contrast, a northeastward inflow of high-salinity water, which turns to flow poleward (southward) on approaching the coast, dominates the summer situation. Andrews (1977) referred to this flow as the West Australian Current. Low-salinity water is also drawn down from the north in summer (Andrews, 1977 and 1983) and the combined flow subsequently turns east towards the Great Australian Bight. Summer satellite images sometimes indicate a narrow poleward flow starting from near North West Cape. Presumably this is a stream of low-salinity tropical water. The summer LUCIE cruise occurred after two months of weak or zero poleward flow (Figure 6), and no low-salinity water was found. In contrast to Andrews (1977), Kitani (1977) and Rochford (1969a) referred to the northward flowing West Australian Current. There is clearly a problem of terminology of the flow in the eastern Indian Ocean; unfortunately, we have been unable to find a clear resolution to this difficulty.

The source of the low-salinity water seems to be a fan-shaped area off northwestern Australia, as depicted in Figure 1. Gentilli (1972) suggested that autumn and winter throughflow from the Pacific Ocean to the Indian Ocean was isolated off northwestern Australia over summer and then became the source for southward flow in the next autumn and winter. Godfrey and Ridgway's (1985) steric height maps are consistent with this interpretation. Thompson (1984) suggested that only a small portion of the Leeuwin Current flux came from the North West Shelf. Recent theoretical work (McCreary et al., 1986; Thompson, 1988) suggests that much of the inflow to the Leeuwin Current is the result of geostrophic near-surface inflow from west of Australia. However, this inflow is of high-salinity water, and there must be a northern source region as well.

The differences between the Leeuwin Current and other eastern boundary currents are illustrated by Figure 11 (Godfrey and Ridgway, 1985). This shows annual mean temperature sections at similar latitudes in the Leeuwin Current, and off California and South Africa (representative of the other four major eastern boundary currents). Even at 500 km offshore, there are substantial differences: inspection of Figure 11 shows that throughout the top 300 m, 500 km offshore, temperatures off western Australia are higher by about 2° to 4°C than those off California,

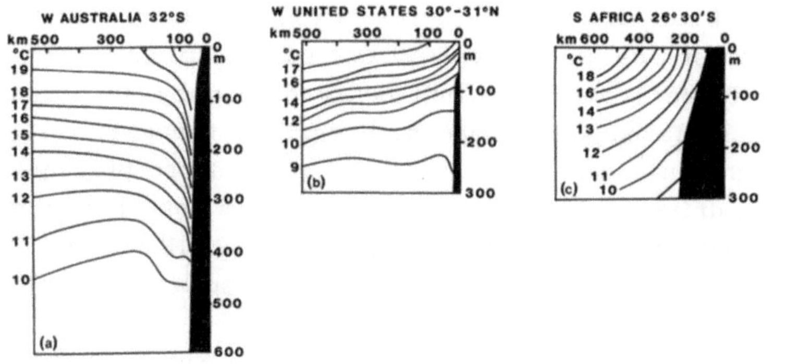

Figure 11: Annual average temperatures, along (a) a section at 32°S off western Australia, (b) a section at 30° to 31°N out from California, and (c) a section at 26°30'S off the west coast of South Africa (from Godfrey and Ridgway, 1985).

and by 1° to 2°C than those off South Africa. At the shelf edge these differences increase greatly because of the southward flow and downwelling associated with the Leeuwin Current, and the upwelling off the other coasts. Thus, at 200 m at the shelf edge, annual mean temperature off western Australia is some 8°C warmer than in the other two current systems!

Some of these differences are compensated by higher salinity, but specific volume anomaly (SVA) is generally higher all along the western Australian coast than along any of the other ocean eastern boundaries. Godfrey (1988) has developed a simple extension of the familiar Sverdrup models, that appears to explain (in a depth integrated sense) these differences between different ocean basins. Specifically, he has shown using Levitus' (1982) annual mean hydrographic data that depth-integrated steric height (or double depth-integrated SVA) in the top 1000 to 2000 m, relative to appropriate "depths of no motion," have roughly characteristic values for each of seven eastern boundaries: eastern Pacific, western New Zealand, western Australia, Indonesia-Malayasia-Burma, western India, western Malagasy, and eastern Atlantic. Furthermore, the quite large differences between these seven characteristic values are quite well modeled by certain integrals of the annual mean wind stress. In particular, western Australian values of double depth-integrated SVA are higher than those in the eastern Pacific because of the action of zonal winds along the equatorial Pacific, which pile up warm, light water near Indonesia. This effect may propagate via baroclinic waves along the western Australian coast.

The result provides a partial explanation for the generally high level of temperature in Figure 11a relative to Figure 11b (there is a similar explanation for the low levels in Figure 11c), but it does not explain the existence of the Leeuwin Current. Godfrey (1988) suggests the Leeuwin Current might be indirectly driven by heat loss to the atmosphere associated with these high surface temperatures. The heat losses at high latitudes create a strong along-shore pressure gradient which in turn drives the Leeuwin Current. These heat losses are unique in eastern boundary regions (see Hsiung, 1986). LUCIE was partly designed in the hope that it might throw some light on such speculations.

Through the year-long use of the CTD and acoustic Doppler current profiler on Franklin, along with the current measurements across the continental shelf and slope and satellite imagery, the Leeuwin Current, during LUCIE, has been observed more precisely than in the past. Data analysis in progress is addressing the relationship between the currents and the pressure and wind stress forcing functions. By combining the various data sets, it is anticipated that a much more accurate and detailed description will become available, leading to a greater understanding of the Leeuwin Current dynamics and the development of useful models.

ACKNOWLEDGEMENTS

We would like to thank everybody who participated in the planning and execution of LUCIE.

From: Joseph L. Reid, Scripps Institution of Oceanography

On: Review and Commentary to paper **THE LEEUWIN CURRENT** by John A. Church, George R. Cresswell, and J. Stuart Godfrey

This is a clear and straightforward description of the present information about the Leeuwin Current. I have only three comments or questions that the authors may wish to consider.

First, the poleward flow in winter, with an equatorward flow offshore, is rather like the California Current. It is the equatorward subsurface flow that is so remarkably different.

The equatorward surface flow in summer seems to carry subtropical water, however, while the equatorward flow off North and South America carries higher-latitude, colder, and less saline waters. Is it the difference in extent of latitude (35°S) of the end of Australia from that of the eastern Pacific continents that accounts for this? South Africa ends at about that latitude. How do the flow and characteristics of the Benguela differ from western Australian waters?

Second, the Leeuwin Current is discussed alone, without reference to the general circulation. For example, is the equatorward undercurrent part of a larger-scale anticyclonic flow in the central South Indian Ocean, with the Leeuwin Current a very interesting aberration from most eastern boundary currents? Is the underlying salinity minimum near 600 m a part of the Intermediate Water, with the oxygen maximum above it, corresponding to the structure seen off South America in the Scorpio data?

Third, is it the narrowness of the Leeuwin Current that has made it so little known until recently? With speeds of 1.5 m s^{-1}, seamen must have been aware of it.

Reply to comments by Professor Reid, sent by J. Stuart Godfrey:

> We have added text and a figure to the original manuscript that address [Prof. Reid's] first and second points. With respect to the third point, I believe the explanation is as follows:
>
> Ships generally crossed the narrow Leeuwin Current en route to South Africa or the Red Sea, and reported ship's drifts from about a day's run. Hence, the Leeuwin Current was missed with nearshore (equatorward) flows and offshore eddies. The

Current does not show up very clearly in current atlases, e.g., the very comprehensive Dutch atlas "Sea Areas Around Australia," published in 1947. Also, early hydrographic studies usually took no stations until they were in water over 1500 m deep, to get a good "depth of no motion"—this was offshore from most of the Leeuwin Current!

PART III:

SPECIAL TOPICS

AN APPLICATION OF TURBULENCE DATA TO THE CALIFORNIA UNDERCURRENT

Rolf Lueck
Hidekatsu Yamazaki

Chesapeake Bay Institute
The Johns Hopkins University
315-711 W 40th Street
Baltimore, MD 21211

ABSTRACT

Persistent and large rates of the dissipation of kinetic energy were observed in the intrusive region of the California Undercurrent. Viscous dissipation can extract the total mechanical energy (potential plus kinetic) of the undercurrent in 6 months and its kinetic energy in only 11 days, suggesting the occurrence of geostrophic adjustments on similar time scales. A simple mechanical energy budget does not show a balance between dissipation and the work done by the pressure field contrary to findings for the Equatorial Undercurrent.

INTRODUCTION

The California Undercurrent is a poleward flow of relatively warm and saline water along the west coast of North America originating from the Equatorial Undercurrent and the subsurface South Equatorial Countercurrent (Tsuchiya, 1981). Wooster and Jones (1970) show that the California Undercurrent "jet" reaches average speeds of 0.3 m s^{-1} off northern Baja California with speeds as high as 0.4 m s^{-1} in August, 1966. They estimated a width of 20 km and a vertical scale of 300 m. Reid (1962) observed a maximum speed of 0.2 m s^{-1} at a depth of 250 m in December, 1961 on a zonal section at 36°. Cannon, Laird, and Ryan (1975) estimate a geostrophic speed of 0.05 m s^{-1} for the undercurrent off Washington State. The undercurrent appears to be a permanent feature throughout the year as far north as 50° (Reed and Halpern, 1976).

Our turbulence measurements, made in April, 1982 over the San Diego trough near the 1000 m isobath (Yamazaki and Lueck, 1987), show persistent and high levels of kinetic energy dissipation in an interleaving region between 150 and 250 meters depth. This region had the T-S signature of the undercurrent, but the presence of the undercurrent over the San Diego trough during our observations is ambiguous. Monthly means of geopotential anomaly (Tsuchiya, 1980) show an undercurrent of 0.05 m s^{-1} at our observation site. Current meter observations by Denbo, Polzin, Allen, Huyer, and Smith (1984) show a mean northward flow of 0.35 m s^{-1} at 139 m off Purisuma Point (34.7°N) between February and May, 1982. Observations by Tsuchiya (1980) show that the California Undercurrent in the San Diego trough is generally northward at 0.05 m s^{-1}. The surface flow is southward at 0.1 m s^{-1} in March and April and weakly to the north during the rest of the year. Osborn and Cox (1972) repeatedly observed enhanced levels of temperature gradient microstructure over the San Diego trough near 90, 180, 200, and 400 meters depth. Gregg and Cox (1972) also observed persistent temperature and conductivity microstructure over the trough and suggested that it is driven by either salt fingering or shear produced by intrusions.

TURBULENT DISSIPATION

The results from 17 microstructure velocity profiles made over the San Diego trough using Camel II (Lueck, Crawford, and Osborn, 1983) have recently been reported by Yamazaki and Lueck (1987). A profile (Figure 1) of two orthogonal components of shear (1-100 cpm) temperature and its gradient shows the typical vertical distribution of turbulence near the 1000 m isobath. Turbulent layers, 1 to 20 m thick, were found between 50 and 70 meters and near 155, 200, and 240 m. The most dissipative layers are associated with the interleaving of water types. A CTD profile (Figure 2), made 35 minutes after the microstructure profile, shows fine structure very similar to that observed by Camel II. However, many of the profiles that were made approximately 30 minutes apart show changes in the vertical scale of temperature inversions, suggesting that the intrusions between 150 and 250 m are evolving rapidly in space and time. The temperature inversion near 215 m in the microstructure profile (220 in the CTD profile) is salinity compensated and occurs at S = 34.12, T = 8.7°C and σ_t = 26.48. This isopycnal is only 0.1 lighter than the one used by Wooster and Jones (1970) and Tsuchiya (1980) to trace the core of the undercurrent. The average dissipation between 150 and 250 meters, where intrusions with water of southern origin were observed, is 1.4×10^{-6} W m^{-3}. For comparison, Lueck et al. (1983) reported 0.53×10^{-6} W m^{-3} as the average dissipation between 200 and 500 m over the continental slope off Vancouver Island.

For the California Undercurrent to maintain its identity as a current system, it cannot dissipate its energy over a period shorter than the time scale of its longitudinal flow. The time that a water parcel is in the current system is $L/U = 4 \times 10^7$ seconds (15 months), where we have used

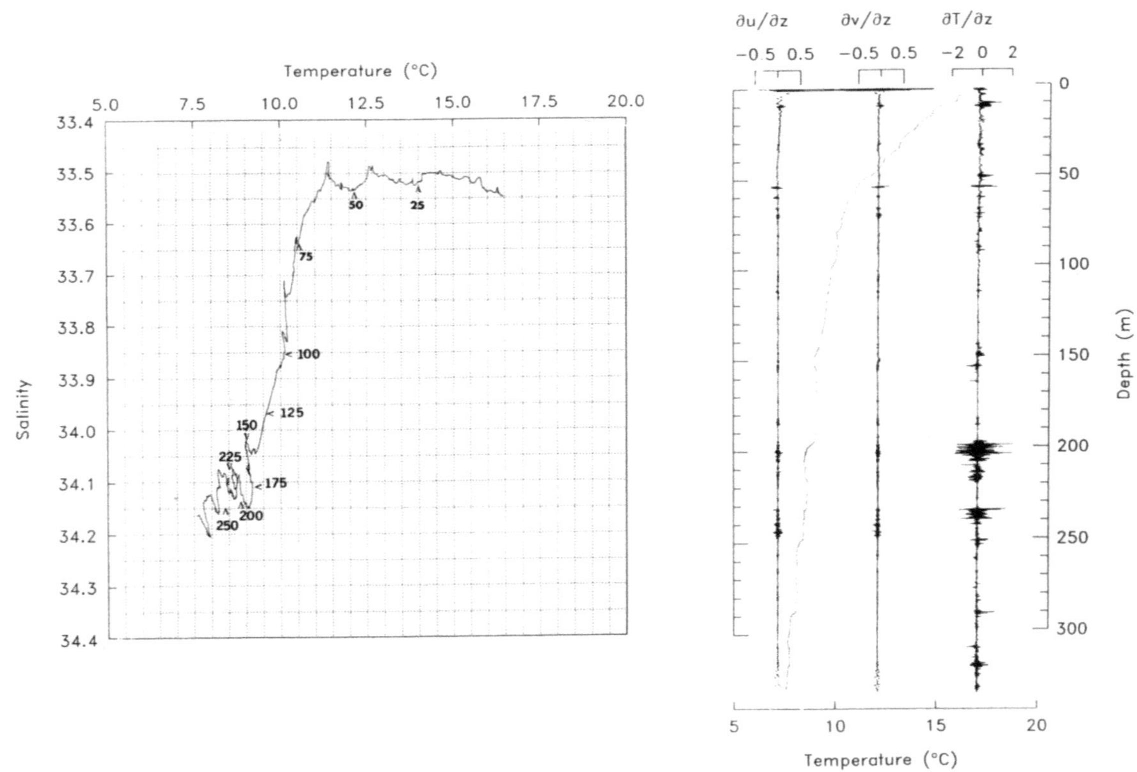

Figure 1: Vertical profile of the microstructure of temperature, its vertical gradient, and two components of the vertical shear of horizontal velocity (right half). The dissipation of kinetic energy is proportional to the variance of the shear. The variance of the temperature gradient is proportional to the dissipation of temperature variance. Bipolar gradients of temperature at microscales indicate overturning motions at such scales even if the fine scale density is salinity compensated. Temperature-salinity diagram from a CTD cast made 35 minutes after the microstructure profile (left half).

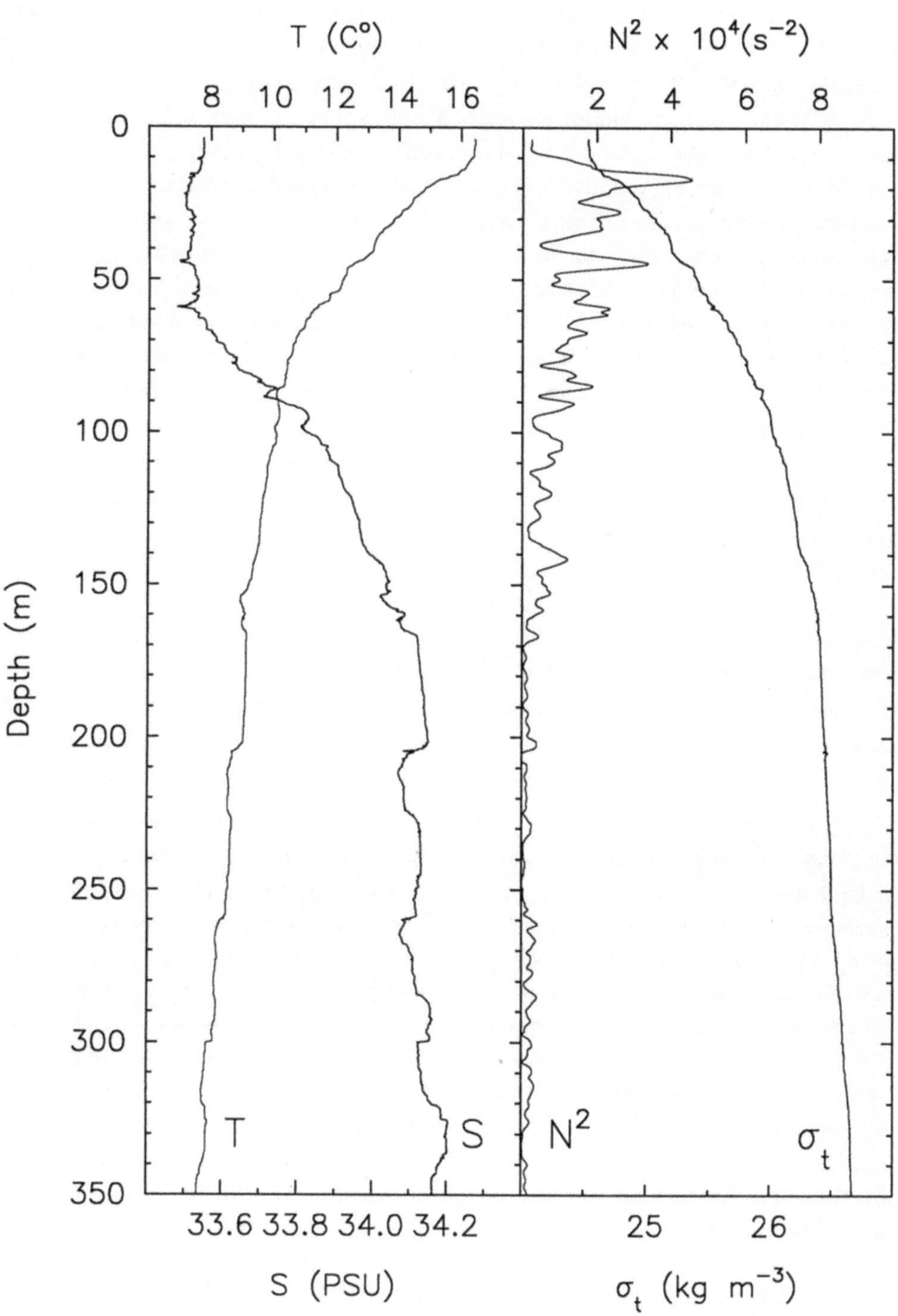

Figure 2: Profile of temperature, salnity, σ_t , and buoyancy frequency squared from CTD cast made 35 minutes after microstructure profile in Figure 1.

the distance separating southern California from mid-Vancouver Island ($L \approx 2 \times 10^6$ m), and a speed scale $U \approx 0.05$ m s^{-1}. The time scale for the dissipation of the current's kinetic energy is only $0.5\rho U^2/\langle\varepsilon\rangle \approx 1 \times 10^6$ s (11 days), where $\langle\varepsilon\rangle = 1.4 \times 10^{-6}$ W m^{-3} is the average measured dissipation. Unless we have underestimated the speed of the undercurrent by a factor of 3.5, the rate of dissipation of kinetic energy will extract the kinetic energy of a fluid parcel off San Diego long before it reaches Vancouver Island. The current speed that would yield a dissipation time scale equal to the longitudinal time scale is 0.18 m s^{-1}, but the undercurrent is weakest in April (Tsuchiya, 1980 and Hickey, 1979). Our work, of course, represents only one position in a current that extends over 2000 km and so does not necessarily give a mean dissipation value for the entire current system. Local effects such as internal tides and basin oscillations may bias our dissipation values upwards. However, a bias, if present, has to selectively favor the intrusion region because that is the only depth range where dissipation rates are anomalously large (Yamazaki and Lueck, 1987).

A dissipation time scale much shorter than the longevity of a current system is, however, not inconsistent with the balance of total energy. Lueck and Osborn (1986) noted that the time scale for the dissipation of the kinetic energy of a Gulf Stream ring was also much shorter than the life time of a typical ring. However, the time scale for the dissipation of total energy (kinetic plus potential) was 2 to 3 years. The ratio of potential to kinetic energy is (Gill, 1982) $\left(k_H N/mf\right)^{-2}$ where k_H is the horizontal wavenumber, N/mf is the Rossby radius of deformation, m is the vertical wavenumber, N is the buoyancy frequency, and f the local Coriolis parameter. To estimate the ratio of potential to kinetic energy we will take $k_H \approx (40000)^{-1}$, $m \approx (200)^{-1}$, $N = 0.004$ s^{-1}, and $f = 8.0 \times 10^{-5}$ s^{-1}. This corresponds closely to Tsuchiya (1980) and gives a ratio of 16. Therefore, the time scale for the dissipation of the total energy of the undercurrent is 16×10^6 s (6 months), and it is not inconsistent with the times scale of the longitudinal flow. Thus, a fluid parcel in the undercurrent dissipates about e^{-1} of its total energy traveling from California to Vancouver Island. To maintain a geostrophic balance, energy must be transferred from the potential into the kinetic field with a time scale comparable to the dissipation time.

The axis of the California Undercurrent is approximately parallel to the coast of North America because of the confines imposed by Coriolis acceleration and the eastern boundary. The axis of the Equatorial Undercurrent is restrained zonally by the reversal of the sign of the Coriolis acceleration. This motivates the notion that the two current systems may be similar. A mechanical energy budget for the Atlantic Equatorial Undercurrent (Crawford and Osborn, 1980) shows a close balance between the work done by the pressure field and the rate of dissipation of kinetic energy with minor contributions from the advection and the local time rate of change of kinetic energy. A similar budget is made here, but the historical mean pressure field in the California Undercurrent is too poorly resolved to permit a definitive mechanical energy budget. Following Gill (1982, equation 4.6.3), the mechanical energy equation is

$$\partial/\partial t \left(1/2\rho\mathbf{u}^2\right) + \nabla \cdot \left\{\left(p + 1/2\ \rho\mathbf{u}^2\right)\mathbf{u} - \mu\nabla\left(1/2\mathbf{u}^2\right)\right\} = -gw\rho' - \varepsilon + p\nabla\cdot\mathbf{u}$$

where p is the perturbation pressure, ρ' the perturbation density, and $\mathbf{u}^2 = \mathbf{u}\cdot\mathbf{u}$. We will assume that the fluid is incompressible, $\nabla\cdot\mathbf{u} = 0$; the viscous transports are negligible, $\mu\nabla\cdot u^2 = 0$; and that the buoyancy flux is small compared to viscous dissipation, $gw\rho' \ll \varepsilon$. Further, we assume that the undercurrent is contained in an imaginary box so that flow normal to its surface vanishes everywhere except at the zonal boundaries. The zonal boundaries are taken orthogonal to the eastern boundary formed by the continental slope. The upper boundary is the interface between the surface current and the undercurrent, where the horizontal velocity vanishes; the bottom boundary is where the along-shore velocity is very small compared to the mean speed of the undercurrent. We assume that the western boundary can be defined by a zero-flow streamline. The cross-sectional area of the southern zonal boundary is A_1, which may be different from the area of the northern boundary, A_2. These assumptions exclude zonal eddy fluxes and internal wave radiation from further consideration. Finally, we assume that the current system is contiguous, a fact not yet established. The volume integral over the box, divided by the contained volume, then yields

$$d/dt\left(1/2\rho\langle\mathbf{u}^2\rangle\right) + v_1A_1(AL)^{-1}\left\{\left(p_2 - p_1\right) + 1/2\rho\left(\mathbf{u}_2^2 - \mathbf{u}_1^2\right)\right\} = -\langle\varepsilon\rangle \quad (1)$$

where $A = 1/2\left(A_1 + A_2\right)$, angled braces denote a volume average. We have used $v_1A_1 = v_2A_2$, and v represents the along-shore (meridional) mean velocity. The intensity of the undercurrent varies seasonally and so a lower bound for the time scale for the rate of change of the kinetic energy is one month. The magnitude of the current is approximately $|\mathbf{u}| = 0.05\ \mathrm{m\ s^{-1}}$ and can change up or down by about the same amount in a month or longer. The first term in equation (1) then is $\pm 5\times10^{-7}\ \mathrm{W\ m^{-3}}$. The historic mean pressure field for January is estimated from the 200/500 dBar dynamics height field of Schartzlose and Reid (1972), which indicated a change of 0.00 to 0.02 dynamic meters over 12 degrees of latitude $\left(L \approx 1.3\times10^6\ \mathrm{m}\right)$. However, the contour interval is only 0.02

dynamic meters making this estimate quite unreliable. The pressure difference over 12 degrees is $p_2 - p_1 = 10\rho\Delta D = 0$ to $-200\ \mathrm{J\ m^{-3}}$ where ΔD is the dynamic height difference in dynamic meters. The kinetic energy difference is maximized when $\mathbf{u}_2^2 \ll \mathbf{u}_1^2$ and is at most $-1\ \mathrm{J\ m^{-3}}$ for a $0.05\ \mathrm{m\ s^{-1}}$ flow making it negligible compared to the pressure difference. The work by the pressure field then is less than $-A_1A^{-1} \times 8 \times 10^{-6}\ \mathrm{W\ m^{-3}}$. The last term in equation (1), the rate of dissipation of kinetic energy, is $1.4 \times 10^{-6}\ \mathrm{W\ m^{-3}}$ based on our observations at one site only. So the magnitude of the terms in equation (1) is

$$\pm 5 \times 10^{-7} + 4 \times 10^{-8} A_1 A^{-1} \left\{ (0 \text{ to } -200) + (-1) \right\} = -1.4 \times 10^{-6}$$

It is not clear by how much the undercurrent broadens towards the north, but a balance between dissipation and pressure work requires a northern boundary 10 times larger than the area of the southern boundary, which probably is unrealistic. Zonal eddy fluxes may be important, and further measurements of both the dissipation and the pressure field are needed before a reliable energy budget can be made. The importance of zonal eddy fluxes is also implied by low values of the vertical eddy diffusivity of density (Yamazaki and Lueck, 1987).

CONCLUSION

The kinetic energy of the California Undercurrent is dissipated on a time scale much shorter than its transit time from San Diego to Vancouver Island. The time scale for the dissipation of the current's total energy is comparable to its transit time. Turbulent dissipation is sufficiently large to cause a geostrophic imbalance if there is not transfer of energy from the potential to the kinetic field. In contrast to the Equatorial Undercurrent, it is not possible to show that the rate of dissipation is balanced by the work done by the pressure field, largely due to uncertainties in the dynamic height anomaly of the California Undercurrent. The available data suggests that the two are not in balance and that lateral fluxes may be important in the mechanical energy budget.

MODEL SIMULATIONS OF A COASTAL JET AND UNDERCURRENT IN THE PRESENCE OF EDDIES AND JETS IN THE CALIFORNIA CURRENT SYSTEM

Mary L. Batteen
Department of Oceanography
Naval Postgraduate School
Monterey, CA 93943

ABSTRACT

A high-resolution, multi-level, primitive equation ocean model is used to examine the response of an initial baroclinic jet in an idealized, flat-bottomed, oceanic regime along an eastern ocean boundary. The initial system, intending to represent the mean California Current System during the upwelling season, consists of an equatorward coastal jet overlying a poleward undercurrent. After ~30 days, eddies and jets are generated, due to the baroclinic instability associated with the vertical shear between the Coastal Upwelling Jet (CUJ) and the California Undercurrent (CUC). An examination of both the near-surface and subsurface flow fields in the presence of the eddies shows a complex structure. In the presence of cyclonic (anticyclonic) eddies close to shore, the near-surface (subsurface) current meanders cyclonically (anticyclonically) around the eddies, thereby extending its offshore domain. In the presence of anticyclonic (cyclonic) eddies, the near-surface (subsurface) flow continues equatorward (poleward) in a narrowly confined nearshore region of ~40 km for the surface flows (~8-20 km for the subsurface flows).

INTRODUCTION

Recent observations have shown that, superimposed on the broad, slow (~5 cm s^{-1}), climatological mean flow in the California Current System (CCS), are highly energetic, mesoscale eddies and meandering jets (Bernstein et al., 1977; Mooers and Robinson, 1984; Rienecker et al., 1985). The eddies have wavelengths of several hundred kilometers, which can intensify over several months and be "cutoff," creating isolated eddies (Bernstein et al., 1977). Strong baroclinic jets with peak velocities of ~50 cm s^{-1} are embedded in this field of cyclonic and anticyclonic eddies. These jets are ~40 km wide, extend from the surface to at least 100 m depth and have offshore extensions of several hundred kilometers (Mooers and Robinson, 1984; Rienecker et al., 1985; Flament et al., 1985).

The dynamical processes responsible for the generation and evolution of these intense and complex eddy and jet patterns in the CCS have yet to be fully identified. One possible generation mechanism currently under investigation is that the eddy and jet fields are the result of either a baroclinic or barotropic instability of the mean California Current System (CCS) which, during the upwelling season (from ~April to October), consists of an equatorward coastal upwelling jet overlying a poleward undercurrent (Ikeda and Emery, 1984).

Both baroclinic and barotropic instabilities have been suggested to generate eddies and jets in the Northeast Pacific (Wright, 1980). Evidence has been shown for baroclinic instability being an important mechanism for eddy generation off Vancouver Island (Emery and Mysak, 1980; Thomson, 1984), and off the coast of Oregon and northern California (Ikeda and Emery, 1984). In particular, Ikeda and Emery (1984) have hypothesized that current meanders are triggered by along-shore variations in the coastline (capes) close to the California coast and grow as a result of the baroclinic instability of the coastal, equatorward, near-surface jet (~40 km wide and associated with coastal upwelling) and the poleward California Undercurrent. As these unstable meanders intensify, they carry the cool, upwelled water offshore and are often cutoff, creating pairs of isolated eddies or "vortex pairs" consisting of a cyclonic and an anticyclonic eddy (Bernstein et al., 1977). Evidence for barotropic instability as a dominant mechanism is still forthcoming, although one case of eddy-eddy-jet interaction observed by OPTOMA in the summer of 1983 showed characteristics of barotropic instability (Robinson et al., 1984).

In this study a high-resolution, multi-level, primitive equation ocean model is used to examine the response of an idealized, flat-bottomed oceanic regime along an eastern boundary to an imposed baroclinic jet. Baroclinic instability, associated with the vertical shear between the Coastal Upwelling Jet (CUJ) and the California Undercurrent (CUC), occurs after ~30 days, resulting in the generation of both cyclonic and anticyclonic eddies and jets. Significant changes from the imposed initial conditions in the CUJ and CUC fields are due to the presence of the eddies and jets.

MODEL DESCRIPTION AND INITIAL CONDITIONS

Model Equations

The numerical model used in this study was developed by Haney (1985) and is a multi-level, primitive equation (PE) model of a baroclinic ocean on an f-plane. The model is based on the hydrostatic and Boussinesq approximations with salinity neglected and the rigid lid approximation made. Depth of the model is variable; however, this study will only consider a flat bottom. In addition, the depth-averaged current is neglected in these experiments. The governing equations, written in sigma coordinates, are as follows:

$$\frac{du}{dt} = \frac{-1}{\rho_o}\frac{\partial p'}{\partial x} + \frac{\sigma}{\rho_o D}\frac{\partial p'}{\partial \sigma}\frac{\partial D}{\partial x} + fv - A_m \nabla^4 u + \frac{K_m}{D^2}\frac{\partial^2 u}{\partial \sigma^2} + \delta_d(u) \qquad (2\text{-}1)$$

$$\frac{dv}{dt} = \frac{-1}{\rho_o}\frac{\partial p'}{\partial y} + \frac{\sigma}{\rho_o D}\frac{\partial p'}{\partial \sigma}\frac{\partial D}{\partial y} - fu - A_m \nabla^4 v + \frac{K_m}{D^2}\frac{\partial^2 v}{\partial \sigma^2} + \delta_d(v) \qquad (2\text{-}2)$$

$$\frac{\partial w}{\partial \sigma} + \frac{\partial u}{\partial x} + \frac{\partial v}{\partial y} = 0 \qquad (2\text{-}3)$$

$$p' = D\int_{\sigma}^{o} B\,d\xi - \int_{-1}^{o}\left[D\int_{\sigma}^{o} B\,d\xi\right]d\sigma \qquad (2\text{-}4)$$

$$B = \alpha g\left(T - T_o\right) \qquad (2\text{-}5)$$

$$\frac{dT}{dt} = -A_H \nabla^4 T + \frac{K_H}{D^2}\frac{\partial^2 T}{\partial \sigma^2} + \delta_d(T) \qquad (2\text{-}6)$$

In the above equations, sigma, denoted by σ, is equal to z/D. All horizontal partial derivatives are on constant sigma surfaces. In addition, the variables used are defined in Table 1, while Table 2 gives other symbols in the model equations along with the corresponding values of constants used throughout this study.

Domain Size and Resolution

The study region extends approximately 500 km offshore from the west coast of North America, and it spans the California coastline from Point Sur in the south to Cape Blanco in the north (Figure 1). The model domain extends 6° in longitude and 6° in latitude (512 km by 640 km) and has 8 km by 10 km horizontal resolution with ten levels in the vertical.

Finite Difference Scheme

In the horizontal, a space-staggered B-scheme (Arakawa and Lamb, 1977) is used while a sigma coordinate system controls the vertical. The noise-free version of the hydrostatic equation, advocated by Arakawa and Suarez (1983), and Batteen (1988), is used.

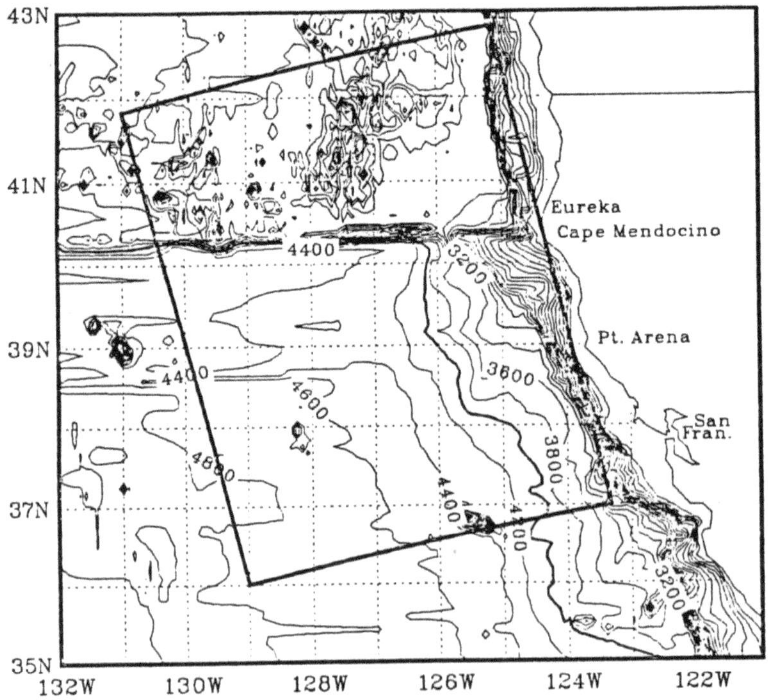

Figure 1: Study domain.

TABLE 1

DEFINITIONS OF VARIABLES USED IN THE MODEL

SYMBOL	DEFINITION
z	height (increasing upwards)
t	time
p'	pressure perturbation from a vertical average
T	temperature
u, v, w	eastward, northward and vertical (sigma) velocity components, respectively
B	buoyancy
δ_d	dynamic adjustment term

TABLE 2

VALUES OF CONSTANTS USED IN THE MODEL

	VALUE	NAME
Ω	$2\pi day^{-1}$	earth rotation rate
T_o	278.2°K	constant reference temperature
ρ_o	1.0276 gm cm^{-3}	density of sea water at T_o
α	2.01×10^{-4} (°K)$^{-1}$	thermal expansion coefficient
K	10	number of levels in the vertical
ΔX	8.0×10^{5} cm	cross-shore grid spacing
ΔY	1.0×10^{6} cm	along-shore grid spacing
D	4.5×10^{5} cm	total ocean depth
$\emptyset_o$	36.5°N	latitude of southern boundary
$\emptyset_m$	42.5°N	latitude of northern boundary
λ_o	124°W	longitude of eastern boundary
λ_m	130°W	longitude of western boundary
Δt	800.0 s	time step
f_o	0.93×10^{-4} s^{-1}	mean Coriolis parameter
g	980.0 cm^{-2}	acceleration of gravity
A_m	2.0×10^{17} cm^4 s^{-1}	biharmonic momentum diffusion coefficient
A_H	2.0×10^{17} cm^4 s^{-1}	biharmonic heat diffusion coefficient
K_m	0.5 cm^2 s^{-1}	vertical eddy viscosity
K_H	0.5 cm^2 s^{-1}	vertical eddy conductivity

Boundary Conditions

The northern, western and southern boundaries are open using a modified version of the radiation boundary conditions of Camerlengo and O'Brien (1980). The eastern boundary, representing the west coast of North America, is modeled as a straight, vertical wall. In this study, a symmetric, free-slip condition on the tangential velocity is invoked at the coastline.

Initial Conditions

The model can either be spun up from rest by a surface wind stress or heat flux, or it can be initialized with a specified current field. In this study, the model is unforced and initialized, following Ikeda and Emery (1984), with representative vertical (Figure 2a) and horizontal (Figure 2b) profiles of mean flow during the upwelling season. The initial flow pattern is based on the type of along-shore baroclinic coastal jet which Hickey (1979) observed in summer after about a week of strong equatorward winds favorable for upwelling. An estimation of the magnitude and sign of terms in the quasigeostrophic potential vorticity equation for the imposed initial conditions was used to determine that the necessary condition for baroclinic instability would be satisfied, i.e., since the potential vorticity gradient changes sign in the domain, the coastal jet is expected to become baroclinically unstable.

The initial mean stratification used in this study is a pure exponential temperature profile with a vertical length-scale of H=450 m. The exact form is

$$T(z) = T_B + \Delta T\, e^{z/450} \,, \qquad (2\text{-}7)$$

where $T_B = 2°C$ is the temperature at great depth, and $\Delta T = 13°C$ is the increase in temperature between the bottom of the ocean and the surface.

This temperature profile was derived from observations used to support the Princeton Dynalysis model (Blumberg and Mellor, 1987) and is representative of the long-term, mean climatological temperature stratification for the California coastal region. Additionally, the

Brunt-Vaisala frequency profile, $N^2(z)$, was calculated analytically using the temperature function in Eq. (2-7) from

$$N^2 = \alpha\, g\, \partial T/\partial z\ . \qquad (2\text{-}8)$$

The model temperature profile and resultant N^2 profile are shown in Figures 2c and 2d, respectively. Using the method of Feliks (1985), the model baroclinic radius of deformation was calculated to be ~30 km, which compares favorably with observed values (Emery et al., 1984).

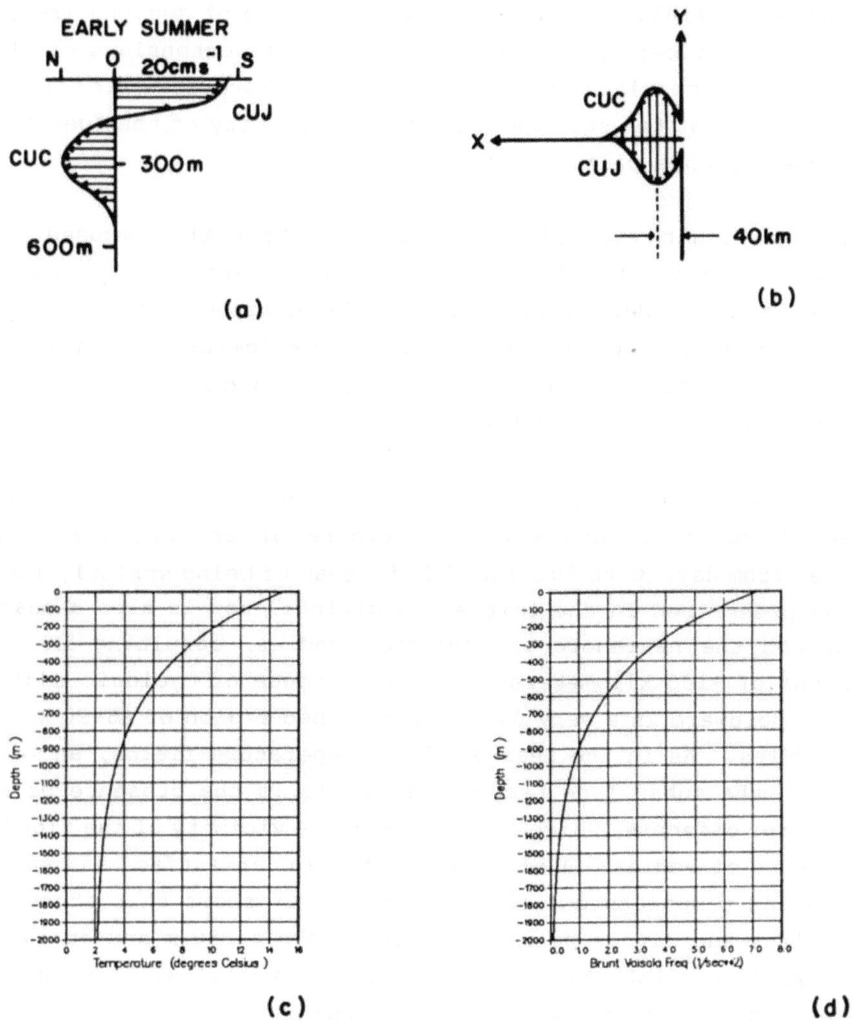

Figure 2: Model initial conditions. (a) Vertical and (b) horizontal profiles of mean flow for the early summer (after Ikeda and Emery, 1984). (c) Model initial temperature profile and (d) initial profile of square Brunt-Vaisala frequency N^2.

RESULTS

Baroclinic instability, associated with the vertical shear between the CUJ and CUC, occurs after ~30 days, resulting in the generation of eddies and jets. Figures 3 and 4 show the near-surface (~50 km depth), instantaneous perturbation pressure (Figure 3) and temperature fields (Figure 4) at five day intervals from days 35 to 50. The CUJ with equatorward flow is discernible in the pressure fields (Figure 3) as is the establishment of anticyclonic vortices, and pinched off cyclonic eddies on the seaward side of the CUJ. Also formed are meanders and dipole pairs of eddies with strong jets between them. An important feature in the temperature fields (Figure 4) is the presence of squirts (regions of intense seaward flow) of cold water with offshore extensions of ~100-200 kilometers. A comparison of the temperature and pressure fields (Figures 3 and 4) shows that the squirts occur in the vicinity of the (geostrophic) flow between dipole pairs of eddies.

A significant evolution of the flow from the imposed initial conditions is that the CUJ, instead of being strictly a southward flow, now is also a meandering flow around the nearshore cyclonic eddies, resulting in offshore extensions of ~100-200 kilometers. In the presence of anticyclonic vortices, the CUJ flows equatorward in a narrowly confined region of ~40 km inshore of the eddies.

The subsurface (~350 m depth), instantaneous perturbation pressure (Figure 5) and temperature fields (Figure 6) are shown at five day sequences from days 35 to 50. The CUC, instead of being strictly a poleward flow, as prescribed in the initial conditions, now is also a meandering flow around the nearshore anticyclonic eddies, resulting in offshore extensions of ~100 kilometers. In the presence of cyclonic eddies, the CUC flows poleward in a very narrowly confined region of ~8-20 km inshore of the eddies. As in the near-surface temperature fields, an important feature in the subsurface temperature field is the presence of squirts of cold water offshore, which also occur in the vicinity of the flow between dipole pairs of eddies. A comparison of the near-surface and subsurface pressure and temperature fields (Figures 3-6) shows that these are the same squirts and dipole pairs of eddies shown at 50 m and at 350 m, and that there is a slight tilt (toward the warm water) with depth of these features, indicative of baroclinic instability.

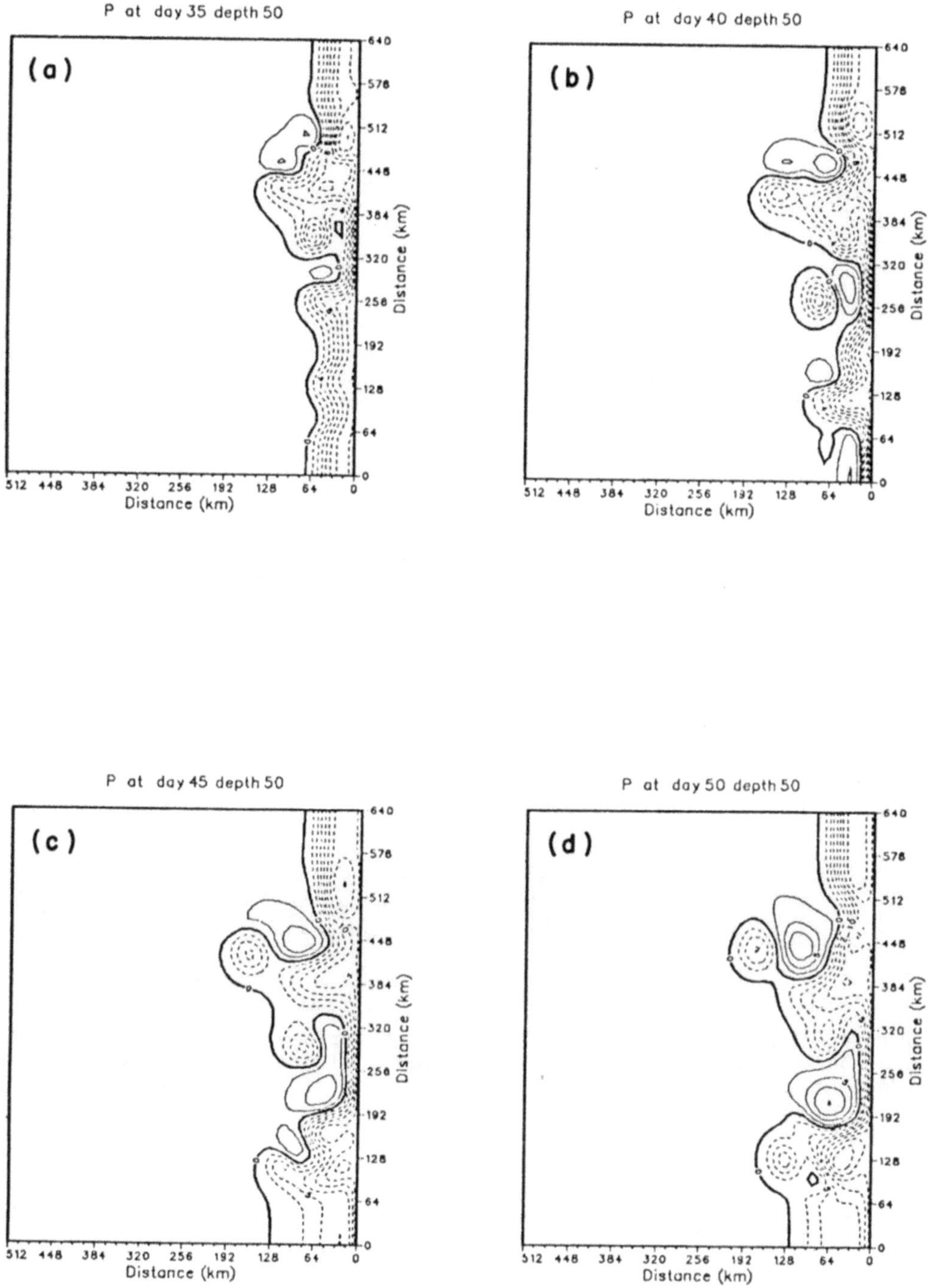

Figure 3: Instantaneous perturbation pressure fields at 50 m depth for days a) 35, b) 40, c) 45, and d) 50. Solid closed contours denote anticyclonic eddies.

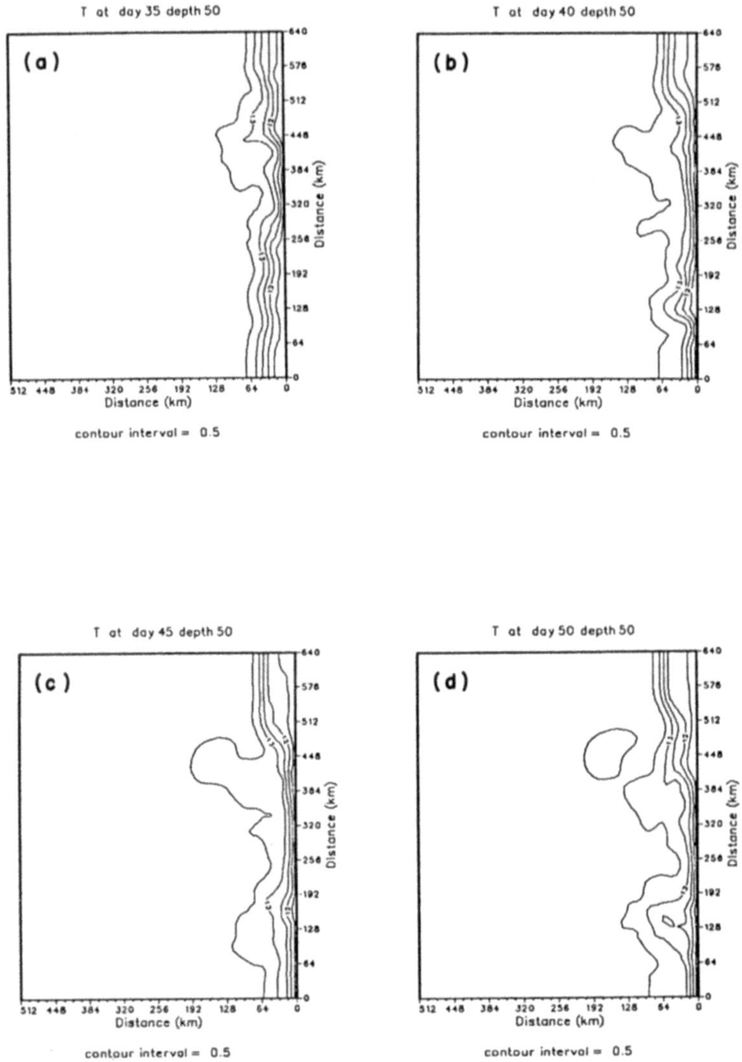

Figure 4: Temperature fields at 50 m depth for days a) 35, b) 40, c) 45, and d) 50. Contour interval is 0.5 degrees Celsius. The temperature is coldest at the coast.

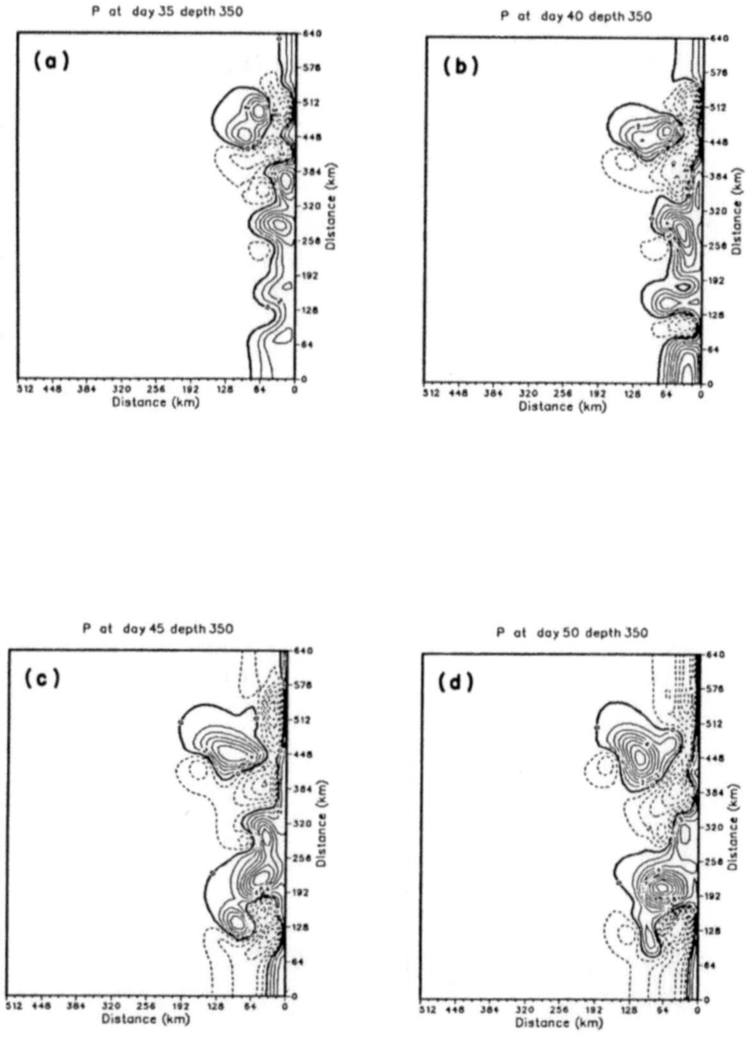

Figure 5: Same as Figure 3 but for 350 m depth.

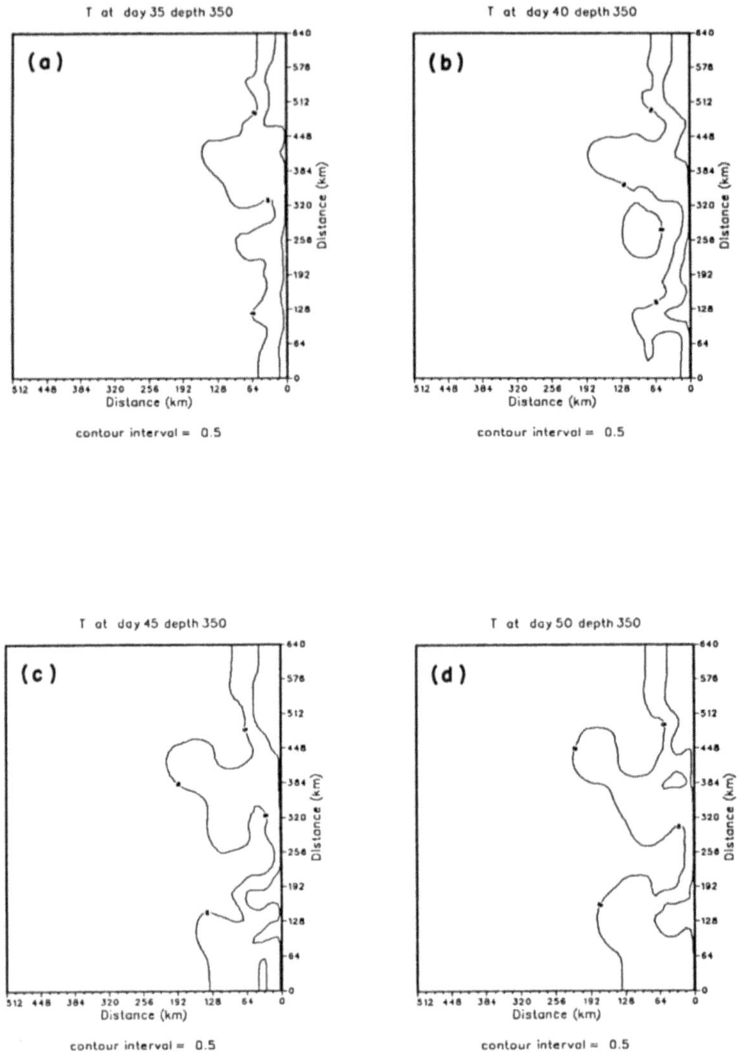

Figure 6: Same as Figure 4 but for 350 m depth.

Figure 7 shows instantaneous horizontal (Figures 7a, b) and vertical cross-sectional (Figure 7c) maps of the northward velocity (v) at day 50. Consistent with the horizontal pressure maps at depths 50 and 350 m (Figures 3d and 5d, respectively), horizontal maps of the v-fields show a cyclone with a meridional velocity of ~20 cm/s at 50 m depth adjacent to an anticyclone with a meridional velocity of ~35 cm/s at 50 m depth. This vortex pair is offshore of the coastal jet with peak velocities greater than 50 cm/s at 50 m depth.

The vertical cross-shore section of v at day 50 at y = 290 km (Figure 7c) shows that the vertical extent of the cyclone is ~1000 m (between ~25-50 km width) with maximum values of ~40 cm/s at the surface (~120 km from the coast). The vertical extent of the anticyclone is ~1200 m depth (~50 km width) with maximum values of ~70 cm/s at ~100 m depth (~50 km from the coast). The vertical extent of the coastal jet is ~400 m (~20 to 40 km width) with maximum values greater than 50 cm/s at the surface. A poleward undercurrent of ~15-20 km width and velocities of ~10 cm/s (maximum velocity at ~450 m depth) is discernible beneath the coastal jet nearshore.

Figure 8 shows vertical cross-sections of the along-shore and time-averaged (ten-day) v-fields before the onset of eddies (Figure 8a) and after the onset of eddies (Figure 8b). A comparison of the cross-sections shows significant changes in the vertical structure of both the coastal jet and the undercurrent. The depth of penetration of the coastal jet has shallowed nearshore, and the core of the equatorward jet has moved 20 km farther offshore. Instead of near-zero surface velocities at the coast, maximum surface CUJ velocities of ~50 cm/s are discernible. The undercurrent, instead of lying between ~15 to 50 km offshore, adjoins the coast.

A comparison of the instantaneous (Figure 7c) and along-shore and time-averaged (Figure 8b) vertical cross-sections of the v-field shows that, in the presence of eddies, the vertical and offshore structures of the coastal jet are readily evident in the time-averaged but not in the instantaneous cross-section of the v-field. This is consistent with recent observations of Kosro (1987) and Huyer and Kosro (1987), who suggest that a classical two-dimensional upwelling-induced coastal jet may not be very representative of actual conditions in the CCS. Their observations show that there is great variability in time and space in the CCS and that the classical coastal jet has not been frequently observed off California. Only one of their six surveys off Point Arena resembled a simple two-dimensional, equatorward, baroclinic coastal jet associated with coastal upwelling, and this was during a period of wind relaxation instead of winds

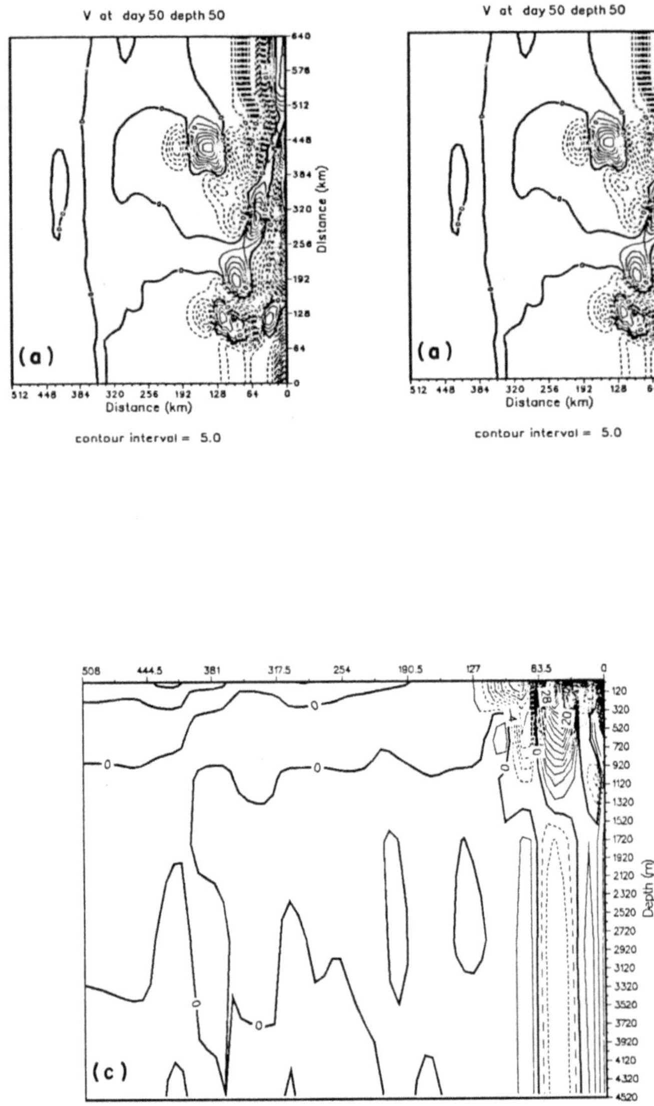

Figure 7: Instantaneous meridional velocity (v) fields at day 50. Solid closed contours denote anticyclonic eddies, while solid contours nearshore denote the undercurrent or poleward flow. The horizontal maps are shown at depths 50 m (a) and 350 m (b). The vertical cross-section (c) of v has been taken at y = 290 km. Contour interval is 5 cm/s.

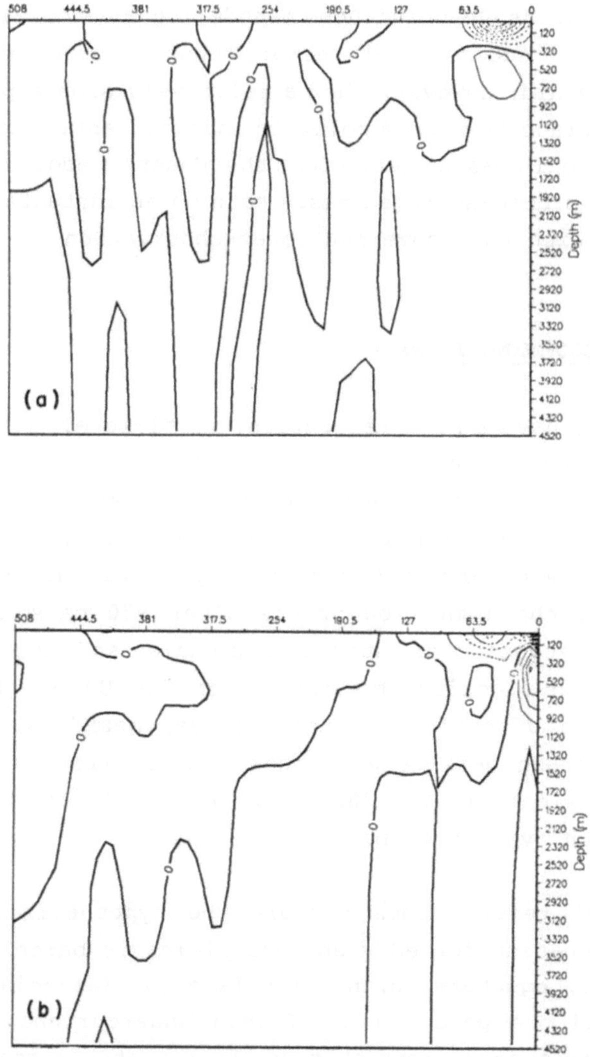

Figure 8: Vertical cross-sections of the v-field which are along-shore and time-averaged (a) over the first 10 days and (b) over days 41-50 of model simulation time. Contour intervals is 5 cm/s. The solid contours nearshore denote the undercurrent.

favorable for upwelling. The other five surveys showed complex circulation patterns which included meandering equatorward flow, cyclonic and anticyclonic eddies, and inshore countercurrents. The overall averaged fields did, however, show a relatively simple pattern: a broad equatorward surface flow and a poleward undercurrent. The present results show that the model coastal jet, after the onset of eddies, is more readily discernible on a time-averaged basis than on an instantaneous one, which is consistent with the above-mentioned observations.

SUMMARY AND RECOMMENDATIONS

This study used a high-resolution, multi-level, primitive equation ocean model to examine the response of an idealized, flat-bottomed oceanic regime along an eastern ocean boundary to an imposed jet. The imposed jet, intended to simulate the mean CCS during the upwelling season, consists of an equatorward CUJ overlying a poleward CUC. Baroclinic instability of the mean flow occurs after ~30 days resulting in the generation of cyclonic and anticyclonic eddies and jets. Examination of the results shows significant evolution of the CUJ and the CUC fields due to the presence of cyclones, anticyclones, meanders, dipole pairs of eddies with strong jets between them, and squirts of cold water with offshore extensions of ~100-200 kilometers. All of these features are consistent with available observational data.

The model results thus support the hypothesis that a possible generation mechanism for eddy and jet fields is baroclinic instability of the coastal, equatorward, near-surface jet (associated with coastal upwelling) and the poleward California Undercurrent. They are also consistent with observations that as these unstable meanders intensify, they carry the cool, upwelled water offshore and are often cutoff, creating vortex pairs of isolated cyclonic and anticyclonic eddies (Bernstein et al., 1977). The model simulations of the CCS demonstrate great variability in time and space and show a classical two-dimensional coastal jet only when the meridional velocity field is time-averaged. This is consistent with recent observations of Kosro (1987) and Huyer and Kosro (1987), who suggest that a classical two-dimensional upwelling-induced coastal jet may not be very representative of instantaneous observations in the CCS.

This study employed twin constraints of a flat-bottom basin and a regular coastline. Future experiments should include bottom topography and/or an irregular coastline which would represent capes and bays.

Future experiments should also include wind forcing instead of initialization of a baroclinic jet representative of the mean CCS during the upwelling season. It is believed that wind forcing of the local ocean domain in the CCS is another important generation process for eddies and jets in the CCS, and may provide important insight into the dynamics, kinematics and energetics of this complex eastern boundary current.

ACKNOWLEDGEMENTS

This work was done in the Department of Oceanography at the Naval Postgraduate School (NPS) while I was holding a National Research Council (NRC) Research Associateship, and continued under the support of the NPS Research Foundation. I wish to thank Dr. C.N.K. Mooers, my NRC sponsor, and Dr. R.L. Haney, who graciously allowed me to use and modify his primitive equation model.

From: R.H. Bourke and A.A. Bird

On: Review and commentary to paper **MODEL SIMULATION OF A COASTAL JET AND UNDERCURRENT IN THE PRESENCE OF EDDIES AND JETS IN THE CCS** by Mary L. Batteen

This is an interesting study which demonstrates, through the use of a high-resolution, multi-level, primitive equation model, the effects on the circulation induced by eddies generated by baroclinic instability of the mean flow. The rotational sense of the eddy determines whether the mean flow is then confined to a narrow inshore region, as in the case of the undercurrent, or forced to meander offshore possibly producing a "squirt." This was also the only model presented during the workshop which showed eddies in the presence of an undercurrent. Since eddies can be observed from remote sensing techniques, knowledge of this interaction of the eddy field and the undercurrent will be useful input for sampling strategies in field programs.

Although the model is formulated for the California Current System, the results have broad implications for other regions as well. In particular, given enough vertical shear to induce baroclinic instability, The same results should be applicable to other eastern boundary current systems.

Of particular importance to observationalists is the finding that the structure of the coastal jet is discernible only after employing a degree of time averaging. Synoptic observations are likely to miss it due to the complex nature of the short-term flow. This points out the need to employ long-term measuring systems if we are to understand the nature of the flow in meandering, boundary-current regimes.

In this study, Batteen used a flat bottom, a regular coastline, and a prescribed initial flow field, including a poleward undercurrent. Future work, as mentioned in the recommendations, should include bottom topography, various coastline geometries, and generation of eddies by wind forcing. Ongoing work by Batteen to incorporate the barotropic component of flow into the model will allow the energetics of the baroclinic and barotropic components of the eddy field to be determined. It is important that each process be included in the model systematically and compared with the results presented here in order to analyze and understand the dynamical processes and their effects. Separation of each of these processes should also help in the interpretation of field observations of eddy and jet fields.

DO NITROGEN TRANSFORMATIONS IN THE POLEWARD UNDERCURRENT OFF PERU AND CHILE HAVE A GLOBALLY SIGNIFICANT INFLUENCE?

L.A. Codispoti, R.T. Barber, and G.E. Friederich

Monterey Bay Aquarium Research Institute

ABSTRACT

Although it represents only about 0.01% of the total oceanic volume, the Poleward Undercurrent off the coasts of Peru and northern Chile is the site of globally significant nitrogen transformations. The co-occurrence of oxygen deficient (<~0.05 ml/l) waters and high fluxes of organic material causes the regional denitrification rate to comprise about 20% of the total marine rate and most of this occurs within the Undercurrent. Theoretical considerations and the existence of extremely low and extremely high nitrous oxide concentrations in and near the Undercurrent suggest that nitrous oxide turnover in this region occurs at much higher rates than in the "average ocean."

The processes of denitrification and nitrous oxide turnover are of more than casual interest. Denitrification converts combined nitrogen into nitrogen gas which is unavailable to most plants. Therefore, this nitrogen transformation can limit the ocean's ability to sequester atmospheric carbon dioxide. Changes in nitrous oxide cycling in regions such as the Poleward Undercurrent could significantly affect the rate at which nitrous oxide is being added to the atmosphere. This gas contributes significantly to the "greenhouse effect" and is the main source of nitrogen oxides in the stratosphere.

The advective field associated with the Poleward Undercurrent exerts a major control on the supply of oxygen to the subsurface waters off the coasts of Peru and Chile, and helps to establish counter-current systems that can create concentrations of organic material. Since the nitrogen transformations described above are sensitive to both oxygen levels and the supply of organic material, understanding these processes requires an understanding of the dynamics of the Poleward Undercurrent.

INTRODUCTION

The Eastern South Pacific Ocean (ESPO) attracts considerable scientific interest because of its high biological productivity and because it displays some of the most intense signals associated with the El Niño-Southern Oscillation phenomenon (e.g., Halpern, 1983). It is generally less well-appreciated that the waters in the upper ~500 m of this region (0-25°S and within 500 km off the coast, Figure 1) have a globally significant impact on nitrogen cycling (e.g., Codispoti and Christensen, 1985) despite the relatively small areas and volumes involved.

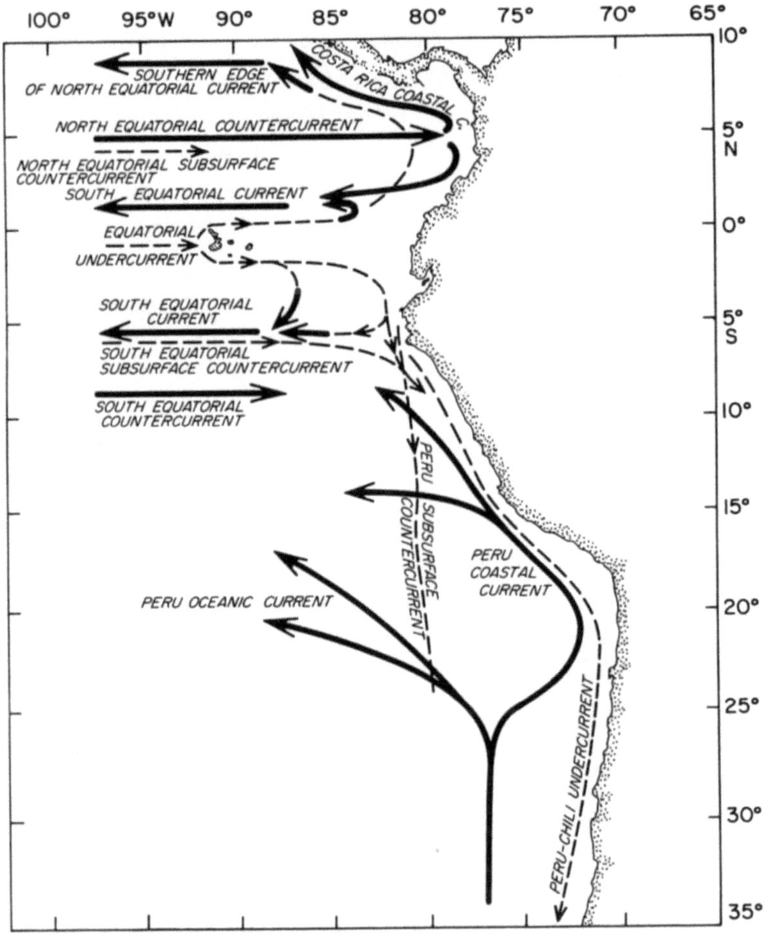

Figure 1: Generalized circulation scheme for the Eastern South Pacific Ocean. (Based on the work of various authors)

In this paper, we will consider the influence of the Poleward Undercurrent in this region on global nitrogen cycling. This current represents less than 0.01% of the total oceanic volume, but because of its position and chemical state, it is a site of particularly intense carbon and nitrogen cycling. We will review data which indicate that the nitrogen transformations that occur within this relatively minor current have globally significant effects that may be important vis-a-vis the issues of climatic change and global habitability.

We think that the most important reasons for studying eastern boundary regions and their associated poleward undercurrents involve providing a better understanding of the factors that govern these extremely interesting systems. Such arguments often fail to convince governments to fund research into these regions, however, and much of the research has been justified because of the relatively rich fishery resources. It should be recognized, however, that the issue of global habitability is becoming increasingly important as mankind's numbers and technological prowess increase. Thus, we believe that the "practical" reasons for studying nitrogen and carbon cycling in the ESPO include the general question of understanding how man is impinging on the environment. This question includes the issues of climatic change and changes in atmospheric chemistry as well as the questions that relate to fishery resources.

METHODS AND DATA

This paper is intended as an overview and review for those who are not intimately familiar with the marine nitrogen cycle. Consequently, some of the data that are presented appear in published papers listed in the references. Other data (including discussions of the methods employed) are presented in the following reports: Friederich et al., 1985; Hafferty et al., 1978; Hafferty et al., 1979.

A parameter called the "nitrate deficit" is used in this paper. This term may be unfamiliar to many readers, so we note here that it estimates the amount of free nitrogen that has been added to a water parcel from the reduction of combined nitrogen during denitrification (see below). Methods for calculating the nitrate deficit are described in Codispoti and Christensen (1985). We used the methods employed by Codispoti and Packard (1980), except for Figure 14 (A) where the values should be multiplied by 1.15 to be comparable.

BACKGROUND

General

Later on, we will show how changes in nitrogen cycling can affect atmospheric carbon dioxide levels, but we note here that both carbon dioxide and nitrous oxide are increasing in the atmosphere, and contribute significantly to the "greenhouse effect" (at present, ~60 and ~10%, of the total, respectively; Lacis et al., 1981; Bolin and Cook, 1983). The present-day rate of increase for atmospheric carbon dioxide is ~3000 Tg C yr^{-1} (Bolin, 1983). For nitrous oxide, the rate of increase is ~2.5 Tg N yr^{-1} (Weiss, 1981; Bolin, 1983; Crutzen, 1983). The consensus view is that increases in these two atmospheric trace gases and in methane and the freons are likely to increase the average temperature of the troposphere by ~5°C within the next ~200 years (Keeling and Bacastow; Lacis et al., 1981). In addition, nitrous oxide plays a significant role in atmospheric chemistry (Crutzen, 1981, 1983) as the main source of NO_x in the stratosphere.

For this discussion, salient features of the ESPO and its Poleward Undercurrent include the following: 1) a high flux of inorganic nutrients into the photic zone (e.g., Codispoti et al., 1982; Codispoti, 1983), 2) high primary production rates that, in turn, support a high downwards flux of organic carbon and nitrogen (e.g, Chavez and Barber, 1987), 3) high subsurface respiration rates arising from "2") (e.g., Packard et al., 1983), and 4) a well-developed oxygen minimum that contains extensive regions between ~75-500 m where oxygen concentrations are almost zero (Codispoti et al., 1986; Wooster et al., 1965; Love, 1970-74; and Zuta and Guillen, 1970).

The well-developed oxygen minimum zone in the eastern South Pacific Ocean is a partial consequence of the ocean-wide fields of advection and biology (e.g., Codispoti et al., 1982), but local processes enhance oxygen-depleted conditions. For example, an upwelling cross-circulation is superimposed on the Poleward Undercurrent. When combined with the high nutrient content of the upwelling waters, this results in high primary production. The resulting flux of organic material means that respiratory rates within the Undercurrent will tend to be high (e.g., Pace et al., 1987; Suess, 1980; Packard et al., 1983). In addition, the counter-current circulations associated with the upwelling circulation and arising in the longshore direction from the presence of the Undercurrent may also lead to the accumulation of nutrients and to enhanced oxygen depletion (Barber and Smith, 1981).

REGIONAL OCEANOGRAPHY

We shall consider the coastal ocean off South America between ~5-25°S (Figure 1). Observations suggest the Poleward Undercurrent in this region exists between depths of ~30 to 200-300 m, and has an average poleward velocity of ~5-10 cm s^{-1} (Brink et al., 1980; Smith, 1981; Fahrbach et al., 1981; Enfield et al., 1978). Its offshore extent is uncertain, but we believe that a choice of 200 km for this dimension is reasonable.

Figures 1 and 2 give highly schematic views of the current system in our study region and show that there are at least two flows that can "feed" the Undercurrent. These are extensions of the Equatorial Undercurrent (EUC) and of the Subsurface South Equatorial Countercurrent (SSECC). Although, oxygen concentrations are relatively low in these two currents (Tsuchiya, 1975 and 1983), they must nevertheless be regarded as sources of oxygen because conditions tend to become even more oxygen depleted within the Undercurrent. This situation arises partially from the high downwards flux of organic material into the Undercurrent. Additional causes are the existence of both cross-shelf and longshore countercurrent systems arising from the upwelling circulation and from the juxtaposition of the shallow equatorward wind drift just above the Poleward Undercurrent (Figures 1 and 2). Such countercurrent circulations form "nutrient traps" that can accumulate more nutrients than originally present in the source waters (Calvert and Price, 1971; Barber and Smith, 1981; Margalef, 1978). Since nutrients accumulate as a result of the breakdown of organic material that is being supplied to the "nutrient trap," these sites are also zones of enhanced respiration and oxygen consumption. Equations (Figure 3), that describe how chemicals can be stripped from an upper layer by the sinking and breakdown of organic matter in either a slower moving lower layer or a lower layer moving in the opposite direction to the surface flow, have been given by Redfield et al. (1963). Accumulation in the lower layer tends to be more extreme when the layers are moving in opposite directions and when the upper layer is highly productive as would be the case for estuarine and upwelling circulations. Such a situation would also apply to the juxtaposition of the Equatorward surface layer and Poleward Undercurrent off Peru and Chile (see Figures 2 and 3).

So far, we have identified two sources of oxygen for the Poleward Undercurrent, namely the waters supplied by the Equatorial Undercurrent between ~3-5°, and the waters suplied by the SSECC at approximately 10°S. A third source of oxygenated and relatively low nutrient water is provided by the Subantarctic waters that move equatorward. The bulk of this water

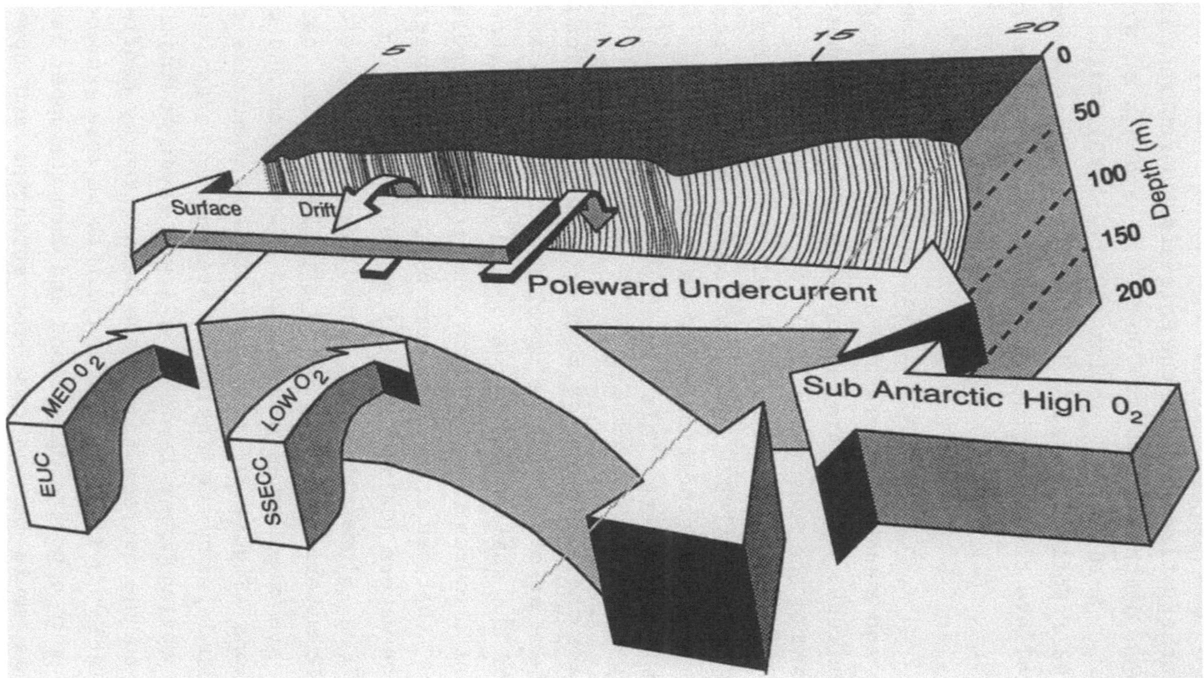

Figure 2: Block diagram showing the Poleward Undercurrent, the surface "wind-drift" and the three advective sources of oxygen for the Undercurrent which are the Equatorial Undercurrent, the Subsurface South Equatorial Countercurrent, and the equatorward flow of Subantarctic waters in the Peru Oceanic Current.

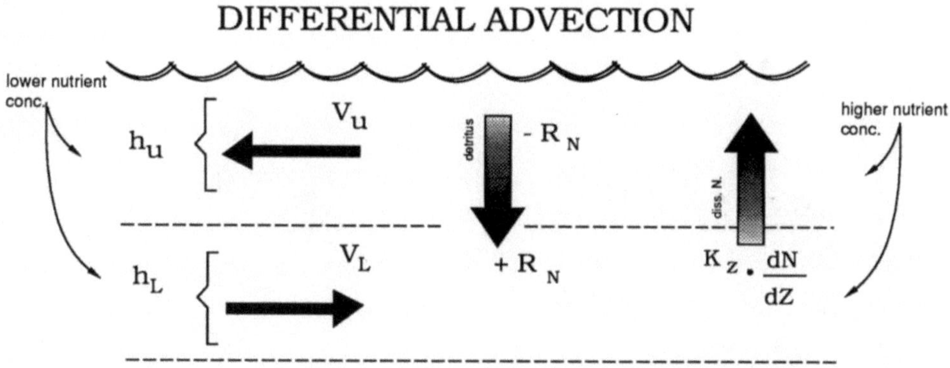

X ⟶ increases in this direction & $\frac{dN_u}{dx} + \frac{dN_L}{dx}$ are positive

$$V_u \cdot \frac{dN_u}{dx} = -R_N + \frac{K_z}{h} \cdot \frac{dN}{dZ}$$ *; at steady state*

$$V_L \cdot \frac{dN_L}{dx} = +R_N - \frac{K_z}{h} \cdot \frac{dN}{dZ}$$ *; at steady state*

$$\therefore \frac{dN_L}{dx} = -\frac{V_u}{V_L} \cdot \frac{dN_u}{dx}$$

Figure 3: Diagram and equations showing how a countercurrent system can lead to enhanced nutrient levels and oxygen depletion that increase in the downstream direction in the lower current. The equations are taken from Redfield et al. (1963) and assume upper and lower layers of equal depth (h), and a steady-state. It is also assumed that the downward loss of some bioactive particulate chemical (N) is ultimately balanced by upwards diffusion of the N that is released into solution by decomposition and respiration in the lower layer. Nutrients increase towards the right in both layers, but the highest values will be in the lower layer. The upwards transport of N could also be accomplished by upwards vertical advection, so for the upwelling case the re-supply of N to the surface layer (per m^3) would more properly be expressed as the sum of vertical diffusive and advective losses.

tends to turn offshore near 25°S (Wyrtki, 1963, 1977), but the influence of the Subantarctic waters which have higher oxygen concentrations than EUC water within the ESPO (e.g., Friederich and Codispoti, 1981) can be felt within the zone that is dominated by the Undercurrent. This influence extended as far north as 10°S during southern winter 1976, and during March-May 1977, a front between Subantarctic and Equatorial Water occurred near 16°S on the 26.0 sigma-t surface (Figure 4). During February 1985, continuous vertical profiles suggested interleaving of equatorial and subantarctic waters near 15°S (Friederich and Codispoti, 1987).

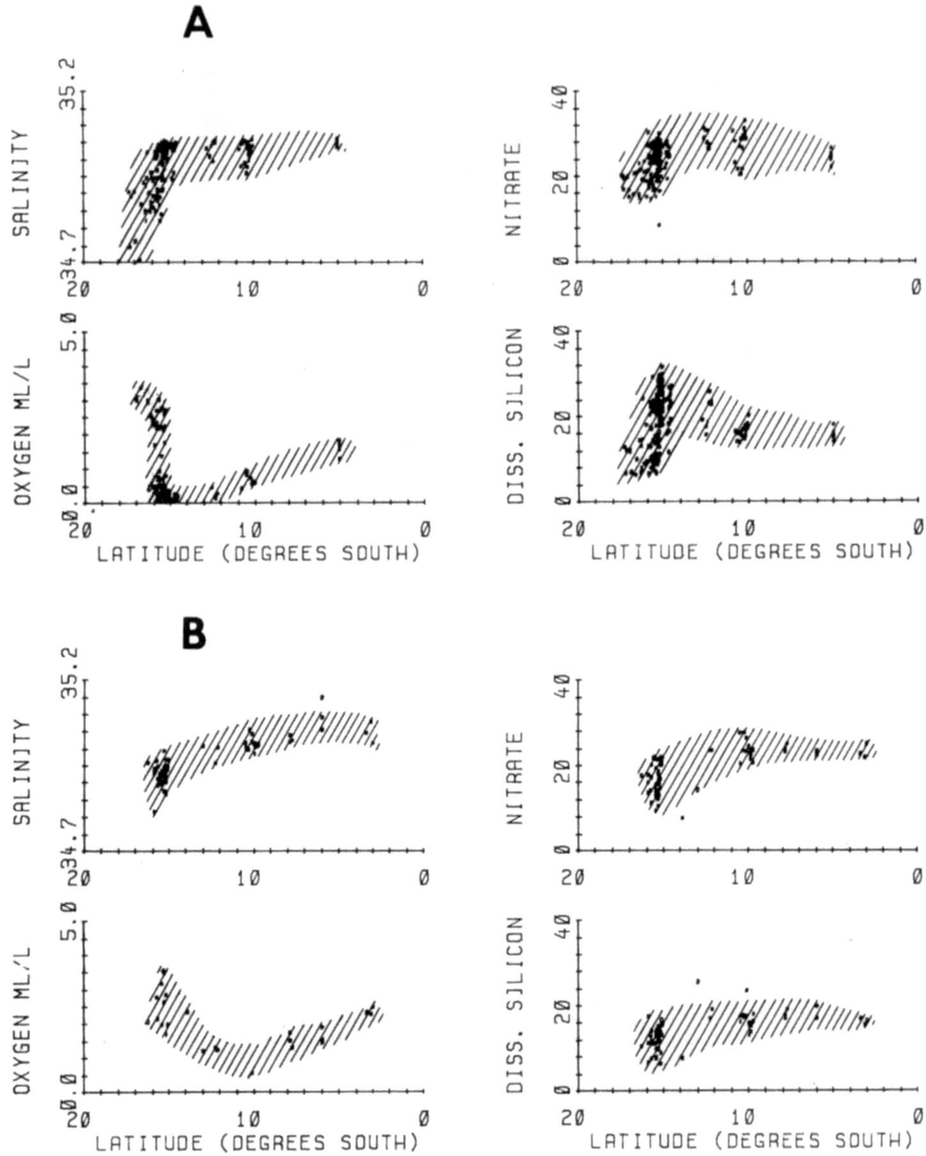

Figure 4: Trends of salinity, dissolved oxygen, nitrate, and dissolved silica on the 26.0 σ_t surface along the coast of Peru during March to May, 1977 (A), and during July to October, 1976 (B). Depth $\geq$ 40m. Nitrate and dissolved silica concentrations are in μM (from Friederich and Codispoti, 1981).

Sections taken along the coasts of Peru and Chile suggest that filaments of low oxygen water penetrate as least as far south as 28°S (Stommel et al., 1973). Some of these low oxygen (<0.5 ml/l) waters are found as far as ~1000 km offshore, but at 28°S the significant nitrate and nitrite consumption appeared to be restricted to the region that one would expect to be influenced by the Poleward Undercurrent (i.e., within 200 km of shore). The offshore zones of low oxygen may arise from the advection of the Peru Countercurrent (Figure 1) which may exist further offshore, but the relationship between the Undercurrent and Countercurrent has not yet been clearly defined. In a similar manner, sections taken near 15°S frequently show horizontal variability in oxygen depletion that could be interpreted as the juxtaposition of equatorward and poleward flows or variations in strength of the poleward flow even though the mean advection may be poleward. In our simplified view (Figure 2) we assume that the Undercurrent and Countercurrent are joined near 5° and "split" near 15°S.

THE NITROGEN CYCLE

Figure 5 outlines the biological nitrogen cycle. Nitrogen is a major constituent of organisms and the supply of available nitrogen is a control on the global carbon cycle. This is because carbon tends to be abundant relative to the needs of organisms, but nitrogen in forms readily available to plants (nitrate, nitrite, ammonium, urea, etc.) is often present in limiting quantities.

Dugdale and Goering (1967) made a distinction between the "new production" supported by the flux of inorganic nitrogen into the photic zone and the total primary production rate which depends on re-cycling of nitrogen within the photic zone. In a steady-state, the export of organic material from the photic zone cannot exceed the rate of "new production." The fundamental distinction is between primary production supported by the transport of available nitrogen into the photic zone vs. the primary production supported by re-cycling of nitrogenous compounds within the photic zone. Although nitrogen fixation (Figure 5) and inputs from the atmosphere and from runoff can support new production, the dominant source of nitrogen arises from the advection and mixing of nutrient-rich subsurface waters into the photic zone. Nitrate is the major form of combined nitrogen supplied by these ascending waters, and Dugdale and Goering (1967) stated that in most cases new production could be estimated by measuring the nitrate uptake rate. They also suggested that the production supported by re-cycling within the photic zone could be estimated by measuring the ammonium uptake rate since ammonium seems to

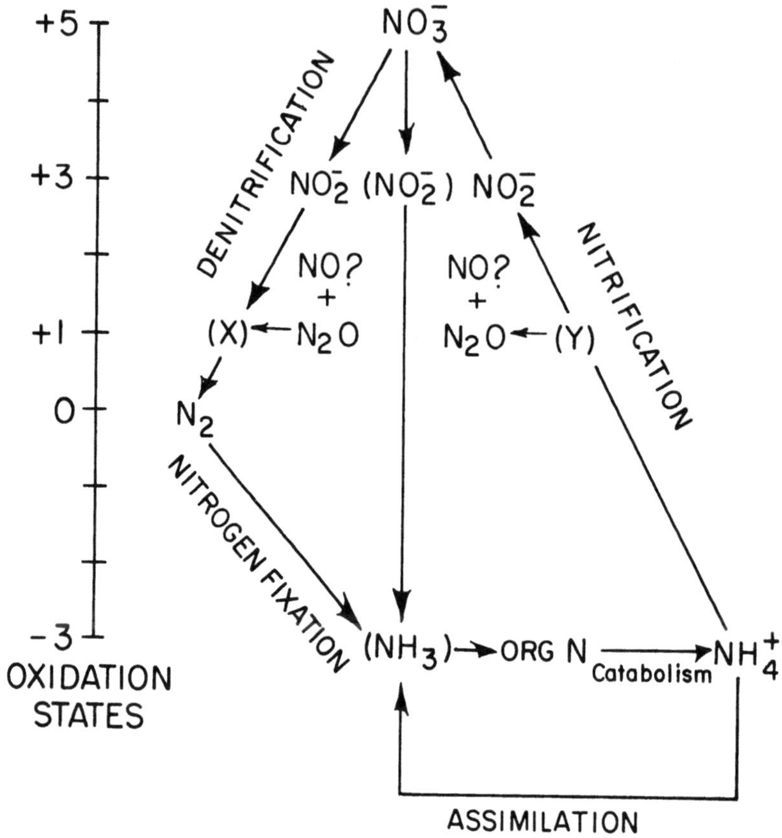

Figure 5: Diagram for the marine nitrogen cycle. This figure is a slightly modified version of a diagram presented by Liu (1979, 1982). (X) and (Y) represent intra-cellular intermediates that do not appear to accumulate in seawater. NO has been added to the original diagram based on recent work (Lipschultz et al., 1981), and the diagram has been modified to suggest that N_2O is produced mainly during nitrification and consumed during denitrification in the sea. Photochemical processes that occur near the sea surface (Zafiriou and True, 1979a,b) are not included in this figure.

be the most abundant form of re-cycled combined nitrogen within the photic zone.

Most new production is exported from the photic zone by a downward flux of particles ($-R_n$ in Figure 3) or dissolved organic matter. Since the supply of nitrate is exhausted by phytoplankton uptake in most of the oceanic photic zone, the downward flux of organic nitrogen and associated biogenic material is essentially equal to the time-averaged upward nitrate transport into the photic zone throughout most of the ocean. The notable exception is at high latitudes where light limitation may be more important than nitrogen limitation. In such cases, waters may leave the surface containing appreciable amounts of unutilized or "preformed nutrients" (Redfield et al., 1963) that behave as conservative tracers. Because of this phenomenon, some models of changing atmospheric carbon dioxide concentration invoke a greater utilization of these preformed nutrients during times when the atmospheric carbon dioxide was lower (e.g., Knox and McElroy, 1984; Sarmiento and Toggweiller, 1984).

The process of nitrification (Figure 5) is of interest to this discussion because it produces nitrate, nitrite and nitrous oxide which are used to oxidize organic matter during denitrification, and because the first nitrification step($NH_4 \rightarrow NO_2^-$) is thought to be the major in situ source for nitrous oxide (Elkins et al., 1978; Yoshinari, 1976). Although nitrification requires oxygen, nitrifying bacteria can thrive at low oxygen concentrations (Goreau et al., 1980; Carlucci and McNally, 1969). The major end-product of the first stage of nitrification is nitrite, but nitrous oxide is produced as a side-product, and yields of nitrous oxide appear to increase as oxygen concentrations decrease (Goreau et al., 1980). It is possible that nitrous oxide is the major end-product at extremely low (<0.05 ml/1) oxygen concentrations (Codispoti and Christensen, 1985; Delwiche, 1981).

When oxygen concentrations fall below ~0.05 ml/1, nitrate reduction ($NH_4 \rightarrow NO_2^-$) and denitrification become important respiratory modes (Figure 5). For brevity, we will consider nitrate reduction to be an initial step in the denitrification process even though more species of bacteria can perform the former than the latter. Denitrification is classically defined as the bacterially mediated reduction of the ionic nitrogen oxides (nitrate and nitrite) to the gaseous oxides (nitric oxide, and nitrous oxide) or to free nitrogen (N_2) during the breakdown of organic material (Knowles, 1982). Free nitrogen appears to be the major end-product of denitrification in the marine water column, and there appears to be a net consumption of nitrous oxide (e.g., Cohen and Gordon, 1978). The important aspects of oceanic denitrification for the purposes of this paper are as follows:

1) This process converts combined nitrogen in a form readily available to plants (nitrate) to free nitrogen which can only be used by nitrogen fixing plants and bacteria
2) Denitrification appears to be the only in situ sink for nitrous oxide; the rest of the nitrous oxide produced during nitrification probably escapes to the atmosphere
3) Denitrification appears to become a major respiratory process only when oxygen concentrations fall below ~1% of the saturation value (e.g., Devol, 1978).

As a consequence, most marine denitrification may be restricted to oxygen deficient portions of the water column (~50%) which represent only ~0.1% of the total oceanic volume and to shallow and hemipelagic sediments (~50%; Codispoti and Christensen, 1985). Codispoti and Packard (1980) estimated that the denitrification rate in the low oxygen environments of the ESPO was ~25 Tg N/yr $\left(Tg = 10^9 g\right)$ and that most (~80%) of this occurred within 175 km of the coast.

Globally significant variability has been observed in the denitrification rate in the waters off Peru (Codispoti and Packard, 1980; Codispoti et al., 1986). McElroy (1983) has suggested that the variability in marine denitrification could be responsible for some of the variability in atmospheric carbon dioxide concentrations that has been inferred from ice-core data. Periods of increased denitrification could cause the combined nitrogen inventory of the ocean to experience transient decreases, and Codispoti and Christensen (1985) have suggested that the present-day ocean may be in deficit by ~70 Tg N/yr. Such an imbalance is possible because increases in denitrification may not be closely coupled to the source terms for combined nitrogen. Piper and Codispoti (1975) argued, for example, that denitrification and nitrogen fixation may not be closely coupled since the latter process appears to be concentrated in the photic zone and may occur at relatively low rates (Capone and Carpenter, 1982; Carpenter and Capone, 1983).

Figure 6 shows how increases in marine denitrification could lead to increases in atmospheric carbon dioxide as suggested by McElroy. During the transient period when increased denitrification is not balanced by decreases in subsurface respiration or by increased source terms, deeper waters rising to the surface will give off carbon dioxide that was sequestered by biological productivity when the nitrate flux to the photic zone was higher. The photic zone, however, will take up less carbon dioxide from the atmosphere because the nitrate supply is reduced. If nitrogen fixation could increase rapidly, the imbalance would be short-lived, but if we must rely on reduced subsurface respiration arising from reduced new production to gradually decrease the denitrification rate, several ocean mixing times might be required to restore a new steady-state. The equations in Figure 6 are simplified and not stoichiometrically balanced, but they do point out one factor that makes it possible for dentrification rates to vary rapidly; during phytoplankton growth the N/C uptake ratio is ~6 (by atoms) but during denitrification, the $\Delta N/\Delta C$ ratio is ~1. Thus, only 18% of the organic material produced by phytoplankton could remove all of the nitrate associated with this production if respiration proceeded by denitrification. The practical consequence of this is that relatively small changes in the organic matter supply could cause large

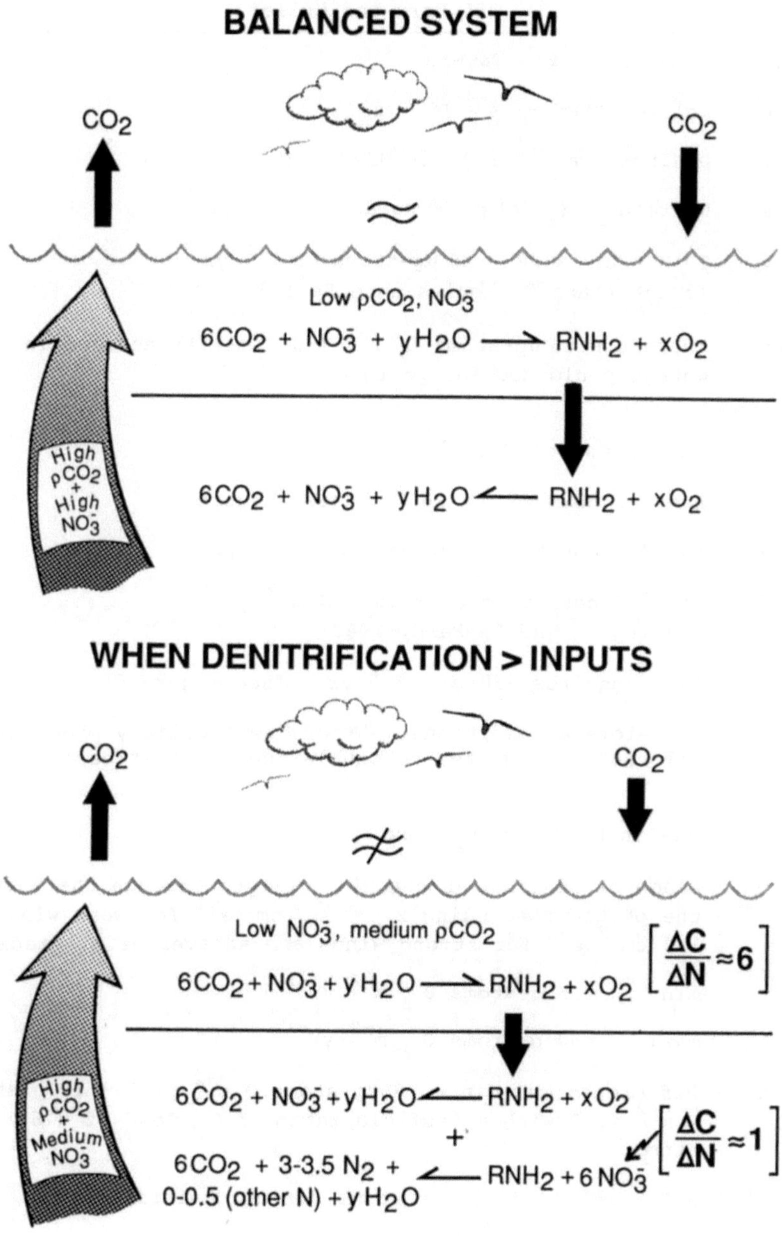

Figure 6: Conceptual models of (upper panel) an ocean with a balanced combined nitrogen budget (combined-N inputs = outputs) and (lower panel) an ocean in which excess denitrification leads to a net of loss of carbon dioxide from the ocean (from Codispoti et al., 1988). The carbon dioxide loss will occur only until a new balance is established for combined nitrogen, and this would eventually have to occur as reduced primary production arising from decreased nitrate levels caused decreases in subsurface respiration (Broecher and Peng, 1982; Piper and Codispoti, 1975). A balance might occur faster if nitrogen fixation rates could increase rapidly in the face of increased denitrification which, at present, is an open question.

Table 1. Mechanisms for Doubling the Denitrification Rate Off Peru and Chile*

A. O_2 Decrease in Source Waters

1. Vol. of MSNM ~1.4×10^{14} m^3
2. Denitrification rate in MSNM ~ 2×10^{13}g N yr^{-1}
3. Nitrate → N_2 Rate ~10 µg-atoms l^{-1} yr^{-1}
4. 10 Nitrate ×5/2 = 25 µg-atoms O_2 l^{-1} yr^{-1} [5/2=Nitrate/0 Equivalence Ratio]
5. Therefore 25 µg-atoms O_2 l^{-1} (~0.3 ml/l) decrease in source waters could double the rate.

B. Increase in Carbon Flux

1. Area of MSNM ~1.0×10^{12} m^2 (see Figure 9)
2. "New" productivity of ~400 g C m^{-2} yr^{-1}
3. Dentrification rate of 20 g N m^{-2} yr^{-1} (Codispoti and Packard, 1980)
4. This requires ~20 g C m^{-2} yr^{-1} (see Figure 6)
5. Therefore an additional ~5% of "new" primary productivity is all that is required to double the denitrification rate.

C. Decrease in Downward O_2 Flux

1. Range of possible downward O_2 fluxes through the bottom of the photic zone using k_z = 0.3 cm^2 s^{-1} for weak winds and 1.0 cm^2 s^{-1} for strong winds and an average O_2 gradient

   min: 15 g-atoms O_2 m^{-2} yr^{-1}

   max: 50 g-atoms O_2 m^{-2} yr^{-1}

2. Difference max-min = 35 g-atoms $O_2 m^{-2}$ yr^{-1} = 13 g-atoms of C m^{-2} yr^{-1} with a Redfield ratio of O_2 to C of 276:106 by atoms
3. A C:N ratio of 1:1 by atoms during denitrification
4. ∴ 13 g-atoms C m^{-2} yr^{-1} ≈ 13 g-atoms N m^{-2} or 182 g N m^{-2} yr^{-1}
5. Therefore changes in downward flux of O_2 only 10% as large as the one estimated above could double the denitrification rate since the area of the MSNM is ~ 1.0×10^{12} m^2.

*These calculations assume a doubling of the rate that Codispoti and Packard (1980) estimated for the main secondary nitrite maximum (MSNM). This table is from Codispoti et al. (1988).

Table 1:

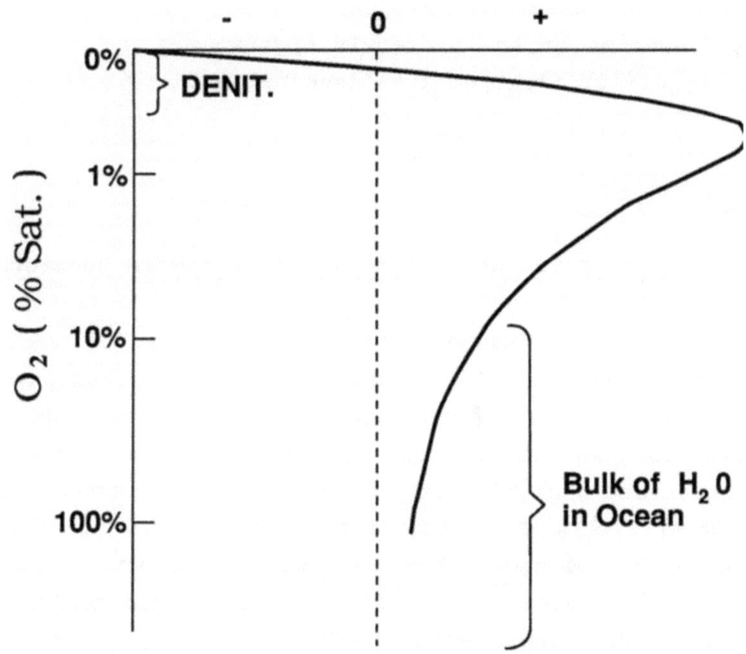

Figure 7: A "cartoon" of nitrous oxide production and consumption rates as a function of dissolved oxygen concentration in the oceanic water column.

changes in denitrification. In addition, since denitrification occurs at low oxygen concentrations and since water column denitrification may be confined largely to 0.1% of the marine volume, very small changes in the global oxygen distribution could also cause large changes in denitrification (see Table 1). Since a large fraction of marine denitrification occurs in eastern boundary regions such as the ESPO, understanding the rates and variability of denitrification in these zones is important.

Because nitrous oxide production is enhanced at low oxygen concentrations and because this gas is consumed by denitrification at even lower concentrations, Codispoti and Christensen (1985) suggested that nitrous oxide turnover in the ocean could be very uneven as shown schematically in Figure 7. Their calculations suggested that nitrous oxide turnover in productive eastern boundary regions and in the Arabian Sea (~1% of the oceanic area) rivaled that in the rest of the ocean and was significant in comparison to the rate at which nitrous oxide is increasing in the atmosphere. Their analysis suggested that small changes in oxygen distribution could have massive changes on nitrous oxide and that a globally significant amount of nitrous oxide was consumed by denitrification in the ESPO. On balance, however, the ESPO may be a net source of N_2O because of the extremely high values occurring at the periphery of the denitrification zone (see below).

DISCUSSION

Advective Controls on Nitrogen Cycling in the Poleward Undercurrent

We have established above that key portions of the nitrogen cycle depend critically on the distribution of dissolved oxygen. In particular, nitrous oxide production is enhanced at lower oxygen concentrations (Goreau et al., 1980 and Figure 7), and both nitrous oxide consumption and denitrification are favored by oxygen concentrations of <~0.05 ml/l (<1% of the saturation value). Scaling calculations (Table 1) suggest, for example, that a 0.3 ml/l decrease in oxygen in the water supplied to the region could cause a doubling of the denitrification rate which in turn would represent a 10-30% increase in the total oceanic rate. The question of the advective supply of oxygen to the Poleward Undercurrent is therefore an important one. Our schematic view of the Undercurrent system suggests two "upstream" sources for the Undercurrent, the EUC and the SSECC (Figure 2). Of these, we believe that the EUC may be a more important source of oxygen. Data from the El Niño Watch experiment (Patzert et al., 1978,), for example suggest that the extension of the SSECC carries waters of lower oxygen content into the region than is carried in by the EUC (Figure 8).

We believe that the relatively greater importance of the EUC is also suggested by the tendency for a monotonic progression of decreasing oxygen followed by increases in nitrite and the nitrate deficit produced by denitrification as the Undercurrent proceeds to the South. In other words, reversals in trend tend to occur only when the Undercurrent travels far enough south to become significantly influenced by Subantarctic waters which can have oxygen concentrations that are higher than those in the EUC within our study area (Figure 4). For example, a high nitrite feature which has been called the Main Secondary Nitrite Maximum (Codispoti and Packard, 1980) is typically found between ~10-25°S, and it has its maximum offshore extent between ~12-15° (Figure 9). This is consistent with the view that total oxygen deficiency (O_2 depletion plus nitrate deficit) in the Undercurrent tends to increase towards the South until influenced by the Subantarctic waters which tend to turn offshore near 25°S.

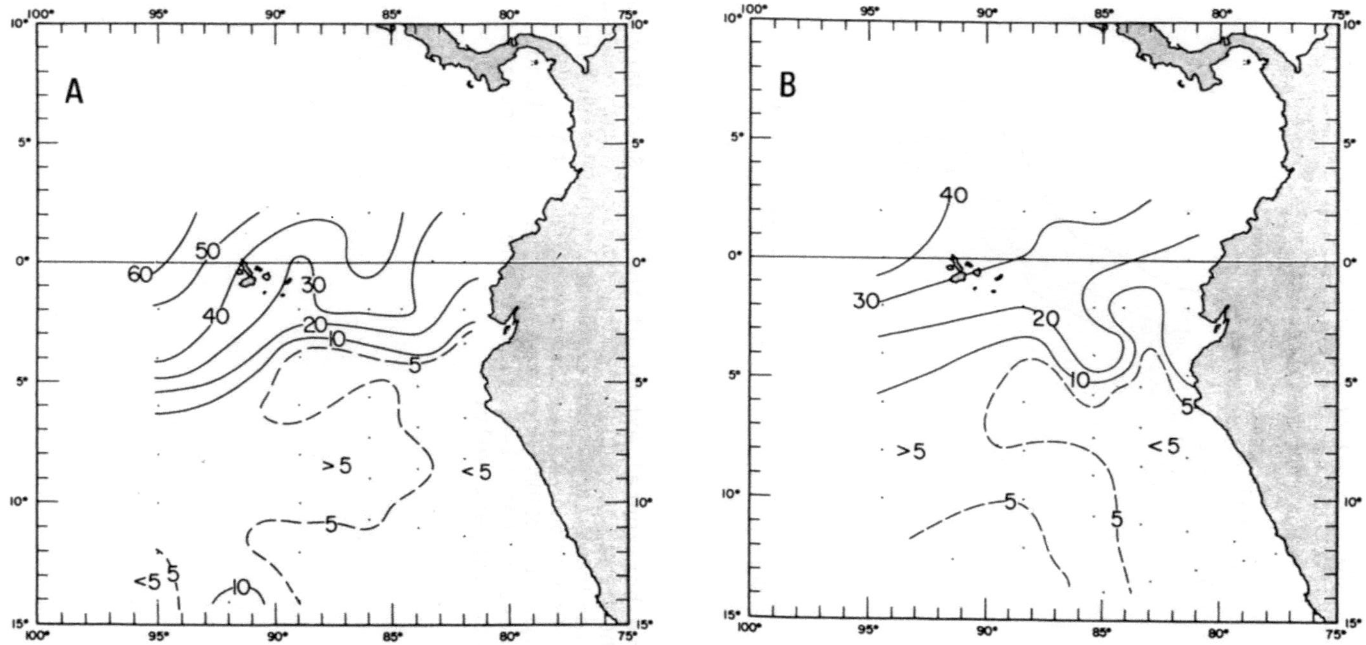

Figure 8: (A) Oxygen (μmoles/kg) on the surface where δ=160 cl/t during 20 February-31 March, 1975, and (B) during 17 April-27 May, 1975 (from Patzert et al., 1978) during the El Niño watch experiment. The 160 cl/t surface is ≈ 26.35 σ_t.

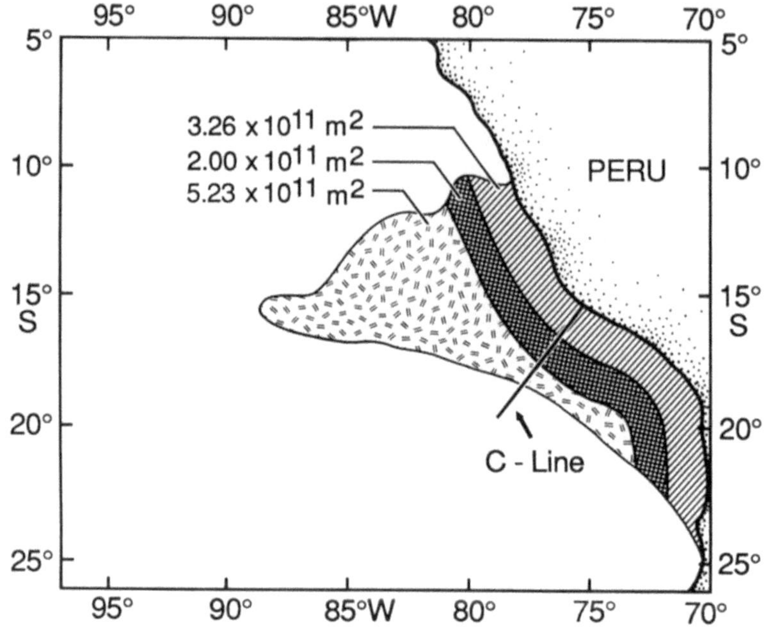

Figure 9: Approximate horizontal extent of the Main Secondary Nitrite Maximum (MSNM) with the position of the "C-line" indicated (From Codispoti and Packard, 1980 as re-drawn from Wooster et al., 1965). The inner band is 175 km wide and Codispoti and Packard suggest that 76% of the denitrification within the MSNM occurred in the inner band. Their calculation also suggested that 9% and 15% of the total occurred in the middle and outer bands, respectively.

A series of sections taken during May-June 1977 (Figures 10-12) also suggests progressively increasing oxygen depletion within the depth range of the Undercurrent (~30 to 250-300 m) as it proceeds from ~5 to 15°S. Similarly, high nitrous oxide concentrations were found within the Undercurrent equatorward of 10° during 1976 (Pierotti and Rasmussen, 1980) while some extremely low values were found poleward of 10° in 1978 (Elkins et al., 1978) suggesting a progression from low oxygen conditions which favor high nitrous oxide production rates to oxygen deficiency which favors net consumption (Figures 7 and 13).

There are times when oxygen deficiency has been observed equatorward of 10° (Sorokin, 1978; Patzert et al., 1978; Zuta and Guillen, 1970), but during such a situation observed in 1985 (Codispoti et al,. 1986) there was still a tendency for increasing oxygen deficiency towards the south at least over the slope and shelf as indicated by nitrate deficit sections taken at 10 and ~15° (Figure 14). Further offshore, a break in high nitrite concentrations near 11° (Codispoti et al., 1986) may have signaled the influence of the SSECC and geostrophic calculations indicated onshore flow in this region. A Coastal Zone Color Scanner Image (supplied by G. Feldman) also indicated an onshore movement in the area of the break in high nitrite concentrations.

The unusual 1985 observations which were made in February and March have been attributed to a region-wide shoaling of the thermocline that may have been an accentuation of a phenomenon that occurs with some frequency during southern summer (Codispoti et al., 1986; Codispoti et al., in press). It would be interesting, however, to know whether episodic denitrification equatorward of 10°S is also influenced by changes in the oxygen supplied by the EUC. An experiment to look at the annual cycle of the EUC extension would be highly instructive since the observations of the oxygen deficiency equatorward of 10°S seem to be concentrated in southern summer.

Minima in Nitrite Concentrations Over the Outer Shelf and Upper Slope

Figures 15 and 16 display minima in nitrite concentrations over the outer shelf and upper slope at 10 and 12°S and a corresponding zone of relatively high oxygen at 12°S. Correlations between light scattering (a measure of the suspended particle load) and poleward velocities using data collected in 1977 near 15°S suggested that the suspended load decreased when poleward velocities increased (G. Kullenberg, personal communication). In addition, Fahrbach et al., (1981) suggest that poleward velocities were strong and reached the surface over the outer

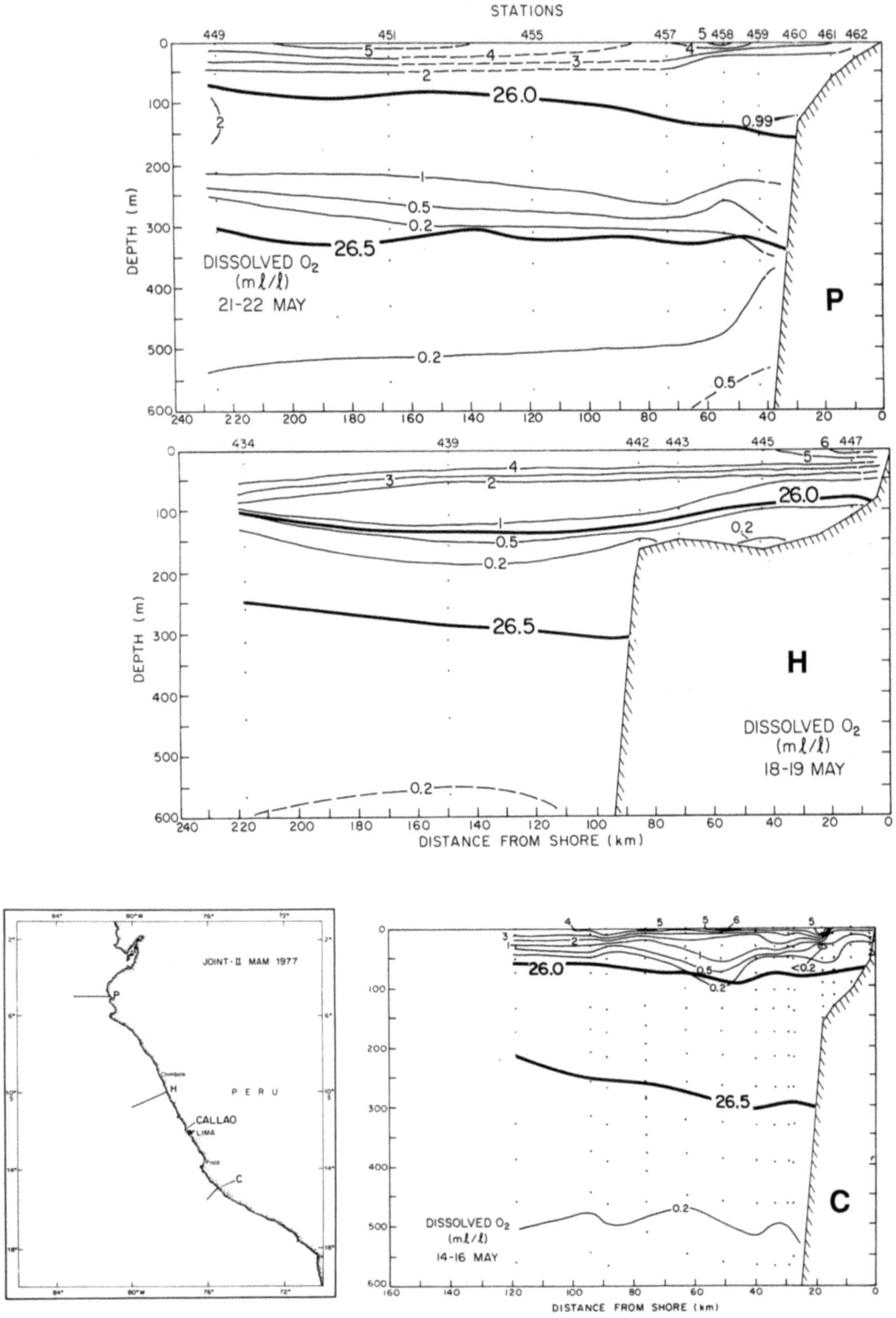

Figure 10: Oxygen distributions at ~5, 10 and 15°S during May 1977 (re-drawn from Hafferty et al., 1979). The 26.0 and 26.5 σ_t surface are indicated by the thick lines.

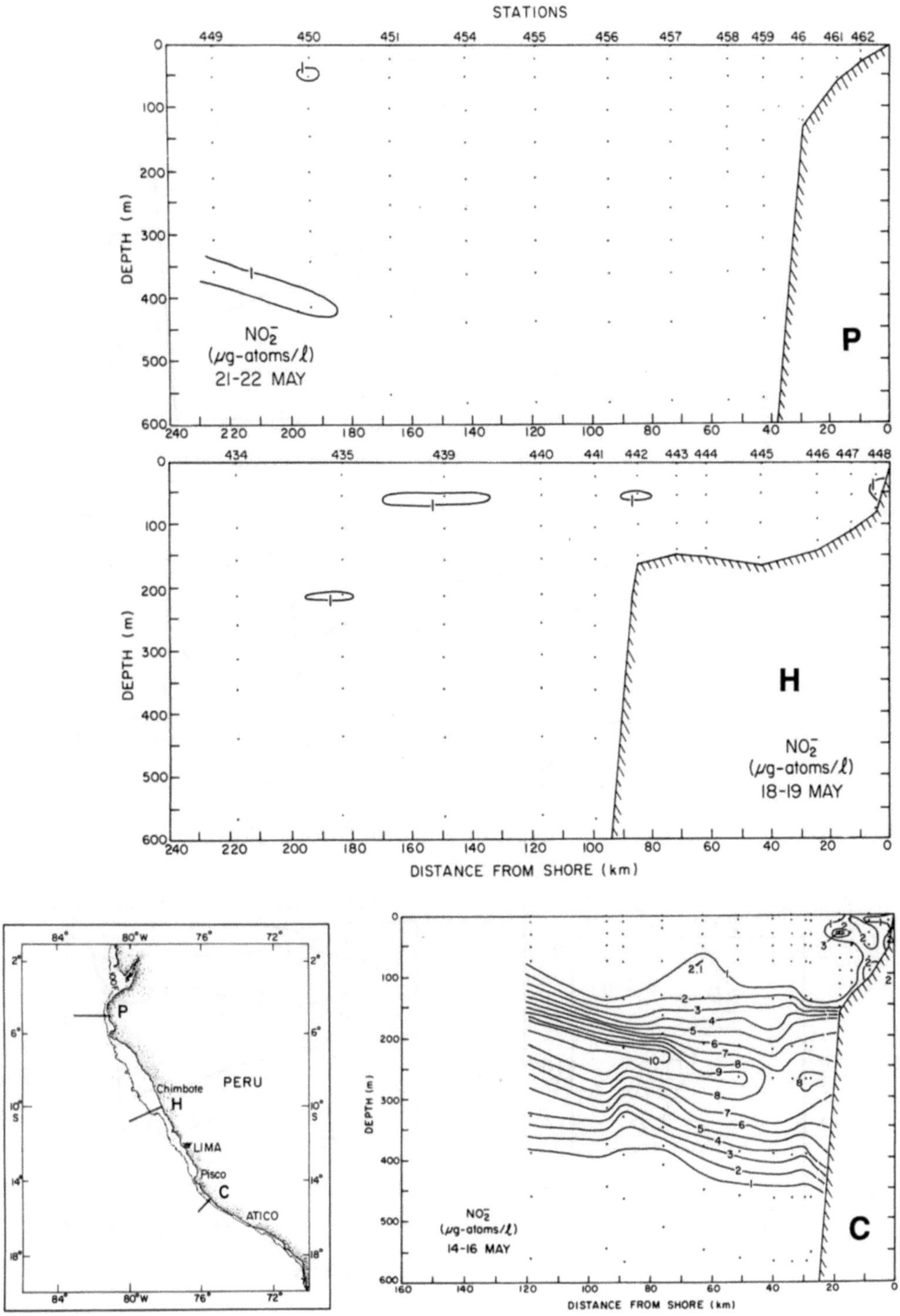

Figure 11: Nitrite distributions at ~5, 10 and 15°S during May 1977 (re-drawn from Hafferty et al., 1979).

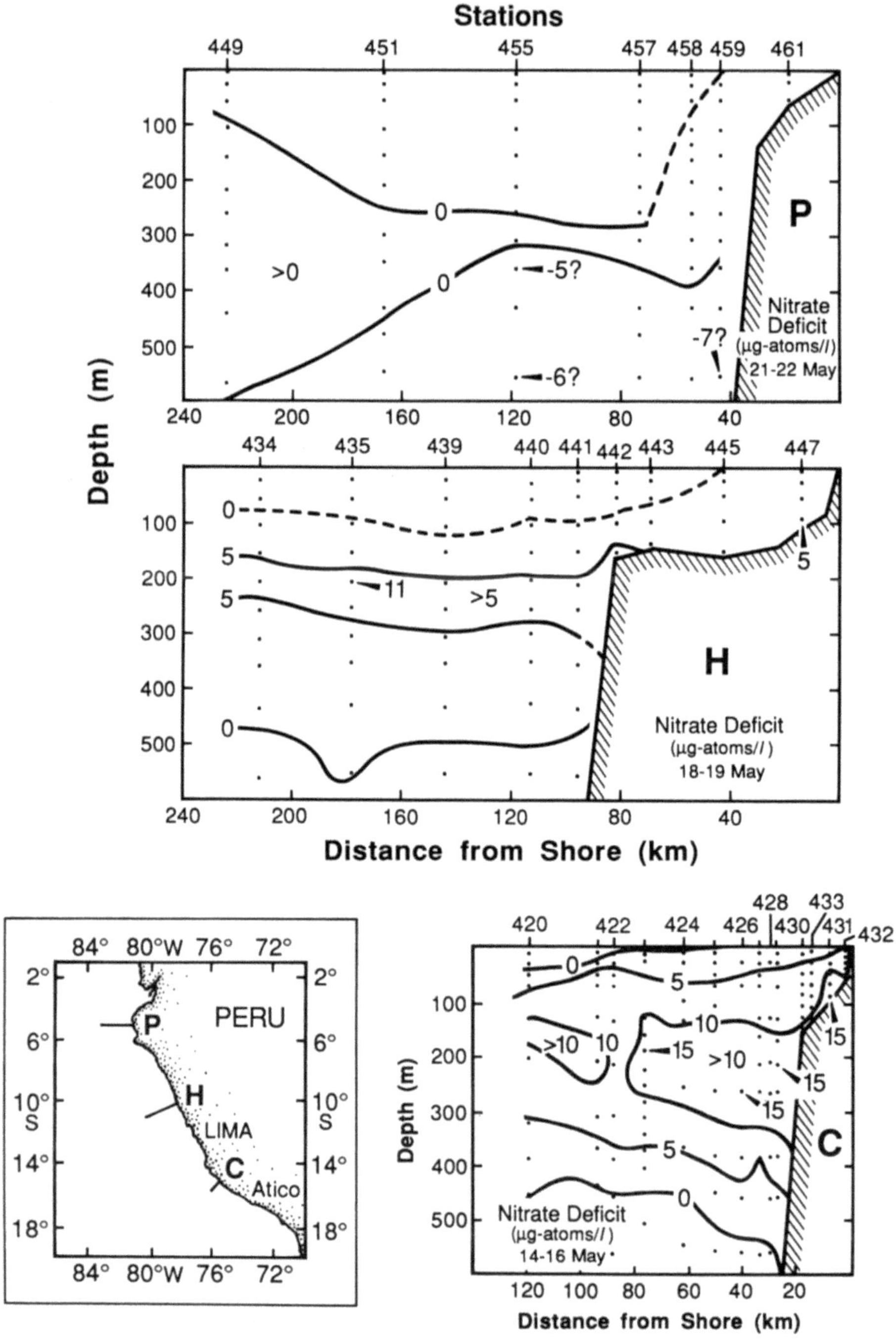

Figure 12: Nitrate deficits at ~5, 10 and 15°S during May 1977 (from data presented in Hafferty et al., 1978).

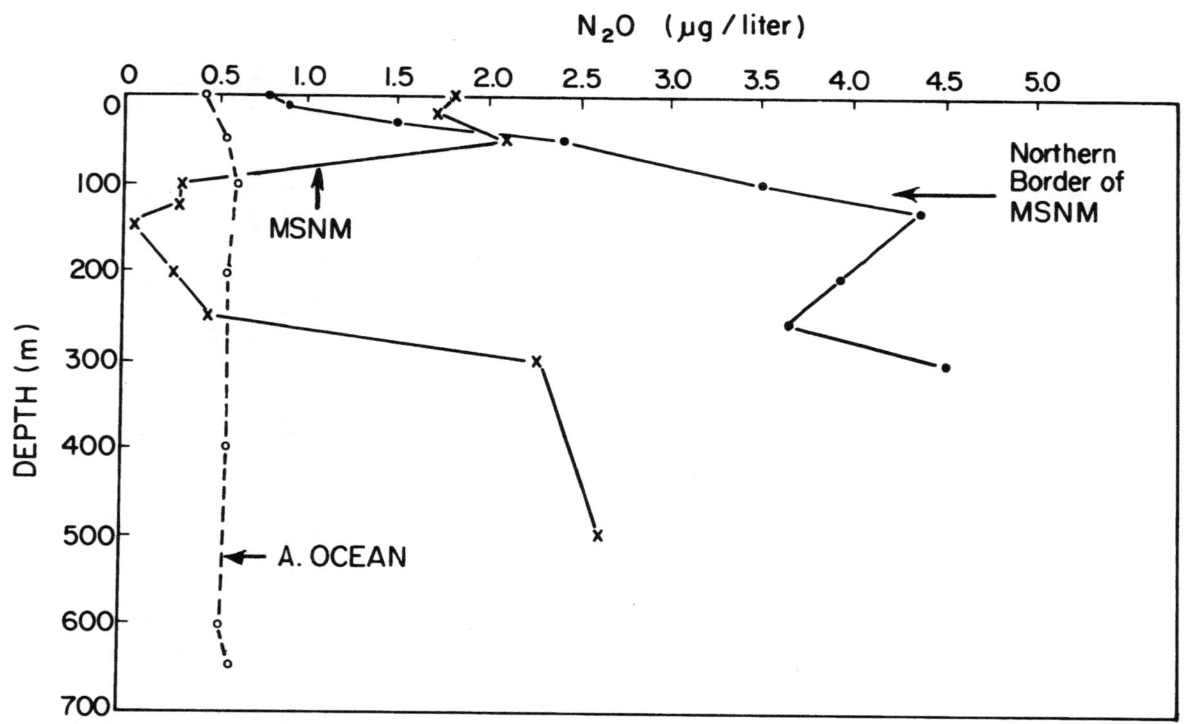

Figure 13: Nitrous oxide distributions in the North Atlantic Ocean (from Hahn, 1981), the Main Secondary Nitrite Maximum that is usually centered at ~200 m (from Elkins, 1978), and just outside of the northern border of the Main Secondary Nitrite Maximum (from Pierotti and Rasmussen, 1980). This figure is taken from Codispoti and Christensen, 1985).

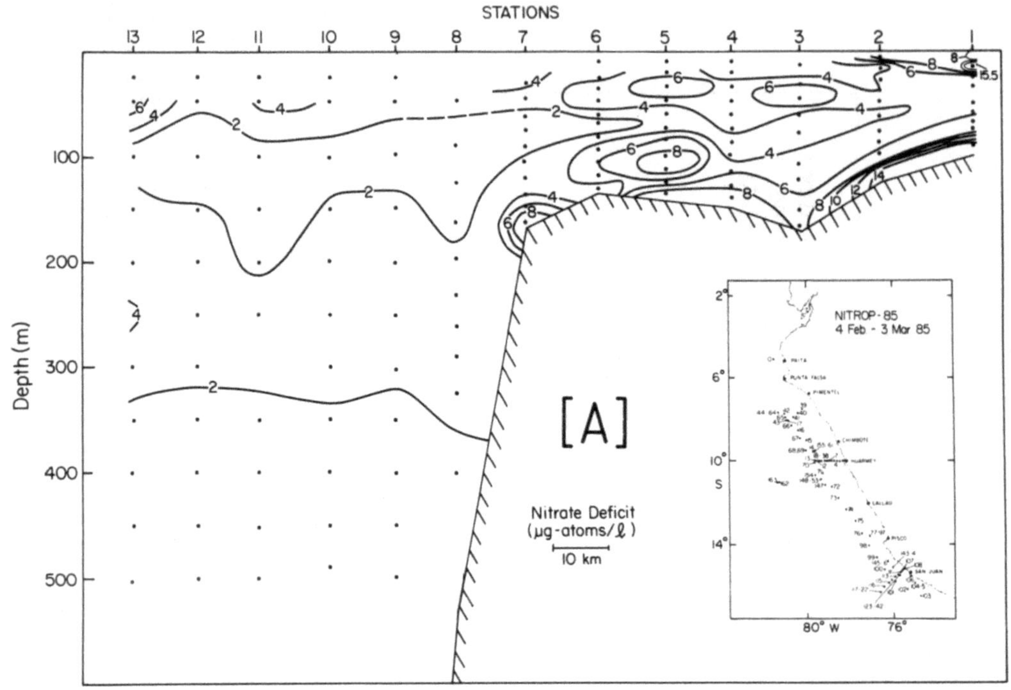

Figure 14a: Nitrate deficit sections at 10° (A) and ~15° (B) during February 1985. The inset shows the station locations. Figure 14B is from Codispoti et al., (1988).

shelf and slope at 5° during observations made in 1977. Consequently, one explanation for these minima is that they signal the "core" of the Undercurrent. Given the three-layered upwelling circulation with onshore flow at mid-depths, nitrite uptake in the photic zone which is about 20m deep, and accumulation of organic material on the bottom, part of this distribution may be attributed to the upwelling circulation. However, one might expect to find higher nitrite concentrations in the bottom layer at the outer shelf if the cross-circulation and local biological processes were the only operative factors. The relationship between poleward velocity and chemical and optical properties is an interesting problem that one can address with existing technology since moored sampling devices exist for acquiring nitrate and nitrite samples (Friederich et al., 1986) as well as current velocity and light transmission data.

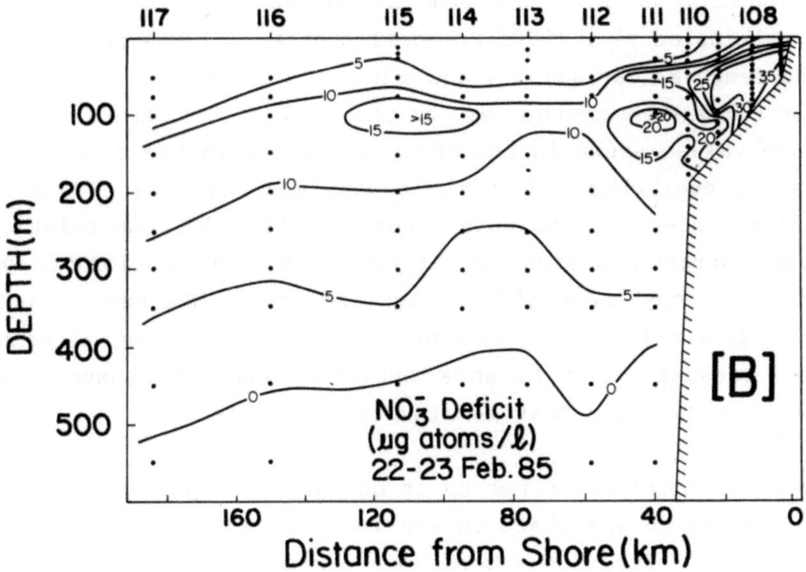

Figure 14b:

Rates of Denitrification and Nitrous Oxide Consumption Within the Undercurrent

If we consider the region between 10 and 25°S, the volume of the Undercurrent, as we have defined it (200 km wide by ~250 m deep), is 8×10^{13} m ^{3} or ~<0.01% of the total oceanic volume. Codispoti and Packard (1980) have estimated the denitrification within 175 km of shore in this region to be about 20 Tg N yr^{-1} and most of this occurs within the depth range of the Undercurrent since respiration rates tend to decrease exponentially with depth (e.g., Packard et al., 1983). Since estimates for the total marine denitrification rate range from ~50 to ~120 Tg N yr^{-1} (Hattori, 1983; Liu, 1979; Codispoti and Christensen, 1985), it is clear that the Undercurrent has an importance to the oceanic combined nitrogen budget that is immense when compared to the small volume of water that is involved. Similarly, Codispoti and Christensen (1985) have suggested that 0.4 Tg N yr^{-1} of nitrous oxide is consumed by denitrification in the ESPO and, once again, most of this consumption might occur within the Undercurrent as we have defined it. Since the rate of atmospheric increase in nitrous oxide is ~2.5 Tg N yr^{-1}, the Poleward Undercurrent may also be an important factor in the global nitrous oxide distribution.

Existing nitrous oxide data from the Poleward Undercurrent (Figures 13 and 15, Elkins et al., 1978; Friederich et al., 1985) reveal extreme spatial and temporal variability and contain both extremely high and low values when compared with the rest of the ocean. For example, the high near-surface values shown in the inset in Figure 15 are amongst the 10 highest values ever observed in the ocean, and less than 100 m below, extremely low concentrations were observed. Clearly, the relationship between Undercurrent dynamics and nitrous oxide distribution off Peru is a question that may have global significance. Whether or not the Undercurrent is a major source or sink for nitrous oxide or, alternately, be neither because of a balance between enhanced consumption and production is a question that remains to be settled.

Since the Undercurrent per se is defined as a poleward advection, it is interesting to speculate on its importance as an exporter of the products of denitrification and nitrous oxide consumption relative to the total exports which would include diffusion and vertical advection. Taking our simplified definition of the Undercurrent (200 km wide, 250 m deep) and assuming an average net poleward advection of 5-10 cm s^{-1} (Enfield et al., 1978; Fahrbach et al., 1981) gives an average poleward transport of ~2.5-5 Sv. Nitrate deficits within the Undercurrent at the "downstream" end of the MSNM are ~10 M (Figures 12 and 14b) (=140 mg N m^{-3}) and multiplying gives an annual export of nitrate deficit by the Undercurrent of 10-20 Tg N yr^{-1}. Since nitrate deficit estimates the free nitrogen produced by denitrification, this is an approximate estimate of the free nitrogen produced by denitrification that is exported by poleward advection in the Undercurrent. Since the total marine denitrification rate is ~50-150 Tg N yr^{-1} (see above), this calculation suggests that horizontal Undercurrent advection may be a globally important term in calculating denitrification rates.

Importance of the Vertical Advection

Since the waters that upwell within the Rossby radius of deformation in the ESPO come from a depth of ~50-100 m (e.g., Wyrtki, 1963), it is generally agreed that the Undercurrent is the major source of upwelling water. The volume transport of upwelling between 6-24°S has been estimated as ~3.3 Sv (Wyrtki, 1963). If we take 3 Sv to be the upwelling rate between 10-24°S and assume an average nitrate deficit of 5 μM in the ascending waters, an annual nitrate deficit loss due to coastal upwelling would be 7 Tg N yr^{-1} in agreement with a similar calculation by Tsunogai (1971).

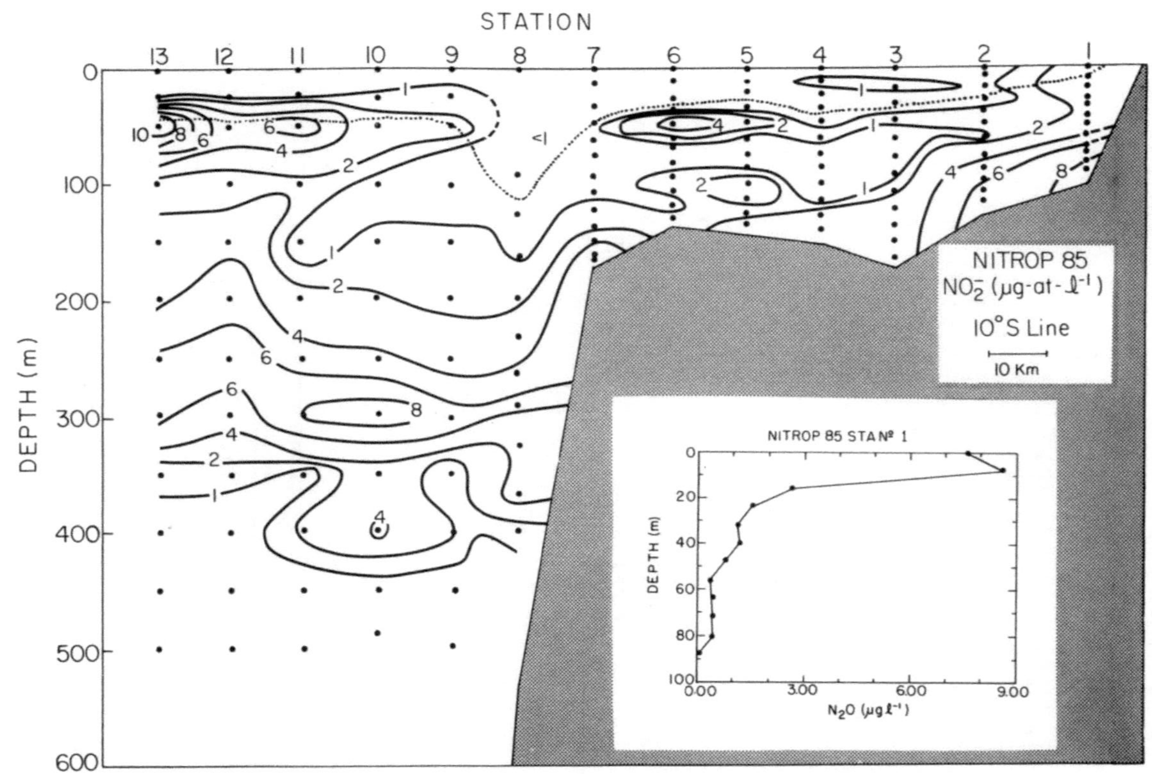

Figure 15: Nitrite concentrations at 10°S during February 1985. The inset shows the nitrous oxide levels distributions at the innermost stations. The high nitrous oxide levels are amoungst the ~10 highest values ever observed in the ocean, but ~50 m below in the oxygen deficient waters the concentrations are extremely low. These data are in line with the concept summarized in Figure 7 (from Codispoti et al., 1986).

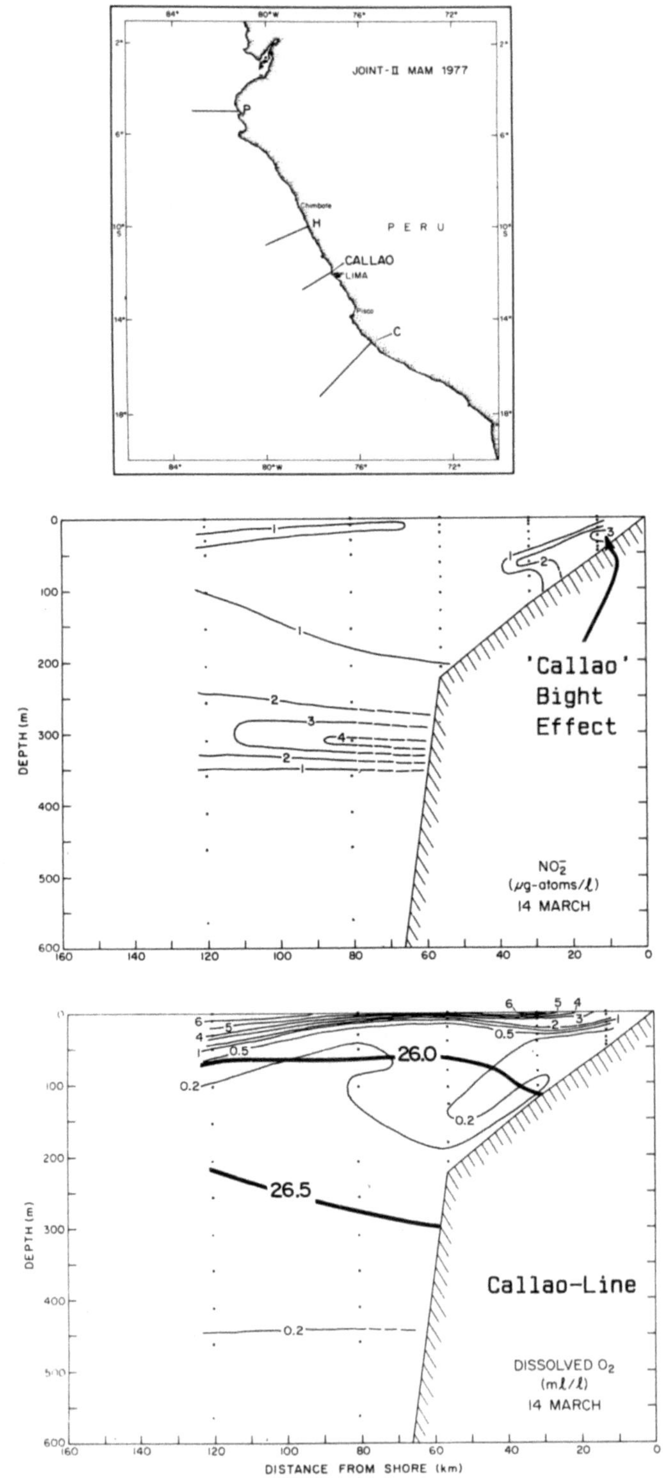

Figure 16: Nitrite and dissolved oxygen concentration on the "Callao Line" during March 1977. The thick lines show the 26.0 and 25.5 σ_t surfaces (re-drawn from Hafferty et al., 1979).

Wyrtki (1963) suggests that the downwelling rate through the 300 m surface is also ~3 Sv, and the nitrate deficits in these waters are also about 5 μM. Thus, the total loss of nitrate-deficit due to vertical advection within the undercurrent regime is ~14 Tg N yr^{-1} . This indicates that vertical advection associated with the Undercurrent waters is another important term for global denitrification.

Carbon Production Supported by Coastal Upwelling Vs. Denitrification

The upwelling waters supplied by the Undercurrent have a potential new production rate of ~200 Tg C yr^{-1} (Chavez and Barber, 1987; Chavez, et al., in press) which is vital to the maintenance of denitrification and nitrous oxide turnover rates close to the coast, but may also be important to the entire regional rate. For example, as indicated by the equations in Figure 6, the ratio of carbon to nitrogen in the denitrification process is approximately 1 by atoms or 0.85 by wt., so the new production supported by coastal upwelling is more than sufficient to support the entire regional denitrification rate. Available data do not permit an evaluation of what percentage of the regional denitrification rate is supported by organic matter production, arising from coastal upwelling (within the Rossby radius) versus production occurring further offshore. It has been suggested, however, that organic material sinking to the bottom over the shelf and outer slope could be transported into the offshore portion of the MSNM because of the net offshore transport in the bottom layer over the shelf and outer slope (Pak et al., 1980; Codispoti and Packard, 1980). This offshore transport should be higher when Undercurrent velocities are higher, and it is interesting to note that data from 1976 during a weak El Niño suggest that Undercurrent velocities were stronger than in 1977 (Codispoti and Packard, 1980; Enfield et al., 1978). It is possible, therefore, that the reduction in the flux of organic material to the Undercurrent due to decreased primary production during an El Niño is compensated to some extent by an increase in the offshore transport of organic matter that may accumulate over the shelf and slope when the Undercurrent is weaker. This is another question that is amenable to investigation with the current generation of moored instruments. Clearly, the organic matter supply and transport processes associated with Undercurrent dynamics is important to the oceanic denitrification rate. This is particularly so since only small changes in the carbon supply can cause large changes in denitrification (Table 1).

CONCLUSION

This paper is speculative and our description of the Undercurrent is certainly over-simplified. Our major intent has been to convince our physical oceanographic colleagues that the Poleward Undercurrent off Peru may be a major "actor" in the global nitrogen cycle, even though it is a dynamic "pipsqueak" in comparison to major currents such as the Gulf Stream. We will be encouraged if this paper engenders a greater appreciation of the relationship between studies of the Poleward Undercurrent in the ESPO and the issue of global habitability.

ACKNOWLEDGEMENTS

Our research in the eastern South Pacific has been supported financially by the National Science Foundation and the Office of Naval Research. We thank Jim Rollins for drafting assistance including the excellent job that he did with the block diagram (Figure 2). We are also indebted to P. Colby and A. Sanico for their skill in turning our rough drafts into readable type.

From: Dr. Theresa Paluszkiewicz, MMS Branch of Environmental Modeling, Reston VA

On: Review and commentary to paper **DO NITROGEN TRANSFORMATIONS IN THE POLEWARD UNDERCURRENT OFF PERU AND CHILE HAVE A GLOBALLY SIGNIFICANT INFLUENCE?"**, by L.A. Codispoti, R.T. Barber, and G.E. Friederich

GENERAL

This paper discusses the nitrogen cycle along the Eastern South Pacific Ocean (ESPO), its possible global significance, and the influence of the circulation characteristic to that region. The results summarized in this paper should be published because they add to the understanding of this complex region and challenge one to consider the importance of considering this region as an ecosystem with globally significant effects. The few comments and suggestions that I offer are meant to clarify the issues for the general reader and strengthen the case for the role of the undercurrent in this ecosystem.

ON THE ISSUE OF THE GLOBAL SIGNIFICANCE OF THE NITROGEN CYCLE IN THIS REGION

As this is a major point in this paper, it probably deserves a little more attention in the introduction. A reader unfamiliar with the nitrogen cycle literature might not be aware of how denitrification can be significant to the carbon dioxide concentrations and just what values are globally significant. The issue of global significance becomes clearer and convincing when it is discussed in more detail [later in the paper].

ON THE POLEWARD UNDERCURRENT AND REGIONAL OCEANOGRAPHY

The description of the undercurrent and the associated dynamics of the region [can be updated]. For example, there are recently published papers which strengthen the case for the influence of the Equatorial Undercurrent on the Peru Undercurrent and are pertinent to the work presented here. These papers are cited below:

Lukas, R. (1986). The Termination of the Equatorial Undercurrent in the Eastern Pacific. Prog. Oceanog., 16, 63-90. See Figure 14.

Rothstein, L.M. (1984). A Model of the Equatorial Sea Surface Temperature Field and Associated Circulation Dynamics. J. Phys. Oceanogr., 14, 1875-1892. This modeling study of the equatorial SST field and associated circulation generates a coastal undercurrent continuous with the EUC.

This section could be improved by a more detailed description of Figure 3 in the context of the ESPO system. For instance, the processes which govern vertical exchange of N are of particular interest to this paper. When is a simple sinking assumed? How are R and Kz parameterized to reflect the various processes which occur in this region? How does a system where there is considerable vertical shear (due to opposing currents in the vertical) effect the magnitude of the vertical exchange? Has this been explored or measured? Can the profiles that exist be used to approximate this term or will new measurements be needed? How will the spatial variability of this term affect the nitrogen cycle in this region?

Consider the difference in the upwelling at 5°S versus that at 10° or 15°S; i.e., regions of less upwelling versus more enhanced upwelling. The paper by Huyer, et al. (JGR, 92, 1987) clearly shows examples of this spatial variability. In fact, Barber and Huyer (Geophysical Research Letters, 6, 1979) indicate that the accumulation of organic material is not a simple passive accumulation of material from sinking from above, but that the variable mixing dynamics of this region are of importance. The interleaving structures found by Codispoti and Friederich may be further indications of the mixing dynamics in the region.

By including issues such as this it may make the "over-simplified" description "less-simplified," but it will point out the dynamics which may be a factor in the uniqueness of regions such as these to the global significance of the nitrogen cycle.

ON THE DISCUSSION

The discussion suggests that there are large temporal variations in the nitrogen cycle and budget that may be related to temporal variations in the strength, persistence or penetration of the EUC and its influence on the ESPO circulation. There is certainly evidence of the waning of the EUC during the 1982-83 El Nino (see Firing, et al., Science, 222, 1983). It would be interesting to see if there were any significant changes in the nitrogen cycle in the ESPO at this time, should you have any coincident data. Dr. Huyer may have hydrographic data during this period. This issue of the temporal and spatial variability of the ESPO circulation is germane also to other points discussed here.

Some useful information may also be gained from the previously cited reference on the variability of the upwelling; namely, that the temporal and spatial variability in the upwelling may be an important factor in the dynamics of this system. This may have added significance when one considers that the most noted case of this variability took place during the 1982-83 El Nino during times of other climatic imbalances.

There are additional references which would aid the discussion of the increased undercurrent velocities. An article by Smith (Science, 221, 1983) describes incidences of higher velocities in the undercurrent associated with the 1982-83 El Nino. In addition, there is a report "Current Measurements Off the West Coast of South America, November 1981-March 1985" (Knoll, et al., 1987, OSU Report 140, 87-31) which shows time series of velocity from current measurements in the undercurrent.

The association of stronger offshore flow associated with stronger alongshore flow may not be entirely correct. Examination of shelf currents at 50 and 100 m from the above mentioned report shows little evidence of the increased offshore flow; this may be due to the insensitivity of the location of the meters and rotation angle, for this particular question. It is difficult to support this statement with data because the location of the measurements may not have been chosen to examine this question. As this is an important question, but one that is not readily answered by available data, it may serve as a worthwhile hypothesis for further study.

AUTHORS' RESPONSE TO THE REVIEWER'S COMMENTS

Because of participation in an extended field program, it was impossible to revise our paper in accordance with the excellent suggestions of the reviewer, and an editorial decision was made to append her remarks to our paper. We agree with her comments, and think that reading the additional references that she cites, and considering some of the complexities that she mentions would be useful for the interested reader. In our defense, we will point out that our main goal was to pique the interest of physical oceanographers in a subject with which most of them are unfamiliar. Consequently, the paper was conceptualized as a lecture to interested non-specialists, and contains a significant amount of review material as well as some previously unpublished material. Glossing over some of the complexities is, of course, a time honored technique when giving introductory lectures, but it can lead to misconceptions which the addendum should help elminate.

TOPOGRAPHIC STRESS IN COASTAL CIRCULATION DYNAMICS

Greg Holloway
Institute of Ocean Sciences
Sidney, B.C., V8L 4B2 Canada

Ken Brink
Woods Hole Oceanographic Institution
Woods Hole, MA 02543

Dale Haidvogel
Chesapeake Bay Institute
Johns Hopkins University
The Rotunda
711 West 40th Street
Baltimore, MD 21211

ABSTRACT

The interaction of nonlinear shelf waves or eddies with longshore variation of topography is addressed. The dynamics are subtle and, perhaps for this reason, have been largely overlooked. Our own results are preliminary; but they point toward an unexpectedly strong influence such that, in many coastal environments, topographic stress may overwhelm other forces. The topographic stress acts to drive undercurrents in the sense of intrinsic shelf wave propagation, hence poleward on eastern boundaries. Potentially, a very important consideration is that a secondary circulation associated with this stress has the sense of upwelling on all boundaries in either hemisphere.

THE PROBLEM

A host of forces act on the coastal ocean, viz. steady and fluctuating wind, windstress curl, buoyancy sources/sinks, longshore pressure gradients, and encounters with open ocean eddies. No doubt this list could be extended. Many of the forces cannot be measured accurately and/or with adequate space-time resolution to obtain confident statistics. Response of the coastal ocean is correspondingly complicated and difficult to observe with sufficient resolution, whatever "sufficient" means! It is an understatement to acknowledge the difficulty to verify by observation a causal relation between forcing and response. Under these circumstances, one may not wish to consider yet another force acting on the coastal ocean. Nonetheless, that is our purpose.

We are concerned with the role of topographic form drag, i.e., the force arising from correlations of pressure fluctuations with bottom topographic slopes. In studies of global atmospheric angular momentum, these are the correlations that support mountain torque (Newton, 1971). A new result is to find that this "drag" really is not a drag. Rather it is a systematic stress with a definite tendency which, among other effects, drives poleward undercurrents on eastern boundaries. More generally the topographic form stress tends to drive currents in the sense of intrinsic shelf wave propagation, i.e., cyclonically with respect to ocean basins.

Not only does topographic stress appear to drive longshore flow, but a secondary circulation has the sense of upwelling on all boundaries in either hemisphere. With significant consequences in terms of biological productivity, regional climate, waste disposal, etc., as well as concern for large scale, longshore circulation, it would seem surprising that an effect such as this has gone unnoticed. The key question is the quantitative one: Is the process so weak as to be ineffective? It is probably too early to answer this question definitely, but our preliminary results are exciting or alarming, depending upon one's predispositions.

In the following sections we trace certain precedents provide a synopsis of a statistical, theoretical approach, then provide results from numerical experimentation.

PREMONITIONS AND EXTREMAL CONJECTURES

We will try to show that there are three necessary ingredients:

a. A large scale gradient of potential vorticity.
b. Topographic variation along the large scale contours of potential vorticity.
c. Nonlinearity.

These ingredients were present in early ocean circulation models under simplifying assumptions of quasigeostrophy and ß-plane with small amplitude bottom roughness. Among the results of such modeling was the observation by Bretherton and Karweit (1975) that topographic interaction tended to propel westward flow. Flow around individual topographic features tended also to mean circulation of anticyclonic sense about elevations, and cyclonic about depressions. Thus, the flow everywhere tended "westward" in the deformed f/H contours, as discussed also by Holloway and Hendershott (1974).

Interpreting these numerical experimental results, two quite different extremal conjectures were advanced. Bretherton and Haidvogel (1976, hereafter BH) proposed that such flows approach a condition of minimum total enstrophy at given total energy. Let the overall depth be written $H = H_o\left(1-h(\underline{x})/f\right)$ where H_o is mean depth, h/f is fractional elevation of the bottom above the mean depth, $\underline{x} = (x,y)$ is spatial coordinate with x east and y north, and f is Coriolis parameter. In a barotropic, quasigeostrophic flow with rigid upper lid, motion is defined by a velocity streamfunction $\underline{u} = z \times \underline{\nabla}\psi$ where $\underline{u} = (u,v)$, z is unit vertical and $\underline{\nabla} = (\partial_x , \partial_y)$. Omitting forcing and dissipation, and taking f constant the equation of motion is

$$\frac{D}{Dt}\left[\nabla^2 \psi + h\right] = 0 \tag{1}$$

with

$$\frac{D}{Dt} = \partial_t + J\left(\psi, \cdot\right)$$

and

$$J\left(A,B\right) = \partial_x A \partial_y B - \partial_y A \partial_x B.$$

In a closed domain or in a domain satisfying periodicity conditions on $\nabla^2\psi$ and h, integral invariants of (1) are energy E and total enstrophy Q, where

$$E = \frac{1}{2}\int dxdy \;|\underline{\nabla}\psi|^2 \tag{2}$$

and

$$Q = \int dxdy \left(\nabla^2\psi + h\right)^2 \tag{3}$$

In brief, the argument of BH is that turbulent shearing will draw out contours of potential vorticity $q = \nabla^2\psi + h$, producing steep gradients of q. If one hypothesizes potential vorticity mixing at short length scales, then q-variance (i.e., Q) will be dissipated. E is at larger scales and, moreover, will be retained at large scales by any reverse cascading tendency in 2D turbulence. The result is to suggest that the flow approach a state of miminum Q for given E. Maps of ψ calculated from this extremal method were found to approximate numerical experimental results also obtained by BH and to agree in sense with Bretherton and Karweit (1975) and Holloway and Hendershott (1974).

Extended to the ß-plane for which $q = \nabla^2\chi + h + \beta y$, BH obtained

$$\psi(\underline{x}) = \Sigma_{\underline{k}} \frac{\hat{h}_k e^{i\underline{k}\cdot\underline{x}}}{k^2 + \mu} + \frac{\beta y}{\mu} \tag{4}$$

where $h(\underset{\sim}{x}) = \sum \hat{h}_{\underset{\sim}{k}} e^{i\underset{\sim}{k}\cdot\underset{\sim}{x}}$ and μ is a positive constant which depends upon E. Interest in (4) is directed to the term $\beta y/\mu$ which provides a westward flow of magnitude $\beta y/\mu$. If f/H contours are dominated by the topography of a continental margin, the "westward" flow of BH is the poleward flow on an eastern boundary.

While the extremal principle of BH was advanced to explain these results, a very different extremal approach was given by Salmon et al. (1976, hereafter SHH). The problem considered was again restricted to quasigeostrophy, with one or two layers in the vertical. For the case of one layer on a ß-plane with bottom topography, $q = \nabla^2\psi + h + \beta y$ as previously. Without forcing or dissipation, integral invariants are E and Q, including contributions from βy in the latter. In a closed basin, boundary conditions may be taken as $\psi = 0$ along the boundary. SHH supposed that $\psi(\underset{\sim}{x}, t)$ is expanded upon a complete, orthogonal set of eigenfunctions

$$\psi = \sum_i \psi_i(t)\, \phi_i(\underset{\sim}{x}) \tag{5}$$

where

$$\nabla^2\phi_i + k_i^2\, \phi_i = 0$$

with $\phi_i = 0$ on the boundary. Truncating the sum in (5) at some large N, the vector $\{\psi_i\}$ describes the motion of a point in N-dimensional phase space. SHH then inquired what is the probability distribution of $\{\psi_i\}$ for fixed expectations of E and Q. The extremal solution is that which maximizes entropy subject to E and Q. For the ith expansion coefficient, SHH (eqn. 4.3) obtains

$$P_i(\psi_i) = \left(\frac{a_i + a_2 k_i^2}{\pi}\right)^{1/2} \exp\left\{-\left(a_1 + a_2 k_i^2\right) k_i \left(\psi_i - \langle\psi_i\rangle\right)^2\right\} \tag{6}$$

with

$$\langle\psi_i\rangle = \frac{a_2 k_i^2}{a_1 + a_2 k_i^2}\, h_i \tag{7}$$

where h_i is the expansion coefficient of $h(\underset{\sim}{x}) + \beta y$ and a_1 and a_2 are determined by E and Q.

The solution (7) in fact contains (4) in the limit that one choose Q at its minimum for given E, in which case a_1 and a_2 collapse to the single parameter μ. In common with (4), (7) expresses a westward flow $\langle\psi\rangle \cong \frac{a_2}{a_1}\, \beta y$ which is valid for all Q including the case where Q is at its minimum. (At unrealistically large Q for given E, the sign of a_2/a_1 reverses to give eastward flow). The closed basin solutions include eastward return flows in narrow boundary currents. More recent investigation by Carnevale and Frederiksen (1987) confirms the maximum entropy result that mean flow is westward, relating this also to nonlinear stability theory. In re-entrant ß-channel geometry, the mean is westward at all latitudes without the eastward return flows required by closed basin geometry.

One is left in a quandary that is not resolved to date. Two quite different extremal conjectures have been advanced: that flows tend either to minimize enstrophy or to maximize entropy. A question remains as to which is "right," with some uncertainty what constitutes "right." For present purposes we may leave this question open, noting only that each conjecture leads to a tendency for westward flow on a ß-plane corresponding to poleward flow on an eastern boundary where topography controls the gradient of f/h.

MECHANISM: ASYMMETRICAL FORM DRAG

In their interpretation of model results, BK recognized the important role of asymmetrical topographic form drag. When a flow was forced toward the east on the ß-plane, large topographic form drag

$$D_x = -\frac{H_o}{f}\int_A dA\; p\partial_x h \qquad (8)$$

arose, where p is pressure at the bottom and $\int_A dA$ is an integral over a fluid region of interest. When a flow was toward the west, BK remarked that the form drag vanished. This asymmetry is due to the possibility of stationary Rossby in case of eastward flow, whereas only freely propagating Rossby waves were possible in the westward flow. Qualitatively, these considerations appeared to support the numerical experimental result that westward flow was "favored."

Analytical results for the form drag due to an isolated topographic bump on a ß-plane were obtained by Johnson (1977) for cases of uniform upstream flow from the west or from the east. In cases of flow from the west, Rossby wave radiation of energy resulted in drag which could be expressed in terms of length and height of the bump and amplitude of the upstream flow. No drag resulted in cases of flow from the east. Johnson's results thus supported the account given by BK. Although the analytical results were available only for cases of steady flow with uniform upstream conditions and for bump heights of such small amplitude that closed recirculation above the bump would not occur, Clarke (1982) has argued that Johnson's drag is broadly consistent with inferred topographic resistance to the eastward flowing Antarctic Circumpolar Current.

Brink (1986) considers the theoretical analysis of topographic drag on a continental shelf and slope, extending previous analyses which were limited to quasi-geostrophic dynamics on the ß-plane. Steady, inviscid, barotropic motion is assumed while the topography is permitted to have small amplitude departure from a mean continental margin which is uniform in the longshore coordinate. Except for a small amplitude restriction, the topography is permitted to have extensive spatial structure with the provision (for analytical convenience) that topographic variations vanish beyond some large distance from the region of interest.

Brink's procedure is to expand perturbatively in the amplitude of the topographic variation, scaled by a small quantity ε . At zeroth order the steady flow is directed alongshelf with arbitrary shear in the cross-shelf direction. To first order in ε, the transport streamfunction perturbation is expanded on eigenfunction bases which depend upon the zeroth order flow and mean continental margin profile. The topographic drag (8) can then be evaluated at order ε^2, resulting in Brink's eqn. 2.26. For details of this calculation, one should refer to Brink's article. Importantly, the drag is expressed in terms of integral statistics of the topographic variations.

Brink identifies three possibilities:

a. Zeroth order flow is in the direction of free shelf wave propagation. Drag vanishes.

b. Zeroth order flow opposes free shelf wave propagation and is slower than the first mode intrinsic phase speed. This is the situation which supports non-zero drag due to standing shelf lee waves.

c. Zeroth order flow opposes free shelf wave propagation with amplitude greater than the maximum intrinsic phase speed. Termed "supersonic" by Brink, drag vanishes in this case since the mean flow is so swift that all perturbations are swept downstream.

Using a gridded topographic data set off Oregon (Peffley and O'Brien, 1976) and assuming a southward zeroth order flow with values from 6.3 cm s^{-1} inshore to 2.2 cm s^{-1} offshore, Brink (1986, Figure 6) obtains the topographic drag shown here as Figure 1. Averaged across the shelf, this stress is 3.36 dyne cm^{-2}. Thus Brink remarks that if one were to ignore questions of validity of the perturbation expansion and use actual, unscaled topography ($\varepsilon = 1$), the apparent northward stress of 3.36 dyne cm^{-2} would overwhelm the coastal momentum balance. However, validity of the perturbation expansion is in doubt and the assumption of steady flow is not satisfied. As well, Brink remarks that the actual near bottom flow is persistently northward (Kundu and Allen, 1976) which, if assumed for the sense of the zeroth order flow, would result in no topographic drag at all, corresponding to case a. The important point here is to realize that a topographic drag can arise with, possibly, quite large values. To go further than Brink's analysis requires numerical simulation such as reported by Haidvogel and Brink (1986, hereafter HB).

HB treats the fully nonlinear, time-dependent problem of barotropic flow above a continental margin with finite amplitude topographic irregularities. A flat rigid upper lid is assumed. In coordinates such that x is alongshelf and y is cross-shelf, HB solves numerically for evolution of transport streamfunction

$$\left(-\partial_y\phi,\ \partial_x\phi\right) = \left(Hu,\ Hv\right) \tag{9}$$

with H given by

$$H = H_o \exp\left\{\alpha y + \sin\frac{\pi y}{W}\cdot\Sigma d_m \sin\left(k_m x + \theta_m\right)\right\} \tag{10}$$

The flow is in a channel of width W bounded by straight free-slip sidewalls. H and ϕ are assumed to be periodic over some x-interval L.

The study by HB is motivated by a thought experiment based upon the topographic drag analysis of Brink. It was seen that mean flow opposed to the sense of shelf wave propagation generates large drag due to lee wave formation, whereas mean flow in the sense of shelf wave propagation encounters no topographic drag. Suppose then that an x-directed wind stress is applied which is periodic in time about zero mean stress. Although the alongshelf wind stress is symmetric with respect to periods of positive and negative x-driving, the resulting flow will encounter strongly asymmetrical topographic drag and result in a time averaged flow

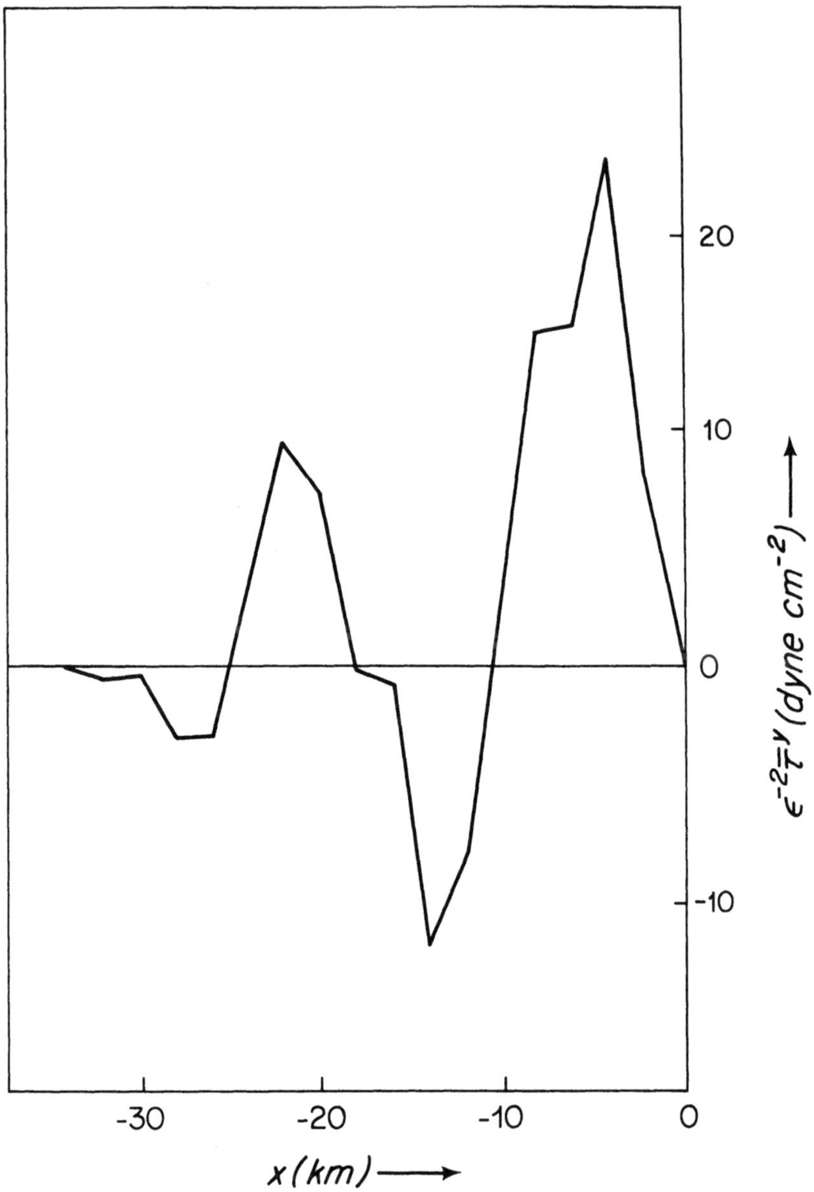

Figure 1: Bottom stress as a function of cross-shelf distance for a case of topography resembling the Oregon shelf.

directed toward positive x (the direction of intrinsic wave propagation in the coordinate scheme of HB).

Numerical simulation results of HB tend to bear out the thought experiment over a range of parameters including wavelength and amplitude

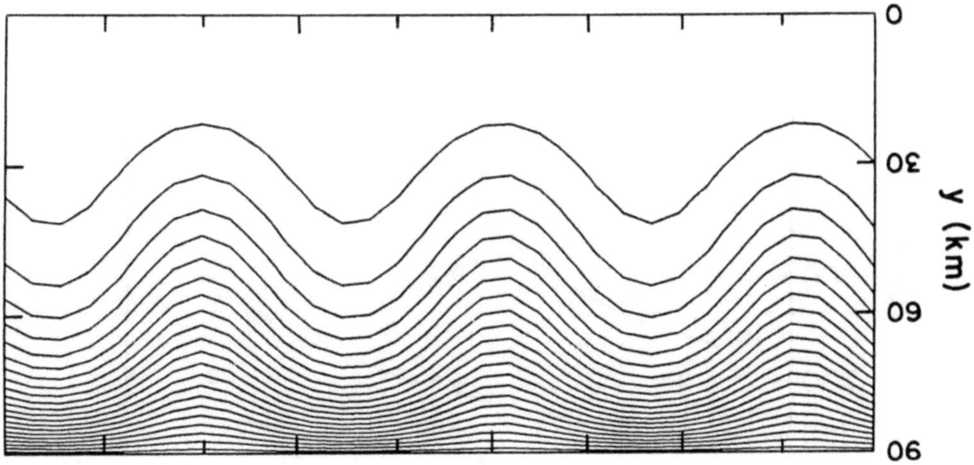

Figure 2a: Bottom topographic contours for the central experiment studied by HB. The contour interval is 100 meters.

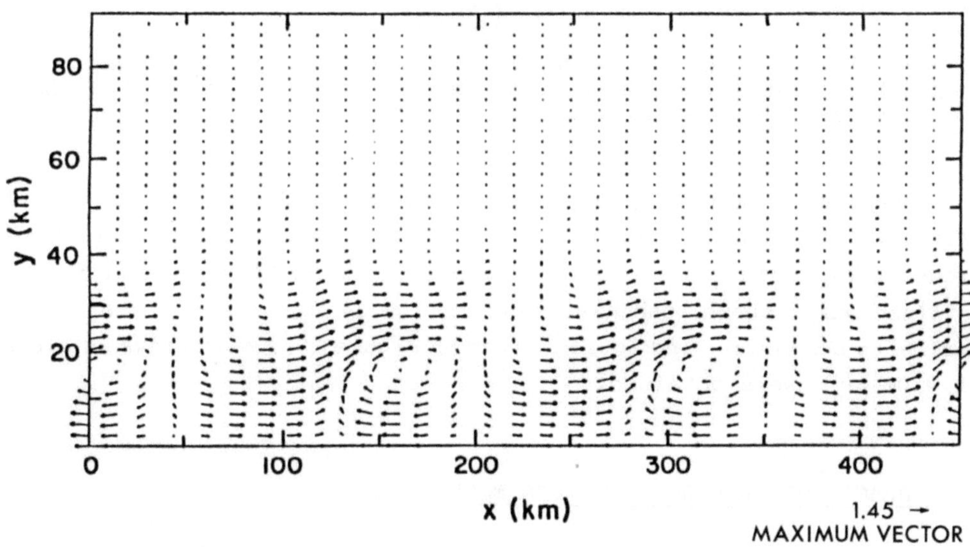

Figure 2b: Time-mean Eulerian Velocity vectors for the central experiment of HB. The maximum vector length is 1.45 cm/sec. The circulation is forced by a spatially broad, along-channel wind stress which oscillates in time with a ten day period and an amplitude of 1 dyne/cm**2.

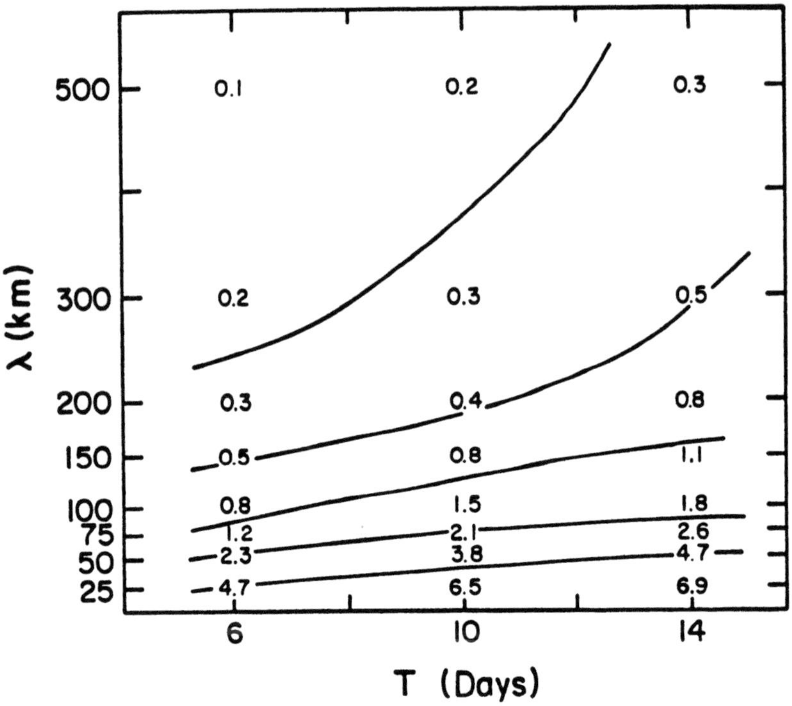

Figure 3: Maximum time and zonally averaged along-channel currents (9 cm/sec) observed by HB, as a function of bottom bump wavelength (km) and wind stress periods (days).

of topographic variation, strength and period of wind stress forcing, and coefficient of linear bottom friction. Isobaths of topography and a time averaged flow field are shown in Figure 2 (from HB Figure 2 and 5). In the illustration shown, the maximum mean flow is 1.45 cm s^{-1}. For other experiments summarized in Figure 3 (from HB Figure 10), mean flows as large as 6.9 cm s^{-1} were obtained.

WHEN A DRAG'S NOT A DRAG: NEPTUNE EFFECT

It is clear from the numerical experiments just described that substantial mean flows can result from zero mean but time-periodic applied stress. What is not so clear is that this is due to asymmetrical topographic drag. In particular, cases where the mean flow is persistently in the sense of shelf wave propagation should be characterized by absence of topographic drag. The question is re-opened: What drives such flows?

A different theoretical approach was developed by Holloway (1986, 1987, hereafter H86 H87) following analytical methods from turbulence theory. The cumbersome calculation is made tractable by returning to barotropic quasigeostrophic flow on a ß-plane. Nonlinear time-dependent motion is considered in the presence of a spectrum of statistically homogeneous topographic variation $H_{\underline{k}} = \left|h_{\underline{k}}\right|^2$. Relative vorticity $\zeta(\underline{x},t) = \hat{z}\cdot\underline{\nabla}\times\underline{u}$ is likewise expanded in a wavenumber spectrum $Z_{\underline{k}} = \left|\zeta_{\underline{k}}\right|^2$. The problem is posed as follows:

Given a spectrum H_k, given a statistical distribution of wind stress torques as well as any mean zonal (along-shelf) wind stress, and given values for ß and for frictional parameters, one seeks the statistical expectation, for vorticity-topography cross-spectra

$$C_{\underline{k}} = R_{\underline{k}} + iI_{\underline{k}} = \xi_{\underline{k}}^* h_{\underline{k}}$$

where asterisk denotes complex conjugation, and for a component of spatially uniform zonal (along-shelf) flow.

Due to nonlinearity of the equation of motion, equations for the evolution of second moment quantities Z_k and C_k involve unknown third order correlations $\zeta\zeta\zeta$, $\zeta\zeta h$, ζhh, while evolution of third and higher order correlations involve successively higher order correlations. This unclosed hierarchy is then "closed" by reference to an entropy argument related to eqn. (6): it is required that in the absence of any external forcing or dissipation, flow statistics tend toward maximum entropy. This is accomplished at the level of fourth cumulant statistics, i.e., the portion of fourth order correlations which represent departures from random phase configurations. It is assumed that fourth cumulants have the effect of providing a linear relaxation of third order correlations. The result (H87, eqn. 16) is

$$C_{\underline{k}} = \frac{ik_x UH_{\underline{k}} - \gamma_{\underline{k}} H_{\underline{k}}}{-i\omega_{\underline{k}} + \eta_{\underline{k}} + \nu_k} \tag{11}$$

where k_x is the x-component wavenumber, $\gamma_{\underline{k}}$ and $\eta_{\underline{k}}$ are expressed (H87, eqn. 12 and 13) in terms of weighted sums over all wavevectors of the vorticity and topography variance of the vorticity and topography variance spectra, ν_k is the wavevector representaton of any explicit dissipation law, and $\omega_{\underline{k}} = k_x\left(U - ß/k^2\right)$ is the Doppler shifted Rossby wave frequency.

The importance of C_k is that the total x-directed force F which the topography exerts upon the flow can be expressed in terms of the imaginary part of C_k as $F = \Sigma_{\underline{k}} F_{\underline{k}} = -\Sigma_{\underline{k}} k_x k^{-2} I_{\underline{k}}$. From (11)

$$F_k = -\frac{\eta_k + \nu_k}{\omega_k^2 + \left(\eta_k + \nu_k\right)^2}\frac{k_x^2}{k^2} UH_{\underline{k}} + \frac{\omega_k k_x}{\omega_k^2 + \left(\eta_k + \nu_k\right)^2}\frac{\gamma_k}{k^2} H_{\underline{k}} \tag{12}$$

The first term on the right side of (12) is an asymmetrical drag, i.e., opposed in sense to U. (Both η_k and γ_k are positive.) However, the derivation is not restricted to steady ideal flow nor to lowest order in an amplitude expansion nor to cases where recirculation above bumps does not occur. More significantly, the result differs from previous analyses insofar as drag arises on either sign of U and has no large $|U|$ cutoff. The drag is asymmetrical in that flow in the sense of intrinsic wave propagation, i.e., $U < 0$ in β-plane geometry, results in large $\omega_k^2 = 0$ thereby suppressing contribution to F_k. Flow opposed to intrinsic propagation, i.e., $U > 0$, admits large contributions near $k^2 = \beta/U$ for which $\omega_k^2 = 0$. It is noteworthy that the drag contributions exist for all values of U even if one assumes inviscid flow for which $\nu_k = 0$.

Although topographic drag according to the first term on the right side of (12) differs from the results of Johnson or of Brink, a qualitative thought experiment such as suggested by HB is not sensitive to these differences. This leaves open the question whether asymmetry of drag is sufficient to account for rectified flow. Especially one should be puzzled if flow is persistently in the sense of intrinsic wave propagation, for which it was previously argued that drag vanishes and it is now argued that the drag does not vanish but in fact propels the flow when eddy activity is present.

An answer, perhaps, and a novel effect is given by the second term on the right side of (12). This term takes the sign of $\omega_k k_x$ and so, in particular, for $U < 0$ this term tends to further accelerate U negatively. Apparently the two terms in (12) may oppose each other. It is suggested in H87 that the net force can be estimated, albeit roughly, if $\eta_k \approx \gamma_k$ and ν_k plays a minor role. Then the first term acts to eliminate the Doppler shift from $\omega_k k_x$ in the second term so that the net force takes the sense of the intrinsic phase propagation. A more quantitative evaluation of η_k and of γ_k is required to determine the net force quantitatively. However, H87 provides estimates that overall F could correspond to stress of order 1 dyne cm^{-2} in many "typical" coastal environments.

The surprise in this section is to see that topographic stress may not be a drag at all but rather tends to be a force of definite sign, namely that which acts in the sense of intrinsic shelf wave propagation. There is no obvious name for the phenomenon leading to the second term in (12). Following a cartoon in H86, we call it the 'Neptune effect.'

A mathematical result (12) which runs against intuition begs for some physical interpretation. H87 offers the following account:

a. There is a tendency for relative vorticity to correlate with topographic variation in the sense (N. hemisphere) for negative vorticity above topographic elevation. This tendency was seen in numerical experiments by BK and by HH and has been discussed theoretically by BH and by SHH. One can also say in a loose way that if fluid columns were to relocate "randomly" then a column which finds itself over a local elevation will be foreshortened and hence, on average, exhibit negative vorticity.

b. A geostrophic tendency, which need not be complete, will associate positive pressure anomaly with negative vorticity (N. hemisphere) and therefore, with a., tend to associate positive pressure anomaly on geopotential surfaces with positive elevation of topographic variation (in either hemisphere).

c. Larger scale gradient of potential vorticity, whether by β or by gradient of f/h on continental margin, will induce a tendency for vorticity and pressure anomalies to propagate in the sense of Rossby or shelf wave propagation. Nonlinearity and topographic variations tend to scatter the propagating waves.

d. Competition between tendency b. to establish pressure anomaly above elevation and c. to propagate as waves until scattered leads to an equilibrium tendency for positive pressure to lie on the side of wave propagation relative to a topographic bump. Negative pressure anomaly tends similarly to lie on the side of wave propagation relative to topographic depression. The resulting pressure-slope correlation (in either hemisphere) supports a transfer of momentum from solid earth to fluid which has the same sense as Rossby or shelf wave propagation.

Finally, in case the physical account is no more compelling than the mathematical one, numerical experiments are also reported by H87 as shown here in Figure 4. A steady mean stress $\hat{U}$ is applied in a number of cases ranging from poleward to equatorward stress (in eastern boundary geometry). As well, random wind stress torques (with zero mean momentum) are applied in two of three cases for each choice of $\hat{U}$. The fluid is of nominal depth 300 m and the bottom is characterized by random topographic variations of rms height 30 m. Dashed and dash-dot curves show the response of mean U to increasing amplitudes of random wind stress

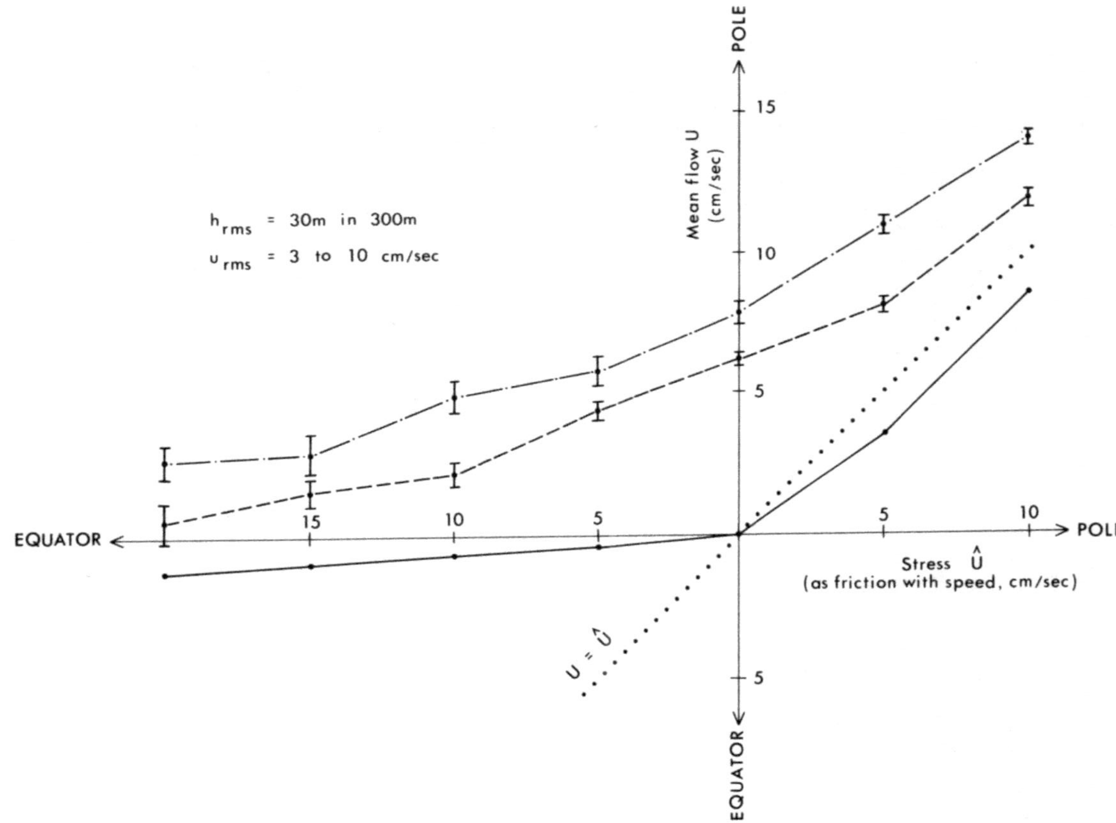

Figure 4: Mean (area averaged) flow U is plotted against mean applied stress $\hat{U}$. A dotted curve shows the purely frictional solution $U = \hat{U}$. When topographic roughness of r.m.s. amplitude 30 m with a reference depth of 300 m is included, U departs from $\hat{U}$. A solid curve shows the response when only the mean stress $\hat{U}$ is applied. Dashed and dash-dot curves show the response when random wind stress torques of different r.m.s. amplitudes are also present. Under the influence of random torques, U is unsteady. Vertical bars show one standard deviation of fluctuating U about its time averaged value.

torques. Of particular interest are experiments in which $\hat{U}$ is equatorward with a strength that would produce equatorward U of 5 or 10 cm s^{-1} in absence of topography. When wind stress torques are sufficient to maintain eddy energy levels corresponding to several cm s^{-1} rms speeds, mean U is poleward at some few to several cm s^{-1}. Values of U fluctuate in time and vertical bars shows one standard deviation from time averaged U. Although these mean currents are persistently poleward, the net topographic stress is also poleward resisting both bottom friction and applied $\hat{U}$. Indeed, eddy energy is being withdrawn by working against the mean applied stress $\hat{U}$.

Results presented in this section and in the preceding section are so recent that they should be regarded tentatively. Indications are that flow interaction with longshore variation of topography may indeed have very substantial consequences. Much closer investigation is warranted.

SECONDARY CIRCULATIONS AND ROLE OF BAROCLINICITY

To this point both theories and numerical experiments have been based upon barotropic models. (Not quite. Johnson (1977) included effects of stratification). In this final section we address two further aspects.

First, even within the context of barotropic theory, we may observe that there is an important secondary recirculation within the vertical planes normal to the coastline. At (11) we calculated a longshore momentum source $F = -\sum k_x k^{-2} I_{\underline{k}}$ due to pressure-bottom slope correlations. Let us now consider the correlation of onshore velocity $v = \partial_x \psi$ (in the notation of H86 and H87) with topographic elevation h. That is

$$\overline{vh} = \mathrm{Re}\sum\left(ik_x \psi_{\underline{k}}\right)^* h_{\underline{k}} = -\sum k_x k^{-2} I_{\underline{k}} \tag{13}$$

Contributions to the sum in (13) are just those already evaluated at (12). Arguments following (12) indicated a strong tendency for negative sign, hence for $\overline{vh} < 0$, i.e., onshore motion tends to occur above negative topography such as canyons or other depressions. If, for example, we suppose a situation with no longshore gradient of mean pressure, then we expect no average off/onshore motion of water in a nearly geostrophic portion of water column. However, within a depth range spanned by the range of topographic variation about the mean depth, longshore pressure differences are supported by the topography. It is within this depth range that negative $\overline{vh}$ supports net onshore (upslope) transport due to Neptune effect.

A return transport is required to avoid net onshore mass transfer—else Coriolis reaction to such transfer would need to be included in the longshore momentum budget. Three return paths are available. There may be an adverse mean pressure gradient $\overline{\partial_x p} \cong F < 0$, supporting net offshore geostrophic transport in the water column above depths spanned by topographic variation. A mean adverse wind stress may drive an offshore transport in a surface Ekman layer, thereby closing an upwelling cell. Thirdly, the mean flow may be driven so strongly in the direction of shelf wave propagation that a benthic Ekman layer supports the offshore transport. The proportions in which these paths are used will depend upon external forcing of wind stress and longshore mean pressure gradients. The key remark here is that the transport associated with the topographic interaction has a definite preferred sense, favoring onshore motion or upwelling of the deeper water.

Discussion thus far has not addressed truly baroclinic effects due to stable stratification. In a qualitative way, one anticipates stratification shielding much of the water column from the influence of topography. Thus equatorward wind stress at an eastern boundary might drive equatorward surface circulation while topographic stress drives a poleward undercurrent. Exploration of these circumstances would appear to be the natural province of numerical modeling. Very preliminary stratified simulations, similar to those described by HB but utilizing a primitive equation coastal model, appear to support these conjectures. We hope to report these calculations in the near future.

ACKNOWLEDGEMENTS

We are grateful to Billie Mathias for typing and to Patricia Kimber for assistance with the figures. We appreciate support received from the Office of Naval Research (contracts N00014-C-85-0440, -87-G-0262, and -87-K-0092.)

PART IV:

EPILOGUE

EPILOGUE

We sent to each workshop participant a set of questions about Eastern Boundary Poleward Undercurrents (EBPUC). From their answers we assembled a consensus of scientific opinion on the nature of flows along eastern ocean boundaries.

1. WHAT DOES YOUR OWN STUDY SUGGEST AS THE APPROPRIATE SPATIAL-TEMPORAL SCALES OF THE EBPUC?

Rationale: Smith posed the thesis that the EBPUC must have some kind of persistence - spatial, temporal or both; Mittelstaedt suggests that the EBPUC of the North Atlantic is "connected" from Africa to Norway, say 15 to 65°N; Neshyba pointed out that the salinity-oxygen characteristic of the Peru-Chile EBPUC is traced through more than a dozen cross-shelf hydrographic sections from 10 to 48°S, from a data set spanning many years; Yamazaki and Lueck state that the time scale of dissipation of total mechanical energy in the California Undercurrent compares to its transit time from 33 to 50°N; Freeland finds the North Pacific EBPUC is persistent at Vancouver Island, and shows a similar generic feature in the Canadian Arctic Basin.

Opinion: **HUTHNANCE.** There are several aspects to the notion of spatial persistence or continuity: (i) transport volume, (ii) advection of identifiable water, (iii) propagation of slope current fluctuations, (iv) cross-slope structure.

For quasi-geostrophic flow along depth contours, (iv) is determined by (i) in detail between individual depth contours so that the persistence is similar. The persistence of (ii) may be much less, however, if there is strong lateral exchange but the volume is maintained by large scale driving. Persistence of (iii) may be different again, being the frictional decay distance for coastal trapped waves (Huthnance 1987) which also corresponds to the along-shelf evolution/persistence scale for establishing (i) or (iv) after a barrier is encountered. From 48 to 60°N in the Atlantic, (i) is quite well conserved along the slope (with values) between 1.2 and 2.2 Sv (Huthnance 1986), but rapidly increases NE of the Wyville-Thomson Ridge as upper layer flows from across the Ridge and Faeroe-Shetland Channel recirculation are added. Thus the implied greater than 1000 km scale south of 60°N is upset by the Ridge. Statistics for (ii) are poor but water characteristics and a few drogue tracks suggest 500 km or so from Rockall Trough northwards. Evidence regarding (iii) comes from our CONtinental SLope EXperiment (CONSLEX) in winter 1982-3; along-shore coherences are significant, but not good in general, up to 500 km at periods of two days, but are upset by the Wyville-Thomson Ridge intersecting the slope; coherences are poorer at shorter periods except actually on the shelf where the scale exceeds 500 km. I hesitate to comment on (iv) with the poor cross-slope resolution hitherto.

McCLIMANS. I can say that the Norwegian Coastal Current (NCC) and the Norwegian Current (NC) have seasonal and mesoscale variability. The NCC exhibits a more direct response to variable winds, primarily in the Skagerrak near its source, but also along the coast, especially with gale or stronger winds and transport less than 1/4 Sv. Pressure fluctuations over Scandinavia make the Baltic Sea behave like a bellows, pressing out large amounts of buoyant water during the fall and winter months.

MITTELSTAEDT. The climatological mean boundary flow along the Eastern Atlantic Margin is permanent and exhibits a pronounced seasonal signal.

HICKEY. Data from both Washington and the Southern California Bight almost always clearly indicate a poleward undercurrent (as defined by a subsurface maximum in poleward flow) over the slope during summer and early fall. I feel confident that the Undercurrent is a very large scale feature of about 4000 km length whose principal time scale is seasonal. If we could remove smaller spatial scales and shorter time scales, the Undercurrent would emerge as a continuous longshore feature.

MOOERS. The seasonal aspects of the EBPUCs need to be determined and related to the seasonal aspects of coastal upwelling; eastern boundary currents per se, including planetary Rossby waves; and related mesoscale variablility.

2. IS THERE A "NORMAL" PATTERN TO THE EBPUC?

Rationale: Can you associate numbers with descriptive characteristics: width? thickness? relation to topography? steady-state? seasonal/ transient aspects?

Opinion: **HUTHNANCE.** The slope current west of Britain, 48 to 60°N, appears to be confined to the steep slope bordering the continental shelf (Booth and Ellett 1983; Huthnance 1986) and, therefore, has a width of 20 to 50 km. It appears to decrease downward but is essentially barotropic in the depths where it occurs, to 2000 m or more at 48°N decreasing to 1000 m or less at 60°N and beyond. It closely follows the depth contours along the slope. Although varying, it is steady in comparison with most obvious local atmospheric forcing, i.e., rarely reversing. There is some evidence for larger transports in winter (Gould, et al. 1985; Huthnance 1986).

BATTEEN. My model results suggest that the Undercurrent can change significantly if mesoscale eddies are present. My paper in this volume shows typical seasonal vertical (Figure 2a) and horizontal (Figure 2b) profiles and vertical cross-sections (Figure 8) of an equatorward coastal jet overlying a poleward undercurrent during the upwelling season and before the onset of eddies. These are used as initial conditions. After about 30 days, eddies and jets are generated, due to the baroclinic instability associated with the vertical shear between the coastal upwelling jet (CUJ) and the California Undercurrent (CU). Results show significant changes from the imposed initial conditions in the CUJ and CU fields due to the presence of eddies and jets. For example, the CU, instead of being a strictly northward flow, as prescribed in the initial conditions, now meanders around nearshore anticyclone eddies, resulting in offshore extensions of about 100 km. In the presence of cyclonic eddies, the CU flows poleward in a very narrowly confined region of 8 to 20 km inshore of the eddies.

McCLIMANS. In our work we rely heavily on models, from very simple dimensional arguments to 1, 2, and 3 dimensional numerical models and laboratory models. Laboratory models are quite valuable for studies of instabilities due to their high spatial resolution, but are limited to regional dynamics over which $\underline{f}$ is reasonably constant, about 1000 km. Remote driving mechanisms must be simulated as boundary conditions. This is satisfactory for the Norwegian Coastal Current and the Norwegian Current, but may prove to be a problem for the undercurrents if it turns out that they are part of the basin-wide dynamics.

HICKEY. Temporal scale: Off Washington and Southern California, the vertical shear associated with the Undercurrent has a seasonal cycle, such that the maximum poleward flow occurs in summer and early fall. The poleward flow itself at a depth of 100 to 200 m appears to have a semi-annual signal with maxima in summer and in winter. I believe that the second maximum, which is generally not associated with a strong subsurface maximum, may have a different driving mechanism than the first. Width: The off Washington appears to be about 20 km or about the width of the continental slope. In the California Bight, an undercurrent occurs across the entire Santa Monica basin, about 40 km, i.e., it does not appear to be trapped to one side or the other as might be expected. Thickness: This is about 200 m to the 5 cm s^{-1} contour. The depth of the subsurface maximum is shallower in the Southern California Bight, about 100 m, than off

Washington, about 200 m. Transients: There is clearly much high frequency variability. The one Washington data set which had four moorings across the undercurrent did not indicate that meandering was significant, in spite of suggestions by Mysak and others.

MITTELSTAEDT. The EBPUC's observed off Northwest Africa and off the Iberian Peninsula represent the "boundary signal" of a much broader and thicker subsurface circulation than say 100 km in the horizontal and a few hundred meters in the vertical. The domain of the poleward subsurface flow in the eastern basin seems to increase with latitude.

When/where the regional/seasonal external driving forces like the winds dominate the large scale pressure gradient, the surface flow is different from the flow at depths. This is the case in winter and spring at latitudes south of about 23°N and throughout the year north thereof up to about 50°N in the Eastern Atlantic. This is about the latitudinal zone where I expect an EBPUC to become disconnected from the surface layer. There are, of course, local and transient perturbations due to boundary eddies and waves.

Topography is important. It locally determines the flow direction and may generate eddies/waves that in turn interfere with the boundary flow.

MOOERS. The instabilities of EBPUCs seem to be linked to the mesoscale variability of eastern boundary current regimes. The interactions between the EBPUCs and the mesoscale eddies and meandering jets observed in those regimes may play an important role in cross-shore exchange processes and, thus, ecosystem dynamics. Similarly, the interactions of the EBPUCs with the coastal upwelling centers found equatorward of coastal capes may also play an important role in cross-shore exchange processes and, thus, ecosystem dynamics.

3. ARE THERE ESSENTIALLY TWO EBPUCs IN EACH BASIN?

Rationale: One EBPUC may be coupled to the eastern boundary of the sub-tropical gyre, while another is coupled to the eastern boundary current of the sub-polar gyre, the two being discontinuous?

Opinion: **HICKEY.** The seasonal variation of the NE Pacific poleward undercurrent of about 50°N suggests that this EBPUC is tied to the sub-tropical gyre. I was not aware that the feature has been observed north of Vancouver Island. By the "feature", I define a flow pattern with a subsurface maximum. Thus, I continue to differentiate in my mind between flows that may be directly driven by local poleward wind stress and flows that generally oppose the local wind stress and hence usually have a subsurface maximum of some sort, even though the depth of the maximum may be a function of wind stress or other factors.

MITTELSTAEDT. I believe there is a connection of the EBPUC off (North) Africa and Europe. The dynamics of the various branches of the boundary flow along the eastern margin of the Atlantic are different. But still, they all might be a common manifestation of the boundary dynamics to the large scale processes in the interior of the ocean.

HUTHNANCE. Confinement over the slope distinguishes the EBPUC west of Britain from a broader drift (hundreds of kilometers) northwards in the upper layers of the Rockall Trough and Plateau, and not tied to topography. To my mind, the most pertinent respective theories are poleward density gradients with the shelf/slope/ocean topography (Huthnance 1984) and a thermal-diffusion eastern boundary layer (McCreary 1981). These account for the two widths. However, the poleward density gradient applies from the tropics to high latitudes and is not directly coupled to the two gyres separately; essentially one (not two) slope currents would be expected in each basin. More observations in Biscay and between Portugal and Africa are required to establish this in view of the aforementioned persistence scale of 1000 km or less. On the other hand, the eastern boundary layer would adjoin two gyres separately. This boundary layer form of the EBPUC will be quasi-steady, but will be transient with season owing to its association with slow baroclinic Rossby waves decaying westwards. In fact, Rockall Trough is so occupied by eddies that consistent estimates of any northward drift are hard to determine.

NESHYBA. Having traced the continuity of the Peru-Chile Undercurrent from 10 to 48°S using its salinity/oxygen characteristic, we noted that this PUC transcends the zone centered on 43°S where the West Wind Drift of the Southeast Pacific impinges the continent. It appears that the PUC proceeds through this the zone of stagnation. To the south of the impingement zone, along the Chilean coast in what constitutes the eastern boundary of a sub-polar flow system (not really a gyre because the Drake Passage removes the eastern boundary south of about 53°S), it gradually loses its subsurface geostrophic signature by blending into the surface Cape Horn Current that gains volume via cumulative runoff from the Cordillera. These data suggest that there is but one EBPUC in the Southeast Pacific.

4. WHAT IDEAS DO YOU HAVE ON HOW TO PARTITION EBPUC FORCING INTO LOCAL AND DISTANT EFFECTS?

Rationale: Some boundary phenomena are remotely driven, e.g., coastal trapped waves; are EBPUCs also?

Opinion: **HUTHNANCE.** My thoughts on this are expounded in Huthnance (1987) for simple barotropic contexts. Any eastern boundary has a set of coastal trapped waves which allow a more or less complete description of the motion on the slope from top to bottom. A forced problem has the solution

Particular Integral
+ Complementary Function (viz a combination of coastal trapped waves to fit initial conditions in the along-shore coordinate and decaying as they propagate along the shelf/slope.)

The slope current is the Particular Integral (corresponding to density gradient forcing, say) but evolves through the decay of coastal trapped waves (e.g., northwards from the equator or a cross-slope barrier). These are initially present in the solution to enable the initial conditions to be satisfied in the presence of the Particular Integral. The evolution/ decay distance is the persistence scale (iii) discussed in the first question above.

HICKEY. I believe that the EBPUC is driven by the seasonal along-shore pressure gradient; some of the pressure gradient signal is remotely driven and some locally driven by the coastal wind field. South of Point Conception, other processes must drive the along-shore pressure gradient. Other driving mechanisms for the poleward flow have been suggested and these are decoupled from the pressure gradient mechanism. I suspect these mechanisms will not be as significant as those related to the pressure gradient field.

MITTELSTAEDT. I believe that the EBPUC we observed off Northwest Africa, as well as the one off Europe, is essentially regionally driven.

BATTEEN. One can gain ideas on how to partition local and distant forcing of EBPUCs by running numerical experiments which use local but no remote forcing, then remote but no local forcing, and comparing the results with observations.

5. HOW CAN YOU BETTER USE THEORETICAL STUDIES?

Rationale: Could you use model diagnostics to interpret results? How are model results used for experimental design? —for setting priorities in field work proposed for large scale science like WOCE? What can concretely be done to bring theoretical and observed mechanisms together?

Opinion: **HICKEY.** Topography and wind stress climates vary enough along the West Coast that I believe the models could be used to partition the variance due to specific forcing mechanisms. The various models should be coupled more tightly with the observations. Perhaps this workshop is a step in the right direction.

BATTEEN. Model diagnostics can be used to interpret results, particularly if you use a model to run a number of controlled experiments, such as varying the type of forcing or topography, and then obtain model diagnostics for each case. Close comparison of the results should contribute to a deeper understanding of physical mechanisms responsible for observed oceanic behavior.

A model's structural detail could be used to help plan field investigations. For example, a model's horizontal and vertical cross-sections of horizontal velocity, temperature, and pressure can provide information on the surface and subsurface structure of jets and undercurrents which would influence the design of observational grids and arrays.

HUTHNANCE. Theory is used, as implicit in the above discussions, in a distinction between wind forcing (fluctuating currents on the shelf), density gradient forcing (steady currents along the slope), and the broader diffusive boundary layer. Because coastal trapped waves underlie the first two, their study in context can confirm their form and elucidate shelf and slope flows generally (Huthnance 1986; Freeland et al. 1986). A knowledge of coastal trapped wave dispersion (for wavelength-frequency relations) and decay (for evolution-persistence scales) can assist experiment design, specifically along-slope spacing. Care is required, however, for the degrading effect of higher modes with shorter scales; they may be more prevalent in chaotic reality than in idealized models.

MOOERS. The capabilities and limitations of oceanic observations and models suggest using them jointly to make optimal estimates of ocean fields in general. Considering the intense spatial and temporal variability of EBPUCs relative to our ability to observe or model them, the methods of four-dimensional data assimilation may be essential for dynamical interpolation of limited observations on one hand, and for observational constraints to be imposed on numerical models on the other hand. Time sequences of optimal field estimates are necessary for diagnostic studies of the EBPUC regime.

5. WHICH TECHNOLOGIES WOULD BE USEFUL IN MODERN EBPUC STUDIES?

Rationale: See Jane Huyer's paper What's Next?

Opinion: **MITTELSTAEDT.** Among others, we require multi-ship programs, satellite data products, extensive ADCP application, and both surface and subsurface Lagrangian devices in depths ranging from 100 to 1500 m for periods of 1 to 2 years.

BATTEEN. All the above plus long-term current meter deployment, and high-resolution multi-level primitive equation models such as developed by Haidvogel, Haney, and Batteen.

HICKEY. Pop-up drifters may be important for investigating the continuity of the EBPUC. Shipboard and in-situ acoustic Doppler current meters would provide additional detail on spatial scales needed to properly examine questions of instability, etc.

HUTHNANCE. We used current meters extensively to good effect for variability in CONSLEX, but they are vulnerable to fishing activity in many productive shelf and slope areas. The proposed five years or more of current meter measurements should be within WOCE Core Project 3.

For many years SMBA have worked a CTD section across the Hebrides Shelf and Rockall Trough to Rockall, at two-monthly intervals. For a narrow slope current (not the previous purpose of this CTD work) the spacing would need to be closer, with several casts within the steep slope area. We have relied on remote wind and atmospheric pressure observations from the many land and island recording stations, together with some information about their relation to more open offshore records.

Bottom pressure and coastal sea level records in CONSLEX gave some information on temporal variations, relating best to the currents in 'events' extending 500 km or more along-shore. Pressure gradients were difficult to estimate, however; motion took place on scales of 100 km or less over which pressure differences approach instrumental noise level, especially along-shore.

The role of EBPUCs in transporting (say) heat can be estimated properly only by budgeting for a closed 'box'. Even a section across a channel between two closed boundaries gives only a difference between transports in opposite directions, with no test of their reliability. Estimates from CONSLEX current meters with thermistors were very scattered. Long records are needed but also some preliminary work on how long and on the spatial representativeness of estimates from one mooring. In CONSLEX, cross-slope current variations were a minimum over the upper slope, which makes this location strategic for cross-slope fluxes.

NESHYBA. New Technology: There is no doubt in my mind that a pop-up Lagrangian drifter, preferably air-launched, is the technological key to reveal along-shore continuity within subsurface EBPUCs. This way we can obtain more answers more quickly and cost-effectively than with other technology now possible.

Old Plus New Technology: As all the above commentators point out, long-term records are needed, particularly to evaluate transports and the coherency links between EBPUC transports and other major gyre features. We might set up long-term sections in certain places where each terminus has a sea-level station, and the section is as short as need be to capture eddies, squirts and jets as well as steady flows. The section between Costa Rica and Cocos Island, about 500 km, could monitor the North Equatorial Countercurrent transport. Cocos Island has already been positioned by the Global Positioning System (GPS). When repeated enough times, this device also yields vertical leveling between section end stations. A similar setup is available between Valparaiso and J. Fernandez Island to monitor differential transport in the Humboldt Current system and the strength of the South Pacific subtropical gyre. We can now see far enough ahead technologically to plan for definitive long-term data sets needed to resolve our basic questions about EBPUCs. All we require is resolve, a green light, and about 1/1000th part of NASA's funding for outer space!

MOOERS. The Lagrangian methods, together with long-term current meter moorings on the slope, seem essential. There may be tracer methods which could be used to study the water mass continuity issues. Adroit use of satellite IR and ocean color imagery could prove useful, though indirectly, in monitoring the state of the EBPUCs. Satellite radar altimetry should help determine the offshore boundary conditions and the spatial structure of the pressure gradient field, especially on the mesoscale.

7. SHOULD WE RECOMMEND THAT AN EBPUC PANEL BE CHARTERED WITHIN THE WOCE, FOR THE PURPOSE OF DEFINING THE PRIORITY OF EBPUCs IN OCEAN SCIENCE FOR THE COMING DECADE?

Rationale: The continuity of EBPUCs can be examined as part of the division of West Wind Drifts as they impinge on eastern boundaries, at which they become an integral part of gyre behavior. Included also are the coherency analyses between the EBPUC and basin-wide phenomena, as well as between EBPUC and EBC.

Opinion: **HUTHNANCE**. EBPUCs are relevant to WOCE specifically. The WOCE Science Plan envisages that sea-level recording will carry on, after the five-year special observational period, as a continuing monitor of ocean circulation. It is therefore important that sea-level circulation relationships be understood, implicating all boundary currents across which sea-level changes occur.

Core Project 3 appears to be the relevant part of WOCE. I support the formation of an EBPUC panel or, even wider, a Lateral Boundary Conditions panel, within Core Project 3, by whatever means are appropriate, presumably through the WOCE Science Steering Committee.

HICKEY. I think WOCE may be the one chance to examine the forcing of the EBPUC relative to the Pacific basin forcing, as well as to resolve questions of continuity within these features of eastern boundary flows. The formation of an EBPUC panel in WOCE should be a high priority.

MITTELSTAEDT. Yes.

NESHYBA. EBPUCs ought to be a definitive target in WOCE gyre experiments because of their role in poleward transports of heat and water. For example, it is difficult to foresee success in budget studies of the South Pacific that do not explicitly account for EBPUC transports.

Beyond the contribution that EBPUC research holds for physical science, it is critical to success in fisheries. Eastern boundaries are sites for four out of five major coastal upwelling-fishery zones of the world ocean: (1) Oregon and vicinal states, (2) Peru-Chile, (3) Namibia-Angola, and (4) North Africa. The fifth, Somalia-Arabia, differs in that it is monsoon driven and not an eastern ocean boundary. We now know empirically that ecosystems evolve with integral dependency on a coherency with these physical systems often in the form of migration patterns of nekton. Physical scientists gain new insight when patterns of biological behavior can be blended into physics, and our results are essential to describing life cycles within eastern boundary zones. Although WOCE itself excludes biology explicitly, physical ocean. Science in general would be remiss if its results are not blended into related coastal research.

PART V

REFERENCES

Aagaard, K. and L.K. Coachman (1968). The East Greenland Current North of the Denmark Strait, II, Arctic, 21, 267-290.

Aagaard, K. and P. Greisman (1975). Toward New Mass and Heat Budgets for the Arctic Ocean, J. Geophys. Res., 80, 3821-3827.

Aagaard, K., C. Darnall, A. Foldvik, and T. Torresen (1985a). Fram Strait Current Measurements 1984-1985, Report No. 63, Department of Oceanography, Geophysical Institute, University of Bergen, Bergen, Norway.

Aagaard, K., J.H. Swift, and E. Carmack (1985b). Thermohaline Circulation in the Arctic Mediterranean Seas, J. Geophys. Res., 90 (C3), 4833-4846.

Aagaard, K., A. Foldvik, and S.R. Hillman (1987). The West Spitsbergen Current: Disposition and Water Mass Transformation. J. Geophys. Res., 92 (C4), 3778-3784.

Aas, E. (1977). The Mean Seasonal Variation in the Transport of Atlantic Water Through the Faeroe-Shetland Channel. Institute of Geophysics Report Series, No. 18, 20 pp. University of Oslo.

Ahumada, R., A. Rudolph, and V. Martinez (1983). Circulation and Fertility of Waters in Concepcion Bay. Coastal and Shelf Science, 16, 95-105.

Allain, Ch. (1970). Observations hydrologiques sur le talus du Banc d'Arguin en Decembre 1962, Rapp. Process-Verbaux Reunion ICES, 159, 86-89.

Allan, R.J. and J.I. Pariwono (1987). Aspects of Large-Scale Ocean-Atmosphere Interactions in Low-Latitude Australia. Tropical Ocean-Atmosphere Newsletter, 38, 6-10.

Allen, J.S. and R.L. Smith (1981). On the Dynamics of Wind-Driven Shelf Currents. Philosophical Transactions, the Royal Society of London, A302, 617-634.

Alvarez, L.G., R. Durazo, J. Perez, L.F. Navarro, and R. Hernandez (1984). Observaciones de corrientes costeras superficiales mediante trazadores Lagrangeanos. II. Erendira, Baja California, 1979-1981. CICESE, Dept. of Oceanography, (Unpublished Report).

Alvarez, L.G. (1986). Estudio de corriente litoral y difusion frente a la costa Noroeste de Baja California, con relacion al transporte de desechos vertidos al mar. Departamento de Oceanografia, CICESE. 96 pp. (Unpublished Report).

Alvarez, L.G., A. Badan-Dangon, and A. Valle (1988). A Note on Coastal Currents off Tehuantepec. Estuarine, Coastal, and Shelf Science in press.

Ambar, I. and M.R. Howe (1979). Observations of the Mediterranean Out-flow 1. Mixing in the Mediterranean Outflow, Deep-Sea Res., 26, 535-554.

Ambar, I. (1983). A Shallow Core of Mediterranean Water Off Western Portugal, Deep-Sea Res., 30, 677-680.

Ambar, I. (1984). Seis meses de medicoes de correntes, temperaturas e salinidades na vertente continental ao largo da costa alentejana, Relatorio tecnico, 1/84, Grupo de Oceanografia, Universidade de Lisboa, 47 pp. (Unpublished Manuscript).

Ambar, I. (1985). Seis meses de medicoes de correntes, temperaturas e salinidades na vertente continental Portuguesa a 40°N, Relatorio tecnico, 1/85, Grupo de Oceanografia, Universidade de Lisboa, 40 pp. (Unpublished Manuscript).

Ambar, I. and A. Fiuza (1987). Observations of a Poleward Current Off Portugal.

Anderson, D.L.T. and A.E. Gill (1975). Spin-up of a Stratified Ocean, with Application to Upwelling. Deep-Sea Res., 22, 583-596.

Andrews, J.C. (1977). Eddy Structure and the West Australian Current. Deep-Sea Res., 24, 1133-1148.

Andrews, J.C. (1983) Ring Structure in the Poleward Boundary Current Off Western Australia in Summer. Aust. J. Mar. Freshwat. Res., 34, 547-561.

Arakawa, A. and V.R. Lamb (1977). Computational design of the basic dynamical processes of the UCLA general circulation model. Methods in Computational Physics. (J. Chang, Ed.), Academic Press, 17, 173-265.

Arakawa, A. and M.J. Suarez (1983). Vertical differencing of the primitive equations in sigma coordinates. J. Atmos. Sci., 40, 34-45.

Arana, P. and A. Nakanishi (1971). La pesqueria del camaron nylon (Heterocarpus reedi), frente a la costa de Valparaiso. Inv. Mar. 2(4), 78-88.

Arfi, R. (1987). Variabilite interannuelle de l'hydrologie d'une region d'upwelling (Cap Blanc, Mauritanie), Oceanologica Acta, 10 (2), 151-159.

Austin, T.S. (1960). Oceanography of the East Central Equatorial Pacific as Observed During Expedition EASTROPIC. Fisheries Bulletin, 60, 257-282.

Badan-Dangon, A. (1972). Interpretacion de los datos obtenidos por medio de botellas de deriva en el sistema de la Corriente de California. Tesis de Licenciatura. UABC.Ensenada, 30 pp.

Badan-Dangon, A., K.H. Brink, and R.L. Smith (1986). On the Dynamical Structure of the Midshelf Water Column Off Northwest Africa. Continental Shelf Research, 5, 629-644.

Bainbridge, A.E. (1981). GEOSECS-Atlantic Expedition. Sections and Profiles. Vol. 2, Natl. Sci. Foundation, Washington, D.C., 198 pp..

Bakun, A. and R.H. Parrish (1980). Environmental inputs to fishery population models for eastern boundary current regions, in Intergov. Oceanogr. Comm., UNESCO, Workshop Rep. 28, G.D. Sharp, ed., 67-104.

Bakun, A. (1988). Applications of Maritime Data to the Study of Surface Forcing of Seasonal and Interannual Variability in Eastern Boundary Regions. PhD. Dissertation, Oregon State University, College of Oceanography.

Bang, N.D. (1973). Characteristics of an Intense Ocean Frontal System in the Upwell Regime West of Cape Town. Tellus, 25(3), 256-265.

Barber, R.T. and R.L. Smith (1981). Coastal Upwelling Ecosystems In: Analysis of Marine Ecosystems, A.R. Longhurst (Ed.), Academic Press, 31-68.

Barton, E.D., R.D. Pillsbury, and R.L. Smith (1975). A Compendium of Physical Observations From JOINT-I, Reference 75-17, School of Oceanography, Oregon State University, 60 pp.

Barton, E.D. and M.L. Argote (1980). Hydrographic Variability in an Upwelling Area Off Northern Baja California in June, 1976. J. Mar. Res., 38, 631-649.

Barton, E.D., J.M. Robles, A. Amador, and C. Morales (1980). A Year of Current and Temperature Observations Off Baja California North. Data Report. CICESE, Ensenada, 162 pp.

Barton, E.D. and P. Hughes (1982). Variability of Water Mass Interleaving Off NW Africa, J. Mar. Res., 40 (4), 963-984.

Barton, E.D. (1985). Low-Frequency Variability of Currents and Temperatures on the Pacific Continental Shelf Off Northern Baja, California, 1978 to 1979. Continental Shelf Research, 4, 425-443.

Barton, E.D. (1987). Meanders, Eddies, and Intrusions in the Thermohaline Front Off Northwest Africa, Oceanologica Acta, 10(3), 267-283.

Barton, E.D., This Volume. The Poleward Undercurrent on the Eastern Boundary of the Subtropical North Atlantic.

Batteen, M.L. (1988). On the use of sigma coordinates in large-scale ocean circulation models. Ocean Modelling, 77, 3-5.

Baumgartner, T.R. and N. Christensen, Jr. (1985). Coupling of the Gulf of California to Large-Scale Interannual Climatic Variability. J. Mar. Res., 43, 825-848.

Beardsley, R.C., C.E. Dorman, C.A. Friehe, L.K. Rosenfeld, and D.D. Winant (1987). Local Atmospheric Forcing During the Coastal Ocean Dynamics Experiment 1, A Description of the Marine Boundary Layer and Atmospheric Conditions Over Northern California.

Bennett, E.B. (1963). An Oceanographic Atlas of the Eastern Tropical Pacific Ocean, Based on Data from the ESTROPIC Expedition, October - December 1955. Inter-American Tropical Tuna Commission, Bulletin 8, 33-165.

Bernal, P.A. (1981). A Review of the Low-Frequency Response of the Pelagic Ecosystem in the California Current. CalCOFI Report XXII, 49-62.

Bernal, P.A. and J.A. McGowan (1981). Advection and Upwelling in the California Current. In: Coastal Upwelling, F.A. Richards (Ed.), American Geophysical Union, Washington, 381-399.

Bernstein, R.L., L.C. Breaker and R. Whritner (1977). California current eddy formation: Ship, air, and satellite results. Science, 195, 353-359.

Blackburn, M. (1962). An Oceanographic Study of the Gulf of Tehuantepec. U.S. Fish and Wildlife Service, Special Scientific Reports, Fisheries, No. 404, 28 pp.

Blanco, J. (1984). Caracteristicas de la Circulacion sobre la plataforma conteinental de Talcahuano. Thesis. Esc de Ciencias del Mar U.C.V., 41 pp.

Blumberg, A.F. and G.L. Mellor (1987). A description of a three-dimensional coastal ocean circulation model. Three-dimensional Coastal Ocean Models, (N. Heaps, Ed.), American Geophysical Union, 4, 1-16.

Bolin, B. (1983). C, N, P, and S Cycles: Major Reservoirs and Fluxes. In: The Major Biogeochemical Cycles and Their Interactions, Bolin, B. and R.B. Cook Eds., Wiley, N.Y., 41-45.

Bolin, B. and R.B. Cook (1983). The Major Biogeochemical Cycles and Their Interactions, Wiley, New York, 532 pp.

Booth, D.A. and D.J. Ellett (1983). The Scottish Continental Slope Current. Continental Shelf Research, 2, 127-146.

Booth, D.A. and D.T. Meldrum (1987a). Drifting Buoys in the Northeast Atlantic. Journal due Conseil International pour l'Exploration de la Mer, 43, 261-267.

Booth, D.A. and D.T. Meldrum (1987b). Northeast Atlantic Satellite-Tracked Buoy Drifts. Offshore Technology Report OTH 87 270, HMSO London, 95 pp.

Bourke, R.H., J.L. Newton, R.G. Paquette, and M.D. Tunnicliffe (1987). Circulation and Water Masses of the East Greenland Shelf. J. Geophys. Res., 92, (C7), 6729-6740.

Braathen, B.R. and R. Saetre (1973). Farming Salmon in Norwegian Coastal Waters. Environment and Equipment. Fisken og Havet. Ser. B. (9) (in Norwegian).

Brandhorst, W. (1963). Descripcion de las condiciones oceanograficas de las aguas costeras entre Valparaiso y el golfo de Arauco, con especial referencia al contenido de oxigeno y su relacion con la pesca. (Resultados de la expedicion AGRIMAR, 1959). Min. Agricultura Dir. Agr. y Pesca, pp. 3-55, Dantiago, Chile.

Brandhorst, W. (1971). Condiciones oceanograficas estivales frente a la costa de Chile, Rev. Bio. Mar., Valparaiso, 14(3), 45-84.

Bretherton, F.P. and M.J. Karweit (1975). Mid-Ocean Mesoscale Modeling. In: Numerical Models of the Ocean Circulation, Ocean Affairs Board, National Research Council, National Academy of Sciences, Washington, DC, 237-249.

Bretherton, F.P. and D.B. Haidvogel (1976). Two-dimensional Turbulence above Topography. J. Fluid Mech., 78, 129-154.

Brink, K.H. and J.S. Allen (1978). On the Effect of Bottom Friction on Barotropic Motion Over the Continental Shelf. J. Phys. Oceanogr., 8, 919-922.

Brink, K.H., D. Halpern, and R.L. Smith (1980). Circulation in the Peruvian Upwelling System Near 15°S. J. Geophys. Res., 85, 4036-4048.

Brink, K.H., D. Halpern, A. Huyer, and R.L. Smith (1983). The Physical Environment of the Peruvian Upwelling System. Progress in Oceanography, 12, 285-305.

Brink, K.H. and D.C. Chapman (1985). Programs for Computing Properties of Coastal-Trapped Waves and Wind-Driven Motions Over the Continental Shelf and Slope. WHOI Tech. Rep., 85-17.

Brink, K.H. (1986). Topographic Drag Due to Barotropic Flow Over the Continental Shelf and Slope. J. Phys. Oceanogr., 16, 2150-2158.

Brink, K.H. and R.D. Muench (1986). Circulation in the Point Conception-Santa Barbara Channel Region. J. Geophys. Res., 91 (C1), 877-895.

Brockmann, C., E. Fahrgach, A. Huyer and R.L. Smith (1980). The Poleward Undercurrent Along the Peru Coast: 5 to 15°S. Deep-Sea Res., 27, 847-856.

Broecker, W.S. and T.H. Peng (1982). Tracers in the Sea. Eldigio Press, New York, 690 pp.

Brown, P.C. and L. Hutchings (1985) - Phytoplankton Distribution and Dynamics in the Southern Benguela Current. In: International Symposium on the Most Important Upwelling Areas Off Western Africa (Cape Blanco and Benguela){Barcelona, 1983.} Bas, C., Margelef, R. and P. Rubias (Eds.) Barcelona, Institute de Investigaciones Pesqueras, 319-344.

Brown, W.S., J.D. Irish, and M.R. Erdman (1983). CODE-1: Bottom Pressure Observations. Chapter V, CODE Technical Report No. 21, WHOI Tech. Report, 83-23, 117-138.

Brundrit, G.B. (1984). Monthly Mean Sea Level Variability Along the West Coast of Southern Africa. S. Afr. J. Mar. Sci., 2, 195-203.

Brundrit G.B., B.A. De Cuevas, and A.M. Shipley (1987). Long-Term Sea-Level Variability in the Eastern South Atlantic and a Comparison With That in the Eastern Pacific. In: The Benguela and Comparable Ecosystems. Payne, A.I.L., J.A. Gulland, and K.H. Brink (Eds.). S. Afr. J. Mar. Sci., 5, 73-78.

Calvert, S.E. and N.B. Price (1971). Upwelling and Nutrient Regeneration in the Benguela Current, October 1968. Deep-Sea Res., 18, 505-523.

Camerlengo, A.L. and J.J. O'Brien (1980). Open boundary conditions in rotating fluids. J. Comput. Physics, 35, 12-35.

Cannon, G.A., N.P. Laird, and T.V. Ryan (1975). Flow Along the Continental Slope Off Washington, Autumn 1971. J. Mar. Res., 33, Suppl., 97-107.

Capone, D.G. and E.J. Carpenter (1982). Nitrogen Fixation in the Marine Environment. Science, 217, 1140-1142.

Carlucci, A.F. and P.M. McNally (1969). Nitrification by Marine Bacteria in Low Concentrations of Substrate and Oxygen. Limnol. Oceanogr., 14, 736-739.

Carmack, E. (1972). On the Hydrography of the Greenland Sea, PhD Thesis, University of Washington, Seattle, WA.

Carmack, E. and K. Aagaard (1973). On the Deep Water of the Greenland Sea, Deep-Sea Res., 20, 687-715.

Carnevale, G.F. and J.S. Frederiksen (1987). Nonlinear Stability and Statistical Mechanics of Flow Over Topography. J. Fluid Mech., 175, 157-181.

Carpenter, E.J. and D.G. Capone (1983). Nitrogen in the Marine Environment. Academic Press, N.Y., 900 pp.

Chapman D.C. and K.H. Brink (1987). Shelf and Slope Circulation Induced by Fluctuating Offshore Forcing. J. Geophys. Res., 92, 741-759.

Chapman, P. and L.V. Shannon (1985). The Benguela Ecosystem. II Chemistry and Related Processes. In: Oceanography and Marine Biology. An Annual Review. 23, Barnes, M. (Ed)., Aberdeen; University Press, 183-251.

Chavez, F.P. and R.T. Barber (1987). An Estimate of New Production in the Equatorial Pacific. Deep-Sea Res., 34(7), 1229-1243.

Chavez, F.P., R.T. Barber and M.P. Sanderson. The Potential Primary Production of the Peruvian Upwelling Ecosystem: 1953-1984. In: ICLARM Conference Proceedings Series. Pauly, D., H. Salzwedel, P. Muck, and J. Mendo, (Eds.), ICLARM/PROCOPA/IMARPE. In Press.

Chelton, D.B. (1981). Interannual Variability of the California Current - Physical Factors. CalCOFI Reports, XXII, 34-48.

Chelton, D.B. (1982). Large-Scale Response of the California Current. CalCOFI Report XXIII, 130-148.

Chelton, D.B., P.A. Bernal, and J.A. McGowan (1982). Large-Scale Interannual Physical and Biological Interaction in the California Current. J. Mar. Res., 40, 1095-1125.

Chelton, D.B. and R.E. Davis (1982). Monthly Mean Sea Level Variability Along the West Coast of North America. J. Phys. Oceanogr., 12, 757-784.

Chelton, D.B. (1984). Seasonal Variability of Alongshore Geostrophic Velocity Off Central California. J. Geophys. Res., 89, 3473-3486.

Chelton, D.B., R.L. Bernstein, A. Bratkovitch, and P.M. Kosro (1987). The Central California Coastal Circulation Study. EOS, 68, 1-13.

Chelton, D.B., A.W. Bratkovitch, R.L. Bernstein, and P.M. Kosro (1988). Poleward Flow off Central California During the Spring and Summer of 1981 and 1984. J. Geophys. Res., 93(10), 604-10,620.

Christensen, N., Jr. and N. Rodriguez (1979). A Study of the Sea Level Variations and Currents Off Baja California. J. Phys. Oceanogr., 9, 631-638.

Christensen, N., Jr., R. de la Paz, and G. Gutierrez (1983). A Study of Sub-Inertial Waves Off the West Coast of Mexico. Deep-Sea Res., 30, 835-850.

Clarke, A.J. (1977). Observational and Numerical Evidence for Wind-Forced Coastal-Trapped Long Waves. J. Phys. Oceanogr., 7, 231-247.

Clarke, A.J. (1982). The Dynamics of Large-Scale, Wind-Driven Variations in the Antarctic Circumpolar Current. J. Phys. Oceanogr., 12, 1092-1105.

Clarke, A.J. and S. Van Gorder (1986). A Method for Estimating Wind-Driven Frictional, Time-Dependent, Stratified Shelf and Slope Water Flow. J. Phys. Oceanogr., 16, 1013-1028.

Coachman, L.K. and K. Aagaard (1974). Physical Oceanography of Arctic and Subarctic Seas, In: Marine Geology and Oceanography of the Arctic Seas, Y. Herman (Ed.), Springer-Verlag, New York, 1-72.

Codispoti, L.A. and F.A. Richards (1976). An Analysis of the Horizontal Regime of Denitrification in the Eastern Tropical North Pacific. Limnol. Oceanogr., 21, 379-383.

Codispoti, L.A. and T.T. Packard (1980). On the Denitrification Rate in the Eastern Tropical South Pacific. J. Mar. Res., 38, 453-477.

Codispoti, L.A. (1981). Temporal Nutrient Variability in Three Different Upwelling Regions. In: Coastal Upwelling, F.A. Richards (Ed.), American Geophysical Union, Washington, 209-220.

Codispoti, L.A., R.C. Dugdale, and H.J. Minas (1982). A Comparison of the Nutrient Regimes Off Northwest Africa, Peru, and Baja, California. Rapports et Proces-Verbaux Reunions, International Council for the Exploration of the Sea, 180, 184-201.

Codispoti, L.A. (1983). Nitrogen in Upwelling System. In: Nitrogen in the Marine Environment, E. Carpenter and D. Capone (Eds.), Academic Press, N.Y.

Codispoti, L.A. and J.C. Christensen (1985). Nitrification, Denitrification and Nitrous Oxide Cycling in the Eastern Tropical South Pacific Ocean. Mar. Chem., 16, 277-300.

Codispoti, L.A., G.E. Friederich, T.T. Packard, H.E. Glover, P.J. Kelley, R.W. Spinrad, R.T. Barbver, J.W. Elkins, B.B. Ward, F. Lipschultz, and N. Lostaunau (1986). High Nitrite Levels Off Northern Peru: A Signal of Instability in the Marine Denitrification Rate. Science, 233, 1200-1202.

Codispoti, L.A., G.E. Friederich, T.T. Packard, and R.T. Barber (1988). Remotely Driven Thermocline Oscillations and Denitrification in the Eastern South Pacific: The Potential for High Denitrification Rates During Weak Coastal Upwelling. Science of the Total Environment, 75, 301-318.

Cohen, Y. and L.I. Gordon (1978). Nitrous Oxide in the Oxygen Minimum of the Eastern Tropical North Pacific. Deep-Sea Res., 25, 509-524.

Colborn, J.G. (1975). The Thermal Structure of the Indian Ocean. International Indian Ocean Expedition Oceanographic Monograph No. 2.

Connary, S.D. (1972). Investigations of the Walvis Ridge and Environs. Ph.D. Thesis, Columbia University, 228 pp.

Crawford, W.R. and T.R. Osborn (1980). Energetics of the Atlantic Equatorial Undercurrent. Deep-Sea Res., 26, Suppl. II, 309-323.

Cresswell, G.R. (1977). The Trapping of Two Drifting Buoys by an Ocean Eddy. Deep-Sea Res., 24, 1203-1209.

Cresswell, G.R. and T.J. Golding (1980). Observations of a South Flowing Current in the Southeastern Indian Ocean. Deep-Sea Res., 27A, 449-466.

Cresswell, G.R. (1986). The Role of the Leeuwin Current in the Life Cycles of Several Marine Creatures. Proceedings International Conference on Pelagic Biogeography, Amsterdam, 1985.

Cromwell, T. and E.B. Bennett (1959). Surface drift Charts for the Eastern Tropical Pacific. Inter-American Tropical Tuna Commission Bulletin, 3, 217-237.

Crowe, F.J. and R.S. Schwartzlose (1972). Release and Recovery Records of Drift Bottles in the California Current Region 1955 through 1971. CalCOFI Atlas No. 16, 140 pp.

Crutzen, P.J. (1981). Atmospheric Chemical Processes of the Oxides of Nitrogen, Including Nitrous Oxide. In: Denitrification Nitrification and Atmospheric Nitrous Oxide, C.C. Delwiche (Ed.), Wiley, N.Y., 17-44.

Crutzen, P.J. (1983). Atmospheric Interactions - Homogeneous Gas Reactions of C, N, and S Containing Compounds. In: The Major Biogeochemical Cycles and Their Interactions, B. Bolin and R.B. Cook (Eds.), 67-112.

Csanady, G.T. (1978). The Arrested Topographic Wave. J. Phys. Oceanogr., 8, 47-62.

Csanady, G.T. (1980). Longshore Pressure Gradients Caused by Offshore Winds. J. Geophys. Res., 85, 1076-1084.

Dakin, W.J. (1919). The Percy Sladen Trust Expeditions to the Abrolhos Islands (Indian Ocean). Report I. Journal of the Linnean Society, 34, 127-180.

Davis, R.E. (1985). Drifter Observations of Coastal Surface Currents During CODE: The Method and Descriptive View. J. Geophys. Res., 90, 4741-4755.

De Decker, A.H.B. (1970). Notes on an Oxygen-depleted Subsurface Current Off the West Coast of South Africa. Investl. Rep. Div. Sea Fish. S. Afr., 84, 24 pp.

Defant, A. (1941). Die Relative Topographie Einzelner Druckflachen im Atlantischen Ozean. "Meteor" Rep., 6(2:4), Teil., X-XVIII, 183-190 +.

Defant, A. (1941). Die absolute Topographie des physikalischen Meeresniveaus und der Druckflachen im Atlantischen Ozean, Wiss. Ergebn. Dt. Atlant. Exped. "Meteor" 1925-7, VI, 2, 191-260.

Delwiche, C.C. (1981). The Nitrogen Cycle and Nitrous Oxide. In: Denitrification, Nitrification and Nitrous Oxide, C.C. Delwiche (Ed.), Wiley, N.Y., 1-16.

Denbo, D.W. and J.S. Allen (1983). Mean Flow Generation on a Continental Margin by a Periodic Wind Forcing. J. Phys. Oceanogr., 13, 78-92.

Denbo, D.W., K. Polzin, J.S. Allen, A. Huyer, and R.L. Smith (1984). Current Meter Observations Over the Continental Shelf Off Oregon and California, February 1981-January 1984. College of Oceanography, Oregon State University, Data Rep. 112, Ref. 84-12.

Devol. A.H. (1978). Bacterial Oxygen Uptake Kinetics as Related to Biological Processes in Oxygen Deficient Zones of the Oceans. Deep-Sea Res., 25, 137-146.

Dickson, R.R. (1972). Variability and Continuity within the Atlantic Current of the Norwegian Sea, Rapp. Cons. Explor. Mer., 162, 167-183.

Dietrich, G., K. Kalle, W. Drauss, and G. Siedler (1980). General Oceanography: An Introduction. InterScience.

Dooley, H.D. and J. Meincke (1981). Circulation and Water Masses in the Faroese Channels During Overflow '73. Deutsche Hydrographische Zeitschrift, 34, 41-54.

Dugdale, R.C. and J.J. Goering (1967). Uptake of New and Regenerated Forms of Nitrogen in Primary Productivity. Limnol. Oceanogr., 12, 196-206.

Dunlop, J.N. and R.N. Wooller (1986). Range Extensions and the Breeding Seasons of Seabirds in Southwestern Australia. Rec. West. Aust. Mus., 12, 389-394.

Elkins, J.W., S.C. Wofsy, M.B. McElroy, C.E. Kolb and W.A. Kaplan (1978). Aquatic Sources and Sinks for Nitrous Oxide. Nature, 275, 602-606.

Ellett, D.J. and D.A. Booth (1983). Some Oceanographic Applications of ODAS Data. COST-43 Technical Document 100, 83-97.

Ellett, D.J., A. Edwards and R. Bowers (1986). The Hydrography of the Rockall Channel--An Overview. Proceedings of the Royal Society of Edinburgh, 88B, 61-81.

Embley, R.W. and J.J. Morley (1980). Quaternary Sedimentation and Paleoenvironmental Studies Off Namibia (South-West Africa). Mar. Geol., 36, 183-204.

Emery, W.J. and L.A. Mysak (1980). Dynamical Interpretations of Satellite-sensed Thermal Features Off Vancouver Island. J. Phys. Oceanogra., 10, 961-970.

Emery, W.J., W.G. Lee and L. Magaard (1984). Geographic and seasonal distributions of Brunt-Vaisala frequency and Rossby radii in the North Pacific and North Atlantic. J. Phys. Oceanogr., 14, 294-317.

Enfield, D.B., R.L. Smith and A. Huyer (1978). A Compilation of Observations from Moored Current Meters. Data Report 70, Oregon State University.

Enfield, D.B. and J.S. Allen (1980). On the Structure and Dynamics of Monthly Mean Sea Level Anomalies Along the Pacific Coast of North and South America. J. Phys. Oceanogr., 10, 557-578.

Enfield, D.B. and J.S. Allen (1983). The Generation and Propagation of Sea Level Variability Along the Pacific Coast of Mexico. J. Phys. Oceanogr., 13, 1012-1033.

Espinoza, R., S. Neshyba and Z. Maoxiang (1983). Surface Water Motion off Chile with Satellite Images of Surface Chlorophyll and Temperature. In: Recursos Marinos del Pacifico. Arana (Ed): 41-57.

Fahrbach, E. (1976). Einige Beobachtungen zur Erzeugung und Ausbreitung interner Gezeitenwellen am Kontinentalabhang vor Sierra Leone, "Meteor" Forsch.-Ergebn., A, 18, 64-77.

Fahrbach, E., C. Brockmann, N. Lostaunau and W. Urquizo (1981). The Northern Peruvian Upwelling System During the ESACAN Experiment. In: Coastal Upwelling, F.A. Richards (Ed.), Am. Geophys. Union, Washington, D.C., 134-145.

Feliks, Y. (1985). Notes and Correspondence on the Rossby Radius of Deformation in the Ocean. J. Phys. Oceanogr., 15, 1605-1607.

Fiuza, A. (1982). The Portuguese Coastal Upwelling System, In: Present Problems of Oceanography in Portugal, Junta Nacional de Investigacao Cientifica e Tecnologica, Lisbon, 45-71 (Unpublished Manuscript).

Fiuza, A. and D. Halpern (1982). Hydrographic Observations of the Canary Current between 21°N and 25.5°N During March-April 1974, Rapp. Proces-Verbaux Reunion ICES, 180.

Flament, P., L. Armi and L. Washburn (1985). The Evolving Structure of an Upwelling Filament. J. Geophys. Res., 90, 11765-11778.

Fonseca, T.R. (1977). Proceso de surgencia en Punta Curaumilla con especial referencia a la Circulacion-Thesis. Esc. de Pesquerias y Alimentos U.C.V., 90 pp.

Fonseca, T. (1985). Fisica de las aguas costeras de la zona Central de Chile. Tralka Vol. (4), 337-354.

Fraga, F. (1974). Distribution des masses d'eau dans l'upwelling de Mauretanie, Tethys, 6, (2), 5-10.

Freeland, H.J. and K.L. Denman (1982). A Topographically Controlled Upwelling Center of Southern Vancouver Island. J. Mar. Res., 40(4), 1069-1093.

Freeland, H.J. (1983). Low Frequency Currents Observed Off Southern Vancouver Island, 1979-1981. Can. Data Rep. Hydrogr. Ocean Sci., 7, 80 pp.

Freeland, H.J., W.R. Crawford, and R.E. Thomson (1984). Currents along the Pacific Coast of Canada. Atmos.-Oc., 22(2), 151-172.

Freitag H.P. and D. Halpern (1981). Hydrographic Observations Off Northern California During May 1977. J. Geophys. Res., 86, 4248-4252.

Friederich, G.E. and L.A. Codispoti (1981). The Effects of Mixing and Regeneration on the Nutrient Content of Upwelling Waters Off Peru. In: Coastal Upwelling, F.A. Richards (Ed.), Am. Geophys. Union, Washington, D.C. 221-227.

Friederich, G.E., P.M. Kelley, L.A. Codispoti, R.W. Spinrad, G. Kullenberg, J.W. Elkins, J. Kogelschatz, T.T. Packard, F. Lipschultz, H.E. Glover, A.E. Smith, and B.B. Ward (1985). Microbial Nitrogen Transformations in the Oxygen Minimum Zone Off Peru. NITROP-85 Data Report, R/V Wecoma Cruise, Feb 1-Mar 5, 1985, Bigelow Lab Tech. Rep. No. 59.

Friederich, G.E., P.J. Kelly, and L.A. Codispoti (1986). An Inexpensive Moored Water Sampler for Investigating Chemical Variability. In: Tidal Mixing and Phytoplankton Dynamics, M. Bowman, C.M. Yentsch, and W.T. Peterson, (Eds.), Springer-Verlag, N.Y.

Friederich, G.E. and L.A. Codispoti (1987). An Analysis of Continuous Vertical Nutrient Profiles Taken During a Cold-Anomaly Off Peru. Deep-Sea Res., 34(5/6), 1049-1065.

Garcia, J. (1983). Variaciones Hidrograficas y eventos de surgencia frente a Punta Colonet, B.C., en junio de 1980. Tesis de licenciatura. UABC. 116 pp.

Gardner, D. (1977). Nutrients as Tracers of Water Mass Structure in the Coastal Upwelling of Northwest Africa, In: A Voyage of Discovery, M. Angel (Ed.), Supplement to Deep-Sea Res., 24, 305-325.

Gentilli, J. (1972). Thermal Anomalies in the Eastern Indian Ocean. Nature, 238, Physical Sciences, 93-95.

Gill A.E. and E.H. Schumann (1974). The Generation of Long Shelf Waves by the Wind. J. Phys. Oceanogr., 4,(1), 83-90.

Gill, A.E. (1982). Atmosphere-Ocean Dynamics, International Geophysics Series, 30, Academic Press, 662 pp.

Gladfelter, W.H. (1964). Oceanography of the Greenland Sea, USS ATKA (AGB-3) Survey Summer 1962, Informal Report 0-64-63, U.S. Naval Oceanographic Office, Washington, D.C., 154 pp.

Godfrey, J.S. and K.R. Ridgway (1985). The Large-Scale Environment of the Poleward-Flowing Leeuwin Current, Western Australia: Longshore Steric Height Gradients, Wind Stresses, and Geostrophic Flow. J. Phys. Oceanogr., 15, 481-495.

Godfrey, J.S., D.J. Vaudrey, and S.D. Hahn (1986). Observations of the Shelf-Edge Current South of Australia, Winter 1982. J. Phys. Oceanogr., 16, 668-679.

Gomez Valdez, J. (1984). Estructura hidrografica promedio frente a Baja California. Ciencias Marinas, 9, 75-85.

Gordon, R.L. (1978). Internal Wave Climate Near the Coast of Northwest Africa During JOINT-1. Deep-Sea Res., 25, 625-643.

Goreau, T.J., W.A. Kaplan, J.C. Wofsy, M.B. McElroy, F.W. Valois, and S.W. Watson (1980). Production of NO2 and N2O by Nitrifying Bacteria at Reduced Concentrations of Oxygen. Appl. Environ. Microbiol., 40, 526-532.

Gould, W.J., J. Loynes, and J. Backhaus (1985). Seasonality in Slope Current Transports NW of Shetland. ICES Hydrography Committee, C.M. 1985/C:7.

Gregg, M.C. and C.S. Cox (1972). The Vertical Microstructure of Temperature and Salinity. Deep-Sea Res., 19, 355-376.

Griesman, P. (1976). Current Measurements in the Eastern Greenland Sea, PhD Thesis, University of Washington, Seattle, WA.

Greisman, P. and K. Aagaard (1979). Seasonal Variability of the West Spitsbergen Current, Ocean Modelling, 19, 3-5.

Griffiths, R.C. (1965). A Study of Ocean Fronts Off Cape San Lucas, Lower, California. U.S. Dept. of the Interior. Special Scientific Report. Fisheries. No. 499.

Griffiths, R.C. (1968). Physical, Chemical, and Biological Oceanography of the Entrance to the Gulf of California, Spring 1960. Special Scientific Report. U.S. Fish and Wildlife Service. Fisheries, 573, 47 pp.

Griffiths, R.W. and A.F. Pearce (1985). Instability and Eddy Pairs on the Leeuwin Current South of Australia. Deep-Sea Res., 32, 1511-1534.

Gunn, J.T., P. Hamilton, H.J. Herring, L.H. Kantha, G.S.E. Lagerloef, G.L. Mellor, R.D. Muench, and G.R. Stegan (1986). Santa Barbara Channel Circulation Model and Field Study, Appendix C: Observational Data. Report Number 92 (C2), Dynalysis of Princeton, 221 pp.

Gunther, E.R. (1936). A Report on Oceanographical Investigations in the Peru Coastal Current. Discovery Rep., 13, 107-276.

Hafferty, A.J., L.A. Codispoti, and A. Huyer (1978). JOINT-II R/V Melville Legs I, II, and IV, R/V Iselin Leg II Bottle Data, March 1977-May 1977. Coastal Upwelling Ecosystem Analysis Data Rep. 45, 779 pp.

Hafferty, A.J., D. Lowman, and L.A. Codispoti (1979). JOINT-II Melville and Iselin Bottle Data Sections, March-May 1977. Coastal Upwelling Ecosystems Analysis Data Rep. 38, 130 pp.

Hahn, J. (1981). Nitrous Oxide in the Oceans. In: Denitrification, Nitrification, and Nitrous Oxide, C.C. Delwiche (Ed.), Wiley, N.Y., 191-240.

Haidvogel, D.B. and K.H. Brink (1986). Mean Currents Driven by Topographic Drag over the Continental Shelf and Slope. J. Phys. Oceanogr., 16, 2159-2172.

Halpern, D. (1983). Special Issue III; Update 1982/83 Equatorial Pacific Warm Event. Tropical Ocean-Atmosphere Newsletter, 21, 1-34.
Halpern, D., R.L. Smith, and R.K. Reed (1978). On the California Undercurrent Over the Continental Slope Off Oregon. J. Geophys. Res., 83, 1366-1372.

Hamann, I., H.-Ch. John, and E. Mittelstaedt (1981). Hydrography and its Effect on Fish Larvae in the Mauretanian Upwelling Area, Deep-Sea Res., 28A, 561-575.

Hamilton, P. and M. Rattray, Jr. (1978). A Numerical Model of the Depth-Dependent, Wind-Driven Upwelling Circulation on a Continental Shelf. J. Phys. Oceanogr., 8, 437-457.

Hamon, B.V. (1965). Geostrophic Currents in the South-Eastern Indian Ocean. Aust. J. Mar. Freshwat. Res., 16, 255-271.

Haney, R.L. (1985). Midlatitude Sea Surface Temperature Anomalies: A Numerical Hindcast. J. Phys. Oceanogr., 15, 787-799.

Hanzlick, D.J. (1983). The West Spitsbergen Current: Transport, Forcing, and Variability, PhD Thesis, University of Washington, Seattle, WA.

Harris, T.F.W. and L.V. Shannon (1979). Satellite-Tracked Drifter in the Benguela Current System. S. Afr. J. Sci., 75(7), 316-317.

Hart, T.J. and R.I. Currie (1960). The Benguela Current. Discovery Reports, 31, 123-182.

Hastenrath, S. and P.J. Lamb (1977). Climatic Atlas of the Tropical Atlantic and Eastern Pacific Oceans. University of Wisconsin Press, Madison.

Hattori, A. (1983). Denitrification and Dissimilatory Nitrate Reduction. In: Nitrogen in the Marine Environment, E. Carpenter and D. Capone (Eds.), Academic Press, N.Y., 191-232.

Helland-Hansen, B. and F. Nansen (1909). The Norwegian Sea, Its Physical Oceanography. Rep. Norweg. Fish. Invest., 2.

Hickey, B.M. (1979). The California Current System-Hypotheses and Facts. Prog. Oceanogr., 8, 191-279.

Hickey, B.M. and Hamilton (1980). A Spin-Up Model as a Diagnostic Tool for Interpretation of Current and Density Measurements on the Continental Shelf of the Pacific Northwest. J. Phys. Oceanogr., 10, 12-24.

Hickey, B.M. (1981). Along-Shore Coherence on the Pacific Northwest Continental Shelf (January-April 1975). J. Phys. Oceanogr., 11, 822-835.

Hickey, B.M. and N. Pola (1983). The Seasonal Along-Shore Pressure Gradient on the West Coast of the United States. J. Geophys. Res., 88, 7823-7633.

Hickey, B.M. (1984). The Fluctuating Longshore Pressure Gradient on the Pacific Northwest Shelf: A Dynamical Analysis. J. Phys. Oceanogr., 14, 276-293.

Hickey, B.M. (1988). Poleward Flow Near the Northern and Southern Boundaries of the U.S. West Coast. This Volume.

Hickey, B.M. (1989a). Patterns and Processes of Circulation In and Over the Continental Shelf Off Washington. In: Coastal Dynamics of the Pacific Northwest. M. Landry and B. Hickey, Wiley.

Hickey, B.M. (1989b). Seasonal Variability Over the Santa Monica/San Pedro Basin and Shelf. J. Geophys. Res. (to be submitted).

Hill, J.W. and A.J. Lee (1957). The Effect of Wind on Water Transport in the Region of the Bear Island Fishery, Proc. Roy. Soc., B, 148, 104-116.

Holden, C.J. (1987). Observations of Low-frequency Currents and Continental Shelf Waves Along the West Coast of South Africa. In: The Benguela and Comparable Ecosystems. Payne, A.I.L., J.A. Gulland and K.H. Brink (Eds). S. Afr. J. Mar. Sci., 5, 197-208.

Holloway, G. and M.C. Hendershott (1974). The Effects of Bottom Relief on Barotropic Eddy Fields. MODE Hotline News, No. 65 (Unpublished Manuscript).

Holloway, G. (1986). A Shelf Wave/Topographic Pump Drives Mean Coastal Circulation. Parts I and II. Ocean Modeling, No. 68 and 69 (Unpublished Manuscripts).

Holloway, G. (1987). Systematic Forcing of Large Scale Geophysical Flows by Eddy-Topographic Interaction. J. Fluid Mech. (To Appear).

Holloway, G., K. Brink and D. Haidvogel (1988). Topographic Stress in Coastal Circulation Dynamics. (This volume.)

Holloway, P.E. and H.C. Nye (1985). Leeuwin Current and Wind Distributions on the Southern Part of the Australian North West Shelf between January 1982 and July 1983. Aust. J. Mar. Freshwat. Res., 36, 123-137.

Horn, W. and J. Meincke (1976). Note on the Tidal Current Field in the Continental Slope Area Off Northwest Africa. Memoires de la Societe' Royale des Sciences Liege, 6(10), 31-42.

Houghton, R. (1976). Circulation and Hydrographic Structure Over the Ghana Continental Shelf During the 1974 Upwelling, J. Phys. Oceanogr., 6, 909-924.

Hsiung, J. (1985). Estimates of the Global Oceanic Meridional Heat Transport. J. Phys. Oceanogr., 15, 1405-1413.

Huber, K., E. Mittelstaedt and G. Weichart (1977). Zur Hydrographie der Gewasser vor Marokko. Physikalische und chemische Daten. Meereskundliche Beobachtungen und Ergebnisse, Nr. 46, Deutches Hydrographisches Institut, Hamburg, 131 pp.

Hughes, P. and E.D. Barton (1974). Stratification and Water Mass Structure in the Upwelling Area Off NW Africa in April/May 1969, Deep-Sea Res., 21, 611-628.

Hurlburt, H.E. and J.D. Thompson (1973). Coastal Upwelling on a B-Plane. J. Phys. Oceanogr., 3, 16-32.

Huthnance, J.M. (1984). Slope Currents and "JEBAR." J. Phys. Oceanogr., 14, 795-816.

Huthnance, J.M. (1986). The Rockall Slope Current and Shelf-Edge Processes. Proceedings of the Royal Society of Edinburgh, 88B, 83-101.

Huthnance, J.M. (1987). Along-Shelf Evolution and Sea Levels Across the Continental Slope. Continental Shelf Research, (To Appear).

Huyer, A., B.M. Hickey, J.D. Smith, R.L. Smith and R.D. Pillsbury (1975a). Along-Shore Coherence at Low Frequencies in Currents Observed Over the Continental Shelf Off Oregon and Washington. J. Geophys. Res., 80 (24), 3495-3505.

Huyer, A., R.D. Pillsbury and R.L. Smith (1975b). Seasonal Variation of the Along-Shore Velocity Field Over the Continental Shelf Off Oregon. Limnol. Oceanogr., 20(1), 90-95.

Huyer, A. (1976). A Comparison of Upwelling Events in Two Locations: Oregon and Northwest Africa, J. Mar. Res., 34, 531-546.

Huyer, A. and R.L. Smith (1976). Observations of a Poleward Undercurrent Over the Continental Slope Off Oregon May-June 1975. Transactions of the Amer. Geophys. Union, 57, 263 pp.

Huyer, A. (1977). Seasonal Variation in Temperature, Salinity, and Density Continental Shelf Off Oregon. Limnol. Oceanogr., 22, 442-453.

Huyer, A., R.L. Smith and E.J.C. Sobey (1978). Seasonal Differences in Low-Frequency Current Fluctuations Over the Oregon Continental Shelf. J. Geophys. Res., 83 (C10), 5077-5089.

Huyer, A., E.J.C. Sobey and R.L. Smith (1979). The Spring Transition in Currents Over the Oregon Continental Shelf. J. Geophys. Res., 84, 6995.

Huyer, A. (1983). Upwelling in the California Current System. Prog. in Oceanog., 12, 259-284.

Huyer, A., R.L. Smith and B.M. Hickey (1984). Observations of a Warm-Core Eddy Off Oregon, January to March 1978. Deep Sea Res., 31 (2), 97-117.

Huyer, A. and R.L. Smith (1985). The Signature of El Nino Off Oregon, 1982-83. J. Geophys. Res., 90, 7133-7142.

Huyer, A. and P.M. Kosro (1987). Mesoscale Surveys Over the Shelf and Slope in the Upwelling Region Near Point Arena, California. J. Geophys. Res., 92, 1655-1681.

Huyer, A., P.M. Kosro, S.J. Lentz and R.C. Beardsley, This Volume. Poleward Flow in the California Current System.

Ikeda, M. and W.J. Emery (1984). Satellite Observation and Modeling of Meanders in the California Current System Off Oregon and Northern California. J. Phys. Oceanogr., 7, 1434-1450.

Inostroza, H. (1972). Some Oceanographic Features of Northern Chilean Waters in July, 1962. In: Oceanography of the South Pacific, New Zealand, 37-46.

Johannessen, O.M., J.A. Johannessen, J. Morison, B.A. Farrelly and E.A.S. Svendsen (1983). Oceanographic Conditions in the Marginal Ice Zone North of Svalbard in Early Fall 1979 with an Emphasis on Mesoscale Processes, J. Geophys. Res., 88 (C5), 2755-276

Johnson, D. (1980). The Gunther Undercurrent and Upwelling Along the North and Central Coast of Chile (Unpublished Manuscript).

Johnson, D., T. Fonseca and H. Sievers (1980). Upwelling in the Humboldt Coastal Current Near Valparaiso, Chile. J. of Marine Res., 38(1), 1-16.

Johnson, D.R., E.D. Barton, P. Hughes and C.N.K. Mooers (1975). Circulation in the Canary Current Region Off Cabo Bojador in August, 1972, Deep-Sea Res., 22, 547-557.

Johnson, D.R. and C.N.K. Mooers (1981). Internal Cross-Shelf Flow Reversals During Coastal Upwelling. In: Coastal Upwelling, F.A. Richards, (Ed.), American Geophysical Union, Washington, D.C.

Johnson, E.R. (1977). Stratified Taylor Columns on a Beta-plane. Geophys. Astrophys. Fluid Dyn., 9, 159-177.

Jones, E.L. (1918). The Neglected Waters of the Pacific Coast. Department of Commerce, U.S. Coast and Geodetic Survey Special Publication No. 48, Washington, D.C.

Jury, M. (1985). Case Studies of Along-Shore Variations in Wind-Driven Upwelling in the Southern Region. In: South Africa Ocean Colour and Upwelling Experiment. L. Shannon (Ed): 29-47.

Kamstra F. (1985). Environmental Features of the Southern Benguela With Special Reference to the Wind Stress. In: South African Ocean Colour and Upwelling Experiment. L.V. Shannon (Ed). Sea Fisheries Research Institute, Cape Town, p. 13-27.

Kamstra F. (1987). Interannual Variability in the Spectra of the Daily Surface Pressure at Four Stations in the Southern Hemisphere. Tellus 39A, 509-514.

Keeling, C.D. and R.B. Bacastow (1977). Impact of Industrial Gases on Climate, Energy and Climate. Geophysics Research Board, National Academy of Science National Research Council, Washington, D.C. 110-160.

Kislyakov, A.G. (1960). Fluctuations in the Regime of the Spitsbergen Current, Translation: Soviet Fisheries Investigations in North European Seas, 39-49.

Kitani, K. (1977). The Movements and Physical Characteristics of the Water Off Western Australia in November 1975. Bulletin of Far Seas Fisheries Research Laboratory, 15, 13-19.

Knowles, R. (1982). Denitrification. Microbiol. Rev., 46, 43-70.

Knox, F. and M.B. McElroy (1984). Changes in Atmospheric CO2: Influence of the Marine Biota at High Latitude. J. Geophys. Res., 89, 4629-4637.

Kosro, P.M. (1987). Structure of the Coastal Current Field off Northern California During the Coastal Ocean Dynamics Experiment. J. Geophys. Res., 92, 1637-1654.

Kremling, K. and A. Wenck (1984). Chemical Data From the NW African Upwelling Region. Berichte aus dem Institut fur Meereskunde, Kiel, Nr. 124, 130 pp.

Krey, J. and B. Babenerd (1976). Phytoplankton Production: Atlas of the International Indian Ocean Expedition. (Kiel: Institut fur Meereskunde).

Kundu, P.J. and J.P. McCreary (1987). A Model of the Throughflow from the Pacific to the Indian Ocean. J. Mar. Res., 45, In press.

Kundu, P.K. and J.S. Allen (1976). Some Three-Dimensional Characteristics of Low-Frequency Current Fluctuations Near the Oregon Coast. J. Phys. Oceanogr., 181-199.

Lacis, A., J. Hansen, P. Lee, T. Mitchell and S. Lebedeff (1981). Greenhouse Effect of Trace Gases, 1970-1980. Geophys. Res. Lett., 8, 1035-1038.

Lamberth, R. and G. Nelson (1987). Field and Analytical Drogue Studies Applicable to the St. Helena Bay Area Off South Africa's West Coast. In: The Benguela and Comparable Ecosystems. Payne, A.I.L., Gulland, J.A. and K.H. Brink (Eds). S. Afr. J. Mar. Sci.

LeFloch, J. (1973). Hydrologie et dynamique de Secteur ouest Marocain. Resultat de la campagne CINECA-Charcot I, Publications du CNEXO, Nr. 06, 27-63.

LeFloch, J. (1974). Quelques aspects de la dynamique et de l'hydrologie des couches superficielles dans l'ouest Marocain Campagnes CINECA-I et III. Tethys. 6, 53-68.

Legeckis, R. and G.R. Cresswell (1981). Satellite Observations of Sea Surface Temperature Fronts Off the Coast of Western and Southern Australia. Deep-Sea Res., 28A, 297-306.

Lenz, J., G. Schneider, M. Elbrachter, P. Fritsche, H. Johannsen and T. Weibe (1985). Hydrographic, chemical and planktological data from the Northwest African Upwelling area obtained from February to April 1983. Berichte aus dem Institut fur Meereskunde, Kiel, Nr. 140, 105 pp.

Lentz, S.J. (1987). A Description of the 1981 and 1982 Spring Transitions Over the Northern California Shelf. J. Geophys. Res., 92, 1545-1567.

Lipschultz, F., O.C. Zafiriou, S.C. Wofsy, M.B. McElroy, F.W. Valois and S.W. Watson (1981). Production of NO and N2O by Soil Nitrifying Bacteria. Nature (London), 294, 642-643.

Liu, K.-K. (1979) Geochemistry of Inorganic Nitrogen Compounds in Two Marine Environments: The Santa Barbara Basin and the Ocean Off Peru. Ph.D. Thesis, University of California, Los Angeles, 354 pp.

Liu, K.-K and I.R. Kaplan (1982). Nitrous Oxide in the Sea Off Southern California. In: The Environment of the Deep Sea, W.G. Ernst and J.G. Marin (Eds.), Prentice Hall, N.Y., 74-92.

Loder, J. (1980). Topographic Rectification of Tidal Currents on the Sides of Georges Bank. J. Phys. Oceanogr., 10, 1399-1416.

Love, C.M. (1970-1974). EASTROPAC Atlas, Cruise Data February, 1967 to March 1968. Circular 330, Volumes 1-11, U.S. Dept. Commerce, Washington, D.C.

Lueck, R.G., W.R. Crawford and T.R. Osborn, (1983). Turbulent Dissipation Over the Continental Slope Off Vancouver Island. J. Phys. Oceanogr., 13, 1809-1818.

Lueck, R.G. and T.R. Osborn, (1986). The Dissipation of Kinetic Energy in a Warm-Core Ring. J. Geophys. Res., 91, 803-818.

Lukas, R. (1986). The Termination of the Equatorial Undercurrent in the Eastern Pacific. Prog. Oceanogr., 16, 63-90.

Lynn, R.J. (1983). The 1982-83 Warm Episode in the California Current. Geophys. Res. Lett., 10, 1093-1095.

Lynn, R.J. and J.J. Simpson (1987). The California Current System: The Seasonal Variability of its Physical Characteristics. J. Geophys. Res., 92, 12,947-12,966.

Madelain, F. (1970). Influence de la topographie du fond sur l'ecoulement Mediterraneen entre le detroit de Gibraltar et le Cap Saint-Vincent, Cahiers Oceanographiques, 22, 43-61.

Margalef, R. (1978). What is an Upwelling Ecosystem? In: Upwelling Ecosystems, R.Boje and M. Tomczak (Eds.), Springer-Verlag, N.Y., 12-14.

Markina, N.P. (1976). Biogeographic Regionalization of Australian Waters of the Indian Ocean. Oceanology, 15, 602-604.

Marmen, H.A. (1926). Coastal Currents Along the Pacific Coast of the United States. Spec. Publ. U.S. Cst. Geodetic. Surv., 121, 91 pp.

Martell, C.M. and J.S. Allen (1979). The Generation of Continental Shelf Waves by Along-Shore Variations in Bottom Topography. J. Phys. Oceanogr., 9, 696-711.

Maxwell, J.G.H. and G.R. Cresswell (1981). Dispersal of Tropical Marine Fauna to the Great Australian Bight by the Leeuwin Current. Aust. J. Mar. Freshwat. Res., 32, 493-500.

McCartney, M.S. and L.D. Talley (1984). Warm-to-Cold Water Conversion in the Northern North Atlantic Ocean. J. Phys. Oceanogr., 14, 922-935.

McCreary, J.P. (1976). Eastern Tropical Ocean Response to Changing Wind Systems, with Application to El Nino. J. Phys. Oceanogr., 6, 632-645.

McCreary, J.P. (1981). A Linear Stratified Ocean Model of the Coastal Undercurrent. Philosophical Transactions, the Royal Society of London, A302: 385-413.

McCreary, J.P., Jr. and S.F. Chao (1985). Three-Dimensional Shelf Circulation Along an Eastern Ocean Boundary. J. Mar. Res., 43, 13-36.

McCreary, J.P., S.R. Shetye and P.J. Kundu (1986). Thermohaline Forcing of Eastern Boundary Currents: With Application to the Circulation Off the West Coast of Australia. J. Mar. Res., 44, 71-92.

McCreary, J.P., Jr., P.K. Kundu and S-Y. Chao (1987). On the Dynamics of the California Current System. J. Mar. Res., 45, 1-32.

McElroy, M.B. (1983). Marine Biological Controls on Atmospheric CO2 Climate. Nature, 302, 328-329.

McLain, D. and D. Thomas (1983). Year-to-Year Fluctuations of the California Counter Current and Effects on Marine Organisms. CalCOFI Rep., Vol. XXIV.

McLain, D.R. and D.H. Thomas (1983). Year-to-Year Fluctuations of the California Countercurrent and Effects on Marine Organisms. CalCOFI Reports, XXIV, 165-181.

McNider, R.T. and J.J. O'Brien (1973). A Multi-Layer Transient Model of Coastal Upwelling. J. Phys. Oceanogr., 3, 258-273.

Meincke, J., G. Siedler and W. Zenk (1975). Some Current Observations Near the Continental Slope Off Portugal, "Meteor" Forsch.-Ergebn., A, 16, 15-22.

Meincke, J. (1978). On the Distribution of Low Salinity Intermediate Waters Around the Faeroes. Dt. Hydrogr. Z., 31, 50-64.

Mejia, A. (1985). Efecto de una cuenca de forma arbitraria en la propagacion de una onda de Kelvin costera. MS Thesis, CICESE, Ensenada, 82 pp.

Midtun, L. (1985). Formation of Dense Bottom Water in the Barents Sea. Deep Sea Res., 32, 1233-1241.

Mittelstaedt, E. (1972). Der hydrographische Aufbau und die zeitliche variabilitat der Schichtung und Stromung im nordwest Afrikanischen Auftriebsgebeit im Frujahr 1968, "Meteor" Forsch.-Ergebn., A, 11, 1-57.

Mittelstaedt, E., R.D. Pillsbury and R.L. Smith (1975). Flow Patterns in the Northwest African Upwelling Area, Dt. Hydrogr. Z., 28, 145-167.

Mittelstaedt, E. (1976). On the Currents Along the Northwest African Coast South of 22°N, Dt. Hydrog. Z., 29, 97-117.

Mittelstaedt, E., D. Halpern, R.L. Smith and J.S. Bottero (1980). Tidal Currents Off Manritania at 21°40'N. Dt. Hydrogr. Z., 33(6), 223-235.

Mittelstaedt, E. and I. Hamann (1981). The Coastal Circulation Off Mauretania, Dt. Hydrogr. Z., 34, 81-118.

Mittelstaedt, E. (1983). The Upwelling Area Off Northwest Africa--A Description of Phenomena Related to Coastal Upwelling, Prog. Oceanogr., 12, 307-331.

Mittelstaedt, E., This volume. On the Subsurface Circulation Along the Moroccan Slope.

Montgomery, R.B. (1938). Circulation in Upper Layers of the Southern North Atlantic Deduced With Use of Isentropic Analysis, Pap. Phys. Oc. Met. MIT, 6, 2.

Montgomery, R.B. (1941). Transport of the Florida Current Off Habana. J. Mar. Res., 4, 198-220.

Mooers, C.N.K. (1975). Several Effects of a Baroclinic Current on the Cross-Stream Propagation of Inertial-Internal Waves. Geophys. Fluid Dyn., 6, 245-275.

Mooers, C.N.K., C.A. Collins and R.L. Smith (1976). The Dynamic Structure of the Frontal Zone in the Coastal Upwelling Region Off Oregon. J. Phys. Oceanogr., 6, 3-21.

Mooers, C.N.K. and A.R. Robinson (1984). Turbulent Jets and Eddies in the California Current and Inferred Cross-Shore Transports. Science, 223, 51-53.

Morales, C., M.L. Argote, A. Amador and E.D. Barton (1978). Mediciones de vientos, corrientes e hidrografia frente a Punta Colonet, B.C. en junio de 1976. Data Report, CICESE, Ensenada. 236 pp.

Munk, W.H. (1950). On the Wind-Driven Ocean Circulation. Journal of Meteorology, 7(2), 79-93.

Nederlandsch Meteorologisch Institut (1949). Seas Around Australia: Oceanographic and Meteorological Data. Koninklijk Nederlands Meteorologisch Institut Publication No. 124, 79 pp. De Bilt.

Nelson C.S. (1977). Wind Stress and Wind Stress Curl Over the California Current. NOAA Technical Report NMFS SSRF-714, U.S. Department of Commerce, Washington, D.C., 89 pp.

Nelson, G. and L. Hutchings (1983). The Benguela Upwelling Area. Prog. in Oceanog., 12(3), 333-356.

Nelson, G. (1985). Notes on the Physical Oceanography of the Cape Peninsula Upwelling System. In: South African Ocean Colour and Upwelling Experiment. Shannon, L.V. (Ed.). Cape Town; Sea Fisheries Research Institute, 63-95.

Nelson G. and A. Polito (1987). Information on Currents in the Cape Peninsula Area, South Africa. In: The Benguela and Comparable Ecosystems. Payne, A.I.L., Gulland, J.A. and K.H. Brink (Eds.). S. Afr. J. Mar. Sci., 5, 287-304.

Neshyba, S. and R. Mendez (1976). Analisis de temperaturas superficiales del mar como indicadores de movimientos de aguas superficiales en el Pacifico Sur-Este. Rev. Com. Perm. Pacifico sur, 5: 139-146.

Newton, C.W. (1971). Global Angular Momentum Balance: Earth Torques and Atmospheric Fluxes. J. Atmos. Sci., 28, 1329-1341.

O'Brien, J.J. and H.E. Hurlburt (1972). A Numerical Model of Coastal Upwelling. J. Phys. Oceanogr., 2, 14-26.

Osborn, T.R. and C.S. Cox, (1972). Oceanic Fine Structure. Geophys. Fluid Dyn., 3, 321-345.

Pace, M.L., G.A. Knauer, D.M. Karl and J.H. Martin (1987). Primary Production, New Production and Vertical Flux in the Eastern Pacific Ocean. Nature, (London), 325, 803-804.

Packard, T.T., P.C. Garfield and L.A. Codispoti (1983). Oxygen Consumption and Denitrification Below the Peruvian Upwelling. In: Coastal Upwelling, Pt. A., E. Suess and J. Theide (Eds.), Plenum Press, N.Y., 147-173.

Pak, H., L.A. Codispoti and R.V. Zaneveld (1980). On the Intermediate Particle Maxima Associated with Oxygen-Poor Water Off Western South America. Deep-Sea Res., 27A, 783-797.

Paquette, R.G., R.H. Bourke, J.L. Newton and W.F. Perdue (1985). The East Greenland Polar Front in Autumn, J. Geophys. Res., 90 (C3), 4866-4882.

Parrish, R.H., A. Bakun, D.M. Husby and C.S. Nelson (1983). Comparative Climatology of Selected Environmental Processes in Relation to Eastern Boundary Current Fish Reproduction. In: Proceedings of the Expert Consultation to Examine Changes in Abundance and Species Composition of Neritic Fish Resources. San Jose, Costa Rica, April 1983. Sharp G. D. and J. Csirke (Eds). F.A.O. Fish Rep. 291 (3), 731-777.

Patzert, W.C., T.J. Cowles and C.S. Ramage (1978). El Nino Watch Atlas of Physical, Chemical, and Biological Oceanographic and Meteorological Data. Ref. Ser. No. 78-7, Scripps Inst. Oceanography, 322 pp.

Pearce, A.F. and B.F. Phillips (1988). ENSO Events, the Leeuwin Current and Larval Recruitment of the Western Rock Lobster. J. Du Conseil., accepted.

Pedlosky, J. (1974). Longshore Currents and the Onset of Upwelling Over Bottom Slope. J. Phys. Oceanogr., 4, 310-320.

Peffley, M.B. and J.J. O'Brien (1976). A Three-Dimensional Simulation of Coastal Upwelling Off Oregon. J. Phys. Oceanogr., 6, 164-180.

Pelaez, J. and J.A. McGowan (1986). Phytoplankton Pigment Patterns in the California Current as Determined by Satellite. Limnol. Oceanogr., 31, 927-950.

Perkin, R.G. and E.L. Lewis (1984). Mixing in the West Spitsbergen Current. J. Phys. Oceanogr., 14, 1315-1325.

Philander, S.G.H. and J-H. Yoon (1982). Eastern Boundary Currents and Coastal Upwelling. J. Phys. Oceanogr., 12, 862-897.

Pierotti, D. and R.A. Rasmussen (1980). Nitrous Oxide Measurements in the Eastern Tropical Pacific Ocean. Tellus, 32, 56-72.

Piper, D.Z. and L.A. Codispoti (1975). Marine Phosphorite Deposits and the Nitrogen Cycle. Science, 188, 15-18.

Poulain, P.M., J.D. Illeman and P.P. Niiler (1987). Drifter Observations in the California Current System (1985-1986). SIO Ref. 87-27, 72 pp.

Preston-Whyte R.A. and P.D. Tyson (1973). Note on Pressure Oscillations Over South Africa. Mon. Wea. Rev. 101, 650-653.

Puls, C. (1895). Oberflachentemperaturen und stromungsverhaltnisse des Aequatorialgurtels des Stillen Ozeans. Doctoral Dissertation, Marburg.

Quadfasel D., J.-C. Gascard and K.-P. Koltermann (1987). Large-Scale Oceanography in Fram Strait During the 1984 Marginal Ice Zone Experiment. J. Geophys. Res., 92 (C7), 6719-6728.

Redfield, A.C., B.H. Ketchum and F.A. Richards (1963). The Influence of Organisms on the Composition of Seawater. In: M.N. Hill (Ed.), The Sea, Vol. 2, Interscience, N.Y., 26-77.

Reed, R.K. and D. Halpern, (1976). Observations of the California Undercurrent Off Washington and Vancouver Island. Limnol. Oceanogr., 21, 389-398.

Reid, B.J. (1987). The Fall Transition of Oregon Shelf Waters. M.S. Thesis, College of Oceanography, Oregon State University, Corvallis, Oregon, 103 pp.

Reid, J.L., G.I. Roden and J.G. Wyllie (1958). Studies of the California Current System. Calif. Coop. Oceanic Ris. Invest. Rep., July 1, 1956-January 1, 1958, 28-56.

Reid, J.L., Jr., G.I. Roden and J.G. Wyllie (1958). Studies of the California Current System. CalCOFI Prog. Report 1, 1956 - 1. 29-57.

Reid, J.L., Jr. (1960). Oceanography of the Northeastern Pacific During the Last Ten Years. CalCOFI Prog. Report 7, 77-90.

Reid, J.L., Jr. (1961). On the Geostrophic Flow at the Surface of the Pacific Ocean with Respect to the 1000-Decibar Surface. Tellus, 13, 489-502.

Reid, J.L., Jr. (1962). Measurements of the California Countercurrent at a Depth of 250 Meters. J. Mar. Res., 20, 134-137.

Reid, J.L., Jr. and R.A. Schwartzlose (1962). Direct Measurements of the Davidson Current Off Central California. J. Geophys. Res., 67, 2491-2497.

Reid, J.L., Jr. (1963). Measurement of the California Countercurrent Off Baja California. J. Geophys. Res., 68, 4819-4822.

Reid, J.L., Jr., R.A. Schwartzlose, and D.M. Brown (1963). Direct Measurements of a Small Surface Eddy Off Northern Baja California. J. Mar. Res., 21, 205-218.

Reid, J.L., Jr. and A.W. Mantyla (1976). The Effect of the Geostrophic Flow Upon Coastal Elevations in the Northern Pacific Ocean. J. Geophys. Res., 81, 3100-3110.

Reid, J.L., Jr. and A.W. Mantyla (1978). On the Mid-Depth Circulation of the North Pacific Ocean. J. Phys. Oceanogr., 8, 946-951.

Reid, R.O. (1948). The Equatorial Currents of the Eastern Pacific as Maintained by the Stress of the Wind. J. Mar. Res., 7, 74-99.

Ridgway, K.R. and R.G. Loch (1987). Mean Temperature-Salinity Relationships in Australian Waters and Their Use Is Water Mass Analysis. Aust. J. Mar. Freshwat. Res., 38, 553-567.

Rienecker, M.M., C.N.K. Mooers, D.E. Hagan and A.R. Robinson (1985). A Cool Anomaly Off Northern California: An Investigation Using IR Imagery and in situ Data. J. Geophys. Res., 90, 4807-4818.

Rienecker, M.M. and C.N.K. Mooers (1989a). A Summary of the OPTOMA Program's Mesoscale Ocean Prediction Studies in the California Current System. In: Three Dimensional Ocean Models of Marine and Estuarine Dynamics (J.C.J. Nihoul and B.M. Jamant, Eds.). Elsevier Scientific Publishing, Amsterdam. In press.

Rienecker, M.M. and C.N.K. Mooers (1989b). Mesoscale Eddies, Jets and Fronts off Pt. Arena, CA, July 1986. J. Geophys. Res. In press.

Robinson, A.R., J.A. Carton, C.N.K. Mooers, L.J. Walstad, E.F. Carter, M.M. Rienecker, J.A. Smith and W.G. Leslie (1984). A Real-time Dynamical Forecast of Ocean Synoptic/Mesoscale Eddies. Nature, 309, 781-783.

Robles, F. (1976). Descripcion general de las condiciones oceanograficas en Aguas Chileans. Instituto de Fomento Pesquero. Santiago de Chile, 90 pp.

Robles, J.M., C. Morales, J. Garcia and C. Flores (1981). Informe de datos hidrograficos de la region de Cabo Colonet, B.C. Crucero Subac 2, Julio de 1980. CICESE, 165 pp.

Rochford, D.J. (1969a). Seasonal Variations in the Indian Ocean along 110°E.I Hydrological Structure of the Upper 500 m. Aust. J. Mar. Freshwat. Res., 20, 1-50.

Rochford, D.J. (1969b). Seasonal Interchange of High and Low Salinity Surface Waters Off South-West Australia. Division of Fisheries and Oceanography Technical Paper No. 29.

Roden, G.I. (1962). Oceanographical Aspects of Eastern Equatorial Pacific. Geofisica Internacional, 2, 77-92.

Roden, G.I. (1970). Aspects of the Mid-Pacific Transition Zone. J. Geophys. Res., 75, 1097-1109.

Roden, G.I. (1971). Aspects of the Transition Zone in the Northeastern Pacific. J. Geophys. Res., 76, 3462-3475.

Roden, G.I. (1972). Thermohaline Structure and Baroclinic Flow Across the Gulf of California Entrance and in the Revilla Gigedo Islands Region. J. Phys. Oceanogr., 2, 177-183.

Roemmich, D. and C. Wunsch (1985). Two Translantic Sections: Meridional Circulation and Heat Flux in the Subtropical North Atlantic Ocean. Deep Sea Res., 32(6), 619-664.

Rogers, J. (1987). Seismic, Bathymetric and Photographic Evidence of Widespread Erosion and a Manganese-Nodule Pavement Along the Continental Rise of the Southeast Cape Basin. Marine Geol., 78, 57-76.

Rojas, O., A. Mujica, G. Ledermann and H. Miles (1983). Estimacion de abundancia relativa de huevos y larvas de peces. CORFO, Gerencia de Desarrollo (AP 83-31). IFOP, Santiago, 98 pp.

Salmon, R., G. Holloway and M.C. Hendershott (1976). The Equilibrium Statistical Mechanics of Simple Quasi-geostrophic Models. J. Fluid Mech., 75, 375-386.

Samelson, R.M. and J.S. Allen (1987). Quasi-geostrophic Topographically Generated Mean Flow over the Continental Margin. J. Phys. Oceanogr., 17, 2043-2064.

Sarmiento, J.L. and J.R. Toggweiler (1984). A New Model for the Role of the Oceans in Determining Atmospheric pCO2. Nature (London), 308, 621-624.

Saville-Kent, W. (1897). The Naturalist in Australia. Chapman & Hall, London, 338 pp.

Schemainda, R., D. Nehrin, S. Schulz (1975). Ozeanologische Untersuchungen zum Produktionspotential der nordwestafrikanischen Wasserauftriebsregion 1970-1973. Geod. Geoph. Veroff R IV, 16, 1-85.

Schwartzlose, R.A. (1963). Nearshore Currents of the Western United States and Baja California as Measured by Drift Bottles. CalCOFI. Reports, 9, 15-22.

Schwartzlose, R.A. and J.L. Reid, (1972). Near-shore Circulation in the California Current. CalCOFI Rept., 16, 57-65.

Send, U., R.C. Beardsley and C.D. Winant (1987). Relaxation from Upwelling in the Coastal Ocean Dynamics Experiment. J. Geophys. Res., 92, 1683-1698.

Shannon, L.V. (1985). The Benguela Ecosystem I. Evolution of the Benguela, Physical Features and Processes. In: Oceanography and Marine Biology. An Annual Review 23. Barnes, M. (Ed.). Aberdeen; University Press, 105-182.

Shannon, L.V. N.M. Walters and S.A. Mostert (1985). Satellite Observations of Surface Temperature and Near-surface Chloroplhyll in the Southern Benguela Region. In: South African Ocean Colour and Upwelling Experiment. Shannon L.V. (Ed.). Sea Fisheries Research Institute, Cape Town, 183-210.

Shannon, L.V. and D. Hunter (1988). Notes on Antarctic Intermediate Water Around Southern Africa. S. Afr. J. Mar. Sci., 6, 107-117.

Shelton, P.A. and L. Hutchings (1982). Transport of Anchovy, Engraulis Capensis Gilchrist, Eggs and Early Larvae by a Frontal Jet Current. J. Cons. Perm. Int. Explor. Mer. 40 (2), 185-198.

Sievers, S., H.A. and N. Silva (1975). Masas de agua y circulacion en el Oceano Pacifico Sudoriental. Latitudes 18°S to 33°S (Operacion Oceanograficas "MARCHILE VIII"). Cienc. y tec. del Mar. Contrib. CONA No 1, 7-67.

Silva, N. and T. Fonseca (1983). Geostrophic Component of the Oceanic Flow Off Northern Chile. Conferencia Internacional Sobre Recursos Marinos del Pacifico. P. Arana (Ed.), 59-70.

Silva, S., N. and S. Neshyba (1979). On the Southernmost Extension of the Peru-Chile Undercurrent. Deep-Sea Res., 26, 1387-1393.

Simons, T.J. (1983). Resonant Topographic Response of Nearshore Currents to Wind Forcing. J. Phys. Oceanogr., 13, 512-523.

Simpson, J.J. (1983). Large-Scale Thermal Anomalies in the California Current During the 1982-1983 El Nino. Geophys. Res. Lett., 10, 937-940.

Simpson, J.J. (1984). El Nino-Induced Onshore Transport in the California Current During 1982-1983. Geophys. Res. Lett., 11, 241-242.

Smith, J.D., B.M. Hickey and J. Beck (1976). Observations from Moored Current Meters on the Washington Continental Shelf from February 1971 to February 1974. Special Report Number 65, University of Washington, Department of Oceanography, 347 pp.

Smith, R.L. (1968). Upwelling. Ocean. Mar. Biol. Ann. Rev., 6, 11-46.

Smith, R.L. (1981). A Comparison of the Structure and Variability of the Flow-Field in Three Coastal Upwelling Regions: Oregon, Northwest Africa, and Peru. In: Coastal Upwelling, F.A. Richard (Ed.), American Geophysical Union, Washington, D.C., 107-118

Smith, R.L. (1983). Peru Coastal Currents During El Nino: 1976 and 1982. Science, 221, 1397-1399.

Sorokin, Y.I. (1978). Description of Primary Production and of the Heterotrophic Microplankton in the Peruvian Upwelling Region. Oceanology, 18, 62-71.

Stabeno, P.J. and R.L. Smith (1987). Deep-Sea Currents Off Northern California, J. Geophys. Res., 92, 755-771.

Stevenson, M.R., J.G. Pattullo and B. Wyatt (1969). Subsurface Currents Off the Oregon Coast as Measured by Parachute Drogues. Deep-Sea Res., 16, 449-461.

Stevenson, M.R., R.W. Garvine and B. Wyatt (1974). Lagrangian Measurements in a Coastal Upwelling Zone Off Oregon. J. Phys. Oceanogr., 4, 321-336.

Stommel, H., E.D. Stroup, J.L. Reid and B.A. Warren (1973). Transpacific Hydrographic Sections at Lats 43°S and 281°S: The Scorpio Expedition..I. Preface. Deep-Sea Res., 20, 1-7.

Stramma, L. (1974). Geostrophic Transport in the Warm Water Sphere of the Eastern Subtropical North Atlantic, J. Mar. Res., 42, 537-558.

Stramma, L. (1984a). Geostrophic Transport in the Warm Water Sphere of the Subtropical North Atlantic. J. Mar. Res. 42, 537-558.

Stramma, L. (1984b). Potential Vorticity and Volume Transport in the Eastern North Atlantic From Two Long CTD Sections. Deutche Hydrographische Zeitschrift, 37, (4), 147-155.

Strub, P.T., J.S. Allen, A. Huyer, R.L. Smith and R.C. Beardsley, (1987a). Seasonal Cycles of Currents, Temperatures, Winds, and Sea Level Over the Northeast Pacific Continental Shelf. J. Geophys. Res., 92, 1507-1526.

Strub, P.T., J.S. Allen, A. Huyer and R.L. Smith (1987b). Large-Scale Structure of the Spring Transition in the Coastal Ocean Off Western North America. J. Geophys. Res., 92, 1527-1544.

Stumpf, H.G. and R. Legeckis (1977). Satellite Observations Mesoscale Eddy Dynamics in the Eastern Tropical Pacific Ocean. J. Phys. Oceanogr., 7, 648-658.

Sturges, W. (1974). Sea Level Slope Along Continental Boundaries. J. Geophys. Res., 79, 825-830.

Suess, E. (1980). Particulate Organic Carbon Flux in the Oceans: Surface Productivity and Oxygen Utilization. Nature (London), 288, 260-263.

Suginohara, N. (1974). Onset of Coastal Upwelling in a Two Layer Ocean by Wind-Stress with Longshore Variation. J. Oceanogr. Soc. Japan, 30, 23-33.

Suginohara, N. (1982). Coastal Upwelling: Onshore-Offshore Circulation, Equatorward Coastal Jet and Poleward Undercurrent over a Continental Shelf Slope. J. Phys. Oceanogr., 12, 272-284.

Suginohara, N. and Y. Kitamura (1984). Long-Term Coastal Upwelling over Continental Shelf-Slope. J. Phys. Oceanogr., 14, 1095-1104.

Sverdrup, H.U., M.W. Johnsen and R.H. Fleming (1942). The Oceans. Prentice-Hall.

Sverdrup, H.V. and R.H. Fleming (1941). The Waters Off the Coast of Southern California, March to July, 1937. Bulletin, Scripps Institution of Oceanography, 4, 261-378.

Swallow, J.C., W.J. Gould and P.M. Saunders (1977). Evidence for a Poleward Eastern Boundary Current in the North Atlantic Ocean, Int. Counc. Explor. Sea, C.M. 1977/C:32, Hydrogr. Comm., 11 pp.

Swift, J.H. (1986). The Arctic Waters, In: The Nordic Seas, ed. B.G. Hurdle, Springer-Verlag, New York, 129-153.

Thompson, R.O.R.Y. and G. Veronis (1983). Poleward Boundary Current Off Western Australia. Aust. J. Mar. Freshwat. Res., 34, 173-185.

Thompson, R.O.R.Y. (1984). Observations of the Leeuwin Current Off Western Australia. J. Phys. Oceanogr., 14, 623-628.

Thompson, R.O.R.Y. (1987). Continental-Shelf-Scale Model of the Leeuwin Current. J. Mar. Res., 45, 813-827.

Thomson, R.E. (1984). A Cyclonic Eddy Over the Continental Margin off Vancouver Island: Evidence for Baroclinic Instability. J. Phys. Oceanogr., 14, 1326-1348.

Thomson, R.E., W.R. Crawford, H.J. Freeland and W.S. Huggett (1985). Low-Pass Filtered Current Meter Records for the West Coast of Vancouver Island: Coastal Oceanic Dynamics Experiment, 1979-81. Can. Data Rep. Hydrogr. Ocean Sci., 40, 102 pp.

Thomson, R.E. and R.E. Wilson (1987). Coastal Countercurrent and Mesoscale Eddy Formation by Tidal Rectification Near an Oceanic Cape. J. Phys. Oceanogr., 17, 2096-2126.

Thorpe, S.A. (1976). Variability of the Mediterranean Undercurrent in the Gulf of Cadiz, Deep-Sea Res., 23, 711-727.

Tibby, R.B. (1941). The Water Masses Off the West Coast of North America. J. Mar. Res., 4, 112-121.

Timofeyev, V.T. (1962). The Movement of Atlantic Water and Heat into the Arctic Basin, Deep-Sea Res., 9, 358-361.

Tomczak, M. (1973). An Investigation into the Occurrence and Development of Cold Water Patches in the Upwelling Region of NW Africa, "Meteor" Forsch.-Ergebn., A, 13, 1-42.

Tomczak, M. 1979. The CINECA Experience. Mar. Policy 3, 59-65.

Tomczak, M. and P. Hughes (1980). Three Dimensional Variability of Water Masses and Currents in the Canary Current Upwelling Region, "Meteor" Forsch.-Ergebn., A, 21, 1-24.

Tomczak, M. (1981). Coastal Upwelling Systems and Eastern Boundary Currents: A Review of Terminology. Geoforum, 12(2), 179-191.

Tont, S.A. (1981). Upwelling: Effects on Air Temperatures and Solar Irradiance. In: Coastal Upwelling, F.A. Richards (Ed.). American Geophysical Union, Washington: 57-62.

Torres Moye, G. and Acosta Ruiz, M.J. (1986). Some Chemical Properties Indicating Coastal Upwelling Events and Subsurface Countercurrent in an Area Punta Colonet, Baja California. Ciencias Marina, 12, 10-25.

Tsuchiya, M. (1974). Variation of the Surface Geostrophic Flow in the Eastern Intertropical Pacific Ocean. Fishery Bulletin, 72, 1075-1086.

Tsuchiya, M. (1975). Subsurface Countercurrents in the Eastern Equatorial Pacific Ocean. J. Mar. Res., 33, (Suppl.), 145-175.

Tsuchiya, M. (1980). Inshore Circulation in the Southern California Bight, 1974-1977. Deep-Sea Res., 27, 99-118.

Tsuchiya, M. 1981. The Origin of the Pacific Equatorial 130°C Water. J. Phys. Oceanogr., 11, 755-770.

Tsuchiya, M. (1983). Use of Nutrient Concentrations for Monitoring the Equatorial Undercurrent. Trop. Ocean-Atmos. Newsl., 20, 3-5.

Tsunogai, S. (1971). Ammonia in the Oceanic Atmosphere and the Cycle of Nitrogen Compounds Through the Atmosphere and the Hydrosphere. Geochem. J., 5, 57-67.

Tucholke, B.E. and R.W. Embley (1984). Cenozoic Regional Erosion of the Abyssal Sea Floor Off South Africa. Memoir American Association. Petroleum Geologists, 36, 145-164.

Tyson, P.D. (1986). Climatic Change and Variability in Southern Africa. Oxford University Press, 220 pp.

Ulrych, T.J., D.E. Smylie, O.G. Jensen and G.K.C. Clarke (1973). Predictive Filtering and Smoothing of Short Records by Using Maximum Entropy. J. Geophys. Res., 78, 4959-4964.

Urrutia-Fucugauchi, J. (1988). Is Mexico Part of North America? EOS, 69, 610.

Wang, D.-P. (1982). Development of a Three-Dimensional, Limited-Area (Island) Shelf Circulation Model. J. Phys. Oceanogr., 12, 605-617.

Warren, B.A. (1981). Transindian Hydrographic Section at Lat. 18°S: Property Distributions and Circulation in the South Indian Ocean. Deep-Sea Res., 28A, 759-788.

Wattenberg, H. (1939). Die Verteilung des Sauerstoffs im Atlantischen Ozean. Atlas. Wissenschaftliche Ergebnisse der Deutschen Atlantischen Exp. "Meteor" 1925-1927, IX.

Webster, I., T.J. Golding and N. Dyson (1979). Hydrological Features of the Near-Shelf Waters Off Fremantle, Western Australia, During 1974. CSIRO Australia Division of Fisheries and Oceanography Report No. 106, 30 pp.

Weichart, G. (1974). Meereschemische Untersuchungen im nordwestafrikanischen Auftriebsgebiet 1968. "Meteor" Forsch.-Ergebn, A, No. 14, 33-70.

Weikert, H. (1986). Occurrence of Calanoides curinatus (Copepoda, Calanoida) Along the Continental Slope Off Morocco-Evidence of a Large-Scale Transport by a Northward Boundary Flow. (Unpublished).

Weiss, R.F. (1981). The Temporal and Spatial Distribution of Trophospheric Nitrous Oxide. J. Geophys. Res., 86, 7185-7195.

Wells, F.E. (1985). Zoogeographical Importance of Tropical Marine Mollusc Species at Rottnest Island, Western Australia. W.A. Naturaliste, 16, 40-45.

Werner, F. and B.M. Hickey (1983). The Role of a Longshore Pressure Gradient in Pacific Northwest Dynamics. J. Phys. Oceanogr., 13, 395-410.

Wickham, J.B. (1975). Observations of the California Countercurrent. J. Mar. Res., 33, 325-340.

Wickham, J.B., A.A. Bird and C.N.K. Mooers (1987). Mean and Variable Flow Over the Central California Continental Margin, 1978 to 1980. Continental Shelf Res., 7, 827-849.

Winant, C.D. (1980). Downwelling over the Southern California Shelf During the Summer. J. Phys. Oceanogr., 10, 791-799.

Winant, C.D. and A.W. Bratkovitch (1981). Temperature and Currents on the Southern California Shelf: A Description of the variability. J. Phys. Oceanogr., 11, 71-86.

Winant, C.D. (1983). Longshore Coherence of Currents on the Southern California Shelf During the Summer. J. Phys. Oceanogr., 13, 54-64.

Winant, C.D., R.C. Beardsley and R.E. Davis (1987). Moored Wind Temperature, and Current Observations Made During Coastal Ocean Dynamics Experiments 1 and 2 over the Northern California Continental Shelf and Upper Slope. J. Geophys. Res., 92, 1569-1604.

Wolf, G., Kaiser, W. (1978). Uber den Jahreszyklus der T-S-Eigenschaften quasipermanenter Wasserarten und Variationen produktionsbiologischer Parameter auf dem Schelf vor Cap Blanc. Geod. Geoph. Veroff, R. IV, 24, 1-81.

Wood, E.J.F. (1954). Dinoflagellates in the Australian Region. Aust. J. Mar. Freshwat. Res., 5, 171-351.

Wooster, W. and M. Gilmartin (1961). The Peru-Chile Undercurrent. J. Mar. Res., 19(3): 97-122.

Wooster, W.S. and J.L. Reid, Jr. (1963). Eastern Boundary Currents, p. 253-280. In: The Sea, Vol. II, M.N. Hill, (Ed.). Interscience Publ., New York, 554 pp.

Wooster, W.S., T.J. Chow and I. Barrett (1965). Nitrite Distribution in the Peru Current Waters. J. Mar. Res., 23, 210-221.

Wooster, W.S. (1970). Eastern Boundary Currents in the South Pacific. In: Scientific Exploration of the South Pacific, W.S. Wooster, (Ed.), National Academy of Sciences, Washington, D.C.

Wooster, W.S. and J.H. Jones (1970). California Undercurrent Off Northern Baja, California. J. Marine Res., 28, 235-250.

Wooster, W., A. Bakun and D.R. McClain (1976). The Seasonal Upwelling Cycle Along the Eastern Boundary of the North Atlantic, J. Mar. Res., 34, 131-141.

Worthington, L.V. (1976). On the North Atlantic Circulation. The Johns Hopkins Oceanographic Studies, No. 6, 110 pp.

Wright, D.G. (1980). On the stability of a fluid with specialized density stratification. Part II. Mixed baroclinic-barotropic instability with application to the Northeast Pacific. J. Phys. Oceanogr., 10, 1307-1322.

Wust, G. (1935). Die Stratosphare des Atlantischen Ozeans. Wissenschaftliche Ergebnisse der Deutschen Atlantischen Expedition "Meteor", 1925-1927, VI, 1, 288 pp.

Wust, G. (1957). Quantitative Untersuchungen zur Statik und Dynamik des Atlantischen Ozeans. Stromgeschwindigkeiten und Strommengen in den Tiefen des Atlantischen Ozeans. Wissenschaftliche Ergebnisse der Deutschen Atlantischen Expedition "Meteor", 1925-1927, VI, 2(6), 420 pp.

Wyllie, J.G. (1966). Geostrophic Flow of the California Current at the Surface and at 200 meters. California Cooperative Oceanic Fisheries Investigations Atlas 4. State of California Marine Research Committee.

Wyrtki, K. (1963). The Horizontal and Vertical Field of Motion in the Peru Current. Bull. Scripps Inst. Oceanogr., 8, 313-346.

Wyrtki, K. (1964). The Thermal Structure of the Eastern Pacific Ocean. Deutsche Hydrographische Zeitschrift. Erganzungsheft, A(6), 84 pp.

Wyrtki, K. (1964). Upwelling in the Costa Rica Dome. U.S. Fish and Wildlife Service, Fisheries Bulletin, 63, 355-372.

Wyrtki, K. (1965). Summary of the Physical Oceanography of the Eastern Pacific Ocean. Institute of Marine Resources, University of California, San Diego, Ref. 65-10. 78 pp.

Wyrtki, K. (1966). Oceanography of the Eastern Equatorial Pacific Ocean. Oceanogr. Mar. Biol. Ann. Rev., 4, 33-68.

Wyrtki, K. (1967). Circulation and Water Masses in the Eastern Pacific Ocean. Journal of Oceanology and Limnology, 1, 117-147.

Wyrtki, K. (1977). Advection in the Peru Current as Observed by Satellite. J. Geophys. Res., 82, 3939-3944.

Yamazaki, H. and R.G. Lueck (1987). Turbulence in the California Undercurrent. J. Phys. Oceanogr., 17, 1378-1396.

Yoon, J.H. and S.G.H. Philander (1982). The Generation of Coastal Undercurrents. J. Ocean. Soc. Japan, 38, 215-224.

Yoshida, K. and H.L. Mao (1957). A Theory of Upwelling of Large Horizontal Extend. J. Marine Res., 16, 40-57.

Yoshida, K. and M. Tsuchiya (1957). Northward Flow in Lower Layers as Indicator of Coastal Upwelling. Records of Oceanographic Works in Japan, 4, 14-22.

Yoshida, K. (1967). Circulation in the Eastern Tropical Oceans With Special Reference to Undercurrents and Upwelling. Japan Journal of Geophysics, 4,1-75.

Yoshinari, T. (1976). Nitrous Oxide in the Sea. Mar. Chem., 4, 189-202.

Zafiriou, O.C. and M.C. True (1979a). Nitrite photolysis in seawater by sunlight. Mar. Chem., 8, 9-32.

Zafiriou, O.C. and M.C. True (1979b). Nitrite Photolysis in Seawater by Sunlight. Mar. Chem., 8, 33-42.

Zenk, W. (1975). On the Mediterranean Outflow West of Gibraltar, "Meteor" Forsch.-Ergebn., A, 16, 23-34.

Zuta, S. and O. Guillen (1970). Oceanografia de las aquas costeras del Peru. Bol. Inst. Mar. Peru, 2, 157-324.

APPENDIX A: MAILING AND PARTICIPANT LIST FOR THE WORKSHOP ON POLEWARD FLOWS ALONG EASTERN OCEAN BOUNDARIES

Monterey, California, December 1986
[*] indicates attendance at the workshop.

David Adamec
Inst. for Naval Oceanography
Bldg. 1103, Rm. 233
Stennis Space Center, MS
39529-5005

Dr. John Allen
College of Oceanography
Oregon State University
Corvallis, OR 97331

W.R.H. Andrews
Wayside
Main Road
Betley, Crewe
Cheshire CW3 9AD, UNITED KINGDOM

Andrew Bakun
Pacific Environmental Group
P.O. Box 831
Monterey, CA 93942

Dr. Richard T. Barber
Monterey Bay Aquarium Res. Ctr.
160 Central Avenue
Pacific Grove, CA 93950

Dr. E.D. Barton *
Dept. of Physical Oceanography
Marine Science Labs.
Menai Bridge
Anglesey
Gwynedd, UNITED KINGDOM

Mary Batteen *
Naval Postgraduate School
Monterey, CA 93940

Dr. Robert Beardsley
Woods Hole Oceanographic Inst.
Woods Hole, MA 02543

Dr. David Behringer
AOML-NOAA
15 Rickenbacker Cswy.
Miami, FL 33149

Dr. Patricio Bernal
Universidad Catolica de Chile
Sede Regional de Talcahuano
4 Casilla 127
Talcahuano, CHILE

Arlene Bird *
Naval Postgraduate School
Dept. Ocean Code 68
Monterey, CA 93940

Dr. Robert Bourke *
Naval Postgraduate School
Monterey, CA 93940

Dr. Alan Bratkovich *
Dept. of Geological Sciences
USC - University Park
Los Angeles, CA 90089-0741

Dr. Larry Breaker *
Moss Landing Marine Lab
Moss Landing, CA 95039

Dr. Kenneth Brink *
Woods Hole Oceanographic Inst.
Woods Hole, MA 02543

Dr. Shenn-yu Chao
Horn Point Laboratory
University of Maryland
Cambridge, MD 21613-0775

Dr. David Chapman
Woods Hole Oceanographic Inst.
Woods Hole, MA 02543

Dr. Dudley Chelton
College of Oceanography
Oregon State University
Corvallis, OR 97331

Dr. John Church *
CSIRO-Marine Laboratories
Division of Oceanography
GPO Box 1538
Hobart, Tasmania 7001, AUSTRALIA

Dr. Allan Clarke *
Dept. of Oceanography
Florida State University
Tallahassee, FL 32306

Dr. Lou Codispoti *
Bigelow Ocean Sciences Lab.
W. Boothbay Harbor, ME 04575

Dr. Curt Collins
Naval Postgraduate School
Monterey, CA 93940

Dr. William Crawford
Institute of Ocean Sciences
Patricia Bay
P.O. Box 6000
9860 W. Saanich Road
Sidney, B.C. V8L 4B2

A. Badan-Dangon *
CICESE
P.O. Box 4844
San Ysidro, CA 92073

Dr. Don Denbo
Ctr. for Earth & Planetary Physics
Harvard University
Cambridge, MA 02138

Dr. Russ Davis
Mail Code A-030
Scripps Inst. of Oceanography
La Jolla, CA 92093

Prof. Richard C. Dugdale
Laboratoire Physique et Chimie Marin
LaDarse, 06230 Villefranche sur Mer
FRANCE

D.J. Ellett
SMBA
P.O. Box 3
Oban, Argyll, PA 34 4AD
UNITED KINGDOM

Felix Espinoza
Inst. Hidrografico de la Armada
Casilla 324
Valparaiso, CHILE

Dr. Armando Fiuza
Oceanography Grp. Physics Dept.
University of Lisbon
Rua Escola Politecnica, 58
1200 Lisbon, PORTUGAL

Tomas Fonseca *
INGEMAR, Condell 1190
Casilla 1443
Valparaiso, CHILE

Dr. Howard Freeland *
Institute of Ocean Sciences
Patricia Bay
P.O. Box 6000
9860 W. Saanich Road
Sidney, B.C. V8L 4B2

Robert Frouis
Mail Code A-021
Scripps Inst. of Oceanography
La Jolla, CA 92093

R.S. Gardiner-Garden
Geophysical Fluid Dynamics Prog.
James Forrestal Campus
P.O. Box 308
Princeton, NJ 08542

Dr. Stuart Godfrey
CSIRO-Marine Laboratories
Division of Oceanography
GPO Box 1538
Hobart, Tasmania 7001, AUSTRALIA

Dr. John Gould
National Institute of Oceanography
Wormley, Godalming
Surrey, ENGLAND

Dr. Dale Haidvogel
The Chesapeake Bay Institute
The Johns Hopkins University
The Rotunda, Suite 340
711 West 40th Street
Baltimore, MD 21211

Dr. David Halpern
J.P.L. M/S 169/236 CAL TECH
4800 Oak Grove Drive
Pasadena, CA 91109

R. Haney *
Naval Postgraduate School
Monterey, CA 93940

Dr. Donald Hansen
NOAA/AOML
15 Rickenbacker Cswy.
Miami, FL 33149

Loren Haury
Scripps Inst. of Oceanography
A-018
La Jolla, CA 92093

Tom Hayward
Scripps Inst. of Oceanography
La Jolla, CA 92093

Dr. Barbara Hickey
Dept. of Oceanography
University of Washington
Seattle, WA 98195

Dr. Greg Holloway *
Institute of Ocean Sciences
P.O. Box 6000
Sidney, B.C. 48L 4B2

Dr. Robert Houghton
Lamont-Doherty Geological Observ.
Columbia University
Palisades, NY 10964

L. Hutchings
Sea Fisheries Branch
Beach Road
Sea Point
Cape Town, SOUTH AFRICA

Dr. John Huthnance
Institute of Ocean Sciences
Bidston, Birkenhead
Merseyside L43 7RA, UNITED KINGDOM

Dr. Adriana Huyer *
College of Oceanography
Oregon State University
Corvallis, OR 97331

Dr. Motoyoshi Ikeda
Bedford Inst. of Oceanography
Dartmouth NS
Vancouver, Nova Scotia
Canada B2Y 4A2

Dr. Donald Johnson
Naval Ocean R&D Activity
Stennis Space Center, MS 39529-5004

Dr. Burton Jones
Dept. of Biological Sciences
Univ. of Southern California
Los Angeles, CA 90007

Dr. P. Michael Kosro *
College of Oceanography
Oregon State University
Corvallis, OR 97331

Dr. Pijush Kundu
Nova Univ., Physical Ocean. Lab
8000 N. Ocean
Dania, FL 33004

Dr. Miguel Lavin
Dept. of Oceanography
CICESE
San Ysidro, CA

Dr. Ants Leetmaa
NOAA/AOML
15 Rickenbacker Cswy
Miami, FL 33149

Steve Lentz *
Woods Hole Oceanographic Inst.
Woods Hole, MA 02543

Rolf G. Lueck
Chesapeake Bay Institute
Johns Hopkins University
315-711 W. 40th Street
Baltimore, MD 21211

Dr. Roger Lukas
Joint Inst. for Marine & Atms. Res.
University of Hawaii at Manoa
1000 Pope Road
Honolulu, HI 96822

R.J. Lynn
Southwest Fisheries Center
P.O. Box 271
La Jolla, CA 92038

T.A. McClimans
Norwegian Hydrotechnical Laboratory
Norwegian Inst. of Technology
Trondheim, Norway

Dr. Julian McCreary *
Nova University
8000 N. Ocean Drive
Dania, FL 33004

Dr. John McGowan
Mail Code A-028
Scripps Inst. of Oceanography
La Jolla, CA 92093

Prof. George L. Mellor
Geophysical Fluid Dynamics Prog.
Princeton University
P.O. Box 308
Princeton, NJ 08540

Dr. Ekkehard Mittelstaedt
Deutsches Hydrographisches Institut
Bernhard-Nocht-Strasse 78
2 Hamburg 4, GERMANY

Dr. C.N.K. Mooers *
Inst. for Naval Oceanography
Bldg. 1100, Rm. 311
Stennis Space Center, MS 39529-5005

Dr. Robin Muench
10533 Ravenna Avenue NE
Seattle, WA 98125

Greville Nelson
Sea Fisheries Research Inst.
Private Bag X2
Roggebaai 8012, SOUTH AFRICA

Dr. Stephen Neshyba *
College of Oceanography
Oregon State University
Corvallis, OR 97331-5503

Dr. H.J. Niebauer
Institute of Marine Science
University of Alaska
Fairbanks, AK 99755

Peter Niiler
Scripps Inst. of Oceanography
Mail Code A-030
La Jolla, CA 92093

Marlene Noble
USGS
Menlo Park, CA

Dr. James J. O'Brien
Dept. of Meteorology Annex
Florida State University
930 Wildwood
Tallahassee, FL 32306

Theresa Paluszkiewicz
Minerals Management Service
U.S. Dept. of Interior
Washington, DC 20240

A. Pares-Kiema
Dept. of Meteorology Annex FSU
930 Wildwood
Tallahassee, FL 32306

Richard Parrish
Pacific Environmental Grp.
P.O. Box 831
Monterey, CA 93942

Dr. Joseph Pedlosky
Dept. of Physical Oceanography
Woods Hole Oceanographic Inst.
Woods Hole, MA 02543

Dr. George Philander
NOAA Geophys. Fluid Dynamics Lab
Box 308
Princeton, NJ 08540

Prof. Steve Ramp *
Naval Postgraduate School
Monterey, CA 93940

Prof. Joseph Reid
Mail Code A-030
Scripps Inst. of Oceanography
La Jolla, CA 92093

Michele Reinicker
Inst. for Naval Oceanography
Bldg. 1103, Rm. 233
Stennis Space Center, MS
39529-5005

Dr. Peter Rhines
School of Oceanography
University of Washington
Seattle, WA 98195

Steve Riser *
School of Oceanography
University of Washington
Seattle, WA 98195

Dr. Lew Rothstein *
School of Oceanography
University of Washington
Seattle, WA 98195

Prof. Tom Royer
Institute of Marine Science
University of Alaska
Fairbanks, AK 99701

Eckart Schumann
Universiteit van Port Elizabeth
Posbus 1600 Port Elizabeth 6000
Republeik van Suid-Afrika

Mr. Nelson Silva
Centro de Investigaciones del Mar
Universidad Catolica de Valparaiso
Casilla 1020
Valparaiso, CHILE

Dr. James Simpson
Mail Code A-030
Scripps Inst. of Oceanography
La Jolla, CA 92093

Prof. D.C. Smith
Naval Postgraduate School
Monterey, CA 93940

Dr. Robert L. Smith *
College of Oceanography
Oregon State University
Corvallis, OR 97331-5503

CPCB Hector SOLDI Soldi
Direccion de Hidrografia y
Navegacion de la Marina
Casilla Postal 80
Chucuito, Callao, PERU

Dr. Raymond K. Steedman
72 Watkins Road
Dalkeith, Western Australia
AUSTRALIA 6009

Lothar Stramma
Institut fur Meereskunde
an der Universitat Kiel
Dusternbrooker Weg 20
D 2300 Kiel 1 FRG

Dr. Rick Thompson
Institute of Ocean Sciences
9860 W. Saanich Road
Sidney, BC V8L 4B2
CANADA

M. Tomczak
Dept. of Geology and Geophysics
University of Sydney
Sydney 2006, NSW AUSTRALIA

Dr. John Whitehead *
Woods Hole Oceanographic Inst.
Woods Hole, MA 02543

Jacob Wickham, Professor Emeritus
Naval Postgraduate School
Monterey, CA 93940

H. Yamazaki
Chesapeake Bay Institute
Johns Hopkins University
315-711 W. 40th Street
Baltimore, MD 21211

Prof. Salvador Zuta
Instituto del Mar IMARPE
Direccion Postal
Casilla 3734
Lima, PERU

Coastal and Estuarine Studies

(formerly Lecture Notes on Coastal and Estuarine Studies)

Vol. 1: J. Sündermann, K.-P. Holz (Eds.), Mathematical Modelling of Estuarine Physics. Proceedings, 1978. 265 pages. 1980.

Vol. 2: D.P. Finn, Managing the Ocean Resources of the United States: The Role of the Federal Marine Sanctuaries Program. 193 pages. 1982.

Vol. 3: M. Tomczak Jr., W. Cuff (Eds.), Synthesis and Modelling of Intermittent Estuaries. 302 pages. 1983.

Vol. 4: H.R. Gordon, A.Y. Morel, Remote Assessment of Ocean Color for Interpretation of Satellite Visible Imagery. 114 pages. 1983.

Vol. 5: D.C.L. Lam, C.R. Murthy, R.B. Simpson, Effluent Transport and Diffusion Models for the Coastal Zone. 168 pages. 1984.

Vol. 6: M.J. Kennish, R.A. Lutz (Eds.), Ecology of Barnegat Bay, New Jersey. 396 pages. 1984.

Vol. 7: W.R. Edeson, J.-F. Pulvenis, The Legal Regime of Fisheries in the Caribbean Region. 204 pages. 1983.

Vol. 8: O. Holm-Hansen, L. Bolis, R. Gilles (Eds.), Marine Phytoplankton and Productivity. 175 pages. 1984.

Vol. 9: A. Pequeux, R. Gilles, L. Bolis (Eds.), Osmoregulation in Estuarine and Marine Animals. 221 pages. 1984.

Vol. 10: J.L. McHugh, Fishery Management. 207 pages. 1984.

Vol. 11: J.D. Davis, D. Merriman (Eds.), Observations on the Ecology and Biology of Western Cape Cod Bay, Massachusetts. 289 pages. 1984.

Vol. 12: P.P.G. Dyke, A.O. Moscardini, E.H. Robson (Eds.), Offshore and Coastal Modelling. 399 pages. 1985.

Vol. 13: J. Rumohr, E. Walger, B. Zeitzschel (Eds.), Seawater-Sediment Interactions in Coastal Waters. An Interdisciplinary Approach. 338 pages. 1987.

Vol. 14: A.J. Mehta (Ed.), Estuarine Cohesive Sediment Dynamics. 473 pages. 1986.

Vol. 15: R.W. Eppley (Ed.), Plankton Dynamics of the Southern California Bight. 373 pages. 1986.

Vol. 16: J. van de Kreeke (Ed.), Physics of Shallow Estuaries and Bays. 280 pages. 1986.

Vol. 17: M.J. Bowman, C.M. Yentsch, W.T. Peterson (Eds.), Tidal Mixing and Plankton Dynamics. 502 pages. 1986.

Vol. 18: F. Bo Pedersen, Environmental Hydraulics: Stratified Flows. 278 pages. 1986.

Vol. 19: K.N. Fedorov, The Physical Nature and Structure of Oceanic Fronts. 333 pages. 1986.

Vol. 20: A. Rieser, J. Spiller, D. VanderZwaag (Eds.), Environmental Decisionmaking in a Transboundary Region. 209 pages. 1986.

Vol. 21: Th. Stocker, K. Hutter, Topographic Waves in Channels and Lakes on the f-Plane. 176 pages. 1987.

Vol. 22: B.-O. Jansson (Ed.), Coastal Offshore Ecosystem Interactions. 367 pages. 1988.

Vol. 23: K. Heck, Jr. (Ed.), Ecological Studies in the Middle Reach of Chesapeake Bay. 287 pages. 1987.

Vol. 24: D.G. Shaw, M.J. Hameedi (Eds.), Environmental Studies in Port Valdez, Alaska. 423 pages. 1988.

Vol. 25: C.M. Yentsch, F.C. Mague, P.K. Horan (Eds.), Immunochemical Approaches to Coastal, Estuarine and Oceanographic Questions. 399 pages. 1988.

Vol. 26: E.H. Schumann (Ed.), Coastal Ocean Studies off Natal, South Africa. 271 pages. 1988.

Vol. 27: E. Gold (Ed.), A Law of the Sea for the Caribbean: An Examination of Marine Law and Policy Issues in the Lesser Antilles. 507 pages. 1988.

Vol. 28: W.S. Wooster (Ed.), Fishery Science and Management. 339 pages. 1988.

Vol.29: D.G. Aubrey, L. Weishar (Eds.), Hydrodynamics and Sediment Dynamics of Tidal Inlets. 456 pages. 1988.

Vol. 30: P.B. Crean, T.S. Murty, J.A. Stronach, Mathematical Modelling of Tides and Estuarine Circulation. 471 pages. 1988.

Vol. 31: G. Lopez, G. Taghon, J. Levinton (Eds.), Ecology of Marine Deposit Feeders. 322 pages. 1989.

Vol. 32: F. Wulff, J.G. Field, K.H. Mann (Eds.), Network Analysis in Marine Ecology. 284 pages. 1989.

Vol. 33: M.L. Khandekar, Operational Analysis and Prediction of Ocean Wind Waves. 214 pages. 1989.

Vol. 34: S.J. Neshyba, Ch.N.K. Mooers, R.L. Smith, R.T. Barber (Eds.), Poleward Flows Along Eastern Ocean Boundaries. 374 pages. 1989.

MIX
Papier aus verantwortungsvollen Quellen
Paper from responsible sources
FSC® C105338

If you have any concerns about our products,
you can contact us on
ProductSafety@springernature.com

In case Publisher is established outside the EU,
the EU authorized representative is:
Springer Nature Customer Service Center GmbH
Europaplatz 3, 69115 Heidelberg, Germany

Printed by Libri Plureos GmbH
in Hamburg, Germany

The Theory of Chaotic Attractors

Springer
New York
Berlin
Heidelberg
Hong Kong
London
Milan
Paris
Tokyo

Brian R. Hunt
Judy A. Kennedy
Tien-Yien Li
Helena E. Nusse
Editors

The Theory of Chaotic Attractors

Springer

Brian R. Hunt
Institute for Physical Science and Technology
University of Maryland
College Park, MD 20742
USA
bhunt@ipst.umd.edu

Judy A. Kennedy
Department of Mathematics
University of Delaware
Newark, Delaware 19716
USA
kennedy@math.udel.edu

Tien-Yien Li
Department of Mathematics
Michigan State University
East Lansing, Michigan 48824
USA
li@math.msu.edu

Helena E. Nusse
Department of Econometrics
University of Groningen
NL-9700 AV Groningen
The Netherlands
h.e.nusse@eco.rug.nl

Mathematics Subject Classification (2000): 37D45, 37A40, 37C40, 37Exx, 28D05, 01A75

Library of Congress Cataloging-in-Publication Data
The theory of chaotic attractors / [edited by] Brian R. Hunt, Judy A. Kennedy, Tien-Yien Li, Helena E. Nusse.
p. cm.
Includes bibliographical references
ISBN 978-1-4419-2330-1 ISBN 978-0-387-21830-4 (eBook)
DOI 10.1007/978-0-387-21830-4
1. Chaotic behavior in systems. I. Nusse, Helena Engelina, 1952— II. Hunt, Brian R.
Q172.5.C45T44 2003
033'.857—dc21 2003053003

ISBN 0-387-40349-3 Printed on acid-free paper.

9 8 7 6 5 4 3 2 1 SPIN 10936931

Springer-Verlag is a part of *Springer Science+Business Media*

springeronline.com

This volume is dedicated to James A. Yorke in commemoration of his 60th birthday.

Preface

The editors felt that the time was right for a book on an important topic, the history and development of the notions of chaotic attractors and their "natural" invariant measures. We wanted to bring together a coherent collection of readable, interesting, outstanding papers for detailed study and comparison. We hope that this book will allow serious graduate students to hold seminars to study how the research in this field developed. Limitation of space forced us painfully to exclude many excellent, relevant papers, and the resulting choice reflects the interests of the editors.

Since James Alan Yorke was born August 3, 1941, we chose to have this book commemorate his sixtieth birthday, honoring his research in this field. The editors are four of his collaborators.

We would particularly like to thank Achi Dosanjh (senior editor mathematics), Elizabeth Young (assistant editor mathematics), Joel Ariaratnam (mathematics editorial), and Yong-Soon Hwang (book production editor) from Springer Verlag in New York for their efforts in publishing this book.

College Park, MD
June 2003

Brian R. Hunt
Judy A. Kennedy
Tien-Yien Li
Helena E. Nusse

Contents

Introduction

Part I: Who is James Yorke?

Recently, the Science and Technology Foundation of Japan announced that James A. Yorke had been named a winner of the 2003 Japan Prize for his work in the field of chaos theory. This foundation has awarded Japan Prizes since 1985 under the auspices of the Japanese prime minister. It is a Japanese version of the Nobel Prize. One of the most esteemed of science and technology prizes, Japan Prizes are given to scientists whose "original and outstanding achievements in science and technology are recognized as having advanced the frontiers of knowledge and served the cause of peace and prosperity for mankind." The Science and Technology Foundation of Japan announced that Yorke and Benoit Mandelbrot, of Yale University, will share $415,000 (50 million yen) for the "creation of universal concepts in complex systems: chaos and fractals". The other 2003 Japan Prize was awarded for discovering the principle behind magnetic resonance imaging (MRI). Past Japan prizes have been awarded for achievements such as the discovery of the HIV virus, the invention of the World Wide Web, the development of artificial intelligence, and the eradication of smallpox. "For a scientist to be awarded the Japan Prize is a distinction as great as any in the world," said University of Maryland President C.D. Mote, Jr. "Jim Yorke has now been officially recognized for his original achievements in nonlinear dynamics that have monumentally advanced the frontiers of science and technology and served the cause of peace and prosperity for mankind. I am so pleased for him and for the inspiration his recognition will provide for others who by pursuing their passions can hope to have an impact as great as Jim's." The presentation ceremony for the 2003 Japan Prize laureates is scheduled to be held in Tokyo in April 2003.

James A. Yorke came to the University of Maryland for graduate studies in 1963 at the age of 22, in part because of the interdisciplinary opportunities offered by the faculty of an interdisciplinary research institute now called the Institute for Physical Science and Technology (IPST). He has spent his entire career at the University of Maryland (College Park). From 1985 through 2001 he was the director of IPST. Currently he is a Distinguished University Professor in the Departments of Mathematics and Physics, and in IPST at the University of Maryland. He has earned an international reputation for his pioneering work in several different areas of mathematics and its applications. The famous papers entitled "On the existence of invariant measures for piecewise monotonic transformations" by A. Lasota and J.A. Yorke, which appeared in 1973 in volume 186 of the *Transactions of the American Mathematical Society*, and "Period three implies chaos" by T.-Y. Li and J.A. Yorke, which appeared in 1975 in volume 82 of the *American Mathematical Monthly* are just two examples. In particular, he has been uniquely innovative in his significant scientific contributions to applied topology, bifurcation theory, numerical methods, biology (including epidemiological modeling), theoretical nonlinear dynamics, and controlling chaos. He is coauthor of over 250 scientific papers and three books.

Yorke is a mathematician who likes to think of himself as a philosopher, but he is also the kind of mathematician who feels compelled to put his ideas of reality to some use, as the following examples demonstrate. In joint work with H.W. Hethcote and A. Nold, Yorke studied the transmission dynamics and control of gonorrhea and produced a report that persuaded the federal government to alter its national strategies for controlling the disease. During the gasoline crisis of the 1970s, Yorke gave testimony to the State of Maryland arguing correctly (but unpersuasively) that the even–odd system of limiting gasoline sales would only make lines longer. In the era of antiwar demonstrations, the government released a spy-plane photograph purporting to show sparse crowds around the Washington Monument at the height of the rally. However, Yorke proved that the photograph had actually been taken later when the rally was breaking up (about a half-hour later), by analyzing the monument's shadow in that photograph. Now he is working on a diverse collection of projects including understanding transmission patterns of HIV/AIDS, how to improve weather prediction, how better to determine genomes (the DNA sequences of plants and animals), and diverse mathematics projects. A recent example of his lively lecturing style was demonstrated during a lecture in which he jumped on a table and stamped a grapefruit flat just to illustrate the idea of a map from three-dimensional to two-dimensional phase space. Those in the audience will remember that lecture.

At the institute, Yorke enjoyed unusual freedom to work on a variety of problems and he had frequent contact with experts in a wide range of disciplines. One of these experts, a fluid dynamicist, encountered Lorenz's 1963 paper "Deterministic Nonperiodic Flow" in 1972 and gave a copy to Yorke. Lorenz's paper was the kind of scientific work that Yorke had been looking for without even knowing it. It was a vivid physical model, a picture of a fluid in motion, and Yorke realized that it had ideas he wanted physicists to see and understand. To a physicist, a "legitimate" example was a differential equation that could be written down in a simple form. When Yorke saw Lorenz's paper, he knew that it was an example that physicists would embrace. Yorke felt that physicists had learned not to see chaos. In daily life, the Lorenzian quality of sensitive dependence on initial conditions lurks everywhere. Small perturbations in one's daily trajectory can have large consequences. When trying to explain this idea, he may pose the question, "How did your parents meet?" Our very existence is the result of improbable events. Yorke understood that there is some disorder virtually everywhere. Physicists and mathematicians want to discover regularities, but one has to know about disorder if one is going to deal with it. Yorke believed that scientists and nonscientists alike can easily mislead themselves about complexity if they are not properly attuned to it and that many scientists must have seen chaotic behavior in innumerable circumstances for many years. In the past, when they were running a physical experiment, and the experiment behaved in an erratic manner, they either tried to fix it or gave up. They explained the erratic behavior by saying that it was due to the noise, or that the experiment was poorly designed. Yorke says that scientists were the last to find out about chaos. He wonders whether there are other phenomena that we are trained to ignore. His goal is to tell scientists what to look for. Some mathematicians prefer to prove results that clarify issues scientists have been discussing, but scientists, Yorke says, generally don't appreciate such rigorous work. They cannot read it and are likely to say "we already knew that." He believes that it is better to try to lead an elephant than to follow.

More than twenty-five years ago, Tien-Yien Li and James A. Yorke published their famous paper "Period three implies chaos." In that paper, "chaos" means (1) infinitely many periods, and (2) sensitivity to initial data. Most of the lines of proof in the paper were devoted to establishing the latter property. The evolution to the final version of the paper provides some insight into the working habits of Yorke. At that time, Li was a graduate student working with Yorke and had finished a draft of the paper. It remained untouched on Yorke's desk for a few months. Li worried that perhaps Yorke did not have a high opinion of the draft and that he was no longer interested, but Yorke was convinced that the results were important, since they illuminated and clarified the message in the works of Lorenz and Smale that physicists were not hearing. However, he did not think that there was need to rush; substantial revision was needed to make the paper more readable for the *Monthly*. Robert May received a copy of the draft when he gave a series of lectures at the University of Maryland in early 1974. May told Yorke about period doubling, and Yorke told May about chaos. (Later in 1974, May published a paper entitled "Biological populations with nonoverlapping generations, stable points, stable cycles and chaos" in volume 186 of the prestigious journal Science mentioning the Li–Yorke "period three implies chaos" theorem, referring to their paper as "in press." He credited Li and Yorke with coining the term "chaos.") An hour after May departed from the University of Maryland, Li remembers Yorke calling to say that they had to rush to finish the revisions so they could submit it the next Monday. This paper was important on its merits, though part of it turned out to be implied by an earlier result of Sarkovskii. Perhaps more than any previous work in the area, it grabbed the attention of nonmathematicians. In the end its most influential feature was its mysterious and mischievous title: "Period three implies chaos." Some colleagues advised them to choose something more sober, but Li and Yorke stuck with a word that came to stand for the whole growing business of deterministic disorder. Soon work of many authors from Maxwell, Poincaré, and Birkhoff to Arnol'd, Sinai, and Smale could be viewed as fitting under the umbrella term "chaos theory."

Jim Yorke always wears red socks. When asked why he says, "Red socks are a trivial reminder to try to think differently."

A Bird's-Eye View on the Variety of Yorke's Important Scientific Contributions

James A. Yorke's principal scientific contributions have been in the area of chaotic dynamical systems. Virtually all his work is with collaborators, and the results described here are equally due to them, but for brevity they are not mentioned. Although he has worked extensively in differential delay equations and in mathematical epidemiology, his interests turned mainly to the area of chaos. Chaos in dynamical systems refers to the fact that even simple deterministic systems can have time evolution that shares many of the properties of randomness. It has become increasingly clear that chaotic dynamics has very important implications for a broad variety of applications (e.g., weather prediction, celestial mechanics, plasmas, mechanical systems, solid state physics). The development of the theory of chaotic dynamics and its evident wide applicability in science and technology is an extremely important achievement of modern mathematics. Indeed, it is the opinion of many scientists that chaotic dynamics is one of the most significant new developments in mathematics in the last few decades. Yorke has played a key role in initiating investigations in fundamental and important questions in chaotic dynamics, always searching for results that scientists need. He has done this consistently over a period of years, starting with the by now famous papers "On the existence of invariant measures for piecewise monotonic transformations" by A. Lasota and J.A. Yorke in 1973, and "Period three implies chaos" by T.-Y. Li and J.A. Yorke in 1975. It is characteristic how differently these two papers approach chaos: one from the view of measure and operators on a Banach space, the other as topology of the real line. This latter paper first used the word "chaos" as a mathematical term to describe this type of dynamics, and thereby named this field of research. This paper introduced the ideas of chaos and sensitivity to initial data to a generation of mathematicians, physicists, engineers, and scientists from a variety of other disciplines. Yorke has been one of the most significant and consistently innovative contributors to the field of chaotic dynamical systems, and he is the driving force behind several of the major innovations. In addition, the theoretical and mathematical work of Yorke and his collaborators led to many applications in a variety of sciences, including physics, astrophysics, biology, chemistry, and economics.

Yorke's pioneering work has helped to bridge the gap between mathematics and the phenomenological approach of physical scientists. One of his goals has been to bring mathematical ideas to scientists by reformulating established concepts into new patterns that they might find useful, for example, the OGY method (Ott–Grebogi–Yorke method) for controlling chaos. He helped to establish a common vocabulary for approaching such problems, as we now illustrate. At the end of the 1970s, Yorke was the first to point out and explain the phenomenon now called "transient chaos," whereby orbits can behave in a chaotic manner before suddenly settling into a more regular steady state, giving both a mathematical theory and numerical experiments to explain how transient chaos phenomena manifest themselves in practice. These studies on transient chaos have paved the way for a host of significant subsequent work by other researchers and, more generally, have demonstrated the practical importance of chaotic sets that are not attractors.

The fractal dimension of strange attractors is a basic attribute of these attractors. Yorke's early work on fractal dimension managed to integrate diverse approaches and expedite measurements of dimensions for chaotic attractors. The Kaplan–Yorke conjecture concerns the idea of determining the fractal dimension of attractors from Lyapunov numbers. These investigations played a major role in initiating the now extensive developments with respect to this important question, to the point that fractal dimensions are now routinely measured in experiments.

Yorke and his collaborators developed methods for analyzing discontinuous changes in the structure of chaotic attractors and the fractal basin boundaries defined by them; the resulting terminology of "crisis" and "metamorphosis" is in common use today. Yorke has also made important and innovative contributions to bifurcation theory, including both local bifurcations and global bifurcations. For example, with collaborators he proved why cascades of period-doubling bifurcations occur in two-dimensional dynamical systems and why there are concurrent creation and annihilation of periodic orbits in any neighborhood of homoclinic tangencies. He has been influential in developing the view that to analyze systems that exhibit irregular behavior, the concept of mathematical modeling of systems needs to be expanded to include algorithms that make direct use of measured time series from the system and that these must be given a firm mathematical foundation. These ideas, which include observation, analysis and control of chaotic systems, have been widely implemented by experimental scientists.

Yorke has done pathbreaking work on the reliability of computer simulations of dynamical systems. He and his collaborators have developed a variety of reliable numerical methods, resulting in the software package "Dynamics." This software is a useful tool for exploring dynamical systems numerically in a way that leads to new fundamental properties and mathematical results of the dynamical system under investigation. The phenomenon of Wada basins in dynamical systems demonstrates that interaction between numerical investigations and development of new mathematical results is of crucial importance. (A Wada basin is a basin of attraction such that every boundary point of that basin is a boundary point of at least three different basins of attraction simultaneously.) Concerning higher-dimensional dynamics, although many mathematical results are known, many more phenomena and results are awaiting to be discovered. Indeed, it is the opinion of several leading mathematicians in dynamical systems that computer explorations are essential to theoretical development and understanding dynamical systems. As of this writing, he is working with collaborators on understanding AIDS in Africa, the variability of chaos in weather prediction, how better to assemble a genome, and a number of more strictly mathematical topics.

A selection of James A. Yorke's publications

Books

H.W. Hethcote and J.A. Yorke,
Gonorrhea Transmission Dynamics and Control,
Lecture Notes in Biomathematics #56,
Springer-Verlag, Berlin, 1984

E. Ott, T. Sauer, and J.A. Yorke (editors)
Coping with Chaos,
Wiley Series in Nonlinear Science
John Wiley & Sons Inc., New York, 1994

K. Alligood, T. Sauer, and J.A. Yorke,
Chaos: An Introduction to Dynamical Systems,
Textbooks in Mathematical Sciences
Springer-Verlag, New York, 1997

H.E. Nusse and J.A. Yorke,
Dynamics: Numerical Explorations,
Second Edition, accompanied by software program Dynamics 2, an interactive program for IBM compatible PC's and Unix computers, (the Unix version of the program is by B.R. Hunt and E.J. Kostelich),
Applied Mathematical Sciences 101,
Springer-Verlag, New York, 1998

Journal Papers (chronologically ordered; a partial but representative list selected by Yorke)

J.A. Yorke, A continuous differential equation in Hilbert space without existence, Funkcialaj Ekvacioj **13** (1970), 19–21

F.W. Wilson, Jr. and J.A. Yorke, Lyapunov functions and isolating blocks, J. Differential Equations **13** (1973), 106–123

W. London, M.D. and J.A. Yorke, Recurrent outbreaks of measles, chicken pox, and mumps, I. Seasonal variation in contact rates, and II. Systematic differences in contact rates and stochastic effects, Amer. J. Epidemiology **98** (1973), 453–468 and 469–482

A. Lasota and J.A. Yorke, The generic property of existence of solutions of differential equations in Banach space, J. Differential Equations **13** (1973), 1–12

A. Lasota and J.A. Yorke, On the existence of invariant measures for piecewise monotonic transformations, Trans. Amer. Math. Soc. **186** (1973), 481–488

T.-Y. Li and J.A. Yorke, Period three implies chaos, Amer. Math. Monthly **82** (1975), 985-992

J.C. Alexander and J.A. Yorke, The implicit function theorem and the global methods of cohomology, J. Functional Analysis, **21** (1976), 330–339

A. Lajmanovich and J.A. Yorke, A deterministic model for gonorrhea in a nonhomogeneous population, Math. Biosci. **28** (1976), 221–236

R.B. Kellogg, T.-Y. Li, and J.A. Yorke, A constructive proof of the Brouwer fixed-point theorem and computational results, SIAM J. Numer. Anal. **13** (1976), 473–383

J.L. Kaplan and J.A. Yorke, On the nonlinear differential delay equation $dx/dt = -f(x(t), x(t-1))$, J. Differential Equations **23** (1977), 293–314

T.Y. Li and J.A. Yorke, Ergodic transformations from an interval into itself, Trans. Amer. Math. Soc. **235** (1978), 183–192

J.C. Alexander and J.A. Yorke, Global bifurcation of periodic orbits, Amer. J. Math. **100** (1978), 263–292

S.N. Chow, J. Mallet-Paret, and J.A. Yorke, Finding zeroes of maps: homotopy methods that are constructive with probability one, Math. of Comp. **32** (1978), 887–899

J.A. Yorke, H.W. Hethcote, and A. Nold, Dynamics and control of the transmission of gonorrhea, Sexually Transmitted Diseases **5** (1978), 51–56

J.A. Yorke, N. Nathanson, G. Pianigiani, and J. Martin, Seasonality and the requirements for perpetuation and eradication of viruses in populations, Amer. J. Epidemiology **109** (1979), 103–123

G. Pianigiani and J.A. Yorke, Expanding maps on sets which are almost invariant: decay and chaos, Trans. Amer. Math. Soc. **252** (1979), 351–366

J.L. Kaplan and J.A. Yorke, Preturbulence: a regime observed in a fluid flow model of Lorenz, Comm. Math. Phys. **67** (1979), 93–108

J.A. Yorke and E.D. Yorke, Metastable chaos: the transition to sustained chaotic oscillations in the Lorenz model, J. Stat. Phys. **21** (1979), 263–277

J. Mallet-Paret and J.A. Yorke, Snakes: oriented families of periodic orbits, their sources, sinks, and continuation, J. Differential Equations **43** (1982), 419–450

C. Grebogi, E. Ott, and J.A. Yorke, Chaotic attractors in crisis, Phys. Rev. Lett. **48** (1982), 1507–1510

P. Frederickson, J.L. Kaplan, E.D. Yorke, and J.A. Yorke, The Lyapunov dimension of strange attractors, J. Differential Equations **49** (1983), 185–207

C. Grebogi, E. Ott, and J.A. Yorke, Crises, sudden changes in chaotic attractors, and transient chaos, Physica D **7** (1983), 181–200

J.D. Farmer, E. Ott, and J.A. Yorke, The dimension of chaotic attractors, Physica D **7** (1983), 153–180

J.A. Yorke and K.T. Alligood, Cascades of period doubling bifurcations: a prerequisite for horseshoes, Bull. Amer. Math. Soc. **9** (1983), 319–322

C. Grebogi, S.W. McDonald, E. Ott, and J.A. Yorke, Final state sensitivity: an obstruction to predictability, Phys. Lett. A **99** (1983), 415–418

J.L. Kaplan, J. Mallet-Paret, and J.A. Yorke, The Lyapunov dimension of a nowhere differentiable attracting torus, Ergodic Theory and Dyn. Sys. **4** (1984), 261–281

J.C. Alexander and J.A. Yorke, Fat baker's transformations, Ergodic Theory and Dyn. Sys. **4** (1984), 1–23

C. Grebogi, E. Ott, S. Pelikan, and J.A. Yorke, Strange attractors that are not chaotic, Physica D **13** (1984), 261–268

E. Ott, W.D. Withers, and J.A. Yorke, Is the dimension of chaotic attractors invariant under coordinate changes?, J. Stat. Phys. **36** (1984), 687–697

C. Grebogi, E. Ott, and J.A. Yorke, Attractors on an N-torus: quasiperiodicity versus chaos, Physica D **15** (1985), 354–373

S.W. McDonald, C. Grebogi, E. Ott, and J.A. Yorke, Fractal basin boundaries, Physica D **17** (1985), 125–153

S.W. McDonald, C. Grebogi, E. Ott, and J.A. Yorke, Structure and crises of fractal basin boundaries, Phys. Lett. A **107** (1985), 51–54

J.A. Yorke and K.T. Alligood, Period doubling cascades of attractors: a prerequisite for horseshoes, Comm. Math. Phys. **101** (1985), 305–321

C. Grebogi, S. W. McDonald, E. Ott, and J. A. Yorke, The exterior dimension of fat fractals, Phys. Lett. A **110** (1985), 1–4; **113** (1986), 495

C. Grebogi, E. Ott, and J.A. Yorke, Comment on "Sensitive dependence on parameters in nonlinear dynamics" and on "Fat fractals on the energy surface" Phys. Rev. Lett. **56** (1986), 266

C. Grebogi, E. Ott, and J.A. Yorke, Metamorphoses of basin boundaries in nonlinear dynamical systems, Phys. Rev. Lett. **56** (1986), 1011–1014

J.A. Yorke, E.D. Yorke, and J. Mallet-Paret, Lorenz-like chaos in a partial differential equation for a heated fluid loop, Physica D **24** (1987), 279–291

C. Grebogi, E. Ott, and J.A. Yorke, Basin boundary metamorphoses: changes in accessible boundary orbits, Physica D **24** (1987), 243–262

C. Grebogi, E. Ott, and J.A. Yorke, Chaos, strange attractors, and fractal basin boundaries in nonlinear dynamics, Science **238** (1987), 632–638

C. Grebogi, E. Ott, and J.A. Yorke, Unstable periodic orbits and the dimension of chaotic attractors, Phys. Rev. A **36** (1987), 3522–3524

S.M. Hammel, J.A. Yorke, and C. Grebogi, Do numerical orbits of chaotic dynamical processes represent true orbits?, J. of Complexity **3** (1987), 136–145

C. Grebogi, E. Ott, F. Romeiras, and J.A. Yorke, Critical exponents for crisis induced intermittency, Phys. Rev. A **36** (1987), 5365–5380

H.E. Nusse and J.A. Yorke, Is every approximate trajectory of some process near an exact trajectory of a nearby process?, Comm. Math. Phys. **114** (1988), 363–379

E.J. Kostelich and J.A. Yorke, Noise reduction in dynamical systems, Phys. Rev. A **38** (1988), 1649–1652

S.M. Hammel, J.A. Yorke, and C. Grebogi, Numerical orbits of chaotic processes represent true orbits, Bull. Amer. Math. Soc. **19** (1988), 465–469

C. Grebogi, E. Ott, and J.A. Yorke, Roundoff-induced periodicity and the correlation dimension of chaotic attractors, Phys. Rev. A **38** (1988), 3688–3692

T.-Y. Li, T. Sauer, and J.A. Yorke, Numerically determining solutions of systems of polynomial equations, Bull. Amer. Math. Soc. **18** (1988), 173–177

I. Kramer, E. D. Yorke, and J. A. Yorke, The AIDS epidemic's influence on the gay contact rate from analysis of gonorrhea incidence, Math. Comput. Modelling **12** (1989), 129–137

T.-Y. Li, T. Sauer, and J.A. Yorke, The cheater's homotopy: an efficient procedure for solving systems of polynomial equations, SIAM J. Numer. Anal. **26** (1989), 1241–1251

H.E. Nusse and J.A. Yorke, A procedure for finding numerical trajectories on chaotic saddles, Physica D **36** (1989), 137–156

W.L. Ditto, S. Rauseo, R. Cawley, C. Grebogi, G.-H. Hsu, E. Kostelich, E. Ott, H.T. Savage, R. Segnan, M. Spano, and J.A. Yorke, Experimental observation of crisis-induced intermittency and its critical exponent, Phys. Rev. Lett. **63** (1989), 923–926

E. Kostelich and J.A. Yorke, Noise reduction: finding the simplest dynamical system consistent with the data, Physica D **41** (1990), 183–196

C. Grebogi, S. M. Hammel, J.A. Yorke, and T. Sauer, Shadowing of physical trajectories in chaotic dynamics: containment and refinement, Phys. Rev. Lett. **65** (1990), 1527–1530

T. Shinbrot, E. Ott, C. Grebogi, and J.A. Yorke, Using chaos to direct trajectories to targets, Phys. Rev. Lett. **65** (1990), 3215–3218

E. Ott, C. Grebogi, and J.A. Yorke, Controlling chaos, Phys. Rev. Lett. **64** (1990), 1196–1199

M. Ding, C. Grebogi, E. Ott, and J.A. Yorke, Massive bifurcation of chaotic scattering, Phys. Lett. A **153** (1991), 21–26

J. Kennedy and J.A. Yorke, Basins of Wada, Physica D **51** (1991), 213–225

H.E. Nusse and J.A. Yorke, Analysis of a procedure for finding numerical trajectories close to chaotic saddle hyperbolic sets, Ergodic Theory and Dyn. Sys. **11** (1991), 189–208

B.R. Hunt and J.A. Yorke, Smooth dynamics on Weierstrass nowhere differentiable curves, Trans. Amer. Math. Soc. **325** (1991), 141–154

T. Sauer and J.A. Yorke, Rigorous verification of trajectories for the computer simulation of dynamical systems, Nonlinearity **4** (1991), 961–979

T. Sauer, J.A. Yorke, and M. Casdagli, Embedology, J. Stat. Phys. **65** (1991), 579–616

Z.-P. You, E.J. Kostelich, and J.A. Yorke, Calculating stable and unstable manifolds, Int. J. Bifurcation and Chaos **1** (1991), 605–623

K.T. Alligood, L. Tedeschini, and J.A. Yorke, Metamorphoses: sudden jumps in basin boundaries, Comm. Math. Phys. **141** (1991), 1–8

H.E. Nusse and J.A. Yorke, A numerical procedure for finding accessible trajectories on basin boundaries, Nonlinearity **4** (1991), 1183–1212

I. Kan, H. Koçak, and J.A. Yorke, Antimonotonicity: concurrent creation and annihilation of periodic orbits, Annals of Mathematics **136** (1992), 219–252

H.E. Nusse and J.A. Yorke, Border collision bifurcations including period two to period three bifurcation for piecewise smooth systems, Physica D **57** (1992), 39–57

T. Shinbrot, W. Ditto, C. Grebogi, E. Ott, M. Spano, and J.A. Yorke, Using the sensitive dependence of chaos (the Butterfly Effect) to direct orbits to targets in an experimental chaotic system, Phys. Rev. Lett. **68** (1992), 2863–2866

K.T. Alligood and J.A. Yorke, Accessible saddles on fractal basin boundaries, Ergodic Theory and Dyn. Sys. **12** (1992), 377–400

J.C. Alexander, J.A. Yorke, Z-P. You, and I. Kan, Riddled basins, Int. J. Bifurcation & Chaos **2** (1992), 795–813

B.R. Hunt, T. Sauer, and J.A. Yorke, Prevalence: a translation-invariant "almost every" on infinite dimensional spaces, Bull. Amer. Math. Soc. **27** (1992), 217–238

T. Shinbrot, C. Grebogi, E. Ott, and J.A. Yorke, Using small perturbations to control chaos, Nature **363** (1993), 411–417

M. Ding, C. Grebogi, E. Ott, T. Sauer, and J.A. Yorke, Plateau onset for correlation dimension: when does it occur?, Phys. Rev. Lett. **70** (1993), 3872–3873

B.R. Hunt and J.A. Yorke, Maxwell on chaos, Nonlinear Science Today **3** (1993), 2–4

J. Kennedy and J. A. Yorke, Pseudocircles in dynamical systems, Trans. Amer. Math. Soc. **343** (1994), 349–366

S.P. Dawson, C. Grebogi, T. Sauer, and J.A. Yorke, Obstructions to shadowing when a Lyapunov exponent fluctuates about zero, Phys. Rev. Lett. **73** (1994), 1927–1930

H.E. Nusse and J.A. Yorke, Border-collision bifurcations for piecewise smooth one-dimensional maps, Int. J. Bifurcation and Chaos **5** (1995), 189–207

J. Kennedy and J.A. Yorke, Bizarre Topology is Natural in Dynamical Systems, Bull. Amer. Math. Soc. **32** (1995), 309–316

H.E. Nusse, E. Ott, and J.A. Yorke, Saddle-node bifurcations on fractal basin boundaries, Phys. Rev. Lett. **75** (1995), 2482–2485

H.E. Nusse and J.A. Yorke, Wada basin boundaries and basin cells, Physica D **90** (1996), 242–261

H.E. Nusse and J.A. Yorke, Basins of attraction, Science **271** (1996), 1376–1380

U. Feudel, C. Grebogi, B. Hunt, and J.A. Yorke, A map with more than 100 coexisting low-period, periodic attractors, Phys. Rev. E **54** (1996), 71–81

E.J. Kostelich, J.A. Yorke, and Z. You, Plotting stable manifolds: error estimates and noninvertible maps, Physica D **93** (1996), 210–222

J. Kennedy and J.A. Yorke, Pseudocircles, diffeomorphisms, and perturbable dynamical systems, Ergodic Theory and Dyn. Sys. **16** (1996), 1031–1057

M. Sanjuan, J. Kennedy, C. Grebogi, and J.A. Yorke, Indecomposable continua in dynamical systems with noise: fluid flow past an array of cylinders, Int. J. Bifurcation & Chaos **7** (1997), 125–138

H.E. Nusse and J.A. Yorke, The structure of basins of attraction and their trapping regions, Ergodic Theory and Dyn. Sys. **17** (1997),463–482

T. Sauer, C. Grebogi,, and J.A. Yorke, How long do numerical chaotic solutions remain valid?, Phys. Rev. Lett. **79** (1997), 59–62

J. Kennedy and J.A. Yorke, The topology of stirred fluids, Topology and Its Applications **80** (1997), 201–238

T. Sauer and J.A. Yorke, Are the dimensions of a set and its image equal under typical smooth functions?, Ergodic Theory and Dyn. Sys. **17** (1997), 941–956

E. Kostelich, I. Kan, C. Grebogi, E. Ott, and J. A. Yorke, Unstable dimension variability: a source of nonhyperbolicity in chaotic systems, Physica D **109** (1997), 81–90

U. Feudel, C. Grebogi, L. Poon, and J.A. Yorke, Dynamical properties of a simple mechanical system with a large number of coexisting periodic attractors, Chaos, Solitons and Fractals **9** (1998), 171–180

C. Robert, K.T. Alligood, E. Ott, and J.A. Yorke, Outer tangency bifurcations of chaotic sets, Phys. Rev. Lett. **80** (1998), 4867–4870

G.-H. Yuan, S. Banerjee, E. Ott, and J.A. Yorke, Border-collision bifurcations in the Buck Converter, IEEE Trans. Circuits and Systems I: Fund. Theory and Appl. **45** (1998), 707–716

T. Sauer, J. Tempkin, and J.A. Yorke, Spurious Lyapunov exponents in attractor reconstruction, Phys. Rev. Lett. **81** (1998), 4341–4344

J. Kennedy, M.A.F. Sanjuan, J.A. Yorke, and C. Grebogi, The Topology of Fluid Flow Past a Sequence of Cylinders, Topology and Its Applications **94** (1999), 207–242

G.-C. Yuan and J.A. Yorke, An open set of maps for which every point is absolutely nonshadowable, Proc. Amer. Math. Soc. **128** (1999), 909–918

J. Miller and J.A. Yorke, Finding all periodic orbits of maps using Newton methods: sizes of basins, Physica D **135** (2000), 195–211

G.-C. Yuan and J.A. Yorke, Collapsing of chaos in one dimensional maps, Physica D **136** (2000), 18–30

C. Robert, K.T. Alligood, E. Ott, and J.A. Yorke, Explosions of Chaotic Sets, Physica D **144** (2000), 44–61

G.-C. Yuan, J.A. Yorke, T.L. Caroll, E. Ott, L.M. Pecora, Testing whether two chaotic one dimensional processes are dynamically identical, Phys. Rev. Lett. **85** (2000), 4265–4268

J. Kennedy and J.A. Yorke, Topological horseshoes, Trans. Amer. Math. Soc. **353** (2001), 2513–2530

D. Sweet, H.E. Nusse, and J.A. Yorke, Stagger and step method: detecting and computing chaotic saddles in higher dimensions, Phys. Rev. Lett. **86** (2001), 2261–2264

D.J. Patil, B.R. Hunt, E. Kalnay, J.A. Yorke, and E. Ott, Local low dimensionality of atmospheric dynamics Phys. Rev. Lett. **86** (2001), 5878–5881

J. Kennedy, S. Koçak, and J.A. Yorke, A chaos lemma, Am. Math. Monthly **108** (2001), 411–423

K. Alligood, E. Sander, and J.A. Yorke, Explosions: global bifurcations at heteroclinic tangencies, Ergodic Theory and Dyn. Sys. **22** (2002), 953–972

Original Contributions in Symposium Proceedings

J.A. Yorke, Spaces of solutions, and Invariance of contingent equations, both in Mathematical Systems Theory and Economics II, Springer-Verlag Lecture Notes in Operations Res. and Math. Econ. #12, 383–403 and 379–381: The Proceedings of International Conference for Mathematical Systems Theory and Economics in Varenna, Italy, June 1967

J.L. Kaplan and J.A. Yorke, Toward a unification of ordinary differential equations with nonlinear semi-group theory, International Conference on Ordinary Differential Equations, H. Antosiewicz, ed., Academic Press (1975), 424–433: The proceedings of a conference in Los Angeles, September 1974

J. Curry and J.A. Yorke, A transition from Hopf bifurcation to chaos: computer experiments with maps in R^2, in The Structure of Attractors in Dynamical Systems, Springer Lecture Notes in Mathematics #668, 48–66: The proceedings of the NSF regional conference in Fargo, ND, June 1977

J.L. Kaplan and J.A. Yorke, Chaotic behavior of multidimensional difference equations, in Functional Differential Equations and Approximation of Fixed Points, H. O. Peitgen and H. O. Walther, eds., Springer Lecture Notes in Mathematics #730 (1979), 204–227

S. N. Chow, J. Mallet-Paret, and J.A. Yorke, A homotopy method for locating all zeroes of a system of polynomials, ibid, 77–78

C. Grebogi, H.E. Nusse, E. Ott, and J.A. Yorke, Basic Sets: Sets that determine the dimension of basin boundaries, in Dynamical Systems, Proc. of Special Year at the University of Maryland, ed. J. Alexander, Springer Lecture Notes in Mathematics #1342 (1988), 220–250

G.-H. Yuan, B.R. Hunt, C. Grebogi, E. Kostelich, E. Ott, and J.A. Yorke, Design and control of shipboard cranes, in Proc. of the 16th ASME Biennial Conference on Mechanical Vibration and Noise, September 1997, Sacramento, CA

Part II: Chaos and SLYRB Measures: The Development of the Theory of Chaotic Attractors

This book is a collection of articles devoted to the theoretical development of chaotic attractors with an emphasis on measures on attractors. The term "SLYRB" in the title of this section is pronounced as the single word "slurb." This acronym stands for "Sinai, Lasota, Yorke, Ruelle, Bowen" and will be explained below in the text. (An asterisk following a citation means that the article appears in this book.) We relate the topics of "chaotic attractors" and the corresponding "natural invariant measures." Limitation of space has forced us to exclude many outstanding papers. Our goal is to have a coherent collection of articles that reflect the theoretical development of chaotic attractors, are interesting to read, and are appropriate for a graduate student seminar. All too often, graduate students preparing to write a thesis have no knowledge of how a field has developed and little knowledge of how individuals have built the field. The courses they take and books they read present the subject as a fully constructed entity. However, we assume that the reader is familiar with a few basic notions from nonlinear dynamics such as "Lyapunov exponents" and "invariant measures." If this is not the case, the reader is advised to consult any good book in nonlinear dynamics covering these notions. The order of the articles is based on the date of publication.

In the 1970s, Lorenz's work became known. Lorenz [L]* discovered sensitivity to initial data in a simple system of three differential equations that served as a simple model for weather prediction. He described the structure of a set that is now called the *Lorenz attractor*. To explain Lorenz's result, Li and Yorke [LiY1]* proved a theorem on sensitivity to initial data in one-dimensional maps, which is now known as the "period three implies chaos" theorem. (This result has considerable overlap with the Sarkovskii theorem [Sa] on periodic orbits, which appeared in 1964 but was virtually unknown in the West until it was featured in a 1977 paper by Stefan [St].) From a different perspective, Rössler [Ro] showed that the invariant set for the Lorenz equations contains horseshoes, so the dynamics display chaos. In 1976, May [Ma]* published a review article on simple models with complicated dynamics, and this article explains the period-doubling phenomena and the Li–Yorke chaos that occur in these models.

In the 1980s, considerable attention was given to attractors in dynamical systems, and conferences devoted to this single topic were held. The notion of "strange attractor" was introduced by Ruelle and Takens [RT], referring to an attractor that is not a fixed-point attractor, a periodic orbit, or a smooth manifold without boundary such as a circle or torus. Nowadays the terms "chaotic attractor" and "strange attractor" are quite common. For some early review articles on this topic, see the review articles by Ott [O1]*, Eckmann [E], and Eckmann and Ruelle [ER]*. For the purpose of this book, for any continuous map from a Euclidean space into itself, we call a compact set A a *Milnor attractor* if (1) A is mapped onto itself, and (2) every open neighborhood of A

contains a set of positive Lebesgue measure such that every initial condition in that set has the property that the distance between the nth iterate of that initial condition and A converges to zero as n goes to infinity; see [Mil]*. The second condition says that there is a positive probability that the trajectory of a randomly chosen point will be attracted to A. (In the literature there exist several (inequivalent) definitions of the notion "attractor.")

We now define "chaotic trajectory" and "chaotic attractor". The trajectory of a point x is called a *chaotic trajectory* if (1) the trajectory of x is bounded and x is not asymptotically periodic, and (2) the trajectory of x has a positive Lyapunov exponent. Therefore, a chaotic trajectory must have infinitely many distinct points. We call a compact set A a *chaotic attractor* if (a) A is a Milnor attractor, and (b) A contains a chaotic trajectory that is dense in A. If the trajectory of some initial point is a chaotic trajectory, may we conclude that there is a chaotic attractor? To illustrate the problem, consider the example of the Hénon map depending on parameters a and b. For specified parameter values (namely, $a = 1.4$ and $b = 0.3$), Hénon [H]* gave the coordinates of the four corner points of a quadrilateral having the property that it is mapped into itself when the map (the Hénon map) is applied once. Hence, the quadrilateral contains an attractor, referred to as the *Hénon attractor*. The Hénon attractor (for these parameter values) is believed to be a chaotic attractor because (according to detailed numerical calculations) it appears to contain a chaotic trajectory that is dense in the Hénon attractor and has a positive Lyapunov exponent. However, there exists no proof of such a fact. One has to prove that there exists a chaotic trajectory that is dense in the closure of the unstable manifold of the fixed point both of whose coordinates are positive. Numerical simulations give the strong impression that this is true. Later in this introduction we discuss the situation for other parameter values.

Generally, a chaotic attractor is a Milnor attractor on which the dynamics are chaotic; that is, the system has at least one positive Lyapunov exponent. Frequently, the two notions of chaotic attractor and "strange attractor" are used as synonyms. Indeed, a chaotic attractor is often a strange attractor in the sense of Ruelle and Takens. However, an example of a strange attractor that is not a chaotic attractor is the so-called *Feigenbaum attractor* (that is, a Milnor attractor that is a Cantor set and for which the Lyapunov exponent of Lebesgue almost every trajectory is zero), which occurs in the one-parameter family of quadratic maps. Also, one encounters in the literature the following. A *strange attractor* is a Milnor attractor that has a Cantor-like structure. There is a large literature on "strange nonchaotic attractors" in systems that have terms that are quasi-periodic of time; see, for example, Ott [O2]. A common way to characterize strange attractors is through their fractal (that is, not necessarily integer-valued) dimension. The articles by Farmer, Ott and Yorke [FOY]*, Grassberger and Procaccia [GP]*, and Grebogi, Ott and Yorke [GOY2]* are recommended. In an earlier paper, Kaplan and Yorke [KY] conjectured a connection between the fractal dimension of an attractor and its

Lyapunov exponents. Their formula is now often used to compute the "Lyapunov dimension" of experimental time series.

At the end of the 1970s, Rössler [Ro] gave an example and Yorke and Yorke [YY] investigated and explained the phenomenon now called "transient chaos", in which trajectories behave chaotically for a finite period of time before settling down to more regular motion. Based on numerical studies of the Lorenz equations, they gave a theory that has found application in many systems, and illustrated the importance of nonattracting chaotic sets. For work on sudden changes in chaotic attractors and the appearance of transient chaos, see [GOY1].

In many applications it is useful to consider not only the set that constitutes a chaotic attractor but also (if it exists) the asymptotic distribution of a typical trajectory converging to the attractor. Indeed, in the physics literature such a distribution is often assumed to exist. When it exists, it is called a "natural invariant measure" [FOY]*. We will discuss some aspects of natural invariant measures. The goal of the paper [HKLN]* is to relate the "Lasota–Yorke measure" for chaotic attractors in one-dimensional maps and the "Sinai–Ruelle–Bowen measure" for chaotic attractors in higher-dimensional dynamical systems, and to introduce the notion of "SLYRB measure," which is a natural invariant measure. Both approaches deal with the problem that attractors are proper subsets of the space and that their ergodic theory should reflect the average behavior of initial points, where "average" is taken with respect to Lebesgue measure. We now discuss briefly the two approaches to the invariant measures derived from Lebesgue measure. Let M be a piecewise smooth map from the space X into itself. One approach follows ideas of Sinai, who since the beginning of the 1970s systematically developed a Markov partition method for dynamical systems. In a 1972 paper [Si], he proved the existence of invariant measures for certain systems having a Markov partition. In fact, he proved this result for Anosov diffeomorphisms. Bowen and Ruelle [BR]* extended these techniques to Axiom A diffeomorphisms and flows in several papers that were published during the period 1973–1976. The resulting measure is often called an *SRB measure*. To describe such an invariant measure, let U be a set of positive Lebesgue measure in X that is mapped into itself and let A be an attractor that is contained in U such that for every initial condition x in U, the distance between the nth iterate of x and A converges to zero as n goes to infinity. They consider the average value along a trajectory of any continuous bounded function. They are interested in cases in which Lebesgue almost all initial conditions x in U yield the same measure. Hence, the significance of such an SRB measure is that it describes the asymptotic distribution of the trajectory of initial points in a set of positive Lebesgue measure in the phase space. The second approach (based on the Perron–Frobenius techniques) was used, for example, by Lasota and Yorke [LaY]*. The Perron–Frobenius operator for M is very useful for studying the evolution of densities under iteration of M. The construction of Lasota and Yorke is an example of a method that is now commonly used for constructing natural invariant measures. The main result of Lasota and Yorke [LaY]* says that piecewise-smooth expanding one-dimensional maps admit an absolutely continuous invariant measure. Lasota and Yorke prove this theorem by establishing convergence for functions of bounded variation and using the fact that these functions are dense in the space of integrable functions. Their original idea of considering the space of functions of bounded variation and the properties of the Perron–Frobenius operator on this space has been used many times since. We refer to the measure in their theorem as a *Lasota–Yorke measure*. This measure is an absolutely continuous invariant measure. It is not difficult to show that each chaotic attractor (with a dense trajectory) for one-dimensional maps has a unique absolutely continuous invariant measure and each Lasota–Yorke measure is a linear combination of these [LiY2]. Yorke reports privately that the goal of [LaY]* initially was only to handle tent maps, but their techniques allowed them to prove the result for more general scalar maps. The theorem of Krzyzewski and Szlenk [KS]* had considered only maps for which there is in fact a smooth invariant density. That is not the case for most tent maps.

The way the two approaches describe the measures appear quite different. However, if the system has one attractor, then the two measures resulting from the two approaches are equal. The theoretical construction of a global SRB measure consists in averaging the iterates of a local measure on an arbitrary unstable manifold. This construction resembles the construction by Lasota and Yorke of pushing forward Lebesgue measure. Hofbauer and Keller [HK]* provide a general setting in which to define the Perron–Frobenius operator of a map and use it to obtain results about invariant measures and equilibrium states. Gora and Boyarsky [GB]* generalized Lasota and Yorke's result to n-dimensional space, and Tsujii [T2]* proved that for every piecewise real-analytic map of a bounded region in the plane there exists a probabilistic absolutely continuous invariant measure. Dellnitz and Junge [DJ]* presented efficient techniques for the numerical approximation of SRB measures based on appropriate discretization of the Perron–Frobenius operator. If m denotes the SRB measure, then it means that a Lebesgue typical initial condition chosen near the attractor A generates a trajectory that is asymptotically distributed according to m; that is, m is a natural invariant measure. Invariant measures with this property have been called by a variety of names in the literature, perhaps putting different emphases on the contributions of Sinai, Ruelle, and Bowen with minor variations in the properties of these measures. They were called Bowen–Ruelle measures by Young [Y1]* and a Bowen–Ruelle–Sinai (or BRS) measures by Tsujii [T1]. On the other hand, Benedicks and Young [BY]* define a Sinai–Bowen–Ruelle (or SBR) measure as an invariant measure with a positive Lyapunov exponent and absolutely continuous conditional measures along unstable manifolds. In a recent paper, Young [Y2] limits the definition of SRB measures to the context of no zero Lyapunov exponents and states a definition of Sinai–Ruelle–Bowen measure. She notes that this definition is a generalization of the ideas in Sinai [Si] and that it first appeared in a paper by Ledrappier [Le]. See also Pesin's book [P] for another variation. The above discussion on the variety of terms and notions of invariant measures demonstrates that the term "SRB measure" is not yet stable.

Notice that all of these concepts require that the measure in question be unique. Generally, this hypothesis is quite difficult to verify. Below, we argue that the term SLYRB, introduced in [HKLN]* and described briefly below, is perhaps better.

We now discuss briefly natural invariant measures for one-dimensional quadratic maps. The family of quadratic maps is not uniformly expanding for any of its members. In fact, near the critical point the quadratic map transforms any bounded density in one step to a density with inverse square-root singularity. A very important result is the famous theorem by Jakobson [J] that says the following. The set of parameters for which the corresponding quadratic map admits an absolutely continuous invariant measure, has positive Lebesgue measure. For an alternative proof, see, for example [Ry1]*. The argument used in the proof of Jakobson's result is that the orbit of the critical point does not approach the critical point too closely too quickly, and the collection of those parameters for which this approach is sufficiently controlled has positive Lebesgue measure. In describing the distribution of the orbit of x, Tsujii [T1] uses a sequence of probability measures. If this sequence converges to a probability measure as n goes to infinity, he calls that measure the *asymptotic distribution* for the orbit of x. Tsujii calls a probability measure m on the interval the *Bowen–Ruelle–Sinai measure* if the asymptotic distribution of the orbit exists and equals m for almost every point in the interval with respect to Lebesgue measure. He studies how the BRS measure depends on the parameter in the family of quadratic maps and considers two cases. One of these cases is that the map admits an absolutely continuous invariant probability measure. In this case the measure is unique and it is the BRS measure for the quadratic map.

We now turn to natural invariant measures for Lozi maps and Hénon maps. For the Hénon attractor $H*$ (which is the attractor for the Hénon map with parameter values $a = 1.4$ and $b = 0.3$ chosen by Hénon [H]*), one may ask the following question: Does there exist a natural invariant Borel probability measure m such that (1) the support of the measure m is $H*$, and (2) the orbit of Lebesgue almost every initial condition in the region defined by Hénon has a well-defined distribution and that this distribution is independent of the choice of the initial condition? Unfortunately, the answer is still unknown. In 1978 Lozi [Lo] introduced a piecewise-linear map as an analogue of the Hénon map. The dynamics of the Lozi map may be easier to understand than those of the Hénon map. It appears that the Lozi map is an intermediate stage between the Axiom A systems and more complicated systems like the Hénon map. Misiurewicz [Mis] showed that for certain parameter values, the Lozi map has a chaotic attractor. The attractor is the closure of the unstable "manifold" of a fixed point P, that is, the collection of points whose orbits converge to P as the map is iterated backwards. The maximal smooth component of the unstable "manifold" of P containing P is called the *local unstable manifold* of P. Collet and Levy [CL]* constructed an invariant measure on the attractor as follows. They started off with Lebesgue measure on the local unstable manifold of P. Then they considered the iterates of this measure under the Lozi map. Since the attractor is compact, one can extract a subsequence that converges to an invariant Borel probability measure m. They showed that m is unique, that it is absolutely continuous with respect to the Lebesgue measure in the unstable direction, and that it is in fact a Bowen–Ruelle measure (that is, a measure generated by the orbits of almost all points). This approach resembles Lasota and Yorke's construction of pushing forward Lebesgue measure.

We continue with an example combining the LY and SRB approaches. Bowen and Ruelle [BR]* assumed that the maps were smooth in the expanding directions. The measure they described, however, was typically not smooth, because the map had contracting directions. In contrast, Lasota and Yorke [LaY]* studied maps that were expanding. Their measure was typically not smooth because the maps were only piecewise smooth, and the successive iterates of the nonsmooth points (points at which the map is not smooth) could be dense in a collection of intervals. The first papers to deal with maps having both properties were those by Rychlik [Ry2], [Ry3]*, Young [Y1]* and Collet and Levy [CL]*. They dealt with maps that had the properties of having a contracting direction and being nonsmooth in the expanding direction. Young [Y1]* introduces and investigates a class of piecewise smooth maps on the unit square R in the plane that are continuous, one-to-one, and map the region R into itself. These maps are hyperbolic piecewise smooth maps and she calls them "generalized Lozi maps." Recall that the above result of Lasota and Yorke states that piecewise expanding maps of the interval have absolutely continuous invariant measures (Lasota–Yorke measure). Young proves an analogous statement for her generalized Lozi maps. A part of the proof follows a combination of the methods used by Sinai, and by Lasota and Yorke. Young's result is that the generalized Lozi maps admit certain invariant measures that imply that the maps have what she calls *Bowen–Ruelle measures*, that is, measures generated by the orbits of almost all initial points.

The existence of invariant measures for the Lozi maps was shown independently by Rychlik [Ry2], [Ry3]* by a method independent of Young [Y1]* and Collet and Levy [CL]*. The results were reported in [Ry2]. The paper [Ry2] contains only sketches of the arguments, and the full proofs can be found in Rychlik's dissertation [Ry3]*. The paper [Ry2] and the dissertation [Ry3]* present a hybrid approach to the "Lasota–Yorke" measures and the "Sinai–Ruelle–Bowen" measures.

We now return to the Hénon map (with parameter values $a = 1.4$ and $b = 0.3$ chosen by Hénon) [H]*. Numerical simulations suggest that for almost every initial condition in a certain region, the orbit has a well-defined distribution as n goes to infinity, and this distribution is independent of the choice of initial condition. However, in this case, as well as for other parameter values for which there seems to be a strange attractor, the situation is very delicate. The question is whether there exists an invariant measure that reflects the statistical behavior of Lebesgue almost every point in a much larger set. But why should such an invariant measure exist? On the other hand, if (one can show that) such a measure exists, it is of great significance. In the proofs of the results by Sinai [Si] and by Bowen and Ruelle [BR]*, Markov partitions were used to connect the dynamics of F to certain one-dimensional lattice systems in statistical mechanics, and m was realized as a Gibbs state or equilibrium state. However, Markov partitions are not available for systems more general than Axiom A, such as the Hénon map. We now explain why it is so difficult to prove the existence of a natural invariant measure for the Hénon map. Let P be the fixed point of the Hénon map that has two positive coordinates. The unstable manifold of P is the collection of points whose orbits converge to P as the map is iterated backwards. In analogy to the construction of an invariant measure by Collet and Levy [CL]* for the Lozi map, suppose we start with the local unstable manifold of the fixed point P and push forward Lebesgue measure on the local unstable manifold. The definition of a local unstable manifold says that the backward nth iterates of points on the local unstable manifold converge to P as n goes to infinity, but generally, one has no idea what happens to points on the local unstable manifold under forward iteration of the map. For example, it is easy to imagine that parts of the unstable manifold belong to basins of attraction of periodic orbits, so the orbits of certain points of the local unstable manifold converge to a periodic orbit attractor when the Hénon map is applied repeatedly. Hence, the question of existence of natural invariant (SRB) measures is very delicate. It is in some sense analogous to the problem of existence of absolutely continuous invariant measures for nonuniformly expanding maps (Jakobson theorem), except that it is more complicated because one has to consider not only rates of expansion, but their directions as well. Despite these difficulties, there have been some major breakthroughs. Benedicks and Carleson [BC] developed an elaborate machinery for analyzing the dynamics of the Hénon map for a positive measure set of parameters. However, this set of parameters does not include the parameter values $a = 1.4$ and $b = 0.3$ chosen by Hénon [H]*. Subsequently, Benedicks and Young [BY]* constructed SRB measures for these attractors. More recently, Young [Y2] and Alves, Bonatti, and Viana [ABV]* investigated diffeomorphisms that are not hyperbolic, but partially hyperbolic. These papers contain breakthrough results. Under quite general assumptions, [Y2] proves that correlations for hyperbolic systems with singularities decay exponentially. If Lyapunov exponents are bounded away from zero, [ABV]* prove that there are only finitely many natural invariant (Sinai–Ruelle–Bowen) measures for such maps and that Lebesgue almost every point is in one of the corresponding basins.

We recall that Lasota–Yorke measures are absolutely continuous invariant measures, while Bowen–Ruelle–Sinai measures are not. Hence, from the point of view of measure theory applied to dynamics, those invariant measures are considered to be different. However, from the point of view of dynamics using measure theory, the existence of a chaotic attractor implies in certain systems the existence of the special invariant measures as discussed above. After comparing the construction of SRB measures and their properties with the construction and properties of Lasota–Yorke measure, in [HKLN]* it is suggested that the natural invariant measure supported on any chaotic attractor might be called a "SLYRB measure." We quote from [HKLN]*: "The SRB concept of measure can be motivated by asking how a trajectory from a typical initial point is distributed asymptotically. Similarly, the SLYRB concept of measure can be motivated by asking what the average distribution is for trajectories of a large collection of initial points in some region not necessarily restricted to a single basin. The latter is analogous to asking where all the raindrops from a rainstorm go, and the former asks about where a single raindrop goes, perhaps winding up distributed throughout a particular lake." In [HKLN]* the following is proposed. A measure m for a dynamical system is called a *SLYRB measure* if m is a weighted sum (with positive weights) of invariant measures such that each of these invariant measures is the asymptotic distribution of a trajectory for a set of initial points of positive measure. Consider a system that has exactly one attractor A and for which almost all initial points chosen near the attractor A generate trajectories that are asymptotically distributed according to a single measure m. Then m would be both an SRB measure (by some standards) and a SLYRB measure. An invariant measure with this latter property for certain two-dimensional systems has been called a Bowen–Ruelle measure by Young [Y1]* and for certain one-dimensional systems a Bowen–Ruelle–Sinai (or BRS) measure by Tsujii [T1], so these measures are special cases of SLYRB measures. Furthermore, the SBR measure defined by Benedicks and Young [BY]*, being an invariant measure with a positive Lyapunov exponent and absolutely continuous conditional measures along unstable manifolds, is also an example of a SLYRB measure. Obviously, the Lasota–Yorke measure is an example of a SLYRB measure. Finally, [HKLN]* proposes that a dynamical system displays *SLYRB chaos* whenever there exists a chaotic attractor that supports a SLYRB measure. Enjoy!

References

[ABV] J.F. Alves, C. Bonatti and M. Viana, SRB measures for partially hyperbolic systems whose central direction is mostly expanding, Invent. Math. **140** (2000), 351–398

[BC] M. Benedicks and L. Carleson, The dynamics of the Hénon map, Annals of Math. **133** (1991), 73–169

[BY] M. Benedicks and L.-S. Young, Sinai-Bowen-Ruelle measures for certain Hénon maps, Invent. Math. **112** (1993), 541–576

[BR] R. Bowen and D. Ruelle, The ergodic theory of Axiom A flows, Invent. Math. **29** (1975), 181–202

[CL] P. Collet and Y. Levy, Ergodic properties of the Lozi mappings, Commun. Math. Phys. **93** (1984), 461–481

[DJ] M. Dellnitz and O. Junge, On the approximation of complicated dynamical behavior, SIAM J. Numer. Anal. **36** (1999), 491–515

[E] J.-P. Eckmann, Roads to turbulence in dissipative dynamical systems, Rev. Modern Phys. **53** (1981), 643–654

[ER] J.-P. Eckmann and D. Ruelle, Ergodic theory of chaos and strange attractors, Rev. Modern Phys. **57** (1985), 617–656

[FOY] J.D. Farmer, E. Ott and J.A. Yorke, The dimension of chaotic attractors, Physica D **7** (1983), 153–180

[GB] P. Gora and A. Boyarsky, Absolutely continuous invariant measures for piecewise expanding C^2 transformation in R^N, Israel J. Math. **67** (1989), 272–286

[GP] P. Grassberger and I. Procaccia, Measuring the strangeness of strange attractors, Physica D **9** (1983), 189–208

[GOY1] C. Grebogi, E. Ott and J.A. Yorke, Crisis, sudden changes in chaotic attractors, and transient chaos, Physica D **7** (1983), 181–200

[GOY2] C. Grebogi, E. Ott, and J.A. Yorke, Unstable periodic orbits and the dimensions of multifractal chaotic attractors, Phys. Rev. A **37** (1988), 1711–1724

[H] M. Hénon, A two-dimensional mapping with a strange attractor, Commun. Math. Phys. **50** (1976), 69–77

[HK] F. Hofbauer and G. Keller, Ergodic properties of invariant measures for piecewise monotonic transformations, Math. Z. **180** (1982), 119–140

[HKLN] B.R. Hunt, J.A. Kennedy, T.-Y. Li and H.E. Nusse, SLYRB measures: natural invariant measures for chaotic systems, Physica D **170**(2002), 50–71

[J] M. Jakobson, Absolutely continuous invariant measures for one-parameter families of one-dimensional maps, Commun. Math. Phys. **81** (1981), 39–88

[KY] J.L. Kaplan and J.A. Yorke, Chaotic behavior of multidimensional difference equations, in Functional Differential Equations and Approximation of Fixed Points, H.O. Peitgen and H.O. Walther, eds., Springer Lecture Notes in Mathematics #730 (1979), 204–227

[KS] K. Krzyzewski and W. Szlenk, On invariant measures for expanding differentiable mappings, Studia Math. **33** (1969), 83–92

[LaY] A. Lasota and J.A. Yorke, On the existence of invariant measures for piecewise monotonic transformations, Trans. Amer. Math. Soc. **186** (1973), 481–488

[Le] F. Ledrappier, Proprietes ergodiques des mesures de Sinai, I.H.E.S. Publ. Math. **59** (1984), 163–188

[LiY1] T.-Y. Li and J.A. Yorke, Period three implies chaos, Amer. Math. Monthly **82** (1975), 985–992

[LiY2] T.-Y. Li and J.A. Yorke, Ergodic transformations from an interval into itself, Trans. Amer. Math. Soc. **235** (1978), 183–192

[L] E.N. Lorenz, Deterministic nonperiodic flow, J. Atm. Sc. **20** (1963), 130–

141
[Lo] R. Lozi, Un attracteur etrange (?) du type attracteur de Hénon, J. Physique (Paris) **39** (Coll. C5) (1978), 9–10
[Ma] R.M. May, Simple mathematical models with very complicated dynamics, Nature **261** (1976), 459–467
[Mil] J. Milnor, On the concept of attractor, Commun. Math. Phys. **99** (1985), 177–195; Comments "On the concept of attractor": corrections and remarks, Commun. Math. Phys. **102** (1985), 517–519
[Mis] M. Misiurewicz, Strange attractors for the Lozi mappings, in Nonlinear Dynamics, R.H.G. Helleman, ed., Ann. New York Acad. Sci. #357 (1980), 348–358
[O1] E. Ott, Strange attractors and chaotic motions of dynamical systems, Rev. Modern Phys. **53** (1981), 655–671
[O2] E. Ott, Chaos in Dynamical Systems, Cambridge University Press, 1993
[P] Ya.B. Pesin, Dimension Theory in Dynamical Systems, Contemporary Views and Applications, University of Chicago Press, 1997
[Ro] O. Rössler, Horseshoe map chaos in the Lorenz equation, Phys. Lett. A **60** (1977), 392–394
[RT] D. Ruelle and F. Takens, On the nature of turbulence, Commun. Math. Phys. **20** (1971), 167–192
[Ry1] M.R. Rychlik, Another proof of Jakobson's theorem and related results, Ergodic Theory Dynamical Systems 8 (1988), 93–109
[Ry2] M. Rychlik, Mesures invariantes et principe variationel pour les applications de Lozi, C.R. Acad. Sc. Paris, t. 296 (1983), 19–22
[Ry3] M. Rychlik, Invariant Measures and Variational Principle for Lozi Applications, Ph.D. dissertation, University of California at Berkeley (1983),
[Sa] A.N. Sarkovskii, Coexistence of cycles of a continuous map of the line into itself, Ukr. Mat. Z. **16** (1964), 61–71
[Si] Ya.G. Sinai, Gibbs measures in ergodic theory, Russian Mathematical Surveys **27** (1972), 21–69
[St] P. Stefan, A theorem of Sarkovskii on the existence of periodic orbits of continuous endomorphisms of the real line, Comm. Math. Phys. **54** (1977), 237–248
[T1] M. Tsujii, On continuity of Bowen-Ruelle-Sinai measures in families of one dimensional maps, Comm. Math. Phys. **177** (1996), 1–11
[T2] M. Tsujii, Absolutely continuous invariant measures for piecewise real-analytic expanding maps on the plane, Comm. Math. Phys. **208** (2000), 605–622
[YY] J.A. Yorke and E.D. Yorke, Metastable chaos: the transition to sustained chaotic behavior in the Lorenz model, J. Stat. Phys. **21** (1979), 263–277
[Y1] L.-S. Young, Bowen-Ruelle measures for certain piecewise hyperbolic maps, Trans. Amer. Math. Soc. **287** (1985), 41–48
[Y2] L.-S. Young, Statistical properties of dynamical systems with some hyperbolicity, Ann. of Math. (2) **147** (1998), 585–650

Reprinted from Journal of the Atmospheric Sciences, Vol. 20, No. 2, March, 1963, pp. 130–141

Deterministic Nonperiodic Flow[1]

Edward N. Lorenz

Massachusetts Institute of Technology

(Manuscript received 18 November 1962, in revised form 7 January 1963)

Abstract

Finite systems of deterministic ordinary nonlinear differential equations may be designed to represent forced dissipative hydrodynamic flow. Solutions of these equations can be identified with trajectories in phase space. For those systems with bounded solutions, it is found that nonperiodic solutions are ordinarily unstable with respect to small modifications, so that slightly differing initial states can evolve into considerably different states. Systems with bounded solutions are shown to possess bounded numerical solutions.

A simple system representing cellular convection is solved numerically. All of the solutions are found to be unstable, and almost all of them are nonperiodic.

The feasibility of very-long-range weather prediction is examined in the light of these results.

1. Introduction

Certain hydrodynamical systems exhibit steady-state flow patterns, while others oscillate in a regular periodic fashion. Still others vary in an irregular, seemingly haphazard manner, and, even when observed for long periods of time, do not appear to repeat their previous history.

These modes of behavior may all be observed in the familiar rotating-basin experiments, described by Fultz, *et al.* (1959) and Hide (1958). In these experiments, a cylindrical vessel containing water is rotated about its axis, and is heated near its rim and cooled near its center in a steady symmetrical fashion. Under certain conditions the resulting flow is as symmetric and steady as the heating which gives rise to it. Under different conditions a system of regularly spaced waves develops, and progresses at a uniform speed without changing its shape. Under still different conditions an irregular flow pattern forms, and moves and changes its shape in an irregular nonperiodic manner.

Lack of periodicity is very common in natural systems, and is one of the distinguishing features of turbulent flow. Because instantaneous turbulent flow patterns are so irregular, attention is often confined to the statistics of turbulence, which, in contrast to the details of turbulence, often behave in a regular well-organized manner. The short-range weather forecaster, however, is forced willy-nilly to predict the details of the large-scale turbulent eddies—the cyclones and anticyclones—which continually arrange themselves into new patterns. Thus there are occasions when more than the statistics of irregular flow are of very real concern.

In this study we shall work with systems of deterministic equations which are idealizations of hydrodynamical systems. We shall be interested principally in nonperiodic solutions, i.e., solutions which never repeat their past history exactly, and where all approximate repetitions are of finite duration. Thus we shall be involved with the ultimate behavior of the solutions, as opposed to the transient behavior associated with arbitrary initial conditions.

A closed hydrodynamical system of finite mass may ostensibly be treated mathematically as a finite collection of molecules—usually a very large finite collection—in which case the governing laws are expressible as a finite set of ordinary differential equations. These equations are generally highly intractable, and the set of molecules is usually approximated by a continuous distribution of mass. The governing laws are then expressed as a set of partial differential equations, containing such quantities as velocity, density, and pressure as dependent variables.

It is sometimes possible to obtain particular solutions of these equations analytically, especially when the solutions are periodic or invariant with time, and, indeed, much work has been devoted to obtaining such solutions by one scheme or another. Ordinarily, however, nonperiodic solutions cannot readily be determined except by numerical procedures. Such procedures involve replacing the continuous variables by a new finite set of functions of time, which may perhaps be the values of the continuous variables at a chosen grid of points, or the coefficients in the expansions of these variables in series of orthogonal functions. The governing laws then become a finite set of ordinary differential

[1] The research reported in this work has been sponsored by the Geophysics Research Directorate of the Air Force Cambridge Research Center, under Contract No. AF 19(604)-4969.

equations again, although a far simpler set than the one which governs individual molecular motions.

In any real hydrodynamical system, viscous dissipation is always occurring, unless the system is moving as a solid, and thermal dissipation is always occurring, unless the system is at constant temperature. For certain purposes many systems may be treated as conservative systems, in which the total energy, or some other quantity, does not vary with time. In seeking the ultimate behavior of a system, the use of conservative equations is unsatisfactory, since the ultimate value of any conservative quantity would then have to equal the arbitrarily chosen initial value. This difficulty may be obviated by including the dissipative processes, thereby making the equations nonconservative, and also including external mechanical or thermal forcing, thus preventing the system from ultimately reaching a state of rest. If the system is to be deterministic, the forcing functions, if not constant with time, must themselves vary according to some deterministic rule.

In this work, then, we shall deal specifically with finite systems of deterministic ordinary differential equations, designed to represent forced dissipative hydrodynamical systems. We shall study the properties of nonperiodic solutions of these equations.

It is not obvious that such solutions can exist at all. Indeed, in dissipative systems governed by finite sets of *linear* equations, a constant forcing leads ultimately to a constant response, while a periodic forcing leads to a periodic response. Hence, nonperiodic flow has sometimes been regarded as the result of nonperiodic or random forcing.

The reasoning leading to these concludions is not applicable when the governing equations are nonlinear. If the equations contain terms representing advection—the transport of some property of a fluid by the motion of the fluid itself—a constant forcing can lead to a variable response. In the rotating-basin experiments already mentioned, both periodic and nonperiodic flow result from thermal forcing which, within the limits of experimental control, is constant. Exact periodic solutions of simplified systems of equations, representing dissipative flow with constant thermal forcing, have been obtained analytically by the writer (1962a). The writer (1962b) has also found nonperiodic solutions of similar systems of equations by numerical means.

2. Phase space

Consider a system whose state may be described by M variables $X_1, \cdots, X_M$. Let the system be governed by the set of equations

$$dX_i/dt=F_i(X_1,\cdots X_M), \quad i=1,\cdots,M, \tag{1}$$

where time t is the single independent variable, and the functions F_i possess continuous first partial derivatives. Such a system may be studied by means of *phase space*—an M-dimensional Euclidean space Γ whose coordinates are $X_1, \cdots, X_M$. Each *point* in phase space represents a possible instantaneous state of the system. A state which is varying in accordance with (1) is represented by a moving *particle* in phase space, traveling along a *trajectory* in phase space. For completeness, the position of a stationary particle, representing a steady state, is included as a trajectory.

Phase space has been a useful concept in treating finite systems, and has been used by such mathematicians as Gibbs (1902) in his development of statistical mechanics, Poincaré (1881) in his treatment of the solutions of differential equations, and Birkhoff (1927) in his treatise on dynamical systems.

From the theory of differential equations (e,g., Ford 1933, ch. 6), it follows, since the partial derivatives $\partial F_i/\partial X_j$ are continuous, that if t_0 is any time, and if $X_{10}, \cdots X_{M0}$ is any point in Γ, equations (1) possess a unique solution

$$X_i=f_i(X_{10},\cdots,X_{M0},t), \quad i=1,\cdots,M, \tag{2}$$

valid throughout some time interval containing t_0, and satisfying the condition

$$f_i(X_{10},\cdots,X_{M0},t_0)=X_{i0}, \quad i=1,\cdots,M. \tag{3}$$

The functions f_i are continuous in $X_{10}, \cdots, X_{M0}$ and t. Hence there is a unique trajectory through each point of Γ. Two or more trajectories may, however, approach the same point or the same curve asymptotically as $t\rightarrow\infty$ or as $t\rightarrow-\infty$. Moreover, since the functions f_i are continuous, the passage of time defines a continuous deformation of any region of Γ into another region.

In the familiar case of a conservative system, where some positive definite quantity Q, which may represent some form of energy, is invariant with time, each trajectory is confined to one or another of the surfaces of constant Q. These surfaces may take the form of closed concentric shells.

If, on the other hand, there is dissipation and forcing, and if, whenever Q equals or exceeds some fixed value Q_1, the dissipation acts to diminish Q more rapidly then the forcing can increase Q, then $(-dQ/dt)$ has a positive lower bound where $Q\geqq Q_1$, and each trajectory must ultimately become trapped in the region where $Q<Q_1$. Trajectories representing forced dissipative flow may therefore differ considerably from those representing conservative flow.

Forced dissipative systems of this sort are typified by the system

$$dX_i/dt=\sum_{j,k} a_{ijk}X_jX_k-\sum_j b_{ij}X_j+c_i, \tag{4}$$

where $\sum a_{ijk}X_iX_jX_k$ vanishes identically, $\sum b_{ij}X_iX_j$ is positive definite, and $c_1, \cdots, c_M$ are constants. If

$$Q=\tfrac{1}{2}\sum_i X_i^2, \tag{5}$$

and if $e_1, \cdots, e_M$ are the roots of the equations

$$\sum_j (b_{ij}+b_{ji})e_j=c_i, \qquad (6)$$

it follows from (4) that

$$dQ/dt=\sum_{i,j} b_{ij}e_ie_j-\sum_{i,j} b_{ij}(X_i-e_i)(X_j-e_j). \qquad (7)$$

The right side of (7) vanishes only on the surface of an ellipsoid E, and is positive only in the interior of E. The surfaces of constant Q are concentric spheres. If S denotes a particular one of these spheres whose interior R contains the ellipsoid E, it is evident that each trajectory eventually becomes trapped within R.

3. The instability of nonperiodic flow

In this section we shall establish one of the most important properties of deterministic nonperiodic flow, namely, its instability with respect to modifications of small amplitude. We shall find it convenient to do this by identifying the solutions of the governing equations with trajectories in phase space. We shall use such symbols as $P(t)$ (variable argument) to denote trajectories, and such symbols as P or $P(t_0)$ (no argument or constant argument) to denote points, the latter symbol denoting the specific point through which $P(t)$ passes at time t_0.

We shall deal with a phase space Γ in which a unique trajectory passes through each point, and where the passage of time defines a continuous deformation of any region of Γ into another region, so that if the points $P_1(t_0)$, $P_2(t_0)$, $\cdots$ approach $P_0(t_0)$ as a limit, the points $P_1(t_0+\tau)$, $P_2(t_0+\tau)$, $\cdots$ must approach $P_0(t_0+\tau)$ as a limit. We shall furthermore require that the trajectories be uniformly bounded as $t\rightarrow\infty$; that is, there must be a bounded region R, such that every trajectory ultimately remains with R. Our procedure is influenced by the work of Birkhoff (1927) on dynamical systems, but differs in that Birkhoff was concerned mainly with conservative systems. A rather detailed treatment of dynamical systems has been given by Nemytskii and Stepanov (1960), and rigorous proofs of some of the theorems which we shall present are to be found in that source.

We shall first classify the trajectories in three different manners, namely, according to the absence or presence of transient properties, according to the stability or instability of the trajectories with respect to small modifications, and according to the presence or absence of periodic behavior.

Since any trajectory $P(t)$ is bounded, it must possess at least one *limit point* P_0, a point which it approaches arbitrarily closely arbitrarily often. More precisely, P_0 is a limit point of $P(t)$ if for any $\epsilon>0$ and any time t_1 there exists a time $t_2(\epsilon,t_1)>t_1$ such that $|P(t_2)-P_0|<\epsilon$. Here absolute-value signs denote distance in phase space. Because Γ is continuously deformed as t varies, every point on the trajectory through P_0 is also a limit point of $P(t)$, and the set of limit points of $P(t)$ forms a trajectory, or a set of trajectories, called the *limiting trajectories* of $P(t)$. A limiting trajectory is obviously contained within R in its entirety.

If a trajectory is contained among its own limiting trajectories, it will be called *central*; otherwise it will be called *noncentral*. A central trajectory passes arbitrarily closely arbitrarily often to any point through which it has previously passed, and, in this sense at least, separate sufficiently long segments of a central trajectory are statistically similar. A noncentral trajectory remains a certain distance away from any point through which it has previously passed. It must approach its entire set of limit points asymptotically, although it need not approach any particular limiting trajectory asymptotically. Its instantaneous distance from its closest limit point is therefore a transient quantity, which becomes arbitrarily small as $t\rightarrow\infty$.

A trajectory $P(t)$ will be called *stable at a point* $P(t_1)$ if any other trajectory passing sufficiently close to $P(t_1)$ at time t_1 remains close to $P(t)$ as $t\rightarrow\infty$; i.e., $P(t)$ is stable at $P(t_1)$ if for any $\epsilon>0$ there exists a $\delta(\epsilon,t_1)>0$ such that if $|P_1(t_1)-P(t_1)|<\delta$ and $t_2>t_1$, $|P_1(t_2)-P(t_2)|<\epsilon$. Otherwise $P(t)$ will be called *unstable* at $P(t_1)$. Because Γ is continuously deformed as t varies, a trajectory which is stable at one point is stable at every point, and will be called a *stable* trajectory. A trajectory unstable at one point is unstable at every point, and will be called an *unstable* trajectory. In the special case that $P(t)$ is confined to one point, this definition of stability coincides with the familiar concept of stability of steady flow.

A stable trajectory $P(t)$ will be called uniformly stable if the distance within which a neighboring trajectory must approach a point $P(t_1)$, in order to be certain of remaining close to $P(t)$ as $t\rightarrow\infty$, itself possesses a positive lower bound as $t_1\rightarrow\infty$; i.e., $P(t)$ is uniformly stable if for any $\epsilon>0$ there exists a $\delta(\epsilon)>0$ and a time $t_0(\epsilon)$ such that if $t_1>t_0$ and $|P_1(t_1)-P(t_1)|<\delta$ and $t_2>t_1$, $|P_1(t_2)-P(t_2)|<\epsilon$. A limiting trajectory $P_0(t)$ of a uniformly stable trajectory $P(t)$ must be uniformly stable itself, since all trajectories passing sufficiently close to $P_0(t)$ must pass arbitrarily close to some point of $P(t)$ and so must remain close to $P(t)$, and hence to $P_0(t)$, as $t\rightarrow\infty$.

Since each point lies on a unique trajectory, any trajectory passing through a point through which it has previously passed must continue to repeat its past behavior, and so must be *periodic*. A trajectory $P(t)$ will be called *quasi-periodic* if for some arbitrarily large time interval τ, $P(t+\tau)$ ultimately remains arbitrarily close to $P(t)$, i.e., $P(t)$ is quasi-periodic if for any $\epsilon>0$ and for any time interval τ_0, there exists a $\tau(\epsilon,\tau_0)>\tau_0$ and a time $t_1(\epsilon,\tau_0)$ such that if $t_2>t_1$, $|P(t_2+\tau)-P(t_2)|$

$<\epsilon$. Periodic trajectories are special cases of quasi-periodic trajectories.

A trajectory which is not quasi-periodic will be called *nonperiodic*. If $P(t)$ is nonperiodic, $P(t_1+\tau)$ may be arbitrarily close to $P(t_1)$ for some time t_1 and some arbitrarily large time interval τ, but, if this is so, $P(t+\tau)$ cannot remain arbitrarily close to $P(t)$ as $t\rightarrow\infty$. Nonperiodic trajectories are of course representations of deterministic nonperiodic flow, and form the principal subject of this paper.

Periodic trajectories are obviously central. Quasi-periodic central trajectories include multiple periodic trajectories with incommensurable periods, while quasi-periodic noncentral trajectories include those which approach periodic trajectories asymptotically. Nonperiodic trajectories may be central or noncentral.

We can now establish the theorem that a trajectory with a stable limiting trajectory is quasi-periodic. For if $P_0(t)$ is a limiting trajectory of $P(t)$, two distinct points $P(t_1)$ and $P(t_1+\tau)$, with τ arbitrarily large, may be found arbitrary close to any point $P_0(t_0)$. Since $P_0(t)$ is stable, $P(t)$ and $P(t+\tau)$ must remain arbitrarily close to $P_0(t+t_0-t_1)$, and hence to each other, as $t\rightarrow\infty$, and $P(t)$ is quasi-periodic.

It follows immediately that a stable central trajectory is quasi-periodic, or, equivalently, that a nonperiodic central trajectory is unstable.

The result has far-reaching consequences when the system being considered is an observable nonperiodic system whose future state we may desire to predict. It implies that two states differing by imperceptible amounts may eventually evolve into two considerably different states. If, then, there is any error whatever in observing the present state—and in any real system such errors seem inevitable—an acceptable prediction of an instantaneous state in the distant future may well be impossible.

As for noncentral trajectories, it follows that a uniformly stable noncentral trajectory is quasi-periodic, or, equivalently, a nonperiodic noncentral trajectory is not uniformly stable. The possibility of a nonperiodic noncentral trajectory which is stable but not uniformly stable still exists. To the writer, at least, such trajectories, although possible on paper, do not seem characteristic of real hydrodynamical phenomena. Any claim that atmospheric flow, for example, is represented by a trajectory of this sort would lead to the improbable conclusion that we ought to master long-range forecasting as soon as possible, because, the longer we wait, the more difficult our task will become.

In summary, we have shown that, subject to the conditions of uniqueness, continuity, and boundedness prescribed at the beginning of this section, a central trajectory, which in a certain sense is free of transient properties, is unstable if it is nonperiodic. A noncentral trajectory, which is characterized by transient properties, is not uniformly stable if it is nonperiodic, and, if it is stable at all, its very stability is one of its transient properties, which tends to die out as time progresses. In view of the impossibility of measuring initial conditions precisely, and thereby distinguishing between a central trajectory and a nearby noncentral trajectory, all nonperiodic trajectories are effectively unstable from the point of view of practical prediction.

4. Numerical integration of nonconservative systems

The theorems of the last section can be of importance only if nonperiodic solutions of equations of the type considered actually exist. Since statistically stationary nonperiodic functions of time are not easily described analytically, particular nonperiodic solutions can probably be found most readily by numerical procedures. In this section we shall examine a numerical-integration procedure which is especially applicable to systems of equations of the form (4). In a later section we shall use this procedure to determine a nonperiodic solution of a simple set of equations.

To solve (1) numerically we may choose an initial time t_0 and a time increment Δt, and let

$$X_{i,n}=X_i(t_0+n\Delta t). \tag{8}$$

We then introduce the auxiliary approximations

$$X_{i(n+1)}=X_{i,n}+F_i(P_n)\Delta t, \tag{9}$$

$$X_{i((n+2))}=X_{i(n+1)}+F_i(P_{(n+1)})\Delta t, \tag{10}$$

where P_n and $P_{(n+1)}$ are the points whose coordinates are

$$(X_{1,n},\cdots,X_{M,n}) \quad\text{and}\quad (X_{1(n+1)},\cdots,X_{M(n+1)}).$$

The simplest numerical procedure for obtaining approximate solutions of (1) is the forward-difference procedure,

$$X_{i,n+1}=X_{i(n+1)}. \tag{11}$$

In many instances better approximations to the solutions of (1) may be obtained by a centered-difference procedure

$$X_{i,n+1}=X_{i,n-1}+2F_i(P_n)\Delta t. \tag{12}$$

This procedure is unsuitable, however, when the deterministic nature of (1) is a matter of concern, since the values of $X_{1,n},\cdots,X_{M,n}$ do not uniquely determine the values of $X_{1,n+1},\cdots,X_{M,n+1}$.

A procedure which largely overcomes the disadvantages of both the forward-difference and centered-difference procedures is the double-approximation procedure, defined by the relation

$$X_{i,n+1}=X_{i,n}+\tfrac{1}{2}[F_i(P_n)+F_i(P_{(n+1)})]\Delta t. \tag{13}$$

Here the coefficient of Δt is an approximation to the time derivative of X_i at time $t_0+(n+\frac{1}{2})\Delta t$. From (9) and (10), it follows that (13) may be rewritten

$$X_{i,n+1}=\tfrac{1}{2}(X_{i,n}+X_{i((n+2))}). \tag{14}$$

A convenient scheme for automatic computation is the successive evaluation of $X_{i(n+1)}$, $X_{i((n+2))}$, and $X_{i,n+1}$ according to (9), (10) and (14). We have used this procedure in all the computations described in this study.

In phase space a numerical solution of (1) must be represented by a jumping particle rather than a continuously moving particle. Moreover, if a digital computer is instructed to represent each number in its memory by a preassigned fixed number of bits, only certain discrete points in phase space will ever be occupied. If the numerical solution is bounded, repetitions must eventually occur, so that, strictly speaking, every numerical solution is periodic. In practice this consideration may be disregarded, if the number of different possible states is far greater than the number of iterations ever likely to be performed. The necessity for repetition could be avoided altogether by the somewhat uneconomical procedure of letting the precision of computation increase as n increases.

Consider now numerical solutions of equations (4), obtained by the forward-difference procedure (11). For such solutions,

$$Q_{n+1}=Q_n+(dQ/dt)_n\Delta t+\tfrac{1}{2}\sum_i F_i^2(P_n)\Delta t^2. \qquad (15)$$

Let S' be any surface of constant Q whose interior R' contains the ellipsoid E where dQ/dt vanishes, and let S be any surface of constant Q whose interior R contains S'.

Since $\sum F_i^2$ and dQ/dt both possess upper bounds in R', we may choose Δt so small that P_{n+1} lies in R if P_n lies in R'. Likewise, since $\sum F_i^2$ possesses an upper bound and dQ/dt possesses a *negative* upper bound in $R-R'$, we may choose Δt so small that $Q_{n+1}<Q_n$ if P_n lies in $R-R'$. Hence Δt may be chosen so small that any jumping particle which has entered R remains trapped within R, and the numerical solution does not blow up. A blow-up may still occur, however, if initially the particle is exterior to R.

Consider now the double-approximation procedure (14). The previous arguments imply not only that $P_{(n+1)}$ lies within R if P_n lies within R, but also that $P_{((n+2))}$ lies within R if $P_{(n+1)}$ lies within R. Since the region R is convex, it follows that P_{n+1}, as given by (14), lies within R if P_n lies within R. Hence if Δt is chosen so small that the forward-difference procedure does not blow up, the double-approximation procedure also does not blow up.

We note in passing that if we apply the forward-difference procedure to a conservative system where $dQ/dt=0$ everywhere,

$$Q_{n+1}=Q_n+\tfrac{1}{2}\sum_i F_i^2(P_n)\Delta t^2. \qquad (16)$$

In this case, for any fixed choice of Δt the numerical solution ultimately goes to infinity, unless it is asymptotically approaching a steady state. A similar result holds when the double-approximation procedure (14) is applied to a conservative system.

5. The convection equations of Saltzman

In this section we shall introduce a system of three ordinary differential equations whose solutions afford the simplest example of deterministic nonperiodic flow of which the writer is aware. The system is a simplification of one derived by Saltzman (1962) to study finite-amplitude convection. Although our present interest is in the nonperiodic nature of its solutions, rather than in its contributions to the convection problem, we shall describe its physical background briefly.

Rayleigh (1916) studied the flow occurring in a layer of fluid of uniform depth H, when the temperature difference between the upper and lower surfaces is maintained at a constant value ΔT. Such a system possesses a steady-state solution in which there is no motion, and the temperature varies linearly with depth, If this solution is unstable, convection should develop.

In the case where all motions are parallel to the $x-z$-plane, and no variations in the direction of the y-axis occur, the governing equations may be written (see Saltzman, 1962)

$$\frac{\partial}{\partial t}\nabla^2\psi=-\frac{\partial(\psi,\nabla^2\psi)}{\partial(x,z)}+\nu\nabla^4\psi+g\alpha\frac{\partial\theta}{\partial x}, \qquad (17)$$

$$\frac{\partial}{\partial t}\theta=-\frac{\partial(\psi,\theta)}{\partial(x,z)}+\frac{\Delta T}{H}\frac{\partial\psi}{\partial x}+\kappa\nabla^2\theta. \qquad (18)$$

Here ψ is a stream function for the two-dimensional motion, θ is the departure of temperature from that occurring in the state of no convection, and the constants g, α, ν, and κ denote, respectively, the acceleration of gravity, the coefficient of thermal expansion, the kinematic viscosity, and the thermal conductivity. The problem is most tractable when both the upper and lower boundaries are taken to be free, in which case ψ and $\nabla^2\psi$ vanish at both boundaries.

Rayleigh found that fields of motion of the form

$$\psi=\psi_0\sin(\pi aH^{-1}x)\sin(\pi H^{-1}z), \qquad (19)$$

$$\theta=\theta_0\cos(\pi aH^{-1}x)\sin(\pi H^{-1}z), \qquad (20)$$

would develop if the quantity

$$R_a=g\alpha H^3\Delta T\nu^{-1}\kappa^{-1}, \qquad (21)$$

now called the *Rayleigh number*, exceeded a critical value

$$R_c=\pi^4a^{-2}(1+a^2)^3. \qquad (22)$$

The minimum value of R_c, namely $27\pi^4/4$, occurs when $a^2=\tfrac{1}{2}$.

Saltzman (1962) derived a set of ordinary differential equations by expanding ψ and θ in double Fourier series in x and z, with functions of t alone for coefficients, and substituting these series into (17) and (18). He arranged the right-hand sides of the resulting equations in double-Fourier-series form, by replacing products of trigonometric functions of x (or z) by sums of trigonometric functions, and then equated coefficients of similar functions of x and z. He then reduced the resulting infinite system to a finite system by omitting reference to all but a specified finite set of functions of t, in the manner proposed by the writer (1960).

He then obtained time-dependent solutions by numerical integration. In certain cases all except three of the dependent variables eventually tended to zero, and these three variables underwent irregular, apparently nonperiodic fluctuations.

These same solutions would have been obtained if the series had at the start been truncated to include a total of three terms. Accordingly, in this study we shall let

$$a(1+a^2)^{-1}\kappa^{-1}\psi = X\sqrt{2}\sin(\pi a H^{-1}x)\sin(\pi H^{-1}z), \tag{23}$$

$$\pi R_c^{-1}R_a\Delta T^{-1}\theta = Y\sqrt{2}\cos(\pi a H^{-1}x)\sin(\pi H^{-1}z) - Z\sin(2\pi H^{-1}z), \tag{24}$$

where X, Y, and Z are functions of time alone. When expressions (23) and (24) are substituted into (17) and (18), and trigonometric terms other than those occurring in (23) and (24) are omitted, we obtain the equations

$$X^{\cdot} = -\sigma X + \sigma Y, \tag{25}$$

$$Y^{\cdot} = -XZ + rX - Y, \tag{26}$$

$$Z^{\cdot} = XY - bZ. \tag{27}$$

Here a dot denotes a derivative with respect to the dimensionless time $\tau = \pi^2 H^{-2}(1+a^2)\kappa t$, while $\sigma = \kappa^{-1}\nu$ is the *Prandtl number*, $r = R_c^{-1}R_a$, and $b = 4(1+a^2)^{-1}$. Except for multiplicative constants, our variables X, Y, and Z are the same as Saltzman's variables A, D, and G. Equations (25), (26), and (27) are the convection equations whose solutions we shall study.

In these equations X is proportional to the intensity of the convective motion, while Y is proportional to the temperature difference between the ascending and descending currents, similar signs of X and Y denoting that warm fluid is rising and cold fluid is descending. The variable Z is proportional to the distortion of the vertical temperature profile from linearity, a positive value indicating that the strongest gradients occur near the boundaries.

Equations (25)–(27) may give realistic results when the Rayleigh number is slightly supercritical, but their solutions cannot be expected to resemble those of (17) and (18) when strong convection occurs, in view of the extreme truncation.

6. Applications of linear theory

Although equations (25)–(27), as they stand, do not have the form of (4), a number of linear transformations will convert them to this form. One of the simplest of these is the transformation

$$X' = X, \quad Y' = Y, \quad Z' = Z - r - \sigma. \tag{28}$$

Solutions of (25)–(27) therefore remain bounded within a region R as $\tau \to \infty$, and the general results of Sections 2, 3 and 4 apply to these equations.

The stability of a solution $X(\tau)$, $Y(\tau)$, $Z(\tau)$ may be formally investigated by considering the behavior of small superposed perturbations $x_0(\tau)$, $y_0(\tau)$, $z_0(\tau)$. Such perturbations are temporarily governed by the linearized equations

$$\begin{bmatrix} x_0 \\ y_0 \\ z_0 \end{bmatrix}^{\cdot} = \begin{bmatrix} -\sigma & \sigma & 0 \\ (r-Z) & -1 & -X \\ Y & X & -b \end{bmatrix} \begin{bmatrix} x_0 \\ y_0 \\ z_0 \end{bmatrix}. \tag{29}$$

Since the coefficients in (29) vary with time, unless the basic state X, Y, Z is a steady-state solution of (25)–(27), a general solution of (29) is not feasible. However, the variation of the volume V_0 of a small region in phase space, as each point in the region is displaced in accordance with (25)–(27), is determined by the diagonal sum of the matrix of coefficients; specifically

$$V_0^{\cdot} = -(\sigma + b + 1)V_0. \tag{30}$$

This is perhaps most readily seen by visualizing the motion in phase space as the flow of a fluid, whose divergence is

$$\frac{\partial X^{\cdot}}{\partial X} + \frac{\partial Y^{\cdot}}{\partial Y} + \frac{\partial Z^{\cdot}}{\partial Z} = -(\sigma + b + 1). \tag{31}$$

Hence each small volume shrinks to zero as $\tau \to \infty$, at a rate independent of X, Y, and Z. This does not imply that each small volume shrinks to a point; it may simply become flattened into a surface. It follows that the volume of the region initially enclosed by the surface S shrinks to zero at this same rate, so that all trajectories ultimately become confined to a specific subspace having zero volume. This subspace contains all those trajectories which lie entirely within R, and so contains all central trajectories.

Equations (25)–(27) possess the steady-state solution $X = Y = Z = 0$, representing the state of no convection. With this basic solution, the characteristic equation of the matrix in (29) is

$$[\lambda + b][\lambda^2 + (\sigma+1)\lambda + \sigma(1-r)] = 0. \tag{32}$$

This equation has three real roots when $r > 0$; all are negative when $r < 1$, but one is positive when $r > 1$. The criterion for the onset of convection is therefore $r = 1$, or $R_a = R_c$, in agreement with Rayleigh's result.

When $r > 1$, equations (25)–(27) possess two additional steady-state solutions $X = Y = \pm\sqrt{b(r-1)}$, $Z = r - 1$.

For either of these solutions, the characteristic equation of the matrix in (29) is

$$\lambda^3+(\sigma+b+1)\lambda^2+(r+\sigma)b\lambda+2\sigma b(r-1)=0. \quad (33)$$

This equation possesses one real negative root and two complex conjugate roots when $r>1$; the complex conjugate roots are pure imaginary if the product of the coefficients of λ^2 and λ equals the constant term, or

$$r=\sigma(\sigma+b+3)(\sigma-b-1)^{-1}. \quad (34)$$

This is the critical value of r for the instability of steady convection. Thus if $\sigma<b+1$, no positive value of r satisfies (34), and steady convection is always stable, but if $\sigma>b+1$, steady convection is unstable for sufficiently high Rayleigh numbers. This result of course applies only to idealized convection governed by (25)–(27), and not to the solutions of the partial differential equations (17) and (18).

The presence of complex roots of (34) shows that if unstable steady convection is disturbed, the motion will oscillate in intensity. What happens when the disturbances become large is not revealed by linear theory. To investigate finite-amplitude convection, and to study the subspace to which trajectories are ultimately confined, we turn to numerical integration.

TABLE 1. Numerical solution of the convection equations. Values of X, Y, Z are given at every fifth iteration N, for the first 160 iterations.

N	*X*	*Y*	*Z*
0000	0000	0010	0000
0005	0004	0012	0000
0010	0009	0020	0000
0015	0016	0036	0002
0020	0030	0066	0007
0025	0054	0115	0024
0030	0093	0192	0074
0035	0150	0268	0201
0040	0195	0234	0397
0045	0174	0055	0483
0050	0097	−0067	0415
0055	0025	−0093	0340
0060	−0020	−0089	0298
0065	−0046	−0084	0275
0070	−0061	−0083	0262
0075	−0070	−0086	0256
0080	−0077	−0091	0255
0085	−0084	−0095	0258
0090	−0089	−0098	0266
0095	−0093	−0098	0275
0100	−0094	−0093	0283
0105	−0092	−0086	0297
0110	−0088	−0079	0286
0115	−0083	−0073	0281
0120	−0078	−0070	0273
0125	−0075	−0071	0264
0130	−0074	−0075	0257
0135	−0076	−0080	0252
0140	−0079	−0087	0251
0145	−0083	−0093	0254
0150	−0088	−0098	0262
0155	−0092	−0099	0271
0160	−0094	−0096	0281

7. Numerical integration of the convection equations

To obtain numerical solutions of the convection equations, we must choose numerical values for the constants. Following Saltzman (1962), we shall let $\sigma=10$ and $a^2=\frac{1}{2}$, so that $b=8/3$. The critical Rayleigh number for instability of steady convection then occurs when $r=470/19=24.74$.

We shall choose the slightly supercritical value $r=28$. The states of steady convection are then represented by the points $(6\sqrt{2}, 6\sqrt{2}, 27)$ and $(-6\sqrt{2}, -6\sqrt{2}, 27)$ in phase space, while the state of no convection corresponds to the origin $(0,0,0)$.

We have used the double-approximation procedure for numerical integration, defined by (9), (10), and (14). The value $\Delta\tau=0.01$ has been chosen for the dimensionless time increment. The computations have been performed on a Royal McBee LGP-30 electronic com-

TABLE 2. Numerical solution of the convection equations Values of X, Y, Z are given at every iteration N for which Z possesses a relative maximum, for the first 6000 iterations.

N	*X*	*Y*	*Z*	*N*	*X*	*Y*	*Z*
0045	0174	0055	0483	3029	0117	0075	0352
0107	−0091	−0083	0287	3098	0123	0076	0365
0168	−0092	−0084	0288	3171	0134	0082	0383
0230	−0092	−0084	0289	3268	0155	0069	0435
0292	−0092	−0083	0290	3333	−0114	−0079	0342
0354	−0093	−0083	0292	3400	−0117	−0077	0350
0416	−0093	−0083	0293	3468	−0125	−0083	0361
0478	−0094	−0082	0295	3541	−0129	−0073	0378
0540	−0094	−0082	0296	3625	−0146	−0074	0413
0602	−0095	−0082	0298	3695	0127	0079	0370
0664	−0096	−0083	0300	3772	0136	0072	0394
0726	−0097	−0083	0302	3853	−0144	−0077	0407
0789	−0097	−0081	0304	3926	0129	0072	0380
0851	−0099	−0083	0307	4014	0148	0068	0421
0914	−0100	−0081	0309	4082	−0120	−0074	0359
0977	−0100	−0080	0312	4153	−0129	−0078	0375
1040	−0102	−0080	0315	4233	−0144	−0082	0404
1103	−0104	−0081	0319	4307	0135	0081	0385
1167	−0105	−0079	0323	4417	−0162	−0069	0450
1231	−0107	−0079	0328	4480	0106	0081	0324
1295	−0111	−0082	0333	4544	0109	0082	0329
1361	−0111	−0077	0339	4609	0110	0080	0334
1427	−0116	−0079	0347	4675	0112	0076	0341
1495	−0120	−0077	0357	4741	0118	0081	0349
1566	−0125	−0072	0371	4810	0120	0074	0360
1643	−0139	−0077	0396	4881	0130	0081	0376
1722	0140	0075	0401	4963	0141	0068	0406
1798	−0135	−0072	0391	5035	−0133	−0081	0381
1882	0146	0074	0413	5124	−0151	−0076	0422
1952	−0127	−0078	0370	5192	0119	0075	0358
2029	−0135	−0070	0393	5262	0129	0083	0372
2110	0146	0083	0408	5340	0140	0079	0397
2183	−0128	−0070	0379	5419	−0137	−0067	0399
2268	−0144	−0066	0415	5495	0140	0081	0394
2337	0126	0079	0368	5576	−0141	−0072	0405
2412	0137	0081	0389	5649	0135	0082	0384
2501	−0153	−0080	0423	5752	0160	0074	0443
2569	0119	0076	0357	5816	−0110	−0081	0332
2639	0129	0082	0371	5881	−0113	−0082	0339
2717	0136	0070	0395	5948	−0114	−0075	0346
2796	−0143	−0079	0402				
2871	0134	0076	0388				
2962	−0152	−0072	0426				

puting machine. Approximately one second per iteration, aside from output time, is required.

For initial conditions we have chosen a slight departure from the state of no convection, namely (0,1,0). Table 1 has been prepared by the computer. It gives the values of N (the number of iterations), X, Y, and Z at every fifth iteration for the first 160 iterations. In the printed output (but not in the computations) the values of X, Y, and Z are multiplied by ten, and then only those figures to the left of the decimal point are printed. Thus the states of steady convection would appear as 0084, 0084, 0270 and −0084, −0084, 0270, while the state of no convection would appear as 0000, 0000, 0000.

The initial instability of the state of rest is evident. All three variables grow rapidly, as the sinking cold fluid is replaced by even colder fluid from above, and the rising warm fluid by warmer fluid from below, so that by step 35 the strength of the convection far exceeds that of steady convection. Then Y diminishes as the warm fluid is carried over the top of the convective cells, so that by step 50, when X and Y have opposite signs, warm fluid is descending and cold fluid is ascending. The motion thereupon ceases and reverses its direction, as indicated by the negative values of X following step 60. By step 85 the system has reached a state not far from that of steady convection. Between steps 85 and 150 it executes a complete oscillation in its intensity, the slight amplification being almost indetectable.

The subsequent behavior of the system is illustrated in Fig. 1, which shows the behavior of Y for the first 3000 iterations. After reaching its early peak near step 35 and then approaching equilibrium near step 85, it undergoes systematic amplified oscillations until near step 1650. At this point a critical state is reached, and thereafter Y changes sign at seemingly irregular intervals, reaching sometimes one, sometimes two, and sometimes three or more extremes of one sign before changing sign again.

Fig. 2 shows the projections on the X-Y- and Y-Z-planes in phase space of the portion of the trajectory corresponding to iterations 1400–1900. The states of steady convection are denoted by C and C'. The first portion of the trajectory spirals outward from the vicinity of C', as the oscillations about the state of steady convection, which have been occurring since step 85, continue to grow. Eventually, near step 1650, it crosses the X-Z-plane, and is then deflected toward the neighborhood of C. It temporarily spirals about C, but crosses the X-Z-plane after one circuit, and returns to the neighborhood of C', where it soon joins the spiral over which it has previously traveled. Thereafter it crosses from one spiral to the other at irregular intervals.

Fig. 3, in which the coordinates are Y and Z, is based upon the printed values of X, Y, and Z at every fifth iteration for the first 6000 iterations. These values determine X as a smooth single-valued function of Y and Z over much of the range of Y and Z; they determine X

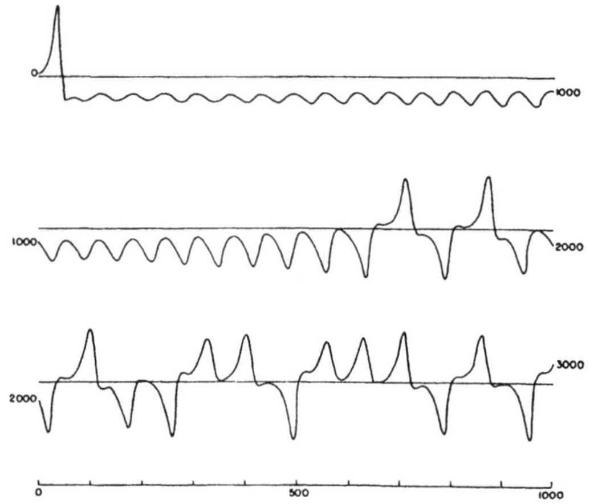

FIG. 1. Numerical solution of the convection equations. Graph of Y as a function of time for the first 1000 iterations (upper curve), second 1000 iterations (middle curve), and third 1000 iterations (lower curve).

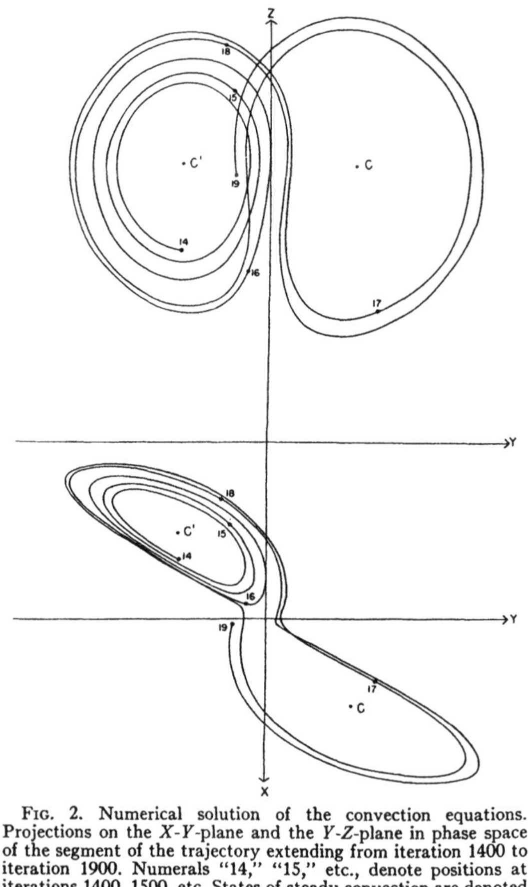

FIG. 2. Numerical solution of the convection equations. Projections on the X-Y-plane and the Y-Z-plane in phase space of the segment of the trajectory extending from iteration 1400 to iteration 1900. Numerals "14," "15," etc., denote positions at iterations 1400, 1500, etc. States of steady convection are denoted by C and C'.

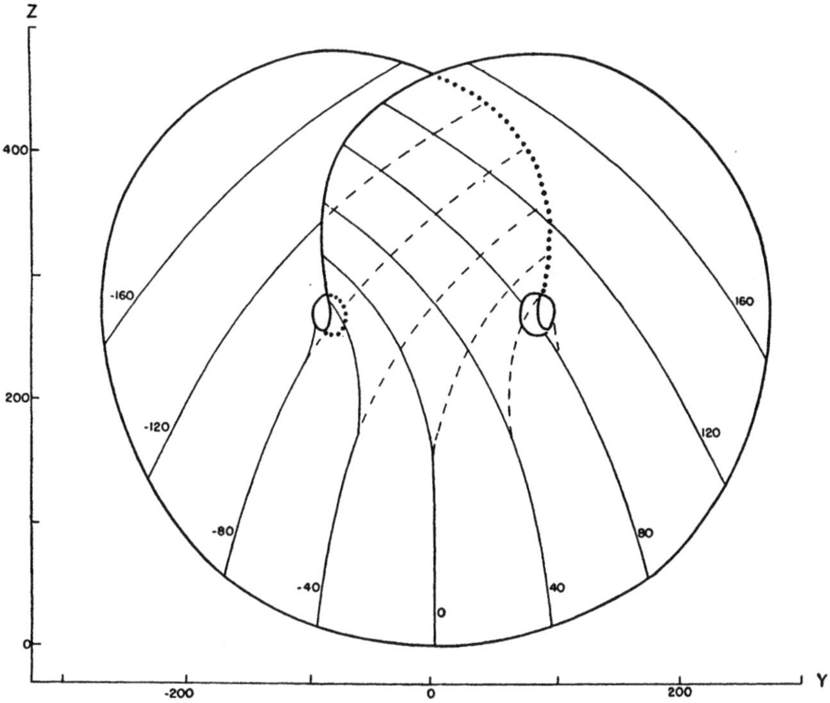

FIG. 3. Isopleths of X as a function of Y and Z (thin solid curves), and isopleths of the lower of two values of X, where two values occur (dashed curves), for approximate surfaces formed by all points on limiting trajectories. Heavy solid curve, and extensions as dotted curves, indicate natural boundaries of surfaces.

as one of two smooth single-valued functions over the remainder of the range. In Fig. 3 the thin solid lines are isopleths of X, and where two values of X exist, the dashed lines are isopleths of the lower value. Thus, within the limits of accuracy of the printed values, the trajectory is confined to a pair of surfaces which appear to merge in the lower portion of Fig. 3. The spiral about C lies in the upper surface, while the spiral about C' lies in the lower surface. Thus it is possible for the trajectory to pass back and forth from one spiral to the other without intersecting itself.

Additional numerical solutions indicate that other trajectories, originating at points well removed from these surfaces, soon meet these surfaces. The surfaces therefore appear to be composed of all points lying on limiting trajectories.

Because the origin represents a steady state, no trajectory can pass through it. However, two trajectories emanate from it, i.e., approach it asymptotically as $\tau \rightarrow -\infty$. The heavy solid curve in Fig. 3, and its extensions as dotted curves, are formed by these two trajectories. Trajectories passing close to the origin will tend to follow the heavy curve, but will not cross it, so that the heavy curve forms a natural boundary to the region which a trajectory can ultimately occupy. The holes near C and C' also represent regions which cannot be occupied after they have once been abandoned.

Returning to Fig. 2, we find that the trajectory apparently leaves one spiral only after exceeding some critical distance from the center. Moreover, the extent to which this distance is exceeded appears to determine the point at which the next spiral is entered; this in turn seems to determine the number of circuits to be executed before changing spirals again.

It therefore seems that some single feature of a given circuit should predict the same feature of the following circuit. A suitable feature of this sort is the maximum value of Z, which occurs when a circuit is nearly completed. Table 2 has again been prepared by the computer, and shows the values of X, Y, and Z at only those iterations N for which Z has a relative maximum. The succession of circuits about C and C' is indicated by the succession of positive and negative values of X and Y. Evidently X and Y change signs following a maximum which exceeds some critical value printed as about 385.

Fig. 4 has been prepared from Table 2. The abscissa is M_n, the value of the nth maximum of Z, while the ordinate is M_{n+1}, the value of the following maximum. Each point represents a pair of successive values of Z taken from Table 2. Within the limits of the round-off in tabulating Z, there is a precise two-to-one relation between M_n and M_{n+1}. The initial maximum $M_1=483$ is shown as if it had followed a maximum $M_0=385$, since maxima near 385 are followed by close approaches to the origin, and then by exceptionally large maxima.

It follows that an investigator, unaware of the nature of the governing equations, could formulate an empirical prediction scheme from the "data" pictured in Figs. 2 and 4. From the value of the most recent maximum of Z, values at future maxima may be obtained by repeated applications of Fig. 4. Values of X, Y, and Z between maxima of Z may be found from Fig. 2, by interpolating between neighboring curves. Of course, the accuracy of predictions made by this method is limited by the exactness of Figs. 2 and 4, and, as we shall see, by the accuracy with which the initial values of X, Y, and Z are observed.

Some of the implications of Fig. 4 are revealed by considering an idealized two-to-one correspondence between successive members of sequences M_0, M_1, $\cdots$, consisting of numbers between zero and one. These sequences satisfy the relations

$$\begin{aligned} M_{n+1}&=2M_n && \text{if } M_n<\tfrac{1}{2} \\ M_{n+1} & \text{ is undefined} && \text{if } M_n=\tfrac{1}{2} \\ M_{n+1}&=2-2M_n && \text{if } M_n>\tfrac{1}{2}. \end{aligned} \tag{35}$$

The correspondence defined by (35) is shown in Fig. 5, which is an idealization of Fig. 4. It follows from repeated applications of (35) that in any particular sequence,

$$M_n=m_n\pm 2^n M_0, \tag{36}$$

where m_n is an even integer.

Consider first a sequence where $M_0=u/2^p$, where u is odd. In this case $M_{p-1}=\tfrac{1}{2}$, and the sequence terminates. These sequences form a denumerable set, and correspond to the trajectories which score direct hits upon the state of no convection.

Next consider a sequence where $M_0=u/2^p v$, where u and v are relatively prime odd numbers. Then if $k>0$, $M_{p+1+k}=u_k/v$, where u_k and v are relatively prime and u_k is even. Since for any v the number of proper fractions u_k/v is finite, repetitions must occur, and the sequence is periodic. These sequences also form a denumerable set, and correspond to periodic trajectories.

The periodic sequences having a given number of distinct values, or phases, are readily tabulated. In particular there are a single one-phase, a single two-phase, and two three-phase sequences, namely,

2/3, $\cdots$,
2/5, 4/5, $\cdots$,
2/7, 4/7, 6/7, $\cdots$,
2/9, 4/9, 8/9, $\cdots$.

The two three-phase sequences differ qualitatively in that the former possesses two numbers, and the latter only one number, exceeding $\tfrac{1}{2}$. Thus the trajectory corresponding to the former makes two circuits about C, followed by one about C' (or vice versa). The trajectory corresponding to the latter makes three circuits about C, followed by three about C', so that actually only Z varies in three phases, while X and Y vary in six.

Now consider a sequence where M_0 is not a rational fraction. In this case (36) shows that M_{n+k} cannot equal

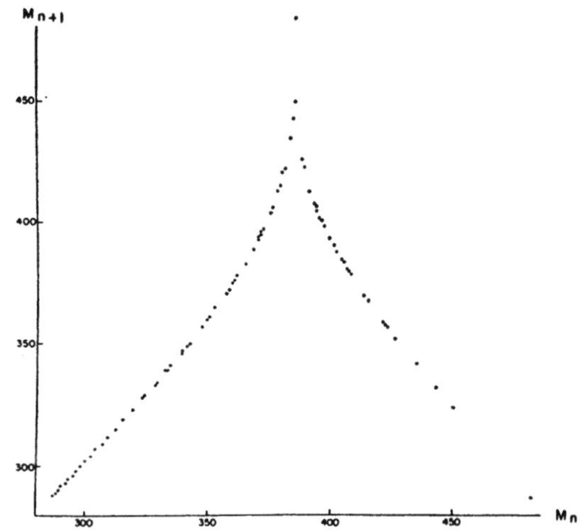

FIG. 4. Corresponding values of relative maximum of Z (abscissa) and subsequent relative maximum of Z (ordinate) occurring during the first 6000 iterations.

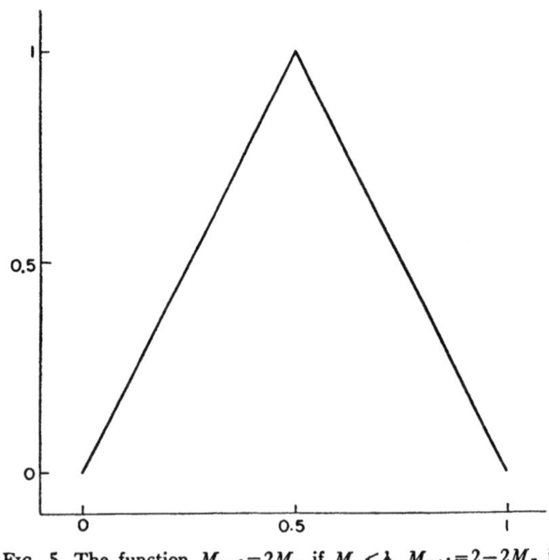

FIG. 5. The function $M_{n+1}=2M_n$ if $M_n<\tfrac{1}{2}$, $M_{n+1}=2-2M_n$ if $M_n>\tfrac{1}{2}$, serving as an idealization of the locus of points in Fig. 4.

M_n if $k>0$, so that no repetitions occur. These sequences, which form a nondenumerable set, may conceivably approach periodic sequences asymptotically and be quasi-periodic, or they may be nonperiodic.

Finally, consider two sequences M_0, M_1, $\cdots$ and M_0', M_1', $\cdots$, where $M_0'=M_0+\epsilon$. Then for a given k, if ϵ is sufficiently small, $M_k'=M_k\pm 2^k\epsilon$. All sequences are therefore unstable with respect to small modifications. In particular, all periodic sequences are unstable, and no other sequences can approach them asymptotically. All sequences except a set of measure zero are therefore nonperiodic, and correspond to nonperiodic trajectories.

Returning to Fig. 4, we see that periodic sequences analogous to those tabulated above can be found. They are given approximately by

$$\begin{array}{l} 398, \cdots, \\ 377, 410, \cdots, \\ 369, 391, 414, \cdots, \\ 362, 380, 419, \cdots. \end{array}$$

The trajectories possessing these or other periodic sequences of maxima are presumably periodic or quasi-periodic themselves.

The above sequences are temporarily approached in the numerical solution by sequences beginning at iterations 5340, 4881, 3625, and 3926. Since the numerical solution eventually departs from each of these sequences, each is presumably unstable.

More generally, if $M_n'=M_n+\epsilon$, and if ϵ is sufficiently small, $M_{n+k}'=M_{n+k}+\Lambda\epsilon$, where Λ is the product of the slopes of the curve in Fig. 4 at the points whose abscissas are M_n, $\cdots$, M_{n+k-1}. Since the curve apparently has a slope whose magnitude exceeds unity everywhere, all sequences of maxima, and hence all trajectories, are unstable. In particular, the periodic trajectories, whose sequences of maxima form a denumerable set, are unstable, and only exceptional trajectories, having the same sequences of maxima, can approach them asymptotically. The remaining trajectories, whose sequences of maxima form a nondenumerable set, therefore represent deterministic nonperiodic flow.

These conclusions have been based upon a finite segment of a numerically determined solution. They cannot be regarded as mathematically proven, even though the evidence for them is strong. One apparent contradiction requires further examination.

It is difficult to reconcile the merging of two surfaces, one containing each spiral, with the inability of two trajectories to merge. It is not difficult, however, to explain the *apparent* merging of the surfaces. At two times τ_0 and τ_1, the volumes occupied by a specified set of particles satisfy the relation

$$V_0(\tau_1)=e^{-(\sigma+b+1)(\tau_1-\tau_0)}V_0(\tau_0), \tag{37}$$

according to (30). A typical circuit about C or C' requires about 70 iterations, so that, for such a circuit, $\tau_2=\tau_1+0.7$, and, since $\sigma+b+1=41/3$,

$$V_0(\tau_1)=0.00007V_0(\tau_0). \tag{38}$$

Two particles separated from each other in a suitable direction can therefore come together very rapidly, and appear to merge.

It would seem, then, that the two surfaces merely appear to merge, and remain distinct surfaces. Following these surfaces along a path parallel to a trajectory, and circling C or C', we see that each surface is really a pair of surfaces, so that, where they appear to merge, there are really four surfaces. Continuing this process for another circuit, we see that there are really eight surfaces, etc., and we finally conclude that there is an infinite complex of surfaces, each extremely close to one or the other of two merging surfaces.

The infinite set of values at which a line parallel to the X-axis intersects these surfaces may be likened to the set of all numbers between zero and one whose decimal expansions (or some other expansions besides binary) contain only zeros and ones. This set is plainly nondenumerable, in view of its correspondence to the set of all numbers between zero and one, expressed in binary. Nevertheless it forms a set of measure zero. The sequence of ones and zeros corresponding to a particular surface contains a history of the trajectories lying in that surface, a one or zero immediately to the right of the decimal point indicating that the last circuit was about C or C', respectively, a one or zero in second place giving the same information about the next to the last circuit, etc. Repeating decimal expansions represent periodic or quasi-periodic trajectories, and, since they define rational fractions, they form a denumerable set.

If one first visualizes this infinite complex of surfaces, it should not be difficult to picture nonperiodic deterministic trajectories embedded in these surfaces.

8. Conclusion

Certain mechanically or thermally forced nonconservative hydrodynamical systems may exhibit either periodic or irregular behavior when there is no obviously related periodicity or irregularity in the forcing process. Both periodic and nonperiodic flow are observed in some experimental models when the forcing process is held constant, within the limits of experimental control. Some finite systems of ordinary differential equations designed to represent these hydrodynamical systems possess periodic analytic solutions when the forcing is strictly constant. Other such systems have yielded nonperiodic numerical solutions.

A finite system of ordinary differential equations representing forced dissipative flow often has the property that all of its solutions are ultimately confined within the same bounds. We have studied in detail the properties of solutions of systems of this sort. Our principal results concern the instability of nonperiodic solutions. A nonperiodic solution with no transient component must be unstable, in the sense that solutions temporarily approximating it do not continue to do so. A nonperiodic solution with a transient component is sometimes stable, but in this case its stability is one of its transient properties, which tends to die out.

To verify the existence of deterministic nonperiodic flow, we have obtained numerical solutions of a system of three ordinary differential equations designed to represent a convective process. These equations possess three steady-state solutions and a denumerably infinite set of periodic solutions. All solutions, and in particular the periodic solutions, are found to be unstable. The remaining solutions therefore cannot in general approach the periodic solutions asymptotically, and so are nonperiodic.

When our results concerning the instability of nonperiodic flow are applied to the atmosphere, which is ostensibly nonperiodic, they indicate that prediction of the sufficiently distant future is impossible by any method, unless the present conditions are known exactly. In view of the inevitable inaccuracy and incompleteness of weather observations, precise very-long-range forecasting would seem to be non-existent.

There remains the question as to whether our results really apply to the atmosphere. One does not usually regard the atmosphere as either deterministic or finite, and the lack of periodicity is not a mathematical certainty, since the atmosphere has not been observed forever.

The foundation of our principal result is the eventual necessity for any bounded system of finite dimensionality to come arbitrarily close to acquiring a state which it has previously assumed. If the system is stable, its future development will then remain arbitrarily close to its past history, and it will be quasi-periodic.

In the case of the atmosphere, the crucial point is then whether analogues must have occurred since the state of the atmosphere was first observed. By analogues, we mean specifically two or more states of the atmosphere, together with its environment, which resemble each other so closely that the differences may be ascribed to errors in observation. Thus, to be analogues, two states must be closely alike in regions where observations are accurate and plentiful, while they need not be at all alike in regions where there are no observations at all, whether these be regions of the atmosphere or the environment. If, however, some unobserved features are implicit in a succession of observed states, two successions of states must be nearly alike in order to be analogues.

If it is true that two analogues have occurred since atmospheric observation first began, it follows, since the atmosphere has not been observed to be periodic, that the successions of states following these analogues must eventually have differed, and no forecasting scheme could have given correct results both times. If, instead, analogues have not occurred during this period, some accurate very-long-range prediction scheme, using observations at present available, may exist. But, if it does exist, the atmosphere will acquire a quasi-periodic behavior, never to be lost, once an analogue occurs. This quasi-periodic behavior need not be established, though, even if very-long-range forecasting is feasible, if the variety of possible atmospheric states is so immense that analogues need never occur. It should be noted that these conclusions do not depend upon whether or not the atmosphere is deterministic.

There remains the very important question as to how long is "very-long-range." Our results do not give the answer for the atmosphere; conceivably it could be a few days or a few centuries. In an idealized system, whether it be the simple convective model described here, or a complicated system designed to resemble the atmosphere as closely as possible, the answer may be obtained by comparing pairs of numerical solutions having nearly identical initial conditions. In the case of the real atmosphere, if all other methods fail, we can wait for an analogue.

Acknowledgments. The writer is indebted to Dr. Barry Saltzman for bringing to his attention the existence of nonperiodic solutions of the convection equations. Special thanks are due to Miss Ellen Fetter for handling the many numerical computations and preparing the graphical presentations of the numerical material.

REFERENCES

Birkhoff, G. O., 1927: *Dynamical systems.* New York, Amer. Math. Soc., Colloq. Publ., 295 pp.

Ford, L. R., 1933: *Differential equations.* New York, McGraw-Hill, 264 pp.

Fultz, D., R. R. Long, G. V. Owens, W. Bohan, R. Kaylor and J. Weil, 1959: Studies of thermal convection in a rotating cylinder with some implications for large-scale atmospheric motions. *Meteor. Monog*, **4**(21), Amer. Meteor. Soc., 104 pp.

Gibbs, J. W., 1902: *Elementary principles in statistical mechanics.* New York, Scribner, 207 pp.

Hide, R., 1958: An experimental study of thermal convection in a rotating liquid. *Phil. Trans. Roy. Soc. London*, (A), **250**, 441–478.

Lorenz, E. N., 1960: Maximum simplification of the dynamic equations. *Tellus*, **12**, 243–254.

——, 1962a: Simplified dynamic equations applied to the rotating-basin experiments. *J. atmos. Sci.*, **19**, 39–51.

——, 1962b: The statistical prediction of solutions of dynamic equations. *Proc. Internat. Symposium Numerical Weather Prediction*, Tokyo, 629–635.

Nemytskii, V. V., and V. V. Stepanov, 1960: *Qualitative theory of differential equations.* Princeton, Princeton Univ. Press, 523 pp.

Poincaré, H., 1881: Mémoire sur les courbes définies par une équation différentielle. *J. de Math.*, **7**, 375–442.

Rayleigh, Lord, 1916: On convective currents in a horizontal layer of fluid when the higher temperature is on the under side. *Phil. Mag.*, **32**, 529–546.

Saltzman, B., 1962: Finite amplitude free convection as an initial value problem—I. *J. atmos. Sci.*, **19**, 329–341.

STUDIA MATHEMATICA, T. XXXIII. (1969)

On invariant measures for expanding differentiable mappings

by

K. KRZYŻEWSKI and W. SZLENK (Warszawa)

This note concerns expanding differentiable mappings first studied by M. Shub, see [5] and [6]. These mappings are closely connected with Anosov diffeomorphisms. But while it is not known whether there always exists a finite Lebesgue measure invariant with respect to an Anosov diffeomorphism (see [1] and [6]), it turns out that such a measure always exists for any expanding differentiable mapping. The purpose of this note is to prove this fact. It seems that this may be of some interest and that is why we publish the proof although the arguments used in it have some points of similarity with the proof of Theorem 1 in [3], p. 483.

The authors are very much indebted to Professor J. G. Sinai for his valuable remarks concerning this paper.

In the sequel M will always denote a compact, connected differentiable manifold of class C^∞ unless stated otherwise. If φ is a map of class C^1 of M into itself, then $d\varphi$ will denote the derivative of φ which is the map of the tangent bundle $T(M)$ into itself. We shall say that φ *is expanding if* there exist a Riemannian metric $\|\cdot\|$ on M, a positive real number a and a real number c greater than 1 and such that

$$\|(d\varphi^n)(\mathfrak{a})\| \geqslant ac^n\|\mathfrak{a}\| \tag{1}$$

for each $\mathfrak{a}\in T(M)$ and $n = 1, 2, \ldots$

EXAMPLE. Let φ be a differentiable mapping of the 2-dimensional torus into itself given by the formula

$$\varphi(x, y) = \big(mx+ny+\varepsilon\cdot f(x, y), px+qy+\varepsilon g(x, y)\big)(\mathrm{mod}\,1), \qquad 0\leqslant x<1, 0\leqslant y<1,$$

where

(i) m, n, p, q are integers;

(ii) the eigenvalues of the matrix $\begin{pmatrix} m, n \\ p, q \end{pmatrix}$ are real and their moduli are greater than 1;

(iii) f and g are the real functions of class C^1 on R^2, periodic with period 1 with respect to each variable;

(iv) ε is a real positive number.

It is easy to see that if ε is sufficiently small, then φ is expanding. For more general examples of expanding mappings see [5] and [6].

Let μ be a Borel measure on M. We shall say that μ is the *Lebesgue measure* if μ is equivalent to the Riemannian measure on M induced by a certain Riemannian metric on M. Now we may state the following

THEOREM. *If φ is an expanding map of class C^2 of M into itself, then*

(a) *there exists a normalised Lebesgue measure μ on M invariant with respect to φ;*

(b) *the dynamical system (μ, φ) is exact* [1] *and therefore ergodic;*

(c) *if $\bar{\mu}$ is any normalised Borel measure on M absolutely continuous with respect to a certain Riemannian measure and invariant with respect to φ, then $\bar{\mu} = \mu$.*

For the proof of the theorem the following lemmas will be needed.

LEMMA 1. *If φ is an expanding map of M into itself, there exist a Riemannian metric $\|\cdot\|_1$ of class C^∞ on M and a real number c_1 greater than 1 such that*

$$\|(d\varphi)(\mathfrak{a})\|_1 \geqslant c_1 \|\mathfrak{a}\|_1 \tag{2}$$

for each $\mathfrak{a} \in T(M)$.

Proof. First we shall prove that there exsits a Riemannian metric $\|\cdot\|_1$, not necessarily of class C^∞, such that (2) is satisfied. For this purpose let k be an integer such that $ac^k > 1$ and $k > 1$, where a and c are from (1). Then we may define the Riemannian metric $\|\cdot\|_1$ on M as follows:

$$\|\mathfrak{a}\|_1^2 = \|\mathfrak{a}\|^2 + \ldots + \|(d\varphi^{k-1})(\mathfrak{a})\|^2, \tag{3}$$

where $\mathfrak{a} \in T(M)$ and where $\|\cdot\|$ is from (1). Since M is compact, there exists a finite real number A such that

$$\|(d\varphi)(\mathfrak{a})\|^2 + \ldots + \|(d\varphi^{k-1})(\mathfrak{a})\|^2 \leqslant A\|\mathfrak{a}\|^2 \tag{4}$$

for $\mathfrak{a} \in T(M)$. Now let c_1 be any real number such that $1 < c_1 < ac^k$ and $a^2c^{2k} - c_1^2 \geqslant A(c_1^2 - 1)$. Then it is easy to see that (2) is satisfied. Now we shall show that $\|\cdot\|_1$ may be chosen to be of class C^∞. For this purpose let us assume that (2) is satisfied for the Riemannian metric $\|\cdot\|_1$. If we prove that for any positive number ε there exists a Riemannian metric $\|\cdot\|_2$ of class C^∞ on M such that

$$|\|\mathfrak{a}\|_2^2 - \|\mathfrak{a}\|_1^2| \leqslant \varepsilon\|\mathfrak{a}\|_1^2 \tag{5}$$

for each $\mathfrak{a} \in T(M)$, the proof of the lemma will be completed. In fact, if we choose ε such that $1 + \varepsilon < c_1^2(1 - \varepsilon)$, then $\|\cdot\|_2$ will satisfy (2) with the constant $c_1\sqrt{\dfrac{1-\varepsilon}{1+\varepsilon}}$.

[1] For the definition of exact dynamical systems see [4].

If we apply the standard method based on the C^∞-partition of unity, the proof of (5) may be reduced to the proving of the following:

(6) for any $\varepsilon > 0$ and any point $p \in M$ there exists a Riemannian metric $\|\cdot\|_{2,U}$ of class C^∞ on some neighborhood U of p such that

$$|\|\mathfrak{a}\|_1^2 - \|\mathfrak{a}\|_{2,U}^2| \leqslant \varepsilon\|\mathfrak{a}\|_1^2 \quad \text{for } \mathfrak{a} \in T(M)/U.$$

Condition (6) may easily be proved by a uniform approximation of the coordinates of $\|\cdot\|_1$ in some coordinate system on M containing p by real functions of class C^∞. Therefore the proof of (6) will be omitted. Thus the proof of the lemma is completed.

From now on, φ will denote, unless stated otherwise, an expanding map of class C^2, $\|\cdot\|_1$ will denote the Riemannian metric on M given by Lemma 1, and $d_1(\cdot,\cdot)$, μ_1 will denote the natural metric and the Riemannian measure induced by $\|\cdot\|_1$ on M respectively.

The following two lemmas may be proved in the standard way and therefore we shall omit the proofs:

LEMMA 2. *Each expanding map is an N-fold covering, where $1 < N < \infty$.*

LEMMA 3. *If $f(\cdot)$ is a real function of class C^1 on M, then there exists a finite real number L such that*

$$|f(x)-f(y)| \leqslant Ld_1(x,y) \quad \textit{for } x, y \in M.$$

Let φ be a map of class C^1 of a Riemannian manifold $(X, \|\cdot\|)$ of class C^∞ into a Riemannian manifold $(X_*, \|\cdot\|_*)$ of class C^∞, where $\dim X = \dim X_*$. Then we may define on X the function $D\varphi$ as

$$(D\varphi)(x) = \mu_*^{\varphi(x)}\big((d\varphi)(A_x)\big)$$

for $x \in X$, where A_x is any Borel set in $T_x(X)$ of measure μ^x equal to 1 and μ^x, μ_*^y are the natural measures induced by the Riemannian metrics $\|\cdot\|$ and $\|\cdot\|_*$ on $T_x(X)$ and $T_y(X_*)$, respectively. This function, as is easy to see, is well-defined and will be termed the *scalar derivative* of φ. If φ is of class C^2, then $D\varphi$ is of class C^1. Further, if φ is a diffeomorphism X onto X_*, then

$$\mu_*\big(\varphi(A)\big) = \int_A (D\varphi)(x)\,d\mu(x)$$

for each Borel set $A \subset X$, where μ, μ_* are the Riemannian measures induced by $\|\cdot\|$, $\|\cdot\|_*$ on X, X_*, respectively.

Now we may prove the following

LEMMA 4. *There exists a real finite number a such that*

(7) $$\mu_1\big(\varphi^{-n}(A)\big) \leqslant a\mu_1(A).$$

for each Borel set $A \subset M$ and $n = 1, 2, \ldots$

Proof. In view of Lemma 2 and the well-known property of the Riemannian metric of class C^∞, there exists an open cover $\{U_i\}_{1\leqslant i\leqslant p}$ of M such that

(8) φ is a diffeomorphism of each component of $\varphi^{-1}(U_i)$ onto U_i for $i = 1, 2, \ldots, p$;

(9) for each pair of points x, y belonging to U_i, there exists a regular curve joining x and y, contained in U_i and such that $d_1(x, y)$ is equal to its length, $i = 1, \ldots, p$.

Let δ be the Lebesgue number of the cover $\{U_i\}_{1\leqslant i\leqslant p}$; then there exist a positive integer k and open sets $A_{i_0,\ldots,i_n}(1\leqslant i_0\leqslant k, 1\leqslant i_1\leqslant N, \ldots, 1\leqslant i_n\leqslant N)$ for $n = 0, 1, \ldots$, where N is from Lemma 2, such that

(10) $\mu_1\big(M - \bigcup_{i_0=1}^{k} A_{i_0}\big) = 0$ and $A_{i_0} \neq \emptyset$ for $i_0 = 1, \ldots, k$

(11) $A_{i_0,\ldots,i_n} \cap A_{i'_0,\ldots,i'_n} = \emptyset$ for each pair $(i_0, \ldots, i_n)$, $(i'_0, \ldots, i'_n)$ of different admissible $(n+1)$-tuples of indices for $n = 0, 1, \ldots$;

(12) $\varphi^{-1}(A_{i_0,\ldots,i_n}) = \bigcup_{i_{n+1}=1}^{N} A_{i_0,\ldots,i_n,i_{n+1}}$;

(13) φ is a diffeomorphism of $A_{i_0,\ldots,i_{n+1}}$ onto $A_{i_0,\ldots,i_n}$;

(14) $\operatorname{diam}(A_{i_0,\ldots,i_n}) < \delta/c_1^n$;

(15) $d_1\big(\varphi(x), \varphi(y)\big) \geqslant c_1 d_1(x, y)$ for each pair of points x, y belonging to $A_{i_0,\ldots,i_{n+1}}$ where $n = 0, 1, \ldots$

The above sets will be defined by induction. First we shall define A_{i_0} for $i_0 = 1, 2, \ldots, k$. For this purpose let us remark that there exists a cover $\{B_i\}_{1\leqslant i\leqslant k}$ of M such that B_i are open balls of radii not greater than $\delta/2$ and such that $\mu_1(\operatorname{Fr}(B_i)) = 0$ for $i = 1, 2, \ldots, k$. Then the open sets defined as

$$A_1 = B_1 \text{ and } A_s = B_s - \bigcup_{j=1}^{s-1} \bar{B}_j \quad \text{for } s = 2, 3, \ldots, k$$

have the required properties if one rejects empty sets. Let us now assume that the sets $A_{i_0,\ldots,i_j}$ have been defined for $1\leqslant i_0\leqslant k, 1\leqslant i_1\leqslant N, \ldots, 1\leqslant i_j\leqslant N$, where $j = 0, 1, \ldots, n$. If $(i_0, \ldots, i_n)$ is any admissible $(n+1)$-tuple of indices, then in view of (14) there exists an $r_{i_0,\ldots,i_n}$ $(1\leqslant r_{i_0,\ldots,i_n}\leqslant p)$ such that $A_{i_0,\ldots,i_n} \subset U_{r_{i_0,\ldots,i_n}}$. On account of (8) $\varphi^{-1}(A_{i_0,\ldots,i_n})$ is equal to $\bigcup_{i_{n+1}=1}^{N} A_{i_0,\ldots,i_n,i_{n+1}}$ where $A_{i_0,\ldots,i_n,i_{n+1}}$ $(i_{n+1} = 1, \ldots, N)$ are the intersections of $\varphi^{-1}(A_{i_0,\ldots,i_n})$ with the components of $\varphi^{-1}(U_{r_{i_0,\ldots,i_n}})$. It is easy to see that (11) and (13) are satisfied if one replaces n by $n+1$, and in view of (9) one obtains (15) and therefore (14) for $A_{i_0,\dots,i_n,i_{n+1}}(i_{n+1} = 1, \dots, N)$. Now we shall prove that there exists a finite real number β such that

(16) $$\frac{(D\varphi^n)(y)}{(D\varphi^n)(x)} \leqslant \beta$$

for $x, y \in A_{i_0,\dots,i_n}$, $n = 1, 2, \dots$ For this purpose let us assume that $x, y \in A_{i_0,\dots,i_n}$, where n is any positive integer. Then, in view of the chain rule for scalar derivatives, one obtains

(17) $$\frac{(D\varphi^n)(y)}{(D\varphi^n)(x)} = \prod_{i=0}^{n-1} \frac{(D\varphi)(\varphi^i(y))}{(D\varphi)(\varphi^i(x))} \leqslant \prod_{i=0}^{n-1}\left(1+\frac{|(D\varphi)(\varphi^i(x))-(D\varphi)(\varphi^i(y))|}{(D\varphi)(\varphi^i(x))}\right) \leqslant \exp\left\{\sum_{i=0}^{n-1}\frac{|(D\varphi)(\varphi^i(x))-(D\varphi)(\varphi^i(y))|}{(D\varphi)(\varphi^i(x))}\right\}.$$

On account of Lemma 3 inequality (17) implies that

$$\frac{(D\varphi^n)(y)}{(D\varphi^n)(x)} \leqslant \exp\left\{\frac{L}{\gamma}\sum_{i=0}^{n-1} d_1(\varphi^i(x), \varphi^i(y))\right\},$$

where $\gamma = \inf_{x\in M}(D\varphi)(x)$, and L is the constant given by Lemma 3. From (15) it follows that

$$d_1(\varphi^{i+1}(x), \varphi^{i+1}(y)) \geqslant c_1 d_1(\varphi^i(x), \varphi^i(y))$$

for $i = 0, 1, \dots, n-1$. This implies that

(18) $$d_1(\varphi^{n-1}(x), \varphi^{n-1}(y)) \geqslant c_1^{n-1-i} d_1(\varphi^i(x), \varphi^i(y))$$

for $i = 0, 1, \dots, n-1$. From (18) it follows that (16) is satisfied with the constant β equal to

$$\exp\left\{\frac{L}{\gamma}\operatorname{diam}(M)\frac{c_1}{c_1-1}\right\}.$$

Now we may prove (7). For this purpose let us remark that in view of (11), (12) and (13) it follows that

(19) $$\mu_1(\varphi^{-n}(A)) = \sum_{\substack{1\leqslant i_1\leqslant N\\ \cdots\cdots\\ 1\leqslant i_n\leqslant N}} \int_A (D\varphi_{i_0,\dots,i_n}^{-n})(y)\,d\mu_1(y),$$

where A is any Borel set contained in A_{i_0} $(1 \leqslant i_0 \leqslant k)$, and $D\varphi_{i_0,\dots,i_n}^{-n}$ denotes the scalar derivative of the inverse map $\varphi_{i_0,\dots,i_n}^{-n}$ to $\varphi^n/A_{i_0,\dots,i_n}$.

But (16) implies

$$\frac{\sup\limits_{x\epsilon A_{i_0}} (D\varphi_{i_0,\ldots,i_n}^{-n})(x)}{\inf\limits_{x\epsilon A_{i_0}} (D\varphi_{i_0,\ldots,i_n}^{-n})(y)} \leqslant \beta. \tag{20}$$

From (19) and (20) one obtains

$$\mu_1(\varphi^{-n}(A)) \leqslant \beta\cdot\mu_1(A) \sum_{\substack{1\leqslant i_1\leqslant N\\ \cdots\cdots\\ 1\leqslant i_n\leqslant N}} \inf_{x\epsilon A_{i_0}} (D\varphi_{i_0,\ldots,i_n}^{-n})(x).$$

Since, in view of (24),

$$\mu_1(\varphi^{-n}(A_{i_0})) \geqslant \mu_1(A_{i_0}) \sum_{\substack{1\leqslant i_1\leqslant N\\ \cdots\cdots\\ 1\leqslant i_n\leqslant N}} \inf_{x\epsilon A_{i_0}} (D\varphi_{i_0,\ldots,i_n}^{-n})(x),$$

taking into account (10) we find that (8) is satisfied with a constant a equal to $\beta\mu_1(M)/\min\limits_{1\leqslant i_0\leqslant K}\mu_1(A_{i_0})$. Thus the lemma is completely proved.

LEMMA 5. *For each open non-empty set $U \subset M$ there exists a positive integer n_0 such that $\varphi^{n_0}(U) = M$.*

Proof. Let (M_*, π) denote the universal covering of M. Since π is a regular map of class C^∞, the Riemannian metric $\|\cdot\|_*$, defined on M_* as $\|\mathfrak{a}\|_* = \|(d\pi)(\mathfrak{a})\|_1$ for $\mathfrak{a}\,\epsilon\, T(M_*)$ is of class C^∞. Let Γ denote the group of cover transformations of the covering (M_*, π). It is well known that Γ is the group of isometries of the Riemannian manifold $(M_*, \|\cdot\|_*)$. Now, since M is compact, there exsists an open set $Z \subset M_*$ such that

$$\pi(Z) = M \quad\text{and}\quad \mathrm{diam}(Z) = \delta_* < \infty \tag{21}$$

(the diameter of Z is in the metric $d_*(\cdot,\cdot)$ induced on M_* by $\|\cdot\|_*$). Now let us remark that

$$\pi(K_*(x_0, \delta_*)) = M, \tag{22}$$

where $K_*(x_0, \delta_*)$ is any closed ball of radius equal to δ_*. Indeed, in view of (21) there exists an $\bar{x}_0\epsilon Z$ such that $\pi(x_0) = \pi(\bar{x}_0)$. Further, since Γ is transitive on each fibre of the covering (M_*, π), there exists a $g\,\epsilon\,\Gamma$ such that $x_0 = g(\bar{x}_0)$. This implies that $g(K_*(\bar{x}_0, \delta_*)) = K_*(x_0, \delta_*)$ and in view of (21) we infer that (22) is satisfied.

Since (M_*, π) is universal, there exists a continuous map $\varphi_*\colon M_* \to M_*$ such that

$$\varphi\pi = \pi\varphi_*. \tag{23}$$

It is easy to see that (M_*, φ_*) is the covering of M_* and φ_* is a regular map of class C^2. Since M_* is simply connected, it follows that φ_* is a homeomorphism, and therefore a diffeomorphism of class C^2 of M_* onto itself. In view of the definition of $\|\cdot\|_*$, we obtain

$$\|(d\varphi_*)(\mathfrak{a})\|_* \geqslant c_1\|\mathfrak{a}\|_* \tag{24}$$

for $\mathfrak{a} \in T(M_*)$. Since φ_* is a diffeomorphism, (24) implies

$$d_*\big(\varphi_*(x), \varphi_*(y)\big) \geqslant c_1 d_*(x, y) \quad \text{for } x, y \in M_*. \tag{25}$$

Now let us assume that U is any open non-empty set, $U \subset M$. There exists a closed ball $K_*(x_0, r)$ in M_* such that $\pi\big(K_*(x_0, r)\big) \subset U$, $K_*(x_0, r) \subset Z$. Now it suffices to show that

$$\varphi^{n_0}\Big(\pi\big(K_*(x_0, r)\big)\Big) = M \tag{26}$$

for a certain positive integer n_0. In view of (23), (26) is equivalent to

$$\pi\Big(\varphi_*^{n_0}\big(K_*(x_0, r)\big)\Big) = M. \tag{27}$$

To prove (27), let n_0 be such a positive integer that

$$c_1^{n_0} r > \delta_*. \tag{28}$$

Then, on account of (22) and (27) it suffices to show that

$$K_*\big(\varphi_*^{n_0}(x_0), \delta_*\big) \subset \varphi_*^{n_0}\big(K_*(x_0, r)\big). \tag{29}$$

To prove (29) let us assume that, on the contrary, there exists a $y \in K_*\big(\varphi_*^{n_0}(x_0), \delta_*\big) - \varphi_*^{n_0}\big(K_*(x_0, r)\big)$. Then there exists a regular curve $k: \langle 0, 1\rangle \to M_*$ such that $k(0) = \varphi_*^{n_0}(x_0)$, $k(1) = y$ and $L_*(k|_0^1) < \delta_* + \varepsilon$, where $L_*(k|_0^1)$ denote the length of k and ε is such that $c_1^{n_0} r \geqslant \delta_* + \varepsilon$, $\varepsilon > 0$. Further, it is easy to see that there exists a t_0, $0 < t_0 < 1$, such that $k(t_0) \in \mathrm{Fr}\Big(\varphi_*^{n_0}\big(K_*(x_0, r)\big)\Big)$. This implies that $L_*(k|_0^{t_0}) < \delta_* + \varepsilon$ and therefore $d_*\big(\varphi_*^{n_0}(x_0), k(t_0)\big) < \delta_* + \varepsilon$, but in view of formula (25) we have $d_*\big(\varphi_*^{n_0}(x_0), k(t_0)\big) \geqslant \delta_* + \varepsilon$. Thus the proof of the lemma is completed.

The following lemma is, in fact, the theorem on p. 525 in [4], suitably modified for our purposes. Therefore its proof will be omitted.

LEMMA 6. *Let φ be an endomorphism of a Lebesgue space (M, μ) with the measure vanishing on points. If*

(i) *there exists a family $\mathscr{A}$ of measurable sets of positive measure such that finite sums of disjoint sets belonging to $\mathscr{A}$ are dense in the space of all measurable sets;*

(ii) *there exist real finite numbers L_1, L_2 and for each $A \in \mathscr{A}$ there exist positive integers n_1, n_2, $n_1 \leqslant n_2$, such that*

($\mathrm{ii_a}$) $\mu\big(\varphi^{n_1}(Z)\big) \leqslant L_1 \dfrac{\mu(Z)}{\mu(A)}$ *for each measurable set $Z \subset A$,*

($\mathrm{ii_b}$) $\mu\big(\varphi^{n_2}(A)\big) = 1$,

(ii$_c$) *$\mu\big(\varphi^{n_2-n_1}(Z)\big) \leqslant L_2\mu(Z)$ for each measurable set $Z \subset \varphi^{n_1}(A)$; then φ is exact.*

Proof of the theorem. We may assume that the measure μ_1 is normalised. From Lemma 4 it follows that

$$\frac{\mu_1(A)+\mu_1\big(\varphi^{-1}(A)\big)+\ldots+\mu_1\big(\varphi^{-n+1}(A)\big)}{n} \leqslant a\cdot\mu_1(A) \tag{30}$$

for each Borel set A and $n = 1, 2, \ldots$ From (30), in view of Theorem 9 in [2] on p. 667, it easily follows that the sequence

$$\frac{1}{n}\sum_{i=1}^{n}\mu_1\big(\varphi^{-i+1}(A)\big)$$

is convergent (see the proof of Theorem 1 on p. 483 in [3]). Let us put

$$\mu(A) = \lim_{n\to\infty}\frac{\mu_1(A)+\ldots+\mu_1\big(\varphi^{-n+1}(A)\big)}{n}.$$

From Corollary 4 in [2] on p. 160 it follows that μ is a normalised Borel measure on M. It is evident that μ is absolutely continuous with respect to μ_1. It remains to show that μ_1 is absolutely continuous with respect to μ. For this purpose we shall prove that

(31) if $\mu_1\big(\varphi^{-n}(A_{i_0})\big) \nrightarrow 0$ for some A_{i_0}, then the condition $\mu(A) = 0$ implies $\mu_1(A) = 0$, where A is a Borel set contained in A_{i_0}.

For this purpose let us assume that $\mu_1\big(\varphi^{-n}(A_{i_0})\big) \nrightarrow 0$. Then, in view of Lemma 4, it follows that

$$\inf_n \mu_1\big(\varphi^{-n}(A_{i_0})\big) = a_{i_0} > 0. \tag{32}$$

Further, keeping in mind the notation from the proof of Lemma 4, one obtains

$$\begin{aligned}\mu_1\big(\varphi^{-n}(A)\big) &\geqslant \mu_1(A)\sum_{\substack{1\leqslant i_1\leqslant N\\ \cdots\cdots\\ 1\leqslant i_n\leqslant N}}\inf_{x\in A_{i_0}}(D\varphi^{-n}_{i_0,\ldots,i_n})(x)\\ &\geqslant \frac{1}{\beta}\mu_1(A)\sum_{\substack{1\leqslant i_1\leqslant N\\ \cdots\cdots\\ 1\leqslant i_n\leqslant N}}\sup_{x\in A_{i_0}}(D\varphi^{-n}_{i_0,\ldots,i_n})(x).\end{aligned} \tag{33}$$

From (32) it follows that

$$\sum_{\substack{1\leqslant i_1\leqslant N\\ \cdots\cdots\\ 1\leqslant i_n\leqslant N}}\sup_{x\in A_{i_0}}(D\varphi^{-n}_{i_0,\ldots,i_n})(x) \geqslant \frac{a_{i_0}}{\mu_1(A_{i_0})}\mu_1(A). \tag{34}$$

From (33) and (34) we find that

$$\mu_1(\varphi^{-n}(A)) \geqslant \frac{1}{\beta} \frac{a_{i_0}}{\mu_1(A_{i_0})} \mu_1(A)$$

for $n = 1, 2, \ldots$ It is easy to see that this completes the proof of (31).

Part (a) of the theorem will be completely proved if we show that (31) is satisfied for each i_0, $1 \leqslant i_0 \leqslant k$. Suppose, on the contrary, that there exists an i_0', $1 \leqslant i_0' \leqslant k$, such that (31) is not satisfied for $A_{i_0'}$. Then, since

$$\mu_1(M) = \sum_{i_0=1}^{k} \mu_1(\varphi^{-n}(A_{i_0})),$$

there exists an i_0'' such that (31) is satisfied for $A_{i_0''}$. Then in view of Lemma 5, there exists a positive integer n_0 such that $\varphi^{n_0}(A_{i_0''}) \supset A_{i_0''}$. Let us put $B = \varphi^{-n_0}(A_{i_0'}) \cap A_{i_0''}$. The set B is open and non-empty; therefore in view of (31)

$$\mu_1(\varphi^{-n}(B)) \nrightarrow 0.$$

But $\varphi^{-n}(B) \subset \varphi^{-n-n_0}(A_{i_0'})$ for $n = 1, 2, \ldots$ This implies that $\mu_1(\varphi^{-n}(B)) \to 0$; thus we obtain a contradiction. This completes the proof of part (a) of the theorem.

Now we shall proceed to part (b). For this purpose it suffices to show that our dynamical system satisfies the hypothesis of Lemma 6 ([2]). To do so let $\mathscr{A}$ be the family of all sets $A_{i_0,\ldots,i_n}$ $(n \geqslant 1)$, from the proof of Lemma 4. Then, to prove that (i) is satisfied, it suffices to show that for each $\varepsilon > 0$ and each open set G there exists a set G_ε, $G_\varepsilon \subset G$, equal to the finite sum of disjoint set belonging to $\mathscr{A}$ such that

$$\mu(G - G_\varepsilon) < \varepsilon. \tag{35}$$

For this purpose let us put $G_k = \{x: x \in G,\ d_1(\operatorname{Fr} G, x) > 1/k\}$, $k = 1, 2, \ldots$ Then $G_k \subset G_{k+1}$ for $k = 1, 2, \ldots$ and $\bigcup_{k=1}^{\infty} G_k = G$. Therefore there exists a k_0 such that $\mu(G - G_{k_0}) < \varepsilon$. In view of (14) there exists an n_0 such that

$$d_1(A_{i_0,\ldots,i_n}) < \frac{1}{k_0} \tag{36}$$

for each admissible (n_0+1)-tuple of indices. Now let us put $G_\varepsilon = \bigcup A_{i_0,\ldots,i_n}$, where the sum is over such admissible (n_0+1)-tuples of indices $(i_0, \ldots, i_n)$ that $G_{k_0} \cap A_{i_0,\ldots,i_{n_0}} \neq \emptyset$. Since

$$\mu(M - \bigcup_{(i_0,\ldots,i_{n_0})} A_{i_0,\ldots,i_{n_0}}) = 1$$

([2]) In fact, to apply Lemma 6 one has to complete the measure.

and in view of (36), it is easy to see that (35) is satisfied. Now we shall prove that (ii) is also satisfied. To do so, let us remark that from the proof of part (a) it follows that there exist two real finite numbers a, $\bar{a}$ such that $\mu(A) \leqslant a\mu_1(A)$ and $\mu_1(A) \leqslant \bar{a}\mu(A)$ for each Borel set A. Therefore it suffices to show that (ii) is satisfied when we replace the measure μ by μ_1. For this purpose let us put $L_1 = \beta$, where β is from (16). Further, in view of Lemma 5, it follows that there exists a positive integer s_0 such that $\varphi^{s_0}(A_{i_0}) = M$ for $i_0 = 1, 2, \ldots, k$, where A_{i_0} is from the proof of Lemma 4. Since φ^{s_0} is a local diffeomorphism and in view of the remarks preceding Lemma 4, it easily follows that there exists a finite real number L_2 such that $\mu_1(\varphi^{s_0}(Z)) \leqslant L_2\mu_1(Z)$ for each Borel set Z. Now let us put for each set $A_{i_0,\ldots,i_n} (n \geqslant 1)$ $n_1 = n$ and $n_2 = n + s_0$. Then (ii_b) and (ii_c) are satisfied; to prove (ii_a) it suffices to apply (21) and act as in the proofs of Lemma 4 and (31) (see the proof of the theorem on p. 525 in [4]). Thus the proof of part (b) is completed.

It remains to prove (c). In fact, in view of (a) and (b), the part (c) follows from the well-known theorem which states that any invariant normalized measure absolutely continuous with respect to a normalized invariant ergodic measure is equal to this measure.

Added in proof. After the paper was submitted for publication, the paper [7] came to our attention. As we understand, there is announced the following result: for each expanding mapping of the compact manifold M into itself there exists an invariant regular Borel measure μ positive on each open set in M.

References

[1] Д. В. Аносов, *Геодезические потоки на замкнутых римановых многообразиях отрицательной кривизны*, Труды математ. инст. им. В. А. Стеклова 90 (1967).

[2] N. Dunford and T. Schwartz, *Linear operators (I)*, 1958.

[3] A. Rényi, *Representations for real numbers and their ergodic properties*, Acta Math. Acad. Scient. Hung. 8 (1957), p. 477-493.

[4] В. А. Рохлин, *Точные эндоморфизмы пространств Лебега*, Изв. Акад. Наук СССР, Серия мат., 25 (1961), 499-530.

[5] M. Shub, Thesis, University of California, Berkeley 1967.

[6] S. Smale, *Differentiable dynamical systems*, Bull. Amer. Math. Soc. 73 (1967), p. 747-817.

[7] A. Avez, *Propriétés ergodiques des endomorphismes dilatants des variétés compactes*, C. R. Acad. Sc. Paris 266 (1968), sér. A, p. 610-612.

Reçu par la Rédaction le 28. 5. 1968

TRANSACTIONS OF THE
AMERICAN MATHEMATICAL SOCIETY
Volume 186, December 1973

ON THE EXISTENCE OF INVARIANT MEASURES FOR PIECEWISE MONOTONIC TRANSFORMATIONS(1)

BY

A. LASOTA AND JAMES A. YORKE

ABSTRACT. A class of piecewise continuous, piecewise C^1 transformations on the interval $[0,1]$ is shown to have absolutely continuous invariant measures.

1. Introduction. The purpose of this note is to prove the existence of absolutely continuous invariant measures for a class of point-transformations of the unit interval [0, 1] into itself. Our main result is Theorem 1 which generalizes some previous results of A. Rényi [5], A. O. Gel'fond [2], W. Parry [4] and A. Lasota [3]. It gives, also, a positive answer to a conjecture of S. Ulam [7, p. 74]. Theorem 1 is stated for a piecewise monotonic function with a finite number of discontinuities but it can be easily extended to some piecewise monotonic functions with infinite number of discontinuities.

Our method is different from the methods of the above mentioned authors. Firstly we explore the fact that the Frobenius-Perron operator corresponding to the point-transformation under consideration has the property of sometimes shrinking the variation of the function. Secondly to prove the existence of invariant measures we use the abstract ergodic theorem which enables us to make our proofs constructive. The advantage of this method is that we do not require that our mappings be local homeomorphisms nor that they generate an exact endomorphism in the sense of Rohlin [6], a property that has been the typical requirement for previous work. §4 describes some extensions, including an extension to higher dimensions.

2. Existence theorem. Denote by $(L_1, \| \|)$ the space of all integrable functions defined on the interval [0, 1]. Lebesgue measure on [0, 1] will be denoted by m. Let τ: [0, 1] → [0, 1] be a measurable nonsingular transformation.

Received by the editors November 9, 1972 and, in revised form, February 13, 1973.

AMS (MOS) subject classifications (1970). Primary 28A65.

Key words and phrases. Frobenius-Perron operator, invariant measures.

(1) The research of both authors was partially supported by the National Science Foundation under grant GP-31386x.

"Nonsingularity" means that $m(\tau^{-1}(A)) = 0$ whenever $m(A) = 0$ for a measurable set A. Given τ we define the Frobenius-Perron operator $P_\tau: L_1 \to L_1$ by the formula

$$P_\tau f(x) = \frac{d}{dx} \int_{\tau^{-1}([0,x])} f(s)\, ds.$$

It is well known that the operator P_τ is linear and continuous and satisfies the following conditions:

(a) P_τ is positive: $f \geq 0 \Rightarrow P_\tau f \geq 0$;

(b) P_τ preserves integrals

$$\int_0^1 P_\tau f\, dm = \int_0^1 f\, dm, \qquad f \in L_1;$$

(c) $P_{\tau^n} = P_\tau^n$ (τ^n denotes the nth iterate of τ);

(d) $P_\tau f = f$ if and only if the measure $d\mu = f\,dm$ is invariant under τ, that is $\mu(\tau^{-1}(A)) = \mu(A)$ for each measurable A.

A transformation $\tau: [0, 1] \to R$ will be called *piecewise* C^2, if there exists a partition $0 = a_0 < a_1 < \cdots < a_p = 1$ of the unit interval such that for each integer i ($i = 1, \ldots, p$) the restriction τ_i of τ to the open interval (a_{i-1}, a_i) is a C^2 function which can be extended to the closed interval $[a_{i-1}, a_i]$ as a C^2 function. τ need not be continuous at the points a_i.

Theorem 1. *Let $\tau: [0, 1] \to [0, 1]$ be a piecewise C^2 function such that $\inf|\tau'| > 1$. Then for any $f \in L_1$ the sequence*

$$\frac{1}{n}\sum_{k=0}^{n-1} P_\tau^k f$$

is convergent in norm to a function $f^* \in L_1$. The limit function has the following properties:

(1) $f \geq 0 \Rightarrow f^* \geq 0$.

(2) $\int_0^1 f^* dm = \int_0^1 f\, dm$.

(3) $P_\tau f^* = f^*$ *and consequently the measure $d\mu^* = f^* dm$ is invariant under τ.*

(4) *The function f^* is of bounded variation; moreover, there exists a constant c independent of the choice of initial f such that the variation of the limiting f^* satisfies the inequality*(2)

$$\bigvee_0^1 f^* \leq c\|f\|.$$

(2) Here and in what follows the symbol $\bigvee_a^b f$ as well as $\bigvee_{[a,b]} f$ denote the varition of f over the closed interval $[a, b]$.

We point out in Theorem 3 that it is sufficient to assume just that some iterate of τ satisfy the derivative condition.

Proof. Write $s = \inf|\tau'|$ and choose a number N such that $s^N > 2$. It is easy to see that the function $\phi = \tau^N$ is piecewise C^2. Denote by $b_0, \cdots, b_q$ the corresponding partition for ϕ. Writing ϕ_i for the corresponding C^2 functions we have

$$(5) \qquad |\phi_i'(x)| \geq s^N, \qquad x \in [b_{i-1}, b_i],\ i = 1, \cdots, q.$$

Computing the Frobenius-Perron operator for ϕ we obtain

$$(6) \qquad P_\phi f(x) = \sum_{i=1}^{q} f(\psi_i(x))\sigma_i(x)\chi_i(x)$$

where $\psi_i = \phi_i^{-1}$, $\sigma_i(x) = |\psi_i'(x)|$ and χ_i is the characteristic function of the interval $J_i = \phi_i([b_{i-1}, b_i])$. From (5) it follows that

$$(7) \qquad |\sigma_i(x)| \leq s^{-N}, \qquad x \in J_i,\ i = 1, \cdots, q.$$

By its very definition the operator P_ϕ is defined as a mapping from L_1 into L_1 but the formula (6) enables us to consider P_ϕ as a map from the space of functions defined on $[0, 1]$ into itself.

Let f be a given function of bounded variation over $[0, 1]$. From (6) and (7) it follows that

$$(8) \qquad \bigvee_0^1 P_\phi f \leq \sum_{i=1}^{q} \bigvee_{J_i} (f \circ \psi_i)\sigma_i + s^{-N} \sum_{i=1}^{q} (|f(b_{i-1})| + |f(b_i)|).$$

In order to evaluate the first sum we write

$$\bigvee_{J_i} (f \circ \psi_i)\sigma_i = \int_{J_i} |d((f \circ \psi_i)\sigma_i)|$$

$$\leq \int_{J_i} |f \circ \psi_i||\sigma_i'|\, dm + \int_{J_i} \sigma_i |d(f \circ \psi_i)|$$

$$\leq K \int_{J_i} |f \circ \psi_i| \sigma_i\, dm + s^{-N} \int_{J_i} |d(f \circ \psi_i)|$$

where $K = \max|\sigma_i'|/\min(\sigma_i)$. Changing the variables we obtain

$$(9) \qquad \bigvee_{J_i} (f \circ \psi_i)\sigma_i \leq K \int_{b_{i-1}}^{b_i} |f|\, dm + s^{-N} \int_{b_{i-1}}^{b_i} |df|.$$

In order to evaluate the second term in (8) we write

$$(10) \qquad |f(b_{i-1})| + |f(b_i)| \leq \bigvee_{b_{i-1}}^{b_i} f + 2d_i$$

where $d_i = \inf\{|f(x)| : x \in [b_{i-1}, b_i]\}$. On the other hand we have an obvious inequality

(11) $$d_i \leq b^{-1} \int_{b_{i-1}}^{b_i} |f|\, dm$$

where $b = \min_i(b_i - b_{i-1})$. From (10), (11) it follows that

(12) $$\sum_{i=1}^{q} (|f(b_{i-1})| + |f(b_i)|) \leq \bigvee_0^1 f + 2b^{-1}\|f\|.$$

Applying (12) and (9) to (8) we obtain $\bigvee_0^1 P_\phi f \leq \alpha\|f\| + \beta \bigvee_0^1 f$ where $\alpha = (K+2b^{-1})$ and $\beta = 2s^{-N} < 1$.

Now, for the same function f, let us write $f_k = P_\tau^k f$. Since $P_\tau^N = P_\phi$ we have

$$\bigvee_0^1 f_{Nk} \leq \alpha\|f_{N(k-1)}\| + \beta \bigvee_0^1 f_{N(k-1)} \leq \alpha\|f\| + \beta \bigvee_0^1 f_{N(k-1)}$$

and consequently

(13) $$\limsup_{k\to\infty} \bigvee_0^1 f_{Nk} \leq \alpha(1-\beta)^{-1}\|f\|.$$

The last inequality and the condition $\|f_k\| \leq \|f\|$ (which follows from (a) and (b)) prove that the set $C = \{f_{Nk}\}_{k=0}^\infty$ is relatively compact in L_1. Since $\{f_k\}_{k=0}^\infty \subset \bigcup_{k=0}^{N-1} P_\tau^k C$, the whole sequence $\{f_k\}_{k=0}^\infty$ is relatively compact, too. By Mazur's theorem the same is true for the sequence

(14) $$\left\{\frac{1}{n}\sum_{k=0}^{n-1} P_\tau^k f\right\}.$$

The set of functions of bounded variation is dense in L_1. We have proved that for any such function f the sequence (14) is relatively compact. Therefore, we are in a position to use the Kakutani-Yosida Theorem (see [1, VIII.5.3]) which says that for any $f \in L_1$ the sequence (14) converges strongly to a function f^* which is invariant under P_τ. From (a) and (b) it follows that f^* satisfies (1) and (2). Therefore it remains only to prove (4). Since the operator P_τ is given by a formula analogous to (6) it is easy to derive the inequality $\bigvee_0^1 P_\tau f \leq c_1 \bigvee_0^1 f + c_2\|f\|$ with some constants c_1 and c_2. This and (13) imply the inequality

$$\limsup_{k\to\infty} \bigvee_0^1 P_\tau^k f \leq c\|f\|$$

(with a positive constant c) which is valid for any f with bounded variation. Consequently for any such f we have also

$$\limsup_{k\to\infty} \bigvee_0^1 \left(\frac{1}{n}\sum_{k=1}^{n-1} P_\tau^k f\right) \le c\|f\|.$$

Writing $Q = \lim_n (1/n)\sum_{k=1}^{n-1} P_\tau^k$ and using Helly's theorem we have $\bigvee_0^1 Qf \le c\|f\|$, for f of bounded variation. The operator Q is linear and contractive. We may therefore apply Helly's theorem once more to extend this inequality for the closure of the set of functions of bounded variation, that is to all of L_1. This finishes the proof.

3. A counterexample. Now we shall show that our assumption $\inf|\tau'| > 1$ is essential. Consider the transformation

$$\gamma(x) = \begin{cases} x/(1-x) & \text{for } 0 \le x < \tfrac{1}{2}, \\ 2x - 1 & \text{for } \tfrac{1}{2} \le x \le 1 \end{cases}$$

for which the assumption $|\gamma'(x)| > 1$ is violated only at $x = 0$. We are going to prove that for any $f \in L^1$ the sequence $P_\gamma^n f$ converges in measure to zero. Therefore the equation $P_\gamma f = f$ has only the trivial solution and there is no absolutely continuous nontrivial measure invariant under γ.

The proof will be given in a few steps. First we prove that for $f_0 \equiv 1$ the sequence $g_n(x) = xf_n(x)$, where $f_n = P_\gamma^n f_0$, converges to a constant c_0. Then using the condition $\|f_n\| = 1$ we derive easily that $c_0 = 0$, and consequently $f_n \to 0$. Finally by an approximation argument we may extend this result to an arbitrary sequence $P_\gamma^n f$ with $f \in L_1$.

The Frobenius-Perron operator P_γ may be written in the form

$$P_\gamma f(x) = \frac{1}{(1+x)^2} f\left(\frac{x}{1+x}\right) + \frac{1}{2} f\left(\frac{1}{2} + \frac{x}{2}\right).$$

Thus for g_n we have the following recursive formula:

$$g_{n+1}(x) = \frac{1}{1+x} g_n\left(\frac{x}{1+x}\right) + \frac{x}{1+x} g_n\left(\frac{1}{2} + \frac{x}{2}\right), \qquad g_0(x) \equiv x. \tag{15}$$

By an induction argument it is easy to check that $g_n' \ge 0$ for each n. Therefore all the functions g_n are positive and increasing. According to (15) we have

$$g_{n+1}(1) = \tfrac{1}{2} g_n(\tfrac{1}{2}) + \tfrac{1}{2} g_n(1) \le g_n(1).$$

This proves the existence of a limit $\lim_n g_n(1) \overset{\text{df}}{=} c_0$. Write $z_0 = 1$ and $z_{k+1} = z_k/(1 + z_k)$. According to (15) we obtain

$$g_{n+1}(z_k) = \frac{1}{1+z_k} g_n(z_{k+1}) + \frac{z_k}{1+z_k} g_n\left(\frac{1}{2} + \frac{z_k}{2}\right).$$

Fix k and suppose that $\lim_n g_n(x) = C_0$ for $z_k \le x \le 1$. (This is certainly true for $k = 0$.) Since $z_k \le \tfrac{1}{2} + \tfrac{1}{2} z_k$, we obtain at the limit as $n \to \infty$

$$C_0 = \frac{1}{1+z_k} \lim_n g_n(z_{k+1}) + \frac{z_k}{1+z_k} C_0.$$

Thus $\lim_n g_n(z_{k+1}) = C_0$. Since g_n are increasing, this proves that $\lim_n g_n(x) = C_0$ uniformly for all $x \in [z_{k+1}, 1]$. Therefore by an induction argument it follows that $\lim_n g_n(x) = C_0$ in any interval $[z_k, 1]$ and consequently, since $\lim_k z_k = 0$, we have $\lim_n g_n(x) = C_0$ for all $0 < x \le 1$. Hence, $\lim_n f_n(x) = c_0/x$. We claim that $c_0 = 0$. If not there would exist $\epsilon > 0$ such that $\int_\epsilon^1 c_0/x\, dx > 1$ and

$$\lim_n \int_\epsilon^1 f_n(x)\, dx = \int_\epsilon^1 \frac{c_0}{x}\, dx > 1$$

which is impossible since $\|f_n\| = 1$ for each n. It can be easily proved by induction that each of the functions f_n is decreasing. Thus the convergence of f_n to zero is uniform on any interval $[\epsilon, 1]$ with $\epsilon > 0$.

Now let f be an arbitrary function. We may write $f = f^+ - f^-$ where $f^+ = \max(0, f)$ and $f^- = \max(0, -f)$. Given $\epsilon > 0$ consider a constant r such that

$$\int_0^1 (f^- - r)^+ dm + \int_0^1 (f^+ - r)^+ dm \le \epsilon.$$

We have

$$\int_\epsilon^1 |P_\gamma^n f|\, dm = \int_\epsilon^1 P_\gamma^n f^+ dm + \int_\epsilon^1 P_\gamma^n f^- dm$$

$$\le 2 \int_\epsilon^1 P_\gamma^n r\, dm + \int_\epsilon^1 P^n(f^+ - r)\, dm + \int_\epsilon^1 P^n(f^- - r)\, dm$$

$$\le 2r \int_\epsilon^1 P_\gamma^n 1\, dm + \epsilon.$$

Since $P_\gamma^n 1$ converges on $[\epsilon, 1]$ uniformly to zero we have

$$\lim_n \int_\epsilon^1 |P_\gamma^n f|\, dm = 0 \qquad \text{for } \epsilon > 0$$

which proves that the sequence $P_\gamma^n f$ converges in measure to zero.

4. Final remarks. Now we want to discuss some extensions of our method to other transformations. First of all we may prove an analogue of Theorem 1 for piecewise C^2 transformations with a countable number of pieces.

Let $r_i: \Delta_i \to [0, 1]$ be a countable sequence of C^2 functions where Δ_i is a sequence of closed intervals such that $\Sigma_i\, m(\Delta_i) = 1$, $m([0, 1] - \bigcup_i \Delta_i) = 0$. The function r defined by the condition

$$r(x) = r_i(x), \qquad x \in \text{interior of } \Delta_i,$$

will be called countably piecewise C^2. Note that the values of r on the set $[0, 1] \setminus \bigcup_i \operatorname{int} \Delta_i$ are arbitrary.

Theorem 2. *Let τ be a countably piecewise C^2 function such that*

$$\inf |\tau'(x)| > 2, \qquad \sup |\tau''(x)| < \infty, \tag{16}$$

$$\tau_i(\Delta_i) = [0, 1] \quad \textit{except for a finite number of intervals.} \tag{17}$$

Then for each $f \in L_1$ the sequence $(1/n)\sum_{k=0}^{n-1} P_\tau^k f$ is convergent in norm to a function f^* which satisfies conditions (1), (2), (3) and (4).

The proof of Theorem 2 is basically the same as the proof of Theorem 1. Thus it can be omitted. Let us only note that the condition (17) is essential. In fact it is easy to construct a countably piecewise linear function with the slope $\tau' > 3$ such that $\inf_{\epsilon \le x \le 1-\epsilon} \tau(x) - x$ is a positive number for each ϵ in $(0, 1/2)$. (The graph of τ lies over the diagonal.) It can be proved by elementary calculation that for any such function τ and $f \in L_1$, $P_\tau^n f \to 0$ in measure as $n \to \infty$.

A close look at the proof of Theorem 1 shows that we have used only the fact that $\sup|(\tau^N)'| > 2$. Therefore, in fact, we have proved the following result.

Theorem 3. *Let τ: $[0, 1] \to [0, 1]$ be a piecewise C^2 function such that $\inf|(\tau^{n_0})'| > 1$ for a positive integer n_0. Then for any $f \in L_1$ the sequence $(1/n)\sum_{k=0}^{n-1} P_\tau^k f$ is convergent in norm to a function f^* which satisfies conditions* (1), (2) *and* (3). *If, in addition,* $\inf|\tau'| > 0$ *then condition* (4) *is also satisfied.*

Observe that in our counterexample the function γ has the property that $(\gamma^n)'_{x=0} = 0$ for each n. This is because the point $(0, \gamma(0))$ lies on the diagonal.

Our techniques can be easily used to obtain new proofs of known results in higher dimensions. See [8], [9], [10] for such results. In this case τ: $M \to M$ is assumed C^1 on a compact manifold M and the variation of a C^1 function f is defined as $\int_M |\text{grad} f(m)|\, dm$. Hence in this case we do not allow discontinuities in τ, or more generally if f is C^1 on $M \backslash \partial M$, we must make assumptions on τ guaranteeing $P_\tau(f)$ is C^1 on $M \backslash \partial M$. The techniques in [8], [9], [10] are quite different from the "bounded variation" approach of this paper.

The study of the functions τ described arose while investigating the design of more durable high speed oil well drilling bits. The invariant measure $f(x)dx$ describes the distribution of impacts on the surface of the bit. The durability and efficiency of the tool depends strongly on f. The first author is part of a team that has obtained patents in Poland for superior bits by slightly altering the bit shape to one with a better impact distribution f.

REFERENCES

1. N. Dunford and J. T. Schwartz, *Linear operators. I. General theory*, Pure and Appl. Math., vol. 7, Interscience, New York, 1958. MR 22 #8302.

2. A. O. Gel'fond, *A common property of number systems*, Izv. Akad. Nauk Ser. Mat. SSSR 23 (1959), 809–814. (Russian) MR 22 #702.

3. A. Lasota, *Invariant measures and functional equations*, Aequationes Math. (in press).

4. W. Parry, *On the β-expansion of real numbers*, Acta Math. Acad. Sci. Hungar. 11 (1960), 401–416. MR 26 #288.

5. A. Rényi, *Representation for real numbers and their ergodic properties*, Acta Math. Acad. Sci. Hungar. 8 (1957), 477–493. MR 20 #3843.

6. V. A. Rohlin, *Exact endomorphisms of Lebesgue spaces*, Izv. Akad. Nauk Ser. Mat. SSSR 25 (1961), 499–530; English transl., Amer. Math. Soc. Transl. (2) 39 (1964), 1–36. MR 26 #1423.

7. S. M. Ulam, *A collection of mathematical problems*, Interscience Tracts in Pure and Appl. Math., no. 8, Interscience, New York, 1960. MR 22 #10884.

8. M. S. Waterman, *Some ergodic properties of multidimensional F-expansions*, Z. Wahrscheinlichkeitstheorie und Verw. Gebiete 16 (1970), 77–103. MR 44 #173.

9. A. Avez, *Propriétés ergodiques des endomorphismes dilatants des variétés compactes*, C. R. Acad. Sci. Paris Sér. A–B 266 (1968), A610–A612. MR 37 #6944.

10. K. Krzyzewski and W. Szlenk, *On invariant measures for expanding differentiable mappings*, Studia Math. 33 (1969), 83–92. MR 39 #7067.

DEPARTMENT OF MATHEMATICS, JAGELLONIAN UNIVERSITY, CRACOW, POLAND

INSTITUTE FOR FLUID DYNAMICS AND APPLIED MATHEMATICS, UNIVERSITY OF MARYLAND, COLLEGE PARK, MARYLAND 20742

Inventiones math. 29, 181–202 (1975)

The Ergodic Theory of Axiom A Flows

Rufus Bowen* (Berkeley) and David Ruelle** (Bures-sur-Yvette)

1. Introduction

Let M be a compact (Riemann) manifold and (f^t): $M \to M$ a differentiable flow. A closed (f^t)-invariant set $\Lambda \subset M$ containing no fixed points is hyperbolic if the tangent bundle restricted to Λ can be written as the Whitney sum of three (Tf^t)-invariant continuous subbundles

$$T_\Lambda M = E + E^s + E^u$$

where E is the one-dimensional bundle tangent to the flow, and there are constants $c, \lambda > 0$ so that

(a) $\|Tf^t(v)\| \leqq c\, e^{-\lambda t} \|v\|$ for $v \in E^s$, $t \geqq 0$ and

(b) $\|Tf^{-t}(v)\| \leqq c\, e^{-\lambda t} \|v\|$ for $v \in E^u$, $t \geqq 0$.

We can choose $t_0 > 0$ and change λ so that the above conditions hold with $c=1$ when $t \geqq t_0$. We can also assume that, for such t, Tf^t (resp. Tf^{-t}) expands E at a smaller rate than it expands any element of E^u (resp. E^s). It is then said that the metric is *adapted* (see [14]) to f^{t_0}. We will always assume that $t_0 \leqq 1$ – this can be achieved by a rescaling of t $(t \to t' = t/t_0)$ which does not affect our main results.

A closed invariant set Λ is a *basic hyperbolic* set if

(a) Λ contains no fixed points and is hyperbolic;

(b) the periodic orbits of $f^t|\Lambda$ are dense in Λ;

(c) $f^t|\Lambda$ is a topologically transitive flow; and

(d) there is an open set $U \supset \Lambda$ with $\Lambda = \bigcap_{t \in \mathbb{R}} f^t U$.

These sets are the building blocks of the Axiom A flows of Smale [27]. We will especially be interested in *attractors*, basic hyperbolic sets Λ for which the U in (d) can be found satisfying $f^t U \subset U$ for all $t \geqq T_0$ (T_0 fixed) and hence $\Lambda = \bigcap_{t \geqq 0} f^t U$. This paper will study the average asymptotic behavior of orbits of points in the neighborhood U of a C^2-attractor.

Precisely we will find an ergodic probability measure μ_φ on a C^2 attractor Λ so that for almost all $x \in U$ w.r.t. Lebesgue measure and all continuous g: $U \to \mathbb{R}$ one has

$$\lim_{T \to \infty} \frac{1}{T} \int_0^T g(f^t x)\, dt = \int g\, d\mu_\varphi \tag{1}$$

* Math. Dept. U.C. Berkeley. Partially supported by NSF GP 14519.

** Institut des Hautes Etudes Scientifiques, Bures-sur-Yvette (France).

(see Theorem 5.1). The measure μ_φ will be described as the unique equilibrium state for a certain function $\varphi=\varphi^{(u)}$ (defined by (2), Section IV) on Λ, i.e. the unique f^t-invariant probability measure μ on Λ which maximizes the expression

$$h_\mu(f^1)+\int \varphi\, d\mu$$

where $h_\mu(f^1)$ is measure theoretic entropy. This variational principle (which is formally identical with one in statistical mechanics [21]) is useful because it gives a description of μ_φ which persists when one lifts μ_φ to a symbol space for closer study.

This paper carries over to flows results previously obtained for diffeomorphisms with regard to equilibrium states [6, 7, 24] and attractors [24]. For Anosov flows ($\Lambda=M$) the measure μ_φ has been studied in [9, 16, 17, 20, 25, 26] and the theory of Gibbs states (a slightly different formalism from equilibrium states which yields the same measures for basic hyperbolic sets) has been developed in [26]. Some results obtained here for flows are new even for diffeomorphisms; this is the case of Theorem 5.6. Results for diffeomorphisms can be obtained from those for flows via suspension (or directly by simplification of the proofs).

The determination of the asymptotic behavior of orbits is a significant problem in the study of differentiable dynamical systems. In particular the asymptotic behavior of solutions of a differential equation is of central interest in physical applications. Here we consider only the case of Axiom A flows. In that case it is known that $f^t x$ often depends in a very sensitive or "unstable" manner on the initial condition x, and (1)—which describes the time-average of an "observable" g—is probably the best way of expressing the asymptotic behavior of $f^t x$. It is a natural problem to extend (1) to non Axiom A situations.

We shall show that $\mu_{\varphi^{(u)}}$ depends continuously on the flow (f^t) (Proposition 5.4). In the same direction, Sinai [26] has proved the stability of μ_φ under small stochastic perturbations for Anosov flows[1]. (1) holds almost everywhere for x in the basin of an Axiom A attractor; one can prove that, for a C^2 Axiom A flow, these basins (and those of point attractors) cover M up to a set of Lebesgue measure zero. Equivalently: if a basic set is not an attractor, its stable manifold has measure zero (Theorem 5.6).

It can be seen that, unless Λ is a periodic orbit, the entropy of μ_φ does not vanish; this indicates "strong ergodic properties" of the system (μ_φ, f^t). In fact, if (f^t) restricted to Λ is C-dense, (μ_φ, f^t) is a Bernoulli flow (see Remark 3.5). The correlation functions

$$\rho_{gg'}(t)=\int (g\circ f^t)\cdot g'\, d\mu_\varphi-\int g\, d\mu_\varphi\cdot\int g'\, d\mu_\varphi$$

are interesting to consider in physical applications. In the C-dense case we have $\lim_{t\to\infty}\rho_{gg'}(t)=0$ if $g, g'\in L^2(\mu_\varphi)$ (Remark 3.5). Assuming that g, g' are C^1, does $\rho_{gg'}(t)$ tend to zero exponentially when $t\to\infty$? The methods of the present paper do not seem capable of answering this question. A positive answer has been obtained for diffeomorphisms ([24, 26]).

[1] The corresponding problem for attractors for Axiom A diffeomorphisms has been treated by Kifer (Sinai, private communication).

Terminology

The manifold M and the Riemann metric on M are C^∞. The flow (f^t) is called C^r $(r \geqq 1)$ if it corresponds to a C^r vector field on M; a basic hyperbolic set Λ for (f^t) is then called a C^r basic hyperbolic set. The flow (f^t) restricted to Λ is topologically transitive if it has a dense orbit.

For easy reference, we collect here the definitions of stable manifolds

$$W_x^s = \{y \in M: \lim_{t \to \infty} d(f^t x, f^t y) = 0\}$$

$$W_x^{cs} = \bigcup_{t \in \mathbb{R}} W_{f^t x}^s .$$

A distance on M is defined by

$$\delta_T(x, y) = \sup_{0 \leqq t \leqq T} d(f^t x, f^t y)$$

when $0 \leqq T < \infty$; $B_x(\varepsilon, T)$ is the closed ε-neighbourhood of x for that distance; also

$$W_x^s(\varepsilon) = W_x^s \cap B_x(\varepsilon, \infty).$$

Replacing t by $-t$ and s by u we obtain the definition of unstable manifolds. We also write

$$W_\Lambda^s(\varepsilon) = \bigcup_{x \in \Lambda} W_x^s(\varepsilon), \qquad \text{etc.}$$

The basic hyperbolic set Λ is C-dense if $W_x^s \cap \Lambda$ is dense in Λ for some (hence for all) $x \in \Lambda$.

In general we write $f^* \mu$ the image of a measure μ by a continuous map f.

2. Symbolic Dynamics

Let us recall the symbolic dynamics of a basic hyperbolic set Λ [4]. For $A = [A_{ij}]$ an $n \times n$ matrix of 0's and 1's we define

$$\Sigma_A = \{\mathbf{x} = (x_i)_{i=-\infty}^{+\infty} \in \{1, \ldots, n\}^{\mathbb{Z}}: A_{x_i x_{i+1}} = 1 \; \forall i \in \mathbb{Z}\}$$

and $\sigma_A: \Sigma_A \to \Sigma_A$ by $\sigma_A(\mathbf{x}) = (x_i')_{i=-\infty}^{\infty}$ where $x_i' = x_{i+1}$. If we give $\{1, \ldots, n\}$ the discrete topology and $\{1, \ldots, n\}^{\mathbb{Z}}$ the product topology, then Σ_A becomes a compact metrizable space and σ_A a homeomorphism. σ_A (or Σ_A) is called a *subshift of finite type* if $\sigma_A: \Sigma_A \to \Sigma_A$ is topologically transitive (i.e. for U, V non-empty open sets there is an $n > 0$ with $f^n U \cap V \neq \emptyset$).

For $\psi: \Sigma_A \to \mathbb{R}$ a positive continuous function one can define a special (or suspension) flow as follows. Let

$$Y = \{(\mathbf{x}, s): s \in [0, \psi(\mathbf{x})], \mathbf{x} \in \Sigma_A\} \subset \Sigma_A \times \mathbb{R}.$$

Identify the points $(\mathbf{x}, \psi(\mathbf{x}))$ and $(\sigma_A(\mathbf{x}), 0)$ for all $\mathbf{x} \in \Sigma_A$ to get a new space $\Lambda(A, \psi)$. Then $\Lambda(A, \psi)$ is a compact metric space (see [8] for a metric) and one can define a flow g^t on $\Lambda(A, \psi)$ by

$$g^t(\mathbf{x}, s) = (\mathbf{x}, s + t) \qquad \text{for } s + t \in [0, \psi(\mathbf{x})]$$

and remembering identifications. More precisely, if $z=q(\mathbf{x}, s)$ where $q: Y \to \Lambda(A, \psi)$ is the quotient map, then $g^t(z)=q(\sigma_A^k \mathbf{x}, v)$ where k is chosen so that

$$v=t+s-\sum_{j=0}^{k-1} \psi(\sigma_A^j \mathbf{x}) \in [0, \psi(\sigma_A^k \mathbf{x})].$$

The flow g^t on $\Lambda(A, \psi)$ will be important to us with ψ satisfying an additional condition. For $\psi: \Sigma_A \to \mathbb{R}$ let

$$\operatorname{var}_n \psi = \sup\{|\psi(\mathbf{x})-\psi(\mathbf{y})|: \mathbf{x}, \mathbf{y} \in \Sigma_A,\ x_i=y_i \forall |i| \leqq n\}.$$

Let

$$\mathscr{F}_A=\{\psi \in C(\Sigma_A): \exists b>0,\ \alpha \in (0,1) \text{ so that } \operatorname{var}_n \psi \leqq b\alpha^n \text{ for all } n \geqq 0\}.$$

2.1. **Lemma.** *Let Λ be a basic hyperbolic set. Then there is a topologically mixing*[2] *subshift of finite type $\sigma_A: \Sigma_A \to \Sigma_A$, a positive $\psi \in \mathscr{F}_A$ and a continuous surjection $\rho: \Lambda(A, \psi) \to \Lambda$ so that*

$$\begin{array}{ccc} \Lambda(A,\psi) & \xrightarrow{g^t} & \Lambda(A,\psi) \\ \downarrow{\scriptstyle \rho} & & \downarrow{\scriptstyle \rho} \\ \Lambda & \xrightarrow{f^t} & \Lambda \end{array}$$

commutes.

This is from [4], Section 2, except for the mixing condition on σ_A. If $\sigma_A: \Sigma_A \to \Sigma_A$ is *not* mixing, then for some $m>0$ $\Sigma_A = X_1 \cup \cdots \cup X_m$, a disjoint union of closed sets with $\sigma_A(X_i)=X_{i+1}$ and $\sigma_A^m | X_i: X_i \to X_i$ conjugate to a mixing subshift of finite type (see e.g. 2.7 of [2]).

Identifying $\sigma_A^m: X_1 \to X_1$ with some $\sigma_B: \Sigma_B \to \Sigma_B$ and defining $\psi': \Sigma_B \to \mathbb{R}$ by

$$\psi'(x)=\psi(x)+\psi(\sigma_A x)+\cdots+\psi(\sigma_A^{m-1} x)$$

one can see that $\Lambda(B, \psi')$ is homeomorphic to $\Lambda(A, \psi)$ in a natural way and $\psi' \in \mathscr{F}_B$.

There are other properties of the map ρ which we shall recall as we need them. Throughout the remainder of the paper ψ will always denote a positive function in $\mathscr{F}_A$ and σ_A a mixing subshift of finite type.

For any homeomorphism f the set of f-invariant Borel probability measures will be denoted $M(f)$. If $F=(f^t)_{t\in\mathbb{R}}$ is a continuous flow we will write $M(F)= \bigcap_{t\in\mathbb{R}} M(f^t)$.

3. Equilibrium States

Let us review the definition of topological pressure for a homeomorphism $f: X \to X$ of a compact metric space and a continuous function $\varphi: X \to \mathbb{R}$ [23, 28]. For given $\varepsilon>0$ and $n>0$, a subset $E \subset X$ is called *(ε, n)-separated* if

$$x, y \in E, \quad x \neq y \Rightarrow d(f^k x, f^k y)>\varepsilon \quad \text{for some } k \in [0, n].$$

[2] A homeomorphism $F: X \to X$ is topologically mixing if, for U, V open nonempty in X, $U \cap F^n V \neq \emptyset$ for all sufficiently large n.

One defines

$$Z_n(f, \varphi, \varepsilon)=\sup\left\{\sum_{x\in E}\exp\sum_{k=0}^{n-1}\varphi(T^k x)\colon E \text{ is } (\varepsilon, n)\text{-separated}\right\}$$

$$P(f, \varphi, \varepsilon)=\limsup_{n\to\infty}\frac{1}{n}\log Z_n(f, \varphi, \varepsilon)$$

and

$$P(f, \varphi)=\lim_{\varepsilon\to 0} P(f, \varphi, \varepsilon).$$

For $\varphi=0$ the number $P(f, \varphi)$ is just the topological entropy $h(f)$ of f; the theory of topological pressure generalizes that of topological entropy. The main general result is that

$$P(f, \varphi)=\sup_{\mu\in M(f)}\left(h_\mu(f)+\int\varphi\, d\mu\right).$$

This was proved by Walters [28]; for f expansive there is a $\mu\in M(f)$ with $h_\mu(f)+\int\varphi\, d\mu=P(f, \varphi)$.

An *equilibrium state* for $\varphi\colon X\to\mathbb{R}$ with respect to $f\colon X\to X$ is a $\mu\in M(f)$ with $h_\mu(f)+\int\varphi\, d\mu=P(f, \varphi)$, i.e. a $\mu\in M(f)$ maximizing the quantity $h_\mu(f)+\int\varphi\, d\mu$

Now we will consider the case of a flow $F=(f^t\colon X\to X)$ and $\varphi\colon X\to\mathbb{R}$. A set $E\subset X$ is (ε, T)-*separated* if

$$x, y\in E, \quad x\neq y \;\Rightarrow\; d(f^t x, f^t y)>\varepsilon \quad \text{for some } t\in[0, T].$$

Then we define

$$Z_T(F, \varphi, \varepsilon)=\sup\left\{\sum_{x\in E}\exp\int_0^T\varphi(f^t x)\, dt\colon E \text{ is } (\varepsilon, T)\text{-separated}\right\}$$

$$P(F, \varphi, \varepsilon)=\limsup_{T\to\infty}\frac{1}{T}\log Z_T(F, \varphi, \varepsilon)$$

and

$$P(F, \varphi)=\lim_{\varepsilon\to 0} P(F, \varphi, \varepsilon).$$

The definition of $P(F, \varphi)$ is independent of the choice of metric on M. It is a straightforward exercise to check that if one lets $\varphi^1(x)=\int_0^1\varphi(f^t x)\, dt$, then $P(F, \varphi)=P(f^1, \varphi^1)$; also, for $\mu\in M(F)$ one has $\int\varphi\, d\mu=\int\varphi^1\, d\mu$. As $M(F)\subset M(f^1)$ one has

$$h_\mu(f^1)+\int\varphi\, d\mu=h_\mu(f^1)+\int\varphi^1\, d\mu\leqq P(f^1, \varphi^1)=P(F, \varphi)$$

for $\mu\in M(F)$. There is an argument ([28], see also [10], p. 359–360) to show that for any $\mu'\in M(f^1)$ one can find a $\mu\in M(F)$ with

$$h_\mu(f^1)+\int\varphi^1\, d\mu\geqq h_{\mu'}(f^1)+\int\varphi^1\, d\mu'.$$

Hence it follows that

$$P(F, \varphi)=\sup_{\mu\in M(F)}\left(h_\mu(f^1)+\int\varphi\, d\mu\right).$$

By an *equilibrium state* for φ (with respect to F) we mean a $\mu\in M(F)$ with

$$h_\mu(f^1)+\int\varphi\, d\mu=P(F, \varphi).$$

For $G=\{g^t\}$ the special flow on $\Lambda(A,\psi)$ there is a well-known bijection between $M(G)$ and $M(\sigma_A)$. For $\nu\in M(\sigma_A)$ and m Lebesgue measure, $\nu\times m$ gives measure 0 to the identifications on $Y\to\Lambda(A,\psi)$ and so $\mu_\nu=(\nu\times m(Y))^{-1}\,\nu\times m|Y$ gives a probability measure on $\Lambda(A,\psi)$. One can check that $\nu\in M(\sigma_A)$ implies $\mu_\nu\in M(G)$ and that $\nu\to\mu_\nu$ defines a bijection $M(\sigma_A)\to M(G)$.

It is known that any function $\gamma\in\mathscr{F}_A$ has a unique equilibrium state ν w.r.t. σ_A [6, 11, 22, 24] and that ν depends continuously on γ (weak topology for ν, uniform topology for γ). We will now state the corresponding condition on φ: $\Lambda(A,\psi)\to\mathbb{R}$ which guarantees a unique equilibrium state.

3.1 **Proposition**[3]. *Let* φ: $\Lambda(P,\psi)\to\mathbb{R}$ *be continuous,* $\Phi(\mathbf{x})=\int_0^{\psi(\mathbf{x})}\varphi(\mathbf{x},t)\,dt$ *and* $c=P(G,\varphi)$. *Assume that* $\Phi\in\mathscr{F}_A$. *Then there is a measure* $\mu_\varphi\in M(G)$ *so that*

(a) μ_φ *is the unique equilibrium state for* φ *with respect to* G.

(b) $\mu_\varphi=\mu_{\nu_0}$ *where* ν_0 *is the unique equilibrium state for* $\Phi-c\psi$ *on* Σ_A.

(c) μ_φ *is ergodic and positive on non-empty open sets and*

(d) *for* $\varepsilon>0$ *there is a* $C_\varepsilon>0$ *so that*

$$\mu_\varphi(B_{x,G}(\varepsilon,T))\geqq C_\varepsilon\exp\left(-cT+\int_0^T\varphi(g^t x)\,dt\right)$$

for all $x\in\Lambda(A,\psi)$, $T\geqq 0$ *where*

$$B_{x,G}(\varepsilon,T)=\{y\in\Lambda(A,\psi):\ d(g^t y,g^t x)\leqq\varepsilon\ \text{ for all }\ t\in[0,T]\}.$$

Let $\gamma=\Phi-c\psi$. As $\Phi,\psi\in\mathscr{F}_A$ we have $\gamma\in\mathscr{F}_A$. This guarantees that γ has a unique equilibrium state ν_0. By Fubini's theorem, for any $\nu\in M(\sigma_A)$, $(\nu\times m)(Y)=\int\psi\,d\nu$ and $\int\varphi\,d\mu_\nu=\dfrac{\int\Phi\,d\nu}{\int\psi\,d\nu}$. A theorem of Abramov [1] states that

$$h_{\mu_\nu}(g^1)=\frac{h_\nu(\sigma_A)}{\int\psi\,d\nu}.$$

Hence

$$\begin{aligned}c=P(G,\varphi)&=\sup_{\mu\in M(G)}\left(h_\mu(g^1)+\int\varphi\,d\mu\right)\\&=\sup_{\nu\in M(\sigma_A)}\frac{h_\nu(\sigma_A)+\int\Phi\,d\nu}{\int\psi\,d\nu}.\end{aligned}$$

Thus $P(\sigma_A,\gamma)=\sup_\nu\left(h_\nu(\sigma_A)+\int(\Phi-c\psi)\,d\nu\right)=0$ with ν attaining the supremum (i.e. $\nu=\nu_0$) precisely if μ_ν is the unique equilibrium state for φ. This shows that $\mu_\varphi=\mu_{\nu_0}$ satisfies (a) and (b); (c) is true because ν_0 has these same properties ([6] or [24], Appendix B).

We now verify (d). Let $x\in\Lambda(A,\psi)$ be represented by $x=(\mathbf{x},t_1)$, $t_1\in[0,\psi(\mathbf{x}))$ and $g^T x=(\sigma_A^n\mathbf{x},t_2)$, where $n=n(x)$ is such that

$$t_2=T+t_1-\sum_{k=0}^{n-1}\psi(\sigma_A^k\mathbf{x})\in[0,\psi(\sigma_A^n\mathbf{x})).$$

[3] Another proof of (a), (b), (d) has now been obtained by E. Franco-Sanchez, Berkeley thesis, 1974.

Given $\varepsilon>0$ one can find $\delta_\varepsilon>0$ and $s_\varepsilon>0$ (not depending on x or T) so that[4]

$$B_{x,G}(\varepsilon,T)\supset\{(\mathbf{y},t)\colon |t-t_1|\leqq s_\varepsilon \text{ and } d(\sigma_A^k\mathbf{y},\sigma_A^k\mathbf{x})\leqq\delta_\varepsilon \text{ for all } k\in[0,n-1]\}.$$

We do not go through the details but do point out that, for any $\alpha>0$, $\psi\in\mathscr{F}_A$ implies that for δ small enough

$$d(\sigma_A^k\mathbf{x},\sigma_A^k\mathbf{y})\leqq\delta\,\forall k\in[0,n-1] \;\Rightarrow\; \sum_{k=0}^{n-1}|\psi(\sigma_A^k\mathbf{x})-\psi(\sigma_A^n\mathbf{y})|<\alpha.$$

Then

$$\mu_\varphi B_{x,G}(\varepsilon,T)\geqq\frac{s_\varepsilon}{\int\psi\,d\nu_0}\nu_0\{\mathbf{y}\colon d(\sigma_A^k\mathbf{y},\sigma_A^k\mathbf{x})\leqq\delta_\varepsilon\forall k\in[0,n-1]\}.$$

By [6] Lemma 5, this right side is at least

$$a_\varepsilon\exp\left(\sum_{k=0}^{n-1}\gamma(\sigma_A^k\mathbf{x})-nP(\sigma_A,\gamma)\right)=a_\varepsilon\exp\sum_{k=0}^{n-1}\gamma(\sigma_A^k\mathbf{x})$$

for some $a_\varepsilon>0$. Since

$$\sum_{k=0}^{n-1}\psi(\sigma_A^k\mathbf{x})+t_2=T+t_1$$

and

$$\int_0^{t_1}\varphi(\mathbf{x},t)\,dt+\int_0^T\varphi(g^t x)\,dt-\int_0^{t_2}\varphi(\sigma_A^n\mathbf{x},t)\,dt=\sum_{k=0}^{n-1}\Phi(\sigma_A^k\mathbf{x}),$$

one sees that $\sum_{k=0}^{n}\gamma(\sigma_A^k\mathbf{x})$ differs from $-cT+\int_0^T\varphi(g^t x)\,dt$ by at most $2\,\|\psi\|(|c|+\|\varphi\|)$.

This proves (d).

3.2. *Remark.* Special flows are simple enough that parts (a) and (b) above could have been derived without appealing so much to general (and harder) results on topological pressure.

3.3 **Theorem.** *Assume that Λ is a basic hyperbolic set for F and that $\varphi\colon\Lambda\to\mathbb{R}$ satisfies a Hölder condition of positive exponent. Then φ has a unique equilibrium state μ_φ. Furthermore, μ_φ is ergodic and positive on non-empty open sets of Λ, and for any $\varepsilon>0$ there is a $C_\varepsilon>0$ so that*

$$\mu_\varphi(B_{x,F|\Lambda}(\varepsilon,T))\geqq C_\varepsilon\exp\left(-P(F|\Lambda,\varphi)\,T+\int_0^T\varphi(f^t x)\,dt\right)$$

for all $x\in\Lambda$, $T\geqq 0$.

We apply the preceding proposition to the function $\varphi^*=\varphi\circ\rho$ on $\Lambda(A,\psi)$. There are $b_1>0$ and $\tau\in(0,1)$ so that

$$d(\rho(\mathbf{x},0),\rho(\mathbf{y},0))\leqq b_1\tau^N$$

if $x_i=y_i$ for all $|i|\leqq N$ (see [4], Lemma 2.2.(i)).

[4] In this formula $(\mathbf{y},t)$ has to be replaced by $(\sigma_A^{-1}\mathbf{y},\psi(\sigma_A^{-1}\mathbf{y})+t)$ resp. by $(\sigma_A\mathbf{y},t-\psi(\mathbf{y}))$ when $t<0$ resp. $t>\psi(\mathbf{y})$.

Since F is a differentiable flow, there is a constant b_2 so that

$$d(f^t x, f^t y) \leqq b_2\, d(x, y) \quad \text{provided } t \in [0, \|\psi\|].$$

The Hölder condition on φ states that

$$|\varphi(x) - \varphi(y)| \leqq b_3\, d(x, y)^\alpha \quad \text{with} \quad \alpha > 0.$$

Combining these estimates, when $x_i = y_i \forall i \in [-N, N]$ we have

$$\begin{aligned} &\left| \int_0^{\psi(\mathbf{x})} \varphi^*(\mathbf{x}, t)\, dt - \int_0^{\psi(\mathbf{y})} \varphi^*(\mathbf{y}, t)\, dt \right| \\ &\quad \leqq \|\varphi\|\, |\psi(\mathbf{x}) - \psi(\mathbf{y})| + \int_0^{\psi(\mathbf{x})} |\varphi(f^t \rho(\mathbf{x}, 0)) - \varphi(f^t \rho(\mathbf{y}, 0))|\, dt \\ &\quad \leqq \|\varphi\|\, |\psi(\mathbf{x}) - \psi(\mathbf{y})| + \|\psi\|\, b_3 (b_2 b_1)^\alpha (\tau^\alpha)^N. \end{aligned}$$

Since $\psi \in \mathscr{F}$, this gives $\Phi^* \in \mathscr{F}$ where $\Phi^*(\mathbf{x}) = \int_0^{\psi(\mathbf{x})} \varphi^*(\mathbf{x}, t)\, dt$. So φ^* has a unique equilibrium state μ_{φ^*} as in the preceding proposition.

We recall that there are closed subsets $A_s = \rho^{-1}(\Delta^s \mathscr{M})$ and $A_u = \rho^{-1}(\Delta^u \mathscr{M})$ of $\Lambda(A, \psi)$ so that (see [4])

(a) $A_s \neq \Lambda(P, \psi) \neq A_u$

(b) $g^t A_s \subset A_s$, $g^{-t} A_u \subset A_u \forall t \geqq 0$ and

(c) ρ is one-to-one off $\bigcup_{t \in \mathbb{R}} g^t (A_u \cup A_s) = \bigcup_{n \in \mathbb{Z}} g^n (A_u \cup A_s)$.

Because μ_{φ^*} is positive on non-empty open sets $\mu_{\varphi^*}(A_s) \neq 1 \neq \mu_{\varphi^*}(A_u)$; since μ_{φ^*} is ergodic and each of these sets is invariant under one direction of time, $\mu_{\varphi^*}(A_s) = 0 = \mu_{\varphi^*}(A_u)$. By (c) then ρ gives a conjugacy of the measurable flows (G, μ_{φ^*}) and $(F(\Lambda, \mu_\varphi))$ where $\mu_\varphi = \rho^* \mu_{\varphi^*}$. In particular $h_{\mu_\varphi}(f^1) = h_{\mu_{\varphi^*}}(g^1)$ and

$$h_{\mu_\varphi}(f^1) + \int \varphi\, d\mu_\varphi = h_{\mu_{\varphi^*}}(g^1) + \int \varphi^*\, d\mu_{\varphi^*} = P(G, \varphi^*).$$

As $F|\Lambda$ is the quotient of G and $\varphi^* = \varphi \circ \rho$, one has $P(F|\Lambda, \varphi) \leqq P(G, \varphi^*)$ (see Walters [28], Theorem 2.2); because $h_{\mu_\varphi}(f^1) + \int \varphi\, d\mu_\varphi = P(G, \varphi^*)$ one has $P(F|\Lambda, \varphi) = P(G, \varphi^*)$ and μ_φ is an equilibrium state for φ. If μ were another equilibrium state for φ, then $\mu = \rho^* \mu'$ for some $\mu' \in M(G)$ (by an easy application of the Hahn-Banach and Markov-Kakutani theorems) and

$$h_{\mu'}(g^1) + \int \varphi^*\, d\mu' \geqq h_\mu(f^1) + \int \varphi\, d\mu = P(F|\Lambda, \varphi) = P(G, \varphi^*).$$

So μ' is an equilibrium state for φ^*, $\mu' = \mu_{\varphi^*}$ and $\mu = \mu_\varphi$. Thus μ_φ is the unique equilibrium state. The remaining properties for μ_φ follow from the corresponding ones for μ_{φ^*} in Proposition 3.1.

3.4 *Remark.* For $\varphi = 0$ the uniqueness of equilibrium state just says that $F|\Lambda$ has a unique invariant measure maximizing entropy. This was proved earlier in [5].

3.5. *Remark.* It was proved in [2] and [24] that (σ_A, ν_0) is isomorphic to a Bernoulli shift where ν_0 is the equilibrium state of $\gamma \in \mathscr{F}_A$. The corresponding result for $(F|\Lambda, \mu_\varphi)$ follows from results proved elsewhere. If $F|\Lambda$ is C-dense (i.e. $W^u(x) \cap \Lambda$ is dense in Λ for every $x \in \Lambda$ where $W^u(x) = \{y\colon d(f^{-t} x, f^{-t} y) \to 0$ as $t \to +\infty\}$), then G is also C-dense and Sinai [26] p. 48–9, applied a theorem of

Gurevič [12] to show that $(G, \mu_{\varphi*})$ is a K-flow. We mention that, although Sinai uses the formalism of Gibbs states instead of equilibrium states, the measure $\mu_{\varphi*}$ is the same as the one he constructs. M. Ratner [20] (also Bunimovič [9]) has proved in this C-dense case that $(G, \mu_{\varphi*})$ is actually Bernoulli (i.e. $(g^t, \mu_{\varphi*})$ is isomorphic to a Bernoulli shift for each $t \neq 0$). Since $(F|\Lambda, \mu_\varphi) \approx (G, \mu_{\varphi*})$, $(F|\Lambda, \mu_\varphi)$ is Bernoulli when $F|\Lambda$ is C-dense and φ is Hölder continuous. In that case we have

$$\lim_{t\to\infty} \int (g \circ f^t) \cdot g' \, d\mu_\varphi = \int g \, d\mu_\varphi \cdot \int g' \, d\mu_\varphi$$

for all $g, g' \in L^2(\mu_\varphi)$. [This follows from the fact that (μ_φ, f^t) is equivalent to a Bernoulli shift for each t, and the continuity of the flow (f^t)].

4. Attractors

Now assume Λ is a C^2 basic hyperbolic set. For $x \in \Lambda$ let $\lambda_t(x)$ be the Jacobian of the linear map $D f^t \colon E_x^u \to E_{f^t x}^u$ using inner products induced by the Riemannian metric.

Define

$$\varphi^{(u)}(x) = -\left.\frac{d \ln \lambda_t(x)}{dt}\right|_{t=0} = -\left.\frac{d\lambda_t(x)}{dt}\right|_{t=0} \tag{2}$$

which exists and depends differentiably on E_x^u (hence continuously on x) as f^t is a C^2 flow. Since $\lambda_{T+t}(x) = \lambda_t(f^T x)\, \lambda_T(x)$ one has

$$-\ln \lambda_t(f^T x) = -\ln \lambda_{T+t}(x) + \ln \lambda_T(x)$$

and so

$$\varphi^{(u)}(f^T x) = -\left.\frac{d \ln \lambda_s(x)}{ds}\right|_{s=T}.$$

This implies that

$$\int_0^T \varphi^{(u)}(f^t x)\, dt = -\ln \lambda_T(x).$$

This integral is the one appearing in Theorem 3.3 for $\varphi = \varphi^{(u)}$.

4.1. **Lemma.** *For Λ a C^2 basic hyperbolic set and $\varphi^{(u)} \colon \Lambda \to \mathbb{R}$ as above, $\varphi^{(u)}$ satisfies a Hölder condition of positive exponent.*

$x \to E_x^u$ is Hölder continuous (3.1 of [19]) and $E_x^u \to \varphi^{(u)}(x)$ is differentiable, so the composition $x \to \varphi^{(u)}(x)$ is Hölder.

4.2. **Lemma** (Volume lemma). *Let Λ be a C^2 basic hyperbolic set and define*

$$B_x(\varepsilon, T) = \{y \in M \colon d(f^t x, f^t y) \leqq \varepsilon \text{ for all } t \in [0, T]\}.$$

For small $\varepsilon > 0$ there is a constant $c_\varepsilon > 1$ so that

$$m(B_x(\varepsilon, T))\, \lambda_T(x) \in [c_\varepsilon^{-1}, c_\varepsilon]$$

for all $x \in \Lambda$ and $T \geqq 0$, where m is the measure on M derived from the Riemann metric.

4.3. **Lemma** (Second volume lemma). *For small* ε, $\delta > 0$ *there is* $d = d(\varepsilon, \delta) > 0$ (d independent of n) *so that*

$$m(B_y(\delta, n)) \geqq d \cdot m(B_x(\varepsilon, n))$$

whenever $x \in \Lambda$ *and* $y \in B_x(\varepsilon, n)$.

These two lemmas are proved in the Appendix.

4.4. **Proposition.** (a) *Let* Λ *be a* C^2 *basic hyperbolic set and define*

$$B_\Lambda(\varepsilon, T) = \bigcup_{x \in \Lambda} B_x(\varepsilon, T).$$

Then (for sufficiently small ε*)*

$$P(F|\Lambda, \varphi^{(u)}) = \limsup_{T \to \infty} \frac{1}{T} \log m(B_\Lambda(\varepsilon, T)) \leqq 0. \tag{3}$$

(b) *Define*

$$W_x^s(\varepsilon) = \{y \in M : \lim_{t \to \infty} d(f^t y, f^t x) = 0 \text{ and } d(f^t y, f^t x) \leqq \varepsilon \, \forall t \geqq 0\}$$

when $x \in \Lambda$, *and let* $W_\Lambda^s(\varepsilon) = \bigcup_{x \in \Lambda} W_x^s(\varepsilon)$. *If* $m(W_\Lambda^s(\varepsilon)) > 0$, *then* $P(F|\Lambda, \varphi^{(u)}) = 0$ *and*

$$h_{\mu_{\varphi^{(u)}}}(f^1) = -\int \varphi^{(u)} \, d\mu_{\varphi^{(u)}}. \tag{4}$$

This is true in particular if Λ *is an attractor.*

Let $0 < \delta \leqq \varepsilon$. If E is a maximal (δ, T)-separated set for $F|\Lambda$, then

$$\bigcup_{x \in E} B_x(\delta/2, T) \subset B_\Lambda(\varepsilon, T) \subset \bigcup_{x \in E} B_x(\delta + \varepsilon, T)$$

where the $B_x(\delta/2, T)$ are disjoint, and the second inclusion follows from $\Lambda \subset \bigcup_{x \in E} B_x(\delta, T)$. Thus, assuming ε small enough and using the volume lemma,

$$c_{\delta/2}^{-1} \sum_{x \in E} \lambda_T(x)^{-1} \leqq m(B_\Lambda(\varepsilon, T)) \leqq c_{\delta+\varepsilon} \sum_{x \in E} \lambda_T(x)^{-1}.$$

Therefore

$$c_{\delta/2}^{-1} Z_T(F|\Lambda, \varphi^{(u)}, \delta) \leqq m(B_\Lambda(\varepsilon, T)) \leqq c_{\delta+\varepsilon} Z_T(F|\Lambda, \varphi^{(u)}, \delta)$$

and

$$P(F|\Lambda, \varphi^{(u)}, \delta) = \limsup_{T \to \infty} \frac{1}{T} \log m(B_\Lambda(\varepsilon, T)).$$

The limit $\delta \to 0$ yields (3), proving (a).

By Theorem 3.3 and Lemma 4.1, $\varphi = \varphi^{(u)}$ has a unique equilibrium state $\mu_{\varphi^{(u)}}$. By the definition of equilibrium states, (4) is equivalent to $P(F|\Lambda, \varphi^{(u)}) = 0$. The latter statement follows from (3) with

$$m(B_\Lambda(\varepsilon, T)) \geqq m(W_\Lambda^s(\varepsilon)) > 0.$$

There is a neighbourhood V of Λ so that

$$W_\Lambda^s(\varepsilon) \supset \{y \in M : f^t y \in V \text{ for all } t \geqq 0\}.$$

Indeed 5.1 of [13] gives this for diffeomorphisms and [13] indicates how to do the proof for flows. If Λ is an attractor, one can find a small neighbourhood U' of Λ so that $f^t y \in V$ for all $t \geqq 0$ whenever $y \in U'$. Then $U' \subset W_\Lambda^s(\varepsilon)$ and therefore $m(W_\Lambda^s(\varepsilon)) > 0$.

4.5. *Remark.* A rescaling of t: $t \to t' = t/t_0$ does not change the invariant measures, it replaces $h_\mu(f^1)$ by $h_\mu(f'^{t_0}) = t_0\, h_\mu(f^1)$ (see [1]) and $\varphi^{(u)}$ by $\varphi'^{(u)} = t_0\, \varphi^{(u)}$. Therefore – as indicated in the Introduction – rescaling of t does not change $\mu_{\varphi^{(u)}}$ or the main results below. A change of Riemann metric on M changes $\varphi^{(u)}$ but not $\mu_{\varphi^{(u)}}$, or $P(F|\Lambda, \varphi^{(u)})$ as one readily sees.

4.6. **Corollary.** *Let Λ be a C^2 basic hyperbolic set. For sufficiently small $\varepsilon > 0$ there is a constant c'_ε such that*

$$m(B_x(2\varepsilon, T)) \leqq c'_\varepsilon\, \mu_{\varphi^{(u)}}(B_x(\varepsilon, T))$$

for all $x \in \Lambda$, $T \geqq 0$.

This follows from Theorem 3.3, Lemma 4.2, and Proposition 4.4(a), with $c'_\varepsilon = c_{2\varepsilon}/C_\varepsilon$.

5. Main Results

5.1. **Theorem.** *Let Λ be a C^2 hyperbolic attractor, W^s_Λ its basin. Then for m-almost all points $x \in W^s_\Lambda$ one has*

$$\lim_{T\to\infty} \frac{1}{T} \int_0^T g(f^t x)\, dt = \int g\, d\mu_{\varphi^{(u)}}$$

for all continuous g: $M \to \mathbb{R}$ (i.e. x is a generic point for $\mu_{\varphi^{(u)}}$).

We can replace W^s_Λ by a neighbourhood U of Λ such that $f^t U \subset U$ for all $t \geqq t_0$ and $\bigcap_{t \geqq 0} f^t U = \Lambda$.

Let us write

$$\bar{g}(T, x) = \frac{1}{T} \int_0^T g(f^t x)\, dt \quad \text{and} \quad \bar{g} = \int g\, d\mu_\varphi.$$

Let

$$E(g, \delta) = \{x \in U: \limsup_{T\to\infty} |\bar{g}(T, x) - \bar{g}| \geqq \delta\}.$$

Choose $\varepsilon > 0$ so small that $|g(f^t x) - g(f^t x')| < \delta/4$ whenever $d(x, x') \leqq \varepsilon$ and $0 \leqq t \leqq 1$. If we set $C_n(g, \delta') = \{x \in U: |\bar{g}(n, x) - \bar{g}| > \delta'\}$, then

$$E(g, \delta) \subset \bigcap_{N=0}^{\infty} \bigcup_{n=N}^{\infty} C_n\left(g, \frac{3\delta}{4}\right) \subset E\left(g, \frac{3\delta}{4}\right).$$

Now fix $N > 0$ and choose finite subsets $S_N, S_{N+1}, \ldots,$ of Λ successively as follows. Let $S_n (n \geqq N)$ be a maximal subset of $C_n(g, \delta/2) \cap \Lambda$ satisfying the conditions:

(a) $B_x(\varepsilon, n) \cap B_y(\varepsilon, k) = \emptyset$ for $x \in S_n$, $y \in S_k$, $N \leqq k < n$ and

(b) $B_x(\varepsilon, n) \cap B_{x'}(\varepsilon, n) = \emptyset$ for $x, x' \in S_n$, $x \neq x'$.

(Notice that each S_n is finite.) Choose $\alpha > 0$ so that

$$B_\Lambda(\alpha) \subset W^s_\Lambda(\varepsilon) = \bigcup_{z \in \Lambda} W^s_z(\varepsilon)$$

($B_\Lambda(\alpha)$ is the closed α-neighbourhood of Λ). If

$$y \in B_\Lambda(\alpha) \cap C_n\left(g, \frac{3\delta}{4}\right) \quad (n \geqq N)$$

and $y \in W_z^s(\varepsilon)$ with $z \in \Lambda$, then $z \in C_n(g, \delta/4)$ because of the way ε was chosen. By the maximality of S_n one has

$$B_z(\varepsilon, n) \cap B_x(\varepsilon, k) \neq \emptyset \quad \text{for some } x \in S_k, \ N \leqq k \leqq n$$

and then $B_x(2\varepsilon, k) \supset B_z(\varepsilon, n) \supset W_z^s(\varepsilon) \ni y$. Thus one has

$$B_\Lambda(\alpha) \cap \bigcup_{n=N}^{\infty} C_n\left(g, \frac{3\delta}{4}\right) \subset \bigcup_{k=N}^{\infty} \bigcup_{x \in S_k} B_x(2\varepsilon, k).$$

Using Lemma 4.2 one has

$$m\left(B_\Lambda(\alpha) \cap \bigcup_{n=N}^{\infty} C_n\left(g, \frac{3\delta}{4}\right)\right) \leqq A'_{2\varepsilon} \sum_{k=N}^{\infty} \sum_{x \in S_k} \exp \int_0^k \varphi(f^t x)\, dt. \tag{5}$$

The definition of S_n implies that $V_N = \bigcup_{k=N}^{\infty} \bigcup_{x \in S_k}^{\infty} B_x(\varepsilon, k)$ is a disjoint union. The choice of ε gives that $B_\varepsilon(x, k) \subset C_k(g, \delta/4)$ for $x \in S_k \subset C_k(g, \delta/2)$ and so $V_N \subset \bigcup_{k=N}^{\infty} C_k(g, \delta/4)$. Because $\mu_{\varphi^{(u)}}$ is ergodic,

$$0 = \mu_{\varphi^{(u)}}\left(E\left(g, \frac{\delta}{4}\right)\right) \geqq \mu_{\varphi^{(u)}}\left(\bigcap_{N=0}^{\infty} \bigcup_{k=N}^{\infty} C_k\left(g, \frac{\delta}{4}\right)\right) = \lim_{N \to \infty} \mu_{\varphi^{(u)}}\left(\bigcup_{k=N}^{\infty} C_k\left(g, \frac{\delta}{4}\right)\right)$$

and $\lim_{N \to \infty} \mu_{\varphi^{(u)}}(V_N) = 0$. By Theorem 3.3 (and $P(F|\Lambda, \varphi^{(u)}) = 0$, confer Proposition 4.4)

$$\mu_{\varphi^{(u)}}(V_N) \geqq C_\varepsilon \sum_{k=N}^{\infty} \sum_{x \in S_k} \exp \int_0^k \varphi(f^t x)\, dt.$$

Hence the sum on the right converges to 0 as $N \to \infty$ and using (5) above we get

$$\lim_{N \to \infty} m\left(B_\Lambda(\alpha) \cap \bigcup_{n=N}^{\infty} C_n\left(g, \frac{3\delta}{4}\right)\right) = 0.$$

This in turn gives $m(B_\Lambda(\alpha) \cap E(g, \delta)) = 0$.

Now $f^t E(g, \delta) \subset E(g, \delta)$ for all $t \geqq 0$ and $f^t(U) \subset B_\Lambda(\alpha)$ for some $t > 0$. As f^t is a diffeomorphism, $m(f^t E(g, \delta)) \leqq m(B_\Lambda(\alpha) \cap E(g, \delta)) = 0$ implies $m(E(g, \delta)) = 0$. Letting $\{g_k\}_{k=1}^{\infty}$ be a dense sequence of continuous functions $\bar{U} \to \mathbb{R}$, we get that for x outside the m-null set

$$\bigcup_{k, m \geqq 1} E\left(g_k, \frac{1}{m}\right)$$

one has $\lim_{T \to \infty} \bar{g}_k(T, x) = \bar{g}_k$; as the g_k are dense, it follows that $\lim_{T \to \infty} \bar{g}(T, x) = \bar{g}$ for all continuous $g: \bar{U} \to \mathbb{R}$.

5.2. *Remark.* For the special case of an Anosov flow ($\Lambda = M$) with an invariant (probability) measure μ' absolutely continuous w.r.t. m, this theorem implies the known fact that $\mu' = \mu_{\varphi^{(u)}}$.

5.3. **Theorem.** *Let Λ be a C^2 attractor, W_Λ^s its basin, and let v be a probability measure absolutely continuous with respect to m and with support in W_Λ^s. If the flow F restricted to Λ is C-dense, then*

$$\lim_{t \to \infty} \int (g \circ f^t)\, dv = \int g\, d\mu_{\varphi^{(u)}}$$

for all continuous $g: M \to \mathbb{R}$.

We may choose U as in the proof of Theorem 5.1, and assume that $\operatorname{supp} \nu \subset U$. Define $f^{*t} \nu$ by

$$(f^{*t} \nu)(g) = \nu(g \circ f^t)$$

and write $\mu_{\varphi^{(u)}} = \mu$. We have to prove that weak $\lim_{t \to \infty} f^{*t} \nu = \mu$. In showing this we may assume that $\nu = r \cdot m$ where $r \geqq 0$ is bounded (by density of bounded functions in L^1).

Given $\varepsilon > 0$ we find as in the proof of Proposition 4.4 that if U' is a sufficiently small neighbourhood of Λ, then $U' \subset W^s_\Lambda(\varepsilon)$. We can choose $t(\varepsilon) > 0$ so that $f^{t(\varepsilon)} U \subset U'$ and therefore

$$\operatorname{supp} f^{*t(\varepsilon)} \nu \subset W^s_\Lambda(\varepsilon).$$

Let $E \subset \Lambda$ be a maximal (T, ε)-separated set for $F|\Lambda$, we have thus

$$\operatorname{supp} f^{*t(\varepsilon)} \nu \subset \bigcup_{x \in E} B_x(2\varepsilon, T).$$

Let $(\psi_x)_{x \in E}$ be a non-negative measurable partition of unity on $\operatorname{supp} f^{*t(\varepsilon)} \nu$ subordinate to the covering by the $B_x(2\varepsilon, T)$ and let χ_x be the characteristic function of $B_x(\varepsilon, T)$. We write

$$\nu_{\varepsilon, T} = \sum_{x \in E} \left(\frac{\int \psi_x \, d(f^{*t(\varepsilon)} \nu)}{\int \chi_x \, d\mu} \right) \cdot \chi_x \mu.$$

The measure $\nu_{\varepsilon, T}$ is a probability measure absolutely continuous with respect to μ, with density bounded independently of T. This is because

$$\frac{\int \psi_x \, d(f^{*t(\varepsilon)} \nu)}{\int \chi_x \, d\mu} \leqq \|r_{t(\varepsilon)}\|_\infty \frac{m(B_x(2\varepsilon, T))}{\mu(B_x(\varepsilon, T))} \leqq C'_\varepsilon \|r_{t(\varepsilon)}\|_\infty$$

by Corollary 4.6 ($r_{t(\varepsilon)}$ denotes the density of $f^{*t(\varepsilon)} \nu$ with respect to m).

Notice that $\nu_{\varepsilon, T}$ is obtained by redistributing the mass of $f^{*t(\varepsilon)} \nu$ in such a manner that all that goes to $B_x(\varepsilon, T)$ comes from $B_x(2\varepsilon, T)$. Therefore also $f^{*t} \nu_{\varepsilon, T}$ is obtained by redistributing the mass of $f^{*(t + t(\varepsilon))} \nu$ in such a manner that all that goes to $f^t B_x(\varepsilon, T)$ comes from $f^t B_x(2\varepsilon, T)$. The diameter of $f^t B_x(2\varepsilon, T)$ is at most 4ε when $0 \leqq t \leqq T$; therefore if $\mathcal{N}$ is a closed weak neighbourhood of the origin in the space of real measures on M we have

$$f^{*(t + t(\varepsilon))} \nu - f^{*t} \nu_{\varepsilon, T} \in \mathcal{N}$$

when $0 \leqq t \leqq T$, provided ε has been chosen sufficiently small.

Remember now that $\nu_{\varepsilon, T} = s_{\varepsilon, T} \cdot \mu$ where $s_{\varepsilon, T} \in L^\infty(\mu)$ and $\|s_{\varepsilon, T}\|_\infty$ is bounded independently of T. We can thus choose $T_n \to \infty$ such that $s_{\varepsilon, T_n} \to s_\varepsilon$ in $L^\infty(\mu)$ with its topology of weak dual of $L^1(\mu)$. We have thus

$$f^{*(t + t(\varepsilon))} \nu - f^{*t}(s_\varepsilon \cdot \mu) \in \mathcal{N} \tag{6}$$

for all $t \geqq 0$. We use now Remark 3.5: (μ, f^t) is a Bernoulli flow and

$$\lim_{t \to \infty} \int s_\varepsilon \cdot (g \circ f^t) \, d\mu = \mu(g)$$

for all continuous $g: \Lambda \to \mathbb{R}$. There is thus $t_{\mathcal{N}}$ such that

$$f^{*t}(s_\varepsilon \cdot \mu) - \mu \in \mathcal{N} \tag{7}$$

for $t \geqq t_{\mathcal{N}}$. From (6) and (7) we obtain

$$f^{*t} \nu - \mu \in 2\mathcal{N}$$

when $t \geqq t(\varepsilon) + t_{\mathcal{N}}$. Therefore $f^{*t} \nu$ tends weakly to μ when $t \to \infty$.

5.4. **Proposition.** *Let Λ be a C^2 basic hyperbolic set. The measure $\mu_{\varphi^{(u)}}$ depends continuously on the C^2 flow F for the weak topology on measures and the C^1 topology on flows. Also the pressure of $\varphi^{(u)}$ and the entropy of $\mu_{\varphi^{(u)}}$ with respect to F depend continuously on F for the C^1 topology on flows.*

Let $\mu = \mu_{\varphi^{(u)}}$, and μ' be the corresponding measure for a flow F'. We have to show that $\mu' \to \mu$ weakly when $F' \to F$ in the C^1 sense; μ' is a measure carried by the F'-basic set Λ' close to Λ.

Going back to Lemma 2.1 and using [4] we have a special flow G' on $\Lambda'(A, \psi')$ and a continuous surjection $\rho': \Lambda'(A, \psi') \to \Lambda'$. By [4] and the Ω-stability theorem [5], we can construct $\Lambda'(A, \psi')$ from the same subshift $\sigma_A: \Sigma_A \to \Sigma_A$ which was used for $\Lambda(P, \psi)$. When $F' \to F$, we have $\psi' \to \psi$ uniformly, and $\rho'(\mathbf{x}, t) \to \rho(\mathbf{x}, t)$ (uniformly in $(\mathbf{x}, t)$ for $0 \leqq t \leqq \min\{\psi(\mathbf{x}), \psi'(\mathbf{x})\}$). Furthermore $E'^u_{\rho'(\mathbf{x}, t)} \to E^u_{\rho(\mathbf{x}, t)}$ where E'^u is the unstable subbundle for F' (by Theorem (6.1) of [15]).

We know that

$$\begin{aligned} \Phi(\mathbf{x}) &= \int_0^{\psi(\mathbf{x})} \varphi^{(u)}(\rho(\mathbf{x}, t))\, dt = \int_0^{\psi(\mathbf{x})} \varphi^{(u)}(f^t \rho(\mathbf{x}, 0))\, dt \\ &= -\ln \lambda_{\psi(\mathbf{x})}(\rho(\mathbf{x}, 0)) \end{aligned}$$

and correspondingly

$$\Phi'(\mathbf{x}) = -\ln \lambda'_{\psi'(\mathbf{x})}(\rho'(\mathbf{x}, 0)).$$

Therefore when $F' \to F$, we have $\Phi' \to \Phi$ uniformly.

According to Section 3 we have

$$P(F|\Lambda, \varphi^{(u)}) = P(G, \varphi^{(u)} \circ \rho) = \sup_{\nu \in M(\sigma_A)} \frac{h_\nu(\sigma_A) + \int \Phi\, d\nu}{\int \psi\, d\nu}.$$

Therefore $P(F|\Lambda, \varphi^{(u)})$ depends continuously on F. Using the notation of Proposition 3.1, we let ν_0 be the unique equilibrium state for $\Phi - P(F|\Lambda, \varphi^{(u)}) \cdot \psi$. By the continuity of the equilibrium state indicated just before Proposition 3.1, $\nu_0' \to \nu_0$ when $F' \to F$. Thus $\mu_{\nu_0'} \to \mu_{\nu_0}$ and $\mu' = \rho'^* \mu_{\nu_0'} \to \mu = \rho^* \mu_{\nu_0}$.

Finally, the entropy of $\mu_{\varphi^{(u)}}$ is

$$P(F|\Lambda, \varphi^{(u)}) - \int \varphi^{(u)}\, d\mu_{\varphi^{(u)}}$$

where

$$\int \varphi^{(u)}\, d\mu_{\varphi^{(u)}} = \int (\varphi^{(u)} \circ \rho)\, d\mu_{\nu_0} = \int \Phi\, d\nu_0 / \int \psi\, d\nu_0$$

depends continuously on F.

5.5. **Proposition.** *Let Λ be a C^1 basic hyperbolic set and let $\varepsilon>0$.*

(a) *If $W_x^u(\varepsilon)\subset\Lambda$ for some $x\in\Lambda$, then Λ is an attractor.*

(b) *If Λ is not an attractor, there exists $\gamma>0$ such that for all $x\in\Lambda$, there is $y\in W_x^u(\varepsilon)$ with $d(y,\Lambda)>\gamma$.*

If $W_x^u(\varepsilon)\subset\Lambda$, and $u>0$, the set

$$U_x=\bigcup\{W_y^s(\varepsilon):\ y\in\bigcup_{|t|\leqq u} f^t\,W_x^u(\varepsilon)\}\tag{8}$$

is a neighbourhood of x in M (see [13], Lemma 4.1). Choose a periodic point $p\in U_x\cap\Lambda$ and let t_0 be its period. For some $\beta\in(0,\varepsilon]$, we have $W_p^u(\beta)\subset U_x$, hence

$$W_p^u(\beta)\subset W_\Lambda^s(\varepsilon)\cap W_\Lambda^u(\varepsilon)=\Lambda$$

(by [18], Theorem 3.2). Now

$$W_p^u=\bigcup_{n=1}^{\infty} f^{-nt_0}\,W_p^u(\beta)\subset\Lambda$$

and

$$W_p^{cu}=\bigcup_{0\leqq t\leqq t_0} W_{f^tp}^u$$

is dense in Λ (see for instance [3], p. 11–13).

For each $x\in W_p^{cu}$, the set U_x defined by (8) is a neighbourhood of x in M. As $W_x^s(\varepsilon)$, $W_x^u(\varepsilon)$ depend continuously on $x\in\Lambda$, one can find $\delta>0$ independent of x such that $U_x\supset B_x(2\delta)$ for all $x\in W_p^{cu}$ (see [13], Lemma 4.1). In view of this, and the density of W_p^{cu} in Λ,

$$B_\Lambda(\delta)\subset\bigcup\{U_x:\ x\in W_p^{cu}\}.$$

Therefore, if $z\in B_\Lambda(\delta)$ there exist $x\in W_p^{cu}$ and $y\in\bigcup_{|t|\leqq u} f^t\,W_x^u(\varepsilon)\subset W_p^{cu}\subset\Lambda$ such that $z\in W_y^s(\varepsilon)$. When $t\to\infty$ then $d(f^tz,f^ty)\to0$ uniformly in z. Therefore

$$\bigcap_{t\geqq0} f^t\,B_\Lambda(\delta)=\Lambda,$$

which shows that Λ is an attractor and proves (a).

To prove (b), notice that the set

$$V_\gamma=\{x\in\Lambda:\ d(y,\Lambda)>\gamma \text{ for some } y\in W_x^u(\varepsilon)\}$$

is open in Λ since $W_x^u(\varepsilon)$ varies continuously with x. Also V_γ increases as γ decreases and, by part (a) of the present proposition, $\bigcup_{\gamma>0}V_\gamma=\Lambda$. Therefore, by compactness, $V_\gamma=\Lambda$ for some $\gamma>0$.

5.6. **Theorem**[5]. *Let Λ be a C^2 basic hyperbolic set. The following conditions are equivalent:*

(a) *Λ is an attractor;*

(b) *$m(W_\Lambda^s)>0$;*

(c) *$P(F|\Lambda,\varphi^{(u)})=0$.*

[5] Some of the ideas in the proof of this theorem were earlier discovered by J. Franks and R.F. Williams.

Since $W^s_\Lambda = \bigcup_{n=0}^{\infty} f^{-n} W^s_\Lambda(\varepsilon)$, (b) can be replaced by $m(W^s_\Lambda(\varepsilon))>0$ for any small $\varepsilon>0$. In Proposition 4.4 (b) we have seen that (a) $\Rightarrow$ (b) $\Rightarrow$ (c). To complete the proof we assume that Λ is not an attractor, and show that $P(F|\Lambda, \varphi^{(u)})<0$.

Given a small $\varepsilon>0$, choose γ as in Proposition 5.5(b). There is $t>0$ such that if $x\in\Lambda$, $f^t W^u_x(\gamma/4) \supset W^u_{f^t x}(\varepsilon)$. Let $E\subset\Lambda$ be (γ, T)-separated. For any $x\in E$ and $T>0$, there is $y(x, T)\in B_x(\gamma/4, T)$ such that $d(f^{t+T}(y(x, T), \Lambda)>\gamma$ [because $f^T B_x(\gamma/4, T)\supset W^u_x(\gamma/4)$, hence $f^{T+t} B_x(\gamma/4, T)\supset W^u_{f^t x}(\varepsilon)$]. Choose $\delta\in(0, \gamma/4]$ such that $d(f^t z, f^t y)<\gamma/2$ whenever $d(z, y)\leqq\delta$. Then

$$B_{y(x,T)}(\delta, T)\subset B_x(\gamma/2, T)$$

$$f^{T+t} B_{y(x,T)}(\delta, T)\cap B_\Lambda(\gamma/2)=\emptyset,$$

hence

$$B_{y(x,T)}(\delta, T)\cap B_\Lambda(\gamma/2, T+t)=\emptyset.$$

Using the second volume lemma we have thus

$$\begin{aligned} m(B_\Lambda(\gamma/2, T))-m(B_\Lambda(\gamma/2, T+t)) &\geqq \sum_{x\in E} m(B_{y(x,T)}(\delta, T)) \\ &\geqq d(3\gamma/2, \delta)\sum_{x\in E} m(B_x(3\gamma/2, T))\geqq d(3\gamma/2, \delta)\, m(B_\Lambda(\gamma/2, T)) \end{aligned}$$

and therefore

$$m(B_\Lambda(\gamma/2, T+t))\leqq(1-d(3\gamma/2, \delta))\, m(B_\Lambda(\gamma/2, T))$$

so that by Proposition 4.4(a)

$$P(F|\Lambda, \varphi^{(u)})\leqq\frac{1}{t}\log(1-d(3\gamma/2, \delta))<0.$$

5.7. **Corollary.** *Let F be a C^2 Axiom A flow on the compact manifold M.*

(a) *The closures of the basins of the attractors cover M.*

(b) *If Λ is a basic hyperbolic set and $m(\Lambda)>0$, then Λ is a connected component of M and $F|\Lambda$ is an Anosov flow.*

Since F satisfies Axiom A, $M=\bigcup\{W^s_\Lambda$: Λ is a basic hyperbolic set$\}$.

The complement of the basins of the attractors is $\bigcup\{W^s_\Lambda$: Λ is not an attractor$\}$, it has measure 0, and therefore contains no open set. This proves (a).

If Λ is a basic set and $m(\Lambda)>0$, then $m(W^s_\Lambda)>0$ and $m(W^u_\Lambda)>0$ so that Λ is an attractor for both F and the opposite flow F^{-1}. Since Λ is an attractor for F, then $W^u_\Lambda=\Lambda$. Since Λ is an attractor for F^{-1}, then W^u_Λ is open. Therefore Λ is open and closed. Also, Λ is connected since W^{cu}_p is dense in Λ for periodic p (see [3], p. 11–13). This proves (b).

Appendix

Throughout what follows Λ will be a hyperbolic set for the C^r flow (f^t) on the manifold M $(r\geqq 1)$.

We recall that, by assumption, the Riemann metric on M is adapted to f^1 (see Introduction). Notice also that there is $K>0$ such that $\|Tf^n|E\|\leqq K$ for all $n\geqq 0$.

Denote by indices 0, 1, 2 the components in E_x, E_x^s, E_x^u of a vector in T_xM. We shall use in this Appendix a new scalar product in T_xM defined by

$$\langle u, v\rangle_x = \sum_0^2 (u_i, v_i). \tag{A.1}$$

We write $\|u\| = \|u\|_x = (\langle u, u\rangle_x)^{1/2}$. If E' is a subspace of T_xM, $E'(\varepsilon)$ will denote the closed ε-ball centered at the origin of E' for this metric.

A.1. **Convenient Charts.** *For sufficiently small $\varepsilon > 0$ and each $x \in \Lambda$, let us define a C^r chart $\varphi_x: T_xM(\varepsilon) \to M$ such that*[6]

$$\varphi_x(E_x + E_x^s)(\varepsilon) \subset W_x^{cs}, \qquad \varphi_x(E_x + E_x^u)(\varepsilon) \subset W_x^{cu}$$

and the map $F = \varphi_{fx}^{-1} \circ f^1 \circ \varphi_x$ is tangent to $T_x f^1$ at the origin of T_xM.

If $\exp_x^{-1} W_x^{cs}$ is (in a neighbourhood of the origin of T_xM) the graph of a function $\psi': E_x \times E_x^s \to E_x^u$ we set

$$\varphi'(u) = u_0 + u_1 + \big(u_2 + \psi'(u_0, u_1)\big).$$

If $\varphi'^{-1} \exp_x^{-1} W_x^{cu}$ is the graph of $\psi'': E_x \times E_x^u \to E_x^s$ we set

$$\varphi''(u) = u_0 + \big(u_1 + \psi''(u_0, u_2)\big) + u_2.$$

Then $\varphi_x = \exp_x \circ \varphi' \circ \varphi''$ has the desired properties.

If $u \in T_xM(\varepsilon)$, ε sufficiently small, Taylor's formula yields

$$\|F_2(u)\| = \|F_2(u_0 + u_1 + u_2) - F_2(u_0 + u_1)\| \geqq \gamma^{-1} \|u_2\| \tag{A.2}$$

for some $\gamma \in (0, 1)$ independent of x. Similarly if $0 < \omega \leqq 1$, ε can be chosen so small that

$$\|F_0(u) + F_1(u) - F_0(v) - F_1(v)\| \leqq \omega \|F_2(u) - F_2(v)\| \tag{A.3}$$

whenever

$$\|u_0 + u_1 - v_0 - v_1\| \leqq \omega \|u_2 - v_2\| \quad \text{and} \quad u, v \in T_xM(\varepsilon).$$

We define

$$D_x(\varepsilon, n) = \{u \in T_xM: \|F^k u\|_{f^k x} \leqq \varepsilon \text{ for } k = 0, 1, \ldots, n\}. \tag{A.4}$$

Let $u \in D_x(\varepsilon, n)$, then (A.2) yields

$$\|(F^k)_2(u)\| \leqq \gamma^{n-k} \|(F^n)_2(u)\| \leqq \varepsilon \gamma^{n-k}$$

for $k = 0, 1, \ldots, n$. Let $v = u_0 + u_1$. We may assume that $v \in D_x\big((K+1)\varepsilon, n\big)$ and, for ε suitably small, apply (A.3) with $\omega = 1$.

We obtain

$$\|(F^k)_0(u) + (F^k)_1(u) - (F^k)_0(v) - (F^k)_1(v)\| \leqq \|(F^k)_2(u)\| \leqq \varepsilon \gamma^{n-k}$$

and therefore

$$\|F^k(u) - F^k(v)\| \leqq 2\varepsilon \gamma^{n-k} \tag{A.5}$$

for $k = 0, 1, \ldots, n$. (In particular $v \in D_x(3\varepsilon, n)$.)

[6] $W_x^{cs} = \bigcup\{W_y^s: y \in \text{orbit of } x\}$, $W_x^{cu} = \bigcup\{W_y^u: y \in \text{orbit of } x\}$.

A.2. **Lemma.** *Let $r=1$ and $\psi: M \to \mathbb{R}$ be C^1. Given $\theta>0$ there is $\delta>0$ such that: if $x\in\Lambda$, $y\in M$, $n>0$ and $d(f^k y, f^k x)\leqq\delta$ for $k=0,1,\dots,n$, then*

$$\left|\int_0^n \psi(f^t y)\,dt - \int_0^n \psi(f^t x)\,dt\right| < \theta.$$

Because of the C^1 assumptions there exists $C>0$ such that

$$\left|\int_0^1 \psi(f^t p)\,dt - \int_0^1 \psi(f^t q)\,dt\right| \leqq C\,d(p,q).$$

Given ε, one can choose δ so small that $d(x,y)<\delta$ implies $u=\varphi_x^{-1} y\in T_x M(\varepsilon/2)$ for $x\in\Lambda$ and $y\in M$. Define $v=u_0+u_1$ and $z=\varphi_x v\in W_x^{cs}$. We may then assume that $z\in W_{f^s x}^s(\varepsilon)$ with $|s|<C_0\varepsilon$ (C_0 independent of ε). Then

$$d(f^k z, f^{k+s} x) < C_1 \varepsilon \gamma'^k$$

for some $\gamma'\in(0,1)$. If

$$d(f^k x, f^k y)\leqq\delta \quad \text{for } k=0,1,\dots,n,$$

(A.5) yields

$$d(f^k y, f^k z) < C_2 \varepsilon \gamma^{n-k}$$

with $C_2>0$. Thus

$$\begin{aligned}\left|\int_0^n \psi(f^t y) - \int_0^n \psi(f^t x)\right| &\leqq \sum_{k=0}^{n-1}\left|\int_0^1 \psi(f^t f^k y) - \int_0^1 \psi(f^t f^k z)\right| \\ &\quad + \sum_{k=0}^{n-1}\left|\int_0^1 \psi(f^t f^k z) - \int_0^1 \psi(f^t f^{k+s} x)\right| \\ &\quad + \left|\int_0^n \psi(f^{t+s} x) - \int_0^n \psi(f^t x)\right| \\ &\leqq C\left[\frac{C_2\varepsilon}{1-\gamma} + \frac{C_1\varepsilon}{1-\gamma'}\right] + 2C_0\varepsilon\|\psi\|.\end{aligned}$$

A.3. **Lemma.** *Let $\pi: G^q M \to M$ be the Grassmannian bundle of q-dimensional subspaces in TM and $G^q f^t: G^q M \to G^q M$ be the diffeomorphism induced by Tf^t. We assume that $r=2$ so that $G^q f$ is a C^1 flow on $G^q M$. If $q=\dim E_x^u$ and $\Lambda^*=\{E_x^u : x\in\Lambda\}$, then Λ^* is a hyperbolic set for the flow $G^q f$.*

For $x\in\Lambda$, define the manifolds

$$V_x^{*s} = \{E\in\pi^{-1}(W_x^s(\varepsilon)) : d(E-E_x^u)<\varepsilon\}; \qquad V_x^{*u} = TW_x^u(\varepsilon)$$

in $G^q M$. The manifold $\pi^{-1}(x)$ contains E_x^u, and $G^q f^t[\pi^{-1}(x)] = \pi^{-1}(f^t x)$. It is known (and easily seen) that $G^q f^1$ contracts a neighbourhood of E_x^u in $\pi^{-1}(x)$ (see B.1 of [19]). Since f^1 contracts $W_x^s(\varepsilon)$, it follows that, when ε is sufficiently small, Gf^1 contracts V_x^{*s}. Thus TGf^1 is a contraction of $E^{*s}=TV_x^{*s}$. Clearly $G^q f^{-1}$ contracts V^{*u} and therefore $TG^q f^{-1}$ is a contraction of $E^{*u}=TV_x^{*u}$. We have

$$\begin{aligned}\dim E^{*s} + \dim E^{*u} &= \dim W_x^s(\varepsilon) + \dim\pi^{-1}(x) + \dim W_x^u(\varepsilon) \\ &= \dim M - 1 + \dim\pi^{-1}(x) = \dim G^q M - 1\end{aligned}$$

which concludes the proof.

A.4. **Lemma.** *Let $r=2$. Given $\theta>0$, there is $\varepsilon>0$ so that the following holds.*

If $x\in\Lambda$ and $v\in D_x(\varepsilon,n)$ (defined by (A.4)), *let $E_v^*\in G^qM$ be the tangent at φ_x^v to the manifold $\varphi_x(v+E_x^u(\varepsilon))$. Then*

$$e^{-\theta}\leqq\frac{\operatorname{Jac}Tf^n|E_v^*}{\operatorname{Jac}Tf^n|E_x^u}\leqq e^{\theta}\tag{A.6}$$

and

$$e^{-\theta}\leqq\frac{\operatorname{Jac}D_v(F^n|v+E_x^u)}{\operatorname{Jac}D_0(F^n|E_x^u)}\leqq e^{\theta}.\tag{A.7}$$

(D_0, D_v are derivatives in charts.)

The Jacobian in (A.7) is computed with respect to the scalar product (A.1). Clearly the estimate (A.7) differs from (A.6) only by bounded factors and it suffices to prove (A.6). To do this we apply Lemma A.2 to the hyperbolic set Λ^* for the C^1 flow G^qf (cf. Lemma A.3), with the replacement $x\to E_x^u$, $y\to E_v^*$. We have to check that (for sufficiently small ε),

$$d(Tf^kE_x^u,\,Tf^kE_v^*)\leqq\delta$$

for $k=0,1,\ldots,n$. Using the charts φ_x, this results from (A.4) and (A.3). To conclude the proof it suffices to define

$$\psi(E)=\frac{d}{dt}\ln(\operatorname{Jac}Tf^t|E)\Big|_{t=0}$$

and remark that

$$\int_0^n\psi(Tf^tE_v)\,dt=\ln\operatorname{Jac}Tf^n|E_v.$$

A.5. **Proof of the Volume Lemma.** We shall show that for sufficiently small $\varepsilon>0$ there exist $b_\varepsilon, b'_\varepsilon>0$ so that

$$b_\varepsilon\leqq m_x(D_x(\varepsilon,n))\cdot\operatorname{Jac}D_0(F^n|E_x^u)\leqq b'_\varepsilon\tag{A.8}$$

for all $x\in\Lambda$, $n>0$. Here m_x denotes the measure on T_xM associated with the scalar product (A.1). This will prove the volume lemma because the use of the charts φ_x multiplies all distances, measures and Jacobians (see Lemma A.4) by positive factors bounded away from 0 and ∞.

If $v\in(E_x+E_x^s)(\varepsilon)$ define

$$N_v(\varepsilon,n)=\{u\in T_xM\colon u_0+u_1=v\ \text{and}\ (F^k)_2(u)\in E^u_{f^kx}(\varepsilon)\ \text{for}\ k=0,1,\ldots,n\}.$$

If ε is sufficiently small, $\|F^kv\|\leqq(K+1)\,\varepsilon$ for all $k\geqq0$.

Also, using (A.3) with $\omega\leqq1$ and induction on n, we find that $F^nN_v(\varepsilon,n)$ is the graph of a C^1 function $g\colon E^u_{f^nx}(\varepsilon)\to(E_{f^nx}+E^s_{f^nx})((K+2)\,\varepsilon)$ such that $\|Dg\|\leqq\omega$[7]. In particular we obtain the second inclusion of:

$$D_x(\varepsilon,n)\subset\bigcup_v N_v(\varepsilon,n)\subset D_x((K+3)\,\varepsilon,n).$$

[7] This is an easy adaptation of the first part of the proof of Theorem 2.3 of [14].

To prove (A.8) it suffices thus to show that

$$c_\varepsilon \leqq m_x\left(\bigcup_v N_v(\varepsilon, n)\right) \cdot \operatorname{Jac} D_0(F^n | E_x^u) \leqq c'_\varepsilon \tag{A.9}$$

for some $c_\varepsilon, c'_\varepsilon > 0$.

Since $\|Dg\| \leqq \omega$, the measure of $F^n N_v(\varepsilon, n)$ (induced on the manifold by the metric (A.1) on $T_{f^n x}M$) is contained between bounds $d_\varepsilon, d'_\varepsilon > 0$. In view of (A.7), the measure of $N_v(\varepsilon, n)$ multiplied by $\operatorname{Jac} D_0(F^n | E_x^u)$ is contained between $d_\varepsilon e^{-\theta}$ and $d'_\varepsilon e^{\theta}$.

Finally, using Fubini's theorem to integrate over $v \in (E_x + E_x^s)(\varepsilon)$ yields (A.9).

A.6. **Proof of the Second Volume Lemma.** Let $w \in D_x(\varepsilon, n)$, and define

$$D_{xw}(\delta, n) = \{u \in T_x M : \|F^k u - F^k w\|_{f^k x} \leqq \delta \text{ for } 0, 1, \ldots, n\}.$$

It will suffice to show that there is $b_\delta > 0$ so that

$$m_x(D_{xw}(\delta, n)) \cdot \operatorname{Jac} D_0(F^n | E_x^u) \geqq b_\delta \tag{A.10}$$

for all $x \in \Lambda$, $w \in D_x(\varepsilon, n)$, $n > 0$. Furthermore it suffices to prove (A.10) for $\delta < \varepsilon$.

Let

$$\Delta = \{u \in T_{f^n x} M : u_0 = (F^n w)_0,\ u_2 = (F^n w)_2,\ u_1 \in E^s_{f^n x}(3\varepsilon)\}.$$

For each $v \in \Delta$, let $\Gamma_v = \{f^t \cdot \varphi_{f^n x} v : |t| < \alpha\}$ so that $\varphi_{f^n x}^{-1} \Gamma_v$ is a "piece of trajectory" through v. Then, for small ε and suitable α,

$$W = \bigcup_{v \in \Delta} (\varphi_{f^n x}^{-1} \Gamma_v)$$

is a C^2 manifold in $T_{f^n x} M$, which is the graph of a function ψ defined on a subset of $E_{f^n x} + E^s_{f^n x}$ with values in $E^u_{f^n x}$ and such that $\|D\psi\| \leqq \omega$ with $\omega \in (0, 1)$.

We may assume that the domain of ψ contains $E_{f^n x}(2\varepsilon) + E^s_{f^n x}(2\varepsilon)$ and let W' be the graph of the restriction of ψ to $E_{f^n x}(2\varepsilon) + E^s_{f^n x}(2\varepsilon)$. We write

$$W' = \bigcup_{\tau : |\tau| < 2\varepsilon} W'_\tau$$

where $W'_\tau \subset \{u : u_0 = \tau\}$. Let

$$W''_\tau = \{u \in T_x M : F^n u \in W'_\tau \text{ and } (F^k)_1(u) \in E^s_{f^k x}(2\varepsilon) \text{ for } k = 0, 1, \ldots, n\}$$

$$W'' = \bigcup_{\tau : |\tau| < 2\varepsilon} W''_\tau.$$

Applying (A.3) and the argument in A.5 to $F^n W''_\tau$, F^{-1} instead of $N_v(\varepsilon, n)$, F we find that W''_τ is the graph of a function $g_\tau : E_x^s(2\varepsilon) \to (E_x + E_x^u)(2(K+1)\varepsilon)$ such that $\|Dg_\tau\| \leqq \omega$. On the other hand W'' is a union of "pieces of trajectories" which are graphs of maps $E_x \to E_x^s + E_x^u$ with derivative $\leqq \omega$ (for sufficiently small ε). The C^2 manifold W'' is thus the graph of a function ψ'' defined on a subset of $E_x + E_x^s$ with values in E^u and such that $\|D\psi''\| \leqq 1$ (for small ε, hence small ω). The manifold W'' imitates a piece of center-stable manifold through w.

Notice that F^k contracts or expands a "piece of trajectory" in W'' by a factor bounded away from 0 and ∞ (contained between $(K+1)^{-1}$ and $(K+1)$, say). From this and the above properties of W'' it follows that the domain of ψ'' contains a ball B of radius β around $w_0 + w_1$, in $E_x + E_x^s$ for sufficiently small β

and (taking $\beta<\delta/4(K+1)$) we have

$$d(F^k\psi'' v, F^k w)\leqq\frac{\delta}{2}$$

for $k=0,1,\ldots,n$ whenever $v\in B$.

For each $v\in B$, define

$$N_v^*\left(\frac{\delta}{2},n\right)^{\text{l}} =\{u\in T_x M:\ u_0+u_1=v \text{ and } \|(F^k)_2(u)-(F^k)_2(\psi'' v)\|\leqq\frac{\delta}{2} \text{ for } k=0,1,\ldots,n\}.$$

We have

$$\bigcup_{v\in B} N_v^*\left(\frac{\delta}{2},n\right)^{\text{l}}\subset D_{xw}(\delta,n)$$

and proceeding as in A.5 we find that

$$m_x\left(\bigcup_v N_v^*\left(\frac{\delta}{2},m\right)\right)\cdot \operatorname{Jac} D_0(F^n\,|\,E_x^u)\geqq c_\delta$$

for some $c_\delta>0$. This proves (A.10) and therefore the second volume lemma.

References

1. Abramov, L.M.: On the entropy of a flow. A.M.S. Translations **49**, 167–170 (1966)
2. Bowen, R.: Periodic points and measures for Axiom A diffeomorphisms. Trans. A.M.S. **154**, 377–397 (1971)
3. Bowen, R.: Periodic orbits for hyperbolic flows. Amer. J. Math. **94**, 1–30 (1972)
4. Bowen, R.: Symbolic dynamics for hyperbolic flows. Amer. J. Math. **95**, 429–459 (1973)
5. Bowen, R.: Maximizing entropy for a hyperbolic flow. Math. Systems Theory. To appear
6. Bowen, R.: Some systems with unique equilibrium states. Math. Systems Theory. To appear
7. Bowen, R.: Bernoulli equilibrium states for Axiom A diffeomorphisms. Math. Systems Theory. To appear
8. Bowen, R., Walters, P.: Expansive one-parameter flows. J. Diff. Equs. **12**, 180–193 (1972)
9. Bunimovič, L.A.: Imbedding of Bernoulli shifts in certain special flows (in Russian). Uspehi mat. Nauk **28**, 171–172 (1973)
10. Dinaburg, E.I.: On the relations among various entropy characteristics of dynamical systems. Math. USSR Izvestia **5**, 337–378 (1971)
11. Dobrušin, R.L.: Analyticity of correlation functions in one-dimensional classical systems with slowly decreasing potentials. Commun. math. Phys. **32**, 269–289 (1973)
12. Gurevič, B.M.: Some existence conditions for K-decompositions for special flows. Trans. Moscow Math. Soc. **17**, 99–128 (1967)
13. Hirsch, M., Palis, J., Pugh, C., Shub, M.: Neighborhoods of hyperbolic sets. Inventiones Math. **9**, 121–134 (1970)
14. Hirsch, M., Pugh, C.: Stable manifolds and hyperbolic sets. Proc. Symp. in Pure Math. **14**, 133–163 (1970)
15. Hirsch, M., Pugh, C., Shub, M.: Invariant manifolds. To appear
16. Livšic, A.N., Sinai, Ya.G.: On invariant measures compatible with the smooth structure for transitive U-systems. Soviet Math. Dokl. **13**, 1656–1659 (1972)
17. Margulis, G.A.: Certain measures associated with U-flows on compact manifolds. Func. Anal. and its Appl. **4**, 55–67 (1970)
18. Pugh, C., Shub, M.: The Ω-stability theorem for flows. Inventiones math. **11**, 150–158 (1970)
19. Pugh, C.: Ergodicity of Anosov Actions. Inventiones math. **15**, 1–23 (1972)

20. Ratner, M.: Anosov flows with Gibbs measures are also Bernoullian. Israel J. Math. To appear
21. Ruelle, D.: Statistical Mechanics. New York: Benjamin 1969
22. Ruelle, D.: Statistical mechanics of a one-dimensional lattice gas. Commun. Math. Phys. **9**, 267–278 (1968)
23. Ruelle, D.: Statistical mechanics on a compact set with Z^ν action satisfying expansiveness and specification. Trans. A.M.S. To appear
24. Ruelle, D.: A measure associated with Axiom *A* attractors. Amer. J. Math. To appear
25. Sinai, Ya. G.: Markov partitions and *Y*-diffeomorphisms. Func. Anal. and its Appl. **2**, 64–69 (1968)
26. Sinai, Ya. G.: Gibbs measures in ergodic theory. Russ. Math. Surveys **166**, 21–69 (1972)
27. Smale, S.: Differentiable dynamical systems. Bull. A.M.S. **73**, 747–817 (1967)
28. Walters, P.: A variational principle for the pressure of continuous transformations. Amer. J. Math. To appear

R. Bowen
University of California
Department of Mathematics
Berkeley, Calif. 94720, USA

David Ruelle
Institut des Hautes Etudes Scientifiques
35, Route de Chartres
F-91449 Bures-sur-Yvette/France

(Received May 18, 1974)

Reprinted from the AMERICAN MATHEMATICAL MONTHLY
Vol. 82, No. 10, December 1975
pp. 985–992

PERIOD THREE IMPLIES CHAOS

TIEN-YIEN LI AND JAMES A. YORKE

1. Introduction. The way phenomena or processes evolve or change in time is often described by differential equations or difference equations. One of the simplest mathematical situations occurs when the phenomenon can be described by a single number as, for example, when the number of children susceptible to some disease at the beginning of a school year can be estimated purely as a function of the number for the previous year. That is, when the number x_{n+1} at the beginning of the $n+1$st year (or time period) can be written

$$x_{n+1} = F(x_n), \tag{1.1}$$

where F maps an interval J into itself. Of course such a model for the year by year progress of the disease would be very simplistic and would contain only a shadow of the more complicated phenomena. For other phenomena this model might be more accurate. This equation has been used successfully to model the distribution of points of impact on a spinning bit for oil well drilling, as mentioned in [8, 11], knowing this distribution is helpful in predicting uneven wear of the bit. For another example, if a population of insects has discrete generations, the size of the $n+1$st generation will be a function of the nth. A reasonable model would then be a generalized logistic equation

$$x_{n+1} = rx_n[1 - x_n/K]. \tag{1.2}$$

A related model for insect populations was discussed by Utida in [10]. See also Oster *et al* [14, 15].

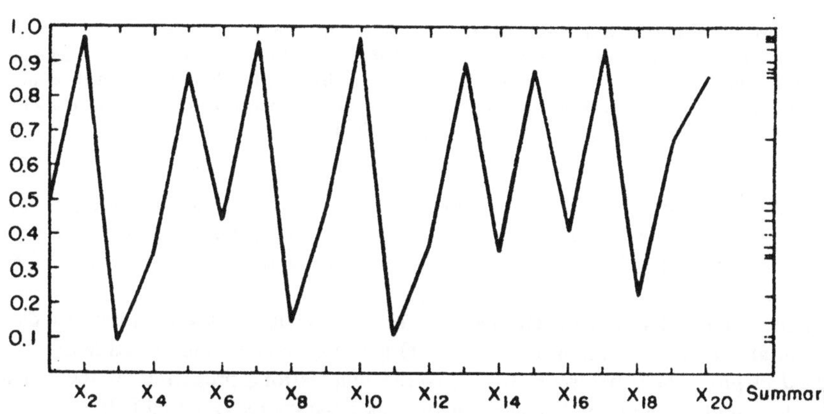

FIG. 1. For $K = 1$, $r = 3.9$, with $x_1 = .5$, the above graph is obtained by iterating Eq. (1.2) 19 times. At right the 20 values are repeated in summary. No value occurs twice. While $x_2 = .975$ and $x_{10} = .973$ are close together, the behavior is not periodic with period 8 since $x_{18} = .222$.

These models are highly simplified, yet even this apparently simple equation (1.2) may have surprisingly complicated dynamic behavior. See Figure 1. We approach these equations with the viewpoint that irregularities and chaotic oscillations of complicated phenomena may sometimes be understood in terms of the simple model, even if that model is not sufficiently sophisticated to allow accurate numerical predictions. Lorenz [1–4] took this point of view in studying turbulent behavior in a fascinating series of papers. He showed that a certain complicated fluid flow could be modelled by such a sequence $x, F(x), F^2(x), \cdots$, which retained some of the chaotic aspects of the original flow. See Figure 2. In this paper we analyze a situation in which the sequence $\{F^n(x)\}$ is non-periodic and might be called "chaotic." Theorem 1 shows that chaotic behavior for (1.1) will result in any situation in which a "population" of size x can grow for two or more successive generations and then having reached an unsustainable height, a population bust follows to the level x or below.

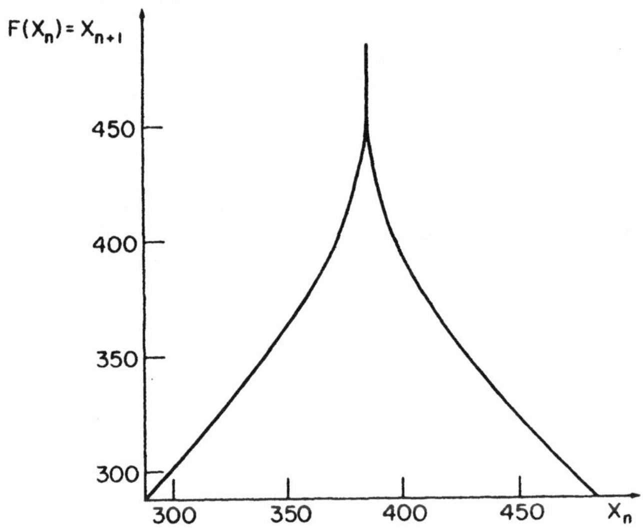

FIG. 2. Lorenz [1] studied the equations for a rotating water-filled vessel which is circularly symmetric about its vertical axis. The vessel is heated near the rim and cooled near its center. When the vessel is annular in shape and the rotation rate high, waves develop and alter their shape irregularly. From a simplified set of equations solved numerically, Lorenz let X_n be in essence the maximum kinetic energy of successive waves. Plotting X_{n+1} against X_n, and connecting the points, the above graph is obtained.

In section 3 we give a well-known simple condition which guarantees that a periodic point is stable and then in section 4 we quote a result applicable when F is like the one in Figure 2. It implies that there is an interval $J_\infty \subset J$ such that for almost every $x \in J$, the set of limit points of the sequence $\{F^n(x)\}$ is J_∞.

A number of questions remain unanswered. For example, is the closure of the periodic points an interval or at least a finite union of intervals? Other questions are mentioned later.

Added in proof. May has recently discovered other strong properties of these maps in his independent study of how the behavior changes as a parameter is varied [17].

2. The main theorem. Let $F: J \to J$. For $x \in J$, $F^0(x)$ denotes x and $F^{n+1}(x)$ denotes $F(F^n(x))$ for $n = 0, 1, \cdots$. We will say p is a **periodic point with period** n if $p \in J$ and $p = F^n(p)$ and $p \neq F^k(p)$ for $1 \leqq k < n$. We say p is **periodic** or is a **periodic point** if p is periodic for some $n \geqq 1$. We say q is **eventually periodic** if for some positive integer m, $p = F^m(q)$ is periodic. Since F need not be one-to-one, there may be points which are eventually periodic but are not periodic. Our objective is to understand the situations in which iterates of a point are very irregular. A special case of our main result says that if there is a periodic point with period 3, then for each integer $n = 1, 2, 3, \cdots$, there is a periodic point with period n. Furthermore, there is an uncountable subset of points x in J which are not even "asymptotically periodic."

THEOREM 1. *Let J be an interval and let $F: J \to J$ be continuous. Assume there is a point $a \in J$ for which the points $b = F(a)$, $c = F^2(a)$ and $d = F^3(a)$, satisfy*

$$d \leqq a < b < c \text{ (or } d \geqq a > b > c).$$

Then

T1: *for every $k = 1, 2, \cdots$ there is a periodic point in J having period k.*

Furthermore,

T2: *there is an uncountable set $S \subset J$ (containing no periodic points), which satisfies the following conditions:*

(A) *For every $p, q \in S$ with $p \neq q$,*

$$\limsup_{n \to \infty} |F^n(p) - F^n(q)| > 0 \tag{2.1}$$

and

$$\liminf_{n \to \infty} |F^n(p) - F^n(q)| = 0. \tag{2.2}$$

(B) *For every $p \in S$ and periodic point $q \in J$,*

$$\limsup_{n \to \infty} |F^n(p) - F^n(q)| > 0.$$

REMARKS. Notice that if there is a periodic point with period 3, then the hypothesis of the theorem will be satisfied.

An example of a function satisfying the hypotheses of the theorem is $F(x) = rx[1 - x/K]$ as in (1.2) for $r \in (3.84, 4]$ with $J = [0, K]$ and for $r > 4$, $F(x) = \max\{0, rx[1 - x/K]\}$ with $J = [0, K]$. See [2] for a detailed description of iterates of this function for $r \in [0, 4)$. The case $r = 4$ is discussed in [6, 7, 12].

While the existence of a point of period 3 implies the existence of one of period 5, the converse is false. (See Appendix 1).

We say $x \in J$ is **asymptotically periodic** if there is a periodic point p for which

$$F^n(x) - F^n(p) \to 0 \quad \text{as} \quad n \to \infty. \tag{2.3}$$

It follows from (B) that the set S contains no asymptotically periodic points. We remark that it is unknown what the infimum of r is for which the equation (1.2) has points which are not asymptotically periodic.

Proof of Theorem 1. The proof of T1 introduces the main ideas for both T1 and T2. We now give the proof of T1 with necessary lemmas and relegate the tedious proof of T2 to Appendix 2.

LEMMA 0. *Let $G: I \to R$ be continuous, where I is an interval. For any compact interval $I_1 \subset G(I)$ there is a compact interval $Q \subset I$ such that $G(Q) = I_1$.*

Proof. Let $I_1 = [G(p), G(q)]$, where $p, q \in I$. If $p < q$, let r be the last point of $[p, q]$ where $G(r) = G(p)$ and let s be the first point after r where $G(s) = G(q)$. Then $G([r, s]) = I_1$. Similar reasoning applies when $p > q$.

LEMMA 1. *Let $F: J \to J$ be continuous and let $\{I_n\}_{n=0}^{\infty}$ be a sequence of compact intervals with $I_n \subset J$ and $I_{n+1} \subset F(I_n)$ for all n. Then there is a sequence of compact intervals Q_n such that $Q_{n+1} \subset Q_n \subset I_0$ and $F^n(Q_n) = I_n$ for $n \geqq 0$. For any $x \in Q = \cap Q_n$ we have $F^n(x) \in I_n$ for all n.*

Proof. Define $Q_0 = I_0$. Then $F^0(Q_0) = I_0$. If Q_{n-1} has been defined so that $F^{n-1}(Q_{n-1}) = I_{n-1}$, then $I_n \subset F(I_{n-1}) = F^n(Q_{n-1})$. By Lemma 0 applied to $G = F^n$ on Q_{n-1} there is a compact interval $Q_n \subset Q_{n-1}$ such that $F^n(Q_n) = I_n$. This completes the induction.

The technique of studying how certain sequences of sets are mapped into or onto each other is often used in studying dynamical systems. For instance, Smale uses this method in his famous "horseshoe example" in which he shows how a homeomorphism on the plane can have infinitely many periodic points [13].

LEMMA 2. *Let $G: J \to R$ be continuous. Let $I \subset J$ be a compact interval. Assume $I \subset G(I)$. Then there is a point $p \in I$ such that $G(p) = p$.*

Proof. Let $I = [\beta_0, \beta_1]$. Choose $\alpha_i (i = 0, 1)$ in I such that $G(\alpha_i) = \beta_i$. It follows $\alpha_0 - G(\alpha_0) \geqq 0$ and $\alpha_1 - G(\alpha_1) \leqq 0$ and so continuity implies $G(\beta) - \beta$ must be 0 for some β in I.

Assume $d \leqq a < b < c$ as in the theorem. The proof for the case $d \geqq a > b > c$ is similar and so is omitted. Write $K = [a, b]$ and $L = [b, c]$.

Proof of T1: Let k be a positive integer. For $k > 1$ let $\{I_n\}$ be the sequence of intervals $I_n = L$ for $n = 0, \cdots, k-2$ and $I_{k-1} = K$, and define I_n to be periodic inductively, $I_{n+k} = I_n$ for $n = 0, 1, 2, \cdots$. If $k = 1$, let $I_n = L$ for all n.

Let Q_n be the sets in the proof of Lemma 1. Then notice that $Q_k \subset Q_0$ and $F^k(Q_k) = Q_0$ and so by Lemma 2, $G = F^k$ has a fixed point p_k in Q_k. It is clear that p_k cannot have period less than k for F; otherwise we would need to have $F^{k-1}(p_k) = b$, contrary to $F^{k+1}(p_k) \in L$. The point p_k is a periodic point of period k for F.

3. Behavior near a periodic point. For some functions F, the asymptotic behavior of iterates of a point can be understood simply by studying the periodic points. For

$$F(x) = ax(1 - x) \tag{3.1}$$

a detailed discussion of the points of period 1 and 2 may be found in [1] for $a \in [0, 4]$ and we now summarize some of those results. For $a \in [0, 4]$, $F: [0, 1] \to [0, 1]$.

For $a \in [0, 1]$, $x = 0$ is the only point of period 1; in fact, for $x \in [0, 1]$, the sequence $F^n(x) \to 0$ as $n \to \infty$.

For $a \in (1, 3]$, there are two points of period 1, namely 0 and $1 - a^{-1}$, and for $x \in (0, 1)$, $F^n(x) \to 1 - a^{-1}$ as $n \to \infty$.

For $a > 3$ there are also two points of period 2 which we may call p and q and of course $F(p) = q$ and $F(q) = p$. For $a \in (3, 1 + \sqrt{6} \approx 3.449)$ and $x \in (0, 1)$, $F^{2n}(x)$ converges to either p or q while $F^{2n+1}(x)$ converges to the other, except for those x for which there is an n for which $F^n(x)$ equals the point $1 - a^{-1}$ of period 1. There are only a countable number of such points so that the behavior of $\{F^n(x)\}$ can be understood by studying the periodic points.

For $a > 1 + \sqrt{6}$, there are 4 points of period 4 and for a slightly greater than $1 + \sqrt{6}$, $F^{4n}(x)$ tends to one of these 4 unless for some n, $F^n(x)$ equals one of the points of period 1 or 2. Therefore we may summarize this situation by saying that each point in [0, 1] is asymptotically periodic.

For those values of a for which each point is asymptotically periodic, it is sufficient to study only the periodic points and their "stability properties." For any function F a point $y \in J$ with period k is said to be **asymptotically stable** if for some interval $I = (y - \delta, y + \delta)$ we have

$$|F^k(x) - y| < |x - y| \quad \text{for all} \quad x \in I.$$

If F is differentiable at the points $y, F(y), \cdots, F^{k-1}(y)$, there is a simple condition that will guarantee this behavior, namely

$$\left| \frac{d}{dx} F^k(x) \right| < 1.$$

By the chain rule

$$\begin{aligned}\frac{d}{dx}F^k(y) &= \frac{d}{dx}F(F^{k-1}(y))\cdot\frac{d}{dx}F^{k-1}(y)\\ &= \frac{d}{dx}F(F^{k-1}(y))\times\frac{d}{dx}F(F^{k-2}(y))\times\cdots\times\frac{d}{dx}F(y)\\ &= \prod_{n=0}^{k-1}\frac{d}{dx}F(y_n),\end{aligned}\tag{3.2}$$

where y_n is the nth iterate, $F^n(y)$. Therefore y is asymptotically stable if

$$\left|\prod_{i=0}^{k-1}\frac{d}{dx}F(y_i)\right| < 1, \qquad \text{where} \qquad y_i = F^i(y).$$

This condition of course guarantees nothing about the limiting behavior of points which do not start "near" the periodic point or one of its iterates. The function in Figure 2 which was studied by Lorenz has the opposite behavior, namely, where the derivative exists we have

$$\left|\frac{d}{dx}F(x)\right| > 1.$$

For such a function every periodic point is "unstable" since for x near a periodic point y of period k, the kth iterate $F^k(x)$ is further from y than x is. To see this, approximate $F^k(x)$ by

$$F^k(y)+\frac{d}{dx}F^k(y)[y-x] = y+\frac{d}{dx}F^k(y)[y-x].$$

Thus for x near y, $|F^k(x)-y|$ is approximately $|x-y|\,|(d/dx)F^k(y)|$. From (3.2) $|(d/dx)F^k(y)|$ is greater than 1. Therefore $F^n(x)$ is further from y than x is.

We do not know when values of a begin to occur for which F in (3.1) has points which are not asymptotically periodic. For $a = 3.627$, F has a periodic point (which is asymptotically stable) of period 6 (approx. $x = .498$). This x is therefore a point of period 3 for F^2 and so Theorem 1 may be applied to F^2. Since F^2 has points which are not asymptotically periodic, the same is true of F.

In order to contrast the situations in this section with other possible situations discussed in the next section, we define the limit set of a point x. The point y is a **limit point** of a sequence $\{x_n\}\subset J$ if there is a subsequence x_{n_i} converging to y. The **limit set** $L(x)$ is defined to be the set of limit points of $\{F^n(x)\}$. If x is asymptotically periodic, then $L(x)$ is the set $\{y, F(y),\cdots,F^{k-1}(y)\}$ for some periodic point y of period k.

4. Statistical properties of $\{F^n(x)\}$. Theorem 1 establishes the irregularity of the behavior of iterates of points. What is also needed is a description of the regular behavior of the sequence $\{F^n(x)\}$ when F is piecewise continuously differentiable (as is Lorenz's function in Figure 2) and

$$\inf_{x\in J_1}\left|\frac{dF}{dx}\right| > 1 \qquad \text{where} \qquad J_1 = \left\{x: \frac{dF}{dx}\text{ exists}\right\}.\tag{4.1}$$

One approach to describing the asymptotic behavior for such functions is to describe $L(x)$, if possible. A second approach, which turns out to be related, is to examine the average behavior of $\{F^n(x)\}$. The fraction of the iterates $\{x,\cdots,F^{N-1}(x)\}$ of x that are in $[a_1,a_2]$ will be denoted by $\phi(x,N,[a_1,a_2])$. The limiting fraction will be denoted

$$\phi(x,[a_1,a_2]) = \lim_{N\to\infty}\phi(x,N,[a_1,a_2])$$

when the limit exists. The subject of ergodic theory, which studies transformation on general spaces, motivates the following definition. We say g is the **density of** x (for F) if the limiting fraction satisfies

$$\phi(x,[a_1,a_2]) = \int_{a_1}^{a_2} g(x)dx \quad \text{for all} \quad a_1, a_2 \in J; \quad a_1 < a_2.$$

The techniques for the study of densities use non-elementary techniques of measure theory and functional analysis, so that we shall only summarize the results. But their value lies in the fact that for certain F almost all $x \in J$ have the same density. Until recently the existence of such densities had not been proved, except for the simplest of functions F. The following result has recently been proved:

THEOREM 2. **[5]**. *Let F: $J \to J$ satisfy the following conditions*:

1) *F is continuous.*
2) *Except at one point $t \in J$, F is twice continuously differentiable.*
3) *F satisfies* (4.1).

Then there exists a function g: $J \to [0,\infty)$, such that for almost all $x \in J$, g is the density of x. Also for almost all $x \in J$, $L(x) = \{y: g(y) > 0\}$ which is an interval. Moreover, the set $J_\infty = \{y: g(y) > 0\}$ is an interval, and $L(x) = J_\infty$ for almost all x.

The proof makes use of results in **[8]**. The problem of computationally finding the density is solved in **[9]**.

A detailed discussion of (3.1) is given in **[16]**, describing how $L(x)$ varies as the parameter a in (3.1) varies between 3.0 and 4.0.

A major question left unsolved is whether (for some nice class of functions F) the existence of a stable periodic point implies that almost every point is asymptotically periodic.

Appendix 1: Period 5 does not imply period 3. In this Appendix we give an example which has a fixed point of period 5 but no fixed point of period 3.

Let F: $[1,5] \to [1,5]$, be defined such that $F(1) = 3$, $F(2) = 5$, $F(3) = 4$, $F(4) = 2$, $F(5) = 1$ and on each interval $[n, n+1]$, $1 \leqq n \leqq 4$, assume F is linear. Then

$$F^3([1,2]) = F^2([3,5]) = F([1,4]) = [2,5].$$

Hence, F^3 has no fixed points in $[1,2]$. Similarly, $F^3([2,3]) = [3,5]$ and $F^3([4,5]) = [1,4]$, so neither of these intervals contains a fixed point of F^3. On the other hand,

$$F^3([3,4]) = F^2([2,4]) = F([2,5]) = [1,5] \supset [3,4].$$

Hence, F^3 must have a fixed point in $[3,4]$. We shall now demonstrate that the fixed point of F^3 is unique and is also a fixed point of F.

Let $p \in [3,4]$ be a fixed point of F^3. Then $F(p) \in [2,4]$. If $F(p) \in [2,3]$, then $F^3(p)$ would be in $[1,2]$ which is impossible since then p could not be a fixed point. Hence $F(p) \in [3,4]$ and $F^2(p) \in [2,4]$. If $F^2(p) \in [2,3]$ we would have $F^3(p) \in [4,5]$, an impossibility. Hence p, $F(p)$, $F^2(p)$ are all in $[3,4]$. On the interval $[3,4]$, F is defined linearly and so $F(x) = 10 - 2x$. It has a fixed point 10/3 and it is easy to see that F^3 has a unique fixed point, which must be 10/3. Hence there is no point of period 3.

Appendix 2. Proof of T2 of Theorem 1. Let $\mathcal{M}$ be the set of sequences $M = \{M_n\}_{n=1}^{\infty}$ of intervals with

(A.1) $$M_n = K \quad \text{or} \quad M_n \subset L, \quad \text{and} \quad F(M_n) \supset M_{n+1}$$

$$\text{if} \quad M_n = K \quad \text{then}$$

(A.2) $$n \text{ is the square of an integer and } M_{n+1}, M_{n+2} \subset L,$$

where $K = [a, b]$ and $L = [b, c]$. Of course if n is the square of an integer, then $n+1$ and $n+2$ are not, so the last requirement in (A.2) is redundant. For $M \in \mathcal{M}$, let $P(M, n)$ denote the number of i's in $\{1, \cdots, n\}$ for which $M_i = K$. For each $r \in (3/4, 1)$ choose $M^r = \{M_n^r\}_{n=1}^{\infty}$ to be a sequence in $\mathcal{M}$ such that

$$\text{(A.3)} \qquad \lim_{n\to\infty} P(M^r, n^2)/n = r.$$

Let $M_0 = \{M^r : r \in (3/4, 1)\} \subset M$. Then $\mathcal{M}_0$ is uncountable since $M^{r_1} \neq M^{r_2}$ for $r_1 \neq r_2$. For each $M^r \in \mathcal{M}_0$, by Lemma 1, there exists a point x_r with $F^n(x_r) \in M_n^r$ for all n. Let $S = \{x_r : r \in (3/4, 1)\}$. Then S is also uncountable. For $x \in S$, let $P(x, n)$ denote the number of i's in $\{1, \cdots, n\}$ for which $F^i(x) \in K$. We can never have $F^k(x_r) = b$, because then x_r would eventually have period 3, contrary to (A.2). Consequently $P(x_r, n) = P(M^r, n)$ for all n, and so

$$\rho(x_r) = \lim_{n\to\infty} P(X_r, n^2) = r$$

for all r. We claim that

(A.4) for $p, q \in S$, with $p \neq q$, there exist infinitely many n's such that $F^n(p) \in K$ and $F^n(q) \in L$ or vice versa.

We may assume $\rho(p) > \rho(q)$. Then $P(p, n) - P(q, n) \to \infty$, and so there must be infintely many n's such that $F^n(p) \in K$ and $F^n(q) \in L$.

Since $F^2(b) = d \leqq a$ and F^2 is continuous, there exists $\delta > 0$ such that $F^2(x) < (b+d)/2$ for all $x \in [b-\delta, b] \subset K$. If $p \in S$ and $F^n(p) \in K$, then (A.2) implies $F^{n+1}(p) \in L$ and $F^{n+2}(p) \in L$. Therefore $F^n(p) < b - \delta$. If $F^n(q) \in L$, then $F^n(q) \geqq b$ so

$$|F^n(p) - F^n(q)| > \delta.$$

By claim (A.4), for any $p, q \in S$, $p \neq q$, it follows

$$\limsup_{n\to\infty} |F^n(p) - F^n(q)| \geqq \delta > 0.$$

Hence (2.1) is proved. This technique may be similarly used to prove (B) is satisfied.

Proof of 2.2. Since $F(b) = c$, $F(c) = d \leqq a$, we may choose intervals $[b^n, c^n]$, $n = 0, 1, 2, \cdots$, such that

(a) $[b, c] = [b^0, c^0] \supset [b^1, c^1] \supset \cdots \supset [b^n, c^n] \supset \cdots$,

(b) $F(x) \in (b^n, c^n)$ for all $x \in (b^{n+1}, c^{n+1})$,

(c) $F(b^{n+1}) = c^n$, $F(c^{n+1}) = b^n$.

Let $A = \bigcap_{n=0}^{\infty} [b^n, c^n]$, $b^* = \inf A$ and $c^* = \sup A$, then $F(b^*) = c^*$ and $F(c^*) = b^*$, because of (c).

In order to prove (2.2) we must be more specific in our choice of the sequences M^r. In addition to our previous requirements on $M \in M$, we will assume that if $M_k = K$ for both $k = n^2$ and $(n+1)^2$ then $M_k = [b^{2n-(2j-1)}, b^*]$ for $k = n^2 + (2j-1)$, $M_k = [c^*, c^{2n-2j}]$ for $k = n^2 + 2j$ where $j = 1, \cdots, n$. For the remaining k's which are not squares of integers, we assume $M_k = L$.

It is easy to check that these requirements are consistent with (A.1) and (A.2), and that we can still choose M^r so as to satisfy (A.3). From the fact that $\rho(x)$ may be thought of as the limit of the fraction of n's for which $F^{n^2}(x) \in K$, it follows that for any $r^*, r \in (3/4, 1)$ there exist infinitely many n such that $M_k^r = M_k^{r^*} = K$ for both $k = n^2$ and $(n+1)^2$. To show (2.2), let $x_r \in S$ and $x_{r^*} \in S$. Since $b^n \to b^*$, $c^n \to c^*$ as $n \to \infty$, for any $\varepsilon > 0$ there exists N with $|b^n - b^*| < \varepsilon/2$, $|c^n - c^*| < \varepsilon/2$ for all $n > N$. Then, for any n with $n > N$ and $M_k^r = M_k^{r^*} = K$ for both $k = n^2$ and $(n+1)^2$, we have

$$F^{n^2+1}(x_r) \in M_k^r = [b^{2n-1}, b^*]$$

with $k = n^2+1$ and $F^{n^2+1}(x_r)$ and $F^{n^2+1}(x_{r^*})$ both belong to $[b^{2n-1}, b^*]$. Therefore, $|F^{n^2+1}(x_r) - F^{n^2+1}(x_{r^*})| < \varepsilon$. Since there are infinitely many n with this property, $\liminf_{n\to\infty}|F^n(x_r) - F^n(x_{r^*})| = 0$. □

REMARK. The theorem can be generalized by assuming that $F\colon J \to R$ without assuming that $F(J) \subset J$ and we leave this proof to the reader. Of course $F(J) \cap J$ would be nonempty since it would contain the points a, b, and c, assuming that b, c, and d, are defined.

Research partially supported by National Science Foundation grant GP-31386X.

References

1. E. N. Lorenz, The problem of deducing the climate from the governing equations, Tellus, 16 (1964) 1–11.
2. ———, Deterministic nonperiodic flows, J. Atmospheric Sci., 20 (1963) 130–141.
3. ———, The mechanics of vacillation, J. Atmospheric Sci., 20 (1963) 448–464.
4. ———, The predictability of hydrodynamic flow, Trans. N.Y. Acad. Sci., Ser. II, 25 (1963) 409–432.
5. T. Y. Li and J. A. Yorke, Ergodic transformations from an interval into itself, (submitted for publication).
6. P. R. Stein and S. M. Ulam, Nonlinear transformation studies on electronic computers, Los Alamos Sci. Lab., Los Alamos, New Mexico, 1963.
7. S. M. Ulam, A Collection of Mathematical Problems, Interscience, New York, 1960, p. 150.
8. A. Lasota and J. A. Yorke, On the existence of invariant measures for piecewise monotonic transformations, Trans. Amer. Math. Soc., 186 (1973) 481–488.
9. T. Y. Li, Finite approximation for the Frobenius-Perron operator—A Solution to Ulam's conjecture, (submitted for publication).
10. Syunro Utida, Population fluctuation, an experimental and theoretical approach, Cold Spring Harbor Symposia on Quantitative Biology, 22 (1957) 139–151.
11. A. Lasota and P. Rusek, Problems of the stability of the motion in the process of rotary drilling with cogged bits, Archium Gornictua, 15 (1970) 205–216 (Polish with Russian and German Summary).
12. N. Metropolis, M. L. Stein and P. R. Stein, On infinite limit sets for transformations on the unit interval, J. Combinatorial Theory Ser. A 15, (1973) 25–44.
13. S. Smale, Differentiable dynamical systems, Bull. A.M.S., 73 (1967) 747–817 (see §1.5).
14. G. Oster and Y. Takahashi, Models for age specific interactions in a periodic environment, Ecology, in press.
15. D. Auslander, G. Oster and C. Huffaker, Dynamics of interacting populations, a preprint.
16. T. Y. Li and James A. Yorke, The "simplest" dynamics system, (to appear).
17. R. M. May, Biological populations obeying difference equations, stable cycles, and chaos, J. Theor. Biol. (to appear).

DEPARTMENT OF MATHEMATICS, UNIVERSITY OF UTAH, SALT LAKE CITY, UT 84112.
INSTITUTE FOR FLUID DYNAMICS AND APPLIED MATHEMATICS, UNIVERSITY OF MARYLAND, COLLEGE PARK, MD 20742.

(Reprinted from Nature, Vol. 261, No. 5560, pp. 459–467, June 10, 1976)

Simple mathematical models with very complicated dynamics

Robert M. May*

First-order difference equations arise in many contexts in the biological, economic and social sciences. Such equations, even though simple and deterministic, can exhibit a surprising array of dynamical behaviour, from stable points, to a bifurcating hierarchy of stable cycles, to apparently random fluctuations. There are consequently many fascinating problems, some concerned with delicate mathematical aspects of the fine structure of the trajectories, and some concerned with the practical implications and applications. This is an interpretive review of them.

THERE are many situations, in many disciplines, which can be described, at least to a crude first approximation, by a simple first-order difference equation. Studies of the dynamical properties of such models usually consist of finding constant equilibrium solutions, and then conducting a linearised analysis to determine their stability with respect to small disturbances: explicitly nonlinear dynamical features are usually not considered.

Recent studies have, however, shown that the very simplest nonlinear difference equations can possess an extraordinarily rich spectrum of dynamical behaviour, from stable points, through cascades of stable cycles, to a regime in which the behaviour (although fully deterministic) is in many respects "chaotic", or indistinguishable from the sample function of a random process.

This review article has several aims.

First, although the main features of these nonlinear phenomena have been discovered and independently rediscovered by several people, I know of no source where all the main results are collected together. I have therefore tried to give such a synoptic account. This is done in a brief and descriptive way, and includes some new material: the detailed mathematical proofs are to be found in the technical literature, to which signposts are given.

Second, I indicate some of the interesting mathematical questions which do not seem to be fully resolved. Some of these problems are of a practical kind, to do with providing a probabilistic description for trajectories which seem random, even though their underlying structure is deterministic. Other problems are of intrinsic mathematical interest, and treat such things as the pathology of the bifurcation structure, or the truly random behaviour, that can arise when the nonlinear function $F(X)$ of equation (1) is not analytical. One aim here is to stimulate research on these questions, particularly on the empirical questions which relate to processing data.

Third, consideration is given to some fields where these notions may find practical application. Such applications range from the abstractly metaphorical (where, for example, the transition from a stable point to "chaos" serves as a metaphor for the onset of turbulence in a fluid), to models for the dynamic behaviour of biological populations (where one can seek to use field or laboratory data to estimate the values of the parameters in the difference equation).

*King's College Research Centre, Cambridge CB2 1ST; on leave from Biology Department, Princeton University, Princeton 08540.

Fourth, there is a very brief review of the literature pertaining to the way this spectrum of behaviour—stable points, stable cycles, chaos—can arise in second or higher order difference equations (that is, two or more dimensions; two or more interacting species), where the onset of chaos usually requires less severe nonlinearities. Differential equations are also surveyed in this light; it seems that a three-dimensional system of first-order ordinary differential equations is required for the manifestation of chaotic behaviour.

The review ends with an evangelical plea for the introduction of these difference equations into elementary mathematics courses, so that students' intuition may be enriched by seeing the wild things that simple nonlinear equations can do.

First-order difference equations

One of the simplest systems an ecologist can study is a seasonally breeding population in which generations do not overlap[1–4]. Many natural populations, particularly among temperate zone insects (including many economically important crop and orchard pests), are of this kind. In this situation, the observational data will usually consist of information about the maximum, or the average, or the total population in each generation. The theoretician seeks to understand how the magnitude of the population in generation $t+1$, X_{t+1}, is related to the magnitude of the population in the preceding generation t, X_t: such a relationship may be expressed in the general form

$$X_{t+1} = F(X_t) \tag{1}$$

The function $F(X)$ will usually be what a biologist calls "density dependent", and a mathematician calls nonlinear; equation (1) is then a first-order, nonlinear difference equation.

Although I shall henceforth adopt the habit of referring to the variable X as "the population", there are countless situations outside population biology where the basic equation (1), applies. There are other examples in biology, as, for example in genetics[5,6] (where the equation describes the change in gene frequency in time) or in epidemiology[7] (with X the fraction of the population infected at time t). Examples in economics include models for the relationship between commodity quantity and price[8], for the theory of business cycles[9], and for the temporal sequences generated by various other economic quantities[10]. The general equation (1) also is germane to the social sciences[11], where it arises, for example, in theories of learning (where X may be the number of bits of information that can be remembered after an interval t), or in the propagation of rumours in variously structured societies (where X is the number of people to have heard the rumour after time t). The imaginative reader will be able to invent other contexts for equation (1).

In many of these contexts, and for biological populations in particular, there is a tendency for the variable X to increase from one generation to the next when it is small, and for it to decrease when it is large. That is, the nonlinear function $F(X)$ often has the following properties: $F(0)=0$; $F(X)$ increases monotonically as X increases through the range $0<X<A$ (with $F(X)$ attaining its maximum value at $X=A$); and $F(X)$ decreases monotonically as X increases beyond $X=A$. Moreover, $F(X)$ will usually contain one or more parameters which "tune" the severity of this nonlinear behaviour; parameters which tune the steepness of the hump in the $F(X)$ curve. These parameters will typically have some biological or economic or sociological significance.

A specific example is afforded by the equation[1,4,12–23]

$$N_{t+1} = N_t(a-bN_t) \qquad (2)$$

This is sometimes called the "logistic" difference equation. In the limit $b=0$, it describes a population growing purely exponentially (for $a>1$); for $b\neq0$, the quadratic nonlinearity produces a growth curve with a hump, the steepness of which is tuned by the parameter a. By writing $X=bN/a$, the equation may be brought into canonical form[1,4,12–23]

$$X_{t+1} = aX_t(1-X_t) \qquad (3)$$

In this form, which is illustrated in Fig. 1, it is arguably the simplest nonlinear difference equation. I shall use equation (3) for most of the numerical examples and illustrations in this article. Although attractive to mathematicians by virtue of its extreme simplicity, in practical applications equation (3) has the disadvantage that it requires X to remain on the interval $0<X<1$; if X ever exceeds unity, subsequent iterations diverge towards $-\infty$ (which means the population becomes extinct). Furthermore, $F(X)$ in equation (3) attains a maximum value of $a/4$ (at $X=\frac{1}{2}$); the equation therefore possesses non-trivial dynamical behaviour only if $a<4$. On the other hand, all trajectories are attracted to $X=0$ if $a<1$. Thus for non-trivial dynamical behaviour we require $1<a<4$; failing this, the population becomes extinct.

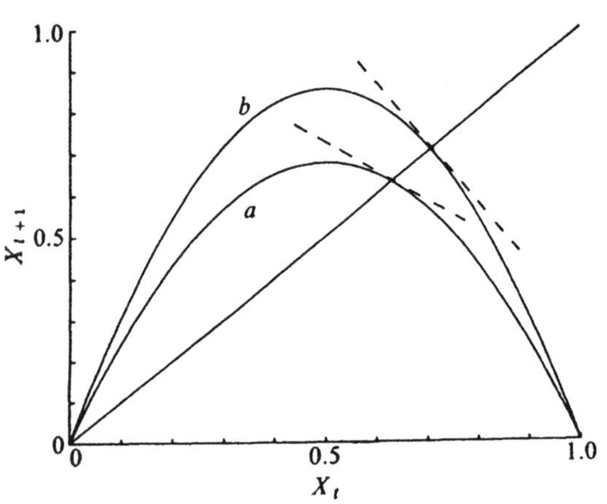

Fig. 1 A typical form for the relationship between X_{t+1} and X_t described by equation (1). The curves are for equation (3), with $a = 2.707$ (*a*); and $a = 3.414$ (*b*). The dashed lines indicate the slope at the "fixed points" where $F(X)$ intersects the 45° line: for the case *a* this slope is less steep than −45° and the fixed point is stable; for *b* the slope is steeper than −45°, and the point is unstable.

Another example, with a more secure provenance in the biological literature[1,23–27], is the equation

$$X_{t+1} = X_t \exp[r(1-X_t)] \qquad (4)$$

This again describes a population with a propensity to simple exponential growth at low densities, and a tendency to decrease at high densities. The steepness of this nonlinear behaviour is tuned by the parameter r. The model is plausible for a single species population which is regulated by an epidemic disease at high density[28]. The function $F(X)$ of equation (4) is slightly more complicated than that of equation (3), but has the compensating advantage that local stability implies global stability[1] for all $X>0$.

The forms (3) and (4) by no means exhaust the list of single-humped functions $F(X)$ for equation (1) which can be culled from the ecological literature. A fairly full such catalogue is given, complete with references, by May and Oster[1]. Other similar mathematical functions are given by Metropolis *et al.*[16]. Yet other forms for $F(X)$ are discussed under the heading of "mathematical curiosities" below.

Dynamic properties of equation (1)

Possible constant, equilibrium values (or "fixed points") of X in equation (1) may be found algebraically by putting $X_{t+1}=X_t=X^*$, and solving the resulting equation

$$X^* = F(X^*) \qquad (5)$$

An equivalent graphical method is to find the points where the curve $F(X)$ that maps X_t into X_{t+1} intersects the 45° line, $X_{t+1}=X_t$, which corresponds to the ideal nirvana of zero population growth; see Fig. 1. For the single-hump curves discussed above, and exemplified by equations (3) and (4), there are two such points: the trivial solution $X=0$, and a non-trivial solution X^* (which for equation (3) is $X^*=1-[1/a]$).

The next question concerns the stability of the equilibrium point X^*. This can be seen[24,25,19–21,1,4] to depend on the slope of the $F(X)$ curve at X^*. This slope, which is illustrated by the dashed lines in Fig. 1, can be designated

$$\lambda^{(1)}(X^*) = [dF/dX]_{X=X^*} \qquad (6)$$

So long as this slope lies between 45° and −45° (that is, $\lambda^{(1)}$ between +1 and −1), making an acute angle with the 45° ZPG line, the equilibrium point X^* will be at least locally stable, attracting all trajectories in its neighbourhood. In equation (3), for example, this slope is $\lambda^{(1)}=2-a$: the equilibrium point is therefore stable, and attracts all trajectories originating in the interval $0<X<1$, if and only if $1<a<3$.

As the relevant parameters are tuned so that the curve $F(X)$ becomes more and more steeply humped, this stability-determining slope at X^* may eventually steepen beyond −45° (that is, $\lambda^{(1)}<-1$), whereupon the equilibrium point X^* is no longer stable.

What happens next? What happens, for example, for $a>3$ in equation (3)?

To answer this question, it is helpful to look at the map which relates the populations at successive intervals 2 generations apart; that is, to look at the function which relates X_{t+2} to X_t. This second iterate of equation (1) can be written

$$X_{t+2} = F[F(X_t)] \qquad (7)$$

or, introducing an obvious piece of notation,

$$X_{t+2} = F^{(2)}(X_t) \qquad (8)$$

The map so derived from equation (3) is illustrated in Figs 2 and 3.

Population values which recur every second generation (that

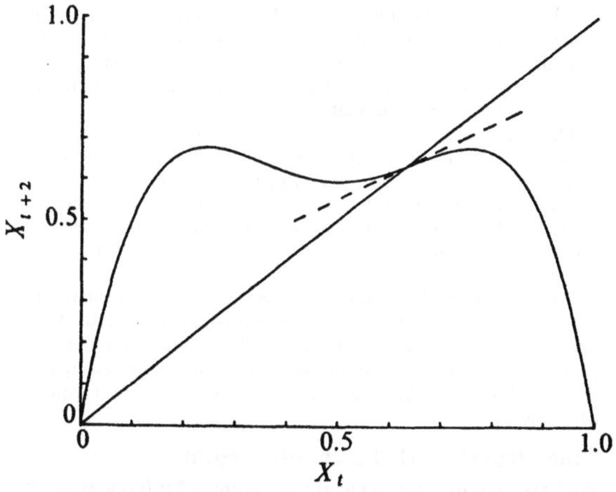

Fig. 2 The map relating X_{t+2} to X_t, obtained by two iterations of equation (3). This figure is for the case (*a*) of Fig. 1, $a = 2.707$: the basic fixed point is stable, and it is the only point at which $F^{(2)}(X)$ intersects the 45° line (where its slope, shown by the dashed line, is less steep than 45°).

is, fixed points with period 2) may now be written as X^*_2, and found either algebraically from

$$X^*_2 = F^{(2)}(X^*_2) \tag{9}$$

or graphically from the intersection between the map $F^{(2)}(X)$ and the 45° line, as shown in Figs 2 and 3. Clearly the equilibrium point X^* of equation (5) is a solution of equation (9); the basic fixed point of period 1 is a degenerate case of a period 2 solution. We now make a simple, but crucial, observation[1]: the slope of the curve $F^{(2)}(X)$ at the point X^*, defined as $\lambda^{(2)}(X^*)$ and illustrated by the dashed lines in Figs 2 and 3, is the square of the corresponding slope of $F(X)$

$$\lambda^{(2)}(X^*) = [\lambda^{(1)}(X^*)]^2 \tag{10}$$

This fact can now be used to make plain what happens when the fixed point X^* becomes unstable. If the slope of $F(X)$ is less than $-45°$ (that is, $|\lambda^{(1)}| < 1$), as illustrated by curve *a* in Fig. 1, then X^* is stable. Also, from equation (10), this implies $0 < \lambda^{(2)} < 1$ corresponding to the slope of $F^{(2)}$ at X^* lying between 0° and 45°, as shown in Fig. 2. As long as the fixed point X^* is stable, it provides the only non-trivial solution to equation (9). On the other hand, when $\lambda^{(1)}$ steepens beyond $-45°$ (that is, $|\lambda^{(1)}| > 1$), as illustrated by curve *b* in Fig 1, X^* becomes unstable. At the same time, from equation (10) this implies $\lambda^{(2)} > 1$, corresponding to the slope of $F^{(2)}$ at X^* steepening beyond 45°, as shown in Fig. 3. As this happens, the curve $F^{(2)}(X)$ must develop a "loop", and two new fixed points of period 2 appear, as illustrated in Fig. 3.

In short, as the nonlinear function $F(X)$ in equation (1) becomes more steeply humped, the basic fixed point X^* may become unstable. At exactly the stage when this occurs, there are born two new and initially stable fixed points of period 2, between which the system alternates in a stable cycle of period 2. The sort of graphical analysis indicated by Figs 1, 2 and 3, along with the equation (10), is all that is needed to establish this generic result[1,4].

As before, the stability of this period 2 cycle depends on the slope of the curve $F^{(2)}(X)$ at the 2 points. (This slope is easily shown to be the same at both points[1,20], and more generally to be the same at all *k* points on a period *k* cycle.) Furthermore, as is clear by imagining the intermediate stages between Figs 2 and 3, this stability-determining slope has the value $\lambda = +1$ at the birth of the 2-point cycle, and then decreases through zero towards $\lambda = -1$ as the hump in $F(X)$ continues to steepen. Beyond this point the period 2 points will in turn become unstable, and bifurcate to give an initially stable cycle of period 4. This in turn gives way to a cycle of period 8, and thence to a hierarchy of bifurcating stable cycles of periods 16, 32, 64, . . ., 2^n. In each case, the way in which a stable cycle of period *k* becomes unstable, simultaneously bifurcating to produce a new and initially stable cycle of period 2*k*, is basically similar to the process just adumbrated for $k=1$. A more full and rigorous account of the material covered so far is in ref. 1.

This "very beautiful bifurcation phenomenon"[22] is depicted in Fig. 4, for the example equation (3). It cannot be too strongly emphasised that the process is generic to most functions $F(X)$ with a hump of tunable steepness. Metropolis *et al.*[18] refer to this hierarchy of cycles of periods 2^n as the harmonics of the fixed point X^*.

Although this process produces an infinite sequence of cycles with periods 2^n ($n \to \infty$), the "window" of parameter values wherein any one cycle is stable progressively diminishes, so that the entire process is a convergent one, being bounded above by some critical parameter value. (This is true for most, but not all, functions $F(X)$: see equation (17) below.) This critical parameter value is a point of accumulation of period 2^n cycles. For equation (3) it is denoted a_c: $a_c = 3.5700\ldots$

Beyond this point of accumulation (for example, for $a > a_c$ in equation (3)) there are an infinite number of fixed points with different periodicities, and an infinite number of different periodic cycles. There are also an uncountable number of initial points X_0 which give totally aperiodic (although bounded) trajectories; no matter how long the time series generated by $F(X)$ is run out, the pattern never repeats. These facts may be established by a variety of methods[1,4,20,22,29]. Such a situation, where an infinite number of different orbits can occur, has been christened "chaotic" by Li and Yorke[20].

As the parameter increases beyond the critical value, at first all these cycles have even periods, with X_t alternating up and down between values above, and values below, the fixed point X^*. Although these cycles may in fact be very complicated (having a non-degenerate period of, say, 5,726 points before repeating), they will seem to the casual observer to be rather like a somewhat "noisy" cycle of period 2. As the parameter value continues to increase, there comes a stage (at $a=3.6786\ldots$ for equation (3)) at which the first odd period cycle appears. At first these odd cycles have very long periods, but as the parameter value continues to increase cycles with smaller and smaller odd periods are picked up, until at last the three-point

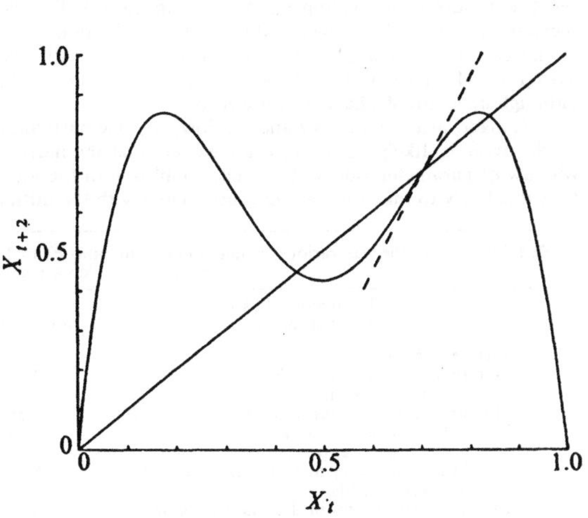

Fig. 3 As for Fig. 2, except that here $a = 3.414$, as in Fig. 1*b*. The basic fixed point is now unstable: the slope of $F^{(2)}(X)$ at this point steepens beyond 45°, leading to the appearance of two new solutions of period 2.

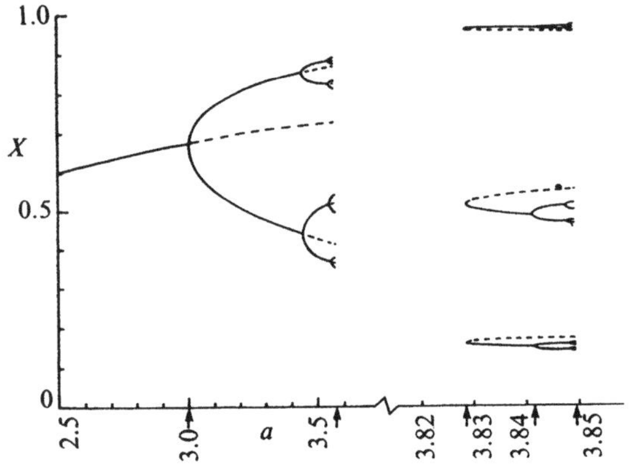

Fig. 4 This figure illustrates some of the stable (———) and unstable (— — — —) fixed points of various periods that can arise by bifurcation processes in equation (1) in general, and equation (3) in particular. To the left, the basic stable fixed point becomes unstable and gives rise by a succession of pitchfork bifurcations to stable harmonics of period 2^n; none of these cycles is stable beyond $a = 3.5700$. To the right, the two period 3 cycles appear by tangent bifurcation: one is initially unstable; the other is initially stable, but becomes unstable and gives way to stable harmonics of period 3×2^n, which have a point of accumulation at $a = 3.8495$. Note the change in scale on the a axis, needed to put both examples on the same figure. There are infinitely many other such windows, based on cycles of higher periods.

cycle appears (at a=3.8284 . . for equation (3)). Beyond this point, there are cycles with every integer period, as well as an uncountable number of asymptotically aperiodic trajectories: Li and Yorke[20] entitle their original proof of this result "Period Three Implies Chaos".

The term "chaos" evokes an image of dynamical trajectories which are indistinguishable from some stochastic process. Numerical simulations[12,15,21,23,25] of the dynamics of equation (3), (4) and other similar equations tend to confirm this impression. But, for smooth and "sensible" functions $F(X)$ such as in equations (3) and (4), the underlying mathematical fact is that for any specified parameter value there is one unique cycle that is stable, and that attracts essentially all initial points[22,29] (see ref. 4, appendix A, for a simple and lucid exposition). That is, there is one cycle that "owns" almost all initial points; the remaining infinite number of other cycles, along with the asymptotically aperiodic trajectories, own a set of points which, although uncountable, have measure zero.

As is made clear by Tables 3 and 4 below, any one particular stable cycle is likely to occupy an extraordinarily narrow window of parameter values. This fact, coupled with the long time it is likely to take for transients associated with the initial conditions to damp out, means that in practice the unique cycle is unlikely to be unmasked, and that a stochastic description of the dynamics is likely to be appropriate, in spite of the underlying deterministic structure. This point is pursued further under the heading "practical applications", below.

The main messages of this section are summarised in Table 1, which sets out the various domains of dynamical behaviour of the equations (3) and (4) as functions of the parameters, a and r respectively, that determine the severity of the nonlinear response. These properties can be understood qualitatively in a graphical way, and are generic to any well behaved $F(X)$ in equation (1).

We now proceed to a more detailed discussion of the mathematical structure of the chaotic regime for analytical functions, and then to the practical problems alluded to above and to a consideration of the behavioural peculiarities exhibited by non-analytical functions (such as those in the two right hand columns of Table 1).

Fine structure of the chaotic regime

We have seen how the original fixed point X^* bifurcates to give harmonics of period 2^n. But how do new cycles of period k arise?

The general process is illustrated in Fig. 5, which shows how period 3 cycles originate. By an obvious extension of the notation introduced in equation (8), populations three generations apart are related by

$$X_{t+3} = F^{(3)}(X_t) \tag{11}$$

If the hump in $F(X)$ is sufficiently steep, the threefold iteration will produce a function $F^{(3)}(X)$ with 4 humps, as shown in Fig. 5 for the $F(X)$ of equation (3). At first (for $a<3.8284$ . . in equation 3) the 45° line intersects this curve only at the single point X^* (and at X=0), as shown by the solid curve in Fig. 5. As the hump in $F(X)$ steepens, the hills and valleys in $F^{(3)}(X)$ become more pronounced, until simultaneously the first two valleys sink and the final hill rises to touch the 45° line, and then to intercept it at 6 new points, as shown by the dashed curve in Fig. 5. These 6 points divide into two distinct three-point cycles. As can be made plausible by imagining the intermediate stages in Fig. 5, it can be shown that the stability-determining slope of $F^{(3)}(X)$ at three of these points has a common value, which is $\lambda^{(3)}=+1$ at their birth, and thereafter steepens beyond $+1$: this period 3 cycle is never stable. The slope of $F^{(3)}(X)$ at the other three points begins at $\lambda^{(3)}=+1$, and then decreases towards zero, resulting in a stable cycle of period 3. As $F(X)$ continues to steepen, the slope $\lambda^{(3)}$ for this intially stable three-point cycle decreases beyond -1; the cycle becomes unstable, and gives rise by the bifurcation process discussed in the previous section to stable cycles of period 6, 12, 24, . . ., 3×2^n. This birth of a stable and unstable pair of period 3 cycles, and the subsequent harmonics which arise as the initially stable cycle becomes unstable, are illustrated to the right of Fig. 4.

Table 1 Summary of the way various "single-hump" functions $F(X)$, from equation (1), behave in the chaotic region, distinguishing the dynamical properties which are generic from those which are not

The function $F(X)$ of equation (1)	$aX(1-X)$	$X \exp[r(1-X)]$	aX; if $X<\frac{1}{2}$ $a(1-X)$; if $X>\frac{1}{2}$	λX; if $X<1$ λX^{1-b}; if $X>1$
Tunable parameter	a	r	a	b
Fixed point becomes unstable	3.0000	2.0000	1.0000*	2.0000
"Chaotic" region begins [point of accumulation of cycles of period 2^n]	3.5700	2.6924	1.0000	2.0000
First odd-period cycle appears	3.6786	2.8332	1.4142	2.6180
Cycle with period 3 appears [and therefore every integer period present]	3.8284	3.1024	1.6180	3.0000
"Chaotic" region ends	4.0000†	∞‡	2.000†	∞‡
Are there stable cycles in the chaotic region?	Yes	Yes	No	No

* Below this a value, $X = 0$ is stable.
† All solutions are attracted to $-\infty$ for a values beyond this.
‡ In practice, as r or b becomes large enough, X will eventually be carried so low as to be effectively zero, thus producing extinction in models of biological populations.

Table 2 Catalogue of the number of periodic points, and of the various cycles (with periods $k = 1$ up to 12), arising from equation (1) with a single-humped function $F(X)$

k	1	2	3	4	5	6	7	8	9	10	11	12
Possible total number of points with period k	2	4	8	16	32	64	128	256	512	1,024	2,048	4,096
Possible total number of points with non-degenerate period k	2	2	6	12	30	54	126	240	504	990	2,046	4,020
Total number of cycles of period k, including those which are degenerate and/or harmonics and/or never locally stable	2	3	4	6	8	14	20	36	60	108	188	352
Total number of non-degenerate cycles (including harmonics and unstable cycles)	2	1	2	3	6	9	18	30	56	99	186	335
Total number of non-degenerate, stable cycles (including harmonics)	1	1	1	2	3	5	9	16	28	51	93	170
Total number of non-degenerate, stable cycles whose basic period is k (that is, excluding harmonics)	1	–	1	1	3	4	9	14	28	48	93	165

There are, therefore, two basic kinds of bifurcation processes[1,4] for first order difference equations. Truly new cycles of period k arise in pairs (one stable, one unstable) as the hills and valleys of higher iterates of $F(X)$ move, respectively, up and down to intercept the 45° line, as typified by Fig. 5. Such cycles are born at the moment when the hills and valleys become tangent to the 45° line, and the initial slope of the curve $F^{(k)}$ at the points is thus $\lambda^{(k)} = +1$: this type of bifurcation may be called[1,4] a tangent bifurcation or a $\lambda = +1$ bifurcation. Conversely, an originally stable cycle of period k may become unstable as $F(X)$ steepens. This happens when the slope of $F^{(k)}$ at these period k points steepens beyond $\lambda^{(k)} = -1$, whereupon a new and initially stable cycle of period $2k$ is born in the way typified by Figs 2 and 3. This type of bifurcation may be called a pitchfork bifurcation (borrowing an image from the left hand side of Fig. 4) or a $\lambda = -1$ bifurcation[1,4].

Putting all this together, we conclude that as the parameters in $F(X)$ are varied the fundamental, stable dynamical units are cycles of basic period k, which arise by tangent bifurcation, along with their associated cascade of harmonics of periods $k2^n$, which arise by pitchfork bifurcation. On this basis, the constant equilibrium solution X^* and the subsequent hierarchy of stable cycles of periods 2^n is merely a special case, albeit a conspicuously important one (namely $k=1$), of a general phenomenon. In addition, remember[1,4,22,29] that for sensible, analytical functions (such as, for example, those in equations (3) and (4)) there is a unique stable cycle for each value of the parameter in $F(X)$. The entire range of parameter values ($1 < a < 4$ in equation (3), $0 < r$ in equation (4)) may thus be regarded as made up of infinitely many windows of parameter

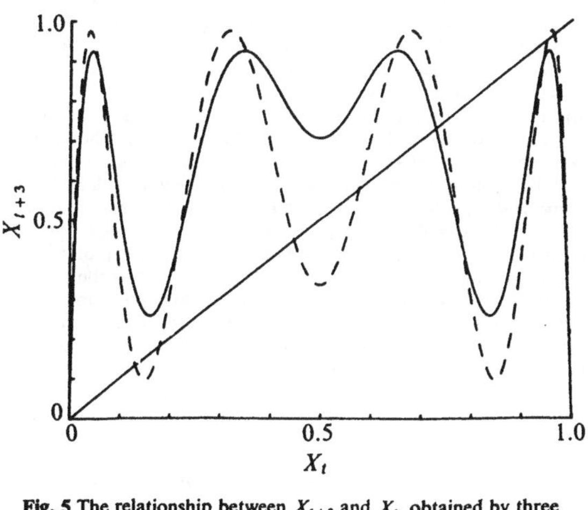

Fig. 5 The relationship between X_{t+3} and X_t, obtained by three iterations of equation (3). The solid curve is for $a = 3.7$, and only intersects the 45° line once. As a increases, the hills and valleys become more pronounced. The dashed curve is for $a = 3.9$, and six new period 3 points have appeared (arranged as two cycles, each of period 3).

values—some large, some unimaginably small—each corresponding to a single one of these basic dynamical units. Tables 3 and 4, below, illustrate this notion. These windows are divided from each other by points (the points of accumulation of the harmonics of period $k2^n$) at which the system is truly chaotic, with no attractive cycle: although there are infinitely many such special parameter values, they have measure zero on the interval of all values.

How are these various cycles arranged along the interval of relevant parameter values? This question has to my knowledge been answered independently by at least 6 groups of people, who have seen the problem in the context of combinatorial theory[16,30], numerical analysis[13,14], population biology[1], and dynamical systems theory[22,31] (broadly defined).

A simple-minded approach (which has the advantage of requiring little technical apparatus, and the disadvantage of being rather clumsy) consists of first answering the question, how many period k points can there be? That is, how many distinct solutions can there be to the equation

$$X^*_k = F^{(k)}(X^*_k)? \qquad (12)$$

If the function $F(X)$ is sufficiently steeply humped, as it will be once the parameter values are sufficiently large, each successive iteration doubles the number of humps, so that $F^{(k)}(X)$ has 2^{k-1} humps. For large enough parameter values, all these hills and valleys will intersect the 45° line, producing 2^k fixed points of period k. These are listed for $k \leq 12$ in the top row of Table 2. Such a list includes degenerate points of period k, whose period is a submultiple of k; in particular, the two period 1 points ($X=0$ and X^*) are degenerate solutions of equation (12) for all k. By working from left to right across Table 2, these degenerate points can be subtracted out, to leave the total number of non-degenerate points of basic period k, as listed in the second row of Table 2. More sophisticated ways of arriving at this result are given elsewhere[13,14,16,22,30,31].

For example, there eventually are $2^6 = 64$ points with period 6. These include the two points of period 1, the period 2 "harmonic" cycle, and the stable and unstable pair of triplets of points with period 3, for a total of 10 points whose basic period is a submultiple of 6; this leaves 54 points whose basic period is 6.

The 2^k period k points are arranged into various cycles of period k, or submultiples thereof, which appear in succession by either tangent or pitchfork bifurcation as the parameters in $F(X)$ are varied. The third row in Table 2 catalogues the total number of distinct cycles of period k which so appear. In the fourth row[14], the degenerate cycles are subtracted out, to give the total number of non-degenerate cycles of period k: these numbers must equal those of the second row divided by k. This fourth row includes the (stable) harmonics which arise by pitchfork bifurcation, and the pairs of stable–unstable cycles arising by tangent bifurcation. By subtracting out the cycles which are unstable from birth, the total number of possible stable cycles is given in row five; these figures can also be obtained by less pedestrian methods[13,16,30]. Finally we may subtract out the stable cycles which arise by pitchfork bifurcation, as harmonics of some simpler cycle, to arrive at the final

Table 3 A catalogue of the stable cycles (with basic periods up to 6) for the equation $X_{t+1} = aX_t(1-X_t)$

Period of basic cycle	a value at which: Basic cycle first appears	Basic cycle becomes unstable	Subsequent cascade of "harmonics" with period $k2^n$ all become unstable	Width of the range of a values over which the basic cycle, or one of its harmonics, is attractive
1	1.0000	3.0000	3.5700	2.5700
3	3.8284	3.8415	3.8495	0.0211
4	3.9601	3.9608	3.9612	0.0011
5(*a*)	3.7382	3.7411	3.7430	0.0048
5(*b*)	3.9056	3.9061	3.9065	0.0009
5(*c*)	3.99026	3.99030	3.99032	0.00006
6(*a*)	3.6265	3.6304	3.6327	0.0062
6(*b*)	3.937516	3.937596	3.937649	0.000133
6(*c*)	3.977760	3.977784	3.977800	0.000040
6(*d*)	3.997583	3.997585	3.997586	0.000003

row in Table 2, which lists the number of stable cycles whose basic period is k.

Returning to the example of period 6, we have already noted the five degenerate cycles whose periods are submultiples of 6. The remaining 54 points are parcelled out into one cycle of period 6 which arises as the harmonic of the only stable three-point cycle, and four distinct pairs of period 6 cycles (that is, four initially stable ones and four unstable ones) which arise by successive tangent bifurcations. Thus, reading from the foot of the column for period 6 in Table 2, we get the numbers 4, 5, 9, 14.

Using various labelling tricks, or techniques from combinatorial theory, it is also possible to give a generic list of the order in which the various cycles appear[1,13,16,22]. For example, the basic stable cycles of periods 3, 5, 6 (of which there are respectively 1, 3, 4) must appear in the order 6, 5, 3, 5, 6, 6, 5, 6: compare Tables 3 and 4. Metropolis *et al.*[16] give the explicit such generic list for all cycles of period $k \leq 11$.

As a corollary it follows that, given the most recent cycle to appear, it is possible (at least in principle) to catalogue all the cycles which have appeared up to this point. An especially elegant way of doing this is given by Smale and Williams[22], who show, for example, that when the stable cycle of period 3 first originates, the total number of other points with periods k, N_k, which have appeared by this stage satisfy the Fibonacci series, N_k=2, 4, 5, 8, 12, 19, 30, 48, 77, 124, 200, 323 for $k=1, 2, \ldots, 12$: this is to be contrasted with the total number of points of period k which will eventually appear (the top row of Table 2) as $F(X)$ continues to steepen.

Such catalogues of the total number of fixed points, and of their order of appearance, are relatively easy to construct. For any particular function $F(X)$, the numerical task of finding the windows of parameter values wherein any one cycle or its harmonics is stable is, in contrast, relatively tedious and inelegant. Before giving such results, two critical parameter values of special significance should be mentioned.

Hoppensteadt and Hyman[21] have given a simple graphical method for locating the parameter value in the chaotic regime at which the first odd period cycle appears. Their analytic recipe is as follows. Let α be the parameter which tunes the steepness of $F(X)$ (for example, $\alpha = a$ for equation (3), $\alpha = r$ for equation (4)), $X^*(\alpha)$ be the fixed point of period 1 (the non-trivial solution of equation (5)), and $X_{max}(\alpha)$ the maximum value attainable from iterations of equation (1) (that is, the value of $F(X)$ at its hump or stationary point). The first odd period cycle appears for that value of α which satisfies[21,31]

$$X^*(\alpha) = F^{(2)}(X_{max}(\alpha)) \tag{13}$$

As mentioned above, another critical value is that where the period 3 cycle first appears. This parameter value may be found numerically from the solutions of the third iterate of equation (1): for equation (3) it is[14] $a = 1 + \sqrt{8}$.

Myrberg[13] (for all $k \leq 10$) and Metropolis *et al.*[16]. (for all $k \leq 7$) have given numerical information about the stable cycles in equation (3). They do not give the windows of parameter values, but only the single value at which a given cycle is maximally stable; that is, the value of a for which the stability-determining slope of $F^{(k)}(X)$ is zero, $\lambda^{(k)}=0$. Since the slope of the k-times iterated map $F^{(k)}$ at any point on a period k cycle is simply equal to the product of the slopes of $F(X)$ at each of the points X^*_k on this cycle[1,8,20], the requirement $\lambda^{(k)}=0$ implies that $X = A$ (the stationary point of $F(X)$, where $\lambda^{(1)}=0$) is one of the periodic points in question, which considerably simplifies the numerical calculations.

For each basic cycle of period k (as catalogued in the last row of Table 2), it is more interesting to know the parameter values at which: (1) the cycle first appears (by tangent bifurcation); (2) the basic cycle becomes unstable (giving rise by successive pitchfork bifurcations to a cascade of harmonics of periods $k2^n$); (3) all the harmonics become unstable (the point of accumulation of the period $k2^n$ cycles). Tables 3 and 4 extend the work of May and Oster[1], to give this numerical information for equations (3) and (4), respectively. (The points of accumulation are not ground out mindlessly, but are calculated by a rapidly convergent iterative procedure, see ref. 1, appendix A.) Some of these results have also been obtained by Gumowski and Mira[32].

Practical problems

Referring to the paradigmatic example of equation (3), we can now see that the parameter interval $1 < a < 4$ is made up of a one-dimensional mosaic of infinitely many windows of a-values, in each of which a unique cycle of period k, or one of its harmonics, attracts essentially all initial points. Of these windows, that for $1 < a < 3.5700\ldots$ corresponding to $k=1$ and its harmonics is by far the widest and most conspicuous. Beyond the first point of accumulation, it can be seen from Table 3 that these windows are narrow, even for cycles of quite low periods, and the windows rapidly become very tiny as k increases.

As a result, there develops a dichotomy between the underlying mathematical behaviour (which is exactly determinable) and the "commonsense" conclusions that one would draw from numerical simulations. If the parameter a is held constant at one value in the chaotic region, and equation (3) iterated for an arbitrarily large number of generations, a density plot of the observed values of X_t on the interval 0 to 1 will settle into k equal spikes (more precisely, delta functions) corresponding to the k points on the stable cycle appropriate to this a-value. But for most a-values this cycle will have a fairly large period, and moreover it will typically take many thousands of generations before the transients associated with the initial conditions are damped out: thus the density plot produced by numerical simulations usually looks like a sample of points taken from some continuous distribution.

An especially interesting set of numerical computations are due to Hoppensteadt (personal communication) who has combined many iterations to produce a density plot of X_t for each one of a sequence of a-values, gradually increasing from 3.5700 . . to 4. These results are displayed as a movie. As can be expected from Table 3, some of the more conspicuous cycles do show up as sets of delta functions: the 3-cycle and its first few harmonics; the first 5-cycle; the first 6-cycle. But for most values of a the density plot looks like the sample function of a random process. This is particularly true in the neighbourhood of the a-value where the first odd cycle appears ($a=3.6786\ldots$), and again in the neighbourhood of $a=4$: this is not surprising, because each of these locations is a point of accumulation of points of accumulation. Despite the underlying discontinuous changes in the periodicities of the stable cycles, the observed density pattern tends to vary smoothly. For example, as a increases toward the value at which the 3-cycle appears, the density plot tends to concentrate around three points, and it smoothly diffuses away from these three points after the 3-cycle and all its harmonics become unstable.

I think the most interesting mathematical problem lies in designing a way to construct some approximate and "effectively continuous" density spectrum, despite the fact that the exact density function is determinable and is always a set of delta functions. Perhaps such techniques have already been developed in ergodic theory[33] (which lies at the foundations of statistical mechanics), as for example in the use of "coarse-grained observers". I do not know.

Such an effectively stochastic description of the dynamical properties of equation (4) for large r has been provided[28], albeit by tactical tricks peculiar to that equation rather than by any general method. As r increases beyond about 3, the trajectories generated by this equation are, to an increasingly good approximation, almost periodic with period $(1/r)\exp(r-1)$.

The opinion I am airing in this section is that although the exquisite fine structure of the chaotic regime is mathematically fascinating, it is irrelevant for most practical purposes. What seems called for is some effectively stochastic description of the deterministic dynamics. Whereas the various statements about the different cycles and their order of appearance can be made in generic fashion, such stochastic description of the actual dynamics will be quite different for different $F(X)$: witness the difference between the behaviour of equation (4), which for large r is almost periodic "outbreaks" spaced many generations apart, versus the behaviour of equation (3), which for $a\to4$ is not very different from a series of Bernoulli coin flips.

Mathematical curiosities

As discussed above, the essential reason for the existence of a succession of stable cycles throughout the "chaotic" regime is that as each new pair of cycles is born by tangent bifurcation (see Fig. 5), one of them is at first stable, by virtue of the way the smoothly rounded hills and valleys intercept the 45° line. For analytical functions $F(X)$, the only parameter values for which the density plot or "invariant measure" is continuous and truly ergodic are at the points of accumulation of harmonics, which divide one stable cycle from the next. Such exceptional parameter values have found applications, for example, in the use of equation (3) with $a=4$ as a random number generator[34,35]: it has a continuous density function proportional to $[X(1-X)]^{-\frac{1}{2}}$ in the interval $0<X<1$.

Non-analytical functions $F(X)$ in which the hump is in fact a spike provide an interesting special case. Here we may imagine spikey hills and valleys moving to intercept the 45° line in Fig. 5, and it may be that both the cycles born by tangent bifurcation are unstable from the outset (one having $\lambda^{(k)}>1$, the other $\lambda^{(k)}<-1$), for all $k>1$. There are then no stable cycles in the chaotic regime, which is therefore literally chaotic with a continuous and truly ergodic density distribution function.

One simple example is provided by

$$\begin{aligned} X_{t+1} &= aX_t;\ \text{if } X_t<\tfrac{1}{2} \\ X_{t+1} &= a(1-X_t);\ \text{if } X_t>\tfrac{1}{2} \end{aligned} \quad (14)$$

defined on the interval $0<X<1$. For $0<a<1$, all trajectories are attracted to $X=0$; for $1<a<2$, there are infinitely many periodic orbits, along with an uncountable number of aperiodic trajectories, none of which are locally stable. The first odd period cycle appears at $a=\sqrt{2}$, and all integer periods are represented beyond $a=(1+\sqrt{5})/2$. Kac[36] has given a careful discussion of the case $a=2$. Another example, this time with an extensive biological pedigree[1-8], is the equation

$$\begin{aligned} X_{t+1} &= \lambda X_t;\ \text{if } X_t<1 \\ X_{t+1} &= \lambda X_t^{1-b};\ \text{if } X_t>1 \end{aligned} \quad (15)$$

If $\lambda>1$ this possesses a globally stable equilibrium point for $b<2$. For $b>2$ there is again true chaos, with no stable cycles: the first odd cycle appears at $b=(3+\sqrt{5})/2$, and all integer periods are present beyond $b=3$. The dynamical properties of equations (14) and (15) are summarised to the right of Table 2.

The absence of analyticity is a necessary, but not a sufficient, condition for truly random behaviour[31]. Consider, for example,

$$\begin{aligned} X_{t+1} &= (a/2)X_t;\ \text{if } X_t<\tfrac{1}{2} \\ X_{t+1} &= aX_t(1-X_t)\ ;\ \text{if } X_t>\tfrac{1}{2} \end{aligned} \quad (16)$$

This is the parabola of equation (3) and Fig. 1, but with the left hand half of $F(X)$ flattened into a straight line. This equation does possess windows of a values, each with its own stable cycle, as described generically above. The stability-determining slopes $\lambda^{(k)}$ vary, however, discontinuously with the parameter a, and the widths of the simpler stable regions are narrower than for equation (3): the fixed point becomes unstable at $a=3$; the point of accumulation of the subsequent harmonics is at $a=3.27\ldots$; the first odd cycle appears at $a=3.44\ldots$; the 3-point cycle at $a=3.67\ldots$ (compare the first column in Table 1).

These eccentricities of behaviour manifested by non-analytical functions may be of interest for exploring formal questions in ergodic theory. I think, however, that they have no relevance to models in the biological and social sciences, where functions such as $F(X)$ should be analytical. This view is elaborated elsewhere[37].

As a final curiosity, consider the equation

$$X_{t+1} = \lambda X_t[1+X_t]^{-\beta} \quad (17)$$

Table 4 Catalogue of the stable cycles (with basic periods up to 6) for the equation $X_{t+1} = X_t\exp[r(1-X_t)]$

Period of basic cycle	*r* value at which: Basic cycle first appears	Basic cycle becomes unstable	Subsequent cascade of "harmonics" with period $k2^n$ all become unstable	Width of the range of *r* values over with the basic cycle, or one of its harmonics, is attractive
1	0.0000	2.0000	2.6924	2.6924
3	3.1024	3.1596	3.1957	0.0933
4	3.5855	3.6043	3.6153	0.0298
5(*a*)	2.9161	2.9222	2.9256	0.0095
5(*b*)	3.3632	3.3664	3.3682	0.0050
5(*c*)	3.9206	3.9295	3.9347	0.0141
6(*a*)	2.7714	2.7761	2.7789	0.0075
6(*b*)	3.4558	3.4563	3.4567	0.0009
6(*c*)	3.7736	3.7745	3.7750	0.0014
6(*d*)	4.1797	4.1848	4.1880	0.0083

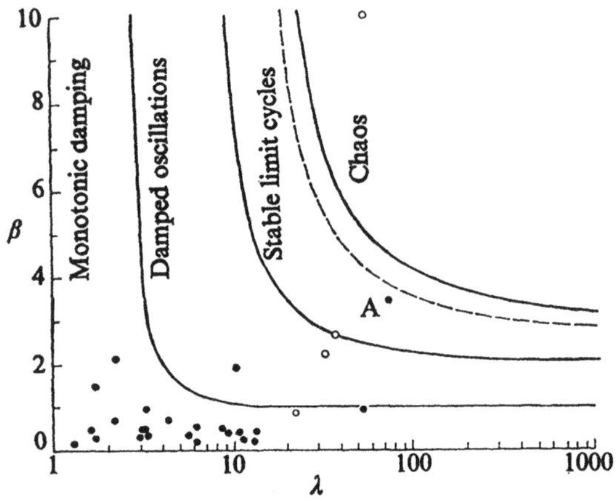

Fig. 6 The solid lines demarcate the stability domains for the density dependence parameter, β, and the population growth rate, λ, in equation (17); the dashed line shows where 2-point cycles give way to higher cycles of period 2^n. The solid circles come from analyses of life table data on field populations, and the open circles from laboratory populations (from ref. 3, after ref. 39).

This has been used to fit a considerable amount of data on insect populations[38,39]. Its stability behaviour, as a function of the two parameters λ and β, is illustrated in Fig. 6. Notice that for $\lambda<7.39$ . . there is a globally stable equilibrium point for all β; for $7.39 \ldots < \lambda < 12.50$ . . this fixed point becomes unstable for sufficiently large β, bifurcating to a stable 2-point cycle which is the solution for all larger β; as λ increases through the range $12.50 \ldots < \lambda < 14.77$ . . various other harmonics of period 2^n appear in turn. The hierarchy of bifurcating cycles of period 2^n is thus truncated, and the point of accumulation and subsequent regime of chaos is not achieved (even for arbitrarily large β) until $\lambda > 14.77 \ldots$

Applications

The fact that the simple and deterministic equation (1) can possess dynamical trajectories which look like some sort of random noise has disturbing practical implications. It means, for example, that apparently erratic fluctuations in the census data for an animal population need not necessarily betoken either the vagaries of an unpredictable environment or sampling errors: they may simply derive from a rigidly deterministic population growth relationship such as equation (1). This point is discussed more fully and carefully elsewhere[1].

Alternatively, it may be observed that in the chaotic regime arbitrarily close initial conditions can lead to trajectories which, after a sufficiently long time, diverge widely. This means that, even if we have a simple model in which all the parameters are determined exactly, long term prediction is nevertheless impossible. In a meteorological context, Lorenz[15] has called this general phenomenon the "butterfly effect": even if the atmosphere could be described by a deterministic model in which all parameters were known, the fluttering of a butterfly's wings could alter the initial conditions, and thus (in the chaotic regime) alter the long term prediction.

Fluid turbulence provides a classic example where, as a parameter (the Reynolds number) is tuned in a set of deterministic equations (the Navier–Stokes equations), the motion can undergo an abrupt transition from some stable configuration (for example, laminar flow) into an apparently stochastic, chaotic regime. Various models, based on the Navier–Stokes differential equations, have been proposed as mathematical metaphors for this process[15,40,41]. In a recent review of the theory of turbulence, Martin[42] has observed that the one-dimensional difference equation (1) may be useful in this context. Compared with the earlier models[15,40,41], it has the disadvantage of being even more abstractly metaphorical, and the advantage of having a spectrum of dynamical behaviour which is more richly complicated yet more amenable to analytical investigation.

A more down-to-earth application is possible in the use of equation (1) to fit data[1,2,3,38,39,43] on biological populations with discrete, non-overlapping generations, as is the case for many temperate zone arthropods. Figure 6 shows the parameter values λ and β that are estimated[39] for 24 natural populations and 4 laboratory populations when equation (17) is fitted to the available data. The figure also shows the theoretical stability domains: a stable point; its stable harmonics (stable cycles of period 2^n); chaos. The natural populations tend to have stable equilibrium point behaviour. The laboratory populations tend to show oscillatory or chaotic behaviour; their behaviour may be exaggeratedly nonlinear because of the absence, in a laboratory setting, of many natural mortality factors. It is perhaps suggestive that the most oscillatory natural population (labelled *A* in Fig. 6) is the Colorado potato beetle, whose present relationship with its host plant lacks an evolutionary pedigree. These remarks are only tentative, and must be treated with caution for several reasons. Two of the main caveats are that there are technical difficulties in selecting and reducing the data, and that there are no single species populations in the natural world: to obtain a one-dimensional difference equation by replacing a population's interactions with its biological and physical environment by passive parameters (such as λ and β) may do great violence to the reality.

Some of the many other areas where these ideas have found applications were alluded to in the second section, above[5–11]. One aim of this review article is to provoke applications in yet other fields.

Related phenomena in higher dimensions

Pairs of coupled, first-order difference equations (equivalent to a single second-order equation) have been investigated in several contexts[4,44–46], particularly in the study of temperate zone arthropod prey–predator systems[2–4,23,47]. In these two-dimensional systems, the complications in the dynamical behaviour are further compounded by such facts as: (1) even for analytical functions, there can be truly chaotic behaviour (as for equations (14) and (15)), corresponding to so-called "strange attractors"; and (2) two or more different stable states (for example, a stable point and a stable cycle of period 3) can occur together for the same parameter values[4]. In addition, the manifestation of these phenomena usually requires less severe nonlinearities (less steeply humped $F(X)$) than for the one-dimensional case.

Similar systems of first-order ordinary differential equations, or two coupled first-order differential equations, have much simpler dynamical behaviour, made up of stable and unstable points and limit cycles[48]. This is basically because in continuous two-dimensional systems the inside and outside of closed curves can be distinguished; dynamic trajectories cannot cross each other. The situation becomes qualitatively more complicated, and in many ways analogous to first-order difference equations, when one moves to systems of three or more coupled, first-order ordinary differential equations (that is, three-dimensional systems of ordinary differential equations). Scanlon (personal communication) has argued that chaotic behaviour and "strange attractors", that is solutions which are neither points nor periodic orbits[48], are typical of such systems. Some well studied examples arise in models for reaction–diffusion systems in chemistry and biology[49], and in the models of Lorenz[15] (three dimensions) and Ruelle and Takens[40] (four dimensions) referred to above. The analysis of these systems is, by virtue of their higher dimensionality, much less transparent than for equation (1).

An explicit and rather surprising example of a system which

has recently been studied from this viewpoint is the ordinary differential equations used in ecology to describe competing species. For one or two species these systems are very tame: dynamic trajectories will converge on some stable equilibrium point (which may represent coexistence, or one or both species becoming extinct). As Smale[50] has recently shown, however, for 3 or more species these general equations can, in a certain reasonable and well-defined sense, be compatible with any dynamical behaviour. Smale's[50] discussion is generic and abstract: a specific study of the very peculiar dynamics which can be exhibited by the familiar Lotka-Volterra equations once there are 3 competitors is given by May and Leonard[51].

Conclusion

In spite of the practical problems which remain to be solved, the ideas developed in this review have obvious applications in many areas.

The most important applications, however, may be pedagogical.

The elegant body of mathematical theory pertaining to linear systems (Fourier analysis, orthogonal functions, and so on), and its successful application to many fundamentally linear problems in the physical sciences, tends to dominate even moderately advanced University courses in mathematics and theoretical physics. The mathematical intuition so developed ill equips the student to confront the bizarre behaviour exhibited by the simplest of discrete nonlinear systems, such as equation (3). Yet such nonlinear systems are surely the rule, not the exception, outside the physical sciences.

I would therefore urge that people be introduced to, say, equation (3) early in their mathematical education. This equation can be studied phenomenologically by iterating it on a calculator, or even by hand. Its study does not involve as much conceptual sophistication as does elementary calculus. Such study would greatly enrich the student's intuition about nonlinear systems.

Not only in research, but also in the everyday world of politics and economics, we would all be better off if more people realised that simple nonlinear systems do not necessarily possess simple dynamical properties.

I have received much help from F. C. Hoppensteadt, H. E. Huppert, A. I. Mees, C. J. Preston, S. Smale, J. A. Yorke, and particularly from G. F. Oster. This work was supported in part by the NSF.

1 May, R. M., and Oster, G. F., *Am. Nat.*, **110** (in the press).
2 Varley, G. C., Gradwell, G. R., and Hassell, M. P., *Insect Population Ecology* (Blackwell, Oxford, 1973).
3 May, R. M. (ed.), *Theoretical Ecology: Principles and Applications* (Blackwell, Oxford, 1976).
4 Guckenheimer, J., Oster, G. F., and Ipaktchi, A., *Theor. Pop. Biol.* (in the press).
5 Oster, G. F., Ipaktchi, A., and Rocklin, I., *Theor. Pop. Biol.* (in the press).
6 Asmussen, M. A., and Feldman, M. W., *J. theor. Biol.* (in the press).
7 Hoppensteadt, F. C., *Mathematical Theories of Populations: Demographics, Genetics and Epidemics* (SIAM, Philadelphia, 1975).
8 Samuelson, P. A., *Foundations of Economic Analysis* (Harvard University Press, Cambridge, Massachusetts, 1947).
9 Goodwin, R. E., *Econometrica*, **19**, 1–17 (1951).
10 Baumol, W. J., *Economic Dynamics*, 3rd ed. (Macmillan, New York, 1970).
11 See, for example, Kemeny, J., and Snell, J. L., *Mathematical Models in the Social Sciences* (MIT Press, Cambridge, Massachusetts, 1972).
12 Chaundy, T. W., and Phillips, E., *Q. Jl Math. Oxford*, **7**, 74–80 (1936).
13 Myrberg, P. J., *Ann. Akad. Sc. Fennicae*, *A*, I, No. 336/3 (1963).
14 Myrberg, P. J., *Ann. Akad. Sc. Fennicae*, *A*, I, No. 259 (1958).
15 Lorenz, E. N., *J. Atmos. Sci.*, **20**, 130–141 (1963); *Tellus*, **16**, 1–11 (1964).
16 Metropolis, N., Stein, M. L., and Stein, P. R., *J. Combinatorial Theory*, **15(A)**, 25–44 (1973).
17 Maynard Smith, J., *Mathematical Ideas in Biology* (Cambridge University Press, Cambridge, 1968).
18 Krebs, C. J., *Ecology* (Harper and Row, New York, 1972).
19 May, R. M., *Am. Nat.*, **107**, 46–57 (1972).
20 Li, T-Y., and Yorke, J. A., *Am. Math. Monthly*, **82**, 985–992 (1975).
21 Hoppensteadt, F. C., and Hyman, J. M. (Courant Institute, New York University: preprint, 1975).
22 Smale, S., and Williams, R. (Department of Mathematics, Berkeley: preprint, 1976).
23 May, R. M., *Science*, **186**, 645–647 (1974).
24 Moran, P. A. P., *Biometrics*, **6**, 250–258 (1950).
25 Ricker, W. E., *J. Fish. Res. Bd. Can.*, **11**, 559–623 (1954).
26 Cook, L. M., *Nature*, **207**, 316 (1965).
27 Macfadyen, A., *Animal Ecology: Aims and Methods* (Pitman, London, 1963).
28 May, R. M., *J. theor. Biol.*, **51**, 511–524 (1975).
29 Guckenheimer, J., *Proc. AMS Symposia in Pure Math.*, *XIV*, 95–124 (1970).
30 Gilbert, E. N., and Riordan, J., *Illinois J. Math.*, **5**, 657–667 (1961).
31 Preston, C. J. (King's College, Cambridge: preprint, 1976).
32 Gumowski, I., and Mira, C., *C. r. hebd. Séanc. Acad. Sci., Paris*, **281a**, 45–48 (1975); **282a**, 219–222 (1976).
33 Layzer, D., *Sci. Am.*, **233**(6), 56–69 (1975).
34 Ulam, S. M., *Proc. Int. Congr. Math.1950, Cambridge, Mass.; Vol. II*, pp. 264–273 (AMS, Providence R.I., 1950).
35 Ulam, S. M., and von Neumann, J., *Bull. Am. math. Soc.* (abstr.), **53**, 1120 (1947).
36 Kac, M., *Ann. Math.*, **47**, 33–49 (1946).
37 May, R. M., *Science*, **181**, 1074 (1973).
38 Hassell, M. P., *J. Anim. Ecol.*, **44**, 283–296 (1974).
39 Hassell, M. P., Lawton, J. H., and May, R. M., *J. Anim. Ecol.* (in the press).
40 Ruelle, D., and Takens, F., *Comm. math. Phys.*, **20**, 167–192 (1971).
41 Landau, L. D., and Lifshitz, E. M., *Fluid Mechanics* (Pergamon, London, 1959).
42 Martin, P. C., *Proc. Int. Conf. on Statistical Physics, 1975, Budapest* (Hungarian Acad. Sci., Budapest, in the press).
43 Southwood, T. R. E., in *Insects, Science and Society* (edit. by Pimentel, D.), 151–199 (Academic, New York, 1975).
44 Metropolis, N., Stein, M. L., and Stein, P. R., *Numer. Math.*, **10**, 1–19 (1967).
45 Gumowski, I., and Mira, C., *Automatica*, **5**, 303–317 (1969).
46 Stein, P. R., and Ulam, S. M., *Rosprawy Mat.*, **39**, 1–66 (1964).
47 Beddington, J. R., Free, C. A., and Lawton, J. H., *Nature*, **255**, 58–60 (1975).
48 Hirsch, M. W., and Smale, S., *Differential Equations, Dynamical Systems and Linear Algebra* (Academic, New York, 1974).
49 Kolata, G. B., *Science*, **189**, 984–985 (1975).
50 Smale. S. (Department of Mathematics, Berkeley: preprint, 1976).
51 May, R. M., and Leonard, W. J., *SIAM J. Appl. Math.*, **29**, 243–253 (1975).

Commun. math. Phys. 50, 69—77 (1976)

Communications in
Mathematical
Physics

A Two-dimensional Mapping with a Strange Attractor

M. Hénon

Observatoire de Nice, F-06300 Nice, France

Abstract. Lorenz (1963) has investigated a system of three first-order differential equations, whose solutions tend toward a "strange attractor". We show that the same properties can be observed in a simple mapping of the plane defined by: $x_{i+1}=y_i+1-ax_i^2$, $y_{i+1}=bx_i$. Numerical experiments are carried out for $a=1.4$, $b=0.3$. Depending on the initial point (x_0, y_0), the sequence of points obtained by iteration of the mapping either diverges to infinity or tends to a strange attractor, which appears to be the product of a one-dimensional manifold by a Cantor set.

1. Introduction

Lorenz (1963) proposed and studied a remarkable system of three coupled first-order differential equations, representing a flow in three-dimensional space. The divergence of the flow has a constant negative value, so that any volume shrinks exponentially with time. Moreover, there exists a bounded region R into which every trajectory becomes eventually trapped. Therefore, all trajectories tend to a set of measure zero, called *attractor*. In some cases the attractor is simply a point (which is then a stable equilibrium point) or a closed curve (known as a limit cycle). But in other cases the attractor has a much more complex structure; it appears to be locally the product of a two-dimensional manifold by a Cantor set. This is known as a *strange attractor*. Inside the attractor, trajectories wander in an apparently erratic manner. Moreover, they are highly sensitive to initial conditions. These phenomena are of interest for weather prediction (Lorenz, 1963) and more generally for turbulence theory (Ruelle and Takens, 1971; Ruelle, 1975). Further numerical explorations of the Lorenz system have been made by Lanford (1975) and Pomeau (1976).

We present her a "reductionist" approach in which we try to find a model problem which is as simple as possible, yet exhibits the same essential properties as the Lorenz system. Our aim is (i) to make the numerical exploration faster and more accurate, so that solutions can be followed for a longer time, more

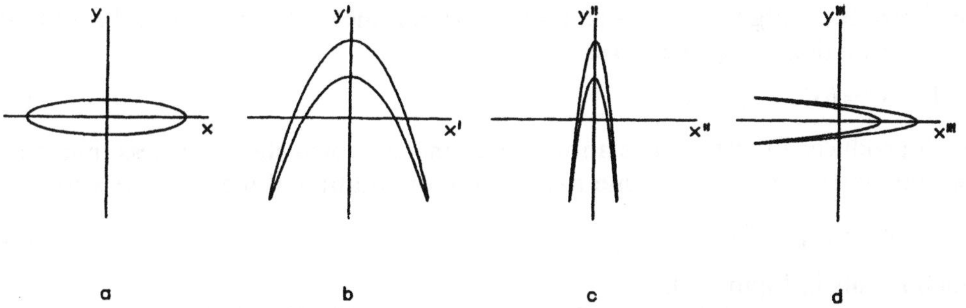

Fig. 1. The initial area a is mapped by T' into b, then by T'' into c, and finally by T''' into d

detailed explorations can be conducted, etc.; (ii) to provide a model which might lend itself more easily to mathematical analysis.

2. The Model

Our first step is classical (Birkhoff, 1917) and consists in considering not the whole trajectories in the three-dimensional space, but only their successive intersections with a two-dimensional *surface of section S*. We define a mapping T of S into itself as follows: given a point A of S, we follow the trajectory which originates from A until it intersects S again; this new point is $T(A)$. This mapping is sometimes called a *Poincaré map*. A trajectory is thus replaced by an infinite set of points in S, obtained by repeated application of the mapping T. The essential properties of the trajectory are reflected into corresponding properties of the set of points. We have thus formally reduced the problem to the study of a two-dimensional mapping.

At this point, however, the only advantage really gained is in clarity of presentation of the results; the actual computation of the mapping still requires the numerical integration of the differential equations. Now comes the second and decisive step: we forget about the differential system, and we define a mapping T by explicit equations, giving directly $T(A)$ when A is known. This of course simplifies the computation drastically. The new mapping T does not any more correspond to the Lorenz system; however, by choosing it carefully we may hope to retain the essential properties which we wish to study. Past experience in the measure-preserving case (see Hénon, 1969, and references therein) has shown indeed that the same features are found in dynamical systems defined by differential equations and in mappings defined as such.

The third step consists in specifying T. Here we have been inspired by the numerical results of Pomeau (1976) on the Lorenz system, which show clearly how a volume is stretched in one direction, and at the same time folded over itself, in the course of one revolution. This folding effect has been also described by Ruelle (1975, Fig. 5 and 6). We simulate it by the following chain of three mappings of the (x, y) plane onto itself. Consider a region elongated along the x axis (Fig. 1a). We begin the folding by

$$T' : x' = x, \quad y' = y + 1 - ax^2, \tag{1}$$

which produces Figure 1b; a is an adjustable parameter. We complete the folding by a contraction along the x axis:

$$T'' : x'' = bx', \quad y'' = y', \tag{2}$$

which produces Figure 1c; b is another parameter, which should be less than 1 in absolute value. Finally we come back to the orientation along the x axis by

$$T'' : x''' = y'', \quad y''' = x'', \tag{3}$$

which results in Figure 1d.

Our mapping will be defined as the product $T = T'''T''T'$. We write now (x_i, y_i) for (x, y) and (x_{i+1}, y_{i+1}) for (x''', y''') (as a reminder that the mapping will be iterated) and we have

$$T: x_{i+1} = y_i + 1 - ax_i^2, \quad y_{i+1} = bx_i. \tag{4}$$

This mapping has some interesting properties. Its Jacobian is a constant:

$$\frac{\partial(x_{i+1}, y_{i+1})}{\partial(x_i, y_i)} = -b. \tag{5}$$

The geometrical interpretation is quite simple: T' preserves areas; T''' also preserves areas but reverses the sign; and T'' contracts areas, multiplying them by the constant factor b. The property (5) is welcome because it is the natural counterpart of the constant negative divergence in the Lorenz system.

A polynomial mapping satisfying (5) is known as an *entire Cremona transformation*, and the inverse mapping is also given by polynomials (Engel, 1955, 1958). Indeed we have here

$$T^{-1}: x_i = b^{-1} y_{i+1}, \quad y_i = x_{i+1} - 1 + ab^{-2} y_{i+1}^2. \tag{6}$$

Thus T is a one-to-one mapping of the plane onto itself. This is also a welcome property, because it is the natural counterpart of the fact that in the Lorenz system there is a unique trajectory through any given point.

The selection of T could have been approached in a different way, by looking for the "simplest" non-trivial mapping. It is natural then to consider polynomial mappings of progressively increasing order. Linear mappings are trivial, so the polynomials must be at least of degree 2. The most general quadratic mapping is

$$\begin{aligned} x_{i+1} &= f + ax_i + by_i + cx_i^2 + dx_iy_i + ey_i^2, \\ y_{i+1} &= f' + a'x_i + b'y_i + c'x_i^2 + d'x_iy_i + e'y_i^2 \end{aligned} \tag{7}$$

and depends on 12 parameters. But if we impose the condition that the Jacobian is a constant, some relations must be satisfied by these parameters. We can further reduce the number of parameters by an appropriate linear change of coordinates in the plane. In this way, by a slight extension of the results of Engel (1958), it can be shown that the general form (7) is reducible to a "canonical form" depending on two parameters only. This is a generalization of our earlier result (Hénon, 1969) that a quadratic *area-preserving* mapping can be brought into a form depending on one parameter only. The canonical form can be written in several different ways; and one of them turns out to be identical with (4), which is thus reached by an entirely different road! The mapping (4), which was initially constructed in empirical fashion, is in fact the most general quadratic mapping with constant Jacobian.

One difference with the Lorenz problem is that the successive points obtained by repeated application of T do not always converge towards an attractor; sometimes they "escape" to infinity. This is because the quadratic term in (4) dominates when the distance from the origin becomes large. However, for particular values of a and b it is still possible to prove the existence of a bounded "trapping region" R, from which the points can never escape once they have entered it (see below Section 5).

T has two invariant points, given by

$$x=(2a)^{-1}[-(1-b)\pm\sqrt{(1-b)^2+4a}], \quad y=bx. \tag{8}$$

These points are real for

$$a>a_0=(1-b)^2/4. \tag{9}$$

When this is the case, one of the points is always linearly unstable, while the other is unstable for

$$a>a_1=3(1-b)^2/4. \tag{10}$$

3. Choice of Parameters

We select now particular values of a and b for a numerical study. b should be small enough for the folding described by Figure 1 to occur really, yet not too small if one wishes to observe the fine structure of the attractor. The value $b=0.3$ was found to be adequate. A good value of a was found only after some experimenting. For $a<a_0$ or $a>a_3$, where a_0 is given by (9) and a_3 is of the order of 1.55 for $b=0.3$, the points always escape to infinity: apparently there exists no attractor in these cases. For $a_0<a<a_3$, depending on the initial values (x_0, y_0), either the points escape to infinity or they converge towards an attractor, which appears to be unique for a given value of a. We concentrate now on this attractor. For $a_0<a<a_1$, where a_1 is given by (10), the attractor is the stable invariant point. When a is increased over a_1, at first the attractor is still simple and consists of a periodic set of p points. (An equivalent attractor in the Lorenz problem would be a limit cycle intersecting the surface of section p times). The value of p increases through successive "bifurcations" as a increases, and appears to tend to infinity as a approaches a critical value a_2, of the order of 1.06 for $b=0.3$. For $a_2<a<a_3$, the attractor is no more simple, and the behaviour of the points becomes erratic. This is the case in which we are interested. We adopt the following values:

$$a=1.4, \quad b=0.3. \tag{11}$$

4. Numerical Results

Figure 2 shows the result of plotting 10000 successive points, obtained by iteration of T, starting from the arbitrarily chosen initial point $x_0=0$, $y_0=0$; the vertical scale is enlarged to give a better picture. Figure 3 shows the result of 10000

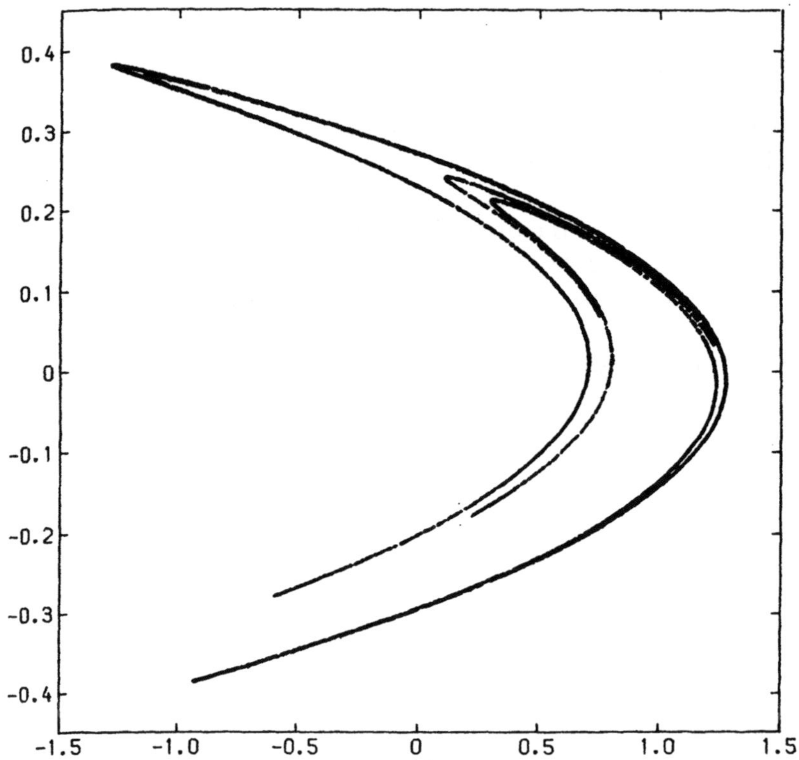

Fig. 2. 10000 successive points obtained by iteration of the mapping T starting from $x_0=0$, $y_0=0$

iterations of T again, starting from a different point: $x_0=0.63135448$, $y_0=0.18940634$ (this choice will be explained below). The two figures are seen to be almost identical. This suggests strongly that what we see in both figures is essentially the attractor itself: the successive points quickly approach the attractor and soon become undistinguishable from it at the scale of the figure. This is confirmed if one looks at the first few points on Figure 2. The initial point at $x_0=0$, $y_0=0$ and the first iterate at $x_1=1$, $y_1=0$ are clearly visible; the second iterate is still visible at $x_2=-0.4$, $y_2=0.3$; the third iterate can barely be distinguished at $x_3=1.076$, $y_3=-0.12$; and the fourth iterate at $x_4=-0.7408864$, $y_4=0.3228$ is already lost inside the attractor at the resolution of Figure 2. The following points then wander over the attractor in an apparently erratic manner.

One of the two unstable invariant points has the coordinates, given by (8):

$$x=0.63135448\ldots, \qquad y=0.18940634\ldots . \tag{12}$$

This point appears to belong to the attractor. The two eigenvalues λ_1, λ_2 and the slopes p_1, p_2 of the corresponding eigenvectors are

$$\lambda_1=0.15594632\ldots, \qquad p_1=1.92373886\ldots,$$
$$\lambda_2=-1.92373886\ldots, \quad p_2=-0.15594632\ldots . \tag{13}$$

The instability is due to λ_2. The corresponding slope p_2 appears to be tangent to the "curves" in Figure 2.

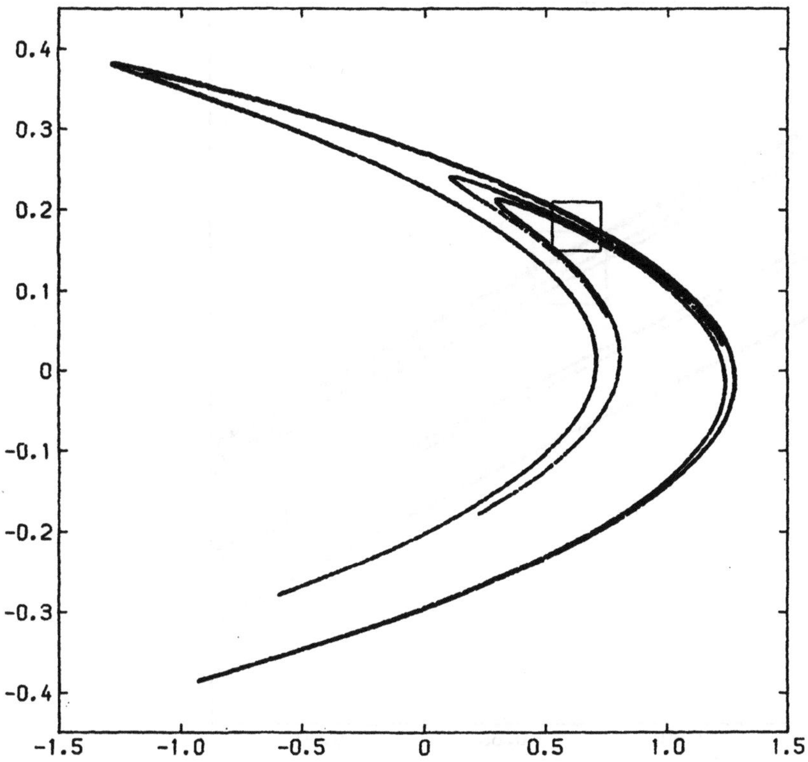

Fig. 3. Same as Figure 2, but starting from $x_0=0.63135448$, $y_0=0.18940634$

These properties allow us to eliminate the "transient regime" in which the points approach the attractor, and which is not of much interest: we simply start from the close vicinity of the unstable point (12), by rounding off its coordinates to 8 digits. This is done in Figure 3 and in the following figures. The points quickly move away along the line of slope p_2 since $|\lambda_2|$ is appreciably larger than 1.

The attractor appears to consist of a number of more or less parallel "curves"; the points tend to distribute themselves densely over these curves. The few gaps that can still be seen on Figures 2 and 3 have probably no particular significance. Their locations are not the same on the two figures. They are simply due to statistical fluctuations in the quasi-random distribution of points, and they would disappear if more moints were plotted. Thus, the *longitudinal structure* of the attractor (along the curves) appears to be simple, each curve being essentially a one-dimensional manifold.

The *transversal structure* (across the curves) appears to be entirely different, and much more complex. Already on Figures 2 and 3 a number of curves can be seen, and the visible thickness of some of them suggests that they have in fact an underlying structure. Figure 4 is a magnified view of the small square of Figure 3: some of the previous "curves" are indeed resolved now into two or more components. The number n of iterations has been increased to 10^5, in order to have a sufficient number of points in the small region examined. The small square in Figure 4 is again magnified to produce Figure 5, with n increased to 10^6: again the

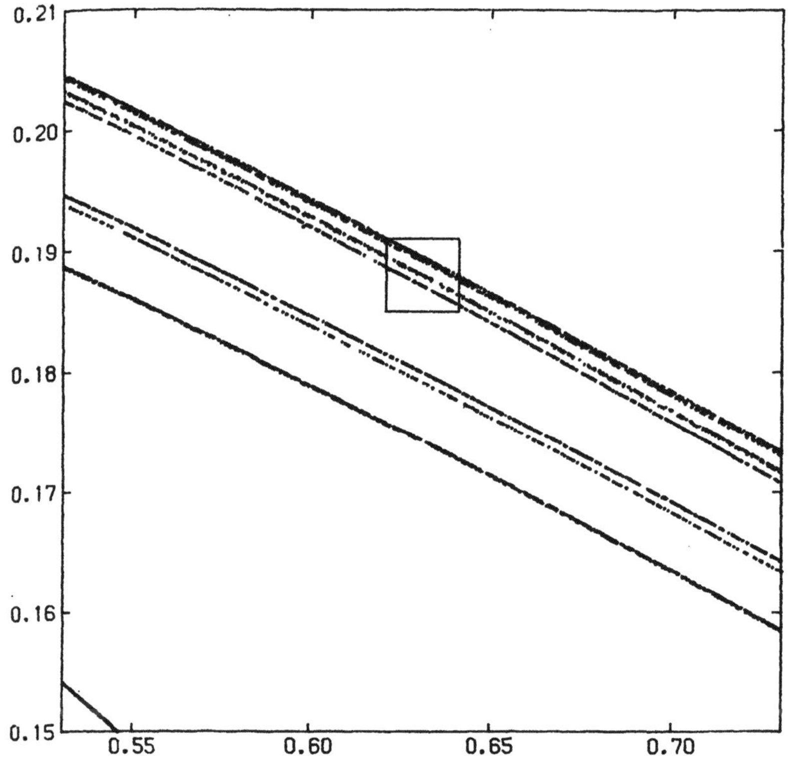

Fig. 4. Enlargement of the squared region of Figure 3. The number of computed points is increased to $n = 10^5$

number of visible "curves" increases. One more enlargement results in Fig. 6, with $n = 5 \times 10^6$: the points become sparse but new curves can still easily be traced.

These figures strongly suggest that the process of multiplication of "curves" will continue indefinitely, and that each apparent "curve" is in fact made of an infinity of quasi-parallel curves. Moreover, Figures 4 to 6 indicate the existence of a hierarchical sequence of "levels", the structure being practically identical at each level save for a scale factor. This is exactly the structure of a Cantor set.

The frames of Figures 4 to 6 have been chosen so as to contain the invariant point (12). This point appears to lie on the upper boundary of the attractor. Surprisingly, its presence is completely invisible on the figures; this contrasts with the area-preserving case, were stable and unstable invariant points play a very conspicuous role (see for instance Hénon, 1969). On the other hand, the presence of the invariant point explains, locally at least, the hierarchy of similar structures: at each application of the mapping, the scale of the transversal structure is multiplied by λ_1 given by (13). At the same time, the points spread out along the curves, as dictated by the value of λ_2.

5. A Trapping Region

The fact that even after 5×10^6 iterations the points have not diverged to infinity suggests that there is a region of the plane from which the points cannot escape.

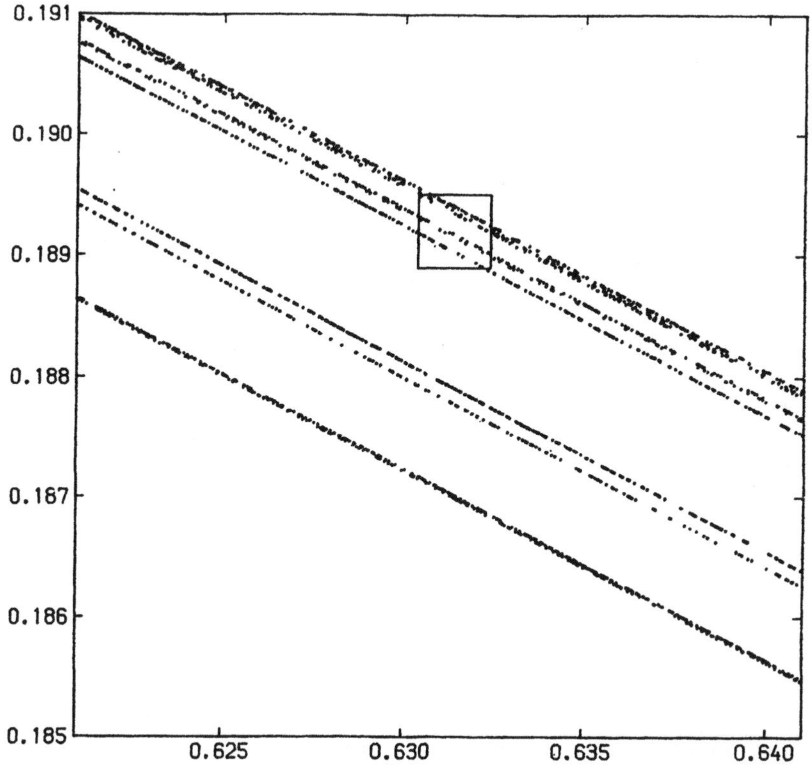

Fig. 5. Enlargement of the squared region of Figure 4; $n=10^6$

This can be actually proved by finding a region R which is mapped inside itself. An example of such a region is the quadrilateral $ABCD$ defined by

$$x_A=-1.33, \quad y_A=0.42, \quad x_B=1.32, \quad y_B=0.133,$$
$$x_C=1.245, \quad y_C=-0.14, \quad x_D=-1.06, \quad y_D=-0.5. \tag{14}$$

The image of $ABCD$ is a region bounded by four arcs of parabola, and it can be shown by elementary algebra that this image lies inside $ABCD$. Plotting the quadrilateral on Figure 2 or 3, one can verify that it encloses the observed attractor.

6. Conclusions

The simple mapping (4) appears to have the same basic properties as the Lorenz system. Its numerical exploration is much simpler: in fact most of the exploratory work for the present paper was carried out with a programmable pocket computer (HP-65). For the more extensive computations of Figures 2 to 6, we used a IBM 7040 computer, with 16-digit accuracy. The solutions can be followed over a much longer time than in the case of a system of differential equations. The accuracy is also increased since there are no integration errors.

Lorenz (1963) inferred the Cantor-set structure of the attractor from reasoning, but could not observe it directly because the contracting ratio after one "circuit"

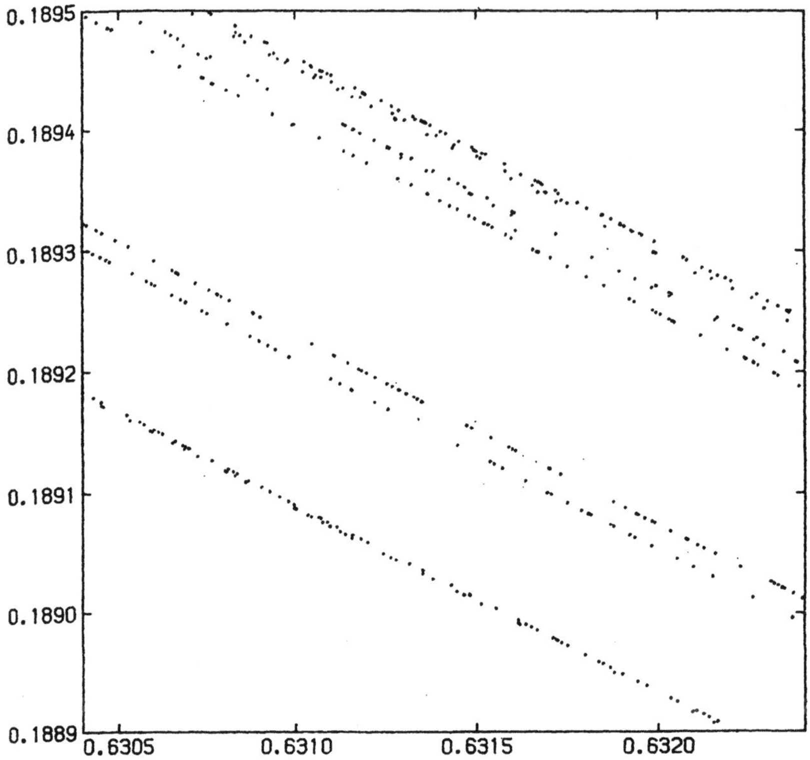

Fig. 6. Enlargement of the squared region of Figure 5; $n = 5 \times 10^6$

was too small: 7×10^{-5}. A similar experience was reported by Pomeau (1976). In the present mapping, the contracting ratio after one iteration is 0.3, and one can easily observe a number of successive levels in the hierarchy. This is also facilitated by the larger number of points.

Finally, for mathematical studies the mapping (4) might also be easier to handle than a system of differential equations.

References

Birkhoff, G. D.: Trans. Amer. Math. Soc. **18**, 199 (1917)
Engel, W.: Math. Annalen **130**, 11 (1955)
Engel, W.: Math. Annalen **136**, 319 (1958)
Hénon, M.: Quart. Appl. Math. **27**, 291 (1969)
Lanford, O.: Work cited by Ruelle, 1975
Lorenz, E. N.: J. atmos. Sci. **20**, 130 (1963)
Pomeau, Y.: to appear (1976)
Ruelle, D., Takens, F.: Comm. math. Phys. **20**, 167; **23**, 343 (1971)
Ruelle, D.: Report at the Conference on "Quantum Dynamics Models and Mathematics" in Bielefeld, September 1975

Communicated by K. Hepp

Received March 25, 1976

Strange attractors and chaotic motions of dynamical systems

Edward Ott

Department of Physics and Astronomy, University of Maryland, College Park, Maryland 20742
and Department of Electrical Engineering, University of Maryland, College Park, Maryland 20742

A review is presented of recent work related to strange attractors and chaotic motions of dynamical systems. First, simple systems capable of displaying chaotic behavior are discussed. In order of increasing dimensionality of the system, they are one-dimensional noninvertible maps, two-dimensional invertible maps, and autonomous systems of three coupled ordinary differential equations. The concept of fractional dimension of the strange attractor is stressed. Several physical examples well be reviewed, along with the possible relevance to turbulence in systems, such as fluids or plasmas, that are described by partial differential equations.

CONTENTS

I. INTRODUCTION

In this report, a review will be presented of recent developments related to strange attractors and chaotic motions of dynamical systems. Emphasis will be placed on those aspects that may prove useful for physical scientists.

At this point it might be appropriate to discuss some of the terms used in the title. A *dynamical system* may be thought of as any set of equations giving the time evolution of the state of a system from a knowledge of its previous history. Examples are Maxwell's equations, the Navier-Stokes equations, and Newton's equations of motion for a particle with suitably specified forces. The adjective *chaotic* is used here to describe a type of time evolution resulting from a dynamical system. In particular, it describes motions which are commonly thought of as "turbulent," i.e., motions whose time evolution appears, on detailed examination, to be very complex. For such motions, one often has the feeling that a statistical description may be of more use than actual knowledge of the true evolution. (A more precise definition of the term *chaotic* appears in Sec. II.) At this point I shall not attempt to define a *strange attractor*. Rather, we only note that its presence can lead to chaotic motion. Thus, the topic under discussion in this review is related to the occurrence of turbulent-type motions in physical systems. More specifically, we shall be concerned with nonconservative systems. [Chaotic motion in Hamiltonian (conservative) systems is not within the scope of the present review.]

It is becoming increasingly clear that the topic of strange attractors is one that will find abundant applications in a wide variety of physical situations. The list of such applications is already large, including problems in the onset of turbulence in fluids (Lorenz, 1963; McLaughlin and Martin, 1975; Ruelle and Takens, 1971), chemically reacting systems (Tomita and Kai, 1979), buckling beams (Holmes, 1980), nonlinear wave interactions in plasmas (Adam, Bussac, and Laval, 1980; Wersinger, Finn, and Ott, 1980a, 1980b, 1980c; Vyshkind, 1978; Vyshkind and Rabinovich, 1976; Russell and Ott, 1980; Wang, 1980; Masui and Wang, 1980), solid-state physics (Huberman and Crutchfield, 1979), lasers (Haken, 1975), self-generation of the earth's magnetic field (Robbins, 1977), magnetohydrodynamic flow (Maschke and Saramito, 1980; Treve and Manley, 1980), etc. (The above is only a partial listing and many other relevant references exist.)

As can be surmised from the dates of the references just mentioned, virtually all the activity in this field (at least when restricted to problems in the physical sciences) has occurred since 1975. The notable exceptions to this statement are the papers of Lorenz (1963) and of Ruelle and Takens (1971). These two papers, independently and from quite different points of view, originally suggested the relevance of strange attractors to the onset of turbulence in fluid flows. Lorenz was interested in explaining the presence of chaotic behavior in numerical solutions of a model system of three coupled, first-order, nonlinear, ordinary differential equations which modeled the nonlinear evolution of the Benard instability, i.e., the instability which results when a fluid layer subjected to gravity is heated sufficiently strongly from below. By a combination of careful analysis of the computer generated solutions and analytical reasoning, Lorenz was able to deduce that the solution of his equations was eventually trapped in a region of the phase space of the system which had a very intricate (strange) geometric structure. The now-recognized general implications of Lorenz's paper were not widely appreciated until many years after its publication. In 1971, Ruelle and Takens, making use of then recent developments in mathematics, offered a possible mechanism by which turbulent solutions to the Navier-Stokes equations could appear as a parameter is varied (e.g., as the Reynolds number is increased). In particular, they showed on the basis of quite general arguments that a strange attractor could appear. It is to be anticipated that the application of the type of considerations initiated by these two papers will lead to new insights in a variety of fields of physics. The purpose of the present paper is to facilitate this process by providing an elementary introduction designed for researchers and students in the physical sciences.

Some previous related reviews from various points of view are those of Treve (1978), Swinney and Gollub (1978), Rabinovich (1978), Ruelle (1977), Sinai (1977), Holmes (1977), Helleman (1980), Shaw (1981), and Yorke and Yorke (1981).

We shall proceed by discussing dynamical systems of progressively higher dimension (all of which can display chaotic behavior) [Ruelle (1977)]:

(1) One-dimensional noninvertible maps,

(2) Two-dimensional invertible maps,

(3) First-order systems of autonomous ordinary differential equations,

$$\frac{dx_i(t)}{dt} = f_i[x_1(t), x_2(t), \ldots, x_n(t)] \quad i=1,2,\ldots,n, \text{ with } n>3.$$

(4) Partial differential equations.

One-dimensional maps will be discussed in the next section (Sec. II). These are relations of the form $x_{n+1} = F(x_n)$; thus, for some initial x_0, a sequence, $x_0, x_1, x_2, \ldots$, is generated. Surprisingly, very simple one-dimensional maps will turn out to yield rather good qualitative models for behavior in two-dimensional maps (Sec. III), ordinary differential equations (Sec. IV), and partial differential equations (Sec. V).

II. ONE-DIMENSIONAL NONINVERTIBLE MAPS

To begin, we consider one-dimensional maps

$$x_{n+1} = F(x_n), \tag{2.1}$$

where $F(x)$ is a scalar function. [Some relevant references on one-dimensional maps are Li and Yorke (1975); May (1976); Guckenheimer *et al.* (1977); Feigenbaum (1978); and Guckenheimer (1979). In addition, a monograph by Coullet and Eckmann (1980) has been published on the subject.] We only consider the case in which the sequence $x_0, x_1, x_2, \ldots$, generated by F is bounded, $P<x_n<Q$, for all n. We will often say that such a sequence is "chaotic" or "turbulent," by which we mean that it has the following properties: (1) sensitive dependence on initial conditions (if two initial points x_0^a and x_0^b are chosen very close to each other, the distance between their successive images under F initially diverges exponentially); (2) the average correlation function for a given sequence satisfies $C(m) \to 0$ as $m \to \infty$, where

$$C(m) = \lim_{N\to\infty} \frac{1}{N}\sum_{m=1}^{N} (x_n - \langle x\rangle)(x_{n+m} - \langle x\rangle),$$

$$\langle x\rangle = \lim_{N\to\infty}\frac{1}{N}\sum_{n=1}^{N} x_n;$$

(3) The sequence is nonperiodic.

We now introduce the following two one-dimensional maps, each of which typifies a broad class,

$$F_1(x) \equiv a(1-2|x-\tfrac{1}{2}|), \quad 0<a\le 1 \tag{2.2}$$

$$F_2(x) = 4bx(1-x), \quad 0<b\le 1 \tag{2.3}$$

where a and b are constants. The map F_1 has a sharp peak, while the map F_2 has a rounded smooth maximum. For $0<(a,b)<1$, both F_1 and F_2 map the interval 0 to 1 into itself, and we shall only consider this range of x (i.e., initial conditions are always assumed to satisfy $1>x_0>0$). Both F_1 and F_2 are noninvertible, since, given x_{n+1}, one cannot solve $x_{n+1} = F(x_n)$ for x_n. (From Fig. 1 it is seen that for each value of x_{n+1} there are *two* possible values of x_n.) Thus one may say that it is possible to go forward in time but not backward in time. This represents a basic difference with ordinary differential equations, $\dot{\mathbf{x}} = \mathbf{f}(\mathbf{x})$, which may, in principal, be integrated either forward or backward in time.

First consider F_1. There are essentially two cases of interest, $0<a<\frac{1}{2}$ and $\frac{1}{2}<a\le 1$, both illustrated in Fig. 1(a). In Fig. 1, the dashed line represents $x_n = x_{n+1}$. For $0<a<\frac{1}{2}$, Fig. 1(a) shows that $x_{n+1}<x_n$, since, over

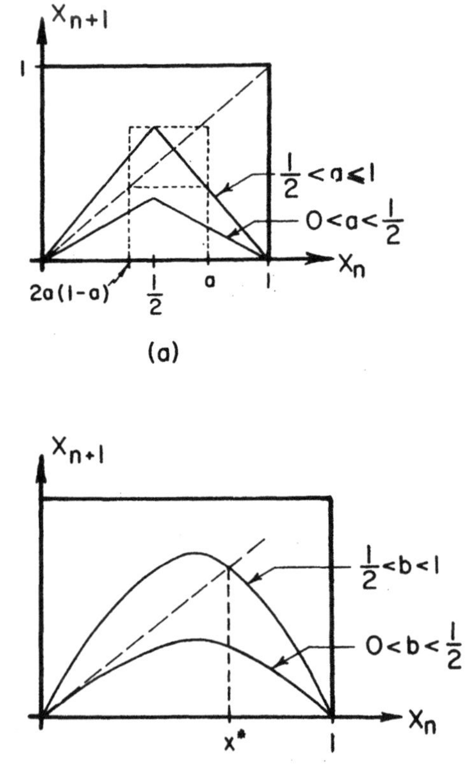

FIG. 1. (a) Map F_1 [Eq. (2.2)], and (b) the map F_2 [Eq. (2.3)].

the entire range, $F_1(x_n)$ lies below the line $x_n = x_{n+1}$. In this case it is clear that x_n converges to zero as n increases. Now, consider $\frac{1}{2} < a \leq 1$. For any initial x_0 between 0 and 1, the sequence eventually becomes trapped in the interval $2a(1-a) < x < a$ through which it will typically wander chaotically. To illustrate this, consider the particularly simple case $a = 1$ for which the chaotic interval becomes $0 < x < 1$. In this case, we may consider the map to represent two steps: (1) a uniform stretching of the interval 0 to 1 to twice its original length, and (2) a folding in half of the stretched interval so that it now has its original length. These steps are illustrated in Fig. 2. The stretching property leads to exponential separation of nearby points and hence, sensitive dependence on initial conditions. The folding property keeps the generated sequence bounded, but also causes the map to be noninvertible, since it causes two different x_n points to be mapped into one x_{n+1} point.

Conversely, for a general one-dimensional map, in order to have the distance between nearby points separate exponentially, it is necessary for the map to be, on the average, stretching. On the other hand, to have the sequence remain bounded (confined between 0 and 1 in the case of F_1), folding must take place. Thus we conclude that in order for a one-dimensional map to exhibit chaotic behavior, it must be noninvertible. Figure 3 illustrates the stretching and folding properties of F_1 for a value of a less than one ($\frac{1}{2} < a < 1$). From Fig. 3(a) we see that after one application of F_1, there are no points in $a < x < 1$. From Fig. 3(b), we see that the interval 0 to $2a(1-a)$ is stretched but that no points are folded back onto it. Thus any point in $0 < x < 2a(1-a)$ will eventually leave that interval and never return. Thus the generated sequence is eventually trapped in $2a(1-a) < x < a$. An alternative way of seeing the fact that F_1 has sensitive dependence on initial conditions is illustrated in Fig. 4 for the case $a = 1$. Figure 4(a) shows x_{n+2} versus x_n obtained from two applications of F_1; that is, $x_{n+2} = F_1[F_1(x_n)]$. We will use the notation $x_{n+2} = F_1^{(2)}(x_n)$ to denote $F_1[F_1(x_n)]$. For x_{n+m} versus x_n, Fig. 4(a) generalizes to Fig. 4(b), $x_{n+m} = F_1^{(m)}(x_n)$. From Fig. 4(b), one can see that if the initial condition has an uncertainty $\pm\varepsilon$, then after $m \sim \ln_2(1/\varepsilon)$ iterations of F_1, we will have essentially no clue as to where x lies in the interval 0 to 1 (example: for $\varepsilon \cong 10^{-12}$, $m \cong 40$).

We now turn to a consideration of the map F_2 [Eq.

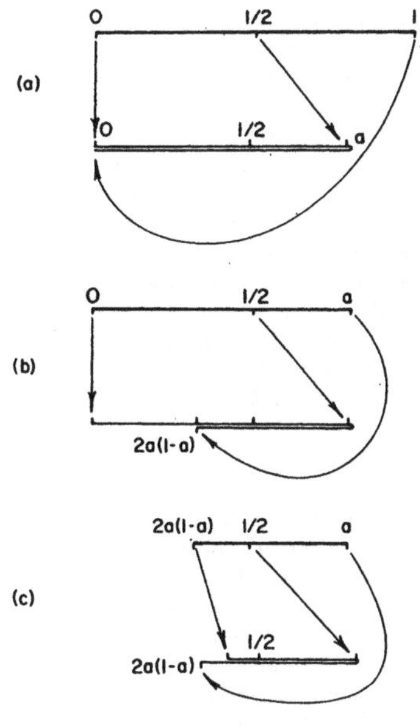

FIG. 3. Map F_1 for $a = 0.8$; (a) mapping of the interval 0 to 1, (b) 0 to a, and (c) $2a(1-a)$ to a.

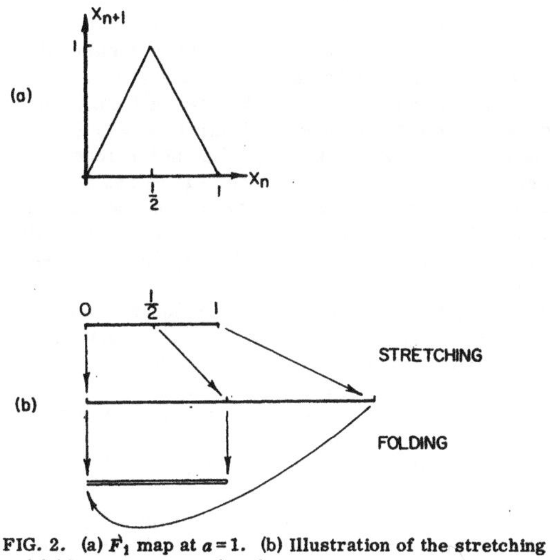

FIG. 2. (a) F_1 map at $a = 1$. (b) Illustration of the stretching and folding properties of F_1 for $a = 1$.

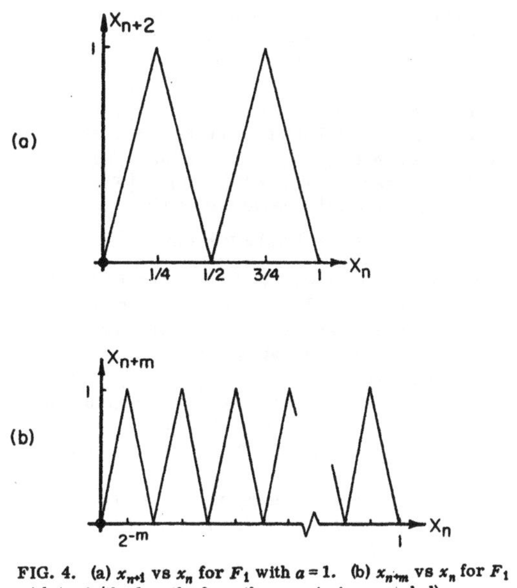

FIG. 4. (a) x_{n+1} vs x_n for F_1 with $a = 1$. (b) x_{n+m} vs x_n for F_1 with $a = 1$ (the length along the x_n axis is expanded).

(2.3) and Fig. 1(b)], which has a smooth rounded maximum at $x_n=\frac{1}{2}$ as opposed to the sharp peak in F_1 [Fig. 1(a)]. From Fig. 1(b) we note that the line $x_n=x_{n+1}$ intersects the map at $x_n=0$ for $0<b<\frac{1}{4}$ and at two points, $x_n=0$ and $x_n=x^*$, for $\frac{1}{4}<b<1$, where, from (2.3), $x^*=1-(4b)^{-1}$. These points of intersection are fixed points of the map; that is, if the initial point is chosen to be a fixed point, then successive applications of the map leave it unmoved. It is important to discover whether the fixed points are stable to small perturbations. Let $\bar{x}$ be a fixed point, $\bar{x}=F(\bar{x})$, and consider a perturbation from it, $x_n=\bar{x}+\delta_n$. From (2.1), $\bar{x}+\delta_{n+1}=F(\bar{x}+\delta_n)$. For δ_n small, we can Taylor-series expand $F(\bar{x}+\delta_n)$ as $F(\bar{x})+F'(\bar{x})\delta_n=\bar{x}+F'(\bar{x})\delta_n$, from which we obtain

$$\delta_{n+1}/\delta_n=F'(\bar{x}), \tag{2.4}$$

where $F'\equiv dF/dx$. Thus, if $|F'(\bar{x})|>1$, images under F of points near $\bar{x}$ successively move farther away from it, and $\bar{x}$ is unstable. For $|F'(\bar{x})|<1$, points near $\bar{x}$ converge to it, and $\bar{x}$ is stable. (For example, for F_1 given by Eq. (2.2), $|F_1'(x)|=2a$ for $x\neq\frac{1}{2}$, and thus $\bar{x}=0$ is stable for $0<a<\frac{1}{2}$, while zero and the second intersection [cf. Fig. 1(a)] are both unstable for $\frac{1}{2}<a\leq 1$.) For F_2, Eq. (2.3) shows that $F'(0)=4b$, and $F'(x^*)=2(1-2b)$. Thus, zero is stable for $b<\frac{1}{4}$. The fixed point $x=x^*$ first appears at $b=\frac{1}{4}$, and, simultaneously with its appearance, the zero fixed point loses its stability. The fixed point $x^*=1-(4b)^{-1}$ is stable in $\frac{3}{4}>b>\frac{1}{4}$, since $|F'(x^*)|<1$ in this range. Corresponding to these results, it can be shown that the sequence generated by F_2 converges to zero for $0\leq b<\frac{1}{4}$ and to x^* for $\frac{1}{4}<b<\frac{3}{4}$. $F'(x^*)=1$ at $b=\frac{1}{4}$ and decreases as b increases, becoming zero at $b=\frac{1}{2}$, minus one at $b=\frac{3}{4}$, and less than minus one (unstable) for $b>\frac{3}{4}$. The question which then arises is what happens in the range $\frac{3}{4}<b<1$ for which both 0 and x^* are unstable. To begin answering this question it is instructive to examine the map $x_{n+2}=F_2^{(2)}(x_n)$ shown in Fig. 5 for b slightly below $\frac{3}{4}$ and for b slightly above $\frac{3}{4}$. Values of x which recur every second iteration, i.e., in a series $e,f,e,f,e,f,\ldots$ are fixed points of $F^{(2)}$, $e=F^{(2)}(e)$, and $f=F^{(2)}(f)$, with $e=F(f)$ and $f=F(e)$. For the special case $e=f$, e will also be a fixed point of F itself. Now consider the stability of the sequence $e,f,e,f,\ldots$. Taking $x_n=e+\delta_n$, $x_{n+2}=e+\delta_{n+2}$, we have $e+\delta_{n+2}=F^{(2)}(e+\delta_n)=F[F(e+\delta_n)]$, which, when Taylor-series expanded, yields

$$\delta_{n+2}/\delta_n=F'(e)F'(f)=F^{(2)\prime}(e)=F^{(2)\prime}(f). \tag{2.5}$$

Applying (2.5) to a fixed point of F, we have $F^{(2)\prime}(x^*)=[F'(x^*)]^2$. Thus, when x^* loses stability, the slope of $F^{(2)}$ at x^* becomes greater than one. Referring, now, to Fig. 5 for F_2, it is seen that, when this occurs, two new intersections of $F_2^{(2)}$ with $x_{n+2}=x_n$, e and f, simultaneously appear. Furthermore, when these points appear, they initially have $F_2^{(2)\prime}(e)=F_2^{(2)\prime}(f)=1$, and the slopes decrease as b is raised. Thus, the new fixed points of $F^{(2)}$ are initially stable, and it is found that, when b slightly exceeds $\frac{3}{4}$, the generated sequence converges to an alternating one, $e,f,e,f,\ldots$. As b increases further, however, $F_2^{(2)\prime}(e)=F_2^{(2)\prime}(f)$ decreases; eventually, past $b=0.862\ldots$, $F_2^{(2)\prime}(f)$ becomes less than minus one, and the e,f cycle becomes unstable.

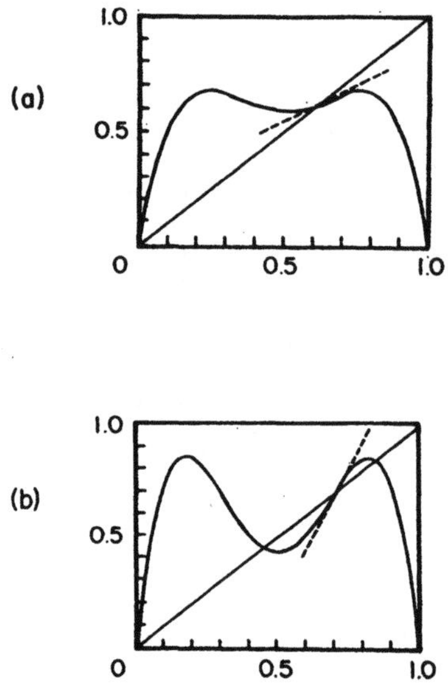

FIG. 5. x_{n+2} vs x_n for F_2 with (a) $b=0.678$ and (b) $b=0.854$. The dashed lines are the slope at $x=x^*$. The intersections of $x_{n+2}=x_n$ with $x_{n+2}-F_2^{(2)}(x_n)$ are $e>x^*>f$.

What happens next can be deduced, in an analogous way, from the map $x_{n+4}=F_2^{(2)}[F_2^{(2)}(x_n)]\equiv F_2^{(4)}(x_n)$. When the e,f cycle loses stability, a stable four-point periodic cycle simultaneously appears: $g,h,i,j,g,h,i,j,g,h,\ldots$, which then gives way ("bifurcates") to an eight-point cycle, which then gives way to a 16-point cycle, etc. Furthermore, the band of b values over which a given 2^k-point cycle is stable decreases geometrically with k, so that

$$\frac{b_k-b_{k-1}}{b_{k+1}-b_k}\rightarrow 4.669\,201\ldots \tag{2.6}$$

for $k\rightarrow\infty$, where b_k is the value of b at the point where the 2^k-point cycle bifurcates to a 2^{k+1}-point cycle. Also $b_k\rightarrow 0.892\ldots$ for $k\rightarrow\infty$; that is, there is an accumulation point of an infinite number of bifurcations at $b_\infty=0.892\ldots$. Fiegenbaum (1978) has derived Eq. (2.6) using arguments based on scale invariance near b_∞, and has also obtained other properties of the generated sequences for b near b_∞. These properties apply independently of the detailed functional form of the map as functional form of the map as long as it has a quadratic maximum as does F_2. Thus Eq. (2.6) applies to a wide class of maps.

Just past b the orbit generated by F_2 looks like a noisy cycle of periodicity 2^p with $p\rightarrow\infty$ as b approaches b_∞ from above. By a "noisy cycle of periodicity 2^p" we mean that the orbit is confined to 2^p disjoint intervals in $1>x>0$ which it visits in a sequential order. Thus the orbit always comes back to the same interval after 2^p iterations. On the other hand, if one looks at the points generated by $F_2^{(2^p)}$ with an initial condition in one of these intervals, then the orbit looks completely chaotic in this interval. As b increases, these intervals merge in pairs so that a noisy 2^p cycle goes into a noisy 2^{p-1} cycle as b increases past a critical value $\bar{b}_p$. Furthermore, $\bar{b}_p$ obeys the same scaling relation[1] as in Eq. (2.6) (Coullet and Tresser, 1980). As b increases past $\bar{b}_1$, chaotic motion over a single connected band emerges.

In addition to the noisy 2^p cycles, narrow windows in b also exist within $b_\infty < b < 1.0$ for which the generated sequence is exactly periodic. Generally, these periodic sequences first appear with some period N and then go through a sequence of period doubling bifurcations, creating periods $2^k N$, with an accumulation point at $k \to \infty$ ending the particular periodic window. The widest such window is for 3×2^k periodic cycles, $0.9571 < b < 0.9624$. Other periodic windows are exceedingly narrow in b, and most of the range $b_\infty < b < 1$ appears to be chaotic. As an example of how these cycles first appear, consider the onset of the $N=3$ period cycle. Figure 6 shows the map x_{n+3} versus x_n [obtained from $F_2^{(3)}(x_n)$] for values of b just below and just above that for which the $N=3$ cycle first appears. As b increases past its critical value, the minimum of $F_2^{(3)}$ lowers until two new intersections of $F_2^{(3)}$ with $x_{n+3}=x_n$ are created near this minimum, one with slope greater than one (unstable) and one with slope less than one (initially stable). This type of phenomenon, whereby a periodic orbit appears after a region of chaotic motion, is called a *tangent bifurcation*. For values of b where numerically generated sequences appear to be chaotic, it is not, at present, known whether they are truly chaotic, or whether, in fact, they are really periodic, but with exceedingly large periods and very long transients required to settle down. Recent numerical results do, however, strongly suggest that the sequences are truly chaotic (Lorenz, 1979). Figure 7 summarizes some of the previously described results for F_2.

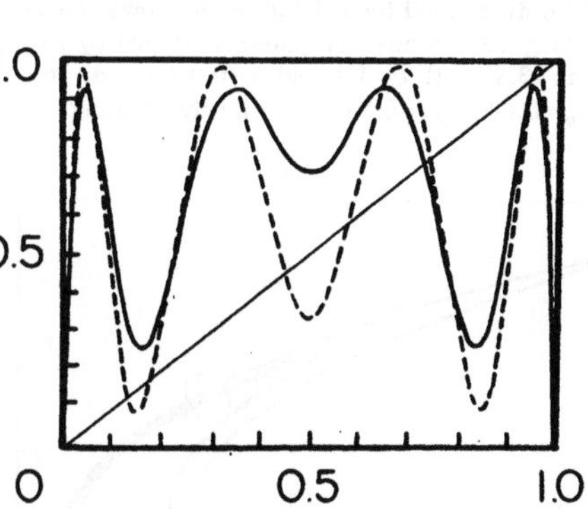

FIG. 6. x_{n+3} vs x_n for $b=0.975$ (dashed curve) and for $b=0.925$ (solid curve).

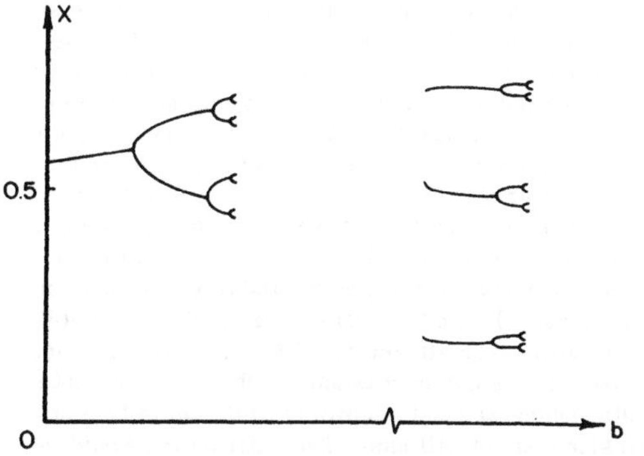

FIG. 7. Summary diagram for some of the bifurcations of $F^{(2)}$. The scale about the period 3×2^k cycles has been greatly expanded.

III. TWO-DIMENSIONAL INVERTIBLE MAPS

A general two-dimensional map can be written as

$$x_{n+1}=f_1(x_n,y_n),\quad y_{n+1}=f_2(x_n,y_n). \tag{3.1}$$

The map is invertible if (3.1) can be solved uniquely for x_n and y_n as functions of x_{n+1} and y_{n+1}, $x_n=g_1(x_{n+1},y_{n+1})$ and $y_n=g_2(x_{n+1},y_{n+1})$. That is, it is possible to go either backwards or forwards in time. A two-dimensional invertible map can easily be constructed from a one-dimensional noninvertible map as follows:

$$x_{n+1}=F(x_n)+y_n, \tag{3.2a}$$

$$y_{n+1}=\beta x_n, \tag{3.2b}$$

where $F(x)$ is noninvertible. For $\beta=0$, $x_{n+1}=F(x_n)$, and the noninvertible one-dimensional map is recovered. However, as long as $\beta\neq 0$, no matter how small it is, the map (3.2) is invertible: $x_n=y_{n+1}/\beta$ and $y_n=x_{n+1}-F(y_{n+1}/\beta)$. On the other hand, if β is sufficiently small, the variation of x is well described by the one-dimensional map, $x_{n+1}=F(x_n)$. Furthermore, for small β, the range of variation of y_n will be small compared to that for x_n [cf. Eq. (3.2b)], and thus, if the points generated by (3.2) are plotted on the xy plane, the generated sequence will appear to lie on a line (the x axis) with some small spread about the line. For very small β, the spread may, in practical terms, be unmeasurable, so that (3.2) becomes indistinguishable from a one-dimensional noninvertible map. (This will, indeed, turn out to be similar to what happens in certain differential equation examples to be discussed in the next section.)

It is of interest to compute the Jacobian of the map (3.2),

$$J\equiv\begin{vmatrix}\dfrac{\partial x_{n+1}}{\partial x_n} & \dfrac{\partial x_{n+1}}{\partial y_n}\\[2ex] \dfrac{\partial y_{n+1}}{\partial x_n} & \dfrac{\partial y_{n+1}}{\partial y_n}\end{vmatrix}=-\beta.$$

Thus, for $|\beta|<1$, we see that areas will contract by the factor $|\beta|$ on each application of the mapping (3.2). Thus, if the generated sequence of pairs (x_n,y_n) re-

[1]The subject of the universal scaling properties of maps with a quadratic maximum is currently a very active research topic. Recent work includes the study of scaling of the Lyapunov number past b_∞ (Chang and Wright, 1981; Huberman and Rudnick, 1980), the noise power spectrum past b_∞ (Huberman and Zisook, 1981; Wolf and Swift, 1981), and scaling behavior with the addition of random noise (Crutchfield *et al.*, 1981; Shraiman *et al.*, 1981). Furthermore, universal scaling properties of conservative systems with period doubling have also recently been examined (for example, Green *et al.*, 1981).

mains in a bounded region of the xy plane, then the sequence must asymptotically approach a subset of the original bounded xy region which has zero area. This subset is called an *attractor*. For example, if the sequence becomes attracted to an N-point periodic cycle, then the attractor would be the N points plotted in the xy plane, clearly a subset of zero area. Another possible subset of zero area that a sequence might asymptote to is a curve. At first sight, these two possibilities, a zero-dimensional subset (points) and a one-dimensional subset (a curve), might appear to exhaust all possibilities for zero-area attractors. This is, however, not the case. There can be attractors which have noninteger dimension (at least according to the definition of dimension that we will use). Such attractors would be termed strange. The relevant definition of dimension is that due to Hausdorf[2] [see, for example, Mandelbrot (1977)]:

$$d=\lim_{\varepsilon\to 0} \ln N(\varepsilon)/\ln\left(\frac{1}{\varepsilon}\right), \tag{3.3}$$

where, if the set in question is a subset of a p-dimensional ordinary space, then $N(\varepsilon)$ is the number of p-dimensional cubes of side ε needed to cover the set. Alternatively, (3.3) implies that for small ε, $N(\varepsilon) \cong K\varepsilon^{-d}$. Thus, if one is content to know where the set lies to within an accuracy ε, then, to specify the location of the set, we need only specify the positions of the $N(\varepsilon)$ cubes covering the set. Hence, the dimension may be viewed as telling us how much information is necessary to specify the location of the set to within a given accuracy. If the set has complicated fine-scale structure, then, as a practical matter, it may be advantageous to introduce some coarse-graining into the description of the set, and then ε may be thought of as specifying the degree of coarse-graining. As an example of the application of Eq. (3.3), if the set in question is a point, then $N(\varepsilon)=1$, and, according to (3.3), the Hausdorf dimension is zero; if the set in question is the section of the xy plane given by $0<x<1$ and $0<y<1$, then $N(\varepsilon)=\varepsilon^{-2}$, and the Hausdorf dimension is two; if the set is a straight line joining (0, 0) and (1.0), then $N(\varepsilon)=\varepsilon^{-1}$, and the Hausdorf dimension is one. These examples all yield the obvious results, so that (3.3) conforms to our intuition in these cases. As an example of a set with a noninteger dimension, consider the following construction of a Cantor set (illustrated in Fig. 8): take a line of unit length, $0\le x\le 1$, and remove the middle third $\frac{1}{3}<x<\frac{2}{3}$; then take the two remaining intervals between 0 and $\frac{1}{3}$ and between $\frac{2}{3}$ and 1, divide them in thirds, and remove the two middle thirds; in the limit as this process is repeated an infinite number of times, what is left is a set that has zero net length and an uncountable number of elements. To apply (3.3) to this set, we note the following which is evident from Fig. 8,

$$\varepsilon=\tfrac{1}{3}, \quad N=2$$

$$\varepsilon=\tfrac{1}{9}, \quad N=4$$

$$\varepsilon=\tfrac{1}{3}^{p}, \quad N=2^{p}.$$

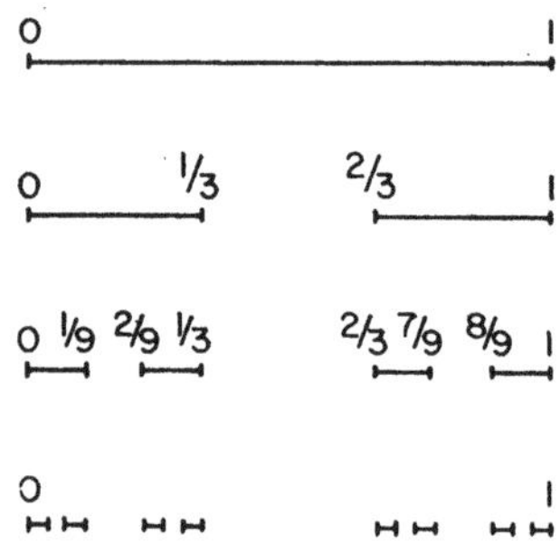

FIG. 8. First few steps in the construction of an example of a Cantor set.

Thus from (3.3)

$$d=(\ln 2)/(\ln 3)=0.630.$$

Note that the Cantor set just constructed has the property of scale invariance. That is, by the nature of the construction, the set between 0 and 1 will look precisely the same as that part of it between 0 and $\frac{1}{3}$, if the latter is examined under a magnifying glass which magnifies by a factor of three.

As a concrete example, we now consider a mapping first studied by Henon (1976):

$$x_{n+1}=1-cx_n^2+y_n, \quad y_{n+1}=\beta x_n. \tag{3.4}$$

This mapping is essentially equivalent to Eq. (3.2) with F given by Eq. (2.3). [A study of the map obtained from (3.2) and (2.2) has been presented by Lozi (1978), but will not be discussed here.] Figure 9 shows the results of plotting 10^4 successive points obtained by iterating the map (3.4) with $c=1.4$ and $\beta=0.3$ from an initial point $x_0=0.631$, $y_0=0.189$. Similarly obtained

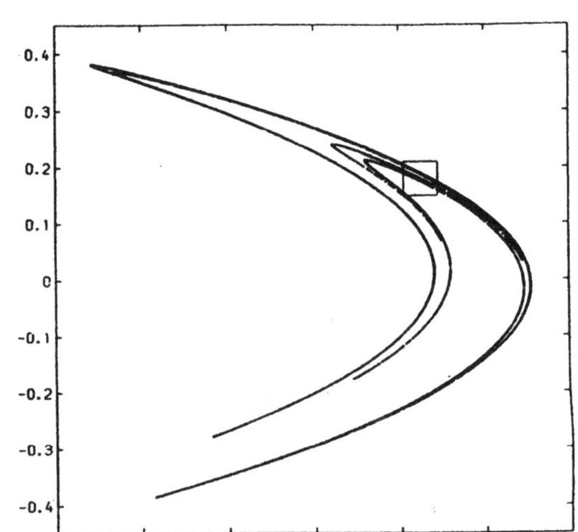

FIG. 9. Iterated points of the map, Eq. (3.4), 10^4 iterations.

[2]Actually, the Hausdorf dimension has a somewhat more involved definition than that given by Eq. (3.3). More precisely, d given by (3.3) defines the "capacity" of the set. For cases of interest to us here, however, Eq. (3.3) probably gives the same result as would the actual Hausdorf dimension definition.

plots starting with other initial values are almost identical (except for an initial transient), suggesting that Fig. 9 is, in fact, essentially a picture of the attractor. That is, as the map is iterated, points come closer and closer to the attractor and eventually become indistinguishable from it on the scale of the figure. Figure 10(a) shows a blow-up of the squared region in Fig. 9; Figs. 10(b) and 10(c) are successive blow-ups of the squared regions in the preceding figure. Scale invariant, Cantor-set-like structure transverse to the linear structure is evident. Thus, the attractor is probably strange with dimension between one and two. In fact, a recent study (Russell *et al.*, 1980) gives $d \cong 1.26$. Further studies of the map (3.4) have been carried out by Feit (1978), Curry (1979), and Simó (1979), who have explicitly verified exponential divergence of initially close points. In addition, Feit (1978) and Simó (1979) have studied how the character of the generated sequences changes as c is varied with β fixed. It is found that, dispersed among intervals of c where the motion is chaotic, there are many small subintervals where the motion is periodic. On each such subinterval, there appear attractors of period $k, 2k, 4k, \ldots, 2^n k, \ldots$, similar to the phenomenon observed for the one-dimensional map, Eq. (3.4) with $\beta = 0$ [which is equivalent to Eq. (2.3)].

Bridges and Rowlands (1977) have given a procedure for investigating maps of the type (3.2) essentially by a power series in β. For example, to lowest order in β (3.2a) gives $x_{n+1} = F(x_n)$ (as previously noted), which, when substituted into (3.2b), yields the result that the attractor lies in the vicinity of the curve

$$x = F(y/\beta). \tag{3.5}$$

Higher-order approximations yield increasing detail to the curve. For Henon's map, Eq. (3.4), with $c = 1.4$ and $\beta = 0.3$, Eq. (3.5) yields a surprisingly good first-order approximation of the attractor.

The reader may be aware that two-dimensional maps have been extensively utilized as models of Hamiltonian systems, and, in particular, to model ergodic behavior in Hamiltonian systems. Since Hamilton's equations conserve phase-space volume (by Liouville's theorem), two-dimensional maps modeling Hamiltonian systems do not cause uniform contraction of areas, and chaotic motions (i.e., ergodicity) generated by such maps generally fill up a two-dimensional area. Thus, no strange (fractional dimensional) set is involved. If dissipation (e.g., friction) is added to a Hamiltonian system, phase-space contraction may be expected to result. Thus it is of interest to consider the relation of ergodicity in Hamiltonian systems to the possible

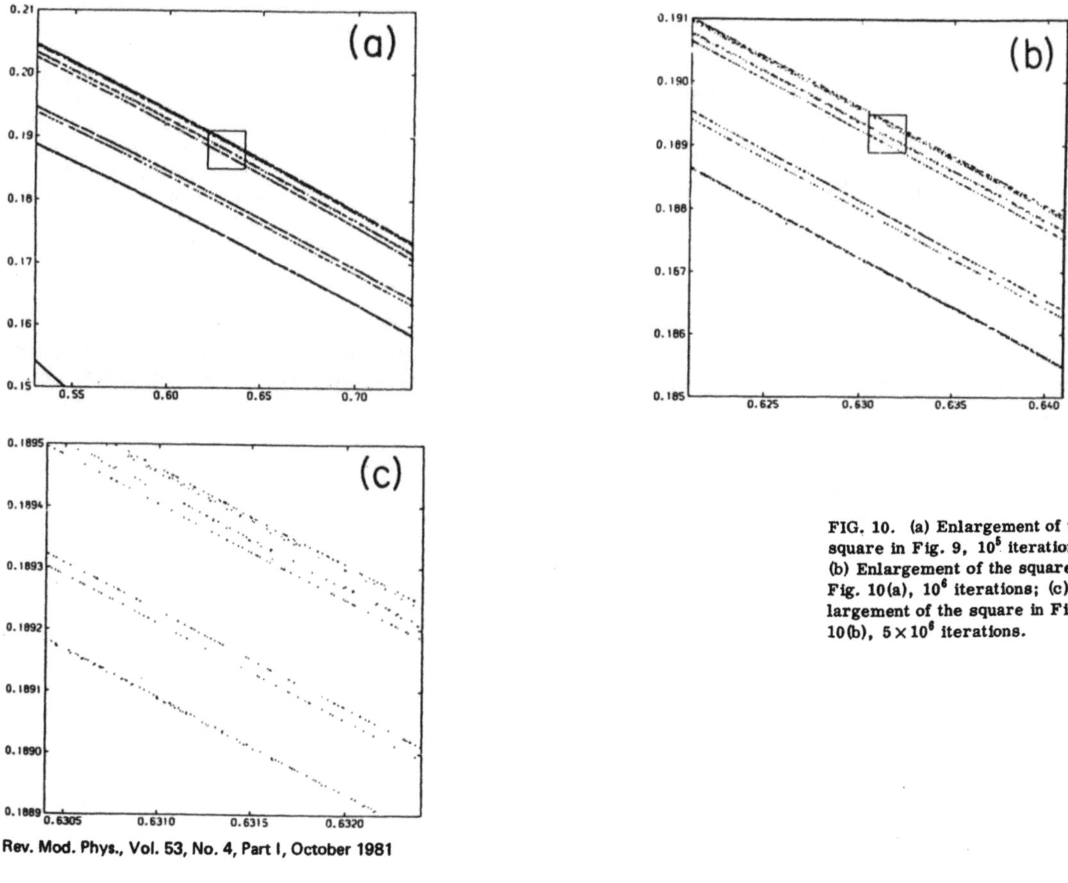

FIG. 10. (a) Enlargement of the square in Fig. 9, 10^5 iterations; (b) Enlargement of the square in Fig. 10(a), 10^6 iterations; (c) Enlargement of the square in Fig. 10(b), 5×10^6 iterations.

appearance of strange attractors when dissipation is added to these systems (McLaughlin, 1979; Zaslavskii, 1978). Recent studies addressed to this question (Zaslavskii, 1978; Zaslavskii and Rachko, 1979) consider the two-dimensional map

$$y_{n+1} = e^{-\Gamma}(y_n + \varepsilon \cos 2\pi x_n), \tag{3.6a}$$

$$x_{n+1} = \left\langle x_n + \frac{\Omega}{2\pi} + \frac{\alpha\Omega}{2\pi\Gamma}(1 - e^{-\Gamma})y_n + \frac{K}{\Gamma}(1 - e^{-\Gamma})\cos 2\pi x_n \right\rangle, \tag{3.6b}$$

where $\langle \cdots \rangle$ denotes the fractional part of the argument and $K = \alpha\varepsilon\Omega/2\pi$. The Jacobian of the map (3.6) is $\exp(-\Gamma)$. Thus, the map contracts areas for $\Gamma > 0$ and is area preserving for $\Gamma = 0$. In the absence of dissipation, $\Gamma = 0$, and (3.6) reduces to

$$y_{n+1} = y_n + \varepsilon \cos 2\pi x_n, \tag{3.7a}$$

$$x_{n+1} = \left\langle x_n + \frac{\Omega}{2\pi} + \frac{\alpha\Omega}{2\pi} y_n + K\cos 2\pi x_n \right\rangle, \tag{3.7b}$$

which has been extensively studied as a basic model of stochasticity in Hamiltonian systems (Rosenbluth *et al.*, 1966; Stix, 1973; Rechester and Stix, 1976; Chirikov, 1979; Greene, 1979) (and also as a model for stochastic magnetic field line topologies in magnetic confinement controlled thermonuclear fusion devices). The onset of ergodic behavior in the map (3.7) can be estimated from Chirikov's island overlap condition which, for small ε, yields ergodicity for $K \simeq 1$. By qualitative arguments based upon the exponential divergence of nearby chaotic orbits Zaslavskii is able to estimate the condition for the appearance of chaotic orbits and a strange attractor for the map (3.6), $K\mu \gtrsim 1$, where $\mu = (1 - e^{-\Gamma})\Gamma^{-1}$. Numerical results confirm this rough estimate.

In connection with chaotic maps a useful notion is that of the Lyapunov numbers. An illustration of the Lyapunov numbers is given in Fig. 11. For a two-dimensional map, the Lyapunov numbers, λ_1 and λ_2, are the average principal stretching factors for a very small circular area; more formally

$$(\lambda_1, \lambda_2) = \lim_{n\to\infty}[\text{magnitude of the eigenvalues of } \underline{J}(x_n, y_n)\underline{J}(x_{n-1}, y_{n-1})\cdots\underline{J}(x_1, y_1)]^{1/n}, \tag{3.8}$$

where $\underline{J}(x,y)$ is the Jacobian matrix of the map:

$$\underline{J}(x,y) = \begin{bmatrix} \dfrac{\partial f_1(x,y)}{\partial x} & \dfrac{\partial f_1(x,y)}{\partial y} \\[2ex] \dfrac{\partial f_2(x,y)}{\partial x} & \dfrac{\partial f_2(x,y)}{\partial y} \end{bmatrix}$$

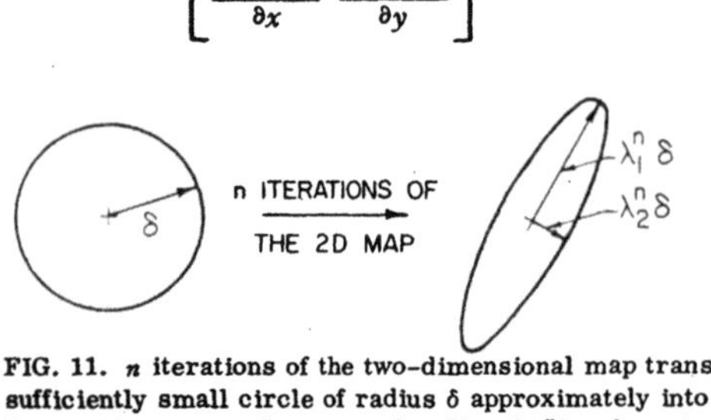

FIG. 11. n iterations of the two-dimensional map transform a sufficiently small circle of radius δ approximately into an ellipse with major and minor radii $\lambda_1^n\delta$ and $\lambda_2^n\delta$, where λ_1 and λ_2 are Lyapunov numbers, for $n \to \infty$.

[f_1 and f_2 are defined by Eq. (2.2)], and (x_1, y_1), (x_2, y_2), $\ldots, (x_n, y_n)$ is a sequence generated by the map. Thus, the Lyapunov numbers specify the average stretching rate of nearby points. Say $\lambda_1 > \lambda_2$. If the map is chaotic, then λ_1 must exceed unity, so that the distance between almost all nearby points increases on successive mappings. If the map is area contracting, $\lambda_1\lambda_2 < 1$; if it is area preserving, $\lambda_1\lambda_2 = 1$. The calculation of λ_1 and λ_2 from Eq. (3.8) is fairly easy using a digital computer and can then be used as a criterion for chaos, $\lambda_1 > 1$. An interesting conjecture concerning Lyapunov numbers has been put forward by Frederickson *et al.* (1980) [see also Kaplan and Yorke (1979a)], namely, that the Hausdorf dimension of a strange attractor of a map with the eigenvalues of $\underline{J}$ independent of x and y is related to its Lyapunov numbers. For a two-dimensional map with Lyapunov numbers $\lambda_1 > 1 > \lambda_2$, $\lambda_1\lambda_2 < 1$, their conjecture states that the dimension is given by

$$d = 1 + (\ln\lambda_1)/(\ln\lambda_2^{-1}). \tag{3.9}$$

For example, for the Henon map, Eq. (3.4), with $c = 1.4$ and $b = 0.3$, $\lambda_1 \cong 0.2$, and Eq. (3.9) gives $d \cong 1.26$. Recently, numerical experiments have been performed which tend to confirm (3.9) (Russell *et al.*, 1980). Furthermore, we note that these numerical experiments also tested maps for which the eigenvalues $\underline{J}$ are not independent of x and y and still obtained excellent agreement with Eq. (3.9). While it is known by counterexample that Eq. (3.9) fails in general if the eigenvalues of $\underline{J}$ depend on x and y, the results of these computer experiments indicate that (3.9) still yields a surprisingly good approximation to the dimension in some cases.

A possible motivation for (3.9) is depicted in Fig. 12, which considers a map of the unit square into itself. The mapping consists of two steps. The first step is a stretching along y by $\lambda_1 > 1$ and a contraction along x by $\lambda_2 < 1$. For the second step, we assume that λ_1 is an integer ($\lambda_1 = 3$ for Fig. 12), and move the stretched and contracted area back into the unit square, as shown. This may be represented analytically by the map

$$y_{n+1} = \lambda_1 y_n \bmod 1,$$

$$x_{n+1} = \lambda_2 x_n + y_n - \lambda_1^{-1}(\lambda_1 y_n \bmod 1).$$

From the construction in Fig. 12, it is clear that λ_1 and λ_2 are the Lyapunov numbers of this map. If the proc-

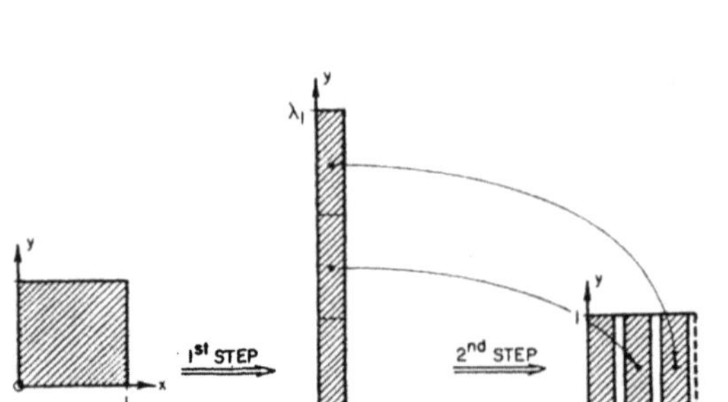

FIG. 12. Map motivating Eq. (3.9). The illustrated map may be represented analytically as $y_{n+1} = 3y_n \bmod 1$, $x_{n+1} = \lambda_2 x_n + y_n - y_{n+1}/3$.

ess in Fig. 12 is repeated p times, λ_1^p vertical strips of width λ_2^p will be created. To apply Eq. (3.3) for the dimension, let $\varepsilon = \lambda_2^p$. Then, the number of squares of side ε needed to cover the strips is $N \cong \lambda_1^p \lambda_2^{-p}$, and Eq. (3.3) yields (3.9). In addition, the conjecture can be used to predict the dimension of strange attractors for systems of ordinary differential equations (Russell *et al.*, 1980). The map of Fig. 12 is also interesting, since it demonstrably yields a strange attractor, while for maps such as (3.4) and (3.6) one can only say that numerical results suggest the presence of a strange attractor.

IV. THREE-DIMENSIONAL SYSTEMS OF ORDINARY DIFFERENTIAL EQUATIONS

A. Background

Here we consider systems of three coupled, nonlinear, autonomous, differential equations,

$$dx_i(t)/dt = f_i[x_1(t), x_2(t), x_3(t)], \quad i = 1, 2, 3. \tag{4.1}$$

The system is "autonomous" because f_i does not depend explicitly on t but only on x_i. Alternatively, a nonautonomous system of two coupled equations, $dx_1/dt = g_1(x_1, x_2, t)$ and $dx_2/dt = g_2(x_1, x_2, t)$, can be written in the form (4.1) by defining $f_1 = g_1(x_1, x_2, x_3)$; $f_2 = g_2(x_1, x_2, x_3)$; $f_3 = 1$.

We now define the Poincaré map of a system like (4.1). The Poincaré map represents a reduction of a system like (4.1) to a two-dimensional map, such as those studied in Sec. III. Figure 13 illustrates the construction of a Poincaré map. Consider a particular solution of (4.1) to generate an orbit in x_1, x_2, x_3 phase space. We now assume that some appropriate surface (the "surface of section") in this space has been chosen, and we study intersections of the orbit with the chosen surface. In Fig. 13 the chosen surface is the plane $x_2 = K$. Every time the orbit crosses the chosen surface in a particular direction ($dx_2/dt < 0$ for Fig. 13) we record the crossing point, e.g., points A and B in Fig. 13. For Fig. 13, it is clear that point A uniquely determines point B, since the solution of (4.1) is unique. Likewise, point B determines point A by time reversal of (4.1). Thus the Poincaré map in this illustration represents an invertible transformation of a point in the plane $x_2 = K$ into another point, i.e., it is an invertible two-dimensional map.

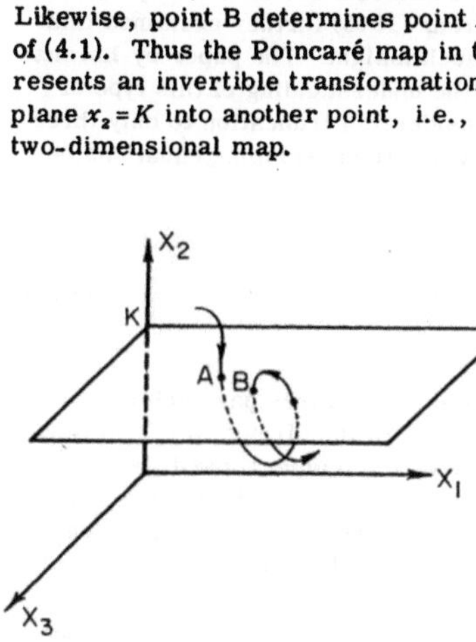

FIG. 13. Poincaré map of Eq. (4.1).

As an example, we consider the following system of two nonautonomous ordinary differential equations:

$$\frac{dp}{dt} = f(q) \sum_{n=-\infty}^{+\infty} \delta(\omega t - 2n\pi) - \nu(p - p_0),$$

$$\frac{dq}{dt} = p,$$

where $\delta(\theta)$ denotes the delta function of θ. This can be written as three autonomous equations

$$\frac{dp}{dt} = f(q) \sum_{n=-\infty}^{+\infty} \delta(\theta - 2n\pi) - \nu(p - p_0), \tag{4.2a}$$

$$\frac{dq}{dt} = p, \tag{4.2b}$$

$$\frac{d\theta}{dt} = \omega. \tag{4.2c}$$

We take the surface of section to be $\theta = 2m\pi - \varepsilon$, where $\varepsilon \to 0^+$. Defining $(p_m, q_m) = \lim_{\varepsilon \to 0^+} [p(t_m - \varepsilon), q(t_m - \varepsilon)]$, and $t_m = 2m\pi/\omega$, Eqs. (4.2) then yield by simple integration, the following two-dimensional map:

$$y_{m+1} = e^{-\Gamma}[y_m + f(q_m)] \tag{4.3a}$$

$$q_{m+1} = q_m + (1 - e^{-\Gamma})\nu^{-1}[y_m + f(q_m)] + 2\pi p_0/\omega, \tag{4.3b}$$

where $\Gamma = 2\pi\nu/\omega$, and $y_m = p_m - p_0$. By a suitable choice of $f(q)$ this two-dimensional map may be reduced to that studied by Zaslavskii (1978) and discussed in the preceding section.

Now, we turn to a discussion of the evolution of phase-space volumes as governed by the system (4.1). That is, we consider the volume (V) enclosed by some closed surface S in the x_1, x_2, x_3 phase space, and let the surface evolve by having each point on the surface follow an orbit generated by (4.1). By the divergence theorem

$$\frac{dV}{dt} = \int_V \left(\sum_{i=1}^{3} \frac{\partial f_i}{\partial x_i} \right) dx_1 dx_2 dx_3. \tag{4.4}$$

In the special case where divergence of the phase-space flow, $\sum_{i=1}^3 \partial f_i/\partial x_i$, is a negative constant, $\sum_{i=1}^3 \partial f_i/\partial x_i = -k$, Eq. (4.4) yields $dV/dt = -kV$, so that

$$V(t) = V(0)\exp(-kt). \tag{4.5}$$

Thus phase-space volumes shrink exponentially in time. Many of the physical examples yielding a system of the form (4.1) that have been investigated for strange attractors also happen to have a constant negative divergence. We shall restrict our discussion in the remainder of this section to this case. [In fact, Eqs. (4.2) fall in this class with the flow divergence being $-\nu$.] In Sec. V, more general cases, without negative divergence, will be discussed.

The special case of three ordinary autonomous differential equations with negative phase-space flow divergence presents a very clear case for the necessity of introducing the concept of a strange attractor. Since phase-space volume contracts to zero in the limit of large time, it follows that any attractor must have zero volume. A natural assumption might then be that the attractor would have to be a surface (two dimensional), a curve (one dimensional), or a point (zero dimensional). However, none of these allows chaotic motion. In particular, not even the highest dimension of the above three possibilities (two) allows chaos. For example, for orbits within a finite section of a plane, the Poincaré-Bendixson theorem shows that the only possible attractor for the orbit must be either a point, a simple closed curve, or a self-intersecting closed curve[3] (e.g., a figure eight (e.g., Hirsch and Smale, 1974). Thus if one observes chaotic motion in the system (4.1), and if (4.1) has negative phase-space flow divergence, then one is faced with something of a paradox. One way out is to realize that attractors with zero volume need not only have dimension zero, one, or two, but can, in fact, have noninteger dimension. In particular, chaotic motion is possible if Eqs. (4.1) have an attractor of dimension greater than two but less than three (the latter so that the volume of the attractor is zero), i.e., a strange attractor. We will now outline some work on physically interesting systems exhibiting strange attractors. The three examples which we will discuss are (1) the so-called Lorentz system, which represents a simple model of the convective motions that result when a temperature difference is maintained across a fluid layer which is subjected to gravity, (2) a simple model for the saturation of an unstable mode by coupling energy through quadratic nonlinearities from the unstable mode to damped modes, and (3) a model similar to (2) but with cubic nonlinearities (resulting, for example, from a nonlinear Schrödinger equation).

B. Examples

The first two examples, (1) and (2), are for parameters such that the phase-space contraction rate is large. This has the consequence that the attractor in the surface of section appears to be one dimensional. Actually, closer examination under magnification would reveal thickness containing structure within the attractor. Furthermore, the dimension in the surface of section must be $1<d<2$, as for any invertible two-dimensional map with a strange attractor. Here, however, d is only slightly larger than one. This leads to the result that the system dynamics can be well approximated by a one-dimensional noninvertible map of the types discussed in Sec. II. [Recall that, as discussed in Sec. III, this also is true in the analogous case of the two-dimensional map, Eq. (3.2) with β small.] The example in (1) yields a one-dimensional map that is like that given by Eq. (2.2); it has a sharp (nondifferentiable) maximum, and, in the notation of Sec. II, $|F'(x)|>1$ for all points on the chaotic orbit generated by $x_{n+1}=F(x_n)$. The example in (2) yields a one-dimensional map that is like that given in Eq. (2.3). As a consequence, example (2) leads to a pattern of period doubling bifurcations, tangent bifurcations, and chaotic orbits that is essentially the same as that for the quadratic map, Eq. (2.3) (cf. Fig. 7). Example (3) will be discussed for a range of parameters for which the contraction rate is not large. Thus, for example (3), non-one-dimensional structure will be readily evident in the surface of section. Furthermore, it will be shown that this structure appears to have approximate scale-invariant properties upon magnification, in analogy to the Cantor set example of Sec. III (cf. Fig. 8) and to the Henon map (cf. Figs. 9 and 10). Thus these three examples serve to illustrate the relevance of essential features of one- and two-dimensional chaotic maps to systems of ordinary differential equations modeling physical systems.

1. Lorenz's treatment of the Benard instability

Consider two rigid plane parallel walls at $z=0, L$ with a fluid occupying the space in between. Gravity is in the negative z direction, and the plate at $z=0$ is maintained at a higher temperature than the plate at $z=L$, $T_0>T_L$. A possible equilibrium of this system is one in which the fluid is at rest and heat is transported from $z=0$ to $z=L$ via thermal conduction. Lord Rayleigh studied the linear stability of this equilibrium and found that if (T_0-T_L) exceeds a critical value, then the system becomes unstable to perturbations in the form of circulating fluid flow. The linear analysis cannot, however, be used to specify the ultimate nonlinear state of the fluid once instability sets in. Considering variations only in two dimensions, Saltzman (1962) derived a set of nonlinear ordinary differential equations by expanding the stream function and the temperature perturbation in double spatial Fourier series, with coefficients functions of t alone. By substituting the series into the original governing set of partial differential equations and truncating the infinite sum to a finite number of terms, he obtained a set of ordinary differential equations. Lorenz (1963) further examined this problem and added much insight. The paper by Lorenz has greatly added to the understanding of this type of problem. Lorenz considered a truncation to only three Fourier modes, for which the describing equations become

$$dX/dt=-\sigma X+\sigma Y\,, \tag{4.6a}$$

$$dY/dt=-XZ+rX-Y\,, \tag{4.6b}$$

$$dZ/dt=XY-bZ\,, \tag{4.6c}$$

where σ, r, and b are dimensionless parameters of the system. X is proportional to the circulatory fluid flow velocity, Y characterizes the temperature difference between ascending and descending fluid elements, and Z is proportional to the distortion of the vertical temperature profile from its equilibrium (which is linear with height).

Setting $dX/dt=dY/dt=dZ/dt=0$, we find that (4.6) possesses steady-state solutions, $X=Y=Z=0$, and if $r>1$,

[3]For motion described by more than three autonomous ordinary differential equations with phase-space contraction, $\sum_i \partial f_i/\partial x_i<0$, or by three autonomous equations where $\sum \partial f_i/\partial x_i$ is not everywhere negative, an attractor which is a toroidal surface is possible. Motion on the toroidal surface is doubly periodic; that is, the solution of (4.1) can be represented as $x_i(t)=l_i(t,t)$, where l_i is periodic in both variables, $l_i(t+T_1,t+T_2)=l_i(t,t)$, and T_1/T_2 is not a rational number. Alternatively, if the orbit is doubly periodic, then the only frequency components of the Fourier spectrum of $x_i(t)$ are $(m/T_1)+(n/T_2)$, where m and n are integers. Doubly periodic motion in three dimensions with phase-space contraction is not possible, because the volume V enclosed by the toroidal attractor would be time independent, contrary to Eq. (4.5). At any rate, doubly periodic orbits are not chaotic, either.

$X = Y = \pm[b(r-1)]^{1/2}$, $Z = r - 1$. The equilibrium $X = Y = Z = 0$ represents the case of no fluid flow, while the two possible equilibria for $r > 1$ represent steady circulating convection. Linearization of Eqs. (4.6) about these equilibria reveals that the $X = Y = Z = 0$ equilibrium looses stability when $r > 1$, while the steadily convecting equilibria become unstable if

$$r > \sigma(\sigma + b + 3)(\sigma - b - 1) \equiv r_c \tag{4.7}$$

and $\sigma > b + 1$.

Lorenz numerically considered the solution of (4.6) for a case in which (4.7) is satisfied: $\sigma = 10$, $b = \frac{8}{3}$, $r = 28$. [Note that from (4.7) $r_c = 24.74$ in this case.] For these values of the parameters he obtained a chaotic solution and examined its properties in detail. It is instructive also briefly to consider the character of the solutions of (4.6) for other parameters. In particular, say $\sigma = 10$, $b = \frac{8}{3}$, and increase r. For $r < 1$ all initial conditions eventually decay to the convection-free equilibrium. For $1 < r \lesssim 24.06$ all initial conditions eventually settle into one of the two stable convective equilibria. For $24.07 \lesssim r \lesssim 24.74$, depending on initial conditions, the solution settles into either a chaotic motion or into one of the two stable convective equilibria. This phenomenon of dependence upon initial conditions is called hysteresis. For $r > 24.74$ the eventual orbit is chaotic for all initial conditions. [For r still larger, say $r \gtrsim 50$, the behavior can again change (Shimzu and Morioda, 1978).] Even in the range $r < 24.06$ interesting behavior arises, in particular, the phenomenon of "chaotic transients" first noted by Yorke and Yorke (1979) and by Kaplan and Yorke (1979b). These authors find that the evolution from a given initial condition initially looks very much like the time dependence of the solution in the chaotic regime, after which the solution quickly settles into one of the two stable equilibria. Furthermore, they find that the time to settle into equilibrium is sensitive to the initial condition, and can be fairly long with its duration increasing with r. Similar phenomena are to be expected in other cases (e.g., Shimzu and Morioda, 1978) and in other systems.

We now return to a description of Lorenz's results for $r = 28$. Figure 14 shows a projection of the phase-space orbit of the system onto the YZ plane. The points labeled C and C' are the steady convection equilibria points. Evidently, the orbit spirals outward from one of the points C or C' until it exceeds some critical distance from the origin, at which point it starts spiraling about the other point. If one were to make a sequential list of the number of circuits the solution makes around one point before it switches to the other point, the sequence would appear to be chaotic. By examination of the solution, Lorenz has deduced that the orbit appears to be confined to a surface. Actually this apparent "surface" must have some small thickness, inside of which is embedded the more complicated structure of the strange attractor. In fact, if one were to pass a line through this surface normal to it, one would find that the intersection of the line with the surface is a set of dimension $0 < d < 1$, i.e., like the Cantor set of Fig. 8. However, since the thickness of the strange attractor is small, the presence of structure in this intersection would only be visible upon magnification, and, unmagnified, it would appear to be a point.

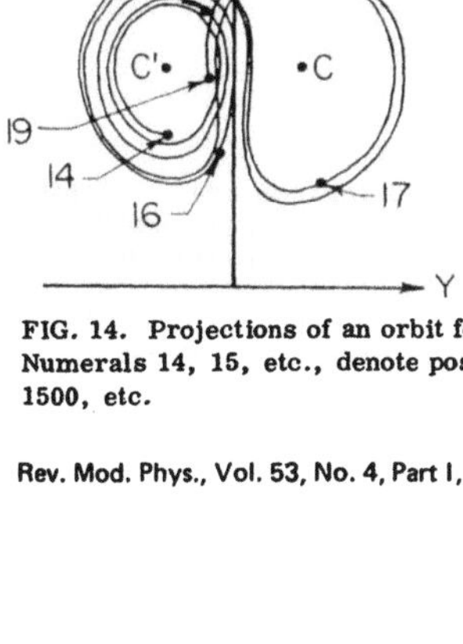

FIG. 14. Projections of an orbit for $r = 28$ onto the YZ plane. Numerals 14, 15, etc., denote positions at iterations 1400, 1500, etc.

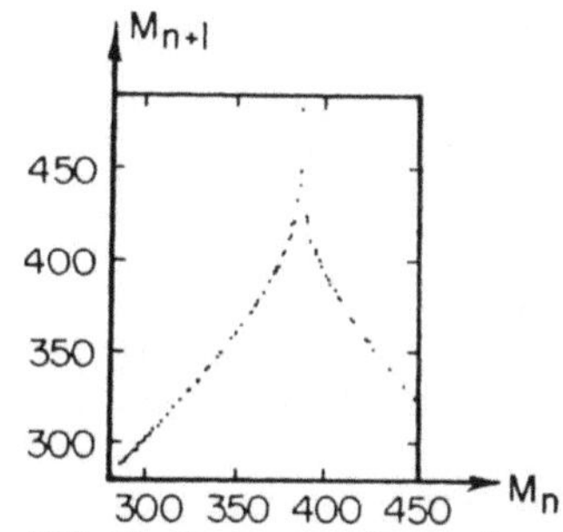

FIG. 15. Maxima vs subsequent maxima of Z occurring during 6000 iterations.

Figure 15 shows a plot, obtained by Lorenz, of M_n, the nth maximum of Z, versus M_{n+1}, the value of the following maximum. It is clear that an (approximate) one-dimensional map is generated. Furthermore, $|dM_{n+1}/dM_n| > 1$, which is similar to the result for Eq. (2.2) with $a > \frac{1}{2}$. Thus, as for Eq. (2.2), we expect this one-dimensional map to generate a chaotic sequence. For later comparison with the results of example (2), note that the maxima of Z may be regarded to lie in a surface of section, $bZ = XY$ [put $dZ/dt = 0$ in (4.6c)]. For further discussion of the Lorenz attractor see Lanford (1976), Afraymovich *et al.* (1977), and Bunimovich and Sinai (1977).

2. Instability saturation by quadratically nonlinear mode coupling

An important problem in plasma physics (and in other fields as well) is that of determining the nonlinear state resulting from a linearly unstable wave. An elementary process by which saturation can occur is that of resonant three-wave mode coupling of energy in the linearly unstable wave to two other waves which are linearly damped. The following normalized system of equations describes this process:

$$dC_1/dt = C_1 + C_2 C_3 \exp(i\delta t),$$

$$dC_{2,3}/dt = -\gamma_{2,3} C_{2,3} - C_1 C_{3,2}^* \exp(i\delta t),$$

where C_i $(i=1,2,3)$ are time-dependent complex wave amplitudes, C_i^* is the complex conjugate of C_i, times have been normalized to the growth rate of wave 1 (the higher frequency wave), δ represents the effect of a mismatch in the frequency resonance, and amplitudes have been normalized so that the coefficients of the nonlinear terms are one. Introducing $a_1 \exp(i\phi_1)=C_1$, $a_{2,3}\exp(i\phi_{2,3})=C_{2,3}\exp(i\delta t/2)$, and $\phi=\phi_1-\phi_2-\phi_3$, where a_i and ϕ_i are real, the previous system gives four real equations for $a_{1,2,3}$ and ϕ. These equations readily yield

$$d(a_2^2-a_3^2)/dt=-2(\gamma_2 a_2^2-\gamma_3 a_3^2).$$

Thus in the special case $\gamma_2=\gamma_3$, $a_2^2-a_3^2$ decreases exponentially. Restricting consideration to $\gamma_2=\gamma_3\equiv\gamma$ and $a_2=a_3$, the basic equations then become (Vyshkind and Rabinovich, 1976)

$$da_1/dt=a_1+a_2^2\cos\phi\,,$$

$$da_2/dt=-a_2(\gamma+a_1\cos\phi)\,,$$

$$d\phi/dt=-\delta+a_1^{-1}(2a_1^2-a_2^2)\sin\phi\,.$$

For definiteness in examining this system, Wersinger *et al.* (1980a, 1980b, 1980c) chose $\delta=2$ and studied the properties of numerical solutions[4] for a range of γ from $\gamma=1$ to $\gamma=25$. For $\gamma\lesssim 3$ the damping was not strong enough to arrest the instability, and the system evolution was apparently unbounded. For $3\leq\gamma\leq 8.5$, all initial conditions led to a simple periodic limit cycle. Using $\phi(t)=\pi/2$ as the surface of section, this limit cycle was manifested as a single point in the surface of section. For γ somewhat larger than 8.5 a limit cycle is still observed, but the single fixed point in the surface of section that previously manifested the limit cycle splits into two points that are visited alternately. Correspondingly, the single-peak-per-period function for $3\lesssim\gamma\lesssim 8.5$ becomes a function with two alternating maxima (cf. Figs. 16 and 17). As γ is increased, the two-point cycle splits into a four-point cycle (for $\gamma \gtrsim 11.9$), which splits into an eight-point cycle (for $\gamma \cong 13.15$), ..., followed by a region of γ ($\gamma\lesssim 13.4$) for

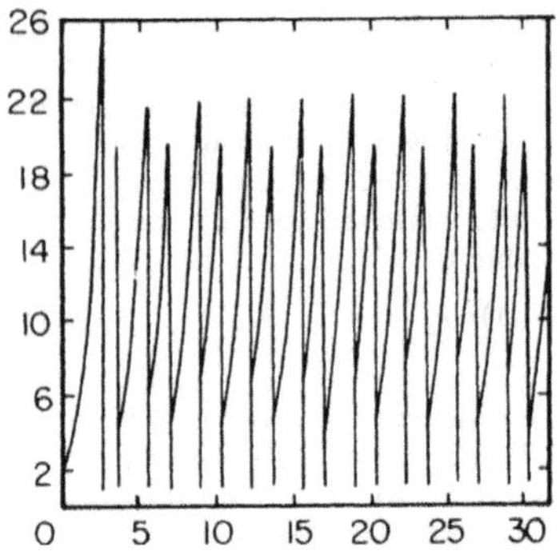

FIG. 17. a_1 vs t for $\gamma=9$.

which the generated sequence appears chaotic[5] (cf. Fig. 18). For still larger γ a sequence from a three-point cycle to a six-point cycle, etc., briefly appears. To summarize, the phenomenology is precisely the same as that depicted in Fig. 7 for the one-dimensional quadratic map,[6] Eq. (2.3)! Similar results for entirely different physical situations have also been obtained [see, for example, Tomita and Kai (1979)]. To make the correspondence with Eq. (2.3) more concrete, first consider results from the numerically generated surface of section shown in Fig. 19 for a value of γ in the chaotic regime, $\gamma=15$. It is seen that the points generated in the surface of section *appear* to lie on an arc. Since this arc has no visible thickness, it is natural to attempt an approximate reduction to a one-dimensional map. This can be done, for example, by plotting $x_{n+1}\equiv a_2(t_{n+1})$ versus $x_n\equiv a_2(t_n)$, where t_n is the nth time at which the system orbit pierces the surface of section. The points so generated (cf. Fig. 20) lie along a curve $x_{n+1}=F(x_n)$. By a change of variables (e.g., $\bar{x}_n\equiv(\text{const})-x_n$) the one-dimensional map in Fig. 20 can be turned upside down so that it has a smooth rounded maximum (rather than

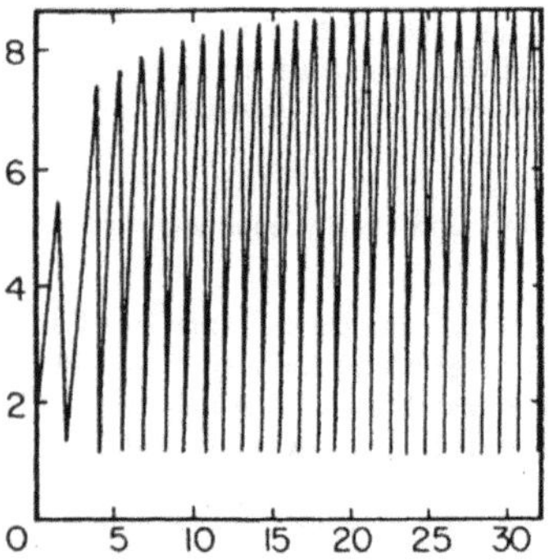

FIG. 16. a_1 vs t for $\gamma=3$. After an initial transient, the solution settles into a limit cycle.

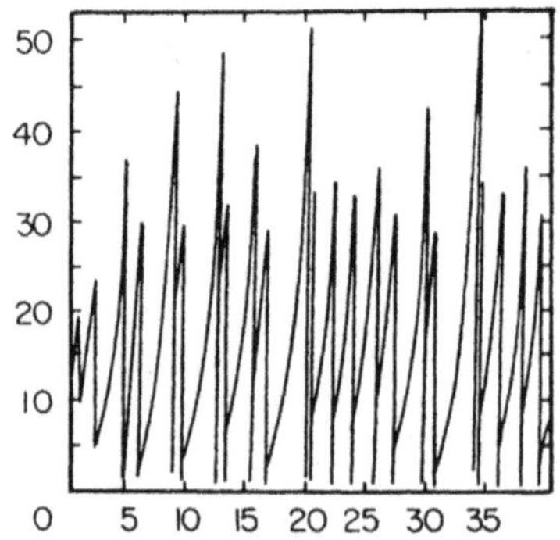

FIG. 18. a_1 vs t for $\gamma=15$.

[4]For larger values of δ time-independent equilibrium solutions of the equations exist.

[5]Feigenbaum (1979) has shown that the frequency power spectrum at the accumulation point of the period doubling bifurcations of a differential equation system has universal properties.

[6]This includes the phenomenon of noisy 2^p cycles studied by Crutchfield *et al.* (1980) by computations on another set of differential equations.

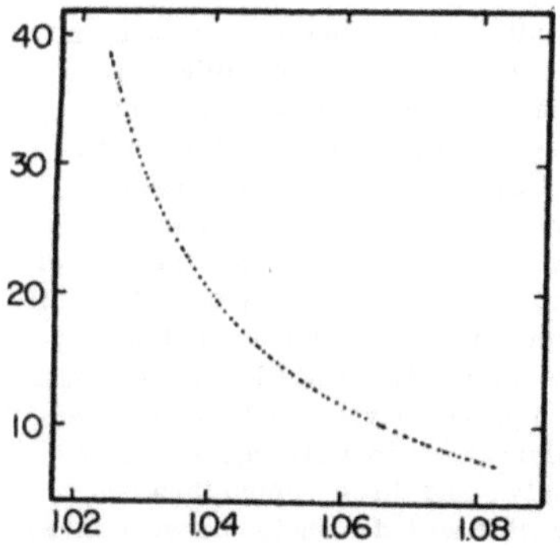

FIG. 19. Surface of section for $\gamma = 15$.

minimum). Thus we see that the one-dimensional map generated by the particular system discussed is of the same general character as the simple quadratic map discussed in Sec. II, Eq. (2.3), i.e., it is concave down with a smooth rounded maximum.

3. Instability saturation by cubicly nonlinear mode coupling

Here we very briefly describe another study of nonlinear mode coupling saturation of an unstable plasma wave. The physical situation, however, is somewhat different from that considered in example (2) in that a resonant three-wave process is precluded by the wave dispersion relation. The physical system is assumed to be well modeled by the following normalized, one-dimensional, nonlinear Schröedinger equation with linear wave growth and damping included,

$$i\left(\frac{\partial E}{\partial t}+\hat{\gamma}E\right)+\frac{\partial^2 E}{\partial x^2}+[\,|E|^2-|E|_0^2]E=0\,,$$

where E is the complex amplitude coefficient of the x-directed electric field, $|E|_0^2$ denotes the spatial average of $|E|^2$, and $\hat{\gamma}$ is a linear growth-damping operator defined so that the Fourier transform of $\hat{\gamma}E(x,t)$ is $\gamma(k)E_k(t)$ with $E_k(t)$ the Fourier coefficient of E and $\gamma(k)$ the linear damping rate of a wave of wave number k [$\gamma(k)<0$ for growth]. This equation may be used to model a situation where an electron beam with a thermal spread is injected into a plasma [cf. Russell and Ott (1980) and references therein]. Under suitable conditions on $\gamma(k)$ it is possible to consider $E(x,t)$ to consist of just three wave number components, k_0, k_1, k_2, with $2k_0=k_1+k_2$. There then results a system of ordinary differential equations for the three complex wave amplitudes corresponding to the three wave numbers,

$$\frac{dE_0}{dt}=-\gamma(k_0)E_0+i[\,|E_1|^2E_0+|E_2|^2E_0+2E_0^*E_1E_2\exp(2i\delta t)]\,,$$

$$\frac{dE_1}{dt}=-\gamma(k_1)E_1+i[\,|E_0|^2E_1+|E_2|^2E_1+E_0^2E_2^*\exp(-2i\delta t)]\,,$$

$$\frac{dE_2}{dt}=-\gamma(k_2)E_2+i[\,|E_0|^2E_2+|E_1|^2E_2+E_0^2E_1^*\exp(-2i\delta t)]\,,$$

where we assume $\gamma(k_0)<0$ and $\gamma(k_{1,2})>0$. As in example (2), if $\gamma(k_1)=\gamma(k_2)$, these three complex equations can be reduced to a system of three real equations:

$$\frac{da_0}{dt}=a_0+2a_0a_1^2\sin\phi\,,$$

$$\frac{da_1}{dt}=-\gamma a_1-a_0^2a_1\sin\phi\,,$$

$$\frac{d\phi}{dt}=-2\delta+2(a_1^2-a_0^2)+2(2a_1^2-a_0^2)\cos\phi\,.$$

The above system has been examined analytically and numerically as a function of the dimensionless parameters of the system (δ and γ). It was found that the model exhibits a wealth of characteristic dynamical behavior, including stationary equilibria, bifurcations from stationary equilibria to periodic orbits, period doubling bifurcations, chaotic solutions on a strange attractor, tangent bifurcations from chaotic to periodic solutions, transient chaos, and hysteresis. It is not our purpose here to detail the behavior of this system. Rather, we wish to use it as an illustration of scale invariant nature in a strange attractor. For a range of the parameters δ and γ, this system exhibits chaotic time dependence with only a moderate phase-space volume contraction rate. Thus structure of the associated strange attractor is evident in the surface of section (Fig. 21). Furthermore, upon magnification, it is evi-

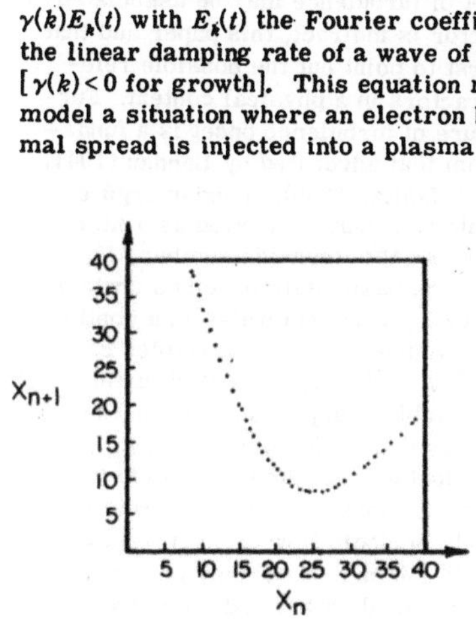

FIG. 20. Points x_{n+1} vs x_n lie approximately on a curve defining a one-dimensional map.

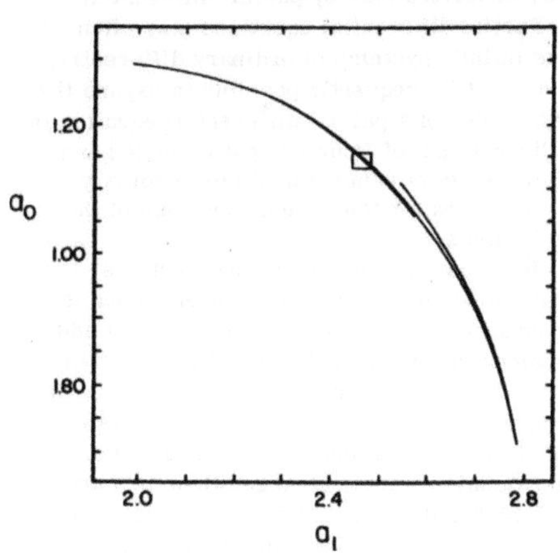

FIG. 21. Surface of section for a particular set of parameters for example (3).

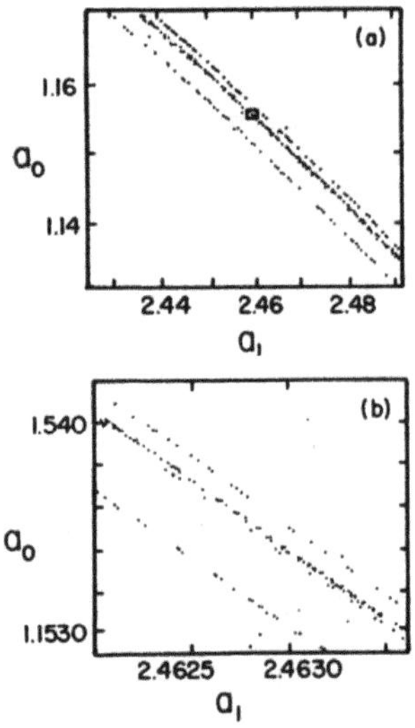

FIG. 22. (a) Magnification of the region in Fig. 21 shown in the rectangle, and (b) magnification of the rectangular region indicated in Fig. 22(a). Scale invariance is evident.

dent that the structure appears to be approximately scale invariant. This is illustrated in Fig. 22 which is to be considered analogous to Fig. 10 for the Henon map.

V. PARTIAL DIFFERENTIAL EQUATIONS, ORDINARY DIFFERENTIAL EQUATIONS, AND TURBULENCE

It is natural to ask whether the phenomena revealed by the examples of the preceding section carry over to more complicated systems. In particular, what happens if the restriction of phase-space volume contraction is lifted and if the dimension of the system is larger than three? Furthermore, the examples in Sec. IV were meant as approximate models for phenomena which are more exactly described only by partial differential equations. Partial differential equations may often be thought of as infinite systems of ordinary differential equations; e.g., it is frequently possible to expand the dependent variables of a partial differential equation in an infinite discrete set of Fourier spatial modes and to derive an infinite set of coupled nonlinear ordinary differential equations for the time dependence of the Fourier coefficients.

It seems clear that such more general systems might display the same characteristic phenomena as those discussed in Sec. IV, but could also reveal additional phenomena ruled out by the specific constraints adopted in Sec. IV. For example, we have found in Sec. IV cases where chaotic solutions occur for a strange attractor of dimension between two and three. For higher dimensional systems, strange attractors are possible with higher dimensionality—e.g., one might have a strange attractor of dimension between five and six if a system of n differential equations with $n \geq 6$ were investigated. (Such higher dimensional strange attractors might be quite difficult to diagnose in actual situations.) At any rate, it seems reasonable to suppose that, as a parameter which characterizes the strength of destabilizing forces in a system described by partial differential equations is cranked up (e.g., the Reynolds number), the general (although not uniform) tendency would be toward motion on attractors of increasing dimension. For example, a stable attracting stationary equilibrium point (zero-dimensional attractor) might bifurcate to a periodic orbit (one-dimensional attractor), which then proceeds via an infinite number of period doubling bifurcations to a strange attractor of dimension between two and three, which then becomes a strange attractor with dimension between three and four, etc. Another possible sequence might be a stationary point (zero dimensions) that bifurcates to a periodic orbit (one dimension), which bifurcates to a doubly periodic orbit (a two-dimensional attractor formed by the surface of a torus), which then bifurcates to a strange attractor (dimension> 2). The former route to a strange attractor [cf. Fig. 23(a)] has been demonstrated in Sec. IV.B.2. The route to a strange attractor via transition from a doubly periodic orbit (motion on a two-dimensional toroidal surface) is illustrated in Fig. 23(b) and was first discussed in a pioneering paper by Ruelle and Takens (1971) [cf. also Newhouse *et al.*, 1978). (Note that this route to a strange attractor [Fig. 23(b)] is specifically ruled out for the examples in Sec. IV, since, as previously mentioned, doubly periodic orbits are not possible in a system of three ordinary differential equations with phase-space contraction.)

Ruelle and Takens (1971) conjectured that small nonlinearities would destroy triply periodic motions. They therefore reasoned that, as the Reynolds number of a fluid flow is increased, the sequence in Fig. 23(b) ought to occur. In particular, they concluded that the last step, (doubly periodic motion)→(strange attractor), is quite likely, since doubly periodic motions cannot bifurcate to triply periodic motions if the latter are unstable. Although the reasoning leading to their conclusion that the onset of turbulence may be associated with a strange attractor is indirect, this paper and that of Lorenz are the first to point out the possible relevance of strange attractors in a physical context. Furthermore, their picture of turbulence onset is a fundamental departure from that advocated by Landau (1941) (cf. also Landau and Lifshitz, 1959). Landau argued that turbulence in fluid flow may be viewed as a hierarchy of instabilities. As the Reynolds number, R, is increased from zero, the basic state becomes unstable to a mode of frequency $\tilde{\omega}_1$ which saturates in a nonlinear periodic state for which dependent variables can be written in the form $\sum_l a_l \exp(-il\omega_1 t)$; as R is further increased, another instability appears at $\tilde{\omega}_2$ and the saturated state becomes doubly periodic, $\sum_{l,m} a_{lm} \times \exp(-il\omega_1 t - im\omega_2 t)$; further increase of R leads to the successive appearance of more and more discrete frequencies so that doubly periodic flow (ω_1, ω_2) transforms to triply periodic, then to quadruply periodic, etc. Thus, as R is increased, more and more frequencies are present, and the flow pattern becomes more and more complicated. Thus in this model the

(A)

{STATIONARY POINT} → {BIFURCATION TO A PERIODIC ORBIT} → {PERIOD DOUBLING BIFURCATIONS} → {ACCUMULATION POINT OF AN INFINITE NUMBER OF PERIOD DOUBLING BIFURCATIONS} → {STRANGE ATTRACTOR}

(B)

{STATIONARY POINT} → {BIFURCATION TO A PERIODIC ORBIT} → {BIFURCATION TO A DOUBLY PERIODIC ORBIT} → {STRANGE ATTRACTOR}

FIG. 23. Two possible routes to a strange attractor.

Fourier spectrum is always discrete and approximates a continuum only in the case where a large number of discrete frequencies are present. Furthermore, the flow is never truly chaotic, since the time correlation functions of multiply periodic functions do not tend to zero for large argument. On the other hand, the appearance of a strange attractor (as, for example, by the sequence of events in Fig. 23) can lead to turbulent motions in a direct way. Recent experiments on the onset of turbulence induced by various instabilities lend support to the idea that turbulence onset may be due to the appearance of a strange attractor (Swinney and Gollub, 1978; Ahlers and Behringer, 1978; Fenstermacher *et al.*, 1979; Walden and Donnelly, 1979; Lashinsky, 1980; Donnelly *et al.*, 1980; Moon and Holmes, 1980).

We now discuss some evidence that the sequence to a strange attractor via a route something like that depicted in Fig. 23(b) can occur. To begin we note the recent study by Curry (1978). In this paper, the author considers the same physical problem as that considered by Lorenz, but includes a greater number of Fourier spatial modes before truncating the series representation of the dependent variables. The result is a system of 14 coupled ordinary differential equations, rather than the three studied by Lorenz. The numerical results of Curry are somewhat different than those of Lorenz. In particular, he finds that chaotic time dependence is preceded by doubly periodic motion on a two-dimensional toroidal surface that is embedded in the full 14-dimensional phase space. The existence of the torus is found by using the surface of section technique: a surface of section $x_1 = \text{const}$ was chosen (where x_1 is one of the 14 variables), and another two of the 14 variables, which we denote by x_2 and x_3, were singled out. Every time the orbit crossed the surface of section the x_2, x_3 coordinates of the crossing point were plotted. Figure 24 shows a typical result for the case where a doubly periodic orbit exists. As can be seen from Fig. 24, the points lie on a closed curve. This is the result to be expected for doubly periodic behavior, since the intersection of a torus with a plane (the surface of section) is a closed curve. Related numerical studies have been performed by Yahata (1978) and Sherman and McLaughlin (1978).

Another relevant study is that of Curry and Yorke (1978), who consider a simple two-dimensional map that models the onset of chaos preceded by doubly periodic motion. The mapping that they study is one which may conveniently be represented as the result of two successive transformations, the first in polar coordinates

$$(\rho, \theta) \to [\varepsilon \ln(1+\rho), \theta + \theta_0] \tag{5.1a}$$

and the second in rectangular coordinates

$$(x, y) \to (x, y + x^2), \tag{5.1b}$$

where $\varepsilon \geq 0$ and $\theta_0 \geq 0$ are parameters to be chosen. Note that for $\varepsilon > 1$, the mapping $\rho \to \varepsilon \ln(1+\rho)$ has two fixed points, one of which is $\rho = 0$ and the other of which (denoted by ρ_ε) is the positive root of $\rho_\varepsilon = \varepsilon \ln(1+\rho_\varepsilon)$. For $1 \geq \varepsilon > 0$, the only fixed point is $\rho = 0$. Consider first the iteration of the polar map, $(\rho, \theta) \to [\varepsilon \ln(1+\rho), \theta + \theta_0]$, alone. For $\varepsilon > 1$, all points tend asymptotically to the circle $\rho = \rho_\varepsilon > 0$. Furthermore, if the rotation angle θ_0 is not 2π times a rational number, then the orbit fills up the entire circle $\rho = \rho_\varepsilon$. As ε approaches one from above, $\rho_\varepsilon \to 0$, so that is may be considered that this map undergoes a bifurcation from a stable fixed point to an attracting orbit on the circle $\rho = \rho_\varepsilon$.

Thus the map $(\rho, \theta) \to [\varepsilon \ln(1+\rho), \theta + \theta_0]$ can be viewed as modeling a surface of section for some differential equation that undergoes a bifurcation of a stable periodic orbit to a doubly periodic orbit on a torus (where $\rho = \rho_\varepsilon$ represents the intersection of this torus with the surface of section and $\rho = 0$ represents the intersection of the periodic orbit with the surface of section).

For ε only slightly larger than one, $\rho_\varepsilon \ll 1$, and the map $(x, y) \to (x, y + x^2)$ is almost the identity map, since

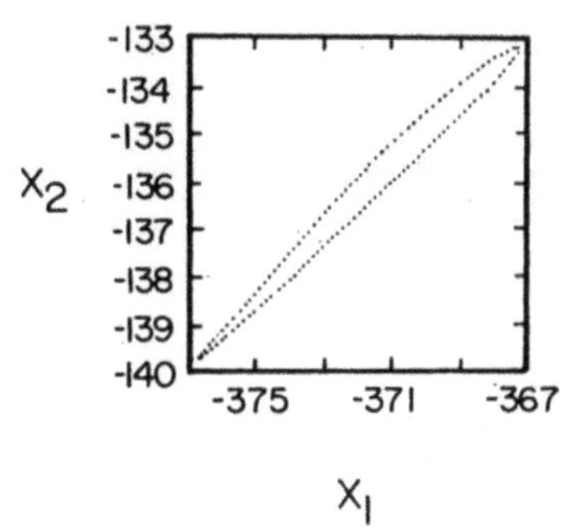

FIG. 24. Orbit intersections with the surface of section, obtained by Curry (1978).

the quadratic term, x^2, is comparatively small. Thus if one considers the map resulting from the successive application of the two above-mentioned transformations, then for $\varepsilon>1$ the map $(x,y)\to(x,y+x^2)$ introduces a nonlinearity that becomes stronger as ε is increased. Curry and Yorke have described a sequence of numerical studies of the attracting set for the composite map, Eqs. (5.1), with $\theta_0=2$. At $\varepsilon=1.01$ a nearly circular loop encircles the origin. As ε increases, the loop distorts and increases in size until $\varepsilon=1.28$. For $1.39\gtrsim\varepsilon>1.28$ an attracting period three orbit is found. Immediately after $\varepsilon\cong1.3953$, an apparently connected loop returns. This apparent loop, however, actually has a complicated structure and is, in fact, a strange attractor.

As ε increases somewhat, the strange aspect of the attractor becomes more apparent, as evidenced by Fig. 25 for $\varepsilon=1.45$. The bump containing the point P_1 in Fig. 25 is mapped to P_2, to P_3, to P_4, ... to P_{11}. The cusplike structure at P_4 shows signs of being flattened against the rest of the set. Each successive cusp is more flattened and more elongated. Soon the cusps are so flattened that they are no longer discernable. This sequence, P_i, is, however, infinite. Note, too, that the successive elongation of the cusps represents a stretching apart of two nearby points. As discussed in Sec. III, such stretching should be expected to lead to chaotic behavior, and this is in fact the case here. Similar behavior has also been observed in a map studied by Coullet *et al.* (1980).

The character of the strange attractor for the example of Curry and Yorke appears to be that of a very complicated curve folded over on itself an infinite number of times and with an infinite length. In this connection, it may be of some interest to give a simple example of an infinite length curve with Hausdorf dimension between one and two (Mandelbrot, 1977). Consider the sequence of operations shown in Fig. 26. We start with an equilateral triangle, divide each of the sides in thirds, and erect smaller equilateral triangles on the middle thirds on each side. The process is repeated twice in the figure. On each application of the process the length of the bounding curve increases by $\frac{4}{3}$. In the limit that the process is repeated an infinite number of times, the length of the bounding curve approaches infinity, although the curve remains in a bounded region of the plane. Furthermore, it may be verified that the Hausdorf dimension of this curve is $\ln 4/\ln 3\cong1.26$.

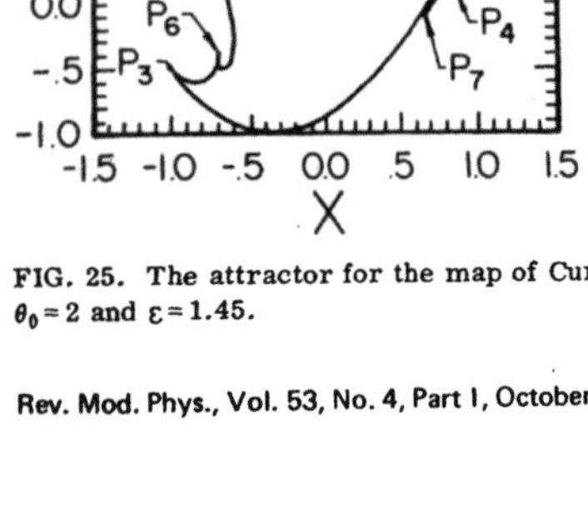

FIG. 25. The attractor for the map of Curry and Yorke with $\theta_0=2$ and $\varepsilon=1.45$.

FIG. 26. Illustration of the construction of a curve of dimension ln4/ln3.

VI. CONCLUSIONS

This subject matter is certain to receive much future attention from researchers in the physical sciences and in mathematics. It seems clear that there is a great need for further work in order that the subject develop to the point that theory can provide answers to many of the most practical questions. For example, given a system of equations, can one predict the occurrence of a strange attractor? To what extent can properties of the strange attractor such as its dimension, associated distribution function, power spectra, and correlation functions be predicted? What is the distribution function on a strange attractor and what is the most efficient way to find it and characterize it? Further work will also certainly be done identifying physical systems which exhibit chaotic motions associated with strange attractors. The list of such systems (partially enumerated in the Introduction) is already impressive, and, as it grows, it is to be expected that interest on the part of physical scientists will also grow.

ACKNOWLEDGMENTS

The author wishes to thank Professor J. A. Yorke for many informative discussions. This work was supported by the U. S. Department of Energy.

REFERENCES

Adam, J. C., M. N. Bussac, and G. Laval, 1980, in *Intrinsic Stochasticity in Plasmas*, edited by G. Laval and D. Gresillon (Les Editions de Physique Courtaboeuf, Orsay, France), p. 415.
Afraymovich, V. S., V. V. Bykov, and L. P. Shilnikov, 1977, Dokl. Adak. Nauk, SSSR 234, 336.
Ahlers, G., and R. P. Behringer, 1978, Phys. Rev. Lett. 40, 712.
Bridges, R., and G. Rowlands, 1977, Phys. Lett. A 63, 189.
Bunimovich, L. A., and Ya. G. Sinai, 1977, in *Proceedings of the Spring School on Nonlinear Waves, Gorki, U.S.S.R.*
Chang, S.-J., and J. Wright, 1981, Phys. Rev. 23, 1419.
Chirikov, B. V., 1979, Phys. Rep. 52, 463.
Coullet, P., and J.-P. Eckmann, 1980, *Iterated Maps on the Interval as Dynamical Systems* (Birkhauser, Boston).
Coullet, P., and C. Tresser, 1980, J. Phys. (Paris), Lett. 41, L255.
Coullet, P., C. Tesser, and A. Arneodo, 1980, Phys. Lett. A 77, 327.
Crutchfield, J., D. Farmer, N. Packard, R. Shaw, G. Jones, and R. J. Donnelly, 1980, Phys. Lett. A 76, 1.
Crutchfield, J., M. Nauenberg, and J. Rudnick, 1981, Phys.

Rev. Lett. **46**, 933.
Curry, J. H., 1979, Commun. Math. Phys. **68**, 129.
Curry, J. H., and J. A. Yorke, 1978, in *The Structure of Attractors in Dynamical Systems*, edited by N. G. Markley, J. C. Martin, and W. Perrizo, Lecture Notes in Mathematics (Springer, Berlin), Vol. 668, p. 48.
Donnelly, R. J., K. Park, R. Shaw, and R. W. Walden, 1980, Phys. Rev. Lett. **44**, 987.
Feigenbaum, M. J., 1978, J. Stat. Phys. **19**, 25.
Feigenbaum, M. J., 1979, Phys. Lett. A **74**, 375.
Feit, S. D., 1978, Commun. Math. Phys. **61**, 249.
Fenstermacher, P. R., H. L. Swinney, and J. P. Gollub, 1979, J. Fluid Mech. **94**, 103.
Frederickson, P., J. L. Kaplan, and J. A. Yorke, 1980, preprint.
Greene, J. M., R. S. MacKay, F. Vivaldi, and M. J. Feigenbaum, 1981, preprint.
Greene, J. M., 1979, J. Math. Phys. **20**, 1183.
Guckenheimer, J., 1979, Commun. Math. Phys. **70**, 133.
Guckenheimer, J., G. Oster, and Ipaktchi, 1977, J. Math. Biol. **4**, 101.
Haken, H., 1975, Phys. Lett. A **53**, 77.
Helleman, R. H. G., 1980, in *Fundamental Problems in Statistical Mechanics 5*, edited by E. G. D. Cohen (North-Holland, Amsterdam), p. 165.
Henon, M., 1976, Commun. Math. Phys. **50**, 69.
Hirsch, M. W., and S. Smale, 1974, *Differential Equations, Dynamical Systems and Linear Algebra* (Academic, New York).
Holmes, P. J., 1977, Appl. Math. Modeling **1**, 362.
Holmes, P. J., 1980, preprint.
Huberman, B. A., and J. P. Cruchfield, 1979, Phys. Rev. Lett. **43**, 1743.
Huberman, B. A., and J. Rudnick, 1980, Phys. Rev. Lett. **45**, 154.
Huberman, B. A., and A. B. Zisook, 1981, Phys. Rev. Lett. **46**, 626.
Kaplan, J. L., and J. A. Yorke, 1979a, in *Functional differential Equations and Approximation of Fixed Points*, edited by H.-O. Peitzen and H.-O. Walther, Lecture Notes in Mathematics (Springer, Berlin), Vol. 730, p. 228.
Kaplan, J. L., and J. A. Yorke, 1979b, Commun. Math. Phys. **67**, 93.
Landau, L. D., 1941, C. R. (Dokl.) Acad. Sci. URSS **44**, 311.
Landau, L. D., and E. M. Lifshitz, 1959, *Fluid Mechanics* (Pergamon, London), Chap. 3.
Lanford, O. E., III, 1976, in *Statistical Mechanics and Dynamical Systems* (Mathematics Department, Duke University, Durham, North Carolina), Chap. 4.
Lashinsky, J., in *Intrinsic Stochasticity in Plasmas*, 1980, edited by G. Laval and D. Gresillon (Les Editions de Physique Courtaboeuf, Orsay, France), p. 425.
Li, T.-Y., and J. A. Yorke, Am. Math. Mon. **82**, 985 (1975).
Lorentz, E. N., 1979, presentation at the International Conference on Nonlinear Dynamics, New York Academy of Sciences (unpublished).
Lorenz, E. N., 1963, J. Atmos. Sci. **20**, 130.
Lozi, R., 1978, J. Phys. (Paris) Suppl. 8, tome 35, p. C5–9.
Mandelbrot, B. B., 1977, *Fractals* (Freeman, San Francisco).
Maschke, E. K., and B. Saramito, 1980, in *Intrinsic Stochasticity in Plasmas*, edited by G. Laval and D. Gresillon (Les Editions de Physique Courtaboeuf, Orsay, France), p. 383.
Masui, K., and P. K. C. Wang, 1980 (unpublished).
May, R. M., 1976, Nature **261**, 459.
McLaughlin, J., 1979, Phys. Rev. A **20**, 2114.
McLaughlin, J. B., and P. C. Martin, 1975, Phys. Rev. A **12**, 186.
Moon, F. C., and P. J. Holmes, 1979, J. Sound Vib. **65**, 275.
Newhouse, S., D. Ruelle, and F. Takens, 1978, Commun. Math. Phys. **64**, 35.
Rabinovitch, M. I., 1978, Sov. Phys.—Usp. **21**, 443.
Rechester, A. B., and T. H. Stix, 1976, Phys. Rev. Lett. **35**, 597.
Robbins, K. A., 1977, Math. Proc. Cambridge Philos. Soc. **82**, 209.
Rosenbluth, M. N., R. Z. Sagdeev, J. B. Taylor, and G. M. Zaslavskii, 1966, Nucl. Fusion **6**, 297.
Ruelle, D., 1977, in *Mathematical Problems in Theoretical Physics*, edited by G. Dell'Antonio, S. Doplicher, and G. Jona-Lasinio, Lecture Notes in Physics (Springer, Berlin), Vol. 80, p. 341.
Ruelle, D., and F. Takens, 1971, Commun. Math. Phys. **20**, 167.
Russell, D. A., J. D. Hanson, and E. Ott, 1980, Phys. Rev. Lett. **45**, 1175.
Russell, D. A., and E. Ott, 1980, preprint.
Saltzman, B., 1962, J. Atmos. Sci. **19**, 329.
Shaw, R., 1981, Z. Naturforsch. A **36**, 80.
Sherman, J., and J. B. McLaughlin, 1978, Commun. Math. Phys. **58**, 9.
Shimzu, T., and N. Morioda, 1978, Phys. Lett. A **69**, 148.
Shraiman, B., C. E. Wayne, and P. C. Martin, 1981, Phys. Rev. Lett. **46**, 935.
Simó, 1979, J. Stat. Phys. **21**, 465.
Sinai, Ya. G., 1977, in *Proceedings of the Spring School on Nonlinear Waves, Gorki, U.S.S.R.*
Stix, T. H., 1973, Phys. Rev. Lett. **30**, 833.
Swinney, H. L., and J. P. Gollub, 1978, Phys. Today 31 (8), 41.
Tomita, K., and T. Kai, 1979, J. Stat. Phys. **21**, 65.
Treve, Y., 1978, in *Topics in Nonlinear Dynamics*, edited by S. Jorna, AIP Conference Proceedings (American Institute of Physics, New York), no. 46, p. 147.
Treve, Y., and O. P. Manley, 1980, in *Intrinsic Stochasticity in Plasmas*, edited by G. Laval and D. Gresillon (Les Editions de Physique Courtaboeuf, Orsay, France), p. 393.
Vyshkind, S. Ya., 1978, Radiofizika **21**, 850 [Sov. Radiophys. **21**, 600].
Vyshkind, S. Ya., and M. I. Rabinovich, 1976, Zh. Eksp. Teor. Fiz. **71**, 557 [Sov. Phys.—JETP **44**, 292 (1976)].
Walden, R. W., and R. J. Donnelly, 1979, Phys. Rev. Lett. **42**, 301.
Wang, P. K. C., 1980, J. Math. Phys. **21**, 398.
Wersinger, J. M., J. M. Finn, and E. Ott, 1980a, in *Intrinsic Stochasticity in Plasmas*, edited by G. Laval and D. Gresillon (Les Editions de Physique Courtaboeuf, Orsay, France), p. 403.
Wersinger, J. M., J. M. Finn, and E. Ott, 1980b, Phys. Rev. Lett. **44**, 453.
Wersinger, J. M., J. M. Finn, and E. Ott, 1980c, Phys. Fluids **23**, 1142.
Wolf, A., and J. Swift, 1981, preprint.
Yahata, H., 1978, Prog. Theor. Phys. Suppl. **64**, 165.
Yorke, J. A., and E. D. Yorke, 1979, J. Stat. Phys. **21**, 263.
Yorke, J. A., and E. D. Yorke, 1981, in *Hydrodynamic Instabilities and the Transition to Turbulence*, edited by H. L. Swinney and J. P. Gollub (Springer, Berlin), p. 77.
Zaslavskii, G. M., 1978, Phys. Lett. A **69**, 145.
Zaslavskii, G. M., and Kh.-R. Ya Rachko, 1979, Zh. Eksp. Teor. Fiz. **76**, 2052 [Sov. Phys.—JETP **49**, 1039 (1979)].

Math. Z. 180, 119 – 140 (1982)

Mathematische Zeitschrift

Ergodic Properties of Invariant Measures for Piecewise Monotonic Transformations

Franz Hofbauer[1,*] and Gerhard Keller[2,**]

[1] Institut für Mathematik, Universität Wien, Strudlhofgasse 4, A-1090 Wien, Austria
[2] Institut für Mathematische Statistik der Universität, Lotzestraße 13, D-3400 Göttingen, Federal Republic of Germany

Introduction

During the last decade a lot of research has been done on onedimensional dynamics. Different kinds of invariant measures for certain classes of piecewise monotonic transformations have been considered. In most of these cases the Perron-Frobenius-operator plays an important role. In this paper we try to unify these different examples, which are discussed in detail below. First we give a discription of the results proved in this paper.

The general setting we propose is the following.

X is a totally ordered, order-complete set and $T: X \to X$ a transformation which is piecewise monotonic and order-continuous. P is a Perron-Frobenius-operator on the space of all bounded measurable functions on X given by

$$Pf(x) = \sum_{y \in T^{-1}x} g(y) f(y),$$

where g is a function of bounded variation on X with $0 < g(x) \leqq d < 1$. Furthermore we assume the existence of a Borel probability measure m on X, which is invariant for P, i.e. $m(Pf) = m(f)$ for all bounded measurable $f: X \to \mathbb{R}$.

Under these assumptions we give a kind of spectral representation for P (Theorem 1) which is fundamental for the subsequent results:

(1) There is a "greatest" T-invariant m-absolutely continuous probability μ (also denoted by $\mu \ll m$), $\mu = hm$, where h is of bounded variation. "Greatest" means that all other T-invariant, m-absolutely continuous probabilities are absolutely continuous with respect to μ (Theorem 1).

(2) The σ-algebra of T-invariant sets is μ-a.s. finite (Theorem 2). This gives the possibility to split up μ into its finitely many ergodic components. Furthermore, there is a k such that the finitely many ergodic components $W_i \subset X$ of

---

* Research for this paper was done, when the first author visited the Institut für mathematische Statistik, Göttingen

** The second author was supported by the Deutsche Forschungsgemeinschaft

(X, μ, T^k) are already weak mixing. We refer to $(W_i, \mu|W_i, T^k|W_i)$ as a weak mixing component of μ (Remark 2).

(3) The asymptotic σ-algebra $\mathfrak{B}_\infty = \bigcap_{j=0}^{\infty} T^{-j}\mathfrak{B}$ of T, where $\mathfrak{B}$ denotes the Borel sets of X, is μ-a.s. finite. It is trivial for a weak mixing component of μ and hence such a component is exact (Theorem 3).

(4) Each finite partition into intervals is a weak Bernoulli partition for every weak mixing component W_i of μ. The corresponding mixing-coefficient decreases exponentially (Theorem 4 and its Corollary). For an $f: W_i \to \mathbb{R}$ of bounded p-variation the stationary stochastic process $\xi_n = f \circ T^{kn}$ $(n \geqq 0)$ satisfies a central limit theorem and an a.s. invariance principle with respect to μ (Theorem 5).

The examples of invariant measures, which fit in our situation, are the following:

(i) Lasota and Yorke [11] consider piecewise monotonic and C^2 transformations T on $[0, 1]$ such that $\inf|T'| > 1$. If one sets $g = |1/T'|$, then the operator P satisfies $\lambda(Pf) = \lambda(f)$ for the Lebesguemeasure λ. In [11] the existence of a function h with $Ph = h$ is proved, which implies the existence of an invariant $\mu \ll \lambda$. In [13] μ is split up into its finitely many ergodic components. Bowen [3] shows that μ is weak Bernoulli, if it is weak mixing. Wong [23] generalizes the results of [11] and [13] to the class of T's, which are piecewise monotonic and C^1, such that T' is of bounded variation and satisfies $\inf|T'| > 1$. Wagner [20] shows that the asymptotic σ-algebra $\mathfrak{B}_\infty$ is μ-a.s. finite. For the class of T considered in [11] the spectral decomposition of P can be found in [10] (cf. also [14]), from which a central limit theorem for stationary processes $f \circ T^n$ on $([0, 1], \mu)$ is deduced, if μ is weak mixing (a special case of this is shown in [24]).

As $g = |1/T'|$, where T is as in [23], satisfies our requirements, all these results are contained in this paper. Furthermore, we show for a weak mixing component of μ, that the mixing coefficients in the weak Bernoulli property decrease at an exponential rate and that the process $f \circ T$ satisfies for certain f a central limit theorem and an almost sure invariance principle.

(ii) Next we consider any piecewise monotonic transformation T on $[0, 1]$ with $h_{\text{top}}(T) > 0$. In [6] and [7] it is shown that every maximal measure can be split up into finitely many ergodic components, which are isomorphic to Markov measures on a finite type subshift with finite or countable alphabet.

If one sets $g(x) = \exp(-h_{\text{top}}(T)) < 1$ for all $x \in [0, 1]$, it is shown in §5 of Chap. III that there is an m with $m(Pf) = m(f)$ and that μ is then a measure with maximal entropy for T. Hence the results of this paper give a decomposition of the maximal measure into its finitely many weak mixing components, each of which is weak Bernoulli and the process $f \circ T^n$ on such a component satisfies for certain f a central limit theorem and an almost sure invariance principle.

(iii) Now let T be the transformation $x \to \beta x \pmod 1$ on $[0, 1]$. Using a limit theorem for the operator P, Walters [22] shows, that there is a unique equilibrium state μ for every Lipschitz continuous function $\varphi: [0, 1] \to \mathbb{R}$, and that μ is weak Bernoulli.

In §4 of Chap. III we consider $\varphi: [0,1] \to \mathbb{R}$ of bounded variation, which satisfy $\sum_{i=1}^{\infty} \operatorname{var}_i \varphi < \infty$, where $\operatorname{var}_i \varphi = \sup\{|\varphi(x) - \varphi(y)|: x$ and y are contained in the same inerval, on which T^i is monotone$\}$. Using the same technique as Walters [22], we show the existence of a probability m, a $\lambda > 0$ and an $n \in \mathbb{N}$ such that $m(Pf) = m(f)$ and $\|g\|_\infty < 1$ is satisfied for $g(x) = \exp\left(\sum_{i=0}^{n-1} \varphi(T^i x)\right) \Big/ \lambda^n$. Furthermore we get that μ is then an equilibrium state for φ.

This means that we generalize the results of Walters to a larger class of functions φ and show besides the weak Bernoulli property also a central limit theorem and an almost sure invariance principle for certain processes $f \circ T^n$.

(iv) Also subshifts X of finite type with finite alphabet are piecewise monotonic, if one introduces the lexicographic ordering in X.

Using a limit theorem for the operator P, Bowen [2] has shown the weak Bernoulli property with exponentially decreasing coefficient and a central limit theorem for equilibrium states of Hölder continuous functions $\varphi: X \to \mathbb{R}$, i.e. $\operatorname{var}_i \varphi \leqq C q^i$ for some $C > 0$ and $0 < q < 1$.

In §3 of Chap. III we show for every $\varphi: X \to \mathbb{R}$ of bounded variation, which satisfies $\sum_{i=1}^{\infty} \operatorname{var}_i \varphi < \infty$, the existence of a probability measure m, a $\lambda > 0$ and an $n \in \mathbb{N}$ such that $m(Pf) = m(f)$ and $\|g\|_\infty < 1$ is satisfied for $g(x) = \exp\left(\sum_{i=0}^{n=1} \varphi(T^i x)\right) \Big/ \lambda^n$. Furthermore μ is then an equilibrium state for φ. Hence for subshifts of finite type we show Bowen's results for a slightly different class of φ's.

Finally we should like to thank M. Denker for encouraging discussions during the preparation of the paper.

I. Spectral Decomposition of the Operator P and the Measure μ

Let X be a totally ordered, complete set. Then the order topology, this is that one generated by open intervals, makes X into a compact space [see e.g. Birkhoff, Lattice Theory, A.M.S. Colloquium Publications 25, Chap. X, §7, Theorem 12]. We call a transformation T on X *piecewise monotonic*, if $X = \bigcup_{i=1}^{N} I_i$, the I_i are disjoint intervals and $T|I_i$ is continuous and preserves or reverses the ordering such that $T(I_i)$ is again an interval. We consider $g: X \to (0, d]$ with

(a) $$d < 1 \quad \text{and} \quad V_X g < \infty$$

where the variation $V_X g$ of g over X is defined by

$$\sup\left\{\sum_{i=1}^{k-1} |g(a_i) - g(a_{i+1})|: k \geqq 2,\ a_1 < \ldots < a_k,\ a_i \in X\right\}.$$

The *Perron-Frobenius-operator* P on the set $\mathfrak{F}$ of all bounded complex-valued functions on X, which are measurable with respect to the Borel-σ-algebra $\mathfrak{B}$, is defined by

$$Pf(x)=\sum_{y\in T^{-1}x} g(y)f(y)=\sum_{i=1}^{N} g(T_i^{-1}x)f(T_i^{-1}x)\,1_{T(I_i)}(x)$$

where $T_i=T\,|\,I_i$. Finally we assume that there is a Borel-probability-measure m on X which satisfies

(b) $$m(Pf)=m(f) \quad \text{for all } f\in\mathfrak{F}.$$

§1. We begin with the investigation of the support of the measure m.

Lemma 1. *For $B\in\mathfrak{B}$, $m(B)=0$ implies that $m(TB)=0$.*

Proof. We can assume that $B\subset I_i$ for some i. Suppose that $m(TB)>0$. We set $A_n=\left\{x\in T(I_i)\colon g(T_i^{-1}x)\geqq\frac{1}{n}\right\}$. As $g(x)>0$ for all $x\in X$, it follows that $\bigcup_{n=1}^{\infty}A_n=T(I_i)$. This implies that there is a k with $m(A_k\cap TB)>0$. Then we have

$$\begin{aligned}m(B)=m(P1_B)&=m\left(\sum_{j=1}^{N} g(T_j^{-1}x)\,1_B(T_j^{-1}x)\,1_{TI_j}(x)\right)\\&=m(g(T_i^{-1}x)\,1_{TB}(x))\geqq\frac{1}{k}\,m(TB\cap A_k)>0.\end{aligned}$$

This contradicts $m(B)=0$. Hence $m(TB)=0$.

Let $\mathfrak{P}=\{I_1,\dots,I_N\}$ denote the partition of X into intervals of continuity and monotonicity of T. Then $\mathfrak{P}_0^k=\bigvee_{i=0}^{k-1}T^{-i}\mathfrak{P}$ is the finite partition of X into intervals, on which T^k is monotone and continuous.

Lemma 2. *$m(A)\leqq d^k$ for all $A\in\mathfrak{P}_0^k$. In particular, $m(\{x\})=0$ for all $x\in X$.*

Proof. This is proved by induction. Let $A\in\mathfrak{P}_0^n$ with $n\geqq1$ be such that $A\subset I_i$. Then

$$m(A)=m(P1_A)=m(g(T_i^{-1}(x))\,1_{TA}(x))\leqq d\,m(TA).$$

One sees easily that TA is contained in an element of $\mathfrak{P}_0^{n-1}$. Hence $m(TA)\leqq d^{n-1}$ by induction hypothesis and $m(A)\leqq d^n$.

Lemma 3. *Let $Y=\bigcap\{F\colon F$ is a closed subset of X with $m(F)=1\}$ be the support of m. T can be changed at a finite number of points in such a way that then $T(Y)\subseteq Y$ and $T^{-1}\{x\}\subseteq Y$ for all $x\in Y$ except finitely many ones.*

Proof. Let J be a subinterval of X, which is contained in I_i for some i. We show that $m(TJ)>0$ if and only if $m(J)>0$. It follows from Lemma 1 that $m(TJ)>0$ implies $m(J)>0$. The other direction follows from

$$m(T^{-1}B)=m(1_B\circ T)=m(P(1_B\circ T))=m(1_B\cdot P1)\leqq N\,d\,m(B)$$

because $P1(x)\leqq Nd$ for all $x\in X$ ($g(x)\leqq d$ and the set $T^{-1}\{x\}$ contains at most N elements).

If x is now in the interior of some I_i then one takes for J open subintervals of I_i containing x and it follows that $x \in Y$ if and only if $Tx \in Y$. In particular, $T^{-1}\{x\} \subseteq Y$ for all $x \in Y$, which are not endpoints of some TI_i. As we shall change T only at endpoints of I_i's, this gives the second assertion. By Lemma 2, Y contains no isolated points. Hence, for an endpoint x of some I_i, we can redefine $T(x)$ as a one-sided limit of $T(t)$ for $t \in Y$. Then $T(Y) \subseteq Y$, because Y is closed and $T(t) \in Y$ for all $t \in Y$, which are in the interior of some I_i. This proves the lemma.

If T is changed on a finite set, $Pf(x)$ changes only for finitely many x. By Lemma 2, we have then $m(Pf)=m(f)$ also for the modified T of Lemma 3. We consider now the modified T, restrict it to the closed set $Y \subseteq X$, which is T-invariant by Lemma 3, and denote it again by T. As the sets $T^{-1}\{x\}$ for the original and the restricted T differ only for finitely many $x \in Y$ by Lemma 3, it follows from Lemma 2 that $m(Pf)=m(f)$ holds also for the restricted T. Y is again a totally ordered set, which is compact and has the advantage, that $m(J)>0$ for every nontrivial interval $J \subseteq Y$ (important for Lemmas 5 and 7 below). T on Y is again piecewise monotonic. Hence we consider from now on only the dynamical system (Y, T).

For $f: Y \to \mathbb{C}$ we define the *variation over a subset J of Y* by

$$V_J f = \sup\left\{\sum_{i=1}^{k-1} |f(a_i)-f(a_{i+1})| : k \geqq 2,\ a_1 < \ldots < a_k,\ a_i \in J\right\}.$$

We shall need a variation $v(f)$ for $f \in L^1_m$, the set of all equivalence classes of complex-valued, m-integrable functions on Y. We define

$$v(f) = \inf\{V_Y \tilde{f} : \tilde{f} \text{ is a version of } f\}.$$

Remark 1. If $\tilde{f}$ is defined on $Y \setminus N$, if $Y \setminus N$ is dense in Y - this is satisfied, if N has measure zero, because every interval has positive m-measure - and if $V_{Y \setminus N} \tilde{f} < \infty$, then one shows easily that for all $x \in Y$ at least one of the limits $\tilde{f}(x^+) = \lim_{z \downarrow x} \tilde{f}(z)$ and $\tilde{f}(x^-) = \lim_{z \uparrow x} \tilde{f}(z)$ exists, since each $x \in Y$ can be approximated from above or below (Y contains no isolated points). Extend $\tilde{f}$ to the whole of Y such that for each $x \in N$ $\tilde{f}(x) = \tilde{f}(x^+)$ or $\tilde{f}(x) = \tilde{f}(x^-)$. Then $V_Y \tilde{f} = V_{Y \setminus N} \tilde{f}$ as is easily seen.

Now set $BV = \{f \in L^1_m : v(f) < \infty\}$, which is a linear subspace of L^1_m, but it is not closed with respect to $\| \ \|_1$. We define for $f \in BV$

$$\|f\|_v = \int |f|\, dm + v(f) = \|f\|_1 + v(f).$$

It is easy to see that $\| \ \|_v$ is a norm on BV. The following lemma is needed for Lemma 5.

Lemma 4. *Let $f \in L^1_m$ satisfy $v(f) \leqq c < \infty$. Then $\int |f - E_m(f \mid \mathfrak{A}_n)|\, dm \leqq c d^n$, where $E_m(f \mid \mathfrak{A}_n)$ denotes the conditional expectation of f with respect to the measure m and the σ-algebra $\mathfrak{A}_n$ generated by $\mathfrak{P}_0^n$.*

Proof.

$$\int |f-E_m(f\,|\,\mathfrak{A}_n)|\,dm \leqq \sum_{A\in\mathfrak{P}_0} m(A)(\operatorname{ess\,sup}_{x\in A} f(x)-\operatorname{ess\,inf}_{x\in A} f(x)) \leqq d^n\cdot v(f)\leqq c d^n.$$

For the second inequality we have used Lemma 2.

The following lemma and Lemmas 6 and 7 will be requirements for the theorem of Ionescu-Tulcea and Marinescu (Theorem 1 below).

Lemma 5. (i) *For every* $c>0$ *the set* $E=\{f\in L^1_m:\ \|f\|_v\leqq c\}$ *is compact in* $(L^1_m, \|\ \|_1)$.

(ii) $(BV, \|\ \|_v)$ *is a Banach space*

(iii) BV *is dense in* $(L^1_m, \|\ \|_1)$.

Proof. (i) Let f_n be in E $(n\geqq 1)$. From Lemma 4 and Theorem IV.8.18 in [4] it follows that there is a subsequence (g_n) of (f_n) and a $f\in L^1_m$ with $\lim_{n\to\infty}\|g_n-f\|_1=0$. We must show that $f\in E$: Let $\varepsilon>0$ be arbitrary. There are versions $\tilde{g}_n$ of g_n such that $V_Y(\tilde{g}_n)\leqq v(g_n)+\varepsilon$ for all n. As g_n converges to f in L^1_m, we can find a subsequence $(\tilde{h}_n)$ of $(\tilde{g}_n)$ and a set $N\subseteq Y$ of measure zero such that $\lim_{n\to\infty}\tilde{h}_n(x)=\tilde{f}(x)$ for all $x\in Y\setminus N$, where $\tilde{f}$ is a version of f defined on $Y\setminus N$. The corresponding subsequence of (g_n) will be called (h_n). We thus have

$$\begin{aligned}
V_{Y\setminus N}(\tilde{f}) &\leqq \varepsilon+\sum_{i=1}^{k}|\tilde{f}(a_i)-\tilde{f}(a_{i-1})| \qquad \text{for suitable } k\in\mathbb{N} \text{ and } a_i\in Y\setminus N,\\
&=\varepsilon+\lim_{n\to\infty}\sum_{i=1}^{k}|\tilde{h}_n(a_i)-\tilde{h}_n(a_{i-1})|\\
&\leqq \varepsilon+\limsup_{n\to\infty} V_Y(\tilde{h}_n)\\
&\leqq 2\varepsilon+\limsup_{n\to\infty} v(h_n).
\end{aligned}$$

Extending $\tilde{f}$ to the whole of Y as in Remark 1 we obtain

$$v(f)\leqq V_Y(\tilde{f})=V_{Y\setminus N}(\tilde{f})\leqq \limsup_{n\to\infty} v(h_n),$$

as $\varepsilon>0$ was arbitrary and $h_n\in E$. Together with $\lim_{n\to\infty}\|h_n-f\|_1=0$ this implies $\|f\|_v\leqq \limsup_{n\to\infty}\|h_n\|_v\leqq c$, i.e. $f\in E$. This proves (i).

(ii) follows immediately from (i) and I.1.6 in [18].

(iii) follows from the fact that the functions constant on the elements of some $\mathfrak{P}_0^n$ are dense in $(L^1_m, \|.\|_1)$, which is a standard result of measure theory, as $\bigcup_{n=1}^{\infty}\mathfrak{P}_0^n$ generates the Borel-σ-algebra m-a.e.

§2. Now we turn to the investigation of the operator P. The following result is an easy consequence of (b) and the formula $\varphi\cdot Pf=P(f\cdot(\varphi\circ T))$, which is easiliy verified.

Lemma 6. (i) P *is an operator on* L^1_m *satisfying* $m(Pf)=m(f)$ *for* $f\in L^1_m$.

(ii) $\|Pf\|_1\leqq\|f\|_1$ *for all* $f\in L^1_m$.

Proof. We show (ii) first for $f\in\mathfrak{F}$ using (b). We have $|Pf(x)|=Pf(x)\cdot\varphi(x)$, where $\varphi(x)=Pf(x)/|Pf(x)|$. Since $|\varphi(x)|=1$ for all $x\in Y$, we have

$$\begin{aligned}\|Pf\|_1&=\int(Pf)\,\varphi\,dm=\int P(f\cdot(\varphi\circ T))\,dm\\&=\int f(\varphi\circ T)\,dm\leqq\int|f|\cdot|\varphi\circ T|\,dm=\|f\|_1.\end{aligned}$$

If one changes f on a set B of measure zero, then Pf is changed on TB, which has measure zero by Lemma 1. Hence P can be considered as an operator on the set of all bounded $f\in L^1_m$. Because of $\|Pf\|_1\leqq\|f\|_1$ for $f\in\mathfrak{F}$ and as $\mathfrak{F}$ is dense in L^1_m, P can be extended continuously to all $f\in L^1_m$. This proves the lemma.

Now we choose an $M\in\mathbb{N}$ such that $\|g_M\|_\infty=\sup\{g_M(x): x\in Y\}<\frac{1}{3}$, where g_n for $n\geqq1$ is given by $g_n(x)=g(x)\,g(Tx)\ldots g(T^{n-1}x)$. Such an M exists, because $\|g\|_\infty\leqq d<1$ (cf. (a)). The following result is a modification of a result in [23] (cf. also [11]).

Lemma 7. *There is an $\alpha\in(0,1)$ and a $\beta\in(0,\infty)$ with*

$$v(P^M f)\leqq\alpha v(f)+\beta\|f\|_1 \quad \textit{for all } f\in L^1_m.$$

Proof. Writing Q for P^M and U for T^M, we have $Qf(x)=\sum_{y\in U^{-1}x} g_M(y)f(y)$. Let $(J_1,\ldots,J_{\tilde N})$ be a partition of Y into intervals such that U/J_i is monotone and continuous and that $V_{J_i}g_M\leqq\|g_M\|_\infty$. As $T(I_i)$ is an interval in Y, also $U(J_i)$ is an interval in Y. For $\varepsilon>0$, let $\tilde f$ be a version of $f\in L^1_m$ such that $V_Y\tilde f\leqq v(f)+\varepsilon$. As $m(J_i)>0$ for $1\leqq i\leqq\tilde N$ and as $U(J_i)$ is an interval in Y, the same computation as in Theorem 1 of [23] works for Q and $\tilde f$. Hence we get $V_Y Q\tilde f\leqq\alpha V_Y\tilde f+\beta\|\tilde f\|_1$, where $\alpha=3\|g_M\|_\infty<1$ and $\beta=3\|g_M\|_\infty/\min\{m(J_i): 1\leqq i\leqq\tilde N\}$. Since $Q\tilde f$ is a version of Qf (cf. the proof of Lemma 6) we have

$$v(Qf)\leqq V_Y Q\tilde f\leqq\alpha V_Y\tilde f+\beta\|\tilde f\|_1\leqq\alpha\cdot v(f)+\beta\|f\|_1+\alpha\varepsilon.$$

Because ε was arbitrary, we get the desired result.

The properties of the space BV and of the operator P, which are listed in Lemma 5, 6 and 7 allow to apply an ergodic theorem of Ionescu-Tulcea and Marinescu [9], stating in an abstract Banach space setting, the following results (they are formulated here for the special situation under consideration).

Theorem 1. *Under the above assumptions on P, T, X and m we have*

(i) *$P: L^1_m\to L^1_m$ only has a finite number of eigenvalues $\lambda_1,\ldots,\lambda_r$ of modulus 1.*
(ii) *Set $E_i=\{f\in L^1_m: Pf=\lambda_i f\}$ for $1\leqq i\leqq r$. Then $E_i\subset BV$ and $\dim(E_i)<\infty$.*
(iii) *The operator P can be represented as*

$$P=\sum_{i=1}^{r}\lambda_i\phi_i+\psi$$

where the ϕ_i are projections onto the eigenspaces E_i, $\|\phi_i\|_1\leqq1$, and ψ is a linear operator on L^1_m with $\sup_{n\geqq1}\|\psi^n\|_1<\infty$. Furthermore $\phi_i\phi_j=0\,(i\neq j)$ and $\phi_i\psi=0$ (all i).

(iv) *$\psi(BV)\subset BV$ and, considered as a linear operator on $(BV, \|\ \|_v)$, ψ satisfies $\|\psi^n\|_v \leqq H q^n$ $(n \geqq 1)$ for some constants H and q with $H>0$ and $0<q<1$.*

(v) *1 is an eigenvalue of P. We assume $\lambda_1 = 1$.*

(vi) *Writing $h := \phi_1(1)$ and $\mu = hm$ one has $h \geqq 0$, $\int h\, dm = 1$ and $Ph = h$. Thus μ is a T-invariant probability on Y and hence on X. If μ' is T-invariant and $\mu' \ll m$, then $\mu' \ll \mu$.*

Proof. (i) to (iv) are immediate consequences of the theorem of Ionescu-Tulcea and Marinescu. It remains to prove (v) and (vi).

Set $h_n = \frac{1}{n} \sum_{k=0}^{n-1} P^k(1)$. Then $h_n \geqq 0$, because P is a positive operator and $\int h_n\, dm = 1$ because of $m(Pf) = m(f)$. Using (iii) and evaluating geometric series it follows that h_n converges in L^1_m. The limit function is invariant under P, nonnegative and has integral 1. Thus 1 is an eigenvalue of P, say $\lambda_1 = 1$, and it follows from (iii) as above that h_n converges to $\phi_1(1) = h$. That μ is T-invariant follows from the following equalities, because $Ph = h$:

$$\mu(f \circ T) = m(h \cdot (f \circ T)) = m(P(h \cdot (f \circ T))) = m(f \cdot Ph)$$
$$\mu(f) = m(fh).$$

If $\mu' = h'm$ is T-invariant, it follows from these equalities that $Ph' = h'$. Thus $h' \in BV$ by (ii) and h' is bounded by some constant c. This implies $h' = \frac{1}{n} \sum_{k=0}^{n-1} P^k(h') \leqq c h_n$, hence $h' \leqq ch$ and $\mu' \ll \mu$.

§3. Once having obtained the invariant measure μ for T, we focus attention to the dynamical system (T, μ). Sometimes we shall adopt the Hilbert-space-viewpoint, interpreting T as a linear operator on the complex Hilbert space L^2_μ and writing $\langle f, g\rangle$ for $\int f\bar{g}\, d\mu$, such that $\langle Tf, Tg\rangle = \langle f, g\rangle$.

The next lemma will provide a description of the adjoint T^* of T in terms of P.

Lemma 8. (i) *$T^* f = \frac{P(f \cdot h)}{h}$ for all $f \in L^2_\mu$.*

(ii) *$T^* f = \lambda f \Leftrightarrow P(fh) = \lambda fh$ for all $f \in L^2_\mu$ and $\lambda \in \mathbb{C}$ (all identities are L^2_μ-identities).*

Proof. (i) for $f, g \in L^2_\mu$ one has

$$\langle T^* f, g\rangle = \langle f, Tg\rangle = \int f \cdot (\bar{g} \circ T)\, h\, dm = \int \frac{P(fh)}{h}\, \bar{g}\, d\mu = \left\langle \frac{P(fh)}{h}, g \right\rangle.$$

(ii) $T^* f = \lambda f \Leftrightarrow P(fh) = \lambda fh$ because of (i).

Lemma 9. (i) *$(T^*)^n \circ T^n = Id$ for each $n \in \mathbb{N}$.*

(ii) *$T^n \circ (T^*)^n = Pr_n$ for each n, where Pr_n denotes the orthogonal projection onto the subspace $T^n(L^2_\mu)$.*

Proof. (i) It suffices to show that $T^* T = Id$. But for $f, g \in L^2_\mu$ one has $\langle T^* Tf, g\rangle = \langle Tf, Tg\rangle = \langle f, g\rangle$.

(ii) Since $(T^n \circ (T^*)^n)^* = T^n \circ (T^*)^n$ and $(T^n \circ (T^*)^n)^2 = T^n \circ (T^*)^n$ (because of (i)), $T^n \circ (T^*)^n$ is an orthogonal projection. On the other hand,

$$(T^n \circ (T^*)^n)(L^2_\mu) \subset T^n(L^2_\mu) \quad \text{and} \quad (T^n \circ (T^*)^n)(T^n f) = T^n f$$

for each $f \in L^2_\mu$. Thus $T^n \circ (T^*)^n = Pr_n$.

Now we can easily clarify the ergodic structure of (T, μ) up to the level of weak mixing (for weak mixing cf. [21]).

Theorem 2. *Let $G \subseteq \{z \in \mathbb{C}: |z| = 1\}$ be the set of eigenvalues of T. Then*

(i) $G = \{\lambda_1, \ldots, \lambda_r\}$

(ii) $Tf = \lambda f \Leftrightarrow T^*\bar{f} = \lambda \bar{f}$ *and* $|\lambda| = 1$ *for all* $f \in L^2_\mu$ *and* $\lambda \in \mathbb{C}$.

(iii) $\dim(E_{(T,\mu)}(\lambda)) < \infty$ *for each* $\lambda \in G$, *where* $E_{(T,\mu)}(\lambda) = \{f \in L^2_\mu: Tf = \lambda f\}$.

Proof. Let us first assume that $Tf = \lambda f$ for some $f \in L^2_\mu$, $f \neq 0$. Then $|\lambda| = 1$, as μ is T-invariant, and $T^*\bar{f} = \lambda T^*(T\bar{f}) = \lambda \bar{f}$. On the other hand, if $T^*f = \lambda f$ for some $f \in L^2_\mu$, $f \neq 0$ and $\lambda \in \mathbb{C}$, $|\lambda| = 1$, we have

$$\langle f, f \rangle = \langle T^*f, \lambda f \rangle = \langle f, \lambda T f \rangle \quad \text{and} \quad \langle \lambda T f, \lambda T f \rangle = |\lambda|^2 \langle T f, T f \rangle = \langle f, f \rangle.$$

This implies $\langle \lambda Tf - f, f \rangle = 0$ and $\langle \lambda Tf - f, \lambda Tf \rangle = 0$, hence $\lambda Tf = f$ or, equivalently, $T\bar{f} = \lambda \bar{f}$. This proves (ii).

If now $Tf = \lambda f$, we get from (ii) and Lemma 8 that $|\lambda| = 1$ and $P(\bar{f}h) = \lambda(\bar{f}h)$. Hence it follows from Theorem 1 that $G \subseteq \{\lambda_1, \ldots, \lambda_r\}$. If $Pg = \lambda_i g$ for some i, then $g \in BV$ by Theorem 1 and $|g| = |P^n g| \leqq P^n |g| \leqq \|g\|_\infty \cdot P^n(1)$. Hence

$$|g| = \frac{1}{n} \sum_{i=0}^{n-1} |P^i g| \leqq \|g\|_\infty \frac{1}{n} \sum_{i=0}^{n-1} P^i(1) \to \|g\|_\infty h \qquad \text{for } n \to \infty.$$

This gives $\frac{|g|}{h} \leqq \|g\|_\infty$ such that $f := \frac{g}{h} \in L^2_\mu$. Furthermore it follows from Lemma 8 and (ii) that $T\bar{f} = \lambda_i \bar{f}$. Hence $G \supseteq \{\lambda_1, \ldots, \lambda_r\}$ proving (i). The same arguments show also that $\dim(E_{(T,\mu)}(\lambda_i)) = \dim(E_i) < \infty$. This is (iii).

The next theorem describes the structure of (T, μ) up to the level of exactness (for exactness see [17]).

Theorem 3. *Let $\mathfrak{B}_\infty(T) := \bigcap_{n=1}^{\infty} T^{-n}\mathfrak{B}$, where $\mathfrak{B}$ denotes the Borel-σ-algebra on Y. Then*

(i) $\{f \in L^2_\mu: f$ *is* $\mathfrak{B}_\infty(T)$*-measurable*$\} = \bigcap_{n=1}^{\infty} T^n(L^2_\mu) = \sum_{i=1}^{r} E_{(T,\mu)}(\lambda_i)$. *In particular,* $\mathfrak{B}_\infty(T)$ *is finite* μ-a.s.

(ii) *If* (T, μ) *is weakly mixing, then* $\mathfrak{B}_\infty(T)$ *is trivial, i.e.* (T, μ) *is exact.*

Proof. (i) The first of the above equalities is generally true. Hence we only have to show the second one. First observe that for $f = \sum_{i=1}^{r} f_i$, $f_i \in E_{(T,\mu)}(\lambda_i)$, we have $f = T^n\left(\sum_{i=1}^{r} \bar{\lambda}_i^n f_i\right)$ such that $\sum_{i=1}^{r} E_{(T,\mu)}(\lambda_i) \subseteq \bigcap_{n=1}^{\infty} T^n(L^2_\mu)$.

The other direction is a bit harder to prove. Let $g \in L^2_\mu \cap BV$. Then $gh \in L^1_m \cap BV$ and since $P(\phi_i(gh)) = \lambda_i \phi_i(gh)$, we have $T^*(\phi_i(gh)/h) = \lambda_i \phi_i(gh)/h$ by Lemma 8. Hence $T(\phi_i(gh)/h) = \bar{\lambda}_i \phi_i(gh)/h$ by Theorem 2.

As $G = \{\lambda_1, \ldots, \lambda_r\}$ generates a finite subgroup of the circle, there is a $p \in \mathbb{N}$ such that $\lambda_i^p = 1$ for $1 \leqq i \leqq r$. This gives for any $n \in \mathbb{N}$

$$\begin{aligned} Pr_{np}(g) &= T^{np}(T)^{*np}(g) = T^{np}\left(\sum_{i=1}^{r} \phi_i(gh)/h + \psi^{np}(gh)/h\right) \\ &= \sum_{i=1}^{r} \phi_i(gh)/h + T^{np}(\psi^{np}(gh)/h). \end{aligned}$$

We have used Lemma 9 and 8 and (iii) of Theorem 1. Now one has for any $f \in \bigcap_{n=1}^{\infty} T^n(L^2_\mu)$, any $g \in L^2_\mu \cap BV$ and any $n \in \mathbb{N}$ that

$$\begin{aligned} \int \left| f - \sum_{i=1}^{r} \phi_i(fh)/h \right| d\mu &\leqq \int |Pr_{np}(f) - Pr_{np}(g)|\, d\mu \\ &\quad + \sum_{i=1}^{r} \int |\phi_i(fh)/h - \phi_i(gh)/h|\, d\mu + \int |\psi^{np}(gh)/h| \circ T^{np}\, d\mu \\ &\leqq \|Pr_{np}(f-g)\|_{2,\mu} + r \int |f-g|\, h\, dm + \int |\psi^{np}(gh)|\, dm \\ &\leqq (r+1)\, \|f-g\|_{2,\mu} + Hq^{np}\, \|gh\|_v. \end{aligned}$$

Since BV is dense in L^2_μ this shows that $f = \sum_{i=1}^{r} \phi_i(fh)/h$ μ-a.e., i.e. $f \in \sum_{i=1}^{r} E_{(T,\mu)}(\lambda_i)$.

(ii) If (T, μ) is weakly mixing, then 1 is the only eigenvalue, and the space of fix-elements of T only contains the constant functions. Hence $\mathfrak{B}_\infty(T)$ is trivial by (i).

Remark 2. Let $W \subset Y$ satisfy $T^{-1} W = W$, i.e. $1_W \circ T = 1_W$. Define

$$m'(A) = m(A \cap W)/m(W).$$

Then

$$\begin{aligned} m'(Pf) &= m(1_W Pf)/m(W) = m(P(f \cdot (1_W \circ T)))/m(W) \\ &= m(f \cdot (1_W \circ T))/m(W) = m(f \cdot 1_W)/m(W) = m'(f). \end{aligned}$$

Similarly one can show that $P(h \cdot 1_W) = h 1_W$. Hence every ergodic component of μ is again of the form $h \cdot m$ for some m and h satisfying $m(Pf) = m(f)$ and $Ph = h$.

If one considers T^k instead of T, where k is such that $\lambda_i^k = 1$ for all i, then an ergodic component of μ is already weak mixing for T^k (cf. Theorem 2). Hence every weak mixing component of μ is again of the form hm for some m and h with $m(Pf) = m(f)$ and $Ph = h$. Here one has to take T^k instead of T and g_k instead of g in the definition of P. Therefore we can assume that μ is weak mixing. We shall do this in the next chapter.

Remark 3. The spectral representation of Theorem 1 together with Theorem 2 allows an easy proof of a central limit theorem for stationary processes of the kind $\xi_n = f \circ T^n$ with $f \in BV$. The proof is completely analogous to the one in [10], where it is shown how to reduce this problem to a central limit theorem of Gordin [5]. As in the next chapter we shall even derive almost sure invariance principles, an exact statement of this theorem is omitted.

II. Weak Bernoullicity and Invariance Principles

In this chapter we shall show that, if (T, μ) is weakly mixing (cf. Remark 2), then the natural partition $\mathfrak{P} = (I_1, \ldots, I_N)$ of Y into intervals of continuity and monotonicity of T (and even more each finite partition of Y into intervals) is a weak-Bernoulli-partition for T. This means that the associated "label-process" (ξ_n), where $\xi_n(x) = i$, if $T^n x \in I_i$, is absolutely regular. The corresponding mixing-coefficients will be shown to decrease at an exponential rate.

§1. We start with the investigation of the weak Bernoulli property. Technically we can follow very closely Ratner's article [16]. Remember that, for all $k \geqq 1$, $\mathfrak{P}_0^k = \bigvee_{i=0}^{k-1} T^{-i}\mathfrak{P}$ is a finite partition of Y into intervals of m-measure $\leqq d^k$ (cf. Lemma 2) such that m-measure plays the same role here as Lebesgue-measure (respectively "length") does in the article of Ratner [16]. In that paper the invariant measure also is called μ, but its density is called p and the function $\varphi(x) = 1/|f'(x)|$ plays the role of our $g(x)$.

The first lemma is proved in Bowen's paper [3] and is easily seen to be true also in this situation.

Lemma 10. *If $A \in \mathfrak{P}_0^k$, and if $\bar{A}$ contains no endpoint of an interval $I_i \in \mathfrak{P}$, then $TA \in \mathfrak{P}_0^{k-1}$.*

The next lemma is the same as Lemma 3.2 in [16].

Lemma 11. *There is a constant $L > 0$ such that, given $\varepsilon > 0$, we can find for all $n > L \cdot \log \frac{1}{\varepsilon}$ a collection $\alpha_n \subseteq \mathfrak{P}_0^n$ with $\mu(\bigcup \alpha_n) > 1 - \varepsilon$ such that for any $x, y \in A \in \alpha_n$*

$$\frac{h(x)}{h(y)} \in [e^{-\varepsilon}, e^{\varepsilon}] \quad \text{and} \quad \frac{g(x)}{g(y)} \in [e^{-\varepsilon}, e^{\varepsilon}].$$

Proof. The proof is quite analogous to the one in [16] except for the estimate of the measure of the set $S_l = \{x \in Y: g(x) < l\}$ $(l > 0)$, which must be replaced by

$$\begin{aligned}
\mu(S_l) &\leqq \|h\|_\infty \cdot m(S_l) = \|h\|_\infty \int P(1_{S_l})\, dm \\
&\leqq \|h\|_\infty \sum_{i=1}^{N} \|(g \circ T_i^{-1}) \cdot 1_{T_i(S_l)}\|_\infty, \quad \text{where } T_i = T/I_i \\
&\leqq \|h\|_\infty \cdot N \cdot l, \quad \text{by definition of } S_l.
\end{aligned}$$

The next lemma, which is basic for our further considerations, can be proved just as the corresponding one (3.3) in [16], too.

Lemma 12. *There is a constant $D>0$ such that for all sufficiently large $M\in\mathbb{N}$ and all $m\geqq 0$ one can find a collection of atoms $\beta=\beta_{m,M}\subseteq\mathfrak{P}_0^{m+M}$ with*

(i) $T^m B\in\mathfrak{P}_0^M$ *for* $B\in\beta$

(ii) $\mu(\bigcup\beta)>1-e^{-M/D}$

(iii) $\left|\frac{\mu(T^m\tilde{B})}{\mu(T^m B)}-\frac{\mu(\tilde{B})}{\mu(B)}\right|<e^{-M/D}\frac{\mu(\tilde{B})}{\mu(B)}$ *for any measurable* $\tilde{B}\subseteq B\in\beta$ *with* $\mu(B)>0$.

Proof. The proof is again as in [16] using Lemmas 10 and 11 instead of the corresponding ones in that paper. One only has to observe that for any measurable $\tilde{B}\subset B\in\mathfrak{P}_0^n$ (n arbitrary) the following holds. Put $q(x):=h(T^n x)/(h(x)\,g_n(x))$, where $g_n(x)=g(x)\,g(Tx)\dots g(T^{n-1}x)$. Then

$$\begin{aligned}\int_{\tilde{B}} q\,d\mu&=\int P^n(1_{\tilde{B}}\cdot(h\circ T^n)/g_n)\,dm\\&=\int\sum_{y\in T^{-n}x} g_n(y)\,1_{\tilde{B}}(y)(h(T^n y)/g_n(y))\,dm(x)\\&=\int\Big(\sum_{y\in T^{-n}x}1_{\tilde{B}}(y)\Big)\,h(x)\,dm(x)\\&=\int 1_{T^n\tilde{B}}(x)\,h(x)\,dm(x)\qquad\text{since }\tilde{B}\subseteq B\in\mathfrak{P}_0^n\\&=\mu(T^n\tilde{B})\end{aligned}$$

and

$$\frac{q(y)}{q(z)}=\frac{h(T^n y)}{h(T^n z)}\cdot\frac{h(z)}{h(y)}\cdot\prod_{k=0}^{n-1}\frac{g(T^k z)}{g(T^k y)}.$$

The next lemma "translates" the information about P, which is contained in the spectral representation, into the "language" of partitions. We introduce some new notation before. Let $\mathfrak{R}$ and $\mathfrak{S}$ be two finite collections of pairwise disjoint measurable subsets of Y. Define

$$D(\mathfrak{R},\mathfrak{S}):=\sum_{R\in\mathfrak{R}}\sum_{S\in\mathfrak{S}}|\mu(R\cap S)-\mu(R)\,\mu(S)|.$$

Lemma 13. *Under the assumption that (T,μ) is weakly mixing there is a constant $C>0$ such that for all M, n, $l\geqq 0$*

$$D(\mathfrak{P}_0^M,\mathfrak{P}_{M+l}^{M+l+n})\leqq C\cdot N^M q^l$$

where q is the constant of (iv) *of Theorem* 1 *and N the number of elements of* $\mathfrak{P}$.

Proof. Set $\mathfrak{R}=\mathfrak{P}_0^M$ and $\mathfrak{S}=\mathfrak{P}_{M+l}^{M+l+n}=T^{-l}(\mathfrak{P}_M^{M+n})$. Then

$$D(\mathfrak{R},\mathfrak{S})=\sum_{R\in\mathfrak{R}}\{|\mu(R\cap S_R^+)-\mu(R)\,\mu(S_R^+)|+|\mu(R\cap S_R^-)-\mu(R)\,\mu(S_R^-)|\},$$

where

$$S_R^+=\cup\{S\in\mathfrak{S}:\mu(R\cap S)>\mu(R)\,\mu(S)\}$$

and

$$S_R^-=\cup\{S\in\mathfrak{S}:\mu(R\cap S)\leqq\mu(R)\,\mu(S)\}=Y\setminus S_R^+.$$

Because of $\mathfrak{S}=T^{-l}(\mathfrak{P}_M^{M+n})$, there are measurable sets $\tilde{S}_R^+$ and $\tilde{S}_R^-$ with $S_R^+=T^{-l}(\tilde{S}_R^+)$ and $S_R^-=T^{-l}(\tilde{S}_R^-)$. Hence

$$D(\mathfrak{R}, \mathfrak{S}) = \sum_{R\in\mathfrak{R}} \{|\int 1_{S_R^+}(1_R-\mu(R))\, h\, dm| + |\int 1_{S_R^-}(1_R-\mu(R))\, h\, dm|\}$$
$$= \sum_{R\in\mathfrak{R}} \{|\int 1_{\bar{S}_R^+} P^l((1_R-\mu(R))\, h)\, dm| + |\int 1_{\bar{S}_R^-} P^l((1_R-\mu(R))\, h)\, dm|\}$$
$$\leqq 2 \sum_{R\in\mathfrak{R}} \|P^l((1_R-\mu(R))\, h)\|_v.$$

Since (T, μ) is weakly mixing, it follows from Theorems 2 and 1 that $P^l = \phi_1 + \psi^l$. Furthermore this implies $\phi_1((1_R-\mu(R))\, h) = (\int (1_R-\mu(R))\, h\, dm)\cdot h \equiv 0$. Thus

$$D(\mathfrak{R}, \mathfrak{S}) \leqq 2 \sum_{R\in\mathfrak{R}} \|\psi^l((1_R-\mu(R))\, h)\|_v$$
$$\leqq 2 \sum_{R\in\mathfrak{R}} H q^l \, \|(1_R-\mu(R))\, h\|_v$$
$$\leqq 2\, |\mathfrak{R}|\, H q^l\, 3\, \|h\|_v$$
$$\leqq 6H\, \|h\|_v\, N^M q^l = C N^M q^l.$$

The third inequality follows because the function $1_R - \mu(R)$ has the value $1-\mu(R)$ on the interval R and the value $-\mu(R)$ on $Y\setminus R$. Hence $v((1_R-\mu(R))\, h) \leqq 3v(h)$. The last inequality follows from $|\mathfrak{R}| \leqq N^M$, which is true, because $\mathfrak{R} = \mathfrak{P}_0^M$.

Now we can derive an estimate of $D(\mathfrak{P}_0^n, \mathfrak{P}_{n+l}^{2n+l})$ independent of n.

Theorem 4. *If (T, μ) is weakly mixing, then there are constants $K>0$ and p with $0<p<1$ such that*

$$D(\mathfrak{P}_0^n, \mathfrak{P}_{n+l}^{n+l+k}) \leqq K p^l$$

for arbitrary n and k.

Proof. Fix $\delta > 0$ such that $\tilde{q} = q N^\delta < 1$ (q as in Lemma 13). It suffices to show Theorem 4 for $n \geqq [\delta l] + 1$. The case $n \leqq [\delta l]$ reduces immediately to $n = [\delta l] + 1$ using the triangle inequality.

Set $M = [\delta l] + 1$ and $m = n - M \geqq 0$. Then

$$D(\mathfrak{P}_0^n, \mathfrak{P}_{n+l}^{n+l+k}) \leqq D(\beta_{m,M}, \mathfrak{P}_{n+l}^{n+l+k}) + 2e^{-M/D}$$

by Lemma 12. We set $\mathfrak{S} = \mathfrak{P}_{n+l}^{n+l+k}$ and $\hat{\mathfrak{S}} = \mathfrak{P}_{M+l}^{M+l+k}$. Because $\mu(T^m S) = \mu(S)$ for every $S\in\mathfrak{S}$, we get $(\beta = \beta_{m,M})$

$$D(\beta, \mathfrak{S}) = \sum_{B\in\beta} \mu(B) \sum_{S\in\mathfrak{S}} \left\{ \left| \frac{\mu(B\cap S)}{\mu(B)} - \frac{\mu(T^m(B\cap S))}{\mu(T^m B)} \right| + \left| \frac{\mu(T^m(B\cap S))}{\mu(T^m B)} - \mu(T^m S) \right| \right\}$$
$$\leqq \sum_{B\in\beta} \sum_{S\in\mathfrak{S}} e^{-M/D} \mu(B\cap S) + \sum_{\hat{B}\in\mathfrak{P}_0^M} \left\{ \sum_{\substack{B\in\beta \\ T^m B = \hat{B}}} \mu(B) \sum_{\hat{S}\in\hat{\mathfrak{S}}} \left| \frac{\mu(\hat{B}\cap\hat{S})}{\mu(\hat{B})} - \mu(\hat{S}) \right| \right\}$$

by Lemma 12

$$\leqq e^{-M/D} + \sum_{\hat{B}\in\mathfrak{P}_0^M} \sum_{\hat{S}\in\hat{\mathfrak{S}}} |\mu(\hat{B}\cap\hat{S}) - \mu(\hat{B})\, \mu(\hat{S})|$$
$$\leqq e^{-M/D} + D(\mathfrak{P}_0^M, \mathfrak{P}_{M+l}^{M+l+k})$$
$$\leqq e^{-M/D} + C N^M q^l \quad \text{by Lemma 13.}$$

Putting together the two estimates and observing that $M=[\delta l]+1$ we get

$$D(\mathfrak{P}_0^n, \mathfrak{P}_{n+l}^{n+l+k}) \leqq 3e^{-l\delta/D} + CN\tilde{q}^l.$$

This gives the desired result if one sets $p=\max\{\tilde{q}, e^{-\delta/D}\}$ and $K=3+CN$.

Corollary. *If (T, μ) is weakly mixing and $\mathfrak{D}$ is an arbitrary finite partition of Y into intervals, then there are constants $\tilde{K}$ and $\tilde{p}$, $0<\tilde{p}<1$ such that $D(\mathfrak{D}_0^n, \mathfrak{D}_{n+l}^{n+l+k}) \leqq \tilde{K}\tilde{p}^l$ for any k and n.*

Proof. The refinement $\mathfrak{P} \vee \mathfrak{D}$ of $\mathfrak{P}$ and $\mathfrak{D}$ is again a finite partition of Y into intervals, on which T is monotone and continuous. Hence Theorem 4 is also true for $\mathfrak{P} \vee \mathfrak{D}$ instead of $\mathfrak{P}$ (with different constants $\tilde{K}$ and $\tilde{p}$). But then it is also true for each coarser partition, in particular for $\mathfrak{D}$.

§2. In order to show central limit theorems, a.s. invariance principles and other limit theorems for stationary processes governed by the dynamical system (T, μ) one only has to observe the following: For $\mathfrak{P}=\{I_1, \ldots, I_N\}$ let $(\xi_n)_{n\geqq 0}$ be the associated label-process, i.e. $\xi_n(x)=i$, if $T^n x \in I_i$. If $\mathfrak{F}_a^b$ denotes the σ-algebra generated by $\{\xi_n: a \leqq n < b\}$, then $\mathfrak{F}_a^b$ is the σ-algebra generated by $\mathfrak{P}_a^b$. Define

$$\beta(\mathrm{n}) := \sup_{t \geqq 1} E_\mu [\sup_{A \in \mathfrak{F}_{t+n}^\infty} |\mu(A \mid \mathfrak{F}_0^t) - \mu(A)|].$$

If $\beta(n) \to 0$, (ξ_n) is called *absolutely regular.* One easily checks that (cf. [19])

$$2\beta(n) = \sup_{t \geqq 1} \sup_{k \geqq 1} D(\mathfrak{P}_0^t, \mathfrak{P}_{t+n}^{t+n+k}).$$

Hence $\beta(n) \leqq Kp^n$ by Theorem 4.

This enables us to apply some well-known limit theorems to the process (ξ_n) and to functionals of it. First define that $f: Y \to \mathbb{R}$ is of *bounded p-variation* on an interval $A \subseteq Y$ if

$$V_A^p f := \sup \sum_{i=1}^n |f(a_i) - f(a_{i-1})|^p < \infty,$$

where the supremum is taken over all finite subsets $a_0 < a_1 < \ldots < a_n$ of A.

Theorem 5. *Assume that (T, μ) is weakly mixing. Let $f: Y \to \mathbb{R}$ be a function of bounded p-variation, $p \geqq 1$, and $\int f\, d\mu = 0$. Define $S(t) := \sum_{i=0}^{t-1} f \circ T^i$, which is a stochastic process on (Y, μ). Then the series*

$$\sigma^2 := \int f^2\, d\mu + 2 \sum_{k=1}^\infty \int f(f \circ T^k)\, d\mu$$

converges absolutely, $\int S(t)^2\, d\mu = t\sigma^2 + O(1)$, and if $\sigma^2 \neq 0$ the following holds

(i) $\sup_{z \in \mathbb{R}} \left| \mu\left\{\frac{1}{\sigma\sqrt{t}} S(t) \leqq z\right\} - \frac{1}{\sqrt{2\pi}} \int_{-\infty}^z e^{-x^2/2}\, dx \right| = O(t^{-\nu})$ *for some* $\nu > 0$.

(ii) *Without changing its distribution one can redefine the process $(S(t))_{t\geqq 0}$ on a richer probability space together with standard Brownian motion $(B(t))_{t\geqq 0}$ such that*

$$|\sigma^{-1} S(t) - B(t)| = O(t^{1/2-\lambda}) \qquad \mu\text{-a.s.}$$

for some $\lambda > 0$.

Proof. We show that theorems in [15] and [1] are applicable here. First observe that the Borel-σ-algebra on Y is generated by the family $(\mathfrak{P}_0^k: k \in \mathbb{N})$, because each interval in Y has positive m-measure and $m(A) \leqq d^k$ for each $A \in \mathfrak{P}_0^k$ (see Lemma 2). Thus f can be considered to be μ-a.e. a functional of the label-process (ξ_n). Furthermore

(i) as a function of bounded p-variation, f is bounded.

(ii) For each $r \geqq p$

$$\begin{aligned}\int |f - E_\mu(f \mid \mathfrak{F}_0^k)|^r \, d\mu &\leqq \sum_{A \in \mathfrak{P}_0^k} \mu(A) \cdot \sup_{x, y \in A} |f(x) - f(y)|^r \\ &\leqq d^k \|h\|_\infty \sum_{A \in \mathfrak{P}_0^k} V_A^p f \cdot (2\|f\|_\infty)^{r-p} \qquad \text{by Lemma 2} \\ &\leqq d^k \|h\|_\infty V_Y^p f \cdot (2\|f\|_\infty)^{r-p}.\end{aligned}$$

(iii) Set $R(k) := \int f(f \circ T^k) \, d\mu$. Then

$$\begin{aligned}|R(k)| &\leqq \int |f - E_\mu(f \mid \mathfrak{F}_0^{[k/2]})| \cdot |f \circ T^k| \, d\mu + \int |E_\mu(f \mid \mathfrak{F}_0^{[k/2]}) \cdot (f \circ T^k)| \, d\mu \\ &\leqq (d^{[k/2]} \|h\|_\infty V_Y^p f)^{1/p} \|f\|_\infty + 4 \|f\|_\infty^2 \cdot \beta([k/2])\end{aligned}$$

by Theorem 17.2.1 of [8]. Hence $|R(k)|$ decreases exponentially with $k \to \infty$ such that $\sigma^2 = R(0) + 2 \sum_{k=1}^{\infty} R(k)$ converges absolutely, and

$$\begin{aligned}\int S(t)^2 \, d\mu &= t \cdot R(0) + 2 \sum_{k=1}^{t-1} (t-k) R(k) \\ &= t\sigma^2 - 2t \sum_{k=t}^{\infty} R(k) - 2 \sum_{k=1}^{t-1} k \cdot R(k) \\ &= t\sigma^2 + O(1)\end{aligned}$$

since $|R(k)|$ decreases exponentially.

Thus Theorem 7.1 in [15] can be applied to obtain (ii), while (i) follows from [1].

Remark 4. (i) The order of approximation in (ii) of Theorem 5 is good enough to imply integral tests, log-log-laws, and weak invariance principles for the process $(f \circ T^i)_{i \geqq 0}$ (see e.g. [15]).

(ii) Each Hölder-continuous function on $[0, 1]$ with exponent $\frac{1}{p}$ is of bounded p-variation.

III. Equilibrium States

In this chapter we show first that μ is an equilibrium state for the function $\log g$, i.e. $h(\mu) + \mu(\log g) = \sup\{h(\nu) + \nu(\log g): \nu$ is a T-invariant probability on

$Y\}$, where $h(\mu)$ denotes the entropy of μ. Then we give a method, how one can show the existence of a measure m and a function g such that (a) and (b) of Chap. I are satisfied. This method is then applied in three different situations.

§1. In order to show that μ is an equilibrium state for $\log g$, we need the following lemmas. For a function $f: Y\to\mathbb{R}$ let f^+ be defined by $f^+(x)=\max\{f(x), 0\}$.

Lemma 14. *If $f: Y\to\mathbb{R}$ is measurable and ν is a T-invariant probability on Y, then $(f-f\circ T)^+\in L^1_\nu$ implies $f-f\circ T\in L^1_\nu$ and $\int(f-f\circ T)\,d\nu=0$.*

Proof. Setting $F:=(f-f\circ T)^+$, we have $0\leqq F+f\circ T-f$, $F\geqq 0$ and $F\in L^1_\nu$. Define $f_n: Y\to\mathbb{R}$ by $f_n(x)=\min\{f(x), n\}$. Then we have $0\leqq F+f_n\circ T-f_n$ and

$$\lim_{n\to\infty}(f_n\circ T-f_n)=f\circ T-f.$$

Hence by Fatou's lemma

$$\begin{aligned}\int(F+f\circ T-f)\,d\nu&=\int\liminf(F+f_n\circ T-f_n)\,d\nu\\&\leqq\liminf\int(F+f_n\circ T-f_n)\,d\nu=\int F\,d\nu<\infty.\end{aligned}$$

This says that $F+f\circ T-f\in L^1_\nu$ and hence $f\circ T-f\in L^1_\nu$. As $|f_n\circ T-f_n|\leqq|f\circ T-f|$, it follows from Lebesgue's theorem of dominated convergence that

$$\int(f\circ T-f)\,d\nu=\lim_{n\to\infty}\int(f_n\circ T-f_n)\,d\nu=0.$$

Lemma 15. *Set $\bar g:=gh/(h\circ T)$. If ν is a T-invariant probability on Y, then*

$$\int\log\bar g\,d\nu=\int\log g\,d\nu.$$

Proof. It follows from $g(x)\leqq d<1$ (cf. (a)) that $\log g(x)<0$ for all $x\in Y$ and from $\bar g(x)\geqq 0$ and $\sum_{y\in T^{-1}x}\bar g(y)=1$, which is implied by $Ph=h$, that $\log\bar g(x)\leqq 0$ for all $x\in Y$.

If $\int\log g\,d\nu=\int\log\bar g\,d\nu=-\infty$, we have nothing to show. If $\int\log g\,d\nu>-\infty$, we have $\log g\in L^1_\nu$. Because of $\log h-\log h\circ T=\log\bar g-\log g$, it follows from $\log\bar g\leqq 0$ that $(\log h-\log h\circ T)^+\leqq|\log g|$ and hence from Lemma 14 that $\int(\log h-\log h\circ T)\,d\nu=0$. This implies $\log\bar g\in L^1_\nu$ and $\int\log g\,d\nu=\int\log\bar g\,d\nu$. If $\int\log\bar g\,d\nu>-\infty$, we apply the same arguments to $\log h^{-1}-\log h^{-1}\circ T=\log g-\log\bar g$ and get again $\int\log g\,d\nu=\int\log\bar g\,d\nu$.

Theorem 6. *μ is an equilibrium state for $\log g$ on Y.*

Proof. Define the operator $\bar P$ by $\bar Pf(x)=\sum_{y\in T^{-1}x}\bar g(y)f(y)$. It follows from $m(Pf)=m(f)$ and $Ph=h$ that $\mu(\bar Pf)=\mu(f)$ and $\bar P1=1$. Now we can use a result of Ledrappier (Theorem 1 of [12]), which says that μ is then an equilibrium state for the function $\log\bar g$, i.e. $h(\mu)+\mu(\log\bar g)=\sup_\nu\{h(\nu)+\nu(\log\bar g)\}$. But by Lemma 15, we have $\nu(\log\bar g)=\nu(\log g)$ for every T-invariant probability ν on Y and hence $h(\mu)+\mu(\log g)=\sup_\nu\{h(\nu)+\nu(\log g)\}$, which says that μ is an equilibrium state for $\log g$.

§2. Let T be a piecewise monotonic transformation on a totally ordered set X, such that the order topology is compact and separable. Let $\varphi \in C(X)$ satisfy $\sum_{i=1}^{\infty} \operatorname{var}_i \varphi < \infty$, where $\operatorname{var}_i \varphi = \sup \{|\varphi(x) - \varphi(y)| : x, y \in A, A \in \mathfrak{P}_0^i\}$. A method, to show that an equilibrium state for φ fits in the situation described at the beginning of Chap. I, is the following one. It is adapted from [22].

We have to modify X a little bit. Set $\tilde{W} = \bigcup_{j=0}^{\infty} T^{-j}\{\lim_{t \uparrow x} T^k(t), \lim_{t \downarrow x} T^k(t) : k \geq 1$ and x is an endpoint of some $I_i \in \mathfrak{P}, 1 \leq i \leq N\}$.

Let W be the set of those elements of $\tilde{W}$, which are not an endpoint of an interval, which is open and closed. Now we consider $X' = X \cup W'$, where W' is a copy of W disjoint from X. We extend the order relation to X'. If $y < x < z$ in X, $x \in W$ and x' is the element of W' corresponding to x, then we define $y < x < x' < z$ or $y < x' < x < z$, such that we can extend T continuously to X'. T is then continuous on X' and if one defines $\tilde{W}$ for X', all elements of $\tilde{W}$ are endpoints of intervals, which are open and closed. The order topology of X' is again compact and separable. We extend also φ continuously to X' and define $\tilde{P}$ by $\tilde{P}f(x) = \sum_{y \in T^{-1}x} e^{\varphi(y)} f(y)$ for $x \in X'$.

Because the intervals $T(I_i)$ for $1 \leq i \leq N$ are open and closed in X', $\tilde{P}$ maps $C(X')$ into $C(X')$. We can apply the Schauder-Tychonoff-theorem (see Dunford and Schwartz, Linear operators I, p. 456) to the continuous map $m \mapsto \tilde{P}^* m / \tilde{P}^* m(1)$ on the set of all Borel-probability measures, which is a compact, convex subset of the dual of $C(X')$ with w^*-topology. $\tilde{P}^*$ denotes the dual operator of $\tilde{P}$. We get a fixed point m for this map, i.e. $m(\tilde{P}f) = \lambda m(f)$ for all $f \in C(X')$, where $\lambda = \tilde{P}^* m(1)$.

Now set $\tilde{g}(x) = e^{\varphi(x)}/\lambda$. The problem is to show that $\|\tilde{g}_n\|_\infty < 1$ for some n ($\tilde{g}_n(x) = \tilde{g}(x)\,\tilde{g}(Tx) \ldots \tilde{g}(T^{n-1}x)$) and that $m(W') = 0$. To this end we prove two lemmas.

Lemma 16. *Let A be a Borel subset of X' such that T^n/A is monotone for all n. Suppose that for every $j \geq 1$ there is a B_j, which is a Borel subset of some element of $\mathfrak{P}_0^{k_j}$ and which satisfies $T^{k_j}(B_j) = A$ for some k_j. Furthermore suppose that $B_i \cap B_j = \emptyset$ for $i \neq j$ and that there is an $l \in \mathbb{N}$ such that B_j and A are contained in the same element of $\mathfrak{P}_0^{k_j - l}$ for all $j \geq 1$. Then $m(A) = 0$.*

Proof. We have $m(B_j) = \lambda^{-k_j} m(\tilde{P}^{k_j} 1_{B_j}) \geq \inf\{\tilde{g}_{k_j}(y) : y \in B_j\} \cdot m(A)$ and similarly $m(A) \leq \sup\{\tilde{g}_{k_j}(y) : y \in A\} \cdot m(T^{k_j} A)$. This implies

$$\begin{aligned} m(B_j)\, m(T^{k_j} A) &\geq (\inf\{\tilde{g}_{k_j}(y) : y \in B_j\} / \sup\{\tilde{g}_{k_j}(y) : y \in A\})\, m(A)^2 \\ &\geq \exp\left(-\sum_{i=0}^{\infty} \operatorname{var}_i \varphi - 2l\, \|\varphi\|_\infty\right) \cdot m(A)^2 \end{aligned}$$

because A and B_j are contained in the same element of $\mathfrak{P}_0^{k_j - l}$. As the B_j are disjoint, we have $m(B_j) \to 0$ as $j \to \infty$. Of course, $m(T^{k_j} A) \leq 1$, hence $m(A) = 0$.

Lemma 17. *Let J be a closed interval, which is a subset of some element of $\mathfrak{P}_0^n$. Then*

$$\sup_J \tilde{g}_n \leqq K m(J)/m(T^n J), \qquad \textit{where} \quad K = \exp\left(\sum_{i=1}^{\infty} \operatorname{var}_i \varphi\right).$$

Proof.

$$\begin{aligned} m(J) &= \lambda^{-n} m(\tilde{P}^{n_1} J) \geqq (\inf_J \tilde{g}_n)\, m(T^n J) \\ &\geqq (\sup_J \tilde{g}_n) \exp\left(-\sum_{i=1}^{n} \operatorname{var}_i \varphi\right) \cdot m(T^n J) \\ &\geqq K^{-1} (\sup_J \tilde{g}_n)\, m(T^n J). \end{aligned}$$

This implies the desired inequality.

We shall apply these lemmas in three special cases in §§ 3-5.

§3. Let $X = \{x_0 x_1 \ldots \in \{1, 2, \ldots, N\}^{\mathbb{N}}: M(x_i x_{i+1}) = 1\}$ be a finite type subshift, which is topologically transitive, i.e. the transition matrix M is irreducible. Furthermore we assume that $h_{\text{top}}(X) > 0$, i.e. X consists not only of a periodic orbit. If we introduce the lexicographic ordering, the shift transformation σ becomes a piecewise monotinic transformation on X. We have here $W = \varnothing$ and $X' = X$. We apply Lemma 16 to $A = \{x\}$ for $x \in X$. Because X is not only a periodic orbit, we can find an inverse image $y_0 y_1 \ldots y_{m-1} x_0 x_1 \ldots$ of $x = x_0 x_1 \ldots$, which is not periodic. As M is irreducible, there is a k_0 such that for any x_j, $x_0 \in \{1, \ldots, N\}$ there are $i_0 = x_j$, $i_1, \ldots, i_k = y_0$ with $k \leqq k_0$ and $M(i_r, i_{r+1}) = 1$ for $0 \leqq r \leqq k = 1$. Then among the points

$$x_0 x_1 \ldots x_j i_1 \ldots i_{k-1} y_0 y_1 \ldots y_{m-1} x_0 x_1 x_2 \ldots \in X \qquad \text{for } j = 1, 2, 3, \ldots$$

there are infinitely many different ones. They can be used in Lemma 16 as B_j $(l = m + k_0)$ to get $m(\{x\}) = 0$.

Set $q_k = \max\{m(A): A \in \mathfrak{P}_0^k\}$. In this case $\mathfrak{P}_0^k$ is the set of all cylinders $[x_0 \ldots x_{k-1}] = \{y_0 y_1 \ldots \in X: y_i = x_i \text{ for } 0 \leqq i \leqq k-1\}$. We have $q_k \to 0$ $(k \to \infty)$, because otherwise one finds an $x \in X$ and an $\varepsilon > 0$, such that each neighbourhood of x contains an $A \in \mathfrak{P}_0^k$ for some k with $m(A) \geqq \varepsilon$. This implies $m(\{x\}) \geqq \varepsilon$, a contradiction.

Next we show that the support Y of m is all of X. There is at least one i with $m([i]) > 0$, because otherwise $m(X) = 0$. Choose some cylinder $[j_0 \ldots j_{n-1}] \subset X$. As above one finds $i_0 = j_{n-1}, i_1, \ldots, i_k = i$ with $M(i_r, i_{r+1}) = 1$. Then

$$m([j_0 \ldots j_{n-1} i_1 \ldots i_k]) \geqq (\inf \tilde{g})^{k+n-1}\, m([i_k]) > 0$$

and hence $m([j_0 \ldots j_{n-1}]) > 0$. This gives $Y = X$. In particular $m([i]) > 0$ for $1 \leqq i \leqq N$.

Now set $J := [x_0 \ldots x_{n-1}]$. Then $T^n(J) \supset [i]$ for some i, hence

$$m(T^n J) \geqq \min\{m([i]): 1 \leqq i \leqq N\} =: c > 0.$$

Lemma 17 implies then that $\sup_J \tilde{g}_n \leqq \frac{K}{c} q_n$. As J was an arbitrary element of $\mathfrak{P}_0^n$, this shows that there is an n with $\|\tilde{g}_n\|_\infty < 1$. Setting $g = \tilde{g}_n$, we have $P = \tilde{P}^n / \lambda^n$ and (a) and (b) are satisfied, if we assume additionally that φ and hence also $\tilde{g}$ and g are of bounded variation. As $Y=X$ and as $h(\mu, T^n) = nh(\mu, T)$, it follows from Theorem 6 that μ is then an equilibrium state for φ. (cf. the results of Chap. 1 of [2]).

§4. We consider the transformation $T: x \mapsto \beta x \pmod 1$ on $[0, 1)$. Let $\chi: [0, 1) \to \{1, \ldots, N\}^{\mathbb{N}}$ (N such that $\beta \leqq N < \beta + 1$) be the β-expansion $x \mapsto i_0 i_1, \ldots, i_j$ such that $i_j - 1 \leqq \beta T^j(x) < i_j$. Set $e_0 e_1 \ldots = \lim_{t \uparrow 1} \chi(t)$ and

$$X = \{x_0 x_1 \ldots \in \{1, \ldots, N\}^{\mathbb{N}}: x_k x_{k+1} \ldots \leqq e_0 e_1 \ldots \text{ for } k \geqq 0\},$$

where $\leqq$ denotes the lexicographic ordering. It follows from Theorem 1 of [6] that $\overline{\chi([0, 1))} = X$. Furthermore $\chi \circ T = \sigma \circ \chi$ and $X \setminus \chi([0, 1))$ is countable. Hence we can consider (X, σ) instead of $([0, 1), T)$, because m will have no atoms. It follows from the definitions, that $e_k e_{k+1} \ldots \neq 111 \ldots$ for all $k \geqq 0$.

We proceed as in §3. Let $x \in X'$ and $y = y_0 \ldots y_{m-1} x_0 x_1 \ldots \in X'$ be an inverse image of x, which is not periodic. We choose a sequence (k_j) and integers z_j with $1 \leqq z_j \leqq x_{k_j}$ as follows. If $\sigma^i(y) \neq e_0 e_1 \ldots$ for all i then choose k_1 such that $x_i = e_i$ for $0 \leqq i \leqq k_1 - 1$ and $x_{k_1} < e_{k_1}$. Having chosen $k_1 < \ldots < k_{n-1}$ choose k_n such that $x_{k_{n-1}+i} = e_i$ for $0 \leqq i \leqq k_n - k_{n-1} - 1$ and $x_{k_n} < e_{k_n - k_{n-1}}$. Using the definition of X, one checks easily that $x_0 \ldots x_{k_i}$ does not end with an initial segment of $e_0 e_1 \ldots$. Set $z_j = x_{k_j}$. If $\sigma^i(y) = e_0 e_1 \ldots$ for some i choose the $k_j \geqq \max\{i, m\}$ such that $e_{k_j - i} = x_{k_j} > 1$. This is possible, because $e_j e_{j+1} \ldots \neq 11 \ldots$ for all j. Set $z_j = x_{k_j} - 1$. One has again that $x_0 \ldots x_{k_j - 1} z_j$ does not end with an initial segment of $e_0 e_1 \ldots$. We set

$$B_j = \{x_0 \ldots x_{k_j - 1} z_j y_0 \ldots y_{m-1} x_0 x_1 \ldots\} \subset X'.$$

Then the requirements of Lemma 16 are satisfied for these B_j, $A = \{x\}$ and $l = m+1$. Hence $m(\{x\}) = 0$. In particular $m(W') = 0$ and we can consider again X instead of X'.

As in §3 we get that $q_k = \sup\{m(A): A \in \mathfrak{P}_0^k\}$ tends to zero as $k \to \infty$. We set $J = [x_0 \ldots x_{n-1}]$. If $x_0 \ldots x_{n-1}$ does not end with an initial segment of $e_0 e_1 \ldots$, then $\sigma^n(J) = X$. Hence by Lemma 17 we get

$$\sup_J \tilde{g}_n \leqq K q_n.$$

If $x_0 \ldots x_{n-1}$ ends with an initial segment of $e_0 e_1 \ldots$, choose k minimal such that $x_{n-k} \ldots x_{n-1} = e_0 \ldots e_{k-1}$. Then $J' = [x_0 \ldots x_{n-k-1} (x_{n-k} - 1) x_{n-k+1} \ldots x_{n-1}]$ satisfies $\sigma^n(J') = X$. Hence

$$\sup_J \tilde{g}_n \leqq (\sup_{J'} \tilde{g}_n) \cdot \exp\left(\sum_{i=1}^{\infty} \operatorname{var}_i \varphi + 2\|\varphi\|_\infty\right) \leqq K q_n \cdot \tilde{K}.$$

Therefore we find again an n such that $\|\tilde{g}_n\|_\infty < 1$. We set $g = \tilde{g}_n$. As in §3, one can show that the support Y of m is all of X. Hence by Theorem 6, μ is an equilibrium state for φ.

This means that we have generalized the results of [22] from the class of all Lipschitz continuous φ to the class of φ satisfying $\sum \operatorname{var}_i \varphi < \infty$ and $V_X \varphi < \infty$.

§5. Now let X be the interval $[0, 1]$ and T any piecewise monotonic transformation on X with positive topological entropy. For φ we take the constant function zero. Then $\tilde{g}$ is the constant function λ^{-1}. We have to show that $\lambda > 1$. Then we can set $g = \tilde{g}$ and (a) and (b) are satisfied.

To this end we need some results of [6] (cf. also [7]). Let D be the following set of subintervals of $I_i \subset X'$ $(1 \leqq i \leqq N)$, on which T is monotone. An interval K is called a successor of an interval $J \subset I_i$, if K is one of the nonempty intervals among $T(J) \cap I_j$ for $1 \leqq j \leqq N$. We define D in the following way: D contains I_i for $1 \leqq i \leqq N$ and with every interval also its successors. Because all images of I_i's are open and closed subsets of X', all elements of D are open and closed. In [6] the elements of D are subsets of a shift space isomorphic to $([0, 1], T)$, but this makes no difference. Furthermore we define a $D \times D$-matrix M with entries 0 and 1 by $M(J, K) = 1$ if and only if K is a successor of J, for $J, K \in D$. It happens that elements of D are successors of more than one element of D.

It is shown in [6] that every irreducible submatrix L of M corresponds to a closed T-invariant subset Ω of $[0, 1]$. If $F \subset D$ is the index set of L, i.e. $L = M/F$, then $I = \cup \{J : J \in \tilde{F}\}$, where $\tilde{F} := \{J \in D$: there are $J_0, J_1, \ldots, J_n = J$ with $J_0 \in F$ and $M(J_i, J_{i+1}) = 1\}$, and $I' = \cup \{J : J \in \tilde{F} \setminus F\}$ are T-invariant sets, which are finite unions of intervals and $\Omega = \bigcap_{i=0}^{\infty} T^{-i}\overline{(I \setminus I')}$. We choose an Ω, such that $r(L) = r(M)$, where r denotes the spectral radius of the l^1-operators L and M. Because I is T-invariant, T/I is again a piecewise monotonic transformation. Hence we can take I as X and assume that m is concentrated on I.

Lemma 18. (i) *If* $M(J_i J_{i+1}) = 1$ *for* $0 \leqq i \leqq k-1$ *then* $T^k(J_0 \cap \ldots \cap T^{-k} J_k) = J_k$.

(ii) *For every* $J \in F$, *we have* $m(J) > 0$.

Proof. (i) is proved by induction. The induction step is as follows

$$\begin{aligned} T^k(J_0 \cap T^{-1} J_1 \cap \ldots \cap T^{-k} J_k) &= T^{k-1}(T J_0 \cap J_1 \cap \ldots \cap T^{-k+1} J_k) \\ &= T^{k-1}(J_1 \cap \ldots \cap T^{-k+1} J_k) \end{aligned}$$

because $T J_0 \supset J_1$, since J_1 is a successor of J_0.

In order to show (ii) note that there is a $K \in \tilde{F}$ with $m(K) > 0$, since $m(I) = 1$. We can find $J_0 = J, J_1, \ldots, J_k = K$ with $M(J_i, J_{i+1}) = 1$ and $J \in F$. As L is irreducible, we can do this for all $J \in F$. Set $Z = J_0 \cap T^{-1} J_1 \cap \ldots \cap T^{-k} J_k \subset J$. Then $m(Z) = \lambda^{-k} m(\tilde{P}^k 1_Z) = \lambda^{-k} m(T^k Z) = \lambda^{-k} m(K) > 0$, using (i). Hence $m(J) > 0$ proving (ii).

It follows from §3 of [6] that $r(M) = \exp h_{\text{top}}(X)$. Thus $r(L) = r(M) > 1$, and L consists not only of a cycle. Therefore we can find $J_0, J_1, \ldots, J_n = J_0 \in F$ and $K_0 = J_0$, $K_1, \ldots, K_k = J_r \in F$ for some $n, k \geqq 0$ and $0 \leqq r \leqq n-1$ such that $M(J_i, J_{i+1}) = M(K_i, K_{i+1}) = 1$. Now set

$$A_k = \bigcap_{i=0}^{k-1} T^{-ni}\left(\bigcap_{j=0}^{n-1} T^{-j} J_j\right), \qquad A = \bigcap_{k=0}^{\infty} A_k,$$

$$B_j = \bigcap_{i=0}^{jn} T^{-i} J_{i(\text{mod } n)} \cap \bigcap_{i=1}^{k} T^{-jn-i} K_i \cap \bigcap_{i=1}^{\infty} T^{-jn-k-i} J_{r+i(\text{mod } n)}.$$

The B_j are disjoint, because $T^{jn} B_j \subset K_1$, $T^{jn} B_{j+m} \subset J_1$ and K_1 and J_1 are disjoint as different successors of J_0. Furthermore $T^{jn+k+n-r} B_j = A$ and A and B_j are contained in the same element of $\mathfrak{P}_0^{jn}$. Hence it follows from Lemma 16 that $m(A)=0$. Now we apply Lemma 17 and get

$$\sup_{A_k} \tilde{g}_{kn} \leqq K m(A_k)/m(T^{kn} A_k).$$

Because of $T^{kn} A_k = T(J_{n-1}) \supset J_0$ ((i) of Lemma 18) and $m(J_0)>0$ ((ii) of Lemma 18), the right hand side of this inequality tends to zero. Hence there is a k with $\sup_{A_k} \tilde{g}_{kn} < 1$. Because of $\tilde{g}_{kn} \equiv \lambda^{-kn}$, we get $\lambda > 1$, the desired result. It follows now from Lemma 2 that $m(\{x\})=0$ for all $x \in X'$. Hence $m(W')=0$.

A similar proof as that of (ii) of Lemma 18 shows that the support Y of m contains Ω. As $h_{\text{top}}(\Omega) = \log r(L) = \log r(M) = h_{\text{top}}(X)$, it follows from Theorem 6, that μ is a measure with maximal entropy.

Acknowledgement. We thank the referee for some comments which simplified the proof of Lemma 5.

References

1. Babbel, B.: Diplomarbeit, Institut für Mathematische Statistik und Wirtschaftsmathematik der Universität Göttingen 1980
2. Bowen, R.: Equilibrium states and the ergodic theory of Anosov diffeomorphisms. Lecture Notes in Mathematics **470**. Berlin-Heidelberg-New York: Springer 1975
3. Bowen, R.: Bernoulli maps of an interval. Israel J. Math. **28**, 298–314 (1978)
4. Dunford, N., Schwartz, J.T.: Linear Operators, Part I. New York: Interscience 1957
5. Gordin, M.I.: The central limit theorem for stationary processes. Soviet Math. Dokl. **10**, 1174–1176 (1969)
6. Hofbauer, F.: On intrinsic ergodicity of piecewise monotonic transformations with positive entropy. Israel J. Math. **34**, 213–237 (1979)
7. Hofbauer, F.: On intrinsic ergodicity of piecewise monotonic transformations with positive entropy II. Israel J. Math. (To appear)
8. Ibragimov, I.A., Linnik, Y.V.: Independent and stationary sequences of random variables. Groningen: Wolters-Noordhoff 1971
9. Ionescu-Tulcea, C., Marinescu, G.: Théorie ergodique pour des classes d'opérations non complètement continues. Ann. of Math. (2) **52**, 140–147 (1950)
10. Keller, G.: Un théorème de la limite centrale pour une classe de transformations monotones per morceaux. C.R. Acad. Sci. Paris Sér. A **291**, 155–158 (1980)
11. Lasota, A., Yorke, J.: On the existence of invariant measures for piecewise monotonic transformations. Trans. Amer. Math. Soc. **186**, 481–488 (1973)
12. Ledrappier, F.: Principe variationnel et systemes dynamiques symboliques. Z. Wahrscheinlichkeitstheorie verw. Gebiete **30**, 185–202 (1974)
13. Li, T., Yorke, J.: Ergodic transformations from an interval into itself. Trans. Amer. Math. Soc. **235**, 183–192 (1978)
14. Li, T., Yorke, J.: Iterating piecewise expanding maps: Asymptotic dynamics of probability densities. Preprint
15. Philipp, W., Stout, W.: Almost sure invariance principles for partial sums of weakly dependent random variables. Mem. Amer. Math. Soc. **161** (1975)
16. Ratner, M.: Bernoulli flows over maps of the interval. Israel J. Math. **31**, 298–314 (1978)
17. Rohlin, V.A.: Exact endomorphisms of Lebesgue spaces. Amer. Math. Soc. Transl. (2) **39**, 1–36 (1964)

18. Schäfer, H.H.: Topological Vector Spaces. New York: MacMillan 1966
19. Volkonski, V.A., Rozanov, Y.A.: Some limit theorems for random functions II. Theor. Probability Appl. **6**, 186–198 (1961)
20. Wagner, G.: The ergodic behavior of piecewise monotonic transformations. Z. Wahrscheinlichkeitstheorie und verw. Gebiete **46**, 317–324 (1979)
21. Walters, P.: Ergodic theory. Lecture Notes in Mathematics **458**. Berlin-Heidelberg-New York: Springer 1975
22. Walters, P.: Equilibrium states for β-transformations and related transformations. Math. Z. **159**, 65–88 (1978)
23. Wong, S.: Some metric properties of piecewise monotonic mappings of the unit interval. Trans. Amer. Math. Soc. **246**, 493–500 (1978)
24. Wong, S.: A central limit theorem for piecewise monotonic mappings of the unit interval. Ann. Probability **7**, 500–514 (1979)

Received March 25, 1981; received in final form December 7, 1981

Physica 7D (1983) 153–180
North-Holland Publishing Company

THE DIMENSION OF CHAOTIC ATTRACTORS

J. Doyne FARMER
Center for Nonlinear Studies and Theoretical Division, MS B258, Los Alamos National Laboratory, Los Alamos, New Mexico 87545, USA

Edward OTT
Laboratory of Plasma and Fusion Energy Studies, University of Maryland, College Park, Maryland, USA

and

James A. YORKE
Institute for Physical Science and Technology and Department of Mathematics, University of Maryland, College Park, Maryland, USA

Dimension is perhaps the most basic property of an attractor. In this paper we discuss a variety of different definitions of dimension, compute their values for a typical example, and review previous work on the dimension of chaotic attractors. The relevant definitions of dimension are of two general types, those that depend only on metric properties, and those that depend on the frequency with which a typical trajectory visits different regions of the attractor. Both our example and the previous work that we review support the conclusion that all of the frequency dependent dimensions take on the same value, which we call the "dimension of the natural measure", and all of the metric dimensions take on a common value, which we call the "fractal dimension". Furthermore, the dimension of the natural measure is typically equal to the Lyapunov dimension, which is defined in terms of Lyapunov numbers, and thus is usually far easier to calculate than any other definition. Because it is computable and more physically relevant, we feel that the dimension of the natural measure is more important than the fractal dimension.

Table of contents

1. Introduction

It is the purpose of this paper to discuss and review questions relating to the dimension of chaotic attractors. Before doing so, however, we should first say what we mean by the work "attractor".

1.1. *Attractors*

In this paper we consider dynamical systems such as maps (discrete time, n)

$$x_{n+1} = F(x_n),$$

or ordinary differential equations (continuous

time, t)

$$\frac{dx(t)}{dt} = G(x(t)),$$

where in both cases x is a vector. Thus, given an initial value of x (at $n = 0$ for the map or $t = 0$ for the differential equations) an orbit is generated ($(x_1, x_2, \ldots, x_n, \ldots)$ for the map and $x(t)$ for the differential equations). We shall be interested in attractors for such systems. Loosely speaking, an attractor is something that "attracts" initial conditions from a region around it once transients have died out. More precisely, an *attractor* is a compact set, A, with the property that there is a neighborhood of A such that for almost every* initial condition the limit set of the orbit as n or $t \to +\infty$ is A. Thus, almost every trajectory in this neighborhood of A passes arbitrarily close to every point of A. The *basin of attraction* of A is the closure of the set of initial conditions that approach A.

We are primarily interested in *chaotic* attractors. We give a definition of chaos in section 3, but the reader may also wish to see the reviews given in references 1–4.

1.2. *Why study dimension?*

The dimension of an attractor is clearly the first level of knowledge necessary to characterize its properties. Generally speaking, we may think of the dimension as giving, in some way, the amount of information necessary to specify the position of a point on the attractor to within a given accuracy (cf. section 2). The dimension is also a lower bound on the number of essential variables needed to model the dynamics. For an extensive discussion of dimension in many contexts, see Mandelbrot [5, 6, 46].

* The phrase "almost every" here signifies that the set of initial conditions in this neighborhood for which the corresponding limit set is not A can be covered by a set of cubes of arbitrarily small volume (i.e. has Lebesgue measure zero).

† Mod 1 means that the values of x and y are truncated to be less than or equal to one and their integer part are discarded, so that the map is defined on the unit square.

For simple attractors, defining and determining the dimension is easy. For example, using any reasonable definition of dimension, a stationary time independent equilibrium (fixed point) has dimension zero, a stable periodic oscillation (limit cycle) has dimension one, and a doubly periodic attractor (2-torus) has dimension two. It is because their structure is very regular that the dimension these simple attractors takes on integer values.

Chaotic (strange) attractors, however, often have a structure that is not simple; they are often not manifolds, and frequently have a highly fractured character. For chaotic attractors, intuition based on properties of regular, smooth examples does not apply. The most useful notions of dimension take on values that are typically not integers.

To fully understand the properties of a chaotic attractor, one must take into account not only the attractor itself, but also the "distribution" or "density" of points on the attractor. This is more precisely discussed in terms of what we shall call the *natural measure* associated with a given attractor. The natural measure provides a notion of the relative frequency with which an orbit visits different regions of the attractor. Just as chaotic attractors can have very complicated properties, the natural measures of chaotic attractors often have complicated properties that make the relevant assignment of a dimension a nontrivial problem.

Precise definitions of such terms as "natural measure" follow, but we would first like to give an example in order to motivate the central questions we are addressing in this paper.

Consider the following two dimensional map†:

$$\begin{aligned} x_{n+1} &= x_n + y_n + \delta \cos 2\pi y_n \mod 1, \\ y_{n+1} &= x_n + 2y_n \mod 1. \end{aligned} \tag{1}$$

For small values of δ, Sinai [7] has shown that the attractor of this map is the entire square, and is thus of dimension 2. Therefore almost every initial condition generates a trajectory that eventually comes arbitrarily close to every point on the square. However, consider the typical trajectory

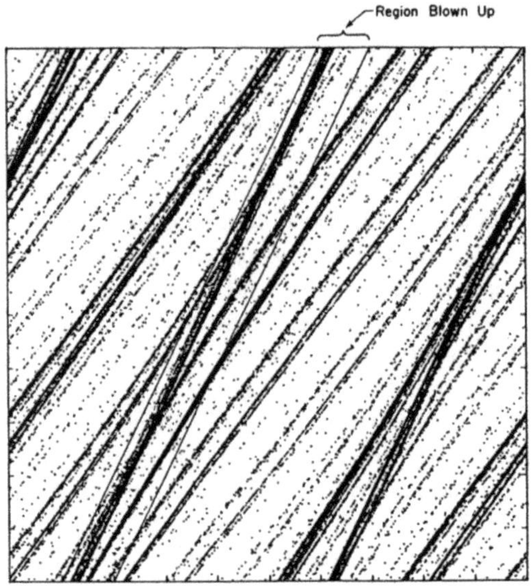

Fig. 1. Successive iterates of the initial point $x_0 = 0.5$, $y_0 = 0.5$ using eq. (1) with $\delta = 0.1$. 80,000 points are shown. Almost any initial condition gives a qualitatively similar plot; the location of the individual points of course changes, but the location of the dark bands does not. The density of these points is described by the *natural measure* of this attractor. (For example, the outlined parallelogram (which is blown up in fig. 2) contains approximately 27% of the points of a typical trajectory, and thus can be said to have a natural measure of approximately 0.27.)

Fig. 2. A blow-up of the strip marked in fig. 1. This strip was chosen in order to follow one of the dark bands; the blow-up was made by expanding the strip in a direction perpendicular to its long sides (roughly horizontally), and the top and bottom were trimmed to make the result square. What appears to be a single band in fig. 1 is now seen as a collection of bands.

shown in fig. 1. Certain regions are visited far more often than others. The natural measure of a given region is proportional to the frequency with which it is visited (see section 2.2.2), in this case the natural measure is highly concentrated in diagonal bands whose density of points is much greater than the average*. Furthermore, as shown in fig. 2, if a small piece of the attractor is magnified, the same sort of structure is still seen.

For this map we do not know if the value of δ chosen to construct fig. 1 is small enough to insure that the dimension of the attractor is two. For practical purposes, though, this may be irrelevant. Even if a trajectory eventually comes arbitrarily close to any given point, the amount of time required for this to happen may be enormous. In order to assign a relevant dimension that will characterize the trajectories on the attractor, the natural measure must be taken into account. For this example the dimension that characterizes properties of the natural measure is between one and two.

These considerations are not as esoteric as they might seem. One may not be as interested in whether the dimension of a given attractor is 3.1 or 3.2 as in whether it is on the order of three or on the order of thirty. As we shall see, a proper understanding of *probabilistic* notions of dimension leads to an efficient method of computing the dimension of chaotic attractors, that provides the best known method of answering such questions.

The main points of this paper can be summarized as follows:

1) Although there are a variety of different definitions of dimension, the relevant definitions

* In fact, for small values of δ, Sinai [7] has shown that for any $\epsilon > 0$, there exists a collection of tiny squares whose total area is less than ϵ, and such that almost every trajectory spends $1 - \epsilon$ of the time inside this collection of squares. These squares cover what is called the *core* of the attractor. (See section 7).

Table I
Current evidence indicates that typically the first two dimensions take on the same value, called the fractal dimension, while the next five dimensions take on another typically smaller value, called the dimension of the measure.

Name of dimension	Symbol	Generic name	Symbol
Capacity	d_C	fractal	d_F
Hausdorff dimension	d_H	dimension	
Information dimension	d_I		
ϑ-capacity	$d_C(\vartheta)$		
ϑ-Hausdorff dimension	$d_H(\vartheta)$	dimension of the	d_μ
Pointwise dimension	d_P	natural measure	
Hausforff dimension of the core	d_H(core)		
Lyapunov dimension	d_L		

are of two types, those which only depend on metric properties, and those which depend on metric and probabilistic properties (i.e., they involve the natural measure of the attractor).

2) Current evidence supports the conclusion that all of the metric dimensions typically take on the same value, and all of the frequency dependent dimensions take on another, typically smaller, common value.

3) Current evidence supports a conjectured relationship whereby the dimension of the natural measure can be found from a knowledge of the stability properties of an orbit on the attractor (i.e., knowledge of the Lyapunov numbers).

4) For typical chaotic attractors we conjecture that the distribution of frequencies with which an orbit visits different regions of the attractor is, in a certain sense, log-normal (section 5).

Points 1–3 are summarized in table I. The first two entries in the table are metric dimensions, while the next five are frequency dependent dimensions. Under the hypothesis that all the metric dimensions yield the same value (point 2), we call this value the *fractal dimension* and denote it d_F. Similarly, if all the probabilistic dimensions yield the same value, we call this value the *dimension of the natural measure*, and denote it d_μ. Although in special cases d_F equals d_μ, typically $d_F > d_\mu$. Finally, the last entry in table I, the Lyapunov dimension, is by definition the predicted value of d_μ obtained from the Lyapunov numbers (cf. Point 3). The Lyapunov dimension is in a different category than the other dimensions listed, since it is defined in terms of dynamical properties of an attractor, rather than metric and natural measure properties.

1.3. *Outline*

This paper is organized as follows: In section 2 we give several definitions of dimension. Section 3 reviews conjectures relating Lyapunov numbers to dimension. These conjectures are particularly important because the Lyapunov numbers provide the only known efficient method to compute dimension. In sections 4, 5, 6, and 7, we compute all the dimensions discussed here for an explicitly soluble example, the generalized baker's transformation. In addition, based on this example, in section 5 we propose a new conjecture concerning the frequency with which different values of the probability occur. Section 7 gives a discussion of the "core" of attractors, and section 8 gives another example supporting the connection between Lyapunov numbers and dimension (an attractor which is topologically a torus but is nowhere differentiable). Section 9 reviews relevant results from numerical computations of the dimension of chaotic attractors. Concluding remarks are given in section 10.

In general terms, this paper has two functions. One is to present a review of the current status of research on the dimension of chaotic attractors. The other purpose is to present new results (sections 4–6).

2. Definitions of dimension

In this section we define and discuss six different concepts of dimension. The first two of these, the capacity and the Hausdorff dimension, require only a metric (i.e., a concept of distance) for their definition, and consequently we refer to them as "metric dimensions". The other dimensions we will discuss in this section are the information dimension, the ϑ-capacity, the ϑ-Hausdorff dimension, and the pointwise dimension. These dimensions require both a metric and a probability measure for their definition, and hence we will refer to them as "probabilistic dimensions".

In this paper we compute the values of these dimensions for an example that we believe is general enough to be "typical" of chaotic attractors, at least regarding the question of dimension. We find that the metric dimensions take on a common value. Whenever this is the case, we will refer to this common value d_F as the *fractal dimension**. For our example we also find that the probabilistic dimensions take on a common value d_μ, which we will refer to as the *dimension of the natural measure*. As we summarize in conjecture 1, we feel that this equality is a general property, true for typical cases.

Conjecture 1. For a typical chaotic attractor the capacity and Hausdorff dimensions have a common value d_F, and the information dimension, ϑ-capacity, ϑ-Hausdorff dimension, and pointwise dimensions have a common value d_μ, i.e., in the notation of table I,

$$d_C = d_H \equiv d_F$$

and

$$d_I = d_C(\vartheta) = d_H(\vartheta) = d_P \equiv d_\mu.$$

Note: For the case of diffeomorphisms† in two dimensions, L.S. Young has rigorously proven that information dimension, pointwise dimension, and the Hausdorff dimension of the core (see section 7) all take on the same value [12].

In addition to the dimensions defined in this section, we will also discuss three others‡. The Lyapunov dimension, the capacity of the core, and the Hausdorff dimension of the core. Lyapunov dimension is discussed in section 3, and the latter two dimensions are discussed in section 7. For our example the Lyapunov dimension and Hausdorff dimension of the core are equal to d_μ, while the capacity of the core is equal to d_F.

2.1. *Metric dimensions*

We begin by discussing two concepts of dimension which apply to sets in spaces on which a concept of distance, i.e., a metric is defined. In particular we begin by discussing the capacity and the Hausforff dimension.

2.1.1. *Capacity*

The capacity of a set was originally defined by Kolmogorov [14]. It is given by

$$d_C = \lim_{\epsilon \to 0} \frac{\log N(\epsilon)}{\log(1/\epsilon)}, \tag{2}$$

* The term *fractal* was originally coined by Mandelbrot [5]. However, he uses "fractal dimension" as a synonym for Hausdorff dimension. We should also mention that in some of our previous papers on this subject [8–11], we used the term "fractal dimension" as a synonym for capacity, rather than our current usage as described in the text.

† A diffeomorphism is a differentiable invertible mapping whose Jacobian has non-zero determinant everywhere.

‡ Note that in this paper we will not discuss the concept of *topological dimension*, since its application to chaotic dynamics is not clear. Its value is an integer and it is generally equal to neither d_F nor d_μ. For discussions of topological dimension, we refer the reader to Hurwicz and Wallman [13].

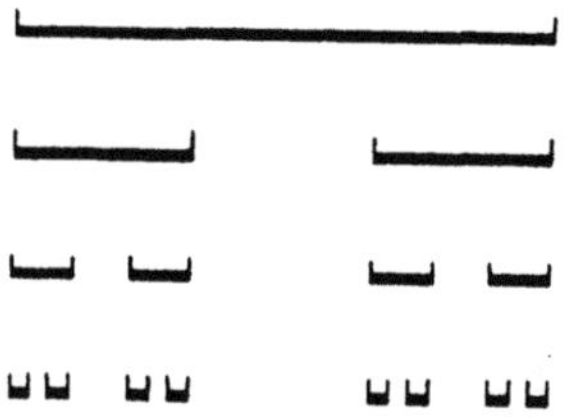

Fig. 3. The first few steps in the construction of the classic example of a Cantor set.

where, if the set in question is a bounded subset of a p-dimensional Euclidean space $\mathbb{R}^p$, then $N(\epsilon)$ is the minimum number of p-dimensional cubes of side ϵ needed to cover the set. For a point, a line, and an area, $N(\epsilon) = 1$, $N(\epsilon) \sim \epsilon^{-1}$, and $N(\epsilon) \sim \epsilon^{-2}$, and eq. (2) yields $d_C = 0$, 1, and 2, as expected. However, for more general sets (dubbed *fractals* by Mandelbrot), d_C can be noninteger*. For example, consider the Cantor set obtained by the limiting process of deleting middle thirds, as, illustrated in fig. 3. If we choose $\epsilon = (1/3)^m$, then $N = 2^m$, and eq. (2) yields

$$d_C = \frac{\log 2}{\log 3} = 0.630\ldots .$$

If one is content to know where the set lies to within an accuracy ϵ, then to specify the location of the set, we need only specify the position of the $N(\epsilon)$ cubes covering the set. Eq. (2) implies that for small ϵ, $\log N(\epsilon) \approx d_C \log(1/\epsilon)$. Hence, the dimension tells us how much information is necessary to specify the location of the set to within a given accuracy. If the set has a very fine-scaled structure (typical of chaotic attractors), then it may be advantageous to introduce some coarse-graining into the description of the set. In this case, ϵ may be thought of as specifying the degree of coarse-graining.

* Sets can be constructed for which the limit of eq. (2) does not exist. We would then say that the capacity is not defined.

† For example, for the set of numbers 1, 1/2, 1/3, 1/4, ..., the Hausdorff dimension is zero while (2) yields $d_C = \frac{1}{2}$.

‡ To show that $d_C \geq d_H$, consider a covering consisting of cubes of equal side $\epsilon_i = \epsilon$. Then due to the infimum in the definition of $l_d(\epsilon)$, we see that $\tilde{l}_d(\epsilon) \equiv \Sigma \epsilon^d = N(\epsilon)\epsilon^d$ satisfies $\tilde{l}_d(\epsilon) \geq l_d(\epsilon)$. Thus taking the limit $\epsilon \to 0$ and making use of eq. (2) we see that $d_H \leq d_C$.

2.1.2. *Hausdorff dimension*

The capacity may be viewed as a simplified version of the Hausdorff dimension, originally introduced by Hausdorff in 1919 [15]. (We have reversed historical order and defined capacity before Hausdorff dimension because the definition of Hausdorff dimension is more involved.) We believe that for attractors these two dimensions are generally equal. While it is possible to construct simple examples of sets where the Hausdorff dimension and the capacity are unequal†, these do not seem to apply to attractors. (Although they may apply to the core of attractors. See section 7.)

To define the Hausdorff dimension of a set lying in a p-dimensional Euclidean space, consider a covering of it with p-dimensional cubes of variable edge length ϵ_i. Define the quantity $l_d(\epsilon)$ by

$$l_d(\epsilon) = \inf \sum_i \epsilon_i^d,$$

where the infimum (i.e. minimum) extends over all possible coverings subject to the constraint that $\epsilon_i \leq \epsilon$. Now let

$$l_d = \lim_{\epsilon \to 0} l_d(\epsilon).$$

Hausdorff showed that there exists a critical value of d above which $l_d = 0$ and below which $l_d = \infty$. This critical value, $d = d_H$, is the Hausdorff dimension. (Precisely at $d = d_H$, l_d may be either 0, ∞, or a positive finite number.) This concept of dimension will be used in sections 4, 6, and 7. It is easy to see that $d_C \geq d_H$‡.

2.2. *Dimensions for the natural measure*

2.2.1. *The natural measure on an attractor*

Note that, in computing d_C from eq. (2), all cubes used in covering the attractor are equally important even though the frequencies with which an orbit on the attractor visits these cubes may be very different. In order to take the frequency with which each cube is visited into account, we need to consider not only the attractor itself, but the relative frequency with which a typical orbit visits different regions of the attractor as well. We can say that some regions of the attractor are more probable than others, or alternatively we may speak of a measure on the attractor*. We define the natural measure of an attractor as follows: For each cube C and initial condition x in the basin of attraction, define $\mu(x, C)$ as the fraction of time that the trajectory originating from x spends in C†. If almost every such x gives the same value of $\mu(x, C)$, we denote this value $\mu(C)$ and call μ the *natural measure* of the attractor [16]. The natural measure gives the relative probability of different regions of the attractor as obtained from time averages, and therefore is the "natural" measure to consider. We will assume throughout that any attractor we consider has a natural measure, at least whenever C is one of the cubes we are using to cover the attractor.

The four definitions discussed in the remainder of this section are defined for attractors with a metric and a natural measure defined on them.

2.2.2. *Information dimension*

The *information dimension*, d_I, is a generalization of the capacity that takes into account the relative probability of the cubes used to cover the set. This dimension was originally introduced by Balatoni and Renyi [17].

The information dimension is given by

$$d_I = \lim_{\epsilon \to 0} \frac{I(\epsilon)}{\log(1/\epsilon)}, \tag{3}$$

where

$$I(\epsilon) = \sum_{i=1}^{N(\epsilon)} P_i \log \frac{1}{P_i}$$

and P_i is the probability contained within the ith cube. Letting the ith cube of side ϵ be C_i, $P_i = \mu(C_i)$. Note that if all cubes have equal probability then $I(\epsilon) = \log N(\epsilon)$, and hence $d_C = d_I$. However, for unequal probabilities $I(\epsilon) < \log N(\epsilon)$. Thus, in general, $d_C \geqslant d_I$.

In information theory the quantity $I(\epsilon)$ defined in eq. (3) has a specific meaning [18]. Namely, it is the amount of information necessary to specify the state of the system to within an accuracy ϵ, or equivalently, it is the information obtained in making a measurement that is uncertain by an amount ϵ. Since for small ϵ, $I(\epsilon) \approx d_I \log(1/\epsilon)$, we may view d_I as telling how fast the information necessary to specify a point on the attractor increases as ϵ decreases. (For a more extensive discussion of the physical meaning of the information dimension, see refs. 9 and 10.)

2.2.3. *ϑ-Capacity*

Another definition of dimension which we shall be interested in is what we will call the ϑ-capacity, $d_C(\vartheta)$. Essentially, this quantity is the capacity of that part of the attractor of highest probability,

$$d_C(\vartheta) = \lim_{\epsilon \to 0} \frac{\log N(\epsilon; \vartheta)}{\log(1/\epsilon)}, \tag{4}$$

where $N(\epsilon; \vartheta)$ is the minimum number of cubes of side ϵ needed to cover at least a fraction ϑ of the natural measure of the attractor. In other words, the cubes must be chosen so that their combined natural measure is at least ϑ. Thus $d_C(1) = d_C$. For the examples we study here, we find that for any value of $\vartheta < 1$, the ϑ-capacity is independent of ϑ, but that $d_C(\vartheta)$ for $\vartheta < 1$ may differ from its value at $\vartheta = 1$. In particular $d_C(\vartheta) = d_\mu$ for $\vartheta < 1$ and

* Although there are many measures possible for a given attractor, we are only interested in one of them, the natural measure.

† $\mu(x, C) = \lim_{\tau \to \infty} \mu_\tau(x, C)$, where $\mu_\tau(X, C)$ is the fraction of time spent in C up to some finite time τ.

$d_C(\vartheta) = d_C$ for $\vartheta = 1$. ϑ-capacity was originally defined by Frederickson et al. [8]. Similar quantities have also been defined by Ledrappier [18], and Mandelbrot [6, 45].

2.2.4. *ϑ-Hausdorff dimension*

In analogy with the relationship between capacity (a metric dimension) and ϑ-capacity (a probability dimension), we introduce here a probability dimension based on the Hausdorff dimension. We call this new dimension the ϑ-Hausdorff dimension and denote it $d_H(\vartheta)$. To define the ϑ-Hausdorff dimension, modify the definition of Hausdorff dimension as follows: Define $l_d(\epsilon, \vartheta)$ by

$$l_d(\epsilon, \vartheta) = \inf \sum_i \epsilon_i^d,$$

where now the infimum extends over all possible $\epsilon_i < \epsilon$ which cover a fraction ϑ of the total probability of the set. We define $d_H(\vartheta)$ as that value of d below which $l_d(\vartheta) = \infty$ and above which $l_d(\vartheta) = 0$, where $l_d(\vartheta) = \lim_{\epsilon\to 0} l_d(\epsilon, \vartheta)$. This concept of dimension will be used in section 6.

2.2.5. *Pointwise dimension*

Roughly speaking, the pointwise dimension d_p is the exponent with which the total probability contained in a ball decreases as the radius of the ball decreases. To make this notion more precise, let μ denote the natural probability measure on the attractor, and let $B_\epsilon(x)$ denote a ball of radius ϵ centered about a point x on the attractor. Roughly speaking, $\mu(B_\epsilon(x)) \sim \epsilon^{d_p}$. More precisely, define this dimension as

$$d_p(x) = \lim_{\epsilon\to 0} \frac{\log \mu(B_\epsilon(x))}{\log \epsilon}. \qquad (5)$$

If $d_p(x)$ is independent of x for almost all x with respect to the measure μ*, we call $d_p(x) = d_p$ the *pointwise* dimension. Similar definitions of dimension have also been given by Takens [20], Billingsley [31], Young [11], and Janssen and Tjon [21].

* By "almost all x with respect to the measure μ" we mean that the set of x which does not satisfy this is a set of μ measure zero.

2.3. *Using a grid of cubes to compute dimension*

Some of the definitions we have used, such as the capacity, allow any location or orientation of the cubes used to cover the attractor. In a numerical experiment, however, it is much more convenient to select the cubes used to cover the attractor out of a fixed grid, as shown in fig. 4. For these dimensions (d_C, d_I, and $d_C(\vartheta)$) it can be shown that selecting from a fixed grid of cubes gives the same value of the dimension as an optimal collection of cubes. For example, for the case of an attractor in a two-dimensional space, using a fixed grid to compute $N(\epsilon)$ in eq. (2) results in an increase of at most a factor of four in $N(\epsilon)$, which has no effect on the value of the dimension. Note that this is *not* true for the Hausdorff dimension, which requires a more general cover.

In principle, the definitions of dimension given in this section and the use of a fixed grid provide specific prescriptions for obtaining capacity, information dimension, and ϑ-capacity. To find approximate values for these dimensions, one can generate an orbit on the attractor using a computer, and then divide the space containing the orbit into cubes of side ϵ in order to estimate the numbers $N(\epsilon)$, $I(\epsilon)$, or $N(\epsilon; \vartheta)$. By examining how

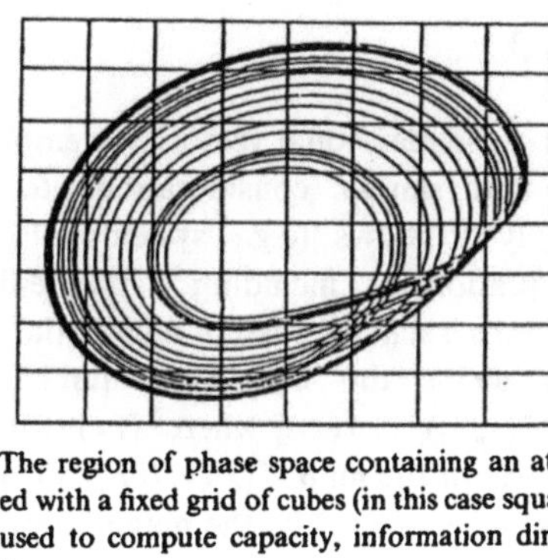

Fig. 4. The region of phase space containing an attractor can be divided with a fixed grid of cubes (in this case squares), which can be used to compute capacity, information dimension, or ϑ-capacity.

$N(\epsilon)$, $I(\epsilon)$, and $N(\epsilon; \vartheta)$ vary as ϵ is decreased the value of these dimensions can be estimated.

As discussed in section 9, however, in practice the agenda described above for computing dimension may be difficult, costly, or impossible. Thus it is of interest to consider other means of obtaining the dimension of chaotic attractors. The next section deals with this question. In particular, we discuss a conjecture that the dimension of chaotic attractors can be determined directly from the dynamics in terms of Lyapunov numbers.

3. Lyapunov numbers and Lyapunov dimension

The Lyapunov numbers quantify the average stability properties of an orbit on an attractor. For a fixed point attractor of a mapping, the Lyapunov numbers are simply the absolute values of the eigenvalues of the Jacobian matrix evaluated at the fixed point. The Lyapunov numbers generalize this notion for more complicated attractors. As we shall see, for a typical attractor there is a connection between average stability properties and dimension. The possibility of such a connection was first pointed out by Kaplan and Yorke [22] and later by Mori [23].

3.1. *Definition of Lyapunov numbers*

For expository purposes, for most of this paper we shall consider p-dimensional maps,

$$x_{n+1} = F(x_n),$$

where x is a p-dimensional vector. We emphasize, however, that similar considerations to those below apply to flows (e.g., systems of differential equations), including infinite-dimensional systems such as partial differential equations. To define the Lyapunov numbers, let $J_n = [J(x_n)J(x_{n-1})\ldots J(x_1)]$ where $J(x)$ is the Jacobian matrix of the map, $J(x) = (\partial F/\partial x)$, and let $j_1(n) \geqslant j_2(n) \geqslant \cdots \geqslant j_p(n)$ be the magnitudes of the eigenvalues of J_n. The Lyapunov numbers are

$$\lambda_i = \lim_{n\to\infty} [j_i(n)]^{1/n}, \quad i = 1, 2, \ldots, p, \tag{6}$$

where the positive real nth root is taken. The Lyapunov numbers generally depend on the choice of the initial condition x_1. The Lyapunov numbers were originally defined by Oseledec [24]. We have the convention

$$\lambda_1 \geqslant \lambda_2 \geqslant \cdots \geqslant \lambda_p.$$

For a two-dimensional map, for example, λ_1 and λ_2 are the average principal stretching factors of an infinitesimal circular are (cf. fig. 5). For a chaotic attractor on the average nearby points initially diverge at an exponential rate, and hence at least one of the Lyapunov numbers is greater than one. This makes quantitative the notion of "sensitive dependence onititial conditions". We will take $\lambda_1 > 1$ as our definition of *chaos*. (Note that many authors refer to *Lyapunov exponents* rather than Lyapunov numbers. The Lyapunov exponents are simply the logarithms of the Lyapunov numbers.)

In this paper we assume that *almost every* initial condition in the basin of any attractor that we consider has the same Lyapunov numbers. Thus, the spectrum of Lyapunov numbers may be considered to be a property of an attractor. This assumption is supported by numerical experiments [25]. Exceptional trajectories, such as unstable fixed points on the attractor, typically do not sample the whole attractor and thus typically have Lyapunov numbers that are different from those of the attractor. Those points in the basin of attraction that have different Lyapunov numbers or for

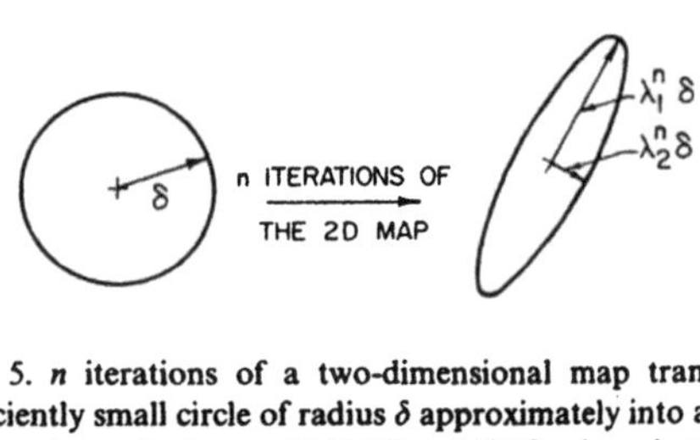

Fig. 5. n iterations of a two-dimensional map transform a sufficiently small circle of radius δ approximately into an ellipse with major and minor radii $(\lambda_1)^n\delta$ and $(\lambda_2^n)\delta$, where λ_1 and λ_2 are the Lyapunov numbers.

which Lyapunov numbers do not exist are here assumed to be of measure zero. (In other words, they may be covered by a collection of cubes of varying size having arbitrarily small total volume).

3.2. *Definition of Lyapunov dimension*

The following discussion contains a heuristic argument that motivates a connection between Lyapunov numbers and dimension. Consider a two-dimensional map. Suppose we wish to compute the capacity of a chaotic attractor, for which $\lambda_1 > 1 > \lambda_2$. Cover the attractor with $N(\epsilon)$ squares of side ϵ. Now, iterate the map q times. For q fixed and ϵ small enough, the action of the mapping is roughly linear over the square, and each square will be stretched into a long thin parallelogram. From the definition of the Lyapunov numbers, the average length of these parallelograms is $(\lambda_1)^q\epsilon$, and the average width is $(\lambda_2)^q\epsilon$. Now, suppose we had used a finer cover of squares of side $(\lambda_2)^q\epsilon$. (See fig. 6.) To cover each parallelogram takes about $(\lambda_1/\lambda_2)^q$ smaller squares. Thus, *if it is supposed* that all squares on the attractor behave in this typical way, then one is lead to the estimate

$$N(\lambda_2^q\epsilon) \approx \left(\frac{\lambda_1}{\lambda_2}\right)^q N(\epsilon). \tag{7}$$

Motivated by eq. (2), assume $N(\epsilon) \approx k(1/\epsilon)^{d_C}$, and

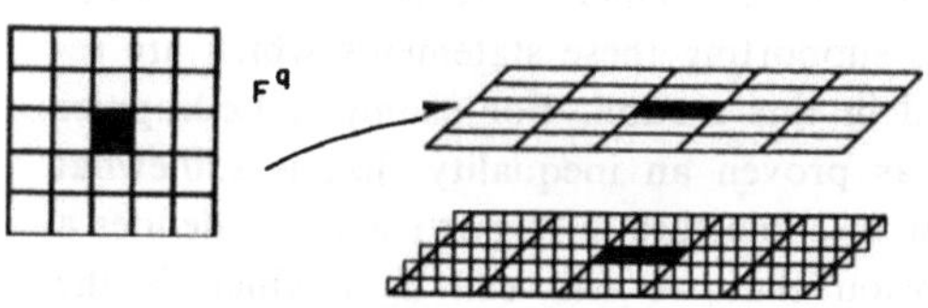

Fig. 6. A schematic illustration of the heuristic argument for the Lyapunov dimension. The image of each small square in (a) is approximately a parallelogram which has been stretched horizontally be a factor of λ_1^q and contracted vertically by a factor λ_2^q. The images in (b) thus have a smaller cover of squares as shown in (c).

substitute into both sides of eq. (7). This gives

$$k\left(\frac{1}{\lambda_2^q\epsilon}\right)^{d_C} \approx k\left(\frac{\lambda_1}{\lambda_2}\right)^q\left(\frac{1}{\epsilon}\right)^{d_C}.$$

Collecting terms, taking logarithms, and solving for d_C gives

$$d_C = 1 + \frac{\log\lambda_1}{\log(1/\lambda_2)}.$$

We will see that this expression is often meaningful even when this heuristic derivation is invalid, so we will call it the *Lyapunov dimension* d_L.

$$d_L = 1 + \frac{\log\lambda_1}{\log(1/\lambda_2)}. \tag{8}$$

Generalization of the above heuristic argument to p-dimensional maps gives (cf. ref 7)

$$d_L = k + \frac{\log(\lambda_1\lambda_2\ldots\lambda_k)}{\log(1/\lambda_{k+1})}, \tag{9}$$

where k is the largest value for which $\lambda_1\lambda_2\ldots\lambda_k \geqslant 1$. If $\lambda_1 < 1$, define $d_L = 0$; if $\lambda_1\lambda_2\ldots\lambda_p \geqslant 1$, define $d_L = p$. We shall refer to d_L as the *Lyapunov dimension*. This quantity was originally defined by Kaplan and Yorke [22], who originally gave it as a lower bound on the fractal dimension.

From the above argument one might be tempted to guess that $d_C = d_L$. The Lyapunov numbers are *average* quantities, however, and to compute an average, each cube must be weighted according to its probability. The capacity does not distinguish between probable and improbable cubes. To understand how some cubes might have vastly different probabilities than others, consider an atypical square of a two-dimensional map. If the area of the images of this square decreases half as fast as the average for k iterations, then its kth image will be 2^k times larger than the image of a typical square, and the number of squares needed to cover it will be 2^k times greater than the typical value. In fact, as will be evident from considerations of explicit examples (cf. section 5), it is commonly the case that the vast majority of cubes needed to cover the attractor are atypical, and do not represent the properties of time averages. By this we mean that all the atypical cubes taken together contain an extremely small fraction of the total probability on the attractor yet account for almost all of $N(\epsilon)$. Furthermore, this tendency increases as ϵ decreases. The behavior of the atypical cubes under iteration is in general not described by the Lyapunov numbers. It is clear, then, that in order for this estimate to be valid, we must consider only the more probable cubes, i.e., the estimate should be in terms of the dimension of the natural measure rather than the capacity. Assuming the equality of probabilistic dimensions (conjecture 1), we are led to the following conjecture:

Conjecture 2. For a typical* attractor $d_\mu = d_L$.

In the following six sections we present evidence supporting this conjecture. Also, L.S. Young has proved some rigorous results along these lines, which are reviewed in the next subsection.

In the special case that *every* initial condition on the attractor generates the same Lyapunov numbers, we will say that the attractor has *absolute* Lyapunov numbers. In this case it is not necessary to distinguish probable from improbable cubes, and the above conjecture can be made in terms of the fractal dimension rather than the dimension of the natural measure. We call this conjecture 3,

Conjecture 3. If *every* (not just almost every) initial condition generates the same set of p Lyapunov numbers $\lambda_1, \lambda_2, \ldots \lambda_p$, and if $\lambda_1 > 1$, then for a typical attractor of this type $d_F = d_L = d_\mu$.

The requirement of conjecture 3 that every initial condition on the attractor generate the same Lyapunov numbers is very restrictive and only holds for special cases. For example, it holds if the Jacobian matrix of the map is independent of x. In more general cases, the requirement of conjecture 3 would be expected to fail because of the existence of unstable fixed and periodic points on the attractor. For example, if x_1 is chosen to be precisely on an unstable fixed point, the Lyapunov numbers generated will simply be the eigenvalues of $J(x_1)$. These will typically be different from those generated by a chaotic orbit on the attractor. Examples for which conjecture 3 is valid will be special cases of the more general example presented in the following section. In addition, an example for which conjecture 3 can be proven to hold is given in section 8.

3.3. *Review of rigorous results concerning Lyapunov dimension*

In addition to the analytic and numerical evidence we will give for conjectures 1–3 in the remainder of this paper, there are several rigorous results supporting these statements which are reviewed in this section. For example, Ledrappier [19] has proven an inequality that is somewhat similar to conjecture 2. In particular, he defines a dimension that we will call d_{Led}, which is the ϑ-capacity in the limit as ϑ goes to one, i.e.

$$d_{Led} = \lim_{\vartheta \to 1} d_C(\vartheta).$$

For C^2 diffeomorphisms he has shown that

$$d_L \geqslant d_{Led}.$$

* The reason for the word "*typical*" is that there exist examples of maps that do not satisfy $d_\mu = d_L$. These maps are exceptional, however, in that arbitrarily small perturbations of them restore the conjectured equality of d_μ and d_L. An example of such an atypical case is where a point x_0 is attracting and yet has $\lambda_1 = 1$ (i.e., the Jacobian matrix $\partial F/\partial x$ has an eigenvalue $+1$ at x_0). The attraction here is due to higher order terms. The attractor is a point and so has dimension zero, yet $d_L \geqslant 1$. Small perturbations, however, will destroy this delicate balance. For example, the one-dimensional map $x_{i+1} = F(x_i) \equiv \alpha x_i - x_i^3$ has a fixed point at $x = 0$ with $\lambda_1 = 1$ for $\alpha = 1$ yet $x = 0$ is attracting. This situation is changed, however, as soon as $\alpha \neq 1$. When $|\alpha| < 1$, $d_L = 0$, and when $|\alpha| > 1$, $x = 0$ is no longer attracting.

The proof is a rigorous version of the heuristic argument that we have given (fig. 6). Also, Douady and Oesterle [26] have proven that an upper bound for the fractal dimension can be obtained yielding an expression like eq. (8), where the numbers they use are basically upper bounds for the Lyapunov numbers.

L.S. Young [12] has proven several results that strongly support conjectures 1 and 2. Particularly relevant are the following two theorems*.

1. If d_p exists then

$$d_p = d_I = d_H(\text{core}) = d_{Led}. \tag{10}$$

2. For two-dimensional C^2 diffeomorphisms with $\lambda_1 > 1 > \lambda_2$, d_p exists, and

$$d_p = \frac{h_\mu}{\log \lambda_1}\left(1 + \frac{\log \lambda_1}{\log(1/\lambda_2)}\right). \tag{11}$$

(See section 7 for a definition of d_H(core).) h_μ denotes the Kolmogorov entropy† of the attractor taken with respect to the measure μ, and λ_1 and λ_2 are the Lyapunov numbers with respect to μ. (More precisely, almost every initial condition x with respect to μ give λ_1 and λ_2 as the Lyapunov numbers.)

For Axiom-A attractors Bowen and Ruelle [16] have shown that there is a natural measure such that h_μ with respect to this measure is the sum of the logarithms of the Lyapunov numbers that are greater than one. For attractors with only one Lyapunov number greater than one, this implies that $h_\mu = \log \lambda_1$. Thus, for axiom-A attractors of two-dimensional maps, eqs. (9)–(11) yield $d_\mu = d_L$. Therefore Young has shown that conjecture 2 holds for this case. (It has been conjectured that the relationship between h_μ and the positive λ_i holds for non-axiom-A attractors that have a natural measure.) This result for the case of axiom-A attractors of two-dimensional maps has also been obtained independently by Pelikan [30].

* For these results Young does not require the existence of a natural measure, but rather assumes simply the existence of some invariant measure μ. In this case the Lyapunov numbers are those obtained when starting at almost every initial point with respect to μ.

† The Kolmogorov entropy, originally defined by Shannon [18] and applied to dynamical systems by Kolmogorov [27] and Sinai [28], puts a quantitative value on the average amount of new information obtained from a sequence of measurements. See [10] or [29] for physically motivated reviews. Note that this is also called metric entropy. The name *metric* entropy derives from the invariance properties of this quantity; in fact, the definition of metric entropy does not require a metric (but does require a measure).

‡ Except for the ϑ-Hausdorff dimension, for which we only obtain an upper bound.

4. Generalized baker's transformation: scaling

4.1. *Definition of generalized baker's transformation*

In this section we define the example which we will study in detail in this and the following four sections. Although we feel that this example is general enough to be typical of low-dimensional chaotic attractors (at least concerning its dimensional properties), it is also simple enough that all of the dimensions discussed in this paper can be analytically calculated‡. Thus, for this example, we shall be able to verify conjectures 1–3 in a case where generally $d_F \neq d_\mu$. As we shall show in section 5, another nice property of this map is that it allows us to investigate certain properties of the natural probability distribution in detail.

The map to be considered is

$$x_{n+1} = \begin{cases} \lambda_a x_n, & \text{if } y_n < \alpha, \\ \tfrac{1}{2} + \lambda_b x_n, & \text{if } y_n > \alpha; \end{cases} \tag{12a}$$

$$y_{n+1} = \begin{cases} \dfrac{1}{\alpha} y_n, & \text{if } y_n < \alpha, \\ \dfrac{1}{1-\alpha}(y_n - \alpha), & \text{if } y_n > \alpha; \end{cases} \tag{12b}$$

where we shall assume $0 \leqslant x_n \leqslant 1$ and $0 \leqslant y_n \leqslant 1$. If this condition is satisfied initially it is also satisfied

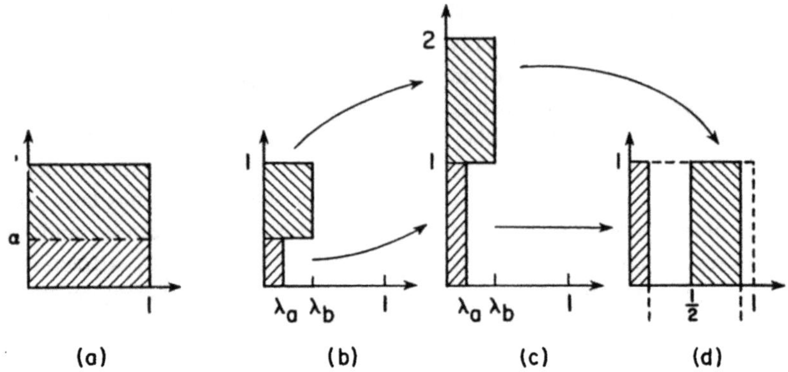

Fig. 7. The generalized baker's transformation. One iteration of the map takes us from (a) to (d). Steps (b) and (c) are conceptual intermediate stages.

at all subsequent iterates. Fig. 7 illustrates the action of this map on the unit square. As shown in fig. 7, we take α, λ_a, $\lambda_b \leqslant \frac{1}{2}$, and $\lambda_b \geqslant \lambda_a$. Fig. 8 shows the result of applying the map two times to the unit square. From fig. 8 it is seen that, if the x interval $[0, \lambda_a]$ is magnified by a factor $1/\lambda_a$, it becomes a precise replica of fig. 7d. Similarly, if the x interval $[\frac{1}{2}, \frac{1}{2} + \lambda_b]$ is magnified by $1/\lambda_b$, a replica of fig. 7d again results. This self similarity property of eq. (12) will subsequently be used to obtain d_C, d_I, d_H, and d_p.

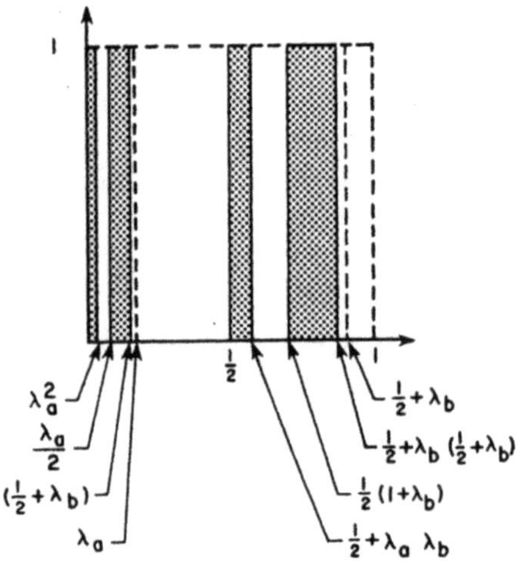

Fig. 8.

4.2. *Lyapunov numbers of generalized baker's transformation*

Now we consider the Lyapunov numbers. Eq. (12b) involves y alone and consists of a linear stretching on each of the y intervals $[0, \alpha]$ and $[\alpha, 1]$. Thus almost every y initial condition in $[0, 1]$ will generate an ergodic orbit in y with uniform density in $[0, 1]$. The Jacobian of eq. (12) is diagonal and depends only on y.

$$J = \begin{pmatrix} L_2(y) & 0 \\ 0 & L_1(y) \end{pmatrix},$$

where

$$L_2(y) = \begin{cases} \lambda_a, & \text{if } y < \alpha, \\ \lambda_b, & \text{if } y > \alpha, \end{cases}$$

and

$$L_1(y) = \begin{cases} \dfrac{1}{\alpha}, & \text{if } y < \alpha, \\ \dfrac{1}{1-\alpha}, & \text{if } y > \alpha. \end{cases}$$

Thus applying eq. (6) we have

$$\lambda_1 = \lim_{n\to\infty} [L_1(y_n) \ldots L_1(y_1)]^{1/n},$$

or

$$\log \lambda_1 = \lim_{n\to\infty} \left\{ \frac{n_\alpha}{n} \log \frac{1}{\alpha} + \frac{n_\beta}{n} \log \frac{1}{\beta} \right\},$$

where $\beta = 1 - \alpha$. n_α is the number of times the orbit has been in the set $y < \alpha$, and n_β is the number of times the orbit has been in the set $y > \alpha$. Since for *almost* any y_1, the orbit in y is ergodic with uniform density in [0, 1], $\lim_{n\to\infty} n_\alpha/n = \alpha$, and similarly $\lim_{n\to\infty} n_\alpha/n = \beta$. Thus

$$\log \lambda_1 = \alpha \log \frac{1}{\alpha} + \beta \log \frac{1}{\beta}. \tag{13}$$

Similarly, we obtain for λ_2

$$\log \lambda_2 = \alpha \log \lambda_a + \beta \log \lambda_b. \tag{14}$$

To simplify notation in this and subsequent expressions, let

$$H(\alpha) = \alpha \log \frac{1}{\alpha} + (1-\alpha) \log \frac{1}{1-\alpha}. \tag{15}$$

$H(\alpha)$ is called the *binary entropy function* and is the amount of information contained in a coin-toss where heads has a probability α.

The Lyapunov dimension of the attractor of the generalized Baker's transformation (eq. (12)) is

$$d_L = 1 + \frac{H(\alpha)}{\alpha \log(1/\lambda_a) + \beta \log(1 + \lambda_b)}. \tag{16}$$

In the following sections we compute the values of the dimensions defined in this paper, and show that all the probabilistic dimensions take on the value given in eq. (16).

For all but special values of λ_a, λ_b, and α, there exist unstable periodic orbits whose Lyapunov numbers are different from those given in eqs. (13) and (14)*. Thus, in general we expect that conjecture 2 rather than conjecture 3 applies and $d_F \neq d_\mu$.

* To see that for almost all parameter values the Lyapunov numbers of the generalized baker's transformation are not absolute, consider the special initial condition on the attractor with y-value $y_1 = y_a$ where $y_a = \alpha^2(1-\alpha+\alpha^2)^{-1}$. This initial condition corresponds to one of the points on the unstable period 2 orbit, $(y_a, y_b, y_a, y_b, \ldots)$, where $y_b = \alpha^{-1} y_a$. Since $0 < y_a < \alpha < y_b < 1$, we have $n_a = n_\beta = \frac{1}{2}$, and the Lyapunov numbers generated by this initial condition are $\lambda_1 = (\alpha\beta)^{-1/2}$ and $\lambda_2 = (\lambda_a \lambda_b)^{1/2}$, rather than those given by eq. (13) and (14).

4.3. *Capacity of generalized Baker's transformation*

To calculate d_C we first note that the attractor is a product of a Cantor set along x and the interval [0, 1] along y. Thus the capacity, or any of the other dimensions, are in the form $d_C \equiv 1 + \bar{d}_C$, where $\bar{d}_C$ is the dimension of the attractor in the x-direction. We will generally use a bar over a dimension to refer to the dimension along the x-direction.

We now obtain $\bar{d}_C$ by making use of the scaling property of the generalized Baker's transformation, discussed at the end of section 4.1. We write $N(\epsilon)$ as

$$N(\epsilon) = N_a(\epsilon) + N_b(\epsilon),$$

where $N_a(\epsilon)$ is the number of x-intervals of length ϵ needed to cover that part of the attractor which lies in the x-interval $[0, \lambda_a]$, and $N_b(\epsilon)$ is the analogous quantity for the x-interval $[\frac{1}{2}, \frac{1}{2} + \lambda_b]$. From the scaling property, $N_a(\epsilon) = N(\epsilon/\lambda_a)$, and similarly $N_b(\epsilon) = N(\epsilon/\lambda_b)$. Thus

$$N(\epsilon) = N(\epsilon/\lambda_a) + N(\epsilon/\lambda_b). \tag{17}$$

Assuming heuristically that $N(\epsilon) \approx k\epsilon^{-\bar{d}_C}$ for small ϵ, substituting into eq. (17) gives

$$k\left(\frac{1}{\epsilon}\right)^{\bar{d}_C} = k\left(\frac{\lambda_a}{\epsilon}\right)^{\bar{d}_C} + k\left(\frac{\lambda_b}{\epsilon}\right)^{\bar{d}_C},$$

implying that

$$1 = \lambda_a^{\bar{d}_C} + \lambda_b^{\bar{d}_C}, \tag{18}$$

which is a transcendental equation for $\bar{d}_C$. As

expected, eqs. (16) and (18) show that, in general, $1+\bar{d}_C = d_C \neq d_L$. However, for the special choice $\lambda_a = \lambda_b$, $\alpha = \frac{1}{2}$, corresponding to eq. (12) with $\lambda_a = \lambda_1 = 2$, the two agree. Note that for this case the Jacobian matrix is constant, the Lyapunov numbers are therefore absolute, and conjecture 3 applies.

In obtaining eq. (18), in order to keep the argument simple, we have made the strong assumption that $N(\epsilon) \approx k\epsilon^{-\bar{d}_C}$ for small ϵ, which implies the existence of the limit given in the definition of capacity, eq. (2). We can, however, show that the limit given in eq. (2) exists and $\bar{d}_C$ must satisfy eq. (18) in a rigorous manner, as follows:

Define $E_C(\epsilon)$ by

$$N(\epsilon) = E_C(\epsilon)\epsilon^{-\bar{d}},$$

where $\bar{d}$ is defined by $1 = \lambda_a^{\bar{d}} + \lambda_b^{\bar{d}}$. Substituting this into eq. (17) then yields

$$E_C(\epsilon) = \bar{\alpha} E_C\left(\frac{\epsilon}{\lambda_a}\right) + \bar{\beta} E_C\left(\frac{\epsilon}{\lambda_b}\right), \tag{19}$$

where $\bar{\alpha} = \lambda_a^{\bar{d}}$ and $\bar{\beta} = \lambda_b^{\bar{d}}$, and are independent of ϵ. Notice that by definition $\bar{\alpha} + \bar{\beta} = 1$, so the above expression says that $E_C(\epsilon)$ is a weighted average of its values at ϵ/λ_a and ϵ/λ_b. Choose ϵ_1 and ϵ_2 so that $\epsilon_1 > \epsilon_2 > 0$. Since $N(\epsilon)$ and hence $E_C(\epsilon)$ are finite and positive for any finite ϵ, there exist finite non-zero numbers $B_1 > B_2 > 0$ such that $B_2 < E_C(\epsilon) < B_1$ for $\epsilon_1 > \epsilon > \epsilon_2$. We can assume that ϵ_1 and ϵ_2 are chosen so that ϵ_1/ϵ_2 is large. Since $\bar{\alpha} + \bar{\beta} = 1$, eq. (19) implies that $B_2 < E_C(\epsilon) < B_1$ also applies to the wider interval $\epsilon_1 > \epsilon > \lambda_b\epsilon_2$. Repeating this argument increases the domain of validity of the bound to $\epsilon_1 > \epsilon > \lambda_b^2\epsilon_2$, and so on. Hence $E_C(\epsilon)$ is bounded uniformly from above and below for arbitrarily small ϵ. Thus the limit of eq. (2) exists and $\bar{d}_C = \bar{d}$. (In fact it can be shown that eq. (19) implies that $\lim_{\epsilon\to 0} E_C(\epsilon)$ is a constant if $\log \lambda_a / \log \lambda_b$ is an irrational number.) Note that in eq. (18), since both terms on the right-hand side are monotonically decreasing, d_C obtained from solving this equation is unique.

4.4. *Computation of Hausdorff dimension*

The Hausdorff dimension d_H can be calculated by an argument that is very similar to the one used above in computing the capacity. Let $\bar{d}_H \equiv d_H - 1$, the Hausdorff dimension along x. Applying the scaling property of the map to the quantity $l_d(\epsilon)$ (defined in section 2), we obtain

$$l_d(\epsilon) = (\lambda_a)^d l_d\left(\frac{\epsilon}{\lambda_a}\right) + (\lambda_b)^d l_d\left(\frac{\epsilon}{\lambda_b}\right).$$

Substituting $l_d(\epsilon) = E_H(\epsilon)\epsilon^{-(\bar{d}-d)}$ into the above equation, we again find that $E_H(\epsilon)$ satisfies eq. (19). Thus the limit $\epsilon \to 0$ yields $l_d = \infty$ or $l_d = 0$ for $d < \bar{d}_C$ or $d > \bar{d}_C$, respectively. Hence, as predicted in section 2, the Hausdorff dimension and capacity are equal, $d_H = d_C$.

4.5. *Calculation of information dimension*

The information dimension d_I can also be calculated by a scaling argument similar to that used above in computing the capacity. Once again, let $d_I = 1 + \bar{d}_I$ and express the summation for $I(\epsilon)$ in eq. (3) as the sum of contributions from the two strips in fig. 7d,

$$I(\epsilon) = I_a(\epsilon) + I_b(\epsilon). \tag{20}$$

The total probability contained in strip $[0, \lambda_a]$ is α, and that in strip $[\frac{1}{2}, \lambda_b + \frac{1}{2}]$ is β. Assuming that it takes $N(\epsilon)$ strips of width ϵ to cover the whole attractor, then from the scaling property of eq. (12), covering the strip $[0, \lambda_a]$ at resolution $\epsilon\lambda_a$ also requires $N(\epsilon)$ strips. Thus

$$I_a(\epsilon\lambda_a) = \sum_{i=1}^{N(\epsilon)} \alpha P_i \log \frac{1}{\alpha P_i}$$

$$= \alpha\left[\log \frac{1}{\alpha} + I(\epsilon)\right].$$

Hence, replacing $\epsilon\lambda_a$ by ϵ in the above,

$$I_a(\epsilon) = \alpha \log\frac{1}{\alpha} + \alpha I\left(\frac{\epsilon}{\lambda_a}\right),$$

$$I_b(\epsilon) = \beta \log\frac{1}{\beta} + \beta I\left(\frac{\epsilon}{\lambda_b}\right).$$

Thus

$$I(\epsilon) = \alpha I\left(\frac{\epsilon}{\lambda_a}\right) + \beta I\left(\frac{\epsilon}{\lambda_b}\right) + H(\alpha), \tag{21}$$

where $H(\alpha)$ is given by eq. (15). Motivated by eq. (3), if we assume that $I(\epsilon) = \bar{d}_I \log(1/\epsilon)$ for small ϵ, and substitute for $I(\epsilon)$, $I(\epsilon/\lambda_a)$, and $I(\epsilon/\lambda_b)$ in the above equation we obtain

$$\bar{d}_I = \frac{H(\alpha)}{\alpha \log(1/\lambda_a) + \beta \log(1/\lambda_b)},$$

which is in turn equal to $\bar{d}_L$. The assumption that $I(\epsilon) = \bar{d}_I \log(1/\epsilon)$ can be made rigorous in the limit as $\epsilon \to 0$ using an argument that is completely analogous to that used in deriving the capacity in the last part of subsection 4.3.

We should mention that Alexander and Yorke [11] have computed the Lyapunov and information dimensions of the generalized baker's transformation for the special case $\alpha = \frac{1}{2}$, $\lambda = \lambda_a = \lambda_b$, where $\lambda > \frac{1}{2}$. In this case $d_L = 2$. For uncountably many values of λ they find that also $d_I = 2$, although there are certain special values of λ for which $d_I < 2$.

In order to calculate the other probability dimensions listed in table I more information concerning the probability distribution is required. This is dealt with in section 5, and we therefore defer calculation of the remaining dimensions to the sections following section 5.

5. Distribution of probability

In this section we derive the form of the probability distribution $\{P_i(\epsilon)\}$ associated with the natural measure μ of the generalized baker's transformation. Here P_i denotes the probability of the ith cube C_i of edge ϵ, i.e., $P_i = \mu(C_i)$. The collection of numbers $\{P_i(\epsilon)\}$ may be also be thought of as the result of coarse graining the natural measure. This probability distribution is interesting both for its own sake, and because it is needed to compute some of the dimensions that we are interested in. In what follows we restrict ourselves to the case in which $\lambda_a = \lambda_b \equiv \lambda_2$, which keeps the width of all the strips the same. Thus a particularly convenient partition for computing $\{P_i\}$ is the set of 2^n non-empty strips obtained by iterating the unit square n times.

Starting with a uniform probability distribution, on one application of the map two strips are produced, one with total probability α and the other with total probability β. (See fig. 7d.) If the map is applied again (fig. 8), there results one strip of probability α^2, one of probability β^2, and two of probability $\alpha\beta$. In general, after n applications of the map, there result 2^n strips of width $(\lambda_2)^n$ and probabilities $\alpha^m\beta^{(n-m)}$, $m = 0, 1, 2, \ldots, n$. The number of strips with probability $\alpha^m\beta^{(n-m)}$ is

$$Z(n,m) = \frac{n!}{(n-m)!m!}, \tag{22}$$

i.e., the binomial coefficient. Since we take $\alpha < \frac{1}{2} < \beta$, lower m corresponds to more probable strips, i.e. strips of greater natural measure. The total probability contained in these $Z(n,m)$ strips is

$$W(n,m) \equiv \alpha^m\beta^{(n-m)}Z(n,m). \tag{23}$$

Note the similarity to a sequence of coin tosses. Using a coin with probability α of heads and β of tails, for a sequence of n flips the total number of sequences with m occurrences of heads is given by eq. (22), and the likelihood of all such sequences is given by eq. (23).

For large n (small ϵ) it is convenient to have smooth estimates for $Z(n,m)$ and $W(n,m)$. Using

Sterling's approximation, i.e.

$$\log n! = (n+\tfrac{1}{2})\log(n+1) - (n+1) + \log(2\pi)^{1/2} + \mathcal{O}(n^{-1}),$$

we obtain from eq. (22)

$$\log Z \approx (n+\tfrac{1}{2})\log(m+1) - \log(2\pi)^{1/2} + 1.$$

Expanding this expression in a Taylor series about its maximum value, $m = n/2$, yields

$$Z(n,m) \approx \frac{2^n}{\sqrt{2\pi}}\sqrt{\frac{4}{n}}\exp\left\{-\frac{1}{2}\left[\frac{4}{n}\left(m-\frac{n}{2}\right)^2\right]\right\}. \quad (24)$$

Similarly, from eq. (23), $W(n,m)$ is

$$W(n,m) \approx \frac{1}{\sqrt{2\pi n\alpha\beta}}\exp\left\{-\frac{(m-n\alpha)^2}{2n\alpha\beta}\right\}. \quad (25)$$

Note that, since these expressions were obtained by Taylor series expansion, eq. (24) is only valid for $|m/n - \frac{1}{2}| \ll 1$, and eq. (25) is only valid for $|m/n - \alpha| \ll 1$. However, since the width of these Gaussians is $\mathcal{O}(1/n^{1/2})$, eq. (24) is valid for most of the strips, and eq. (25) is valid for most of the probability.

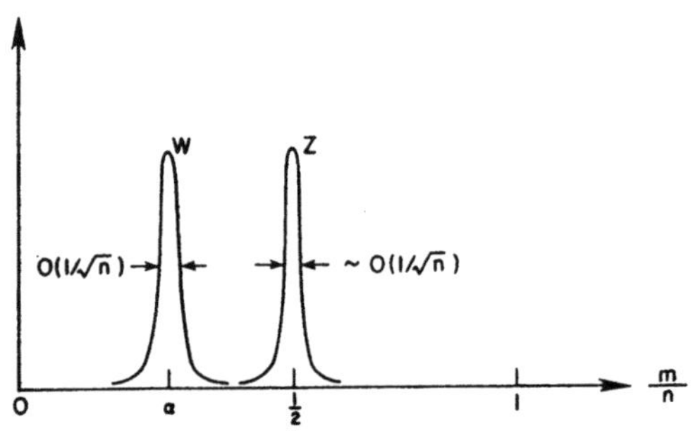

Fig. 9. A schematic representation of the distribution of probabilities on the attractor. $Z(n,m)$ is the number of cubes with probability $p = \alpha^m\beta^{(n-m)}$, and $W(n,m)$ is the sum of the probability contained in cubes of probability p. For large n and m/n close to its mean value, these are both approximately Gaussian distributions in m/n whose width is proportional to n. In the limit as $n \to \infty$, W and Z become delta functions, and no longer overlap.

Fig. 9 shows a schematic plot of Z and W. It is clear from this figure that, for large n, almost all of the probability is contained in a very small fraction of the total number of strips. Furthermore, the situation is accentuated as ϵ gets smaller (n gets larger), since the width of the Gaussians given in eqs. (24) and (25) decreases according to $n^{1/2}$. In the limit as $\epsilon \to 0$ these Gaussians approach delta functions, and they do not overlap. We feel that the above properties are typical features of chaotic attractors.

5.1. *Log-normal distribution of probabilities*

It is instructive to rewrite eq. (25) in another form. Let $p = \alpha^m\beta^{(n-m)}$ denote the probability of a strip, and reexpress eq. (25) in terms of $u = \log(1/p)$ rather than m. Noting that m is proportional to u, and letting $\epsilon = \lambda_2^n$ $W(n,m)$ becomes

$$F(u) = \frac{1}{\sqrt{2\pi\sigma}}\, e^{-(u-u_0)^2/2\sigma^2}, \quad (26)$$

where

$$\sigma^2 = \frac{[\alpha\beta(\log(\beta/\alpha))^2\log(1/\epsilon)]}{\log(1/\lambda_2)}$$

and

$$u_0 = \bar{d}_L \log\frac{1}{\epsilon}, \quad (27)$$

with d_L given by eq. (16). Eq. (26) is only valid if

$$\frac{(u-u_0)^2}{\sigma^2} \ll \log\frac{1}{\epsilon}, \quad (28)$$

corresponding to $|m/n - \alpha| \ll 1$. $F(u)\mathrm{d}u$ is the total probability contained in strips whose values of $u = \log(1/p)$ fall between u and $u + \mathrm{d}u$. Thus we see that the values the logarithm of p asymptotically have a Gaussian distribution, or in other words, the values of p asymptotically have a log-normal distribution. We believe that this is typically true of chaotic attractors. In particular, we offer the following conjecture*:

Conjecture 4. Let A be a chaotic attractor of a p-dimensional *invertible* dynamical system, and assume that this attractor has a natural measure μ. Cover A with a fixed grid of p-dimensional cubes of side length ϵ. Assign each nonempty cube C_i probability $P_i = \mu(C_i)$, and let $U_i = \log(1/P_i)$. Let u_0 be the mean of the numbers U_i, and let σ^2 be the variance. For typical chaotic attractors, in the limit as $\epsilon \to 0$, values of U_i sufficiently close to the mean (in the sense of eq. (28)) approach a Gaussian distribution. In other words, the corresponding values of P_i approach a log-normal distribution.

Note that U_i is the information obtained in a measurement that finds the orbit inside of the ith cube [1, 9, 10]. Thus, conjecture 4 states that for chaotic attractors the information is approximately normally distributed for small ϵ.

The function $Z(n, m)$ given in eq. (24), can also be reexpressed in terms of p rather than m. When this is done, with similar restrictions to those of eq. (28), the result is also a Gaussian in terms of $\mu = \log(1/p)$. When recast in the more general setting of conjecture 4, this says that the number of cubes C_i whose values U_i lie between u and $u + du$ are given by a Gaussian distribution. (Similar restrictions to those given in conjecture 4 apply.)

6. Computation for the natural measure dimensions

In this section we verify conjectures 1 and 2 for the generalized baker's transformation by explicitly computing all of the probability dimensions defined in section 2. In order to simplify the computations, for all but the ϑ-Hausdorff dimension we restrict ourselves to the case in which $\lambda_a = \lambda_b \equiv \lambda_2$. For the ϑ-Hausdorff dimension we treat the most general case in which $\lambda_a \neq \lambda_b$, but are only able to obtain an upper bound for the dimension.

* The form of this conjecture was developed in collaboration with Erica Jen.

6.1. *Alternate derivation of information dimension*

Now that we know the probability distribution for the generalized baker's transformation for $\lambda_a = \lambda_b \equiv \lambda_2$, we can obtain the information dimension directly from its definition. From eq. (3) and eq. (26), $I(\epsilon)$ is the average value of $\log(1/P_i)$ or

$$I(\epsilon) = \int uF(u)\,du = u_0.$$

Since from eq. (27) $u_0 = d_L \log(1/\epsilon)$, eq. (3) yields $d_I = d_L$ (previously shown in section 4 for the more general case $\lambda_a \neq \lambda_b$). Thus the mean value of the log-normal distribution is simply the information contained in the probability distribution, and its scaling rate is the dimension of the nature measure, i.e., $I(\epsilon) \approx d_\mu \log(1/\epsilon)$.

6.2. *Determination of ϑ-capacity*

Here we calculate $d_C(\vartheta)$ for $\lambda_a = \lambda_b = \lambda_2$. We choose ϵ equal to the width of a strip, $\epsilon = \lambda_2^n$. As usual, for convenience we compute the ϑ-capacity of the attractor projected onto the x-axis, i.e. $\tilde{d}_C(\vartheta) = d_C(\vartheta) - 1$. The ϑ-capacity $\tilde{d}_C(\vartheta)$ is defined in terms of the minimum number of intervals $N(\epsilon; \vartheta)$ of width ϵ that have total natural measure at least ϑ,

$$N(\epsilon; \vartheta) = \sum_{m=0}^{m_\vartheta} Z(n, m), \tag{29}$$

where m_ϑ is the largest integer such that

$$\sum_{m=0}^{m_\vartheta - 1} W(n, m) \leqslant \vartheta, \tag{30}$$

To find m_ϑ we use eq. (25) and approximate the sum in eq. (30) by an integral,

$$\vartheta \approx \frac{1}{\sqrt{2\pi\alpha\beta n}} \int_{-\infty}^{m_\vartheta} e^{-(m - n\alpha)^2/2n\alpha\beta}\,dm.$$

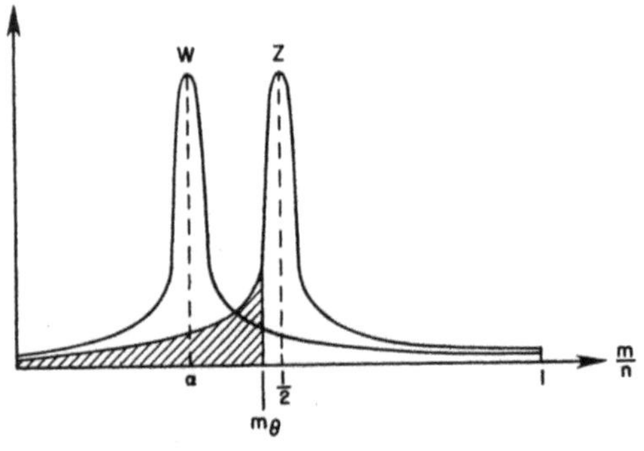

Fig. 10. The principal contribution to the sum needed to compute $N(\epsilon; \vartheta)$ (eq. (29)) comes from values of m near m_ϑ.

Thus for fixed ϑ we obtain

$$\frac{m_\vartheta}{n} \approx \alpha + \operatorname{erfc}^{-1}(\vartheta)\sqrt{\frac{\alpha\beta}{n}}, \tag{31}$$

where $\operatorname{erfc}(x) = (1/\sqrt{2\pi})\int_{-\infty}^{x} e^{-x^2/2}\,dx$. Now, consider eq. (29). The principal contribution to the sum will come from m values very close to m_ϑ, as depicted in fig. 10. Thus we use eqs. (23) and (25) to approximate $Z(n, m)$ as

$$Z(n, m) \approx \frac{\beta^{-n}(\beta/\alpha)^m}{\sqrt{2\pi n\alpha\beta}}\, e^{-(m-n\alpha)^2/2n\alpha\beta}. \tag{32}$$

The term $(\beta/\alpha)^m$ decreases as m decreases away from m_ϑ, and this decrease is very rapid compared to the variation of $e^{-(m-n\alpha)^2/2n\alpha\beta}$. Thus in performing the sum in eq. (29), we may approximate $e^{-(m-n\alpha)^2/2n\alpha\beta}$ as being constant and equal to its value at $m = m_\vartheta$. Hence the only m dependent term in the sum is $(\beta/\alpha)^m$. Since

$$\sum_{m=0}^{m_\vartheta}\left(\frac{\beta}{\alpha}\right)^m \approx \left(\frac{\beta}{\alpha}\right)^{m_\vartheta}\frac{\beta}{(\beta-\alpha)},$$

we find that

$$N(\epsilon; \vartheta) \sim \beta^{-(n-m_\vartheta)}\alpha^{-m_\vartheta}n^{-1/2}.$$

From eq. (4) and $n = \log(1/\epsilon)/\log(1/\lambda_2)$, the above estimate of $N(\epsilon; \vartheta)$ yields $d_C(\vartheta) = d_L$, in agreement with conjecture 2.

6.3. *Computation of ϑ-Hausdorff dimension*

In this section we obtain an upper bound on the ϑ-Hausdorff dimension of the generalized baker's transformation with $\lambda_a \neq \lambda_b$. (Recall that for our work in the previous section we took $\lambda_a = \lambda_b$.) We obtain an inequality for the ϑ-Hausdorff dimension by using a specific covering along x to compute the sum

$$l_d^*(\epsilon, \vartheta) = \sum_i \epsilon_i^d,$$

where the $\epsilon_i < \epsilon$ cover a fraction ϑ of the natural measure of the attractor. Our choice for the ϵ_i is specified below. Taking the limit as $\epsilon \to 0$, we find that there is a value of d at which $l_d^*(\epsilon, \vartheta)$ crosses over from ∞ to 0. For the partition we have chosen, we find that crossover occurs at $d = \bar{d}_L$. We believe that the value we obtain is in fact the true ϑ-Hausdorff dimension. However, we cannot be sure that the particular covering we have chosen gives the lowest possible value of d, and thus we can only say that we have obtained an upper limit.

After n iterations of the map, an initially uniform probability distribution in the unit square is transformed to 2^n strips with widths $\lambda_a^m\lambda_b^{(n-m)}$ and probabilities $\alpha^m\beta^{(n-m)}$, $m = 0, 1, 2, \ldots, n$. As shown in eq. (22), the number of such strips is $Z(n, m)$. We shall choose the ϵ_i to cover the most probable strips so that ϑ of the total probability is covered. ϵ for our covering is equal to the width of the widest strip, which is either $(\lambda_a)^n$ or $(\lambda_b)^n$, whichever is larger. Letting $U_d(n, m)$ be

$$U_d(n, m) = (\lambda_b^{m-n}\lambda_a^m)^d Z(n, m),$$

we have that

$$l_d^*(\epsilon, \vartheta) = \sum_i \epsilon_i^d = \sum_m U_d(n, m), \tag{33}$$

We still have yet to specify which m values are to be included in the sum. To do this, we expand $U_d(n,m)$ about its maximum value (as done for Z and W in section 5), and obtain

$$U_d(n,m) \approx \frac{[\lambda_a^d + \lambda_b^d]^n}{\sqrt{2\pi n \frac{\lambda_a^d \lambda_b^d}{(\lambda_a^d+\lambda_b^d)(\lambda_a^d+\lambda_b^d)}}} \times \exp -\frac{1}{2}\left\{\frac{\left[\frac{m}{n} - \frac{\lambda_a^d}{\lambda_a^d+\lambda_b^d}\right]^2}{\frac{\lambda_a^d\lambda_b^d}{n(\lambda_a^d+\lambda_b^d)(\lambda_a^d+\lambda_b^d)}}\right\} \quad (34)$$

In order to compute $l_d^*(\epsilon, \vartheta)$, we must consider the natural measure as well as $U_d(n,m)$. Note that for the general case we are considering now with $\lambda_a \neq \lambda_b$, $W(n,m)$ obtained in eq. (25) continues to be the correct expression for the distribution of probabilities in each strip. Depending on the values of α, d, λ_a, and λ_b, W may peak at a value of m that is smaller, larger, or equal to the value of m at the peak of U_d. Comparing the location of the peaks of the Gaussians in eq. (34) (for U) and in eq. (25) (for W), we see that there are three cases:

Case 1: $\alpha < \frac{\lambda_a^d}{(\lambda_a^d+\lambda_b^d)}$,

Case 2: $\alpha > \frac{\lambda_a^d}{(\lambda_a^d+\lambda_b^d)}$,

Case 3: $\alpha = \frac{\lambda_a^d}{(\lambda_a^d+\lambda_b^d)}$.

Cases 1 and 2 may be shown to be equivalent as follows. From the case 2 inequality and the fact that $\alpha+\beta=1$, we obtain $\beta < \lambda_b^d/(\lambda_a^d+\lambda_b^d)$. But, if we define m' by $m = n - m'$, and change the sums over m to sums over m', then the roles of (α, λ_a) and (β, λ_b) are interchanged, and case 2 is converted to case 1. We shall not consider case 3 here; suffice it to say that it does not alter the results obtained from consideration of cascs 1 and 2. Therefore it is sufficient to compute the ϑ-Hausdorff dimension for case 1.

For case 1, selecting the best covering of intervals that contain ϑ of the total probability is easy. Since W remains valid, we get a covering that includes ϑ of the total natural measure by including intervals whose value of m is less than m_ϑ, just as we did for the computation of ϑ-capacity. Furthermore, since U_d peaks at a larger value of m/n than W does, this selection gives the smallest value of l_d^*. The situation is analogous to the computation of ϑ-capacity, except that here the role of Z is played by U_d (cf. fig. 10). To evaluate

$$l_d^* = \sum_{m=0}^{m_\vartheta} U_d(n,m), \quad (35)$$

we note that, as for the analogous evaluation for ϑ-capacity in the previous subsetion, the principal contribution to the sum comes from m-values close to m_ϑ. Thus we approximate $U_d(n,m)$ as

$$U_d(n,m) \approx \frac{\lambda_a^m \lambda_b^{n-m}}{\alpha^m \beta^{(n-m)}} W(n,m),$$

with W approximated by eq. (25). Proceeding as in section 6.2 we obtian an estimate for $l_d^*(\epsilon, \vartheta)$,

$$l_d^*(\epsilon, \vartheta) \sim n^{-1/2}\left(\frac{\beta}{\lambda_b^d}\right)^{m_\vartheta - n}\left(\frac{\alpha}{\lambda_a^d}\right)^{-m_\vartheta},$$

or

$$\log[l_d^*(\epsilon, \vartheta)] \approx -n[d - (\bar{d}_L)] \log\left(\frac{1}{\lambda_2}\right).$$

For $\epsilon \to 0$ (i.e., $n \to \infty$) we obtain $l_d^*(\vartheta) = 0$ for $d > \bar{d}_L$ and $l_d^*(\vartheta) = \infty$ for $d < \bar{d}_L$. Thus remembering that $d_H(\vartheta) = \bar{d}_H(\vartheta) + 1$,

$$d_H(\vartheta) \leqslant d_L. \quad (36)$$

As already mentioned, we expect that the above inequality is really an equality. This expectation is reinforced by the fact that when $\vartheta = 1$ we recover the exact expression for the Hausdorff dimension computed in eq. (18). To see that this is true, replace m_ϑ in eq. (35) by n. From the form of U_d, this sum is simply the binomial expansion of $(\lambda_a^d + \lambda_b^d)^n$. As $n \to \infty$, this quantity is 0 or ∞ for $d > \bar{d}_H$ or $d < \bar{d}_H$, where $\bar{d}_H$ satisfies $\lambda_a^{\bar{d}_H} + \lambda_b^{\bar{d}_H} = 1$, which is the same as eq. (18). That is, for the specific choice of ϵ_i that we have used, we obtain the correct value of d_H. Since the same choice of the ϵ_i was used in obtaining $d_H(\vartheta)$, it seems plausible that the equality might apply in eq. (36).

6.4. *Computation of the pointwise dimension*

We now consider the pointwise dimension for the generalized baker's transformation with $\lambda_a = \lambda_b < \frac{1}{2}$, and we show that d_p exists and is equal to d_L.

As previously noted in section 5, application of the map n times to the unit square produces 2^n strips of widths $(\lambda_a)^n$. (Recall that we are assuming $\lambda_a = \lambda_b$.) In order to compute the pointwise dimension, we choose a point x at random with respect to the natural measure μ, compute the natural measure contained in an ϵ ball centered about x, (i.e. $\mu(B_\epsilon(x))$), and compute the ratio of $\log \mu(B_\epsilon(x))$ to $\log \epsilon$ in the limit as ϵ goes to zero (cf. eq. (5)). The simplest case for this computation occurs when $\lambda_a < \frac{1}{4}$, so that the gaps between strips are bigger than the strips themselves, as pictured in fig. 11a. Choosing a point x at random with respect to the natural measure μ, let S_n denote the nth order strip of width $(\lambda_a)^n$ that the point x lies in. Letting $\epsilon = (\lambda_a)^n$, the natural measure contained in a ball of radius ϵ around x (i.e., the x-interval $[x - (\lambda_a)^n, x + (\lambda_a)^n]$) will be equal to the natural measure of the strip S_n, regardless of where in the strip x lies. (See fig. 11a.) The natural measure contained in a given strip is $\alpha^m \beta^{(n-m)}$, where $n \geq m \geq 0$, where m depends on the particular strip that x happens to lie in. (See section 5.) Thus, we have

$$\lim_{n\to\infty} \frac{\mu(B_\epsilon(x))}{\log \epsilon} = \lim_{n\to\infty} \frac{\log \mu(S_n)}{n \log \lambda_a}$$

$$= \lim_{n\to\infty} \frac{m \log \alpha + (n-m) \log \beta}{n \log \lambda_a}. \tag{37}$$

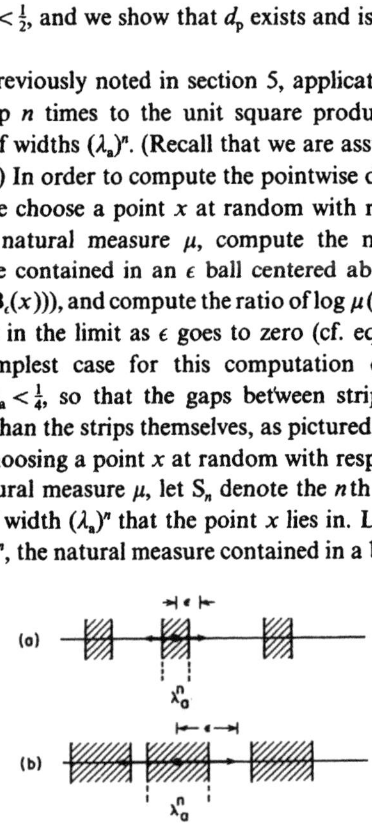

Fig. 11. (a) Computing the pointwise dimension for the case that $\lambda_a < \frac{1}{4}$. (b) The case $\lambda_a > \frac{1}{4}$, in which the computation is a little more complicated.

In the limit as n grows large, as shown in section 5 (see fig. 9), the total probability $W(n,m)$ contained in strips of a given m value is distributed as a Gaussian centered about $m/n = \alpha$. Thus, in the limit as $n \to \infty$ it becomes overwhelmingly likely that $m/n = \alpha$. Thus for almost every x with respect to the natural measure μ, $\lim_{n\to\infty} m/n = \alpha$. (This is just a statement of the law of large numbers.) Putting this into eq. (37) gives

$$d_p = \lim_{n\to\infty} \frac{\mu(B_\epsilon(x))}{\log \epsilon} = \frac{\alpha \log \alpha + \beta \log \beta}{\log \lambda_a}$$

$$= \frac{H(\alpha)}{\log(1/\lambda_d)} = \bar{d}_L. \tag{38}$$

(See eqs. (15) and (16).)

To extend this computation of the pointwise dimension to the case that $\frac{1}{2} > \lambda_a > \frac{1}{4}$, for any $\lambda_a < \frac{1}{2}$ choose a k such that $\lambda_a^{k+1} \leq \frac{1}{2} - \lambda_a$ (e.g., for $\lambda_a \leq \frac{1}{4}$ this relation is satisfied for any $k \geq 0$; for $\lambda_a \leq 0.365\ldots$, for any $k \geq 1$; etc.). Then we can show $\mu(B_\epsilon(x)) \leq \alpha^{-k} \mu(S_n)$, where without loss of generality, we have assumed $\alpha \leq \beta$. Since $B_\epsilon(x) \supset S_n$, we have also $\mu(B_\epsilon(x)) \geq \mu(S_n)$. Thus $\mu(S_n) \leq \mu(B_\epsilon(x)) \leq \alpha^{-k} \mu(S_n)$ which with eq. (37) yields $d_p = \bar{d}_L$. (Our evaluation of eq. (37) holds not for $\epsilon \to 0$ but rather holds for the restricted set of $\epsilon = \lambda_a^n$, $n = 1, 2, \ldots$, however, it is not hard to show that that in fact implies eq. (38) for every sequence of ϵ going to 0.)

Thus we have shown that for the generalized baker's transformation the pointwise dimension is equal to the dimension of the natural measure. (Although we have only shown this for $\lambda_a = \lambda_b$, it is not hard to extend this result to $\lambda_a \neq \lambda_b$.)

7. The core of attractors

As shown in section 5, for the generalized baker's transformation, typically almost all of the probability is contained in a very small fraction of the total number of cubes needed to cover the attractor. In the limit as ϵ goes to zero, this fraction goes to zero. Thus, the natural measure of the attractor is *concentrated* on a subset of the attractor. We will call this subset the *core* of the attractor.

To get a better feel for why this comes about, and to see how the properties of the core are related to those of the attractor and its natural measure, consider the special case of the generalized baker's transformation where $\lambda_a = \lambda_b = \frac{1}{2}$. As we have already seen, at the nth level of approximation the natural measure consists of 2^n vertical strips of probability $\alpha^m\beta^{n-m}$. For large n and $\beta > \alpha$, a small fraction of the strips whose m values are close to αn contain much more of the natural measure than all other strips. Fig. 12 shows a plot of the nth level approximation to the probability distribution as a function of x with $\lambda_a = \lambda_b = \frac{1}{2}$, and $\alpha < \frac{1}{2}$ and $n = 10$. The probability distribution looks as though it were made up of spikes, showing that already at $n = 10$ the natural measure has become quite concentrated in certain cubes (in this case intervals).

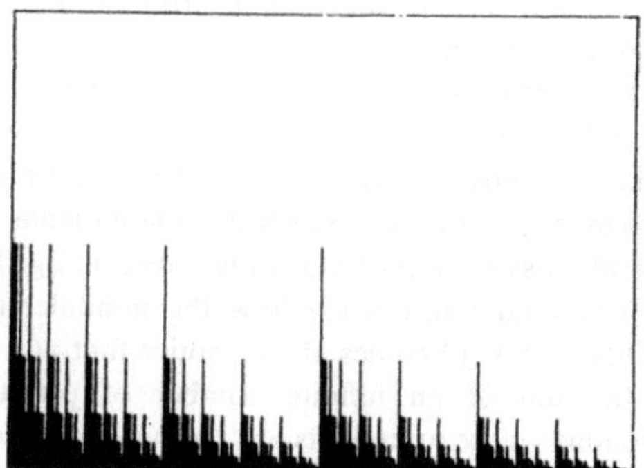

Fig. 12. The natural probability distribution of the generalized baker's transformation projected onto the x-axis, and coarse grained using intervals of width $\epsilon = 2^{-10}$. In this case $\alpha < \frac{1}{2}$, and $\lambda_a = \lambda_b = \frac{1}{2}$.

To understand the form of this probability distribution, it is instructive to represent the probability distribution of these strips in terms of x rather than m. To do this, approximate x using its first n binary digits, i.e. as a binary decimal truncated after n digits. Let m be the number of ones contained in the first n digits of the binary expansion of x. The natural measure of the strip $S_n(x)$ containing x is then $\mu(S_n(x)) = \alpha^m\beta^{(n-m)}$. (See the discussion at the beginning of section 5.) As we have already shown (see fig. 9), when written in terms of m, for large n the natural measure is approximately a Gaussian centered about αn, and in the limit where n is large almost all the measure is contained in strips with $m \approx \alpha n$. In other words, the natural measure of the generalized baker's transformation for $\lambda_a = \lambda_b = \frac{1}{2}$ is concentrated on those values of x that have 1's in their binary expansions in the fraction α, or equivalently, 0's in the fraction β. In the limit $n \to \infty$, *all* the natural measure is contained in this set, which we will call the core of this attractor.

For this case ($\lambda_a = \lambda_b = \frac{1}{2}$) the attractor is the entire unit square. The core of this attractor is dense on the attractor. In other words, any point of the attractor has points of the core arbitrarily close to it. Hence any covering of the core must also be a covering of the attractor, and vice versa. Thus the capacity of the core is the same as that of the attractor. The Hausdorff dimension, in contrast, is more subtle, and in fact, computing the Hausdorff dimension of the set of numbers whose binary expansions have a given fraction of ones is a classic problem in the study of Hausdorff dimension [31]. The Hausdorff dimension of this set is

$$d_H = \frac{H(\alpha)}{\log 2}.$$

(See eq. (15).) This result was conjectured by Good in 1941 [32] and proved by Eggleston in 1949 [33]. Also, the Hausdorff dimension of a very similar example (involving ternary rather than binary expansions) was proven by Besicovitch in 1931 [34].

Thus, for this example we see that the Hausdorff dimension of the core is equal to the dimension of the natural measure, and the capacity of the core is equal to the fractal dimension of the attractor (cf. eq. (16)). For the case of diffeomorphisms of the plane, the former result has been proven by Young [12]. We suspect that this is a property of typical attractors.

8. An attractor that is a nowhere differentiable torus

This section contains a review of the work of Kaplan, Mallet-Paret, and Yorke [35] on the dimension of a chaotic attractor in a setting that is quite different from that of the generalized baker's transformation. The attractor described below has the same topological form as a torus, and yet is nowhere differentiable, thus providing an interesting example of the nonanalytic forms that can be produced by chaotic dynamics.

Consider the following map:

$$\begin{aligned} x_{n+1} &= 2x_n + y_n \mod 1, \\ y_{n+1} &= x_n + y_n \mod 1, \\ z_{n+1} &= \lambda z_n + p(x_n, y_n). \end{aligned} \tag{39}$$

where x and y are taken mod 1, z can be any real number, and p is periodic in x and y with period 1 and is at least five times differentiable. (For example, $p(x, y) = \cos 2\pi x$.) In order to keep z bounded, λ is chosen between 0 and 1. Note that the eigenvalues and eigenvectors of the Jacobian matrix of eq. (39) are independent of x, y, and z. Thus *every* initial condition has the same Lyapunov numbers, i.e., the Lyapunov numbers are absolute, so that in this case conjecture 3 is relevant, and we expect that the fractal dimension and the dimension of the natural measure should be equal.

The equations for x and y are independent of z, and in fact are the classic Anasov or "cat" map [36],

$$\begin{pmatrix} x_{n+1} \\ y_{n+1} \end{pmatrix} = A \begin{pmatrix} x_n \\ y_n \end{pmatrix} \mod 1,$$

where

$$A = \begin{pmatrix} 2 & 1 \\ 1 & 1 \end{pmatrix}.$$

Thus, the x–y dynamics are chaotic, and are unaffected by the value of z.

To understand the shape of the attractor in the z-direction, put a sample initial condition into eq. (39). For example, consider $(x_0, y_0, 0)$. z_n takes on the form

$$z_n = \sum_{k=1}^{n} \lambda^{k-1} p(x_{n-k}, y_{n-k}).$$

Making use of the fact that $\binom{x_{n-k}}{y_{n-k}} = A^{-k}\binom{x_n}{y_n}$, and letting n go to infinity, it can be shown that the surface given by

$$z(x, y) = \sum_{k=1}^{\infty} \lambda^{k-1} p\left(A^{-k}\begin{pmatrix} x \\ y \end{pmatrix}\right),$$

is invariant and is the unique attractor of this dynamical system.

$z(x, y)$ has some very interesting properties. For $\lambda < 1/R$, where $R = (3 + \sqrt{5})/2$, $z(x, y)$ is smooth and has dimension 2. If $\lambda > 1/R$, however, for most choices of p, $z(x, y)$ is nowhere differentiable. A typical cross section of $z(x, y)$ is shown in fig. 13. To understand intuitively how the nondifferentiability of $z(x, y)$ comes about, notice that $z(x, y)$ is the sum of an infinite number of periodic functions whose arguments are the successive iterates of the cat map. Unless λ is small enough to diminish the effect of higher order iterates, the value of the sum can swing wildly as x or y change.

The Lyapunov numbers of the map given in eq.

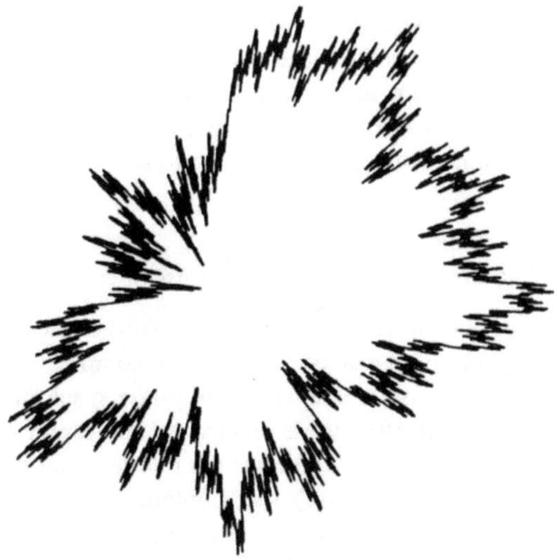

Fig. 13. A cross section of a nowhere differentiable torus, made using eq. (39) with $p(x, y) = \cos 2\pi x$ and λ chosen so that $d_C = 2\frac{1}{2}$.

(39) are $\lambda_1 = R$, $\lambda_2 = \lambda$, and $\lambda_3 = 1/R$, where $R = (3 + \sqrt{5})/2$, as given above. Kaplan, Mallet-Paret, and Yorke [35] have shown that there are two possibilities for the dimension of $z(x, y)$: Either

(i) $z(x, y)$ is nowhere differentiable and

$$d_C = d_L$$

or

(ii) $z(x, y)$ is differentiable and $d_C = 2$.

For given p, the nowhere differentiable case occurs for nearly every choice of λ. Thus we see that conjecture 3 is satisfied for this example.

9. Numerical computations

In this section we discuss some aspects of the numerical computation of dimension. First we will discuss the basic ideas behind numerical computations of dimension, secondly we will discuss some of the problems encountered in such computations, and finally we will review some previous numerical work.

The methods to compute dimension vary considerably depending on the dimension that one wishes to compute. Thus far, we are aware of numerical computations only of capacity [37–41], Lyapunov dimension [37–40], and Hausdorff dimension [42]. Of these, only the studies involving the capacity and the Lyapunov dimension were applied to attractors of dynamical systems. In each case, the computations follow from the definitions. As we shall see, the capacity is (in principle) straightforward to compute, but is in practice unfeasible to compute for all but very low dimensional attractors. The Lyapunov dimension, in contrast, is much more feasible to compute. We will begin the discussion with a description of the computation of Lyapunov dimension, and then go on to discuss the computation of capacity.

9.1. *Numerical computation of Lyapunov dimension*

The Lyapunov dimension is defined in terms of the Lyapunov numbers. (See section 3.) Thus, the work involved in computing Lyapunov dimension is in computing the Lyapunov numbers. Numerical methods for doing this have been discussed by Bennetin et al. [43], Shimada and Nagashima [44], and in infinite dimensions by Farmer [38]. With appropriate numerical caution, the largest k Lyapunov numbers can be computed by following the evolution of k nearby trajectories simultaneously and measuring their rate of separation. There are various numerical problems with this method, however, and a better method is to follow only one trajectory, but also follow k trajectories of the associated equations for the evolution of vectors in the tangent space. These methods have been successfully used in a variety of numerical studies.

For low-dimensional cases, such as two-dimensional maps or systems of three autonomous ordinary differential equations, with a modern computer and plenty of computer time, numerical computation of the dimensions we discuss here directly from their definitions is feasible, as discussed in the next subsection. Even in such low-dimensional cases, however, the computation of Lyapunov dimension is by far less costly in terms of computer time and memory than the computation of other dimensions. For higher dimensional attractors it appears that only the Lyapunov dimension is computationally feasible. The key reason that the Lyapunov dimension is feasible to compute numerically even for attractors of rather high dimension (e.g. $d_L \approx 10$) is that the difficulty of the computation scales linearly with the dimension of the attractor times the dimension of the space it lies in, rather than exponentially as it does for a computation of the fractal dimension, or any of the other dimensions discussed in this paper. The memory needed to compute the largest j Lyapunov numbers is equal to the memory needed to numerically integrate the equations under study, multiplied by $j+1$. (Memory requirements are usually a problem only in computations involving partial differential equations.) The computer time needed is the time needed to compute a time average to the desired accuracy (which depends, among other things, on the irregularity of the natural measure of the attractor), multiplied by $j+1$. Fortunately it is only necessary to compute the largest Lyapunov numbers, and the number of these needed depends on the dimension of the attractor rather than the dimension of the phase space. (See eq. (9).) This linear dependence on the dimension of the attractor has allowed computation of the Lyapunov dimension for attractors of dimension as large as twenty [38].

We should mention one disadvantage concerning Lyapunov dimension. Namely, it is not presently known how the Lyapunov dimension can be determined directly from a physical experiment. The difficulty comes about because, in some sense, in order to determine Lyapunov numbers it is necessary to be able to follow adjacent trajectories. To determine all the necessary Lyapunov numbers, it is necessary to follow some trajectories (at least one) which are not on the attractor. Thus it is not possible to compute the Lyapunov dimension by simply observing behavior on the attractor; one must perturb the system from the attractor, and do so in a very well defined way. This poses a very severe problem in the computation of dimension from experimental data, one that is not present in the computation of other dimensions.

9.2. *Computation of fractal dimension*

In principle, it is quite straightforward to use the definition of capacity, eq. (2), to compute the fractal dimension. The region of phase space surrounding the attractor is divided up into a grid of cubes of size ϵ, the equations are iterated, and the number of cubes $N(\epsilon)$ that contain part of the attractor are counted. ϵ is decreased and the process is repeated. If $\log N(\epsilon)$ is plotted against $\log \epsilon$, in the limit as ϵ goes to zero the slope is the fractal dimension.

The difficulty with this method is that one must use values of ϵ small enough to insure that the asymptotic scaling has been reached. (See Froehling et al. [40] and Greenside et al. [39].) The total number of cubes containing part of the attractor scales roughly as

$$N(\epsilon) \sim \epsilon^{-d_C}. \tag{40}$$

Thus, the number of cubes increases *exponentially* with the fractal dimension of the attractor. To get a feel for the seriousness of this problem, plug in some typical numbers: If $\epsilon = 0.01$ and $d_C = 3$, then $N \approx 10^6$, exceeding the core memory of all but the biggest current computers. Thus, computations of fractal dimension are currently not feasible for attractors of dimension significantly greater than three.

In addition, there is another potential problem involved in computing capacity. In counting cubes, how can one be sure that all the nonempty cubes have been counted? This problem is compounded by the highly nonuniform distribution of probability on an attractor. In particular, if our hypothesis that the probability is distributed log-normally is correct, in order to count the highly improbable cubes present in the wings of the distribution requires that a large number of points on the attractor must be generated. Furthermore, this number increases rapidly as ϵ decreases.

The conclusion is that a great deal of care must be taken in the computation of fractal dimension, and in particular, a sufficiently large number of points on the attractor must be generated to insure that low probability cubes are not left out in the determination of $N(\epsilon)$.

Although there are as yet no extensive results on direct computations of the dimension of the natural measure, it may be easier to reliably compute than the fractal dimension. The reason for this is that very improbable cubes are irrelevant for a computation of the dimension of the natural measure. Numerical experiments on this topic are currently in progress.

9.3. *Summary of past numerical experiments*

In this section we summarize previous numerical experiments on dimension computation. The two studies most relevant to the topic under discussion are those of Russel et al. [37] and Farmer [38]. Both of these were made in an attempt to test the Kaplan–Yorke conjecture [8, 22]. (See section 3.) In both of these studies, the capacity of chaotic attractors was computed directly from the definition. The Lyapunov dimension was also computed, and compared to the capacity.

In the study of Russel et al., five examples were examined. In each case, the computed capacity agree with the computed Lyapunov dimension to within experimental accuracy. These computations were done on the Crayl, a state of the art mainframe computer; at the smallest value of $\epsilon = 2^{-14}$, more than 10^5 cubes were counted.

The numerical experiments of Farmer were done using high-dimensional approximations to an infinite dimensional dynamical system. Because the equations under study were more time consuming to integrate, and because the capacity computations were done on a minicomputer, it was only possible to achieve about two significant figures of accuracy. The computed capacity and Lyapunov dimension agreed to this accuracy at the two parameter values tested.

In 1980, Mori [23] conjectured an alternate formula relating the fractal dimension to the spectrum of Lyapunov numbers. For attractors in a low-dimensional phase space, such as those studied by Russel et al. [37], Mori's formula and the Kaplan–Yorke formula (eq. (9)) predict the same value. For higher dimensional phase spaces, however, the two formulas no longer agree. Farmer's results support the Kaplan–Yorke formula.

One puzzling aspect of both of these numerical experiments is the striking agreement between the computed value of capacity and the Lyapunov dimension. The Kaplan–Yorke conjecture equates the Lyapunov dimension to the dimension of the natural measure, and therefore only gives a lower bound on the fractal dimension. Why, then, was such good agreement obtained between the computed capacity and the computed Lyapunov dimension? We do not yet understand the answer to this question, though further numerical experiments may resolve the question.

10. Conclusions

We have given several different definitions of dimension. These divide into two types, those that require a probability measure for their definition, and those that do not. (Refer back to table I.) For an example that we believe is typical of chaotic attractors, i.e., the generalized baker's transformation, our computations of dimension show that all of the probabilistic definitions take on one value, which we call the dimension of the natural measure, while the definitions that do not require a probability measure take on another value, which we call the fractal dimension of the attractor. We believe that this is true for typical attractors.

If the probability distribution on the attractor is "coarse grained" by covering the attractor with cubes, for the generalized baker's transformation we find that the probability contained in these cubes is distributed nearly log-normally when the cubes are sufficiently small. In other words, the total probability contained in cubes whose natural measure is between $u = \log p_i$ and $u + du$ has a distribution that is nearly Gaussian, and as the size of the cubes is decreased, it becomes more nearly Gaussian. Furthermore, the number of cubes in a given interval of u also has a Gaussian distribution, but with a different mean and variance. (See fig. 9.) As ϵ decreases, both of these distributions become narrower in a relative sense, in that the ratio of their variance to their mean decreases. In the limit as ϵ goes to zero, both distributions approach delta functions; since their means are different, in this limit the two distributions typically do not overlap. Thus, almost all of the natural measure is contained in almost none of the cubes, and the natural measure is concentrated on a core set. The capacity of the core is the fractal dimension of the attractor, while the Hausdorff dimension of the core is the dimension of the natural measure. Once again, although we have demonstrated the results memtioned in this paragraph only for the generalized baker's transformation, we feel that they are true for typical chaotic attractors.

Most of the dimensions that we have defined are difficult to compute numerically. The Lyapunov dimension, however, is much easier to compute numerically than any of the other dimensions. We compute the Lyapunov dimension for the generalized baker's transformation, and show that it is equal to the dimension of the natural measure obtained from any of the other probabilistic dimensions that we have investigated. This supports the conjecture of Kaplan and Yorke [22].

References

[1] R. Shaw, "Strange Attractors, Chaotic Behavior, and Information Flow", Z. Naturforsch. 36a (1981) 80.

[2] E. Ott, "Strange Attractors and Chaotic Motions of Dynamical Systems", Rev. Mod. Phys. 53 (1981) 655.

[3] R. Helleman, "Self-Generated Chaotic Behavior in Nonlinear Mechanics", Fundamental Problems in Stat. Mech. 5, E.G.D. Cohen, ed. (North-Holland, Amsterdam and New York, 1980) 165–233.

[4] J.A. Yorke and E.D. Yorke, "Chaotic Bahavior and Fluid Dynamics", Hydrodynamic Instabilities and the Transition to Turbulence, H.L. Swinney and J.P. Gollub eds., Topics in Applied Physics 45 (Springer, Berlin, 1981) pp. 77–95.

[5] B. Mandelbrot, Fractals: Form, Chance, and Dimension (Freeman, San Francisco, 1977).

[6] B. Mandelbrot, The Fractal Geometry of Nature (Freeman, San Francisco, 1982.)

[7] Ja. Sinai, "Gibbs Measure in Ergodic Theory", Russ. Math. Surveys 4 (1972) 21–64.

[8] P. Frederickson, J. Kaplan, E. Yorke, and J. Yorke, "The Lyapunov Dimension of Strange Attractors", J. Diff. Eqns. in press.

[9] J.D. Farmer, "Dimension, Fractal Measures, and Chaotic Dynamics", Evolution of Order and Chaos, H. Haken, ed., (Springer, Berlin, 1982), p. 228.

[10] J.D. Farmer, "Information Dimension and the Probabilistic Structure of Chaos", first chapter of UCSC doctoral disseration (1981) and Z. Naturforsch. 37a (1982) 1304–1325.

[11] J. Alexander and J. Yorke, "The Fat Baker's Transformation", U. of Maryland preprint (1982).

[12] L.S. Young, "Dimension, Entropy, and Lyapunov Exponents", to appear in Ergodic Theory and Dynamical Systems.

[13] W. Hurwicz and H. Wallman, Dimension Theory (Princeton Univ. Press, Princeton, 1948).

[14] A.N. Kolmogroov, "A New Invariant for Transitive Dynamical Systems", Dokl. Akad. Nauk SSSR 119 (1958) 861–864.

[15] Hausdorff, "Dimension und Auβeres Maβ", Math. Annalen. 79 (1918) 157.

[16] R. Bowen and D. Ruelle, "The Ergodic Theory of Axiom-A Flows", Inv. Math. 29 (1975) 181–202.

[17] J. Balatoni and A. Renyi, Publ. Math. Inst. of the Hungarian Acad. of Sci. 1 (1956) 9 (Hungarian). English translation, Selected Papers of A. Renyi, vol. 1 (Akademiai Budapest, Budapest, 1976), p. 558. See also A. Renyi, Acta Mathematica (Hungary) 10 (1959) 193.

[18] C. Shannon, "A Mathematical Theory of Communication", Bell Tech. Jour. 27 (1948) 379–423, 623–656.

[19] F. Ledrappier, "Some Relations Between Dimension and Lyapunov Exponents" Comm. Math. Phys. 81 (1981) 229–238.

[20] F. Takens, "Invariants Related to Dimension and Entropy", to appear in Atas do 13 Colognio Brasiliero de Mathematica.

[21] T. Janssen and J. Tjon, "Bifurcations of Lattice Structure", Univ. of Utrecht preprint (1982).

[22] J. Kaplan and J. Yorke, Functional Differential Equations and the Approximation of Fixed Points, Proceedings, Bonn, July 1978, Lecture Notes in Math. 730, H.O. Peitgen and H.O. Walther, eds., (Springer, Berlin, 1978), p. 228.

[23] H. Mori, "Prog. Theor. Phys. 63 (1980) 3.

[24] V.I. Oseledec, "A Multiplicative Ergodic Theorem. Lyapunov Characteristic Numbers for Dynamical Systems", Trans. Moscow Math. Soc. 19 (1968) 197.

[25] C. Grebogi, E. Ott, and J. Yorke, "Chaotic Attractors in Crisis", in this volume.

[26] A. Douady and J. Oesterle, "Dimension de Hausdorff des Attracteurs, Comptes Rendus des Seances de L'academie des Sciences 24 (1980) 1135–38.

[27] A.N. Kolmogorov, Dolk. Akad. Nauk SSSR 124 (1959) 754. English summary in MR 21, 2035.

[28] Ya. G. Sinai, Dolk. Akad. Nauk SSSR 124 (1959) 768. English summary in MR 21, 2036.

[29] J. Crutchfield and N. Packard, "Symbolic Dynamics of One-Dimensional Maps: Entropies, Finite Precision, and Noise", Int'l. J. Theo. Phys. 21 (1982) 433.

[30] S. Pelikan, private communication.

[31] P. Billingsley, Ergodic Theory and Information, New York, 1965).

[32] I.J. Good, "The Fractional Dimensional Theory of Continued Fractions", Proc. Camb. Phil. Soc. 37 (1941) 199–228.

[33] H.G. Eggleston, "The Fractional Dimension of a Set Defined by Decimal Properties", Quart. J. Math. Oxford Ser. 20 (1949) 31–36.

[34] A. Besicovitch, "On the Sum of Digits of Real Numbers Represented in the Dyadic System", Math. Annalen. 110 (1934) 321.

[35] J.L. Kaplan, J. Mallet-Paret, and J.A. Yorke, "The Lyapunov Dimension of a Nowhere Differentiable Attracting Torus", Univ. of Maryland Preprint (1982).

[36] V.I. Arnold and Avez, Ergodic Theory in Classical Mechanics (New York, 1968).

[37] D. Russel, J. Hansen, and E. Ott, "Dimensionality and Lyapunov Numbers of Strange Attractors", Phys. Rev. Lett. 45 (1980) 1175.

[38] J.D. Farmer, "Chaotic Attractors of an Infinite Dimensional Dynamical System", Physica 4D (1982) 366–393.

[39] H. Greenside, A. Wolf, J. Swift, and T. Pignataro, "The Impracticality of a Box Counting Algorithm for Calculating the Dimensionality of Strange Attractors", Phys. Rev. A25 (1982) 3453.

[40] H. Froehling, J. Crutchfield, J.D. Farmer, N. Packard, and R. Shaw, "On Determining the Dimension of Chaotic Flows", Physica 3D (1981) 605.

[41] R. Kautz, private communication.

[42] A.J. Chorin, "The Evolution of a Turbulent Vortex", Comm. Math. Phys. 83 (1982) 517–535.

[43] G. Bennètin, L. Galgani and J. Strelcyn, Phys. Rev. A 14 (1976) 2338; also see G. Benettin, L. Galgani, A. Giorgilli and J. Strelcyn, Meccanica 15 (1980) 9.

[44] I. Shimada and T. Nagashima, Prog. Theor. Phys. 61 (1979) 228.

[45] B.B. Mandelbrot, "Intermittent turbulence in self-similar cascades: Divergence of high moments and dimension of the carrier". J. Fluid Mechanics 62 (1974) 331–358. See 351, 354 for a discussion of a frequency dependent dimension.

[46] B.B. Mandelbrot, "Fractals and turbulence: attractors and dispersion". Turbulence Seminar, Berkeley 1976/7. Lecture Notes in Mathematics 615, 83–93 (Springer, New York).

Physica 9D (1983) 189–208
North-Holland Publishing Company

MEASURING THE STRANGENESS OF STRANGE ATTRACTORS

Peter GRASSBERGER† and Itamar PROCACCIA
Department of Chemical Physics, Weizmann Institute of Science, Rehovot 76100, Israel

Received 16 November 1982
Revised 26 May 1983

We study the correlation exponent ν introduced recently as a characteristic measure of strange attractors which allows one to distinguish between deterministic chaos and random noise. The exponent ν is closely related to the fractal dimension and the information dimension, but its computation is considerably easier. Its usefulness in characterizing experimental data which stem from very high dimensional systems is stressed. Algorithms for extracting ν from the time series of a single variable are proposed. The relations between the various measures of strange attractors and between them and the Lyapunov exponents are discussed. It is shown that the conjecture of Kaplan and Yorke for the dimension gives an upper bound for ν. Various examples of finite and infinite dimensional systems are treated, both numerically and analytically.

1. Introduction

It is already an accepted notion that many nonlinear dissipative dynamical systems do not approach stationary or periodic states asymptotically. Instead, with appropriate values of their parameters, they tend towards strange attractors on which the motion is chaotic, i.e. not (multiply) periodic and unpredictable over long times, being extremely sensitive on the initial conditions [1–4].

A natural question is by which observables this situation is most efficiently characterized. Even more basically, when observing a seemingly strange behaviour, one would like to have clear-cut procedures which could exclude that the attractor is indeed multiply periodic, or that the irregularities are e.g. caused by external noise [5].

The first possibility can be ruled out by making a Fourier analysis, but for the second one has to turn to some other measures. These measures should be sensitive to the *local* structure, in order to distinguish the blurred tori of a noisy (multi-) periodic motion from the strictly deterministic motion on a fractal. Also, they should be able to distinguish between different strange attractors.

In this paper we shall propose such a measure. Before doing so we shall discuss however the existing approaches to the subject.

In a system with F degrees of freedom, an attractor is a subset of F-dimensional phase space towards which almost all sufficiently close trajectories get "attracted" asymptotically. Since volume is contracted in dissipative flows, the volume of an attractor is always zero, but this leaves still room for extremely complex structures.

Typically, a strange attractor arises when the flow does not contract a volume element in *all* directions, but stretches it in some. In order to remain confined to a bounded domain, the volume element gets folded at the same time, so that it has after some time a multisheeted structure. A closer study shows that it finally becomes (locally) Cantor-set like in some directions, and is accordingly a fractal in the sense of Mandelbrot [6].

Ever since the notion of strange attractors has been introduced, it has been clear that the Lyapunov exponents [7, 8] might be employed in studying them. Consider an infinitesimally small F-dimensional ball in phase space. During its

† Permanent address: Department of Physics, University of Wuppertal, W. Germany.

evolution it will become distorted, but being infinitesimal, it will remain an ellipsoid. Denote the principal axes of this ellipsoid by $\epsilon_i(t)$ $(i = 1, \ldots, F)$. The Lyapunov exponents λ_i are then determined by

$$\epsilon_i(t) \approx \epsilon_i(0)\, e^{\lambda_i t}. \tag{1.1}$$

The sum of the λ_i, describing the contraction of volume, has of course to be negative. But since a strange attractor results from a stretching and folding process, it requires at least one of the λ_i to be positive. Inversely, a positive Lyapunov exponent implies sensitive dependence on initial conditions and therefore chaotic behaviour.

One drawback of the λ_i's is that they are not easily measured in experimental situations. Another limitation is that while they describe the *stretching* needed to generate a strange attractor, they don't say much about the *folding*.

That these two are at least partially independent is best seen by looking at a horshoe-like map† embedded in 3-dimensional space (fig. 1). Assume that each step of the evolution consists of (i) stretching in the x-direction by a factor of 2, (ii) squeezing in the y- and z-direction by different factors $\mu_z < \mu_y < \frac{1}{2}$, and (iii) folding in the (x, y) plane (fig. 1a) or in the (x, z) plane (fig. 1b). From fig. 1 one realizes already that the attractor will in both cases be a Cantorian set of lines, being more "plane-filling" in the first case than in the second case. Indeed, using the results of Section 7, one finds easily that the fractal dimensions are $D_a = 1 + \ln 2/|\ln \mu_y|$ and $D_b = 1 + \ln 2/|\ln \mu_z|$, respectively.

It is this fractal (or Hausdorff–Besikovich) dimension which has until now attracted most attention [9–14] as a measure of the local structure of fractal attractors. In order to define it [5], one first covers the attractor by F-dimensional hypercubes of side length l and considers the limit $l \to 0$. If the

† Notice that this is not a Smale's horseshoe. We also neglect in the following the bent parts of the horseshoe, in comparison to the parallel parts (i.e. we assume $L_x \gg L_y$, L_z; see fig. 1).

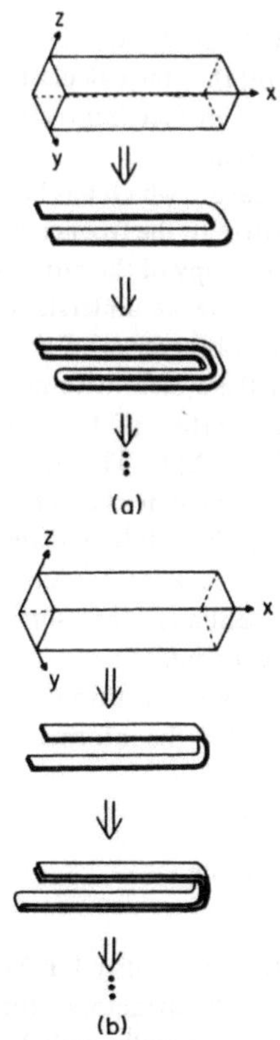

Fig. 1. Shape of an originally rectangular volume element after two iterations, each consisting of stretching, squeezing and folding. In fig. 1a (1b), the folding is in the $y(z)$-direction, which is the direction of lesser (stronger) squeezing.

minimal number of cubes needed for the covering grows like

$$M(l) \underset{l \to 0}{\simeq} l^{-D}, \tag{1.2}$$

the exponent D is called the Hansdorff dimension of the attractor [5].

Being a purely geometric measure, D is independent of the frequency with which a typical trajectory visits the various parts of the attractor.

Even if these frequencies are very inequal, developing maybe even singularities somewhere, all parts contribute to D equally. It has been documented [12, 14] that the calculation of D is exceedingly hard and in fact impractical for higher dimensional systems.

Another measure which has been considered and which is sensitive to the frequency of visiting, is the information entropy of the attractor. By "information entropy" here we understand the information gained by an observer who measures the actual state $X(t)$ of the system with accuracy l, and who knows all properties of the system but not the initial condition $X(0)$. This is very similar to the entropy in statistical mechanics if we relate $X(t)$ to the microstate ($F \approx 10^{23}$), and the "system" to the macrostate. It is *not* the Kolmogorov entropy which is essentially the sum of all positive Lyapunov exponents.

Using the above partition of phase space into cells with length l, the information entropy can be written as

$$S(l) = -\sum_{i=1}^{M(l)} p_i \ln p_i \,, \tag{1.3}$$

where p_i is the probability for $X(t)$ to fall into the ith cell. For all attractors studied so far, $S(l)$ increases logarithmically with $1/l$ as $l \to 0$, and we shall accordingly make the ansatz

$$S(l) \simeq S_0 - \sigma \ln l \,. \tag{1.4}$$

The constant σ will be called, following ref. 8, the information dimension. It is always a lower bound to the Hausdorff dimension, and in most cases they are almost the same within numerical errors.

The measure on which we shall concentrate mostly in this paper, has been recently introduced by the present authors [15]. It is obtained from the correlations between random points on the attractor. Consider the set $\{X_i, i = 1 \cdots N\}$ of points on the attractor, obtained e.g. from a time series, i.e. $X_i \equiv X(t + i\tau)$ with a fixed time increment τ between successive measurements. Due to the exponential divergence of trajectories, most pairs (X_i, X_j) with $i \neq j$ will be *dynamically* uncorrelated pairs of essentially random points. The points lie however on the attractor. Therefore they will be spatially correlated. We measure this spatial correlation with the correlation integral $C(l)$, defined according to

$$C(l) = \lim_{N\to\infty} \frac{1}{N^2} \times \{\text{number of pairs } (i,j) \text{ whose distance } |X_i - X_j| \text{ is less than } l\} \,. \tag{1.5}$$

The correlation integral is related to the standard correlation function

$$c(\mathbf{r}) = \lim_{N\to\infty} \frac{1}{N^2} \sum_{\substack{i,j=1 \\ i\neq j}}^{N} \delta^F(X_i - X_j - \mathbf{r}) \tag{1.6}$$

by

$$C(l) = \int_0^l \mathrm{d}^F r \, c(\mathbf{r}) \,. \tag{1.7}$$

One of the central aims of this paper is to establish that for small l's $C(l)$ grows like a power

$$C(l) \sim l^{\nu} \,, \tag{1.8}$$

and that this "correlation exponent" can be taken as a most useful measure of the local structure of a strange attractor. It seems that ν is more relevant, in this respect, than D. In any case, its calculation yields also an estimate of σ and D, since we shall argue that in general one has

$$\nu \leqslant \sigma \leqslant D \,. \tag{1.9}$$

We found that the inequalities are rather tight in most cases, but not in all. Given an experimental signal, if one finds eq. (1.8) with $\nu < F$, one knows that the signal stems from deterministic chaos rather than random noise, since random noise will always result in $C(l) \sim l^F$. Explicit algorithms will be proposed below.

One of the main advantages of ν is that it can easily be measured, at least more easily than either σ or D. This is particularly true for cases where the fractal dimension is large ($\gtrsim 3$) and a covering by small cells becomes virtually impossible. We thus expect that the measure ν will be used in experimental situations, where typically high dimensional systems exist.

In theoretical cases, when the evolution law is known analytically, the easiest quantities to evaluate are the Lyapunov exponents. General formulae expressing D in terms of the λ_i have been proposed by Mori [9] and by Kaplan and Yorke [10]. If they were correct, they would obviously be very useful. They have been verified in simple cases [11, 14]. But Mori's formula was shown to be wrong in one case by Farmer [8], and the above example shown in fig. 1 shows that also the Kaplan–Yorke formula

$$D = D_{KY} \equiv j + \frac{\lambda_1 + \lambda_2 + \cdots + \lambda_j}{|\lambda_{j+1}|} \tag{1.10}$$

does not hold even in all those cases where $\nu = \sigma = D$. Here, the exponents are ordered in descending order $\lambda_1 \geq \lambda_2 \geq \cdots \geq \lambda_F$, and j is the largest integer for which $\lambda_1 + \lambda_1 + \ldots + \lambda_j \geq 0$.

In section 7 we shall take up this question again. We shall show that the counterexample in fig. 1b is not generic. We shall however claim that eq. (1.10) cannot generally be expected to be correct, and that in fact D_{KY} is an upper bound, if $\nu = \sigma = D$.

In the next section, we shall present numerical results for several simple models, for which the fractal dimensions are known from the literature. This will serve to illustrate the scaling law (1.8), and to verify the inequality $\nu \leqslant D$. This inequality and its stronger version, eq. (1.9), will be derived in section 3. The case of one-dimensional maps at infinite bifurcation (Feigenbaum [16]) points is special in that there the information dimension σ and the exponent ν can be calculated exactly, with the result $\nu \neq \sigma \neq D$. It is treated in section 4. Section 5 is dedicated to an important modification which allows to extract ν from a time series of one single variable, instead of from the series $\{X_i\}$. This is of course most important for infinite-dimensional systems, but it is also very useful in low-dimensional cases where it diminishes systematic errors. Among others, we shall apply this method in section 6 to the Mackey–Glass [17] delay equation studied in great detail in ref. 8.

In section 7 we discuss the relation of ν to the Lyapunov exponents, and establish the result

$$\nu \leqslant D_{KY}\,. \tag{1.11}$$

A summary and a discussion of the actual method of treating experimental signals is offered in section 8.

2. Case studies of low-dimensional systems

In this section we shall establish that $C(l)$ can be very well represented by a power law l^ν, by exhibiting numerical results for a number of low dimensional systems. These results are summarized in table I. In section 5 we shall show that this is the case also in high (and infinite) dimensional systems. Details of the numerical algorithms are discussed in appendix A.

2.1. *One-dimensional maps*

The simplest cases of chaotic system are represented by maps of some interval into itself, as e.g. the logistic map [2]

$$x_{n+1} = ax_n(1 - x_n)\,. \tag{2.1}$$

We shall study this map both at the point of onset of chaos via period doubling bifurcations, i.e. when $a = a_\infty = 3.5699456\ldots$ and for the case $a = 4.0$. In fig. 2 we show the result for the first case. It is well known [2, 16] that for this map the attractor* is

* Note that the term "attractor" would not be universally accepted here due to the fact that in any neighbourhood there exist trajectories which do not tend towards it asymptotically.

Table I

	ν	No. of iterations, time increment τ	D	σ
Hénon map $a = 1.4$, $b = 0.3$	1.21 ± 0.01[d]) 1.25 ± 0.02[e])	15000	1.26 (ref. 11)	-
Kaplan–Yorke map $\alpha = 0.2$	1.42 ± 0.02	15000	1.431(ref. 11)	-
Logistic eq., $b = 3.5699456\cdots$	0.500 ± 0.005 $0.4926 < \nu < 0.5024$[f])	25000	0.538(ref. 13)	0.5170976
Lorenz eq.[a])	2.05 ± 0.01	15000; $\tau = 0.25$	2.06 ± 0.01	-
Rabinovich–[b]) Fabrikant eq.	2.19 ± 0.01	15000; $\tau = 0.25$	-	-
Zaslavskii map[c])	(≈ 1.5)	25000	1.39(ref. 11)	-

[a])Parameters as in refs. 7 and 11.
[b])Parameters as in section 3 of ref. 20.
[c])Parameters as in ref. 11.
[d])From eqs. (1.5) and (1.8).
[e])From single variable time series, with $f = 3$.
[f])Exact analytic bound.

Cantor-like with a fractal dimension satisfying the exact bound [13] $0.5376 < D < 0.5386$. In section 4 we shall prove exactly that $\sigma = 0.517097\ldots$, and that $0.4926 < \nu < 0.5024$ while from Fig. 2 we find $\nu = 0.500 \pm 0.005$. For very small distances, the data for $C(l)$ deviate from a power law, but that was to be expected: the behaviour at $a = a_\infty$ is not yet chaotic, and therefore the values x_n are strongly correlated. We verified that indeed the powerlaw holds down to smaller values of l if we increase N or use only values $x_i, x_{i+p}, x_{i+2p}, x_{i+2p}, \ldots$ with p being a large odd number.

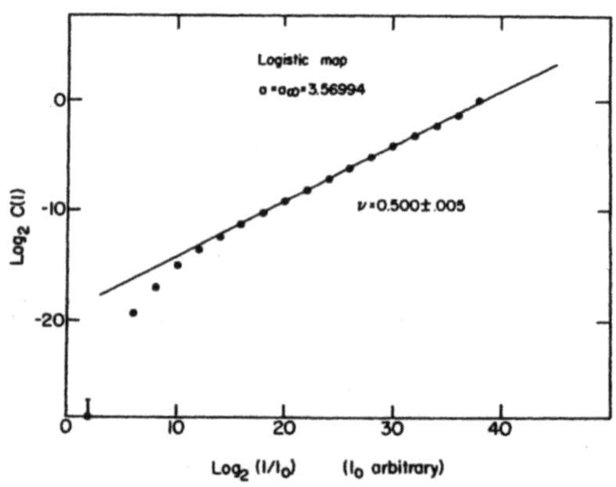

Fig. 2. Correlation integral for the logistic map (2.1) at the infinite bifurcation point $a = a_\infty = 3.699\ldots$ The starting point was $x_0 = \frac{1}{2}$, the number of points was $N = 30.000$.

The same map can be used also to introduce the important issue of corrections to scaling. These are found for the parameter value $a = 4$. It is well known that in this the attractor* consists of the interval [0, 1], and that the invariant probability density is equal to

$$p(x) \equiv \lim_{N\to\infty} \frac{1}{N} \sum_{i=1}^{N} \delta(x_i - x) \tag{2.2}$$

$$= \frac{1}{\pi} [x(1-x)]^{-1/2} \,. \tag{2.3}$$

From this, one finds easily

$$\nu = \sigma = D = 1 \,. \tag{2.4}$$

Notice, however, that while the scaling laws (1.2) and (1.4) are exact, the scaling law (1.8) for $C(l)$

* Again, the term is questionable, as no point outside the interval [0, 1] gets attracted towards it. We shall ignore this irrelevant point, which could be avoided by using $a = 4 - \epsilon$.

requires logarithmic corrections, due to the singular behaviour of $p(x)$:

$$C(l) = \int_0^1\int dx\, dy p(x)p(y)\theta(|x-y|-l) \underset{l\to 0}{\simeq} \frac{4}{\pi^2} l \ln 1/l . \qquad (2.5)$$

Thus, a numerical calculation of ν is expected to converge very slowly. This problem and a remedy for it are discussed further in section 5.

2.2. *Maps of the plane*

Here we examined the Hénon [18] map

$$x_{n+1} = y_n + 1 - ax_n^2 , \quad y_{n+1} = bx_n , \qquad (2.6)$$

with $a = 1.4$ and $b = 0.3$, the Kaplan–Yorke [10] map

$$x_{n+1} = 2x_n \pmod 1 , \quad y_{n+1} = \alpha y_n + \cos 4\pi x_n \qquad (2.7)$$

with $\alpha = 0.2$, and the Zaslavskii [19] map

$$x_{n+1} = [x_n + \nu(1+\mu y_n) + \epsilon\nu\mu \cos 2\pi x_n] \pmod 1, \quad y_{n+1} = e^{-\Gamma}(y_n + \epsilon \cos 2\pi x_n) , \qquad (2.8)$$

with the parameters

$$\mu = \frac{1-e^{-\Gamma}}{\Gamma} \qquad (2.9)$$

and $\Gamma = 3.0$, $\nu = 400/3$, and $\epsilon = 0.3$ taken from ref. 11.

Figs. 3–5 exhibit the results for the correlation integrals. In the first two cases, we find excellent agreement with a power law; while for the Kaplan–Yorke map we find $\nu = 1.42 \pm .02$ in agreement with the published [11] value of D, a fit to the Hénon map yields $\nu_{eff} = 1.21$, smaller than

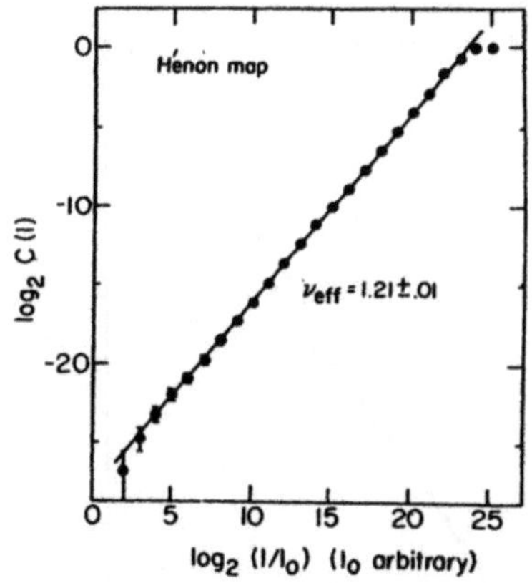

Fig. 3. Correlation integral for the Hénon map (2.6) with $a = 1.4$, $b = 0.03$ and $N = 15.000$.

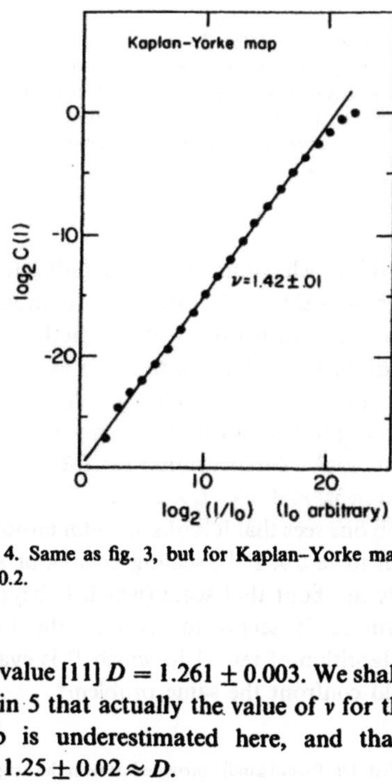

Fig. 4. Same as fig. 3, but for Kaplan-Yorke map (2.7) with $\alpha = 0.2$.

the value [11] $D = 1.261 \pm 0.003$. We shall argue in sectin 5 that actually the value of ν for the Hénon map is underestimated here, and that instead $\nu = 1.25 \pm 0.02 \approx D$.

The case of the Zaslavskii map is exceptional as it was the only system for which we did not find

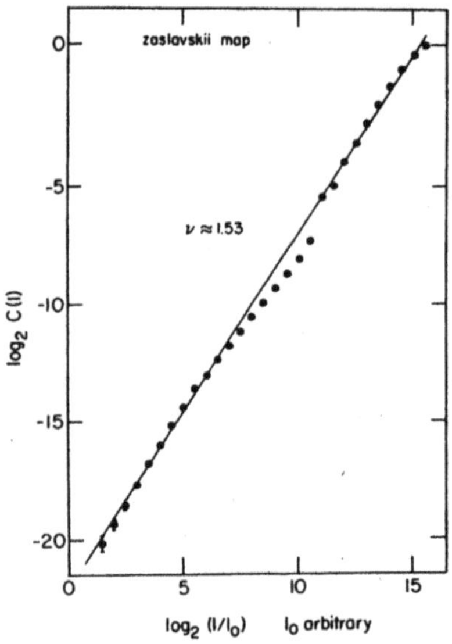

Fig. 5. Correlation integral for Zaslavskii map (eqs. (2.8), (2.9)); $N = 25.000$, parameters as in the text. For faster scaling, the y-coordinate was blown up by a factor of 25, rendering the attractor square-like at low resolution (see fig. 6; without this, the attractor would have looked effectively 1-dimensional for $l \gtrsim l_{max}/25$).

clear-cut power behaviour. Also, an (admittedly poor) fit would yield $\nu \approx 1.5$, in clear violation of the bound $\nu < D$. The reasons why our method has to fail for this map – with the parameters as quoted above – becomes clear when looking at fig. 6. Call l_0 the outer length scale. From fig. 6a one sees that the attractor looks 2-dimensional for $l \gtrsim l_0 \times 2^{-5}$ and $\approx$ 1-dimensional for $l_0 \times 2^{-5} \gtrsim l \gtrsim l_0 \times 2^{-9}$. From fig. 6b one sees that it looks $\approx$ 2-dimensional again down to $\approx l_0 \times 2^{-14}$, scaling behaviour setting in only at about that scale (which is beyond our resolution). It seems to us that the box-counting algorithm of ref. 11 in which D is evaluated, should confront the same problem†.

† Note added: Dr. Russel kindly provided us with the original data of $M(\epsilon)$ versus ϵ. From these, it seems that indeed a similar phenomenon occurs and that accordingly a value $D \approx 1.5$ cannot be excluded.

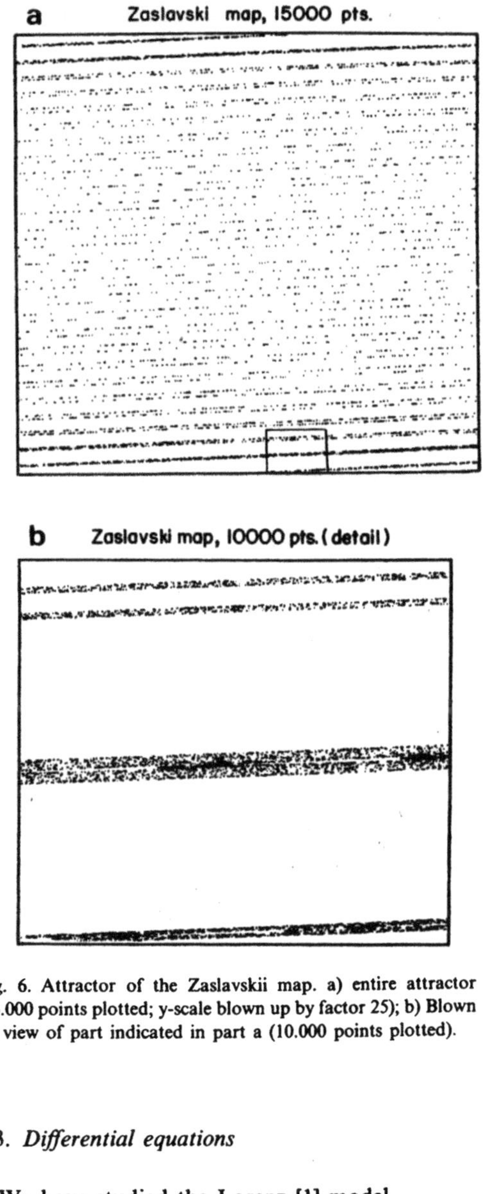

Fig. 6. Attractor of the Zaslavskii map. a) entire attractor (15.000 points plotted; y-scale blown up by factor 25); b) Blown up view of part indicated in part a (10.000 points plotted).

2.3. *Differential equations*

We have studied the Lorenz [1] model

$$\dot{x} = \sigma(y - x),$$
$$\dot{y} = -y - xz + Rx, \qquad (2.10)$$
$$\dot{z} = xy - bz,$$

with $R = 28$, $\sigma = 10$, and $b = 8/3$, and the Rabinovich–Fabrikant [20] equations

$$\begin{aligned} \dot{x} &= y(z - 1 + x^2) + \gamma x, \\ \dot{y} &= x(3z + 1 - x^2) + \gamma y, \\ \dot{z} &= -2z(\alpha + xy), \end{aligned} \tag{2.11}$$

with $\gamma = 0.87$ and $\alpha = 1.1$.

As seen in fig. 7 we get adequate power laws for $C(l)$, and in the case of the Lorenz model, where D is known [11], we obtain $\nu \simeq D$.

Further examples will be studied in section 6, in the context of higher dimensional systems.

It should be stressed that the algorithm used to calculate ν converged quite rapidly. Although each entry in table I and figs. 2–7 were based on $\approx$15.000–25.000 points each, reasonable results (i.e. results for ν within $\pm$ 5%) were obtained in most cases already with only a few thousand points. This should be contrasted with the difficulties associated with estimating D in box-counting algorithms [11, 14].

Summarizing this section, we can say that except for the logistic map at $a = a_\infty$ ("Feigenbaum attractor") we found in all cases that $\nu \approx D$ within the limits of accuracy. We now turn to a theoretical analysis of the relations between ν, σ and D.

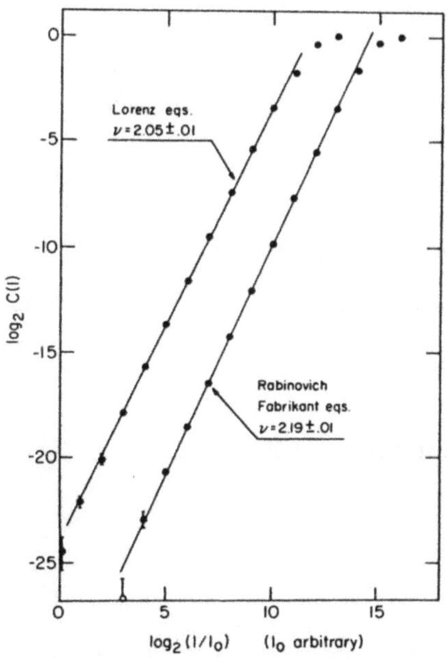

Fig. 7. Correlation integrals for the Lorenz equations (eq. (2.10); dots) and for the Rabinovich–Fabrikant equation (eq. (2.11); open circles). In both cases, $N = 15.000$ and $\tau = 0.25$.

3. Relations between ν, σ and D

In this section we shall establish the inequalities (1.9). We shall do this in 3 steps.

a) The easiest inequality to prove is $\sigma \leqslant D$. Consider a covering of the attractor by hypercubes ("cells") of edge length l, and a time series $\{X_k; k = 1, \ldots, N\}$. The probabilities p_i for an arbitrary X_k to fall into cell i are simply

$$p_i = \lim_{N \to \infty} \frac{1}{N} \mu_i . \tag{3.1}$$

where μ_i is the number of points X_k which fall into cell i.

If the coverage of the attractor is uniform, one has,

$$p_i = \frac{1}{M(l)}, \tag{3.2}$$

where $M(l)$ is the number of cells needed to cover the attractor, and one finds from eqs. (1.3) and (1.2)

$$S(l) = S^{(0)}(l) = \ln M(l) = \text{const} - D \ln l . \tag{3.3}$$

In the general case, one uses the convexity of $x \ln x$ in the usual way to prove that $S(l) \leqslant S^{(0)}(l)$. Invoking the ansatz $S(l) = \text{const} - \sigma \ln l$, we find $\sigma \leqslant D$.

b) Instead of showing immediately $\nu \leqslant \sigma$, let us proceed slowly and show first that $\nu \leqslant D$.

From the definition of $C(l)$, we get up to a factor

of order unity

$$C(l) \simeq \lim_{N\to\infty} \frac{1}{N^2} \sum_{i=1}^{M(l)} \mu_i^2 = \sum_{i=1}^{M(l)} p_i^2 . \tag{3.4}$$

Here, we have replaced the number of pairs with distance $< l$ by the number of pairs which fall into the same cell of length l. The error committed should be independent on l, and thus should not affect the estimation of ν. Using the Schwartz inequality we get

$$C(l) = M(l)\langle p_i^2\rangle \geqslant M(l)\langle p_i\rangle^2 = \frac{1}{M(l)} \sim l^D . \tag{3.5}$$

In this equation square brackets denote average over all cells. Comparing eqs. (3.5) and (1.8) we find immediately $\nu \leqslant D$.

c) In order to derive $\nu \leqslant \sigma$, consider two nested coverings with cubes of lengths l and $2l$. The numbers of cubes that contain a piece of the attractor are then related by

$$M(l) = 2^D M(2l) . \tag{3.6}$$

Denote by p_i the probability to fall in cube i of the finer coverage, and by P_j the probability to fall in cube j of the coarser. Define $\omega_i (i = 1, \ldots M(l))$ by

$$p_i = \omega_i P_j \quad (i \in j) . \tag{3.7}$$

Evidently we have

$$P_j = \sum_{i\in j} p_i, \quad \sum_{i\in j} \omega_i = 1 . \tag{3.8}$$

We can then write the correlation integral as

$$C(l) \simeq \sum_{i=1}^{M(l)} p_i^2 = \sum_{j=1}^{M(2l)} P_j^2 \sum_{i\in j} \omega_i^2 . \tag{3.9}$$

Consider now the ratio

$$\frac{C(l)}{C(2l)} = \frac{\sum_j P_j^2 \sum_{i\in j} \omega_i^2}{\sum_j P_j^2} , \tag{3.10}$$

and compare it to the entropy difference

$$S(2l) - S(l) = \sum_{i=1}^{M(l)} p_i \ln p_i - \sum_{j=1}^{M(2l)} P_j \ln P_j = \sum_{j=1}^{M(2l)} P_j \sum_{i\in j} \omega_i \ln \omega_i . \tag{3.11}$$

In order to estimate eq. (3.10) in terms of eq. (3.11), we have to introduce a new assumption. We assume that the ω_i's are distributed independently of the P_j. This means essentially that locally the attractor looks the same in regions where it is rather dense (P_j large) as in regions where P_j is small. Although we cannot further justify this assumption, it seems to us very natural. It leads immediately to

$$\frac{C(l)}{C(2l)} = \frac{\langle \omega^2\rangle}{\langle \omega\rangle} = 2^D \langle \omega^2 \rangle , \tag{3.12}$$

and to

$$S(2l) - S(l) = 2^D \langle \omega \ln \omega \rangle . \tag{3.13}$$

Define now a normalized variable W by

$$W = \frac{\omega}{\langle \omega \rangle} = 2^D \omega . \tag{3.14}$$

Using the inequality [21]

$$\langle W^2 \rangle > \exp\langle W \ln W \rangle , \tag{3.15}$$

we establish

$$\frac{C(l)}{C(2l)} \geqslant \exp[S(2l) - S(l)] \tag{3.16}$$

and thus

$$\nu \leqslant \sigma . \tag{3.17}$$

Remarks. From the proofs it is clear that if the attractor is uniformly covered, one has equalities

$$\nu = \sigma = D . \tag{3.18}$$

It is an interesting question how non-uniform the coverage must be in order to break them. With the exception of the Feigenbaum map (logistic map with $a = a_\infty$), which is however not generic, all examples of the last section were compatible with eq. (3.18).

In cases where $\nu \neq D$, we claim that indeed ν is the more relevant observable. In these cases, the neighbourhoods of certain points have higher "seniority" in the sense that they are visited more often than others. The fractal dimension is ignorant of seniority, being a purely geometric concept. But both the correlation integral and the entropy dimension weight regions according to their seniority.

Eqs. (1.9) and (3.18) have been used previously in the context of fully developed homogeneous turbulence [22]. The connection

$$c(l) \propto l^{D-F}, \quad l \in R^f$$

following from $\nu = D$ has been used previously also in percolation theory [23] and in a model for dendritic growth [24].

4. Information entropy and ν of the Feigenbaum attractor

In this section we shall compute exactly the information dimension and ν of one-dimensional maps

$$x_{n+1} = F(x_n) \tag{4.1}$$

at the onset of chaos. The method follows closely the one of ref. 13.

It is well known that such maps – provided they have a unique quadratic maximum – have universal scaling features, studied in most detail by Feigenbaum [16]. This behaviour is most easily described by observing that the iterations

$$F^{(2^n)}(x) = \underbrace{F(F(\dots F(x)\dots))}_{2^n \text{ times}} \tag{4.2}$$

tend after a suitable rescaling towards a universal function

$$g(x) = \lim_{n\to\infty} \frac{1}{F^{(2^n)}(0)} F^{(2^n)}(xF^{(2^n)}(0)). \tag{4.3}$$

This "Feigenbaum function" $g(x)$ satisfies the exact scaling relation

$$g(g(x)) = -\frac{1}{\alpha} g(\alpha x), \tag{4.4}$$

with $\alpha = 2.50290\dots$, and the normalization condition $g(0) = 1$. We have here assumed that the maximum of $F(x)$ is at $x = 0$, which can always be achieved by a change of variables. In order to obtain the information dimension of the logistic map at $a = a_\infty = 3.5699345\dots$, it is thus sufficient to compute σ for the Feigenbaum map.

The "attractor" (see the reservations in section 2) of $g(x)$ consists of the sequence $\{\xi_n, n = 0, 1, 2, \dots\}$ with

$$\xi_0 = 0 \tag{4.5}$$

and

$$\xi_{n+1} = g(\xi_n). \tag{4.6}$$

The first few ξ_k's are shown in fig. 8. There, it is also indicated how they build up the Cantorian structure of the attractor: the points $\xi_1, \xi_2, \xi_3, \dots, \xi_{2^k+1}$ form the end-points of 2^k intervals, and the following ξ_k's fall all into these intervals. Furthermore, any sequence $\{\xi_n, \xi_{n+1} \dots \xi_{n+2^k-1}\}$ of 2^k successive points visits each of these intervals exactly once. Thus, the a priori probabilities $p_i(i = 1, \dots, 2^k)$ for an arbitrary x_n to fall into the ith interval are all equal to $p_i = 2^{-k}$.

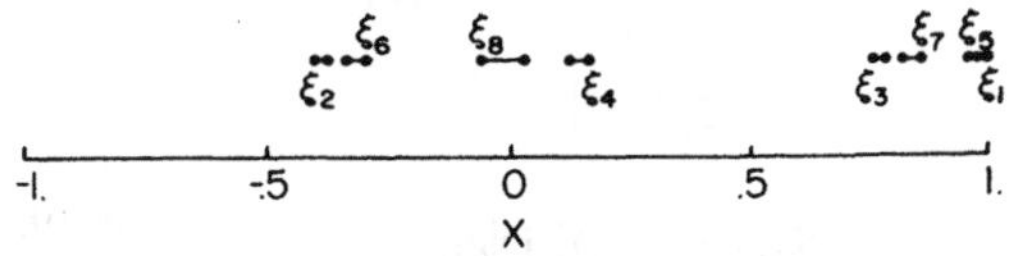

Fig. 8. First 16 points $\xi_1, \dots, \xi_{16}$ of the attractor of the Feigenbaum equation describing the onset of chaos in 1-dimensional systems.

By the grouping axiom, we can first write the information entropy as

$$S(l) = \tfrac{1}{2}[S_{[2,4]}(l) + S_{[3,1]}(l)] + \ln 2\,, \tag{4.7}$$

where we denote by $S_{[i,j]}$ the information needed to specify the point on the interval $[\xi_i, \xi_j]$, and where we have used the fact that an arbitrary x_n has equal probability to be on $[\xi_2, \xi_4]$ or on $[\xi_3, \xi_1]$. From eq. (4.4) we find, however, that

$$\xi_{2n} = -\frac{1}{\alpha}\xi_n\,. \tag{4.8}$$

Thus, the interval $[\xi_2, \xi_4]$ is a down-scaled image of the whole attractor, and we have

$$S_{[2,4]}(l) = S(\alpha l) \approx S(l) - \sigma \ln \alpha\,, \tag{4.9}$$

where we have used the scaling ansatz (1.4).

In order to estimate $S_{[3,1]}(l)$, we decompose the interval [3, 1] into the 2^{k-1} subintervals discussed above, defined by the ξ_n with odd n's:

$$S_{[3,1]}(l) = (k-1)\ln 2 + 2^{-k+1}\sum_{i=1}^{2^{k-1}} S_i(l)\,.$$

Again, we have applied the grouping axiom, using that $p_i = 2^{-k}$. The $S_i(l)$ are the informations needed to pin down x_n provided one knows that it falls into the ith subinterval. Since each subinterval maps onto one on the left-hand piece $[\xi_2, \xi_4]$, each $S_i(l)$ is equal to the information $\tilde{S}_i(|g_i'|l)$ needed to pin x_{n+1} on the corresponding interval on the left-hand side. Here, g_i' is some average derivative of $g(x)$ in the ith subinterval. Using that $\tilde{S}_i(|g_i'|l) \simeq \tilde{S}_i(l) - \sigma \ln|g_i'|$, we obtain

$$\begin{aligned} S_{[3,1]}(l) &= (k-1)\ln 2 + 2^{-k+1}\sum_{i=1}^{2^{k-1}} \tilde{S}_i(l) \\ &\quad - \sigma \sum_{i=1}^{2^{k-1}} \ln|g_i'| \\ &= S_{[2,4]}(l) - \sigma \sum_{i=1}^{2^{k-1}} \ln|g_i'|\,. \end{aligned} \tag{4.10}$$

Inserting this and eq. (4.9) into eq. (4.7), we find after a few manipulations and after taking the limit $k \to \infty$

$$\sigma = \lim_{k\to\infty} \frac{\ln 2}{\ln\alpha + \frac{1}{2^{k+1}}\sum_{i=1}^{2^k} |g'(\xi_{2i-1})|}\,. \tag{4.11}$$

The limit converges very quickly, leading (for $k > 7$) to

$$\sigma = 0.5170976\,. \tag{4.12}$$

The calculation of the correlation exponent, or rather of the exponent of the Renyi entropy (see eq. (3.4))

$$R(l) = \sum_{i=1}^{M(l)} p_i^2 \tag{4.13}$$

follows even more closely the one in ref. 13.

As in that paper, we obtain a nested set of bounds. The first (and least stringent) is obtained by writing

$$R(l) = \tfrac{1}{4}\{R_{[2,4]}(l) + R_{[3,1]}(l)\} \tag{4.14}$$

and using $R_{[2,4]}(l) = R(\alpha.l)$ and $R_{[3,1]}(l) = R(\alpha g' l)$ with

$$|g'(\xi_3)| < g' < |g'(\xi_1)|\,. \tag{4.15}$$

Assuming $R(l) \sim l^\nu$, we obtain

$$1 + |g'(\xi_3)|^\nu < \frac{4}{\alpha^\nu} < 1 + |g'(\xi_1)|^\nu\,, \tag{4.16}$$

leading to $0.4857 < \nu < 0.5235$.

For the next more stringent bounds, we write further

$$R_{[3,1]}(l) = \tfrac{1}{4}\{R_{[3,7]}(l) + R_{[5,1]}(l)\}\,, \tag{4.17}$$

with

$$R_{[3,7]}(l) + R(\alpha^2 g^{(1)}.l), \quad |g'(\xi_3)| < g^{(1)} < |g'(\xi_7)| \tag{4.18}$$

and

$$R_{[5,1]}(l) + R_{[3,1]}(\alpha g^{(2)} l), \quad |g'(\xi_5)| < g^{(2)} < |g'(\xi_1)| . \tag{4.19}$$

Some algebra leads then to

$$|g'(\xi_5)|^\nu + \frac{\alpha^\nu}{4-\alpha^\nu} |g'(\xi_3)|^\nu < \frac{4}{\alpha^\nu}$$
$$< |g'(\xi_1)|^\nu + \frac{\alpha^\nu}{4-\alpha^\nu} |g'(\xi_7)|^\nu , \tag{4.20}$$

with the result

$$0.4926 < \nu < 0.5024 , \tag{4.21}$$

in agreement with the numerical value $\nu = 0.500 \pm 0.005$.

5. Using a single-variable time series

Very often one does not have access to a time series $\{X_n\}$ of F-dimensional vectors. Instead one follows only one or at most a few components of X_n. This is particularly relevant for real (as opposed to computer) experiments where the number of degrees of freedom often is very high if not infinite. Such systems nevertheless can have low-dimensional attractors. It would be very desirable to have a reliable method which allows a characterization of this attractor from a single-variable time series. $\{x_i,\ i = 1, \ldots, N;\ x_i \in R\}$.

The essential idea [25, 26] consists in constructing d-dimensional vectors

$$\xi_i = (x_i, x_{i+1}, \ldots, x_{i+d-1}) \tag{5.1}$$

and using ξ-space instead of X-space. The correlation integral would e.g. be

$$C(l) = \lim_{N\to\infty} \frac{1}{N^2} \sum_{i,j=1}^{N} \theta(l - |\xi_i - \xi_j|) . \tag{5.2}$$

More generally, one can use

$$\xi_i = (x(t_i), x(t_i + \tau) \ldots x(t_i + (d-1)\tau)) , \tag{5.3}$$

with τ some fixed interval. The magnitude of τ should not be chosen too small since otherwise $x_i \approx x_{i+\tau} \approx x_{i+2\tau} \approx \cdots$ so that the attractor in ξ-space would be stretched along the diagonal and thus difficult to disentangle. On the other hand, τ should not be chosen too large since distant values in the time series are not strongly correlated (due to the exponential divergence of trajectories and unavoidable small errors).

A similar compromise must be chosen for the dimension d. Clearly, d must be larger than the Hausdorff dimension D of the attractor (otherwise, $C(l) \sim l^d$). If the attractor is Cantorian in more than one dimension, this might however not be sufficient. Also, it might be that, when looked at in d dimension, the density

$$p(\xi) = \lim_{N\to\infty} \frac{1}{N} \sum_{i=1}^{N} \delta(\xi_i - \xi) \tag{5.4}$$

develops singularities which are absent in more than d dimensions (such singularities occur e.g. when one projects a sphere with constant density, $p(\xi) = p\delta(x^2 + y^2 + z^2 - R^2)$, onto the x–y plane: the new density $\tilde{p}(x, y)$ is infinite at $x^2 + y^2 = R^2$).

On the other hand, one cannot make d too large without getting lost in experimental errors and lack of statistics.

In the next section, we shall study an infinite-dimensional system from this point of view. In the remainder of the present section, we shall apply these considerations to the logistic map with $a = 4$, and to the Hénon map.

In the logistic map, we have seen that there are logarithmic corrections to the power law $C(l) \sim l^\nu$. They result precisely from singularities of $p(x)$, at $x = 0$ and $x = 1$. While embedding the attractor in a higher dimensional space does not completely remove these singularities, it substantially reduces their influence. The reason is that embedding in higher dimensional space always results in stretching the attractor. However, the portions which are most strongly stretched are those which are most densely populated at the lower dimension. For example in the logistic map with $a = 4$ the "attractor" is the interval [0, 1] in 1d but is the parabola in 2d. The parabola has highest slopes at the end points, exhibiting the stronger stretching associated with regions of singular distributions at a lower dimension. A similar effect appears when going from $d = 2$ to $d = 3$. We thus expect that the importance of the singularities in the distribution would be reduced in higher dimensions.

In order to check this, we have calculated for the logistic map at $a = 4$ the original correlation integral and the modified integral obtained by embedding in a 2- and 3-dimensional space. The results are shown in fig. 9. We observe indeed the expected decrease of systematic error when increasing d, accompanied by an increase of the statistical error.

Analogous results for the Hénon map are shown in fig. 10. There, we used as time series the series $\{x_n, x_{n+2}, x_{n+4}, \ldots\}$. While the 2-dimensional correlation integral gives an effective ν in agreement with the result of section 2, the 3-dimensional embedding gives a larger values $\nu = 1.25 \pm 0.02$ which agrees with the value of D found in refs. 11 and 14.

No such effects were observed in the Lorenz model, where both the originally defined $C(l)$ and

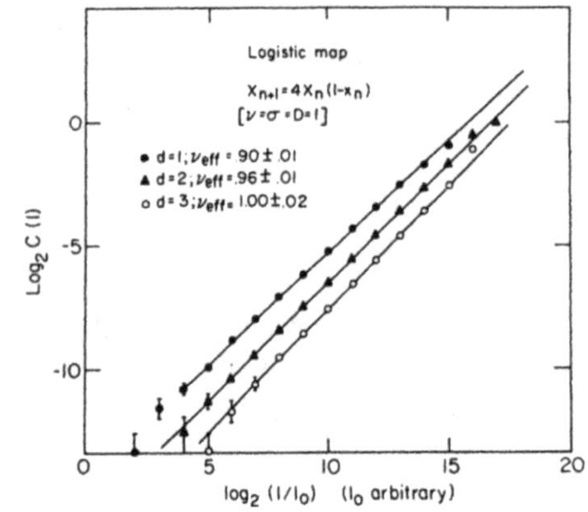

Fig. 9. Modified correlation integrals for the logistic map (2.1) with $a = 4$. The distance l between 2 points ξ_n and ξ_m on the attractor is defined as $l^2 = (\xi_n - \xi_m)^2 = (x_n - x_m)^2 + \ldots + (x_{n+d-1} - x_{m+d-1})^2$. For each value of d, we took $N = 15.000$.

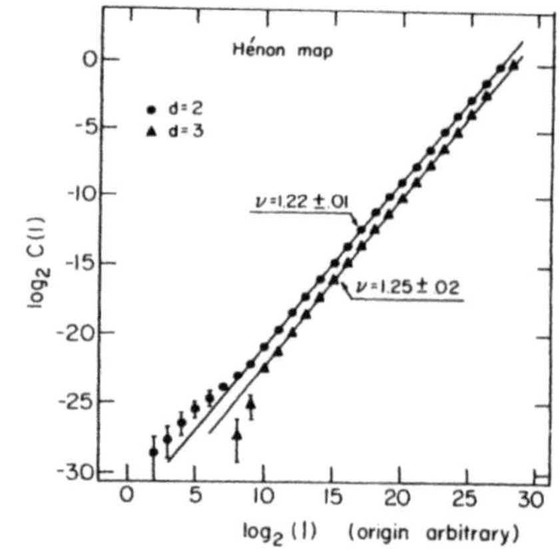

Fig. 10. Modified correlation integrals for the Hénon map (2.6). The time series consisted of coordinates x_n, x_{n+2}, $x_{n+4}, \ldots$, and $\xi = (x_n, x_{n+2}, \ldots, x_{n+2(d-1)})$ for each d. For $d = 2$, we took $N = 30.000$; for $d = 3$, we took $N = 20.000$.

the modified correlation integral using only a single coordinate time series gave values of ν which agreed with D [15].

The conclusion drawn from these examples is that it is often useful to represent the attractor in a higher dimensional space than absolutely necessary, in order to reduce systematic errors. These errors result from a strongly non-uniform coverage of the attractor, provided this non-uniformity is not so strong as to make $\nu \neq D$.

6. Infinite-dimensional systems: an example

An extremely convenient way of generating very high dimensional systems is to consider delay differential equations of the type

$$\frac{dx(t)}{dt} = F(x(t), x(t-\tau)), \tag{6.1}$$

where τ is a given time delay. Such a delay equation is in fact infinite dimensional, as is most easily seen from the initial conditions necessary to solve eq. (6.1): they consist of the function $x(t)$ over a whole interval of length τ.

Following ref. 8, we shall study a particular example, introduced by Mackey and Glass [17] as a model for regeneration of blood cells in patients with leukemia. It is

$$\dot{x}(t) = \frac{ax(t-\tau)}{1+[x(t-\tau)]^{10}} - bx(t). \tag{6.2}$$

As in ref. 8, we shall keep $a = 0.2$ and $b = 0.1$ fixed, and study the dependence on the delay time τ.

For the numerical investigation, eq. (6.2) is turned into an n-dimensional set of difference equations, with $n = 600$–1200. Details are described in the appendix. The time series was always chosen as $\{x(t), x(t+\tau), x(t+2\tau), \ldots\}$ except for some runs with $\tau = 100$, where we took points at times t, $t+\tau/2$, $t+2\tau/2, \ldots$.

The results for the correlation integral are shown in figs. 11–14. Estimated values of ν are given in table II, together with values of D obtained in ref. 8 by applying the defining eq. (1.2) to a Poincaré return map. Also shown in table II are the

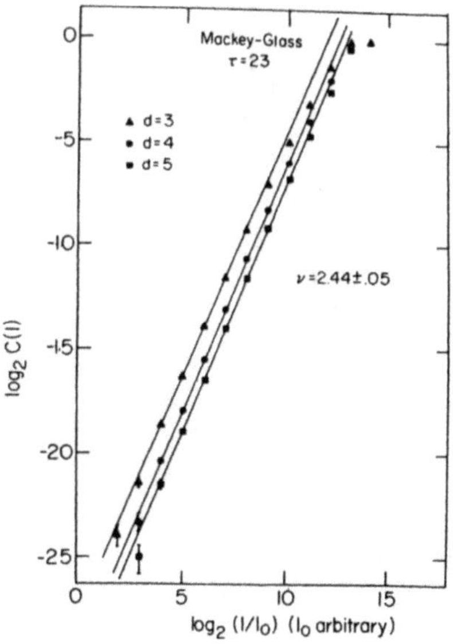

Fig. 12. Same as fig. 11, but for $\tau = 23$.

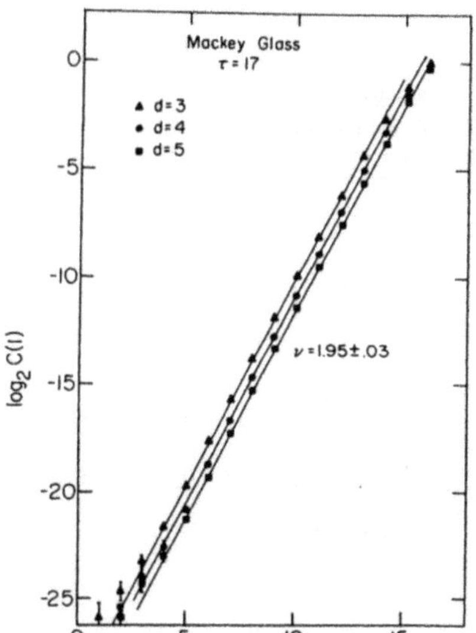

Fig. 11. Modified correlation integrals for the Mackey–Glass delay equation (6.2), with delay $\tau = 17$. The time series consisted of $\{X(t+i\tau);\ i = 1, \ldots, 25.000\}$.

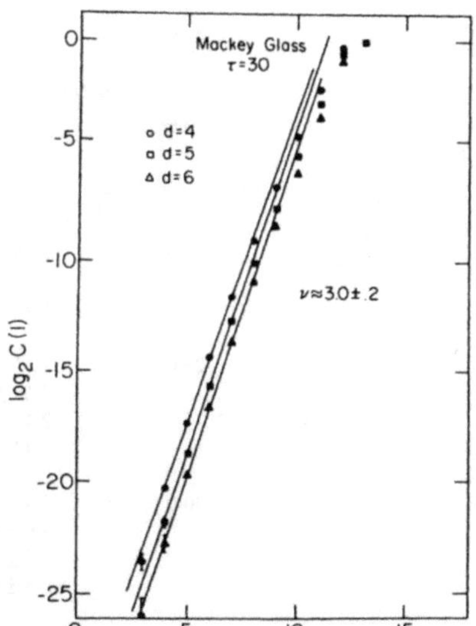

Fig. 13. Same as fig. 11, but for $\tau = 30$.

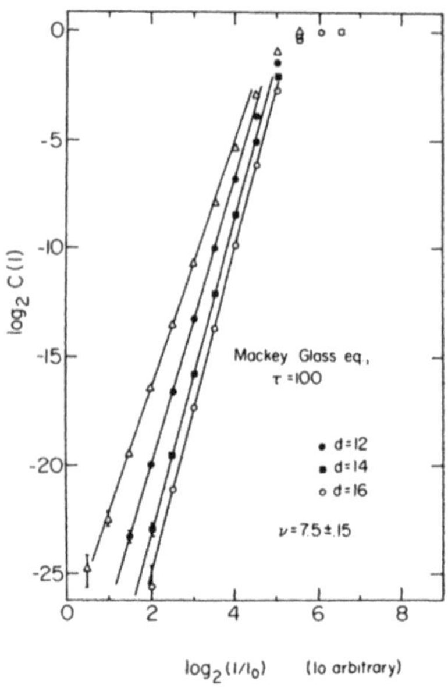

Fig. 14. Same as fig. 11, but for $\tau = 100$. For $d = 16$, the time series consisted of points $\{X(t + i\tau/2);\ i = 1, \ldots, 25.000\}$.

Kaplan–Yorke dimension D_{KY} (see eq. (1.10)) which will in the next section be shown to be an upper bound to ν, and the number of positive Lyapunov exponents, both taken from ref. 8. It is obvious that this latter number, called D_{LB}, is a lower bound to D. If the density of trajectories on the attractor is not too non-uniform, we expect that D_{LB} yields also a lower bound to ν.

From table II we see that indeed in all cases

$$D_{LB} \leqslant \nu \leqslant D \leqslant D_{KY}, \tag{6.3}$$

except for $\tau = 17$ where ν is slightly less than D_{LB}. However, for those small values of τ for which box-counting according to the definition of D had been feasible, our values of ν are considerably smaller than the values of D found in ref. 8, while the values of D were fairly close to D_{KY}.

In all cases, the linearity of the plot of log $C(l)$ versus log l improved substantially when increasing d above its minimal required value. For increasing values of d, the effective exponent at first also

Table II
Estimates of the correlation exponent ν for the Mackey–Glass equation (6.2) with $a = 0.2$, $b = 0.1$. Values for D_{LB}, D and D_{KY} are from ref. 8. For $\tau = 100$ the value of ν saturated at $d = 16$

τ	D_{LB}	ν	D	D_{KY}
17.0	2	1.95 ± 0.03 ($d = 3$)	2.13 ± 0.03	2.10 ± 0.02
		1.35 ± 0.03 ($d = 4$)		
		1.95 ± 0.03 ($d = 5$)		
23.0	2	2.38 ± 0.15 ($d = 3$)	2.76 ± 0.06	2.92 ± 0.03
		2.43 ± 0.05 ($d = 4$)		
		2.44 ± 0.05 ($d = 5$)		
		2.42 ± 0.1 ($d = 6$)		
30.0	3	2.87 ± 0.3 ($d = 4$)	> 2.94	3.58 ± 0.04
		3.0 ± 0.2 ($d = 5$)		
		3.0 ± 0.2 ($d = 6$)		
		2.8 ± 0.3 ($d = 7$)		
100.0	6	5.8 ± 0.3 ($d = 10$)		≈ 10.0
		6.6 ± 0.2 ($d = 12$)		
		7.2 ± 0.2 ($d = 14$)		
		7.5 ± 0.15 ($d = 16$)		

increases, but settles at a value which we assume to be the true value of ν. We must stress that we have no *proof* that the values of ν obtained with the highest chosen d represent the "true" exponent. We feel however that they surely represent reasonable estimates even for attractors with dimensions as high as ≈ 7.

In real experiments, where Lyapunov exponents are not available and thus D_{LB} and D_{KY} not easily obtained, our method seems the only one which could distinguish such an attractor from a system where the stochasticity is due to random noise. In that case, one would expect $C(l) \sim l^d$ as the trajectory is space-filling, in clear distinction from what we observe.

7. Relation to Lyapunov exponents and the Kaplan–Yorke conjecture

As we already mentioned in the introduction, the Lyapunov exponents are related to the evolution of the shape of an infinitesimal F-dimensional ball in phase space: being infinitesimal, it depends only on the linearized part of the flow, and thus becomes an ellipsoid with exponentially shrinking or growing axes. Denoting the principal axes by $\epsilon_i(t)$, the Lyapunov exponents are given by

$$\lambda_i = \lim_{t\to\infty} \lim_{\epsilon_i(0)\to 0} \frac{1}{t} \ln \frac{\epsilon_i(t)}{\epsilon_i(0)} . \tag{7.1}$$

Directions associated with positive Lyapunov exponents are called "unstable", those associated with negative exponents are called "stable".

Originally [10], Kaplan and Yorke had conjectured that D_{KY} is equal to D. In a recent preprint [27], they claim that D_{KY} is generically equal to a "probabilistic dimension", which seems to be the same as σ.

This latter claim has been partially supported in ref. 28, where essentially D_{KY} is proven to be an upper bound to the probabilistic dimension.

As shown by the counter example mentioned in the introduction, there are (possibly exceptional) cases where this bound is not saturated. In this section, we shall elucidate this question by giving a heuristic proof for the inequality $\nu \leqslant D$. From this, we see necessary conditions for the Kaplan–Yorke conjecture to hold, and which do not seem to be met generally.

Consider two infinitesimally close-by trajectories $X(t)$ and $X'(t) = X(t) + \Delta(t)$, where the latter could indeed be $X'(t) = X(t+T)$, which for sufficiently large T is essentially independent of $X(t)$. We assume that $\Delta_i(t)$ increase exponentially, without any fluctuations, as

$$\Delta_i(t) = \Delta_i(0)\, e^{\lambda_i t} , \tag{7.2}$$

where the components are along the principal axes discussed above. This is of course a strong assumption which would imply, in particular, that $\nu = \sigma = D$. Corrections to it will be treated in a forthcoming paper, but our main conclusion will remain unchanged. Conservation of the number of trajectories implies that the correlation function increases like

$$c(\Delta(t)) = \left| \frac{\partial(\Delta(0))}{\partial(\Delta(t))} \right| c(\Delta(0)) = e^{-t\Sigma_{i=1}^{F}\lambda_i} c(\Delta(0)) . \tag{7.3}$$

To proceed further, we need a scaling assumption which generalizes the scaling ansatz

$$c(|\Delta|) \sim |\Delta|^{\nu - F} . \tag{7.4}$$

Observing that the attractor is locally a topological product of an R^n with Cantor sets, and that the relevant axes are the principal axes, we associate with each axis an exponent ν_i, $0 < \nu_i < 1$, and make the ansatz

$$c(\Delta) \approx \prod_{i=1}^{F} c_i(\Delta_i) , \tag{7.5}$$

with

$$c_i(x) \propto \begin{cases} x^{\nu_i - 1}, & \text{if} \quad 0 < \nu_i \leqslant 1 , \\ \delta(x), & \text{if} \quad \nu_i = 0 . \end{cases} \tag{7.6}$$

If $\nu_i = 0$, this means that the motion along this axis dies asymptotically (example: directions normal to a limit cycle). Directions with $\nu_i = 1$ are the unstable directions, with the continuous density. Directions with $0 < \nu_i < 1$, finally, are either Cantorian or, in exceptional cases, directions along which the distribution is continuous but singular at $\Delta_i = 0$. Notice that $\nu_i > 1$ is impossible.

Substituting eq. (7.5) into (7.3), we find

$$\prod_i (\Delta_i(0)^{\nu_i - 1} e^{t\lambda_i(\nu_i - 1)}) = e^{-t\Sigma_i\lambda_i} \prod_i \Delta_i(0)^{\nu_i - 1}, \qquad (7.7)$$

or

$$\sum_{i=1}^{F} \lambda_i \nu_i = 0. \qquad (7.8)$$

In addition we have, from eqs. (7.5) and (7.4),

$$\sum_{i=1}^{F} \nu_i = \nu, \qquad (7.9)$$

and

$$0 \leqslant \nu_i \leqslant 1. \qquad (7.10)$$

It is now easy to find the maximum of ν subject to the constraints (7.8)–(7.10). It is obtained when

$$\nu_i = \begin{cases} 1, & \text{for } i \leqslant j, \\ 0, & \text{for } i \geqslant j+2, \end{cases} \qquad (7.11a)$$

and

$$\nu_{j+1} = \frac{1}{|\lambda_{j+1}|} \sum_{i<j} \lambda_i. \qquad (7.11b)$$

Here, we have used that $\lambda_1 \geqslant \lambda_2 \geqslant \ldots$, and that $\Sigma_j \lambda_i < 0$. Expressed in words, the distribution (7.11) means that the attractor is the most extended along the most unstable directions. Inserting it into eq. (7.9), we obtain

$$\nu \leq j + \frac{\sum_{i \leq j} \lambda_i}{|\lambda_{j+1}|} \equiv D_{KY}. \qquad (7.12)$$

as we had claimed.

From the derivation it is clear that the Kaplan–Yorke conjectures $\sigma = D_{KY}$ or $D = D_{KY}$ cannot be expected to hold when *either the attractor is Cantorian in more than one dimension, or if the folding occurs in a direction which is not the minimally contracting one.* The latter was indeed the case for example b in fig. 1. But example b of fig. 1 is not generic, the generic case being the one where the folding is in a plane which encloses an arbitrary angle ϕ with the z-axis (see fig. 15). It is easy to convince oneself that $D = D_{KY}$ whenever $\phi \neq 0$, i.e. nearly always. A still more general case is obtained if we fold in each $(2n)$th iteration in a plane characterized by ϕ_1, and each $(2n+1)$st iteration in a different plane. Again, it seems that $D = D_{KY}$ is generic.

The examples might suggest that indeed $D = D_{KY}$ in all those generic cases in which $\nu = D$, but we consider it as not very likely in high-dimensional cases. For invertible two-dimensional maps, the above conditions are of course satisfied, and thus $\sigma = D_{KY}$ if $\nu = \sigma = D$ (see ref. 29).

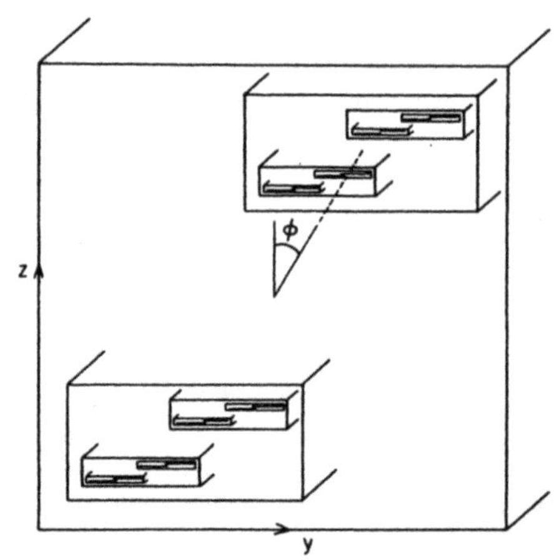

Fig. 15. Cross section through a rectangular volume element and its first 4 iterations under a map which stretches in x-direction, contracts in y- and z-directions (factors $\frac{1}{2}$ and $\frac{1}{4}$, respectively), and folds back under an angle ϕ with respect to the z-direction.

8. Conclusions

The theoretical arguments of section 3 and 6 of this paper have shown (though not with mathematical rigour) that the correlation exponent ν introduced in this paper is closely related to other quantities measuring the local structure of strange attractors.

The numerical results presented in section 2, 5 and 7 have yielded proof that ν can indeed be calculated with reasonable efforts. While all results presented in this paper were based on time series of 10.000–30.000 points, reasonable estimates of ν can already be obtained with series of a few thousand points, in most cases. Surely, for higher dimensional attractors one needs longer time series. However, rather than taking longer time series, we found it in general more important to embed the attractor in higher dimensional spaces, and to choose this embedding dimension judiciously. Compared to box-counting algorithms used previously by other authors, our method has two advantages: First, our storage requirements are drastically reduced. Secondly, in a box-counting algorithm one should iterate until *all* non-empty boxes of a given size l have been visited. This is clearly impractical, in particular if l is very small. Thus, one has systematic errors even if the number of iterations N is excessively large. In our method, there is no such problem. In particular, the finiteness of N induces no systematic errors beyond the corrections to the scaling law $C(l) \sim l^\nu$.

We found that in most cases ν was very close to the Hausdorff dimensions D and to the information dimension σ, with two notable exceptions. One was the Feigenbaum map, corresponding to the onset of chaos in 1 dimension. In that case, we were able to compute σ exactly in an analytic way, with the result $\sigma \neq D$, supporting the numerical evidence for $\nu < \sigma$.

The other exception was the Mackey–Glass delay equation, where we found numerically $\nu < D$. The information dimension has not been calculated directly in this case. Accepting the claim made in ref. 8 that the Kaplan–Yorke formula (1.10) predicts correctly σ, we would have $\nu < \sigma = D = D_{KY}$. This seems somewhat surprising, since we argued in section 7 that a rather direct connection (as an inequality $\nu \leqslant D_{KY}$) exists between ν and D_{KY}, while a connection between σ and D_{KY} seems less evident to us.

The main conclusion of this paper, as far as experiments are concerned, is that one can distinguish deterministic chaos from random noise. By analyzing the signal as explained in section 5, and embedding the attractor in an increasingly high dimensional space, one finds whether $C(l)$ scales like l^ν or l^d. With a random noise the slope of log $C(l)$ vs. log l will increase indefinitely as d is increased. For a signal that comes from a strange attractor the slope will reach a value of ν and will then become d independent.

An issue of experimental importance is the effect of random noise *on top* of the deterministic chaos. The treatment of this question is beyond the scope of this paper and is treated elsewhere [30]. Here we just remark that when there is an external noise of a given mean square magnitude, a plot of log $C(l)$ vs. log l has two regions. For length scales above those on which the random component blurrs the fractal structure, $C(l)$ continues to scale like l^ν. On length scales below those that are affected by the random jitter of the trajectory, $C(l)$ scales like l^d. The analysis of experimental signals along these lines can therefore yield simultaneously a characterization of the strange attractor *and* and estimate of the size of the random component. For more details see ref. 30.

It is thus our hope that the correlation exponent will indeed be measured in experiments whose dynamics is governed by strange attractors.

Acknowledgements

This work has been supported in part by the Israel Comission for Basic Research. P.G. thanks the Minerva Foundation for financial support. We thanks Drs. H.G.E. Hentschel and R.M. Mazo for a number of useful discussions.

Appendix A

All numerical calculations were performed in double precision arithmetic on an IBM 370/165 at the Weizmann Institute.

The integrations of the Lorenz and Rabinovich–Fabrikant equations were done using a standard Merson–Runge–Kutta subroutine of the NAG library.

In order to integrate the Mackey–Glass delay equation we approximated it by a N-dimensional set of difference equations by introducing a time step

$$\Delta t = \tau / n \,, \tag{A.1}$$

with n being some large integer, and writing

$$x(t + \Delta t) \approx x(t) + \frac{\Delta t}{2}(\dot{x}(t) + \dot{x}(t + \Delta t)) \,. \tag{A.2}$$

Notice that this, being the optimal second-order approximation, is a very efficient algorithm – provided we can compute $\dot{x}(t + \Delta t)$. In the present case we can, due to the special form

$$\dot{x}(t) = f(x(t - \tau)) - bx(t) \,. \tag{A.3}$$

Inserting this in eq. (A.2) and rearranging terms, we arrive at

$$x(t + \Delta t) = \frac{2 - b\Delta t}{2 + b\Delta t} x(t) + \frac{\Delta t}{2 + b\Delta t} \times \{ f(x(t - \tau)) + f(x(t - \tau + \Delta t)) \} \,. \tag{A.4}$$

In all runs shown in this paper, we used $n = 600$ (corresponding to $0.03 \lesssim \Delta t \lesssim 0.15$), except for the runs with $\tau = 100$, where we used $n = 1200$ and with $n = 600$, finding no appreciable differences.

We also performed control runs with a fourth-order approximation instead of eq. (A.2). The correlation integral was unchanged within statistical errors, and the stability of the solutions did not seem to improve much. This could result from the very large higher derivatives of x, resulting from the tenth power in eq. (6.2).

In order to ensure that all x_i are on the attractor, the first 100–200 iterations were discarded.

Generating the time series $\{X_i\}_{i=1}^{N}$ was indeed the less time-consuming part of our computation, the more important part consisting of calculating the $N(N-1)/2 \gtrsim 10^8$ pairs of distances $r_{ij} = |X_i - X_j|$ and summing them up to get the correlation integral.

In particular, we found that an efficient algorithm for the latter was instrumental in applying the method advocated in this paper.

Such a fast algorithm was found using the fact that floating-point numbers are stored in a computer in the form

$$r = \pm \text{ mantissa} \cdot \text{base}^{+\text{exp}} \,. \tag{A.5}$$

with base = 16 in our case 1/base < mantissa < 1, and exp being an integer. If one can extract the exponent, one can bin the r_{ij}'s in bins of widths increasing geometrically. By extracting the exponent of an arbitrary power r^p of r, one can furthermore choose the width of this binning arbitrarily. Access to the exponent is made very easy and fast by using the shifting and masking operations available e.g. in extended IBM and in CDC Fortran. After having computed the numbers N_K of pairs (i, j) in the interval $2^{k-1} < r_{ij} < 2^k$, the correlation integrals are obtained by

$$c(r = 2^k) = \frac{1}{N^2} \sum_{k'=-\infty}^{k} N_{k'} \,. \tag{A.6}$$

We found this method to be nearly an order of magnitude faster than computing e.g. the logarithmics of r_{ij} directly, and binning by taking their integer parts. A typical run with 20.000 points took – depending on the model studied – between 15 and 30 minutes CPU time.

References

[1] E.N. Lorenz, J. Atmos. Sci. 20 (1963) 130.
[2] R.M. May, Nature 261 (1976) 459.
[3] D. Ruelle and F. Takens, Commun. Math. Phys. 20 (1971) 167.

[4] E. Ott, Rev. Mod. Phys. 53 (1981) 655.
[5] J. Guckenheimer, Nature 298 (1982) 358.
[6] B. Mandelbrot, *Fractals – Form, Chance and Dimension* (Freeman, San Francisco, 1977).
[7] V.I. Oseledec, Trans. Moscow Math. Soc. 19 (1968) 197. D. Ruelle, Proc. N.Y. Acad. Sci. 357 (1980) 1 (R.H.G. Helleman, ed.).
[8] J.D. Farmer, Physica 4D (1982) 366.
[9] H. Mori, Progr. Theor. Phys. 63 (1980) 1044.
[10] J.L. Kaplan and J.A. Yorke, in: Functional Differential Equations and Approximations of Fixed Points, H.-O. Peitgen and H.-O. Walther, eds. Lecture Notes in Math. 730 (Springer, Berlin, 1979) p. 204.
[11] D.A. Russel, J.D. Hanson and E. Ott, Phys. Rev. Lett. 45 (1980) 1175.
[12] H. Froehling, J.P. Crutchfield, D. Farmer, N.H. Packard and R. Shaw, Physica 3D (1981) 605.
[13] P. Grassberger, J. Stat. Phys. 26 (1981) 173.
[14] H.S. Greenside, A. Wolf, J. Swift and T. Pignataro, Phys. Rev. A25 (1982) 3453.
[15] P. Grassberger and I. Procaccia, Phys. Rev. Lett. 50 (1983) 346. Related discussions can be found in a preprint by F. Takens "Invariants Related to Dimensions and Entropy".
[16] M. Feigenbaum, J. Stat. Phys. 19 (1978) 25; 21 (1979) 669.
[17] M.C. Mackey and L. Glass, Science 197 (1977) 287.
[18] M. Hénon, Commun. Math. Phys. 50 (1976) 69.
[19] G.M. Zaslavskii, Phys. Lett. 69A (1978) 145.
[20] M.I. Rabinovich and A.L. Fabrikant, Sov. Phys. JETP 50 (1979) 311. (Zh. Exp. Theor. Fiz. 77 (1979) 617).
[21] W. Feller, An Introduction to Probability Theory and its Applications, vol. 2, 2nd ed. (Wiley, New York, 1971) p. 155.
[22] B.B. Mandelbrot, in: *Turbulence and the Navier–Stokes Equations*, R. Teman, ed., Lecture Notes in Math. 565 (Springer, Berlin, 1975). H.G.E. Hentschel and I. Procaccia, Phys. Rev. A., in press.
[23] D. Stauffer, Phys. Rep. 54C (1979) 1.
[24] T.A. Witten, Jr., and L.M. Sander, Phys. Rev. Lett. 47 (1981) 1400.
[25] N.H. Packard, J.P. Crutchfield, J.D. Farmer and R.S. Shaw, Phys. Rev. Lett. 45 (1980) 712.
[26] F. Takens, in: Proc. Warwick Symp. 1980, D. Rand and B.S. Young, eds, Lectures Notes in Math. 898 (Springer, Berlin, 1981).
[27] P. Frederickson, J.L. Kaplan, E.D. Yorke and J.A. Yorke, "The Lyapunov Dimension of Strange Attractors" (revised), to appear in J. Diff. Eq.
[28] F. Ledrappier, Commun. Math. Phys. 81 (1981) 229.
[29] L.S. Young, "Dimension, Entropy, and Lyapunov Exponents" preprint.
[30] A. Ben-Mizrachi, I. Procaccia and P. Grassberger, Phys. Rev. A, submitted.

Invariant Measures and the Variational Principle for Lozi Mappings

Marek Rychlik

Department of Mathematics
University of Arizona
Tucson, AZ 85721 USA
phone: (602) 621-6864, FAX: (602) 621-8322
rychlik@math.arizona.edu

Summary. This article contains a proof of the existence of SBR measures for the family of maps of the plane known as Lozi maps: $L(x,y) = (1 - a|x| + y,\ bx)$. We also prove that the number of SBR measures is finite. Our approach also yields invariant measures for other classes of dynamical systems like piecewise expanding mappings of an interval with infinitely many pieces and Hölder derivative. Although known to specialists from my 1983 dissertation (University of California, Berkeley), the results presented in this article have never been published. I am delighted that the article found its place in the current volume, as the ideas of several papers that Jim Yorke co-authored or inspired played an important role in my approach.

1 Introduction

1.1 Preliminary Remarks

The main goal of this paper is the study of Lozi mappings introduced in (cf. [9]). Lozi noticed the chaotic behavior of the orbits of these mappings, similar to that of the Hénon mappings (cf. [5]). The formulas for Lozi mappings

$$L(x,y) = (1 - a|x| + y,\ bx) \tag{1}$$

and for Hénon mappings

$$H(x,y) = (1 - ax^2 + y,\ bx) \tag{2}$$

show some qualitative similarity. However, Lozi mappings are much easier to study because they admit uniform hyperbolic structure (cf. [10]). The only complication is the presence of the singularity $x = 0$. It makes Lozi mappings similar to the dispersed billiards of Bunimovich and Sinai. Using this analogy, Collet (cf. [3]) proved similar results to ours by constructing Markov partitions for Lozi mappings, repeating the method of Bunimovich and Sinai (cf. [2]) (however for a smaller range of parameters).

We use a different analogy. Lozi mappings are similar to maps of an interval. More precisely, if we build a factor map, utilizing the partition into stable intervals, then this factor map has properties formulated as conditions I–IV in section 2 of the paper, which are satisfied for piecewise expanding, $C^{1+\epsilon}$ mappings of an interval.

Pursuing this analogy, in section 2 we prove the existence of absolutely continuous invariant measures for a map satisfying conditions I–IV. In general, the map has more than one such measure (and this occurs for certain cases of iterations of Lozi mappings). We need a spectral decomposition theorem (Theorem 2) which gives all *ergodic* absolutely continuous measures. Actually, without much trouble (using the Pinsker partition) we could prove that the natural extension of this factor map (which for Lozi mappings coincides with L itself) has a finite number of K-components. From Pesin theory and from section 3, which gives a detailed description of invariant measures for Lozi mappings, it follows that L is Bernoulli on the K-components.

Misiurewicz in (cf. [10]) proved that L has a strange attractor for certain parameters a and b. We repeat his considerations in section 3, slightly modifying his trapping region and the methods of the proofs. In section 3 we also prove that a large class of measures (including the Lebesgue measure restricted to the trapping region) satisfy an ergodic theorem (Theorem 4).

We also prove a version of Bowen's theorem for functions (Theorem 5).

In section 4 we prove a variational principle for the measures found in section 2. This result holds in the form suggested by Bowen and Ruelle (cf. [1]). We could also say the same in Pesin's terms: For every invariant measure $\mu, h_\mu(L)$ is not larger than the positive Lyapunov exponent; $h_\mu(L)$ equals the Lyapunov exponent iff μ is one of the measures constructed in this paper.

The author hopes that the methods developed in this paper will be useful in studying other dynamical systems, in particular Hénon mappings.

There is one generalization of the results which is almost obvious: instead of L we can consider a small perturbation $L + f$ where $f \in C^{1+\epsilon}(\mathbb{R}^2)$. The results remain valid. What makes it possible is that the iterations of the singularities under the new map behave similarly to those for L.

We notice that section 2 creates a new class of examples of piecewise expanding mappings of class $C^{1+\epsilon}$ on the pieces and with infinite number of pieces. The case of finitely many pieces was considered in (cf. [7] and in one unpublished paper of the author).

2 Abstract Theorems

2.1 Abstract Reduction Theorem

Let (X, Σ, m) be a Lebesgue space with a σ-algebra Σ and a probability measure m. Let $T : X \to X$ be a measurable, nonsingular mapping. The latter means that $T_* m$ is absolutely continuous with respect to m (abbr. $T_* m \ll m$).

We define the Perron-Frobenius operator of T. We have $P_T : L^1(X, \Sigma, m) \to L^1(X, \Sigma, m)$ and:

$$P_T f = \frac{d(T_*(fm))}{dm} \qquad \text{(Radon-Nikodym derivative).} \tag{3}$$

It is easy to see that P_T is determined by the following condition:

$$\forall h \in L^\infty(X, \Sigma, m) : \int_X (h \circ T) \cdot f dm = \int_X h \cdot P_T f dm. \tag{4}$$

This means that $T : L^\infty(X, \Sigma, m) \to L^\infty(X, \Sigma, m)$ defined by $Th = h \circ T$ is the adjoint operator of P_T.

A measurable, countable partition α of X is called regular (or T-regular) iff for every $A \in \alpha$, $T(A)$ is Σ-measurable and $T|A$ maps $(A, \Sigma|A)$ onto $(T(A), \Sigma|T(A))$ isomorphically (spaces with fixed σ-algebras, where $\Sigma|A$ and $\Sigma|T(A)$ are σ-algebras defined by the restriction of Σ to A and $T(A)$ respectively).

For any regular partition we define $g_T : X \to \mathbb{R}_+$ as follows:

$$g_T(x) = \frac{d(T_*(\chi_A \cdot m))}{dm}(Tx) \quad \text{as } x \in A \in \alpha. \tag{5}$$

In other words, $g_T = \sum_{A \in \alpha} (TP)(\chi_A)\chi_A$. One can check easily that g_T does not depend on the choice of α. Moreover, g_T is the reciprocal of the Jacobian of T. Using g_T, we can express P_T in the following way:

$$P_T f(x) = \sum_{y \in T^{-1}(x)} g_T(y) \cdot f(y), \quad x \in X. \tag{6}$$

Of course, g_T is determined up to a set of measure 0 and (6) holds m-a. e.

We study in detail the case where T is a factor of another mapping $S : Y \to Y$, where (Y, β, ν) is a Lebesgue space. We assume that S is nonsingular. By ξ we denote the measurable partition of Y which is S-invariant ($S^{-1}\xi \le \xi$) and $X = Y/\xi$ and $T = S_\xi$ (factor map). Let $\pi : Y \to X$ be the natural projection. Let $C(x)$ denote the element $\pi^{-1}(x) \in \xi$. We have $S(C(x)) \subset C(Tx)$. The situation is illustrated by following commutative diagram:

$$\begin{array}{ccc} Y & \xrightarrow{S} & Y \\ \downarrow{\scriptstyle\pi} & & \downarrow{\scriptstyle\pi} \\ X & \xrightarrow{T} & X \end{array} \tag{7}$$

Our aim is to find the connection between P_T and P_S.

Proposition 1. *For every $f \in L^1(Y,\beta,\nu)$:*

$$P_T(f|\xi) = E_\nu(P_S f|\xi) \tag{8}$$

where $E_\nu(\cdot|\xi)$ is the operator of taking conditional expectation with respect to the partition (and meant as an operator from $L^1(Y,\beta,\nu)$ to $L^1(X,\Sigma,m)$.

$$\begin{array}{ccc} L^1(\nu) & \xrightarrow{P_S} & L^1(\nu) \\ \downarrow{\scriptstyle E_\nu(\cdot|\xi)} & & \downarrow{\scriptstyle E_\nu(\cdot|\xi)} \\ L^1(m) & \xrightarrow{P_T} & L^1(m) \end{array} \tag{9}$$

Proof. Let $h \in L^\infty(m)$. Then:

$$\begin{aligned} &\int_X h \cdot P_T E(f|\xi)\,dm = \int_X (h\circ T)\cdot E_\nu(f|\xi)\,dm \\ &= \int_Y (h\circ T\circ\pi)(E_\nu(f|\xi)\circ\pi)\,d\nu = \int_Y (h\circ\pi\circ S)\cdot f\,d\nu \\ &= \int_Y (h\circ\pi)\cdot P_S f\,d\nu = \int_Y (h\circ\pi)(E_\nu(P_S f|\xi)\circ\pi)\,d\nu \\ &= \int_X h\cdot E_\nu(P_S f|\xi)\,dm. \end{aligned}$$

We assume that S has a regular partition α with the following property:

$$S^{-1}\xi \vee \alpha = \xi. \tag{10}$$

Usually we will have $\xi = \alpha^- = \bigvee_{k=0}^{\infty} S^{-k}(\alpha)$ and (10) will hold automatically.

Lemma 1. *The family $\beta = \{\pi(A)\}_{A\in\alpha}$ is a T-regular partition of X.*

Proof. By (10) it follows that $\alpha \leq \xi$ and $\pi^{-1}(\pi A) = A$ for every $A \in \alpha$. Thus, β is a partition. We also check easily that $\pi^{-1}(T(\pi A)) = S(A)$. Then, $T(\pi A)$ is also measurable. Obviously $T|\pi(A) : \pi(A) \to T(\pi(A))$ is a factor of $S|A : A \to S(A)$. Moreover, (10) shows that $T|\pi(A)$ is 1:1 and $(T|\pi(A))^{-1}$ is a factor of $(S|A)^{-1}$. So, $T|\pi(A)$ is an isomorphism of $(\pi(A), \Sigma|\pi(A))$ and $(T(\pi(A)), \Sigma|T(\pi(A)))$.

By $(\nu_C)_{C\in\xi}$ we denote the family of all conditional measures of ν with respect to ξ. The following proposition joins g_S, g_T and $(\nu_C)_{C\in\xi}$ together in one formula.

Proposition 2. *For almost every $x \in X$ and for almost every $y \in C(x)$ (in the sense of $\nu_{C(x)}$):*

$$g_T(x) = g_S(y)\frac{d\left((S|A)_*^{-1}\nu_{C(Tx)}\right)}{d\nu_{C(x)}}(y) \tag{11}$$

where A is the only element of α which contains $C(x)$ (we notice that $C(x) = S^{-1}C(Tx)\cap A$). In particular, $(S|A)_*^{-1}\nu_{C(Tx)}$ is equivalent to $\nu_{C(x)}$ for almost every $x \in X$.

Proof. Let us fix some $f \in L^\infty(\nu)$. Then $E_\nu(P_S f|\xi)(x)$ can be written as $\int_{C(x)} \sum_{A\in\alpha}(f \cdot g_S) \circ (S|A)^{-1} d\nu_{C(x)}$ or $\int f\, d\sigma_{1,x}$, and $P_T E_\nu(f|\xi)(x)$ can be written as $\sum_{z\in T^{-1}(x)} \int_{C(z)} f d\nu_{C(z)} \cdot g_T(z)$ or $\int f\, d\sigma_{2,x}$, where

$$\sigma_{1,x} = \sum_{z\in T^{-1}(x)} g_S \cdot \left((S|C(z))_*^{-1} \nu_{C(x)}\right), \qquad \sigma_{2,x} = \sum_{z\in T^{-1}(x)} g_T(z) \cdot d\nu_{C(z)}.$$

Since f is arbitrary, these formulas and Proposition 1 give us $\sigma_{1,x} = \sigma_{2,x}$ for almost every $x \in X$. Since $\nu_{C(z)}$ have disjoint supports and since g_T and g_S are positive almost everywhere, this implies (11).

In many examples the elements of ξ are endowed with a measure which is different from the conditional one. We assume that $(l_C)_{C\in\xi}$ is a family of measures such that for every $C \in \xi$, l_C is equivalent to ν_C and the function

$$\rho = \frac{d\nu_C}{dl_C}$$

defined on Y is β-measurable. We notice that for a. e. $x \in X$, $(S|C(x))_*^{-1} l_{C(Tx)}$ is also equivalent to l_C. Besides, function λ, defined by

$$\lambda(y) = \frac{d\left((S|C(x))_*^{-1}\, l_{C(Tx)}\right)}{dl_{C(x)}}(y), \quad \text{as} \quad y \in C(x) \tag{12}$$

is also β-measurable.

Example 1. Let $S : Y \to Y$ be an Anosov diffeomorphism of a Riemannian manifold Y. Let ξ be an invariant measurable partition which consists of pieces of stable manifolds (one can easily construct such ξ using a Markov partition). Every element $C \in \xi$ is endowed with the induced Riemannian metric and, hence, with the volume l_C. If ν is the volume measure on the entire manifold, then (if S is at least of class $C^{1+\epsilon}$) conditional measures are equivalent to l_C.

We observe that Proposition 2 gives

$$g_T \circ \pi = g_S \frac{\rho \circ S}{\rho} \lambda. \tag{13}$$

If y and y' belong to the same C then (13) yields

$$\frac{\rho(y)}{\rho(y')} = \frac{g_S(y)}{g_S(y')} \frac{\rho(Sy)}{\rho(Sy')} \frac{\lambda(y)}{\lambda(y')}. \tag{14}$$

By induction we prove that

$$\frac{\rho(y)}{\rho(y')} = \left(\prod_{k=0}^{n-1} \frac{g_S(S^k y)\lambda(S^k y)}{g_S(S^k y')\lambda(S^k y')}\right) \frac{\rho(S^n y)}{\rho(S^n y')}. \tag{15}$$

This observation gives the following

Proposition 3. *For a. e. $x \in X$ and for $\nu_{C(x)}$ a. e. $y, y' \in C(x)$, the following conditions are equivalent:*

1.
$$\lim_{n\to\infty} \frac{\rho(S^n y)}{\rho(S^n y')} = 1$$

2.
$$\frac{\rho(y)}{\rho(y')} = \prod_{k=0}^{\infty} \frac{g_S(S^k y)\lambda(S^k y)}{g_S(S^k y')\lambda(S^k y')}$$

and the product converges.

Remark 1. If (1) holds a. e. then (2) and the condition $\int \rho dl_C = 1$ determine ρ completely.

2.2 The Existence of Invariant Measures and the Spectral Decomposition

As in the previous subsection, T is a nonsingular map of X. X is a Lebesgue space. Let β be a regular partition of X which is a generator for T. A generator means one-sided generator, i. e. $\beta^- = \bigvee_{k=0}^{\infty} T^{-k}(\beta) = \epsilon$, where ϵ is a partition into points.

We ask if there exist T-invariant measures absolutely continuous with respect to m. This problem is equivalent to that of the existence of an eigenvector of P_T corresponding to eigenvalue 1. We notice that 1 is always in the spectrum of P_T, but it may happen that it is not an eigenvalue (see cf. [8]). However, in particular situations we are able to give a positive answer to this question.

In this section we write g_n instead of g_{T^n}, g instead of $g_T = g_1$ and P instead of P_T (g_n is well defined since β^n is a regular partition for T^n); supremum (infimum) means essential supremum (infimum). For every $A \in \Sigma$ we define $\beta(A) = \{B \in \beta : m(B \cap A) > 0\}$.

The following conditions allow us to prove the existence of an eigenvector of P corresponding to eigenvalue 1.

(I) Distorsion condition: $\exists d \in (0,\infty)\forall_{n\geq 1}\forall_{B\in\beta^n} : \sup_B g_n \leq d \cdot \inf_B g_n$;

(II) Localization condition: $\exists \epsilon > 0, r \in (0,1)\ \forall_{n\geq 1}\ \forall_{B\in\beta^n}$:
$$m(T^n B) < \epsilon \implies \sum_{B'\in\beta(T^n B)} \sup_B g \leq r;$$

(III) Bounded variation condition: $\sum_{B\in\beta} \sup_B g < +\infty$.

The reader will note the order of quantifiers in these statements. Conditions (I) and (III) are often used in the theory of the existence of absolutely continuous measures (see [13]). Condition (II) is crucial in our method of treating situations with no Markov partition available.

Remark 2. Condition (II) holds automatically if for some $\epsilon' > 0$, $m(T^n B) > \epsilon'$ for all n and $B \in \beta^n$ (this is true if, for instance, β is related to a finite Markov partition). It says that if $T^n B$ is small in the sense of measure then it cannot intersect too many elements of β with a large value of g.

Let us note that if (I) and (III) hold for T, β and g, then they hold for T^N, β^n and g_N for every $N \geq 1$. Condition (I) holds with the same value of d. Besides,

$$\sum_{B\in\beta^n} \sup_B g_N \leq \left(\sum_{B\in\beta} \sup_B g \right)^N .$$

Thus we know that conditions (I)–(III) for iterations of T can be verified in the following way: we check (I) and (III) for T and then check only (II) for some iteration of T.

Theorem 1. *Let (I)–(III) be satisfied. Then the sequence $(P^n 1)_{n\geq 1}$ is bounded in $L^\infty(m)$ and the averages*

$$\frac{1}{n}\sum_{k=0}^{n-1} P^k 1 \tag{18}$$

converge in $L^1(m)$ to some $\phi \in L^\infty(m)$ such that $P\phi = \phi$.

Proof. If the averages (18) are bounded in $L^\infty(m)$ then they form a weakly sequentially conditionally compact sequence and hence, by Yosida's ergodic theorem [4] they converge to $\phi \in L^\infty(m)$ with the property $P\phi = \phi$. Therefore, we only have to check that $(P^n 1)_{n\geq 1}$ is bounded in $L^\infty(m)$.

Let $\gamma_n = \sum_{B\in\beta^n} \sup_B g_n$. It is obvious that $\|P^n 1\|_\infty \leq \gamma_n$ for every $n \geq 1$. If we prove that γ_n is bounded then the theorem would follow. We will prove by induction that

$$\gamma_{n+1} \leq r\gamma_n + C_1, \tag{19}$$

where $C_1 = (\gamma_1 \cdot d)/\epsilon$ ($\gamma_n < \infty$ because of (III)). In view of $g_{n+1} = g_n(g \circ T^n)$ we have for every $B' \in \beta^n$:

$$\sum_{\{B\in\beta^{n+1}:B\subset B'\}} \sup_B g_{n+1} \leq \sup_{B'} g_n \sum_{\{B''\in\beta(T^n B')\}} \sup_{B''} g. \tag{20}$$

Let $\beta_1 = \{B' \in \beta^n : m(T^n B') < \epsilon\}$ and $\beta_2 = \{B' \in \beta^n : m(T^n B') \geq \epsilon\}$. If $B' \in \beta_1$ then the right-hand side of (20) is not greater than $r \cdot \sup_{B'} g_n$, by (II). If $B' \in \beta_2$ then by (I) $\sup_{B'} g_n \leq d \cdot \inf_{B'} g_n \leq d\frac{m(B')}{m(T^n B')} \leq \frac{d}{\epsilon} m(B')$.

For every $B \in \beta^{n+1}$ there is a unique $B' \in \beta^n$ containing B. We obtain $\gamma_{n+1} = \sum_{B\in\beta^{n+1}} \sup_B g_{n+1} = \sum_{B'\in\beta^n} \sum_{\{B\in\beta^{n+1}:\, B\subset B'\}} \sup_B g_{n+1} \leq \sum_{B'\in\beta_1} r \cdot \sup_{B'} g_n + \sum_{B'\in\beta_2} \gamma_1 \frac{d}{\epsilon} m(B') \leq r\gamma_n + C_1$. Hence, we have proved (19). Since $0 < r < 1$, this implies that $(\gamma_n)_{n\geq 1}$ is bounded.

Remark 3. The inequality (19) implies that $\|\phi\|_\infty \leq \frac{C_1}{1-r} = \frac{\gamma_1 \cdot d}{\epsilon(1-r)}$.

Theorem 1 states the existence of an invariant, absolutely continuous with respect to m measure $\phi \cdot m$, but it gives no information about the number and properties of such measures. One can improve this result by adding the following condition:

(IV) Expanding condition: $\exists r \in (0,1) : \sup_X g \leq r$.

We can always assume that r is chosen to satisfy both (II) and (IV).

Theorem 2. *We assume (I)–(IV). Then there exists a bounded, finite dimensional projection $Q : L^1(m) \to L^\infty(m)$ such that*

1. *$Q(L^1(m)) \subset L^\infty(m)$ and Q is bounded as an operator from L^1 to L^∞.*
2. *For every $f \in L^1(m)$ the averages*
$$\frac{1}{n}\sum_{k=0}^{n-1} P^k f \tag{21}$$
converge in $L^1(m)$ to Qf, as $n \to \infty$.
3. *The range of Q (denoted by $\mathcal{R}(Q)$) consists of all eigenvectors of P corresponding to the eigenvalue 1 (and of 0).*
4. *There exist non-negative functions $\phi_1, \phi_2, \ldots, \phi_s \in \mathcal{R}(Q)$ which span $\mathcal{R}(Q)$ and $\phi_i \wedge \phi_j = 0$ as $i \neq j$, where $\phi \wedge \psi \stackrel{\text{def}}{=} \min(\phi, \psi)$. Moreover, $m(\phi_i) = 1$ for $i = 1, 2, \ldots, s$ and if*
$$C_i = \bigcup_{n=0}^{\infty} T^{-n}\{x : \phi_i(x) > 0\}$$
is the basin of the measure $\phi_i m$ then Q can be represented as
$$Qf = \sum_{i=1}^{s} \int_{C_i} f\, dm \cdot \phi_i. \tag{22}$$

Moreover, $\bigcup_{i=1}^{s} C_i = X$ (up to a set of measure 0) and $(\phi_i)_{i=1}^{s}$ are the only functions $\phi \in L^1(m)$ such that $\phi \cdot m$ is a T-invariant, ergodic, probabilistic measure.

Proof. Let $\mathcal{A}$ be the set of all pairs (k, B) where k is a non-negative integer and $B \in \beta^n$ for some $n \geq k$. If $a \in \mathcal{A}$, $a = (k, B)$ then for every $B' \in \beta$, aB' denotes the pair $(k+1, B \cap T^{-k}B') \in \mathcal{A}$. We notice that for $k < n$ the set $B \cap T^{-k}B'$ has positive measure for the unique B' determined by the condition $T^kB \subset B'$ and $B \cap T^{-k}B' = B$. We denote this element of $\mathcal{A}$ by $a+1$. By f_a we denote the function $P^k(\chi_B) \in L^1(m)$ (as $a = (k, B)$). Using this notation, we can write
$$Pf_a = \sum_{B' \in \beta} f_{aB'} = \begin{cases} f_{a+1} & \text{if } n > k, \\ \sum_{B' \in \beta(T^nB)} f_{aB'} & \text{if } n = k. \end{cases} \tag{23}$$

Hence, the subspace of all linear combinations of the elements of the family $(f_a)_{a\in\mathcal{A}}$ is P-invariant (but not closed in $L^1(m)$, in general). We wish to know how the coefficients in such linear combinations change under P. Therefore, we introduce the vector space V of all families $(x_a)_{a\in\mathcal{A}}$ of real numbers, only a finite number of them being different from 0. Let $(e_a)_{a\in\mathcal{A}}$ be the standard basis of V (e_a is the family of reals that has the unit on the a-th place and zeros elsewhere). We define a linear map $\Phi : V \to L^1(m)$ by putting $\Phi e_a = f_a$ for every $a \in \mathcal{A}$. If $\mathbf{x} = (x_a)_{a\in\mathcal{A}} \in V$ then

$$\Phi\mathbf{x} = \sum_{a\in\mathcal{A}} x_a f_a. \tag{24}$$

We also define the linear operator $\bar{P} : V \to V$ by

$$\bar{P}e_a = \begin{cases} e_{a+1} & \text{if } n > k \\ \sum_{B'\in\beta(T^n B)} e_{aB'} & \text{if } n = k, \end{cases} \tag{25}$$

where $a = (k, B)$ and $B \in \beta^n$, so that the following diagram commutes:

$$\begin{array}{ccc} V & \xrightarrow{\Phi} & L^1(m) \\ \big\downarrow \bar{P} & & \big\downarrow P \\ V & \xrightarrow{\Phi} & L^1(m) \end{array} \tag{26}$$

One can say that $\bar{P}$ acts on the coefficients of linear combinations while P acts on those combinations themselves.

By V_+ we denote the convex cone of all families of non-negative numbers. It is obvious that $W_0 = \mathcal{R}(\Phi)$ is dense in $L^1(m)$ because β is a generator. We also introduce a norm in V by

$$\|\mathbf{x}\| = \sum_{a\in\mathcal{A}} |x_a| \cdot w_a \tag{27}$$

where $(w_a)_{a\in\mathcal{A}}$ is the family of reals given by $w_a = \|f_a\|_\infty$, as $a \in \mathcal{A}$.

Let $|x|$ denote the vector $(|x_a|)_{a\in\mathcal{A}} \in V_+$ First we prove the following lemma:

Lemma 2. *There exists* $R \in (0, \infty)$ *such that for every* $\mathbf{x} \in V_+$*:*

$$\|\bar{P}\mathbf{x}\| \le r \cdot \|\mathbf{x}\| + R\|\Phi\mathbf{x}\|_1. \tag{28}$$

Proof. Given $a = (k, B) \in \mathcal{A}$ we notice that $w_a = \sup_B g_k$. We put $C_2 = d/\epsilon$. We claim that if $\|f_a\|_\infty \ge C_2\|f_a\|_1$ then

$$\sum_{B\in\beta} w_{aB} \le r \cdot w_a. \tag{29}$$

In fact, $\|f_a\|_1 = m(B)$ and $\|f_a\|_\infty \ge C_2\|f_a\|_1$ imply that $\sup_B g_k \ge C_2 m(B)$. But $m(B) = (m(B)/m(T^k B)) \cdot m(T^k B) \ge \inf_B g_k \cdot m(T^k B) \ge 1/d \cdot \sup_B g_k \cdot$

$m(T^kB)$. Therefore $m(T^kB) \le C_2^{-1}d = \epsilon$. Our assertion follows by applying inequality (20) and reasoning as in the proof of Theorem 1.

Let $\mathcal{A}'$ be the set of all $a \in \mathcal{A}$ such that $\|f_a\|_\infty \ge C_2 \cdot \|f_a\|_1$ and $\mathcal{A}'' = \mathcal{A}\backslash\mathcal{A}'$. Then $\|\bar{P}\mathbf{x}\| \le \sum_{a\in\mathcal{A}} x_a \|\bar{P}e_a\| \le \sum_{a\in\mathcal{A}} x_a \sum_{B'\in\beta} w_{aB'} \le \sum_{a\in\mathcal{A}'} x_a r w_a + C_2\gamma_1 \sum_{a\in\mathcal{A}''} x_a \|f_a\|_1 \le r\|\mathbf{x}\| + C_2\gamma_1 \|\Phi\mathbf{x}\|_1$. Putting $R = C_2\gamma_1$, we obtain the lemma.

Now we continue with the proof of Theorem 2. Let us fix some $x \in V_+$ and take $f = \Phi x$. We notice that $\|P^n f\|_\infty \le \|P^n x\|$, since $\|\Phi x\|_\infty \le \|x\|$ (see formula 26). Lemma 2 gives us $\|\bar{P}^n x\| \le R(1+r+\ldots+r^{n+1})\|f\|_1 + r^n\|x\| \le \frac{R}{1-r}\|f\|_1 + r^n\|x\|$. Hence the sequence $(P^n f)_{n\ge1}$ is bounded in $L^\infty(m)$. So, we can apply the Yosida ergodic theorem again and we get that the limit of (21) exists for at least $f \in \Phi(V_+)$. Moreover, for such f we can use the previous inequality and obtain that for every $f \in L^1(m)$, $f \ge 0$ $\|Qf\|_\infty \le \frac{R}{1-r}\|f\|_1$. This is because $\Phi(V_+)$ is dense in the set of all nonnegative functions f, so Qf is well defined on this set (by virtue of $\|P\|_1 = 1$). Moreover, since every $f \in L^1(m)$ can be represented as $f_+ - f_-$, where $f_+ = f \vee 0$ and $f_- = (-f)\vee 0$, then this limit exists for every $f \in L^1(m)$ and $\|Qf\|_\infty \le \|Qf_+\|_\infty + \|Qf_-\|_\infty \le \frac{R}{1-r}\|f_+\|_1 + \frac{R}{1-r}\|f_-\|_1 = \frac{R}{1-r}\|f\|_1$. Other conclusions of Theorem 2 can be proved easily if one uses the fact that the set $\{f \in \mathcal{R}(Q) : f \ge 0, m(f) = 1\}$ has only finitely many extreme points. We define $\phi_1, \phi_2, \ldots, \phi_s$ to be the only extreme points of this set. For details, see [13].

Remark 4. In [13] it was noted that from the fact that $Q : L^1(m) \to L^\infty(m)$ is bounded with norm $\le \frac{R}{1-r}$ follows an explicit bound on s, the number of ergodic components. If $f = \chi_{C_i}$ for some i then $\|f\|_1 = m(C_i)$ and $\|Qf\|_\infty \le \|Q1\|_\infty = \|1\|_\infty = 1$. Hence, $1 \le \frac{R}{1-r} m(C_i)$, and thus $m(C_i) \ge \frac{1-r}{R}$. But $\sum_{i=1}^{s} m(C_i) = 1$, and therefore $s\frac{1-r}{R} \ge 1$ and $s \le \frac{R}{1-r}$.

3 Applications to Lozi Mappings

3.1 Geometrical Properties of Lozi Mappings

We consider a mapping $L : \mathbb{R}^2 \to \mathbb{R}^2$ defined by

$$L(x,y) = (1 - a|x| + by, x) \tag{30}$$

where a and b are real parameters.

For the set of parameters, which we describe later, this mapping has a strange attractor and some hyperbolicity property. These facts were proved in the paper of M. Misiurewicz [10] (we extend the set of parameters described in that paper to some negative b).

L is differentiable except for the y-axis. L is continuous for all parameters. If $p \in \mathbb{R}^2\backslash\{x = 0\}$ then

$$DL(p) = \begin{bmatrix} \mp a & b \\ 1 & 0 \end{bmatrix} \stackrel{\text{def}}{=} A_\pm . \tag{31}$$

("+" for $x(p) > 0$ and "-" for $x(p) < 0$).

The iteration L^n is differentiable at $p \in \mathbb{R}^2 \setminus \bigcup_{k=0}^{n-1} L^k(\{x = 0\})$ and

$$DL^n(p) = DL(L^{n-1}p)\, DL(L^{n-2}p) \ldots DL(p) = A_{\epsilon_{n-1}} A_{\epsilon_{n-2}} \cdots A_{\epsilon_0} \tag{32}$$

where $\epsilon_i = sgn(x(L^i p))$, $i = 0, 1, \ldots, n-1$.

We want to describe some fields of invariant directions for L, or, more precisely, for DL. A convenient way to describe them is to use the projective coordinates in the tangent space of the plane. The computation $A_\pm \begin{bmatrix} 1 \\ u \end{bmatrix} = \begin{bmatrix} \mp a + bu \\ 1 \end{bmatrix}$ $= (\mp a + bu) \begin{bmatrix} 1 \\ \dfrac{1}{\mp a + bu} \end{bmatrix}$ shows that the corresponding transformation on directions is $\alpha_\pm(u) = \frac{1}{\mp a + bu}$. (Each line is described by parameter u of the only point of this line which has the form $(1, u)$, except for the y-axis.)

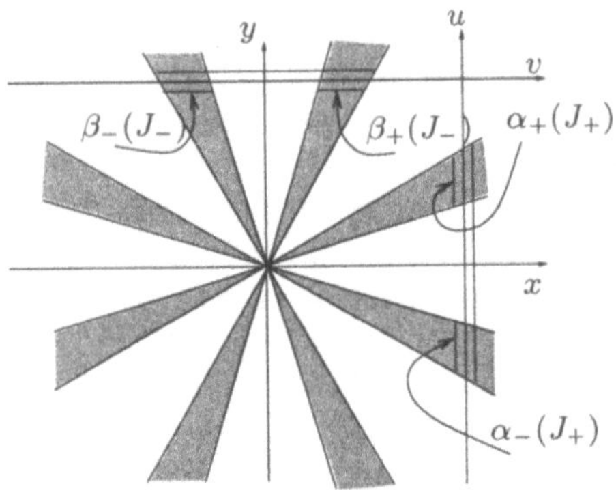

Fig. 1. Invariant cones.

If $b \neq 0$ then L^{-1} is a homeomorphism and is differentiable except for the points p on the y-axis, and

$$DL^{-1}(p) = \begin{bmatrix} 0 & 1 \\ b^{-1} & \pm ab^{-1} \end{bmatrix} \stackrel{\text{def}}{=} B_\pm . \tag{33}$$

This time "+" corresponds to $y(p) > 0$ and "−" to $y(p) < 0$. The corresponding map on directions is $\beta_\pm(v) = \frac{b}{\pm a + v}$ where any line going through 0 and point $(v, 1)$ is described by its unique point of the form $(v, 1)$.

In the sequel we always assume that $|b| < \frac{a^2}{4}$ and $a > 0$.

Lemma 3. *Let* $\theta_0 = \frac{a}{2} - \sqrt{\left(\frac{a}{2}\right)^2 - |b|}$. *Then the intervals* $J_+ = \{u \in \mathbb{R} : |bu| \leq \theta_0\}$ *and* $J_- = \{v \in \mathbb{R} : |v| \leq \theta_0\}$ *are* $\alpha_\pm$- *and* $\beta_\pm$-*invariant, respectively. Moreover, if* $\theta_1 = \frac{|b|}{a + \theta_0}$ *then* $\alpha_\pm(J_+) \subseteq \{u \in \mathbb{R} : |bu| \geq \theta_1\}$ *and* $\beta_\pm(J_-) \subseteq \{v \in \mathbb{R} : |v| \geq \theta_1\}$. *See Figure 1.*

Lemma 4. *Let* $\kappa = \frac{\frac{a}{2} - \sqrt{\left(\frac{a}{2}\right)^2 - |b|}}{\frac{a}{2} + \sqrt{\left(\frac{a}{2}\right)^2 - |b|}}$. *Then* $\sup_{J_+} |\alpha'_\pm(u)| = \sup_{J_-} |\beta'_\pm(v)| = \kappa <$

Lemmas 3 and 4 allow an easy proof of the existence of two fields of invariant directions.

For any sequence $\epsilon_0, \epsilon_1, \ldots \in \{+,-\}$ by the Mean Value Theorem we have

$$|\beta_{\epsilon_0} \circ \beta_{\epsilon_1} \circ \ldots \circ \beta_{\epsilon_{n-1}}(J)| \leq \kappa^n |J_-| \tag{34}$$

(where $|\cdot|$ stands for the length of an interval). Therefore, the set

$$\bigcap_{n=1}^{\infty} \beta_{\epsilon_0} \circ \ldots \circ \beta_{\epsilon_{n-1}}(J_-) \tag{35}$$

consists of exactly one point. This point might also be expressed as a continued fraction

$$\cfrac{b}{\epsilon_0 a + \cfrac{b}{\epsilon_1 a + \cfrac{b}{\epsilon_2 a + \cdots}}} \tag{36}$$

which converges by (34).Similarly, for any sequence $\eta_0, \eta_1, \eta_2, \ldots \in \{+,-\}$, the set $\bigcap_{n=1}^{\infty} \alpha_{\eta_0} \circ \ldots \circ \alpha_{\eta_{n-1}}(J_+)$ consists of exactly one point which can be represented as a continued fraction

$$\cfrac{1}{-\eta_0 a + \cfrac{b}{-\eta_1 a + \cfrac{b}{-\eta_2 a + \cdots}}} \tag{37}$$

The above construction gives us invariant directions for L. For example, if $p \in \mathbb{R}^2 \setminus \bigcup_{n=0}^{\infty} L^{-n}(\{x=0\})$ then we can define direction $E^s(p)$ by taking $\epsilon_i = sgn\, x(L^i p)$ for $i = 0, 1, \ldots$. Similarly, if $p \in \mathbb{R}^2 \setminus \bigcup_{n=0}^{\infty} L^n(\{y=0\})$ then by putting $\eta_i = sgn\, y(L^{-i}p)$ for $i = 1, 2, \ldots$, we can define $E^u(p)$. We also introduce two functions

$$u(p) = \cfrac{-1}{\eta_0 a + \cfrac{b}{\eta_1 a + \cfrac{b}{\eta_2 a + \cdots}}} \quad \text{and} \quad v(p) = \cfrac{b}{\epsilon_0 a + \cfrac{b}{\epsilon_1 a + \cfrac{b}{\epsilon_2 a + \cdots}}} \tag{38}$$

which correspond to these fields of directions.

These functions are discontinuous, as defined on the plane, but become Hölder if we consider them as functions of the sequences of $\pm$ only. For example, if the orbits of two points $p_1, p_2 \in \mathbb{R}^2 \setminus \bigcup_{k=0}^{n-1} L^{-k}(\{x=0\})$ fulfill the condition $sgn\, x(L^k p_1) = sgn\, x(L^k p_2)$ for $k = 0, 1, \ldots, n-1$ then $|u(p_1) - u(p_2)| \leq \kappa^n |J_+|$. A similar inequality holds for v $|v(L^n p_1) - v(L^n p_2)| \leq \kappa^n |J_-|$. We summarize these remarks in

Proposition 4. *If $a > 0$ and $|b| < (a/2)^2$ then there are two fields of directions E^s and E^u defined on $D^s = \mathbb{R}^2 \setminus \bigcup_{n=0}^{\infty} L^{-n}(\{x=0\})$ and $D^u = \mathbb{R}^2 \setminus \bigcup_{n=0}^{\infty} L^{n}(\{y=0\})$, respectively. This means that $DL(p)E^s(p) \subset E^s(Lp)$ for $p \in D^s$ and $DL^{-1}(p)E^u(p) \subset E^u(L^{-1}p)$ for $p \in D^u$.*

Remark 5. As we can see, $L(D^s) \subset D^s$ and $L^{-1}(D^u) \subset D^u$. The sets D^s and D^u are dense in the plane and have full Lebesgue measure. They are of type G_δ.

Let $\lambda^s(p)$ and $\lambda^u(p)$ be the rates of change of the length in the direction of $E^s(p)$ and $E^u(p)$ respectively (the orientation change is not reflected in the sign) under L and L^{-1}, respectively.

Lemma 5. *The following formulas hold*

$$\lambda^s(p) = |v(p)| \cdot \frac{h_1(p)}{h_1(Lp)}, \qquad \lambda^u(p) = |u(p)| \cdot \frac{h_2(p)}{h_2(L^{-1}p)} \tag{39}$$

where

$$h_1 = \frac{1}{\sqrt{1+v^2}}, \qquad h_2 = \frac{1}{\sqrt{1+u^2}}. \tag{40}$$

Proof. We have $DL(p)\begin{bmatrix} v(p) \\ 1 \end{bmatrix} = A_\pm \begin{bmatrix} v(p) \\ 1 \end{bmatrix} = \begin{bmatrix} \mp a v(p) + b \\ v(p) \end{bmatrix} = v(p)\begin{bmatrix} \mp a + b/v(p) \\ 1 \end{bmatrix} = v(p) \cdot \begin{bmatrix} v(Lp) \\ 1 \end{bmatrix}$. Hence, $\lambda^s(p) = |v(p)| \cdot \frac{\|(v(Lp),1)\|}{\|(v(p),1)\|} = |v(p)| \frac{h_1(p)}{h_1(Lp)}$. The proof of the other formula is identical.

Let $\lambda_n^s(p)$ and $\lambda_n^u(p)$ be analogous functions defined for the iterations L^n and L^{-n} respectively. We have $\lambda_n^s(p) = \lambda^s(L^{n-1}p) \cdots \lambda^s(Lp)\lambda^s(p) = |v(L^{n-1}p)| \cdots |v(Lp)| \cdot |v(p)| \frac{h_1(p)}{h_1(L^n p)}$ due to telescoping cancellations. There are constants $c_1, c_2 \in (0, \infty)$ such that $c_1 \le h_1 \le c_2$. By Lemma 3 $|v(p)| \le \theta_0$. Hence $\lambda_n^s \le c_3 \theta_0^n$ where $c_3 = c_2 c_1^{-1}$. We want L to be hyperbolic. This requires $\theta_0/|b| < 1$. After easy calculations we find that this is equivalent to $a > |b| + 1$. Similarly, $\lambda_n^u \le \left| u(L^{-(n-1)}p) \cdots u(L^{-1}p) u(p) \frac{h_2(p)}{h_2(L^{-n}p)} \right| < c_4 \left(\frac{\theta_0}{|b|} \right)^n$ where c_4 is some constant $\in (0, \infty)$. We consider $|b| \le 1$ only. Thus if $a > |b| + 1$ then $\theta_0 < |b|$ and we obtain

Proposition 5. *If $a < 0$, $0 < |b| \le 1$, $a > |b| + 1$, and $|b| < (a/2)^2$ then L is hyperbolic in the following sense: let $\lambda_+ = \theta_0/|b|$ and $\lambda_- = \theta_0$; then $\lambda_+, \lambda_- \in (0,1)$ and there exists a constant $C \ge 0$ such that $|\lambda_n^s(p)| \le C\lambda_-^n$ as $p \in D^s$, and $|\lambda_n^u(p)| \le C\lambda_+^n$ as $p \in D^u$.*

Now we are going to describe certain invariant sets for L.

Lemma 6. *Let θ_2 be the positive root of the equation $\theta(a + |b|\theta) = 1$ (i.e. $\theta_2 = \frac{1}{|b|}\left(-\frac{a}{2} + \sqrt{\frac{a^2}{4} + |b|} \right)$. Then under the assumptions of Proposition 5 the set*

$$\Delta_0 = \{(x, y) \in \mathbb{R}^2 : \ |y| \leq \theta_2(p - x)\},$$

where $q = (1 - |b|\theta_2)^{-1}$, *is L-invariant, i.e.* $L(\Delta_0) \subset \Delta_0$.

Proof. Δ_0 is a union of rays starting from $p_0 = (q, 0)$ and with directions, for which $u \in [-\theta_0, \theta_0]$. (see Figure 2(a)). Let Z be one of the rays and let u be the corresponding parameter. Hence $Z = \{(q - t, t \cdot u) \in \mathbb{R}^2 : \ t \in \mathbb{R}_+\}$. Z intersects the line $\{x = 0\}$ at exactly one point $p_1 = (0, qu)$. We have $L(p_1) = (1 + bqu, 0) \stackrel{\text{def}}{=} p_2$. The calculation $1 + qbu \leq 1 + \frac{|b|\theta_2}{1-|b|\theta_2} = \frac{1}{1-|b|\theta_2} = q$, implies that $p_2 \in \Delta_0$. We can see that $L(Z)$ is a union of a ray Z that starts at p_2 with the direction

$$\alpha_-(u) = (a + bu)^{-1} \in \left[\frac{1}{a + |b|\theta_0}, \frac{1}{a - |b|\theta_0}\right] \subseteq [-\theta_0, \theta_0]$$

and the interval Z_2 joining p_2 with $Lp = (1 - qa, q)$. We have $Lp_0 \in \Delta_0$ because

$$\begin{aligned} \frac{q}{(1 - qa) - q} &= \frac{1}{q^{-1} - (a + 1)} = \frac{1}{1 - |b|\theta_2 - (a + 1)} \\ &= \frac{1}{a + |b|\theta_2} = -\theta_2. \end{aligned} \tag{41}$$

Hence $Lp_0 \in \partial\Delta_0$. Thus $Z_1, Z_2 \subset \Delta_0$ and $L(Z) \subset \Delta_0$. In view of $|b| < 1$, L has two fixed points p_+ and p_- with $x(p_+) > 0$ and $x(p_-) < 0$. Both are hyperbolic by Proposition 5. Moreover, $p_- = (-(a+b+1)^{-1}, -(a+b-1)^{-1})$.

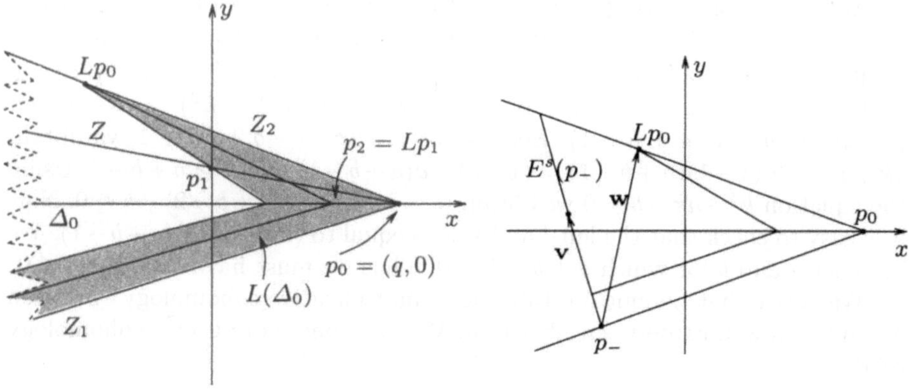

(a) The infinite trapping region.

(b) The triangular trapping region.

Fig. 2. The trapping regions for L.

Let $\mathbf{v} = (v(p_-), 1)$ and $\mathbf{w} = Lp_0 - p_-$. The condition that $L(p_0)$ is to the right of $E^s(p_-)$ is equivalent to $\mathbf{v} \times \mathbf{w} \leq 0$. This condition guarantees that the triangle

$$\Delta = \{p \in \Delta_0 : \ p \text{ is to the right of } E^s(p_-)\} \tag{42}$$

is invariant. The direction of $E^s(p)$ is defined by the following value of the parameter v:

$$\cfrac{b}{-a + \cfrac{b}{-a + \dots}}. \tag{43}$$

This expression is equal to $-k$ where k is the positive root of the equation $b = k(a+k)$. From now on we always assume that the hypothesis of Proposition 5 is satisfied. We can put $\mathbf{v} = (-k, 1)$ and $\mathbf{w} = Lp_0 - p_- = (1 - aq + e, q + e)$ where $e = 1/(a + b - 1)$. Hence,

$$\mathbf{v} \times \mathbf{w} = \begin{vmatrix} -k & 1 - aq + e \\ 1 & q + e \end{vmatrix} = -k(q+e) - (1 - aq + e) \leq 0. \tag{44}$$

This condition for $b > 0$ may be rewritten as $2a + b - 4 \leq 0$, as a lengthy computation shows. In the case of $b < 0$, simplifying of this condition is more complicated.

Proposition 6. *Let a, b fulfill the conditions $a > 0$, $0 < |b| \leq 1$, $a > |b| + 1$ and (44) (as we stated, in the case of $b > 0$ this reduces to $2a + b \leq 4$). Then $L(\Delta) \subset \Delta$.*

Remark 6. If we require that $L(\Delta) \subset \operatorname{int} \Delta$ then we may use Δ shifted to the right by a sufficiently small distance. This works provided that $2a + b < 4$ ($Lp_0 \notin E^s(p_-)$).

Now we are going to give a proof of the equivalence of (44) and $2a + b - 4 \leq 0$ for $b > 1$. We notice that $k = b\theta_2$. We have a sequence of equivalent inequalities: (44) and then $1 - aq + e + k(q+e) \geq 0$ and $1 - \frac{a}{1-k} + e + \frac{k}{1-k} + ke \geq 0$. Thus $1 - k - a + e - ek + k + ke - k^2 e \geq 0$, $e(1 - k^2) \geq a - 1$ and $1 - k^2 \geq (a-1)(a+b-1)$. Because of $1 - k^2 = -b + ak + 1$ we obtain $ak + 1 - b \geq (a-1)(a+b-1) = (1-b) + a(a+b-1)$ and $k \geq a + b - 2$. Using the equation $k^2 + ak - b = 0$ we obtain $(a+b-2)^2 + a(a+b-2) - b \leq 0$. Now it is easy to check that the left-hand side is equal to $(2a+b-4)(a+b-1)$. We consider a and b for which $a + b - 1 > 0$. Thus we must have $2a + b - 4 \leq 0$.

We notice that Lemma 5 establishes "multiplicative cohomology" between λ^s and v as well as between λ^u and u. We have one more useful cohomology relation.

Lemma 7. *There is a function h_3, bounded and separated from 0, such that for all $p \in D^s \cap L^{-1}(D^u)$ we have*

$$\lambda^s(p) = \lambda^u(Lp) \cdot |b| \cdot \frac{h_3(p)}{h_3(Lp)}.$$

Proof. Let $\mathbf{v}(p) = (v(p), 1)$ and $\mathbf{u}(p) = (1, u(p))$. The linear transformation $DL(p)$ with respect to the pair of bases

$$\left(\frac{\mathbf{v}(p)}{\|\mathbf{v}(p)\|}, \frac{\mathbf{w}(p)}{\|\mathbf{u}(p)\|}\right) \text{ and } \left(\frac{\mathbf{v}(Lp)}{\|\mathbf{v}(Lp)\|}, \frac{\mathbf{w}(Lp)}{\|\mathbf{u}(Lp)\|}\right)$$

has a diagonal matrix

$$\Lambda(p) = \begin{bmatrix} \pm\lambda^s(p) & 0 \\ 0 & \pm\lambda^u(Lp)^{-1} \end{bmatrix}. \tag{45}$$

Let $S(p)$ be the transition matrix from the first basis to the standard basis. Then $\Lambda(p) = S(Lp)^{-1}DL(p)S(p)$. Taking the determinants of both sides we get $\lambda^s(p)(\lambda^u(Lp))^{-1} = |b| \cdot \frac{h_3(p)}{h_3(Lp)}$, where $h_3(p) = |\det S(p)|$.

3.2 Invariant Partitions for Lozi Mappings

The results of Misiurewicz listed in the previous subsection give the existence of stable and unstable invariant directions on sets D^s and D^u respectively.

The idea applied by Misiurewicz and coming from billiard theory, that the trajectory of almost every point does not approach the singularity too fast, gives the existence of local stable and unstable manifolds (intervals).

We use the main geometrical idea of the paper, illustrated by Lemma 8 and Proposition 7 in order to prove some stronger estimates than those used by Misiurewicz ([10]).

Let $\tilde{\alpha}$ be the cover of the plane with the half-planes $\{x \leq 0\}$ and $\{x \geq 0\}$. It is easy to see that the cover $\tilde{\alpha}^n$ consists of convex sets bounded by broken lines. Let us set $\alpha = \tilde{\alpha}|\Delta$, where Δ is the trapping region constructed in the previous subsection. Hence α^n consists of convex polygons. We will call α a partition, because it becomes a genuine partition after discarding singularities of the iterates of L from the plane (which form a set of measure 0).

Lemma 8. *For every $N \geq 1$ there is an open cover $\mathcal{U}_N$ of Δ such that every element of $\mathcal{U}_N$ intersects not more than $2N$ elements of α^n.*

Proof. We will prove by induction that not more than $2N$ elements of α^n can have a nonempty intersection. The case of $N = 1$ is obvious. Let us assume that our assertion is true for some $N \geq 1$. From the definition $\alpha^{N+1} = \alpha \vee L^{-1}(\alpha^n)$. The partition $L^{-1}(\alpha^N)$ is clearly a partition into polygons (although they may not be convex). By the induction hypothesis, not more than $2N$ segments from the boundaries of the elements of $L^{-1}(\alpha^N)$ meet at one point. Because α^{N+1} is obtained by adding the line $\{x = 0\}$ to the boundaries, the number of segments meeting at one point can increase by 2, at most. The assertion has been proved. The Lemma follows easily.

Proposition 7. *There exist constants $F \in \mathbb{R}_+$ and $r \in (0,1)$ such that for every segment I with the direction from the unstable cone J_+ we have*

$$\sum_{J\in\alpha^N|I} \frac{|J|}{|L^n(J)|} \leq F(r^n + |I|), \tag{46}$$

where $|\cdot|$ stands for the length of a segment.

Proof. We choose N in such a way that $r_0 \overset{\text{def}}{=} (2N)\cdot(C\lambda_+^N) < 1$, where C, λ_+ are as in Proposition 5.

Let ϵ_0 be the Lebesgue constant of the cover $\mathcal{U}_n$ given by Lemma 8. Let for $n = 1, 2, \ldots$

$$\gamma_n = \sum_{J\in\alpha^{nN}|I} \frac{|J|}{|L^{nN}(J)|}. \tag{47}$$

We are going to show that

$$\gamma_{n+1} \leq r_0\gamma_n + \epsilon_0^{-1}R_0|I|, \tag{48}$$

where $R_0 = \sup_I \gamma_1$, and I is any segment with the direction from the unstable cone.

Let $J \in \alpha^{nN}|I$. Either $|J| < \epsilon_0$ or $|J| > \epsilon_0$. In the first case $\alpha^{(n+1)N}|J$ consists of not more than $2N$ elements (Lemma 8) and

$$\begin{aligned}\sum_{J\in\alpha^{(n+1)N}|I} \frac{|J|}{|L^{(n+1)N}(J)|} &\leq 2N\cdot\max_J \frac{|J|}{|L^{(n+1)N}(J)|}\\ &\leq 2N\frac{|J|}{|L^{nN}(J)|}\cdot\max\frac{|L^{nN}(J)|}{|L^{(n+1)N}(J)|}\\ &\leq 2N\cdot\frac{|J|}{|L^{nN}(J)|}\cdot C\lambda_+^N = r_0\frac{|J|}{|L^{nN}(J)|}.\end{aligned} \tag{49}$$

In the second case we get

$$\sum_{J'\in\alpha^{(n+1)N}|J} \frac{|J'|}{|L^{(n+1)N}(J')|} \leq R_0\cdot\frac{|J|}{|L^{nN}(J)|} \leq R_0\epsilon_0^{-1}|J|. \tag{50}$$

Summation over all J gives (48) easily.

To derive (46) from (48) we notice that $n = k\cdot N + l$, $0 \leq l \leq N-1$. In fact, (48) implies that $\gamma_n \leq r_0^n + \frac{\epsilon_0^{-1}R_0}{(1-r_0)\cdot|I|}$. So, we check easily that the left-hand side of (46) is not larger than $r_0^k\cdot\gamma_l + \frac{\epsilon_0^{-1}R_0}{(1-r_0)\cdot|I|}$. Let $r = r_0^{1/N}$ and let

$$F = \max\left(\max_{0\leq l\leq N-1}\frac{\gamma_l}{r^{N-1}}, \epsilon_0^{-1}R_0(1-r_0)^{-1}\right).$$

We check that F does not exceed the right-hand side of (46).

Remark 7. This proof was similar to that of Theorem 1.

Let $\xi = \alpha^- = \bigvee_{n=0}^{\infty} L^{-n}(\alpha)$. This partition is obviously into convex sets and the hyperbolicity of L gives that all elements of ξ are either segments with the direction from the unstable cone or points (otherwise some element of ξ would be expanded without breaking on the singularity, which contradicts the fact that Δ is bounded).

We will prove that almost every element of ξ is a segemnt of positive length.

Lemma 9. *Let*

$$D^S(\delta) = \{p \in \Delta : \text{ dist } (L^n p, \{x=0\}) \geq \delta \lambda_n^S(p) \text{ for } n = 0,1,2,\dots\}. \quad (51)$$

For every $p \in D^S(\delta)$ the distance from p to the endpoints of $\xi(p)$ is not smaller than δ (in particular, $|\xi(p)| \geq 2\delta$).

Proof. See ([10]).

Lemma 10. *Let $\lambda = (\lambda_n)$ be a sequence of positive numbers such that $S \stackrel{\text{def}}{=} \sum_n \lambda_n < +\infty$. Let:*

$$D^S(\delta, \lambda) = \{p \in \Delta : \ n \in \mathbb{Z}_+ : \ dist(L^n p, \{x=0\}) \leq \delta \lambda_n\}. \quad (52)$$

Let I be a segment as in Proposition 7. Then there is a constant A_1 such that

$$|I \backslash D^S(\delta, \lambda)| \leq A_1 S \cdot \delta. \quad (53)$$

Proof. See Figure 3. Let $p \in I \backslash D^S(\delta, \lambda)$.

For some $n \in \mathbb{Z}_+$, $\text{dist}(L^n p, \{x=0\}) < \delta\lambda_n$. Let $J \in \alpha^n | I$ be the element containing p. Thus, the image $L^n p$ belongs to an interval in $L^n J$ of length $\leq A_0 \cdot \lambda_n$.

This means that p itself belongs to the interval $J \cap L^{-n}(\{|x| \leq \delta\lambda_n\})$ which has length not larger than

$$|J \cap L^{-n}(\{|x| \leq \delta\lambda_n\})| \leq \frac{A_0 \delta \lambda_n}{|L^n J|} \cdot |J|.$$

This gives $|L^{-n}(\{|x| \leq \delta\lambda_n\}) \cap I| \leq A_0 \delta \lambda_n \cdot F(r^n + |I|) \leq A_0 F(1 + \text{diam}(\Delta)) \delta \lambda_n$. Thus, by summation over all n,

$$|I \backslash D^S(\delta, \lambda)| \leq A_0 F(1 + \text{diam}(\Delta)) S \delta.$$

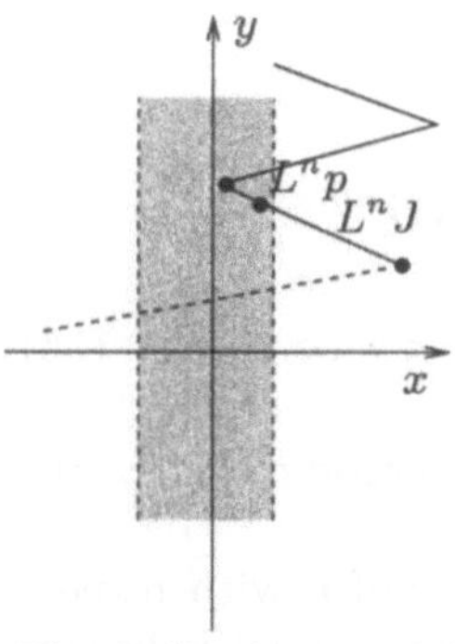

Fig. 3. Trimming of the images of J.

So, we can put $A_1 = A_0 F(1 + \text{diam}(\Delta))$.

Corollary 1. *Using the above notation,* $|I \backslash D^S(\delta)| \leq A_2 \cdot \delta$ *where* $A_2 = A_1 \sum_{n=0}^{\infty} C\lambda_-^n = A_1 C/(1-\lambda_-)$.

Proof. Let $\lambda_n = C\lambda_-^n, n \in \mathbb{Z}_+$. Because $\lambda_S^n(p) \leq \lambda_n$, we have $D^S(\delta) \supset D^S(\delta, \lambda)$. This gives the result.

Corollary 2. *The set* $\tilde{D}^S \stackrel{\text{def}}{=} \bigcup_{\delta>0} D^S(\delta)$ *has full Lebesgue measure in* Δ. *Moreover, there is a constant* A_2 *such that* $\nu(\Delta \backslash D^S(\delta)) \leq A_2\delta$ *where* ν *is the Lebesgue measure restricted to* Δ.

We need integrability of the function $1/D$ where $D(p) = |\xi(p)|$. This cannot be derived from the above estimate, but we have:

Proposition 8. *There is a constant* $A_3 \in \mathbb{R}_+$ *such that for an arbitrary* $\delta > 0$,

$$\nu(\{p \in \Delta : \ D(p) < \delta\}) \leq A_3 \delta^2. \tag{54}$$

Proof. This set is contained in the set of points p such that $\mathrm{dist}(L^n p, \{x = 0\}) < \delta\lambda_n(p)$ for at least *two* different $n \in \mathbb{Z}_+$ (because each end of $\xi(p)$ has to be trimmed). Let us consider the family of strips $\Pi_\alpha = \{p \in \mathbb{R}^2 : y(p) < \alpha\}$ $(\alpha > 0)$. The last condition means that $L^n p \in \Pi_\alpha \backslash D^S(\alpha)$ where $\alpha = \delta\lambda_n^S(p)$. By Lemma 10 and Fubini theorem, $\nu(\Pi_\alpha \backslash D^S(\alpha)) \leq A_2\alpha^2 \leq A_2(C\lambda_-^n)^2$. Because of the Jacobian of L is $|b|$, we have

$$\nu\left(L^{-n}(\Pi_\alpha \backslash D^S(\alpha))\right) \leq A_2 C^2 \lambda_-^{2n} |b|^{-n}. \tag{55}$$

We also have $\lambda_- = \theta_0$ and $\theta_0 \leq |b|$. Hence, $\lambda_-^{2n}|b|^{-n} \leq \lambda_-^n$ and summing (55) over all n we get $\nu(\{p \in \Delta : \ D(p) < \delta\}) \leq \sum_{n=0}^{\infty} A_2 C^2 \lambda_-^{2n} |b|^{-n} = A_2 C^2 \sum_{n=0}^{\infty} \lambda_-^u \frac{A_2 C^2}{1-\lambda_-}$. So we can put $A_3 = A_2 C^2/(1-\lambda_-)$.

Corollary 3. *The function* $\Delta \ni p \to 1/D(p) \in \mathbb{R}_+$ *is integrable with any power less than* 2.

Proof. Standard computation. Let $\beta \in (1,2)$. Then

$$\begin{aligned} \int_\Delta \left(\frac{1}{D}\right)^\beta d\nu &= \int_0^\infty \nu(\{D^{-\beta} > \gamma\}) d\gamma \leq 1 + \int_1^\infty \nu(\{D^{-\beta} > \gamma\}) d\gamma \\ &\leq 1 + \int_1^\infty \nu(\{D < \gamma^{-1/\beta}\}) d\gamma \leq 1 + \int_1^\infty B\gamma^{-2/\beta} d\gamma < +\infty. \end{aligned} \tag{56}$$

Now we are going to prove that from the point of view of ergodic theory, the elements of ξ are parallel.

Let $(\nu_C)_{C\in\xi}$ denote the family of conditional measures of ν with respect to the partition ξ and let $(l_C)_{C\in\xi}$ be the family of one-dimensional Lebesgue measures on the elements of ξ.

Proposition 9. *For almost every* $C \in \xi$, ν_C *is absolutely continuous with respect to* l_C *and the Radon-Nikodym derivative* $\frac{d\nu_C}{dl_C}$ *is constant on* C *(equal to* $1/|C|$*).*

Proof. Let A_n be the polygon of the partition α^n, $n \in \mathbb{Z}_+$, containing $C \in \xi$. Because A_n is convex, the projection of the measure $\frac{1}{\nu(A_n)}\nu|A_n$ onto the y-axis is a measure absolutely continuous with respect to the Lebesgue measure and the density ρ_n is a concave function on some interval (a_n, b_n) and $\rho_n \equiv 0$ outside of this interval. We have $(a_{n+1}, b_{n+1}) \subset (a_n, b_n)$ and $a_n \to a$, $b_n \to b$, where a and b are the ends of the projection of C onto the y-axis. (See Figures 4(a) and 4(b).)

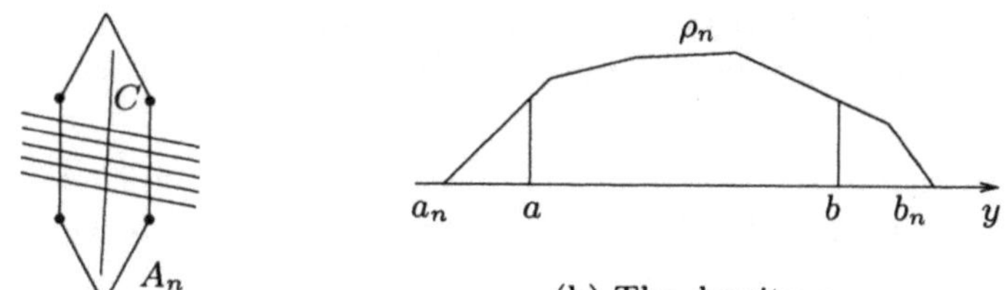

(a) A polygon of the partition α_n.

(b) The density ρ_n.

Fig. 4. Figures for the proof of Proposition 9.

Hence, ρ_{n_k} converges for some sequence $n_k \nearrow \infty$ (at least pointwise), because $\int \rho_n = 1$. Let $\rho = \lim_{n\to\infty} \rho_{n_k}$. It is easy to see that projecting ρ back onto C we get $\frac{d\nu_C}{dl_C}$ (which is concave, too). (As M. Wojtkowski noticed, even for any partition into intervals with measurable ends $\frac{d\nu_C}{dl_C}$ is linear.) Now the proposition can be derived easily from Proposition 3 and Lemma 10. Indeed, if $\rho = \frac{d\nu_C}{dl_C}$, it is enough to check that for almost every $x \in \Delta$ and almost every $y, y' \in C(x) = C$:

$$\lim_{n\to\infty} \frac{\rho(L^n y)}{\rho(L^n y')} = 1. \tag{57}$$

By Lemma 10 we assume that $\mathrm{dist}(L^n C, \{x = 0\}) > \delta\lambda^n$, where $\delta > 0$ and $\lambda \in (\lambda_-, 1)$. By concavity (see Figure 5(a)) of $\rho|\xi(L^n x)$:

$$\frac{\rho(L^n y)}{\rho(L^n y')} \leq 1 + \frac{\mathrm{dist}(L^n y, L^n y')}{\mathrm{dist}(L^n C, \partial\xi(L^n x))} \leq 1 + \frac{C\lambda_-^n}{\delta\lambda^n} \tag{58}$$

which goes to 1 as $n \to \infty$.

So $\overline{\lim}_{n\to\infty} \frac{\rho(L^n y)}{\rho(L^n y')} \leq 1$. By symmetry we get (57).

3.3 The Reduction for Lozi Mappings

In the previous subsection we proved the basic results which allow us to apply section 2 of this paper to Lozi mappings. The purpose of this subsection is to check the assumptions of section 2 for this case.

We use the notation of subsection 2.1. So, for example, $Y = \Delta$ and β is the σ-algebra of Borel sets. First we consider the case when ν is the Lebesgue measure restricted to Δ. We put $S = L|\Delta$.

We already know that $\rho = 1/D$. Applying formula (13) and the fact that $g_S = |b|^{-1}$ we get

Lemma 11. *Under the assumptions formulated above,*

$$g_T = |b|^{-1}\lambda^S \frac{D}{D \circ T}. \tag{59}$$

What we do not like about this formula is the presence of D, which is "very discontinuous". Fortunately, it can be eliminated by replacing ν with $(1/D)\cdot\nu$ (the elimination is possible because formula (59) has a form of a cohomology relation).

Proposition 10. *If we apply the results of subsection 2.1 to the measure $\nu = 1/D\hat{\nu}$, where $\hat{\nu}$ is the Lebesgue measure restricted to Δ, then*

$$g_T = |b|^{-1}\lambda^S. \tag{60}$$

Proof. Using the definition 5 and the fact that $m = 1/D \cdot \hat{m}$ where m is the factor of the Lebesgue measure on $X = \Delta/\xi$, we get on $A \in \alpha$:

$$\begin{aligned} g_T &= \frac{d\,(T_*(\chi_A \cdot m))}{dm} \circ T = \frac{d\left(T_*(\chi_A \cdot D^{-1} \cdot \hat{m})\right)}{d(D^{-1}\hat{m})} \circ T \\ &= \frac{D^{-1} \circ (T|A)^{-1} \cdot dT_*(\chi_A \hat{m})}{D^{-1} d\hat{m}} \circ T \\ &= \frac{D^{-1}}{D^{-1} \circ T} \cdot |b|^{-1}\lambda^S \cdot \frac{D}{D \circ T} = |b|^{-1}\lambda^S. \end{aligned} \tag{61}$$

(We used Lemma 11.)

This function g_T has nice properties and we can apply the results of subsection 2.2. Now using the notation of subsection 2.2 we are going to verify assumptions I–IV.

Lemma 12. *Condition I is satisfied.*

Proof. Let $n \in \mathbb{Z}_+$ and $B \in \beta^n$. Let $x_1, x_2 \in B$ ($x_1, x_2 \in X$ are also in ξ). Thus, $L^k_{x_1}$ and $L^k_{x_2}$ are on the same side of the line $x = 0$ for $k = 0, 1, \ldots, n-1$. Using formula 39 and $|v(L^k x_1) - v(L^k x_2)| \le \kappa^{n-k}|J_-|$ for $k = 0, \ldots, n-1$ we can see that for some constant $d_0 \in \mathbb{R}_+$, $\exp(-d_0\kappa^{n-k}) \le \frac{g_T(L^k x_1)}{g_T(L^k x_2)} \le \exp(d_0\kappa^{n-k})$. Because $g_n = g_T \cdot g_T \circ T \cdot \ldots \cdot g_T \circ T^{n-1}$, we get

$$\exp\left(-\sum_{k=0}^{n-1} d_0\kappa^{n-k}\right) \le \frac{g_n(x_1)}{g_n(x_2)} \le \exp\left(\sum_{k=0}^{n-1} d_0\kappa^{n-k}\right). \tag{62}$$

So, we might put $d = \exp\left(\frac{d_0}{1-\kappa}\right)$, to satisfy condition I of subsection 2.2.

Lemma 13. *Conditions II and IV are satisfied for some iteration $T^N, N \in \mathbb{Z}_+$.*

Proof. IV is satisfied because $g_n = |b|^{-n} \cdot \lambda_n^S$, so $|g_n| \leq |b|^{-n}(C\theta_0^n) = C(\theta_0/|b|)^n$. For large n, $|g_n| < 1$, since $\theta_0/|b| < 1$.

The argument that shows that II is satisfied for some iteration is similar to the proof of Proposition 7. Let us choose N, r_0 and ϵ_0 as in the proof of that proposition. We notice that $\lambda_+ = \theta_0/|b|$ so $|g_n| \leq C \cdot \lambda_+^n$ as $n \in \mathbb{Z}_+$.

Let $B \in \beta$ and let $A = \pi^{-1}(B)$. So $T^n(B) = \pi(S^n A)$. $S^n A$ is a convex polygon (see Figure 5(b)) such that

$$\partial(S^n A) \subset \bigcup_{k=0}^{n} L^k(\{x = 0\}). \tag{63}$$

So, the sides of $\partial(S^n A)$ are segments with the direction belonging to J_+. We have two possibilities:

1. $\operatorname{diam}(S^n A) > \epsilon_0$. Then it contains a segment I with the unstable direction (from J_+) of length $> \epsilon_0$.
   By Corollary 1, for arbitrary $\delta > 0$, $|I \backslash D^S(\delta)| \leq A_2\delta$. The set $\bar{A} = \pi^{-1}\pi(S^n A) = \pi^{-1}(T^n B)$ has measure larger that $A_4^{-1}(1 - A_2\delta)$, where A_4 is a constant from the next lemma. So, we put $\delta = \frac{1}{2}A_2^{-1}$ and $\epsilon = A_4^{-1}(1 - A_2\delta) = \frac{1}{2}A_4^{-1}$. So $\nu(\bar{A}) > \epsilon$ and since $m = \pi_*\nu$, $m(T^n B) = \nu(\bar{A}) > \epsilon$.
2. $\operatorname{diam}(S^n A) < \epsilon_0$. Then $T^n B$ is contained in not more than $2N$ elements of β^N and
$$\sum_{\beta' \in \beta^N(T^n B)} \sup_B g_N \leq (2 \cdot N) \cdot (C\lambda_+^N) = r_0. \tag{64}$$

So, in particular II holds for T^N with $r = r_0$.

Lemma 14. *Let I be a segment with the direction from J_+ and let l_I be the Lebesgue measure on I. Then the measure $\pi_*(l_I)$ is absolutely continuous with respect to m and*

$$\frac{d\pi_*(l_I)}{dm}(x) = \csc \angle(I, x) \tag{65}$$

for $x \in X$. In particular, for some $A_4 \in \mathbb{R}_+$,

$$A_4^{-1} \leq \frac{d\pi_*(l_T)}{dm} \leq A_4. \tag{66}$$

Proof. Fix some small $\delta > 0$. Let I_δ be a strip of width containing I. We notice that if $x \in D^S(\delta) \cap I$ and $\operatorname{dist}(x, \partial I) > \delta$, then $x \cap I_\delta$ is an interval of length $\delta \cdot \csc\omega(x)$, where $\omega(x) = \angle(I, x)$ (see Figure 5(c)). So $\nu_x(I_\delta) = \delta \csc\omega(x)/D(x)$. Let E be a subinterval of I. Thus, if $\hat{\nu}$ is the restriction of the Lebesgue measure to Δ, $\hat{\nu}(\pi^{-1}(\pi E) \cap I_\delta) = \int_{\pi(E)} \nu_x(x \cap I_\delta) d\hat{m} = \delta \int_{\pi(E)} \csc\omega(x) dm$. On the other hand, using Corollary 1 we get $\hat{\nu}(\pi^{-1}(\pi E)) = l_I(E) \cdot \delta + o(\delta)$. This implies (65) immediately.

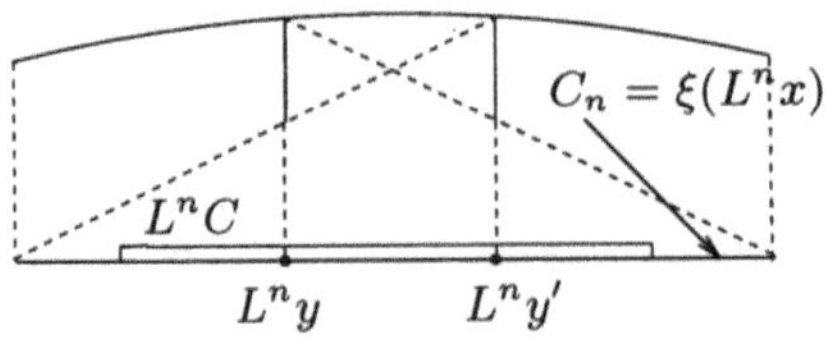

(a) Concavity of $\rho|\xi(L^n x)$.

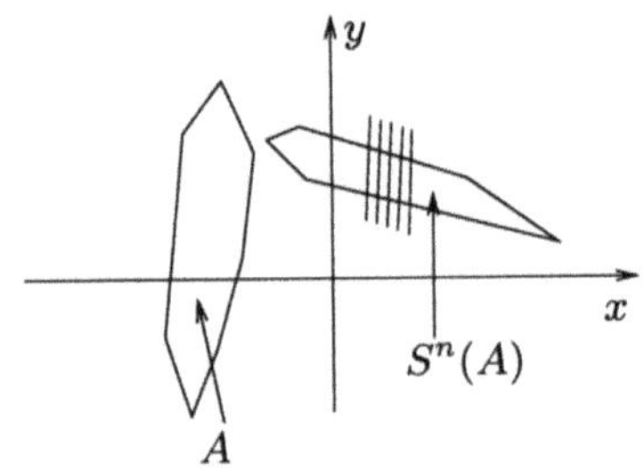

(b) The images of A.

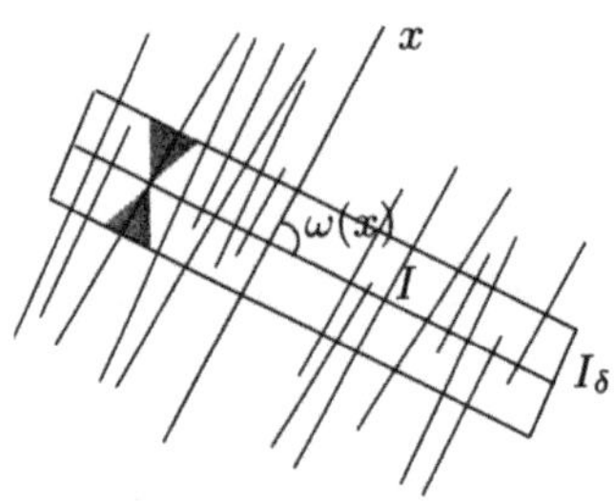

(c) The strip I_δ.

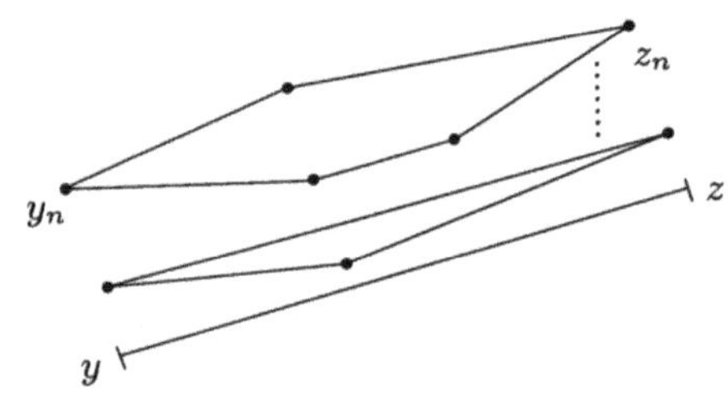

(d) An arbitrary sequence in $\mathcal{F}$.

Fig. 5. Figures illustrating convergence of partitions and densities.

Condition III of subsection 2.2 holds since β is finite.

Theorem 3. *The results of section 2, in particular Theorems 1 and 2 apply to Lozi mappings.*

3.4 Invariant Measures for Lozi Mappings

Using the general theory presented in section 2, we produced invariant measures of form $\phi \cdot m$ for the factor map T. We are now going to describe two methods of constructing the L-invariant measure μ, whose projection $\pi_*(\mu)$ onto X coincides with $\phi{\cdot}m$, $P\phi = \phi$. The first method is based on the following construction.

Let $f : \Delta \to \mathbb{R}$ be a continuous function. We have to define $\mu(f)$. Let us first define $f^<(p) = \inf_{\xi(p)} f$ and $f^>(p) = \sup_{\xi(p)} f$ where $p \in \Delta$. We put

$$\mu(f) = \lim_{n\to\infty} \tilde{\mu}\left((f \circ S^n)^<\right) \tag{67}$$

where $\tilde{\mu} = \phi \cdot m$. (We notice that $f^<$ and $f^>$ are always ξ-measurable.)

Lemma 15. *The limits* $\lim_{n\to\infty} \tilde{\mu}\left((f \circ S^n)^<\right)$ *and* $\lim_{n\to\infty} \tilde{\mu}\left((f \circ S^n)^>\right)$ *exist and are equal.*

Proof. Let $f_n^< = (f \circ S^n)^< \circ S^{-n}$ and $f_n^> = (f \circ S^n)^> \circ S^{-n}$. We have $f \geq f_n^<$ for every $n \in \mathbb{Z}_+$ and $f_1^< \leq f_2^< \leq \ldots \leq f_n^< \leq \ldots$. Similarly, $f \leq f_n^>$ and $f_1^> \geq f_2^> \geq \ldots \geq f_n^> \geq \ldots$. Besides, if $\xi_n = S^n\xi$ is a partition of $S^n\Delta$, then $f_n^>(p) - f_n^<(p) = \sup_{\xi_n(p)} f - \inf_{\xi_n(p)} f \leq \omega_{\delta_n}(f)$ where $\delta_n = \sup_p \operatorname{diam}(\xi_n(p))$ and

$$\omega_\delta(f) = \sup_{dist(x,y)<\delta} |f(x) - f(y)|. \tag{68}$$

So, $f_n^> - f_n^< \to 0$ as $n \to \infty$ and, consequently, $f_n^> \searrow f$, $f_n^< \nearrow f$ uniformly as $n \to \infty$. Since we have $(f \circ S^n)^< = f_n^< \circ S^n$, then $|\tilde{\mu}\left((f \circ S^n)^>\right) - \tilde{\mu}\left((f \circ S^n)^<\right)| \leq \sup|(f \circ S^n)^> - (f \circ S^n)^<| = \sup|f_n^> - f_n^<| \leq \omega_{\delta_n}(p)$ which goes to 0 as $n \to \infty$.

Using Theorem 2, we can construct measures $\mu_1, \ldots, \mu_S$ such that $\pi_*(\mu_i) = \phi_i m$.

Proposition 11. *Let* $\tilde{\mu}$ *be an arbitrary measure on* Λ *which is* T*-invariant and such that the sets of* Σ *are* $\tilde{\mu}$*-measurable. Then there exists a unique measure* μ *on* Δ *such that* μ *is* S*-invariant and* $\pi_*(\mu) = \tilde{\mu}$.

Proof. Let μ be constructed as the above measure and let $\bar{\mu}$ be some other measure for which (1) and (2) hold. Obviously, for every $f \in C(\Delta)$ $\bar{\mu}(f^<) \leq \bar{\mu}(f) \leq \bar{\mu}(f^>)$ and, since $\bar{\mu}(f^<) = (\pi_*\bar{\mu})(f^<) = \tilde{\mu}(f^<)$ for every f (in particular for $f \circ S^n$), we get $\tilde{\mu}\left((f \circ S^n)^<\right) \leq (\bar{\mu})(f) \leq \tilde{\mu}\left((f \circ S^n)^>\right)$ because $\bar{\mu}(f) = \bar{\mu}(f \circ S^n)$. A simple limit passage gives the proposition.

Theorem 4. *Let* μ *be Borel, regular measure on* Δ *such that* $\pi_*\mu$ *is absolutely continuous with respect to* m. *Then*

$$\frac{1}{n}\sum_{k=0}^{n-1} L_*^k\mu \overset{n\to\infty}{\longrightarrow} \sum_{i=1}^{S} \mu(\bar{C}_i)\cdot\mu_i \tag{69}$$

where $\bar{C}_i = \pi^{-1}(C_i)$ *and* $\mu_1, \ldots, \mu_s$ *are as above,* $C_1, \ldots, C_n$ *is as in Theorem 2, and the convergence is in weak-*-topology.*

Proof. Let $\rho = \frac{d(\pi_*\mu)}{dm}$. We check easily that

$$\frac{d(\pi_*(L^k\mu))}{dm} = P^k\rho, \qquad k \in \mathbb{Z}_+. \tag{70}$$

This implies

$$\pi_*\left(\frac{1}{n}\sum_{k=0}^{n-1} L_*^k\mu\right) = \left(\frac{1}{n}\sum_{k=0}^{n-1} P^k\rho\right)\cdot m. \tag{71}$$

Let $f \in C(\Delta)$ and let $N \in \mathbb{Z}_+$ be fixed. For $k \geq N$, $f_N \circ L^k$ is ξ-measurable and

$$\mu(f_N^> \circ L^k) = \int f_N^> \circ L^k \rho\, dm = \int (f_N^> \circ L^N) \circ T^{k-N} \rho\, dm$$
$$= \int (f_N^> \circ L^N) \cdot P^{k-N} \rho\, dm. \tag{72}$$

Thus by Theorem 2,

$$\frac{1}{n} \sum_{k=0}^{n-1} \mu(f_N^> \circ L^k) = \frac{1}{n} \sum_{k=0}^{N-1} \mu(f_N^> \circ L^k)$$
$$+ \int \left(\frac{1}{n} \sum_{k=N}^{n-1} P^{k-N} \rho \right) \cdot (f_N^> \circ L^N) dm$$
$$\to \int \left(\sum_{i=1}^{S} \mu(C_i) \cdot \phi_i \right) \cdot (f_N^> \circ L^N) dm$$
$$= \sum \mu(C_i) \cdot \mu_i(f_N^> \circ L^N), \quad \text{as} \quad n \to \infty. \tag{73}$$

Since $f_N^> \to f$ as $N \to \infty$ and $\mu_i(f_N^> \circ L^N) = \mu_i(f_N^>)$, we get

$$\frac{1}{n} \sum_{k=0}^{n-1} \mu(f \circ L^k) \to \sum \mu(\bar{C}_i) \cdot \mu_i(f) \quad \text{as} \quad n \to \infty. \tag{74}$$

This completes the proof of the theorem.

Remark 8. The class of measures satisfying the hypothesis of Theorem 4 includes measures absolutely continuous with respect to the Lebesgue measure and measures l_I defined in Lemma 14 (as well as the convex hull of these measures).

Now we are going to describe the second construction of invariant measures. From this construction it will follow that unstable invariant intervals for L exist almost everywhere in the sense of measure $\mu_1 + \ldots + \mu_S$.

Lemma 16. *The family* $\mathcal{F} = \bigcup_{n\in\mathbb{Z}_+} \bigcup_{k\leq n} L^k \alpha^n$ *is precompact in the Hausdorff topology of all compact subsets of* Δ. *Its cluster points are segments* I *which are contained in some element of the partition*

$$(\alpha|\Lambda)^+ = \bigvee_{k=0}^{\infty} L^k(\alpha|\Lambda). \tag{75}$$

Proof. All compact subsets of Δ form a compact space in the Hausdorff topology. So every set, in particular $\mathcal{F}$, is precompact.

Let $A_n \in \mathcal{F}$ be an arbitrary sequence (see Figure 5(d)). We can see easily that A_n is a polygon and $\partial A_n \subset \bigcup_{j=-m(n)+k(n)}^{k(n)} L^j(\{x = 0\})$ if $A_n \in L^{k(n)}\alpha^{m(n)}$, where $k(n)$, $m(n) \in \mathbb{Z}_+$, $k(n) \leq m(n)$. We can assume that $\operatorname{diam}(A_n) \not\to 0$ (otherwise A_n has a subsequence converging to a point of Λ). This implies that the difference $m(n) - k(n)$ is bounded by some integer N. Taking a subsequence, we may assume $m(n) - k(n) = N$ for all n. Eventually, taking a sequence $L^N A_n$ instead of A_n we may assume that $A_n \in L^{m(n)}\alpha^{m(n)}$. In this case we have $\partial A_n \subset \bigcup_{j=1}^{m(n)} L^j(\{x=0\})$. We can also assume that $m(n) \to \infty$ as $n \to \infty$. Hence, A_n becomes "thinner and thinner". More precisely, ∂A_n is a sum of two broken lines joining the points of A_n with the largest and smallest values of the x-coordinate, which we denote by z_n and y_n respectively. Passing to a subsequence, we may assume that $y_n \to y$ and $z_n \to z_n$. Let I be the segment joining y and z. Then both the upper and the lower broken lines in ∂A_n converge to I (after a suitable reparameterization even in the Lipschitz topology). From the continuity of L it follows that $L^{-n}I \subset \Delta$ and is always contained in one of the half-planes $x \leq 0$ and $x \geq 0$ for all $n \in \mathbb{Z}_+$.

Let us use the ideas of the proof of Theorem 2.

Lemma 17. *Let $\mathcal{D} = \{$all segments $\subset \Delta$with the directions $\in J_+\}$. Then*

$$\mathcal{K} \stackrel{\text{def}}{=} \left\{\frac{d\pi_*(l_I)}{dm}\right\}_{I\in\mathcal{D}} \tag{76}$$

is precompact in $L^1(m)$. Moreover, if $f_I = \frac{d\pi_*(l_I)}{dm}$, then

$$Pf_I = \sum_{J\in L(\alpha|I)} \lambda^u(J) f_J \tag{77}$$

where $\lambda^u(J)$ is equal to the ratio $|L^{-1}J|/|J|$.

Proof. This is an easy consequence of the fact that

$$L_*(l_I) = \sum_{J'\in\alpha|J} L_*(l_{J'}) = \sum_{J'\in\alpha|J} \frac{|J'|}{|LJ'|} LJ' \tag{78}$$

and the definition of P.

Using Proposition 1 and formula (77) we can easily derive

Lemma 18. *Let $\{t_I\}_{I\in\mathcal{D}}$ be any family of nonnegative numbers. If $f = \sum_{I\in\mathcal{D}} t_I f_I$, then $P^N f = \sum_{I\in\mathcal{D}} t_I' f_I$, where*

$$\left|\sum_{I\in\mathcal{D}} t_I'\right| \leq r_0 \left|\sum t_I\right| + R_0 \cdot \|f\|_1 \tag{79}$$

where N, r_0, R_0 are as in the proof of Proposition 1.

Corollary 4. *Let $\mathcal{L}_\gamma = (k \cdot \overline{\mathrm{conv}(\mathcal{K} \cup \{0\})}) \cap \{\|f\|_1 \leq \gamma\}$. $\mathcal{L}_\gamma$ is a compact set in $L^1(m)$ and $P^N(\mathcal{L}_\gamma) \subset (r_0 \cdot k + R_0\gamma) \cdot \mathcal{L}_\gamma$ where N, r_0, R_0 are as in the proof of Proposition 1 and γ is an arbitrary nonnegative number. In particular, for every $k > R_0(1-r_0)^{-1}\gamma$, $P(k\mathcal{L}_\gamma) \subset k\mathcal{L}_\gamma$.*

Thus, P^N preserves compact set $k \cdot \mathcal{L}_\gamma$ and, by Szauder fixed point theorem, it has a fixed point ϕ. This gives another proof of the existence of an invariant measure for T^N absolutely continuous with respect to m.

In the sequel we put $\gamma = 1$ and $k = R_0(1 - r_0)^{-1}$ and $\mathcal{L} = k \cdot \mathcal{L}_1$.

Lemma 19. *$\bigcap_{n=0}^{\infty} P^n(\mathcal{L}) \subset k\bar{\mathcal{L}}$, where*

$$\bar{\mathcal{L}} = \left\{ \frac{d\pi_*(l_I)}{dm} \right\}_{I \in \mathcal{D}_\infty} \tag{80}$$

and $\mathcal{D}_\infty$ is the set of segments contained in some element of $(\alpha|\Lambda)^+$.

Proof. We notice that if I_n is a sequence of points of $\mathcal{D}$ and $\delta_n \stackrel{\text{def}}{=} dist(I_n, I) \to 0$ for some segment I, then by Corollary 1, $m(\pi(I_n) \div \pi(I)) \leq A_2\delta_n + 2\delta_n \to 0$. This implies $\|f_{I_n} - f_I\|_1 \leq (\sup f_{I_n}) \cdot m(I_n \backslash I) + \sup f_I \cdot m(I \backslash I_n) + \sup_{E_n} |f_{I_n}|E_n - f_I|E_n| \cdot m(E_n)$ where $E_n = \pi(I_n) \cap \pi(I) \subset X$. We may assume $|I| > 0$. By Lemma 14, since $\angle(I_n, x) \to \angle(I, x)$ for $x \in E_n$, we can see that the last summand tends to 0 as $n \to \infty$.

Let us fix some $N \in \mathbb{Z}_+ \cup \{\infty\}$. The set $\mathcal{L}(N) = \overline{\bigcap_{n=0}^{N} P^n(\mathcal{L})}$ is convex and compact. Moreover, if $N < +\infty$, then $\mathcal{L}(N) = \overline{\text{conv}\{f_I : I \in \mathcal{D}_N\}}_N$ where $\mathcal{D}_N = \{I \subset L(\Delta) :\ x$ has the same sign on $L^{-i}(I)$ for $i = 1, \ldots, N\}$. Since we proved that the map $\mathcal{D}_N \ni I \to f_I \in L^1(m)$ is continuous (and even Lipschitz), then by known theorems of functional analysis for every function $\phi \in \mathcal{L}(N)$ there is a regular, Borel, probabilistic measure τ_N on $\mathcal{D}_N$ such that $\phi = \int_{\mathcal{D}_N} f_I \cdot d\tau_N(I)$. If $\phi \in \mathcal{L}(\infty)$ then we choose a subsequence of $\{\tau_N\}_{N \in \mathbb{Z}_+}$ converging in the weak-$*$-topology to some τ. Thus, $\text{supp}(\tau) \subset \cap_{N=0}^{\infty} \mathcal{D}_N = \mathcal{D}_\infty$ and

$$\phi = \int_{\mathcal{D}_\infty} f_I d\tau(I). \tag{81}$$

Corollary 5. *For each invariant density ϕ_i, $1 \leq i \leq s$, there is a measure τ_i on $\mathcal{D}_\infty$ such that*

$$\phi_i = \int_{\mathcal{D}_\infty} f_I d\tau_i(I). \tag{82}$$

Let μ_i be the corresponding L-invariant measure. Then

$$\mu_i = \int_{\mathcal{D}_\infty} l_I d\tau_i(I). \tag{83}$$

(The integral makes sense, since $\mathcal{D}_\infty \ni I \to l_I \in C(\Delta)^*$ is continuous, if $\mathcal{D}$ is endowed with the Hausdorff topology and $C(\Delta)^*$ is endowed with the weak-$*$-topology set and the image of this map is a weak-$*$-compact set.)

For a while let μ_i be defined via (83).

Proof. We have $\pi_*(\mu_i) = \int_{\mathcal{D}_\infty} \pi_*(l_I) d\tau_i(I)$. Hence,

$$\frac{d\tau_*(\mu_i)}{dm} = \int_{\mathcal{D}_\infty} \frac{d\pi_*(l_I)}{dm} d\tau_i(I) = \int_{\mathcal{D}_\infty} f_I d\tau_i(I) = \phi_i. \tag{84}$$

By Proposition 11, both definitions of μ_i coincide.

Proposition 12. *For every $\phi \in L^\infty(m)_+$ such that $P\phi = \phi$, there exists a measure τ such that* $\mathrm{supp}(\tau) \subset (\alpha|\Lambda)^+$ *and*

$$\phi = \int_{\mathcal{D}_\infty} f_I d\tau(I). \tag{85}$$

Moreover, if we define a measure μ on Δ via

$$\mu(f) = \int_{\mathcal{D}_\infty} l_I(f) d\tau(I), \qquad f \in C(\Delta) \tag{86}$$

then μ is the measure satisfying $\pi_*(\mu) = \phi$ and the conditional measures μ_I, where $I \in (\alpha|\Lambda)^+$ are given by

$$\mu_I = \csc\omega \cdot l_I. \tag{87}$$

Proof. The proposition is an easy consequence of formula 85. We notice that $I \in \mathcal{D}_\infty$ belongs to $(\alpha|\Lambda)^+$ iff $\partial I \subset \bigcup_{n=0}^\infty L^n(\{x=0\})$.

The idea of the proof is to iterate the measure in formula 81 in order to get more desired intervals. This is possible since $I \in \mathcal{D}_\infty$ is expanded by L, so I breaks on the singularity $x = 0$ and produces intervals $J \in (\alpha|\Lambda)^+$.

For each n and $A \in \alpha^n$, let $\Psi^{A,n} : \mathcal{D} \to \mathcal{D}$ be defined by $\Psi^{A,n}(I) = L^n(I \cap A)$. We write Ψ^A instead of $\Psi^{A,1}$. Let $U : C(\Delta)^* \to C(\Delta)^*$ be an operator defined by $U_\tau = \sum_{A\in\alpha} \lambda^u \cdot \Psi^A_*(\tau)$. We check easily that for each ϕ of the form

$$\phi = \int_{\mathcal{D}} f_I d\tau(I) \tag{88}$$

we have $P\phi = \int_{\mathcal{D}} f_I d(U\tau)(I)$. This implies $P^n\phi = \int_{\mathcal{D}_n} f_I d(U^n\tau)(I)$, $n \in \mathbb{Z}_+$, where

$$U^n\tau = \sum_{A\in\alpha^n} \lambda^u_n \Psi^{A,n}_*(\tau). \tag{89}$$

We notice that $\alpha^n|I$ consists of intervals J such that all except these containing the ends of I have the property $\partial I \subset \cup_{k=o}^{n-1} L^{-k}(\{x=0\})$. So, $L^n J \in (\alpha|\Lambda)^+$ except perhaps these two (at most) intervals. This implies that if

$$\mathcal{E}_n = \{I : \partial I \subset \cup_{k=1}^n L^k(\{x=0\})\} \tag{90}$$

then $U^n_\tau(\mathcal{D}\backslash\mathcal{E}_n) \le 2 \cdot \sup \lambda^n_u \cdot \|\tau\| \le 2\lambda^n_+ \cdot \|\tau\|$. Lemma 18 says that for every $\tau \in C(\Delta)^*$, $\|U^{n+N}\tau\| \le r_0\|\tau\| + R_0\tau(|I|)$. So, by considerations similar to those of Proposition 7, we get $\|U^n\tau\| \le F(r^n\|\tau\| + \tau(|I|))$ for some $F \ge 0$ and $r \in (0,1)$. In particular, $\|U^n\tau\| \le A_5\|\tau\|$ for some $A_5 \ge 0$.

If we apply (90) to $U^{m-n}\tau$ instead of τ, we get for every $m \geq n$, $U^m\tau(\mathcal{D}\backslash\mathcal{E}_n) \leq 2A_5\lambda_+^n\|\tau\|$. Let τ be a measure from (88) and let $\bar{\tau}$ be a weak accumulation point of the sequence $\tau^n = \frac{1}{n}\sum_{k=0}^{n} U^k\tau$. Then $\frac{1}{n}\sum_{k=0}^{n-1} P^k\phi \to \int_{\mathcal{D}} f_I d\bar{\tau}(I)$ at least for a subsequence. We have $\tau^m(\mathcal{E}_n) \geq 1 - 2\lambda_+^n A_5$ for each n and $m \geq n$. Because $\mathcal{E}_n$ is closed in $\mathcal{D}$, we get $\bar{\tau}(\mathcal{E}_n) \geq 1 - 2A_z\lambda_+^n \to 1$ as $n \to \infty$. This implies $\bar{\tau}\left(\bigcup_{n=1}^{\infty}\mathcal{E}_n\right) = 1$. Therefore, $\bar{\tau}\left((\alpha|\Lambda)^+\right) = 1$. If $P\phi = \phi$, we get $\phi = \int_{(\alpha|\Lambda)^+} f_I d\bar{\tau}(I)$.

Theorem 5. *There exist sets* $\mathcal{M}_1, \ldots, \mathcal{M}_s \subset \Delta$ *such that* $l_I(\bar{C}_i\backslash\mathcal{M}_i) = 0$ *for every* $I \in \mathcal{D}$ *and* $\mathcal{M}_i \subset \bar{C}_i$ $(i = 1, \ldots, s)$, *and for every* $p \in \mathcal{M}_i$ *and* $f \in C(\Delta)$ *we have* $\lim_{n\to\infty}\frac{1}{n}\sum_{k=0}^{n-1} f(L^k p) = \mu_i(f)$.

Proof. By the ergodic theorem, the statement about the limit holds on the set $\mathcal{M}_i^0$ of generic points of μ_i. It is well known that $\mu_i(\mathcal{M}_i^0) = 1$. Let $\mathcal{M}_i = \{p \in \bar{C}_i : \exists p' \in \mathcal{M}_i,\ j \in \mathbb{Z}_+ : L^j p \in \xi(p')\}$. We can see that the limit statement holds on $\mathcal{M}_i$, too. We want to prove that $l_I(\bar{C}_I\backslash\mathcal{M}_I) = 0$. Let us assume that it is not true and let us define $\mu = \chi_{\bar{C}_i\backslash\mathcal{M}_i}\cdot l_I$. By Theorem 4, $\frac{1}{n}\sum_{k=0}^{n-1} L^k(\mu) \to l_I(\bar{C}_i\backslash\mathcal{M}_i)\cdot\mu_i$. Projecting onto X, we obtain $\frac{1}{n}\sum_{k=0}^{n-1} P^k(\chi_{\bar{C}_i\backslash\mathcal{M}_i}\cdot f_I) \to l_I(\bar{C}_i\backslash\mathcal{M}_i)\cdot\phi_i$ in $L^1(m)$. The left-hand side restricted to $\pi(\mathcal{M}_i^0)$ is equal to 0, while the right-hand is not, unless $l_I(\bar{C}_I\backslash\mathcal{M}_i) = 0$.

4 The Variational Principle and the Convergence of the Iterations of the Perron-Frobenius Operator

4.1 Basic Estimates

We are going to prove that in general the convergence of the iterations of the Perron-Frobenius operator (as defined in section 2) is a sufficient condition for the measures $\mu_1, \ldots, \mu_s$ from Theorem 2 to satisfy a maximum principle, which is analogous to the variational principle of Ruelle and Bowen [1].

In order to formulate the results we have to modify the assumptions of section 2 a little.

Let (X, Σ) be a measurable space and let $T : X \to X$ be a measurable transformation. We assume that there exists a measurable partition β of X which is countable and is a one-sided generator. We also assume that for every $B \in \beta$, $T|B : (B, \Sigma|B) \to (TB, \Sigma|TB)$ is an isomorphism. (These assumptions have been copied from section 2).

We also assume that a Σ-measurable function $g : X \to \mathbb{R}_+$ is given and a measure m on X such that (X, Σ, m) is a Lebesgue space and $g = g_T$ almost everywhere. We notice that g_T is defined a. e. , while g is defined *everywhere.* We assume that $\sup_X g < +\infty$ and $\inf_X g > 0$. We also assume that the "g-operator" $Pf(x) = \sum_{y\in T^{-1}(x)} g(y)f(y)$ is well defined on the space of all bounded measurable functions $f : X \to \mathbb{R}$. For example, $\sum_{B\in\beta}\sup_B g < +\infty$ is a sufficient condition.

Let $\mathcal{M}$ be a family of T-invariant probability measures defined on Σ. We assume that for every $\mu \in \mathcal{M}$ the entropy $H_\mu(\beta) < +\infty$. Hence (see [11] and [14]), $h_\mu(T) = h_\mu(T, \beta) = \int_X \ln J\, d\mu < +\infty$. (we notice that (X, Σ, μ) is always a Lebesgue space). J, as usual, denotes the Jacobian of T.

The pressure functional $\mathcal{P}: \mathcal{M} \to \mathbb{R} \cup \{-\infty\}$ is defined by

$$\mathcal{P}(\mu) = h_\mu(T) + \int_X \ln g\, d = \int_X \ln(Jg)\, d. \tag{91}$$

We will call a measure $\mu \in \mathcal{M}$ an equilibrium state (with respect to g) iff μ realizes the maximum of $\mathcal{P}$ and $\mathcal{P}(\mu) = 0$. We sometimes write $\mathcal{P}(T, \mu)$, if it is not clear what transformation we are considering.

Lemma 20. *For every* $\mu \in \mathcal{M}$

$$\mathcal{P}(\mu) \leq \inf_{\phi, n \geq 1} \frac{1}{n} \ln \int_X \frac{P^u \phi}{\phi} d\mu \tag{92}$$

where the infimum is over all positive, measurable functions $\phi: X \to \mathbb{R}_+$.

Proof. For each point $x \in X$ the set $T^{-1}(x)$ is countable, so there is a function $h: X \to \mathbb{R}_+$ such that $\mu_{T^{-1}(x)} = \sum_{y \in T^{-1}(x)} h(y)\delta_y$ where $\{\mu_{T^{-1}(x)}\}_{x \in X}$ is the family of the conditional measures with respect to the partition $\{T^{-1}(x)\}_{x \in X}$ ($= T^{-1}\epsilon$, where ϵ is the partition into points).

The condition $T_* \mu = \mu$ implies $h(y) = 1/J(y)$ almost everywhere on X (see [14]). Thus, $\mathcal{P}(\mu) = \int_X \ln(J \cdot g) d\mu = \int_X \mu_{T^{-1}(x)}(\ln(J \cdot g)) \cdot d\mu(x) = \int \sum_{y \in T^{-1}(Tx)} J^{-1}(y) \cdot \ln(Jg)(y) d\mu(x)$. By Jensen inequality

$$\sum_{y \in T^{-1}(Tx)} J^{-1}(y) \cdot \ln(J \cdot g)(y) \leq \ln \left(\sum_{y \in T^{-1}(Tx)} g(y) \right) = \ln(P1 \circ T)(x).$$

Hence $\mathcal{P}(\mu) \leq \int \ln(P1 \circ T) d\mu \leq \ln \int P1 \circ T d\mu = \ln \int P1 d\mu$. Let $\phi: X \to \mathbb{R}_+$ be any bounded and bounded away from 0 (Σ-measurable) function. If we apply the previous inequality and definition 91 to $g \cdot \phi/\phi \circ T$ instead of g, we obtain $h_\mu(T) + \int \log(g \cdot \phi/\phi \circ T) d\mu \leq \ln \int \frac{P\phi}{\phi} d\mu$. Since $\int \ln(g\phi/\phi \circ T) = \int \ln g\, d\mu + \int \ln \phi d\mu - \int \ln \phi \circ T\, d\mu = \int \ln g\, d\mu$, these last two formulas yield $\mathcal{P}(\mu) \leq \ln \int \frac{P(\phi)}{\phi} d\mu$. Replacing T by T^n and using $\mathcal{P}(\mu) = \frac{1}{n}\mathcal{P}(T^n, \mu)$, we get the lemma.

Lemma 21. *Let* T, β *and* m *satisfy the assumptions of section 2. Let* $\phi \in L^\infty(m)$ *be nonnegative and* $P\phi = \phi$, $\int \phi\, dm = 1$. *Then* $\mathcal{P}(\mu) = 0$, *where* $\mu = \phi \cdot m$.

Proof. We use the following fact from [6] : Let $f: X \to \mathbb{R}$ be Σ-measurable and let $(f - f \circ T)_+$ be μ-integrable. Then $f - f \circ T$ is integrable and $\int_X (f - f \circ T) d\mu = 0$. We notice that $\phi(Tx) = P\phi(Tx) = \sum_{y \in T^{-1}(Tx)} g(y)\phi(y) \geq$

$g(x)\phi(x)$. Hence, $\phi/\phi \circ T \leq g^{-1} = J$ and $\sup J = 1/\inf g$ yields $\int(\ln\phi - \ln\phi\circ T)_+ d\mu < +\infty$. Thus, $\int_X \ln\frac{\phi}{\phi\circ T}d\mu = 0$. It is not difficult to see that the Jacobian J_μ of μ is equal to $J\cdot\phi/\phi\circ T$, where $J = 1/g$. Hence $\mathcal{P}(\mu) = \int_X \ln(J_\mu\cdot g)d\mu = \int_X \ln(J\cdot g\cdot\phi/\phi\circ T)d\mu = \int_X \ln\phi/\phi\circ T d\mu = 0$.

From now on we assume that the following condition is satisfied:

(*) There is a constant $C \in \mathbb{R}_+$ such that for every $\epsilon > 0$ and $A \in \Sigma$ and for every $\mu \in \mathcal{M}$ there is a β^k-measurable set $A_1 \in \Sigma$ such that $\mu(A \div A_1) < \epsilon$ and for sufficiently large $n \in \mathbb{Z}_+$ $P^n(\chi_{A_1}) \leq C\cdot m(A_1)$.

Obviously, if $\mathcal{M}$ is the family of measures $\pi_*(\nu)$, where ν is any L-invariant, Borel, probability, non-atomic measure, then this condition is satisfied by virtue of $\|P^{n^N}(\chi_{A_1})\|_\infty \leq r^n \ \#\{B \in \beta^k : \ B \subset A_1\} + \frac{R}{1-r}m(A_1)$ following from the estimates of the proof of Theorem 2. We notice that only non-atomic measures have to be considered because g is not defined on the set $\bigcup_{k=0}^{\infty} L^{-k}(\{x = 0\})$ which could contain some atoms of μ and the integral $\int \ln g\, d\mu$ would not make sense. For nonatomic μ we have $\mu\left(\bigcup_{k=0}^{+\infty} L^k(\{x=0\})\right) = 0$ because this set is countable.

Theorem 6. *Each equilibrium state μ is of the form $\phi\cdot m$ where $\phi \geq 0$ $\int\phi\, dm = 1$ and $P\phi = \phi$.*

Proof. Let $\mu \in \mathcal{M}$ and $\mathcal{P}(\mu) \geq 0$. We can assume that μ is ergodic, because otherwise we could work with an ergodic component of μ. We want to prove that $\mu \ll m$. Let us suppose that on the contrary, $\mu \not\ll m$. There exists a set $A \in \Sigma$ such that $m(A) = 0$ and $\mu(A) = 1$. Let $\epsilon = 1/9C$ and let $A_1 \in \Sigma$ be such that $\mu(A \div A) < \epsilon$ and $m(A_1) < \epsilon$. Let us put $\phi = t_1\chi_{A_1} + t_2\chi_{X\setminus A_1}$ where $t_1 = 1/3\epsilon^{-1}$ and $t_2 = \frac{1-t_1 m(A_1)}{1-m(A_1)}$. By condition (*) we get that for some $n \in \mathbb{Z}_+$ we have $\|P^n\phi\|_\infty \leq C\,(t_1 m(A_1) + t_2(1-m(A_1))) \leq C$. Hence,

$$P^n\phi/\phi \leq \begin{cases} 1/3 & \text{on } A_1 \\ 3/2C & \text{on } X\setminus A_1. \end{cases} \tag{93}$$

(We observe that $t_2 \geq \frac{1-1/3}{1-\epsilon} \geq 2/3$). Finally, $\mu\left(\frac{P^n\phi}{\phi}\right) \leq 1/3\mu(A_1) + 3/2C\mu(X\setminus A_1) \leq 1/3 + 3/2C\cdot\epsilon = 1/2 < 1$. So, $\mathcal{P}(\mu) \leq \frac{1}{n}\ln\mu\left(\frac{P^n\phi}{\phi}\right) \leq 0$. This contradicts the assumption $\mu \ll m$.

4.2 The Variational Principle for Lozi Mappings

For every L-invariant, non-atomic, probabilistic measure μ on Δ we have $\mu\left(\bigcup_{k=-\infty}^{+\infty} L^k(\{x=0\})\right) = 0$. This implies that both λ^u and λ^s are well-defined μ-a. e. Let $\phi^u = \log\lambda^u \circ L$. The geometrical meaning of $\lambda^u(Lp)$ is the reciprocal of the expansion coefficient along the unstable direction $E^u(p)$. So, ϕ^u is analogous to the function considered by Bowen and Ruelle [1].

Because of the cohomology relation 59 we have

$$h\mu(L) + \int_\Lambda \phi^u d\mu = h_\mu(L) + \int_\Lambda \log(\lambda^s|b|^{-1})d\mu$$
$$+ \int \log h_3 \circ L d\mu - \int \log h_3 d\mu = h_\mu(L) + \int \log g d\mu. \qquad (94)$$

The last expression is equal to $P(\pi_* \mu)$. The previous section yields:

Theorem 7. *The only measures maximizing the pressure functional* $h_\mu(L) + \int_\Lambda \phi^u d\mu$ *are of the form* $t_1\mu_1 + \ldots + t_s\mu_s$, $\sum_i t_i = 1$, $t_i \geq 0$. *Moreover, for such measures the value of the pressure functional is equal to* 0.

References

1. Bowen, R. *Equilibrium states and ergodic theory of Anosov diffeomorphisms* Lecture Notes in Math., Vol. **470**, Addison-Wesley, Berlin, Heidelberg, New York, p. 108, 1978.
2. Bunimovich, L. A., Sinai, Ya. *Markov partitions for dispersed billiards* Commun. Math. Phys. **78**, 247–280 (1980).
3. Collet, P., Levy, Y. *Ergodic properties of the Lozi Mappings* Commun. Math. Phys. **93**, 461–481 (1984).
4. Dunford, N., Schwartz, J. T. *Linear Operators*, Part 1, Wiley, New York (1959).
5. Hénon, M. *A two-dimensional mapping with a strange attractor* Commun. Math. Phys. **50**, 69–77 (1976).
6. Hofbauer, F., Keller, G. *Ergodic properties of invariant measures for piecewise monotonic transformations* Math. -Z. **180**, no. 1, 119-140 (1982).
7. Keller, G. *Generalized bounded variation and its applications to piecewise monotonic transformations* Z. Wahrsch. Verw. Gebiete **69**, no. 3, 461–478 (1985).
8. Lasota, A., Yorke, J. *On the existence of invariant measures for piecewise monotonic transformations* Trans. AMS **186**, 481–488 (1973).
9. Lozi, R. *Un attracteur étrange du type attracteur de Hénon* J. Phys. (Paris) **39**, (Coll. C5), 9–10 (1978).
10. Misiurewicz, M. *Hyperbolicity of the Lozi mappings* Annals of the New York Acad. of Sci., Vol. **357** (Nonlinear Dynamics), 348–358 (1980).
11. Rokhlin, V. A. *Lectures on the entropic theory of transformations with an invariant measure* UMN **12**, No. 5, 3–56 (1967) in Russian.
12. Rychlik, M. *Mesures invariantes et principe variationel pour les applications de Lozi* Compte Rend. Acad. Sci. Paris, t. 296 (1983).
13. Rychlik, M. *Bounded variation and invariant measures* Studia Math., t. 76, 69–80 (1983).
14. Walter, P. *Invariant measures and equilibrium states for some mappings which extend distances* Trans. AMS **236**, 121–153 (1978).

Commun. Math. Phys. 93, 461–481 (1984)

Communications in
Mathematical
Physics

Ergodic Properties of the Lozi Mappings

P. Collet and Y. Levy

Centre de Physique Théorique, Ecole Polytechnique, F-91128 Palaiseau Cédex, France

Abstract. In this paper, we construct the Bowen-Ruelle measure for the Lozi mapping, an almost everywhere hyperbolic diffeomorphism of the plane. We also derive some of its properties which are similar to those of an axiom A system.

I. Introduction

The Lozi mapping T is a homeomorphism of $\mathbb{R}^2$ given by

$$\begin{pmatrix} x \\ y \end{pmatrix} \to \begin{pmatrix} 1+y-a|x| \\ bx \end{pmatrix}.$$

For some values of a and b, Lozi [Lo] observed complicated behaviour for the trajectories of this system. For $b=0.5$ and $a=1.7$ one observes numerically a strange attractor, which is very similar to the attractor of the Henon map [He]. The main advantage of the Lozi map over the Henon map is that one can prove hyperbolicity without much effort. This is the main reason why so little is known for the Henon map, where hyperbolicity is believed to occur only on Cantor-like sets of parameters. Our opinion is that the Lozi mapping is an intermediate stage between the Axiom A dynamical systems and more complicated systems like the Henon map. As we shall see below, its dynamical structure is more complicated than in the Axiom A systems, although some detailed ergodic properties are the same. The Lozi map is rather similar to Sinai's billiards, and in this article, we shall use this analogy. In particular, the discontinuity of the differential allows the uniform hyperbolicity as in the billiards case. A proof of hyperbolicity for the Lozi map was first given by Misiurewicz [M]. He also derived many important consequences which will be described below.

This article is devoted to the investigation of the metric properties of the Lozi map. In the next paragraph, we briefly describe some properties of the map which will be needed later on. Most of them were known before. In the third paragraph we construct an invariant measure; its ergodic properties (absolute continuity with respect to the Lebesgue measure in the unstable direction, ergodicity, K-property, Bowen-Ruelle property, Bernoulli character) are derived in Sect. IV.

Similar results were obtained by Rychlik [Ry] using a different proof. We learned that Young [Y 1] has proven similar results for piecewise C^2 hyperbolic maps.

II. Notations – General Properties

We set some notations which will be of constant use in this paper. $\mathcal{M}$ is the Borel σ-algebra of $\mathbb{R}$. If A is a measurable subset of $\mathbb{R}^2$, $\mathcal{M}_A$ is the corresponding factor sub σ-algebra. l and m are the 1-dim and 2-dim Lebesgue measure. d is the Euclidean distance in $\mathbb{R}^2$. We shall sometimes use the notations $\pm$ and $\mp$. We adopt the values of [M] for a and small values for b. For a partition ξ of the space X, $\xi(x)$, $x \in X$ is the atom of ξ containing x. Let $S_0^- = Oy$, $S_0^+ = Ox = TS_0^-$; $S_n^\pm = T^{\pm n} S_0^\pm$, $n \in \mathbb{N}$, are finite broken lines. Note that $T^{\pm n}$ is singular on $S_n^\mp$, $n \in \mathbb{N}$.

The fields of stable (unstable) directions $E^s(\cdot)$ $[E^u(\cdot)]$ are defined outside $\bigcup_{n \geqq 0} S_n^- \left(\bigcup_{n \geqq 0} S_n^+ \right)$ (cf. [M]). If $\theta^s(x)$ and $\theta^u(x)$ are their angles with respect to the x axis, we have continuous fraction expansions for $\tan\theta^s(x)$ and $\tan\theta^u(x)$ which are given by

$$\tan\theta^u(x) = \frac{-b}{a \cdot \varepsilon(T^{-1}x) - \tan\theta^u(T^{-1}x)}, \qquad \tan\theta^s(x) = a \cdot \varepsilon(x) + \frac{b}{\tan\theta^s(Tx)},$$

where $\varepsilon(x)$ is the sign of x. These formulae express the fact that $E^s(\cdot)$ and $E^u(\cdot)$ are invariant fields. If b/a is small enough, the above continuous fraction expansions are convergent. We shall denote by $\lambda(x)$ the expansion factor in the direction $E^u(x)$, i.e. this is the length of the image by DT_x of the unit vector in the direction $E^u(x)$. It is easy to verify that

$$\lambda(x) = \frac{[a^2 + b^2 + \tan^2(\theta^u(x)) - 2a\varepsilon(x)\tan\theta^u(x)]^{1/2}}{(1 + \tan^2\theta^u(x))^{1/2}}.$$

Let $W^s(x)$ [respectively $W^u(x)$] be the global stable (respectively unstable) manifold of x. The maximal smooth component $W^s_{\text{loc}}(x)$ $[W^u_{\text{loc}}(x)]$ of $W^s(x)$ $[W^u(x)]$ containing x will be called the local stable (local unstable) manifold of x. x splits $W^u_{\text{loc}}(x)$ into two "semi local unstable manifolds."

Let X be the fixed point of T with positive coordinates. Z is the intersection of the positive x axis with $\overline{W^u_{\text{loc}}(X)}$. We shall denote by F the triangle defined by the points Z, $T(Z)$, $T^2(Z)$ (see [M]). Ω will denote the strange attractor of T which is equal to $\bigcap_0^\infty T^n(F)$ [M]. We shall denote by λ and λ_+ the infimum and the supremum of $\lambda(x)$ for x in the set $F \setminus \bigcup_{-\infty}^{+\infty} S_i^\pm : \lambda > 0$.

It is now easy, using the continuous fraction expansion, to verify the following lemma

Lemma II.1. *Let* x *and* y *in* F *be such that for* $0 \leqq j \leqq q$, $T^j(x)$ *and* $T^j(y)$ *are on the same side of* S_0. *Then*

i) *The angle between* $W^s_{loc}(x)$ *and* $W^s_{loc}(y)$ *(if they are defined) is bounded by* $2\left(\frac{b}{\lambda}\right)^{q-1}$.

ii) $\left|\frac{\lambda(x)}{\lambda(y)}-1\right| \leq 2\left(\frac{b}{\lambda}\right)^{q-1}$.

We also observe that the angle between two local stable manifolds is between $\frac{2\pi}{5}$ and $-\frac{2\pi}{5}$ if $a<2$, and the angle between a local stable and a local unstable manifold is greater than $\frac{\pi}{5}$.

We now introduce a set H which is more convenient for our purposes than the set F. This set was already used by Misiurewicz but we recall here its properties.

Lemma II.2. *For b small enough, there is a polygon H such that*

i) $TH \subset H$,
ii) $\Omega \subset H$,
iii) *The boundary of H is contained in* $T^q W^u_{loc}(X) \cup W^s_{loc}(X)$ *for some* $q>0$.
iv) $T^r F \subset H$ *for some positive* r.

Proof. We adopt the notations of [M]. H_0 is the triangle XZP, where $P = W^s_{loc}(X) \cap (Z, T(Z))$ and $H = \bigcup_{n=0}^{\infty} T^n(H_0)$. ($T$ is denoted f in [M].)

In Proposition 2 of [M], it is shown that there is a positive integer p such that

$$H = \bigcup_{n=0}^{p} T^n(H_0).$$

Moreover, in the same Proposition 2, it is established that

$$T^p F \subset H \quad \text{and} \quad TH \subset H. \quad \text{Q.E.D.}$$

We shall use the following mixing property derived by [M].

Proposition II.3. (Ω, T) *is topologically mixing, i.e.: if* $A, B \subset \mathbb{R}^2$ *are open, then*

$$(A \cap \Omega \neq \emptyset,\ B \cap \Omega \neq \emptyset) \Rightarrow (\exists N, n>N \Rightarrow T^n A \cap B \cap \Omega \neq \emptyset).$$

We investigate now the absolute continuity of the unstable foliation. Let W^1 and W^2 be two local unstable manifolds in Ω. We define a map $P = P_{W^1 W^2}$ from W^1 to W^2 by

$$x \in W^1 \to P(x) = W^s_{loc}(x) \cap W^2 \quad \text{if} \quad W^s_{loc}(x) \cap W^2 \neq \emptyset.$$

P is defined on $\mathcal{D}(P) = \{x \in W^1 \mid W^s_{loc}(x) \cap W^2 \neq \emptyset\}$.

Proposition II.4. *Given W^1 and W^2 as above there is a constant $L_1>0$ which is $\mathcal{O}(1)$ such that for any Borel subset A of W^1, $A \subset \mathcal{D}(P)$,*

$$(1 - L_1(d(W^1, W^2))^{1/3}) l(A) \leq l(P(A)) \leq (1 + L_1(d(W^1, W^2))^{1/3}) l(A).$$

The proof is given in the appendix.

III. Construction of Invariant Measures

Let $L_0 = W^u_{\mathrm{loc}}(X)$. We know from Misiurewicz's work [M] that Ω is the closure of $\bigcup_{n \geq 0} T^n L_0$. Therefore, it is natural to try to obtain an invariant measure by iterating the Lebesgue measure supported by L_0. We define a sequence $(\mu'_n)_{n \in \mathbb{Z}}$ of probability measures on $\mathcal{M}_\Omega$ (μ'_n has support in $T^n L_0$) by $\mu'_n(A) = l(T^{-n} A \cap L_0)/l(L_0)$ for $A \in \mathcal{M}_\Omega$. The sequence $(\mu_n)_{n \in \mathbb{N}}$ defined by

$$\mu_n = \frac{1}{n} \sum_{j=0}^{n-1} \mu'_j$$

is a sequence of probability measures on $\mathcal{M}_\Omega$. As Ω is compact one can extract a subsequence which vaguely converges to an invariant probability measure μ (cf. [B]). Let $A^\pm_\varepsilon$ denote the stripe of width ε around $S^\pm_0$.

In order to obtain some properties of the measure μ, we shall first estimate the μ_n-measures of $A^\pm_\varepsilon$.

Proposition III.1. *For any positive number* $\tau < \left(1 - \frac{1}{K \log \lambda}\right)$ *(K is the integer appearing in Lemma III.2), there is a positive real number* ε_0*, such that for* $0 < \varepsilon \leqq \varepsilon_0$*, we have* $\mu_n(A^\pm_\varepsilon) \leqq \varepsilon^\tau$, $\forall n \in \mathbb{N}$.

We give the proof for $A_\varepsilon = A^-_\varepsilon$. We shall first give some geometrical considerations. For $n \in \mathbb{N}$, $T^n L_0$ is a segment or a broken line. Let J_n be the set of maximal smooth components contained in $T^n L_0$. For $M \in J_n$, the endpoints of M belong to $S_p \cup S_q$ for some p, q $1 \leqq p, q \leqq n+2$. We define $k(M)$ by $k(M) = \inf(p, q)$. For M belonging to J_n, we shall denote by $R_p(M)$ the element of J_{n-p} containing $T^{-p} M$ for $p \in \mathbb{Z}$. We now define recursively a finite sequence of integers $k_i(M)$ by

$$k_0(M) = 0,$$

$$k_1(M) = k(M),$$

$$k_{i+1}(M) = k_i(M) + k(R_{k_i(M)}(M)) \quad \text{as long as} \quad R_{k_i(M)}(M) \neq L_0.$$

We shall write k_i instead of $k_i(M)$ when there is no ambiguity. We now prove a lower bound on $k(\cdot)$.

Lemma III.2. *Assume b is small enough, there is an integer* $K > 4$ *and* $\theta > 0$ *such that if* $M \in J_n$, $l(M) < \theta$*, and* $M \cap S^-_0 \neq \emptyset$*, then* $k(M) > K$.

Proof. For b small enough, there is an integer $K > 2$ such that $0 < p \leqq K$ implies $(S^-_0 \cap F) \cap S^-_p = \emptyset$.

Let $\theta = \inf_{0 < p \leqq K} d(S^-_0 \cap F, S^-_p \cap F)$, then if $l(M) < \theta$, we have $k(M) > K$. Q.E.D.

We now come to the basic estimate. It is enough to prove the assertion for μ'_n, $n \in \mathbb{Z}$. The proof is recursive. Let τ be a number such that $0 < \tau < 1 - \frac{1}{K \log \lambda}$. The estimate is obvious for μ'_{-1}, since $l(L_0) > 1$; for μ'_n, $n \leqq -2$ and ε small enough, $\mu'_n(A_\varepsilon) = 0$. From now on, we assume the bound has been already proven for μ'_p, $-1 \leqq p < n$.

Let ϱ, $0<\varrho<1$ be such that

$$(1-\varrho)\left(1-\frac{1}{K\log\lambda}\right)>\tau .$$

Note that such a ϱ exists since $\tau<1-\dfrac{1}{K\log\lambda}$.

Let $E_1=\{M\in J_n | l(M)>4\varepsilon^\varrho\}$. If $M\in E_1$, M is a straight segment of T^nL_0, therefore

$$\frac{l(T^{-n}(A_\varepsilon\cap M))}{l(T^{-n}M)}=\frac{l(A_\varepsilon\cap M)}{l(M)}\leqq\frac{4\varepsilon}{4\varepsilon^\varrho}=\varepsilon^{1-\varrho},$$

since $l(A_\varepsilon\cap M)\leqq 4\varepsilon$ (we have used the fact that the contraction coefficient by T^{-n} is constant along M). Therefore,

$$\sum_{M\in E_1} l(T^{-n}(A_\varepsilon\cap M))\leqq\varepsilon^{1-\varrho}\sum_{M\in E_1} l(T^{-n}M)\leqq\varepsilon^{1-\varrho}l(L_0).$$

We now define a subset E_2 of J_n by

$$E_2=\{M\in J_n|\exists i\in\mathbb{N},\ k_i(M)<c|\log\varepsilon|,\ l(R_{k_i}(M))>4\varepsilon^\varrho\lambda^{-k_i(M)}\},$$

where c is a fixed number satisfying

$$K(1-\varrho)/\log 2>c>(1-\varrho)/\log\lambda \quad \left(\text{note that } \frac{K\log\lambda}{\log 2}>1\right).$$

For $M\in E_2$ let $\sigma(M)$ be the smallest integer i such that

$$l(R_{k_i}(M))>4\varepsilon^\varrho\lambda^{-k_i(M)}.$$

Let p be a positive integer, and let $\tilde{M}$ be an element of J_{n-p} such that $\tilde{M}\cap S_0^-\neq\emptyset$. We shall denote by $N(\tilde{M},p)$ the number of elements M of J_n such that $T^{-p}M$ is included in $\tilde{M}$, and $k_{\sigma(M)}\geqq p$.

Let

$$L_p=\sup_{0\leqq k\leqq p}\ \sup_{\substack{\tilde{M}\in J_{n-k}\\ \tilde{M}\cap S_0\neq\emptyset}} N(\tilde{M},k).$$

Note that L_p is a non-decreasing sequence. We shall now give an upper bound for L_p. For $\tilde{M}$ as above let $\tilde{M}_L$ (respectively $\tilde{M}_R$) be the segment $\tilde{M}\cap\{(x,y)|x\leqq 0\}$ [respectively $\tilde{M}\cap\{(x,y)|x\geqq 0\}$]. Assume moreover that the subset $E_L(p)$ of E_2 defined by $E_L(p)=\{M\in E_2|T^{-p}M\subset\tilde{M}_L,\ k_{\sigma(M)}\geqq p\}$ is non-empty. [If $E_L(p)$ is empty, then $E_R(p)$ is not empty or $L_p=0$.] Let q be the smallest positive integer such that $T^q\tilde{M}_L\cap S_0^-\neq\emptyset$. There are two cases.

Case 1. $q<p$. Then $T^q\tilde{M}_L\in J_{n-p+q}$, and for any $M\in E_L(p)$, we have $R_{k_i}(M)=T^q\tilde{M}_L$ for some integer i. Moreover, $k_i(M)=p-q<p\leqq k_{\sigma(M)}$. This implies $i<\sigma(M)$, and

$$l(T^q\tilde{M}_L)\leqq 4\varepsilon^\varrho\lambda^{-k_i(M)}.$$

If ε_0 is small enough, we have $4\varepsilon^{\varrho} < \theta$, and we can apply Lemma III.2 to conclude that $q > K$. Therefore,

$$\begin{aligned}\mathrm{card}(E_L(p)) &= \mathrm{card}\{M \in E_2 | T^{-p+q}M \subset T^q\tilde{M}_L,\ k_{\sigma(M)} \geqq p\}\\ &\leqq \mathrm{card}\{M \in E_L(p) | T^{-p+q}M \subset T^q\tilde{M}_L,\ k_{\sigma(M)} \geqq p-K\}\\ &\leqq \sup_{0 \leqq k \leqq p-K} \sup_{\substack{\tilde{M}' \in J_{n-k}\\ \tilde{M}' \cap S_0 \neq \emptyset}} N(\tilde{M}', k) = L_{p-K}\,.\end{aligned}$$

Case 2. $q \geqq p$. In this case we have $T^p\tilde{M}_L \in J_n$, and $\mathrm{card}\,E_L(p) = 1$. Similarly, we define $E_R(p)$ by

$$E_R(p) = \{M \in E_2 | T^{-p}M \subset \tilde{M}_R,\ k_{\sigma(M)} \geqq p\},$$

and we obtain as before $\mathrm{card}\,E_R(p) = 1$ or $\mathrm{card}\,E_R(p) \leqq L_{p-K}$. Therefore, $N(\tilde{M}, p) = \mathrm{card}\,E_L(p) + \mathrm{card}\,E_R(p) \leqq \sup(2, 2L_{p-K})$, and we obtain the bound

$$L_p \leqq 2^{1+p/K}.$$

Let now p be an integer such that $0 \leqq p \leqq c|\log\varepsilon|$. Let $\tilde{M} \in J_{n-p}$. We note that if there is an $M \in J_n$ for which $T^{-p}M \subset \tilde{M}$, and $k_{\sigma(M)} = p$, we have $\tilde{M} \cap S_0 \neq \emptyset$. For such an M, we have

$$l(T^{-p}(A_\varepsilon \cap M)) \leqq \lambda^{-p}, \qquad l(A_\varepsilon \cap M) < 4\varepsilon\lambda^{-p},$$

and from $l(\tilde{M}) > 4\varepsilon^{\varrho}\lambda^{-p}$ [since $p = k_{\sigma(M)}(M)$], we deduce

$$\begin{aligned}\sum_{\substack{M \in E_2\\ T^{-p}M \subset \tilde{M}\\ k_{\sigma(M)}(M) = p}} l(T^{-p}(A_\varepsilon \cap M)) &< 4\varepsilon\lambda^{-p}L_p \leqq 4\varepsilon\lambda^{-p}2^{1+p/K}\\ &< 2\varepsilon^{1-\varrho-\frac{c}{K}\log 2}\,l(\tilde{M})\,.\end{aligned}$$

If $M \in J_n$ and $T^{-p}M \subset \tilde{M}$, $\tilde{M} \in J_{n-p}$, we have

$$\frac{l(T^{-n}(A_\varepsilon \cap M))}{l(T^{-n+p}\tilde{M})} = \frac{l(T^{-p}(A_\varepsilon \cap M))}{l(\tilde{M})}\,.$$

Therefore,

$$\sum_{\substack{M \in E_2\\ T^{-p}M \subset \tilde{M}\\ k_{\sigma(M)}(M) = p}} l(T^{-n}(A_\varepsilon \cap M)) < 2\varepsilon^{1-\varrho-\frac{c}{K}\log 2}\,l(T^{-(n-p)}\tilde{M}),$$

and

$$\begin{aligned}\sum_{M \in E_2} l(T^{-n}(A_\varepsilon \cap M)) &= \sum_{p=0}^{E(c|\log\varepsilon|)} \sum_{\tilde{M} \in J_{n-p}} \sum_{\substack{M \in E_2\\ T^{-p}M \subset \tilde{M}\\ k_{\sigma(M)}(M) = p}} l(T^{-n}(A_\varepsilon \cap M))\\ &< 2\varepsilon^{1-\varrho-\frac{c}{K}\log 2} \sum_{p=0}^{E(c|\log\varepsilon|)} \sum_{\tilde{M} \in J_{n-p}} l(T^{-(n-p)}\tilde{M})\\ &< 4\varepsilon^{1-\varrho-\frac{c}{K}\log 2}\, c|\log\varepsilon|\, l(L_0)\,.\end{aligned}$$

Let now E_2 be defined by $E_3 = J_n \backslash (E_1 \cup E_2)$. We shall assume $E_3 \neq \emptyset$, otherwise, the proof is finished. This implies $n > c|\log\varepsilon|$. For $M \in E_3$, let i be the unique integer such that

$$k_i(M) < c|\log\varepsilon| \leqq k_{i+1}(M).$$

Note that $k_{i+1}(M)$ exists if ε_0 is small enough since $l(L_0) > 1$. We now observe that $T^{-(k_{i+1}-k_i)}R_{k_i}(M) \cap S_0^- \neq \emptyset$, and

$$l(T^{-(k_{i+1}-k_i)}R_{k_i}(M)) < \lambda^{-(k_{i+1}-k_i)} l(R_{k_i}(M)) \leqq 4\varepsilon^{\varrho}\lambda^{-k_{i+1}}.$$

Therefore, $T^{-(k_{i+1}-k_i)}R_{k_i}(M) \subset A_{4\varepsilon^{\varrho}\lambda^{-k_{i+1}}}$. We shall now use the recursive assumption. We have

$$\sum_{\substack{M\in E_3\\ k_{i+1}(M)=p}} l(T^{-n}(A_\varepsilon \cap M)) = \mu'_{n-p}\Bigg(\bigcup_{\substack{M\in E_3\\ k_{i+1}(M)=p}} T^{-p}(A_\varepsilon \cap M)\Bigg)\cdot l(L_0)$$

$$\leqq \mu'_{n-p}(A_{4\varepsilon^{\varrho}\lambda^{-p}})\cdot l(L_0).$$

Therefore,

$$\sum_{M\in E_3} l(T^{-n}(A_\varepsilon \cap M)) \leqq \sum_{p=c|\log\varepsilon|}^{n} \mu'_{n-p}(A_{4\varepsilon^{\varrho}\lambda^{-p}})$$

$$\leqq \frac{4^{\tau}\varepsilon^{\varrho\tau}\lambda^{-c\tau|\log\varepsilon|}}{1-\lambda^{-\tau}}.$$

From $\varrho + c\log\lambda > 1$, $1-\varrho > \tau$, and $1-\varrho-\frac{c}{K}\log 2 > \tau$, we obtain $\mu'_n(A_\varepsilon) \leqq \varepsilon^{\tau}$ if $\varepsilon < \varepsilon_0$ for ε_0 small enough but independent of n. Q.E.D.

We fix now τ as in Proposition III.1 and denote by $(\mu_n)_{n\in\mathbb{N}}$ a subsequence of the previous sequence which converges weakly to μ.

Corollary III.3. i) *For ε small, $n\in\mathbb{N}$, $k\in\mathbb{Z}$, $\mu_n(T^kA_\varepsilon) < \varepsilon^{\tau}$, $\mu(T^kA_\varepsilon) < \varepsilon^{\tau}$.*

ii) *For $N\in\mathbb{N}$, we define $H^N = \Big\{x\in\Omega$: one (at least) of the endpoints of $W^u_{\mathrm{loc}}(x)$ does not lie on $\bigcup_{k=0}^{N} S_k^+\Big\}$. There exist $c>0$ and α, $0<\alpha<1$, such that for $n\in\mathbb{N}$, $\mu_n(H^N) < c\alpha^N$, and $\mu(H^N) < c\alpha^N$.*

iii) *For $\varepsilon\in\mathbb{R}^+_*$, we define $H_\varepsilon = \{x\in\Omega$: one (at least) of the semi-loc unstable manifolds is shorter than $\varepsilon\}$. For ε small, $\mu_n(H_\varepsilon) < \varepsilon^{\tau}$, $n\in\mathbb{N}$, and $\mu(H_\varepsilon) < \varepsilon^{\tau}$.*

iv) *For $x\in\Omega$, $\mu(\{x\}) = 0$.*

v) *For $x\in\Omega$, $\mu(\{W^u(x)\}) = \mu(\{W^s(x)\}) = 0$.*

Proof. The first part of i) follows from Proposition III.1, by the definition of μ_n; the second part is a consequence of the weak convergence of (μ_n), for A_ε has smooth boundary.

To prove ii) consider the line Δ passing through $x\in\Omega$ in the unstable direction, and let y be one of the endpoints of $W^u_{\mathrm{loc}}(x)$.

By construction, either $y\in S_k^+$ for some $k\in\mathbb{N}$ or $y\notin \bigcup_{k\geqq 0} S_k^+$, but y is an accumulation point of $\bigcup_{k\geqq 0} S_k^+ \cap \Delta$. In the latter case, one can construct a sequence $(y_n)_{n\in\mathbb{N}}$, $y_n\in\Delta$, such that, for some strictly increasing function $k(\cdot)$, $y_n\in S^+_{k(n)}$, $n\in\mathbb{N}$, and moreover, $\bigcup_{k=0}^{k(n)} S_k^+$ does not cross Δ between y_n and x. Then, $T^{-k(n)}$ is linear on the segment $[x, y_n] \subset \Delta$. As $\lim_{n\to\infty} y_n = y \in \Omega$ and Ω is bounded, we can suppose $d(x, y_n) < 10$, so that $d(T^{-k(n)}x, S_0^+) < 10/\lambda^{k(n)}$, $\forall n \in \mathbb{N}$. Thus, $x \in \bigcap_{n\geq 0} T^{k(n)} A_{10/\lambda^{k(n)}}$, which is a set of μ-measure zero, by i): almost surely, the endpoints of $W^u_{\rm loc}(x)$, $x \in \Omega$, lie in $\bigcup_{k\geq 0} S_k^+$. Suppose now $y \in S_k^+$, $k \geq N$. Then $x \in T^k A_{10/\lambda^k}$ whence ii) by i) with $\alpha = \lambda^{-\tau}$.

With the same notations, suppose $x \in H_\varepsilon$; then if $y \in S_k^+$, $k \in \mathbb{N}$, $x \in T^k A_{\varepsilon/\lambda^k}$, whence iii) by i) and the convergence of $\sum_{k\geq 0} \lambda^{-k\tau}$.

The proof of iv) is similar to the proof of Proposition III.1: Let $x \in \Omega$, and let ε be a positive number. Let B be the ball of radius ε centered at x. Let J_n be as above. We have

$$\mu_n(B) = \sum_{M \in J_n} l(T^{-n}(B \cap M))/l(L_0).$$

Let $J_n^+ = \left\{M \in J_n \mid l(M) > \varepsilon^{\frac{1}{1+\tau}}\right\}$, $J_n^- = J_n \backslash J_n^+$. If $M \in J_n^+$, from $l(B \cap M) \leqq 2\varepsilon$, we obtain

$$l(T^{-n}(B \cap M)) = l(T^{-n}M)\frac{l(B \cap M)}{l(M)} \leqq \varepsilon^{\frac{\tau}{1+\tau}} l(T^{-n}M).$$

Therefore,

$$\sum_{M \in J_n^+} l(T^{-n}(B \cap M)) \leqq \varepsilon^{\frac{\tau}{1+\tau}} \sum_{M \in J_n^+} l(T^{-n}M) \leqq \varepsilon^{\frac{\tau}{1+\tau}} l(L_0).$$

For J_n^-, we have

$$\sum_{M \in J_n^-} l(T^{-n}(B \cap M)) \leqq \sum_{M \in J_n^-} l(T^{-n}M) = \mu_n\Big(\bigcup_{M \in J_n^-} M\Big) \cdot l(L_0)$$

$$\leqq \mu_n(H\varepsilon^{1/1+\tau}) \cdot l(L_0) \leqq c \cdot l(L_0) \cdot \varepsilon^{\frac{\tau}{1+\tau}} \quad \text{by iii)}.$$

We obtain

$$\mu(B) \leqq l(L_0)(1+c) \cdot \varepsilon^{\frac{\tau}{1+\tau}}$$

and get iv) if we let $\varepsilon \to 0$.

We shall prove v) for $W^s(x)$, the proof for $W^u(x)$ is similar. It is enough to prove that $(W^s_{\rm loc}(x)) = 0$, since $W^s(x) = \bigcup_{j=0}^{\infty} T^{-j} W^s_{\rm loc}(T^j x)$. Assume $x \in \Omega$ satisfies $\mu(W^s_{\rm loc}(x)) > \beta > 0$. Then, there is a couple of integers i and j, $i \neq j$ such that $T^i W^s_{\rm loc}(x) \cap T^j W^s_{\rm loc}(x) \neq \emptyset$, otherwise, we get a contradiction from $\mu(T^l W^s_{\rm loc}(x)) > \beta$, $\forall l \geqq 0$. Assume $i > j$ and let $k = i - j$. Since T is a bijection, we have $T^k W^s_{\rm loc}(x) \cap W^s_{\rm loc}(x) \neq \emptyset$, and therefore, $T^k W^s_{\rm loc}(x) \subset W^s_{\rm loc}(x)$. Since $T^k|_{W^s_{\rm loc}(x)}$ is a contraction, this means that $W^s_{\rm loc}(x)$ contains a k periodic point P. Moreover, $P = \bigcap_{l=0}^{\infty} T^{kl} W^s_{\rm loc}(x)$, and therefore,

$$\mu(\{P\}) = \inf_l \mu(T^{kl} W^s_{\rm loc}(x)) = \mu(W^s_{\rm loc}(x)) > \beta.$$

a contradiction with iv).

IV. Ergodic Properties of the Invariant Measure

So far, the measure μ is not unique. The uniqueness will be proven by showing that μ is the Bowen-Ruelle measure. We first investigate the properties of the conditional expectations of μ on the unstable foliation, making use of the sequence $(\mu_n)_{n\in\mathbb{N}}$. In order to investigate and use these properties, we define two countable partitions α and β which decompose Ω.

Let ζ^+ (ζ^-) be the decomposition of μ.a.a. Ω into local unstable (stable) manifolds. As ζ^+ is a partition generated by $\bigcup_{n\geq 0} S_n^+$, it is measurable. We can define the restriction μ^+ of μ to the sub σ algebra $\mathcal{M}^+\subset\mathcal{M}_\Omega$ of the sets ζ^+-saturate. By Corollary III.3ii), for μ.a.e. $W_0\in\zeta^+$, there are two maximal smooth components I_0 and J_0, contained in $\bigcup_{k=0}^{N_0} S_k^+$ for some $N_0\geqq 0$ such that the endpoints of W_0 lie on I_0 and J_0. We then define the partition α by $\alpha(W_0)=\{W\in\zeta^+$ with endpoints on I_0 and $J_0\}$. Let $\zeta_N^+\subset\zeta^+$ be the union of the elements of ζ^+ with endpoints on $\bigcup_{k=0}^{N} S_k^+$, $N\in\mathbb{N}$. ζ_N^+ is a finite union of atoms of α; since, by Corollary III.3ii), $\lim_{N\to\infty}\mu^+(\zeta_N^+)=1$, α is a countable partition of ζ^+.

We now look for a partition of Ω into parallelograms. Let P_N^+ (P_N^-) be the partitions generated by $\bigcup_{n=0}^{N} S_n^+\left(\bigcup_{n=0}^{N} S_n^-\right)$. As S_n^+ is a broken line, folded only on $\bigcup_{i=0}^{n-1} S_i^+$, a simple recursion argument shows that the $P_N^\pm(x)$, $x\in\Omega$, are convex sets. Note that $\zeta^\pm=\lim_{N\to\infty} P_N^\pm$. Let $x\in\Omega$ such that $W^+=W_{\text{loc}}^u(x)$ and $W^-=W_{\text{loc}}^s(x)$ have positive length. Let $N_\pm(x)$ be the smallest integers such that the endpoints of W^- (W^+) lie outside $P_{N^+(x)}^+(x)$ $[P_{N^-(x)}^-(x)]$. It is easy to see that $N_\pm$ are finite; $N_\pm$ are obviously measurable functions, so that we can define, for such an element $x\in\Omega$ the atom $\beta(x)$ by:

$$\beta(x)=N_x^{-1}(N_+(x))\cap N_-^{-1}(N_-(x))\cap P_{N^+(x)}^+(x)\cap P_{N^-(x)}^-(x)\,.$$

As usual, we define the unstable and stable fibers $\gamma^+(x)$, $\gamma^-(x)$ by

$$\gamma^+(x)=\beta(x)\cap W_{\text{loc}}^u(x)\,,$$

$$\gamma^-(x)=\beta(x)\cap W_{\text{loc}}^s(x)\,,\quad \text{i.e.}\quad \gamma^\pm=\beta\vee\zeta^\pm\,.$$

The set $\beta(y)$ is a parallelogram in the following sense

$$x\in\beta(y)\;\Rightarrow\;\begin{cases}\exists!\,z\in\beta(y)\,, & z=\gamma^+(x)\cap\gamma^-(y)\,,\\ \exists!\,z'\in\beta(y)\,, & z'=\gamma^-(x)\cap\gamma^+(y)\,.\end{cases}$$

This is easily checked by using the convexity of $P_{N^\pm(x)}^\pm(x)$ and the fact that $N^\pm$ are constant on atoms of $P_{N^\pm(x)}^\pm\vee\zeta^\mp$. By definition β is a countable partition of Ω into parallelograms.

We now come back to the properties of μ. As ζ^+ is a measurable partition, we can apply the usual theorem on disintegration (see [Ro] e.g.).

Namely, there is a μ-a.s. unique family $\{\mu_W,\,W\in\zeta^+\}$ of probability measures on Ω such that:

a) μ_W has μ-a.s. support on W,

b) for $A \in \mathcal{M}_\Omega$ the map $W \to \mu_W(A)$ is in $L^1(\zeta^+, d\mu^+)$, and $\mu(A) = \int_{\zeta^+} \mu_W(A) d\mu^+(W)$, denoted $\mu^+(\mu.(A))$.

For $W \in \zeta^+$, let l_W denote the normalized $1-d$ Lebesgue measure on W.

Proposition IV.1. *The conditional expectations of μ on the local unstable manifolds are the corresponding $1-d$ Lebesgue probabilities, i.e.: $\mu_W = l_W$ for μ.a.e. $W \in \zeta^+$.*

Proof. As stated before, some subsequence $(\mu_{n_i})_{i \in \mathbb{N}}$ of $(\mu_n)_{n \in \mathbb{N}}$ converges weakly to μ; we shall still denote it $(\mu_n)_{n \in \mathbb{N}}$. Let μ_n^+ denote the restriction of μ_n to $\mathcal{M}^+$, $n \in \mathbb{N}$. By the geometrical properties of T, it is easy to show that for $A \in \mathcal{M}_\Omega$, $\mu_n(A) = \int_{\zeta^+} l_W(A) d\mu_n^+(W) = \mu_n^+(l.(A))$. It is enough to prove $\lim_{n \to \infty} \mu_n^+(l.(f)) = \mu^+(l.(f))$ for $f \in C^0(\Omega)$. Let $f \in C^0(\Omega)$. With respect to the Hausdorff topology τ_H on ζ^+, the map $\zeta^+ \to \mathbb{R} : W \to l_W(f)$ is continuous. We shall show that μ_n^+ converges weakly to μ^+ in the sense of τ_H.

By Corollary III.3ii), $\mu_k^+(\zeta^+ \backslash \zeta_N^+)$, $k \in \mathbb{N}$ and $\mu^+(\zeta^+ \backslash \zeta_N^+)$ are simultaneously bounded by $C \cdot \varrho^N$ for some $\varrho \in]0, 1[$ and some positive constant C. Let $Q \subset \zeta^+$ be an atom of α. Since ζ_N^+ is a finite union of atoms of α the compactness of ζ_N^+ would follow from the compactness of Q. As two elements of ζ^+ cannot intersect, the elements of Q depend continuously on, e.g. the vertical coordinate of their rightmost endpoint. Since Ω is compact, the limit of a convergent sequence of elements of Q is a segment contained in Ω, thus, it belongs to Q : Q is compact. As Ω is totally regular, the hypothesis of Prokhorov's theorem are fulfilled (cf. [B]), so that $(\mu_n^+)_{n \in \mathbb{N}}$ converges weakly to some probability measure on ζ^+, which has to be the restriction μ^+ of μ. Q.E.D.

We are now able to derive the ergodic properties of T.

Proposition IV.2. *(Ω, T, μ) is ergodic.*

Proof. Take $f \in C^0(\Omega)$. By Birkhoff's ergodic theorem, there is a set $B \subset \Omega$, $\mu(B) = 0$ and a function $\bar{f} \in L^1(d\mu)$ such that, if $x \in \Omega \backslash B$ the limits $f^\pm(x) = \lim_{N \to \infty} \frac{1}{N} \sum_{n=0}^{N-1} f(T^{\pm n}x)$ exist and $f^+(x) = f^-(x) = \bar{f}(x)$. What we shall prove is that $\bar{f}$ is almost surely constant. Let $x, y \in \Omega \backslash B$. As $\mu(B) = \mu\left(\bigcap_\varepsilon H_\varepsilon\right) = 0$, the conditional measures of $B \cup \left(\bigcap_\varepsilon H_\varepsilon\right)$ (i.e. the corresponding normalized lengths) are zero on $W_{\text{loc}}^u(x)$ and $W_{\text{loc}}^u(y)$ for $\mu \times \mu$-a.e. (x, y). Thus, by Proposition II.4, it is enough to find a subset $A \subset W_{\text{loc}}^u(x)$ of positive length and an integer N such that for $z \in A$, $W_{\text{loc}}^s(z)$ crosses $T^N W_{\text{loc}}^u(y)$, because f^+ (f^-) is obviously constant on stable (unstable) manifolds. Consider now $\beta(x)$. Almost surely $\mu(\beta(x)) > 0$, and thus, by Proposition II.4, $l(\gamma^+(x)) > 0$. By Corollary III.3v), we can define $x_1, x_2 \in \beta(x)$ by demanding that the quadrilateral Q defined by $\zeta^\pm(x_1)$, $\zeta^\pm(x_2)$ be the smallest such that $\mu(Q \cap \beta(x)) = \mu(\beta(x))$.

For some N, by Proposition II.3, $T^N W_{\text{loc}}^u(y)$ "enters" Q, and one of its smooth components, that we shall call $\tilde{W}$, crosses $\zeta^-(x_1)$ or $\zeta^-(x_2)$ or both . Thus, we can define the canonical isomorphism P along stable fibers from $W^u_{loc}(x)$ into $\tilde{W} \subset T^N W^u_{loc}(y)$, with a domain $\mathscr{D}(P)$ of positive length. The proof is completed by taking $A = \mathscr{D}(P)$.

Proposition IV.3. i) *For* $n \in \mathbb{N}^*$, (Ω, T^n, μ) *is ergodic.*
ii) (Ω, T, μ) *is a K-system.*

Proof. The proof of i) is similar to the proof of Proposition IV.2 ii) follows from the fact the Pinsker σ-algebra Π of T is smaller than the σ-algebra of the measurable sets saturate by stable and unstable manifolds (cf. [P]). The proof of Proposition IV.2 shows that if $f \in L^1(d\mu)$ is constant along stable and unstable manifolds then f is μ-a.s. constant; thus, Π is trivial for μ, whence ii).

In order to prove the Bernoullian property, we introduce some notations. We define decreasing sequences $(\mathscr{M}_n^{\pm})_{n \in \mathbb{N}}$, where $\mathscr{M}_n^{\pm}$ is the sub σ-algebra of the elements of $\mathscr{M}_\Omega$ which are $T^{\pm n}\zeta^{\pm}$-saturate. By Proposition IV.3ii) the σ-algebras $\lim_n \mathscr{M}_n^{\pm} = \bigcap_n \mathscr{M}_n^{\pm}$ are both trivial for μ. We note $\mu_{\pm}$ the restriction of μ to $\mathscr{M}_0^{\pm}$.

Proposition IV.1 about the conditional expectation of μ with respect to $\mathscr{M}_0^+$ will allow us to prove the following:

Proposition IV.4. (Ω, T, μ) *is isomorphic to a Bernoulli shift.*

Proof. The proof is similar to [L1].

The canonical map $(\Omega, \mathscr{M}_0^+ \vee \mathscr{M}_0^-) \xrightarrow{P} (\zeta^+ \times \zeta^-, \mathscr{M}_0^+ \times \mathscr{M}_0^-)$ given by $P(x) = (\zeta^+(x), \zeta^-(x))$ is a.e. defined. Let $\nu = \mu \circ P^{-1}$ denote the image of μ through P. We have:

Lemma IV.5. ν *is absolutely continuous with respect to* $\mu^+ \otimes \mu^-$, i.e. $\nu \ll \mu^+ \otimes \mu^-$.

Proof. Let $A \in \mathscr{M}_0^+ \vee \mathscr{M}_0^-$ such that $P(A) \in \mathscr{M}_0^+ \times \mathscr{M}_0^-$. By definition, we have:

$$\mu^+ \otimes \mu^-(P(A)) = \int_{W \in \zeta^+} d\mu^+(W) \cdot \mu^-\Big(\bigcup_{x \in A \cap W} \zeta^-(x)\Big).$$

Suppose $\mu^+ \otimes \mu^-(P(A)) = 0$. Then, for some $\mathscr{W}_1 \subset \zeta^+$ of full measure:

$$W \in \mathscr{W}_1 \Rightarrow \mu^-\Big(\bigcup_{x \in A \cap W} \zeta^-(x)\Big) = 0.$$

Suppose that, for some $W' \in \gamma^+$, included in an element C of β of positive measure, we have $\mu(A|W') = 0$. Then, by Proposition II.4 and Proposition IV.1, we have:

$$\mu\Big(C \cap \bigcup_{x \in A \cap W'} \zeta^-(x)\Big) > 0,$$

so that if W is the ζ^+-saturate of W', we have:

$$\mu^-\Big(\bigcup_{x \in A \cap W} \zeta^-(x)\Big) > 0,$$

so that $W \notin \mathscr{W}_1$. Thus, $\mu(A|W') = 0$ almost surely and $\mu(A) = 0$. Q.E.D.

We shall denote the measure $\mu^+ \otimes \mu^-$ by $\mu^\times$. By Radon-Nikodym's theorem, there is a $\mu^\times$-integrable function $h: \zeta^+ \times \zeta^- \to \mathbb{R}^+$, such that $d\nu(x) = h(x) \cdot d\mu^\times(x)$. If $A \in \mathscr{M}_0^+ \times \mathscr{M}_0^-$ and $\nu(A) > 0$, the conditional probability $\nu_A = \nu(\cdot|A)$ is given, for a function f, by:

$$\nu_A(f)=\nu(f\cdot\chi_A)/\nu(\chi_A)=\mu^\times(f\cdot h\cdot\chi_A)/\mu^\times(h\cdot\chi_A)$$
$$=\mu^\times(f\cdot h)/\mu_A^\times(h)$$

or, more briefly:

$$\nu_A=\mu_A^\times(h\cdot)/\mu_A^\times(h).$$

We are now able to prove the weak Bernoulli property.

Lemma IV.5. $(\Omega,\mathcal{M},\mu,T)$ *is weak Bernoulli, that is* μ *and* $\mu^\times\circ P$ *coïncide on* $\bigwedge_n(\mathcal{M}_n^+\vee\mathcal{M}_n^-)$.

Proof. As $(\mathcal{M}_n^+\vee\mathcal{M}_n^-)_{n\in\mathbb{N}}$ is a decreasing sequence, the conditional expectation on A of μ with respect to $\bigwedge_n(\mathcal{M}_n^+\vee\mathcal{M}_n^-)$ is given almost everywhere by:

$$\mu_A^{\bigwedge_n(\mathcal{M}_n^+\vee\mathcal{M}_n^-)}=\lim_{n\to\infty}\mu_A^{\mathcal{M}_n^+\vee\mathcal{M}_n^-}$$
$$=\lim_{n\to\infty}\nu_{P(A)}^{\mathcal{M}_n^+\times\mathcal{M}_n^-},$$

by definition of ν. Therefore,

$$\mu_A^{\bigwedge_n(\mathcal{M}_n^+\vee\mathcal{M}_n^-)}=\lim_{n\to\infty}(\mu^\times)_{P(A)}^{\mathcal{M}_n^+\times\mathcal{M}_n^-}(h\cdot)/(\mu^\times)_{P(A)}^{\mathcal{M}_n^+\times\mathcal{M}_n^-}(h).$$

As $(\Omega,\mathcal{M}_\Omega,\mu,T)$ is a K-system, the σ-algebras $\bigwedge_{n\in\mathbb{N}}\mathcal{M}_n^+$ and $\bigwedge_{n\in\mathbb{N}}\mathcal{M}_n^-$ both coïncide with the trivial algebra mod μ: the conditional expectations

$$\mu^{\bigwedge_{n\in\mathbb{N}}\mathcal{M}_n^\pm}=\lim_{n\to\infty}\mu_\pm^{\mathcal{M}_n^\pm}$$

are constant μ-almost everywhere.

Thus, both $\mu^{\bigwedge_n(\mathcal{M}_n^+\vee\mathcal{M}_n^-)}$ and $(\mu^\times\circ P)^{\bigwedge_n(\mathcal{M}_n^+\vee\mathcal{M}_n^-)}$ are μ-a.e. constant and thus coïncide. Q.E.D.

The statement of Proposition IV.4 follows from Lemma IV.5 (cf. [L2] for instance).

We prove now that μ is the (unique) Bowen-Ruelle measure, id est:

Proposition IV.6. *For* $g\in C^0(F)$ *and* m*-almost any* $x\in F$,

$$\lim_{n\to\infty}\frac{1}{n}\sum_{k=0}^{n-1}g(T^kx)=\mu(g).$$

The proof follows three steps: We consider the points x of $\bigcup_{n\in\mathbb{N}}T^nL_0$ such that, for some $g\in C^0(F)$, we have not $\lim_{N\to\infty}\frac{1}{N}\sum_{n=0}^{N-1}g(T^nx)=\mu(g)$, and prove that the length of this set is zero (Lemma IV.7). Then we notice that, for m-a.e. $x\in F$ one can find positive integers p,n such that $W_{\text{loc}}^s(T^nx)$ crosses T^pL_0 (Lemma IV.8). Thus, what we have to show is that, m-almost surely in F, this intersection does not fall in the exceptional set estimated in Lemma IV.7.

Let $\mathscr{A}=\left\{x\in\bigcup_{n\geqq 0}T^nL_0 \mid \exists g\in C^0(F)\text{ and }\lim_{N\to+\infty}\frac{1}{N}\sum_{n=0}^{N-1}g(T^nx)\text{ does not exist or is not equal to }\mu(g)\right\}$.

Lemma IV.7. $\forall n\in\mathbb{N}$, $\mu_n(\mathscr{A})=l(\mathscr{A}\cap L_0)=0$.

Proof. We first observe that $\mathscr{A}=T\mathscr{A}$, and if $x,y\in T^nL_0$, $y\in W^s_{\text{loc}}(x)$, then $x\in\mathscr{A}$ is equivalent to $y\in\mathscr{A}$. This last property implies using Propositions II.4 and IV.1, that $W\to\mu(\mathscr{A}|W)$ is a continuous function on (ζ^+,τ_H). From $\mu\left(\bigcup_{n\geqq 0}T^nL_0\right)=0$, we deduce $\mu(\mathscr{A})=0$. Therefore, using the Birkhoff ergodic Theorem, we have

$$0=\mu(\mathscr{A})=\int_{\zeta^+}d\nu(W)\mu(\mathscr{A}|W)=\lim_{n\to+\infty}\int_{\zeta^+}d\nu_n(W)\mu(\mathscr{A}|W)=\lim_{n\to+\infty}\mu_n(\mathscr{A}).$$

However, $\mu_n(\mathscr{A})=\dfrac{l(T^{-n}\mathscr{A}\cap L_0)}{l(L_0)}=\dfrac{l(\mathscr{A}\cap L_0)}{l(L_0)}$, hence $l(\mathscr{A}\cap L_0)=0$. Q.E.D.

Let $\mathscr{L}=\{x\in F \mid \forall n,p\in\mathbb{N},\ W^s_{\text{loc}}(T^nx)\cap T^pL_0=\emptyset\}$.

Lemma IV.8. $m(\mathscr{L})=0$.

Proof. Let H and r be as in Lemma II.2. This lemma implies $\mathscr{L}\subset\bigcup_{n>0}T^{-n}(\mathscr{L}\cap H)$. Therefore, it is enough to show that $m(\mathscr{L}\cap H)=0$ since T^{-1} is absolutely continuous. For $\alpha>0$ and $n\in\mathbb{N}$, let $\mathscr{M}^\alpha_n=\{x\in H \mid l(W^s_{\text{loc}}(T^nx))<\alpha^n\}$. We have

$$\mathscr{M}^\alpha_n=T^{-n}\{y\in T^nH \mid l(W^s_{\text{loc}}(y))<\alpha^n\}$$

and using [M], we derive

$$m(\mathscr{M}^\alpha_n)\leqq\delta\left(\frac{\alpha}{b}\right)^n,$$

where δ is a positive constant. Therefore, if $\alpha<b$, we have with $\mathscr{M}^\alpha=\bigcap_{n=0}^{\infty}\mathscr{M}^\alpha_n$,

$$m(\mathscr{M}^\alpha)=0.$$

Let $x\in(H\backslash\mathscr{M}^\alpha)\cap\mathscr{L}$. We have $x\notin\mathscr{M}^\alpha_n$ for n large enough, but

$$W^s_{\text{loc}}(T^nx)\cap T^{n+r}L_0=\emptyset,$$

therefore,

$$W^s_{\text{loc}}(T^nx)\subset T^nH.$$

$T^{-n}W^s_{\text{loc}}(T^nx)$ is a broken line of total length bounded below by $\left(\dfrac{\lambda\alpha}{b}\right)^n$. However, for b small enough, it is easy to verify that for $y\in H$ such that $W^s_{\text{loc}}(y)\subset H$, then $\overset{0}{\widehat{W^s_{\text{loc}}(y)}}$ does not intersect both S_0^- and S_0^+. This implies that $T^{-n}W^s_{\text{loc}}(T^nx)$ is composed of at most $2^{[n/2]+1}$ straight segments. Therefore, since each segment has a length at most one, we have

$$l(T^{-n}(W^s_{\mathrm{loc}}(T^n x))) \leqq 2^{[n/2]+1}.$$

We now choose $\alpha < b$, such that $\frac{\sqrt{2}}{\lambda}\frac{b}{\alpha} < 1$ (this is possible since $\lambda > \sqrt{2}$). We have

$$2^{[n/2]+1} \geqq l(T^{-n}(W^s_{\mathrm{loc}}(T^n x))) \geqq \left(\frac{\lambda\alpha}{b}\right)^n,$$

which is a contradiction if n is large enough. Therefore, $\mathscr{L} \cap H \subset \mathscr{M}^\alpha$, and we have

$$m(\mathscr{L} \cap H) \leqq m(\mathscr{M}^\alpha) = 0. \quad \text{Q.E.D.}$$

We now come to the proof of Proposition IV.6.

We first observe that if $x \in F \backslash \mathscr{L}$, one can find an integer n, and $p \geqq n$ such that $W^s_{\mathrm{loc}}(T^n x) \cap T^p L_0 \neq \emptyset$. We define a new set $\mathscr{B}$ by

$$\mathscr{B} = \{x \in F \backslash \mathscr{L} \mid \forall n, p, p \geqq n,\ W^s_{\mathrm{loc}}(T^n x) \cap T^p L_0 \subset \mathscr{A}\}.$$

If $x \in F \backslash (\mathscr{L} \cup \mathscr{B})$, there is an $n \in \mathbb{N}$, and an integer $p \geqq n$ such that some point y of $W^s_{\mathrm{loc}}(T^n x) \cap T^p L_0$ does not belong to $\mathscr{A}$. Therefore, if g belongs to $C^0(\mathbb{R}^2)$, we have

$$\begin{aligned}\lim_{m \to +\infty} \frac{1}{m} \sum_{j=0}^{m-1} g(T^j x) &= \lim_{m \to \infty} \frac{1}{m} \sum_{j=0}^{m-1} g(T^{j+n} x) \\ &= \lim_{m \to \infty} \frac{1}{m} \sum_{j=0}^{m-1} g(T^j y) = \mu(g).\end{aligned}$$

We shall now show that $m(\mathscr{B}) = 0$. Let $\mathscr{B}_p$ be defined by

$$\mathscr{B}_p = \{y \in F \backslash \mathscr{L} \mid W^s_{\mathrm{loc}}(y) \cap T^p L_0 \subset \mathscr{A}\}.$$

From $\mathscr{B} = \bigcap_{p > n \geqq 0} T^{-n} \mathscr{B}_p$, it is enough to show that $m(\mathscr{B}_p) = 0$ for every integer p. From the definition of $\mathscr{B}_p$, we have

$$\mathscr{B}_p \subset \bigcup_{x \in \mathscr{B} \cap T^p L_0} (W^s_{\mathrm{loc}}(x) \cap F).$$

Thus, an unstable segment W being chosen, we consider $\mathscr{B}_W = \bigcup_{x \in \mathscr{A} \cap W} W^s_{\mathrm{loc}}(x)$, and prove $m(\mathscr{B}_W) = 0$, using $l(\mathscr{A} \cap W) = 0$.

Let $\mathscr{A}_W = \{x \in \mathscr{A} \cap W \mid W^s_{\mathrm{loc}}(x) \neq \{x\}\}$. It is enough to prove $m\left(\bigcup_{x \in \mathscr{A}_W} W^s_{\mathrm{loc}}(x)\right) = 0$. If the W^s_{loc}'s were depending smoothly on x, this would be a consequence of Fubini's theorem. Instead we use Lemma II.1. Let $\varepsilon > 0$ be given sufficiently small. As $l(\mathscr{A} \cap W) = 0$, $\mathscr{A} \cap W$ can be covered by a countable union of open disjoint intervals of total length smaller than ε. Consider one of these intervals, say I; let $\varepsilon' < \varepsilon$ be its length. Let now $n \in \mathbb{N}^*$; $\bigcup_0^{n+1} S_k^-$ splits I into at most 2^{n+2} segments; if J is such a segment, by Lemma II.1, the dispersion of the angles of $\{W^s_{\mathrm{loc}}(x), x \in J\}$ is bounded by $2(b/\lambda)^n$, so we obtain

$$m\left(\bigcup_{x \in J} W^s_{\mathrm{loc}}(x)\right) \leqq 4(b/\lambda)^n + l(J).$$

Thus, $m\left(\bigcup_{x\in I} W^s_{\text{loc}}(x)\right) \leqq 16(2b/\lambda)^n + l(I) \leqq 16(2b/\lambda)^n + \varepsilon'$. As n is arbitrary and $\lambda > 2b$, we get

$$m\left(\bigcup_{x\in I} W^s_{\text{loc}}(x)\right) \leqq 2\varepsilon',$$

so that

$$m\left(\bigcup_{x\in \mathscr{A}W} W^s_{\text{loc}}(x)\right) \leqq 2\varepsilon,$$

which proves that $m(\mathscr{B}_W)=0$.

V. The Hausdorff Dimension

Let us denote by χ_+ and χ_- the characteristic exponents of (Ω, T, μ): $\chi_\pm = \lim_{n\to\pm\infty} \frac{1}{n}\log\|D_x T^n\|$ for μ.a.e. $x\in\Omega$. Let h denote the μ-entropy of T. In [Y2] Young proves that, if M is a compact surface, (M, f, m) an ergodic C^2 dynamical system with characteristic exponents $\chi_1 \geqq 0 \geqq \chi_2$, the Hausdorff dimension of m is given by $HD(m) = h_m(f)\left(\frac{1}{\chi_1} - \frac{1}{\chi_2}\right)$, where $h_m(f)$ is the m-entropy of f. If the limit exists almost everywhere, $HD(m)$ is defined by $HD(m) = \lim_{\varepsilon\to 0} \log m(\mathscr{B}(x,\varepsilon))/\log\varepsilon$, where $\mathscr{B}(x,\varepsilon) = \{y\in M | d(x,y) < \varepsilon\}$. As we shall show this result remains valid in our case, despite the fact that T is only almost everywhere C^∞. We have the following theorem.

Proposition V.1. *For μ-a.e. $x\in\Omega$,*

$$\lim_{\alpha\to 0} \log\mu(\mathscr{B}(x,\alpha))/\log\alpha = h\left(\frac{1}{\chi_+} - \frac{1}{\chi_-}\right).$$

In order to prove this, we merely adapt ideas of Ledrappier [L3] to our case. We prove separately two inequalities which lead to Proposition V.1.

Lemma V.2. $\liminf_{\alpha\to 0} \log m(\mathscr{B}(x,\alpha))/\log\alpha \geqq h\left(\frac{1}{\chi_+} - \frac{1}{\chi_-}\right)$ *μ-a.e.*

Proof. Let P denote the partition of Ω defined by $S = S_0^+ \cup S_0^-$. For simplicity, let $F_\pm(x)$ denote the quantities $\|D_x T^{\pm 1}\|$, for $x\in\Omega$. Note that $F_\pm$ is constant on each side of $S_0^\mp$.

We first prove that, $\varepsilon > 0$ being given, we can find for μ-a.e. $x\in\Omega$ an integer $N(x)$ and a number $C(x,\varepsilon) > 0$ such that if $n > N(x)$ and $d(x,y) \leqq C(x,\varepsilon)\cdot e^{-n}$, then x and y belong to the same atom of $\bigvee_{-n_-}^{n_+} T^k P$, where $n_\pm = [n/(\mu(F_\pm) + 2\varepsilon)]$. Suppose $d(x,y) \leqq d(x,S)$. Then x and y belong to the same atom of P, and $d(Tx, Ty) \leqq F_+(x) \cdot d(x,y)$. If, moreover, $d(x,y) \leqq d(Tx,S)/F_+(x)$, we deduce $d(Tx,Ty) \leqq d(Tx,S)$: x and y belong to the same atom of $T^{-1}P$, and $d(T^2x\cdot T^2y) \leqq F_+(Tx)\cdot F_+(x) \cdot d(x,y)$. Similar arguments are valid for T^{-1} and F_-. Thus, we see that a sufficient set of conditions to insure that x and y lie in the same atom of $\bigvee_{-n_-}^{n_+} T^k P$ is that

$$\begin{cases} d(x,y) \leqq d(T^k x, S) \Big/ \prod_{j=0}^{k-1} F_+(T^j x), & 0 \leqq k \leqq n_-, \\ d(x,y) \leqq d(T^{-k} x, S) \Big/ \prod_{j=0}^{k-1} F_-(T^{-j} x), & 0 \leqq k \leqq n_+ \end{cases}$$

(where the product is taken equal to 1 if $k=0$).

We have the two following estimates:

1) Let $B(\alpha^k)$ be the neighborhood of S of diameter α^k. We recall that $\mu(B(\alpha^k)) < K \cdot \alpha^{\tau k}$, where K, τ are two positive constants. Thus, if $\alpha < 1$, the series $\sum_{k=0}^{\infty} \mu(B(\alpha^k))$ converges. This allows us, through a measure theoretic result (see [L3] e.g.) to derive the existence a.e. of a measurable function $C(x, \varepsilon)$, $0 < C(x,\varepsilon) \leqq 1$ for μ-a.e. $x \in \Omega$ such that for $k \in \mathbb{Z}$ $d(T^k x, S) \geqq C(x,\varepsilon) e^{-|k|\varepsilon}$. In particular, we get

$$\begin{cases} d(T^k x, S) \geqq C(x,\varepsilon) e^{-n_- \cdot \varepsilon}, & 0 \leqq k \leqq n_-, \\ d(T^{-k} x, S) \geqq C(x,\varepsilon) e^{-n_+ \cdot \varepsilon}, & 0 \leqq k \leqq n_+ . \end{cases}$$

2) By the ergodic theorem, for μ-a.e. $x \in \Omega$, $\exists \bar{N}(x)$ such that if $n_- > \bar{N}(x)$, $n_+ > \bar{N}(x)$, we have

$$\begin{cases} \prod_{j=0}^{k-1} F_+(T^j x) \leqq \exp(n_-(\mu(F_+) + \varepsilon)), & 0 \leqq k \leqq n_-, \\ \prod_{j=0}^{k-1} F_-(T^{-j} x) \leqq \exp(n_+(\mu(F_-) + \varepsilon)), & 0 \leqq k \leqq n_+ . \end{cases}$$

Assuming ε is sufficiently small and setting $N(x) = \bar{N}(x) \cdot \mu(F_+ + F_-)$, we get, for $n > N(x)$,

$$\begin{cases} d(T^k x, S) \Big/ \prod_{j=0}^{k-1} F_+(T^j x) \geqq C(x,\varepsilon) \cdot \exp[-n_-(\mu(F_+) + 2\varepsilon)] = C(x,\varepsilon) \cdot e^{-n}, \\ \qquad 0 \leqq k \leqq n_-, \\ d(T^{-k} x, S) \Big/ \prod_{j=0}^{k-1} F_-(T^{-j} x) \geqq C(x,\varepsilon) \cdot \exp[-n_+(\mu(F_-) + 2\varepsilon)] = C(x,\varepsilon) \cdot e^{-n}, \\ \qquad 0 \leqq k \leqq n_+ . \end{cases}$$

Then, $B(x, C(x,\varepsilon) \cdot e^{-n}) \subset \left(\bigvee_{-n_-}^{n_+} T^k P \right)(x)$ for $n > N(x)$.

Using Shannon-Mac Millan-Breimann's theorem (see [Bi]) we obtain for μ-a.e. $x \in \Omega$,

$$\liminf_{\alpha \to 0} \log \mu(B(x,\alpha)) / \log \alpha \geqq h \cdot \left(\frac{1}{\mu(F_+) + 2\varepsilon} + \frac{1}{\mu(F_-) + 2\varepsilon} \right).$$

As ε is arbitrarily small, we have μ-a.e.:

$$\liminf_{\alpha \to 0} \log \mu(B(x,\alpha)) / \log \alpha \geqq h \cdot \left(\frac{1}{\mu(F_+)} + \frac{1}{\mu(F_-)} \right).$$

We can reproduce the same arguments replacing T by T^n, $S_0^\pm$ by $\bigcup_{k=0}^{n-1} T^{\pm k}S_0^\pm$, P by $\bigvee_{|k|<n} T^kP$, $F_\pm$ by $\|DT^{\pm n}\|$.

As $\lim_{n\to\infty} \int \|D_xT^{\pm n}\| d\mu(x) = \chi_\pm$, we get

$$\liminf_{\alpha\to 0} \log\mu(B(x,\alpha))/\log\alpha \geqq h\cdot\left(\frac{1}{\chi_+}-\frac{1}{\chi_-}\right).$$

Lemma V.3. $\limsup_{\alpha\to 0} \log\mu(B(x,\alpha))/\log\alpha \leqq h\cdot\left(\frac{1}{\chi_+}-\frac{1}{\chi_-}\right).$

Proof. Let P still denote the partition defined by S. We first prove that, $\varepsilon>0$ being given, there is a constant K such that, for μ-a.e. $x\in\Omega$, we can find $N(x)$ such that

$$n>N(x) \Rightarrow \left(\bigvee_{-n_-}^{n_+} T^kP\right)(x)\subset B(x, Ke^{-n}),$$

where $n_+ = n/[-\chi_- - \varepsilon]$, $n_- = n/[\chi_+ - \varepsilon]$.

Let $x\in\Omega$ and let $Q=\left(\bigvee_{-n_-}^{n_+} T^kP\right)(x)$.

We first exhibit "unstable and stable widths" $w_\pm$ of Q. For $y\in Q$, we draw the line $\Delta=\Delta(y)$ passing through y in the unstable direction, and note $w_+(y)$ the length of $\Delta\cap Q$; we recall that Q is a convex set. We set $w_+ = \sup_{y\in Q} w_+(y)$ and define w_- similarly. As the unstable and stable directions are transverse, and as the S_p^+, $p\in\mathbb{N}$, and the S_q^-, $q\in\mathbb{N}$, are also transverse, the minimum ball constaining Q has a radius smaller than $C\cdot(w_+ + w_-)$, where C is a constant. We now estimate w_+; we fix $y\in Q$ such that $w_+(y)>w_+/2$. T^{n_-} is linear on $\Delta(y)\cap Q$. Thus, the usual argument shows that $10>l(T^{n_-}(\Delta(y)\cap Q)) = w_+(y)\cdot \prod_{i=0}^{n_- -1} J^+(T^iy)$, where J^+ is the jacobian in the unstable direction. By the ergodic theorem, $\prod_{i=0}^{n_- -1} J^+(T^iy) = e^{n_-(\chi_+ - \varepsilon(n_-))}$, where $\varepsilon(\cdot)$ is a function depending on y such that $\lim_{m\to\infty}\varepsilon(m)=0$. As y is here fixed, for n large enough, n_- is large enough and $\varepsilon(n_-)<\varepsilon$. We get $10>w_+(y)\cdot e^n$; similarly, $w_-(y)<10\cdot e^{-n}$, and we get $\left(\bigvee_{-n_-}^{n_+} T^kP\right)(x)\subset B(x,Ke^{-n})$ with $K=20\cdot C$. Letting n go to infinity, we get

$$\limsup_{\alpha\to 0}\frac{\log\mu(B(x,\alpha))}{\log\alpha} \leqq h\left(\frac{1}{\chi_+ - \varepsilon}+\frac{1}{-\chi_- - \varepsilon}\right).$$

Lemma V.3 now follows from the fact that ε is arbitrarily small, Lemmata V.2 and V.3 end the proof of Proposition V.1.

We add the following result, which is the mere consequence of [L, S].

Proposition V.4. *With the above notations,* $h=\chi_+$.

Proof. One can check that hypothesis of [K, S] are fulfilled, so that the result of [L, S] applies.

Appendix. Proof of Proposition II.4

Lemma A.1. *There is a positive constant L'_1 such that for any pair W_1, W_2 of local unstable manifolds in Ω, there is an ε_0, $0<\varepsilon_0<1$ such that if $0<\varepsilon<\varepsilon_0$, one can find an open subset $\mathscr{A}_\varepsilon$ of Ω with $l(\mathscr{A}_\varepsilon \cap W_1) < \frac{2\varepsilon}{1-4\sqrt{b}/\lambda}$, and if $x, x' \in \mathscr{D}(P_{W^1W^2}) \setminus \mathscr{A}_\varepsilon$, $d(x,x')<\varepsilon^2$, then*

$$(1-L'_1 D^{1/3}) d(x,x') \leqq d(P_{W^1W^2}(x), P_{W^1W^2}(x')) \leqq d(x,x')(1+L'_1 D^{1/3}),$$

where $D = d(W_1, W_2)$.

Proof. Let K be a real number such that $\frac{2}{\lambda} < K < \frac{1}{\sqrt{b}}$. We define $\mathscr{A}_\varepsilon$ by

$$\mathscr{A}_\varepsilon = \bigcup_{j=0}^{\infty} (A^{-j}_{2\varepsilon/K^j} \cap \Omega), \quad \text{where} \quad A^{-j}_\alpha = T^{-j} A^-_\alpha .$$

Every $A^{-j}_{2\varepsilon/K^j}$ crosses W_1 at most 2^j times, therefore,

$$\begin{aligned} l(\mathscr{A}_\varepsilon \cap W_1) &\leqq \sum_{j=0}^{\infty} l(A_{2\varepsilon/K^j} \cap W_1) \\ &\leqq 2 \sum_{j=0}^{\infty} 2^j \frac{1}{(K\lambda)^j} 2\varepsilon = \frac{4\varepsilon}{1-2/K\lambda}, \end{aligned}$$

because the stable and unstable manifolds are transverse. We note $p = P_{W^1W^2}$. Assume $x, x' \in \mathscr{D}(p)$, and $x \notin \mathscr{A}_\varepsilon$ and $d(x,x') \leqq \varepsilon^2$. Let $\delta = d(x,x')$. Let N be the integer such that $2\varepsilon \left(\frac{1}{K\lambda_+}\right)^{N+1} \leqq \delta < 2\varepsilon \left(\frac{1}{K\lambda_+}\right)^N$. Assume that for some k, $0 \leqq k \leqq N$, S_k^- crosses W_1 at some point y between x and x'. From $d(x,y)<\delta$, we deduce $d(T^k x, T^k y) \leqq \delta \lambda_+^k$. Therefore,

$$d(T^k x, S_0) \leqq \delta \lambda_+^k \leqq \delta \lambda_+^N \leqq 2\varepsilon K^{-N} \leqq 2\varepsilon K^{-k},$$

which contradicts $x \notin \mathscr{A}_\varepsilon$. Assume now that S_k^- crosses W_2 for some k, $0 \leqq k \leqq N$, at some point which is on the segment $(p(x), p(x'))$. Since S_k^- cannot cross a local stable manifold, S_k^- must have a corner inside the parallelogram $Q = (x, x', p(x), p(x'))$. This implies that there is a point z belonging to this parallelogram such that $z \in S_{k'}^-$, for some k', $0 \leqq k' < k$. By induction we obtain that Q must intersect S_0^-, a contradiction since S_0^- is a straight line. Let j be an integer such that $0 \leqq j \leqq N$. The above argument implies that $T^{N-j}Q$ is a parallelogram. Moreover, we have

$$d(T^{N-j}x, T^{N-j}x') \geqq \delta \lambda^{N-j}, d(T^{N-j}x, T^{N-j}(p(x))) < 2\left(\frac{b}{\lambda}\right)^{N-j} D,$$

and

$$d(T^{N-j}x', T^{N-j}(p(x'))) \leqq 2\left(\frac{b}{\lambda}\right)^{N-j} D,$$

since the angle between a local stable and a local unstable manifold is greater than $\frac{\pi}{5}$. We have, therefore,

$$\left|\frac{d(T^{N-j}(p(x)), T^{N-j}(p(x')))}{d(T^{N-j}(x), T^{N-j}(x'))} - 1\right| < 10\left(\frac{b}{\lambda^2}\right)^{N-j}\frac{D}{\delta},$$

which implies

$$\left|\frac{d(T^{N-j}(p(x)), T^{N-j}(p(x')))}{d(T^{N-j}(x), T^{N-j}(x'))} - 1\right| < \frac{1}{10}\sqrt{D} \quad \text{by our choice of } N,$$

provided ε_0 is chosen small enough (independently of N and δ), and $j \leqq \frac{1}{2}\frac{|\log D|}{\log(\lambda^2/b)}$. We also have

$$\frac{d(T^{N-j}(p(x)), T^{N-j}(p(x')))}{d(T^{N-j}(x), T^{N-j}(x'))} \Big/ \frac{d(p(x), p(x'))}{d(x, x')} = \prod_{l=1}^{N-j} \Delta_l,$$

where

$$\Delta_l = \frac{d(T^l(p(x)), T^l(p(x')))}{d(T^l(x), T^l(x'))} \Big/ \frac{d(T^{l-1}(p(x)), T^{l-1}(p(x')))}{d(T^{l-1}(x), T^{l-1}(x'))}$$

satisfies

$$1 - 2\left(\frac{b}{\lambda}\right)^{N-l} \leqq \Delta_l \leqq 1 + 2\left(\frac{b}{\lambda}\right)^{N-l},$$

according to Lemma I.1. Therefore, if $j = \left[\frac{1}{2}\frac{|\log D|}{\log(\lambda^2/b)}\right] - \log 8$, we have

$$\left|\prod_{l=1}^{N-j} \Delta_l - 1\right| \leqq \frac{\sqrt{D}}{2}, \quad \text{if } D \text{ is smaller than } \frac{b}{64\lambda^2}.$$

Combining the two estimates, we obtain the result for $D < \frac{b}{64\lambda^2}$ with $L_1' = 1$. For $D > \frac{b}{64\lambda^2}$, we apply T one time [this is enough since $D \leqq \mathcal{O}(1)b$],and we can apply the estimate unless W^1 and W^2 cross S_0, in which case the estimation is performed with respect to S_0. Q.E.D.

We still note $p = P_{W^1W^2}$.

Proof of Proposition II.4. It is enough to prove the proposition with $A = \bar{A} \cap \mathcal{D}(p)$, $\bar{A}$ a closed interval in W^1 and $l(A) > 0$. Let ε_0 be as in Lemma A.1, and choose ε such that $0 < \varepsilon < \varepsilon_0$ and $\varepsilon < l(A)D^{1/3}$. We observe that $W_1 \backslash \mathcal{D}(p)$ is an open subset. Therefore, there is a sequence $(U_k)_{k \in \mathbb{N}}$ of disjoint open intervals such that

$$W_1 \backslash \mathcal{D}(p) = \bigcup_{k=0}^{\infty} U_k.$$

In the following, for $a, b \in \mathscr{D}(p)$ we shall denote by $\tilde{p}(]a,b[)$ the segment $]p(a), p(b)[$ (although p is not everywhere defined on $]a,b[$, this definition makes sense because p is order preserving).

Let $V_k = \tilde{p}(U_k)$. From $l(W_1) < \infty$ and $l(W_2) < \infty$, we deduce that there is an integer N_1 such that

$$\sum_{k=N_1}^{\infty} l(U_k) < \varepsilon\,, \qquad \sum_{k=N_1}^{\infty} l(V_k) < \varepsilon\,, \qquad l(V_k) < \varepsilon^2 \quad \text{if} \quad k \geqq N_1\,.$$

Let $A' = \bar{A} \Big\backslash \bigcup_{k=0}^{N_1-1} U_k$. We have $l(A) \leqq l(A') \leqq l(A) + \varepsilon$ and $l(p(A)) \leqq l(\tilde{p}(A'))$. The set $\mathscr{A}' = W_2 \cap \mathscr{A}_\varepsilon \cup \left(\bigcup_{k=N_1}^{\infty} V_k \right)$ is an open subset of W_2 such that

$$l(\mathscr{A}') \leqq \left(1 + \frac{2}{1 - 4\sqrt{b/\lambda}} \right) \varepsilon \leqq 4\varepsilon \quad \text{if } b \text{ is small enough}\,.$$

We can find a sequence $(V_k')_{k \in \mathbb{N}}$ of disjoint open intervals of W_2 such that

$$\mathscr{A}' = \bigcup_{k=0}^{\infty} V_k'\,.$$

Let N_2 be an integer such that

$$l(V_k') < \varepsilon^2 \quad \text{if} \quad k \geqq N_2 \quad \text{and} \quad \sum_{k=N_2}^{\infty} l(V_k') < \varepsilon\,.$$

Let

$$A'' = \tilde{P}^{-1} \left(\tilde{P}(A') \Big\backslash \bigcup_{k=0}^{N_2-1} V_k' \right),$$

we have $l(A'') \leqq l(A')$ and $l(\tilde{p}(A'')) \leqq l(\tilde{p}(A')) \leqq l(\tilde{p}(A'')) + 4\varepsilon$. $\tilde{p}(A'')$ is a finite union of closed intervals of W_2 whose endpoints are in $\mathscr{D}(p^{-1}) \backslash \mathscr{A}_\varepsilon$. Let $I = [u, v]$ be such an interval. We claim that there is a finite sequence $(u_s)_{s=0,\ldots,q}$ such that

i) $u_0 = u$, $u_q = v$,

ii) $u_j \in \mathscr{D}(p^{-1} \backslash \mathscr{A}_\varepsilon)$,

iii) $d(u_j, u_{j+1}) \leqq \varepsilon^2$.

This is obvious from the above construction. From Lemma A.2.1, we have

$$l(\tilde{p}^{-1}[u_j, u_{j+1}]) \leqq (1 + L_1' D^{1/3}) l[u_j, u_{j+1}] \forall j, \qquad 0 \leqq j \leqq q-1\,,$$

which implies

$$l(\tilde{p}^{-1}(I)) \leqq (1 + L_1' D^{1/3}) l(I) \quad \text{and} \quad l(A'') \leqq (1 + L_1' D^{1/3}) l(\tilde{p}(A''))\,.$$

Therefore,

$$\begin{aligned} l(p(A)) &\leqq l(\tilde{p}(A')) \leqq l(\tilde{p}(A'')) + 4\varepsilon \leqq (1 + L_1' D^{1/3}) l(A'') + 4\varepsilon \\ &\leqq (1 + L_1' D^{1/3}) l(A') + 4\varepsilon \leqq (1 + L_1' D^{1/3}) l(A) + (5 + L_1' D^{1/3}) \varepsilon \\ &\leqq (1 + L_1 D^{1/3}) l(A)\,, \quad \text{where} \quad L_1 = 2L_1' + 5\,. \end{aligned}$$

The inequality in the other direction is obtained by interchanging W_1 and W_2. Q.E.D.

Acknowledgement. We would like to thank J. Lascoux and F. Ledrappier for many helpful discussions.

References

[B] Bourbaki, M.: Eléments de Mathématiques, Integration chapter IX. Paris: Hermann 1969

[Bi] Billingsley, P.: Ergodic theory and information. New York: Wiley 1965

[He] Henon, M.: A two-dimensional mapping with a strange attractor. Commun. Math. Phys. **50**, 69–77 (1976)

[K, S] Katok, A., Strelcyn, J.M.: Invariant manifolds for smooth maps with singularities, Part I: Existence. Preprint (1980)

[L1] Ledrappier, F.: Propriétés ergodiques des mesures de Sinaï. Preprint (1982)

[L2] Ledrappier, F.: Sur la condition de Bernoulli faible et ses applications. Théorie ergodique Rennes 73/74. In: Lecture Notes in Mathematics, Vol. 532. Berlin, Heidelberg, New York: Springer 1975

[L3] Ledrappier, F.: Quelques propriétés des exposants caractéristiques. Ecole d'été de probabilités de St. Flour 1982. In: Lecture Notes in Mathematics. Berlin, Heidelberg, New York: Springer 1983 (to appear)

[L, S] Ledrappier, F., Strelcyn, J.M.: A proof of the estimation from below in Pesin entropy formula. Preprint. Paris-Nord (1982)

[Lo] Lozi, R.: Un attracteur étrange du type attracteur de Henon. J. Phys. (Paris) **39** (Coll. C5), 9–10 (1978)

[M] Misiurewicz, M.: Strange attractors for the Lozi mappings. In: Nonlinear dynamics, R. G. Helleman (ed.). New York: The New York Academy of Sciences 1980

[P] Parry, W.: Entropy and generators in ergodic theory. New York: Benjamin 1969

[Ro] Rohlin, V.A.: On the fundamental ideas of measure theory. Math. Sbornik **25** (67), 107–150 (1949)

[Ry] Rychlik, M.: Théorie ergodique, mesures invariantes et principe variationnel pour les applications de Lozi. C. R. Acad. Sci. Paris (1983) (to appear)

[Y1] Young, Lai-Sang: Bowen-Ruelle measures for certain piecewise hyperbolic maps. Preprint, University of Wisconsin (1982)

[Y2] Young, Lai-Sang: Dimension, entropy, and Lyapunov exponents. Ergod. Theor. Dynam. Syst. **2**, 109–124 (1982)

Communicated by J. Lascoux

Received June 13, 1983; in revised form November 1, 1983

Commun. Math. Phys. 99, 177–195 (1985)

Communications in Mathematical Physics

On the Concept of Attractor

John Milnor

Institute for Advanced Study, Princeton, NJ 08540, USA

Abstract. This note proposes a definition for the concept of "attractor," based on the probable asymptotic behavior of orbits. The definition is sufficiently broad so that every smooth compact dynamical system has at least one attractor.

Attractors have played an increasingly important role in thinking about dynamical systems since their introduction some twenty years ago; yet there is no agreement as to the most useful definition. Section 1 of this note compares several definitions from the literature, and Sect. 2 proposes an alternative definition based on asymptotic behavior for almost every choice of initial point. The remaining two sections illustrate this definition by a number of examples, and discuss the stability and robustness of these attractors. There are three appendices. The first compares a closely related purely topological definition, the second studies real quadratic maps of their interval as a test case, and the third discusses strange attractors.

1. History

The following is quoted from Auslander, Bhatia, and Seibert (1964):

> "In the study of topological properties of ordinary differential equations, the stability theory of compact invariant sets (which may be regarded as generalizations of critical points and limit cycles) plays a central role. ... By Liapunov stability (or just stability) of the compact invariant set M, we mean that every orbit starting sufficiently close to M will remain in a given neighborhood of M. The set M is asymptotically stable if it is stable and is also an 'attractor' – that is, all orbits in a neighborhood ... of M approach M."

(Compare La Salle and Lefschetz, p. 31.) The word attractor, applied to a single invariant point for a smooth flow, had been used earlier by Coddington and

Levinson (1955) and by Mendelson (1960). However, attractors consisting of more than one point seem to have been first studied in this Auslander-Bhatia-Seibert paper. They explicitly considered the unstable case, in which an orbit which starts arbitrarily close to the attractor may wander far away before converging back towards the attractor. Compare Example 1 in Sect. 4 below. Although their definition is occasionally used by other authors (see La Salle, Hirsch, and perhaps Smale 1977), it has not been widely accepted, and most subsequent authors have required some form of stability as part of the definition. Smale (1967), in a widely read survey article, defined a more complicated object which might better be called an axiom A attractor. Smale's attractors, for a smooth map f from a compact manifold to itself, were "hyperbolic sets," containing a dense orbit, with periodic points everywhere dense, and satisfying the following rather awkward stability condition: The attractor A must have a neighborhood U so that A is equal to the intersection of the images $f^m(U)$ for $m>0$. Williams (1968) gave a related but simpler definition:

> "A subset A of $\Omega(f)$ is an *attractor* of f, provided it is indecomposable and has a neighborhood U such that $f(U)\subset U$ and $\bigcap_{i>0} f^i(U)=A$",

where $\Omega(f)$ is the non-wandering set of f, and where a closed f-invariant set is *indecomposable* if is not the union of two disjoint closed invariant subsets. The concept of attractor acquired great interest when Ruelle and Takens (1971) suggested that turbulent behavior in fluids might be caused by the presence of "strange" attractors. In fact, an explicit example in support of this idea had been worked out much earlier by Lorenz (1963). The definition of attractor used by Ruelle and Takens, like that of Smale, used an awkward form of stability:

> "A closed subset A of the non-wandering set Ω is an attractor if it has a neighborhood U such that $\bigcap_{t>0} D_{X,t}(U)=A$",

where the notation $D_{X,t}$ stands for the flow on a smooth manifold generated by a vector field X. (Note that this condition does not imply asymptotic stability. For example, according to Besicovitch, there is a homeomorphism of the plane, fixing the origin, such that the intersection of the successive images of the unit disk is the origin, even though every nonzero orbit is everywhere dense. See also Anosov and Katok.)

It is my contention that all of these definitions are too restrictive, since they exclude many interesting examples. Furthermore, in many cases they lead to an awkward situation in which there is no convenient language to describe where most points actually go when one follows a flow or iterates a mapping. In order to illustrate this point, Sect. 3 will describe a quadratic map of the interval which has no attractor at all according to the above definitions, although almost every orbit converges to a single uniquely defined compact set. A less restrictive definition has been given by Guckenheimer (1976), who requires only that an attractor must have:

> "... a fundamental system of neighborhoods, each of which is forward invariant under the flow generated by X".

In the language of Auslander, Bhatia, and Seibert, this is the condition of Liapunov stability (cf. Sect. 4). Sets having this property are important and useful; however, I believe that not every Liapunov stable set should be called an attractor. Here is an example. Consider a diffeomorphism or a flow on the plane which reduces to the identity map on a collection of concentric circles converging to the origin, but which pushes points slightly away from the origin otherwise. Then the origin is Liapunov stable, and hence would be called an attractor by Guckenheimer's definition, although it does not attract any other point.

Many other definitions of attractor can be found in the literature. The author's favorite reference is Collet and Eckmann (1980), which informally defines the attractor of a map f as

> "the set of points to which most points evolve under iterates of f".

This idea forms the basis for the present paper. For further discussion and other definitions the reader is referred to Conley, Guckenheimer and Holmes (Sect. 5.4), Kan, Ruelle (1981, 1983), as well as Zeeman.

2. Attractors, Minimal Attractors, and the Likely Limit Set

Let M be a smooth compact manifold, possibly with boundary, and let f be a continuous map from M into itself. The notation $f^n = f \circ \ldots \circ f$ will stand for the nth iterate of f. Recall that the *omega limit set* $\omega(x)$ of a point $x \in M$ is the collection of all accumulation points for the sequence $x, f(x), f^2(x), \ldots$ of successive images of x. If we choose some metric for the topological space M, then $\omega(x)$ can also be described as the smallest closed set S such that the distance from $f^n(x)$ to the nearest point of S tends to zero as $n \to \infty$. The definition of omega limit set in the case of a continuous flow on M is completely analogous. Note that $\omega(x)$ is always closed and nonvacuous, with $f(\omega(x)) = \omega(x)$. Furthermore, $\omega(x)$ is always contained in the nonwandering set $\Omega(f)$.

Choose some measure μ on M which is equivalent to Lebesgue measure when restricted to any coordinate neighborhood. This can be constructed using a partition of unity, or using the volume form associated with a Riemannian metric. It doesn't really matter which particular measure we use, since we will usually only distinguish between sets of measure zero and sets of positive measure.

Definition. A closed subset $A \subset M$ will be called an *attractor* if it satisfies two conditions:

(1) the *realm of attraction* $\varrho(A)$, consisting of all points $x \in M$ for which $\omega(x) \subset A$, must have strictly positive measure; and

(2) there is no strictly smaller closed set $A' \subset A$ so that $\varrho(A')$ coincides with $\varrho(A)$ up to a set of measure zero.

The first condition says that there is some positive possibility that a randomly chosen point will be attracted to A, and the second says that every part of A plays an essential role.

Note. In the literature, the set $\varrho(A)$ is usually called the "basin of attraction" if it is an open set, and the "stable manifold" if it is a lower dimensional smooth manifold. I have avoided both terminologies since our sets $\varrho(A)$ are not open in general (cf. Sect. 3), and are certainly not lower dimensional manifolds. For any closed set A, it is not difficult to check that $\varrho(A)$ is necessarily a Borel set (or more precisely a countable intersection of σ-compact sets), and hence is measurable.

Basic properties of attractors are the following. An attractor A is necessarily closed, nonvacuous, and contained in the nonwandering set $\Omega(f)$, with $f(A)=A$. Any finite union of attractors is again an attractor; and more generally the closure of an arbitrary union of attractors is an attractor. Proofs are easily supplied. In order to show that attractors always exist, we will first construct one particularly important attractor.

Definition. The *likely limit set* $\Lambda=\Lambda(f)$ is the smallest closed subset of M with the property that $\omega(x)\subset\Lambda$ for every point $x\in M$ outside of a set of measure zero.

Lemma 1. *This likely limit set Λ is well defined and is an attractor for f. In fact, Λ is the unique maximal attractor, which contains all others.*

Sketch of Proof. Let $\{U_i\}$ be a countable basis for the open subsets of M, and let U be the union of those U_i such that $U_i\cap\omega(x)=\emptyset$ for almost every x. Then it follows that $U\cap\omega(x)=\emptyset$ for almost every x. The complement of U is the required likely limit set Λ. Further details of the argument are straightforward. □

If f is a measure preserving transformation, that is if $\mu(S)=\mu(f^{-1}(S))$ for every measurable set S, then $\Lambda(f)$ will be equal to the entire manifold M. Furthermore, every attractor will coincide with its realm of attraction, up to a set of measure zero. In this case, these constructions are probably not too interesting. However, when Λ is a proper subset of M, it provides a usefull tool for studying asymptotic behavior for almost all orbits.

Here is a more general construction for attractors. If $S\subset M$ is any subset of positive measure, then we define $\Lambda(f,S)$ to be the smallest closed subset of M which contains $\omega(x)$ for almost every point x of S. It is easy to check that $\Lambda(f,S)$ is well defined, and is an attractor. Note that its realm of attraction necessarily contains S, up to a set of measure zero. Evidently every attractor can be obtained by this construction. One special case is of particular interest.

Lemma 2. *If S is a compact set of positive measure with the property that $f(S)\subset S$, then S necessarily contains at least one attractor.*

For $\Lambda(f,S)$ is an attractor, and is clearly contained in S. □

We will be particularly interested in *minimal attractors*, that is, attractors for which no proper subset is an attractor. Evidently a closed set $A\subset M$ is a minimal attractor if and only if

(1′) its realm of attraction $\varrho(A)$ has positive measure, and

(2′) there is no strictly smaller closed set $A'\subset A$ for which $\varrho(A')$ has positive measure.

The number of distinct minimal attractors for f is at most countably infinite. If A is a minimal attractor, note that $\omega(x)$ is precisely equal to A for almost every x in $\varrho(A)$. There are many interesting cases in which the union of the minimal attractors for f is equal to the entire likely limit set Λ, or at least is everywhere dense in Λ.

Lemma 3. *Suppose that the likely limit set Λ is a union of finitely many disjoint minimal attractors $A_1, \ldots, A_n$. Then the corresponding realms of attraction $\varrho(A_i)$ form a partition of M into disjoint sets of positive measure, up to a set of measure zero. Every attractor is a union of minimal attractors, and for almost every point x of M the limit set $\omega(x)$ is precisely equal to some A_i.*

Proof. Choose neighborhoods U_i of the A_i so that $f(U_i)$ is disjoint from U_j for $i \neq j$, and let U be the union of the U_i. Then almost every orbit $x_0 \mapsto x_1 \mapsto \ldots$ will satisfy $x_t \in U$ for large t; hence x_t must belong to just one U_i for large t. The proof is now straightforward. □

In cases where this lemma applies, the collection of minimal attractors forms a very useful tool. However, the minimality condition may be very hard to verify, and there is no guarantee that a map f has any minimal attractors at all. Counterexamples are provided by the identity map, the concentric circle example of Sect. 1, or by more interesting examples such as the polynomial map

$$(c, y) \mapsto (c, y^2 - c)$$

in two variables (see Fig. 1). In such cases where there are not enough minimal attractors, or in cases where the minimal attractors are not known, it is necessary to fall back on the more general concept of attractor.

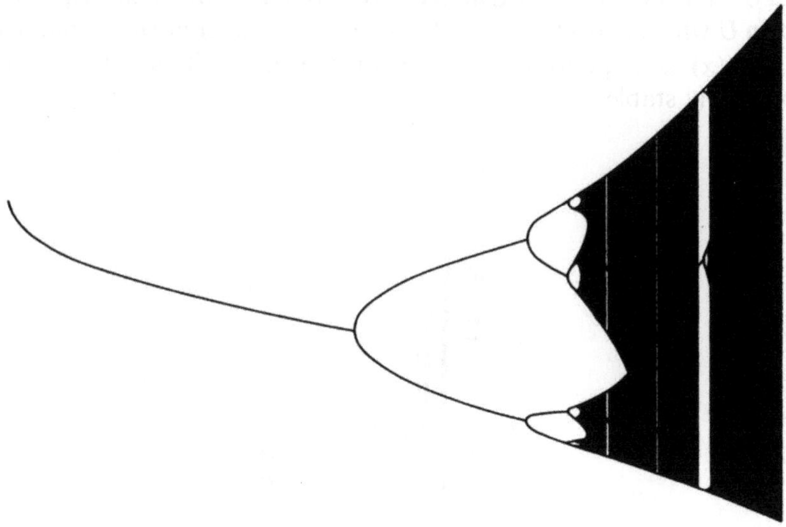

Fig. 1. Likely limit set (in the finite plane) for $(c, y) \mapsto (c, y^2 - c)$

3. The Feigenbaum Attractor

This section will illustrate these ideas by describing a smooth map which has a unique attractor according to the definitions in Sect. 2, but no attractor at all according to most of the definitions in Sect. 1. Let $c=1.401155189\ldots$ be the smallest real number for which the real quadratic map $x \mapsto x^2-c$ has infinitely many distinct periodic orbits. This map $f(x)=x^2-c$ has been studied by Feigenbaum, since it represents the point of transition from stable periodic behavior to chaotic behavior. In order to have a compact domain of definition, let us choose some interval I containing the origin with $f(I)\subset I$, and think of f as a map from I into itself (Fig. 2).

The orbit of zero under f is an almost periodic sequence which can be described as follows (cf. Misiurewicz). The numbers $0=a_0 \mapsto a_1 \mapsto a_2 \ldots$ are all distinct. However, the difference a_m-a_n is very small whenever $m-n$ is divisible by a high power of two. Thus the closure A of this orbit is a Cantor set, homeomorphic to the ring $\varprojlim(Z/2^kZ)$ of 2-adic integers in such a way that each $a_n=f^n(0)$ corresponds to the 2-adic integer n. Note that f restricted to A corresponds to the homeomorphism which adds one to each 2-adic integer. It follows that there are no periodic points in A. However, it can be shown that there are periodic points arbitrarily close to every point of A. These periodic points are all unstable, with period equal to a power of two.

The dynamic structure of $f: I\to I$ is as follows. For almost every initial point x_0 in I the successive images $x^n=f^n(x_0)$ converge towards the Cantor set A. Hence $A=\Lambda(f)$ is the unique attractor for f. But there are a countable infinity of exceptional points whose successive images do not converge towards A. Furthermore, these exceptional points are everywhere dense in I. For according to Guckenheimer (1979), since f has no periodic attractor, any arbitrarily small open interval $U\subset I$ has some forward image $f^n(U)$ which contains the origin. Choosing a point x in U which maps to a periodic point in this neighborhood of the origin, we see that $\omega(x)$ is a periodic orbit disjoint from A. Thus A is certainly not asymptotically stable.

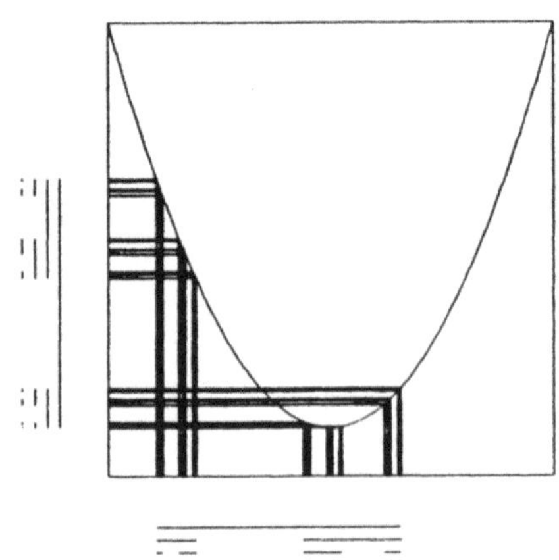

Fig. 2. The unique attractor for the map $x\mapsto x^2-1.4011\ldots$ on an appropriate interval is the indicated Cantor set

However, A is Liapunov stable, and in fact is equal to the intersection of a nested sequence of asymptotically stable sets. In spite of the dense set of exceptional points, the tendency to converge towards A is extremely strong. If U is an arbitrarily small neighborhood of A, let us choose a smaller neighborhood V so that $f(V) \subset V$. Then if we start at any point x_0 in I and follow the successive images $x_n = f^n(x_0)$ there are just two possibilities. Either this orbit eventually hits the open set V and remains trapped within $V \subset U$ forever after, or else it manages to precisely hit one of the finitely many periodic points which lie outside of V, and remains trapped in a finite periodic orbit outside of V thereafter. The first possibility occurs for a dense open set of initial points x_0, while the second possibility occurs only for a countable nowhere dense set of x_0.

Further details of this argument will be given in Appendix 2. For readers who prefer to work with flows, Kan has described somewhat analogous examples of smooth flows on the 3-dimensional sphere so that almost every orbit converges towards a "solenoid," locally homeomorphic to the product of a Cantor set and a 1-manifold, which contains no periodic orbits, but can be approximated arbitrarily closely by periodic orbits. These examples are also Liapunov stable, but not asymptotically stable. (See also Grebogi, Ott, Pelikan, and Yorke. Related but less differentiable examples have been described by Bowen and Franks, and by Franks and Young.) It seems likely that this lack of asymptotic stability may be shared by many other fractal attractors.

4. Stability and Robustness

Let $f: M \to M$ be a fixed smooth map. As in previous sections, we will say that a closed subset $A \subset M$ with $f(A) = A$ is *Liapunov stable* (also called "orbitally stable") if it has arbitrarily small neighborhoods U with $f(U) \subset U$; and *asymptotically stable* if it is Liapunov stable and also satisfies the Auslander-Bhatia-Seibert condition that its realm of attraction $\varrho(A)$ is an open set. In the asymptotically stable case, if we choose U with closure contained in $\varrho(A)$, then it follows easily that A is equal to the intersection of the sequence of forward images $U \supset f(U) \supset f^2(U) \supset \ldots$.

Note that we have not imposed any stability requirement at all as part of the definition of attractor. Of course, many interesting attractors are asymptotically stable, or at least Liapunov stable. However, if we required some form of stability as part of the definition, then it would be awkward to talk about transitional case, and it would no longer be true that every smooth map on a compact manifold has at least one attractor. Here is a simple minded example to illustrate this point.

Example 1. Let M be the circle of real numbers modulo 2π, and let

$$f(\theta) \equiv \theta + 1 - \cos(\theta) \pmod{2\pi}.$$

Then the fixed point $\theta = 0$ is the unique attractor. This attractor is not Liapunov stable, since no nontrivial neighborhood is mapped into itself by f. (Compare Fig. 3. We will describe such a fixed point as "one-sided stable.") A completely analogous example on the real projective line $R \cup \infty$, is the map $f(x) = x + 1$, with

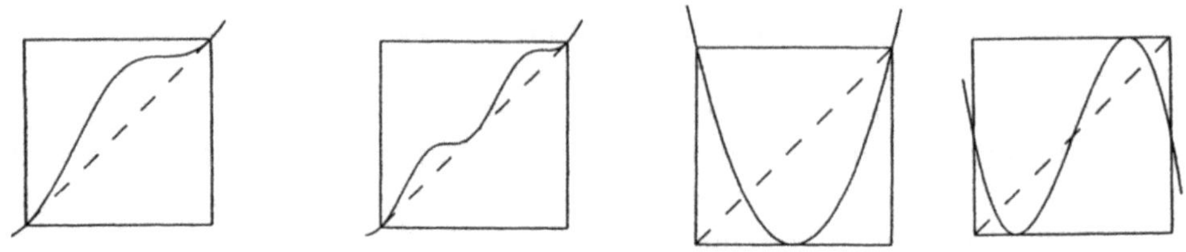

Fig. 3. Graphs of Examples 1–4

the point at infinity as unique attractor. There is a well known classical example of a fractal attractor with similar stability properties, namely the Denjoy C^1-diffeomorphism of the circle, which has a Cantor set as unique attractor (cf. Schweitzer). Here is a really pathological example. Let $f(x)=4x(1-x^2)^2$ for $|x|\leqq 1$, with $f(x)=0$ otherwise. Then surely the only attractor is the origin, which is a strictly repulsive fixed point.

Here is a simple variant of Example 1.

Example 2. On the circle of real numbers modulo 2π, let $f(\theta)=\theta+\sin^2(\theta)$. Then A consists of the two points 0 and π. Each is a minimal attractor, but neither is stable since there are points arbitrarily close to either one whose successive images converge to the other.

Note that an attractor which has positive measure need not attract anything outside of itself. For example an irrational rotation of the circle has the entire circle as minimal attractor.

Example 3. The Ulam von Neumann Attractor. Consider the quadratic map $f(x)=2x^2-1$ from $R\cup\infty$ to itself. Then for almost every point x in the interval $I=[-1,1]$ the limit set $\omega(x)$ is equal to the entire interval I. A proof can be based on the corresponding property for the squaring map $z\mapsto z^2$ on the unit circle, making use of the observation that f restricted to I is equal to the correspondence $\mathrm{Re}(z)\mapsto\mathrm{Re}(z^2)$, where z ranges over the unit circle. On the other hand, for every point x outside of I the orbit of x diverges to infinity. It follows that $\Lambda(f)$ consists of the interval I, which is an unstable minimal attractor, together with the point at infinity which is a stable attractor. Every point outside of I is repelled by I.

By a linear change of coordinate, this map can be put into the form $x\mapsto x^2-2$. More generally, for any parameter c we can consider the map $x\mapsto x^2-c$. If c belongs to the interval $[-1/4,2]$, this map has at least one bounded attractor. It may be true that it has only one bounded attractor, which is either a periodic orbit, a finite union of intervals, or (in uncountably many exceptional cases) an almost periodic Cantor set as in Sect. 3 (cf. Lemma 4 in Appendix 2). Presumably, for almost all parameter values this bounded attractor is asymptotically stable. However, in the almost periodic case it is only Liapunov stable, and there are at least a countable infinity of cases which are not even Liapunov stable. These include the two extreme values $c=-1/4$ and $c=2$, and the value $c=7/4$ which corresponds to a one-sided stable orbit of period 3.

Note that two distinct minimal attractors need not be disjoint from each other. However, the intersection of two minimal attractors must be too small to attract points from any set of positive measure.

Example 4. The likely limit set for the cubic map $f(x)=3\sqrt{3}(x-x^3)/2$ consists of two unstable minimal attractors $[-1,0]$ and $[0,1]$, together with the stable attractor $\{\infty\}$. Here the coefficient $3\sqrt{3}/2$ is chosen so that each of these intervals will map precisely onto itself. In this case, it is interesting to note that the union $[-1,1]$ of the two unstable attractors is asymptotically stable. (The proof that $[0,1]$ is a minimal attractor can be outlined as follows. One can first check that the change of variable $x=\sin^2(\theta)$ converts f, restricted to this interval, to an expanding map. It then follows from Li and Yorke that it has an ergodic invariant measure which is equivalent to Lebesgue measure.)

It is often convenient to describe instability properties of such examples by constructing an associated graph.

Definition. The *stability diagram* associated with a map f is the 1-dimensional complex with one vertex a_i for each minimal attractor A_i, and with edges as follows. There is a directed edge from vertex a_i to the distinct vertex a_j whenever there exist points arbitrarily close to A_i whose limit set $\omega(x)$ intersects A_j; and also there is a loop joining a_i to itself whenever there exist points arbitrarily close to A_i whose successive images temporarily wander away, outside of some fixed neighborhood of A_i, and yet have limit set $\omega(x)$ which intersects A_i.

If A_i is Liapunov stable, note that no edge can lead away from the corresponding vertex. The stability diagrams corresponding to the four examples are shown in Fig. 4.

Next let us study the behavior of the set $\Lambda=\Lambda(f)$ as we perturb the map f. Recall that the *Hausdorff distance* between two closed sets is the smallest number δ such that each closed ball of radius δ centered at a point of either set necessarily contains a point of the other set.

Definition. We will say that the likely limit set Λ for a given map f is *robust* if the Hausdorff distance between $\Lambda(f)$ and $\Lambda(g)$ tends to zero whenever g tends to f in the C^∞-topology (roughly speaking, whenever the kth derivative of g converges uniformly to the kth derivative of f for every $k\geqq 0$). Similarly, an attractor A for f is *robust* if any g which is C^∞-close to f has an attractor A' which is Hausdorff-close to A.

Note that this attractor A' may be qualitatively quite different from A. The condition of Hausdorff closeness guarantees only that these two sets will be rather hard to distinguish by a computer experiment. For example, it follows from Lemma 2 that any asymptotically stable periodic orbit is necessarily robust.

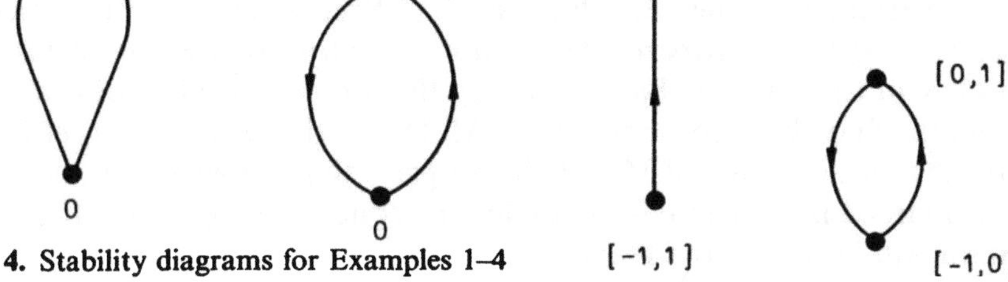

Fig. 4. Stability diagrams for Examples 1–4

However, the attractor A' which Hausdorff approximates this orbit A may have period equal to some multiple of the period of A, or may even be infinite. The most common possibility is bifurcation. For example, each of the maps $x \mapsto x - x^3$ and $x \mapsto x^2 - 3/4$ has a unique stable finite fixed point, which bifurcates under perturbation. The unique attractor $\Lambda(f)$ for the Feigenbaum map f of Sect. 3 is also robust. Again, for g close to f the likely limit set $\Lambda(g)$ may be qualitatively quite different from $\Lambda(f)$. This Cantor set may simplify into a stable periodic orbit, or complicate into a chaotic attractor.

It is conjectured that any robust attractor must be Liapunov stable. To illustrate this conjecture, we have noted that Examples 1, 2, 3, and 4 are not stable, and it is easy to show that they also are not robust.

An attractor may fail to be robust in at least three different ways. It may simply evaporate, or explode into a much bigger set, or implode into a much smaller set. (In the case of an asymptotically stable attractor, only the last possibility can occur, by Lemma 2.) Here are examples to illustrate these three possibilities. The map $x \mapsto x + x^2$, with a one-sided stable attractor at the origin, can be approximated arbitrarily closely by a map for which the only attractor is the point at infinity. The translation $x \mapsto x + 1$ of Example 1, with the infinite point as unique attractor, can be approximated arbitrarily closely by a fractional linear transformation $x \mapsto (1+x)/(1-\varepsilon x)$, for which every orbit is dense, so that the only attractor is the entire set $R \cup \infty$. Finally, let $f(x) = x^2 - c$, where $c = 1.543689\ldots$ is the real root of the equation $c^3 - 2c^2 + 2c - 2 = 0$, so that $f^4(0) = f^3(0) = -f^2(0)$. Then f has a unique finite attractor $A = [f(0), f^2(0)]$, which is asymptotically stable. (The proof is similar to that in Example 4.) Yet f is equal to the limit of a sequence $f_3, f_4, \ldots$ of quadratic maps, where the unique finite attractor A_n for f_n is a superstable orbit of period n consisting of zero and $n-2$ negative numbers, but only one positive number $f_n^2(0)$. Thus A_n is substantially smaller than A for all n.

Note that the likely limit set may fail to be robust even if every attractor is robust, since new attractors may suddenly appear. As an example, the unique attractor for the map $f(x) = x + x^3$ is the stable fixed point at infinity; yet maps arbitrarily close to f have an additional stable fixed point at the origin.

Even when an attractor disappears under perturbation of a map, it leaves behind a "resonance," that is a region where successive images of a point may get trapped for many iterations before escaping (cf. Grebogi, Ott, and Yorke). In practical applications, if the perturbation is very small so that the expected escape time is very large, it may be difficult to tell the difference between such a resonance and an actual attractor.

Appendix 1. The Generic Limit Set

The following variation on the likely limit set Λ of Sect. 2 is sometimes easier to work with. Let us say that a property of a point $x \in M$ is true for *generic* x if it is true for almost all x in the sense of Baire category theory, i.e., for all x outside of a countable union of nowhere dense subsets of M. Define the *generic limit set* $\Gamma \subset M$ to be the smallest closed subset of M with the property that, for generic x, the limits set $\omega(x)$ is contained in Γ. There is an analogous concept of "generic-attractor." The definition will be left to the reader.

In many cases the two limit sets Γ and Λ are equal to each other. However, this is not always true, and I do not know any good criterion for equality. The set Γ is likely to be easier to compute since its definition is purely topological, whereas the definition of Λ uses both topology and measure theory. On the other hand, the set Λ is more closely related to the probable asymptotic behavior which one would like to study in applications. Also, the set Λ is closer to something which one can actually draw pictures of in numerical experiments.

The only general relation between these two sets which is known to me is the statement that the intersection $\Gamma \cap \Lambda$ is always non-vacuous. This follows from the more precise statement that, for generic x, the intersection $\omega(x) \cap \Lambda$ is non-vacuous. In fact, for any neighborhood U of Λ the set of x whose full orbit is disjoint from U is closed and of measure zero, hence nowhere dense. Therefore, the orbit of a generic x intersects every such neighborhood, and hence satisfies $\omega(x) \cap \Lambda \neq \emptyset$.

Following are two rather pathological examples.

Example 5. The likely limit set Λ can be strictly larger than Γ. Consider the gradient flow associated with a smooth real valued function on the 2-dimensional sphere which takes its minimum along a Cantor set of positive measure, but has only non-degenerate critical points otherwise, with just one local maximum x_0. Then $\Gamma = \{x_0\}$, yet Λ contains an entire Cantor set. A similar example involving a C^1-horseshoe has been given by Bowen (1975).

Example 6. The set Λ can be strictly smaller than Γ, at least in the case of a non-differentiable map f. More precisely, there is a continuous map of the interval for which a generic orbit is dense, so that Γ is the entire interval, and for which Λ is a Cantor set. In fact, with suitable modification of the example, Λ will split up into two or more minimal attractors, which may be disjoint or not according to taste. I don't know whether there exists a smooth map with these properties.

Let us start with the continuous measure-preserving map $F(x) = 2 \cdot |x| - 1$ on the interval $[-1, 1]$. Then it is not difficult to show that, for generic x, the limit set $\omega(x)$ is equal to the entire interval $[-1, 1]$ (cf. Appendix 2). Hence $\Gamma = [-1, 1]$. However, we will construct a Borel probability measure μ on M so that, with μ-probability equal to one, the limit set $\omega(x)$ avoids some fixed open subset of $[-1, 1]$. Let C be the closed set consisting of all points x whose full orbit is disjoint from the interval $(1/2, 1]$. It is not difficult to check that C is a Cantor set. In fact, the "kneading sequence" $\operatorname{sgn}(x_0)$, $\operatorname{sgn}(x_1), \ldots$ associated with an orbit $x_0 \mapsto x_1 \mapsto \ldots$ in C can be any sequence of signs which does not have two $+1$'s in a row. Therefore, we can choose a probability measure μ_0 with support equal to C, and so that points have measure zero. Now define a sequence of probability measures inductively by setting

$$2\mu_{n+1}(S) = \mu_n(F(S \cap [-1, 0])) + \mu_n(F(S \cap [0, 1])),$$

so that the support of μ_n is equal to the full t-fold inverse image $F^{-n}(C)$. Then the measure $\sum \mu_n / 2^{n+1}$ has support equal to the entire interval $[-1, 1]$. Yet, with μ-probability equal to one, the limit set $\omega(x)$ is contained in C. [If μ_0 is constructed with more care, then $\omega(x)$ will actually be equal to C with probability one.] Note that the continuous change of coordinate $y = \int_{-1}^{x} d\mu$ transform μ to the standard

Lebesgue measure on the unit interval. Therefore, F is topologically conjugate to a continuous map for which Γ is strictly larger than Λ.

Similarly, let $\hat{C}$ be the Cantor set consisting of all points whose orbits are disjoint from the interval $(-1/2, 0)$, and let $\hat{\mu}$ be an associated measure, so that with $\hat{\mu}$-probability equal to one the limit set $\omega(x)$ must be equal to $\hat{C}$. If we combine these two constructions by using the probability measure $(\mu+\hat{\mu})/2$, then $\omega(x)$ will be equal to C with probability 1/2, and to $\hat{C}$ with probability 1/2. Thus in this case there are two minimal attractors, which intersect only along the periodic orbit $C \cap \hat{C} = \{-3/5, 1/5\}$. We can modify the example so that the two minimal attractors will be disjoint, by using the slightly larger interval $(-3/4, 0)$ for the construction of $\hat{C}$.

Appendix 2. Maps of the Interval

The following discussion will depend strongly on Guckenheimer [17].

Let $x \mapsto f(x)$ be a real quadratic mapping which carries some finite interval I into itself. If we choose this interval I to be maximal, then $f: I \to I$ will belong to the class of maps studied by Guckenheimer. That is: f is smooth of class C^3, with a single interior critical point, with negative Schwarzian derivative

$$2(Df)(D^3 f) - 3(D^2 f)^2 < 0,$$

and f maps both endpoints of I to a single endpoint x_0 at which the derivative satisfies $Df(x_0) \geqq 1$. (This last condition is essential, but was inadvertently omitted in [17].)

Lemma 4. *For $f: I \to I$ satisfying these conditions, the generic limit set Γ is either a finite periodic orbit, a Cantor set equal to the closure of the orbit of the critical point, or a finite union of say n intervals, bounded by the first $2n$ forward images of the critical point, and containing the critical point in its interior. Furthermore, for generic x, the omega limit set $\omega(x)$ is precisely equal to Γ.*

In the first two cases, for every x outside of a set of measure zero we have $\omega(x) = \Gamma$. Hence the likely limit set Λ is equal to Γ, and is a minimal attractor. It is quite possible that this statement is true in the third case also, but I have not been able to decide this question (cf. Example 6 above).

Proof. First suppose that f has a periodic attractor, that is either a stable or a one-sided stable periodic orbit A. Then there exists a neighborhood (or one-sided neighborhood) U consisting of points whose orbits converge uniformly to A. According to [17, p. 135 and 2.8, 3.1], for every point x outside of a closed set of measure zero the orbit of x eventually falls into this neighborhood U, and hence converges to A. Clearly, it follows that $\Gamma = \Lambda = A$.

Henceforth, let us assume that f has no stable or one-sided stable periodic orbit. In order to handle this case we will need the following.

Definition. We will say that the map f is *reducible* if there exists a closed interval J about the origin and an integer $n \geqq 2$ so that the successive images $J, f(J), f^2(J), \ldots, f^{n-1}(J)$ have disjoint interiors, and so that $f^n(J) \subset J$.

The idea is that whenever this is the case we can reduce the study of f to the study of the n-fold iterate $g=f^n$, considered a map from the subinterval J into itself. In fact, if f is reducible, [17, 2.6 and 3.1] proves that, for every x outside of a closed set of measure zero, the orbit of x eventually meets the interior of J. It follows that the likely limit set $\Lambda=\Lambda(f)$ is equal to the n-fold union $\Lambda(g)\cup f(\Lambda(g)) \cup \ldots f^{n-1}(\Lambda(g))$; with a similar description for $\Gamma(f)$.

We may always assume that the interval J is chosen so as to be maximal. It then follows easily that $g=f^n$ maps the boundary of J to one boundary point. Since there are no stable periodic orbits, the derivative of g at this boundary fixed point, must be at least one. Since the condition of negative Schwarzian derivative is preserved under composition, we see that this new map $g:J\to J$ satisfies all of Guckenheimer's conditions.

If this new map $g:J\to J$ is also reducible, then we iterate the construction. We must distinguish between the "finitely reducible case" in which the reduction process terminates with a non-reducible map after finitely many steps, and the "infinitely reducible case" in which it continues ad infinitum (Fig. 2). Evidently, in order to understand the finitely reducible case, we need only understand the case of a non-reducible map.

Suppose then that f is not reducible. We will prove that f is topologically conjugate to a piecewise linear map of the form

$$F(y)=s\cdot|y|-1 \quad \text{for} \quad |y|\leqq 1/(s-1),$$

where $\log(s)>0$ is the topological entropy [17, 4.5]). Since we have excluded the trivial case where f has a stable or one-sided stable fixed point, it is not difficult to check that f must have strictly positive topological entropy. Hence, by Milnor and Thurston, f is topologically semi-conjugate to F, say $h\circ f=F\circ h$, where h is a monotone map. Thus every inverse image $h^{-1}(y)$ is either a point or an interval J. If some $h^{-1}(y)$ were an interval, then the orbit of y would have to contain the critical point 0, say $F^m(y)=0$. For otherwise any two points of J would have the same kneading sequence, which is impossible by [17, 2.6]. In fact, this orbit would have to contain 0 more than once, since the interval $f^{m+1}(J)$ also maps to a single point under h. Hence $F^n(0)$ would be 0 for some n, necessarily greater than one. Therefore, f^n would map the interval $h^{-1}(0)$ into itself, and hence f would be reducible, contradicting our hypothesis. This contradiction proves that h must in fact be a homeomorphism, so that f is topologically conjugate to F.

Thus we are reduced to studying the piecewise linear map F. We may assume that the growth number s is greater than $\sqrt{2}$, since otherwise it is easy to check that F is reducible. Then we will prove that the generic limit set of F is equal to the interval

$$A=[-1,s-1]=[F(0),F^2(0)].$$

In fact, it is easy to check that $F(A)=A$, and that the orbit of any y in the open interval $|y|<1/(s-1)$ eventually gets trapped in this subinterval A.

The map F from A to itself has a strong transitivity property as follows. *For any subinterval $J\subset A$ there exists a positive integer n so that $F^n(J)$ is equal to the entire interval A.* For if J is small enough so that the two sets J and $F(J)$ do not both contain zero, then $F^2(J)$ is longer than J by a factor of at least $s^2/2>1$. Thus the successive images of J expand exponentially until say $F^{m-1}(J)$ and $F^m(J)$ both contain zero; and hence $F^{m+1}(J)=A$.

It follows from this property that, for a generic point y, the limit set $\omega(y)$ is equal to the entire set A. For if $\{U_i\}$ is a basis for the open subsets of A, then the set of y whose orbit intersects U_i is dense and open. Thus the orbit of a generic y intersects every U_i, and hence is dense in A. Therefore, the generic limit set $\Gamma(F)$ equals A, and it follows that the generic limit set $\Gamma(f)$ for the smooth map f is the corresponding interval $h^{-1}(A)$. More explicitly, if c is the critical point of the non-reducible map f, then $\Gamma(f)$ is equal to the closed interval bounded by $f(c)$ and $f^2(c)$. Similarly, the generic limit set for a finitely reducible map f is a finite union of intervals, bounded by appropriate forward images of the critical point.

The likely limit set Λ seems much harder to determine. It follows easily from this discussion that Λ is a subset of Γ, but I have not been able to prove that $\Lambda=\Gamma$ in the non-reducible or finitely reducible case.

Finally, we must consider the case where the reduction process does not terminate after finitely many steps, but rather continues ad infinitum. In this infinitely reducible case it is not difficult to check that there is a unique attractor $A=\Lambda=\Gamma$ which is a Cantor set, homeomorphic to an inverse limit of finite cyclic groups Z/nZ. Furthermore, the map f is almost periodic on A, and corresponds to the adding machine map $\alpha\mapsto\alpha+1$ from this limit group to itself. Any such inverse limit of finite cyclic groups can occur, so there are uncountably many distinct examples. Details will be omitted. □

It follows easily from this discussion that f has sensitive dependence on initial conditions (as defined in [17]) if and only if Γ is a finite union of intervals. For the class of maps studied by Guckenheimer, there are just two qualitatively different forms of behavior. If f has a periodic or almost periodic attractor A, then $A=\Lambda(f)=\Gamma(f)$ is a set of measure zero, f does not have sensitive dependence on initial conditions, and the topological entropy of $f|A$ is zero. In all other cases, the generic limit set is an interval or union of intervals, with positive measure, with sensitive dependence on initial conditions, and the entropy of $f|\Gamma(f)$ is strictly positive. In the latter case, the dynamics of f must be described as *chaotic*.

Concluding Remarks. There are many unsolved problems, even for real quadratic maps of the interval. For example, in the chaotic case, I don't know how to decide whether $\Lambda=\Gamma$, or to decide whether there can be more than one attractor. Equivalently, is it true that almost every orbit in the union of intervals Γ is everywhere dense in Γ? More generally, I don't know how to answer the following basic question for maps of $R\cup\infty$ to itself. Suppose either that $f(x)$ is a rational function of degree at least two, or that f has negative Schwarzian derivative. Does it follow that the likely limit set $\Lambda(f)$ is equal to $\Gamma(f)$, and can be expressed as the union of finitely many minimal attractors? Similarly, I have no idea how to decide when the set $\Gamma(f)$ [much less $\Lambda(f)$] is robust. Here is an explicit guess for the class of maps considered by Guckenheimer: The interval $[f(0), f^2(0)]$ is conjectured to be a robust necessarily minimal attractor if and only if the orbit of the critical point is everywhere dense in this interval. (Compare the discussion of $x\mapsto x^2-1.543689\ldots$ in Sect. 4.)

Appendix 3. Strange Attractors

The concept of a *strange attractor*, as described by Ruelle and Takens, is perhaps intentionally not precisely defined. The original discussion emphasized wild topology. An excellent introduction to this subject can be found in Ruelle's survey article (1979–80), which rather emphasizes chaotic dynamical behavior. Guckenheimer and Holmes define a strange attractor as one which has a transversal homoclinic point, and hence has chaotic behavior.

It is important to note that there is no necessary connection between wild topology or fractal geometry and strange dynamical behavior. For example, the fractal attractor of Sect. 3 is topologically wild, but has very sedate almost periodic dynamics, with topological entropy equal to zero. On the other hand, the map $(\theta, t) \mapsto (2\theta, t/2)$ from the cylinder $S^1 \times [-1, 1]$ to itself has a smooth manifold $S^1 \times 0$ as unique attractor; yet the dynamical behavior is completely chaotic. In fact, this map is ergodic and mixing with respect to the standard invariant measure on the circle, and has positive entropy. The behavior of this example under perturbation will be studied below. For an analogous example using a diffeomorphism or flow in place of an iterated map, we could substitute an Anosov diffeomorphism of the torus, or the geodesic flow on the unit tangent bundle of a manifold of negative curvature, in place of the squaring map on the circle.

Here is an example of an attractor which is chaotic and topologically wild, yet simple enough to analyze almost completely (cf. Kaplan and Yorke). Let us start with the map $f_0(e^{i\theta}, x) = (e^{2i\theta}, cx)$ from the cylinder $S^1 \times R$ to itself, where c is a constant in the range $0 < c < 1/2$. Clearly, f_0 has a unique attractor which is asymptotically stable, since all orbits tend towards the circle $S^1 \times 0$, and since almost very orbit on the circle is everywhere dense. This attractor $S^1 \times 0$ is a smooth manifold, but is dynamically chaotic. Now make a small perturbation of f_0 to the map

$$f_\varepsilon(e^{i\theta}, x) = (e^{2i\theta}, cx + \varepsilon\cos\theta).$$

Then the smooth attractor $S^1 \times 0$ will be replaced by a fractal attractor (Fig. 5).

To see this, first note that a scale change in the x-coordinate will replace f_ε by the map

$$f_1(e^{i\theta}, x) = (e^{2i\theta}, cx + \cos\theta).$$

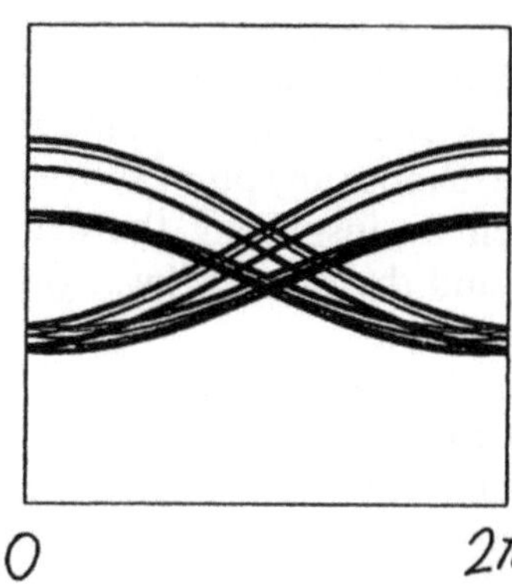

Fig. 5. The unique attractor for the map $(\theta, x) \mapsto (2\theta, 0.4x + \cos\theta)$

We can further simplify by introducing a third coordinate y, and extending f_1 to the smooth map

$$(e^{i\theta}, x, y) \mapsto (e^{2i\theta}, cx + \cos\theta, cy + \sin\theta)$$

from $S^1 \times R \times R$ to itself. Setting $w = e^{i\theta}$ and $z = x + iy$, we can write this map as

$$F(w, z) = (w^2, cz + w)$$

from the 3-dimensional manifold $S^1 \times C$ to itself.

Lemma 5. *There is a unique attractor $A \subset S^1 \times C$, homeomorphic to the dyadic solenoid, which is asymptotically stable, and in fact, attracts every orbit. More precisely, for any compact neighborhood K of A the successive images $F^n(K)$ converge towards A, and for almost every point (w, z) of $S^1 \times C$ the omega limit set $\omega(x)$ is equal to A. This set A supports a measure so that F restricted to A is measure theoretically equivalent to the full Bernoulli shift of entropy* $\log 2$.

Here by the dyadic solenoid $\hat{S}$ we mean the inverse limit of the sequence of squaring homomorphisms

$$S^1 \leftarrow S^1 \leftarrow S^1 \leftarrow \ldots .$$

Thus, by definition, a point of $\hat{S}$ is a sequence $\hat{w} = (w_0, w_1, w_2, \ldots)$ of points on the unit circle so that

$$w_1 = \pm\sqrt{w_0}, \quad w_2 = \pm\sqrt{w_1}, \ldots .$$

Define $A = g(\hat{S})$ to be the image of $\hat{S}$ under the continuous map

$$g(\hat{w}) = (w_0, w_1 + cw_2 + c^2 w_3 + \ldots).$$

Since we have assumed that $|c| < 1/2$, it is quite easy to check that g is a topological embedding. (With more work, one could check this for somewhat larger c.) Note that F maps A homeomorphically onto itself. In fact,

$$F(w_0, w_1 + cw_2 + c^2 w_3) = (w_0^2, w_0 + cw_1 + c^2 w_2 \ldots),$$

so F restricted to A corresponds to the "shift automorphism"

$$(w_0, w_1, w_2, \ldots) \mapsto (w_0^2, w_0, w_1, \ldots)$$

from the topological group $\hat{S}$ to itself.

Next we will see that, for any starting point (w, z) in $S^1 \times C$, the distance between $F^n(w, z)$ and the set A converges geometrically (and uniformly on any compact set) to zero. In fact, choose any point $\hat{w} = (w_0, w_1, \ldots)$ in $\hat{S}$ with $w_0 = w$, and let $g(\hat{w}) = (w_0, z_0) \in A$. Then by inspecting the definition of F we see that the distance between $F^n(w, z)$ and the point $F^n(w_0, z_0) \in A$ is exactly $c^n|z - z_0|$, which converges to zero as $n \to \infty$.

More geometrically, let k be some constant in the open interval from $1/(1-c)$ to $1/c$, and let T be the solid torus consisting of all (w, z) with $|z| \leqq k$. Then it is not hard to see that F maps T homeomorphically into its own interior, with degree two, and that A is equal to the intersection of the forward images $F^n(T)$. In this form, the example has been studied by Smale (1967), by Ruelle and Takens, and by Ruelle (1979–80).

To study the ergodic behavior for F restricted to A, or equivalently for the shift automorphism of $\hat{S}$, write each point of $\hat{S}$ as $(w_0, w_1, \ldots)$, with $w_n = \exp(\pi i x_n)$. Expressing the number x_0 in binary notation as $a_0 + a_1/2 + a_2/4 + \ldots$, it follows that x_1 has the form $a_{-1} + a_0/2 + a_1/4 \ldots$, with similar expressions for the higher x_n. Thus the shift automorphism of $\hat{S}$ corresponds to the Bernoulli shift on a doubly infinite sequence $\ldots a_{-1}, a_0, a_1, a_2, \ldots$ of zeros and ones.

The Hausdorff dimension of $A \subset S^1 \times C$ can be computed as $1 - \log(2)/\log(c)$. This lies strictly between 1 and 2, so A is a fractal set, as defined by Mandelbrodt.

Now let A' be the image of A under the projection map from $S^1 \times C$ to $S^1 \times R$. Then it follows easily from the discussion above that A' is an attractor for the map f. In fact, for any compact neighborhood N of A', the successive images $f^k(N)$ converge to A'. Furthermore, A' supports an invariant measure, with entropy $\log 2$, so that almost every orbit in A' is dense. This attractor A' is nowhere dense, since it is a compact set of Hausdorff dimension $d \leqq 1 - \log(2)/\log(c) < 2$. In fact, A' can be described as the closure of the union of the graphs of the smooth almost periodic functions

$$g_n(\theta) = \cos((\theta + 2\pi n)/2) + c\cos((\theta + 2\pi n)/4) + c^2\cos((\theta + 2\pi n)/8) + \ldots,$$

where n ranges over all integers. Since g_n is uniformly close to g_m whenever n is close to m in the 2-adic topology, the function g_N is also defined and real analytic for any 2-adic integer N. With this convention, we can say simply that A' is the union of the graphs of the uncountably many distinct functions g_N. Thus A' is certainly not a manifold. It seems likely that its Hausdorff dimension is equal to $1 - \log(2)/\log(c) > 1$. (Compare Farmer, Ott, and Yorke.) On the other hand, if we modify this example by choosing c sufficiently close to 1, then it can be shown that the corresponding attractor A' contains an entire open subset of the cylinder. (See Frederickson, Kaplan, Yorke, and Yorke.)

If we carry out an analogous construction starting out with the Anosov diffeomorphism $(\phi, \psi) \mapsto (2\phi + \psi, \phi + \psi)$ of the torus in place of the squaring map on the circle, then we obtain an attractor which is topologically a torus $S^1 \times S^1$. Let $\lambda = (\sqrt{5} - 1)/2$ be the smaller eigenvalue of the associated integer matrix. Then this torus is embedded in a nowhere differentiable manner into the manifold $S^1 \times S^1 \times R$ if $\lambda < c < 1$, but is n times continuously differentiable if $0 < c < \lambda^n$ (see Kaplan, Mallet-Paret, and Yorke).

Acknowledgements. I am grateful to R. Butler, J. Guckenheimer, W. Thurston, R. F. Williams, and J. Yorke for a number of helpful suggestions.

References

Anosov, D.V., Katok, A.B.: New examples in smooth ergodic theory. Trans. Moscow Math. Soc. **23**, 1–35 (1970)

Auslander, J., Bhatia, N.P., Seibert, P.: Attractors in dynamical systems. Bol. Soc. Mat. Mex. **9**, 55–66 (1964)

Besicovitch, A.S.: A problem on topological transformation of the plane. Fundam. Math. **27**, 61–65 (1937)

Bowen, R.: A horseshoe with positive measure. Invent. Math. **28**, 203–204 (1975)

Bowen, R.: Invariant measures for Markov maps of the interval. Commun. Math. Phys. **69**, 1–17 (1979)

Bowen, R., Franks, J.: The periodic points of maps of the disk and the interval. Topology **15**, 337–342 (1976)

Coddington, E., Levinson, N.: Theory of ordinary differential equations. New York: McGraw-Hill 1955

Collet, P., Eckmann, J.-P.: Iterated maps on the interval as dynamical systems. Boston: Birkhäuser 1980

Conley, C.: Isolated invariant sets and the morse index. C.B.M.S. Regional Lect. **38**, A.M.S. 1978

Farmer, J.D., Ott, E., Yorke, J.A.: The dimension of chaotic attractors. Physica **7** D, 153–180 (1983)

Feigenbaum, M.: Quantitative universality for a class of nonlinear transformations. J. Stat. Phys. **19**, 25–52 (1978); **21**, 669–706 (1979)

Franks, J., Young, L.-S.: A C^2 Kupka-Smale diffeomorphism of the disk with no sources or sinks (preprint)

Frederickson, P., Kaplan, J., Yorke, E., Yorke, J.: The Liapunov dimension of strange attractors. J. Diff. Eq. **49**, 185–207 (1983)

Grebogi, C., Ott, E., Pelikan, S., Yorke, J.: Strange attractors that are not chaotic. Physica **13** D, 261–268 (1984)

Grebogi, C., Ott, E., Yorke, J.: Crises, sudden changes in chaotic attractors, and transient chaos. Physica **7** D, 181–200 (1983)

Guckenheimer, J.: A strange attractor, pp. 368–381. In: The Hopf bifurcation and applications. Marsden and McCracken (eds.). Berlin, Heidelberg, New York: Springer 1976

Guckenheimer, J.: Sensitive dependence on initial conditions for one-dimensional maps. Commun. Math. Phys. **70**, 133–160 (1979)

Guckenheimer, J., Williams, R.F.: Structural stability of Lorenz attractors. Pub. Math. I.H.E.S. **50**, 59–72 (1979)

Guckenheimer, J., Holmes, P.: Nonlinear oscillations, dynamical systems and bifurcation of vector fields. Berlin, Heidelberg, New York: Springer 1983

Hénon, M.: A two-dimensional mapping with a strange attractor. Commun. Math. Phys. **50**, 69–77 (1976)

Hirsch, M.: The dynamical systems approach to differential equations. Bull. Am. Math. Soc. **11**, 1–64 (1984)

Jakobson, M.V.: Absolutely continuous invariant measures for one-parameter families of one-dimensional maps. Commun. Math. Phys. **81**, 39–88 (1981)

Jonker, L., Rand, D.: Bifurcations in one dimension. Invent. Math. **62**, 347–365 (1981); **63**, 1–15 (1981)

Kan, I.: Strange attractors of uniform flows. Thesis, Univ. Ill. Urbana 1984

Kaplan, J.L., Mallet-Paret, J., Yorke, J.A.: The Lyapunov dimension of a nowhere differentiable attracting torus. Erg. Th. Dyn. Syst. (to appear)

Kaplan, J.L., Yorke, J.A.: Chaotic behavior of multidimensional difference equations, pp. 204–227. In: Functional differential equations and the approximation of fixed points. Peitgen and Walther (eds.). Lecture Notes in Mathematics, Vol. 730. Berlin, Heidelberg, New York: Springer 1970

Lanford, O.: A computer assisted proof of the Feigenbaum conjecture. Bull. Am. Math. Soc. **6**, 427–434 (1982)

La Salle, J.P.: The stability of dynamical systems. Reg. Conf. App. Math. **25**, S.I.A.M. 1976

La Salle, Lefschetz, S.: Stability by Liapunov's direct method. New York: Academic Press 1961

Li, T., Yorke, J.: Ergodic transformations from an interval into itself. Trans. Am. Math. Soc. **235**, 183–192 (1978)

Liapunoff, M.A.: Problème général de la stabilité du mouvement. Annal. Math. Stud. **17**, Princeton Univ. Press 1947 (1907 translation from the 1892 Russian original)

Lorenz, E.N.: Deterministic nonperiodic flow. J. Atmosph. Sci. **20**, 130–141 (1963)

Mandelbrodt, B.: The Fractal Geometry of Nature. San Francisco: Freeman 1982
Mendelson, P.: On unstable attractors. Bol. Soc. Mat. Mex. **5**, 270–276 (1960)
Milnor, J., Thurston, W.: On iterated maps of the interval (in preparation)
Misiurewicz, M.: Invariant measures for continuous transformations of [0, 1] with zero topological entropy, pp. 144–152 of Ergodic Theory, Denker and Jacobs, Lecture Notes in Mathematics, Vol. 729. Berlin, Heidelberg, New York: Springer 1979
Misiurewicz, M.: Absolutely continuous measures for certain maps of an interval. Pub. Math. I.H.E.S. **53**, 17–51 (1981)
Nusse, L.: Chaos, yet no chance to get lost. Thesis, Utrecht 1983
Ruelle, D., Takens, F.: On the nature of turbulence. Commun. Math. Phys. **20**, 343–344 (1971)
Ruelle, D.: Strange attractors. Math. Intell. **2**, 126–137 (1979–80)
Ruelle, D.: Small random perturbations of dynamical systems and the definition of attractor. Commun. Math. Phys. **82**, 137–151 (1981)
Ruelle, D.: Small random perturbations and the definition of attractor, pp. 663–676 of Geometric Dynamics, Palis (ed.). Lecture Notes in Mathematics, Vol. 1007. Berlin, Heidelberg, New York: Springer 1983
Schweitzer, P.A.: Counterexamples to the Seifert conjecture and opening closed leaves of foliations. Annal. Math. **100**, 386–400 (1974)
Shaw, R.: Strange attractors, chaotic behavior and information flow. Z. Naturforsch. A **36**, 80–112 (1981)
Smale, S.: Differential dynamical systems. Bull. Am. Math. Soc. **73**, 747–817 (1967)
Smale, S.: Dynamical systems and turbulence, pp. 71–82 of Turbulence seminar, Chorin, Marsden, Smale (eds.). Lecture Notes in Mathematics, Vol. 615. Berlin, Heidelberg, New York: Springer 1977
Ulam, S., v. Neumann, J.: On combination of stochastic and deterministic processes. Bull. Am. Math. Soc. **53**, 1120 (1947)
Williams, R.F.: The zeta function of an attractor, pp. 155–161 of Conference on the Topology of Manifolds, J. Hocking (ed.). Boston: Prindle, Weber & Schmidt 1968
Williams, R.F.: Expanding attractors. Pub. Math. I.H.E.S. **43**, 169–203 (1974)
Williams, R.F.: The structure of Lorenz attractors, pp. 94–112 of Turbulence Seminar, Chorin, Marsden, Smale (eds.). Lecture Notes in Mathematics, Vol. 615. Berlin, Heidelberg, New York: Springer 1977
Williams, R.F.: The structure of Lorenz attractors. Pub. Math. I.H.E.S. **50**, 73–100 (1979)
Williams, R.F.: Attractors, strange and perverse (to appear)
Zeeman, E.C.: Catastrophe theory and applications. Summer school on dynamical systems. I.C.T.P. Trieste 1983

Communicated by O. E. Lanford

Received December 5, 1984; in revised form December 26, 1984

Commun. Math. Phys. 102, 517–519 (1985)

Communications in Mathematical Physics

Comments

On the Concept of Attractor: Correction and Remarks

John Milnor
Institute for Advanced Study, Princeton, NJ 08540, USA

The following consists of three unrelated comments on the author's paper [1].

1. Correction

Let f be a continuous map from a compact metric space X to itself, with n^{th} iterate denoted by f^n, and let $A \subset X$ be a closed non-vacuous subset with $f(A) = A$. Consider the following two properties of A.

(I) *For any sufficiently small neighborhood U of A, the intersection of the images $f^n(U)$ for $n \geqq 0$ is equal to A (compare Smale* [2, p. 786]*).*

(II) (Asymptotic stability) *For any sufficiently small neighborhood U, the successive images $f^n(U)$ converge to A, in the sense that for any neighborhood V there exists n_0 so that $f^n(U) \subset V$, for $n \geqq n_0$.*

In [1, Sect. 1] the author mistakenly described an example satisfying (I) but not (II). (The example was based on a remark of Besicovitch [3], which was corrected in a later paper [4].) In fact, (I) implies (II). The following proof is a minor modification of Hurley [8, Lemma 1.6], which demonstrates a corresponding statement for flows on a compact manifold. The proof shows also that (I) implies the existence of arbitrarily small neighborhoods $W \supset A$ with $f(W) \subset W$.

Proof that (I) *implies* (II). Let U be an open neighborhood which is small enough so that the intersection of the forward images of the closure $\bar{U}$ is equal to A. Let U_n be the open neighborhood consisting of all points x such that $f^i(x) \in U$ for $0 \leqq i \leqq n$. Thus $U = U_0 \supset U_1 \supset \ldots \supset A$ and $f(U_n) \subset U_{n-1}$. Hence the intersection W of the U_n satisfies $f(W) \subset W$. We will show that W is equal to U_n for n sufficiently large, and hence that W is an open set. Otherwise, for infinitely many integers n there must exist a point x_n which belongs to U_n but not U_{n+1}. Let $y_n = f^n(x_n) \in U$. Then we can choose some subsequence of these points y_n which converges to a point $y \in \bar{U}$. Since y_n belongs to the intersection of the sets $f^i(\bar{U})$ for $0 \leqq i \leqq n$, it follows that y belongs to the intersection of all of the $f^i(\bar{U})$, which is equal to A by hypothesis. But $f(y_n) \notin U$, hence $f(y) \notin U$, contradicting the hypothesis that $f(A) = A \subset U$. This proves that W is open. Hence the compact set $\bar{W} \subset \bar{U}$ is a neighborhood of A with $f(\bar{W}) \subset \bar{W}$. It follows easily from compactness that the successive images $\bar{W} \supset f(\bar{W}) \supset f^2(\bar{W}) \supset \ldots$ with intersection A actually converge to A in the sense described in (II). □

Here is an example. Let $U \subset X$ be any open set satisfying $f^n(\bar{U}) \subset U$ for some n, and suppose that the intersection $B = \bar{U} \cap f(\bar{U}) \cap \ldots \cap f^{n-1}(\bar{U})$ is non-vacuous. Then the intersection of the forward images of B is a non-vacuous, compact, f-invariant set which satisfies (I) and (II). The proof is not difficult.

In the context of a smooth flow, Conley [5] uses the term *attractor* for a compact, non-vacuous invariant set satisfying the analogues of conditions (I) and (II), while Auslander et al. [6] call such a set an [asymptotically] *stable attractor*. Note, however, that the word attractor is used in a quite different sense in [1]. Here is an example: Let f be the map of the unit square given by $f(x,y) = (1-x)(x,y)$. Then the edge $x=0$ is the unique compact invariant set satisfying (I) and (II). However, the omega limit set $\omega(x,y)$ is equal to the origin for almost every point (x,y) in the square, hence by the definitions of [1] the origin is the unique attractor.

2. Remark

In order to know that the concept of "minimal attractor," as defined in [1], is reasonable and useful, one would at least like an affirmative answer to the following. *For a C^k-generic map or flow on a compact manifold, is it true that almost every point belongs to the realm of attraction of some "minimal attractor"?* Using standard definitions, a related question would be the following: *Does the chain-recurrent set of a C^k-generic map or flow have at most a countable number of chain components?* Whenever this is true, one can at least say that almost every point belongs to the realm of attraction $\varrho(A)$ for some chain component A which is "attractive" in the sense that $\varrho(A)$ has strictly positive measure.

For some of the known generic and non-generic properties of maps or flows, see [7–12] below.

3. Addendum

The discussion of an attractor related to the solenoid in [1, Appendix 3] should have included a reference to Mayer and Roepstorff [13], which contains a detailed discussion of a similar example.

References

1. Milnor, J.: On the concept of attractor. Commun. Math. Phys. **99**, 177–195 (1985)
2. Smale, S.: Differentiable dynamical systems. Bull. Am. Math. Soc. **73**, 747–817 (1967)
3. Besicovitch, A.S.: A problem on topological transformation of the plane. Fundam. Math. **27**, 61–65 (1937) (compare [4])
4. Besicovitch, A.S.: A problem on topological transformations of the plane. II. Proc. Camb. Phil. Soc. **47**, 38–45 (1951)
5. Conley, C.: Isolated invariant sets and the Morse index. CBMS-NSF Reg. Conf. **38**, Am. Math. Soc. 1978
6. Auslander, J., Bhatia, N.P., Seibert, P.: Attractors in dynamical systems. Bol. Soc. Mat. Mex. **9**, 55–66 (1964)

7. Dobrynskii, V.A., Sarkovskii, A.N.: Genericity of dynamical systems almost all orbits of which are stable under sustained perturbations. Sov. Math. Dokl. **14**, 997–1005 (1973)
8. Hurley, M.: Attractors: persistence, and density of their basins. Trans. Am. Math. Soc. **269**, 247–271 (1982) (compare [9])
9. Hurley, M.: Bifurcation and chain recurrence. Ergodic Theory Dyn. Syst. **3**, 231–240 (1983)
10. Newhouse, S.: Lectures on dynamical systems, pp. 1–114 of "Dynamical Systems". Guckenheimer, Moser, Newhouse (ed.). Boston, Basel, Stuttgart: Birkhäuser 1980
11. Robinson, R.C., Williams, R.F.: Finite stability is not generic, pp. 451–462 of "Dynamical Systems". Peixoto, M. (ed.). New York: Academic Press 1973
12. Takens, F.: Tolerance stability, pp. 293–304 of "Dynamical Systems". Lecture Notes in Mathematics, Vol. 468, Manning, A. (ed.). Berlin, Heidelberg, New York: Springer 1975
13. Mayer, D., Roepstorff, G.: Strange attractors and asymptotic measures of discrete-time dissipative systems. J. Stat. Phys. **31**, 309–326 (1983)

Communicated by A. Jaffe

Received September 3, 1985

TRANSACTIONS OF THE
AMERICAN MATHEMATICAL SOCIETY
Volume 287, Number 1, January 1985

BOWEN-RUELLE MEASURES FOR CERTAIN PIECEWISE HYPERBOLIC MAPS

BY
LAI-SANG YOUNG[1]

ABSTRACT. We consider a class of piecewise C^2 Lozi-like maps and prove the existence of invariant measures with absolutely continuous conditional measures on unstable manifolds

This work is a small step forward in the following program: Suppose a compact neighborhood is mapped into itself and the map displays some chaotic behavior. Is there a strange attractor, or more specifically, is there a Bowen-Ruelle measure? Axiom A systems aside, current techniques have hardly begun to provide answers to these questions. One of the purposes of this note is to demonstate, in a very limited way, how certain 1-dimensional results can sometimes be useful in handling dissipative systems in 2-dimension.

The 1-dimensional result we alluded to says that piecewise expanding endomorphisms of the unit interval have smooth invariant measures [LY]. What we prove here is an analogous statement for certain piecewise hyperbolic attractors in 2-dimension. The prime examples that motivate this study are the Lozi maps, though our analysis has little to do with the precise nature of the equations studied by Lozi or Misiurewicz [M]. Our main result is that these "generalized Lozi maps" have invariant measures with absolutely continuous conditional measures on unstable manifolds. As a consequence they have Bowen-Ruelle measures.

We begin by isolating several properties of the Lozi maps. They are essentially the properties upon which our proof depends. Maps satisfying these hypotheses will henceforth be called "generalized Lozi maps". (See Figure 1.)

Let $R = [0,1]\times[0,1]$ and let $f\colon R \to R$ be a continuous injective map. We assume that f or some iterate of f takes R into its interior. Let $0 < a_1 < \cdots < a_q < 1$ and let $S = \{a_1, \ldots, a_q\} \times [0,1]$. We assume that $f|(R-S)$ is a C^2 diffeomorphism onto its image with $|\mathrm{Jac}(f)| < 1$ and that both $f|(R-S)$ and $f^{-1}|f(R-S)$ have bounded second derivative. We further impose the following conditions on $f|(R-S)$ (geometric interpretations are given in parentheses).

(H1) $$\operatorname{Inf}\left\{\left(\left|\frac{\partial f_1}{\partial x}\right| - \left|\frac{\partial f_1}{\partial y}\right|\right) - \left(\left|\frac{\partial f_2}{\partial x}\right| + \left|\frac{\partial f_2}{\partial y}\right|\right)\right\} \geq 0$$

(Df preserves cones making $< 45°$ with the x-axis),

(H2) $$\operatorname{Inf}\left\{\left|\frac{\partial f_1}{\partial x}\right| - \left|\frac{\partial f_1}{\partial y}\right|\right\} = u > 1$$

Received by the editors November 15, 1982.
1980 *Mathematics Subject Classification.* Primary 58F15.
Key words and phrases. Hyperbolicity, unstable manifolds, invariant measures.
[1]Research partially supported by NSF.

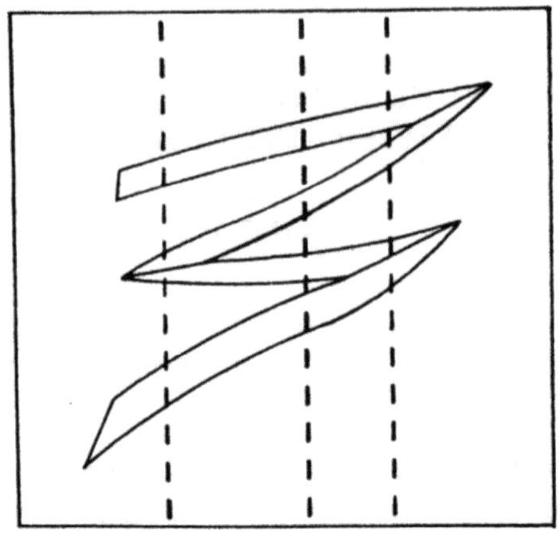

FIGURE 1. Example of a "generalized Lozi map"

(restricted to these cones the action of Df when projected onto the x-axis is uniformly expanding),

(H3) $$\operatorname{Sup}\left\{\frac{|\partial f_1/\partial y| + |\partial f_2/\partial y|}{(|\partial f_1/\partial x| - |\partial f_1/\partial y|)^2}\right\} < 1$$

(horizontal expansion dominates the action of Df on vertical vectors), and

(H4) $$\exists N \in \mathbf{Z}^+ \text{ s.t. } u^N > 2 \text{ and } f^k S \cap S = \emptyset \text{ for } 1 \leq k \leq N$$

(f expands horizontally more than it folds).

Note that (H4) is vacuous when $u > 2$. (Hence Figure 1 is legitimate.) Note also that these hypotheses are indeed satisfied by many Lozi maps. Lozi maps are usually given by

$$L(x, y) = (1 + y - a|x|, bx).$$

Changing coordinates this can be written as

$$f(x, y) = (1 + by - a|x|, x).$$

It is easy to verify that for open intervals of a and b, f takes some square $[c, c] \times [c, c]$ into itself and satisfies (H1)–(H4).[2]

DEFINITION 1. A Borel probability measure μ on R is said to have absolutely continuous conditional measures on unstable manifolds if there exist measurable partitions $\mathcal{P}_1 \subset \mathcal{P}_2 \subset \cdots$ of R and measurable sets $V_1 \subset V_2 \subset \cdots$ s.t.

1. $\mu V_n \uparrow 1$ as $n \to \infty$,
2. each element of $\mathcal{P}_n | V_n$ is an open subset of some unstable manifold and
3. if $\{\mu_c\colon c \in \mathcal{P}_n | V_n\}$ denotes the system of conditional measures on elements of $\mathcal{P}_n | V_n$, and m_c denotes Riemannian measure on c, then for almost every $c \in \mathcal{P}_n | V_n$, we have $\mu_c \ll m_c$.

We now state our main result.

[2] As this manuscript was being written I learned that P. Collet and Y. Levy had jointly obtained a result similar to ours for certain parameter values of the Lozi map [CL].

THEOREM. *If $f\colon R \to R$ is a generalized Lozi map, then f has an invariant Borel probability measure μ s.t.*

1. *Local unstable manifolds exist at μ-a.e. point and*
2. *μ has absolutely continuous conditional measures on unstable manifolds.*

The idea of our proof is as follows: Observe that unstable manifolds (when they exist) are piecewise smooth curves zigzagging across R (turning around at random places. Our strategy is first to construct an invariant measure μ that behaves nicely on neighborhoods of the singularity set S. This is done following a combination of the methods used by Sinai [S] and Lasota and Yorke [LY]. Since all turns are created by passing through S, we now have control over their impact as well. In particular, this allows us to construct a noninvariant measure $\tilde{\mu}$ equivalent to μ on an arbitrarily large set and having the property that its conditional measures on unstable manifolds are (obviously) absolutely continuous. This finishes the proof.

The following notations will be used: The map $p\colon R \to [0,1]$ denotes projection onto the first factor. Lebesgue measure on $[0,1]$ is denoted by m. For $g\colon [a,b] \to \mathbf{R}$, $\bigvee_a^b g$ denotes the total variation of g on $[a,b]$. If μ is a measure on R, then $f_*\mu$ is given by $f_*\mu(E) = \mu(f^{-1}E)$. If $J \subset [0,1]$ is a closed interval and $\alpha\colon J \to [0,1]$ is a C^2 function with $|\alpha'| \le 1$, then $f(\operatorname{graph}(\alpha))$ is a union of finitely many smooth curves (H1). We denote them by $\{L_i(\alpha)\}$, dropping the α whenever no ambiguity arises. For $k > 1$, denote the smooth segments of $f^k\operatorname{graph}(\alpha)$ by $\{L_{i_1\cdots i_k}\}$, where the indices are chosen so that $fL_{i_1\cdots i_k} = \bigcup_j L_{i_1\cdots i_k j}$. Let μ_0 be the measure on $\operatorname{graph}(\alpha)$ s.t. $p_*\mu_0 =$ normalized Lebesgue measure on J. For $k = 1,2,\ldots$, define $\mu_k = (f^k)_*\mu_0$. From (H1) and (H2) we know that for each $i_1\cdots i_k$, $p_*\mu_k|L_{i_1\cdots i_k}$ is absolutely continuous to m. We denote its density by $g_{i_1\cdots i_k}$ and the density of $p_*\mu_k$ by $\hat{g}_k$. That is, we have $\sum_{i_1\cdots i_k} g_{i_1\cdots i_k} = \hat{g}_k$. When it is convenient to consider f^N instead of f, we will write

$$f^{Nk}\operatorname{graph}(\alpha) = \bigcup_{i_1\cdots i_k} L^{(N)}_{i_1\cdots i_k}$$

and

$$d(p_*\mu_{Nk}|L^{(N)}_{i_1\cdots i_k}) = g^{(N)}_{i_1\cdots i_k}\,dm,$$

etc.

LEMMA. *Under the hypotheses of the theorem, there exists an invariant Borel probability measure μ and a function $g\colon [0,1] \to [0,\infty)$ of bounded variation s.t. $d(p_*\mu) = g\,dm$.*

PROOF. Fix $J \subset [0,1]$ and a C^2 function $\alpha\colon J \to [0,1]$ with $|\alpha'| \le 1$. We will show that $\exists M$ s.t. $\bigvee_0^1 \hat{g}_k \le M$ for all k. This will imply that $\bigvee_0^1 (n^{-1}\sum_{k=1}^n \hat{g}_k) \le M$, and since $\int_0^1 (n^{-1}\sum_{k=1}^n \hat{g}_k)\,dm = 1$ the sequence $\{n^{-1}\sum_{k=1}^n \hat{g}_k\}_{n=1,2,\cdots}$ is precompact in $L^1([0,1],m)$. Choose a subsequence $\{n_i\}$ s.t. as $i \to \infty$, $(n_i)^{-1}\sum_{k=1}^{n_i} \mu_k$ converges in the weak star topology to a Borel probability measure μ and

$$\frac{1}{n_i}\sum_{k=1}^{n_i} \hat{g}_k \xrightarrow{L^1} \text{some function } g.$$

It follows immediately that μ is invariant, $d(p_*\mu) = g\,dm$ and that $\bigvee_0^1 g \le M$.

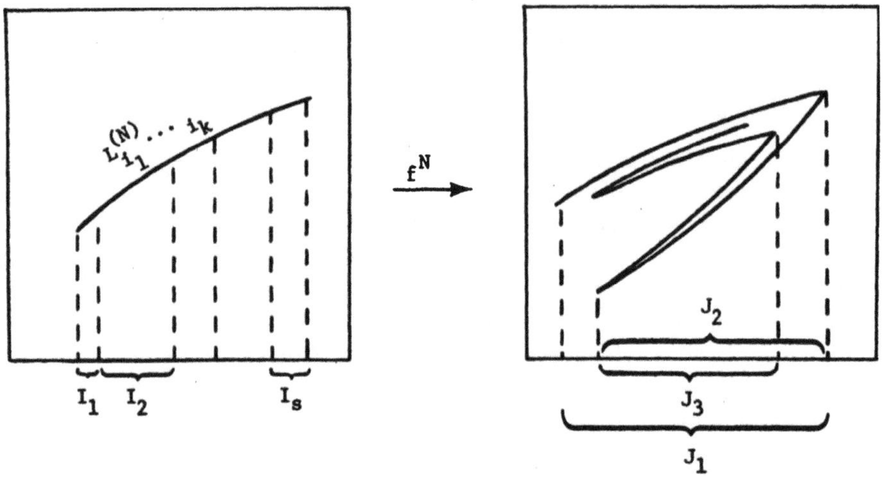

FIGURE 2

Let $N \in \mathbf{Z}^+$ satisfy (H4). It suffices to show the uniform boundedness of $\bigvee_0^1 \hat{g}_{Nk}$ for $k = 1, 2, \ldots$. We will prove that $\exists A$ s.t. for every $i_1 \cdots i_k$,

$$(*) \qquad \sum_j \bigvee_0^1 g^{(N)}_{i_1\cdots i_k j} \le A \int_0^1 g^{(N)}_{i_1\cdots i_k}\, dm + \frac{2}{u^N} \bigvee_0^1 g^{(N)}_{i_1\cdots i_k}.$$

Then if $\beta_k = \sum_{i_1\cdots i_k} \bigvee_0^1 g^{(N)}_{i_1\cdots i_k}$, we would have

$$\beta_{k+1} = \sum_{i_1\cdots i_k} \left(\sum_j \bigvee_0^1 g^{(N)}_{i_1\cdots i_k j} \right) \le A + \frac{2}{u^N}\beta_k$$

so that for all k,

$$\bigvee_0^1 g_{Nk} \le \beta_k \le A \sum_{i=0}^{\infty} \left(\frac{2}{u^N}\right)^i < \infty.$$

We now fix $i_1 \cdots i_k$ and prove $(*)$. Let $pL^{(N)}_{i_1\cdots i_k} = \bigcup_{j=1}^s I_j$, where $\{f^N(x,y) \colon x \in I_j\} = L^{(N)}_{i_1\cdots i_k j}$ and $J_j = pL^{(N)}_{i_1\cdots i_k j}$. Define $\tilde{f}_j \colon I_j \to J_j$ by $\tilde{f}_j(x) = p \circ f^N(x,y)$ for $(x,y) \in L^{(N)}_{i_1\cdots i_k}$. Then each $\tilde{f}_j$ is a C^2 diffeomorphism between I_j and J_j with $|\tilde{f}_j'| \ge u^N$ ((H1), (H2)) and $|(\tilde{f}_j^{-1})''| \le Q$ for some universal Q. (This follows from (H3) by a direct computation.) Now

$$\sum_j \bigvee_0^1 g^{(N)}_{i_1\cdots i_k j} = \sum_j \underbrace{\bigvee_{J_j} \left[(g^{(N)}_{i_1\cdots i_k} \circ \tilde{f}_j^{-1}) |(\tilde{f}_j^{-1})'| \right]}_{①}$$
$$+ \sum_j \underbrace{\left[|\tilde{f}_j'(l_j)|^{-1} g^{(N)}_{i_1\cdots i_k}(l_j) + |\tilde{f}_j'(r_j)|^{-1} g^{(N)}_{i_1\cdots i_k}(r_j) \right]}_{②},$$

where l_j and r_j denote the left and right end points of I_j respectively. Consider one j at a time.

$$\begin{aligned}\bigvee_{J_j}(g^{(N)}_{i_1\cdots i_k}\circ \tilde{f}_j^{-1})|(\tilde{f}_j^{-1})'| &= \int_{J_j} |((g^{(N)}_{i_1\cdots i_k}\circ \tilde{f}_j^{-1})(\tilde{f}_j^{-1})')'|\,dm \\ &= \int_{J_j} |(g^{(N)}_{i_1\cdots i_k}\circ \tilde{f}_j^{-1})'|\,|(\tilde{f}_j^{-1})'|\,dm + \int_{J_j} (g^{(N)}_{i_1\cdots i_k}\circ \tilde{f}_j^{-1})|(\tilde{f}_j^{-1})''|\,dm \\ &\le \frac{1}{u^N}\bigvee_{I_j} g^{(N)}_{i_1\cdots i_k} + \frac{Q}{u^N}\int_{I_j} g^{(N)}_{i_1\cdots i_k}\,dm\end{aligned}$$

so that

$$(**)\qquad \begin{aligned}&①+|\tilde{f}_1'(l_1)|^{-1}g^{(N)}_{i_1\cdots i_k(l_1)} + |\tilde{f}_s'(r_s)|^{-1}g^{(N)}_{i_1\cdots i_k}(r_s) \\ &\le \frac{1}{u^N}\bigvee_0^1 g^{(N)}_{i_1\cdots i_k} + \frac{Q}{u^N}\int_0^1 g^{(N)}_{i_1\cdots i_k}\,dm.\end{aligned}$$

Now we claim that there is a universal $d > 0$ s.t. the points $0, l_2, \ldots, l_s, 1$ are pairwise at least d apart. We have $|0 - l_2|, |l_s - 1| \ge d$ because we may assume that $f^{Nk}R \subset \operatorname{int} R$. That $|r_j - l_j| \ge$ some d for $j = 2, \ldots, s-1$ follows from (H4). Thus if $I = [l, r]$ is either $[0, l_1]$ or $[l_s, 1]$ or $[l_j, r_j]$, $j = 2, \ldots, s-1$, then

$$\begin{aligned}g^{(N)}_{i_1\cdots i_k}(l) + g^{(N)}_{i_1\cdots i_k}(r) &\le 2\left(\min_I g^{(N)}_{i_1\cdots i_k}\right) + \bigvee_I g^{(N)}_{i_1\cdots i_k} \\ &\le \frac{2}{d}\int_I g^{(N)}_{i_1\cdots i_k}\,dm + \bigvee_I g^{(N)}_{i_1\cdots i_k}.\end{aligned}$$

Note that $g^{(N)}_{i_1\cdots i_k}(0) = g^{(N)}_{i_1\cdots i_k}(1) = 0$, so that the terms in ② not accounted for in (**)

$$\le \frac{1}{u^N}\frac{2}{d}\int_0^1 g(N)_{i_1\cdots i_k}\,dm + \frac{1}{u^N}\bigvee_0^1 g^{(N)}_{i_1\cdots i_k}.$$

Thus

$$①+②\le \left(Q + \frac{2}{d}\right)\frac{1}{u^N}\int_0^1 g^{(N)}_{i_1\cdots i_k}\,dm + \frac{2}{u^N}\bigvee_0^1 g^{(N)}_{i_1\cdots i_k},$$

which completes the proof of (*) and the lemma. □

Let $W^u_\delta(x)$ denote the local unstable manifold at x (assuming it exists) s.t. $pW^u_\delta(x) = [p(x) - \delta, p(x) + \delta]$. Implicit in this notation is the assertion that $W^u_\delta(x)$ contains no cusps. Let $D(S, \delta)$ denote the δ-neighborhood of S.

PROOF OF THEOREM. If μ is any invariant probability measure with $d(p_*\mu) = g\,dm$ for some bounded g, say $g \le M_0$, then for $\delta > 0$,

$$\sum_{k=0}^\infty \mu(f^k D(S, \delta u^{-k})) = \sum_{k=0}^\infty \mu D(S, \delta u^{-k}) \le 2\delta M_0 \sum_{k=0}^\infty u^{-k} < \infty.$$

By the Borel-Cantelli lemma, μ-a.e. x is in $f^k D(S, \delta u^{-k})$ for at most finitely many k. That is, for μ-a.e. x, $\exists \delta(x) > 0$ s.t. $f^{-k}x \notin D(S, \delta(x)u^{-k})$ for all $k > 0$. This implies the existence of $W^u_{\delta(x)}(x)$. (For more details see **[KS]**.) Let us fix one measure μ constructed as in the lemma with $\operatorname{graph}(\alpha) = W^u_{\delta(x)}$ for some x. This guarantees that the $L_{i_1\cdots i_k}(\alpha)$ will not cross other unstable manifolds. For $\delta > 0$, let $\Lambda_\delta = \{x \in R\colon d(f^{-k}x, S) \geq \delta u^{-k}\ \forall k \geq 0\}$. Then each Λ_0 is closed and $\lim_{\delta\to 0} \mu\Lambda_\delta = 1$.

We now define a sequence of measurable partitions $\mathcal{P}_1 \subset \mathcal{P}_2 \subset \cdots$. All notations will be as in Definition 1. For $n \in \mathbf{Z}^+$, let $\{U_1, \ldots, U_{2^n}\}$ be the partition of R into 2^n vertical columns of width 2^{-n}. For $x \in U_i \cap \Lambda_{2^{-n}}$, let $c(x) = W^u_{2^{-n}}(x) \cap U_i$. Let $V_n = \bigcup_{x\in\Lambda_{2^{-n}}} c(x)$ and $\mathcal{P}_n = \{c(x)\colon x \in V_n\} \cup \{R - V_n\}$. It suffices to show that for every $\varepsilon > 0$ and every n, there is a set $\tilde{V}_n \subset V_n$ s.t. $\mu\tilde{V}_n > \mu V_n - \varepsilon$ and condition 3 in Definition 1 is satisfied when μ is replaced by $\chi_{\tilde{V}_n}\mu$. In fact, for given ε and n, we will construct a noninvariant measure $\tilde{\mu}$ and a set $\tilde{V}$ with $\mu\tilde{V} > 1 - \varepsilon$ s.t. $\tilde{\mu} \ll \mu$, is equivalent to μ on $\tilde{V}$, and satisfies condition 3 with respect to $\mathcal{P}_{n'}$ for some $n' \geq n$. It is straightforward to verify that this implies the desired result.

Now let $n \in \mathbf{Z}^+$ and $\varepsilon > 0$ be given. Let n' be a large number to be determined later and let $\tilde{U}_1, \ldots, \tilde{U}_{2^{n'}}$ be pairwise disjoint vertical columns of width $2^{-n'}$. We define a sequence of measures $\{\tilde{\mu}_k\}_{k=1,2,\cdots}$ as follows: Recall that in the definition of μ_k, μ_k is carried by $f^k(\operatorname{graph}(\alpha))$, where $f^k(\operatorname{graph}(\alpha))$ is a finite union of smooth curve segments. Let $\tilde{\mu}_k$ be μ_k annihilated on those parts of its support that only partially cross some $\tilde{U}_i$ (See Figure 3.) $\tilde{\mu}_k(R)$ is probably < 1.

We claim that given $\delta > 0$, $\exists n'$ such that for sufficiently large k, $\tilde{\mu}_k(R) > 1 - \delta$. Recall that $\exists M_0$ s.t. $\hat{g}_k \leq M_0\ \forall k$. If $x \in (\operatorname{supp}\mu_k - \operatorname{supp}\tilde{\mu}_k)$, then either x lies in one of the two end pieces of $f^k(\operatorname{graph}(\alpha))$ that only partially crosses some $\tilde{U}_i$, or the horizontal distance between x and a cusp in $f^k(\operatorname{graph}(\alpha))$ is $< 2^{-n'}$, which says that $d(f^{-i}x, S) \leq 2^{-n'}u^{-i}$ for some $1 \leq i \leq k$. Thus

$$\begin{aligned} 1 - \tilde{\mu}_k(R) &= \mu_k(\operatorname{supp}\mu_k - \operatorname{supp}\tilde{\mu}_k) \\ &= \sum_{i=1}^{k} \mu_k\{x\colon\ d(f^{-i}x, S) \leq 2^{-n'}u^{-i}\} + \mu_k\{2 \text{ end pieces}\} \\ &= \sum_{i=1}^{k} \mu_i\{x\colon\ d(x, S) \leq 2^{-n'}u^{-i}\} + \mu_k\{2 \text{ end pieces}\} \\ &\leq 2M_0 q 2^{-n'} \sum_{i=1}^{k} u^{-i} + \mu_k\{2 \text{ end pieces}\}. \end{aligned}$$

The second term $\to 0$ as $k \to \infty$. The first term becomes arbitrarily small as $n' \uparrow \infty$. Recall also that $(n_i)^{-1}\sum_{k=1}^{n_i}\mu_k \to \mu$. Choose a subsequence $\{n_i'\}$ of $\{n_i\}$ s.t. $(n_i')^{-1}\sum_{k=1}^{n_i'}\tilde{\mu}_k \to \tilde{\mu}$ for some $\tilde{\mu}$. It is easy to verify that $\tilde{\mu}E \leq \mu E$ for every Borel set E and hence we have $\tilde{\mu} \ll \mu$ with $0 \leq d\tilde{\mu}/d\mu \leq 1$. But since $\tilde{\mu}(R)$ can be made arbitrarily near 1, we can choose $\tilde{\mu}$ s.t. $\tilde{\mu}$ is equivalent to μ except on a set of μ-measure $< \varepsilon$.

It remains to show that if $\tilde{\mu}_T$ is the transverse measure on $\mathcal{P}_{n'}$ induced by $\tilde{\mu}$, then for $\tilde{\mu}_T$-a.e. $c \in \mathcal{P}_{n'}|V_{n'}$, $\tilde{\mu}_c \ll m_c$. Recall that $g_{i_1\cdots i_k}$ = density of $p_*(\mu_k|L_{i_1\cdots i_k})$. Let $\tilde{g}_{i_1\cdots i_k}$ = density of $p_*(\tilde{\mu}_k|L_{i_1\cdots i_k})$. We claim that $\exists B > 0$ s.t. for any $1 \leq i \leq 2^{n'}$

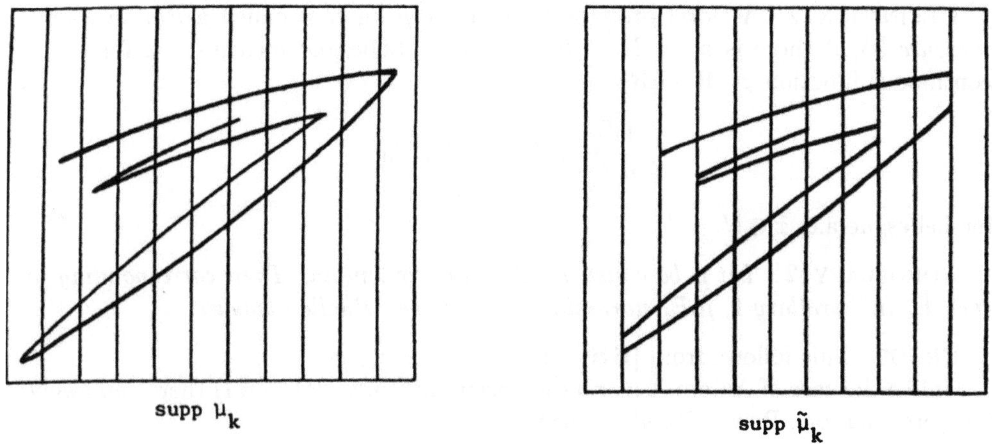

FIGURE 3

and any $i_1 \cdots i_k$, we have either

$$\tilde{g}_{i_1 \cdots i_k} \equiv 0 \quad \text{on } p\tilde{U}_i$$

or

$$\frac{\tilde{g}_{i_1 \cdots i_k}(x)}{\tilde{g}_{i_1 \cdots i_k}(y)} \leq B \quad \forall x, y \in p\tilde{U}_i.$$

This would prove that for almost every $c \in \mathcal{P}_{n'}|V_{n'}$, $d\tilde{\mu}_c = g_c dm_c$ where $\forall x, y \in c$, $g_c(x)/g_c(y) \leq B'$ for some B'. To prove the claim fix $1 \leq i \leq 2^{n'}$ and $i_1 \cdots i_k$. If $L_{i_1 \cdots i_k}$ does not cross the full width of $\tilde{U}_i$, then $g_{i_1 \cdots i_k}|p\tilde{U}_i \equiv 0$. Otherwise there are intervals $E_0, E_1, \ldots, E_k \subset [0,1]$, $E_k = p\tilde{U}_i$, and C^2 diffeomorphisms $h_j\colon E_{j-1} \to E_j$ s.t. $\tilde{g}_{i_1 \cdots i_k}|p\tilde{U}_i = |((h_k \circ \cdots \circ h_1)^{-1})'|$ (see lemma). While these h_j's depend strictly on i and $i_1 \cdots i_k$, they satisfy $|h_j'| \geq u > 1$ and $|h_j''| \leq Q$, where u and Q are universal constants.

The reader can easily verify that

$$|\log|((h_k \circ \cdots \circ h_1)^{-1})'(x)| - \log|((h_k \cdots h_1)^{-1})'(y)||$$
$$\leq \text{some constant depending only on } u,\ Q \text{ and } n'.$$

This completes the proof. □

We have proved our result for a specific class of maps of the square into itself. It is obvious that our hypotheses are more stringent than necessary and that it is easy to make slight generalizations. We do not attempt to do that here, because we do not know what a natural general statement ought to be.

We mention a few corollaries. Since the proofs are standard, we will provide only references. The standing hypothesis for the rest of this article is that f is a generalized Lozi map.

COROLLARY 1. *Let μ be constructed as in our lemma. Then there are measurable sets $E_1, E_2, \cdots \subset R$ s.t. $f^{-1}E_i = E_i$, $\mu E_i > 0$ $\forall i$, $\mu(\bigcup E_i) = 1$ and for each i, $f|E_i\colon (E_i, \mu|E_i) \to (E_i, \mu|E_i)$ is ergodic.*

PROOF. The proof follows that in [**P**]. It uses the absolute continuity of stable manifolds [**KS**] and the fact that on most $c \in \mathcal{P}_n|V_n$, μ_c is equivalent to m_c. □

DEFINITION 2. A Borel probability measure μ on R is called a *Bowen-Ruelle measure* [**B**] if there is a set $U \subset R$ of positive Lebesgue measure s.t. for every continuous function φ: $\mathbf{R} \to \mathbf{R}$,

$$\frac{1}{n}\sum_{i=0}^{n-1} \varphi f^i x \to \int \varphi \, d\mu$$

for Lebesgue-a.e. $x \in U$.

COROLLARY 2. *Let μ be constructed as in our lemma. Then corresponding to each E_i in Corollary 1, $\mu | E_i$ normalized is a Bowen-Ruelle measure.*

PROOF. This follows from [**KS**]. □

Unlike the case of Axiom A attractors, assuming only (H1)–(H4) there can easily be more than one Bowen-Ruelle measure.

COROLLARY 3. *Let f: $R \to R$ be a generalized Lozi map. Then f has an invariant Borel probability measure μ s.t.*

$$h_\mu(f) = \int \lambda_1(x)\, d\mu(x),$$

where $\lambda_1(x)$ is the positive μ-exponent of f at x.

PROOF. See [**LS**]. □

REFERENCES

[**B**] R. Bowen, *Equilibrium states and the ergodic theory of Anosov diffeomorphisms,* Lecture Notes in Math., vol. 47, Springer, Berlin and New York, 1975.

[**CL**] P. Collet and Y. Levy, *Ergodic properties of the Lozi mappings,* Comm. Math. Phys. **93** (1984), 461–482.

[**KS**] A. Katok and J. M. Strelcyn, *Invariant manifolds for smooth maps with singularities, Part* I: *Existence* (preprint).

[**LS**] F. Ledrappier and J. M. Strelcyn, *A proof of the estimation from below in Pesin entropy formula,* J. Ergodic Theory and Dynam. Syst. (to appear).

[**LY**] A. Lasota and J. Yorke, *On the existence of invariant measures for piecewise monotonic transformations,* Trans. Amer. Math. Soc. **186** (1973), 481–488.

[**M**] M. Misiurewicz, *Strange attractors for the Lozi mappings,* Proc. N.Y. Acad. Sci. (1980), 348–358.

[**P**] Ja. Pesin, *Characteristic Lyapunov exponents and smooth ergodic theory,* Russian Math. Surveys **32** (1977), 55–114.

[**S**] Ya. G. Sinai, *Markov partitions and C-diffeomorphisms,* Functional Anal. Appl. **2** (1968), 61–82.

DEPARTMENT OF MATHEMATICS, MICHIGAN STATE UNIVERSITY, EAST LANSING, MICHIGAN 48824 (Current address)

DEPARTMENT OF MATHEMATICS, UNIVERSITY OF NORTH CAROLINA, CHAPEL HILL, NORTH CAROLINA 27514

Ergodic theory of chaos and strange attractors

J.-P. Eckmann

Université de Genève, 1211 Genève 4, Switzerland

D. Ruelle

Institut des Hautes Etudes Scientifiques, 91440 Bures-sur-Yvette, France

Physical and numerical experiments show that deterministic noise, or chaos, is ubiquitous. While a good understanding of the onset of chaos has been achieved, using as a mathematical tool the geometric theory of differentiable dynamical systems, moderately excited chaotic systems require new tools, which are provided by the *ergodic* theory of dynamical systems. This theory has reached a stage where fruitful contact and exchange with physical experiments has become widespread. The present review is an account of the main mathematical ideas and their concrete implementation in analyzing experiments. The main subjects are the theory of *dimensions* (number of excited degrees of freedom), *entropy* (production of information), and *characteristic exponents* (describing sensitivity to initial conditions). The relations between these quantities, as well as their experimental determination, are discussed. The systematic investigation of these quantities provides us for the first time with a reasonable understanding of dynamical systems, excited well beyond the quasiperiodic regimes. This is another step towards understanding highly turbulent fluids.

CONTENTS

*Sections marked with * contain supplementary material which can be omitted at first reading.

I. INTRODUCTION

In recent years, the ideas of differentiable dynamics have considerably improved our understanding of irregular behavior of physical, chemical, and other natural phenomena. In particular, these ideas have helped us to understand the onset of turbulence in fluid mechanics. There is now ample experimental and theoretical evidence that the qualitative features of the time evolution of many physical systems are the same as those of the solution of a typical evolution equation of the form

$$\dot{x}(t)=F_\mu(x(t)),\quad x\in\mathbf{R}^m \tag{1.1}$$

in a space of small dimension m. Here, x is a set of coordinates describing the system (typically, mode amplitudes, concentrations, etc.), and F_μ determines the nonlinear time evolution of these modes. The subscript μ corresponds to an experimental *control parameter,* which is kept constant in each run of the experiment. (Typically, μ is the intensity of the force driving the system.) We write

$$x(t)=f_\mu^t(x(0))\ . \tag{1.2}$$

We usually *assume* that there is a parameter value, say $\mu=0$, for which the equation is well understood and leads to a motion in phase space which, after some transients, settles down to be stationary or periodic.

As the parameter μ is varied, the nature of the asymptotic motion may change.[1] The values μ for which this change of asymptotic regime happens are called bifurcation points. As the parameter increases through successive bifurcations, the asymptotic motion of the system typically gets more complicated. For special sequences of these bifurcations a lot is known, and even quantitative features are predicted, as in the case of the period-doubling cascades ("Feigenbaum scenario"). We do not, however, possess a complete classification of the possible transitions to more complicated behavior, leading eventually to turbulence. *Geometrically,* the asymptotic motion follows an *attractor* in phase space, which will become more and more complicated as μ increases.

The aim of the present review is to describe the current state of the theory of *statistical* properties of dynamical systems. This theory becomes relevant as soon as the system is "excited" beyond the simplest bifurcations, so that precise geometrical information about the shape of the attractor or the motion on it is no longer available. See Eckmann (1981) for a review of the *geometrical* aspects of dynamical systems. The statistical theory is still capable of distinguishing different degrees of complexity of attractors and motions, and presents thus a further step in bridging the gap between simple systems and fully developed turbulence. In particular, the present treatment does not exclude the description of space-time patterns.

After introducing precise dynamical concepts in Sec. II, we address the theory of characteristic exponents in Sec. III and the theory of entropy and information dimension in Sec. IV. In Sec. V we discuss the extraction of dynamical quantities from experimental time series.

It is necessary at this point to clarify the role of the physical concept of *mode,* which appears naturally in simple theories (for instance, Hamiltonian theories with quadratic Hamiltonians), but which loses its importance in nonlinear dynamical systems. The usual idea is to represent a physical system by an appropriate change of variables as a collection of independent oscillators or modes. Each mode is periodic, and its state is represented by an angular variable. The global system is *quasiperiodic* (i.e., a superposition of periodic motions). From this perspective, a dissipative system becomes more and more turbulent as the number of *excited modes* grows, that is, as the number of independent oscillators needed to describe the system progressively increases. This point of view is very widespread; it has been extremely useful in physics and can be formulated quite coherently (see, for example, Haken, 1983). However, this philosophy and the corresponding intuition about the use of Fourier modes have to be completely modified when nonlinearities are important: *even a finite-dimensional motion need not be quasiperiodic in general.* In particular, the concept of "number of excited modes" will have to be replaced by new concepts, such as "number of non-negative characteristic exponents" or "information dimension." These new concepts come from a statistical analysis of the motion and will be discussed in detail below.

In order to talk about a statistical theory, one needs to say what is being averaged and in which sample space the measurements are being made. The theory we are about to describe treats *time averages.* This implies and has the advantage that *transients* become irrelevant. (Of course, there may be formidable experimental problems if the transients become too long.) Once transients are over, the motion of the solution x of Eq. (1.1) settles typically near a subset of $\mathbf{R}^m$, called an *attractor* (mathematical definitions will be given later). In particular, in the case of dissipative systems, on which we focus our attention, the volume occupied by the attractor is in general very small relative to the volume of phase space. We shall not talk about attractors for conservative systems, where the volume in phase space is conserved. For dissipative systems we may assume that phase-space volumes are contracted by the time evolution (if phase space is finite dimensional). Even if a system contracts volumes, this does not mean that it contracts lengths in *all* directions. Intuitively, some directions may be stretched, provided some others are so much contracted that the final volume is smaller than the initial volume (Fig. 1). This seemingly trivial remark has profound consequences. It implies that, even in a dissipative system, the final motion may be unstable *within the attractor.* This instability usually manifests itself by an exponential separation of orbits (as time goes on) of points which initially are very close to each other (on the attractor). The exponential separation takes place in the direction of stretching, and an attractor having this stretching property will be called a *strange attractor.* We shall also say that a system with a strange attractor is *chaotic* or has *sensitive dependence on initial conditions.* Of course, since the attractor is in general bounded, exponential separation can only hold as long as distances are small.

Fourier analysis of the motion on a strange attractor (say, of one of its coordinate components) in general reveals a *continuous power spectrum.* We are used to interpreting this as corresponding to an infinite number of modes. However, as we have indicated before, this

[1]It is to be understood that the experiment is performed with a *fixed* value of the parameter.

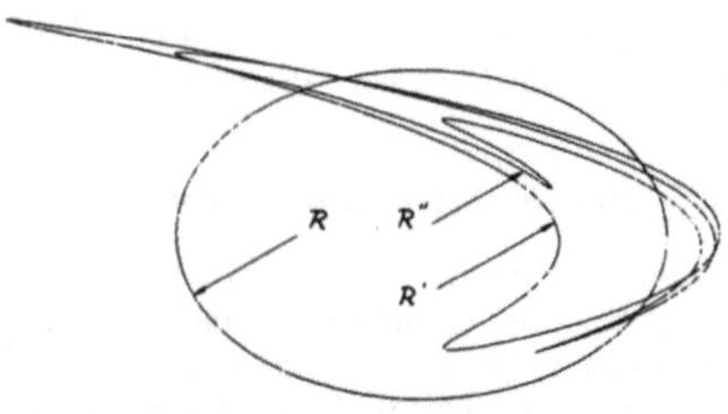

FIG. 1. The Hénon map $x_1'=1-1.4x_1^2+x_2$, $x_2'=0.3x_1$ contracts volumes but stretches distances. Shown are a region R, and its first and second images R' and R'' under the Hénon map.

reasoning is only valid in a "linear" theory, which then has to take place in an infinite-dimensional phase space. Thus, if we are confronted experimentally with a continuous power spectrum, there are two possibilities: We are either in the presence of a system that "explores" an infinite number of dimensions in phase space, or we have a system that evolves nonlinearly on a finite-dimensional attractor. Both alternatives are possible, and the second appears frequently in practice. We shall give below an algorithm which, starting from measurements, gives information on the effective dimension. This algorithm has been successfully used in several experiments, e.g., Malraison *et al.* (1983), Abraham *et al.* (1984), Grassberger and Procaccia (1983b); it has indicated *finite dimensions* in hydrodynamic systems, even though the phase space is infinite dimensional and the system therefore could potentially excite an infinite number of degrees of freedom.

The tool with which we want to measure the dimension and other dynamical quantities of the system is *ergodic theory*. Ergodic theory says that a time average equals a space average. The weight with which the space average has to be taken is an *invariant measure*. An invariant measure ρ satisfies the equation

$$\rho[f^{-t}(E)]=\rho(E),\quad t>0\,, \tag{1.3}$$

where E is a subset of points of $\mathbf{R}^m$ and $f^{-t}(E)$ is the set obtained by evolving each of the points in E backwards during time t. There are in general many invariant measures in a dynamical system, but not all of them are physically relevant. For example, if x is an unstable fixed point of the evolution, then the δ function at x is an invariant measure, but it is not observed. From an experimental point of view, a reasonable measure is obtained according to the following idea of Kolmogorov (see Sec. II.F). Consider Eq. (1.1) with an external noise term added,

$$\dot{x}(t)=F_\mu(x(t))+\varepsilon\omega(t)\,, \tag{1.4}$$

where ω is some noise and $\varepsilon>0$ is a parameter. For suitable noise and $\varepsilon>0$, the stochastic time evolution (1.3) has a unique stationary measure ρ_ε, and the measure we propose as "reasonable" is $\rho=\lim_{\varepsilon\to 0}\rho_\varepsilon$. We shall come back later to the problem of choosing a reasonable measure ρ. For the moment we assume that such a measure exists, and call it the *physical measure*. Physically, we assume that it represents experimental time averages. Mathematically, we only require (for the moment) that it be invariant under time evolution.

A basic virtue of the ergodic theory of dynamical systems is that it allows us to consider only the long-term behavior of a system and not to worry about transients. In this way, the problems are at least somewhat simplified. The physical long-term behavior is on attractors, as we have already noted, but the geometric study of attractors presents great mathematical difficulties. Shifting attention from attractors to invariant measures turns out to make life much simpler.

An invariant probability measure ρ may be *decomposable* into several different pieces, each of which is again invariant. If not, ρ is said to be *indecomposable* or *ergodic*. In general, an invariant measure can be uniquely represented as a superposition of ergodic measures. In view of this, it is natural to assume that the physical measure is not only invariant, but also ergodic. If ρ is ergodic, then the *ergodic theorem* asserts that for every continuous function φ,

$$\lim_{T\to\infty}\frac{1}{T}\int_0^T \varphi[f^t x(0)]dt=\int \rho(dx)\varphi(x) \tag{1.5}$$

for almost all initial conditions $x(0)$ with respect to the measure ρ. Since the measure ρ might be singular, for instance concentrated on a fractal set, it would be better if we could say something for almost all $x(0)$ with respect to the ordinary (Lebesgue) measure on some set $S\subset\mathbf{R}^m$. We shall see below that this is sometimes possible.

One crucial decision in our study of dynamics is to concentrate on the analysis of the separation in time of two infinitely close initial points. Let us illustrate the basic idea with an example in which time is discrete [rather than continuous as in (1.1)]. Consider the evolution equation

$$x(n+1)=f(x(n)),\quad x(i)\in\mathbf{R}\,, \tag{1.6}$$

where n is the discrete time. The separation of two initial points $x(0)$ and $x(0)'$ after time N is then

$$x(N)-x(N)'=f^N(x(0))-f^N(x(0)')$$
$$\approx\left[\frac{d}{dx}(f^N)(x(0))\right][x(0)-x(0)']\,, \tag{1.7}$$

where $f^N(x)=f(f(\cdots f(x)\cdots))$, N times. By the chain rule of differentiation,

$$\frac{d}{dx}(f^N)(x(0))=\frac{d}{dx}f(x(N-1))$$
$$\times\frac{d}{dx}f(x(N-2))\cdots\frac{d}{dx}f(x(0))\,. \tag{1.8}$$

[In the case of m variables, i.e., $x\in\mathbf{R}^m$, we replace the derivative $(d/dx)f$ by the Jacobian matrix, evaluated at x: $D_xf=(\partial f_i/\partial x_j)$.] Assuming that all factors in the above expression are of comparable size, it seems plausible that df^N/dx grows (or decays) exponentially with N.

The same is true for $x(N)-x(N)'$, and we can define the average rate of growth as

$$\lambda = \lim_{N\to\infty} \frac{1}{N} \log | D_{x(0)} f^N \delta x(0) | \ . \tag{1.9}$$

By the theorem of Oseledec (1968), this limit exists for almost all $x(0)$ (with respect to the invariant measure ρ). The average expansion value depends on the direction of the initial perturbation $\delta x(0)$, as well as on $x(0)$. However, if ρ is ergodic, the largest λ [with respect to changes of $\delta x(0)$] is independent of $x(0)$, ρ-almost everywhere. This number λ_1 is called the *largest Liapunov exponent* of the map f with respect to the measure ρ. Most choices of $\delta x(0)$ will produce the largest Liapunov exponent λ_1. However, certain directions will produce smaller exponents $\lambda_2, \lambda_3, \ldots$ with $\lambda_1 \geq \lambda_2 \geq \lambda_3 \geq \cdots$ (see Sec. III.A for details).

In the continuous-time case, one can similarly define

$$\lambda(x,\delta x) = \lim_{T\to\infty} \frac{1}{T} \log | (D_x f^T) \delta x | \ . \tag{1.10}$$

We shall see that the Liapunov exponents (i.e., characteristic exponents) and quantities derived from them give useful bounds on the *dimensions of attractors*, and on the *production of information* by the system (i.e., *entropy* or Kolmogorov-Sinai invariant). It is thus very fortunate that λ and related quantities are experimentally accessible. [We shall see below how they can be estimated. See also the paper by Grassberger and Procaccia (1983a).]

The Liapunov exponents, the entropy, and the Hausdorff dimension associated with an attractor or an ergodic measure ρ all are related to how excited and how chaotic a system is (how many degrees of freedom play a role, and how much sensitivity to initial conditions is present). Let us see by an example how entropy (information production) is related to sensitive dependence on initial conditions.

We consider the dynamical system given by $f(x)=2x \bmod 1$ for $x \in [0,1)$. (This is "left-shift with leading digit truncation" in binary notation.) This map has sensitivity to initial conditions, and $\lambda = \log 2$. Assume now that our measuring apparatus can only distinguish between $x < \frac{1}{2}$ and $x > \frac{1}{2}$. Repeated measurements in time will nevertheless yield eventually all binary digits of the initial point, and it is in this sense that information is produced as "time" (i.e., the number of iterations) goes on. Thus changes of initial condition may be unobservable at time zero, but become observable at some later time. If we denote by ρ the Lebesgue measure on [0,1), then ρ is an invariant measure, and the corresponding mean information produced per unit time is exactly one bit. More generally, the average rate $h(\rho)$ of information production in an ergodic state ρ is related to sensitive dependence on initial conditions. [The quantity $h(\rho)$ is called the entropy of the measure ρ; see Sec. IV.] It may be bounded in terms of the characteristic exponents, and one finds

$$h(\rho) \leq \sum \text{postive characteristic exponents} \ . \tag{1.11}$$

In fact, in many cases (but not all), when a physical measure ρ may be identified, we have Pesin's formula (Pesin, 1977):

$$h(\rho) = \sum \text{positive characteristic exponents} \ . \tag{1.12}$$

Another quantity of interest is the *Hausdorff dimension*. [This quantity has been brought very much to the attention of physicists by Mandelbrot (1982), who uses the term *fractal dimension*. This is also used as a sort of generic name for different mathematical definitions of dimension for "fractal" sets.] The dimension of a set is roughly the amount of information needed to specify points on it accurately. For instance, let S be a compact set and assume that $N(\varepsilon)$ balls of radius ε are needed to cover S. Then a dimension $\dim_K S$, the "capacity" of S, is defined by

$$\dim_K S = \limsup_{\varepsilon\to 0} \log N(\varepsilon) / |\log \varepsilon| \ .$$

[This is a little less than requiring $N(\varepsilon)\varepsilon^d \to$ finite, which means that the "volume" of the set S is finite in dimension d.] Mañé (1981) has shown that the points of S can be parametrized by m real coordinates as soon as $m \geq 2 \dim_K S + 1$.

The definition of the Hausdorff dimension $\dim_H S$ is slightly more complicated than that of $\dim_K S$; it does not assume that S is compact (see Sec. II.J). We next define the *information dimension* $\dim_H \rho$ of a probability measure ρ as the minimum of the Hausdorff dimensions of the sets S for which $\rho(S)=1$. It is not *a priori* clear that sets defined by dynamical systems have *locally* the *same* Hausdorff dimension everywhere, but this follows from the ergodicity of ρ in the case of $\dim_H \rho$. A result of Young [see Eq. (1.13) below] permits in many cases the evaluation of $\dim_H \rho$. Starting from different ideas, Grassberger and Procaccia (1983a,1983b) have arrived at a very similar way of computing the information dimension $\dim_H \rho$ of the measure ρ. Their proposal has been extremely successful, and has been used to measure reproducibly dimensions of the order of 3—10 in hydrodynamical experiments (see, for example, Malraison *et al.*, 1983).

We present some details of the method. Let $\rho[B_x(r)]$ be the mass of the measure ρ contained in a ball of radius r centered at x, and assume that the limit

$$\lim_{r\to 0} \frac{\log \rho[B_x(r)]}{\log r} = \alpha \tag{1.13}$$

exists for ρ-almost all x. The existence of the limit implies that it is constant, by the ergodicity of ρ. Under these conditions, α is equal to the information dimension $\dim_H \rho$, as noted by Young. In an experimental situation, one takes N points $x(i)$, regularly spaced in time, on an orbit of the dynamical system, and estimates $\rho[B_{x(i)}(r)]$ by

$$\frac{1}{N} \sum_{j=1}^{N} \Theta[r - |x(j)-x(i)|] \quad (N \text{ large}) \ , \tag{1.14}$$

where $\Theta(u)=(1+\operatorname{sgn} u)/2$. This permits us in principle to test the existence of the limit. In practice (Grassberger and Procaccia, 1983a,1983b) one defines

$$C(r)=\frac{1}{N^2}\sum_{i,j}\Theta[r-|x(j)-x(i)|] \quad (N \text{ large}), \tag{1.15}$$

$$\text{information dimension}=\lim_{r\to 0}\frac{\log C(r)}{|\log r|}. \tag{1.16}$$

The problem of associating an orbit in $\mathbf{R}^m$ with experimental results will be discussed later. We also postpone discussion of relations between the Hausdorff dimension and characteristic exponents [such relations are described in the work of Frederickson, Kaplan, Yorke, and Yorke (1983); Douady and Oesterlé (1980); and Ledrappier (1981a)].

One may ask to what extent the definition of the above quantities is more than wishful thinking: is there any chance that the dimensions, exponents, and entropies about which we have been talking are finite numbers? For the case of the Navier-Stokes equation,

$$\frac{\partial v_i}{\partial t}=-\sum_j v_j\partial_j v_i+\nu\Delta v_i-\frac{1}{d}\partial_i p+g_i, \tag{1.17}$$

with the incompressibility conditions $\sum\partial_j v_j=0$, one has some comforting results given below. [Note that, in the case of two-dimensional hydrodynamics, one has good existence and uniqueness results for the solutions to Eq. (1.17). Assuming the same to be true in three dimensions (for reasonable physical situations), the conclusions given below for the two-dimensional case will carry over.]

Consider the Navier-Stokes equation in a bounded domain $\Omega\subset\mathbf{R}^d$, where $d=2$ or 3 is the spatial dimension. For *every* invariant measure ρ one has the following relations between the energy dissipation ε (per unit volume and time) and the ergodic quantities described earlier:

$$h(\rho)\le\sum_{\lambda_i\ge 0}\lambda_i\le\frac{B_d}{\nu^{1+d}}\left\langle\int_\Omega \varepsilon^{(2+d)/4}\right\rangle, \tag{1.18}$$

$$\dim_H\rho\le B_d'\frac{|\Omega|^{2/(d+2)}}{\nu^{d/2}}\left\langle\int_\Omega \varepsilon^{(2+d)/4}\right\rangle^{d/(d+2)}, \tag{1.19}$$

where B_d, B_d' are universal constants (see Ruelle, 1982b,1984, and Lieb, 1984, for a detailed discussion of these inequalities). Thus, if some average dissipation is finite, then all of these quantities are finite. In two dimensions, if the average dissipation is finite, i.e., if the *power* pumped into the system is finite, then $h(\rho)$ and $\dim_H\rho$ are also finite. In three dimensions, the situation is less clear because the average of $\int\varepsilon^{5/4}$ occurs instead of the average of $\int\varepsilon$. The lack of an existence and uniqueness theorem is in fact related to this difficulty. Experimentally, however, one finds that $\dim_H\rho$ is finite (implying that there are only finitely many $\lambda_i>0$).

To conclude, let us remark that the dynamical theory of physical systems is a rather mathematical subject, in the sense that it appeals to difficult mathematical theories and results. On the other hand, these mathematical theories still have many loose ends. One might thus be tempted either to disregard rigorous mathematics and go ahead with the physics, or on the contrary to wait until the mathematical situation is sufficiently clarified before going ahead with the physics. Both attitudes would be unfortunate. We believe in the value of the interplay between mathematics and physics, although either discipline offers only incomplete results. A mathematical theorem can prevent us from making "intuitive" assumptions that are already proved to be invalid. On the other hand, the relation between the two disciplines can help us to formulate mathematical conjectures which are made plausible on the basis of our experience as physicists. We are fortunate that the theory of dynamical systems has reached a stage where this kind of attitude seems especially fruitful.

The following are a few general references which are of interest in relation to the topics discussed in the present paper. (These references include books, conference proceedings, and reviews.)

Abraham, Gollub, and Swinney (1984): An overview of the experimental situation.

Bergé, Pomeau, and Vidal (1984): A very nice physics-oriented introduction, to be translated into English.

Bowen (1975): A more advanced introduction, stressing the ergodic theory of hyperbolic systems.

Campbell and Rose (1983): Los Alamos conference.

Collet and Eckmann (1980): A monograph, mostly on maps of the interval.

Cvitanović (1984): A very useful reprint collection.

Eckmann (1981): Review article on the geometric aspects of dynamical systems theory.

Ghil, Benzi, and Parisi (1985): Summer school proceedings on turbulence and predictability in geophysics.

Guckenheimer and Holmes (1983): An easy introduction to differential dynamical systems, oriented towards chaos.

Gurel and Rössler (1979): N.Y. Academy Conference.

Helleman (1980): N.Y. Academy Conference. These two conferences played an important historical role.

Iooss, Helleman, and Stora (1981): Proceedings of a summer school in Les Houches, 1981, with many interesting lectures.

Nobel symposium on chaos (1985).

Shaw (1981): A nice intuitive introduction to the information aspects of chaos.

Vidal and Pacault (1981): Conference proceedings on chemical turbulence.

Young (1984): A brief, but excellent, exposition of the inequalities for entropy and dimension.

II. DIFFERENTIABLE DYNAMICS AND THE RECONSTRUCTION OF DYNAMICS FROM AN EXPERIMENTAL SIGNAL

A. What is a differentiable dynamical system?

A differentiable dynamical system is simply a time evolution defined by an evolution equation

$$\frac{dx}{dt}=F(x) \qquad (2.1)$$

(continuous-time case) or by a map

$$x(n+1)=f(x(n)) \qquad (2.2)$$

(discrete-time case), where f or F are *differentiable* functions. In other words, f or F have continuous first-order derivatives. We may require f or F to be twice differentiable or more, i.e., to have continuous derivatives of second or higher order. Differentiability (possibly of higher order) is also referred to as *smoothness*. The physical justification for the assumed continuity of the derivatives of f or F is simply that physical quantities are usually continuous (small causes produce small effects). This philosophy, however, should, not be adhered to blindly (see Sec. III.D.2).

One introduces the nonlinear time-evolution operators f^t, t real or integer, requiring sometimes $t\geq 0$. They have the property

$$f^0=\text{identity}, \quad f^sf^t=f^{s+t} .$$

The variable x varies over the *phase space* M, which is $\mathbf{R}^m$, or a manifold like a sphere or a torus, or infinite dimensional (Banach spaces, in particular Hilbert spaces, are important in hydrodynamics). If M is a linear space, we define the linear operator D_xf^t (matrix of partial derivatives of f at x, or a bounded operator if M is a Banach space). Writing $f^1=f$, we have

$$D_xf^n=D_{f^{n-1}x}f \cdots D_{fx}fD_xf \qquad (2.3)$$

by the chain rule.

Example.

A viscous fluid in a bounded container $\Omega\subset\mathbf{R}^2$ or $\mathbf{R}^3$ is described by the Navier-Stokes equation

$$\frac{\partial v_i}{\partial t}=-\sum_j v_j\partial_j v_i+\nu\Delta v_i-\frac{1}{d}\partial_i p+g_i , \qquad (2.4)$$

where (v_i) is the velocity field in Ω, ν a constant (the kinematic viscosity), d the (constant) density, p the pressure, and g an external force field. We add to Eq. (2.4) the incompressiblity condition

$$\sum_j \partial_j v_j=0 , \qquad (2.5)$$

which expresses that v_i is divergence free, and we impose $v_i=0$ on $\partial\Omega$ (the fluid sticks to the boundary). Note that the divergence-free vector fields are orthogonal to gradients, so that one can eliminate the pressure from Eq. (2.4) by orthogonal projection of the equation on the divergence-free fields. One obtains thus an equation of the type (2.1) where M is the Hilbert space of square-integrable vector fields which are orthogonal to gradients. In two dimensions (i.e., for $\Omega\subset\mathbf{R}^2$), one has a good existence and uniqueness theorem for solutions of Eqs. (2.4) and (2.5), so that f^t is defined for t real ≥ 0 (Ladyzhenskaya, 1969; Foias and Temam, 1979; Temam, 1979). In three dimensions one has only partial results (Caffarelli, Kohn, and Nirenberg, 1982).

B. Dissipation and attracting sets

For a *conservative system* (Hamiltonian time evolution), Liouville's theorem says that the volume in phase space M is conserved by the time evolution. We shall be mainly interested in *dissipative systems*, for which this is not the case and for which the volume is usually contracted. Let us therefore assume that there is an open set U in M which is contracted by time evolution asymptotically to a compact set A. To be precise, we say that A is an *attracting set* with *fundamental neighborhood* U if (a) for every open set $V\supset A$ we have $f^tU\subset V$ when t is large enough, and (b) $f^tA=A$ for all t. (See Fig. 2.) The open set $\bigcup_{t>0}(f^t)^{-1}U$ is the *basin of attraction* of A. If the basin of attraction of A is the whole of M, we say that A is the *universal attracting set.*

Examples.

(a) If U is an open set in M, and the closure of f^tU is compact and contained in U for all sufficiently large t, then the set $A=\bigcap_{t\geq 0}f^tU$ is a (compact) attracting set with fundamental neighborhood U (see Ruelle, 1981).

(b) The Lorenz time evolution in $\mathbf{R}^3$ is defined by the equation

$$\frac{d}{dt}\begin{bmatrix} x_1 \\ x_2 \\ x_2 \end{bmatrix}=\begin{bmatrix} -\sigma x_1+\sigma x_2 \\ -x_1x_3+rx_1-x_2 \\ x_1x_2-bx_3 \end{bmatrix} , \qquad (2.6)$$

with $\sigma=10$, $b=\frac{8}{3}$, $r=28$ (see Lorenz, 1963). If U is a sufficiently large ball, [i.e., $U=\{(x_1,x_2,x_3):\sum x_i^2\leq R^2\}$ with large R], then U is mapped into itself by time evolution. It contains thus an attracting set A, and A is universal (see Fig. 3).

(c) The Navier-Stokes time evolution in two dimensions also gives rise to a universal attracting set A, because one can again apply (a) to a sufficiently large ball (in a suitable Hilbert space). It can be shown that A has finite dimension (see Mallet-Paret, 1976).

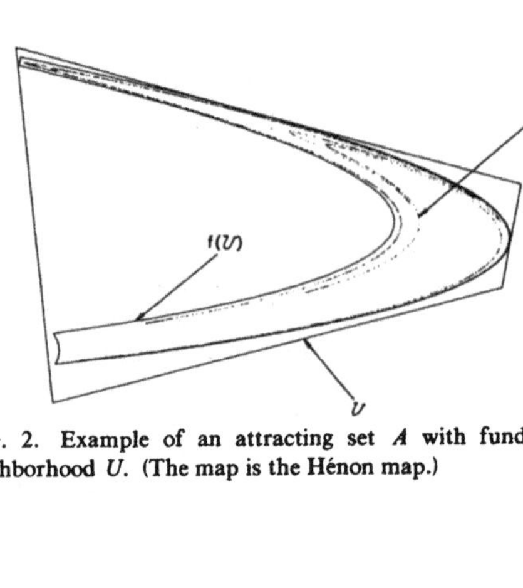

FIG. 2. Example of an attracting set A with fundamental neighborhood U. (The map is the Hénon map.)

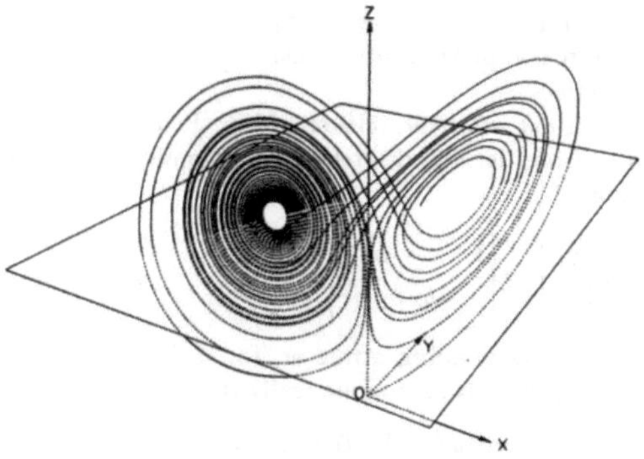

FIG. 3. The Lorenz attractor. From Lanford (1977).

C. Attractors

Physical experiments and computer experiments with dynamical systems usually exhibit transient behavior followed by what seems to be an asymptotic regime. Therefore the point $f^t x$ representing the system should eventually lie on an attracting set (or near it). However, in practice *smaller* sets, which we call *attractors*, will be obtained (they should be carefully distinguished from attracting sets). This is because some parts of an attracting set may not be attracting (Fig. 4).

We should also like to include in the mathematical definition of an attractor A the requirement of irreducibility (i.e., the union of two disjoint attractors is not considered to be an attractor). This (unfortunately) implies that one can no longer impose the requirement that there be an *open* fundamental neighborhood U of A such that $f^t U \to A$ when $t \to \infty$. Instead of trying to give a precise mathematical definition of an attractor, we shall use here the *operational* definition, that it is a set on which *experimental points* $f^t x$ accumulate for large t. We shall come back later to the significance of this operational definition and its relation to more mathematical concepts.

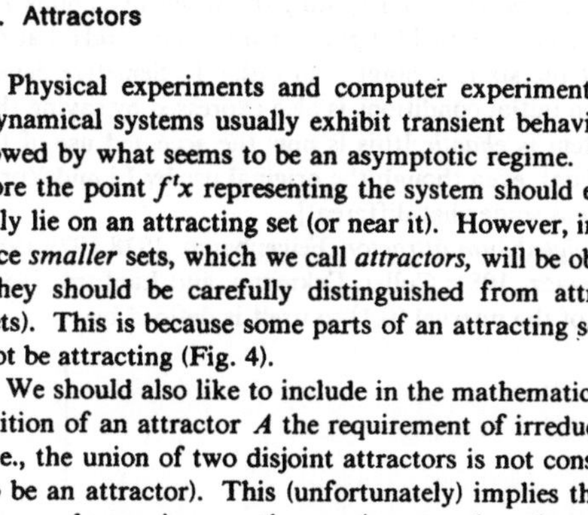

FIG. 4. The dynamical system is $\dot{x}_1 = x_1 - x_1^3$, $\dot{x}_2 = -x_2$. The segment A,B is the universal attracting set, but only the points A,B are attractors. In other words, the whole space is attracted to the segment A,B but only A and B are attractors.

Examples.

(a) Attracting fixed point. Let P be a *fixed point* for our dynamical system, i.e., $f^t P = P$ for all t. The derivative $D_P f^1$ of f^1 (time-one map) at the fixed point is an $m \times m$ matrix or an operator in Hilbert space. If its spectrum is in a disk $\{z : |z| < \alpha\}$ with $\alpha < 1$, then P is an attracting fixed point. It is an attracting set (and an attractor). When the time evolution is defined by the differential equation (1.1) in $\mathbf{R}^m$, the attractiveness condition is that the eigenvalues of $D_P F_\mu$ all have a negative real part. For a discrete-time dynamical system, we say that $(P_1, \ldots, P_n)$ is an attracting periodic orbit, of period n, if $fP_1 = P_2, \ldots, fP_n = P_1$, and P_i is an attracting fixed point for f^n.

(b) Attracting periodic orbit for continuous time. For a continuous-time dynamical system, suppose that there are a point a and a $T > 0$, such that $f^T a = a$ but $f^t a \neq a$ when $0 < t < T$. Then a is a periodic point of period T, and $\Gamma = \{f^t a : 0 \leq t < T\}$ is the corresponding *periodic orbit* (or closed orbit). The derivative $D_a f^T$ has an eigenvalue 1 corresponding to the direction tangent to Γ at a. If the rest of the spectrum is in $\{z : |z| < \alpha\}$ with $\alpha < 1$, then Γ is an attracting periodic orbit. It is again an attracting set and an attractor. The attracting character of a periodic orbit may also be studied with the help of a Poincaré section (see Sec. II.H).

(c) Quasiperiodic attractor. A periodic orbit for a continuous system is really a circle, and the motion on it (by proper choice of coordinate φ) may be written

$$\varphi(t) = \varphi(0) + \omega t \pmod{2\pi}, \tag{2.7}$$

where $\omega = 2\pi/T$. This may be thought of as representing the time evolution of a simple oscillator. Consider now a collection of k oscillators with frequencies $\omega_1, \ldots, \omega_k$ (without rational relations between the ω_i: no linear combination with nonzero integer coefficients vanishes). The motion of the oscillators is described by

$$\varphi_i(t) = \varphi_i(0) + \omega_i t \pmod{2\pi}, \quad i = 1, \ldots, k, \tag{2.8}$$

and this motion takes place on the product of k circles, $(k > 1)$, which is a k-dimensional torus T^k. Suppose that the torus T^k is embedded in $\mathbf{R}^m$, $m \geq k$ (or in Hilbert space), as the periodic orbit Γ was in the previous example; suppose, furthermore, that this torus is an attracting set. Then we say that T^k is a quasiperiodic attractor. Asymptotically, the dynamical system will thus be described by

$$x(t) = f^t x = \Phi[\varphi_1(t), \ldots, \varphi_k(t)] \tag{2.9}$$

$$= \Psi(\omega_1 t, \ldots, \omega_k t), \tag{2.10}$$

where Ψ is periodic, of period 2π, in each argument. A function of the form $t \to \Psi(\omega_1 t, \ldots, \omega_k t)$ is known as a *quasiperiodic function* (with k different periods). Quasiperiodic attractors are a natural generalization of periodic orbits, and they occur fairly frequently in the description of moderately excited physical systems.

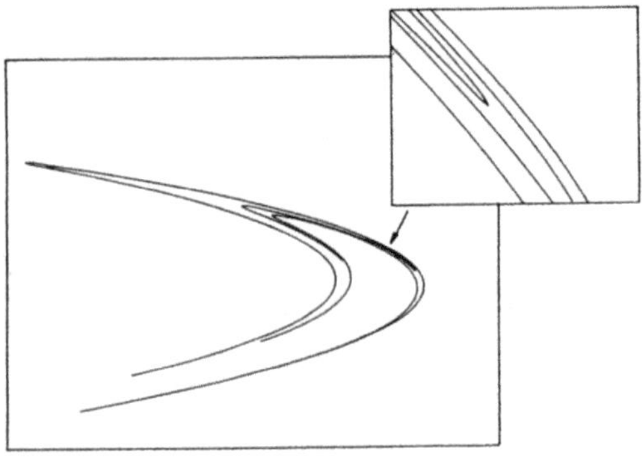

FIG. 5. The Hénon attractor for $a=1.4$, $b=0.3$. The successive iterates f^k of f have been applied to the point (0,0), producing a sequence asymptotic to the attractor. Here, 30000 points of this sequence are plotted, starting with $f^{20}(0,0)$.

D. Strange attractors

The attractors discussed under (a), (b), and (c) above are also attracting sets. They are nice manifolds (point, circle, torus). Notice also that, if a small change $\delta x(0)$ is made to the initial conditions, then $\delta x(t)=(D_x f^t)\delta x(0)$ remains small when $t\to\infty$. [In fact, for a quasiperiodic motion, Eq. (2.7) gives $\delta\varphi(t)=\delta\varphi(0)$.] We shall now discuss more complicated situations.

Examples.

(a) Hénon attractor (Hénon, 1976; Feit, 1978; Curry, 1979). Consider the discrete-time dynamical system defined by

$$f\begin{pmatrix}x_1\\x_2\end{pmatrix}=\begin{pmatrix}1+x_2-ax_1^2\\bx_1\end{pmatrix} \tag{2.11}$$

and the corresponding attractor, for $a=1.4$, $b=0.3$ (see Fig. 5). One finds here numerically that

$$\delta x(t)\approx\delta x(0)e^{\lambda t},\quad \lambda=0.42\ ,$$

i.e., the errors grow exponentially. This is the phenomenon of *sensitive dependence on initial conditions.* In fact (Curry, 1979), computing the successive points $f^n x$ for $n=1,2,\ldots$, with 14 digits' accuracy, one finds that the error of the sixtieth point is of order 1. Sensitive dependence on initial conditions is also expressed by saying that the system is *chaotic* [this is now the accepted use of the word *chaos,* even though the original use by Li and Yorke (1975) was somewhat different].

(b) Feigenbaum attractor (Feigenbaum, 1978,1979,1980; Misiurewicz, 1981; Collet, Eckmann, and Lanford, 1980). A map of the interval [0,1] to itself is defined by

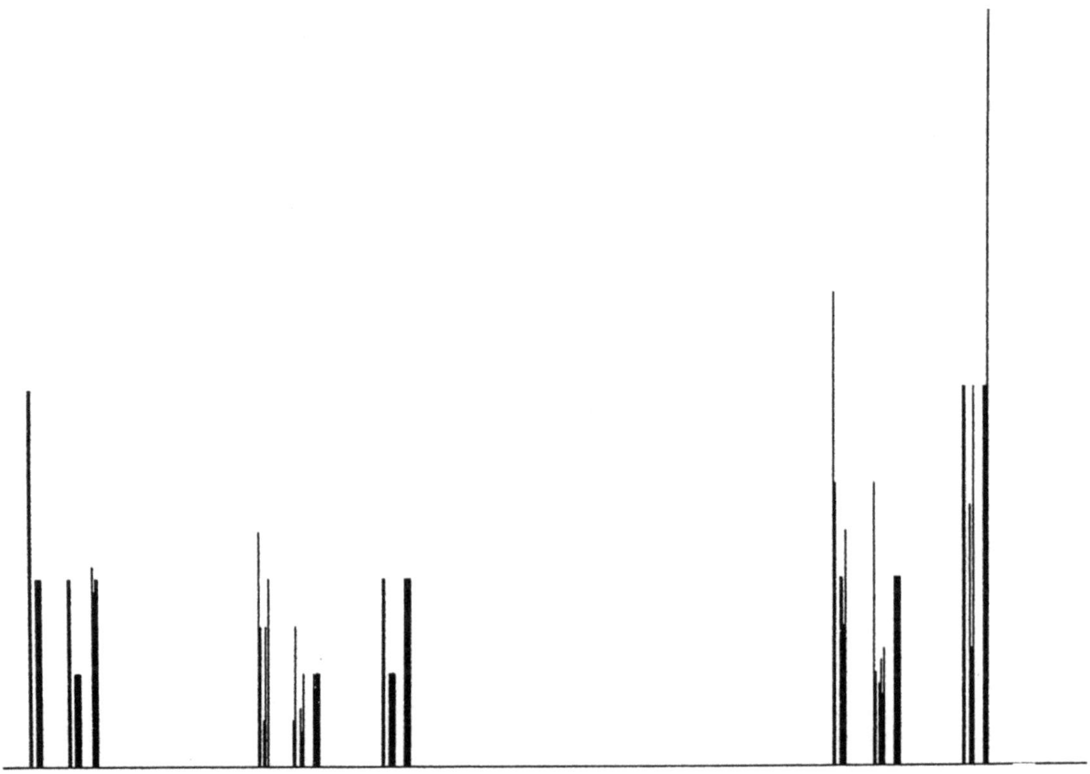

FIG. 6. The Feigenbaum attractor. Histogram of 50000 points in 1024 bins. This histogram shows the unique ergodic measure, which is clearly singular.

$$f_\mu(x)=\mu x(1-x) \tag{2.12}$$

when $\mu\in[0,4]$. It has attracting periodic orbits of period 2^n, with n tending to infinity as μ tends to $3.57\ldots$ through lower values. For the limiting value $\mu=3.57\ldots$, there is a very special attractor A shown in Fig. 6. We shall call it the Feigenbaum attractor (although it was known earlier to many authors). Note that interspersed with this attractor, and arbitrarily close to it, there are repelling periodic orbits of period 2^n, for all n. Therefore the attractor A cannot be an attracting set. One can show, moreover, that, for this very special attractor, there is no sensitive dependence on initial conditions (no exponential growth of errors): the Feigenbaum attractor is not chaotic.

The Hénon and Feigenbaum attractors, as depicted in Figs. 5 and 6, have a complicated aspect typical of *fractal* objects. In general, a fractal set is a set for which the Hausdorff dimension is different from the topological dimension, and usually not an integer. (The exact definition of the Hausdorff dimension is given in Sec. II.J.) The name *fractal* was coined by Mandelbrot. For the rich lore of fractal objects, see Mandelbrot (1982). While many attractors are fractals, and therefore complicated objects, they are by no means featureless. They are unions of *unstable manifolds* (to be defined in Sec. III.E) and often have a Cantor-set structure in the direction transversal to the unstable manifolds. (For the Feigenbaum attractor the unstable manifolds have dimension 0, and only a Cantor set is visible; for the Hénon attractor the unstable manifolds have dimension 1.) An attractor is by definition invariant under a dynamical evolution, and this creates a self-similarity that is often strikingly visible.

In view of both its chaotic and fractal characters, the Hénon attractor deserves to be called a *strange attractor* (this name was introduced by Ruelle and Takens, 1971). The property of being chaotic is actually a more important dynamical concept than that of being fractal, and we shall therefore say that the Feigenbaum attractor is *not* a strange attractor (this differs somewhat from the point of view in Ruelle and Takens). We therefore define a strange attractor to be an attractor with *sensitive dependence on initial conditions.* The notion of strangeness refers thus to the *dynamics* on the attractor, and *not just to its geometry;* it applies whether the time is discrete or continuous. This is again an operational definition rather than a mathematical one. We shall see in Sec. III what should be clarified mathematically. For physics, however, the above operational concept of strange attractors has served well and deserves to be kept.

Example.

(c) Thom's toral automorphisms and Arnold's cat map. Let x_1 (mod1) and x_2 (mod1) be coordinates on the 2-torus T^2; a map $f:T^2\to T^2$ is defined by

$$f\begin{bmatrix}x_1\\x_2\end{bmatrix}=\begin{bmatrix}x_1+x_2\\x_1+2x_2\end{bmatrix}\ (\text{mod}1)\,. \tag{2.13}$$

[Because $\det\binom{11}{12}=1$, the map $\mathbf{R}^2\to\mathbf{R}^2$ defined by the matrix $\binom{11}{12}$ maps Z^2 to Z^2 and therefore, going to the quotient $T^2=\mathbf{R}^2/Z^2$, a map $f:T^2\to T^2$ of the 2-torus to itself is defined. The system is area preserving, and the whole torus is an "attractor."] This is Arnold's celebrated cat map (Arnold and Avez, 1967), well known to be chaotic (see Fig. 7). In fact we have here

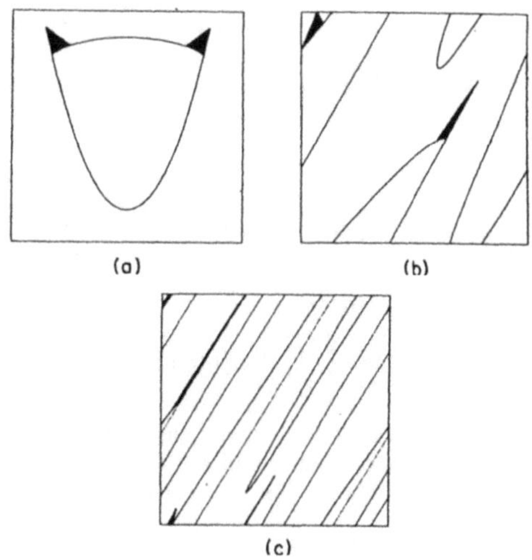

FIG. 7. Arnold's "cat map": (a) the cat; (b) its image under the first iterate; (c) image under the second iterate of the cat map.

$$\delta x(t)\approx\delta x(0)e^{\lambda t}\,, \tag{2.14}$$

with

$$\lambda=\log\frac{3+\sqrt{5}}{2}\,,$$

$(3+\sqrt{5})/2$ being the eigenvalue larger than 1 of the matrix $\binom{11}{12}$.

More generally, if V is an $m\times m$ matrix with integer entries and determinant ±1, it defines a toral automorphism $T^m\to T^m$, and Thom noted that these automorphisms have sensitive dependence on initial conditions if V has an eigenvalue α with $|\alpha|>1$.

Returning to Arnold's cat map, we may imbed T^2 as an attractor A in a higher-dimensional Euclidean space. In this case A is chaotic, but not fractal.

Our examples clearly show that the notions of fractal attractor and chaotic (i.e., strange) attractor are independent. A periodic orbit is neither strange nor chaotic, Arnold's map is strange but not fractal, Feigenbaum's attractor is fractal but not strange, the Lorenz and Hénon attractors are both strange and fractal. [Another strange and fractal attractor with a simple equation has been introduced by Rössler (1976).]

E. Invariant probability measures

An attractor A, be it strange or not, gives a global picture of the long-term behavior of a dynamical system. A more detailed picture is given by the probability measure ρ on A, which describes how frequently various parts of A are visited by the orbit $t \to x(t)$ describing the system (see Fig. 8). Operationally, ρ is defined as the time average of Dirac deltas at the points $x(t)$,

$$\rho = \lim_{t\to\infty} \frac{1}{T} \int_0^T dt\, \delta_{x(t)} . \tag{2.15}$$

Similarly, if a continuous function φ is given, then we define

$$\begin{aligned}\rho(\varphi) &\equiv \int \rho(dx)\varphi(x) \\ &= \lim_{T\to\infty} \frac{1}{T} \int_0^T dt\, \varphi[x(t)] .\end{aligned} \tag{2.16}$$

The measure is *invariant* under the dynamical system, i.e., invariant under time evolution. This invariance may be expressed as follows: For all φ one has

$$\rho(\varphi \circ f^t) = \rho(\varphi) . \tag{2.17}$$

Suppose that the invariant probability measure ρ cannot be written as $\frac{1}{2}\rho_1 + \frac{1}{2}\rho_2$ where ρ_1, ρ_2 are again invariant probability measures and $\rho_1 \neq \rho_2$. Then ρ is called *indecomposable*, or equivalently, *ergodic*.

Theorem. If the compact set A is invariant under the dynamical system (f^t), then there is a probability measure ρ invariant under (f^t) and with support contained in A. One may choose ρ to be ergodic.

[The important assumptions are that the f^t commute and are continuous $A \to A$ (A compact). The theorem results from the Markov-Kakutani fixed-point theorem (see Dunford and Schwartz, 1958, Vol. I).] This is not a very detailed result; it is more in the class referred to as "general nonsense" by mathematicians. But since we shall talk a lot about ergodic measures in what follows, it is good to know that such measures are indeed present.

Theorem (Ergodic theorem). If ρ is ergodic, then for ρ almost all initial $x(0)$ the time averages (2.15) and (2.16) reproduce ρ.

The above theorems show that there are invariant (ergodic) measures defined by time averages. Unfortunately,

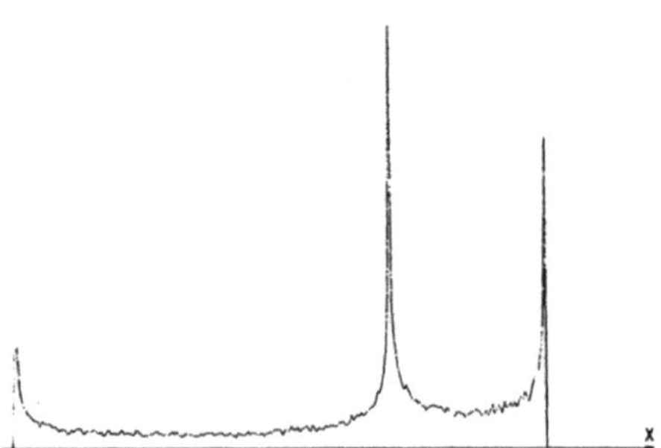

FIG. 8. Histogram of 50 000 iterates of the map $x \to \mu x(1-x)$, in 400 bins. The parameter $\mu = 3.678\,57\ldots$ is the real solution of the equation $(\mu-2)^2(\mu+2) = 16$. It is known that the invariant density is smooth with square-root singularities.

a strange attractor typically carries *uncountably many* distinct ergodic measures. Which one do we choose? We shall propose natural definitions in the next section.

Example.

The points of the circle T^1 may be parametrized by numbers in $[0,1)$, and each such number has a binary expansion $0.a_1a_2a_3\cdots$, where, for each i, $a_i = 0$ or 1 (this coding introduces a little ambiguity, of no importance for what follows). We define a map $f: T^1 \to T^1$ by

$$f(x) = 2(x) \pmod 1 . \tag{2.18}$$

Clearly, f replaces $0.a_1a_2a_3\cdots$ by $0.a_2a_3\cdots$ (an operation called a *shift*). We now choose p between 0 and 1. A probability distribution ρ_p on binary expansions $0.a_1a_2a_3\cdots$ is then defined by requiring that a_i be 0 with probability p, and 1 with probability $1-p$ (independently for each i). One can check that ρ_p is invariant under the shift, and in fact ergodic. It thus defines an ergodic measure for the differentiable dynamical system (2.18), $f: T^1 \to T^1$, and there are uncountably many such measures, corresponding to the different values of p in $(0,1)$.

F. Physical measures

Operationally, it appears that (in many cases, at least) the time evolution of physical systems produces well-defined time averages. The same applies to computer-generated time evolutions. There is thus a selection process of a particular measure ρ which we shall call *physical measure* (another operational definition).

One selection process was discussed by Kolmogorov (we are not aware of a published reference) a long time ago. A physical system will normally have a small level ε of random noise, so that it can be considered as a stochastic process rather than a deterministic one. In a computer study, roundoff errors should play the role of the random noise. Due to sensitive dependence on initial conditions, even a very small level ε of noise has important effects, as we saw in Sec. II.D for the Hénon attractor. On the other hand, a stochastic process such as the one described above normally has only one stationary measure ρ_ε, and we may hope that ρ_ε tends to a specific measure (the *Kolmogorov measure*) when $\varepsilon \to 0$. As we shall see below, this hope is substantiated in the case of *Axiom-A* dynamical systems. However, this approach may have difficulties in general, because an attractor A does not always have an open basin of attraction, and thus the added noise may force the system to jump around on several attractors.

Another possibility is the following: Suppose that M is finite dimensional, and that there is a set $S \subset M$ with Lebesgue measure $\mu(S) > 0$ such that ρ is given by the time averages (2.15) and (2.16) when $x(0) \in S$. This property holds if ρ is an *SRB measure* (to be defined and studied in Sec. IV.C; Sinai, 1972; Bowen and Ruelle, 1975; Ruelle, 1976). For Axiom-A systems, the Kolmogorov and SRB measures coincide, but in general SRB measures are easier to study.

Clearly, Kolmogorov measures and SRB measures are candidates for the description of physical time averages, but they are not always easy to define. Fortunately, many important results hold for an arbitrary invariant measure ρ. Results of this type, which constitute a large part of the ergodic theory of differentiable dynamical systems, will be discussed in Secs. III and IV of this paper.

G. Reconstruction of the dynamics from an experimental signal

In a computer study of a dynamical system in m dimensions, we have an m-dimensional signal $x(t)$, which can be submitted to analysis. By contrast, in a physical experiment one monitors typically only one scalar variable, say $u(t)$, for a system that usually has an infinite-dimensional phase space M. How can we hope to understand the system by analyzing the single scalar signal $u(t)$? The enterprise seems at first impossible, but turns out to be quite doable. This is basically because (a) we restrict our attention to the dynamics on a finite-dimensional attractor A in M, and (b) we can generate several different scalar signals $x_i(t)$ from the original $u(t)$. We have already mentioned that the universal attracting set (which contains all attractors) has finite dimension in two-dimensional hydrodynamics, and we shall come back later to this question of finite dimensionality.

The easiest, and probably the best way of obtaining several signals from a single one is to use *time delays*. One chooses different delays $T_1=0,T_2,\dots,T_N$ and writes $x_k(t)=u(t+T_k)$. In this manner an N-dimensional signal is generated. The experimental points in Fig. 9 below have been obtained by this method. Successive time derivatives of the signal have also been used: $x_{k+1}(t)=d^kx_1(t)/dt^k$, but the numerical differentiations tend to produce high levels of noise. Of course one should measure several experimental signals instead of only one whenever possible.

The reconstruction just outlined will provide an N-dimensional image (or projection) πA of an attractor A which has finite Hausdorff dimension, but lives in a usually infinite-dimensional space M. Depending on the choice of variables (in particular on the time delays), the projection will look different. In particular, if we use fewer variables than the dimension of A, the projection πA will be bad, with trajectories crossing each other. There are some theorems (Takens, 1981; Mañé, 1981) which state that if we use enough variables, typically about twice the Hausdorff dimension, we shall generally get a good projection.

Theorem (Mañé). Let A be a compact set in a Banach space B, and E a subspace of finite dimension such that

$$\dim E > \dim_H(A\times A)+1\ ,$$

or let A be compact and

$$\dim E > 2\dim_K(A)+1\ ,$$

where $\dim_H$ is the Hausdorff dimension and $\dim_K$ is the capacity. Then the set of projections $\pi:B\to E$ such that π restricted to A is injective (i.e., one to one into E) is dense among all projections $B\to E$ with respect to the norm operator topology.

[More precisely, the injective projections are "residual," i.e., contain a countable intersection of dense sets. As noticed by Mañé, his original statement of the theorem needs a slight correction, which is made in the above formulation.]

The choice of variables for the reconstruction of a dynamical system has to be made carefully (by trial and error). This is discussed in Roux, Simoyi, and Swinney (1983).

H. Poincaré sections

The reconstruction process described above yields a line $(f^tx)_0^T$ that may look like a heap of spaghetti and may be difficult to interpret. It is often possible and useful to make a transverse cut through this mess, so that instead of a long curve in N dimensions one now has a set of points S in $N-1$ dimensions (Poincaré section). Figure 9 gives an experimental example corresponding to the Beloussov-Zhabotinski chemical reaction. Given a point x of the Poincaré section, the *first return map* will bring it to Px, which is again in the Poincaré section. When a good model of S and P can be deduced from the experiment, one has essentially understood the dynamical system. This is, however, possible only for low-dimensional attractors.

Notice that the use of a Poincaré section is different from a *stroboscopic* study, where one looks at the system at integer multiples of a *fixed* time interval. By contrast, the time of first return to the Poincaré section is variable

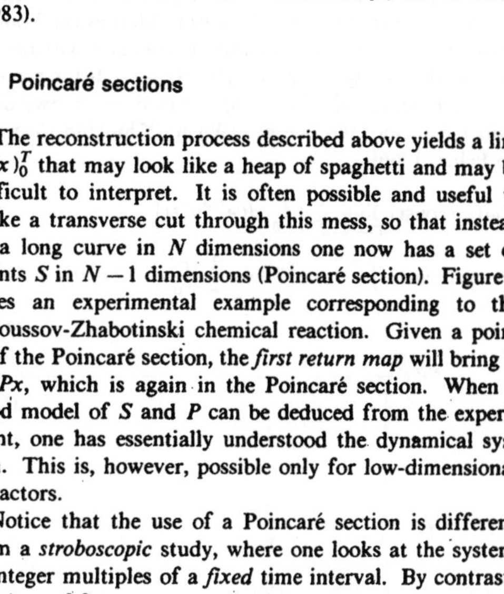

FIG. 9. Experimental plot of a Poincaré section in the Beloussov-Zhabotinski reaction, after Roux and Swinney (1981): (a) the attractor and the plane of Poincaré section; (b) the Poincaré section; (c) the corresponding first return map.

(and has to be determined numerically by interpolation). Sometimes, as for quasiperiodic motions, there is a *natural frequency* (or several) that will stabilize the stroboscopic image. But in general this is not the case, and therefore the stroboscopic study is useless.

I. Power spectra

The power spectrum $S(\omega)$ of a scalar signal $u(t)$ is defined as the square of its Fourier amplitude per unit time. Typically, it measures the amount of energy per unit time (i.e., the power) contained in the signal as a function of the frequency ω. One can also define $S(\omega)$ as the Fourier transform of the time correlation function $\langle u(0)u(t)\rangle$ equal to the average over τ of $u(\tau)u(t+\tau)$. If the correlations of u decay sufficiently rapidly in time, the two definitions coincide, and one has (Wiener-Khinchin theorem; see Feller, 1966)

$$
\begin{aligned}
S(\omega) &= (\text{const}) \lim_{T\to\infty} \frac{1}{T} \left| \int_0^T dt\, e^{i\omega t} u(t) \right|^2 \\
&= (\text{const}) \int_{-\infty}^{\infty} dt\, e^{i\omega t} \lim_{T\to\infty} \frac{1}{T} \int_0^T d\tau' u(\tau') u(t+\tau') \,.
\end{aligned}
\tag{2.19}
$$

Note that the above limit (2.19) makes sense only after averaging over small intervals of ω. Without this averaging, the quantity

$$
\frac{1}{T} \left| \int_0^T dt\, e^{i\omega t} u(t) \right|^2
$$

fluctuates considerably, i.e., it is very noisy. (Instead of averaging over intervals of ω, one may average over many runs).

The power spectrum indicates whether the system is periodic or quasiperiodic. The power spectrum of a periodic system with frequency ω has Dirac δ's at ω and its harmonics $2\omega, 3\omega, \ldots$. A quasiperiodic system with basic frequencies $\omega_1, \ldots, \omega_k$ has δ's at these positions and also at all linear combinations with integer coefficients. (The choice of basic frequencies is somewhat arbitrary, but the number k of independent frequencies is well defined.) In experimental power spectra, the Dirac δ's are not infinitely sharp; they have at least an "instrumental width" $2\pi/T$, where T is the length of the time series used. The linear combinations of the basic frequencies $\omega_1, \ldots, \omega_k$ are dense in the reals if $k > 1$, but the amplitudes corresponding to complicated linear combinations are experimentally found to be small. (A mathematical theory for this does not seem to exist.) A careful experi-

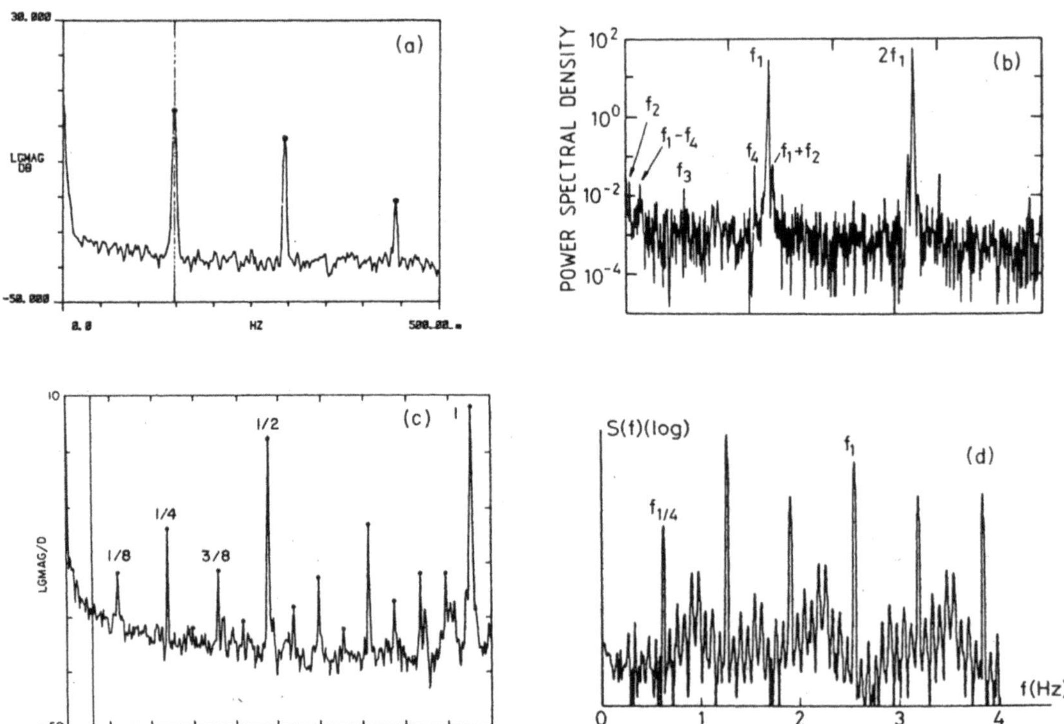

FIG. 10. Some spectra: (a) The power spectrum of a periodic signal shows the fundamental frequency and a few harmonics. Fauve and Libchaber (1982). (b) A quasiperiodic spectrum with four fundamental frequencies. Walden, Kolodner, Passner, and Surko (1984). (c) A spectrum after four period doublings. Libchaber and Maurer (1979). (d) Broadband spectrum invades the subharmonic cascade. The fundamental frequency and the first two subharmonics are still visible. Croquette (1982).

ment may show very convincing examples of quasiperiodic systems with two, three, or more basic frequencies. In fact, $k=2$ is common, and higher k's are increasingly rare, because the nonlinear couplings between the modes corresponding to the different frequencies tend to destroy quasiperiodicity and replace it by chaos (see Ruelle and Takens, 1971; Newhouse, Ruelle, and Takens, 1978). However, for weakly coupled modes, corresponding for instance to oscillators localized in different regions of space, the number of observable frequencies may become large (see, for instance, Grebogi, Ott, and Yorke, 1983a; Walden, Kolodner, Passner, and Surko, 1984). Nonquasiperiodic systems are usually chaotic. Although their power spectra still may contain peaks, those are more or less broadened (they are no longer instrumentally sharp). Furthermore, a noisy background of *broadband spectrum* is present. For this it is *not* necessary that the system be infinite dimensional [Figs. 10(a)–10(d)].

In general, power spectra are very good for the visualization of periodic and quasiperiodic phenomena and their separation from chaotic time evolutions. However, the analysis of the chaotic motions themselves does not benefit much from the power spectra, because (being squares of absolute values) they lose phase information, which is essential for the understanding of what happens on a strange attractor. In the latter case, as already remarked, the dimension of the attractor is no longer related to the number of independent frequencies in the power spectrum, and the notion of "number of modes" has to be replaced by other concepts, which we shall develop below.

J. Hausdorff dimension and related concepts*

Most concepts of dimension make use of a metric. Our applications are to subsets of $\mathbf{R}^m$ or Banach spaces, and the natural metric to use is the one defined by the norm.

Let A be a compact metric space and $N(r,A)$ the minimum number of open balls of radius r needed to cover A. Then we define

$$\dim_K A = \limsup_{r\to 0} \frac{\log N(r,A)}{\log(1/r)} .$$

This is the *capacity* of A (this concept is related to Kolmogorov's ε *entropy and has nothing to do with Newtonian capacity*). *If A and B are compact metric spaces, their product $A\times B$ satisfies*

$$\dim_K(A\times B)\le \dim_K A+\dim_K B . \qquad (2.20)$$

Given a nonempty set A, with a metric, and $r>0$, we denote by σ a covering of A by a (countable) family of sets σ_k with diameter $d_k=\operatorname{diam}\sigma_k\le r$. Given $\alpha\ge 0$, we write

$$m_r^\alpha(A)=\inf_\sigma \sum_k (d_k)^\alpha .$$

When $r\downarrow 0$, $m_r^\alpha(A)$ increases to a (possibly infinite) limit $m^\alpha(A)$ called the *Hausdorff measure of A in dimension α*. We write

$$\dim_H A=\sup\{\alpha : m^\alpha(A)>0\}$$

and call this quantity the *Hausdorff dimension* of A. Note that $m^\alpha(A)=+\infty$ for $\alpha<\dim_H A$, and $m^\alpha(A)=0$ for $\alpha>\dim_H A$. The Hausdorff dimension of a set A is, in general, strictly smaller than the Hausdorff dimension of its closure. Furthermore, the inequality (2.20) on the dimension of a product does not extend to Hausdorff dimensions. It is easily seen that for every compact set A, one has $\dim_H A\le\dim_K A$.

If A and B are compact sets satisfying $\dim_H A=\dim_K A$, $\dim_H B=\dim_K B$, then

$$\dim_H(A\times B)=\dim_K(A\times B)=\dim_H A+\dim_H B .$$

We finally introduce a *topological* dimension $\dim_L A$. It is defined as the smallest integer n (or $+\infty$) for which the following is true: For every finite covering of A by open sets $\sigma_1,\ldots,\sigma_N$ one can find another covering $\sigma_1',\ldots,\sigma_N'$ such that $\sigma_i'\subset\sigma_i$ for $i=1,\ldots,N$ and any $n+2$ of the σ_i' will have an empty intersection:

$$\sigma_{i_0}'\cap\sigma_{i_1}'\cap\cdots\cap\sigma_{i_{n+1}}'=\emptyset .$$

The quantity $\dim_L A$ is also called the *Lebesgue* or *covering dimension* of A.

For more details on dimension theory, see Hurewicz and Wallman (1948) and Billingsley (1965).

III. CHARACTERISTIC EXPONENTS

In this section we review the ergodic theory of differentiable dynamical systems. This means that we study invariant probability measures (corresponding to time averages). Let ρ be such a measure, and assume that it is ergodic (indecomposable). The present section is devoted to the *characteristic exponents of ρ* (also called *Liapunov exponents*) and related questions. We postpone until Sec. V the discussion of how these characteristic exponents can be measured in physical or computer experiments.

A. The multiplicative ergodic theorem of Oseledec

If the initial state of a time evolution is slightly perturbed, the exponential rate at which the perturbation $\delta x(t)$ increases (or decreases) with time is called a *characteristic exponent.* Before defining characteristic exponents for differentiable dynamics, we introduce them in an abstract setting. Therefore, we speak of measurable maps f and T, but the application intended is to continuous maps.

Theorem (multiplicative ergodic theorem of Oseledec). Let ρ be a probability measure on a space M, and $f:M\to M$ a measure preserving map such that ρ is ergodic. Let also $T:M\to$ the $m\times m$ matrices be a measurable map such that

$$\int \rho(dx)\log^+\|T(x)\|<\infty ,$$

where $\log^+u=\max(0,\log u)$. Define the matrix

$T_x^n = T(f^{n-1}x) \cdots T(fx)T(x)$. Then, for ρ-almost all x, the following limit exists:

$$\lim_{n\to\infty} (T_x^{n*} T_x^n)^{1/2n} = \Lambda_x \ . \tag{3.1}$$

(We have denoted by T_x^{n*} the adjoint of T_x^n, and taken the $2n$th root of the positive matrix $T_x^{n*} T_x^n$.)

The logarithms of the eigenvalues of Λ_x are called *characteristic exponents.* We denote them by $\lambda_1 \geq \lambda_2 \geq \cdots$. They are ρ-almost everywhere constant. (This is because we have assumed ρ ergodic. Of course, the λ_i depend on ρ.) Let $\lambda^{(1)} > \lambda^{(2)} > \cdots$ be the characteristic exponents again, but no longer repeated by multiplicity; we call $m^{(i)}$ the multiplicity of $\lambda^{(i)}$. Let $E_x^{(i)}$ be the subspace of $\mathbf{R}^m$ corresponding to the eigenvalues $\leq \exp\lambda^{(i)}$ of Λ_x. Then $\mathbf{R}^m = E_x^{(1)} \supset E_x^{(2)} \supset \cdots$ and the following holds

Theorem. For ρ-almost all x,

$$\lim_{n\to\infty} \frac{1}{n} \log \| T_x^n u \| = \lambda^{(i)} \tag{3.2}$$

if $u \in E_x^{(i)} \setminus E_x^{(i+1)}$. In particular, for all vectors u that are not in the subspace $E_x^{(2)}$ (viz., almost all u), the limit is the largest characteristic exponent $\lambda^{(1)}$.

The above remarkable theorem dates back only to 1968, when the proof of a somewhat different version was published by Oseledec (1968). For different proofs see Raghunathan (1979), Ruelle (1979), Johnson, Palmer, and Sell (1984). What does the theorem say for $m=1$? The 1×1 matrices are just ordinary numbers. Assuming them to be positive and taking the log, the reader will verify that the multiplicative ergodic theorem reduces to the ordinary ergodic theorem of Sec. II.E. The novelty and difficulty of the multiplicative ergodic theorem is that for $m>1$ it deals with *noncommuting* matrices.

In some applications we shall need an extension, where $\mathbf{R}^m$ is replaced by an infinite-dimensional Banach or Hilbert space E and the $T(x)$ are bounded operators. Such an extension has been proved under the condition that the $T(x)$ are *compact* operators. In the Hilbert case this means that the spectrum of $T(x)^* T(x)$ is discrete, that the eigenvalues have finite multiplicities, and that they accumulate only at 0.

Theorem (multiplicative ergodic theorem—compact operators in Hilbert space). All the assertions of the multiplicative ergodic theorem remain true if $\mathbf{R}^m$ is replaced by a separable Hilbert space E, and T maps M to compact operators in E. The characteristic exponents form a sequence tending to $-\infty$ (it may happen that only finitely many characteristic exponents are finite).

See Ruelle (1982a) for a proof. For compact operators on a Banach space, Eq. (3.1) no longer makes sense, but there are subspaces $E_x^{(1)} \supset E_x^{(2)} \supset \cdots$ such that (3.2) holds. This was shown first by Mañé (1983), with an unnecessary injectivity assumption, and then by Thieullen (1985) in full generality. (Thieullen's result applies in fact also to noncompact situations.)

B. Characteristic exponents for differentiable dynamical systems

1. Discrete-time dynamical systems on $\mathbf{R}^m$

We consider the time evolution

$$x(n+1) = f(x(n)) \ , \tag{3.3}$$

where $f: \mathbf{R}^m \to \mathbf{R}^m$ is a differentiable vector function. We denote by $T(x)$ the matrix $(\partial f_i / \partial x_j)$ of partial derivatives of the components f_i at x. For the nth iterate f^n of f, the corresponding matrix of partial derivatives is given by the chain rule:

$$\partial (f^n)_i / \partial x_j = T(f^{n-1}x) \cdots T(fx)T(x) \ . \tag{3.4}$$

Now, if ρ is an ergodic measure for f, with compact support, the conditions of the multiplicative ergodic theorem are all satisfied and the characteristic exponents are thus defined.

In particular, if $\delta x(0)$ is a small change in initial condition (considered as infinitesimally small), the change at time n is given by

$$\begin{aligned}\delta x(n) &= T_x^n \delta x(0) \\ &= T(f^{n-1}x) \cdots T(x) \delta x(0) \ .\end{aligned} \tag{3.5}$$

For most $\delta x(0)$ [i.e., for $\delta x(0) \notin E_{x(0)}^{(2)}$] we have $\delta x(n) \approx \delta x(0) e^{n\lambda_1}$, and sensitive dependence on initial conditions corresponds to $\lambda_1 > 0$. Note that if $\delta x(0)$ is finite rather than infinitely small, the growth of $\delta x(n)$ may not go on indefinitely: if $x(0)$ is in a bounded attractor, $\delta x(n)$ cannot be larger than the diameter of the attractor.

2. Continuous-time dynamical systems on $\mathbf{R}^m$

If the time is continuous, we apply the multiplicative ergodic theorem to the time-one map $f = f^1$. The limits defining the characteristic exponents hold again, with $t \to \infty$ replacing $n \to \infty$ (because of continuity it is not necessary to restrict t to integer values). To be specific, we define

$$T_x^t = \text{matrix } (\partial f_i^t / \partial x_j) \ . \tag{3.6}$$

If ρ is an ergodic measure with compact support for the time evolution, then, for ρ-almost all x, the following limits exist:

$$\lim_{t\to\infty} (T_x^{t*} T_x^t)^{1/2t} = \Lambda_x \ , \tag{3.7}$$

$$\lim_{t\to\infty} \frac{1}{t} \log \| T_x^t u \| = \lambda^{(i)} \ \text{ if } u \in E_x^{(i)} \setminus E_x^{(i+1)} \ , \tag{3.8}$$

where $\lambda^{(1)} > \lambda^{(2)} > \cdots$ are the logarithms of the eigenvalues of Λ_x, and $E_x^{(i)}$ is the subspace of $\mathbf{R}^m$ corresponding to the eigenvalues $\leq \exp\lambda^{(i)}$. Notice, incidentally, that if the Euclidean norm $\| \ \|$ is replaced by some other norm on $\mathbf{R}^m$, *the characteristic exponents and the* $E_x^{(i)}$ *do not change.*

3. Dynamical systems in Hilbert space

We assume that E is a (real) Hilbert space, ρ a probability measure with compact support in E, and f^t a time evolution such that the linear operators $T_x^t = D_x f^t$ (derivative of f^t at x) are compact linear operators for $t>0$. This situation prevails, for instance, for the Navier-Stokes time evolution in two dimensions (as well as in three dimensions, so long as the solution has no singularities). The definition of characteristic exponents is the same here as for dynamical systems in $\mathbf{R}^m$.

4. Dynamical systems on a manifold M

For definiteness, let M be a compact manifold like a sphere or a torus; ρ is a probability measure on M, invariant under the dynamical system. If M is m dimensional, we may cut M into a finite number of pieces which are smoothly parametrized by subsets of $\mathbf{R}^m$ (see Fig. 11). In terms of this new parametrization, the map f is continuous except at the cuts, and so is the matrix of partial derivatives. Since only measurability is needed for the abstract multiplicative ergodic theorem, we can again define characteristic exponents. This definition is independent of the partition of the manifold M that has been used, and of the choice of parametrization for the pieces. The reason is that, for any other choice, the norm used would differ from the original norm by a bounded factor, which disappears in the limit. One could alternatively use a Riemann metric on the manifold and define the characteristic exponents in terms of this metric. If $\mathscr{T}_x M$ denotes the tangent space at x, we now have $\mathscr{T}_x M = E_x^{(1)} \supset E_x^{(2)} \supset \cdots$.

C. Steady, periodic, and quasiperiodic motions

1. Examples and parameter dependence

Before proceeding with the general theory, we pause to discuss illustrations of the preceding results.

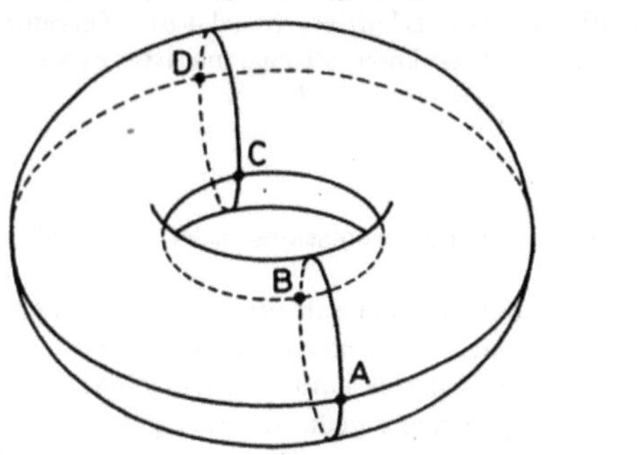

FIG. 11. A two-dimensional torus cut into four rectangular pieces by two horizontal and two vertical circles.

A *steady state* of a physical time evolution is associated with a fixed point P of the corresponding dynamical system. The steady state is thus described by the probability measure $\rho = \delta_P$ (Dirac's delta at P), which is of course invariant and ergodic. We denote by $\alpha_1, \alpha_2, \ldots$, the eigenvalues of the operator $D_P f^1$ (derivative of the time-one map f^1 at P), in decreasing order of absolute values, and repeated according to multiplicity. Then the characteristic exponents are

$$\lambda_1 = \log|\alpha_1|, \quad \lambda_2 = \log|\alpha_2|, \quad \ldots . \tag{3.9}$$

In particular, a stable steady state associated with an attracting fixed point (see Sec. III.C.2) has negative characteristic exponents. If the dynamical system depends continuously on a bifurcation parameter μ, the $\lambda_i = \log|\alpha_i|$ depend continuously on μ, but we shall see in Sec. III.D that this situation is rather exceptional.

A *periodic state* of a physical time evolution is associated with a periodic orbit $\Gamma = \{f^t a : 0 \le t < T\}$ of the corresponding continuous-time dynamical system. It is thus described by the ergodic probability measure

$$\rho = \delta_\Gamma = \frac{1}{T}\int_0^T dt\, \delta_{f^t a} . \tag{3.10}$$

We denote by α_i^T the eigenvalues of $D_a f^T$; then one of these eigenvalues is 1 (corresponding to the direction tangent to Γ at a). The characteristic exponents are the numbers

$$\lambda_i = \frac{1}{T}\log|\alpha_i^T| ,$$

and one of them is thus 0. In particular, a stable periodic state, associated with an attracting periodic orbit (see Sec. II.C.2), has one characteristic exponent equal to zero and the others negative. Here again, if there is a bifurcation parameter μ, the λ_i depend continuously on μ.

Consider now a *quasiperiodic state with k frequencies*, stable for simplicity. This is represented by a quasiperiodic attractor (Sec. II.C.3), i.e., an attracting invariant torus T^k on which the time evolution is described by translations (2.8) in terms of suitable angular variables $\varphi_1, \ldots, \varphi_k$. There is only one invariant probability measure here: the Haar measure ρ on T^k, defined in terms of the angular variables by

$$(2\pi)^{-k} d\varphi_1 \cdots d\varphi_k .$$

Here, k characteristic exponents are equal to zero, and the others are negative. If the dynamical system depends continuously on a bifurcation parameter μ and has a quasiperiodic attractor for $\mu=\mu_0$, it will still have an attracting k torus for μ close to μ_0, but the motion *on* this k torus may no longer be quasiperiodic. For $k \ge 2$, frequency locking may lead to attracting periodic orbits (and negative characteristic exponents). For $k \ge 3$, strange attractors and positive characteristic exponents may be present for μ arbitrarily close to μ_0 (see Ruelle and Takens, 1971; Newhouse, Ruelle, and Takens, 1978). Nevertheless, we have continuity at $\mu=\mu_0$: the characteristic exponents for μ close to μ_0 are close to their values at μ_0.

2. Characteristic exponents as indicators of periodic motion

The examples of the preceding section are typical for the case of negative characteristic exponents. We now point out that, conversely, it is possible to deduce from the negativity of the characteristic exponents that the ergodic measure ρ describes a steady or a period state.

Theorem (continuous-time fixed point). Consider a continuous-time dynamical system and assume that all the characteristic exponents are different from zero. Then $\rho=\delta_P$, where P is a fixed point. (In particular, if all characteristic exponents are negative, P is an attracting fixed point.)

Another formulation: If the support of ρ does not reduce to a fixed point, then one of the characteristic exponents vanishes.

Sketch of proof. One considers the vector function F,

$$F(x)=\left.\frac{d}{d\tau}f^{\tau}x\right|_{\tau=0} . \tag{3.11}$$

If the support of ρ is not reduced to a fixed point, we have $F(x)\neq 0$ for ρ-almost all x. Furthermore, Eq. (3.11) yields

$$T_x^t F(x)=(D_x f^t)F(x)=F(f^t x) .$$

Since ρ is ergodic, $f^t x$ comes close to x again and again, and we find for the limit (3.8)

$$\lim_{t\to\infty}\frac{1}{t}\log\|T_x^t F(x)\|=0 .$$

Thus there is a characteristic exponent equal to 0.

In the next two theorems we assume that the dynamical system is defined by functions that have continuous second-order derivatives. (The proofs use the stable manifolds of Sec. III.E.)

Theorem (discrete-time periodic orbit). Consider a discrete-time dynamical system and assume that all the characteristic exponents of ρ are negative. Then

$$\rho=\frac{1}{N}\sum_{1}^{N}\delta_{f^{k}a} ,$$

where $\{a,fa,\ldots,f^{N-1}a\}$ is an attracting periodic orbit, of period N.

Proof. See Ruelle (1979).

Theorem (continuous-time periodic orbit). Consider a continuous-time dynamical system and assume that all the characteristic exponents of ρ are negative, except λ_1. There are then two possibilities: (a) $\rho=\delta_P$, where P is a fixed point, (b) ρ is the measure (3.9) on an attracting periodic orbit (and $\lambda_1=0$).

Proof. See Campanino (1980).

As an application of these results, consider the time evolution given by a differential equation (2.1) in two dimensions. We have the following possibilities for an ergodic measure ρ:

$\lambda_1=\lambda_2=0$,

$\lambda_1=0$, $\lambda_2<0$: ρ is associated with a fixed point or an attracting period orbit,

$\lambda_1>0$, $\lambda_2=0$: this reduces to the previous case by changing the direction of time, and therefore ρ is associated with a fixed point or a repelling periodic orbit,

λ_1 and λ_2 are nonvanishing: ρ is associated with a fixed point.

None of these possibilities corresponds to an *attractor* with a positive characteristic exponent. Therefore, an evolution (2.1) can be chaotic only in three or more dimensions.

D. General remarks on characteristic exponents

We now fix an ergodic measure ρ, and the characteristic exponents that occur in what follows are with respect to this measure.

1. The growth of volume elements

The rate of exponential growth of an infinitesimal vector $\delta x(t)$ is given in general by the largest characteristic exponent λ_1. The rate of growth of a surface element $\delta\sigma(t)=\delta_1 x(t)\wedge\delta_2 x(t)$ is similarly given in general by the sum of the largest two characteristic exponents $\lambda_1+\lambda_2$. In general for a k-volume element $\delta_1 x(t)\wedge\cdots\wedge\delta_k x(t)$ the rate of growth is $\lambda_1+\cdots+\lambda_k$. (Of course, if this sum is negative, the volume is contracted.) The construction above gives computational access to the lower characteristic exponents (and is used in the proof of the multiplicative ergodic theorem). For instance, for a dynamical system in $\mathbf{R}^m$, the rate of growth of the m-volume element is the rate of growth of the *Jacobian determinant* $|J_x^t|=|\det(\partial f_i^t/\partial x_j)|$, and is given by $\lambda_1+\cdots+\lambda_m$. For a volume-preserving transformation we have thus $\lambda_1+\cdots+\lambda_m=0$. For a map f with constant Jacobian J, we have $\lambda_1+\cdots+\lambda_m=\log|J|$.

Examples.

In the case of the Hénon map [example (a) of Sec. II.D] we have $J=-b=-0.3$, hence $\lambda_2=\log|J|-\lambda_1\approx-1.20-0.42=-1.62$.

In the case of the Lorenz equation [example (b) of Sec. II.B] we have $dJ^t/dt=-(\sigma+1+b)$. Therefore, if we know $\lambda_1>0$ we know all characteristic exponents, since $\lambda_2=0$ and $\lambda_3=-(\sigma+1+b)-\lambda_1$.

2. Lack of explicit expressions, lack of continuity

The ordinary ergodic theorem states that the time average of a function φ tends to a limit (ρ-almost everywhere) and asserts that this limit is $\int\varphi(x)\rho(dx)$. By contrast, the multiplicative ergodic theorem gives no explicit expression for the characteristic exponents. It is true that in the proof of the theorem as given by Johnson *et al.* (1984) there is an integral representation of characteristic exponents in terms of a measure on the space of points (x,Q), where x is a point of our m-dimensional manifold, and Q is an $m\times m$ orthogonal matrix. However, this measure is not constructively given. This situation is similar to that in statistical mechanics where, for example, there is in general no explicit expression for the pressure in terms of the interparticle forces.

For a dynamical problem depending on a bifurcation parameter μ, one would like at least to know some continuity properties of the λ_i as functions of μ. The situation there is unfortunately quite bad (with some exceptions—see Sec. III.C). For each μ there may be several attractors A_α^μ, each having at least one physical measure ρ_α^μ. The dependence of the attractors on μ need not be continuous, because of captures and "explosions," and we do not know that ρ_α^μ depends continuously on A_α^μ. Finally, even if ρ_α^μ depends continuously on μ, it is not true in general that the characteristic exponents do the same. To summarize: the characteristic exponents are in general discontinuous functions of the bifurcation parameter μ.

Example.

The interval [0,1] is mapped into itself by $x\to\mu x(1-x)$ when $0\leq\mu\leq 4$, and Fig. 12 shows λ_1 as a function of μ. There are intervals of values of μ where λ_1 is negative, corresponding to an attracting periodic orbit. It is believed that these intervals are dense in [0,4]. If this is so, λ_1 is necessarily a discontinuous function of μ wherever it is positive. It is believed that $\{\mu\in[0,4]:\lambda_1>0\}$ has positive Lebesgue measure (this result has been announced by Jakobson, but no complete proof has appeared). For some positive results on these difficult problems see Jakobson (1981), Collet and Eckmann (1980a,1983), and Benedicks and Carleson (1984).

The wild discontinuity of characteristic exponents raises a philosophical question: should there not be at least a piecewise continuous dependence of physical quantities on parameters such as one sees, for example, in the solution of the Ising model? Yet we obtain here discontinuous predictions. Part of the resolution of this paradox lies in the fact that our mathematical predictions are *measurable functions* if not continuous, and that measurable functions have much more controllable discontinuities (cf. Luzin's theorem, for instance) than those one could construct with help of the axiom of choice. Another fact is that physical measurements are smoothed by the instrumental procedure. In particular, the definition of characteristic exponents involves a limit $t\to\infty$ [see Eqs. (3.7) and (3.8)], and the great complexity of a curve $\mu\to\lambda_1(\mu)$ will only appear progressively as t is made larger and larger. The presence of noise also smooths out experimental results. At a given level of precision one may find, for instance, that there is one positive characteristic exponent $\lambda_1(\mu)$ in the interval $[\mu_1,\mu_2]$. *This is a meaningful statement,* even though it probably will have to be revised when higher-precision measurements are made; those may introduce *small* subintervals of $[\mu_1,\mu_2]$ where all characteristic exponents are negative. Let us also mention the possibility that for a *large* chaotic system (like a fully turbulent fluid) the distribution of characteristic exponents could again be a smooth function of bifurcation parameters.

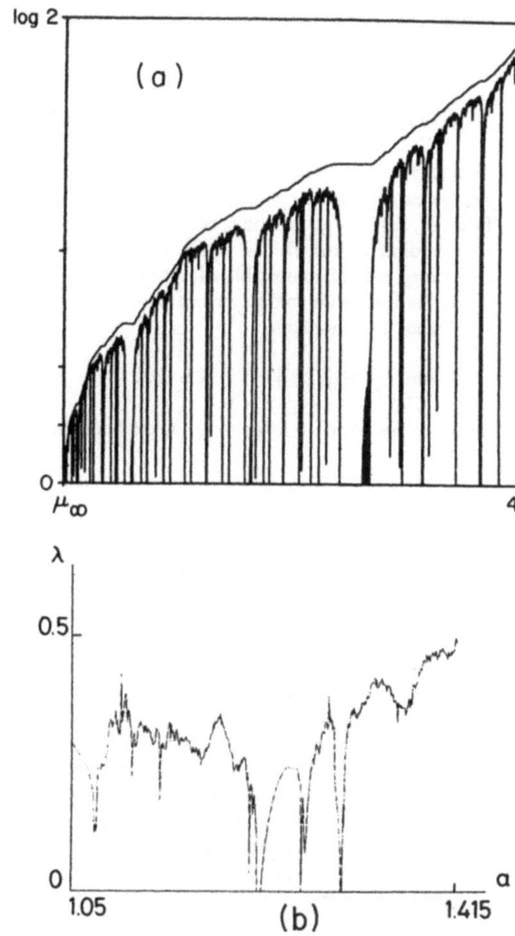

FIG. 12. (a) Topological entropy (upper curve) and characteristic exponent (lower curve) as a function of μ for the family $x\to\mu x(1-x)$. (Graph by J. Crutchfield.) Note the discontinuity of the lower curve. (b) Similar figure for the Hénon map, with $b=0.3$, after Feit (1978).

3. Time reflection

Let us assume that the time-evolution maps f^t are defined for t negative as well as positive. In the discrete-time case this means that f has an inverse f^{-1} which is a smooth map (i.e., f is a *diffeomorphism*). We may consider the time-reversed dynamical system, with time-evolution map $\bar{f}^t=f^{-t}$. If ρ is an invariant (or ergodic) probability measure for the original system, it is also invariant (or ergodic) for the time-reversed system. Furthermore, *the characteristic exponents of an ergodic measure ρ for the time-reversed system are those of the original* *system, but with opposite sign.* We have correspondingly a sequence of subspaces $\bar{E}_x^{(1)} \subset \bar{E}_x^{(2)} \subset \cdots$ for almost all x, such that

$$\lim_{t\to-\infty} \frac{1}{|t|} \log \|T_x^t u\| = -\lambda^{(i)} \quad \text{if } u \in \bar{E}_x^{(i)} \setminus \bar{E}_x^{(i-1)} .$$

Define $F_x^{(i)} = E_x^{(i)} \cap \bar{E}_x^{(i)}$. Then, for ρ-almost all x, the subspaces $F_x^{(i)}$ span $\mathbf{R}^m$ (or the tangent space to the manifold M, as the case may be; compact operators in infinite-dimensional Hilbert space are excluded here because they are not compatible with $t<0$). Furthermore, if T_x^t is the derivative matrix or operator corresponding to $\bar{f}^{|t|}$ when $t<0$, we have

$$\lim_{|t|\to\infty} \frac{1}{t} \log \|T_x^t u\| = \lambda^{(i)} \quad \text{if } u \in F_x^{(i)} ,$$

where t may go to $+\infty$ or $-\infty$. (For details see Ruelle, 1979.)

4. Relations between continuous-time and discrete-time dynamical systems

We have defined the characteristic exponents for a continuous-time dynamical system [see Eqs. (3.7) and (3.8)] so that they are the same as the characteristic exponents for the discrete-time dynamical system generated by the time-one map $f=f^1$.

Given a Poincaré section (see Sec. II.H), we want to relate the characteristic exponents λ_i for a continuous-time dynamical system with the characteristic exponents $\tilde{\lambda}_i$ corresponding to the first return map P. Note that one of the λ_i is zero (first theorem in Sec. III.C); we claim that the other λ_i are given by

$$\lambda_i = \tilde{\lambda}_i / \langle \tau \rangle_\sigma , \tag{3.12}$$

where $\langle \tau \rangle_\sigma$ is the average time between two crossings of the Poincaré section Σ, computed with respect to the probability measure σ on Σ naturally associated with ρ. (The measure σ gives the density of intersections of orbits with Σ.) The proof is not hard and is left to the reader.

5. Hamiltonian systems

Consider a Hamiltonian (i.e., conservative) system with m degrees of freedom. This is a continuous-time dynamical system in $2m$ dimensions. We claim that the set of λ_i's is symmetric with respect to 0. This is readily checked from Eq. (3.7) and the fact that T_x^t is a symplectic matrix. Actually, two of the λ_i vanish; we get rid of one by going to a $(2m-1)$-dimensional energy surface, and one zero characteristic exponent survives in accordance with the first theorem of Sec. III.C.2.

E. Stable and unstable manifolds

The multiplicative ergodic theorem asserts the existence of *linear* spaces $E_x^{(1)} \supset E_x^{(2)} \supset \cdots$ such that

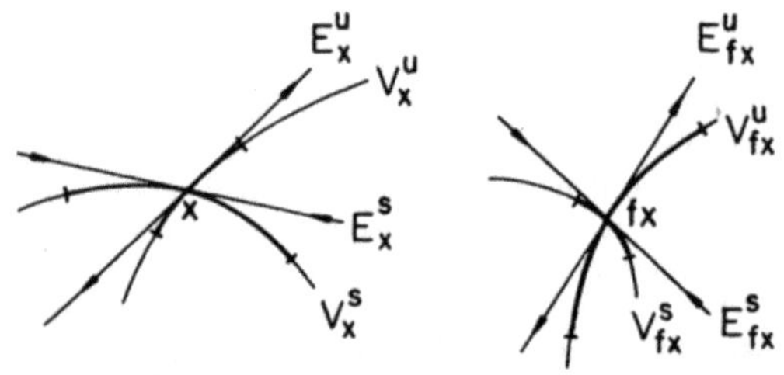

FIG. 13. Stable and unstable manifolds can be defined for points that are neither fixed nor periodic. The stable and unstable directions E_x^s and E_x^u are tangent to the stable and unstable manifolds V_x^s and V_x^u, respectively. They are mapped by f onto the corresponding objects at fx.

$$\lim_{t\to\infty} \frac{1}{t} \log \|T_x^t u\| \le \lambda^{(i)} \quad \text{if } u \in E_x^{(i)} .$$

This means that there exist subspaces $E_x^{(i)}$ such that the vectors in $E_x^{(i)} \setminus E_x^{(i+1)}$ are expanded exponentially by time evolution with the rate $\lambda^{(i)}$. (This expansion is of course a contraction if $\lambda^{(i)}<0$.) See Fig. 13.

One can define a nonlinear analog of those $E_x^{(i)}$ which correspond to negative characteristic exponents. Let $\lambda<0$, $\varepsilon>0$, and write

$$V_x^s(\lambda,\varepsilon) = \{ y : d(f^t x, f^t y) \le \varepsilon e^{\lambda t} \text{ for all } t \ge 0 \} ,$$

where $d(x,y)$ is the distance of x and y (Euclidean distance in $\mathbf{R}^m$, norm distance in Hilbert space, or Riemann distance on a manifold). We shall assume from now on that the time-one map f^1 has continuous derivatives of second as well as first order. If $\lambda^{(i-1)} > \lambda > \lambda^{(i)}$, the set $V_x^s(\lambda,\varepsilon)$ is in fact, for ρ-almost all x and small ε, a piece of *differentiable* manifold, called a *local stable manifold* at x; it is tangent at x to the linear space $E_x^{(i)}$ (and has the

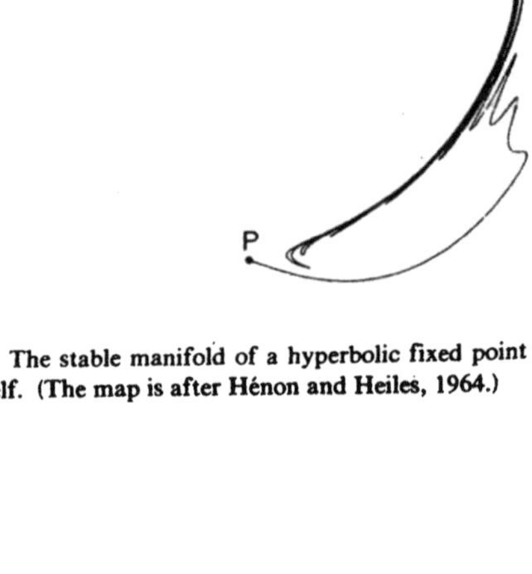

FIG. 14. The stable manifold of a hyperbolic fixed point folds up on itself. (The map is after Hénon and Heiles, 1964.)

same dimension). One shows that $V_x^s(\lambda,\varepsilon)$ is differentiable as many times as f^1.

If we assume that our dynamical system is defined for negative as well as positive times, we can define *global stable manifolds* such that

$$V_x^{(i)s}=\left\{y:\lim_{t\to\infty}\frac{1}{t}\log d(f^tx,f^ty)\le\lambda^{(i)}\right\}$$

$$=\bigcup_{t>0}f^{-t}V_x^s(\lambda,\varepsilon)\,,$$

with negative λ between $\lambda^{(i-1)}$ and $\lambda^{(i)}$ as above.

These global manifolds have the somewhat annoying feature that, while they are locally smooth, they tend to fold and accumulate in a very complicated manner, as suggested by Fig. 14. We can also define *the* stable manifold of x by

$$V_x^s=\left\{y:\lim_{t\to\infty}\frac{1}{t}\log d(f^tx,f^ty)<0\right\}$$

(it is the largest of the stable manifolds, equal to $V_x^{(i)s}$ where $\lambda^{(i)}$ is the largest negative characteristic exponent).

For a dynamical system where negative times are allowed, we obtain *unstable manifolds* V^u instead of stable manifolds simply through replacement of t by $-t$ in the definitions. Instead of assuming that f^t is defined for $t<0$, we find it desirable to make the weaker assumption that f^t and Df^t (defined for $t\ge0$) are *injective*. This means that $f^tx=f^ty$ implies $x=y$ and $D_xf^tu=D_xf^tv$ implies $u=v$. This injectivity assumption is satisfied when the dynamical system is defined for negative as well as positive times, but also in the case of the Navier-Stokes time evolution. The *global unstable manifold* V_x^u is then defined, provided that for every $t>0$ there is x_{-t} such that $f^tx_{-t}=x$; the definition is

$$V_x^u=\left\{y:\text{ there exists }y_{-t}\text{ such that }f^ty_{-t}=y\text{ and }\lim_{t\to\infty}\frac{1}{t}\log d(x_{-t},y_{-t})<0\right\}.$$

If $\lambda>0$ we define similarly

$$V_x^u(\lambda,\varepsilon)=\{y:\text{ there exists }y_{-t}\text{ such that }f^ty_{-t}=y\text{ and }d(x_{-t},y_{-t})\le\varepsilon e^{-\lambda t}\text{ for all }t\ge0\}\,.$$

and if $\lambda>0$ and $\lambda^{(i+1)}<\lambda<\lambda^{(i)}$, we write

$$V_x^{(i)u}=\left\{y:\text{ there exists }y_{-t}\text{ such that }f^ty_{-t}=y\text{ and }\lim_{t\to\infty}\frac{1}{t}d(x_{-t},y_{-t})\le-\lambda^{(i)}\right\}$$

$$=\bigcup_{t>0}f^tV_x^u(\lambda,\varepsilon)\,.$$

The global unstable manifold V_x^u is the largest of the $V_x^{(i)u}$, corresponding to the smallest positive characteristic exponent $\lambda^{(i)}$. Here again one shows that the local unstable manifolds $V_x^u(\lambda,\varepsilon)$ are differentiable (as many times, in fact, as f^1), while the global unstable manifolds $V_x^{(i)u}$ and V_x^u are locally differentiable, but may accumulate on themselves in a complicated manner globally.

The theory of stable and unstable manifolds is part of Pesin theory (for some details, see Sec. III.G).

Examples.

(a) *Fixed points.* If P is a fixed point for a dynamical system (with discrete or continuous time), the characteristic exponents of the δ-measure δ_P at P are called characteristic exponents of the fixed point. They are given explicitly by Eq. (3.9). The fixed point P is said to be *hyperbolic* if all characteristic exponents λ_i are nonzero. When all $\lambda_i<0$, P is *attracting.* When all $\lambda_i>0$, P is *repelling.* When some λ_i are >0 and some <0, P is of *saddle type.* The *stable* and *unstable manifolds* of the hyperbolic fixed point P are defined to be the stable and unstable manifolds of δ_P. One has

$$V_x^s=\left\{y:\lim_{t\to+\infty}f^ty=x\right\},$$

$$V_x^u=\left\{y:\lim_{t\to-\infty}f^ty=x\right\}.$$

(b) *Periodic orbits.* Let Γ be a closed orbit for a continuous-time dynamical system. There is only one invariant measure with support Γ, namely δ_Γ given by Eq. (3.10); it is ergodic. If u is a vector tangent to Γ at x, the corresponding characteristic exponent is zero as one may easily check. If all other characteristic exponents are nonzero, Γ is a *hyperbolic periodic orbit.* The *attracting, repelling,* and *saddle-type* periodic orbits are similarly defined. If $x\in\Gamma$ we have

$$V_x^s=\left\{y:\lim_{t\to+\infty}d(f^tx,f^ty)=0\right\}.$$

This is also called the *strong stable manifold* of x, and a stable manifold of Γ is defined by

$$V_\Gamma^{cs}=\bigcup_{x\in\Gamma}V_x^s$$

$$=\bigcup_{t>0}f^{-t}V_\Gamma^{cs}(\varepsilon)\,,$$

where the local stable manifold $V_\Gamma^{cs}(\varepsilon)$ is defined for small ε by

$$V_\Gamma^{cs}(\varepsilon)=\{y: d(f^t x, f^t y)<\varepsilon \text{ for all } t\geq 0\} .$$

Theorem. If A is an attracting set, and $x\in A$, then $V_x^u\subset A$ i.e., the unstable manifold of x is contained in A.

Proof. If U is a fundamental neighborhood of A, and $y\in V_x^u$, then $f^{-\tau}y\in U$ for sufficiently large τ (because $f^{-\tau}y$ is close to $f^{-\tau}x\in A$). Therefore $y\in \bigcap_{\tau>T} f^\tau U = A$.

Corollary. Let A be an attracting set. The number of characteristic exponents $\lambda_i>0$ for any ergodic measure with support in A is a lower bound to the dimension of A.

Proof. The dimension of A is at least that of V_x^u, which is equal to the dimension of $\overline{E}_x^{(k)}$, where $\lambda^{(k)}$ is the smallest positive characteristic exponent. But $\dim \overline{E}_x^{(k)}$ is the sum of the multiplicities of the positive $\lambda^{(i)}$, i.e., the number of positive characteristic exponents λ_i.

(c) *Visualization of the unstable manifolds.* The Hénon attractor has a characteristic appearance of a line folded over many times (see Fig. 5). A similar picture appears for attractors of other two-dimensional dynamical systems generated by a diffeomorphism (differentiable map with differentiable inverse). The theorem stated above suggests that the convoluted lines seen in such attractors are in fact *unstable manifolds.* This suggestion is confirmed by the fact that in many cases the *physical measure* on an attractor is *absolutely continuous on unstable manifolds,* as we shall discuss below.

In higher dimensions, the unstable manifolds forming an attractor may be lines (one dimension), veils (two dimensions), etc. Attractors corresponding to noninvertible maps in two dimensions often have the characteristic appearance of folded veils or drapes, and it is thus immediately apparent that they do not come from a diffeomorphism (Fig. 15).

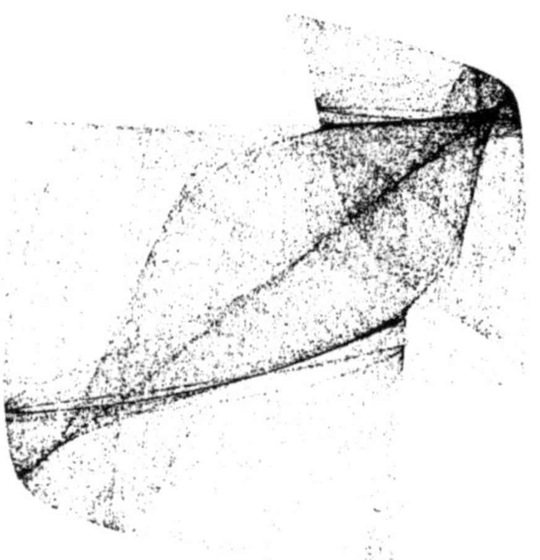

FIG. 15. The map $x'=(A-x-B_1y)x$, $y'=(A-B_2x-y)y$ with $A=3.7$, $B_1=0.1$, $B_2=0.15$ is not invertible. Shown are 50 000 iterates. The map was described in Ushiki *et al.* (1980).

F. Axiom-*A* dynamical systems*

We discuss here some concepts of *hyperbolicity* which will be referred to in Sec. IV. The hurried reader may skip this discussion without too much disadvantage. In this section, M will always be a compact manifold of dimension m. We shall denote by T_xM the tangent space to M at x. If $f:M\to M$ is a differentiable map, we shall denote by $T_xf:T_xM\to T_{fx}M$ the corresponding tangent map. (We refer the reader to standard texts on differential geometry for the definitions.) If a Riemann metric is given on M, the vector spaces T_xM acquire norms $\|\ \|_x$.

1. Diffeomorphisms

Let $f:M\to M$ be a diffeomorphism, i.e., a differentiable map with differentiable inverse f^{-1}.

We say that a point a of M is *wandering* if there is an open set B containing a [say a ball $B_a(\varepsilon)$] such that $B\cap f^kB=\emptyset$ for all $k>0$ (or we might equivalently require this only for all k large enough). The set of points that are not wandering is the *nonwandering set* Ω. It is a closed, f-invariant subset of M.

Let Λ be a closed f-invariant subset of M, and assume that we have linear subspaces E_x^-, E_x^+ of T_xM for each $x\in\Lambda$, depending continuously on x and such that

$$T_xM=E_x^+ + E_x^-,\quad \dim E_x^+ + \dim E_x^- = m .$$

Assume also that $T_xfE_x^-=E_{fx}^-$ and $T_xfE_x^+=E_{fx}^+$ (i.e., E^-,E^+ form a continuous invariant splitting of TM over Λ). One says that Λ is a *hyperbolic set* if one may choose E^- and E^+ as above, and constants $C>0$, $\Theta>1$ such that, for all $n\geq 0$,

$$\|T_xf^n u\|_{f^n x}\leq C\Theta^{-n}\|u\|_x \quad \text{if } u\in E_x^- ,$$

$$\|T_xf^{-n} v\|_{f^{-n} x}\leq C\Theta^{-n}\|v\|_x \quad \text{if } v\in E_x^+ .$$

[Note that, as a consequence, no ergodic measure with support in Λ has characteristic exponents in the interval $(-\Theta^{-1},\Theta^{-1})$.]

If the whole manifold M is hyperbolic, f is called an *Anosov diffeomorphism.* [Arnold's cat map, Sec. II.D, example (c), is an Anosov diffeomorphism.]

If the nonwandering set Ω is hyperbolic, and if the periodic points are dense in Ω, f is called an *Axiom-A diffeomorphism.* (Every Anosov diffeomorphism is an Axiom-A diffeomorphism.)

2. Flows

Consider a continuous-time dynamical system (f^t) on M, where f^t is defined for all $t\in R$; (f^t) is then also called a *flow.*

We say that a point a of M is *wandering* if there is an open set B containing a [say a ball $B_a(\varepsilon)$] such that $B\cap f^tB=\emptyset$ for all sufficiently large t. The set of points that are not wandering is the *nonwandering set* Ω. It is a closed, (f^t)-invariant subset of M.

Let Λ be a closed invariant subset of M containing no fixed point. Assume that we have linear subspaces E_x^-, E_x^0, E_x^+ of T_xM for each $x\in\Lambda$, depending continuously on x, and such that

$$T_xM = E_x^- + E_x^0 + E_x^+ ,$$

$$\dim E_x^0 = 1, \quad \dim E_x^- + \dim E_x^+ = m-1 .$$

Assume also that E_x^0 is spanned by

$$\left.\frac{d}{dt} f^t x\right|_{t=0} ,$$

i.e., E_x^0 is in the direction of the flow, and that

$$T_x f^t E_x^- = E_{f^t x}^- ,$$

$$T_x f^t E_x^+ = E_{f^t x}^+ .$$

One says that Λ is a *hyperbolic set* if one may choose E^-, E^0, and E^+ as above, and constants $C>0$, $\Theta>1$ such that, for all $t\geq 0$

$$\|T_x f^t u\|_{f^t x} \leq C\Theta^{-t}\|u\|_x \quad \text{if } u\in E_x^- ,$$

$$\|T_x f^{-t} v\|_{f^{-t}x} \leq C\Theta^{-t}\|v\|_x \quad \text{if } v\in E_x^+ .$$

More generally, we shall also say that Λ^* is a *hyperbolic set* if Λ^* is the union of Λ as above and of a finite number of hyperbolic fixed points [Sec. III.E, example (a)].

If the whole manifold M is a hyperbolic, (f^t) is called an *Anosov flow.*

If the nonwandering set Ω is hyperbolic, and if the periodic orbits and fixed points are dense in the Ω, then (f^t) is called an *Axiom-A flow.*

3. Properties of Axiom-*A* dynamical systems

Axiom-*A* dynamical systems were introduced by Smale [for reviews, see Smale's original paper (1967) and Bowen (1978)]. Smale proved the following "*spectral theorem*" valid both for diffeomorphism and flows.

Theorem. Ω is the union of finitely many disjoint closed invariant sets $\Omega_1,\ldots,\Omega_s$, and for each Ω_i there is $x\in\Omega_i$ such that the orbit $\{f^t x\}$ is dense in Ω_i. The decomposition $\Omega=\Omega_i\cup\cdots\cup\Omega_s$ is unique with these properties.

The sets Ω_i are called *basic sets,* while those which are attracting sets are called *attractors* (there is always at least one attractor among the basic sets).

Some of the ergodic properties of Axiom-*A* attractors will be discussed in Sec. IV. The great virtue of these systems is that they can be analyzed mathematically in detail, while many properties of a map apparently as simple as the Hénon diffeomorphism [Sec. II.D, example (a)] remain conjectural.

It should be pointed out that there is a vast literature on the Axiom-*A* systems, concerned in particular with *structural stability.*

G. Pesin theory*

We have seen above that the stable and unstable manifolds (defined almost everywhere with respect to an ergodic measure ρ) are differentiable. This is part of a theory developed by Pesin (1976,1977).[2] Pesin assumes that ρ has differentiable density with respect to Lebesgue measure, but this assumption is not necessary for the study of stable and unstable manifolds [see Ruelle (1979), and for the infinite-dimensional case Ruelle (1982a) and Mañé (1983)].

The earlier results on differentiable dynamical systems had been mostly geometric and restricted to *hyperbolic* (Anosov, 1967) or *Axiom-A* systems (Smale, 1967). Pesin's theory extends a good part of these geometric results to arbitrary differentiable dynamical systems, but working now *almost everywhere* with respect to some ergodic measure ρ. (The results are most complete when all characteristic exponents are different from zero.) The original contribution of Pesin has been extended by many workers, notably Katok (1980) and Ledrappier and Young (1984). Many of the results quoted in Sec. IV below depend on Pesin theory, and we shall give an idea of the present aspect of the theory in that section. Here we mention only one of Pesin's original contributions, a striking result concerning area-preserving diffeomorphisms (in two dimensions).

Theorem (Pesin). Let f be an area-preserving diffeomorphism, and f be twice differentiable. Suppose $fS=S$ for some bounded region S, and let S' consist of the points of S which have nonzero characteristic exponents. Then (up to a set of measure 0) S' is a countable union of ergodic components.

In this theorem the area defines an invariant measure on S, which is not ergodic in general, and S can therefore be decomposed into further invariant sets. This may be a continuous decomposition (like that of a disk into circles). The theorem states that *where the characteristic exponents are nonzero, the decomposition is discrete.*

IV. ENTROPY AND INFORMATION DIMENSION

In this section we introduce two more ergodic quantities: the *entropy* (or *Kolmogorov-Sinai invariant*) and the *information dimension.* We discuss how these quantities are related to the characteristic exponents. The measurement of the entropy and information dimension in physical and computer experiments will be discussed in Sec. V.

A. Entropy

As we have noted already in the Introduction, a system with sensitive dependence on initial conditions produces

---

[2]For a systematic exposition see Fathi, Herman, and Yoccoz (1983).

information. This is because two initial conditions that are different but indistinguishable at a certain experimental precision will evolve into distinguishable states after a finite time. If ρ is an ergodic probability measure for a dynamical system, we introduce the concept of *mean rate of creation of information* $h(\rho)$, also known as *measure-theoretic entropy* or the *Kolmogorov-Sinai invariant* or simply *entropy*. When we study the dynamics of a dissipative physicochemical system, it should be noted that the Kolmogorov-Sinai entropy is not the same thing as the thermodynamic entropy of the system. To define $h(\rho)$ we shall assume that the support of ρ is a compact set with a given metric. (More general cases can be dealt with, but in our applications supp ρ is indeed a compact metric space.) Let $\mathscr{A}=(\mathscr{A}_1,\ldots,\mathscr{A}_\alpha)$ be a finite (ρ-measurable) partition of the support of ρ. For every piece $\mathscr{A}_j$ we write $f^{-k}\mathscr{A}_j$ for the set of points mapped by f^k to $\mathscr{A}_j$. We then denote by $f^{-k}\mathscr{A}$ the partition $(f^{-k}\mathscr{A}_1,\ldots,f^{-k}\mathscr{A}_\alpha)$. Finally, $\mathscr{A}^{(n)}$ is defined as

$$\mathscr{A}^{(n)}=\mathscr{A}\vee f^{-1}\mathscr{A}\vee\cdots\vee f^{-n+1}\mathscr{A},$$

which is the partition whose pieces are

$$\mathscr{A}_{i_1}\cap f^{-1}\mathscr{A}_{i_2}\cap\cdots\cap f^{-n+1}\mathscr{A}_{i_n}$$

with $i_j\in\{1,2,\ldots,\alpha\}$. What is the significance of these partitions? The partition $f^{-k}\mathscr{A}$ is deduced from $\mathscr{A}$ by time evolution (note that $f^k\mathscr{A}$ need not be a partition, since f might be many-to-one; this is why we use $f^{-k}\mathscr{A}$). The partition $\mathscr{A}^{(n)}$ is the partition generated by $\mathscr{A}$ in a time interval of length n. We write

$$H(\mathscr{A})=-\sum_{i=1}^{\alpha}\rho(\mathscr{A}_i)\log\rho(\mathscr{A}_i),\tag{4.1}$$

with the understanding that $u\log u=0$ when $u=0$. (We strongly advise using natural logarithms, but $\log_{10}$ and $\log_2$ have their enthusiasts.) Thus $H(\mathscr{A})$ is the information content of the partition $\mathscr{A}$ with respect to the state ρ, and $H(\mathscr{A}^{(n)})$ is the same, over an interval of time of length n. The following limits are asserted to exist, defining $h(\rho,\mathscr{A})$ and $h(\rho)$:

$$h(\rho,\mathscr{A})=\lim_{n\to\infty}[H(\mathscr{A}^{(n+1)})-H(\mathscr{A}^{(n)})]$$

$$=\lim_{n\to\infty}\frac{1}{n}H(\mathscr{A}^{(n)}),\tag{4.2}$$

$$h(\rho)=\lim_{\operatorname{diam}\mathscr{A}\to 0}h(\rho,\mathscr{A}),\tag{4.3}$$

where $\operatorname{diam}\mathscr{A}=\max_i\{\text{diameter of }\mathscr{A}_i\}$. Clearly, $h(\rho,\mathscr{A})$ is the rate of information creation with respect to the partition $\mathscr{A}$, and $h(\rho)$ its limit for finer and finer partitions. This last limit may sometimes be avoided [i.e., $h(\rho,\mathscr{A})=h(\rho)$]; this is the case when $\mathscr{A}$ is a *generating partition*. This holds in particular if $\operatorname{diam}\mathscr{A}^{(n)}\to 0$ when $n\to\infty$, or if f is invertible and $\operatorname{diam}f^n\mathscr{A}^{(2n)}\to 0$ when $n\to\infty$. For example, for the map of Fig. 8, a generating partition is obtained by dividing the interval at the singularity in the middle. For more details we must refer the reader to the literature, for instance the excellent book by Billingsley (1965).

The above definition of the entropy applies to continuous as well as discrete-time systems. In fact, the entropy in the continuous-time case is just the entropy $h(\rho,f^1)$ corresponding to the time-one map. We also have the formula

$$h(\rho,f^T)=|T|\,h(\rho,f^1).$$

Note that the definition of the entropy in the continuous-time case *does not involve a time step t tending to zero*, contrary to what is sometimes found in the literature. Note also that the entropy does not change if f is replaced by f^{-1}.

If (f^t) has a Poincaré section Σ, we let σ be the probability measure on Σ, invariant under the Poincaré map P and corresponding to ρ (i.e., σ is the density of intersection of orbits of the continuous dynamical system with Σ). If we also let τ be the first return time, then we have *Abramov's formula*,

$$h(\rho)=\frac{h(\sigma)}{\langle\tau\rangle_\sigma},$$

which is analogous to Eq. (3.12) for the characteristic exponents.

The relationship of entropy to characteristic exponents is very interesting. First we have a general inequality.

Theorem (Ruelle, 1978). Let f be a differentiable map of a finite-dimensional manifold and ρ an ergodic measure with compact support. Then

$$h(\rho)\le\Sigma\text{ positive }\lambda_i.\tag{4.4}$$

The result is believed to hold in infinite dimensions as well, but no proof has been published yet.

It is of considerable interest that the *equality* corresponding to Eq. (4.4) seems to hold often (but not always) for the *physical measures* (Sec. II.F) in which we are mainly interested. This equality is called the *Pesin identity:*

$$h(\rho)=\Sigma\text{ positive }\lambda_i.$$

Pesin proved that it holds if ρ is invariant under the diffeomorphism f, and ρ has smooth density with respect to Lebesgue measure. More generally, the Pesin identity holds for the *SRB measures* to be studied in Sec. IV.B.

In Sec. V we shall use in addition an entropy concept different from that of Eqs. (4.1)–(4.3). It is given by

$$H_2(\mathscr{A})=-\log\sum_{i=1}^{\alpha}\rho(\mathscr{A}_i)^2,$$
$$K_2(\rho)=\lim_{\operatorname{diam}\mathscr{A}\to 0}\lim_{n\to\infty}\frac{1}{n}H_2(\mathscr{A}^{(n)}),\tag{4.5}$$

if these limits exist (see Grassberger and Procaccia, 1983a). It can be shown that *the K_2 entropy is a lower bound to the entropy $h(\rho)$*:

$$K_2(\rho)\le h(\rho).\tag{4.6}$$

B. SRB measures

We have seen in Sec. III.E that attracting sets are unions of unstable manifolds. Transversally to these, one often finds a discontinuous structure corresponding to the complicated piling up of the unstable manifolds upon themselves. This suggests that invariant measures may have very rough densities in the directions transversal to the foliations of the unstable manifolds. On the other hand, we may expect that—due to stretching in the unstable direction—the measure is smooth when viewed along these directions. We shall call SRB measures (for Sinai, Ruelle, Bowen) those measures that are smooth along unstable directions. They turn out to be a natural and useful tool in the study of physical dynamical systems.

Much of this section is concerned with consequences of the existence of SRB measures. These are mostly relations between entropy, dimensions, and characteristic exponents. To prove the existence of SRB measures for a given system is a hard task, and whether they exist is not known in general. Sometimes no SRB measures exist, but it is unclear how frequently this happens. On the other hand, we do not have much of physical relevance to say about systems without SRB measures.

To repeat, we should like to define, intuitively, SRB measures as measures with smooth density in the stretching, or *unstable*, directions of the dynamical system defined by f. The geometric complexities described above make a rather technical definition necessary. Before going into these technicalities, we discuss the framework in which we shall work.

(a) In the ergodic theory of differentiable dynamical systems, there is no essential difference between discrete-time and continuous-time systems. In fact, if we discretize a continuous-time dynamical system by restricting t to integer values (i.e., use the time-one map $f=f^1$ as a generator), then the characteristic exponents, the stable and unstable manifolds, and the entropy are unchanged. (The information dimension to be defined in Sec. IV.C also remains the same.) We may thus, for simplicity, *consider only discrete-time systems.*

(b) If f is a diffeomorphism (i.e., a differentiable map with differential inverse), then our dynamical system is defined for negative as well as positive times. If, in addition, f is twice differentiable, then the inverse map is also twice differentiable. We shall assume a little less, namely, that f is *twice differentiable* and either a *diffeomorphism* or at least such that f and Df are *injective* (i.e., $fx=fy$ implies $x=y$, and $D_xfu=D_xfv$ implies $u=v$; these conditions hold for the Navier-Stokes time evolution).

Given an ergodic measure ρ (with compact support as usual), unstable manifolds V_x^u are defined for almost all x according to Eq. (3.13). Notice that $y\in V_x^u$ is the same thing as $x\in V_y^u$, so that the unstable manifolds V^u partition the space into equivalence classes. It might seem natural to define SRB measures by using this partition for a decomposition of ρ into pieces ρ_α, carried by different unstable manifolds:

$$\rho=\int \rho_\alpha m(d\alpha)\,, \tag{4.7}$$

where α parametrizes the V^u's, and m is a measure on the "space of equivalence classes." In reality, this space of equivalence classes does not exist in general (as a measurable space) because of the folding and accumulation of the global unstable manifolds [and the existence of a nontrivial decomposition (4.7) would contradict ergodicity].

The correct approach is as follows. Let S be a ρ-measurable set of the form $S=\bigcup_{\alpha\in A}S_\alpha$, where the S_α are disjoint small open pieces of the V^u's (say each S_α is contained in a *local* unstable manifold). If this decomposition is ρ measurable, then one has

$$\rho \text{ restricted to } S=\int \rho_\alpha m(d\alpha)\,,$$

where m is a measure on A, and ρ_α is a probability measure on S_α called the *conditional probability measure* associated with the decomposition $S=\bigcup_{\alpha\in A}S_\alpha$. The ρ_α are defined m-almost everywhere. See Fig. 16. The situation of interest for the definition of SRB measures occurs when *the conditional probabilities ρ_α are absolutely continuous with respect to Lebesgue measure* on the V^u's. This means that

$$\rho_\alpha(d\xi)=\varphi_\alpha(\xi)d\xi \quad \text{on } S_\alpha\,, \tag{4.8}$$

where $d\xi$ denotes the volume element when S_α is smoothly parametrized by a piece of $\mathbf{R}^{m_+}$ and φ_α is an integrable function. The *unstable dimension* m_+ of S_α or V^u is the sum of the multiplicities of the positive characteristic exponents. It is finite even for the case of the Navier-Stokes equation discussed earlier (because $\lambda_i\to-\infty$ when $i\to\infty$, as we have noted).

We say that the ergodic measure ρ is an *SRB measure* if its conditional probabilities ρ_α are absolutely continuous with respect to Lebesgue measure for some choice of S with $\rho(S)>0$, and a decomposition $S=\bigcup_\alpha S_\alpha$ as above. The definition is independent of the choice of S and its decomposition (this is an easy exercise in ergodic theory). We shall also say that ρ *is absolutely continuous along unstable manifolds.*

Theorem (Ledrappier and Young, 1984). Let f be a twice differentiable diffeomorphism of an m-dimensional manifold M and ρ an ergodic measure with compact support. The following conditions are then equivalent: (a) The measure ρ is an SRB measure, i.e., ρ is absolutely

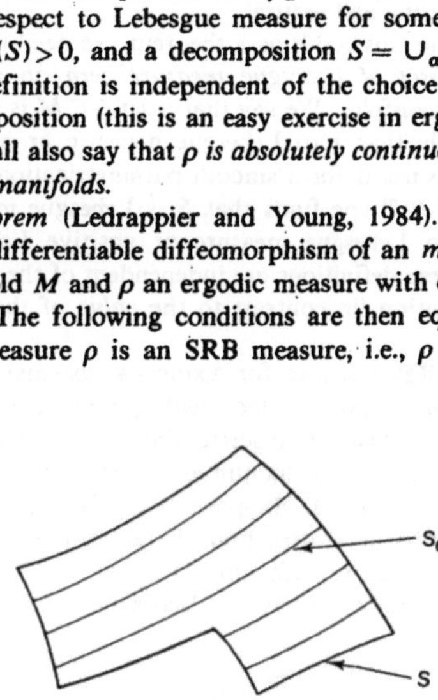

FIG. 16. A decomposition of the set S into smooth leaves S_α, each of which is contained in the unstable manifold.

continuous along unstable manifolds. (b) The measure ρ satisfies Pesin's identity,

$$h(\rho)=\Sigma \text{ positive characteristic exponents} .$$

Furthermore, if these conditions are satisfied, the density functions φ_α in Eq. (4.8) are differentiable.

The theorem says that if ρ is absolutely continuous along unstable manifolds, then the rate of creation of information is the mean rate of expansion of m_+-dimensional volume elements. If, however, ρ is singular along unstable manifolds, then this rate is strictly less than the rate of expansion. These assertions are intuitively quite reasonable, but in fact quite hard to prove. The first proofs have been given for Axiom-A systems (see Sec. III.F) by Sinai (1972; Anosov systems), Ruelle (1976; Axiom-A diffeomorphisms), and Bowen and Ruelle (1975; Axiom-A flows). The general importance of (a) and (b) was stressed by Ruelle (1980).

One hopes that there is an infinite-dimensional extension applying to Navier-Stokes, but such an extension has not yet been proved. Ledrappier (1981b) has obtained a version of the above theorem that is valid for noninvertible maps in one dimension.

The SRB measures are of particular interest for physics because one can show—in a number of cases—that the ergodic averages

$$\frac{1}{n}\sum_{k=0}^{n-1}\delta_{f^k x}$$

tend to the SRB measure ρ when $n\to\infty$, not just for ρ-almost all x, but for x in a set of positive *Lebesgue* measure. Lebesgue measure corresponds to a more *natural notion of sampling* than the measure ρ (which is carried by an attractor and usually singular). The above property is thus both strong and natural.

To formulate this result as a theorem, we need the notion of a *subset of Lebesgue measure zero* on an m-dimensional manifold. We say that a set $S\subset M$ is Lebesgue measurable (has zero Lebesgue measure or positive Lebesgue measure) if for a smooth parametrization of M by patches of $\mathbf{R}^m$ one finds that S is Lebesgue measurable (has zero Lebesgue measure or positive Lebesgue measure). These definitions are independent of the choice of parametrization (in contrast to the *value* of the measure).

Theorem (SRB measures for Axiom-A systems). Consider a dynamical system determined by a twice differentiable diffeomorphism f (discrete time) or a twice differentiable vector field (continuous time) on an m-dimensional manifold M. Suppose that A is an Axiom-A attractor, with basin of attraction U. (a) There is one and only one SRB measure with support in A. (b) There is a set $S\subset U$ such that $U\setminus S$ has zero Lebesgue measure, and

$$\lim_{n\to\infty}\frac{1}{n}\sum_{k=0}^{n-1}\delta_{f^k x}=\rho \quad (\text{discrete time}) ,$$

or

$$\lim_{T\to\infty}\frac{1}{T}\int_0^T dt\,\delta_{f^t x}=\rho \quad (\text{continuous time}) ,$$

whenever $x\in S$.

For a proof, see Sinai (1972; Anosov systems), Ruelle (1976; Axiom-A diffeomorphisms), or Bowen and Ruelle (1975; Axiom-A flows). The "geometric Lorenz attractor" can be treated similarly.

The following theorem shows that the requirement of Axiom A can be replaced by weaker information about the characteristic exponents.

Theorem (Pugh and Shub, 1984). Let f be a twice differentiable diffeomorphism of an m-dimensional manifold M and ρ an SRB measure such that all characteristic exponents are different from zero. Then there is a set $S\subset M$ with positive Lebesgue measure such that

$$\lim_{n\to\infty}\frac{1}{n}\sum_{k=0}^{n-1}\delta_{f^k x}=\rho \tag{4.9}$$

for all $x\in S$.

This theorem is in the spirit of the "absolute continuity" results of Pesin. An infinite-dimensional generalization has been promised by Brin and Nitecki (1985). The theorem fails if 0 is a characteristic exponent, as the following example shows.

Counterexample. A dynamical system is defined by the differential equation

$$\frac{dx}{dt}=x^3$$

on $\mathbf{R}$. Its time-one map has δ_0 as an ergodic measure, with $\lambda_1=0$. However, 0 is (weakly) repelling, so that Eq. (4.9) cannot hold for $x\neq 0$. In fact, if $x\neq 0$, $f^t x$ goes to infinity in a finite time.

We give now an example showing that there is not always an SRB measure lying around, and that there are physical measures that are not SRB.

Counterexample (Bowen, and also Katok, 1980). Consider a continuous-time dynamical system (flow) in $\mathbf{R}^2$ with three fixed points A,B,C where A,C are repelling and B of saddle type, as shown in Fig. 17. The system has an invariant curve in the shape of a "figure 8" (or rather, figure ∞), which is attracting. It can be seen that any point different from A or C yields an ergodic average corresponding to a Dirac δ at B. Therefore δ_B is the physical measure for our system. Clearly it has zero entropy, one strictly positive characteristic exponent and the other strictly negative (and thus, in particular, not zero),

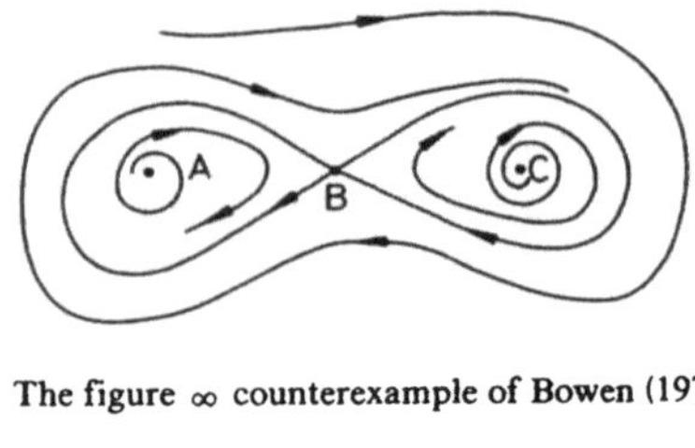

FIG. 17. The figure ∞ counterexample of Bowen (1975).

and is not absolutely continuous with respect to Lebesgue measure on the unstable manifold. (Note that the unstable manifold at B consists of the "figure 8.") On the other hand, it is not hard to see that δ_B is a Kolmogorov measure; i.e., a system perturbed with a little noise ε will spend most of its time near B, and as $\varepsilon \to 0$ the fraction of time spent near B goes to 1.

C. Information dimension

Given a probability measure ρ, we know that its *information dimension* $\dim_H \rho$ is the smallest Hausdorff dimension of a set S of ρ measure 1. Note that the set S is not closed in general, and therefore the Hausdorff dimension $\dim_H(\text{supp}\rho)$ of the support of ρ may be strictly larger than $\dim_H \rho$.

Example. The rational numbers of the interval [0,1], i.e., the fractions p/q with p,q integers, form a countable set. This means that they can be ordered in a sequence $(a_n)_1^\infty$. Consider the probability measure

$$\rho=\frac{1}{e}\sum_{n=1}^{\infty}\frac{1}{n!}\delta_{a_n} ,$$

where δ_x is the δ measure at x. Then ρ is carried by the set S of rational numbers of [0,1], and since this is a countable set we have $\dim_H \rho = 0$. On the other hand, $\text{supp}\rho=[0,1]$, so that $\dim_H(\text{supp}\rho)=1$.

It turns out that the information dimension of a physical measure ρ is a more interesting quantity than the Hausdorff dimension of the attractor or attracting set A which carries ρ. This is both because $\dim_H \rho$ is more accessible experimentally and because it has simple mathematical relations with the characteristic exponents. In any case, we have $\text{supp}\rho \subset A$ and therefore

$$\dim_H \rho \le \dim_H(\text{supp}\rho) \le \dim_H A .$$

The next theorem shows that the information dimension is naturally related to the measure of small balls in phase space.

Theorem (Young, 1982). Let ρ be a probability measure on a finite-dimensional manifold M. Assume that

$$\lim_{r\to 0}\frac{\log\rho[B_x(r)]}{\log r}=\alpha \tag{4.10}$$

for ρ-almost all x. Then $\dim_H \rho = \alpha$.

Young shows that α is also equal to several other "fractal dimensions" (in particular, the "Rényi dimension").

We are of course mostly interested in the case when ρ is ergodic for a differentiable dynamical system. In that situation, the requirement that

$$\lim_{r\to 0}\frac{\log\rho[B_x(r)]}{\log r}$$

exists ρ-almost everywhere already implies that the limit is almost everywhere constant, and therefore equal to $\dim_H \rho$. [The above limit does not always exist, as Ledrappier and Misiurewicz (1984) have shown for certain maps of the interval.]

An interesting relation between $\dim_H \rho$ and the characteristic exponents λ_i has been conjectured by Yorke and others (see references below). We denote by

$$c_\rho(k)=\sum_{i=1}^{k}\lambda_i$$

the sum of the k largest characteristic exponents, and extend this definition by linearity between integers (see Fig. 18):

$$c_\rho(s)=\sum_{i=1}^{k}\lambda_i+(s-k)\lambda_{k+1} \quad \text{if } k\le s<k+1 .$$

The function c_ρ is defined on the interval $[0,+\infty)$ for a dynamical system on a Hilbert space, and on the interval $[0,m]$ for a system on $\mathbf{R}^m$ or an m-dimensional manifold. In the latter cases we write $c_\rho(s)=-\infty$ for $s>m$, so that c_ρ is now in all cases a *concave* function on $[0,+\infty)$, as in Fig. 18. Notice that $c_\rho(0)=0$, that the maximum of $c_\rho(s)$ is the sum of the positive characteristic exponents, and that $c_\rho(s)$ becomes negative for sufficiently large s. (This is because, in the Hilbert case, the λ_i tend to $-\infty$.)

The *Liapunov dimension* of ρ is now defined as

$$\dim_\Lambda \rho=\max\{s: c_\rho(s)\ge 0\} .$$

Notice that when $c_\rho(k)\ge 0$ and $c_\rho(k+1)<0$ we have

$$\dim_\Lambda \rho = k+\frac{c_\rho(k)}{|\lambda_{k+1}|} .$$

The $(k+1)$-volume elements are thus contracted by time evolution, and this suggests that the dimension of ρ must be less than $k+1$, a result made rigorous by Ilyashenko (1983). Yorke and collaborators have gone further and made the following guess.

Conjecture (Kaplan and Yorke, 1979; Frederickson, Kaplan, Yorke, and Yorke, 1983; Alexander and Yorke, 1984). If ρ is an SRB measure, then generically

$$\dim_H \rho = \dim_\Lambda \rho . \tag{4.11}$$

The SRB measures have been defined in Sec. IV.B. and "genericity" means here "in general." What concept of

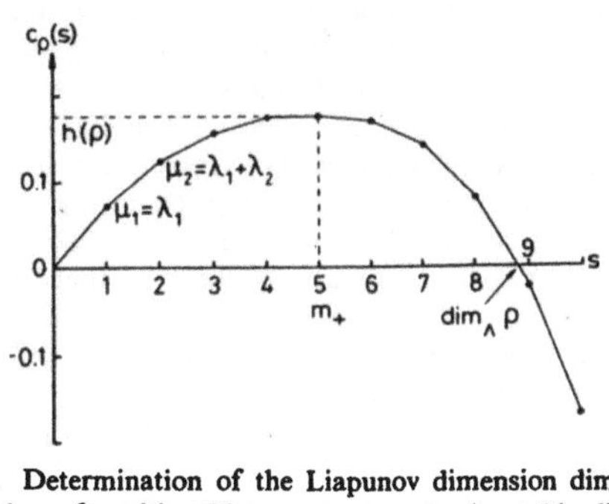

FIG. 18. Determination of the Liapunov dimension $\dim_\Lambda(\rho)$. The number of positive Liapunov exponents (unstable dimension) is m_+. The graph is from Manneville (1985), for the Kuramoto-Sivashinsky model.

genericity is adequate is a very difficult question; we do not know—among other things—how frequently a dynamical system has an SRB measure. An *inequality* is, however, available in full generality.

Theorem. Let f be a twice continuously differentiable map and let ρ be an ergodic measure with compact support. Then

$$\dim_H\rho \le \dim_\Lambda\rho \ . \tag{4.12}$$

The basic result was proved by Douady and Oesterlé (1980), and from this Ledrappier (1981a) derived the theorem as stated. (It holds for Hilbert spaces as well as in finite dimensions.)

An *equality* is also known in special cases, notably the following.

Theorem (Young, 1982). Let f be a twice differentiable diffeomorphism of a two-dimensional manifold, and let ρ be an ergodic measure with compact support. Then the limit (4.10) exists ρ-almost everywhere, and we have

$$\dim_H\rho = h(\rho)\left[\frac{1}{\lambda_1}+\frac{1}{|\lambda_2|}\right] , \tag{4.13}$$

where $\lambda_1>0$ and $\lambda_2<0$ are the characteristic exponents of ρ.

Note, incidentally, that if f is replaced by f^{-1}, the characteristic exponents change sign, the entropy remains the same, and the formula remains correct, as it should. Note also that the cases where λ_1 and λ_2 are not of opposite sign are relatively trivial. From the inequality (4.4) applied for f or f^{-1} we see that $\lambda_1=0$ or $\lambda_2=0$ implies $h(\rho)=0$, so that the right-hand side of Eq. (4.13) becomes indeterminate. If $\lambda_1\ge\lambda_2>0$ or $0>\lambda_1\ge\lambda_2$, a theorem of Sec. III.C.2 applied to f or f^{-1} shows that ρ is carried by a periodic orbit, so that Eq. (4.13) holds with $\dim_H\rho=h(\rho)=0$.

Variants of the above theorem that do not assume the invertibility of f are known (for one dimension see Ledrappier, 1981b, Proposition 4; for holomorphic functions see Manning, 1984).

If ρ is an SRB measure, then Eq. (4.13) becomes $\dim_H\rho=1+\lambda_1/|\lambda_2|$, which is just the conjecture (4.11). The next example shows that the conjecture does not always hold.

Counterexample. Notice first that if a measure ρ has no positive characteristic exponent, then $h(\rho)=0$ by a theorem in Sec. IV.A, and therefore ρ is an SRB measure. If $\dim_H\rho$ is strictly between 0 and 1, then Eq. (4.11) cannot hold (because $\dim_\Lambda\rho$ can only have the value 0 or a value ≥ 1). In particular, the Feigenbaum attractor [example (b) of Sec. II.D] carries a unique probability measure ρ with $\lambda_1=0$ and $\dim_H\rho=0.538\ldots$ so that Eq. (4.11) is violated here. [For the Hausdorff dimension of the Feigenbaum measure see Grassberger (1981); Vul, Sinai, and Khanin (1984); Ledrappier and Misiurewicz (1984).]

Finally, let us mention *lower* bounds on $\dim_H\rho$.

Theorem. If ρ is an SRB measure, then $\dim_H\rho\ge m_+$, where m_+ is the sum of the multiplicities of the positive characteristic exponents (unstable dimension).

This follows readily from the definitions.

D. Partial dimensions

Given an ergodic measure ρ, we can associate with each characteristic exponent $\lambda^{(i)}$ a *partial dimension* $D^{(i)}$. Roughly speaking, $D^{(i)}$ is the Hausdorff dimension in the direction of $\lambda^{(i)}$. The entropy inequality (4.4) and the dimension inequality (4.12) will be natural consequences of the existence of the $D^{(i)}$.

In order to give a precise definition, we assume that f is a twice differentiable diffeomorphism of a compact manifold M (in the case of a continuous-time dynamical system we take for f the time-one map). If $\lambda^{(i)}$ is a positive characteristic exponent and $\lambda^{(i)}>\lambda>\max(0,\lambda^{(i+1)})$, we have defined in Sec. III.E the local *unstable manifolds* $V_x^u(\lambda,\varepsilon)$, which we simply denote here by $V_{\rm loc}^{(i)}$, and $S=\cup_{\alpha\in A}S_\alpha$, where the S_α are open pieces of the $V_{\rm loc}^{(i)}$. Suppose $S=\cup_{\alpha\in A}S_\alpha$, where the S_α are open pieces of the $V_{\rm loc}^{(i)}$, and define *conditional probability measures* $\rho_\alpha^{(i)}$ on S_α such that

$$\rho \text{ restricted to } S=\int \rho_\alpha^{(i)} m(d\alpha) ,$$

where m is some measure on A. This definition is a bit more general than that given in Sec. IV.B. There $\lambda^{(i)}$ was the smallest positive characteristic exponent. We define

$$\delta^{(i)}=\dim_H\rho_\alpha^{(i)}$$

(this is a constant almost everywhere) and write

$$D^{(1)}=\delta^{(1)} \quad \text{if } \lambda^{(1)}>0 ,$$

and

$$D^{(i)}=\delta^{(i)}-\delta^{(i-1)} \quad \text{if } i>1 \text{ and } \lambda^{(i)}>0 .$$

Similarly, if $\lambda^{(i)}<0$, we define conditional probabilities $\rho_\alpha^{(j)}$ on pieces of *stable manifolds* and let

$$\delta^{(j)}=\dim_H\rho_\alpha^{(j)} .$$

We then write

$$D^{(r)}=\delta^{(r)}$$

if the smallest characteristic exponent $\lambda^{(r)}$ is negative, and

$$D^{(j)}=\delta^{(j)}-\delta^{(j+1)} \quad \text{if } j<r,\ \lambda^{(j)}<0 .$$

The definition of $D^{(k)}$ for $\lambda^{(k)}=0$ is somewhat arbitrary [between 0 and the multiplicity $m^{(k)}$ of $\lambda^{(k)}=0$]; we take $D^{(k)}=m^{(k)}$.

Theorem. The partial dimensions $D^{(1)},\ldots,D^{(r)}$ satisfy

$$0\le D^{(i)}\le m^{(i)} \quad \text{for } i=1,\ldots,r ,$$

where $m^{(i)}$ is the multiplicity of $\lambda^{(i)}$. The entropy is given by

$$h(\rho)=\sum_i{}^{+}\lambda(i)D^{(i)}=-\sum_i{}^{-}\lambda^{(i)}D^{(i)} , \tag{4.14}$$

where $\sum^+$ ($\sum^-$) is the sum over positive (negative) characteristic exponents, in particular

$$\sum_i\lambda^{(i)}D^{(i)}=0 . \tag{4.15}$$

The Hausdorff dimension satisfies

$$\dim_H \rho \leq \sum_i D^{(i)} . \tag{4.16}$$

(It is not known if there is equality when no characteristic exponent vanishes.)

The proof of this theorem by Ledrappier and Young (1984) is not easy, but brings further dividends, in particular an interpretation of the numbers $|\lambda^{(i)}|D^{(i)}$ as *partial entropies.*

Some earlier theorems on entropy and Hausdorff dimension are recovered as corollaries of the above theorem, as we now indicate. [We follow Ledrappier and Young (1984); Grassberger (1984); Procaccia (1984).]

(a) First we recover from

$$h(\rho)=\sum_i{}^{+}\lambda^{(i)}D^{(i)}$$

the entropy inequality (4.4),

$$h(\rho)\leq \sum_i{}^{+}\lambda^{(i)}m^{(i)} .$$

(b) The above inequality is in fact an equality (i.e., *ρ is an SRB measure) if and only if $D^{(i)}=m^{(i)}$ for all positive $\lambda^{(i)}$.*

Next we check that we recover the dimension inequality (4.12):

$$\dim_H \rho \leq \dim_\Lambda \rho \tag{4.17}$$

from

$$\dim_H \rho \leq \sum_i D^{(i)} \leq \max\left\{ \sum_i d^{(i)} : 0\leq d^{(i)} \leq m^{(i)} \text{ and } \sum_i d^{(i)}\lambda^{(i)}=0 \right\} .$$

Proof. Let k be such that

$$\sum_1^k \lambda^{(i)} \geq 0 > \sum_i^{k+1} \lambda^{(i)} .$$

We then have $\lambda^{(k+1)}<0$ and

$$\begin{aligned}\dim_H \rho &\leq \sum_i D^{(i)} = \frac{-1}{\lambda^{(k+1)}}\sum_i (-\lambda^{(k+1)})D^{(i)}\\ &= \frac{-1}{\lambda^{(k+1)}}\sum_i (\lambda^{(i)}-\lambda^{(k+1)})D^{(i)} \leq \frac{-1}{\lambda^{(k+1)}}\sum_{i=1}^k (\lambda^{(i)}-\lambda^{(k+1)})D^{(i)}\\ &\leq \frac{-1}{\lambda^{(k+1)}}\sum_{i=1}^k (\lambda^{(i)}-\lambda^{(k+1)})m^{(i)} = \sum_{i=1}^k m^{(i)} + \frac{\sum_{i=1}^k \lambda^{(i)}m^{(i)}}{|\lambda^{(k+1)}|} .\end{aligned}$$

It is easily seen that the right-hand side is just the Liapunov dimension $\dim_\Lambda \rho$, and Eq. (4.17) follows.

(d) Suppose that we have equalities in the proof above, i.e., that the Kaplan-Yorke conjecture holds for ρ. Then we must have $D^{(i)}=m^{(i)}$ for $i=1,\ldots,k$ and $D^{(i)}=0$ for $i=k+2,\ldots,r$ (conversely these properties imply $\sum D^{(i)}=\dim_\Lambda \rho$). In particular, *if the Kaplan-Yorke conjecture holds for ρ then ρ is an SRB measure.*

Remarks.

(a) If $\sum \lambda^{(i)}m^{(i)}>0$ we have $\dim_\Lambda \rho=m$ (the dimension of the manifold), which provides a trivial bound for $\dim_H \rho$. However, if one replaces f by f^{-1}, changing the sign of the $\lambda^{(i)}$, one gets a new Liapunov dimension, which is $<m$ and provides a nontrivial bound on the dimension of ρ.

(b) If there are only two *distinct* characteristic exponents, then $D^{(1)}$ and $D^{(2)}$ can be computed from Eq. (4.14).

(c) Let ρ be an SRB measure with r characteristic exponents such that $\lambda^{(1)}>\cdots>\lambda^{(r-1)}>0>\lambda^{(r)}$ and $\sum_1^r \lambda^{(i)}m^{(i)}<0$. Then what we have said shows that $\sum D^{(i)}=\dim_\Lambda \rho$.

E. Escape from almost attractors

Before asymptotic behavior is reached by a dynamical system, transients of considerable duration are often observed experimentally. This is the case, for instance, for the Lorenz system [Sec. II.B, example (b)] as observed by Kaplan and Yorke (1979): for some values of the parameters *preturbulence* occurs in the form of long chaotic transients, even though the system does not yet have a strange attractor. One may say that the system has an *almost attractor* and try to estimate the escape rate from this set. More generally, one would like to have a precise description of *transient chaos* (see Grebogi, Ott, and Yorke, 1983b).

The situation, as usual, is best understood for the Axiom-*A* systems, where the *basic sets* (see Sec. III.F) are the natural candidates to describe almost attractors. Let Ω_i be a basic set, U a small neighborhood of Ω_i, and μ a measure with positive continuous density with respect to Lebesgue measure on U. Let

$$p(T)=\mu\left[\bigcap_{0\leq t\leq T} f^t U\right]$$

be the amount of mass that has not left U by time T. One finds that $p(T)\approx e^{Pt}$, where

$$P=\max\left\{h(\rho)-\sum \text{positive } \lambda_i(\rho): \rho \text{ ergodic with support in } \Omega_i\right\}. \quad (4.18)$$

Note that P vanishes, as it should, if Ω_i is an attractor; the maximum is given in that case by the SRB measure. If Ω_i is not an attractor, then $P<0$; there is again a unique measure ρ_i realizing the maximum of Eq. (4.18), but it is no longer SRB (see Bowen and Ruelle, 1975).

If our dynamical system is not necessarily Axiom A, the following is a natural guess.

Conjecture. Write

$$P=h(\rho)-\sum \text{positive } \lambda_i(\rho) . \quad (4.19)$$

Then $|P|$ is the rate of escape from the support K of ρ, provided

$$P\geq h(\sigma)-\sum \text{positive } \lambda_i(\sigma)$$

for all ergodic σ with support in K. If $P>h(\sigma)-\sum$ positive $\lambda_i(\sigma)$ when $\sigma\neq\rho$ then ρ describes the time averages over transients near K.

A heuristic argument following the Axiom-A case makes this plausible, but it is unknown how generally the conjecture holds. Some satisfactory experimental verifications have been given by Kantz and Grassberger (1984). They write Eq. (4.19) as follows in terms of the *partial dimensions* $D^{(i)}$ discussed in Sec. IV.D:

$$|P|=-P=\sum_{i:\lambda^{(i)}>0} \lambda^{(i)}(m^{(i)}-D^{(i)}) .$$

F. Topological entropy*

The measure-theoretic entropy of Sec. IV.A gave the rate of information creation with respect to an ergodic measure. A related concept, involving the *topology* rather than a measure, will be discussed here.

Let K be a compact set and $f:K\to K$ a continuous map. If $\mathscr{A}=(\mathscr{A}_1,\ldots,\mathscr{A}_a)$ is a finite open cover of K (i.e., $\cup_i \mathscr{A}_i\supset K$), we write

$$f^{-k}\mathscr{A}=(f^{-k}\mathscr{A}_1,\ldots,f^{-k}\mathscr{A}_k) ,$$

$$\mathscr{A}^{(n)}=\mathscr{A}\vee f^{-1}\mathscr{A}\vee\cdots\vee f^{-n+1}\mathscr{A}$$

$$=(\mathscr{A}_{i_1}\cap f^{-1}\mathscr{A}_{i_2}\cap\cdots\cap f^{-n+1}\mathscr{A}_{i_n}) .$$

Now let $N(\mathscr{A},n)$ be the smallest number of sets in $\mathscr{A}^{(n)}$ that still covers K. The following limit is asserted to exist:

$$h_{\text{top}}(K,\mathscr{A})=\lim_{n\to\infty}\frac{1}{n}\log N(\mathscr{A},n) ,$$

and one defines the *topological entropy* of K by

$$h_{\text{top}}(K)=\sup_{\mathscr{A}} h_{\text{top}}(K,\mathscr{A}) .$$

If we have a metric on K we may write more conveniently

$$h_{\text{top}}(K)=\lim_{\text{diam}\mathscr{A}\to 0} h_{\text{top}}(K,\mathscr{A}) .$$

The following important theorem relates the topological entropy and the measure-theoretic entropies.

Theorem. If K is compact and $f:K\to L$ continuous, then $h_{\text{top}}(K)=\sup\{h(\rho):\rho$ is an ergodic measure with respect to $f\}$.

[This was conjectured by Adler, Konheim, and McAndrew (1965), and proved by Goodwyn, Dinaburg, and Goodman.] For references and more details on topological entropy we must refer the reader to Walters (1975) and Denker, Grillenberger, and Sigmund (1976).

G. Dimension of attractors*

The estimates of $\dim_H\rho$ in Sec. IV.C can be completed by estimates of the dimension of compact invariant sets (like the support of ρ, or attractors and attracting sets).

Theorem. Let A be a compact invariant set for a differentiable map f. Then

$$\dim_H A\leq\sup\{\dim_\Lambda\rho: \rho \text{ is ergodic with support in } A\} . \quad (4.20)$$

This result is due to Ledrappier (1981a), based on Douady and Oesterlé (1980); it is not known whether one can write $\dim_K A$ instead of $\dim_H A$ in Eq. (4.20). Note that, contrary to what Eq. (4.20) might suggest, there are cases where $\dim_H A>\sup\{\dim_H\rho:\rho$ is ergodic with support in $A\}$ (see McCluskey and Manning, 1983).

Lower bounds on $\dim_K A$ are also known. For instance, if a dynamical system has an attracting set A and a fixed point P with unstable dimension $m_+(P)$, then $\dim_K A\geq m_+(P)$ (see the corollary in Sec. III.E). For better estimates see Young (1981).

H. Attractors and small stochastic perturbations*

In this section we discuss how physical measures and attractors are selected by their stability under small stochastic perturbations.

1. Small stochastic perturbations

In Sec. II.F we discussed how the introduction of a small amount of noise in a deterministic system could select a particular invariant measure, the Kolmogorov measure. We can now be more precise. Consider first a discrete-time dynamical system generated by the map $f:M\to M$, where M has finite dimension m. Let $\varepsilon>0$, and for each $x\in M$, let μ_x^ε be a probability measure with support in the ball $\bar{B}_x(\varepsilon)=\{y:d(x,y)\leq\varepsilon\}$.

More specifically, we assume that $\mu_x^\varepsilon(dy)$

$=\varepsilon^{-m}\varphi[x,\varepsilon^{-1}(y-x)]dy$, where φ is continuous, $\varphi\geq 0$, $\varphi(x,0)>0$, $\varphi(x,x-y)=0$ if $d(x,y)\geq 1$, and dy denotes the Lebesgue volume element if $M=\mathbf{R}^m$ (if M is not $\mathbf{R}^m$ this prescription is modified by using a Riemann metric on M; see Kifer, 1974). A *stochastic dynamical system* is a time evolution defined not on M, but at the level of probability measures on M. In our case we replace $f:M\to M$ by the stochastic perturbation

$$\nu\to\int\mu^{\varepsilon}_{fx}\nu(dx) = \left[\int\varepsilon^{-m}\varphi[fx,\varepsilon^{-1}(y-fx)]\nu(dx)\right]dy\ . \tag{4.21}$$

The limit of a small stochastic perturbation corresponds to taking $\varepsilon\to 0$.

In the continuous-time case, the dynamical system defined on $\mathbf{R}^m$ by the equation

$$\frac{dx_i}{dt}=F_i(x)$$

is replaced by an evolution equation for the density Φ of ν. We write $\nu(dx)=\Phi(x)dx$, and

$$\frac{d\Phi(x)}{dt}=\sum_i F_i\partial_i\Phi(x)+\varepsilon\Delta\Phi(x)\ . \tag{4.22}$$

If we have a Riemann manifold, the Laplacian Δ should be replaced by the Laplace-Beltrami operator. The limit of a small stochastic perturbation corresponds again to taking $\varepsilon\to 0$.

Theorem. Let an Axiom-A dynamical system on the compact manifold M be defined by a twice differentiable diffeomorphism f or a twice differentiable vector field F. Let $A_1,\ldots,A_r$ be the attractors, and $\rho_1,\ldots,\rho_r$ the corresponding SRB measures.

(a) In the discrete-time case, for ε small enough, let ρ_i^{ε} be a stationary measure for the process (4.21) with support near A_i. Then $\rho_i^{\varepsilon}\to\rho_i$ when $\varepsilon\to 0$.

(b) In the continuous-time case, there is a unique stationary measure ρ^{ε} for the process (4.22), and any limit of ρ^{ε} when $\varepsilon\to 0$ is a convex combination $\sum\alpha_i\rho_i$ where $\alpha_i\geq 0$, $\sum\alpha_i=1$.

These results have been established by Kifer (1974), following Sinai's work on Anosov systems (1972). Another proof has been announced by Young. The idea behind the theorem is as follows. The noisiness of the stochastic time evolution yields measures which have continuous densities on M. The deterministic part of the time evolution will improve this continuity in the unstable directions by stretching, and roughen it in other directions due to contraction. In the limit one gets measures that are continuous along unstable directions, i.e., SRB measures.

Note the difference between the discrete-time and the continuous-time cases, which is due to the fact that in the discrete-time case, for small ε, a point near one attractor cannot jump out of its basin of attraction.

If we have a general dynamical system (not Axiom A), the stationary states for small stochastic perturbations will again tend to be continuous along unstable directions, but the limit when $\varepsilon\to 0$ need not be SRB (see the counterexample of Sec. IV.C). Moreover, there need not be an open basin of attraction associated with each SRB measure, so that, even in the discrete-time case, the stochastic perturbation may switch from one measure to another, and the limit may be a convex combination of many SRB measures (in particular be nonergodic). That basins of attraction may indeed be a mess is shown by the following result.

Theorem (Newhouse, 1974,1979). There is an open set $S\neq\emptyset$ in the space of twice differentiable diffeomorphisms of a compact two-dimensional manifold, and a dense subset R of S such that each $f\in R$ has an infinite number of attracting periodic orbits.

A variation of this result implies that for the Hénon map [see Sec. II.D, example (a)] the presence of infinitely many attracting periodic orbits is assured for some b and a dense set of values of a in some interval (a_0,a_1). The basins of such attracting periodic orbits are mostly very small and interlock in a ghastly manner.

The study of stochastic perturbations of differentiable dynamical systems is at present quite active; see in particular Carverhill (1984a,1984b) and Kifer (1984).

2. A mathematical definition of attractors

We have defined attractors operationally in Sec. II.C. Here, finally, we discuss a mathematical definition.

If $a,b\in M$, let us write $a\to b$ (a goes to b) provided for arbitrarily small $\varepsilon>0$ there is a chain $a=x_0,x_1,\ldots,x_n=b$ such that $d(x_k,f^{\Theta_k}x_{k-1})<\varepsilon$ with $\Theta_k\geq 1$ for $k=1,\ldots,n$. We accept $a\to a$ (corresponding to a chain of length 0), and it is clear that $a\to b,b\to a$ imply $a\to c$. If for every $\varepsilon>0$ there is a chain $a\to a$ of length ≥ 1 we say that a is *chain recurrent*. If a is chain recurrent, we define its *basic class* $[a]=\{b:a\to b\to a\}$. If $[a]$ consists only of a, then a is a fixed point. Otherwise if $b\in[a]$ then b is chain recurrent and $[b]=[a]$.

We shall say that a basic class $[a]$ is an *attractor* if $a\to x$ implies $x\to a$ (i.e., $x\in[a]$). This definition ensures that $[a]$ is attracting, but in a weaker sense than the definition of attracting sets in Sec. II.B. Here, however, we have irreducibility: an attractor cannot be decomposed into two distinct smaller attractors. (More generally, the set of chain-recurrent points decomposes in a unique way into the union of basic classes.)

It can be shown that any limit when $\varepsilon\to 0$ of a measure stable under small stochastic perturbations of a discrete-time dynamical system is "carried by attractors," at least in a weak sense. More precisely, this can be stated as the following theorem.

Theorem. Let Λ be a compact attracting set for a discrete-time dynamical system, m a probability measure with support close to Λ, and ε sufficiently small. Let also m_{ε}^{k} be obtained at time k from the stochastic evolution (4.22). If $[a]$ is not an attractor, then

$$\lim_{k\to\infty} m_{\varepsilon}^{k}[B_a(\delta)]=0\ ,$$

when δ is sufficiently small. [$B_a(\delta)$ denotes the ball of radius δ centered at a.] In particular, if m_ε^∞ is a limit of m_ε^k when $k\to\infty$, then a does not belong to the support of m_ε^∞. If ρ is a limit when $\varepsilon\to 0$ of m_ε^∞, and if m is ergodic, then its support is contained in an attractor.

Proof. See Ruelle (1981).

The topological definition of an attractor given in this section follows the ideas of Conley (1978) and Ruelle (1981). A rather different definition has been proposed recently by Milnor (1985), based on the privileged role which the Lebesgue measure should play for physical dynamical systems.

I. Systems with singularities and systems depending on time*

The theory of differentiable dynamical systems may to some extent be generalized to differentiable dynamical system *with singularities.* This is of interest, for instance, in Hamiltonian systems with collisions (billiards, hard-sphere problems). On this problem we refer the reader to the considerable work by Katok and Strelcyn (1985).

Another conceptually important extension of differentiable dynamical systems is to systems with *time-dependent forces.* One does not allow here for arbitrary nonautonomous systems, but assumes that

$$x(t+1)=f(x(t),\omega(t)) \quad\text{or}\quad \frac{dx}{dt}=F(x,\omega(t)) ,$$

where ω has a stationary distribution. (For instance, ω is defined by a continuous dynamical system.) It is surprising how many results extend to this more general situation; the extension is without pain, but the formalism more cumbersome. Here again we can only refer to the literature. See Ruelle (1984) for a general discussion, and Carverhill (1984a,1984b) and Kifer (1984) for problems involving stochastic differential equations and random diffeomorphisms.

V. EXPERIMENTAL ASPECTS

Now that we have developed a theoretical background and a language in which to formulate our questions, it is time to discuss their experimental aspects. A basic conceptual problem is that of confronting the limited information that can be obtained in a real experiment with the various limits encountered in the mathematical theory. A similar situation occurs, for instance, in the application of statistical mechanics to the study of phase transitions. Other important problems in the relation between theory and experiment concern numerical efficiency and accuracy. The present section will address those problems.

We shall describe two different fields of experimentation—computer experiments and experiments with real physical systems. There is a quantitative difference between the two fields, since one can study dynamical evolution equations with *fixed* experimental conditions more accurately on a computer than in reality. However, there is also a more important qualitative difference: Since the evolution equations are explicitly known in a computer experiment, it is generally easy to compute directly the "tangent map" Df^t. In a physical experiment, by contrast, only points on a trajectory are directly measurable, and the derivatives (tangents) have to be obtained by a delicate interpolation, to be discussed below.

It must be understood that the information currently being extracted from experiments goes a long way beyond the solid mathematical foundations that we have described in the previous sections. It is a challenge for the mathematical physicist to clarify the relations between the various quantities measured on dynamical systems. Most of them seem indeed very interesting, and very promising, but a lot of work is still necessary to prove the existence of these quantities and establish their relations. Our selection below reflects to some extent our personal taste for measurements based on sound ideas and for which a mathematical foundation can be expected.

A. Dimension

The measurement of dimensions is discussed first because it is most straightforward. We concentrate on the determination of the *information dimension,* using the method advocated by Young (1982), and Grassberger and Procaccia (1983b). The idea is described in the first theorem (by Young) in Sec. IV.C. The method, developed independently by Grassberger and Procaccia, has gained wide acceptance through their work.

We start with an experimental time series $u(1),u(2),\ldots$, corresponding to measurements *regularly spaced in time.* We assume that $u(i)\in\mathbf{R}^\nu$, where $\nu=1$ in the (usual) case of scalar measurements. From the $u(i)$, a sequence of points $x(1),\ldots$, in $\mathbf{R}^{m\nu}$ is obtained by taking $x(i)=[u(i+1),\ldots,u(i+m-1)]$. This construction associates with points $X(i)$ in the phase space of the system (which is, in general, infinite dimensional) their projections $x(i)=\pi_m X(i)$ in $\mathbf{R}^{m\nu}$. In fact, if ρ is the physical measure describing our system (ρ is carried by an attractor in phase space), then the points $x(i)$ are equidistributed with respect to the projected measure $\pi_m\rho$ in $\mathbf{R}^{m\nu}$. [Actually this is not always true: if the time spacing Δt between consecutive measurements $u(i),u(i+1)$ is a "natural period" of the system—for instance, when the system is quasiperiodic—one does not have equidistribution. This exception is easily recognized and handled.] We wish to deduce $\dim_H\rho$ from this information (with the possibility of varying m in the above construction). Before seeing how this is done, a general word of caution is in order. In any given experiment we have only a finite time series, and therefore there are natural limits on what can be extracted from it: some questions are too detailed (or the statistical fluctuations too large) for a reasonable answer to come out. See, for instance, Guckenheimer (1982) for a discussion of such matters.

A serious difficulty seems to arise here from the fact that $\dim_H \pi_m \rho$ need not be equal to the desired $\dim_H \rho$. We remove this objection with the observation that, if $\dim_H \rho \leq M$, then, for most M-dimensional projections p, $\dim_H p\rho = \dim_H \rho$. More precisely, we have the following result.

Theorem. Let $0 < M < n$. If E is a Suslin set in $\mathbf{R}^n$, and $\dim E \leq M$, then there is a Borel set G in the space of orthogonal projections $p:\mathbf{R}^n \to \mathbf{R}^M$ such that its complement has measure zero with respect to the natural rotation-invariant measure on projections, and $\dim pE = \dim E$ for all p in G.

Proof. See Lemma 5.3 in Mattila (1975). We are indebted to C. McCullen for this reference. It is interesting to compare this result with that of Mañé in Sec. II.G, where one obtains (with stronger restrictions) the injectivity of p.

Of course we have not proved that our projection π_m belongs to the good set G of the above theorem (with $M = m\nu$), but this appears to be a reasonable guess, and we shall proceed with the assumption that $\dim H_H \pi_m \rho = \dim_H \rho$ for large enough m.

We now use our sequence $x(1), x(2), \ldots, x(N)$ in $\mathbf{R}^{m\nu}$ to construct functions C_i^m and C^m as follows:

$$C_i^m(r) = N^{-1}\{\text{number of } x(j) \text{ such that } d[x(i),x(j)] \leq r\}, \tag{5.1}$$

$$C^m(r) = N^{-1}\sum_i C_i^m(r)$$
$$= N^{-2}\{\text{number of ordered pairs } [x(i),x(j)] \text{ such that } d[x(i),x(j)] \leq r\}. \tag{5.2}$$

[C_i^m is obtained by sorting the $x(j)$ according to their distances to $x(i)$; C^m is obtained more efficiently directly by sorting pairs than as an average of the C_i^m.] We may use $d(x,x') =$ Euclidean norm of $x'-x$, or any other norm, such as

$$|x'-x| = \max_\alpha |u'(\alpha) - u(\alpha)|,$$

where the $u(\alpha)$ are the m components of x, and $|u'(\alpha)-u(\alpha)|$ is for instance the Euclidean norm in $\mathbf{R}^\nu$ (this will be used in Sec. V.B). Note that when $N \to \infty$, we have

$$\lim C_i^m = (\pi_m \rho)[B_{x(i)}(r)] \tag{5.3}$$

(except perhaps at discontinuity points of the right-hand side). Suppose now that

$$\lim_{r\to 0}\lim_{N\to\infty} \frac{\log C_i^m(r)}{\log r} = \lim_{r\to 0} \frac{\log(\pi_m\rho)[B_{x(i)}(r)]}{\log r} = \alpha_m. \tag{5.4}$$

Then $\dim_H \pi_m \rho = \alpha_m$ (first theorem of Sec. IV.C). Provided the projection π_m is in the "good set" G, we have thus $\alpha_m = m\nu$ if $\dim_H \rho \geq m\nu$ and $\alpha_m = \dim_H \rho$ if $\dim_H \rho \leq m\nu$. Experimentally, α_m may be obtained by plotting $\log C_i^m(r)$ vs $\log r$ and determining the slope of the curve (see below). With a little bit of luck [existence of the limit (5.4), and π_m in the good set] we may thus obtain $\dim_H \rho$ experimentally: *we choose m such that $\alpha_m < m\nu$; then we have* $\dim_H \rho = \alpha_m$. Although we cannot completely verify that π_m is in the good set, we can in principle check (within experimental accuracy) the existence of the limits α_m, and the fact that α_m becomes independent of m when m increases beyond a value such that $\alpha_m \leq m\nu$. Note that α_m should also be independent of the index i in Eq. (5.4).

The information dimension $\dim_H \rho$ may also be obtained by a modification of Eq. (5.4). We describe the method of Grassberger and Procaccia, which has been tested experimentally in a number of cases. This consists in writing

$$\lim_{r\to 0}\lim_{N\to\infty} \frac{\log C^m(r)}{\log r} = \beta_m, \tag{5.5}$$

and asserting that for m sufficiently large, β_m is the information dimension. The only relation that can easily be established rigorously between Eqs. (5.4) and (5.5) is that if both limits exist, then $\alpha_m \geq \beta_m$. However, it seems quite reasonable to assume that in general $\alpha_m = \beta_m$ (i.e., if the C_i^m behave like r^α, then their linear superposition C^m also behaves like r^α). The β_m obtained experimentally do become independent of m for m large enough, as expected. See Fig. 19.

To summarize: the method of Grassberger and Procaccia is a highly successful way of determining the information dimension experimentally. Values between 3 and 10 are obtained reproducibly. The method is not entirely justified mathematically, but nevertheless quite sound. The study of the limit (5.4) is also desirable, even though the statistics there is poorer.

1. Remarks on physical interpretation

a. The meaningful range for $C^m(r)$

Suppose we plot $\log C(r)/\log r$ as a function of $\log r$ (we suppress the superscript m and possibly the subscript i of C). First, for small r, we have a large scatter of points due to poor statistics; then there is a range (r_0, r_1) of near constancy (the constant is the information dimension if m is suitably large). For r larger than r_1 we have deviation from constancy due to nonlinear effects. The "meaningful range" (r_0, r_1) is that in which the distribution of distances between pairs of points is statistically useful.

b. Curves with "knees"

It is not uncommon that the $\log C(r)$ vs $\log r$ plot shows a "knee" (see Fig. 20), so that it has slope α in the range

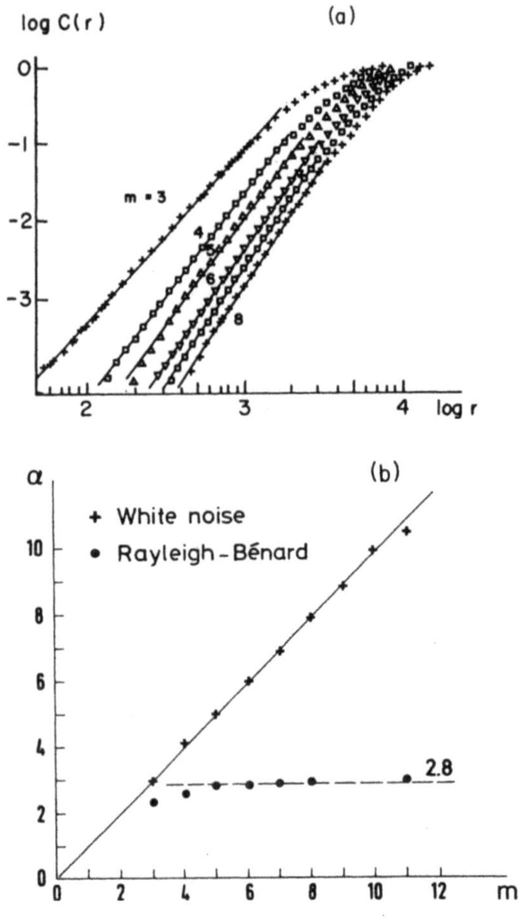

FIG. 19. Experimental results from Malraison *et al.* (1983) and Atten *et al.* (1984); see also Dubois (1982): (a) The plots show $\log C$ vs $\log r$ for different values of the embedding dimension, for the Rayleigh-Bénard experiment. (b) The measured dimension α as a function of the embedding dimension m, both for the Rayleigh-Bénard experiment and for numerical white noise. Note that α becomes nearly constant (but not quite) at $m=3$. The α for white noise is nearly equal to m (but not quite).

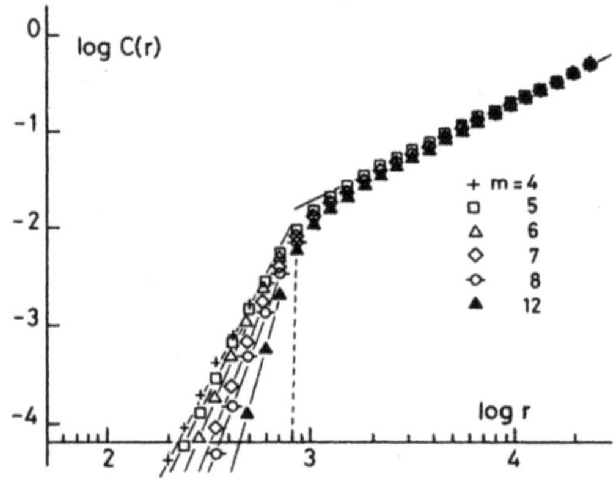

FIG. 20. A $\log C(r)$ vs $\log r$ plot may show a "knee," so that a dimension α appears in the range $(\log r_0, \log r_1)$, and a dimension α' in the range $(\log r_1, \log r_2)$. Here the scalar signal of a deterministic system with information dimension α' is perturbed by the addition of random noise, which yields a dimension α equal to the embedding dimension m (see Atten *et al.*, 1984).

$(\log r_0, \log r_1)$ and a smaller slope α' in the range $(\log r_1, \log r_2)$. The dimension, or "number of degrees of freedom" is thus different for r above and below r_1 (see, for instance, Riste and co-workers, 1985). To see how this situation can arise, let us consider a *product* dynamical system $\mathrm{I}\times\mathrm{II}$ formed of two noninteracting subsystems I and II. Take an observable $u=u_{\mathrm{I}}+u_{\mathrm{II}}$, where u_{I} and u_{II} depend only on the subsystems I and II, respectively, and let the amplitude r_1 of the signal u_{I} be much smaller than that of u_{II}. In the range $r<r_1$ we have statistical information on the complete system $\mathrm{I}\times\mathrm{II}$, giving an information dimension α. In the range $r \gg r_1$ we have statistical information only on the subsystem II, giving an information dimension α'. More generally, suppose that system II evolves independently of I, but that I has a time evolution that may depend on II; then the same conclusions persist for this "semidirect product." (The small-amplitude modes of I are driven here by system II, an example of Haken's "slaving principle.") The above argument makes clear, for instance, how the information dimension found by analysis of a turbulent hydrodynamic system does not take into account small ripples of amplitude less than the discrimination level r_0 of the analysis. (We thank P. C. Martin for useful discussions on this point.) A knee will also appear if the signal from a deterministic chaotic system is perturbed by adding random noise of smaller amplitude (see Fig. 20).

c. Spatially localized degrees of freedom

We have just discussed dynamical systems that have a product structure $\mathrm{I}\times\mathrm{II}$, or where a subsystem II evolves independently and drives other degrees of freedom. Strictly speaking, such decoupling does not seem to occur in realistic situations like that of a turbulent viscous fluid (except for the trivial case where the fluid is in two different uncoupled containers). Normally, in a nonlinear system one may say that "every mode is coupled with all other modes," and *exact* factorization is impossible. An apparent exception is constituted by quasiperiodic motions where factorization is present, but the independent frequencies do not correspond to independent physical subsystems. In other words, if a physical variable $u(t)$ of the system is monitored (for instance a component of the velocity of a viscous fluid at one point), the whole dynamics of the system (on the appropriate attractor) can in principle be reconstructed from the time series $[u(t){:}t$ varying from 0 to $\infty]$. In particular, the information dimension of the system can be obtained indifferently from monitoring u or any other physical variable.

As noted in Secs. V.A.1.a and V.A.1.b, the experimental uncertainties change this situation. At the level of accuracy of an experiment, some degrees of freedom may effectively be driven by others and, having small amplitude, pass unnoticed. (A case in point would be that of eddies of small size in three-dimensional turbulence.) Another frequent and important case occurs when some "oscillators" (possibly complex oscillators) are strongly localized in some region of space.[3] Consider, for instance, the flow between coaxial rotating cylinders in a regime where there is some turbulence superposed with Taylor cells. Some features of the flow are global (like the very existence of the Taylor cells), others seem to be restricted to one Taylor cell, having very little interaction with neighboring cells.

The information dimension d_c, obtained from moderate-precision measurements of one cell, is then likely to be different from the global information dimension d_ν of a column of ν cells (and one expects formulas like $d_c=a+b$, $d_\nu=a+\nu b$). Note that d_ν could be obtained by monitoring a vector signal with ν components, each corresponding to a scalar signal from one cell.

2. Other dimension measurements

The most straightforward way to find the "fractal" dimension of a set A is to cover it with a grid of size r, to count the number $N(r)$ of occupied cells, and to compute

$$\lim_{r\to 0}\frac{\log N(r)}{|\log r|}\ .$$

This box-counting is computationally ineffective (Farmer, 1984). It gives access to the dimension of attractors rather than to the information dimension (the latter seems for the moment to have greater theoretical interest). Another problem with box-counting is that usually the population of boxes is very uneven, so that it may take a considerable amount of time before some "occupied" boxes really become occupied. For all these reasons, the box-counting approach is not used currently.

B. Entropy

The entropy (or Kolmogorov-Sinai invariant) $h(\rho)$ of a physical measure ρ is an important quantity, as we have seen in Sec. IV. Early attempts to measure $h(\rho)$ were based directly on the definitions and used a partition $\mathcal{A}$ (see Sec. IV.A). These attempts (Shimada, 1979; Curry, 1981; Crutchfield, 1981) were interesting but not entirely successful. We describe here another approach due to Grassberger and Procaccia (1983a); see also Cohen and Procaccia (1984). [Similar ideas were developed independently by Takens (1983).] This approach has far greater potential for implementation in experimental situations.

The idea of Grassberger and Procaccia is to exploit the m dependence of the functions $C_i^m(r)$ and $C^m(r)$ defined in Eqs. (5.1) and (5.2). As before, they use $C^m(r)$, which has better statistics, but it is easier to argue with the $C_i^m(r)$, which satisfy

$$C_i^m(r)\approx(\pi_m\rho)[B_{x(i)}(r)] \tag{5.6}$$

for large N [see Eq. (5.3)]. In view of Eq. (5.6), $C_i^m(r)$ is the probability that $x(j)$ satisfies $d[x(j),x(i)]\le r$. Grassberger and Procaccia use the Euclidean norm, but we prefer to follow Takens and to take

$$d[x(j),x(i)]=\max\{\,|u(j)-u(i)|,\dots,|u(j+m-1)-u(i+m-1)|\,\}\ .$$

Usually $\nu=1$ (scalar signal), but the general case is not harder to handle [with $|u(i)-u(j)|$ being the Euclidean norm of $u(i)-u(j)$ in $\mathbf{R}^\nu$]. We may thus interpret $C_i^m(r)$ as the probability that the signal $u(j+k)$ remains in the ball $B^\nu_{u(i+k)}(r)$ for m consecutive units of time [$B^\nu_{u(i)}(r)$ is the ball of radius r in $\mathbf{R}^\nu$ centered at $u(i)$].

With this interpretation, and the fact that $C^m(r)$ is the average of the $C_i^m(r)$, it can be argued that

$$\lim_{r\to 0}\lim_{m\to\infty}\frac{1}{m}\lim_{N\to\infty}[-\log C^m(r)]=\Delta t K_2(\rho)\ , \tag{5.7}$$

where K_2 has been defined at the end of Sec. IV.A, and Δt is the spacing between measurements of the signal u. Since $K_2(\rho)$ is a lower bound to $h(\rho)$, we see that if one obtains $K_2(\rho)>0$ from Eq. (5.7) then one can conclude that $h(\rho)>0$, i.e., that the system is chaotic.

It is, however, also possible to obtain $h(\rho)$ directly as follows. Define

$$\Phi^m(r)=\frac{1}{N}\sum_i \log C_i^m(r)\ .$$

Then

$$\Phi^{m+1}(r)-\Phi^m(r)=\text{average over } i \text{ of } \log[\text{probability that } u(j+m)\in B^\nu_{u(i+m)}(r)$$
$$\text{given that } u(j+k)\in B^\nu_{u(i+k)}(r) \text{ for } k=0,\dots,m-1]\ .$$

Therefore,

[3]For some discussion of the difficult problem of localization in hydrodynamics, see Ruelle (1982b).

$$\lim_{r\to 0}\lim_{m\to\infty}\lim_{N\to\infty}[\Phi^{m+1}(r)-\Phi^m(r)]=\Delta t h(\rho)\ . \qquad (5.8)$$

Remarks.

(a) Like the expressions of Sec. V.A, the identities (5.7) and (5.8) hold "if all goes well." Basically, the condition is that the monitored signal should reveal enough of what is going on in the system.

(b) While the information dimension could be obtained from $C^m(r)$ for one single m (sufficiently large), we have a limit $m\to\infty$ in Eqs. (5.7) and (5.8). In this respect, (5.7) is not optimal (it will contain errors of order $1/m$); it is better to write

$$\lim_{r\to 0}\lim_{m\to\infty}\lim_{N\to\infty}\log\frac{C^m(r)}{C^{m+1}(r)}=\Delta t K_2(\rho)\ ,$$

or to use Eq. (5.8).

C. Characteristic exponents: computer experiments

We recall that the characteristic exponents measure the exponential separation of trajectories in time and are computed from the *derivative* D_xf^t. In computer experiments, the derivative is often directly calculable, whereas in physical experiments it has to be obtained indirectly from the experimental signal. Therefore the methods for evaluating characteristic exponents are somewhat different in the two cases and will be treated separately. In this section, we discuss computed experiments, which have served and still serve an important purpose in the exploration of dynamical systems.

Let us mention here some interesting open problems. What is the distribution of characteristic exponents for a large or a highly excited system? Can one define a density of exponents per unit volume for a spatially extended system? What is the behavior near zero exponent? For a theoretical study in the case of turbulence, see Ruelle (1982b,1984). For an experimental study in the case of the Kuramoto-Sivashinsky model, see Manneville (1985).

In the case of a discrete-time dynamical system defined by a map $f:\mathbf{R}^m\to\mathbf{R}^m$, let

$$T(x)=D_xf\ .$$

This is the matrix of partial derivatives of the m components of $f(x)$ with respect to the m components of x. Write

$$T_x^n=T(f^{n-1}x)\cdots T(fx)T(x) \qquad (5.9)$$

(matrix multiplication on the right-hand side). Then the largest characteristic exponent is given by

$$\lambda_1=\lim_{n\to\infty}\frac{1}{n}\log\|T_x^nu\| \qquad (5.10)$$

for almost any vector u, and this is a very efficient way to obtain λ_1. The other characteristic exponents can in principle be obtained by diagonalizing the positive matrices $(T_x^n)^*T_x^n$ and using the fact that their eigenvalues behave like $e^{2n\lambda_1}, e^{2n\lambda_2},\ldots$. Obviously, for large n, the different eigenvalues have very different orders of magnitude, and this creates a problem if T_x^n is computed without precaution. When $(T_x^n)^*T_x^n$ is diagonalized, the small relative errors on the large eigenvalues might indeed contaminate the smaller ones, causing intolerable inaccuracy. We shall see below how to avoid this difficulty.

Consider next a continuous-time dynamical system defined by a differential equation

$$\frac{dx(t)}{dt}=F(x(t))\ , \qquad (5.11)$$

in $\mathbf{R}^m$. An early proposal to estimate λ_1 (Benettin, Galgani, and Strelcyn, 1976) used solutions $x(t),x'(t),x''(t),\ldots$, chosen as follows. The initial condition $x'(0)$ is chosen very close to $x(0)$, and $x(t)$ remains close to $x'(t)$ up to some time T_1; one then replaces the solution x' by a solution x'' such that $x''(T_1)-x(T_1)=\alpha[x'(T_1)-x(T_1)]$ with α small. Thus $x''(T_1)$ is again very close to $x(T_1)$, and $x''(t)$ remains close to $x(t)$ up to some time $T_2>T_1$, and so on. The rate of deviation of nearby trajectories from $x(\ )$ can thus be determined, yielding λ_1. This simple method has also been applied to physical experiments (Wolf *et al.*, 1984); we shall return to this topic in Sec. V.D.

In the case of (5.11) one can, however, do much better. Namely, one differentiates to obtain

$$\frac{d}{dt}u(t)=(D_{x(t)}F)[u(t)]\ , \qquad (5.12)$$

which is *linear* in u, but with nonconstant coefficients. The solution of (5.11) yields $x(t)=f^t(x(0))$, and the solution of (5.12) yields

$$u(t)=(D_{x(0)}f^t)u(0)\ .$$

Therefore one can readily compute the matrices $T_x^t=D_xf^t$ by integrating Eq. (5.12) with m different initial vectors u. Better yet, one can use the matrix differential equation

$$\frac{d}{dt}T_{x(0)}^t=(D_{x(t)}F)T_{x(0)}^t\ ,$$

with $T_{x(0)}^0$ the identity matrix.

As in the discrete case, it is not advisable to compute T_x^t for large t. We choose a reasonable unit of time τ: not too large, so that the $e^{\lambda_i\tau}$ do not differ too much in their orders of magnitude, but not too small either, because we have to multiply a number of matrices proportional to τ^{-1}. Having chosen τ, we discretize the time (setting $\tilde{f}=f^\tau$) and proceed as in the discrete-time case. If the characteristic exponents for $\tilde{f}$ are $\tilde{\lambda}_i$, then the characteristic exponents for the continuous-time system are $\lambda_i=\tau^{-1}\tilde{\lambda}_i$.

Before discussing the accurate calculation of the λ_i for $i>1$, let us mention that the knowledge of λ_1 [obtained from Eq. (5.10)] is sometimes sufficient to determine *all* characteristic exponents. This is certainly the case for one-dimensional systems, as well as for the Hénon map and the Lorenz equation, as we have seen in the examples of Sec. III.D.1. It is also possible to estimate successively

λ_1 by (5.10), then $\lambda_1+\lambda_2$ as the rate of growth of surface elements, $\lambda_1+\lambda_2+\lambda_3$ as the rate of growth of three-volume elements, etc. This approach was first proposed by Benettin *et al.* (1978). In what follows, we discuss a somewhat different method.

The algorithm we propose for the calculation of the λ_i is very close to the method presented by Johnson *et al.* (1984) for proving the multiplicative ergodic theorem of Oseledec. Remember that we are interested in the product (5.9):

$$T_x^n = T(f^{n-1}x)\cdots T(fx)T(x) .$$

To start the procedure, we write $T(x)$ as

$$T(x)=Q_1R_1 , \tag{5.13}$$

where Q_1 is an orthogonal matrix and R_1 is upper triangular with non-negative diagonal elements. [If $T(x)$ is invertible, this decomposition is unique.] Then for $k=2,3,\ldots$, we successively define

$$T_k' = T(f^{k-1}x)Q_{k-1}$$

and decompose

$$T_k' = Q_kR_k ,$$

where Q_k is orthogonal and R_k upper triangular with non-negative diagonal elements. Clearly, we find

$$T_x^n = Q_nR_n \cdots R_1 .$$

To exploit this decomposition, we shall make use of the results of Johnson *et al.*, but note that those are only proved in the "invertible case" of a dynamical system defined for negative as well as positive times. In the paper referred to, an orthogonal matrix Q is chosen at random (i.e., Q is equidistributed with respect to the *Haar* measure on the orthogonal group), and the initial $T(x)$ is replaced by $T(x)Q$ in Eq. (5.13), the matrices $T(f^{k-1}x)$ for $k>1$ being left unchanged. It is then shown that the diagonal elements $\lambda_{ii}^{(n)}$ of the upper triangular matrix product $R_n\cdots R_1$ obtained from this modified algorithm satisfy

$$\lim_{n\to\infty}\frac{1}{n}\log\lambda_{ii}^{(n)}=\lambda_i \tag{5.14}$$

almost surely with respect to the product of the invariant measure ρ and the Haar measure (corresponding to the choice of Q). *On the right-hand side of Eq. (5.14) we have the characteristic exponents arranged in decreasing order.* For practical purposes, it is clearly legitimate to take Q = identity.

In the case of constant $T(f^kx)$, i.e., $T(f^kx)=A$ for all k, the above algorithm is known as the "Analog of the treppen-iteration using orthogonalization."[4] See Wilkinson (1965, Sec. 9.38, p. 607). The multiplicative ergodic theorem can thus be viewed as the generalization of this algorithm to the case when the $T(f^kx)$ are randomly chosen.

Let us again call $\lambda^{(1)},\ldots,\lambda^{(r)}$ the *distinct* characteristic exponents, and $m^{(1)},\ldots,m^{(r)}$ their multiplicities. The space $E^{(i)}$ associated with the characteristic exponents $\le\lambda^{(i)}$ (see Sec. III.A) is obtained as follows. Consider the last $m_-^{(i)}=m^{(i)}+\cdots+m^{(r)}$ columns of the matrix

$$R_1^{-1}\cdots R_n^{-1}\Delta=(R_n\cdots R_1)^{-1}\Delta ,$$

where Δ is the diagonal matrix equal to the diagonal part of $R_n\cdots R_1$. Let $E^{(i)}(n)$ be the space generated by these $m_-^{(i)}$ column vectors. Then $E^{(i)}=\lim_{n\to\infty}E^{(i)}(n)$.

Note that if we are only interested in the largest s characteristic exponents, $\lambda_1\ge\cdots\ge\lambda_s$, then it suffices to do the decomposition to triangular form only in the upper left $s\times s$ submatrix, leaving the matrices untouched in the lower right $(m-s)\times(m-s)$ corner.

The practical task of decomposing a matrix T_k' as Q_kR_k, as discussed above, is abundantly treated in the literature, and library routines exist for it. According to Wilkinson (1965, Secs. 4.47–4.56), the Householder triangularization is preferable to Schmidt orthogonalization, since it leads to more precisely orthogonal matrices. This algorithm is available in Wilkinson and Reinsch (1971, Algorithm I/8, procedure "decompose"). It exists as part of the packages EISPACK and NAG. This algorithm is numerically very stable, and in fact the size of the eigenvalues should not matter.

D. Characteristic exponents: physical experiments

By contrast with computer experiments, experiments in the laboratory do not normally give direct access to the derivatives D_xf^t. These derivatives must thus be estimated by a detailed analysis of the data. Once the derivatives D_xf^t are known, the problem is analogous to that encountered in computer experiments. The same algorithms can be applied to obtain either the largest characteristic exponent λ_1 or other exponents. Only the positive exponents will be determined, however, or part of them. We have seen above how to restrict the computation to the largest s characteristic exponents, and we shall see below why one can only hope to determine the *positive* λ_i in general.

As in Sec. V.A we start with a time series $u(1),u(2),\ldots$, in $\mathbf{R}^{\nu}$, and from this we construct a sequence $x(1),x(2),\ldots$, in $\mathbf{R}^{m\nu}$, with $x(i)=[u(i),\ldots,u(i+m-1)]$. We shall discuss in remark (c) below how large m should be taken. We shall now try to estimate the derivatives $T_{x(i)}^{\tau}=D_{x(i)}f_{\tau}$. As in Sec. V.C, τ should be such that the $e^{\lambda_k\tau}$ are not too large (we are only interested in positive λ_k); this means that τ should not be larger (and rather smaller) than the "characteristic time" of the system. Also, τ should not be too small, since we have to multiply later a number of matrices $T_{x(i)}^{\tau}$ proportional to τ^{-1}. Of course, τ will be a multiple $p\Delta t$ of the time interval Δt between measurements, so that

[4]We thank G. Wanner for helpful discussions in relation to this problem.

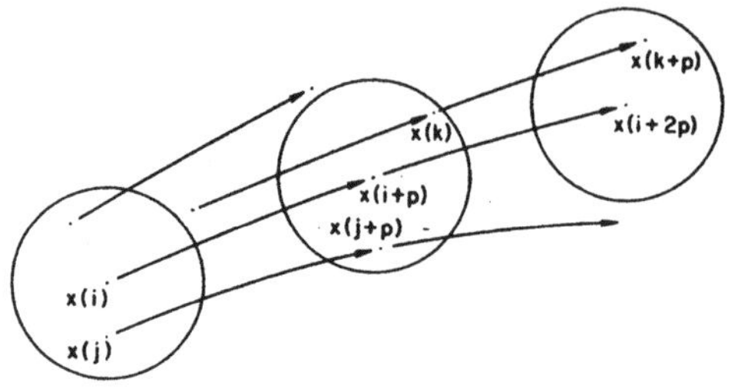

FIG. 21. The balls of radius $\tilde{r}$ centered at $x(i)$, $x(i+p)$, and $x(i+2p)$.

$f_\tau x(i)=x(i+p)$.

The derivatives $T^\tau_{x(i)}$ will be obtained by a best linear fit of the map which, for $x(j)$ close to $x(i)$, sends $x(j)-x(i)$ to $f^\tau(x(j))-f^\tau(x(i))=x(j+p)-x(i+p)$ (see Fig. 21). Close here means that the map should be approximately linear. This will be ensured by choosing $\tilde{r}$ sufficiently small and taking only those $x(j)$ for which

$$d[x(i),x(j)]\leq\tilde{r} \text{ and } d[x(i+p),x(j+p)]\leq\tilde{r}$$

(these conditions imply that $x(j+k)$ is close to $x(i+k)$ for $0<k<p$; one could also require $d[x(i+k), x(j+k)]\leq\tilde{r}$ for each k separately). The choice of $\tilde{r}$ should probably be made by trial and error, monitoring how good a linear fit is obtained. Certainly $\tilde{r}$ should be less than the upper bound r_1 of the "meaningful range for $C^m(r)$," discussed in remark (a) of Sec. V.A. Having chosen $\tilde{r}$, we have to assume that the length N of the original time series is sufficiently long so that there is a fair number of points $x(j)$ in the ball of radius $\tilde{r}$ around $x(i)$, and such that $x(j+p)$ is also in the ball of radius $\tilde{r}$ around $x(i+p)$. In principle, m points are enough to determine a linear map, but we want many points: (a) to overcome the statistical scatter of the $x(j)$, (b) because a symmetric distribution of the $x(j)$ will yield a linear best fit from which the quadratic nonlinear terms have been eliminated. We repeat how $T^\tau_{x(i)}=D_{x(i)}f^\tau$ is obtained, with $\tau=p\Delta t$. Take all $x(j)$ such that $d[x(i),x(j)]\leq\tilde{r}$ and $d[x(i+p),x(j+p)]\leq\tilde{r}$, and determine the $m\nu\times m\nu$ matrix $T^\tau_{x(i)}$ by a least-squares fit[5] such that

$$T^\tau_{x(i)}[x(j)-x(i)]\approx x(j+p)-x(i+p)\ . \qquad (5.15)$$

Note that when we estimate $T^\tau_{x(i+p)}$ we have to start looking again for *all* $x(k)$ such that $d[x(i+p),x(k)]\leq\tilde{r}$ and $d[x(i+2p),x(k+p)]\leq\tilde{r}$, and not just for the $x(k)$ of the form $x(j+p)$!

In general, the points $x(j)$ will not be uniformly distributed in all directions from $x(i)$. In other words, the vectors $x(j)-x(i)$ may not span $\mathbf{R}^{m\nu}$, and therefore the matrix $T^\tau_{x(i)}$ may not be well defined by our prescription. Even if $T^\tau_{x(i)}$ is defined, there will in general be directions in which there are many fewer points than in others, so that the uncertainty on the elements of $T^\tau_{x(i)}$ corresponding to those directions is large. In fact, we can only expect with confidence that the vectors $x(j)-x(i)$ span the *expanding directions* around $x(i)$, i.e., the linear space tangent to the unstable manifold at $x(i)$ (because an SRB measure is absolutely continuous along the unstable manifold or because an attracting set contains the unstable manifolds of the points on it). The fact that the matrix $T^\tau_{x(i)}$ is only known with confidence in the unstable directions need not distress us: It means that we can determine with confidence only the *positive characteristic exponents.* This is done with the method of Sec. V.C, constructing triangular matrices from the T^τ_x for $x=x(i_0)$, $x(i_0+p)$, $x(i_0+2p),\ldots$, computing the characteristic exponents, and discarding those which are ≤ 0. The latter will usually (although not necessarily) be meaningless.

Remarks.

(a) The detailed method presented above for deriving characteristic exponents from experiments seems new. Up to now, attention has been concentrated on obtaining the largest exponent λ_1, using basically the method discussed in Sec. V.C after Eq. (5.11). For a different approach, see Wolf *et al.* (1984).

(b) The example of a time series $u(i)=$const, corresponding to an attracting fixed point, shows that it is not possible in general to obtain the negative characteristic exponents from the long-term behavior of a dynamical system. It is conceivable that our method works up to that k after which the sum of the largest k characteristic exponents becomes negative. If one has access to transients, then negative characteristic exponents are in principle accessible.

(c) We have discussed in this section the determination of the characteristic exponents of a measure ρ from its projection $\pi_m\rho$. How large should one choose m to be? For the determination of the information dimension, it was sufficient to take $m\nu\geq\dim_H\rho$. Here, however, this will usually be insufficient, because we have to reconstruct the *dynamics* in the support of ρ from its projection in $\mathbf{R}^{m\nu}$. We want π_m therefore to be *injective* on the support of ρ. According to Mañé's theorem in Sec. II.G. this may require $m\nu>2\dim_K(\text{support }\rho)+1$. Probably the best evidence that π_m is a "good" projection for the present purposes would be a reasonably good linear fit for Eq. (5.15).

E. Spectrum, rotation numbers

The ergodic quantities that we have discussed in this paper—characteristic exponents, entropy, information dimension—are those which appear at this moment most important and most easily accessible. They are, however, not the only quantities one might consider. For a quasi-periodic system, the *generating frequencies* are of course important. More generally, Frisch and Morf (1981) have drawn attention to the *complex singularities* of the signal

[5] The most convenient algorithms are given in Wilkinson and Reinsch (1971), contribution I/8. (All algorithms in this book are available in the large libraries such as EISPACK and NAG.)

$u(t)$ and their relation to the high-frequency behavior of the power spectrum. One may also look for poles at complex values of the frequency in the power spectrum (*resonances*). Finally one should mention *rotation numbers*, which are not always defined but are interesting quantities when they make sense (see Ruelle, 1985).

VI. OUTLOOK

Our review has led from the definitions of dynamical systems theory to a discussion of those quantities accessible today through the *statistical* analysis of time series for deterministic nonlinear systems. Together with the more geometrical aspects of bifurcation theory, this represents the main body of theoretically and experimentally successful ideas concerning nonlinear dynamics at this time. The purpose of this review is to make this knowledge accessible to a large number of scientists. The results presented here are the combined achievement of many investigators, only incompletely cited. We believe that the next step in the study of dynamical systems should lead to a better understanding of space-time patterns, for which only timid beginnings are now seen. We hope that the present review serves as an encouragement for the undertaking of this difficult problem.

Note added in proof. Another useful reprint collection to be added to the list of Sec. I is Hao Bai-Lin, 1984, *Chaos* (World Scientific, Singapore).

ACKNOWLEDGMENTS

We wish to thank E. Lieb, A. S. Wightman, and especially P. Bergé and F. Ledrappier for their critical reading of the manuscript. This work was partially supported by the Fonds National Suisse.

REFERENCES

Abraham, N. B., J. P. Gollub, and H. L. Swinney, 1984, "Testing nonlinear dynamics," Physica D 11, 252.

Adler, R. L., A. G. Konheim, and M. H. McAndrew, 1965, "Topological entropy," Trans. Am. Math. Soc. 114, 309.

Alexander, J. C., and J. A. Yorke, 1984, "Fat baker's transformations," Ergod. Theory Dynam. Syst. 4, 1.

Anosov, D. V., 1967, "Geodesic flows on a compact Riemann manifold of negative curvature," Trudy Mat. Inst. Steklov 90, [Proc. Steklov Math. Inst. 90 (1967)].

Arnold, V. I., and A. Avez, 1967, *Problèmes Ergodiques de la Mécanique Classique* (Gauthier-Villars, Paris) [*Ergodic Problems of Classical Mechanics* (Benjamin, New York, 1968)].

Atten, P., J. G. Caputo, B. Malraison, and Y. Gagne, 1984, "Détermination de dimension d'attracteurs pour différents écoulements," preprint.

Benedicks, M., and L. Carleson, 1984, "On iterations of $1-ax^2$ on $(-1,1)$," Ann. Math., in press.

Benettin, G., L. Galgani, J. M. Strelcyn, 1976, "Kolmogorov entropy and numerical experiments," Phys. Rev. A 14, 2338.

Benettin, G., L. Galgani, A. Giorgilli, and J. M. Strelcyn, 1978, "Tous les nombres de Liapounov sont effectivement calculables," C. R. Acad. Sci. Paris 286A, 431.

Bergé, P., Y. Pomeau, and C. Vidal, 1984, *L'ordre dans le Chaos* (Herman, Paris).

Billingsley, P., 1965, *Ergodic Theory and Information* (Wiley, New York).

Bowen, R., 1975, *Equilibrium States and the Ergodic Theory of Anosov Diffeomorphisms*, Lecture Notes in Mathematics 470 (Springer, Berlin).

Bowen, R., 1978, *On Axiom A Diffeomorphisms*, Regional conference series in mathematics, No. 35 (Amer. Math. Soc., Providence, R.I.)

Bowen, R., and D. Ruelle, 1975, "The ergodic theory of Axiom *A* flows," Inventiones Math. 29, 181.

Brin, H., and Z. Nitecki, 1985, private communication.

Caffarelli, R., R. Kohn, and L. Nirenberg, 1982, "Partial regularity of suitable weak solutions of the Navier-Stokes equations," Commun. Pure Appl. Math. 35, 771.

Campanino, M., 1980, "Two remarks on the computer study of differentiable dynamical systems," Commun. Math. Phys. 74, 15.

Campbell, D., and H. Rose, 1983, Eds., *Order in Chaos: Proceedings of the International Conference . . . Los Alamos, . . . 1982* (Physica D 7, 1).

Carverhill, A., 1984a, "Flows of stochastic systems: ergodic theory," Stochastics, in press.

Carverhill, A., 1984b, "A 'Markovian' approach to the multiplicative ergodic (Oseledec) theorem for nonlinear dynamical systems," preprint.

Cohen, A., and I. Procaccia, 1984, "On computing the Kolmogorov entropy from the time signals of dissipative and conservative dynamical systems," Phys. Rev. A 31, 1872.

Collet, P., J.-P. Eckmann, and O. E. Lanford, 1980, "Universal properties of maps of an interval," Commun. Math. Phys. 76, 211.

Collet, P., and J.-P. Eckmann, 1980a, "On the abundance of aperiodic behavior for maps on the interval," Commun. Math. Phys. 73, 115.

Collet, P., and J.-P. Eckmann, 1980b, *Iterated Maps on the Interval as Dynamical Systems* (Birkhauser, Cambridge, MA).

Collet, P., and J.-P. Eckmann, 1983, "Positive Liapunov exponents and absolute continuity for maps of the interval," Ergod. Theory Dynam. Syst. 3, 13.

Conley, C., 1978, *Isolated Invariant Sets and the Morse Index*, Regional conference series in mathematics, No. 38 (Amer. Math. Soc., Providence, R.I.)

Croquette, V., 1982, "Déterminisme et chaos," Pour la Science 62, 62.

Crutchfield, J., 1981, private communication.

Curry, J. H., 1979, "On the Hénon transformation," Commun. Math. Phys. 68, 129.

Curry, J. H., 1981, "On computing the entropy of the Hénon attractor," J. Stat. Phys. 26, 683.

Cvitanović, P., 1984, Ed., *Universality in Chaos* (Adam Hilger, Bristol).

Denker, M., C. Grillenberger, and K. Sigmund, 1976, *Ergodic Theory on Compact Spaces*, Lecture Notes in Mathematics 527 (Springer, Berlin).

Douady, A., and J. Oesterlé, 1980, "Dimension de Hausdorff des attracteurs," C. R. Acad. Sci. Paris 290A, 1135.

Dubois, M., 1982, "Experimental aspects of the transition to turbulence in Rayleigh-Bénard convection," in *Stability of Thermodynamic Systems*, Lecture Notes in Physics 164, (Springer, Berlin), pp. 177–191.

Dunford, M., and J. T. Schwartz, 1958, *Linear Operators* (Interscience, New York).

Eckmann, J.-P., 1981, "Roads to turbulence in dissipative dynamical systems," Rev. Mod. Phys. **53**, 643.

Farmer, D., 1984, private communication.

Fathi, A., M. R. Herman, and J.-C. Yoccoz, 1983, "A proof of Pesin's stable manifold theorem," in *Geometric Dynamics*, Lecture Notes in Mathematics 1007 (Springer, Berlin), pp. 177—215.

Fauve, S., and A. Libchaber, 1982, "Rayleigh-Bénard experiment in a low Prandtl number fluid, mercury," in *Chaos and Order in Nature: Proceedings of the International Symposium on Synergetics at Schloss Elmau, 1981*, edited by H. Haken (Springer, Berlin), pp. 25—35.

Feigenbaum, M. J., 1978, "Quantitative universality for a class of nonlinear transformations," J. Stat. Phys. **19**, 25.

Feigenbaum, M. J., 1979, "The universal metric properties of nonlinear transformations," J. Stat. Phys. **21**, 669.

Feigenbaum, M. J., 1980, "The transition to aperiodic behavior in turbulent systems," Commun. Math. Phys. **77**, 65.

Feit, S. D., 1978, "Characteristic exponents and strange attractors," Commun. Math. Phys. **61**, 249.

Feller, W., 1966, *An Introduction to Probability Theory and its Applications*, Vol. II (Wiley, New York).

Foias, C., and R. Temam, 1979, "Some analytic and geometric properties of the solutions of the evolution Navier-Stokes equations," J. Math. Pures Appl. **58**, 339.

Frederickson, P., J. L. Kaplan, E. D. Yorke, and J. A. Yorke, 1983, "The Lyapunov dimension of strange attractors," J. Diff. Equ. **49**, 185.

Frisch, U., and R. Morf, 1981, "Intermittency in nonlinear dynamics and singularities at complex times," Phys. Rev. A **23**, 2673.

Ghil, M., R. Benzi, and G. Parisi, 1985, *Turbulence and Predictability in Geophysical Fluid Dynamics and Climate Dynamics* (North-Holland, Amsterdam).

Grassberger, P., 1981, "On the Hausdorff dimension of fractal attractors," J. Stat. Phys. **26**, 173.

Grassberger, P., 1984, "Information aspects of strange attractors," preprint, Wuppertal.

Grassberger, P., and I. Procaccia, 1983a, "Estimating the Kolmogorov entropy from a chaotic signal," Phys. Rev. A **28**, 2591.

Grassberger, P., and I. Procaccia, 1983b, "Measuring the strangeness of strange attractors," Physica D **9**, 189.

Grebogi, C., E. Ott, and J. A. Yorke, 1983a, "Are three-frequency quasiperiodic orbits to be expected in typical nonlinear dynamical systems?" Phys. Rev. Lett. **53**, 339.

Grebogi, C., E. Ott, and J. A. Yorke, 1983b, "Crises, sudden changes in chaotic attractors and transient chaos," Physica **7D**, 181.

Guckenheimer, J., 1982, "Noise in chaotic systems," Nature **298**, 358.

Guckenheimer, J., and P. Holmes, 1983, *Nonlinear Oscillations, Dynamical Systems, and Bifurcations of Vector Fields* (Springer, New York).

Gurel, O., and O. E. Rössler, 1979, Eds., *Bifurcation Theory and Applications in Scientific Disciplines* (Ann. N. Y. Acad. Sci. **316**).

Haken, H., 1983, *Advanced Synergetics* (Springer, Berlin).

Helleman, R. H. G., 1980, Ed., *International Conference on Nonlinear Dynamics* (Ann. N. Y. Acad. Sci. **357**).

Hénon, M., 1976, "A two-dimensional mapping with a strange attractor," Commun. Math. Phys. **50**, 69.

Hénon, M., and C. Heiles, 1964, "The applicability of the third integral of motion, some numerical experiments," Astron. J. **69**, 73.

Hurewicz, W., and H. Wallman, 1948, *Dimension Theory* (Princeton University, Princeton).

Ilyashenko, Yu. S., 1983, "On the dimension of attractors of k-contracting systems in an infinite dimensional space," Vestn. Mosk. Univ. Ser. 1 Mat. Mekh. 1983 No. 3, 52—58.

Iooss, G., R. Helleman, and R. Stora, 1983, Eds., *Chaotic Behavior of Deterministic Systems* (North-Holland, Amsterdam).

Jakobson, M., 1981, "Absolutely continuous invariant measures for one-parameter families of one-dimensional maps," Commun. Math. Phys. **81**, 39.

Johnson, R. A., K. J. Palmer, and G. Sell, 1984, "Ergodic properties of linear dynamical systems," preprint, Minneapolis.

Kantz, H., and P. Grassberger, 1984, "Repellers, semi-attractors, and long-lived chaotic transients," preprint, Wuppertal.

Kaplan, J. L., and J. A. Yorke, 1979, "Preturbulence: a regime observed in a fluid flow model of Lorenz," Commun. Math. Phys. **67**, 93.

Katok, A., 1980, "Liapunov exponents, entropy and periodic orbits for diffeomorphisms," Publ. Math. IHES **51**, 137.

Katok, A., and J.-M. Strelcyn, 1985, "Smooth maps with singularities, invariant manifolds, entropy and billiards" (in preparation).

Kifer, Yu. I., 1974, "On small random perturbations of some smooth dynamical systems," Izv. Akad. Nauk SSSR Ser. Mat. **38**, No. 5, 1091 [Math. USSR Izv. **8**, 1083 (1974)].

Kifer, Yu. I., 1984, "A multiplicative ergodic theorem for random transformations," preprint.

Ladyzhenskaya, O. A., 1969, *The Mathematical Theory of Viscous Incompressible Flow*, 2nd ed. (Nauka, Moscow, 1970). [2nd English edition Gordon and Breach, New York (1969)].

Lanford, O. E., 1977, in Turbulence Seminar, Proceedings 1976/77, edited by P. Bernard and T. Ratiu, Lecture Notes in Mathematics 615 (Springer, Berlin), pp. 113—116.

Ledrappier, F., 1981a, "Some relations between dimension and Liapunov exponents," Commun. Math. Phys. **81**, 229.

Ledrappier, F., 1981b, "Some properties of absolutely continuous invariant measures on an interval," Ergod. Theory Dynam. Syst. **1**, 77.

Ledrappier, F., and M. Misiurewicz, 1984, "Dimension of invariant measures," preprint.

Ledrappier, F., and L.-S. Young, 1984, "The metric entropy of diffeomorphisms. Part I. Characterization of measures satisfying Pesin's formula. Part II. Relations between entropy, exponents and dimension," preprints, University of California, Berkeley, 1984. "The metric entropy of diffeomorphisms," Bull. Am. Math. Soc. (New Ser.) **11**, 343.

Li, T., and J. A. Yorke, 1975, "Period three implies chaos," Am. Math. Monthly **82**, 985.

Libchaber, A., and J. Maurer, 1979, "Une expérience de Rayleigh-Bénard de géométrie réduite; multiplication, accrochage, et démultiplication de fréquences," J. Phys. (Paris) Colloq. **41**, C3-51.

Lieb, E. H., 1984, "On characteristic exponents in turbulence," Commun. Math. Phys. **92**, 473.

Lorenz, E. N., 1963, "Deterministic nonperiodic flow," J. Atmos. Sci. **20**, 130.

Lundqvist, Stig, 1985, Ed., *The Physics of Chaos and Related Problems: Proceedings of the 59th Nobel Symposium, 1984* (Phys. Scr. **T9**, 1).

Mallet-Paret. J., 1976, "Negatively invariant sets of compact

maps and an extension of a theorem of Cartwright;" J. Diff. Equ. 22, 331.
Malraison, B., P. Atten, P. Bergé, and M. Dubois, 1983, "Dimension of strange attractors: an experimental determination for the chaotic regime of two convective systems," J. Phys. (Paris) Lett. 44, L-897.
Mandelbrot, B., 1982, *The Fractal Geometry of Nature* (Freeman, San Francisco).
Mañé, R., 1981, "On the dimension of the compact invariant sets of certain nonlinear maps," in *Dynamical Systems and Turbulence, Warwick 1980*, Lecture Notes in Mathematics 898 (Springer, New York), pp. 230—242.
Mañé, R., 1983, "Liapunov exponents and stable manifolds for compact transformations," in *Geometrical Dynamics*, Lecture Notes in Mathematics 1007 (Springer, Berlin), pp. 522—577.
Manneville, P., 1985, "Liapunov exponents for the Kuramoto-Sivashinsky model," preprint.
Manning, A., 1984, "The dimension of the maximal measure for a polynomial map," Ann. Math. 119, 425.
Mattila, P., 1975, "Hausdorff dimension, orthogonal projections and intersections with planes," Ann. Acad. Sci. Fenn. Ser. A1, Math., 1, 227.
McCluskey, H., and A. Manning, 1983, "Hausdorff dimension for horseshoes," Ergod. Theory Dynam. Syst. 3, 251.
Milnor, J., 1985, "On the concept of attractor," Commun. Math. Phys., in press.
Misiurewicz, M., 1981, "Structure of mappings of an interval with zero entropy," Publ. Math. IHES 53, 5.
Newhouse, S., 1974, "Diffeomorphisms with infinitely many sinks," Topology 13, 9.
Newhouse, S., 1979, "The abundance of wild hyperbolic sets and non-smooth stable sets for diffeomorphisms," Publ. Math. IHES 50, 102.
Newhouse, S., D. Ruelle, and F. Takens, 1978, "Occurrence of strange axiom A attractors near quasiperiodic flows on T^m, $m \geq 3$," Commun. Math. Phys. 64, 35.
Nobel Symposium, 1984, see Lundqvist, 1985.
Oseledec, V. I., 1968, "A multiplicative ergodic theorem. Lyapunov characteristic numbers for dynamical systems," Trudy Mosk. Mat. Obsc. 19, 179 [Moscow Math. Soc. 19, 197 (1968)].
Pesin, Ya. B., 1976, "Invariant manifold families which correspond to nonvanishing characteristic exponents," Izv. Akad. Nauk SSSR Ser. Mat. 40, No. 6, 1332 [Math. USSR Izv. 10, No. 6, 1261 (1976)].
Pesin, Ya. B., 1977, "Lyapunov characteristic exponents and smooth ergodic theory," Usp. Mat. Nauk 32, No. 4 (196), 55 [Russian Math. Survey 32, No. 4, 55 (1977)].
Procaccia, I., 1984, *The static and dynamic invariants that characterize chaos and the relations between them in theory and experiments*, proceedings of the 59th Nobel Symposium (Phys. Scr. T9, 40).
Pugh, C. C., and M. Shub, 1984, "Ergodic attractors" (in preparation).
Raghunathan, M. S., 1979, "A proof of Oseledec's multiplicative ergodic theorem," Israel J. Math. 32, 356.
Riste, T., and K. Otnes, 1984, *Neutron scattering from a convecting nematic: multicriticality, multistability and chaos*, proceedings of the 59th Nobel Symposium (Phys. Scr. T9, 76).
Rössler, O. E., 1976, "An equation for continuous chaos," Phys. Lett. A 57, 397.
Roux, J.-C., R. H. Simoyi, and H. L. Swinney, 1983, "Observation of a strange attractor," Physica D 8, 257.
Roux, J.-C., and H. L. Swinney, 1981, "Topology of chaos in a chemical reaction," in *Nonlinear Phenomena in Chemical Dynamics*, edited by C. Vidal and A. Pacault (Springer, Berlin), pp. 38—43.
Ruelle, D., 1976, "A measure associated with Axiom *A* attractors," Am. J. Math. 98, 619.
Ruelle, D., 1978, "An inequality for the entropy of differentiable maps," Bol. Soc. Bras. Mat. 9, 83.
Ruelle, D., 1979, "Ergodic theory of differentiable dynamical systems," Phys. Math. IHES 50, 275.
Ruelle, D., 1980, "Measures describing a turbulent flow," Ann. N. Y. Acad. Sci. 357, 1.
Ruelle, D., 1981, "Small random perturbations of dynamical systems and the definition of attractors," Commun. Math. Phys. 82, 137.
Ruelle, D., 1982a, "Characteristic exponents and invariant manifolds in Hilbert space," Ann. Math. 115, 243.
Ruelle, D., 1982b, "Large volume limit of the distribution of characteristic exponents in turbulence," Commun. Math. Phys. 87, 287.
Ruelle, D., 1984, "Characteristic exponents for a viscous fluid subjected to time dependent forces," Commun. Math. Phys. 93, 285.
Ruelle, D., 1985, "Rotation numbers for diffeomorphisms and flows," Ann. Inst. Henri Poincaré, 42, 109.
Ruelle, D., and F. Takens, 1971, "On the nature of turbulence," Commun. Math. Phys. 20, 167. Note concerning our paper "On the nature of turbulence," Commun. Math. Phys. 21, 21.
Shaw, R. S., 1981, "Strange attractors, chaotic behavior and information flow," Z. Naturforsch. 36a, 80.
Shimada I., 1979, "Gibbsian distribution on the Lorenz attractor," Prog. Theor. Phys. 62, 61.
Sinai, Ya. G., 1972, "Gibbs measures in ergodic theory," Usp. Mat. Nauk 27, No. 4, 21 [Russian Math. Surveys 27, No. 4, 21 (1972)].
Smale, S., 1967, "Differentiable dynamical systems," Bull. Am. Math. Soc. 73, 747.
Takens, F., 1981, "Detecting strange attractors in turbulence," in *Dynamical Systems and Turbulence, Warwick 1980*, Lecture Notes in Mathematics 898 (Springer, Berlin), pp. 366—381.
Takens, F., 1983, "Invariants related to dimension and entropy," in *Atas do 13. Col. brasiliero de Matematicas, Rio de Janerio, 1983.*
Temam, R., 1979, *Navier-Stokes Equations*, 2nd ed. (North-Holland, Amsterdam).
Thieullen, P., 1985, "Exposant de Liapounov des fibrés dynamiques pseudocompacts," in preparation.
Ushiki, S., M. Yamaguti, and H. Matano, 1980, "Discrete population models and chaos," Lecture Notes Num. Appl. Anal. 2, 1.
Vidal, C., and A. Pacault, 1981, Eds., *Nonlinear Phenomena in Chemical Dynamics: Proceedings* (Springer, Berlin).
Vul, E. B., Ya. G. Sinai, and K. M. Khanin, 1984, "Universality of Feigenbaum and thermodynamic formalism," Usp. Mat. Nauk 39, No. 3 (237), 3.
Walden, R. W., P. Kolodner, A. Passner, and C. M. Surko, 1984, "Nonchaotic Rayleigh-Bénard convection with four and five incommensurate frequencies," Phys. Rev. Lett. 53, 242.
Walters, P., 1975, *Ergodic Theory—Introductory Lectures*, Lecture Notes in Mathematics 458 (Springer, Berlin).
Wilkinson, J. H., 1965, *The Algebraic Eigenvalue Problem* (Clarendon, Oxford).
Wilkinson, J. H., and C. Reinsch, 1971, *Linear Algebra* (Springer, Berlin).
Wolf, A., J. B. Swift, H. L. Swinney, and J. A. Vastano, 1984,

"Determining Liapunov exponents from a time series," Physica D, in press.
Young, L.-S., 1981, "Capacity of attractors," Ergod. Theory Dynam. Syst. **1**, 381.
Young, L.-S., 1982, "Dimension, entropy and Liapunov exponents," Ergod. Theory Dynam. Syst. **2**, 109.
Young, L.-S., 1984, "Dimension, entropy and Liapunov exponents in differentiable dynamical systems," Physica A **124**, 639.

Another Proof of Jakobson's Theorem and Related Results

Marek Ryszard Rychlik

Department of Mathematics
University of Arizona
Tucson, AZ 85721 USA
phone: (602) 621-6864, FAX: (602) 621-8322
rychlik@math.arizona.edu **

Summary. The author shows that any family C^2-close to $f_\alpha(x) = 1 - ax^2$ $(2 - \epsilon \leq a \leq 2)$ satisfies Jakobson's theorem: For a positive measure set of α the transformation f_α has an absolutely continuous invariant measure. He also indicates some generalizations.

Introduction

In recent years there has been major interest in the following theorem of Jakobson:

Theorem. *Let $f_r(x) = rx(1-x)$, $0 \leq r \leq 4$, be a one parameter family of mappings of the unit interval. There is a positive measure set of those r for which f_r has an absolutely continuous invariant measure (abbreviation: a.c.i.m.).*

The author of the current paper uses his earlier ideas from [5] to give another proof of this theorem. It seems to be less technical than other existing proofs (see [2], [1]) and therefore it yields some interesting generalizations (Section 6). In particular, any family C^2-close to the one above satisfies Jakobson's theorem. Also, the families that contain $f_4(x) = 4x(1-x)$ and do not satisfy Jakobson's theorem form a 'set of codimension ∞' in the set of C^2-families that contain f_4 (or any mapping C^2-close to f_4 with the property that f^2(critical point) = fixed point). In particular, any analytic family of this type satisfies Jakobson's theorem.

** This article was originally published in journal form in *Ergod. The. & Dynam. Sys.* (1988), **8**, 93–109. With kind permission from the publishers it is included in this volume. The author's address at the time of writing of this article was: University of Washington, Department of Mathematics, GN-50, Seattle, Washington 98195 USA and Institute of Mathematics, University of Warsaw, Poland.

One reason to understand Jakobson's result is a possible generalization to higher dimensions. Similar phenomena seem to accompany every period-doubling bifurcation, when we pass the critical value of the parameter (and the 'chaos' is born!). So far M. Rees has found an analogue for rational mappings of the Riemann sphere [4]. (Probably our proof can be modified to work in that case also.)

Let us say a few things about our notation. The Lebesgue measure of a set A is denoted $|A|$. Variables as subscripts mean differentiation. Occasionally we use prime for the derivative over x or when the parameter is the only variable. Sometimes we do not say explicitly that an object depends on the parameter. We also call a set an interval, where obviously the set has two components. For technical reasons we work with the family $f(x,a) = 1 - ax^2$ and interval $[-1,1]$.

The author would like to apologize if some ideas are being used without references. It is difficult, though, to write a proof disjoint with the existing work.

We would like to express our thanks to the referee, whose work allowed us to eliminate many mistakes and other deficiencies of the first version.

The author would like to thank Laura Plaut for her careful typesetting of this paper.[3]

This work was partially supported by NSF Grant DMS 8320356.

1 Sketch of the proof

For $a = 2$ the point $f^2(0) = -1$ is a repelling fixed point.

Let us define $I_n = [-3^{-n}, 3^{-n}]$ for $n = 0, 1, \ldots$. Let us fix $\rho \in (0,1)$ very close to 1. Let $V_n = f(\rho^{-1} I_n)$. It is easy to show (see Appendix) that $f^{n+1}|\rho^{-1}I_n \backslash \rho I_{n+1}$ is expanding with constants $\Lambda_0 = \frac{4}{3}\rho$. It is also easy to show that the set $V_{n+1} \cup f(V_{n+1}) \cup \cdots \cup f^n(V_{n+1}) \subset \{x : |f'(x)| \geq 3\}$. Let $W_n = f^n(V_{n+1})$. We will show that $|W_n| \leq \rho^{-1}\left(\frac{4}{9}\right)^{n-1}$, so $|W_n| \to 0$, as $n \to \infty$.

Let us define $T : I \to I$ by

$$T(x) = f^n(x), \quad \text{as } x \in M_n \stackrel{\text{def}}{=} I_{n-1} \backslash I_n \tag{1}$$

for $n = 1, 2, \ldots$. This transformation is piecewise expanding. From [5] we know that T has an a.c.i.m. ν_0 with bounded density. We can easily verify that the measure

$$\nu = \sum_{n=1}^{\infty} \sum_{k=0}^{n-1} f_*^k(\nu_0 | M_n) \tag{2}$$

[3] The current version was typeset by the author, and all typographical errors are due to him, as well as several corrections of the typographical errors of the original article.

is an a.c.i.m. for f. It is finite, since

$$\nu(I)=\sum_{n=1}^{\infty} n\nu(M_n)\le \text{const.} \sum_{n=1}^{\infty} n\,|M_n|<\infty. \tag{3}$$

We would like to point out that this measure is well known and has density const. $\left(1-x^2\right)^{-1/2}$. The construction we have just presented is a starting point to our construction of a.c.i.m. for $\alpha\neq 2$. It also produces a.c.i.m. for families like $f(x)=1-2x^2+\xi x^2\left(1-x^2\right)$, where ξ is a small parameter. The author does not know any explicit formula for the density of the a.c.i.m. in this case.

Let us start with the observation that given arbitrarily large $N\in\mathbb{Z}^+$ there is $\alpha_N<2$ such that $f^{n+1}|\rho^{-1}I_n\backslash\rho I_{n+1}$ is an expanding for $n\le N$ and $\alpha\in[\alpha_N,2]$. For $\alpha=2$ the set W_n contains -1 for every n. For $\alpha\neq 2$ the set W_n approaches the critical point. The extreme case is when for some n we have $f^n(0)=0$. In this situation 0 attracts a.e. orbit of f and no a.c.i.m. exists. Our hope is that by varying α we can push W_n away from 0. Actually, we will put W_n into a set $\rho^{-1}I_k\backslash\rho I_{k+1}$ for some $k<n$.

Let us fix N sufficiently large and let $\alpha_N<2$ be such that for all $\alpha\in[\alpha_N,2]$ and $n=0,1,\ldots,N$ we have $W_n\subset\{x:|f'(x)|\ge 3\}$. Let $T_{(n)}=f^{n+1}|\rho^{-1}I_n\backslash\rho I_{n+1}$ for $n=0,1,\ldots,N-1$ and let $S_{(n)}=f^n|V_n$ for $n=1,2,\ldots,N$. We have

$$T_{(n)}=S_{(n)}\circ f \quad\text{and}\quad S_{(n+1)}=T_{(0)}\circ S_{(n)} \tag{4}$$

for $n=0,1,\ldots,N-1$.

We will try to extend the definition of expandings $T_{(n)}$ and $S_{(n)}$ on $n>N$ (at least for some parameters). We are going to use induction.

Suppose that I_m, $S_{(m)}$ have been defined for $m\le n$ $(n\ge N)$ and let $T_{(m)}=S_{(m)}\circ f$ for $m<n$. Let us fix $\tau\in\left(\frac12,1\right)$ to the solution of the equation $4^\tau=3$ and let $a=\frac13$.

We define

$$\begin{aligned} I_{n+1}&=\left[-a\left|S'_{(n)}(1)\right|^{-\tau},\ a\left|S'_{(n)}(1)\right|^{-\tau}\right],\\ V_{n+1}&=f\left(\rho^{-1}I_{n+1}\right). \end{aligned} \tag{5}$$

Now we can define $T_{(n)}=S_{(n)}\circ f|\rho^{-1}I_n\backslash\rho I_{n+1}$ and $W_n=S_{(n)}(V_{n+1})$.

Suppose that $W_n\subset\rho^{-1}I_k\backslash\rho I_{k+1}$ for some $k=k(n)<n$. In this case we set

$$S_{(n+1)}=T_{(k)}\circ S_{(n)}|V_{n+1}\quad (k=k(n)). \tag{6}$$

As we have already mentioned, for some parameters there is no k with the property $W_n\subset\rho^{-1}I_k\backslash I_{k+1}$. We need to discard those parameters to proceed with the next step of our induction. Let us describe in detail how we do it.

Let $(s_m)_{m=0}^n$ be a sequence of integers such that $T_{(m)} = f^{s_m}$ (on $\rho^{-1}I_m\backslash\rho^{-1}I_{m+1}$). We impose an extra condition in our construction.

We fix a sufficiently small $\beta > 0$ and require

$$s_{k(m)} \leq \max(\beta m, 1), \quad (m = 0, 1, \dots, n). \tag{7}$$

Along with I_n and $S_{(n)}$ we construct families of intervals $\mathcal{A}_n$ in the space of parameters. This is the corresponding inductive definition:

(i) $\mathcal{A}_N$ consists of a single interval $[\alpha_N, 2]$.

(ii) If $J \in \mathcal{A}_n$ then for $\alpha \in J$ all n steps of our construction work and yield the same sequence $(s_m)_{m=0}^{n-1}$. Let us consider intervals

$$J_k = \left\{\alpha \in J : \ S_{(n)}(1,\alpha) \in I_k\backslash I_{k+1}\right\}. \tag{8}$$

The family $\mathcal{A}_{n+1}$ consists of all intervals J_k for all $J \in \mathcal{A}_n$ and k such that $s_k \leq \beta n$.

We will show that if $\alpha \in J_k$ then $W_n \subset \rho^{-1}I_k\backslash\rho I_{k+1}$. This is possible because the length of $|W_n|$ decays much faster than the length of $I_{k(n)}$.

The set of parameters $J\backslash\bigcup J_k$ is in general nonempty, but we will see that there is $\lambda \in (0,1)$ such that

$$J\backslash\bigcup J_k \subset \left\{\alpha \in J : \ \text{dist}\left(S_{(n)}(1,\alpha), 0\right) \leq a\lambda^{\sqrt{n}}\right\}. \tag{9}$$

Let $A_n = \bigcup \mathcal{A}_n$ and let B_n be the union of the sets on the right-hand side of (9). The crucial part of the proof is to show that $|B_n| \leq \text{const.}\ \lambda^{\sqrt{n}}\,|A_N|$. Once we have obtained this estimate, we can write (in view of $A_n\backslash A_{n+1} \subset B_n$):

$$\begin{aligned} |A_n| &\geq \left|A_N\backslash \bigcup_{m=N}^{n-1} B_m\right| \geq |A_N| - \sum_{m=N}^{n-1} |B_m| \\ &\geq |A_N|\left(1 - \text{const.} \sum_{m=N}^{n-1} \lambda^{\sqrt{m}}\right). \end{aligned} \tag{10}$$

Let $A_\infty = \bigcap_{n=N}^{\infty} |A_n|$. Letting $n \to \infty$ in (10) we get

$$|A_\infty| \geq |A_N|\left(1 - \text{const.} \sum_{m=N}^{\infty} \lambda^{\sqrt{m}}\right). \tag{11}$$

By fixing N sufficiently large we can make the ratio $|A_\infty| / |A_N|$ arbitrarily close to 1, in particular $|A_\infty| > 0$.

From our estimates it easily follows that if $\alpha \in A_\infty$ then f has an a.c.i.m. First we construct a piecewise expanding $T(x) = f^{s_n}(x)$, as $x \in I_n\backslash I_{n+1}$ ($n = 0, 1, \dots$) which has an a.c.i.m. ν_0 with bounded density. Formulas analogous to (2) and (3) yield an a.c.i.m. ν for f.

2 Certain consequences of the Chain Rule

Let u and v be functions of x and α, where α is a parameter. For example, $w = v \circ u$ means that $w(x, \alpha) = v(u(x, \alpha), \alpha)$. We can easily verify:

Lemma 1. *The following formulas hold:*

(i) $w_x = (v_x \circ u)\, u_x$,

(ii) $\dfrac{w_\alpha}{w_x} = \left(\dfrac{v_\alpha}{v_x} \circ u\right) \dfrac{1}{u_x} + \dfrac{u_\alpha}{u_x}$,

(iii) $\dfrac{w_{xx}}{w_x^2} = \dfrac{v_{xx}}{v_x^2} \circ u + \left(\dfrac{1}{v_x} \circ u\right) \dfrac{u_{xx}}{u_x^2}$,

(iv) $\dfrac{w_{\alpha x}}{w_x^2} = \left(\dfrac{v_{\alpha x}}{v_x^2} \circ u\right) \dfrac{1}{u_x} + \left(\dfrac{v_{xx}}{v_x^2} \circ u\right) \dfrac{u_\alpha}{u_x} + \left(\dfrac{1}{v_x} \circ u\right) \dfrac{u_{\alpha x}}{u_x^2}$,

(v) $\dfrac{w_{\alpha\alpha}}{w_x^2} = \left(\dfrac{v_{\alpha\alpha}}{v_x^2} \circ u\right) \dfrac{1}{u_x^2} + 2\left(\dfrac{v_{\alpha x}}{v_x^2} \circ u\right) \dfrac{u_\alpha}{u_x} \dfrac{1}{u_x} + \left(\dfrac{v_{xx}}{v_x^2} \circ u\right) \left(\dfrac{u_\alpha}{u_x}\right)^2 + \left(\dfrac{1}{v_x} \circ u\right) \dfrac{u_{\alpha\alpha}}{u_x^2}$.

Let us introduce the following notations:

$$
\begin{aligned}
\Delta_{xx}(u) &= \frac{|u_{xx}|}{|u_x|^2}, \quad \Delta_{\alpha x}(u) = \frac{|u_{\alpha x}|}{|u_x|^2}, \quad \Delta_{\alpha\alpha}(u) = \frac{|u_{\alpha\alpha}|}{|u_x|^2}, \\
\Delta(u) &= \max\left(\Delta_{xx}(u), \Delta_{\alpha x}(u), \Delta_{\alpha\alpha}(u)\right), \\
\delta(u) &= \frac{u_\alpha}{u_x}, \\
R_{xx}(u) &= 1, \\
R_{\alpha x}(u) &= \left|\frac{1}{u_x}\right| + \left|\frac{u_\alpha}{u_x}\right|, \\
R_{\alpha\alpha}(u) &= \frac{1}{|u_x|^2} + 2\left|\frac{u_\alpha}{u_x}\right| \frac{1}{|u_x|} + \left|\frac{u_\alpha}{u_x}\right|^2, \\
R(u) &= \max\left(R_{xx}(u), R_{\alpha x}(u), R_{\alpha\alpha}(u)\right).
\end{aligned} \tag{12}
$$

Lemma 2. *Let $w = v \circ u$. Then*

$$
\Delta(w) \le \left(\frac{1}{|v_x|} \circ u\right) \Delta(u) + R(u)(\Delta(v) \circ u). \tag{13}
$$

Moreover, this formula holds with $\Delta(w)$, $\Delta(u)$ and $R(u)$ (but not $\Delta(v)$) subscripted with 'xx', 'αx' or '$\alpha\alpha$'.

Proof. By inspection of formulas (iii)-(v) of Lemma 1.

Remark 1. If $|u_x| \ge 1$ then $R(u) \le 4 \max\left(1, |\delta(u)|^2\right)$.

3 Basic concepts

Let f be a C^2-function of x and α. As a function of x, f is a transformation $I \to I$, where $I = [-1, 1]$. We will deal with transformations $T = f^d : \mathcal{D}(T) \to \mathcal{R}(T)$, where $\mathcal{D}(T)$ and $\mathcal{R}(T)$ are subintervals of I. The integer $d \geq 1$ is called the degree of T and denoted by $\deg(T)$.

Let $\Lambda_0 > 1$. We will call T Λ_0-expanding, if $|T'| \geq \Lambda_0$.

Definition 1. *A sequence of expanding maps $(T_i)_{i=1}^n$ is called β-homogeneous ($\beta \geq 0$) if $\deg(T_i) \leq max(\beta i, 1)$ for $i = 1, 2, \ldots, n$.*

This notion proves useful because of the following:

Lemma 3. *Let $(T_i)_{i=1}^n$ be a β-homogeneous sequence of Λ_0-expandings and let $S_i = T_i \circ T_{i-1} \circ \cdots \circ T_1$ for $i = 1, 2, \ldots, n$. Suppose that $\beta \leq 1$. Then*

$$|T_i'| \circ S_{i-1} \leq R_0 \left|S_{i-1}'\right|^{\epsilon}, \tag{14}$$

where R_0 is a constant such that $\sup |f'| \leq R_0$ and $\epsilon = \beta \log R_0 / \log \Lambda_0$.

Proof. $|T_i'| \leq R_0^{\deg(T_i)} \leq R^{\max(\beta i, 1)} \leq \max\left(R_0, R_0^{\beta i}\right)$. Also, $R_0 \left|S_{i-1}'\right|^{\epsilon} \geq R_0 \left(\Lambda_0^{i-1}\right)^{\epsilon} = R_0 R_0^{\beta(i-1)} \geq R_0^{\beta i}$.

Remark 2. (14) holds with T_i replaced by any T verifying $\deg(T) \leq \max(\beta i, 1)$.

Definition 2. *A sequence of functions $(\phi_i)_{i=1}^n$ is called (C, q)-stable ($C \geq 0$, $q \in (0, 1)$) if*

$$|\phi_i - \phi_j| \leq C q^{\min(i,j)} \tag{15}$$

for $i, j = 1, 2, \ldots, n$.

Definition 3. (i) *We say that T has rank (μ, A) if*

$$|\delta(T)| \leq A \left|T_x\right|^{\mu}. \tag{16}$$

(ii) *We say that T has type (σ, B) if*

$$\Delta(T) \leq B \left|T_x\right|^{\sigma}. \tag{17}$$

Theorem 1. *Suppose $(T_i)_{i=1}^n$ is a β-homogeneous sequence of Λ_0-expandings and T_i has rank (μ, A) for $i = 1, 2, \ldots, n$. Let $S_i = T_i \circ \cdots \circ T_1$ for $i \leq n$ and let $S = S_n$, $S_0 = id$. There are constants C_1 and $q \in (0, 1)$ independent of A, μ or n such that for sufficiently small β we have:*

(i) *$\delta(S) \leq C_1$ (equivalently, S has rank $(0, C_1)$).*

(ii) *The sequence of functions $(\delta(S_i))_{i=0}^n$ is (C_1, q)-stable.*

Proof. From Lemma 1 we derive by induction the following formula ($i < j$):

$$\delta(S_j) = \sum_{l=i+1}^{j} (\delta(T_l) \circ S_{l-1}) \left(S'_{l-1}\right)^{-1} + \delta(S_i). \tag{18}$$

This gives

$$|\delta(S_j) - \delta(S_i)| \leq \sum_{l=i+1}^{j} (|\delta(T_l)| \circ S_{l-1}) \left|S'_{l-1}\right|^{-1}. \tag{19}$$

For $l \leq [\beta^{-1}]$ we have $|\delta(T_l)| \leq R_1$ where R_1 depends on f only (note: $\deg(T_l) = 1$). For $l > [\beta^{-1}]$ we have

$$|\delta(T_l)| \circ S_{l-1} \leq A \left|T'_l\right|^{\mu} \circ S_{l-1} \leq A R_0^{\mu} \left|S'_{l-1}\right|^{\epsilon\mu}$$

by Lemma 3. We always have $\left|S'_{l-1}\right| \geq \Lambda_0^{l-1}$. Therefore,

$$|\delta(S_j) - \delta(S_i)| \leq \sum_{1+i \leq l \leq [\beta^{-1}]} R_1 \Lambda_0^{-(l-1)} + \sum_{[\beta^{-1}] < l < j} R_0^{\mu} A \Lambda_0^{-(1-\epsilon\mu)(l-1)}. \tag{20}$$

Suppose $\epsilon\mu < 1/3$ (i.e. $\beta < \log \Lambda_0 / 3\mu \log R_0$). Let $q = \Lambda_0^{-\frac{1}{3}}$ and let β be such that $R_0^{\mu} A \leq R_1 q^{-[\beta^{-1}]}$. The right-hand side of (20) does not exceed

$$\sum_{l=i+1}^{j} R_1 q^{l-1} \leq R_1 (1-q)^{-1} q^i$$

(the terms of the first sum are obviously $\leq R_1 q^{l-1}$ and the second sum $\leq R_1 q^{2(l-1)-[\beta^{-1}]} \leq R_1 q^{l-1}$, since $l - 1 \geq [\beta^{-1}]$). Therefore, the Lemma holds with $C_1 = R_1(1-q)^{-1}$. We notice that (i) can be easily obtained from (ii), since $\delta(S_0) = 0$.

Theorem 2. *Let ν be an arbitrary positive number. Suppose that the assumptions of Theorem 1 are satisfied and, in addition, T_i has type (σ, B) for $i = 1, 2, \ldots, n$. There is a constant C_2 independent of A, B, μ, ν, σ or n such that if β is sufficiently small then S has type (ν, C_2).*

Proof. Lemma 2, Remark 1 and Theorem 1(i) give

$$\Delta(S_n) \leq \Lambda_0^{-1} \Delta(S_{n-1}) + C_3 (\Delta(T_n) \circ S_{n-1}), \tag{21}$$

where $C_3 = 4 \max\left(1, C_1^2\right)$. Therefore, by induction we get

$$\Delta(S_n) \leq C_3 \sum_{i=1}^{n} \Lambda_0^{-(n-1)} \Delta(T_i) \circ S_{i-1} \leq C_3 \left(1 - \Lambda_0^{-1}\right)^{-1} \max_{1 \leq i \leq n} \Delta(T_i) \circ S_{i-1}. \tag{22}$$

We can assume that for $i \leq [\beta^{-1}]$ we have $\Delta(T_i) \leq R_2$ and for $i > [\beta^{-1}]$ we have $\Delta(T_i) \circ S_{i-1} \leq B\,|T_i'|^{\sigma} \circ S_{i-1} \leq R_0^{\sigma} B\,|S'|^{\epsilon\sigma}$. Therefore

$$\Delta(S) \leq \frac{C_3}{1 - \Lambda_0^{-1}} \max\left(R_2, R_0^{\sigma} B\,|S'|^{\epsilon\sigma}\right). \tag{23}$$

Suppose that $\epsilon\sigma < \nu/2$ (i.e. $\beta < \nu \log \Lambda_0 / 2\sigma \log R_0$) and β is such that $R_2 \Lambda_0^{[\beta^{-1}]\nu/2} \geq R_0^{\sigma} B$. Then

$$R_0^{\sigma} B\,|S'|^{\epsilon\sigma} \leq R_0^{\sigma} B\,|S'|^{\nu/2} \leq \left(R_0^{\sigma} B / |S'|^{\nu/2}\right) |S'|^{\nu}.$$

We also have

$$R_0^{\sigma} B / |S'|^{\nu/2} \leq R_0^{\sigma} B \Lambda_0^{-n\nu/2} \leq R_0^{\sigma} B \Lambda_0^{-[\beta^{-1}]\nu/2} \leq R_2.$$

Therefore

$$\Delta(S) \leq \frac{C_3}{1 - \Lambda_0^{-1}} R_2\,|S'|^{\nu}$$

and we can set C_2 to $R_2 C_3 / \left(1 - \Lambda_0^{-1}\right)$.

Definition 4. *Let $V \subset \mathcal{D}(T)$. The number* $\sup_V |T'| / \inf_V |T'|$ *is called the* distortion *of T on V. We omit V, if $V = I$.*

The next lemma is well known in the theory of expandings.

Lemma 4. *Let $S = T_n \circ \cdots \circ T_1$, where T_i is a Λ_0-expanding for $i = 1, 2, \ldots, n$ and $\deg(T_i) = 1$. The distortion of S does not exceed $C_4 = \exp\left(2R_2\left(1 - \Lambda_0^{-1}\right)^{-1}\right)$.*

Proof. In a similar way as in the proof of Theorem 2 we get $\left|S'' / (S')^2\right| \leq C_5 = R_2\left(1 - \Lambda_0^{-1}\right)^{-1}$. Therefore, for $y, z \in \mathcal{D}(S)$ we have

$$\left|\ln \frac{S'(y)}{S'(z)}\right| \leq \int_{[y,z]} \left|\frac{S''}{S'}\right| \leq C_5 \int_{[y,z]} |S'| = C_5\,|S(y) - S(z)| \leq 2C_5. \tag{24}$$

This gives $S'(y)/S'(z) \leq \exp(2C_5) = C_4$ (note: $S'(y)$, $S'(z)$ have the same sign).

Possessing type (σ, B) does not imply bounded distortion on the whole of I, though the distortion is bounded on sufficiently small intervals.

Theorem 3. *Let S be an expanding of type (ν, C_2). For every $\theta_0 > 1$ there is $\eta_0 > 0$ s.t. if for some interval $V \subset \mathcal{D}(S)$ and $y_0 \in V$ we have $|V| \leq \eta_0\,|S'(y_0)|^{-(1+\nu)}$ then the distortion of S on V is bounded by θ_0.*

Proof. Let $y \in V$. We can assume that $S' \geq 0$. We have

$$\left|\frac{1}{S'(y)^{1+\nu}} - \frac{1}{S'(y_0)^{1+\nu}}\right| \leq (1+\nu)\int_{[y,y_0]}\left|\frac{S_{xx}}{S_x^{2+\nu}}\right|$$
$$\leq (1+\nu)C_2\,|V| \leq (1+\nu)\eta_0 C_2\,|S'(y_0)|^{-(1+\nu)} \quad (25)$$

Multiplying (25) by $|S'(y_0)|^{1+\nu}$ yields

$$\left|\left(\frac{S'(y_0)}{S'(y)}\right)^{1+\nu} - 1\right| \leq (1+\nu)\eta_0 C_2. \quad (26)$$

Now it's clear that the distortion is arbitrarily close to 1, if η_0 is sufficiently small.

Now let us go back to the construction of Section 1. We will apply our results to the β-homogeneous sequence $T_i = T_{(k(i))}$ $(i = 1, 2, \ldots, n)$ and $\alpha \in J \in \mathcal{A}_n$. Obviously, $S_i = T_i \circ \cdots \circ T_1 = S_{(i)}$. From now on we will often write S_n instead of $S_{(n)}$, which is consistent with the formula and simplifies our notation.

Theorem 4. *Suppose that $S = S_n$ has rank $(0, C_1)$, type (ν, C_2) and distortion $\leq C_4$. Let $T = T_{(n)} = S \circ f|\rho^{-1}I_n \backslash \rho I_{n+1}$. There exist constants C_7, A, B, μ, σ independent of n such that T has rank (μ, A) and type (σ, B). Moreover, $|T'| \geq C_7\,|S'(1)|^{1-\tau}$.*

Proof. Let us fix C_0 such that

$$\begin{aligned} C_0^{-1}\,|x| \leq |f_x| \leq C_0\,|x|, \\ \max(R(f), \Delta(f)) \leq C_0|x|^{-2}, \\ |\delta(f)| \leq C_0. \end{aligned} \quad (27)$$

We have $|T'| = (|S'| \circ f)\,|f'| \geq C_4^{-1}\,|S'(1)|\,C_0^{-1}\,|x| \geq C_4^{-1}C_0^{-1}a\,|S'(1)|^{1-\tau}$ (see (5)). We set $C_7 = C_4^{-1}C_0^{-1}a$. We notice that

$$|S'(1)| \leq C_8\,|T'|^{1/(1-\tau)}, \quad (28)$$

where $C_8 = C_7^{-1/(1-\tau)}$.

Lemma 2 gives

$$\begin{aligned} \Delta(T) &\leq (|S'|^{-1} \circ f)\Delta(f) + (\Delta(S) \circ f) \cdot R(f) \\ &\leq \max(\Delta(f), R(f)) \times (1 + \Delta(S) \circ f) \\ &\leq C_0\,|x|^{-2}\left(1 + C_2\left(C_4\,|S'(1)|\right)^{\nu}\right) \\ &\leq C_0\left(1 + C_2C_4^{\nu}\right)a^{-2}\,|S'(1)|^{2\tau+\nu} \\ &\leq C_0\left(1 + C_2C_4^{\nu}\right)a^{-2}C_8^{2\tau+\nu}\,|T'|^{(2\tau+\nu)/(1-\tau)} \end{aligned} \quad (29)$$

This yields $\sigma = (2\tau+\nu)/(1-\tau)$ and $B = C_0\left(1+C_2C_4^{\nu}\right)a^{-2}C_8^{2\tau+\nu}$. We have

$$\begin{aligned}|\delta(T)| &= \left|(\delta(S)\circ f)f_x^{-1}+\delta(f)\right| \\ &\le C_1\cdot C_0\,|x|^{-1}+C_0 \le a^{-1}C_1C_0\,|S'(1)|^{\tau}+C_0 \\ &\le \left(1+a^{-1}C_1\right)C_0C_8^{\tau}\,|T'|^{\tau/(1-\tau)}. \end{aligned} \tag{30}$$

This yields $A = \left(1+a^{-1}C_1\right)C_0C_8^{\tau}$, $\mu = \tau/(1-\tau)$.

Remark 3. We can apply the method of the proof to $\Delta_{xx}(T)$, $\Delta_{\alpha x}(T)$, $\Delta_{\alpha\alpha}$ separately. We obtain useful estimates

$$\begin{aligned}\Delta_{xx}(T) &\le C_9\,|S'(1)|^{\max(\nu,2\tau-1)} \\ \Delta_{\alpha x}(T) &\le C_9\,|S'(1)|^{\nu+\tau} \\ \Delta_{\alpha\alpha}(T) &\le C_9\,|S'(1)|^{\nu+2\tau}. \end{aligned} \tag{31}$$

From now on we assume $\nu < 2\tau-1$, which reduces the first exponent to $2\tau-1$.

Remark 4. There exists $\theta_0 > 1$ such that for sufficiently large N, $\alpha\in[\alpha_N,2]$ and all $n\ge 1$ we have $|S_n'(1)|\ge\theta_0^n$ and $\left|T_{(n)}'\right| \ge \max\left(\theta_0, C_7\theta_0^{n(1-\tau)}\right)$. In particular, $\inf\left|T_{(n)}'\right|$ grows exponentially.

Proof. Let us pick $\theta_0\in\left(1,\frac{4}{3}\rho\right)$ arbitrarily and let N be large enough, so that for all $\alpha\in\mathcal{A}_N = [\alpha_N,2]$ we have $\left|T_{(n)}'\right|\ge\theta_0$ for all n satisfying $(*)$ $C_7\theta_0^{n(1-\tau)}<\theta_0$. The number of such n is bounded, so this is possible by the Appendix and C^1-continuity of our family (since $\alpha_N\to 2$, as $N\to\infty$).

We will show by induction that

(i) $\left|S_{(n)}'(1)\right|\ge\theta_0^n$,

(ii) $\left|T_{(k)}'\right|\ge\theta_0$ for all $k<n$.

This is obvious for $n=1$. Using the induction hypothesis and Theorem 4, we get $\left|T_{(n)}'\right|\ge C_7\theta_0^{n(1-\tau)}$, which is $\ge\theta_0$, if n does not satisfy $(*)$. If n does satisfy $(*)$ then (ii) is obvious for $k=n$. Now we apply the definition $S_{(n+1)} = T_{(k)}\circ S_{(n)}$ $(k<n)$ and get $\left|S_{n+1}'(1)\right|\ge\theta_0^{n+1}$, which completes the proof.

Corollary 1. *There exist constants C_1, C_2, C_4, A, B, Λ_0, ν, μ (all positive and $\Lambda_0>1$) s.t. for N sufficiently large and β sufficiently small and for every $n\ge1$ and $\alpha\in A_{\max(n,N)}$ the mapping $S_{(n)}$ has rank $(0,C_1)$, type (ν,C_2), distortion $\le C_4$ and is Λ_0^n-expanding, and $T_{(n)}$ has rank (μ,A), type (σ,B) and is Λ_0-expanding.*

Proof. Induction. Suppose that $T_{(k)}$ has rank (μ,A) and type (σ,B) for all $k<n$. Theorems 1, 2 (applied to the sequence $T_i = T_{(k(i))}$) and Remark 4

imply that $S_{(n)}$ has rank $(0, C_1)$ and type (ν, C_2). If $n \leq N$ then Lemma 4 implies that the distortion of $S_{(n)}$ does not exceed C_4.

Let us notice that $\mathcal{D}(S_{(n)}) = V_n$ and

$$|V_n| \leq \frac{1}{2} C_0 \rho^{-2} |I_n|^2 \leq \frac{1}{2} C_0 \rho^{-2} (2a)^2 \left| S'_{(n)} \right|^{-2\tau}.$$

This means that Theorem 3 applies to $S_{(n)}$ for $n \geq N$ with $\theta_0 = C_4$, if N is large enough (note: $2\tau > 1 + \nu$).

Eventually from Theorem 4 we get that $T_{(n)}$ has rank (μ, A) and type (σ, B). This ends the proof.

4 Families with a prerepelling critical point

Here we consider families a little more general than $f(x, \alpha) = 1 - \alpha x^2$.

Suppose that f has a critical point $c(\alpha)$ for α close to α_0 and suppose that for $\alpha = \alpha_0$ and some $m \in \mathbb{Z}^+$ the point $f^m(c(\alpha_0), \alpha_0) = x_0$ is a repelling point of period κ. We consider the following nondegeneracy condition (cf. [4]). Let $x(\alpha) = f^m(c(\alpha))$ be a differentiable function and let

$$\left. \frac{d}{d\alpha} \right|_{\alpha=\alpha_0} (f^\kappa(x(\alpha)) - x(\alpha)) \neq 0. \tag{32}$$

Let us define a sequence of functions

$$\chi_n(\alpha) = \left(\frac{d}{d\alpha} f^n(x(\alpha), \alpha) \right) \Big/ (f^n)_x (x(\alpha), \alpha). \tag{33}$$

Proposition 1. *The limit $\chi = \lim_{n \to \infty} \chi_n(\alpha_0)$ exists. It is $\neq 0$ iff (32) is satisfied.*

Proof. Let $T = f^\kappa$. Let $n = l\kappa + l_1$, where $0 \leq l_1 \leq \kappa - 1$. We have by Lemma 1(ii):

$$\delta(f^n) = \sum_{i=1}^{l} (\delta(T) \circ T^{i-1}) \frac{1}{(T^{i-1})'} + \frac{1}{(T^l)'} \left(\sum_{j=1}^{l_1} (\delta(f) \circ f^{j-1}) \cdot \frac{1}{(f^{j-1})'} \right) \circ T^l. \tag{34}$$

Substituting (x_0, α_0) and letting $l \to \infty$ yields:

$$\lim_{n \to \infty} \delta(f^n)(x_0, \alpha_0) = \delta(T)(x_0, \alpha_0) \left(1 - T'(x_0)^{-1}\right)^{-1} = \bar{\chi}. \tag{35}$$

Condition (32) is equivalent to

$$(T_x(x_0, \alpha_0) - 1)\, x'(\alpha_0) \neq -T_\alpha (x_0, \alpha_0) \tag{36}$$

or $x'(\alpha_0) \neq -\bar{\chi}$. On the other hand,

$$\left.\frac{d}{d\alpha}\right|_{\alpha=\alpha_0} f^n(x(\alpha),\alpha) \Big/ (f^n)_x(x_0,\alpha_0)$$
$$= x'(\alpha_0) + \delta(f^n)(x_0,\alpha_0) \to x'(\alpha_0) + \bar{\chi} = \chi, \quad \text{as } n \to \infty. \quad (37)$$

This completes the proof.

Theorem 5. *Let $\chi \neq 0$. For every $\eta > 0$ there is $n_0 \in \mathbb{Z}^+$ and $\delta > 0$ such that for every $n \geq n_0$ and $\alpha \in (\alpha_0 - \delta,\ \alpha_0 + \delta)$: if for some interval $U \subset I$, $f(\cdot,\alpha)|U$ is Λ_0-expanding and $\{x(\alpha), f(x(\alpha),\alpha), \dots, f^n(x(\alpha),\alpha)\} \subset U$ then*

$$|\chi_n(\alpha)/\chi - 1| < \eta. \quad (38)$$

Proof. First, we choose n_0 large enough, so that

$$\begin{aligned} |\chi_{n_0}(\alpha_0) - \chi| &< \frac{1}{4}\eta\,|\chi| \\ |\delta(f^{n_0})(x,\alpha) - \delta(f^n)(x,\alpha)| &< C_1 q^{n_0} < \frac{1}{4}\eta\,|\chi| \end{aligned} \quad (39)$$

(we applied Theorem 1).

We pick δ such that if $|\alpha - \alpha_0| < \delta$ then

$$\begin{aligned} |\delta(f^{n_0})(\chi,\alpha) - \delta(f^{n_0})(x_0,\alpha_0)| &< \frac{1}{4}\eta\,|\chi| \\ |x'(\alpha) - x'(\alpha_0)| &< \frac{1}{4}\eta\,|\chi|. \end{aligned} \quad (40)$$

We get Theorem 5 by the triangle inequality.

Remark 5. We will not dwell on the general case, leaving the details to the reader. We notice that in the case of $f(\alpha,x) = 1 - \alpha x^2$ we have $\chi_n(\alpha) = -2\alpha\delta(S_{(n+1)})(1,\alpha)$. It is easy to see that condition (32) holds for this family.

Corollary 2. *Let $D = |\chi/(2\alpha_0)|$. For every $\theta > 1$ we can choose sufficiently large n_1 such that for every $n \geq n_1$ and every α, α_1, $\alpha_2 \in A_n$*

$$\begin{aligned} h \leq \left|\delta(S_{(n)})(1,\alpha)\right| &\leq H, \\ \theta^{-1} \leq \frac{\delta(S_{(n)})(1,\alpha_1)}{\delta(S_{(n)})(1,\alpha_2)} &\leq \theta, \end{aligned} \quad (41)$$

where $h = \theta^{-1}D$, $H = \theta D$ (see Remark 5 and Theorem 1(ii)).

Let us finish this section with one more distortion estimate, this time over α.

Lemma 5. *For every $\theta_1 > 1$ there is $\eta_1 > 0$ such that if $|\alpha_2 - \alpha_1| \leq \eta_1 \left|(S_n)_\alpha(1,\alpha_1)\right|^{-(1+\nu)}$, $\alpha_1, \alpha_2 \in A_n$, then*

$$\theta_1^{-1} \leq \frac{(S_n)_\alpha(1,\alpha_2)}{(S_n)_\alpha(1,\alpha_1)} \leq \theta_1.$$

Proof. We notice that

$$\left|\frac{(S_n)_{\alpha\alpha}}{(S_n)_\alpha^{2+\nu}}\right| = |\delta(S_n)|^{-(2+\nu)}\left|\frac{(S_n)_{\alpha\alpha}}{(S_n)_\alpha^{2+\nu}}\right| \leq h^{-(2+\nu)}C_2. \tag{42}$$

Integrating over α we get

$$\begin{aligned}\left|\frac{1}{(S_n)_\alpha^{1+\nu}(\alpha_2)} - \frac{1}{(S_n)_\alpha^{1+\nu}(\alpha_1)}\right| &\leq (1+\nu)h^{-(2+\nu)}C_2\,|\alpha_2-\alpha_1| \\ &\leq (1+\nu)h^{-(2+\nu)}C_2\eta_1\,|(S_n)_\alpha(1,\alpha_1)|^{-(1+\nu)}.\end{aligned}$$

Multiplying by $|(S_n)_\alpha(1,\alpha_1)|^{1+\nu}$ leads to:

$$\left|\left(\frac{(S_n)_\alpha(1,\alpha_1)}{(S_n)_\alpha(1,\alpha_2)}\right)^{1+\nu} - 1\right| \leq (1+\nu)h^{-(2+\nu)}C_2\eta_1. \tag{43}$$

If η_1 is sufficiently small, (43) yields the lemma.

Corollary 3. *Using the assumptions and notation of Lemma* 5 *we have*

$$\theta_1^{-1}\theta^{-1} \leq \frac{(S_n)_x(1,\alpha_2)}{(S_n)_x(1,\alpha_1)} \leq \theta_1\theta. \tag{44}$$

Remark 6. The sets J_k have not more components than $I_k\backslash I_{k+1}$, since by simple differentiation we can see that the absolute value of the derivative of $S_{(n)}(1,\alpha)$ over α is $>$ absolute value of the derivative of the ends of $I_k\backslash I_{k+1}$.

Proof. Indeed, we ask if

$$\begin{aligned}a\left|\frac{d}{d\alpha}|S_k'(1,\alpha)|^{-\tau}\right| &= a\tau\,|(S_k)_{\alpha x}(1,\alpha)|\cdot|S_k'(1,\alpha)|^{-(\tau+1)} \\ &< |(S_n)_\alpha(1,\alpha)|.\end{aligned} \tag{45}$$

This is equivalent to

$$a\Delta_{\alpha x}(S_k)\cdot\tau\,|S_k'|^{-(\tau-1)} < |\delta(S_n)|\,|S_n'| \tag{46}$$

and because of $\Delta_{\alpha x}(S_k) \leq C_2\,|S_k'|^\nu$ we need

$$a\tau C_2\,|S_k'|^{\nu-\tau+1} < |S_n'|\,|\delta(S_n)|. \tag{47}$$

We have $|S_k'| \leq |S_n'|$. Also $|\delta(S_n)| \geq h > 0$ for large n. Therefore (47) holds for large n, since $\nu < \tau$ (note: $\nu < 2\tau - 1 < \tau$).

5 The measure of A_∞

We use the notation of Section 1.

Proposition 2. (i) *Suppose that $N \geq 2\beta^{-1}$. The numbers s_m, $m = N, N+1, \ldots, n$ satisfy the inequality*

$$s_m \leq \frac{\beta}{2} m^2 + 1. \tag{48}$$

(ii) *If $\alpha \in J_k \in \mathcal{A}_{n+1}$ then $W_n \subset \rho^{-1} I_k \backslash \rho I_{k+1}$.*
(iii) *There is $\lambda \in (0,1)$ such that (9) is satisfied.*

Proof. (i) Induction. We notice that $s_N = N+1 \leq (\beta/2)N^2+1$, if $N \geq 2/\beta$. The induction step:

$$s_{n+1} = s_n + s_{k(n)} \leq \frac{\beta}{2} n^2 + 1 + \beta n \leq \frac{\beta}{2}(n+1)^2 + 1.$$

(ii) We notice that since $S_{(n)}(1) \in I_k \backslash I_{k+1}$:

$$\begin{aligned} \operatorname{dist}\left(S_{(n)}(1), \rho I_{k+1}\right) &\geq \frac{1}{2}(1-\rho)\left|I_{k+1}\right| \\ \operatorname{dist}\left(S_{(n)}(1), \rho^{-1} I_k\right) &\geq \frac{1}{2}\left(\rho^{-1}-1\right)\left|I_k\right|. \end{aligned} \tag{49}$$

It suffices to show that

$$\begin{aligned} 2\left|W_n\right| &\leq \min((1-\rho)\left|I_{k+1}\right|,\ \left(\rho^{-1}-1\right)\left|I_k\right|) \\ &= (1-\rho)\left|I_{k+1}\right|. \end{aligned} \tag{50}$$

Obviously, we have

$$\begin{aligned} \left|W_n\right| &\leq \sup\left|S'_{(n)}(1)\right| \cdot \left|V_{n+1}\right| \leq C_4\left|S'_{(n)}\right| \cdot \frac{1}{2}\rho^{-2} C_0 \left|I_{n+1}\right|^2 \\ &= \frac{1}{2} C_0 \rho^{-2} C_4 a^{-2}\left|S'_{(n)}(1)\right|^{1-2\tau} \leq \frac{1}{2} C_0 \rho^{-2} C_4 a^{-2} \Lambda_0^{n(1-2\tau)}. \end{aligned} \tag{51}$$

On the other hand, $\left|I_{k+1}\right| = 2a\left|S'_{(k)}(1)\right|^{-\tau} \geq 2aR_0^{-\tau s_k}$, since $\deg(S_{(k)}) = s_k - 1 \leq s_k$. Because $s_k \leq \beta n$ for $n \geq N$, we get $R_0^{-\tau s_k} \geq R_0^{-\tau\beta n} = \Lambda_0^{-\epsilon\tau n}$. Suppose that $\epsilon\tau < 2\tau - 1$. For sufficiently large n we have

$$2a(1-\rho)\Lambda_0^{-\epsilon n\tau} \geq C_0\rho^{-2}C_4 a^{-2}\Lambda^{-n(2\tau-1)} \tag{52}$$

and (ii) holds if N is sufficiently large.
(iii) Suppose that $S_{(n)}(1,\alpha) \in I_k$ and $\deg(T_{(k)}) > \beta n$. Then

$$\operatorname{dist}\left(S_n(1,\alpha), 0\right) \leq \frac{1}{2}\left|I_k\right| = a\left|S'_{(k)}(1)\right|^{-\tau} \leq a\Lambda_0^{-k\tau}.$$

From (i) we now have that $\deg(T_{(k)} = s_k \leq (\beta/2)k^2 + 1$. Hence, $\beta n < (\beta/2)k^2 + 1$ or $k > \sqrt{2n - 2/\beta} \geq \sqrt{n}$, since $n \geq N \geq 2/\beta$. This yields (iii) with $\lambda = \Lambda_0^{-\tau}$.

Let us consider the transformation ψ_n defined on every $J \in \mathcal{A}_n$ by $\psi_n(\alpha) = S_n(1,\alpha)$. We assume $N \geq 2\beta^{-1}$.

Theorem 6. *There is a constant C_{10} such that for every $n \geq N$*

$$\sum_{J\in\mathcal{A}_n} \sup_J \frac{1}{|\psi_n'|} \leq C_{10}\,|A_N|\,. \tag{53}$$

Proof. Let π be the partition of I into intervals of equal length d $(d^{-1} \in \mathbb{Z})$. Let $\mathcal{A}_n'$ be the family of intervals defined in a similar way to $\mathcal{A}_n$, except we replace the definition of J_k with

$$J_k(P) = \{\alpha \in J:\ S_n(1,\alpha) \in (I_k \backslash I_{k+1}) \cap P\}\,, \tag{54}$$

where $P \in \pi$. Since every $J \in \mathcal{A}_n$ is a union of element of $\mathcal{A}_n'$, it is sufficient to prove (53) for $\mathcal{A}_n'$ instead of $\mathcal{A}_n$. We will be able to do it, if d is sufficiently small.

Let us introduce two sequences

$$\begin{aligned}\gamma_n &= \sum_{J\in\mathcal{A}_n'} \sup_J \frac{1}{|\psi_n'|}\\ \eta_n &= \sum_{J\in\mathcal{A}_n'} \sup_J \left|\frac{\psi_n''}{(\psi_n')^2}\right| |J|\,.\end{aligned} \tag{55}$$

We are going to derive estimates of $(\gamma_{n+1},\eta_{n+1})$ in terms of (γ_n,η_n).

We fix an arbitrary $\theta > 1$. First, by Corollary 2 we have for n large enough

$$\begin{aligned}\sup_{J_k(P)} \frac{1}{|\psi_{n+1}'|} &\leq \frac{1}{h} \sup_{J_k(P)} \frac{1}{|S_{n+1}'(1)|}\\ &\leq \frac{1}{h} \sup \frac{1}{\left|T_{(k)}'\right|} \sup_{J_k(P)} \frac{1}{|S_n'(1)|}\\ &\leq \frac{1}{h} \sup \frac{1}{\left|T_{(k)}'\right|} H \sup_{J_k(P)} \frac{1}{|\psi_n'|}\\ &= \theta^2 \sup \left|T_{(k)}'\right|^{-1} \sup_{J_k(P)} |\psi_n'|^{-1}\,.\end{aligned} \tag{56}$$

The first supremum is over $\alpha \in J_k(P)$ and $x \in P \cap (I_k \backslash I_{k+1})$ (note: I_k depends on α). There is $k_1 \in \mathbb{Z}^+$ such that

$$\sum_{k\geq k_1} \sup \left|T_{(k)}'\right| \leq \Lambda_0^{-1} \tag{57}$$

(see Remark 4).

Let us choose d small enough, so that for every $P \in \pi$ one of the two alternatives holds: either (1°) $P \subset I_{k_1}$ or (2°) P intersects not more than two of the sets $I_k \backslash I_{k+1}$.

We get:

$$\sum_k \sup_{J_k(P)} \frac{1}{|\psi'_{n+1}|} \le 2\theta^2 \Lambda_0^{-1} \sup_{J(P)} \frac{1}{|\psi'_n|}, \tag{58}$$

where $J(P) = J \cap \psi_n^{-1}(P)$.

We notice that $\psi_n(J(P)) \neq P$ for at most two $P \in \pi$, namely those for which (*) $P \cap \partial\psi_n(J) \neq \emptyset$. Summing up over these P, we get

$$\sum_{k,P}^{*} \sup_{J_k(P)} \frac{1}{|\psi'_{n+1}|} \le 4\theta^2 \Lambda_0^{-1} \sup_{\alpha \in J} \frac{1}{|\psi'_n|}. \tag{59}$$

For those P that (**) $\psi_n(J(P)) = P$ we have

$$\sup_{J(P)} \frac{1}{|\psi'_n|} \le \frac{1}{|\psi_n(J(P))|} \int_{\psi_n(J(P))} \frac{1}{|\psi'_n|} \circ \psi_n^{-1} + \sup_{\alpha \in J(P)} \left| \frac{\psi''_n}{(\psi'_n)^2} \right| |J(P)|, \tag{60}$$

according to the rule 'maximum $\le$ average+max. of derivative$\times$length of the interval' applied to $(1/|\psi'_n|) \circ \psi_n^{-1}$ on $P = \psi_n(J(P))$; we notice that the average is just $d^{-1}|J(P)|$. Totalling over P satisfying (**) we get

$$\sum_{P,k}^{**} \sup_{J_k(P)} \frac{1}{|\psi'_{n+1}|} \le 2\theta^2 \Lambda_0^{-1} \left(\frac{1}{d} |J| + \sup_J \left| \frac{\psi''_n}{(\psi'_n)^2} \right| \cdot |J| \right). \tag{61}$$

Inequalities (59) and (61) and summation over J yield

$$\gamma_{n+1} \le 4\theta^2 \Lambda_0^{-1} \gamma_n + 2\theta^2 \Lambda_0^{-1} \eta_n + \frac{2\theta^2 \Lambda_0^{-1}}{d} |A_N|. \tag{62}$$

Now we intend to estimate η_{n+1}. We have $\psi_{n+1}(\alpha) = T_{(k)}(\psi_n(\alpha), \alpha)$ and this yields

$$\begin{aligned} \psi''_{n+1}(\alpha) = {} & \left(T_{(k)}\right)_{\alpha\alpha} (\psi_n(\alpha), \alpha) \\ & + 2 \left(T_{(k)}\right)_{\alpha x} (\psi_n(\alpha), \alpha)\, \psi'_n(\alpha) \\ & + \left(T_{(k)}\right)_{xx} (\psi_n(\alpha), \alpha) \left(\psi'_n(\alpha)\right)^2 \\ & + \left(T_{(k)}\right)_x (\psi_n(\alpha), \alpha)\, \psi''_n(\alpha). \end{aligned} \tag{63}$$

We also have $\psi'_n(\alpha) = (S_n)_\alpha (1, \alpha)$. Therefore

$$\begin{aligned} \frac{|\psi''_{n+1}|}{(\psi'_{n+1})^2} \le {} & \delta(S_{n+1})^{-2} \left[\Delta_{xx}(T_{(k)}) \delta(S_n)^2 + \frac{2}{|S'_n|} \Delta_{\alpha x}\left(T_{(k)}\right) |\delta(S_n)| \right. \\ & \left. + \frac{1}{(S'_n)^2} \Delta_{\alpha\alpha}(T_{(k)}) \right] + \frac{1}{\left|\left(T_{(k)}\right)_x\right|} \frac{\delta(S_n)^2}{\delta(S_{n+1})^2} \frac{|\psi''_n|}{(\psi'_n)^2}, \end{aligned} \tag{64}$$

where we omitted the arguments for simplicity. This yields

$$\left|\frac{\psi''_{n+1}}{\left(\psi'_{n+1}\right)^2}\right| \leq \Phi_n\left(T_{(k)}\right) + \Lambda_0^{-1}\theta^2 \left|\frac{\psi''_n}{\left(\psi'_n\right)^2}\right| \tag{65}$$

where θ^2 comes from Corollary 2 and

$$\Phi_n(T_{(k)}) = \delta(S_{n+1})^{-2}\left[\Delta_{xx}(T_{(k)})\delta(S_n)^2 + \frac{2}{|S'_n|}\Delta_{\alpha x}(T_{(k)})\delta(S_n) + \frac{1}{\left(S'_n\right)^2}\Delta_{\alpha\alpha}(T_{(k)})\right]. \tag{66}$$

Therefore, we can write

$$\eta_{n+1} = \sum_{J,k,P} \sup_{J_k(P)} \left|\frac{\psi''_{n+1}}{\left(\psi'_{n+1}\right)^2}\right| |J_k(P)| \leq \sum_{J,k,P} \Phi_n(T_{(k)})\,|J_k(P)| + \Lambda_0^{-1}\theta^2\eta_n. \tag{67}$$

Because of the obvious inequality

$$|J_k(P)| \leq \sup_{J_k(P)} \frac{1}{|\psi'_n|} \cdot \left|P \cap \bigcup_{\alpha\in J}(I_k \backslash I_{k+1})\right| \tag{68}$$

we get

$$\eta_{n+1} \leq r\gamma_n + \Lambda_0^{-1}\theta^2\eta_n, \tag{69}$$

where

$$r = \sup_n \max_{P\in\pi} \sum_k \sup_{\alpha,x} \Phi_n(T_{(k)}) \left|P \cap \bigcup_\alpha (I_k\backslash I_{k+1})\right|. \tag{70}$$

We shall soon see that r is finite and even arbitrarily small. Inequalities (62) and (69) imply the theorem, if the eigenvalues of the matrix

$$\mathbf{P} = \begin{bmatrix} 4\theta^2\Lambda_0^{-1} & 2\theta^2\Lambda_0^{-1} \\ r & \theta^2\Lambda_0^{-1} \end{bmatrix} \tag{71}$$

have modulus < 1. As a matter of fact, we can write (62) and (69) as a single matrix inequality

$$\boldsymbol{\xi}_{n+1} \leq \mathbf{P}\boldsymbol{\xi}_n + \mathbf{c} \quad (n \leq N), \tag{72}$$

where $\boldsymbol{\xi}_n = (\gamma_n, \eta_n)$. Here $\mathbf{c}$ is a constant vector and $\|\mathbf{c}\| \leq \text{const.}\,|A_N|$. Here and in what follows, const. means an unspecified constant independent of N. Also, we can easily see that $\|\boldsymbol{\xi}_N\| \leq \text{const.}\,|A_N|$. In fact, it is sufficient to show that $\eta_N \leq \text{const.}\,|A_N|$. The inequality $\gamma_N \leq \text{const.}\,|A_N|$ will follow by the same principle we used to prove (60). First, we notice that

$$\eta_N \leq \sup_{A_N}\left|\frac{\psi''_N}{\left(\psi'_N\right)^2}\right| |A_N|$$

(by definition of Section 1, $\mathcal{A}_N$ has only one element). Let us fix n_1 such that (41) hold for $n \geq n_1$. We can apply (65) for $\alpha \in A_N$ and $n = n_1, n_1+1, \ldots, N$. For these n we have $k = 1$ and $\Phi_n(T_{(1)}) \leq h^{-1}\left(R_2H^2 + 2R_2H + R_2\right) = C_{11}$. If $\Lambda_0^{-1}\theta^2 < 1$, (65) yields

$$\sup_{\mathcal{A}_n} \left| \frac{\psi_N''}{\left(\psi_N'\right)^2} \right| \leq \frac{C_{11}}{1 - \Lambda_0\theta^2} \sup_{A_N} \left| \frac{\psi_{n_1}''}{\left(\psi_{n_1}'\right)^2} \right| . \tag{73}$$

Therefore (72) yields the Theorem if the spectral radius of $\mathbf{P}$ is < 1. We recall that θ is arbitrarily close to 1, as β is sufficiently small. Also, we will see that r is arbitrarily small, as d is sufficiently small.

By Remark 3 we can check that $\Phi_n(T_{(k)}) \leq \text{const.} \left|S_k'(1)\right|^{2\tau-1}$ (note: $\nu < 2\tau - 1$). We also have $|I_k| \leq \text{const.} \left|S_{k-1}'(1)\right|^{-\tau}$, where α on both sides may be different (see Corollary 3). Also, const. $\cdot \left|S_{k-1}'(1)\right| \geq \left|S_k'(1)\right|^{1/(1+\epsilon)}$ by Lemma 3 (indeed, $S_k = T_{(k)} \circ S_{k-1}$ and $|S_k'| = \left(\left|T_{(k)}'\right| \circ S_{k-1}\right) \left|S_{k-1}'\right| \leq R_0 \left|S_{k-1}'\right|^{\epsilon} \left|S_{k-1}'\right| = R_0 \left|S_{k-1}'\right|^{1+\epsilon}$). So, $|I_k| \leq \text{const.} \left|S_k'\right|^{-\tau/(1+\epsilon)}$. Hence

$$\Phi_n(T_{(k)}) \left| P \cap \bigcup_{\alpha \in J} (I_k \backslash I_{k+1}) \right| \leq \text{const.} \left|S_k'(1)\right|^{-\zeta}, \tag{74}$$

where $\zeta = \tau/(1+\epsilon) + 1 - 2\tau$, if β is sufficiently small. We also have $\left|S_k'(1)\right|^{-\zeta} \leq \Lambda_0^{-k\zeta}$. So, the sum in (70) has uniformly exponentially decreasing terms. By fixing sufficiently small d we can make it arbitrarily close to 0.

This proves the theorem, if $4\Lambda_0^{-1} < 1$. When this condition is not satisfied, we pick $p \in \mathbb{Z}^+$ such that $\Lambda_0^p > 4$ and examine the relation between $(\gamma_{n+p}, \eta_{n+p})$ and (γ_n, η_n). The corresponding matrix, as $r \to 0$ looks like

$$\begin{bmatrix} 2M\Lambda_0^{-p} & * \\ 0 & \Lambda_0^{-p} \end{bmatrix}, \tag{75}$$

where M is an integer which is the maximal number of $J' \in \mathcal{A}_{n+p}'$ contained in a fixed $J \in \mathcal{A}_n'$ and such that there is $i \in \{n, n+1, \ldots, n+p-1\}$ with the property: there is $k \leq k_1$ such that $\psi_i(J') \subset I_k \backslash I_{k+1}$. This number can be made $= 2$, if we slightly perturb intervals $I_1, \ldots, I_{k_1}$.

Corollary 4. *We have* $|B_n| \leq 2C_{10}a\lambda^{\sqrt{n}} |A_N|$.

Proof. Clearly, $B_n \subset \psi_n^{-1}\left(\left[-a\lambda^{\sqrt{n}}, a\lambda^{\sqrt{n}}\right]\right)$. Theorem 6 gives

$$\left|\psi_n^{-1}(E)\right| \leq \left(\sum_{J \in \mathcal{A}_n} \sup_J \frac{1}{\left|\psi_n'\right|} \right) |E| \leq C_{10} |E| \, |A_N| \tag{76}$$

for an arbitrary measurable set $E \subset I$.

Remark 7. The existence of an a.c.i.m. for T follows directly from [5], since the function

$$g(x) = \begin{cases} \dfrac{1}{|T'(x)|}, & \text{where } T \text{ is continuous,} \\ 0 & \text{on discontinuities} \end{cases} \tag{77}$$

verifies $\mathrm{Var} g < +\infty$ and $\sup g < 1$ (a sufficient condition for the existence of an a.c.i.m.). One can obtain another proof (and make this paper self-contained) by imitating the proof of Theorem 6 for the sequence of mappings $\psi_n(x) = T^n(x)$ ($\mathcal{A}_n$ becomes the partition into the pieces of continuity of ψ_n). The new functions satisfy the recursive formula $\psi_{n+1}(x) = T_{(k)}(\psi_n(x))$ (if $\psi_n(x) \in I_k \backslash I_{k+1}$). This formula is much easier to differentiate, since x is the only variable.

Proof. We have

$$\mathrm{Var} g < 2 \sum_k \sup \left|T'_{(k)}\right|^{-1} + \sum_k \Delta_{xx}(T_{(k)}) \left|I_k\right|. \tag{78}$$

The first sum converges by Remark 4. The second sum converges by an argument similar to the proof of (74).

In view of Section 1, Proposition 2(iii), Corollary 4 and Remark 7 the main theorem is proved.

6 Some generalizations and final remarks

Our method applies without any changes to families $f(x) = 1 - a|x|^{\xi}$, where ξ is sufficiently close to 2. For $\xi < 2$ we get an example of a family which is not C^2. It is easy to get examples of families with singularities like $|x|^{\xi}$ with any $\xi > 1$.

Another class of examples would be families with a finite or infinite number of singularities like

$$f(x) = \begin{cases} 1 - ax^2 & \text{for } -1 \leq x \leq 0 \\ [10/x] - 1 & \text{for } 1 \geq x \geq 0. \end{cases}$$

Let us discuss a little different result now. Let $f_0(x) = 1 - 2x^2$ and let for every $\eta > 0$ $X(\eta)$ be the η-neighborhood of f_0 in the C^2-topology. If η is sufficiently small then there is a C^1 function $c : X(\eta) \to I$ such that $c(f)$ is the only critical point of $f \in X(\eta)$. Let $M \subset X(\eta)$ be a submanifold of codimension 1 defined as follows:

$$M = \left\{ f \in X(\eta) : \; f^3(c(f)) = f^2(c(f)) \right\}. \tag{79}$$

For $f \in M$, $x(f) \stackrel{\mathrm{def}}{=} f^2(c(f))$ is a hyperbolic fixed point.

Theorem 7. *Suppose that a path $\alpha \to f_\alpha$ intersects M transversally for some α_0. The α_0 is a one-sided Lebesgue density point of the set of those α that f_α has an a.c.i.m.*

Corollary 5. *There is a open set of C^2-families satisfying Jakobson's Theorem.*

One can consider a transversality condition in higher jets

$$\left.\frac{d^k}{d\alpha^k}\right|_{\alpha=\alpha_0} (f_\alpha(x(f_\alpha)) - x(f_\alpha)) \neq 0, \tag{80}$$

where $k \geq 1$ is arbitrary.

Theorem 8. *Theorem 7 remains true if transversality to M is replaced with condition (80).*

The idea of the proof. Let $G : X(\eta) \to \mathbb{R}$ be defined by

$$G(f) = f(x(f)) - x(f). \tag{81}$$

It is easy to check that $dG(f) \neq 0$ for $f \in X(\eta)$, if η is sufficiently small. Therefore, there is a mapping $p : X(\eta) \to M$ such that $X \ni f \mapsto (p(f), G(f)) \in M \times \mathbb{R}$ is a diffeomorphism (one can write down suitable formulas explicitly).

For every $f \in M$ we have a standard family f_γ corresponding to the fiber $\{f\} \times \mathbb{R}$, so that $p(f_\gamma) = f$ and $G(f_\gamma) = \gamma$ for $|\gamma| < \gamma_0$. We repeat our proof simultaneously for all families f_γ. We obtain a family $\mathcal{F}$ of submanifolds $F \subset X(\eta)$ such that if $g \in F \in \mathcal{F}$ then g has an a.c.i.m. Moreover, there is a constant K with the following properties:

(i) Every $F \in \mathcal{F}$ corresponds to a graph of a Lipschitz function $h_F : M \to \mathbb{R}$ with a Lipschitz constant $\leq K$ ($M \in \mathcal{F}$; $h_M \equiv 0$).

(ii) Let $A_\infty(f) = \{\gamma : f_\gamma \in F \text{ for some } F \in \mathcal{F}\}$. The family $\mathcal{F}$ defines a measurable mapping $A_\infty(f_1) \to A_\infty(f_2)$. This mapping has a measure-theoretic Jacobian $\in [K^{-1}, K]$ a.e.

From our estimates it follows easily that for every $f \in M$ and $h > 0$

$$\frac{|A_\infty(f) \cap [0, h]|}{h} \geq 1 - c_1 \exp\left(-c_2 \left(\log \frac{1}{h}\right)^{1/2}\right), \tag{82}$$

where $c_1,\ c_2 > 0$.

These estimates are sufficient to show that every family g_α such that $G(g_{\alpha_0}) = 0$ and $\left.\left(d^k/d\alpha^k\right) G(g_\alpha)\right|_{\alpha=\alpha_0} \neq 0$ for some k intersects the leaves of $\mathcal{F}$ for a positive measure set of parameters α.

Corollary 6. *Any analytic family which contains f_0, satisfies Jakobson's theorem.*

Let us list two more facts concerning our construction.

(i) The density of an a.c.i.m. we constructed $\in L^p(I)$ for every $p \in [1, 2)$. This result is analogous to one of Carleson's results.

(ii) If α_0 is sufficiently close to 2 and $\inf_{n\geq 1} |f^n_{\alpha_0}(0)| > 0$ then α_0 is also a density point of the set of α such that f_α has an a.c.i.m. The condition that α_0 is close to 2 can be replaced with the condition given in [3].

Appendix

In this Appendix $f(x) = 1 - 2x^2$.

Theorem 9. *Let $I_n = [-3^{-n}, 3^{-n}]$ for $n \in \mathbb{Z}^+$. Let $V_n = f(I_n)$. Then for every n and $\rho \in (0,1)$:*

(i) $\left|(f^n)' \,|\, |V_n|\right| \geq \left(3\tfrac{1}{9}\right)^n$ *and* $\sim 4^n$ *for large* n,

(ii) $\left|(f^{n+1})' \,|I_n \backslash \rho I_{n+1}\right| \geq \Lambda = 4\rho/3$ *and* $\geq$ const. $\rho\left(\tfrac{4}{3}\right)^n$,

(iii) $|W_n| = |f^n(V_{n+1})| \leq \left(\tfrac{4}{9}\right)^{n-1} |V_1|$.

Proof. (i) We have $V_n = f(I_n) = [1 - 2\cdot 9^{-n}, 1]$. We have $|f'| \leq 4$. By the Mean Value Theorem we have for $k < n$

$$|f^k(V_n)| \leq 2\cdot 9^{-n}\cdot 4^k = 4^{-(n-k-1)}\cdot\left(\frac{4}{9}\right)^{n-1}|V_1| \leq \left(\frac{4}{9}\right)^{n-1}|V_1|. \tag{83}$$

This implies $f^k(V_n) \subset \pm V_1$ for $n \geq 2$. We also have $\left|f'| \pm V_1\right| \geq 4(1 - |V_1|) = \frac{28}{9} = 3\frac{1}{9}$.

We can write

$$\left|(f^n)'\,|V_n\right| \geq \left|\prod_{k=0}^{n-1} f'|f^k(V_n)\right|. \tag{84}$$

Clearly, it is $\geq \left(3\frac{1}{9}\right)^n$ and also $\sim 4^n$ for large n. Indeed, the product is $\geq$

$$\prod_{k=0}^{n-1} 4\left(1 - |f^k(V_n)|\right) \geq 4^n \prod_{k=0}^{n-1}\left(1 - \left(\frac{4}{9}\right)^{n-1}|V_1|\, 4^{-(n-k-1)}\right). \tag{85}$$

Hence,

$$\left|(f^{n+1})'\,|I_n\backslash\rho I_{n+1}\right| \geq \left|(f^n)'\,|V_n\right|\cdot 4\cdot\frac{1}{2}|\rho I_{n+1}| \geq 3^n 4\rho 3^{-(n+1)} = \frac{4}{3}\rho = \Lambda. \tag{86}$$

Using $\left|(f^n)'\,|V_n\right| \geq \text{const.}\cdot 4^n$ we also get $\left|(f^{n+1})'\,|I_n\backslash\rho I_{n+1}\right| \geq \text{const. } \rho\left(\frac{4}{3}\right)^n$.

If $\rho \geq 27/28$ one gets the same inequalities on $\rho^{-1}I_n\backslash\rho I_n$ with $3\frac{1}{9}$ replaced by 3.

References

1. M. Benedicks & L. Carleson. On iteration of $1 - ax^2$ on $(-1, 1)$. *Ann. of Math.* (2) **122** (1985), no. 1, 1–25.
2. M. V. Jakobson. Absolutely continuous Invariant Measures for one-parameter families of one-dimensional maps. *Comm. in Math. Phys.* **81** (1981), 39–88.
3. M. Misiurewicz. Absolutely continuous measures for certain maps of an interval. *I.H.E.S. Publications Mathématiques*, #**53** (1981), 17–52.
4. M. Rees. Positive measure sets of ergodic rational maps. Preprint. [4]
5. M. Rychlik. Bounded variation and invariant measures. *Studia Math.* t. **76**, (1983), 69-80.

[4] This paper has been published as: M. Rees. Positive measure sets of ergodic rational maps. *Ann. Scient. Ec. Norm. Sup.*, (4) 19 (1986), 383-407.

Unstable periodic orbits and the dimensions of multifractal chaotic attractors

Celso Grebogi
Laboratory for Plasma and Fusion Energy Studies, University of Maryland, College Park, Maryland 20742

Edward Ott
Laboratory for Plasma and Fusion Energy Studies, University of Maryland, College Park, Maryland 20742 and Department of Electrical Engineering and Department of Physics, University of Maryland, College Park, Maryland 20742

James A. Yorke
Institute for Physical Science and Technology and Department of Mathematics, University of Maryland, College Park, Maryland 20742

(Received 28 September 1987)

The probability measure generated by typical chaotic orbits of a dynamical system can have an arbitrarily fine-scaled interwoven structure of points with different singularity scalings. Recent work has characterized such measures via a spectrum of fractal dimension values. In this paper we pursue the idea that the infinite number of unstable periodic orbits embedded in the support of the measure provides the key to an understanding of the structure of the subsets with different singularity scalings. In particular, a formulation relating the spectrum of dimensions to unstable periodic orbits is presented for hyperbolic maps of arbitrary dimensionality. Both chaotic attractors and chaotic repellers are considered.

I. INTRODUCTION

The long time distribution generated by a typical orbit of a chaotic nonconservative dynamical system is generally highly singular. The subset of phase space to which the orbit asymptotes with time, the attractor, can be geometrically fractal. Furthermore, the distribution of orbit points on the attractor can have an arbitrarily fine-scaled interwoven structure of hot and cold spots. Sets with such distributions have been called *multifractals. By hot and cold spots we mean points on* the attractor for which the frequency of close approach of typical orbits is either much greater than typical (a hot spot) or much less than typical (a cold spot). Recently there has been much work developing ways of quantitatively characterizing how such chaotic orbits distribute themselves on attractors.[1–4] In particular, the spectrum of fractal dimensions introduced in Refs. 2–4 are sensitive to the characteristics of the structure of hot and cold spots on the attractor. In this paper we present results which show that, for a large class of chaotic attractors, the infinite number of unstable periodic orbits embedded in the attractor provide the key to an understanding of such issues. (A brief preliminary report of some of this work appears in Grebogi, Ott, and Yorke.[5])

The importance of unstable periodic orbits in determining ergodic properties of chaotic systems has long been recognized in the mathematical literature (e.g., Bowen[6] and Katok[7]). For some more recent work see Refs. 8 and 9 which also illustrate the important point that information about unstable periodic orbits is readily accessible from numerical computation (and perhaps experimentally[9,10]) and can be used for determining ergodic properties. In addition, in the theory of quantum chaos, the distribution of energy levels can be related to unstable periodic orbits of the classical Hamiltonian.[11–13] Another case where unstable periodic orbits appear[10] is in determining the behavior near parameter values where sudden changes in chaotic attractors occur [the argument in connection with our Fig. 1 is similar to that for Eq. (2) of Ref. 10].

The organization of this paper is as follows. Section II presents a discussion of the pointwise dimension for attractors and shows that hot and cold spots occur on the unstable manifolds of saddle periodic orbits in the attractor. Numerical experiments illustrating this are also presented. Section III reviews recent work on the dimensions of attractors including the partition function formalism.[4] Section IV presents our results relating the distribution of typical chaotic orbits on attractors and the associated fractal dimensions to the unstable periodic orbits. Section V illustrates the material of Sec. IV with examples. Arguments yielding the results stated in Sec. IV are presented in Sec. VI for the case of hyperbolic attractors. Section VII treats the case of chaotic sets which are repelling rather than attracting.

The dynamical systems to be discussed throughout this paper are d-dimensional maps of the form $\underline{x}_{n+1}=\underline{F}(\underline{x}_n)$, where $\underline{x}$ is a vector in the d-dimensional phase space of the system. An *attractor* A for such a system is a closed set, invariant under $\underline{F}$, which is the limit set as time goes to $+\infty$ for almost every initial condition in some neighborhood of A. (By "almost every" we mean that the set of initial conditions in the neighborhood that do not approach A can be covered by a set of d-dimensional cubes of arbitrarily small total volume.) The *basin of attraction* for the attractor is the closure of the set of points which asymptote to the attractor as time goes to $+\infty$. In the case of continuous time systems (flows), we can think of $\underline{F}(\underline{x})$ as arising from a Poincaré surface of section.

II. POINTWISE DIMENSION

For most purposes we may think of the *natural measure* of an attractor as follows: For a subset S of the phase space and an initial condition $\underline{x}$ in the basin of attraction of the attractor, we define $\mu(\underline{x},S)$ as the fraction of time the trajectory originating at $\underline{x}$ spends in S in the limit that the length of the trajectory goes to infinity. If $\mu(\underline{x},S)$ is the same for almost every $\underline{x}$ in the basin of attraction, then we denote this value $\mu(S)$ and say that μ is the natural measure of the attractor (cf. Appendix). Henceforth, we assume that the attractor has a natural measure. In particular, this means that the attractor is ergodic (i.e., it cannot be split into two disjoint pieces that each have positive natural measure and are invariant under application of $\underline{F}$).

Let $B(l,\underline{x})$ denote a d-dimensional ball of radius l centered at a point $\underline{x}$ on an attractor embedded in the d-dimensional phase space of the dynamical system being considered. Then the pointwise dimension (at the point $\underline{x}$) of the attractor is defined as

$$D_p(\underline{x})=\lim_{l\to 0}\frac{\log\mu(B(l,\underline{x}))}{\log l} \tag{2.1}$$

or $\mu(B(l,\underline{x}))\sim l^{D_p(\underline{x})}$. For *almost every* point with respect to the natural measure on the attractor, $D_p(\underline{x})$ takes on a common value and is equal to the information dimension (defined in Sec. III). That is, the set of points on the chaotic attractor for which $D_p(\underline{x})$ is not this common value may be covered with a set of d-dimensional cubes of varying sizes which together contain an arbitrarily small amount of the natural measure of the attractor. [Points $\underline{x}$ where $D_p(\underline{x})$ is greater than (less than) the common value it assumes at almost every point with respect to the natural measure are the hot (cold) spots referred to in Sec. I.] For example, a chaotic attractor typically has a dense set of unstable periodic orbits embedded within it, and, as we shall see, $D_p(\underline{x})$ with $\underline{x}$ on one of these periodic orbits does not take on the typical values. The periodic points, however, are countable and so have zero measure. Nevertheless, it is a main point of this paper that this zero measure set is important and leads to interesting properties of the attractor.

To see why $D_p(\underline{x})$ is the same for almost every $\underline{x}$ with respect to the natural measure, assume that it is not. Then we can pick some D_{p0} such that there is a nonzero amount for the natural measure of the attractor for which $D_p(\underline{x})>D_{p0}$ and another nonzero amount for which $D_p(\underline{x})\le D_{p0}$. Thus the attractor is divided into two disjoint sets, $A_>$ and $A_<$, $\mu(A_>)+\mu(A_<)=1$, $\mu(A_{\gtrless})>0$. From the definition (2.1), one can show that $D_p(\underline{x})$ is invariant to smooth changes of coordinates. In particular, for a smooth map $\underline{x}_{n+1}=\underline{F}(\underline{x}_n)$, we can take $\underline{y}=\underline{F}(\underline{x})$ as the change of coordinates. It follows that $D_p(\underline{x})=D_p(\underline{F}(\underline{x}))$. Thus the sets $A_>$ and $A_<$ are invariant under $\underline{F}$. This implies that every orbit on the attractor is confined either to $A_>$ or $A_<$. Hence, contrary to our assumption that the attractor has a natural measure, the attractor has been decomposed into two disjoint $\underline{F}$ invariant sets. We conclude that $D_p(\underline{x})$ must be the same for almost every $\underline{x}$ with respect to the natural measure on the attractor.

Returning now to consideration of the zero measure set of points $\underline{x}$ for which $D_p(\underline{x})$ is not typical (i.e., is not the common value assumed at almost every $\underline{x}$ on the attractor) and taking the map to be two dimensional ($d=2$), we will obtain the following result. Let j be an index labeling the fixed points of the n times iterated map $\underline{F}^n$. (The components of a period n orbit are fixed points of $\underline{F}^n$.) We assume that the Jacobian matrix of $\underline{F}^n$ at fixed point j has one unstable direction and one stable direction. Then for any point $\underline{x}$ *on the unstable manifold* of fixed point j of $\underline{F}^n$,

$$D_p(\underline{x})=1-\frac{\log\lambda_{1j}}{\log\lambda_{2j}}\ , \tag{2.2}$$

where $\lambda_{1j}>1$ and $\lambda_{2j}<1$ are the magnitudes of the unstable and stable eigenvalues of the Jacobian matrix of $\underline{F}^n$. Since points on different periodic orbits typically have different eigenvalues, $D_p(\underline{x})$ will clearly be different for different periodic orbits and hence will not be the typical $D_p(\underline{x})$.

To obtain (2.2) consider a point $\underline{x}_0$ on the unstable manifold of a saddle periodic point and two small circular disks centered at $\underline{x}_0$ with radii l_1 and l_2, where $l_1/l_2=\lambda_{2j}^{-1}$. We iterate the two disks backward a large integral number of periods so that the two disks are now similar ellipses close to the saddle and with their major axes parallel to the stable manifold of the saddle (cf. Fig. 1). We now iterate the l_2 ellipse backward one more period. Since it is close to the saddle, its backward iteration by one period is governed by the linearized map at the saddle (i.e., by the eigenvalues λ_{1j} and λ_{2j}). Thus, since we choose $l_1/l_2=\lambda_{2j}^{-1}$, the major diameter of the l_2 ellipse is now the same as that for the l_1 ellipse, while its minor diameter is smaller than that for the l_1 ellipse by the factor $\lambda_{2j}/\lambda_{1j}$. The inverse images of the disks contain the same natural measure as the original disks. Thus, treating the attractor measure as if it were smooth along the unstable direction, we have $\mu(B(l_2,\underline{x}_0))/\mu(B(l_1,\underline{x}_0))=\lambda_{2j}/\lambda_{1j}$. Setting $\mu(B(l,\underline{x}_0))\sim l^{D_P}$ and $l_2=l_1\lambda_{2j}$, this yields Eq. (2.2), the desired result.

We have tested Eq. (2.2) numerically using the Henon map given by

$$\begin{aligned} x_{n+1}&=1.42-x_n^2+0.3y_n\ ,\\ y_{n+1}&=x_{n+1}\ . \end{aligned} \tag{2.3}$$

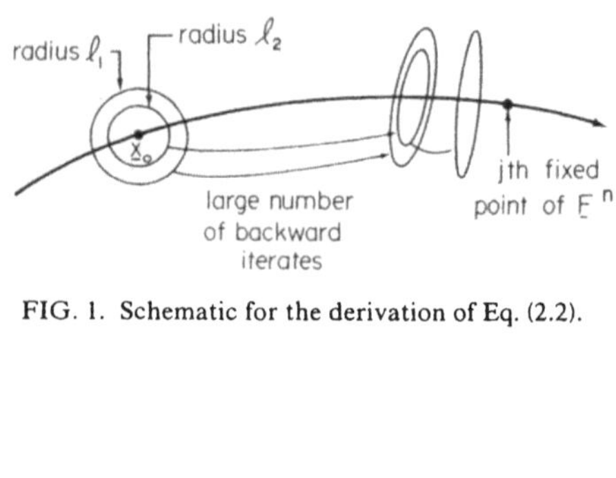

FIG. 1. Schematic for the derivation of Eq. (2.2).

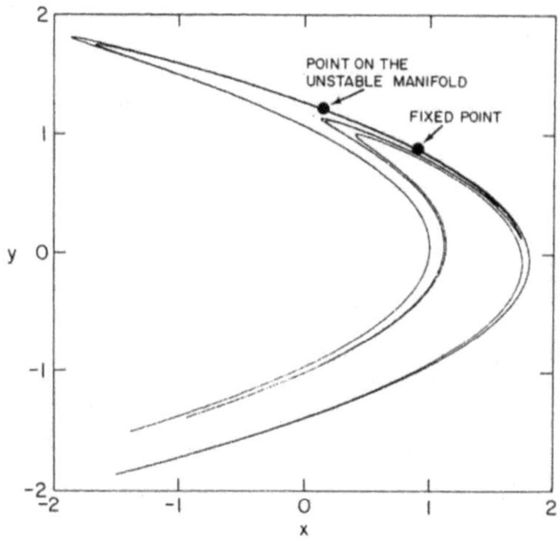

FIG. 2. Plot of the iterates of the Henon map (2.3).

In Fig. 2, we plot iterates of the map (2.3) to show the attractor. In this figure we also indicate the location of a point which we have accurately determined to be on the unstable manifold of an unstable fixed point (i.e., a period one unstable orbit) of the map (2.3). In Fig. 3, we display the result of a calculation of the pointwise dimension $D_p(\underline{x})$ for $\underline{x}$ at this point. In this figure, we plot $\mu(B(l,\underline{x}))$ versus l on a log-log scale. Here $\mu(B(l,\underline{x}))$ is obtained by iterating a randomly chosen initial condition in the basin of the attractor 10^6 times (so that the orbit is essentially on the attractor) and then determining the fraction of subsequent orbit points which fall in $B(l,\underline{x})$ on further iteration. The numerically determined pointwise dimension is the slope of the straight line through the points in Fig. 3; we obtain $D_p(\underline{x}) \cong 1.36$. The magnitudes of the eigenvalues at the fixed point are $1.94\cdots$ and $0.155\cdots$, which, when inserted in Eq. (2.2), yield $D_p(\underline{x}) = 1.36$, in agreement with the data in Fig. 3.

Next we obtain a typical point $\underline{x}$ on the attractor which is within a small distance (1.5×10^{-4} in this case) of the previously chosen point on the unstable manifold of the fixed point. We do this by randomly picking an initial condition, preiterating 2×10^6 times, and then iterating the map until the orbit falls within a circle of radius 2×10^{-4} centered at the previously obtained point on the unstable manifold. Results for $\mu(B(l,\underline{x}))$ versus l for this point are shown in Fig. 4. For $l>1.5\times10^{-4}$ the slope of the straight line fitted to the numerical data is 1.36, which is the same value as obtained in Fig. 3 for the point on the unstable manifold. This agreement is as expected, since for $l \gg 10^{-4}$, the two points are essentially indistinguishable. However, for data points in the range $6\times10^{-6}<l<1.5\times10^{-4}$, we find that the slope of a fitted line is 1.21, which is significantly different from the 1.36 slope for $l>1.5\times10^{-4}$.

There are an infinite number of unstable periodic orbits on the attractor. Thus, although it is true that all typical $\underline{x}$ must have a common value for $D_p(\underline{x})$, one might suspect that there will be significant fluctuations in numerically determined values of D_p, since such calculations are necessarily restricted to a finite range of l. This seems to be the case: We have numerically determined $D_p(\underline{x})$ for the map (2.3) using a range $10^{-1}\le l\le 10^{-5}$ at 20 different typical points, and we find considerable variation in the resulting numerically determined pointwise dimensions. A list of the 20 values obtained appears in Table I. In obtaining these values, a least square fit was used in $10^{-1}\le l\le 1.22\times10^{-5}$, and the root-mean-square deviation of the least square fit is also shown in Table I. This root-mean-square deviation of the fit for $D_p(\underline{x})$ at individual $\underline{x}$ values is small compared with the standard deviation (0.10) about the mean (1.27) obtained by using the twenty $D_p(\underline{x})$ values (given in the first column of Table I). (The 20 typical $\underline{x}$ values used were obtained by choosing a random initial condition in the basin of the attractor, iterating it 20×10^6 times, and selecting every millionth iterate.)

The Kaplan-Yorke formula, discussed in Sec. IV, predicts the typical value of the pointwise dimension in terms of the Lyapunov numbers. For the case considered here, the predicted value is 1.26, which is in good agreement with our mean of the twenty numerically obtained D_p values but is far from the value obtained in Fig. 3.

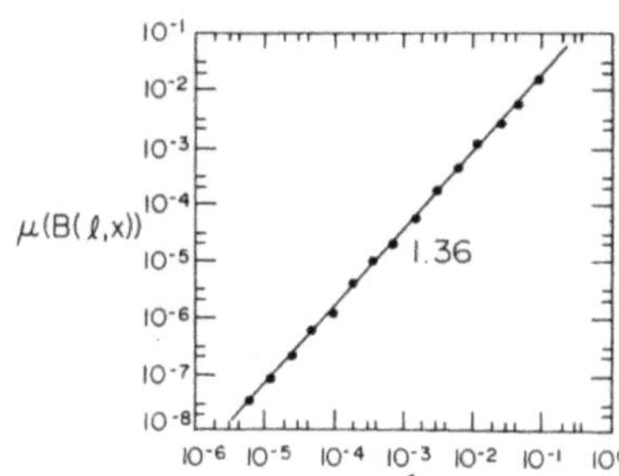

FIG. 3. Log-log plot of $\mu(B(l,x))$ vs l for a point on the unstable manifold of the fixed point.

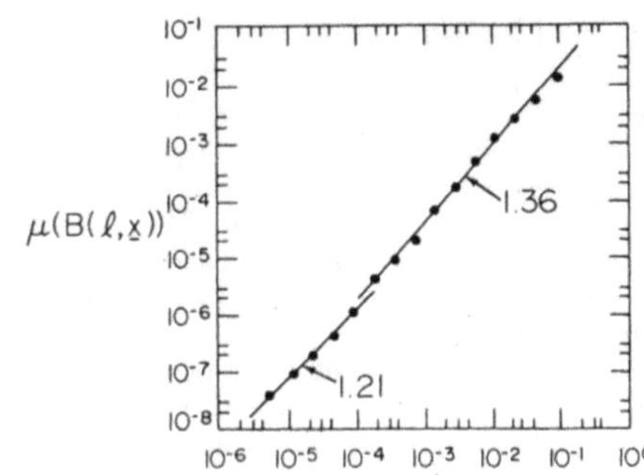

FIG. 4. Log-log plot of $\mu(B(l,\underline{x}))$ vs l for a typical point on the attractor which is within 1.5×10^{-4} of the point on the unstable manifold of the fixed point.

TABLE I. Result of pointwise dimension calculations for 20 typical points. Mean equals 1.27±0.10

$D_p(\underline{x})$	rms deviation of fit
1.29	0.03
1.31	0.03
1.12	0.02
1.34	0.03
1.21	0.02
1.25	0.02
1.21	0.04
1.34	0.03
1.31	0.04
1.38	0.02
1.25	0.04
1.28	0.03
1.26	0.03
1.34	0.03
1.33	0.03
1.15	0.02
1.02	0.01
1.32	0.03
1.41	0.02
1.37	0.02

III. DIMENSIONS OF MULTIFRACTAL CHAOTIC ATTRACTORS

In this section we review past work on the dimensions characterizing multifractal chaotic attractors.[2-4] References 2 and 3 consider the quantity $\tilde{D}_q$ defined by

$$\tilde{D}_q=\frac{1}{q-1}\lim_{l\to 0}\frac{\log\sum_{i=1}^{N(l)}p_i^q}{\log l}\ , \tag{3.1}$$

where the attractor is covered with $N(l)$ d-dimensional cubes from a grid of unit length l, and p_i is the natural measure of the attractor in the ith cube. Taking the limit $q\to 1$, Eq. (3.1) yields[2,3]

$$\tilde{D}_1=\lim_{l\to 0}\left[\sum_i p_i\log p_i\right]\Big/\log l\ , \tag{3.2}$$

which is called the information dimension of the attractor. It is $\tilde{D}_1$ which is the common value assumed by $D_p(\underline{x})$ for almost all $\underline{x}$ with respect to the natural measure on the attractor.[1] We may think of the information dimension as the capacity dimension (cf. below) of the smallest set which contains most of the natural measure of the attractor.[1]

As q is increased past 1, the contribution of the sum $\sum_i p_i^q$ from a relatively few boxes with very little of the total attractor measure but with larger p_i than typical (i.e., hot spots) becomes relatively more important. Similarly, as q is decreased from one, the contribution from low probability boxes begins to be more important. For example, for $q=0$, Eq. (3.1) yields

$$\tilde{D}_0=-\lim_{l\to 0}\log N(l)/\log l\ , \tag{3.3}$$

which is known as the capacity or box-counting dimension of the attractor. Note that all boxes on the attractor contribute democratically to (3.3) no matter what their natural measure p_i is. For typical chaotic attractors, it is to be expected that $\tilde{D}_0>\tilde{D}_1$, since low-probability boxes (cold regions) containing very little of the total natural measure on the attractor may be vastly more numerous than those required to cover most of the natural measure on the attractor.[1]

At this point it is appropriate to discuss another definition of dimension, the Hausdorff dimension,[14] which is, in fact, an older concept than either the capacity[15] or information dimensions.[16] To define the Hausdorff dimension, we cover the attractor with d-dimensional cubes of variable edge length l_i, all of which we restrict to be no bigger than some value l ($l_i\le l$). We then form the quantity

$$\Gamma_H(D,l,\{l_i\})=\sum_i l_i^D\ . \tag{3.4}$$

Next the covering of cubes is optimized so as to make the sum $\sum l_i^D$, minimum,

$$\Gamma_H(D,l)=\inf_{\{l_i\}}\sum_i l_i^D\ , \tag{3.5}$$

where the infimum is taken over all possible collections of cubes that cover the attractor subject to the constraint $l_i\le l$. Finally, the limit $l\to 0$ is taken,

$$\Gamma_H(D)=\lim_{l\to 0}\Gamma_H(D,l)\ . \tag{3.6}$$

The quantity $\Gamma_H(D)$ can be shown to be either zero or infinity except at a critical value of D [cf. Fig. 5(a)]. This critical value defines the Hausdorff dimension which we denote D_H.

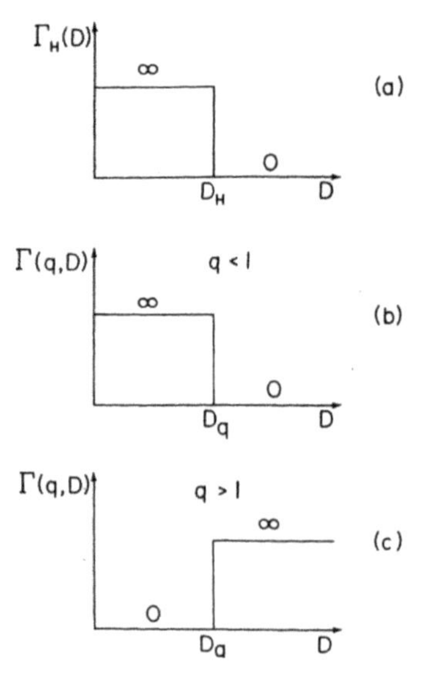

FIG. 5. (a) $\Gamma_H(D)$ vs D_H. (b) $\Gamma(q,D)$ for $q<1$. (c) $\Gamma(q,D)$ for $q>1$.

The capacity dimension $\tilde{D}_0$ and the Hausdorff dimension are closely related. In particular, if we do not optimize over the cube covering, but instead use a cubic grid covering of fixed length l, then all the l_i in (3.4) are equal to l, and we have $\Gamma_H^{(\mathrm{grid})}(D,l)=N(l)l^D$. Equation (3.3) is satisfied by $N(l)\sim l^{-\tilde{D}_0}$, which yields $\Gamma_H^{(\mathrm{grid})}(D,l)\sim l^{D-\tilde{D}_0}$. In the limit $l\to 0$, $\Gamma_H^{(\mathrm{grid})}$ is zero or infinity [as in Fig. 5(a)] with the transition at $D=\tilde{D}_0$. Since the covering using a cubic grid may not be optimal, $\Gamma_H^{(\mathrm{grid})}(D,l)\geq \Gamma_H(D,l)$ and

$$\tilde{D}_0 \geq D_H \ .$$

For examples of ergodic chaotic attractors, where $\tilde{D}_0$ and D_H are analytically calculable, it is found[1] that $\tilde{D}_0=D_H$, and it has been conjectured that this is to be expected, in general, for typical ergodic chaotic attractors (although it is easy to construct sets[1] which are not attractors which have $\tilde{D}_0>D_H$).

Reference 4 gives a formalism which essentially generalizes the Hausdorff procedure and its relation to $\tilde{D}_0$ so that $\tilde{D}_q$ for arbitrary q is treated. In particular, again say we cover the attractor with cubes of variable edge length l_i, but now consider the following quantity, which in analogy with statistical mechanics, has been called the partition function,

$$\Gamma(q,D,l,\{l_i\})=\sum p_i^q/l_i^\tau, \quad \tau=(q-1)D \tag{3.7}$$

where $D\geq 0$. Optimize over covering $\{l_i\}$,

$$\Gamma(q,D,l)=\begin{cases}\sup[\Gamma(q,D,l,\{l_i\})] & \text{for } q>1\ ,\\ \inf[\Gamma(q,D,l,\{l_i\})] & \text{for } q<1\ ,\end{cases} \tag{3.8}$$

and take the limit $l\to 0$,

$$\Gamma(q,D)=\lim_{l\to 0}\Gamma(q,D,l)\ . \tag{3.9}$$

Again $\Gamma(q,D)$ is zero or infinity. The transition point for $\Gamma(q,D)$ is denoted D_q. [$\Gamma(q,D)=0$ for $(q-1)D<(q-1)D_q$, and $\Gamma(q,D)=\infty$ for $(q-1)D>(q-1)D_q$ (cf. Figs. 5(b) and 5(c)).]

Again say that we do not optimize over coverings, but instead use a cubic grid, of basic length l, then $l_i=l$ and we have $\sum_i p_i^q/l_i^\tau=l^{-\tau}\sum_i p_i^q$. From Eq. (3.1), $\sum_i p_i^q\sim l^{(q-1)\tilde{D}_q}$ and hence $\sum_i p_i^q/l_i^\tau\sim l^{(q-1)(\tilde{D}_q-D)}$, which, in the limit $l\to 0$, pass from 0 to ∞ at $D=\tilde{D}_q$. Since in this the optimization over coverings, prescribed in Eq. (3.8), is not done, the quantity D_q defined by Eqs. (3.7)–(3.9) is necessarily less than or equal to $\tilde{D}_q$,

$$D_q\leq \tilde{D}_q\ . \tag{3.10}$$

However, as for the Hausdorff and capacity dimensions, it is to be expected that, in practice, the equality in (3.10) typically holds for chaotic attractors.

A central result of Halsey *et al.*[4] was the demonstration of the way in which D_q is connected to the hot and cold points on the attractor (i.e., those points $\underline{x}$ on the attractor for which the pointwise dimension $(D_p(\underline{x})\neq\tilde{D}_1)$. In particular, they consider the set of $\underline{x}$ values such that $D_p(\underline{x})=\alpha$, and they denote the Hausdorff dimension of this set by $f(\alpha)$. They show that the D_q can be explicitly obtained from the dimensions $f(\alpha)$ via the formulas,

$$df(\alpha)/d\alpha=q\ , \tag{3.11a}$$

$$(q-1)D_q=[q\alpha(q)-f(\alpha(q))]\ . \tag{3.11b}$$

IV. UNSTABLE PERIODIC ORBITS

In this section we state and discuss results obtained in Sec. VI on the relation of unstable periodic orbits on chaotic attractors to the ergodic properties of these attractors. The results to be quoted are for the case of chaotic attractors that are *mixing* and *hyperbolic.* By mixing we mean that for any two sets S_a and S_b in the phase space of the system, we have

$$\lim_{n\to\infty}\mu[S_a\cap \underline{F}^n(S_b)]=\mu(S_a)\mu(S_b)\ ,$$

where μ is the natural measure of the attractor. A chaotic attractor would not be mixing in the commonly encountered situation where the attractor consists of a finite number h of disjoint pieces, and the orbit cycles from piece to piece. In this case, instead of the map $\underline{F}$ one can consider the map $\underline{F}^h$. For $\underline{F}^h$ each of the h pieces of the attractor for $\underline{F}$ is a separate attractor in its own right and is typically mixing. Our result would then apply to the attractors of $\underline{F}^h$. Henceforth, we assume that the attractor is mixing. Hyperbolic attractors are defined in Sec. VI.

We consider a d-dimensional twice differentiable map, $\underline{x}_{m+1}=\underline{F}(\underline{x}_m)$. The magnitudes of the eigenvalues of the Jacobian matrix of the n times iterated map $\underline{F}^n$ at the jth fixed point of $\underline{F}^n$ are denoted $\lambda_{1j},\lambda_{2j},\ldots,\lambda_{uj},\lambda_{(u+1)j},\ldots,\lambda_{dj}$, where we order the eigenvalues as follows: $\lambda_{1j}\geq\lambda_{2j}\geq\cdots\geq\lambda_{uj}>1\geq\lambda_{(u+1)j}\geq\cdots\geq\lambda_{dj}$. Thus in this notation, the number of unstable eigenvalues is u. Let L_j be the product of the unstable eigenvalues at the jth fixed point of $\underline{F}^n$,

$$L_j=\lambda_{1j}\lambda_{2j}\cdots\lambda_{uj}\ . \tag{4.1}$$

Then, as shown in Sec. VI for mixing hyperbolic (axiom A) attractors, the natural probability measure of the attractor contained in some closed subset S of the d-dimensional phase space is the limit as $n\to\infty$ of the sum of the L_j^{-1} over all the fixed points j of $\underline{F}^n$ which lie in S,

$$\mu(S)=\lim_{n\to\infty}\left[\sum_{\substack{\text{fixed points}\\ \text{in } S}} L_j^{-1}\right]\ . \tag{4.2}$$

Thus (4.2) is essentially a representation of the natural measure in terms of the unstable periodic orbits on the attractor. In particular, in the special case where S covers the entire attractor, we have $\mu(S)=1$, and, hence, we obtain a relation amongst the unstable eigenvalues,

$$1=\lim_{n\to\infty}\sum_j L_j^{-1}\ , \tag{4.3}$$

where the sum is over all fixed points of $\underline{F}^n$ on the attractor. Equation (4.3) has been conjectured to apply in general for Hamiltonian chaotic systems[12] and has been used to derive an interesting correspondence between the eigenvalues of a random matrix and the statistics of the semiclassical limit of the energy levels of a bound, time-independent, quantum system whose classical limit is chaotic.[13]

We now give a partition function formulation of the multifractal properties of chaotic attractors. This formulation is in terms of the eigenvalues of the dense set of periodic saddles on the attractor [rather than in terms of the measure of coverings of the attractor, i.e., the p_i in Eqs. (3.7)]. Let

$$D \equiv \Delta + \delta ,$$

where Δ is the integer part of D, and $\delta \equiv D - \Delta = (D \text{ modulo } 1)$ is the fractional part of D. In addition, let

$$S_j(D) \equiv \lambda_{ij}\lambda_{2j} \cdots \lambda_{\Delta j}(\lambda_{(\Delta+1)j})^{\delta} . \tag{4.4}$$

[Note that $S_j(D)$ is a continuous function of D.] In terms of L_j and $S_j(D)$, the result obtained in Sec. VI for the partition function is

$$\hat{\Gamma}(q,D,n) = \sum_j L_j^{-1}[S_j(D)]^{-(q-1)} . \tag{4.5}$$

In the two-dimensional case with $\lambda_{1j} > 1 > \lambda_{2j}$ this reduces to

$$\hat{\Gamma}(q,D,n) = \sum_j \lambda_{1j}^{-q}\lambda_{2j}^{-(D-1)(q-1)} , \tag{4.6}$$

which appears in Grebogi *et al.*[5] and Morita *et al.*[5] and, for the case of the Hausdorff dimension ($q=0$), in Ref. 9. Taking the limit $n \to \infty$ is analogous to taking the limit $l \to 0$ in Eq. (3.9),

$$\hat{\Gamma}(q,D) = \lim_{n \to \infty} \hat{\Gamma}(q,D,n) . \tag{4.7}$$

The quantity $\hat{\Gamma}(q,D)$ is zero or infinity in analogy with the quantity $\Gamma(q,D)$ in Eq. (3.9) [cf. Figs. 5(b) and 5(c)]. We denote the value of D at the transition of $\hat{\Gamma}(q,D)$ from zero to infinity by $\hat{D}_q$ and call it the periodic point dimension.

$$\hat{\Gamma}(q,D) = \begin{cases} 0 & \text{for } (q-1)D > (q-1)\hat{D}_q \\ +\infty & \text{for } (q-1)D < (q-1)\hat{D}_q . \end{cases}$$

In Sec. VI we show that

$$\hat{D}_q \geq D_q . \tag{4.8}$$

We conjecture that for typical chaotic attractors of two-dimensional maps with $\lambda_{1j} > 1 > \lambda_{2j}$

$$\hat{D}_q = D_q . \tag{4.9}$$

Indeed for the two-dimensional case with $\lambda_{1j} > 1 > \lambda_{2j}$, examples are worked out in Sec. V verifying that Eq. (4.9) holds. (For further discussion see Sec. VI.)

Setting $q=1$ and comparing (4.5) with (4.3) we have $\hat{\Gamma}(1,D)=1$. Formally expanding (4.5) around $q=1$ we obtain

$$\hat{\Gamma}(q,D,n) = 1-(q-1)\sum_j \frac{\log S_j(D)}{L_j} + O[(q-1)^2] .$$

Letting $n \to \infty$ the coefficient of the $(q-1)$ term may be expressed using (4.4) as

$$\lim_{n\to\infty} \left[\left(\sum_j L_j^{-1}\log(\lambda_{1j}\lambda_{2j}\cdots\lambda_{\Delta j}) \right) + \delta \left[\sum_j L_j^{-1}\log\lambda_{(\Delta+1)j} \right] \right] .$$

As will become evident from a correspondence with Lyapunov numbers to be discussed subsequently, the first term in large parentheses becomes positive infinite and the second term negative infinite (both sums behave like n for large n). Thus this formal expansion in $(q-1)$ indicates that the transition value $\hat{D}_1 = \Delta_1 + \delta_1$ occurs at

$$\hat{D}_1 = \Delta_1 - \lim_{n\to\infty} \frac{\sum_j L_j^{-1}\log(\lambda_{1j}\cdots\lambda_{\Delta_1 j})}{\sum_j L_j^{-1}\log(\lambda_{(\Delta_1+1)j})} , \tag{4.10}$$

where (to satisfy $0 \leq \delta_1 \leq 1$) we define Δ_1 as the largest integer such that $\sum_j L_j^{-1}\log(\lambda_{1j}\cdots\lambda_{\Delta_1 j})$ is positive. We now compare (4.10) with the Kaplan-Yorke conjecture.[1,17] The Kaplan-Yorke conjecture gives a formula for $\tilde{D}_1$ [defined in (3.2)] in terms of the Lyapunov numbers of typical orbits (i.e., orbits obtained for almost every choice of initial condition in the basin of the attractor). Denoting these (typical) Lyapunov numbers

$$\lambda_1 \geq \lambda_2 \geq \cdots \geq \lambda_d ,$$

the Kaplan-Yorke formula states that for typical systems,

$$\tilde{D}_1 = \Delta_1 - \frac{\ln(\lambda_1\lambda_2\cdots\lambda_{\Delta_1})}{\ln(\lambda_{\Delta_1+1})} , \tag{4.11}$$

where here Δ_1 is the largest integer such that $\lambda_1\lambda_2\cdots\lambda_{\Delta_1} > 1$. Comparing (4.11) and (4.10), see that these equations are the same if we interpret the sums over periodic orbits in (4.10) in terms of Lyapunov numbers,

$$\log\lambda_p = \lim_{n\to\infty} \frac{1}{n} \sum_j L_j^{-1}\log\lambda_{pj} . \tag{4.12}$$

Equation (4.12) is reasonable since our construction, to be discussed in Sec. IV A, shows that, during n iterates of the map, each orbit on the attractor stays close to some orbit of period n, and, furthermore, the natural measure of the orbits which stay close to a given period n orbit is L_j^{-1}. [This latter statement is related to Eq. (4.2).] Thus $\tilde{D}_1 = \hat{D}_1$.

The Kaplan-Yorke conjecture Eq. (4.11) has been shown to apply in a variety of examples and numerical experiments[1] (although it has not yet been proven in general). The correspondence of the $q \to 1$ limit of Eq. (4.5) [i.e., Eq. (4.10)] with the Kaplan-Yorke formula (4.11) supports our conjecture, Eq. (4.9).

Note that the two-dimensional map result for the pointwise dimension on the unstable manifolds of periodic orbits, Eq. (2.2), is the same as (4.11) with $d=2$ and the Lyapunov numbers of a typical orbit replaced by the magnitudes of the eigenvalues of the periodic orbit, $\lambda_{1j} > 1 > \lambda_{2j}$. A generalization of Eq. (2.2) to d-dimensional maps $d > 2$ is the statement that all points on the unstable manifold of the jth fixed point of $\underline{F}^n$ have the same pointwise dimension which is given by (4.11) with the Lyapunov numbers λ_p replaced by the eigenvalues λ_{pj}.

The Hausdorff dimension $f(\alpha)$ of the set of points on the attractor with $D_p(\underline{x})=\alpha$ can also be obtained directly from this formulation. In particular, define $\hat{\Gamma}_\alpha$ to be the same as the expression in (4.6) with $q=0$ (for the Hausdorff dimension) and the sum restricted only to those fixed points j which satisfy $\alpha+\Delta\alpha\geq D_p(\underline{x})\geq\alpha$,

$$\hat{\Gamma}_\alpha(D,\Delta\alpha,n)=\sum_{\substack{j\\ \alpha+\Delta\alpha\geq D_p(\underline{x})\geq\alpha}}\lambda_{2j}^{(D-1)}\ .$$

We then take limits,

$$\hat{\Gamma}_\alpha(D)=\lim_{\Delta\alpha\to 0}\lim_{n\to\infty}\hat{\Gamma}_\alpha(D,\Delta\alpha,n)\ ,$$

and we obtain $f(\alpha)$ as the transition value of D for which $\hat{\Gamma}_\alpha(D)$ goes from infinity to zero as D increases.

The results quoted in this section are for hyperbolic attractors (Sec. VI). For the two-dimensional hyperbolic case with $\lambda_{1j}>1>\lambda_{2j}$, we have $D_p(\underline{x})\geq 1$ for every point on the attractor. For the nonhyperbolic case, $D_p(\underline{x})$ can be less than 1, and we conjecture that (4.6) and (4.9) also hold in the nonhyperbolic case but only for q values corresponding to $\alpha(q)>1$ [cf. (3.11)].

Another result concerning ergodic properties of a map is that for the topological entropy S in terms of N_n, the number of fixed points of the n times iterated map $\underline{F}^n$,

$$S=\lim_{n\to\infty}\frac{1}{n}\log N_n\ . \tag{4.13}$$

For this result for the case of axiom A attractors see the papers by Bowen[6] and by Katok.[7]

V. EXAMPLES

We now illustrate the results on periodic orbits with two analytically tractable two-dimensional, hyperbolic map examples.

A. Example 1: The generalized baker's map

The generalized baker's map was introduced in Ref. 1 as a model for dimension studies which is amenable to analysis yet also has nonconstant stretching and contraction. We divide the square $0\leq(x,y)\leq 1$ into a bottom part, $0\leq y<a$, and a top part, $a<y\leq 1$. This is illustrated in Fig. 6(a) ($b=1-a$ in the figure). We compress the bottom (top) part by a factor λ_a (λ_b) along x, and stretch it in y by a factor a^{-1} (b^{-1}). We then have two rectangles, both of vertical height unity, one of width λ_a and the other of width λ_b [Fig. 6(b)]. We then move the λ_b width strip so that its lower left corner is at $x=\frac{1}{2}$, $y=0$ [Fig. 6(c)]. Thus we have a map of the unit square into itself: $x_{n+1}=\lambda(y_n)x_n+(\frac{1}{2})u(y_n-a)$; $y_{n+1}=\gamma(y_n)[y_n-au(y_n-a)]$; where $\lambda(y)=(\lambda_a,\lambda_b)$ for $y\lessgtr a$, $\gamma(y)=(a^{-1},b^{-1})$ for $y\lessgtr a$, and $u(y)$ is the unit step function.

Using similarity arguments[1] it can be shown directly from the map that the following transcendental equation determines D_q (cf. Refs. 2 and 3),

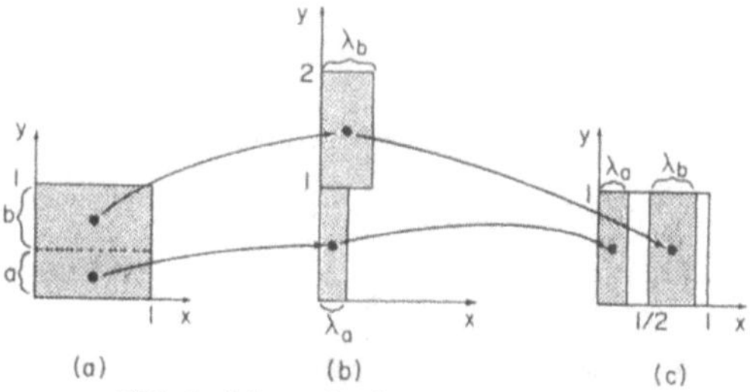

FIG. 6. Schematic of generalized baker's map.

$$1=\bar{A}+\bar{B}\ , \tag{5.1}$$

where $\bar{A}=\lambda_a^{-(q-1)(D_q-1)}a^q$, $\bar{B}=\lambda_b^{-(q-1)(D_q-1)}b^q$.

We now show that the equation determining the periodic point dimension $\hat{D}_q$ is the same as (5.1). We can specify an orbit for the generalized baker's transformation by its symbolic itinerary which specifies whether the orbit's location on successive iterates is in the top (symbolized by a 1) or in the bottom (symbolized by a 0). Thus a periodic orbit of period n which spends $k\leq n$ of its n iterates in $y>a$ is represented by a string of n symbols with k ones and $n-k$ zeros. The eigenvalues associated with such an orbit are $\lambda_2=\lambda_a^{n-k}\lambda_b^k$, $\lambda_1=a^{-(n-k)}b^{-k}$. Equation (4.6) yields

$$\hat{\Gamma}(q,D,n)=\sum_{k=0}^{n}N_{nk}A^{(n-k)}B^k\ , \tag{5.2}$$

where $A=a^q\lambda_a^{-(q-1)(D-1)}$, $B=b^q\lambda_b^{-(q-1)(D-1)}$, and N_{nk} is the number of fixed points of the n times iterated map which belong to periodic orbits which spend k iterates in the top ($y>a$). It can be shown that N_{nk} is the number of ways of arranging k zeros and $n-k$ ones,

$$N_{nk}=\binom{n}{k}\ . \tag{5.3}$$

Hence Eq. (5.2) is just a binomial expansion, $\hat{\Gamma}=(A+B)^n$. Letting $n\to\infty$, we see that the transition of $\hat{\Gamma}(q,D)$ from zero to infinity occurs at $D=D_q$ with D_q given by (5.1). Thus $D_q=\hat{D}_q$ for this example. [This calculation is algebraically equivalent to one in Halsey *et al.*,[4] although the basis for their calculation is Eq. (3.7), while the basis here is the periodic orbit formula, Eq. (4.6).]

To find $f(\alpha)$, we use (2.2) with λ_{1j} set equal to $a^{-(n-k)}b^{-k}$ and λ_{2j} set equal to $\lambda_a^{n-k}\lambda_b^k$ to obtain an equation relating α and $k/n\equiv\kappa$. This gives

$$\alpha=1+\frac{(1-\kappa)\log a+\kappa\log b}{(1-\kappa)\log\lambda_a+\kappa\log\lambda_b}\ .$$

Setting $q=0$ and using only the terms in the sum (5.2) with k/n values near that required by the specification of α and letting $n\to\infty$, we have (again using Stirling's approximation).

$$f(\alpha)=1+\frac{(1-\kappa)\log(1-\kappa)+\kappa\log\kappa}{(1-\kappa)\log\lambda_a+\kappa\log\lambda_b}\ , \tag{5.4}$$

with

$$\kappa=[\log a-(\alpha-1)\log\lambda_a]/[(\alpha-1)\log(\lambda_b/\lambda_a)+\log(a/b)]\ .$$

To obtain the topological entropy we note that $N_n=\sum_{k=0}^{n}N_{nk}$. Since N_{nk} is the binomial coefficient (5.3), this yields $N_n=2^n$, a well-known result for the baker's map. Equation (4.13) then gives $S=\log 2$, also well known.

It is also straightforward to verify Eq. (4.3). In particular, the right-hand side of (4.3) is $\sum\binom{n}{k}a^{n-k}b^k=(a+b)^n=1$ (recall $a+b=1$).

B. Example 2: The baker's apprentice's map

Example 1 has the property that setting $\hat{\Gamma}(q,D,n)=1$ gives precisely the desired result, Eq. (5.1), for all n, rather than only in the $n\to\infty$ limit. The generalized baker's map is exceptional in this regard, and this is due to the exact self-similarity of the attractor. A more typical example, which is still analytically tractable, is illustrated in Figs. 7. Again, we divide the unit square into top $(y>a)$ and bottom $(y<a)$ parts. We again horizontally compress the two parts by λ_a and λ_b. The bottom part is vertically stretched by a^{-1}, as before. The difference is that we now vertically compress the top part by a/b. The parts are then reassembled in the square as shown in Fig. 7(b). [We call this the baker's apprentice's map because the baker squashes and stretches the "dough" just right, so that it is exactly twice its original length, while the less experienced apprentice misses by not stretching enough to make the length double.]

Again we can specify orbits by a string of ones (tops) and zeros (bottoms). In this case, however, an orbit point in the top is always mapped to the bottom. Thus a one is *always* followed by a zero. Replacing B by $\tilde{B}=(b/a)^q\lambda_b^{-(q-1)(D-1)}$ (to account for the compression by a/b as opposed to the stretching by $1/b$ in example 1), we see that Eq. (5.2) still applies. Equation (5.3) for N_{nk}, however, does not.

To find N_{nk} we first note that the number of fixed points of the n times iterated map is the number of possible sequences of length n which contain k ones and $n-k$ zeros, subject to the constraint that a zero always follows a one (except when the last symbol is a one). We consider two cases: (a) the last symbol is a zero, and (b) the last symbol is a one. In case (a), to find the contribution to N_{nk} from such sequences, we regard the sequence (1,0) as a single symbol denoted by a 2. Thus a period n orbit which is located in the top k times is represented by a string of $(n-k)$ symbols of which k are twos and $n-2k$ are zeros (clearly $k\le n/2$). There are $\binom{n-k}{k}$ such symbol

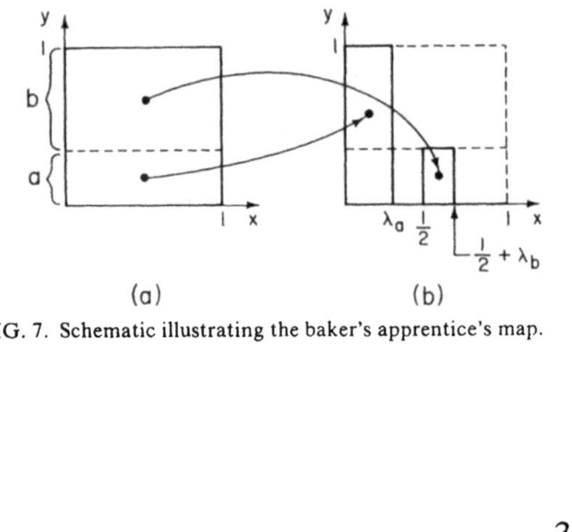

FIG. 7. Schematic illustrating the baker's apprentice's map.

sequences. Sequences ending in the top [case (b)], on the other hand, end in a one. Since the sequence represents a periodic orbit it must also start with a zero. All the rest of the symbols can be thought of as zeros and twos. For this case the zero-two sequence has $n-k-1$ symbols of which $k-1$ are twos. There are $\binom{n-k-1}{k-1}$ sequences of this type. Thus we have

$$N_{nk}=\binom{n-k}{k}+\binom{n-k-1}{k-1}, \tag{5.5}$$

and $N_{nk}=0$ for $k>n/2$.

Using Stirling's approximation to expand $Z(\kappa)\equiv(1/n)\log(N_{nk}A^{(n-k)}\tilde{B}^{k})$ for large n, we have

$$Z(\kappa)\cong(1-\kappa)\log(1-\kappa)-\kappa\log\kappa-(1-2\kappa)\log(1-2\kappa)$$
$$+\kappa\log\tilde{B}+(1-\kappa)\log A\ , \tag{5.6}$$

where $\kappa=k/n$ and $\frac{1}{2}>\kappa>0$. The quantity Z is concave down $(d^2Z/d\kappa^2<0)$ and has one maximum in $\frac{1}{2}>\kappa>0$. The location of this maximum is given by $\kappa_0(1-\kappa_0)=(1-2\kappa_0)^2\tilde{B}/A$. Since the summand in Eq. (5.2) is $\exp[nZ(\kappa)]$, if $n\to\infty$ and $Z(\kappa_0)<0$, then $\hat{\Gamma}(q,D)\to 0$. On the other hand, if $Z(\kappa_0)>0$, then $\hat{\Gamma}(q,D)\to\infty$. Thus at the transition we have the condition $Z(\kappa_0)=0$. This gives a transcendental equation for $\hat{D}_q$, $A+A\tilde{B}=1$, or

$$1=a^q\lambda_a^{(1-q)(\hat{D}_q-1)}+b^q(\lambda_a\lambda_b)^{(1-q)(\hat{D}_q-1)}\ . \tag{5.7}$$

We now show that D_q also satisfies Eq. (5.7). We employ the similarity technique[1-3] [used, for example, in Refs. 2 and 3 to obtain Eq. (5.1)]. We write the sum in the partition function, $\sum_i p_i^q/l_i^\tau$, as a sum over the top region plus a sum over the bottom region, $\Gamma(l)=\Gamma_T(l)+\Gamma_B(l)$. Similarly we write Γ_B as a sum over the bottom left $(x<\frac{1}{2})$, region plus a sum over the bottom right region,

$$\Gamma_B(l)=\Gamma_{BL}(l)+\Gamma_{BR}(l)\ . \tag{5.8}$$

Applying the map to one of the coverings of size l_i in the bottom, we see that it is compressed by λ_a and elongated by $1/a$. Thus this l_i covering can be covered by $(a\lambda_a)^{-1}$ coverings of size $(l_i\lambda_a)$. Each of these new coverings has a probability $(p_ia\lambda_a)$. Inserting this information in the partition function Eq. (3.7) we have that

$$\Gamma_T(l\lambda_a)=\frac{b}{a\lambda_a}\left[\frac{(a\lambda_a)^q}{\lambda_a^\tau}\Gamma_B(l)\right]$$
$$=\frac{b}{a}\frac{a^q}{\lambda_a^{(q-1)(D_q-1)}}\Gamma_B(l)\ .$$

Thus,

$$\Gamma_T(l)=\frac{b}{a}\frac{a^q}{\lambda_a^{(q-1)(D_q-1)}}\Gamma_B(l/\lambda_a)\ .$$

Similarly,

$$\Gamma_{BR}(l)=\frac{(b/a)^{q-1}}{\lambda_b^{(q-1)(D_q-1)}}\Gamma_T(l/\lambda_b)\ .$$

Also, $\Gamma_{BL}=(a/b)\Gamma_T$. Combining these in (5.8) and letting $l \to 0$, we obtain (5.7), and hence $D_q=\hat{D}_q$ for this example as well.

To find $f(\alpha)$ for the apprentice's map, we again use $\alpha=D_p(\underline{x})$ with $D_p(\underline{x})$ given by (2.2). This gives

$$\alpha=1+\frac{(1-\kappa)\log a+\kappa\log(b/a)}{(1-\kappa)\log\lambda_a+\kappa\log\lambda_b} . \qquad (5.9)$$

Including only values of k/n in the sum for $\hat{\Gamma}$ which nearly satisfies this relation and letting $n \to \infty$, we see that the Hausdorff dimension $f(\alpha)$ of the set where $\alpha=D_p(\underline{x})$ is given by setting $Z(\kappa)=0$ in Eq. (5.6) (with $q=0$),

$$f(\alpha)=1+\frac{(1-\kappa)\log(1-\kappa)-\kappa\log\kappa-(1-2\kappa)\log(1-2\kappa)}{(1-\kappa)\log\lambda_a+\kappa\log\lambda_b} \qquad (5.10)$$

where from (5.9)

$$\kappa(\alpha)=\frac{\log a-(\alpha-1)\log\lambda_a}{(\alpha-1)\log(\lambda_b/\lambda_a)+\log(a^2/b)} .$$

To find the topological entropy we perform the sum

$$N_n=\sum_{k=0}^{n/2} N_{nk} ,$$

again by using Stirling's approximation. We obtain $N_n \sim G^n$, where $G=(1+\sqrt{5})/2$ is the golden mean. Thus from (4.13) the topological entropy is $S=\log G$.

Again one may verify Eq. (4.3) by direct calculation. The quantity to be obtained is

$$\lim_{n\to\infty}\sum_{k=0}^{n/2} N_{nk}(b/a)^k a^{n-k} .$$

The computation is somewhat tedious, but straightforward (use Stirling's approximation yet again), and yields 1, as it should.

To conclude this section, we emphasize that these two-dimensional map examples have both been shown to satisfy the conjectured equality of D_q with the periodic point dimension $\hat{D}_q$ [Eq. (4.9)].

VI. DEMONSTRATION OF THE RELATION OF UNSTABLE PERIODIC ORBITS TO THE ERGODIC PROPERTIES OF ATTRACTORS

In this section we obtain the results stated in Sec. IV and illustrated in Sec. V concerning the relationship of periodic orbits to ergodic properties of chaotic attractors. We shall do this only for the case of hyperbolic attractors which have a dense set of periodic orbits (i.e., axiom *A* attractors). Although our arguments are strictly only for the hyperbolic case, we believe that the results may be valid much more generally. For any point $\underline{x}$ in the phase space let $W^s(\underline{x})$ and $W^u(\underline{x})$ denote the stable and unstable manifolds of $\underline{x}$. The stable manifold of $\underline{x}$ is the set of points $\underline{y}$ such that $\|\underline{F}^n(\underline{x})-\underline{F}^n(\underline{y})\| \to 0$ as $n \to +\infty$ (Fig. 8); the unstable manifold is the set of points $\underline{z}$ such that $\|\underline{F}^{-n}(\underline{x})-\underline{F}^{-n}(\underline{z})\| \to 0$ as $n \to +\infty$ (we assume here that

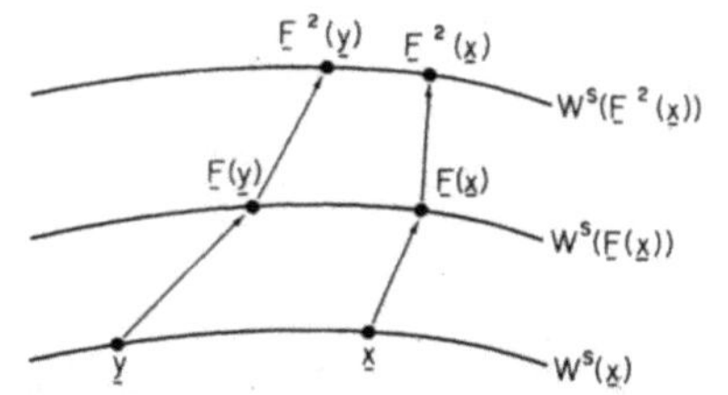

FIG. 8. Schematic illustrating the stable manifold of a point $\underline{x}$.

$\underline{F}$ is invertible). A hyperbolic attractor is one for which the following two conditions are satisfied.

(a) There exist stable and unstable manifolds $W^s(\underline{x})$ and $W^u(\underline{x})$ at each point $\underline{x}$ on the attractor that are not tangent and whose dimensions, d_s and d_u are the same for all $\underline{x}$ on the attractor, with $d_s+d_u=d$, where d is the dimension of the space. [Here $W^s(\underline{x})$ and $W^u(\underline{x})$ are smooth surfaces and d_s and d_u denote their Euclidean dimensions.]

(b) There exists a constant $K>1$ such that for all $\underline{x}$ on the attractor, if a vector $\underline{v}$ is chosen tangent to the unstable manifold, then

$$\|D\underline{F}(\underline{x})\underline{v}\| \geq K\|\underline{v}\| ,$$

and if $\underline{v}$ is chosen tangent to the stable manifold, then

$$\|D\underline{F}(\underline{x})\underline{v}\| \leq \|\underline{v}\|/K .$$

[Here $D\underline{F}(\underline{x})$ denotes the Jacobian matrix of partial derivatives of $\underline{F}(\underline{x})$ evaluated at $\underline{x}$.]

From condition (b) nearby points on the same stable (unstable) manifold approach (separate from) each other with time at least as fast as $\exp(-\kappa n)$ [$\exp(\kappa n)$]. For example, the magnitudes of the eigenvalues of $\underline{F}^n$ at a periodic point of a period n orbit must satisfy $\lambda_{pj} \geq K^n$ for $p \leq u$ and $\lambda_{pj} \leq K^{-n}$ for $p>u$. In particular, there can be no zero eigenvalues. In addition, it is very common for chaotic attractors encountered in practice to not be hyperbolic because they have points where $W^s(\underline{x})$ and $W^u(\underline{x})$ are tangent [in violation of condition (a)]. For example, the Henon attractor is of this type.

We first deal with two-dimensional maps (Secs. VI A–VI B) with $d_u=d_s=1$ and then indicate how the results can be extended to higher dimensions (Sec. VI C).

A. Measure

Imagine that we partition the space into cells C_i, where each cell has as its boundaries stable and unstable manifolds [Fig. 9(a)]. If the cells are very small, the curvature of the boundaries will be slight, and we can regard them as parallelograms [Fig. 9(b)]. Say we consider a given cell C_k and a large number of initial conditions sprinkled within the cell according to the natural probability measure on the attractor. Imagine that we iterate each of these initial conditions n times. After n iterates, a small fraction of the initial conditions may return to the small cell C_k. Since we assume the attractor to be ergodic and

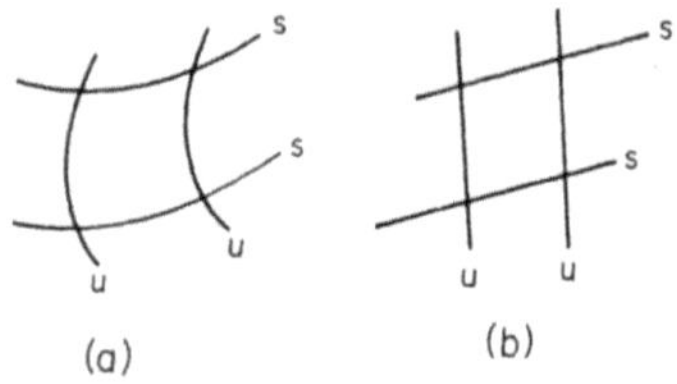

FIG. 9. Cells in a partition of the phase space. The letters s and u label stable and unstable manifold segments bounding the cell.

mixing this fraction is asymptotically (i.e., in the limit $n \to \infty$) equal to the natural measure of the attractor in the cell, $\mu(C_k)$. Let $\underline{x}_0$ be an initial condition that returns and $\underline{x}_n$ its nth iterate. This is illustrated in Fig. 10(a), where we take the stable direction as horizontal and the unstable direction as vertical. The line $ab(c'd')$ through $\underline{x}_0$ ($\underline{x}_n$) is a stable (unstable) manifold segment traversing the cell. Now take the nth forward iterate of ab and the nth backward iterate of $c'd'$. These map to $a'b'$ and cd as shown in Fig. 10(b). Now consider a rectangle constructed by passing unstable manifold segments $e'f'$ through a' and $g'h'$ through b'. By the construction, the nth preimages of these segments are the stable manifold segments ef and gh shown in Fig. 10(c). Thus we have constructed a rectangle $efgh$ in C_k such that all the points in $efgh$ return to C_k in n iterates. That is, $efgh$ maps to $e'f'g'h'$ in n iterates. The intersection of these two rectangles must contain a single saddle fixed point of the n times iterated map [cf. Fig. 10(d)]. Conversely, given a saddle fixed point, we can construct a rectangle of initial conditions $efgh$ which returns to C_k by closely following the periodic orbit which goes through the given fixed point j of $\underline{F}^n$ [the construction is the same as in Figs. 10(a)–10(c) except that $\underline{x}_0=\underline{x}_n$]. Thus all initial conditions which return after n iterates lie in some

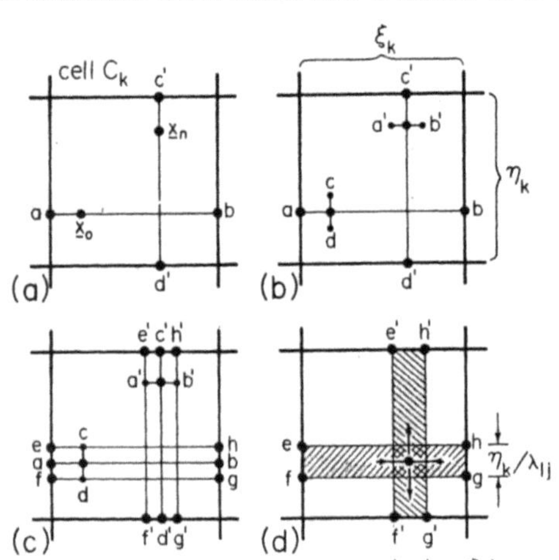

FIG. 10. Schematic of the construction of a rectangle in cell C_k which returns to C_k.

long thin horizontal strip (like $efgh$) which contains a fixed point of the n times iterated map. We label this fixed point j and the magnitudes of its unstable and stable eigenvalues λ_{1j} and λ_{2j}. Denoting the horizontal and vertical lengths of the sides of the cell C_k by ξ_k and η_k [cf. Fig. 10(b)], we see that the initial strip $efgh$ has dimensions ξ_k by (η_k/λ_{1j}) and the final strip has dimensions $\xi_k\lambda_{2j}$ by η_k [cf. Fig. 10(d)]. Since the dynamics is expanding in the vertical direction, the attractor measure varies smoothly in this direction.[18] Since the cell is assumed small, we can treat the attractor as if it were essentially uniform along the vertical direction. Thus the fraction of the measure of C_k occupied by the strip $efgh$ is $1/\lambda_{1j}$. Since, for $n\to\infty$, the fraction of initial conditions starting in C_k which return to it is $\mu(C_k)$, we have

$$\mu(C_k)=\lim_{n\to\infty}\left[\sum_{\substack{\text{fixed points}\\ \text{in } C_k}}\lambda_{1j}^{-1}\right]. \tag{6.1}$$

(Also note that, as n gets larger, λ_{1j}^{-1} and λ_{2j} get exponentially smaller and the number of fixed points in C_k grows exponentially.) Since we imagine that we can make the partition into cells as small as we wish, we can approximate any subset S of the phase space (with reasonably smooth boundaries) by a covering of cells. Thus we obtain the result, Eq. (4.2), of Sec. IV (with $L_j=\lambda_{1j}$ for the case treated here; i.e., $d=2$ with saddle periodic orbits).

In the construction which we used in arriving at Eq. (6.1), we have made two implicit assumptions. Namely, we have assumed that the segment ab maps to a segment $a'b'$ which lies entirely within C_k [i.e., we assume that the situation in Fig. 11(a) does not occur], and we have assumed that the preimage of $c'd'$ is entirely within C_k [i.e., Fig. 11(b) does not occur]. These situations might conceivably occur if $\underline{x}_n$ is too close ($\sim\xi_k\lambda_{2j}$) to a stable boundary or if $\underline{x}_0$ is too close ($\sim\eta_k/\lambda_{1j}$) to an unstable boundary. The point we wish to make here is that, for hyperbolic systems, the partition into cells can be chosen in such a way that the situations depicted in Fig. 11 do not occur. Such partitions are called Markov partitions.[18]

A Markov partition[18] satisfies the following condition. Say we have a Markov partition into cells C_i. Then, if $\underline{x}$ is in cell C_k and $\underline{F}(\underline{x})$ is in cell C_l, we have that

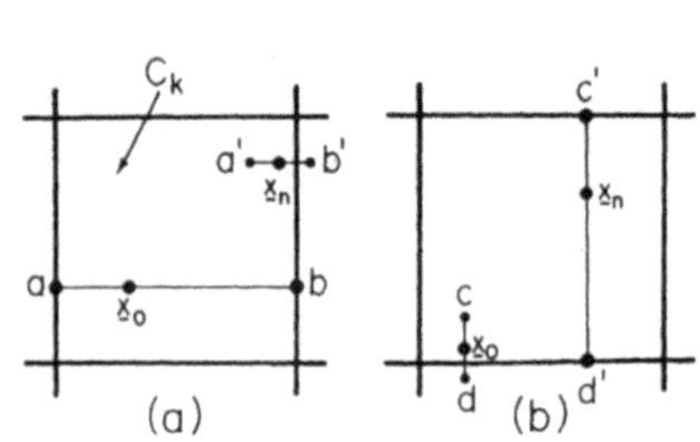

FIG. 11. Schematic depicting what can happen if $\underline{x}_0$ or $\underline{x}_n$ is too close to the boundary of C_k and a Markov partition is not used.

$$\underline{F}(W^u(\underline{x},C_k)) \supset W^u(\underline{F}(\underline{x}),C_l)\,, \tag{6.2a}$$

$$\underline{F}(W^s(\underline{x},C_k)) \subset W^s(\underline{F}(\underline{x}),C_l)\,, \tag{6.2b}$$

where $W^u(\underline{x},C_k)$ denotes a segment of the unstable manifold of $\underline{x}$ which traverses C_k passing through $\underline{x}$, and similarly for $W^s(\underline{x},C_k)$. Equations (6.2) are illustrated in Fig. 12. [Note that, to satisfy (6.2a), the end points of $\underline{F}(W^u(\underline{x},C_k))$ map to cell boundaries.] As an example, Fig. 13 shows a succession of finer Markov partitions for the baker's apprentice's map. The unit square, Fig. 13(a), is mapped into itself and bounded by segments of stable (horizontal) and unstable (vertical) manifolds. The boundaries of the partitions are obtained by taking forward iterates of the vertical (unstable) boundaries and backward iterates of the horizontal (stable) boundaries (as is explained further in the caption to Fig. 13).

Remark. The arguments as presented are not rigorous, but can be made rigorous. In particular, our arguments assume that we can approximate the map on each cell C_k as if it were linear. Also we assume that n is very large, giving the mixing sufficient time to make the probability of returning to C_k nearly equal to the measure $\mu(C_k)$. Detailed calculations based on the second derivative of the map show that our linear estimate (6.1) can be in error by at most a factor of $1\pm\epsilon$, where ϵ depends on the size of C_k and the size of the second-order partial derivatives of the map. Since we can make the size of C_k uniformly small, we can make ϵ uniformly small. Hence Eq. (4.2) holds exactly.

B. Dimension

In Fig. 10 the attractor measure contained in $e'f'g'h'$ is $1/\lambda_{1j}$. Say we cover $e'f'g'h'$ with boxes of edge length ξ_k/λ_{2j} corresponding to the narrow width of $e'f'g'h'$. There are $m_j=\eta_k/(\xi_k\lambda_{2j})$ such boxes. Since the attractor measure in the cell is essentially uniform along the vertical direction, the measure contained within one of the small boxes of edge length $\xi_k\lambda_{2j}$ is $1/(\lambda_{1j}m_j)$. Thus the contribution from $e'f'h'g'$ to the sum $\sum_i p_i^q/l_i^\tau$ in Eq. (3.7) is $m_j(\lambda_{1j}m_j)^{-q}(\xi_k\lambda_{2j})^{-\tau}$, or

$$\chi_k\frac{1}{\lambda_{1j}^q\lambda_{2j}^{(q-1)(D-1)}}\,,$$

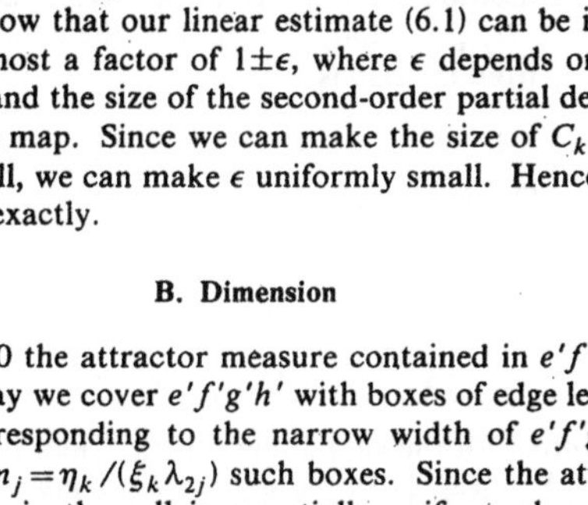

FIG. 12. Schematic illustrating Eqs. (6.2).

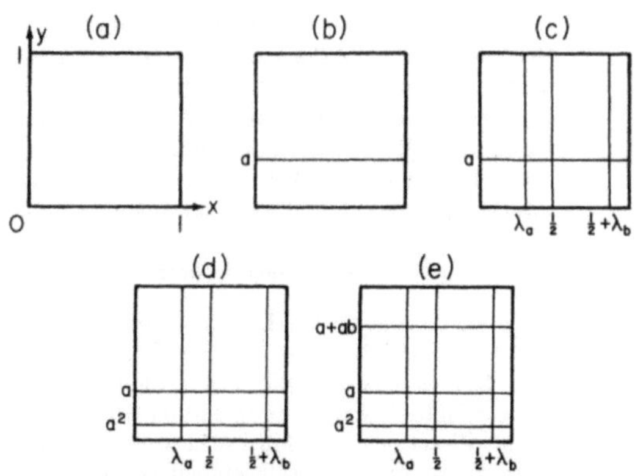

FIG. 13. Several successively finer Markov partitions for the baker's apprentice's map. Starting with (a), (b) is obtained by taking a preimage of the horizontal line $y=1$ (which is a stable manifold segment). (c) is obtained by taking a forward iterate of the vertical (unstable manifold) segments, $x=0$ and $x=1$. (d) is obtained by taking a preiterate of $y=a$ in (b), and (e) is obtained from (d) by preiterating $y=a^2$, etc.

where

$$\chi_k=(\eta_k\xi_k^{D-1})^{-(q-1)}\,.$$

Note that χ_k depends only on the partition and not on n. For a given partition there will be some cell $k=k_M$ where χ_k is the largest and some cell $k=k_m$ where χ_k is the smallest

$$\chi_{k_M}\ge\chi_k\ge\chi_{k_m}\,. \tag{6.3}$$

Now summing over all boxes we have

$$\sum_i p_i^q/l_i^\tau=\sum_j\chi_k\frac{1}{\lambda_{1j}^q\lambda_{2j}^{(q-1)(D-1)}}\,. \tag{6.4}$$

Using the bound (6.3), we have for the quantity in (6.4)

$$\chi_{k_M}\hat{\Gamma}(q,D,n)\ge\sum_j\chi_k\frac{1}{\lambda_{1j}^q\lambda_{2j}^{(q-1)(D-1)}}\ge\chi_{k_m}\hat{\Gamma}(q,D,n)\,, \tag{6.5}$$

where $\hat{\Gamma}(q,D,n)$ is given by (4.6). Letting $l\to 0$ in (3.9) is analogous to letting $n\to\infty$, since the box edge length used for the cover of $e'f'g'h'$ is edge $\xi_k\lambda_{2j}$, and λ_{2j} decreases exponentially with n. Letting n go to infinity, we see that the right-hand side of (6.4) goes through a transition from zero to infinity at the same value of D (denoted $\hat{D}_q$) as does $\sum_i p_i^q/l_i^\tau$. Furthermore, by (6.5) this transition also occurs at $D=\hat{D}_q$ for $\hat{\Gamma}(q,D)$ given by (4.6). This is the desired result for the two-dimensional map case.

In the above we have used a particular covering in obtaining the result (6.4) for $\sum_i p_i^q/l_i^\tau$. In particular, we have used the covering suggested by the dynamics and the Markov partition [cf. Fig. 10(d)]. However, since we have not optimized over all coverings, $\hat{D}_q$ the transition value of $\lim_{l\to 0}\sum_i p_i^q/l_i^\tau$ might overestimate D_q. That is

$\hat{D}_q$ is an upper bound to D_q, Eq. (4.8). It appears, however, that the covering of the attractor that we use is a rather efficient one, and we therefore believe that $D_q = \hat{D}_q$ for typical chaotic attractors. This is confirmed for the examples in Sec. V. [Furthermore, note that our result has turned out to be independent of the particular choice of the Markov partition (i.e., $\hat{\Gamma}$ does not involves χ_k).]

C. Systems of higher dimension

Again we consider a Markov partition, this time in a d-dimensional ($d>2$) space. A fixed point j in a small cell C_k has associated with it a d-dimensional parallelepiped of initial conditions extending across the cell in the stable directions and thin in the unstable directions. The nth iterate of this parallelepiped is a d-dimensional "slab" which extends across the cell in the unstable directions and is thin in the stable directions. This is illustrated for the case $d=3$ and $\lambda_{1j} > \lambda_{2j} > 1 > \lambda_{3j}$ (i.e., two unstable and one stable direction) in Fig. 14. Using this construction, it is readily seen that the derivation given in Sec. VI A for Eq. (6.1) extends to the higher dimensional case, for which we obtain Eq. (4.2). Now we turn to a consideration of the partition function for $d>2$. As shown in Eqs. (6.3)–(6.5) of Sec. VI B, the dimensions of the cells do not affect the final result; thus, for simplicity we set the d edge lengths of the cell C_k equal to 1 (e.g., $\xi_k = \eta_k = \beta_k = 1$ in Fig. 14). Now we cover the slab (cf. Fig. 14) with small d-dimensional cubes. We choose the edge length of the small cubes to be $\lambda_{(k+1)j}$, where we leave k unspecified for the moment, except to say that k is large enough so that $\lambda_{(k+1)j} < 1$ (or $k \geq u$ where u denotes the number of unstable eigenvalues). The number of such small cubes necessary to cover the slab is

$$m_j = \frac{\lambda_{1j}\lambda_{2j}\cdots\lambda_{kj}}{(\lambda_{(k+1)j})^k}\frac{1}{L_j} .$$

The probability measure of the attractor in the slab is L_j^{-1}. Thus, if the measures in each cube used to cover the slab are equal, then

$$p_i = \frac{1}{m_j L_j} \equiv \bar{p} . \tag{6.6}$$

The assumption of equal probabilities was justified in the two-dimensional case (Sec. VI B) by the fact that the measure varies smoothly along unstable manifolds. In the case considered here, however, k can be greater than u. The directions corresponding to subscripts $u+1, u+2, \ldots, k$ are not stretching. Thus the assumption of equal probabilities is not as well founded as it was in Sec. VI B. Nevertheless, it is still useful, as we now show. The contribution to $\sum_i p_i^q / l_i^\tau$ from the slab is $\lambda_{(k+1)j}^{-\tau} \sum_{\text{slab}} p_i^q$. The sum $\sum_{\text{slab}} p_i^q$ subject to the constraint $\sum_{\text{slab}} p_i = L_j^{-1}$ is bounded by the value it assumes when all of the p_i in the slab are equal,

$$\sum_{\text{slab}} p_i^q \leq m_j \bar{p}^{\,q} \quad \text{for } q<1 ,$$

$$\sum_{\text{slab}} p_i^q \geq m_j \bar{p}^{\,q} \quad \text{for } q>1 .$$

Thus using (6.6) for $q<1$ can only increase $\hat{\Gamma}(q,D)$. From Fig. 5(b) we see that this can possibly increase the transition value of D but not decrease it. Similarly, using (6.6) for $q>1$ can only decrease $\hat{\Gamma}(q,D)$, again leading to a possible increase in the transition value of D [cf. Fig. 5(c)]. Thus, we conclude that using (6.6) to obtain $\hat{\Gamma}$ will give a transition value of D for $\hat{\Gamma}$ which is an upper bound on D_q [Eq. (4.8)]. Using (6.6) the contribution to $\sum_i p_i^q / l_i^\tau$ from the slab is

$$m_j \left[\frac{1}{m_j L_j}\right]^q \frac{1}{(\lambda_{(k+1)j})^{(q-1)D}} = L_j^{-1}[S_{kj}(D)]^{-(q-1)} ,$$

where

$$S_{kj}(D) = m_j L_j (\lambda_{(k+1)j})^D$$
$$= \lambda_{1j}\lambda_{2j}\cdots\lambda_{kj}(\lambda_{(k+1)j})^{(D-k)} .$$

In accord with the supremum for $q>1$ and the infimum for $q<1$ in Eq. (3.8), we choose k to make $S_{kj}(D)$ as small as possible. Increasing k decreases $S_{kj}(D)$ so long as $D-k>0$ (recall that $\lambda_{(k+1)j}<1$). Thus $k=\Delta$, where Δ is the integer part of D, and the minimum $S_{kj}(D)$ is $S_j(D)$ given in Eq. (4.4). Now summing over all fixed points and all cells, we obtain the desired result, Eq. (4.5). We note that due to the possible nonuniformity of p_i within the slabs, our conjecture (4.9) is not on as firm ground for $d>2$ as it is for the case $d=2$ treated in Sec. VI B.

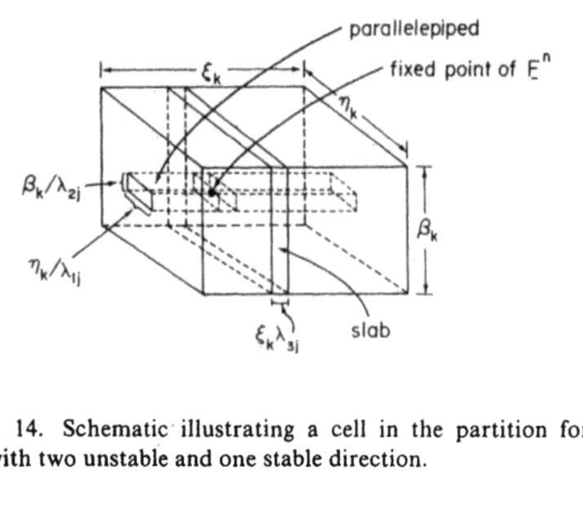

FIG. 14. Schematic illustrating a cell in the partition for $d=3$ with two unstable and one stable direction.

VII. TRANSIENT CHAOS

In many situations there are sets with chaotic dynamics which are not attractors. While such sets do not lead to long-term chaotic behavior of typical initial conditions, they do have an important effect on the gross dynamics.[10,19–21] In particular, they manifest themselves by the presence of chaotic transients. It is the purpose of this section to extend some of the results of the previous section to the case of chaotic sets which are not attracting.

Imagine a region of space Λ which encloses what we shall call a strange saddle. This is illustrated for $d=2$ in Fig. 15. This figure shows the saddle as the intersection of its stable and unstable manifolds, both of which may

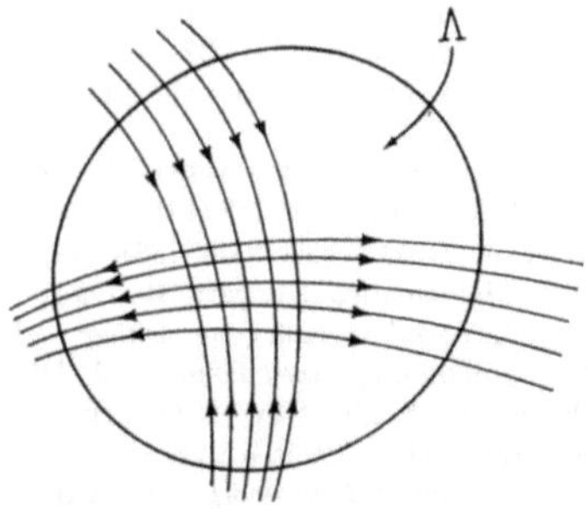

FIG. 15. Strange saddle and its stable and unstable manifolds, and a region Λ containing the strange saddle.

be thought of as a Cantor set of roughly parallel lines. If we sprinkle N_0 initial conditions uniformly in Λ, on subsequent iterates, the orbits leave Λ never to return. Let N_n be the number of orbits that have not left Λ after n iterates. For large n this number will decay exponentially with time,

$$(N_n/N_0)\sim\exp-(n/\tau)\ , \tag{7.1}$$

where we call τ the lifetime of the chaotic transient. We can define a natural measure for this process as follows. Let W be a subset of Λ. The natural measure of W is

$$\mu(W)=\lim_{n\to\infty}\lim_{N_0\to\infty} N_n(W)/N_n\ , \tag{7.2}$$

where $N_n(W)$ is the number of orbit points which fall in W at time n. Equations (7.1) and (7.2) imply that if initial conditions are distributed in accord with the natural measure and evolved in time, then the distribution will decay exponentially at the rate $1/\tau$. (This is not an invariant measure.) Points which leave Λ after a long time do so by being attracted along the stable manifold of the saddle, bouncing around on the saddle in a, perhaps, chaotic way, and then exiting along the unstable manifold.[10,19] The natural measure (7.2) is thus concentrated along the unstable manifold of the strange saddle.

Proceeding as in Sec. VI A, one can show that [compare with (4.3)]

$$\lim_{n\to\infty} e^{n/\tau}\sum_j L_j^{-1}=1\ , \tag{7.3}$$

from which the transient lifetime can be determined,

$$\frac{1}{\tau}=\lim_{n\to\infty}\left[\frac{1}{n}\log\left[\sum_j L_j^{-1}\right]^{-1}\right]\ . \tag{7.4}$$

This equation has been previously conjectured in Ref. 8 where numerical experiments verifying it are also reported. The quantity on the right-hand side of Eq. (7.4) is called the pressure and plays an important role in the ergodic theory of dynamical systems.

One can also ask what the dimensions D_q and $f(\alpha)$ are for the natural measure,[20,21] Eq. (7.2). Proceeding as in Secs. VI B and VI C, we find that the relevant partition function is

$$\hat{\Gamma}(q,D,n)=e^{qn/\tau}\sum_j\frac{1}{L_j}[S_j(D)]^{-(q-1)}\ . \tag{7.5}$$

In (7.5) the sum is over all fixed points of $\underline{F}^n$ which lie on the strange saddle. [The factors $e^{n/\tau}$ and $e^{qn/\tau}$ that appear in Eqs. (7.3) and (7.5) do so essentially in order to compensate for the exponential time decay of orbits in Λ which start with initial conditions distributed in accord with the natural measure (7.2).]

Formally expanding Eq. (7.5) in powers of $q-1$ [as was done in Sec. IV to obtain Eq. (4.10)], we find that the information dimension of the natural measure is

$$D_1=\Delta+\frac{\log(\bar{\lambda}_1\bar{\lambda}_2\cdots\bar{\lambda}_\Delta)-1/\tau}{\log(1/\bar{\lambda}_{\Delta+1})}\ , \tag{7.6}$$

where the Lyapunov numbers $\bar{\lambda}_p$ in this equation are defined in terms of the periodic orbits as follows:

$$\log\bar{\lambda}_p=\lim_{n\to\infty}\frac{e^{n/\tau}}{n}\sum_j\frac{1}{L_j}\log\lambda_{pj}\ . \tag{7.7}$$

Equation (7.6) has been given in Ref. 20 for the two-dimensional case, with the Lyapunov numbers defined in terms of typical orbits rather than periodic orbits [Eq. (7.7)], and we conjecture that the two definitions yield the same Lyapunov numbers. Note that for attractors $\tau\to\infty$ and (7.6) recovers the Kaplan-Yorke formula, Eq. (4.11).

VIII. CONCLUSION

The main result of this paper is a partition function formalism for determining the spectrum of fractal dimensions in terms of unstable periodic orbits. Beyond its conceptual appeal and potential use in further theoretical developments, the utility of this formalism will depend on how easy or difficult it is to extract information on periodic orbits from numerical computations and experimental data.[8-10]

ACKNOWLEDGMENTS

This work was supported by the U.S. Department of Energy (Basic Energy Sciences) and by the U.S. Office of Naval Research.

APPENDIX: NATURAL MEASURE

At the beginning of Sec. II we have given a definition of the natural measure which is adequate for "most purposes." A difficulty with the definition as stated can occur in special cases. For example, if the attractor has zero phase-space volume and we let the set S be the attractor itself, then, for almost every $\underline{x}$ in the basin of attraction, $\mu(\underline{x},S)$ as defined is zero (for finite length trajectories, the orbit approaches S but is not on S). A proper definition should give $\mu(S)=1$ for this S. To correct this, one can define $\mu(\underline{x},S)$ in a slightly different way. Let $\mu_\epsilon(\underline{x},S)$ be the fraction of time the trajectory originating at $\underline{x}$ spends in the ϵ neighborhood of S, and define

$$\mu(\underline{x},S)=\lim_{\epsilon\to 0^+}\mu_\epsilon(\underline{x},S)\ .$$

If this $\mu(\underline{x},S)$ is the same for almost every $\underline{x}$ in the attractor basin, then we denote this value $\mu(S)$ and call it the natural measure of the attractor.

[1]J. D. Farmer, E. Ott, and J. A. Yorke, Physica D 7, 153 (1983).
[2]P. Grassberger, Phys. Lett. **97A**, 227 (1983).
[3]H. G. E. Hentschel and I. Procaccia, Physica D **8**, 435 (1983).
[4]P. Grassberger, Phys. Lett. **107A**, 101 (1985); T. C. Halsey, M. J. Jensen, L. P. Kadanoff, I. Procaccia, and B. I. Shraiman, Phys. Rev. A **33**, 1141 (1986).
[5]C. Grebogi, E. Ott, and J. A. Yorke, Phys. Rev. A **36**, 3522 (1987). Recent related works are the following: T. Morita, H. Hata, H. Mori, T. Horita, nd K. Tomita, Progr. Theor. Phys. **78**, 511 (1987); G. Gunaratne and I. Procaccia, Phys. Rev. Lett. **59**, 1377 (1987); M. H. Jensen (unpublished).
[6]R. Bowen, Trans. Am. Math. Soc. **154**, 377 (1971).
[7]A. B. Katok, Publ. Math. IHES **51**, 137 (1980).
[8]L. P. Kadanoff and C. Tang, Proc. Natl. Acad. Sci. USA **81**, 1276 (1984).
[9]D. Auerbach, P. Cvitanovic, J.-P. Eckmann, G. Gunaratne, and I. Procaccia, Phys. Rev. Lett. **58**, 2387 (1987).
[10]C. Grebogi, E. Ott, and J. A. Yorke, Phys. Rev. Lett. **57**, 1284 (1986); Phys. Rev. A **36**, 5365 (1987).
[11]M. V. Berry, in *Chaotic Behavior of Deterministic Systems*, edited by G. Iooss, R. H. G. Helleman, and R. Stora (North-Holland, Amsterdam, 1983), p. 171–271.
[12]J. H. Hannay and A. M. Ozorio de Almeida, J. Phys. A **17**, 3429 (1984).
[13]M. V. Berry, Proc. R. Soc. London A **400**, 229 (1985).
[14]F. Hausdorff, Math. Annalen **79**, 157 (1918).
[15]A. N. Kolmogorov, Dokl. Akad. Nauk SSSR **119**, 861 (1958).
[16]A. Renyi, Acta Mathematica (Hungary) **10**, 193 (1959).
[17]J. Kaplan and J. Yorke, in *Functional Differential Equations and the Approximation of Fixed Points*, Vol. 730 of *Lecture Notes in Mathematics*, edited by H. O. Peitgen and H. O. Walther (Springer, Berlin, 1978), p. 228.
[18]R. Bowen, *On Axiom A Diffeomorphisms*, CBMS Regional Conference Series in Mathematics (American Mathematical Society, Providence, 1978), Vol. 35.
[19]C. Grebogi, E. Ott, and J. A. Yorke, Physica D **7**, 181 (1983).
[20]H. Kantz and P. Grassberger, Physica D **17**, 75 (1985); G. Hsu, E. Ott, and C. Grebogi, Phys. Lett. A (to be published).
[21]P. Szépfalusy and T. Tél, Phys. Rev. A **34**, 2520 (1986).

ISRAEL JOURNAL OF MATHEMATICS, Vol. 67, No. 3, 1989

ABSOLUTELY CONTINUOUS INVARIANT MEASURES FOR PIECEWISE EXPANDING C^2 TRANSFORMATIONS IN R^N

BY

P. GÓRA[a] AND A. BOYARSKY[b,†]

[a]*Department of Mathematics, Warsaw University, Warsaw, Poland; and*
[b]*Department of Mathematics, Concordia University, 7141 Sherbrooke St. West, Montreal, Canada H4B 1R6*

ABSTRACT

Let S be a bounded region in R^N and let $\mathscr{P} = \{S_i\}_{i=1}^m$ be a partition of S into a finite number of subsets having piecewise C^2 boundaries. We assume that where C^2 segments of the boundaries meet, the angle subtended by tangents to these segments at the point of contact is bounded away from 0. Let $\tau : S \to S$ be piecewise C^2 on $\mathscr{P}$ and expanding in the sense that there exists $0 < \sigma < 1$ such that for any $i = 1, 2, \ldots, m$, $\| D\tau_i^{-1} \| < \sigma$, where $D\tau_i^{-1}$ is the derivative matrix of τ_i^{-1} and $\| \ \|$ is the euclidean matrix norm. The main result provides an upper bound on σ which guarantees the existence of an absolutely continuous invariant measure for τ.

1. Introduction

In 1973 Lasota and Yorke [14] proved a general sufficient condition for the existence of an absolutely continuous invariant measure (a.c.i.m.) for expanding, piecewise C^2 transformations on the interval. In spite of the suggestion at the end of [14] that the "bounded variation" techniques of [14] can be easily used to obtain analogous results in higher dimensions, the generalization of the main result of [14] has taken much longer than expected. This was partly due to the difficulty in finding the right definition of variation in higher dimensions. For smooth maps on boundaryless domains, general results for the existence of a.c.i.m. were known as early as 1969 [12]. For piecewise C^2 maps in R^N, the first major attempt to prove an existence result came in 1979 [11].

† The research of the second author was supported by NSERC and FCAR grants.
Received March 7, 1989

The authors do not use a bounded variation argument but the proof, based on a one-dimensional version [10], is flawed. The first correct, but partial result, appeared in [8]. There, the author considers expanding, piecewise analytic transformations on the unit square partitioned by smooth boundaries. A complicated definition of bounded variation is used and the method cannot be extended beyond dimension 2. For boundaries which are not analytic, the sufficient condition that arises is rather complicated [9].

Working on rectangular partitions and with expanding, piecewise C^2 transformations which are very restrictive (the ith component of the transformation depends only on the ith variable), Jabłoński [7] proved the existence of an a.c.i.m. using the Tonnelli definition of bounded variation. The technique in this special setting is exactly analogous to that in [14].

In [17] a necessary and sufficient condition for the existence of a.c.i.m. is presented, but in most cases it cannot be applied.

With the publication of [3], a major new tool became available. The definition of variation of a function in R^N as the integral of its generalized derivative [3] led to the following partial result [1]: piecewise C^2 transformations on a rectangular partition satisfying a strong expansiveness condition (which depends on the dimension N of the space) have an a.c.i.m. Another partial result was obtained in [6].

In this note we follow through the approach of [1] in a more general setting. With only C^2 restrictions on the boundaries of the partition and with a mild restriction on how these boundaries meet, we prove the existence of an absolutely continuous invariant measure for τ if the slope of τ is sufficiently large.

In this setting, we can invoke the powerful Ionescu Tulcea and Marinescu Theorem [5] to obtain a useful spectral decomposition for the Perron–Frobenius operator of τ and, as a consequence, prove strong ergodic properties of the transformation itself.

Applications of ergodic theory for higher dimensional transformations can be found in [16, 18].

2. Main result

Let S be a bounded region in R^N and let τ be a transformation from S into S. We assume that τ is piecewise C^2 and expanding, i.e.,

(a) there exists a partition $\mathscr{P} = \{S_i\}_{i=1}^m$ of S, where m is a positive integer, and each S_i is a bounded closed domain having a piecewise C^2 boundary of finite $(N-1)$-dimensional measure;

(b) $\tau_i = \tau_{|S_i}$ is a C^2, 1–1 transformation from $\mathrm{int}(S_i)$ onto its image and can be extended as a C^2 transformation onto S_i, $i = 1, 2, \ldots, m$;

(c) there exists $0 < \sigma < 1$ such that for any $i = 1, 2, \ldots, m$,

$$\| D\tau_i^{-1} \| < \sigma, \tag{1}$$

where $D\tau_i^{-1}$ is the derivative matrix of τ_i^{-1} and $\| \quad \|$ is the euclidean matrix norm.

We remark that condition (1) implies, for $\tau_i^{-1}(x)$, $\tau_i^{-1}(y)$ close enough,

$$\rho(\tau_i^{-1}(x), \tau_i^{-1}(y)) < \sigma\rho(x, y),$$

where $x, y \in R_i \equiv \tau(\mathrm{int}(S_i))$ and ρ is the euclidean metric in R^N.

Condition (1) is implied by any of the following equivalent conditions:

(c1) all the eigenvalues of $D\tau_i^{-1}$ are smaller than 1;

(c2) all the eigenvalues of $D\tau_i$ are larger than 1.

If $|\partial\tau_{ij}^{-n}/\partial x_k| < 1$ for some n, where τ_{ij}^{-1} is the jth component of τ_i^{-1}, for $i = 1, \ldots, m$, and $1 \leqq j, k \leqq N$, then condition (c) is true for some iterate τ^l.

Let $Z = \bigcup_{i=1}^m \mathrm{int}(S_i)$. We will consider τ as a transformation from Z into S. Our assumptions imply it is nonsingular, i.e., $\lambda_N(\tau^{-1}(\quad))$ is absolutely continuous with respect to Lebesgue measure λ_N on S. This is enough for τ to induce the Perron–Frobenius operator

$$P_\tau : L_1(S) \rightarrow L_1(S),$$

defined by

$$P_\tau f(x) = \sum_{i=1}^m \frac{f(\tau_i^{-1}(x))}{\mathscr{J}(\tau_i^{-1}(x))} \chi_{R_i}(x),$$

where χ_R is the characteristic function of the set R and $\mathscr{J}(\beta)$ is the absolute value of the Jacobian of β. The properties of P_τ are described in [13], for example. It is well known that f is a τ-invariant density if and only if $P_\tau f = f$.

The main tool of the paper is the multidimensional notion of variation defined using derivatives in the distributional sense [3]:

$$V(f) = \int_{R^N} \| Df \| = \sup\left\{ \int_{R^N} f \operatorname{div}(g) d\lambda_N : g = (g_1, \ldots, g_N) \in C_0^1(R^N, R^N) \right\},$$

where $f \in L_1(R^N)$ has bounded support, Df denotes the gradient of f in the distributional sense, and $C_0^1(R^N, R^N)$ is the space of continuously differentiable functions from R^N into R^N having compact support. We will use the following property of variation which is easily derived from [3, Remark 2.14]:

If $f = 0$ outside a closed domain A whose boundary is Lipschitz continuous, $f_{|A}$ is continuous, $f_{|\mathrm{int}(A)}$ is C^1, then

$$V(f) = \int_{\mathrm{int}(A)} \| Df \| \, d\lambda_N + \int_{\partial A} |f| \, d\lambda_{N-1},$$

where λ_{N-1} is the $(N-1)$-dimensional measure on the boundary of A.

In the sequel we shall consider the Banach space [3, Remark 1.12],

$$BV(S) = \{ f \in L_1(S) : V(f) < +\infty \},$$

with the norm $\| f \|_{BV} = \| f \|_{L_1} + V(f)$.

Before stating the main theorem, we shall need a number of lemmas.

Consider an element $S_i \in \mathscr{P}$. Let x be a point in ∂S_i and $y = \tau(x)$ a point in $\partial(\tau(S_i))$. Let $\mathscr{J}$ be the Jacobian of $\tau_{|S_i}$ at x and $\mathscr{J}_0$ the Jacobian of $\tau_{|\partial(S_i)}$ at x.

LEMMA 1. $\mathscr{J}_0/\mathscr{J} \leqq \sigma$.

PROOF. Let C_n be a neighbourhood of y in $\tau(S_i)$ and $B_n = C_n \cap \partial(\tau(S_i))$, $n = 1, 2, \ldots$. Let γ be a curve perpendicular to $\partial(\tau(S_i))$ at y extending into C_n, and let $\gamma_n = \gamma \cap C_n$. We foliate C_n into hypersurfaces $B_n(t)$, $t \in \gamma_n$, each $B_n(t)$ being perpendicular to γ_n, and thus approximately parallel to B_n. We assume that for any n, $\lambda_{N-1}(B_n(t)) = \lambda_{N-1}(B_n)$ for all $t \in \gamma_n$. Then, if C_n is small enough, we have:

$$\lambda_N(C_n) = (1 + \varepsilon_n) \int_{\gamma_n} \lambda_{N-1}(B_n(t)) d\gamma_n(t) = (1 + \varepsilon_n) \lambda_{N-1}(B_n) \lambda_1(\gamma_n),$$

where $\varepsilon_n \to 0$ as $\mathrm{diam}(C_n) \to 0$. On the other hand, we have

$$1/\mathscr{J} = \lim_{\mathrm{diam}(C_n) \to 0} \lambda_N(\tau^{-1}(C_n))/\lambda_N(C_n).$$

To estimate $\lambda_N(\tau^{-1}(C_n))$, let $\eta_n = \tau^{-1}(\gamma_n)$, $D_n(t) = \tau^{-1}(B_n(t))$, $t \in \gamma_n$. Let $\mathscr{J}_0(t, \zeta)$ be the Jacobian of $\tau_{|D_n(t)}$, where $t \in \gamma_n$, $\zeta \in D_n(t)$. Since $\tau_{|S_i}$ is a C^2-diffeomorphism, $1/\mathscr{J}_0(t, \zeta)$ is a C^1-function. Thus

$$\begin{aligned} \lambda_{N-1}(D_n(t)) &= \int_{B_n(t)} (1/\mathscr{J}_0(t, \zeta)) d\lambda_{N-1}(\zeta) \\ &\leqq \int_{B_n(t)} (1/\mathscr{J}_0 + K \, \mathrm{diam}(C_n)) d\lambda_{N-1}(\zeta) \\ &= (1/\mathscr{J}_0 + K \, \mathrm{diam}(C_n)) \lambda_{N-1}(B_n), \end{aligned}$$

for a constant $K > 0$.

We have

$$\lambda_N(\tau^{-1}(C_n)) \leqq \int_{\eta_n} \lambda_{N-1}(D_n(t))d\eta_n(t)$$

$$= (1/\mathcal{J}_0 + K\operatorname{diam}(C_n))\lambda_{N-1}(B_n)\lambda_1(\eta_n)$$

$$= \sigma(1/\mathcal{J}_0 + K\operatorname{diam}(C_n))\lambda_{N-1}(B_n)\lambda_1(\gamma_n).$$

Thus

$$\lambda_N(\tau^{-1}(C_n))/\lambda_N(C_n) \leqq \sigma(1/\mathcal{J}_0 + K\operatorname{diam}(C_n))(1+\varepsilon_n).$$

Taking the limit as $\operatorname{diam}(C_n) \to 0$, we get $1/\mathcal{J} \leqq \sigma/\mathcal{J}_0$.

We note that this result is a considerable improvement over the condition $\mathcal{J}_0/\mathcal{J} \leqq N\sigma$ derived in [1] for transformations on rectangular partitions.

Let S be a closed domain in R^N with $W = \partial S$, which is piecewise C^2 and of finite $(N-1)$-dimensional measure. Let D denote the set of singular points of W, and let $\nu(x)$ denote the normalized outward normal vector at x (if $x \in D$, there are several possible outward normals). For any $x \in W$, let W_x be a small neighbourhood of x in W, contained completely in one face of W. If $x \in D$, we use a half-neighbourhood which is in one face.

For any $x \in W$, we define an R^N-neighbourhood of x, $\mathcal{U}(\delta, \alpha, x)$, $\delta > 0$, $\pi/2 < \alpha < \pi$, as follows: let $H(y)$, $y \in W_x$ be a C^1 normalized vector field, such that $\angle(H(x), \nu(x)) = \alpha$ ($\angle$ denotes angle). For any point $y \in W_x$, let $L_y = [y, y + \delta H(y)]$, the segment joining y and $y + \delta H(y)$. Now, let $\mathcal{U}(\delta, \alpha, x) = \bigcup_{y \in W_x} L_y$. If W_x and δ are small enough $\mathcal{U}(\delta, \alpha, x)$ lies completely on one side of W_x.

Lemma 2. *For any $x \in W$, and for any $\varepsilon > 0$ sufficiently small, we can choose W_x so that:*

$$\text{(2)} \qquad \frac{(1+\varepsilon)^2}{|\cos\alpha| - \varepsilon} \int_{\mathcal{U}(\delta,\alpha,x)} f\,d\lambda_N \geqq \int_{W_x} \left(\int_{L_y} f(\xi)d\xi \right) d\lambda_{N-1}(y)$$

for any $f \in C^1(R^N)$.

Proof. The inequality (2) obviously holds if W_x is a piece of an $(N-1)$-dimensional hyperplane. The idea of the proof is to convert our situation to that simple case.

By an orthogonal change of variables to the variables $(z_1, \ldots, z_N)$ in R^N, we can ensure that the hyperplane T_x tangent to W_x at x is given by $z_N = 0$, and the angle α between $\nu(x)$ and L_x is contained in the plane $z_1 = z_2 = \cdots = z_{N-2} = 0$. We choose W_x so small that it can be described by the equation: $z_N =$

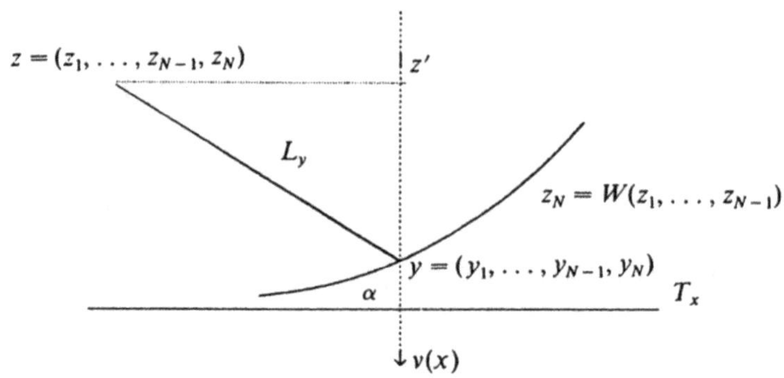

$W(z_1, \ldots, z_N)$; see Fig. 1. First, we straighten out all the segments L_y. We shall do this by keeping the same point y and shifting the point z to z', as shown in Fig. 1. If $z=(z_1, \ldots, z_N)$ is any point on the segment L_y, $y=(y_1, \ldots, y_N)$, then:

$$z=(y_1, \ldots, y_N)+|\operatorname{tg}\alpha|(z_N-y_N)[h_1, \ldots, h_N]$$

and $z'=(y_1, \ldots, y_{N-1}, z_N)$, where $[h_1, \ldots, h_N]=H(y)$, $\alpha=\alpha(y_1, \ldots, y_N)=\angle(H(y), v(x))$. Thus the straightening out is accomplished by the transformation φ_1 defined by:

$$\varphi_1(z_1, \ldots, z_N)=(z_1, \ldots, z_N)-|\operatorname{tg}\alpha|(z_N-y_N)[h_1, \ldots, h_{N-1}, 0],$$

where α, $h_1, \ldots, h_{N-1}$ are C^1 functions of $y_1, \ldots, y_N$, $y_N=W(y_1, \ldots, y_{N-1})$ and $y_1, \ldots, y_{N-1}$ are C^2 functions of $z_1, \ldots, z_N$.

Let

$$\mathscr{J}(\varphi_1)=\left|\det\left(\frac{\partial\varphi_{1,i}}{\partial z_j}\right)_{i,j=1}^{N}\right|$$

be the Jacobian of φ_1.

Notice that for $z=x=y$ we have $h_1=h_2=\cdots=h_{N-2}=\partial W/\partial z_1=\partial W/\partial z_2=\partial W/\partial z_{N-1}=z_N-y_N=0$ and that the derivatives of all involved functions are continuous and bounded. This implies that $\mathscr{J}(\varphi_1)(x)=1$ and that choosing W_x and δ small enough we can ensure $\mathscr{J}(\varphi_1)\leqq 1+\varepsilon$ on $\mathscr{U}(\delta, \alpha, x)$.

We now straighten out the surface W_x. This is done by the transformation

$$\varphi_2(z_1, \ldots, z_N)=(z_1, \ldots, z_{N-1}, z_N-W(z_1, \ldots, z_{N-1})).$$

The Jacobian of φ_2, $\mathcal{J}(\varphi_2)$, is equal to 1. Let $\mathcal{U} = \varphi_2 \circ \varphi_1(\mathcal{U}(\delta, \alpha, x))$ and $W'_x = \varphi_2\varphi_1(W_x)$. Then, we have:

$$\int_{\mathcal{U}(\delta,\alpha,x)} f\,d\lambda_N = \int_{\mathcal{U}} f \frac{1}{J(\varphi_2)} \frac{1}{J(\varphi_1)}\,d\lambda_N \geqq \frac{1}{(1+\varepsilon)} \int_{\mathcal{U}} f\,d\lambda_N$$

$$= \frac{1}{(1+\varepsilon)} \int_{W'_x} \left(\int_{\varphi_2(\varphi_1(L_y))} f(\eta)\,d\eta \right) d\lambda_{N-1}(w),$$

where $\eta = \varphi_2 \circ \varphi_1(\xi)$ and $w = \varphi_2(y)$. To obtain (2), we need $d\eta/d\xi$ and $J(\bar{\varphi}_2)$, where $\bar{\varphi}_2$ *is* φ_2 treated as a transformation from W_x into W'_x. From Fig. 1, it is easy to see that $d\eta/d\xi = |\cos\alpha|$, where $\alpha = \alpha(z_1, \ldots, z_N)$. If W_x is small enough $|\cos\alpha(z_1, \ldots, z_N)| \geqq |\cos\alpha| - \varepsilon$. Also,

$$J(\bar{\varphi}_2)(y) = |\mathrm{Det}(\langle g_i, g_j \rangle)_{i,j=1}^{N-1}|^{-1},$$

where $g(z_1, \ldots, z_{N-1}) = (z_1, \ldots, z_{N-1}, W(z_1, \ldots, z_{N-1}))$ and $g_i = \partial g/\partial z_i$, $i = 1, \ldots, N-1$. Since this determinant is 1 at the point x and all derivatives involved are at least continuous, we can choose W_x so small that

$$J(\bar{\varphi}_2)(y) \geqq \frac{1}{1+\varepsilon}, \qquad \text{for } y \in W_x.$$

We obtain

$$\int_{\mathcal{U}(\delta,\alpha,x)} f\,d\lambda_N \geqq \frac{1}{(1+\varepsilon)} \int_{W_x} \left(\int_{L_y} f(\xi)|\cos\alpha|\,d\xi \right) J(\bar{\varphi}_2)(y)\,d\lambda_{N-1}(y)$$

$$\geqq \frac{|\cos\alpha| - \varepsilon}{(1+\varepsilon)^2} \int_{W_x} \left(\int_{L_y} f(\xi)\,d\xi \right) d\lambda_{N-1}(y). \qquad \blacksquare$$

By a regular cone in R^N we mean a cone whose base is a $(N-1)$-dimensional disk B and such that the central ray L joining the vertex to the center of the disk B is perpendicular to the disk. We define the angle subtended at the vertex of a regular cone to be the angle between L and any line joining the vertex to a point on the boundary of B.

Let S, like before, be a closed domain in R^N having piecewise C^2 boundary of finite $(N-1)$-dimensional measure.

Let us now construct at any singular point $x \in D$, the largest possible regular cone having its vertex at x and which lies completely in S. Let $\theta(x)$ denote the angle subtended at the vertex of this cone. Then define

$$\beta(S) = \min_{x \in D} \theta(x).$$

Since the faces of W meet at angles bounded away from 0, $\beta(S) > 0$. Let $\alpha(S) = \pi/2 + \beta(S)$ and

$$a(S) = |\cos \alpha(S)|.$$

Now we will construct a C^1 field of segments L_y, $y \in W = \partial S$, every L_y being a central ray of a regular cone contained in S, with angle subtended at the vertex y greater than or equal to $\beta(S)$.

We start at the points $y \in D$, where the minimal angle $\beta(S)$ is attained, defining L_y to be central rays of the largest regular cones contained in S. Then we extend this field of segments to the C^1 field we want, making L_y short enough to avoid overlapping. Let $\delta(y)$ be the length of L_y, $y \in W$. By compactness of W we have

$$\delta(S) = \inf_{y \in W} \delta(y) > 0.$$

Now we shorten L_y of our field, making them all of the length $\delta(S)$.

LEMMA 3. *If some S is a closed domain with piecewise C^2 boundary of finite $(N-1)$-dimensional measure, whose smooth faces meet at angles bounded away from zero, and f is a C^1 function on S, then*

$$\int_{\partial S} f(y)d\lambda_{N-1}(y) \leqq \frac{1}{a(S)}\left(\frac{1}{\delta(S)}\int_S f\,d\lambda_N + V(f, \mathrm{int}(S))\right).$$

PROOF. Fix $\varepsilon > 0$, sufficiently small. We partition $W = \partial S$ into sets W_{x_i}, $i = 1, \ldots, M$, for which inequality (2) holds and define sets $\mathcal{U}(\sigma(S), \alpha(S), x_i)$, $i = 1, \ldots, M$, using the field of segments L_y, $y \in W$, constructed above. Now, for any $y \in W \setminus D$, we have:

$$f(y) \leqq \min\{f(x) : x \in L_y\} + V_{L_y}(f),$$

and

$$f(y) \leqq \frac{1}{|L_y|}\int_{L_y} f(\xi)d\xi + V_{L_y}(f) \leqq \frac{1}{\delta(S)}\int_{L_y} f(\xi)d\xi + V_{L_y}(f),$$

where $V_L(f)$ is the variation of f along the line L_y. We also have

$$V_{L_y}(f) \leqq \int_{L_y} \| Df(x) \| \, dx.$$

Integrating over W_{x_i}, $i = 1, \ldots, M$, we get

$$\int_{W_x} f(y)\,d\lambda_{N-1}(y)$$

$$\leqq \frac{(1+\varepsilon)^2}{a(S)-\varepsilon}\frac{1}{\delta(S)}\int_{\mathcal{U}(\delta(S),\alpha(S),x_i)} f\,d\lambda_N + \frac{(1+\varepsilon)^2}{a(S)-\varepsilon}\int_{\mathcal{U}(\delta(S),\alpha(S),x_i)} \| Df \|\,d\lambda_N.$$

Summing up, and noting that $\mathcal{U}(\delta(S), \alpha(S), x_i)$ do not overlap, we get:

$$\int_{\partial S} f(y)\,d\lambda_{N-1}(y) \leqq \frac{(1+\varepsilon)^2}{a(S)-\varepsilon}\left(\frac{1}{\delta(S)}\int_S f\,d\lambda_N + \int_S \| Df \|\,d\lambda_N\right).$$

Since ε is arbitrary, Lemma 3 is proved. ∎

Let $\tau : S \to S$ be a piecewise C^2 expanding transformation. We assume that the sets S_i, $i = 1, \ldots, m$, of its defining partition have piecewise C^2 boundaries of finite $(N-1)$-dimensional measure and that

$$a = \min\{a(S_i) : i = 1, \ldots, m\} > 0.$$

Let

$$\delta = \min\{\delta(S_i) : i = 1, \ldots, m\} > 0.$$

Under these assumptions, we prove the following results.

LEMMA 4. *Let $f \in L_1(S)$. If $V(f) < +\infty$, then*

$$V(P_\tau f) \leqq \sigma(1 + 1/a)V(f) + K \| f \|_{L_1},$$

for some constant $K < +\infty$.

PROOF. First we assume that $f \in C^1(S)$. Then

$$P_\tau f = \sum_{i=1}^{m} \frac{f(\tau_i^{-1})}{\mathcal{J}(\tau_i^{-1})}\chi_{R_i}.$$

Let $F_i = f(\tau_i^{-1})/\mathcal{J}(\tau_i^{-1})$, $i = 1, \ldots, m$. Then,

$$\int_{R^N} \| DP_\tau f \|\,d\lambda_N \leqq \sum_{i=1}^{m}\int_{R^N} \| D(F_i\chi_{R_i}) \|\,d\lambda_N$$

$$\leqq \sum_{i=1}^{m}\left(\int_{R^N} \| (DF_i)\chi_{R_i} \|\,d\lambda_N + \int_{R^N} \| F_i(D\chi_{R_i}) \|\,d\lambda_N\right).$$

We have

$$\int_{R^N} \| (DF_i)\chi_{R_i} \| \, d\lambda_N$$

$$= \int_{R_i} \| (DF_i) \| \, d\lambda_N$$

$$\leqq \int_{R_i} \| (Df(\tau_i^{-1}))/\mathscr{J}(\tau_i^{-1}) \| \, d\lambda_N + \int_{R_i} \| f(\tau_i^{-1}) D(1/\mathscr{J}(\tau_i^{-1})) \| \, d\lambda_N$$

$$\leqq \sum_{j=1}^{N} \int_{R_i} \frac{\left| \frac{\partial f}{\partial x_j}(\tau_i^{-1}) \right| \left| \sum_{k=1}^{N} \frac{\partial \tau_{ij}^{-1}}{\partial x_k} \right|}{\mathscr{J}(\tau_i^{-1})} \, d\lambda_N + \int_{R_i} |f(\tau_i^{-1})| \frac{K_0}{\mathscr{J}(\tau_i^{-1})} \, d\lambda_N$$

$$\leqq \sigma \int_{S_i} \| Df \| \, d\lambda_N + K_0 \int_{S_i} \| f \| \, d\lambda_N,$$

where τ_{ij}^{-1} is the jth component of τ_i^{-1}, $1 \leqq i \leqq m$, $1 \leqq j \leqq N$, and K_0 is an upper bound for $\| D(\mathscr{J}(\tau_i^{-1}))^{-1} \| \mathscr{J}(\tau_i^{-1})$, which exists by the C^2 assumption on τ_i, $1 \leqq i \leqq m$.

Now, using Example 1.4 of [3], we obtain:

$$\int_{R^N} \| F_i(D\chi_{R_i}) \| \, d\lambda_N = \int_{\partial R_i} |F_i| \, d\lambda_{N-1}$$

$$= \int_{\partial R_i} |f(\tau_i^{-1})| \mathscr{J}(\tau_i^{-1})^{-1} d\lambda_{N-1} = \int_{\partial S_i} |f| (\mathscr{J}_0/\mathscr{J}) d\lambda_{N-1},$$

where $\mathscr{J}_0$ is the Jacobian of $\tau_i : \partial S_i \to \partial R_i$.

By Lemma 1, $\mathscr{J}_0/\mathscr{J} \leqq \sigma$. Therefore, using Lemma 3, we obtain:

$$\int_{R^N} \| F_i(D\chi_{R_i}) \| \, d\lambda_N \leqq \sigma \int_{\partial S_i} |f| \, d\lambda_{N-1}$$

$$\leqq \frac{\sigma}{a} V(f, S_i) + \frac{\sigma}{a\delta} \int_{S_i} |f| \, d\lambda_N.$$

Summing up, we get:

$$V(P_\tau f) \leqq \sigma(1 + 1/a) V(f) + (K_0 + \sigma/a\delta) \| f \|_{L_1}.$$

Let $K = K_0 + \sigma/a\delta$.

In general, let $f \in BV(S)$. There exists a sequence of C^1 functions f_n, $n = 1, 2, \ldots$, which approximates f in $BV(S)$. We have:

$$\begin{aligned}
V(P_\tau f) &= \sup_h \int_{R^N} (P_\tau f)\operatorname{div}(h)d\lambda_N \\
&= \sup_h \lim_{n\to+\infty} \int_{R^N} (P_\tau f_n)\operatorname{div}(h)d\lambda_N \\
&= \sup_h \limsup_{n\to+\infty} \int_{R^N} (P_\tau f_n)\operatorname{div}(h)d\lambda_N \\
&\leqq \limsup_{n\to+\infty} V(P_\tau f_n) \\
&\leqq \lim_{n\to+\infty} (\sigma(1+1/a)V(f_n) + K \| f_n \|_{L_1}) \\
&= \sigma(1+1/a)V(f) + K \| f \|_{L_1},
\end{aligned}$$

where $h \in C^1(R^N, R^N)$, $\| h \| \leqq 1$. ■

LEMMA 5. *For any* $f \in BV(S)$

(3) $$\| P_\tau f \|_{BV} \leqq \sigma(1+1/a) \| f \|_{BV} + (K+1) \| f \|_{L_1}.$$

PROOF. Follows directly from Lemma 4 and the definition of $\| \; \|_{BV}$. ■

REMARK. For $N = 1$, the inequality (3) yields the same slope condition as in the original Lasota–Yorke Theorem [14].

We can now state the main result of this paper.

THEOREM 1. *Let* $\tau : S \to S$, $S \subset R^N$, *be a piecewise* C^2, *expanding transformation. If* $\sigma(1+1/a) < 1$, *then* τ *admits an absolutely continuous invariant measure.*

PROOF. From inequality (3) it follows that the set $\{ \| P_\tau^i(1) \|_{BV} \}_{i \geqq 1}$ is uniformly bounded. Hence the set $\{P_\tau^i(1)\}_{i \geqq 1}$ is weakly compact in L_1 (actually it is strongly compact), and it follows from the Kakutani–Yoshida Theorem that P_τ has a nontrivial fixed point f^* which is the density of an a.c.i.m. ■

COROLLARY 1. *Let* $\tau : S \to S$, $S \subset R^N$, *be piecewise* C^2 *and such that some iterate* τ^k *satisfies* $\sigma(1+1/a) < 1$ (σ *and* a *corresponds to* τ^k). *then* τ *admits an a.c.i.m.*

PROOF. Straightforward. ■

EXAMPLE. For a rectangular partition of a rectangular domain in R^N, we have $1/a = \sqrt{N}$, which gives the expansion condition $\sigma(1+\sqrt{N}) < 1$.

REMARK. The expansion condition $\sigma < (1 + 1/a)^{-1}$ depends only on the domain S and not on the transformation τ. Under certain conditions on τ, we can obtain an improved expansion condition such as is done in [9] in dimension 2. These conditions are usually very complex and require accurate knowledge of the transformation.

3. Spectral decomposition

In this section we will use the Ionescu Tulcea and Marinescu Theorem [5]. First we have to check that the assumptions of the theorem are satisfied:

We consider the space BV, $\| \ \|_{BV}$ as included in L_1, $\| \ \|_{L_1}$.

(1) By the semicontinuity property of variation (Theorem 1.9 of [3]), if $\{f_n\} \in BV$, $\| f_n \|_{BV} \leqq D$, for $n = 1, 2, \ldots$ and $f_n \to f$ in L_1, then $f \in BV$ and $\| f \|_{BV} \leqq D$.

(2) The operator norm of the Perron–Frobenius operator P_τ is 1.

(3) There exist constants $R > 0$, $0 < r < 1$ such that

$$\| P_\tau f \|_{BV} \leqq r \| f \|_{BV} + R \| f \|_{L_1}, \qquad \text{for } f \in BV.$$

This follows by Lemma 5 if $\sigma(1 + 1/a) < 1$.

(4) The image of any bounded subset of BV under the Perron–Frobenius operator is relatively compact in L_1. This follows from the compactness Theorem 1.19 of [3].

The Ionescu Tulcea and Marinescu Theorem implies the following result:

THEOREM 2. *Let $\tau : s \to S$, $S \subset R^N$, $N \geqq 1$, be a piecewise C^2 and expanding transformation with $\sigma(1 + 1/a) < 1$ and let $P = P_\tau$ be its Perron–Frobenius operator. Then:*

(a) *P (as an operator from BV into BV) has a finite number of eigenvalues of modulus 1: $\alpha_1, \ldots, \alpha_t$. They are roots of unity and*

$$P = \sum_{i=1}^{t} \alpha_i P_i + T,$$

where $P_i : BV \to BV$ are linear projections with finite dimensional range, and $T : BV \to BV$ is a continuous linear operator;

(b) $P_i^2 = P_i$, $P_i P_j = 0$ $(i \neq j)$, $P_i T = TP_i = 0$, $1 \leqq i, j \leqq t$;

(c) $\| T^n \|_{BV} \leqq M/(1 + h)^n$, $n = 1, 2, \ldots,$ *for some* $M, h > 0$.

REMARK (see [15]). Operators P_i, $i = 1, \ldots, t$ and T have unique extensions onto L_1. Moreover $P_i(L_1) \subset BV$, $\| P_i \|_{L_1} \leqq 1$ and $\sup_n \| T^n \| < +\infty$. For any $f \in L_1$, $T^n f \to 0$ in L_1, as $n \to +\infty$.

The following theorem and corollaries are consequences of the representation of the Perron–Frobenius operator obtained in Theorem 2.

THEOREM 3 (see [15]). *Assume that* 1 *is the only eigenvalue of P with modulus* 1 (*we can consider P^k, where k is the smallest common multiplier of orders of $\alpha_1, \dots, \alpha_t$*). *Let $Ug = g \circ \tau$, for $g \in L_\infty$. Then there exist nonnegative functions $\phi_1, \dots, \phi_s \in BV$ and $\psi_1, \dots, \psi_s \in L_\infty$ such that:*

(a) *For any $f \in L_1$*

$$P_1 f = \sum_{i=1}^{s} \left(\int_{R^N} f \psi_i d\lambda_N \right) \phi_i,$$

(b) $P\phi_i = \phi_i$, $U\psi_i = \psi_i$, $i = 1, \dots, s$.

(c) $\int_{R^N} \phi_i \psi_j d\lambda_N = \delta_{ij}$, $\inf\{\phi_i, \phi_j\} = 0 = \inf\{\psi_i, \psi_j\}$ *as $i \neq j$, and* $\int_{R^N} \phi_i d\lambda_N = 1$, $1 \leqq i, j \leqq s$,

(d) *There exist measurable sets $C_1, \dots, C_s \subset S$ such that $\psi_i = \chi_{C_i}$ a.e., for $i = 1, \dots, s$ and $S = \bigcup_{i=1}^{s} C_i$ a.e.*

(e) $\bigcap_{n=1}^{\infty} U^n(L_1) = \bigcap_{n=1}^{\infty} U^n(L_\infty) = \mathrm{Span}\{\psi_1, \dots, \psi_s\}$.

(f) *For any $f \in L_1$, $U^n f \to P_1^* f$ in $\sigma(L_1, BV)$-topology; for any $f \in L_\infty$, $U^n f \to P_1^* f$ in $\sigma(L_\infty, L_1)$-topology;*

$$P_1^* f = \sum_{i=1}^{s} \left(\int_{R^N} f \phi_i d\lambda_N \right) \psi_i.$$

COROLLARY 2 (see [15]). *For any $1 \leqq i \leqq s$, $\tau^k_{|C_i}$ is an exact transformation.*

COROLLARY 3 (see [15]). *If we assume that τ is mixing* (*or weakly mixing, which is equivalent in this situation*), *and μ is its unique a.c.i.m., then τ has the property of exponential decay of correlation: Let $f \in BV$, $g \in L_\infty$ and $\mu(f) = \int_{R^N} f\, d\mu$, $\mu(g) = \int_{R^N} g\, d\mu$. Then*

$$\int_{R^N} (fg(\tau^n) - \mu(f)\mu(g)) d\mu \leqq \mu(f) V(f) \| g \|_{L_\infty} r^n, \qquad n = 1, 2, \dots,$$

where $0 < r < 1$ is the constant of condition (3).

COROLLARY 4 (see [15]). *If we assume that τ is mixing, then the defining partition $\{S_i\}_{i=1}^{m}$ is weakly Bernoulli for τ, whch implies that the natural extension of the dynamical system (τ, μ) is isomorphic to a Bernoulli shift* (*μ is the τ-invariant absolutely continuous measure*).

COROLLARY 5 (see [4]). *Assume that τ, μ is weakly mixing (it is equivalent to being mixing or exact in our situation). Let $f \in BV$ and $\mu(f) = \int_{R^N} f\,d\mu = 0$. Define*

$$S(t) = \sum_{i=0}^{t-1} f \circ \tau^i,$$

which is a stochastic process on (S, μ). Then the series (σ below has nothing to do with the contraction constant of formula (1), we use it here only for historical reasons)

$$\sigma^2 = \int_S f^2 d\mu + 2 \sum_{k=1}^{\infty} \int_S f(f \circ \tau^k) d\mu$$

converges absolutely, $\int_S S(t)^2 d\mu = t\sigma^2 + O(1)$ and, if $\sigma^2 \neq 0$, the following holds:

(i) $\sup_{z \in R} |\mu((\sigma^2 t)^{-1/2} S(t) \leqq z) - (2\pi)^{-1/2} \int_{-\infty}^{z} \exp(-x^2/2)dx| = O(t^{-\nu})$, *for some $\nu > 0$.*

(ii) *Without changing its distribution, one can redefine the process $(S(t))_{t \geqq 0}$ on a richer probability space together with the standard Brownian motion $(B(t))_{t \geqq 0}$ such that*

$$|\sigma^{-1} S(t) - B(t)| = O(t^{(1/2)-\varepsilon}), \qquad \mu\text{-a.e.},$$

for some $0 < \varepsilon < 1/2$.

(iii) *The process $(S(t))_{t \geqq 0}$ satisfies the iterated log law and other properties of Brownian motion.*

ACKNOWLEDGEMENT

We are grateful to H. Proppe who found errors in earlier drafts of this paper and made many helpful suggestions.

REFERENCES

1. D. Candeloro, *Misure invariante per transformazioni in più dimensioni*, Atti Sem. Mat. Fis. Univ. Modena **XXXV** (1987), 33–42.
2. H. Federer, *Geometric Measure Theory*, Springer-Verlag, New York, 1969.
3. E. Giusti, *Minimal Surfaces and Functions of Bounded Variation*, Birkhauser, 1984.
4. F. Hofbauer and G. Keller, *Ergodic properties of invariant measures for piecewise monotonic transformations*, Math. Z. **180** (1982), 119–140.
5. C. T. Ionescu Tulcea and G. Marinescu, *Théorie ergodique pour des classes d'opérations non completement continues*, Ann. of Math. **52** (1950), 140–147.
6. V. V. Ivanov and A. G. Kachurowski, *Absolutely continuous invariant measures for locally*

expanding transformations, Preprint No. 27, Institute of Mathematics AN USSR, Siberian Section (in Russian).

7. M. Jabłoński, *On invariant measures for piecewise C^2-transformations of the n-dimensional cube*, Ann. Polon. Math. **XLIII** (1983), 185–195.

8. G. Keller, *Ergodicité et mesures invariantes pour les transformations dilatantes par morcaux d'une région bornée du plan*, C.R. Acad. Sci. Paris **289**, Série A (1979), 625–627.

9. G. Keller, *Proprietés ergodiques des endomorphismes dilatants, C^2 par morceaux, des régions bornées du plan*, Thesis, Université de Rennes, 1979.

10. A. A. Kosyakin and E. A. Sandler, *Ergodic properties of a class of piecewise-smooth transformations of an interval*, Izv. VUZ Matematika (3)[118] (1972), 32–40. (English translation from the British Library, Translation Service.)

11. A. A. Kosyakin and E. A. Sandler, *Stochasticity of a certain class of discrete system*, translated from Automatika and Telemekhanika No. 9 (1972), 87–94.

12. K. Krzyżewski and W. Szlenk, *On invariant measures for expanding differentiable mappings*, Studia Math. **33** (1969), 82–92.

13. A. Lasota and M. Mackey, *Probabilistic Properties of Deterministic Systems*, Cambridge University Press, 1985.

14. A. Lasota and J. A. Yorke, *On the existence of invariant measures for piecewise monotonic transformations*, Trans. Am. Math. Soc. **186** (1973), 481–488.

15. M. Rychlik, *Bounded variation and invariant measures*, Studia Math. **LXXVI** (1983), 69–80.

16. F. Schweiger, *Invariant measures and ergodic properties of numbertheoretical endomorphisms*, Banach Center Publications, to appear.

17. E. Straube, *On the existence of invariant absolutely continuous measures*, Commun. Math. Phys. **81** (1981), 27–30.

18. M. Yuri, *On a Bernouilli property for multi-dimensional mappings with finite range structure*, in *Dynamical Systems and Nonlinear Oscillations*, Vol. 1 (Giko Ikegami, ed.), World Scientific, Singapore, 1986.

Invent. math. 112, 541–576 (1993)

Inventiones mathematicae

Sinai–Bowen–Ruelle measures for certain Hénon maps

Michael Benedicks[1] and Lai-Sang Young[2,★]

[1] Department of Mathematics, Royal Institute of Technology, S-100 44 Stockholm, Sweden
[2] Department of Mathematics, UCLA, Los Angeles, CA 90024, USA

Oblatum 20-III-1992

0 Introduction

We study maps $T_{a,b}: \mathbb{R}^2 \to \mathbb{R}^2$ defined by

$$T_{a,b}(x, y) = (1 - ax^2 + y, bx), \quad 0 < a < 2, b > 0.$$

In [BC2] it was proved that for a positive measure set of parameters (a, b), $T_{a,b}$ has a topologically transitive attractor $\Lambda = \Lambda_{a,b}$ on which there is some hyperbolic behavior. The aim of this paper is to study the *statistical* properties of these attractors. Using the machinery developed in [BC2] we prove the following

Theorem. *There is a set $\Delta \subset \mathbb{R}^2$ with* $\mathrm{Leb}(\Delta) > 0$ *such that for all $(a, b) \in \Delta$, $T = T_{a,b}$ admits a unique* SBR *measure λ^*. Moreover,* $\mathrm{supp}(\lambda^*) = \Lambda$ *and (T, λ^*) is Bernoulli.*

A T-invariant Borel probability measure μ is called a Sinai–Bowen–Ruelle (SBR) measure if there is a positive Lyaponov exponent μ-a.e. and the conditional measures of μ on unstable manifolds are absolutely continuous with respect to the Riemannian measure induced on these manifolds. (A more precise definition is given in Sect. 3.4.1.) This notion is due to Sinai [S1, S2]. See also [LS]. The significance of these measures is evident in the following corollary.

Corollary. *For $(a, b) \in \Delta$, $T = T_{a,b}$ has the following property: Let U be a neighborhood of Λ. Then there is a set $\tilde{U} \subset U$ with positive Lebesgue measure such that for all continuous functions $\varphi: U \to \mathbb{R}$,*

$$\frac{1}{n} \sum_{i=0}^{n-1} \varphi \circ T^i(x) \to \int \varphi \, d\lambda^*$$

for every $x \in \tilde{U}$.

★ Benedicks is partially supported by the Swedish Natural Science Research Council (NFR) and the Swedish Board of Technical Development (STU). Young is partially supported by NSF

This corollary follows from our theorem and general nonuniform hyperbolic theory (see [PS, Theorem 1] or Sect. 4.1.2). The property expressed in this corollary is often assumed to be true in physics; it is also taken for granted in numerical experiments. In the case of Axiom A attractors, mathematical justification of this property is provided by the theory of Sinai, Bowen and Ruelle (see e.g. [B], [Ru1] and [S2]). It has been conjectured that many other attractors admit SBR measures, but as far as we know, the Hénon family and similar examples (see next paragraph) are the only nonuniformly hyperbolic attractors for which SBR measures have been constructed.

Mora and Viana [MV] proved recently that in homoclinic bifurcations of surface diffeomorphisms, very small attractors with high periods appear for a positive measure set of parameters. These attractors have the same qualitative estimates as those in [BC2]. Since our proofs rely only on these qualitative estimates, our results apply there also.

This paper is organized as follows. We assume throughout that $T = T_{a,b}$, where (a, b) is a pair of "good" parameters, meaning the ones selected in [BC2]. We do not concern ourselves with how these parameters are selected. In Sect. 1 we give a summary of the results from [BC2] that are relevant to this work. The primary focus of [BC2] is on a certain set of points called "critical points". These points live on a curve W that is the unstable manifold of a fixed point. In Sect. 2 we shift our attention from the critical points to the action of T on all of W. Our main proposition establishes a near complete correspondence between the dynamics of $T|W$ and that of certain interval maps. An SBR measure is constructed in Sect. 3, where Lebesgue measure on a small piece of W is pushed forward. The ergodic properties of this invariant measure are studied in Sect. 4.

1 Dynamics of certain Hénon maps: results from [BC2]

In this section we try to give a self-contained summary of what is known about the "good" Hénon maps. For conceptual simplicity, we will isolate for separate discussion several aspects of their dynamics. The reader should be aware that many of these properties are related, in the sense that their proofs are very delicately intertwined. Some of the ideas in [BC2] will be expressed here in slightly different language. For instance, we will speak of orbits and vectors as being "controlled" (see Sects. 1.4 and 1.5). We will also introduce the notion of "generalized tangential position" (Sect. 1.6).

1.1 The one-dimensional model (Sect. 2 of [BC2])

The b-values considered in [BC2] are very small and the entire analysis there is modeled on that of a certain class of 1-d maps. These are maps of the form $f = f_a: [-1, 1] \circlearrowleft$, where $f_a x = 1 - ax^2$ and a is very near 2. In addition the following conditions are imposed of f:

(1) There is $c > 0$ ($c \approx \log 2$) such that $|Df^n(f0)| \geqq e^{cn}$ for all $n \geqq 0$.

(2) There is a small real number $\alpha > 0$ (e.g. $\alpha = 10^{-6}$) such that $|f^n 0| \geqq e^{-\alpha n}$ for all $n \geqq 1$.

It is proved in [BC2] that the set of parameter values $\mathscr{A} = \{a | f_a$ satisfies (1) and (2)$\}$ has positive Lebesgue measure.

To study the growth of $(f^n)'x$ for $x \in [-1, 1]$, we have three types of derivative estimates given in Sects. 1.1.1–1.1.3. First some notation: Let $\delta > 0$ be a small real number that is nevertheless $\gg 2 - a$. We assume that $\delta = e^{-\mu_0}$ for some $\mu_0 \in \mathbb{Z}^+$. For bookkeeping purposes write

$$(-\delta, \delta) = \dot{\bigcup_{|\mu| \geqq \mu_0}} I_\mu ,$$

where $I_\mu = (e^{-(\mu+1)}, e^{-\mu})$ for $\mu > 0$ and $I_\mu = -I_{-\mu}$ for $\mu < 0$. Each I_μ is further subdivided into μ^2 intervals $\{I_{\mu,j}\}$ of equal length.

For $a \in \mathscr{A}, f = f_a$ has the following properties:

1.1.1 Derivative estimates away from the critical point

There is $c_0 > 0$ and $M_0 \in \mathbb{Z}^+$ such that

(i) if $x, \ldots, f^{j-1}x \notin (-\delta, \delta)$ and $j \geqq M_0$, then $|Df^j(x)| \geqq e^{c_0 j}$;

(ii) if $x, fx, \ldots, f^{k-1}x \notin (-\delta, \delta)$ and $f^k x \in (-\delta, \delta)$, any $k \in \mathbb{Z}^+$, then $|Df^k(x)| \geqq e^{c_0 k}$.

(iii) if $x, fx, \ldots, f^{k-1}x \notin (-\delta, \delta)$, then $|Df^k(x)| \geqq \delta e^{c_0 k}$.

1.1.2 Derivative estimates when bound to the critical orbit

Let $\beta = 14\alpha$. For $x \in (-\delta, \delta)$ define $p(x)$ to be the largest integer p such that

$$|f^j x - f^j 0| < e^{-\beta j} \quad \forall j < p.$$

Then

(i) $\frac{1}{2}|\mu| \leqq p(x) \leqq 5|\mu| \;\; \forall x \in I_\mu$;

(ii) $|Df^p(x)| \geqq e^{c' p}$ for some $c' > 0$.

The orbit of x is said to be *bound* to the critical orbit during the period $j < p$. We may assume that p is constant on each $I_{\mu,j}$.

1.1.3 Distortion of Df^n

Let $\mathscr{P}$ be the partition of $[-1, 1]$ into $[-1, -\delta] \cup [\delta, 1] \cup \bigcup_{\mu,j} I_{\mu,j}$. For $J \in \mathscr{P}$, let J^+ denote the interval consisting of J and its two adjacent intervals in $\mathscr{P}$. Then the following holds: Assume that $\omega \subset [-1, 1]$ is such that for all $k < N$, $f^k \omega \subset J^+$ for some $J \in \mathscr{P}$. Then there is a constant C, independent of N such that

$$\frac{|Df^N(x)|}{|Df^N(y)|} \leqq C \quad \forall x, y \in \omega.$$

1.2 Some geometric properties of Hénon maps

The material in this subsection is quite elementary and is valid for $T = T_{a,b}$ where (a, b) belongs to a set Δ' of the form $\{(a, b): a_1 < a < a_2 < 2, 0 < b < b_0\}$.

1.2.1 The unstable manifold W

T has a unique fixed point $\hat{z}$ in the first quadrant. This fixed point is hyperbolic. Its expanding direction has slope of order $-b/2$ and eigenvalue ≈ -2. Its contractive direction has slope ≈ 2 and eigenvalue $\approx b/2$. The global unstable manifold at $\hat{z}$ is called W. It will play the rôle of the interval $[-1, 1]$ in our 1-d model.

1.2.2 Attracting sets

To guarantee that T has a compact attracting set, we first choose $a_0 < a_1 < 2$ with a_0 sufficiently near 2. Then there exists $b_0 > 0$ sufficiently small compared to $2 - a_1$ such that for all $(a, b) \in \Delta' := [a_0, a_1] \times (0, b_0]$, W stays in a bounded region. Moreover, if $\Lambda = \Lambda_{a,b}$ is the closure of W, then there is an open neighborhood $U = U_{a,b}$ of Λ such that for all $z \in U$, $\operatorname{dist}(T^n z, \Lambda) \to 0$ as $n \to \infty$ ([BC2, Theorem 4]; see also [BM]).

1.2.3 Hyperbolicity and resemblance to 1-d behavior outside of $(-\delta, \delta) \times \mathbb{R}$

Let $\delta > 0$ be at least as small as needed in our one-dimensional analysis and assume that $b_0 \ll 2 - a_0 \ll \delta$. Let $s(v)$ denote the absolute value of slope of the vector v. A simple calculation shows that if $z = (x, y) \notin (-\delta, \delta) \times \mathbb{R}$ and $s(v) \leqq \delta$, then

$$s(DT_z v) \leqq \frac{b}{|x|} \leqq \frac{b}{\delta},$$

which we assume to be $\ll \delta$. ([BC2, Lemma 4.5].) This defines invariant cones outside of $(-\delta, \delta) \times \mathbb{R}$.

For $z = (x, y) \notin (-\delta, \delta) \times \mathbb{R}$ and a unit vector v with $s(v) \leqq \delta$, we have essentially the same estimates as in 1-dimension. That is, there is $c_0 > 0$ and $M_0 \in \mathbb{Z}^+$ such that

(i) if $z, \ldots, T^{j-1} z \notin (-\delta, \delta) \times \mathbb{R}$ and $j \geqq M_0$, then $|DT_z^j v| \geqq e^{c_0 j}$;

(ii) if $z, Tz, \ldots, T^{k-1} z \notin (-\delta, \delta) \times \mathbb{R}$ and $T^k z \in (-\delta, \delta) \times \mathbb{R}$, any $k \in \mathbb{Z}^+$, then $|DT_z^k v| \geqq e^{c_0 k}$;

(iii) if $z, Tz, \ldots, T^{k-1} z \notin (-\delta, \delta) \times \mathbb{R}$ then $|DT^k v| \geqq \delta e^{c_0 k}$.

1.2.4 Curvature estimates

Tangencies between stable and unstable manifolds of T are inevitable. In order to show that some of these tangencies are quadratic, we will need to control the curvature of W. In [BC2], a curve γ in $\mathbb{R}^2$ is called a $C^2(b)$ *curve* if $s(\gamma') \leqq 10b$ and its curvature $\kappa \leqq 10b$. The following lemma is used to verify the $C^2(b)$ property.

Lemma (proved in Section 7.6 of [BC2]). *Let $t \mapsto \gamma(t)$ be a smooth curve. Write $\gamma_n = T^n\gamma$ and let κ_n denote the curvature of γ_n. Assume there is a $c > 0$ such that for all t*

(i) $\kappa_0(t) \leqq 1$;
(ii) $|\gamma'_n(t)|/|\gamma'_{n-j}(t)| \geqq e^{cj}$ $\forall j \leqq n$;
(iii) $\gamma_{n-j}(t) \notin (-\delta, \delta) \times \mathbb{R}$ and $s(\gamma'_{n-j}(t)) < \delta$ *for* $j = 1, \ldots, 4$.

Then $\kappa_n(t) \leqq 10b$ *for all* t.

The discussion from here on applies only to $T = T_{a,b}$, *where* (a, b) *belongs in an inductively constructed set* $\Delta \subset \Delta'$. The set Δ has the property that for each sufficiently small $b > 0$, $\Delta_b := \{a : (a, b) \in \Delta\}$ is a Cantor set and has positive Lebesgue measure.

1.3 The critical set (mostly Sects. 5 and 6 of [BC2])

We use the following notation. For $A \in GL(2, \mathbb{R})$, if $v \mapsto |Av|/|v|$ is not constant, let $e(A)$ denote the unit vector that is contracted the most by A. Write $e_n(z) = e(DT^n_z)$ if it makes sense . For $z \in W$, let $\tau \in T_z\mathbb{R}^2$ denote a unit vector, tangent to W.

A certain subset $\mathscr{C}$ of W called the critical set is singled out in [BC2] to play the rôle of 0 in 1-d. This set is constructed according to the rule in 1.3.1. Each $z_0 \in \mathscr{C}$ has the property that $\tau(z_0)$ is the most contracted direction, i.e. $\lim_{n \to \infty} e_n(z_0) = \tau(z_0)$. This can be thought of as the moral equivalent to $f'(0) = 0$.

1.3.1 Location and rule of construction of $\mathscr{C}$

$\mathscr{C}$ is located in $W \cap ((-10b, 10b) \times \mathbb{R})$. It seems likely that it does not lie on any smooth curve. To give a more precise description of where points of $\mathscr{C}$ are located, we divide W into segments of different "generations": First, there is a unique point $z_0 \in \mathscr{C}$ on the roughly horizontal segment of W containing the fixed point $\hat{z}$. In [BC2], the segment of W between T^2z_0 and Tz_0 is called G_1, the *leaf of generation* 1. Leaves of generation $g \geqq 2$ are defined by

$$G_g := T^{g-1}G_1 - G_{g-1} .$$

We assume (a, b) is sufficiently near (2, 0) so that $\bigcup_{g \leqq 27} G_g$ consists of 2^{26} roughly horizontal segments linked by sharp turns near $x = \pm 1, y = 0$, and that $(\bigcup_{g \leqq 27} G_g) \cap ((-\delta, \delta) \times \mathbb{R})$ consists of 2^{26} $C^2(b)$ curves (see Sect. 1.2.4). If $\mathscr{C}_g := \mathscr{C} \cap \bigcup_{i \leqq g} G_i$, then $\mathscr{C}_{27}$ contains 2^{26} points, one on each of these $C^2(b)$ curves.

For $g > 27$, the following rule is used. Let ρ be a number with $b \ll \rho \ll e^{-72}$, and assume $\mathscr{C}_{g-1}$ is already defined. Consider a maximal piece of $C^2(b)$ curve $\gamma \subset G_g$. If γ contains a segment of length $2\rho^g$ centered at $z'_0 = (x'_0, y'_0)$, and there is a critical point $z_0 = (x_0, y_0) \in \mathscr{C}_{g-1}$ with $x'_0 = x_0$, and $|y'_0 - y_0| \leqq b^{g\sigma}, \sigma = \frac{1}{540}$, then there is a unique point $\hat{z}_0 \in \mathscr{C}_g \cap \gamma$. Moreover, $|\hat{z}_0 - z'_0| \leqq |y'_0 - y_0|^{1/2}$. These are the only points of $\mathscr{C}_g$.

1.3.2 Remarks on the construction of $\mathscr{C}$

Roughly speaking, certain points z on W with $e_n(z) = \tau$ are picked out as approximate critical points. Once a point is designated an approximate critical point, parameters are excluded to ensure that a point in $\mathscr{C}$ is eventually constructed nearby. The construction of $\mathscr{C}$ itself involves an inductive procedure that is guaranteed to work only for the "good" parameters.

As to when to initiate the construction of a critical point, the rules in Sect. 1.3.1 are obviously quite arbitrary. There are two guiding principles. The first is that each "distinguishable" critical orbit of length n causes a certain measure set of parameters to be discarded, so we cannot afford to have too many critical points. On the other hand, as in 1-d, when a critical orbit returns to $(-\delta, \delta) \times \mathbb{R}$, we want it to be "bound" to a suitable critical point of earlier generation. So we must be sure there are enough of these "suitable" critical points. These issues are discussed in Sects. 3 and 6 of [BC2].

1.3.3 Dynamical properties of $\mathscr{C}$

The parameter selection procedure is also designed to guarantee that every $z_0 \in \mathscr{C}$ has the following properties:

(1) For all $n \geqq 0$,

$$|DT^n_{z_0}\binom{0}{1}| \geqq e^{cn} \text{ for some } c \approx \log 2,$$

$$|DT^n_{z_0}\tau| \leqq (Cb)^n \text{ for some } C \text{ independent of } b;$$

(2) there is a small real number α, say $\alpha = 10^{-6}$, such that $T^n z_0$ stays a distance of $\geqq e^{-\alpha n}$ from certain points of $\mathscr{C}$.

The uniform hyperbolicity expressed in (1) is analogous to condition (1) in Sect. 1.1. We will elaborate on (2) and other dynamical properties of $\mathscr{C}$ later on.

1.3.4 Contracting fields in neighborhoods of a critical point

Let $z_0 \in \mathscr{C}$. When an orbit comes near z_0, the course of its interaction with z_0 in the next n iterates is determined to a great extent by the contracting field e_n and expanding field $e_n^\perp$ near z_0. The following facts are proved in Sect. 5 of [BC2][1]:

(i) e_1 is defined everywhere and has slope $= 2ax + \mathcal{O}(b)$.

(ii) There is λ with $b \ll \lambda < 1$ such that for all $z_0 \in \mathscr{C}$, e_n is defined on $B_n(z_0) :=$ the disk of radius $(\lambda/5)^n$ about z_0. Moreover,

$$|DT^n_z e_n| \leqq (Cb)^n, \quad \forall z \in B_n .$$

(iii) There is a constant $C > 0$ such that for all $z_1, z_2 \in B_n$

$$|e_n(z_1) - e_n(z_2)| \leqq C|z_1 - z_2| .$$

[1] The slight difference between the definition of e_n here and that in [BC2] is unimportant

(iv) For $(x_1, y_1), (x_2, y_2) \in B_n$ with $|y_1 - y_2| \leqq |x_1 - x_2|$

$$|e_n(x_1, y_1) - e_n(x_2, y_2)| = (2a + \mathcal{O}(b))|x_1 - x_2| .$$

(v) For $m < n$, $|e_n - e_m| \leqq \mathcal{O}(b^m)$ on $B_n(z_0)$.

Suppose z_0 lies on a $C^2(b)$ curve γ. Then it follows from (i)–(v) that there is a unique z_0^n in $\gamma \cap B_n(z_0)$ at which $e_n = \tau$, and that $|z_0^n - z_0| = \mathcal{O}(b^n)$.

1.4 Controlled orbits

1.4.1 Definitions

(i) Let $z_0 = (x_0, y_0)$ be a critical point lying on a piece of $C^2(b)$ curve $\gamma \subset W$, and let $\zeta_0 = (\xi_0, \eta_0)$ be an arbitrary point in $(-\delta, \delta) \times \mathbb{R}$. We say that ζ_0 is in *tangential position* with respect to z_0 if there is $\hat{z}_0 = (\hat{x}_0, \hat{y}_0) \in \gamma$ with $\hat{x}_0 = \xi_0$ and $|\hat{y}_0 - \eta_0| < |\hat{x}_0 - x_0|^4$.

(ii) Let $\zeta_0 \in W$, and let $\zeta_n = T^n \zeta_0$. We say that *the orbit of* ζ_0 *is controlled on the time interval* $[0, N)$, $N \leqq \infty$, if the following holds. If $\zeta_0 \notin \mathscr{C}$, let $n_1 < n_2 < \ldots$ be the times when ζ_n is in $(-\delta, \delta) \times \mathbb{R}$. If $\zeta_0 \in \mathscr{C}$, take $n_1 > 0$ to be the first time ζ_n returns to $(-\delta, \delta) \times \mathbb{R}$. Then for every $n_i \in [0, N)$, ζ_{n_i} is attached to a critical point with respect to which it is in tangential position. This critical point is denoted by $z(\zeta_{n_i})$ and is called the *binding point* of ζ_{n_i}. See 1.4.3 for a further requirement on the assignment $\zeta_{n_i} \mapsto z(\zeta_{n_i})$.

1.4.2 Control of critical orbits

A key idea in [BC2] is that through parameter exclusion, it is arranged so that for every $z_0 \in \mathscr{C}$

(i) the orbit of z_0 is controlled on the time interval $[0, \infty)$;
(ii) $|z_{n_i} - z(z_{n_i})| \geqq e^{-\alpha n_i}$ $\forall i \geqq 1$;
(iii) $\text{gen}(z(z_{n_i})) < \theta \cdot n_i$ for some small $\theta > 0$.

(ii) is the precise statement of (2) in 1.3.3. To explain the availability of binding points, consider first the following ideal situation. Suppose that z_n is a free return, and that $\bigcup_{g < \theta n} G_g$ contains $2^{[\theta n] - 1}$ $C^2(b)$ curves stretched across $(-\delta, \delta) \times \mathbb{R}$, each containing a critical point. In this case one needs only to consider the curve nearest to z_n. Most locations of z_n are in tangential position relative to the critical point on this curve; bad locations of z_n correspond to deleted parameters. In general, some of the $2^{[\theta n] - 1}$ pieces of W are missing or partially missing. What is true is that at free returns, z_n is always surrounded by a "fairly regular" collection of $C^2(b)$ segments $\{\gamma_j\}$ of W. "Fairly regular" here means that $\text{gen}(\gamma_j) \sim 3^j$, $\text{length}(\gamma_j) \sim \rho^{3^j}$ and $\text{dist}(z_n, \gamma_j) \sim b^{3^j}$ $(3^j < \theta n)$. It is then shown that such a family contains enough critical points for our purposes. (See Sects. 6 and 7 of [BC2].)

1.4.3 Bound periods

Let $z_0 \in \mathscr{C}$, and let ζ_0 be in tangential position with respect to z_0. As a first approximation we define the bound period $\tilde{p}(\zeta_0, z_0)$ to be the largest k such that

$$|T^j\zeta_0 - T^j z_0| < e^{-\beta j} \quad \forall j < k .$$

(Recall that $\beta = 14\alpha$.)

Now consider ζ_0 whose orbit is controlled on $[0, \infty)$, and let $\tilde{p}_i = \tilde{p}(\zeta_{n_i}, z(\zeta_{n_i}))$. Note that it is entirely possible for ζ_n with $n_i < n < n_i + \tilde{p}_i$ to return to $(-\delta, \delta) \times \mathbb{R}$, so that bound periods can be initiated in the middle of bound periods. It is shown in Sect. 6.2 of [BC2] that by modifying slightly the definition of $\tilde{p}$, bound periods can be made "nested". More precisely, one can choose $\{p_i\}_{i=1}^{\infty}$ in such a way that for all i,

(i) $p_i \leqq \tilde{p}_i$ and $|T^{p_i}\zeta_{n_i} - T^{p_i} z(\zeta_{n_i})| \geqq e^{-\beta^* p_i}$ for some $\beta^* \approx \beta$,
(ii) if n_j is such that $n_i < n_j < n_i + p_i$, then $n_j + p_j \leqq n_i + p_i$.

The bound period between ζ_{n_i} and $z(\zeta_{n_i})$ is then defind to be p_i. It is further required that if the bound relation between z_{n_i} and $z(z_{n_i})$ is still in effect at time $n_j > n_i$, then we must have $z(z_{n_j}) = z(T^{n_j - n_i} z(z_{n_i}))$.

1.4.4 Bound of free states

Conceptually, it is convenient to divide a controlled orbit into "free" and "bound" states. Let ζ_0 be as above.

- For $n \leqq n_1$, ζ_n is *free.*
- For $n_1 < n < n_1 + p_1$, ζ_n is said to be in *bound* state. (During this period ζ_n may make multiple returns to $(-\delta, \delta) \times \mathbb{R}$, at which times new bindings are formed. But by our definition of p_i, these "secondary bound periods" expire at or before $n_1 + p_1$.)
- Let n_i be the first return to $(-\delta, \delta) \times \mathbb{R}$ after $n_1 + p_1$. Then ζ_n is free for $n_1 + p_1 \leqq n \leqq n_i$, and ζ_{n_i} is called a *free return.*
- ζ_n is in bound state again during the period $n_i < n < n_i + p_i$, and so on.

1.5 Keeping track of derivatives

Throughout this subsection we assume that the orbit ζ_0 is controlled on $[0, \infty)$, with return times $n_1 < n_2 < \ldots$ and binding points $\zeta_{n_i} \leftrightarrow z(\zeta_{n_i})$ as defined in 1.4.1.

1.5.1 The fold period

For each n_i, the fold period of ζ_{n_i} with respect to $z(\zeta_{n_i})$ is defined to be a number l_i with $2m \leqq l_i \leqq 3m$, where

$$(5b)^m \leqq |z(\zeta_{n_i}) - \zeta_{n_i}| \leqq (5b)^{m-1} .$$

For convenience we will choose l_i such that $n_i + l_i \neq n_j, n_j + 1$ for any j. (The precise definition is given in [BC2, Sect. 6.3] and does not particularly concern us.) We mention three facts about fold periods:

(i) The fold period initiated at a return to $(-\delta, \delta) \times \mathbb{R}$ is very short compared to the bound period initiated at that time. In fact, $l/p \leqq \text{const}/\log(1/b)$, which tends to 0 as $b \to 0$.

(ii) It follows from (i) above and Sect. 1.4.2 that every $z_0 \in \mathscr{C}$ has the following property: for every $m \in \mathbb{Z}^+$, there are integers m_1 and m_2, with $m_1 \leqq m \leqq m_2$ and $m_2 - m_1 \leqq \text{const.}\, m/\log(1/b)$, such that z_{m_1} and z_{m_2} are outside all fold periods. (See Lemma 6.5 in [BC2].)

(iii) If $n_i < n_j \leqq n_i + l_i$ then $n_j + l_j \leqq n_i + l_i$, i.e. a fold period initiated within another fold period does not extend beyond that fold period.

The role of the fold period will become clear in 1.5.4.

1.5.2 The splitting algorithm (Sect. 7.1 of [BC2])

Let $v \in T_{\zeta_0}\mathbb{R}^2$ be a tangent vector. The following algorithm is devised in [BC2] to keep track of $w_n(\zeta_0, v) := DT^n_{\zeta_0} v$. For each n, we decompose $w_n(\zeta_0, v)$ into

$$w_n(\zeta_0, v) = w_n^*(\zeta_0, v) + E_n(\zeta_0, v)\,,$$

where w_n^* and E_n are defined according to the following rules. (We will omit all references to (ζ_0, v) from now on.)

- For $n < n_1$, let $w_n^* = w_n$.
- At time n_1, let $e = e_{l_1}$ be the contractive vector field around $z(\zeta_{n_1})$. Write

  $$w_{n_1} = A_{n_1} e + B_{n_1} \binom{0}{1}$$

  and let

  $$w_{n_1}^* = B_{n_1}\binom{0}{1}, \qquad E_{n_1} = A_{n_1} e\,.$$

- For $n > n_1$, if $n \neq$ any n_i, define

  $$E_n = \sum_{\substack{n_k < n \\ \text{s.t. } n_k + l_k > n}} DT^{n-n_k}_{\zeta_{n_k}} (A_{n_k} e_{l_k})\,.$$

  This of course defines w_n^* as well.
- If $n = n_i$ for some $i > 1$, we let $e = e_{l_i}$, write

  $$DT_{\zeta_{n-1}} w_{n-1}^* = A_n e + B_n \binom{0}{1}$$

  and let

  $$w_n^* = B_n\binom{0}{1}\,.$$

This algorithm has no geometric meaning for arbitrary (ζ_0, v). See 1.5.4 for a discussion of the situation for which it is designed.

1.5.3 Controlled derivatives along controlled orbits

Definition. Let $\zeta_0 \in W$ and let $v \in T_{\zeta_0}\mathbb{R}^2$. We say that the pair (ζ_0, v) is *controlled* on the time interval $[0, N)$ if the orbit of ζ_0 is controlled on $[0, N)$ and whenever $\zeta_n \in (-\delta, \delta) \times \mathbb{R}$, the splitting algorithm in Sect. 1.5 gives

$$(*) \qquad 3|\zeta_n - z(\zeta_n)| \leqq |\bar{B}_n| \leqq 5|\zeta_n - z(\zeta_n)|, \quad 0 \leqq n < N ,$$

where

$$\bar{B}_n = \frac{B_n}{|DT_{\zeta_{n-1}} w^*_{n-1}|} .$$

If $(*)$ holds, we say that the vector $v_n = DT^n_{\zeta_0} v$ *splits correctly*.

One of the most important properties of T proved in [BC2] is that

for every $z_0 \in \mathscr{C}$, $(z_0, \binom{0}{1})$ *is controlled during the time interval* $[0, \infty)$.

It is also proved that if ζ_0 is bound to $z_0 \in \mathscr{C}$ then $(\zeta_0, \binom{0}{1})$ is also controlled during the bound period $[0, p)$.

1.5.4 1-dimensional behavior in 2-dimensions

We now give some idea of how the splitting algorithm is used to study $D^n_{\zeta_0} v$ where (ζ_0, v) is controlled on $[0, \infty)$. For definiteness consider $\zeta_0 \in \mathscr{C}$ and $v = \binom{0}{1}$. We claim that $\{w^*_n\}^{\infty}_{n=0}$ is essentially 1-d in character, and has properties similar to $(f^n)'(f0)$ for $f(x) = 1 - ax^2$.

As before, let $n_1 < n_2 < \dots$ be the return times of ζ_0 to $(-\delta, \delta) \times \mathbb{R}$. First $w^*_1 = \binom{1}{0}$. For $0 < n < n_1$, $\{|w^*_n|\}$ behaves qualitatively like $\{|(f^n)'(f0)|\}$; see 1.2.3. Let us consider $n = n_1$, and let $w_n = A_n e + B_n \binom{0}{1}$ be the decomposition given by the splitting algorithm.

Ignoring the A-term for now, and using the fact that $(\zeta_0, \binom{0}{1})$ is controlled, we see that

$$\begin{aligned} |w^*_{n+1}| = |w^*_n| &= |\bar{B}_n| \cdot |DT_{\zeta_{n-1}} w^*_{n-1}| \\ &\sim 2a|\zeta_n - z(\zeta_n)| \cdot |DT^{n-1}_{\zeta_1} w^*_1|, \end{aligned}$$

which has an obvious resemblance to $|(f^n)'(f0)|$. Moreover, since $w^*_{n+1} = B_n \binom{1}{0}$, the action of DT^j on w^*_{n+1} for $j < n_2 - n_1$ is again essentially 1-d. The vector $DT^{n_2 - n_1} w^*_{n_1}$ is split at time n_2, and the new B-term is treated similarly.

The general philosophy is that the A- terms can, in some sense, be neglected. Let us consider a simple situation, where the fold period l at time $n = n_1$ expires before the next return to $(-\delta, \delta) \times \mathbb{R}$. We claim that $DT^l_{\zeta_n} A_n e$ is extremely short compared to $DT^l_{\zeta_n} B_n \binom{0}{1}$, so that the effect of adding this term at time $n + l$ is negligible. This is because

$$|DT^l_{\zeta_n} A_n e| \leqq |DT_{\zeta_{n-1}} w^*_{n-1}| \cdot (5b)^l ,$$

whereas

$$|DT^l_{\zeta_n} B_n \tbinom{0}{1}| \geqq |DT_{\zeta_{n-1}} w^*_{n-1}| \cdot |\zeta_n - z(\zeta_n)| \cdot |DT^l_{\zeta_n} \tbinom{0}{1}| ,$$

and we know that $|\zeta_n - z(\zeta_n)| \geqq (5b)^{l/2}$ by the definition of l, and also that $|DT^{l-1}_{\zeta_{n+1}}\binom{1}{0}| \geqq \delta e^{c(l-1)}$ by 1.2.3(iii).

In general, the computation is more complicated, but if (ζ_0, v) is controlled one has $|\bar{B}_j| \geqq |\zeta_j - z(\zeta_j)|$ at all returns, and the rejoining of the A-terms of w_n^* at the end of fold periods has negligible effects. (See Sect. 7.3 of [BC2].)

We have made a point of comparing $w_n^*(z_0, \binom{0}{1})$, $z_0 \in \mathscr{C}$, to $(f^n)'(f0)$. This parallel is not complete unless we could guarantee, through parameter exclusion, that for all $z_0 \in \mathscr{C}$, $|w_n^*(z_0, \binom{0}{1})| \geqq e^{cn}$ for some $c > 0$. This is not exactly what we stated in 1.3.3, but it is in fact true.

1.6 Dynamics near the "turns"

1.6.1 Bound state estimates

Assume that ζ_0 is in tangential position with respect to $z_0 \in \mathscr{C}$, and let p and l be its bound and fold periods. We now record some estimates from [BC2] pertaining to the bound states $\{\zeta_j\}_{j=0}^{p}$. These estimates rely on the fact that critical orbits have the properties in Sects. 1.3.1 and 1.5.3.

(i) $|T^j\zeta_0 - T^j z_0| \sim |\zeta_0 - z_0|^2 |DT^j_{z_0}\binom{0}{1}|$ for $l \leqq j \leqq p$.
(For a precise statement, see Sect. 7.5, in particular Lemma 7.4, of [BC2].)

(ii) If $|\zeta_0 - z_0| \approx e^{-\mu}$, $\mu > 0$, then $\frac{1}{2}\mu \leqq p \leqq 5\mu$.
(This follows essentially from (i) and Sect. 1.3.1.)

(iii) $|DT^p_{\zeta_0}\binom{0}{1}| \cdot |\zeta_0 - z_0| \geqq e^{c'p}$ for some $c' > 0$.
(See Lemma 7.5 of [BC2].)

Two points are to be noted here. The first is the striking resemblance between these estimates and those in 1-d. Assertion (i), for instance, says that ζ_0 experiences a quadratic contraction from its interaction with z_0. The second observation we wish to make is that unlike the situation in 1-d, not all the points near z_0 have this "quadratic" behavior. For instance, ζ_0 may lie on a piece of stable manifold through z_0, and be attracted to z_0 forever. Or, if ζ_0 is directly above z_0, then we will have $|T^j\zeta_0 - T^j z_0| \approx |\zeta_0 - z_0| \cdot |DT^j_{z_0}\binom{0}{1}|$ instead of (i) for the first few iterates.

1.6.2 Generalized tangential positions

We wish to say more precisely for which region the estimates in 1.6.1 hold. For $z \in W$, we say that (x', y') is the natural coordinate system at z if $(0, 0)$ is at z, the x'-axis lines up with $\tau(z)$, and the y'-axis with $\tau(z)^\perp$. The following definition is *not* contained in [BC2] but will be useful for us in the next section.

Definition. Let $c > 0$ be a small number $\ll 2a$, say $c = \frac{1}{100}$, and let $z_0 \in \mathscr{C}$. A point ζ_0 near z_0 is said to be in *generalized tangential position* with respect to z_0 if in the natural coordinate system at z_0, $\zeta_0 = (\xi', \eta')$ with $|\eta'| \leqq c\xi'^2$.

Clearly, if ζ_0 is in tangential position wrt z_0, then it is in generalized tangential position, because the segment of W containing z_0 is a $C^2(b)$ curve tangent to the x'-axis at $(0, 0)$. While it is not explicitly stated this way, the proofs in [BC2] show in fact that the bound estimates in 1.6.1 hold for all ζ_0 in generalized tangential position wrt $z_0 \in \mathscr{C}$. Let us recall briefly why this is so:

Let p be the bound period between ζ_0 and z_0. From 1.3.4 we know that there is a contractive direction field defined on a small ball about z_0. The integral curves of e_p are roughly parabolas of the form $y' = \text{const} + a(x' - x'_p)^2$, and have a unique tangency with the x'-axis at $z_0^{(p)} := (x'_p, 0)$. Since $|z_0^{(p)} - z_0| = O(b^p) \ll |z_0 - \zeta_0|$, for our purposes we may as well confuse $z_0^{(p)}$ with z_0. With this simplification we can think of z_0 and ζ_0 as lying on the graph of a function $\varphi : x'\text{-axis} \to y'\text{-axis}$ with $|\varphi'(x')| \leqq 2c|x'|$. The estimates in Sects. 7.4 and 7.5 of [BC2] apply to points on such a curve. What really matters is that all the tangent vectors to graph(φ) split correctly wrt e_p, which a priori is defined only in a small ball about z_0 but is shown in a local induction in [BC2] to be defined on the entire segment of graph(φ) between z_0 and ζ_0. These estimates imply those in 1.6.1.

1.7 Distortion estimates during bound periods

In the course of proving some of the estimates in the last few subsections, the following estimates are used.

(i) For all $z_0 \in \mathscr{C}$, if ζ_0 is bound to z_0 with bound period p then

$$\frac{1}{2} \leqq \frac{|w_j^*(\zeta_0, \binom{0}{1})|}{|w_j^*(z_0, \binom{0}{1})|} \leqq 2 \quad \forall j < p\,.$$

This is Assertion 4(a) in [BC2]. It is proved in Lemmas 7.8, 7.9 and on p. 151.

(ii) There is a constant $C_0 > 0$ such that $\forall z_0 \in \mathscr{C}$, if ζ_0 and ζ'_0 are bound to z_0 during $[0, v]$, then

$$\text{(a)} \quad \frac{|w_v^*(\zeta_0, \binom{0}{1})|}{|w_v^*(\zeta'_0, \binom{0}{1})|} \leqq \exp\left\{ C_0 \sum_{j=0}^{v-1} \frac{\Delta_j(\zeta_0, \zeta'_0)}{d_{\mathscr{C}}(z_j)} \right\},$$

where $\Delta_j(\zeta_0, \zeta'_0) = \max\limits_{0 \leqq i \leqq j} |\zeta_i - \zeta'_i|$ and

$$d_{\mathscr{C}}(z_j) := \begin{cases} |x_j| & \text{if } z_j = (x_j, y_j) \text{ and } |x_j| \geqq \delta, \\ |z_j - z(z_j)| & \text{if } |x_j| \leqq \delta \text{ and } z(z_j) \text{ is the binding point.} \end{cases}$$

(b) $\measuredangle\,(w_v^*(\zeta_0, \binom{0}{1}), w_v^*(\zeta'_0, \binom{0}{1})) < 2b^{\frac{1}{4}}\Delta_v$.

These estimates are proved the same time (i) is proved.

(iii) There is a constant $C'_0 > 0$ such that if z_0, ζ_0 and ζ'_0 are as in (ii) and in addition ζ_0 and ζ'_0 lie on a curve all the tangent vectors of which split correctly with respect to z_0 (compare 1.5.3) then

$$\sum_{j=0}^{p} \frac{\Delta_j(\zeta_0, \zeta'_0)}{d_{\mathscr{C}}(z_j)} \leqq C'_0 \frac{|\zeta_0 - \zeta'_0|}{|\zeta_0 - z_0|}\,.$$

This is proved on pp. 151 and 163 of [BC2].

The constant C_0 and C'_0 above remain uniformly bounded as $\delta \to 0$.

2 Dynamics on the unstable manifold W

2.1 Statements of results

Let $T = T_{a,b}$ be as in Sect. 1. The purpose of this section is to use the dynamics of points in $\mathscr{C}$ to help us understand the dynamics of all points on W. We will prove that the derivative estimates for $DT^n_\zeta \tau$, $\zeta \in W$, are completely parallel to those for $(f^n)'x$ where $f: [-1, 1] \circlearrowleft$ satisfies the conditions in Sect. 1.1.

As usual, $\{\zeta_n\}_{n=-\infty}^{\infty}$ denotes the orbit of ζ_0. Before we begin we need to modify or extend slightly a couple of our definitions from Sect. 1. First we relax our definition of "control" for the orbit of ζ_0 to requiring only that whenever $\zeta_n \in (-\delta, \delta) \times \mathbb{R}$, there is a critical point $z(\zeta_n)$ wrt which ζ_n is in *generalized* tangential position. (See 1.6.2.) Second, we say that the pair (ζ_0, v) is controlled on the time interval $[j, N)$, $-\infty < j < N$, if $(\zeta_j, DT^j_{\zeta_0} v)$ is controlled on $[0, N-j)$, and that (ζ_0, v) is controlled on $(-\infty, N)$ if it is controlled on $[j, N)$ for all $j < N$. Note that for all $\zeta_0 \in W$, ζ_j tends to the unique hyperbolic fixed point $\hat{z}$ as $j \to -\infty$.

2.1.1 Main Proposition and Main Corollary

Proposition 1 (Main proposition) *For all $\zeta_0 \in W$, if $\zeta_i \notin \mathscr{C}$ for all $i < N$, then the pair $(\zeta_0, \tau(\zeta_0))$ is controlled on the time interval $(-\infty, N)$.*

Proposition 1 enables us to describe each ζ_i, $i < N$, as being in a "free state" or "bound state" as discussed in 1.4.4. (Technically one needs to specify a starting point for this to make sense, but since $\zeta_i \to \hat{z}$ as $i \to -\infty$, we can think of our trajectory as starting from ζ_j for some negatively very large j.

Corollary 1 (Main corollary) *Let $\zeta_0 \in W$, and assume that $\zeta_i \notin \mathscr{C}$ for all i. Then ζ_i is in a "free state" infinitely often, and the following holds for $\tau_i := DT^i_{\zeta_0}\tau$:*

I. Expansion outside of $(-\delta, \delta) \times \mathbb{R}$
There is $c_0 > 0$ and $M_0 \in \mathbb{Z}^+$ such that

(i) *if ζ_i is free and $\zeta_i, \ldots, \zeta_{i+M_0} \notin (-\delta, \delta) \times \mathbb{R}$, then*

$$\frac{|\tau_{i+M_0}|}{|\tau_i|} \geqq e^{cM_0};$$

(ii) *if $\zeta_i \notin (-\delta, \delta) \times \mathbb{R}$ is free, and $k > i$ is the first time $\zeta_k \in (-\delta, \delta) \times \mathbb{R}$ then*

$$\frac{|\tau_k|}{|\tau_i|} \geqq e^{c_0(k-i)}.$$

II. Bound period estimates
There is $c_1 \approx \log 2$ such that the following holds: if $\zeta_i \in (-\delta, \delta) \times \mathbb{R}$ is free and becomes bound at this time to $z(\zeta_i) \in \mathscr{C}$ with bound period p, then

(i) *if $e^{-\mu-1} \leqq |\zeta_i - z(\zeta_i)| \leqq e^{-\mu}$, then $\frac{1}{2}\mu \leqq p \leqq 5\mu$;*
(ii) $|\tau_{i+j}|/|\tau_i| \geqq 3|\zeta_i - z(\zeta_i)| \cdot |DT^j_{z(\zeta_i)}\binom{0}{1}| \geqq |\zeta_i - z(\zeta_i)| e^{c_1 j}$, $0 < j < p$;
(iii) $|\tau_{i+p}|/|\tau_i| \geqq e^{c_1 p/3}$.

2.1.2. Global distortion estimates

We will need the following distortion estimate for $DT^n\tau$. Let $\mathscr{P} = \{I_{\mu j}\}$ be the partition of $(-\delta, \delta)$ as described in Sect. 1.1, i.e.

$$(-\delta, \delta) = \dot{\bigcup_{|\mu| \geqq \mu_0}} I_\mu ,$$

where $I_\mu = (e^{-(\mu+1)}, e^{-\mu})$ for $\mu > 0$, $I_\mu = -I_{-\mu}$ for $\mu < 0$, and each I_μ is further subdivided into μ^2 intervals $\{I_{\mu j}\}$ of equal length.

For x_0 with $|x_0| \ll \delta$, we let $\mathscr{P}_{[x_0]}$ denotes a copy of $\mathscr{P}$ with 0 "moved" to x_0. More precisely, let $h:(-\delta, \delta) \to (-\delta, \delta)$ be the piecewise linear homeomorphism taking the points $-\delta, -e^{-(\mu_0+1)}, x_0, e^{-(\mu_0+1)}, \delta$ to $-\delta, -e^{-(\mu_0+1)}, 0, e^{-(\mu_0+1)}, \delta$ respectively and let $\mathscr{P}_{[x_0]} = h^{-1}\mathscr{P}$.

Furthermore, if γ is a roughly horizontal curve and $z_0 = (x_0, y_0)$ is such that $|x_0| \ll \delta$, then $\mathscr{P}_{[z_0]} = \mathscr{P}_{[x_0]}$ is the obvious partition on $\gamma \cap ((-\delta, \delta) \times \mathbb{R})$. Once γ and z_0 are specified, we will speak about $I_{\mu j}$ as though it was a subsegment of γ. Also, for $J = I_{\mu j}$, let J^+ denote the union of J with its adjacent intervals in $\mathscr{P}_{[\cdot]}$

We will use the following notation: if γ_0 is a curve segment then $\gamma_j = T^j\gamma_0$; ζ_j denotes $T^j\zeta_0$ and $\tau_j(\zeta_0) = DT^j_{\zeta_0}\tau$ etc.

Proposition 2 *Let $\gamma_0 \subset W \cap ((-\delta, \delta) \times \mathbb{R})$ be a curve segment, and let $0 = t_0 < t_1 < \ldots < t_q = N$ be its free return times. More precisely, for all $k < q$,*

(1) all points in γ_{t_k} have a common binding point, which we denote by $z^{(k)}$, and $\gamma_{t_k} \subset J^+$ for some $J \in \mathscr{P}_{z^{(k)}}$;

(2) If p_k is the bound period between γ_{t_k} and $z^{(k)}$, then t_{k+1} is the smallest $j \geqq t_k + p_k$ such that $\gamma_j \cap ((-\delta, \delta) \times \mathbb{R}) \neq \emptyset$.
Then for all $\zeta_0, \zeta'_0 \in \gamma_0$,

$$\frac{|\tau_N(\zeta_0)|}{|\tau_N(\zeta'_0)|} \leqq C_1$$

for some C_1 independent of γ_0 or N.

2.1.3 Remarks on the critical set

Recall that in [BC2], the critical set $\mathscr{C}$ is constructed according to some seemingly arbitrary rules. We wish to point out here that $\mathscr{C}$ in fact admits certain intrinsic characterizations. For instance, Corollary 1 gives the following dynamical characterization of $\mathscr{C}$: Let $z_0 \in W$. Then

$$z_0 \text{ lies on a critical orbit} \Leftrightarrow \limsup_{n\to\infty} |DT^n_{z_0}\tau| < \infty \Leftrightarrow \lim_{n\to\infty} |DT^n_{z_0}\tau| = 0 .$$

In fact, $z_0 \in \mathscr{C}$ iff

$$|DT^j_{z_0}\tau| \leqq (5b)^j \ \forall j > 0 \quad \text{and} \quad |DT_{z_{-1}}\tau| > 1 .$$

We mention also that $\mathscr{C}$ admits geometric characterizations. For instance, it is straight-forward to verify using the curvature computations in Sect. 7.6 of [BC2] that $z_0 \in \mathscr{C}$ iff

$$\kappa(z_0) \ll 1 \quad \text{and} \quad \kappa(z_n) > b^{-n} \quad \forall n > 1 .$$

2.2 Some lemmas

The lemmas in Sect. 2.2 are not explicitly stated or proved in [BC2], but their proofs resemble those in [BC2]. We will repeat the shorter arguments and refer the reader to [BC2] for the longer ones.

We will write $w_j(z) := w_j(z, \binom{0}{1})$ and $\tau_j(z) := w_j(z, \tau)$, and let $w_j^*(z)$ and $\tau_j^*(z)$ have the obvious meanings. When we say that a segment $\gamma \subset W$ is "free", it will be assumed implicitly that $\forall z \in \gamma$, (z, τ) is controlled during the time interval $(-\infty, 0)$. The absolute value of the slope of a vector is denoted $s(v)$.

2.2.1 Slope, curvature and derivative estimates

The purpose of this subsection is to prove the following three lemmas:

Lemma 1 *Let γ be a free segment of W. Then*

(i) $\forall z \in \gamma$, $s(\tau(z)) < 2b/\delta$;
(ii) $\forall z \in \gamma \cap ((-\delta, \delta) \times \mathbb{R})$, $s(\tau(z)) < 10b$.

Lemma 2. *Let γ be a free segment of W in $(-\delta, \delta) \times \mathbb{R}$. Then $\kappa(\gamma) < 10b$.*

It follows immediately from these two lemmas that free segments of W in $(-\delta, \delta) \times \mathbb{R}$ are $C^2(b)$ curves. The next lemma is needed in Sect. 3.

Lemma 3 *There exists $c_1 > 0$ such that if $z \in W \cap ((-\delta, \delta) \times \mathbb{R})$ is in a free state, then*

$$|DT_z^{-j}\tau| \leqq e^{-c_1 j} \quad \forall j \geqq 0 .$$

We begin with the following sublemma:

Sublemma 1 *Let z_0 be a point in G_1 near the fixed point $\hat{z}$, and assume that (z_0, τ) is controlled on $[0, j)$. If $z_j \notin (-\delta, \delta) \times \mathbb{R}$ then, $s(\tau_j^*(z_0)) < 2b/\delta$.*

Proof. Proceed by induction exactly as is done in the proof of LI(v')(a) in Sect. 7.3 of [BC2]. □

Proof of Lemma 1 (i) follows immediately from Sublemma 1. To see (ii), note that $T^{-1}z$ is outside of all fold periods. Apply Sublemma 1 and Sect. 1.2.3. □

Sublemma 2 *Let ζ_0 be in generalized tangential position with respect to $z_0 \in \mathscr{C}$. Let p be the bound period between ζ_0 and z_0, and let n be the first free return of ζ_0 to $(-\delta, \delta) \times \mathbb{R}$. Then there is $c > 0$ such that*

(i) $|w_j(\zeta_0)| \geqq e^{cj} \quad \forall j < p$;

(ii) $\dfrac{|w_n(\zeta_0)|}{|w_j(\zeta_0)|} \geqq e^{c(n-j)} \quad \forall j < n$.

Proof. (i) If z_j is outside of all fold periods, then

$$|w_j(\zeta_0)| = |w_j^*(\zeta_0)| \gtrsim |w_j^*(z_0)| \geqq e^{c^* j} .$$

(See 1.7 for "$\gtrsim$" and 1.5.4 for the inequality.) If not, choose $k > j$ such that z_k is outside of all fold periods and $k - j < (C/\log(1/b))j$. (See 1.5.1.) We then have

$$|w_j(\zeta_0)| \geqq 5^{-(C/\log(1/b))j} e^{c^* k} \geqq e^{cj} .$$

(ii) If $j \geqq p$ the estimate follows from 1.2.3(ii) so we only need to consider $j < p$.

Case 1 z_j is outside of all fold periods. We have

$$\frac{|w_n(\zeta_0)|}{|w_j(\zeta_0)|} \gtrsim \frac{|w_n(\zeta_0)|}{|w_p(\zeta_0)|} \cdot \frac{|w_p^*(z_0)|}{|w_j^*(z_0)|} .$$

By 1.2.3 (ii) the first factor is $\geqq e^{c(n-p)}$. To see that the second factor $\geqq e^{c(p-j)}$, let $n_1 < n_2 < \ldots < n_k$ be the times between j and p when z_j returns freely to $(-\delta, \delta) \times \mathbb{R}$. Write $\bar{w}_i(z_0) = DT_{z_{i-1}} w_{i-1}^*(z_0)$. We know from 1.2.3, 1.5.4 and 1.6.1(iii) that every factor in

$$\frac{|w_p^*(z_0)|}{|\bar{w}_{n_k}(z_0)} \cdot \frac{|\bar{w}_{n_k}(z_0)|}{|\bar{w}_{n_{k-1}}(z_0)|} \cdot \ldots \cdot \frac{|\bar{w}_{n_1}(z_0)|}{|w_j^*(z_0)|}$$

is exponential.

Case 2 z_j is in a fold period. Let k be the last time it was outside all fold periods. Because of the rule for construction of the fold periods (see 1.5.1(iii)) ζ_j must still be in fold relation to $z(\zeta_k)$. Let p' and l' be the length of the bound period and fold period resp., initiated at time k. Then $p' < n - k$ since z_n is free and

$$j - k \leqq l' \leqq \frac{Cp'}{\log(1/b)} \leqq \frac{C(n-k)}{\log(1/b)}$$

$$\leqq \frac{C(n-j)}{\log(1/b)} + \frac{C(j-k)}{\log(1/b)} .$$

We move the second term to the left and conclude that

$$j - k \leqq \frac{C'(n-j)}{\log(1/b)} .$$

The same argument as before finishes the proof. □

Proof of Lemma 3 Let $z \in \gamma$. If $T^{-i}z \notin (-\delta, \delta) \times \mathbb{R}$ $\forall i > 0$ our exponential estimate follows directly from (1.2.3)(ii). Otherwise suppose $\zeta_0 = T^{-n}z$ is the previous free return. It is enough to verify

$$\frac{|\tau_n(\zeta_0)|}{|\tau_j(\zeta_0)|} \geqq e^{c_1(n-j)}, \quad 0 \leqq j \leqq n .$$

(If there is more then one free return between $T^{-j}z$ and z, repeat the argument and use the chain rule.)

From Sublemma 2 (i) we know that $e_n(\zeta_0)$ is well-defined. We write

$$\tau(\zeta_0) = Ae_n(\zeta_0) + B\binom{0}{1}$$

and let l be the fold period between ζ_0 and $z_0 = z(\zeta_0)$.

Case 1 $j \geqq l$. Since τ splits correctly at ζ_0, we have

$$|DT^k_{\zeta_0} B\binom{0}{1}| \geqq |B| e^{ck} \gg |DT^k_{\zeta_0} A e_n|$$

$\forall k \geqq l$, and so

$$\frac{|\tau_n(\zeta_0)|}{|\tau_j(\zeta_0)|} \gtrsim \frac{|B||w_n(\zeta_0)|}{|B||w_j(\zeta_0)|} \geqq e^{c(n-j)}$$

by Sublemma 2(ii).

Case 2 $j < l$. At most we have $|\tau_j(\zeta_0)| \leqq 5^l$. Since $l \leqq (C/\log(1/b))n$, we are still guaranteed that

$$\frac{|\tau_n(\zeta_0)|}{|\tau_j(\zeta_0)|} \geqq \frac{e^{cn}}{5^{(C/\log(1/b))n}} \geqq e^{c_1(n-j)} .$$

This completes the proof. □

Proof of Lemma 2 Choose n so that $\zeta_0 = T^{-n}z$ is on the first generation G_1 of W close to the fixed point. Then $\kappa(\zeta_0) \leqq 1$ and the curvature estimate follows immediately from Lemma 3 and 1.2.4. □

2.2.2 Abundance of $C^2(b)$ segments of W near free returns

Lemma 4 *Let $\zeta_0 \in W \cap ((-\delta, \delta) \times \mathbb{R})$ be free. Then for every $m \in \mathbb{Z}^+$ with $3m <$ gen(ζ_0), $\exists m'$ with $m < m' \leqq 3m$ and two $C^2(b)$ curves γ and γ' in W with the following properties*:

(i) *ζ_0 is sandwiched between γ and γ' extending $\geqq 3\rho^{m'}$ to each side of ζ_0*;
(ii) dist(ζ_0, γ), dist$(\zeta_0, \gamma') \leqq (Cb)^{m'}$, $C = 5e^{72}$;
(iii) *if η and η' are the two points in γ and γ' resp. with the same x-coordinate as ζ_0, then $|\tau(\eta) - \tau(\zeta_0)|$, $|\tau(\eta') - \tau(\zeta_0)| \leqq (Cb)^{m'/6}$*;
(iv) gen$(\gamma) = m' + 1$, gen$(\gamma') \leqq m'$.

The proof of Lemma 4 is virtually identical to that of the corresponding result in [BC2] for z_n (instead of ζ_0), where $z_0 \in \mathscr{C}$ and z_n is a free return to $(-\delta, \delta) \times \mathbb{R}$. (See Sect. 1.4.2.) We sketch here the main steps of the proof, referring to [BC2] for more details:

Outline of proof of Lemma 4

Step 1 Show that $\exists m' \in [m, 3m]$ such that $\zeta_{-m'}$ is in a "favorable" position relative to ζ_0. This means that $\zeta_{-m'}$ is outside of all folding periods, and that $\forall k < m'$, $|\zeta_{-m'+k} - z(\zeta_{-m'+k})| \geqq e^{-36k}$ whenever $\zeta_{-m'+k} \in (-\delta, \delta) \times \mathbb{R}$. (See Lemma 6.6 in [BC2] for a proof.)

Step 2 Show that $|DT^k_{\zeta_{-m'}}\tau| \geqq e^{-36k}$ $\forall k < m'$. Idea of proof: Let $t_1 < \ldots < t_j$ be the free return times of $\zeta_{-m'}$. First note that $s(\tau(\zeta_{-m'})) < 2b/\delta$ (Sublemma 1), so that $|DT^{t_1}_{\zeta_{-m'}}\tau| \geqq e^{ct_1}$. Between t_i and t_{i+1}, use the fact that τ splits correctly, and apply Sublemma 2(i) and 1.6.1(iii). (This is also proved in Lemma 7.2 of [BC2].)

Step 3 Show that there is a contractive vector field $e = e_{m'}$ defined on a strip containing $\zeta_{-m'}$ such that

(i) $|DT^j e| \leqq (Cb)^j \quad \forall j < m, C = 5e^{72}$;

(ii) there is an integral curve joining $\zeta_{-m'}$ to two points η_0^1 and η_0^2 on G_1 and G_2 respectively;

(iii) About $\eta_0^r, r = 1, 2$, there is a segment $\gamma^r \subset W$ extending $\geqq 3\rho^{m'}$ to each side of η_0^r such that each point on γ^1(resp. γ^2) is joined to a point on G_2(resp. G_1) by an integral curve of e.

This is proved in Sect. 5.3 of [BC2].

Step 4 Show $T^{m'}\gamma^1$ and $T^{m'}\gamma^2$ contains $C^2(b)$ curves with the properties as claimed. Idea of proof: Again consider the free return times $t_1 < \ldots < t_j$ of $\zeta_{-m'}$. Assuming that γ^r is not so near the "tips", we verify using the criteria in 1.2.4 that $T^{t_1}\gamma^r$ $(r = 1, 2)$ is a $C^2(b)$ curve. Assume now that $T^{t_i}\gamma^r$ contains a $C^2(b)$ curve $\gamma^{r,i}$, trimmed so that it extends by $3\rho^{m'}$ to each side of $\eta_{t_i}^r$. We need to know that τ splits correctly at every point on $\gamma^{r,i}$. Assuming that, we may view $\gamma^{r,i}$ as being tied to $\zeta_{-m'+t_i}$ for the next $t_{i+1} - t_i$ iterates, and the same argument as in Lemma 2 will tell us that $\gamma^{r,i+1}$ is again $C^2(b)$. To see that τ splits correctly on $\gamma^{r,i}$, use the following facts: (i) τ splits correctly at $\zeta_{-m'+t_i}$; (ii) $|\tau(\eta_{t_i}^r) - \tau(\zeta_{-m'+t_i})| < (Cb)^{t_i/6}$ (see Lemma 7.3 in [BC2]), and $(Cb)^{t_i/6} \ll e^{-36t_i} < |\zeta_{-m'+t_i} - z(\zeta_{-m'+t_i})|$; (iii) $\gamma^{r,i}$ is $C^2(b)$ and has length $\ll |\zeta_{-m'+t_i} - z(\zeta_{-m'+t_i})|$. □

Let ζ_0 be as in Lemma 4. Then for every j with $3^{j+1} < \text{gen}(\zeta_0)$, there is m_j with $3^j < m_j \leqq 3^{j+1}$ and $C^2(b)$ curves γ_j and γ_j' with the properties in Lemma 4. In the future we will refer to $\{\gamma_j\}$ and $\{\gamma_j'\}$ as "stacks captured by ζ_0".

2.3 Geometry of the critical set

The purpose of this subsection is to establish some regularity in the structure of the fractal set $\mathscr{C}$.

2.3.1 A basic lemma

Lemma 5 *Let $z_0 \in \mathscr{C}$ be located in a $C^2(b)$ curve $\gamma \subset W$. Assume that γ extends to $\geqq 2d$ to each side of z_0 and let $\zeta_0 \in \gamma$ be such that $|\zeta_0 - z_0| = d$. Then there are no critical points in the disk $B_{d^2}(\zeta_0) := \{z : |z - \zeta_0| \leqq d^2\}$.*

Proof. We will assume that there is a critical point $\tilde{z}_0 \in B_{d^2}(\zeta_0)$ and try to obtain a contradiction.

Let us first assume that $\tilde{z}_0$ lies on a $C^2(b)$ curve $\tilde{\gamma} \subset W$ that extends $\geqq d$ to each side of $\tilde{z}_0$. Using the notation in Sect. 1.3, we choose m with $b^m \ll d^2$ and $2d < (\lambda/5)^m$, and estimate $|\tau(\zeta_0) - e_m(\zeta_0)|$ in 2 different ways:

First note that e_m is defined on all of $B_{2d}(z_0)$, and that for purpose of comparing $e_m(z_0)$ with $e_m(\zeta_0)$, we may assume $e_m(z_0) = \tau(z_0)$ (see 1.3.4). We therefore have

$$|\tau(\zeta_0) - e_m(\zeta_0)| = (2a + \mathcal{O}(b))d\ .$$

Next, we let $\tilde{\zeta}_0$ be the point in $\tilde{\gamma}$ with the same x-coordinate as ζ_0 and write

$$|e_m(\zeta_0) - \tau(\zeta_0)| \leqq |e_m(\zeta_0) - e_m(\tilde{z}_0)| + |e_m(\tilde{z}_0) - \tau(\tilde{\zeta}_0)| + |\tau(\tilde{\zeta}_0) - \tau(\zeta_0)| .$$

The first two terms are $\leqq 5d^2$ (again see 1.3.4). As for the third term, since γ and $\tilde{\gamma}$ are nonintersecting $C^2(b)$ curves of length $\geqq d$ and $|\zeta_0 - \tilde{\zeta}_0| \leqq d^2$, an easy computation gives $|\tau(\zeta_0) - \tau(\tilde{\zeta}_0)| \leqq 2d$. These three terms add up to $2d + 10d^2$, which is $< (2a + \mathcal{O}(b))\cdot d$, and we obtain a contradiction.

Now the $C^2(b)$ curve containing $\tilde{z}_0$ may not be long enough. In that case we use the rules of construction of $\mathscr{C}$ (see 1.3.1) to successively obtain critical points $\tilde{z}_0^{(1)}, \tilde{z}_0^{(2)}, \ldots, \tilde{z}_0^{(k)}$ of lower and lower generation. Let G be such that $\rho^{2G} \leqq d \leqq \rho^G$, and let i be the smallest integer with $\text{gen}(\tilde{z}_0^{(i)}) < G$. Then

$$|\tilde{z}_0^{(i)} - \tilde{z}_0| \leqq \sum_{j = \text{gen}(\tilde{z}_0^{(i-1)})}^{\infty} (b^{\sigma j})^{\frac{1}{2}} \leqq 2b^{\sigma G/2} ,$$

which we may assume is $< d^2$ and $\tilde{z}_0^{(i)}$ lies on a $C^2(b)$ curve extending $\geqq 3\rho^G > d$ to each side on $\tilde{z}_0^{(i)}$. The previous argument can now be repeated with $\tilde{z}_0^{(i)}$ in place of $\tilde{z}_0$. □

2.3.2 View of $\mathscr{C}$ from a point in the free state

Lemma 6 *Let $\zeta \in G_n \cap (-\delta, \delta) \times \mathbb{R}$ be free, and let z_0 and z'_0 be two critical points wrt which ζ is in tangential position. Assume that $|\zeta - z_0|, |\zeta - z'_0| \geqq b^{\frac{n}{100}}$. Then the x-coordinate of ζ cannot lie between those of z_0 and z'_0.*

Proof. We assume this scenario occurs. Let $d = |\zeta - z_0|$, $d' = |\zeta - z'_0|$, and suppose that $d \leqq d'$. Let η be the point on $W^u_{\text{loc}}(z'_0)$ with the same x-coordinate as z_0, and let $D = |\eta - z_0|$. We will show that $D \ll d^2 + d'^2 < (d + d')^2$, which will contradict Lemma 5.

The difficulty in comparing $W^u_{\text{loc}}(z_0)$ and $W^u_{\text{loc}}(z'_0)$ directly is that the two $C^2(b)$ curves may be too far apart and one of them may not be long enough. (We had a similar problem in the proof of Lemma 5.) So again we rely on curves captured by ζ.

First we use ζ to capture a segment W_0 of W of length $> d$ and with $\text{dist}(\zeta, W_0) < d^4$. (This is clearly possible, even with $d \approx b^{\frac{n}{100}}$.) We further require that W_0 be on the opposite side of ζ as $W^u_{\text{loc}}(z_0)$. (See Lemma 4.) Let ζ_{z_0} and ζ_{W_0} be the points on $W^u_{\text{loc}}(z_0)$ and W_0 respectively with the same x-coordinate as ζ. Then $|\zeta_{z_0} - \zeta_{W_0}| < 2d^4$; and $|\tau(\zeta_{z_0}) - \tau(\zeta_{W_0})| < 2d^2$ because $W^u_{\text{loc}}(z_0)$ and W_0 are nonintersecting $C^2(b)$ curves that extend $> d$ to each side of ζ. Putting this together we have

$$D_0 := \text{dist}(z_0, W_0) < 2d^4 + 2d^2\cdot d + 20b\cdot d^2 \ll d^2 .$$

Similarly we let W'_0 be a segment captured by ζ on the opposite side of $W^u_{\text{loc}}(z'_0)$. We require that W'_0 has length $> d'$ and $\text{dist}(\zeta, W'_0) < d'^4$, so that

$$D'_0 := \text{dist}(\eta, W'_0) < 2d'^4 + 2d'^2\cdot d + 20b\cdot d^2 \ll d'^2 .$$

Now D is easily estimated as follows: If $W^u_{\text{loc}}(z_0)$ and $W^u_{\text{loc}}(z'_0)$ are on the same side of ζ, then $D \leqq \max(D_0, D'_0)$. If $W^u_{\text{loc}}(z_0)$ and $W^u_{\text{loc}}(z'_0)$ are on opposite sides of ζ, then $D \leqq D_0 + D'_0$. □

Consider a point $\zeta\in(-\delta,\delta)\times\mathbb{R}$ in the free state. Let $\{\gamma_j\}_{j=1}^k$ be a stack captured by ζ as discussed in 2.2.2. We may assume that there is a critical point on γ_1. Let $j(\zeta)$ be such that $\gamma_{j(\zeta)}$ is the γ_j of highest generation that contains a critical point, and call this critical point $\hat{z}(\zeta)$. We now fix some terminology to describe the location of $\mathscr{C}$ relative to ζ: We say that $\mathscr{C}$ is "*in the middle*" if there exists a stack $\{\gamma_j\}_{j=1}^k$ captured by ζ with the property that $j(\zeta)=k$ and $|\zeta-\hat{z}(\zeta)|<b^{\mathrm{gen}(\zeta)/100}$. If this is not the case, then we say that $\mathscr{C}$ is "*on the left*" (or "*on the right*") if for every stack captured by ζ, $\hat{z}(\zeta)$ lies on the left half (resp. right half) of $\gamma_{j(\zeta)}$.

We remark that if $\mathscr{C}$ is not "in the middle", then it has to be either "on the left" or "on the right" of ζ, because ζ is always in tangential position wrt $\hat{z}(\zeta)$, and Lemma 6 applies.

2.4 Proof of main proposition

2.4.1 General strategy

Let G_n be the leaf of generation n (see 1.3.1 for definition). We assume it has been proved that for all $\zeta_0\in G_n$ the pair (ζ_0,τ) is controlled on the time interval $(-\infty,0)$, and will show that (ζ_0,τ) is controlled on $(-\infty,0]$. Clearly, we need only to consider $\zeta_0\in(-\delta,\delta)\times\mathbb{R}$ and may assume that $\zeta_0\notin T^i\mathscr{C}$ for any $i\geqq 0$.

If ζ_0 is in a bound state, let ζ_{-i} be the last time it was free and let $\tilde{z}_0=z(\zeta_{-i})$. Then ζ_0 is in tangential position wrt $z(\tilde{z}_i)$. Moreover, since $(\zeta_{-i},\binom{0}{1})$ is controlled on the time interval $[0,i]$ (see Sect. 1.5.3) and $\tau(\zeta_{-i})$ splits correctly into $A_{-i}e+B_{-i}\binom{0}{1}$ by our induction hypothesis, the rejoining of the A_{-i}-term (if it has already taken place) has negligible effect. This proves that $\tau(\zeta_0)$ again splits correctly and we are done.

The case where ζ_0 is free is handled in the following lemma:

Lemma 7 *Let γ be a maximal free segment of G_n. If $\gamma\cap((-\delta,\delta)\times\mathbb{R})\neq\emptyset$, then there is a critical point z_0 with respect to which every $\zeta\in\gamma\cap((-\delta,\delta)\times\mathbb{R})$ is in generalized tangential position. Moreover, if $\tau(\zeta)=A(\zeta)e+B(\zeta)\binom{0}{1}$ is the spilitting in* 1.5.2 *with respect to our binding point $z_0:=z(\gamma)$, then*

$$3|\zeta-z_0|\leqq|B(\zeta)|\leqq 5|\zeta-z_0|\,.$$

Let γ_- and γ_+ denote the left and right endpoints of γ respectively. (This makes sense since free segments are roughly horizontal.) If $\gamma_+\in(-\delta,\delta)\times\mathbb{R}$, then it is attached to some $\tilde{z}_+\in\mathscr{C}$ because it is also in a bound state. Similarly, let $\tilde{z}_-=z(\gamma_-)$ if $\gamma_-\in(-\delta,\delta)\times\mathbb{R}$. The following are all the possible geometric configurations:

Case 1 γ is stretched across $(-\delta,\delta)\times\mathbb{R}$, i.e. neither γ_+ nor γ_- is in $(-\delta,\delta)\times\mathbb{R}$. We will show that z_0 lies on γ.

Case 2 γ_+ is in $(-\delta,\delta)\times\mathbb{R}$ and $\tilde{z}_+$ lies to the left of γ_+; $\gamma_-\notin(-\delta,\delta)\times\mathbb{R}$. As in Case 1, we will show that $z_0\in\gamma$.

Case 3 Both γ_- and γ_+ are in $(-\delta,\delta)\times\mathbb{R}$; $\tilde{z}_-$ lies to the right of γ_- and $\tilde{z}_+$ to the left of γ_+. We will also show that $z_0\in\gamma$.

Case 4 $\gamma_+\in(-\delta,\delta)\times\mathbb{R}$ and $\tilde{z}_+$ is to the right of γ_+. In this case we will show that $\tilde{z}_+$ is a viable candidate for z_0.

2.4.2 Proof of Lemma 7

We continue to use the terminology introduced in Sect. 2.3.2. For $n \in \mathbb{Z}^+$, let $\Delta_n > 0$ be sufficiently small that the following holds: Let γ be a free segment in $G_n \cap (-\delta, \delta) \times \mathbb{R}$, and let $\zeta = (\xi, \eta)$ and $\zeta' = (\xi', \eta')$ be two points on γ with $\xi - \Delta_n \leqq \xi' \leqq \xi$. If $\mathscr{C}$ is "on the left" of ζ, then it is either "on the left" or "in the middle" for ζ'. We assume also the analogous statement if $\mathscr{C}$ is "on the right" of ζ'. The existence of Δ_n follows from the proof of Lemma 6. In the following $\hat{z}(\zeta)$ is the "captured critical point" as defined in Sect. 2.3.2.

We now deal with the four cases discussed in Sect. 2.4.1.

Case 1 Since $\gamma \cap (-\delta, \delta) \times \mathbb{R}$ is a $C^2(b)$ curve (Corollary to Lemmas 1 and 2), once we produce a critical point $z_0 \in \gamma$, the rest of the assertion will follow. To product z_0 we start with $\zeta^{(0)}$ on γ with x-coordinate $= \frac{1}{2}\delta$. Then $\hat{z}(\zeta^{(0)})$ must be on the left. We move left along γ by steps of Δ_n to obtain successively $\zeta^{(1)}, \zeta^{(2)}, \ldots$. If for some k, $\mathscr{C}$ is "in the middle", then we are done, because some point on γ "sees" $\hat{z}(\zeta^{(k)})$ (see Sublemma 3 below) and the rules of construction of $\mathscr{C}$ says that a critical point must have been constructed on γ. Clearly we cannot move left indefinitely and continue to have $\mathscr{C}$ "on our left."

Sublemma 3 *Let $\zeta = \zeta^{(k)}$ and γ are as above. Then vert* $\operatorname{dist}(\hat{z}(\zeta), \gamma) < b^{\frac{1}{540}n}$.

Proof. Let η be the point on γ with the same x-coordinate as $\hat{z}(\zeta)$, and let $\hat{\zeta}$ be the point on $\gamma_{j(\zeta)}$ with the same x-coordinate as ζ. Then

$$|\hat{z}(\zeta) - \eta| \leqq |\zeta - \hat{\zeta}| + 20b|\hat{z}(\zeta) - \hat{\zeta}|$$

$$< (Cb)^{\frac{n}{9}} + 20b \cdot b^{\frac{n}{100}} \ll b^{\frac{n}{540}}. \qquad \square$$

Case 2 Let $d_+ = |\gamma_+ - \tilde{z}_+|$. We need the following simple estimate:

Sublemma 4 *Let γ be as in Case 2, and let ζ be a point on γ with $|\zeta - \gamma_+| < d_+$. Then*

$$\operatorname{dist}(\zeta, W^u_{\mathrm{loc}}(\tilde{z}_+)) \ll d_+^{\frac{3}{2}}.$$

Proof. Use γ_+ to capture an unstable leaf γ_j on the opposite side of $W^u_{\mathrm{loc}}(\tilde{z}_+)$ with length$(\gamma_j) \approx d_+$. Let η_j and $\hat{\eta}$ be the points on γ_j and $W^u_{\mathrm{loc}}(\tilde{z}_+)$ resp with the same x-coordinate as γ_+. Then $|\hat{\eta} - \gamma_+| < d_+^4$ because γ_+ is in tangential position wrt $\tilde{z}_+$, and $|\eta_j - \gamma_+| < (Cb)^{m_j} \ll d_+^4$. Also, γ_j and $W^u_{\mathrm{loc}}(\tilde{z}_+)$ are long enough to guarantee that $|\tau(\eta_j) - \tau(\hat{\eta})| < d_+^{\frac{1}{2}}$. So

$$\operatorname{dist}(\zeta, W^u_{\mathrm{loc}}(\tilde{z}_+)) < 2d_+^4 + 2d_+^{\frac{1}{2}} \cdot d_+ + (20b)d_+^{\frac{3}{2}} \ll d_+^{\frac{3}{2}}. \qquad \square$$

Let $R_+ = \{z \in \gamma : |z - \gamma_+| < \frac{1}{2}d_+\}$. We note that if $\zeta \in R_+$ then $\mathscr{C}$ cannot be "in the middle". This is because $d_+ > e^{-\alpha n}$, which is $\gg b^{\frac{n}{100}}$, and if $|\hat{z}(\zeta) - \zeta| < b^{\frac{n}{100}}$ for any stack, then we will have

$$\operatorname{horiz\,dist}(\hat{z}(\zeta), \tilde{z}_+) > \tfrac{1}{3}d_+,$$

and

$$\operatorname{dist}(\hat{z}(\zeta), W^u_{\mathrm{loc}}(\tilde{z}_+)) \ll d_+^{\frac{3}{2}},$$

which contradicts the geometry of the critical set.

We start with $\zeta^{(0)} = \gamma_+$. First we claim that $\mathscr{C}$ is on the left: "middle" has been ruled out in the last paragraph, and "right" is not compatible with the position of $\tilde{z}_+$ (Lemma 6). We move left by steps of Δ_n as before until we reach $\zeta^{(k)}$ with $\mathscr{C}$ in the middle. Now $\zeta^{(k)} \notin R_+$, which guarantees that γ extends $> \frac{1}{2} e^{-\alpha n} \gg 2\rho^n$ to each side of $\zeta^{(k)}$, and so a critical point z_0 on γ is assured. The correct splitting of τ wrt z_0 is automatic as before.

Case 3 We begin as in Case 2, taking $\zeta^{(0)} = \gamma_+$ and moving left in small steps. We must reach some $\zeta^{(k)}$ with $\hat{z}(\zeta^{(k)})$ in the middle before arriving at γ_-, otherwise $\hat{z}(\gamma_-)$ would be on the left, contradicting the fact that $\tilde{z}_-$ is to the right of γ_-. Now $\zeta^{(k)} \notin R_+ \cup R_-$, where R_- has the obvious definition. So again a critical point on γ is guaranteed.

Case 4 Let $d = |\gamma_+ - \tilde{z}_+|$, and let η be the point on $W^u_{\text{loc}}(\tilde{z}_+)$ with the same x-coordinate as γ_+. Then $|\eta - \gamma_+| < 2d^4$ and $|\tau(\eta) - \tau(\tau_+)| < d^2$ (an exercise: cf. the proof of Sublemma 4). We need to show that every point in γ is in a generalized tangential position with respect to $\tilde{z}_+$. Let (x', y') be the natural coordinate system at $\tilde{z}_+$ (see 1.6.2) and let φ be the function whose graph is the curve γ. Then

$$|\varphi(-d)| < 2d^4 + (10b)d^2 \ll \tfrac{1}{100} d^2 ,$$

$$|\varphi'(-d)| < d^2 + (10b)d \ll \tfrac{2}{100} d ,$$

and $|\varphi''| \leqq 10b$. This proves that $|\varphi(x')| < \frac{1}{100} x'^2$ for all the relevant x'.

We also need to know that τ splits "correctly" at every point on γ. This is true because $\tau(\gamma_+)$ splits correctly (it is in a bound state) and γ is a $C^2(b)$ curve.

This completes the proof of Lemma 7. □

We remark that an alternate proof of Lemma 7 is to try to carry out the "capture" argument in Lemma 4 simultaneously for all points on γ.

2.4 Proof of Corollary 1

Part I of Corollary 1 follows from Lemma 1 and Sect. 1.2. Part II follows from the correct splitting guaranteed in Proposition 1, the properties of $\mathscr{C}$ as discussed in 1.3.3, the distortion estimates in 1.7.1, and the bound period estimates in 1.6.

2.5 Proof of Proposition 2

A proof of Proposition 2 is given in Sects. 2 and 8 of [BC2]; see Lemma 8.9 in particular. Because the arguments there are a bit sketchy, and this estimate is of central importance in the construction of SBR measures, we are going to fill in some details. We will assume the distortion estimate for $DT^j_{(\cdot)}\binom{0}{1}$ during bound periods (see 1.7.1), and try to explain how these estimates can be used to control the distortion of τ_j over arbitrarily long periods of time. The fact that (ζ, τ) is controlled for all ζ in question is used implicitly throughout.

We write

$$\log \frac{|\tau_N(\zeta_0)|}{|\tau_N(\zeta_0')|} = \sum_{k<q} S_k' + \sum_{k<q} S_k'',$$

where

$$S'_k = \log \frac{|\tau_{p_k}(\zeta_{t_k})|}{|\tau_{p_k}(\zeta'_{t_k})|} \quad \text{and} \quad S''_k = \log \frac{|\tau_{t_{k+1}-p_k}(\zeta_{t_k+p_k})|}{|\tau_{t_{k+1}-p_k}(\zeta'_{t_k+p_k})|}\,.$$

First we show that $\sum S''_k <$ some C''_1. Consider j with $t_k + p_k \leqq j < t_{k+1}$. Since γ_j is free and free segments have uniformly bounded slopes and curvatures, it follows that

$$|DT_{\zeta_j}\tau - DT_{\zeta'_j}\tau| \leqq \text{const}\,|\gamma_j|\,.$$

So

$$S''_k \lesssim \sum_{j=t_k+p_k}^{t_{k+1}-1} \frac{|DT_{\zeta_j}\tau - DT_{\zeta'_j}\tau|}{|DT_{\zeta'_j}\tau|} \leqq \frac{1}{\delta}\cdot \text{const} \sum_j |\gamma_j|\,,$$

which is $< \text{const}\,|\gamma_{t_{k+1}}|$ because Corollary 1 (part I) tells us that

$$|\gamma_{t_{k+1}}| \geqq e^{c_0(t_{k+1}-j)}|\gamma_j|\,.$$

Using Corollary 1 again (both parts I and II) we conclude that

$$|\gamma_{t_{k+1}}| \geqq |\gamma_{t_k+p_k}| \geqq 2|\gamma_{t_k}|\,.$$

Hence

$$\sum S''_k < \text{const}\,|\gamma_N| < \text{some } C''_1\,.$$

To estimate S'_k we first prove the following

Sublemma 5 *Let $\eta_0, \eta'_0 \in W$ lie in the same $I_{\mu j}$ with respect to some $\tilde{z}_0 \in \mathscr{C}$, and let p be their common bound period. Then there is $C > 0$ not depending on η_0, η'_0 or z_0 such that*

$$\log \frac{|\tau_p(\eta_0)|}{|\tau_p(\eta'_0)|} \leqq C \frac{|\eta_0 - \eta'_0|}{e^{-\mu}}\,.$$

Throughout the proof we will use c to denote a generic constant that is positive and very small.

Proof of Sublemma 5. Split

$$\tau(\eta_0) = Ae + B\binom{0}{1}\ ,$$
$$\tau(\eta'_0) = A'e' + B'\binom{0}{1}\,,$$

where e and e' are the contractive (unit) vectors for the period $[0, p]$. We will use the notation

$$e_j := DT^j_{\eta_0}e, \qquad e'_j := DT^j_{\eta'_0}e'\,,$$
$$w_j := DT^j_{\eta_0}\binom{0}{1}, \qquad w'_j := DT^j_{\eta'_0}\binom{0}{1}\,.$$

Let R_θ be rotation by θ, and let θ be chosen such that R_θ carries w_p to a positive multiple of w'_p. We will write

$$\frac{|\tau_p(\eta_0)|}{|\tau_p(\eta'_0)|} \leqq \text{(I)} + \text{(II)} + \text{(III)} + \text{(IV)} + \text{(V)}\,,$$

where (I)–(V) are defined and estimated as follows:

$$\text{(I)} = \frac{\left| R_\theta B w_p + \dfrac{|Bw_p|}{|B'w'_p|} A' e'_p \right|}{|B'w'_p + A'e'_p|} = \frac{|Bw_p|}{|B'w'_p|},$$

where

$$\frac{|B|}{|B'|} \leqq 1 + 5\frac{|\eta_0 - \eta'_0|}{e^{-\mu}}$$

because of the way τ and e change with z, and

$$\frac{|w_p|}{|w'_p|} \leqq 1 + 2C_0 C'_0 \frac{|\eta_0 - \eta'_0|}{e^{-\mu}}$$

by 1.7.1(ii), (iii). (Note that the constants C_0 and C'_0 are purely numerical so

$$C_0 C'_0 \frac{|\eta_0 - \eta'_0|}{e^{-\mu}}$$

is small if μ_0 is chosen sufficiently large.)

$$\text{(II)} = \frac{1}{|\tau_p(\eta'_0)|} \cdot \left| \frac{|Bw_p|}{|B'w'_p|} A' e_p - \frac{|Bw_p|}{|B'w'_p|} A' e'_p \right| \leqq |e_p - e'_p|,$$

which is $< c|\eta_0 - \eta'_0|$ by Lemma 5.5 of [BC2].

$$\text{(III)} = \frac{1}{|\tau_p(\eta'_0)|} \cdot \left| A' e_p - \frac{|Bw_p|}{|B'w'_p|} A' e_p \right| \leqq \left| 1 - \frac{|Bw_p|}{|B'w'_p|} \right|,$$

which is $< \text{const}\,|\eta_0 - \eta'_0| e^{\mu}$ (same as (I)).

$$\text{(IV)} = \frac{1}{|\tau_p(\eta'_0)|} \cdot |R_\theta A' e_p - A' e_p| \leqq |\theta|,$$

which is $< c\Delta_p \leqq cC'_0 |\eta_0 - \eta'_0| e^{\mu}$ by 1.7. (ii) and (iii).

$$\text{(V)} = \frac{1}{|\tau_p(\eta'_0)|} \cdot |R_\theta A e_p - R_\theta A' e_p| \leqq c|A' - A| \leqq c|\eta_0 - \eta'_0|.$$

Together these estimates prove the sublemma. □

Applying this sublemma to each S'_k, we obtain

$$\sum_{k=0}^{q-1} S'_k \leqq C \sum_{k=0}^{q-1} \frac{|\gamma_{t_k}|}{e^{-\mu_k}},$$

where $\gamma_{t_k} \subset I_{\mu_k j_k}$. To estimate this sum, we let $m(\mu) = \max\{t_k : \mu_k = \mu\}$ for each μ, and use the fact that $|\gamma_{t_{k+1}}| \geqq 2|\gamma_{t_k}|$ to conclude that

$$\sum_{k<q} \leqq \frac{|\gamma_{t_k}|}{e^{-\mu_k}} \leqq \text{const} \sum_{\mu} \frac{|\gamma_{m(\mu)}|}{e^{-\mu}} \leqq \text{const} \sum_{\mu} \frac{1}{\mu^2}.$$

This completes the proof of Proposition 2. □

3 Construction of SBR-measures

We continue to assume that $T = T_{a,b}$, where (a, b) is one of the "good" parameters. Our strategy is as follows. Put Lebesgue measure m on a piece of W. Transport m forward by T, and take the ergodic averages of these measures. We will show that any limit point of these ergoidic averages contains at least one component that has absolutely continuous conditional measures on unstable manifolds. This construction is standard for Axiom A attractors. The piecewise uniformly hyperbolic case is dealt within e.g. [Y2], which contains a simple version of what is done here.

3.1 Bookkeeping on the unstable manifold

We showed in Lemma 7, Sect. 2, that every maximal free segment γ in $W \cap ((-\delta, \delta) \times \mathbb{R})$ is assigned to a critical point with respect to which it is in generalized tangential position. This critical point will be denoted by $z(\gamma)$. On γ, it is natural to consider the partition $\mathscr{P}_{[z(\gamma)]}$, where $\mathscr{P}_{[\cdot]}$ is the partition defined in Sect. 2.1.2. In the construction below we will often speak of $I_{\mu j}$ on γ without explicit mention of $\mathscr{P}_{[z(\gamma)]}$.

We select a piece of W on which to begin our construction. Let, for instance, z_0 be the critical point on G_1, and let $\Delta \subset G_1 \cap ((-\delta, \delta) \times \mathbb{R})$ correspond to $(e^{-(\mu_0+1)}, e^{-\mu_0})$. Let $\mathscr{P}_0 = \mathscr{P}_{[z_0]}|\Delta$. We will describe in the next few paragraphs a sequence of partitions $\mathscr{P}_0 \prec \mathscr{P}_1 \prec \mathscr{P}_2 \ldots$, such that each $\mathscr{P}_n$ divides Δ into a countable number of intervals – or curve segments rather. Points in $\mathscr{P}_n(z)$ can be regarded as having trajectories "indistinguishable" from that of z up to time n.

We consider one element ω of $\mathscr{P}_0$ at a time. Regarding ω as bound to $z(\omega) = z_0$, we let $j_1 > 0$ be the first time when $T^j\omega$ is free and intersects $(-\delta, \delta)$. (We may assume that all points in $T^j\omega$ become free simultaneously.) If $T^{j_1}\omega$ contains some $I_{\mu j}$ then we let $k_1 = j_1$ and go to the next paragraph. If not, then we consider $T^{j_1}\omega$ as bound to $z(T^{j_1}\omega)$ and wait for it to return again in a free state, say at time j_2. From Corollary 1 in Sect. 2 it is clear that $|T^{j_2}\omega| \geqq 2|T^{j_1}\omega|$, so repeating this process for at most a finite number of times, there will be a free return at time j_l, when $T^{j_l}\omega \supseteq$ some $I_{\mu j}$. Set $k_1 = j_l$.

We now define $\mathscr{P}_n|\omega$ for $n \leqq k_1$. Let $\mathscr{P}_n = \mathscr{P}_0$ for $n < k_1$. For $n = k_1$, we first let $\mathscr{P}'_{k_1}|\omega = T^{-k_1}(\mathscr{P}_{[z(T^{k_1}\omega)]} \cup \{(-1, -\delta), (\delta, 1)\})$, and obtain $\mathscr{P}_{k_1}$ from $\mathscr{P}'_{k_1}$ by adjoining each of the two end intervals on $\mathscr{P}'_{k_1}|\omega$ to its neighbor unless the T^{k_1}-image of this end interval lies outside of $(-\delta, \delta)$ and has length $\geqq |I_{\mu_0 j}|$.

We then repeat the argument in the last two paragraphs for each element ω' of $\mathscr{P}_{k_1}$. That is, if k_2 is the first time after k_1 when part of $T^{k_2}\omega'$ returns freely to $(-\delta, \delta)$ and $T^{k_2}\omega' \supset$ some $I_{\mu j}$ then we cut up ω' again at this time according to the locations of $T^{k_2}z$.

Next we introduce a sequence of stopping times $t_0 < t_1 \ldots$ on Δ. Let Δ^+ and Δ^- be the rightmost and leftmost intervals in the partition $\mathscr{P}$ of $(-\delta, \delta)$. (We may assume that Δ^+ and Δ^- are fixed intervals not depending on the location of critical points.) Let $t_0 \equiv 0$. For $z \in \Delta$, we define $t_1(z)$ to be the smallest $k > 0$ such that $T^k(\mathscr{P}_{k-1}(z))$ contains either Δ^+ or Δ^-, $t_2(z)$ to be the smallest $k > t_1(z)$ when $T^k(\mathscr{P}_{k-1}(z))$ contains either Δ^+ or Δ^-, and so on. Note that $t_n(z)$ could take on the value ∞, since it is possible for a point to keep returning to the shorter intervals which get cut before they get a chance to grow long. We will prove in Sect. 3.3, however, that this is an extremely improbable event.

Our construction of $\mathscr{P}_n$ is virtually identical to that of a similar partition in Sect. 1 of [BC2] – except of course that our construction takes place on W whereas the one in [BC2] is carried out in parameter space. Our t_n's correspond essentially (though not exactly) to the "escape times" in [BC2].

3.2 Derivative estimates

Let m denote Lebesgue measure on W, i.e. if $\gamma \subset W$ is a curve segment, then $m(\gamma)$ is equal to the arc length of γ. Let $T^j_*(m|\Delta)$ denote the measure with $T^j_*(m|\Delta)(E) = m(T^{-j}E \cap \Delta)$. Clearly the density of $T^j_*(m|\Delta)$ on $T^j\Delta$ is given by

$$\frac{dT^j_*(m|\Delta)}{dm}(z) = |DT_z^{-j}\tau| .$$

Our first task is to study how $T^j_*(m|\Delta)$ is distributed along W for $j = 1, 2, \ldots$ For this we use the derivative estimates in Corollary 1. The distortion estimate in Proposition 2 is crucial for controlling local fluctuations in densities along certain segments of W. This will be important in our construction. Lemma 3 will also be used.

3.3 Frequency of returns

The aim of this section is to show that a positive measure set of points in Δ return with positive frequency to $\Delta^+ \cup \Delta^-$.

3.3.1 First escape time estimates

Consider first the 1-d map $f:[-1, 1]\circlearrowleft$ satisfying the conditions in Sect. 1.1. Let γ be an interval in $[-1, 1]$. We assume that γ is either $\approx I_{rj}$ for some r, j, or $\gamma \cap (-\delta, \delta) = \emptyset$. (The notation $\gamma \approx I_{rj}$ means that $I_{rj} \subset \gamma \subset I^+_{rj}$, where I^+_{rj} is defined to be the union of I_{rj} and its two adjacent intervals.) We define the *first escape time function* $t|_\gamma$ exactly as t_1 is defined in Section 3.1 – except that we start from γ. This definition is related to our earlier definition of stopping times $t_1 < t_2 < \ldots$ on Δ (had we defined them for our interval map) by

$$t_{i+1}(x) - t_i(x) = (t|_{f^{t_i}(\mathscr{P}_{t_i}(x))})(x) .$$

Note that if $\gamma = f^{t_i}(\mathscr{P}_{t_i}(x))$ for some x and does not intersect $(-\delta, \delta)$, then it has δ or $-\delta$ as one of its end points and has length $\geqq |\Delta^+|$. We claim that for such an interval γ, $t|_\gamma$ is constant on γ and is $\leqq M$ for some M independent of γ. To see this let k_1 be the first time $f^{k_1}\gamma \supset$ some $I_{\mu j}$. Using bound estimates similar to those for $\Delta^\pm$ it is easy to verify that $k_1 \leqq M = C\log(1/\delta)$ and that $f^j\gamma \geqq C/\log(1/\delta) \gg 2\delta$. Therefore $f^{k_1}\gamma$ must contain either Δ^+ or Δ^- and $t|_\gamma = k_1$.

When $\gamma \approx I_{rj}$, the following large deviation estimate for $t|_\gamma$ is proved in Sect. 2.2 of [BC2].

Lemma 8 [BC2] *For $\gamma \approx I_{rj}$ and $n \geqq 6r$*

$$m\{x \in \gamma : t|_\gamma(x) \geqq n\} \leqq e^{-n/20} m(\gamma) .$$

Because of the close correspondence between the derivative estimates in 1 and 2 dimensions (see Corollary 1), these first escape time estimates apply without change to free segments $\gamma \subset T^k \Delta$. We now derive from these estimates the main lemma of Sect. 3.3.

3.3.2 A lemma

Lemma 9 *There is a constant C^* such that for all $i \geqq 0$,*

$$\int_\Delta (t_{i+1} - t_i)\, dm \leqq C^* .$$

Proof. For $i \geqq 1$, let $\hat{\mathscr{P}}_i$ be the partition of Δ into "distinguishable orbits" up to time $t_i - 1$, i.e. $\hat{\mathscr{P}}_i$ refines the partition of Δ by values of t_i and

$$\hat{\mathscr{P}}_i|\{t_i = k\} = \mathscr{P}_{k-1}|\{t_i = k\} \,.$$

This means in particular that for each $\omega \in \hat{\mathscr{P}}_i$, $T^{t_i}\omega \supset \Delta^+$ or Δ^-. Also, since $m\{t_i = \infty\} = 0$, $\hat{\mathscr{P}}_i$ is a genuine partition of Δ up to a set of measure 0. It suffices to give a uniform upper bound for

$$\frac{1}{m(\omega)} \int_\omega (t_{i+1} - t_i)\, dm, \quad \omega \in \hat{\mathscr{P}}_i \,.$$

We fix $\omega \in \hat{\mathscr{P}}_i$ and let $\gamma = T^{t_i}\omega$ Then

$$\int_\gamma (t_{i+1} - t_i) \circ T^{-t_i}\, dm \leqq \sum_{r,j} \int_{I_{r,j}} t|_{I_{rj}} + \int_{\gamma_-} t|_{\gamma_-} + \int_{\gamma_+} t|_{\gamma_+} \,,$$

where γ_- and γ_+ are those parts of γ in $(-1, -\delta) \times \mathbb{R}$ and $(\delta, 1) \times \mathbb{R}$ resp. From Lemma 8, we see that

$$\begin{aligned} \sum_{r,j} \int_{I_{r,j}} t|_{I_{r,j}}\, dm &\geqq \sum_{r,j} \sum_{n=0}^{\infty} m\{z \in I_{r,j} : t|_{I_{r,j}}(z) \geqq n\} \\ &\leqq \sum_{n=0}^{\infty} \left\{ m\left(\bigcup_{|r| \geqq n/6} I_{r,j} \right) + e^{-\frac{n}{20}} m\left(\bigcup_{|r| \leqq n/6} I_{r,j} \right) \right\}, \end{aligned}$$

so that

$$\begin{aligned} \frac{1}{m(\gamma)} \int_\gamma (t_{i+1} - t_i) \circ T^{-t_i}\, dm &\leqq \frac{1}{|\Delta^+|} \left\{ \sum_{n=0}^{\infty} (2e^{-\frac{n}{6}} + e^{-\frac{n}{20}}(2\delta)) + 2M \right\} \\ &\leqq \text{some } C \,. \end{aligned}$$

Our desired estimate then is given by

$$\begin{aligned} \frac{1}{m(\omega)} \int_\omega (t_{i+1} - t_i)\, dm &= \frac{1}{m(\omega)} \int_\gamma (t_{i+1} - t_i) \circ T^{-t_i} d(T^{t_i}(m|\omega)) \,, \\ &\leqq C_1 C \,, \end{aligned}$$

where C_1 is the distortion constant in Proposition 2. The case $i = 0$ is simple. □

3.3.3 A lower bound on the frequency of returns to $\Delta^\pm$

Let $\mathscr{P}_{n,\Delta^\pm} = \bigcup\{\omega \in \mathscr{P}_n : T^n\omega = \Delta^+ \text{ or } \Delta^-\}$.

Lemma 10 *There exists a constant $\alpha^* > 0$ such that*

$$\varliminf_{N\to\infty} \frac{1}{N}\sum_{n=0}^{N} m\mathscr{P}_{n,\Delta^\pm} \geqq \alpha^* > 0 .$$

Proof. From Lemma 9 we know that

$$\int_\Delta t_n\, dm \leqq C^* n$$

and so

$$m\{z\in\Delta : t_n(z) \leqq 2nC^*\} \geqq \tfrac{1}{2}m(\Delta) .$$

This means that

$$\sum_{k=1}^{2nC^*} m\{z\in\Delta : k = t_i(z) \text{ some } i\} \geqq n\cdot\tfrac{1}{2}m(\Delta) .$$

Now for each $\omega\in\mathscr{P}_{k-1}$ with $k = t_i(\omega)$, there is $\omega'\in\mathscr{P}_k$, $\omega'\subset\omega$, such that $T^k\omega \approx \Delta^+$ or Δ^-. Moreover

$$m(\omega') \geqq m(\omega)\cdot\frac{|\Delta^+|}{2}\cdot\frac{1}{C_1} .$$

These estimates together give the desired result. □

3.4 SBR measures as limits of Lebesgue measure on W

3.4.1 Definition of SBR measures

In this subsection we define precisely what we mean by SBR-measures. Let $F: M \circlearrowleft$ be an arbitrary C^2 diffeomorphism of a finite dimensional manifold and let μ be an F-invariant Borel probability measure on M with compact support. We will assume throughout that at μ-a.e. point, there is a strictly positive Lyaponov exponent. Under these conditions, the unstable manifold theorem of Pesin [P1] or Ruelle [Ru2] tells us that passing through μ-a.e. x there is an *unstable manifold* which we denote by $W^u(x)$.

A measurable partition $\mathscr{Q}$ of M is said to be *subordinate to W* (with respect to the measure μ) if at μ-a.e. x, $\mathscr{Q}(x)$ is contained in $W^u(x)$ and contains an open neighborhood of x in $W^u(x)$. On each $\mathscr{Q}(x)$, there are two measures that are of interest to us. One is the restriction to $\mathscr{Q}(x)$ of the Riemann measure induced on $W^u(x)$; let us call this $m_x^{\mathscr{Q}}$. The other is $\mu_x^{\mathscr{Q}}$, where $\{\mu_x^{\mathscr{Q}}\}$ is a canonical family of conditional measures of μ with respect to the partition $\mathscr{Q}$. (For a reference see e.g. Rohlin [Ro].)

Definition. Let $F:(M,\mu)\circlearrowleft$ be as above. We say that μ has *absolutely continuous conditional measures on unstable manifolds* if for every measurable partition $\mathscr{Q}$ subordinate to W^u, $\mu_x^{\mathscr{Q}}$ is absolutely continuous with respect to $m_x^{\mathscr{Q}}$ for μ-a.e. x.

For ease of reference, we will in this paper refer to invariant probability measures with absolutely continuous conditional measures on unstable manifolds as *Sinai–Bowen–Ruelle measures* or simply *SBR-measures*[2].

3.4.2 Pushing forward Lebesgue measure on W

Let

$$m_0 = m|_\Delta$$

and

$$m_n = \frac{1}{n}\sum_{k=0}^{n-1} T_*^k m_0 .$$

We define $\hat{m}_n^+$ to be the restriction to $\Delta^+ \times \mathbb{R}$ of

$$\frac{1}{n}\sum_{\substack{\omega\in\hat{\mathscr{P}}_i \\ t_i(\omega) < n}} T_*^{t_i}(m_0|\omega) ,$$

and let $\hat{m}_n^-$ be defined similarly. From Lemma 10 we know that for either $\{\hat{m}_n^+\}$ or $\{\hat{m}_n^-\}$ – let us say $\{\hat{m}_n^+\}$ – there is a sequence $N_1 < N_2 < \ldots$ such that

$$\hat{m}_{N_i}^+(\mathbb{R}^2) \geqq \frac{\alpha^*}{3} \quad \text{for all } i .$$

Passing to a subsequence if necessary, we may assume that

$$\hat{m}_{N_i}^+ \xrightarrow{\text{weakly}} \text{some } \hat{\lambda}$$

and

$$m_{N_i} \xrightarrow{\text{weakly}} \text{some } \lambda .$$

The following are immediate:

(1) λ is a T-invariant Borel measure, whose support is contained in the attractor $\Lambda = \bar{W}$;

(2) the total mass of $\hat{\lambda}$ is $\geqq \alpha^*/3$;

(3) λ is the sum of $\hat{\lambda}$ and another Borel measure.

Our plan is to use the geometric properties of $\hat{\lambda}$ to show that λ has at least one component with absolutely continuous conditional measures on unstable manifolds.

3.4.3 Geometric properties of $\hat{\lambda}$

Let c_1 be the constant in Lemma 3, and let

$$\Gamma = \{z\in\Lambda\cap(\Delta^+\times\mathbb{R}) : \exists v \neq 0 \in T_z\mathbb{R}^2 \text{ with } |DT_z^{-j}v| \leqq e^{-c_1 j}\ \forall j \geqq 0\} .$$

[2] SBR measures are sometimes defined differently. All the definitions are equivalent for Axiom A attractors, but not all of them have been shown to be equivalent in the nonuniformly hyperbolic setting

Since $|\det DT^{-1}| = b^{-1}$ and the contraction above is uniform, it follows easily that Γ is compact, and that for $z \in \Gamma$, if v_z is the direction contracted by DT^{-j} in the definition of Γ, then $z \mapsto v_z$ is continuous. Moreover, by Lemma 3 we know that $\operatorname{supp} \hat{m}_n^+ \subset \Gamma$ for all n and that $\tau(z) = v_z$.

Sublemma 6 *Let z be an accumulation point of $\bigcup_n \operatorname{supp}(\hat{m}_n^+)$. Then there is a C^1-curve $\gamma(z)$ passing through z such that*

(1) $\gamma(z) = \operatorname{graph}(\varphi)$ *for some* $\varphi : \Delta^+ \to \mathbb{R}$;
(2) $\gamma(z) \subset \Gamma$ *and its tangent vector at* z' *is* $v_{z'}$.

Proof. Let $z_i \in \bigcup_n \operatorname{supp}(\hat{m}_n^+)$ be such that $z_i \to z$. For each z_i, let $\varphi_i : \Delta^+ \to \mathbb{R}$ be the function whose graph is the component of $W \cap (\Delta^+ \times \mathbb{R})$ containing z_i. By the $C^2(b)$-property of free segments the sequence of second derivatives $\{\varphi_i''\}$ is uniformly bounded. Hence a subsequence $\{\varphi_{i_k}'\}$ of $\{\varphi_i'\}$ converges uniformly. It follows immediately that φ_{i_k} converges in the C^1 sense to some φ with the properties in Sublemma 6. □

From Oseledec's theorem we know that Lyapunov exponents are well defined $\hat{\lambda}$-a.e. Sublemma 6 tells us that one of the exponents is positive. We also know from general theory (see [P1] or [Ru2]) that for a.e. point, its local unstable manifold is the unique curve passing through that point that contracts exponentially in backward time. Thus for a typical z, $\gamma(z)$ in Sublemma 6 must be the component of $W^u(z) \cap (\Delta^+ \times \mathbb{R})$ containing z. Let us call it $W^u_{\Delta^+}(z)$.

Let $X \subset \Delta^+ \times \mathbb{R}$ be a measurable set with the following properties: (i) $\hat{\lambda}(\mathbb{R}^2 - X) = 0$, and (ii) X is the disjoint union of $W^u_{\Delta^+}$-curves. Let $\mathcal{Q}$ be the partition of X into $W^u_{\Delta^+}$-leaves, and let $\{\hat{\lambda}_z^{\mathcal{Q}}\}$ be a canonical family of conditional measures of $\hat{\lambda}$. (See Sect. 3.4.1 for notations.)

Sublemma 7 *The measures $\hat{\lambda}_z^{\mathcal{Q}}$ and $m_z^{\mathcal{Q}}$ are equivalent for $\hat{\lambda}$-a.e. z.*

Proof. We claim that there is $C > 0$ such that for all intervals $J \subset \Delta^+$, one has

$$\frac{1}{C}|J| \leqq \hat{\lambda}_z^{\mathcal{Q}}(J \times \mathbb{R}) \leqq C|J|$$

for a.e. z. To see this let $\mathcal{Q}_n$ be a sequence of finite partitions of X such that $\forall z \in X$, $\mathcal{Q}_n(z) \supset W^u_{\Delta^+}(z)$ for all n, and $\bigcap_n \mathcal{Q}_n(z) = W^u_{\Delta^+}(z)$. If some $W^u_{\Delta^+}$-curve γ is contained in the support of $\hat{m}_n^+$, let ρ_n denote the density of $\hat{m}_n^+ | \gamma$. Proposition 2 tells us that for all $z_1, z_2 \in \gamma$,

$$\frac{\rho_n(z_1)}{\rho_n(z_2)} \leqq C_1 .$$

Integrating over $W^u_{\Delta^+}$-curves in each element of $\mathcal{Q}_n$, we obtain

$$\frac{1}{C}|J| \leqq E_{\hat{\lambda}}((J \times \mathbb{R})|\mathcal{Q}_n) \leqq C|J|$$

for some C independent of J. Our assertion follows from the martingale convergence theorem. □

3.4.4 *Completing the construction*

First we observe that $(\lambda \mid X)_z^{\mathscr{Q}}$ is equivalent to $m_z^{\mathscr{Q}}$ for $\hat{\lambda}$-a.e. z. This is true because the σ-algebra of T-invariant measurable sets is contained in the σ-algebra of measurable set made up of entire W^u-leaves, and that for every ergodic measure ν, the conditional measures of ν on local W^u-manifolds are either equivalent to m on a.e. leaf, or they are singular to m on a.e. leaf.

Let

$$X' = \left\{ z : \frac{d\hat{\lambda}}{d\lambda} > 0 \right\}.$$

and let λ' be the saturation of $\lambda|_{X'}$ under T. That is, let $r: X' \to \mathbb{Z}^+$ be the first return time to X' under T and let

$$\lambda' = \sum_{n=0}^{\infty} T_*^n(\lambda \mid (X' \cap \{r > n\}))\,.$$

We noted above that $\lambda \mid X'$ has absolutely continuous conditional measures on W^u-leaves. Hence the same is true for λ'. Moreover, an invariant measure with absolutely continuous conditional measures on W^u-leaves is the sum of at most a countable number of ergodic measures with the same property (see e.g. [L]). We may therefore assume for the rest of this paper that λ^* is one of these ergodic components, normalized to give $\lambda^*(\Lambda) = 1$. This is our SBR-measure.

4 Properties of SBR measures on Λ

Let λ^* be the T-invariant ergodic probability measure we constructed in Sect. 3. The purpose of this section is to prove

(1) the support of λ^* is the entire attractor Λ (Sect. 4.2);
(2) T does not admit any other SBR measures (Sect. 4.3);
(3) (T, λ^*) is Bernoulli (Sect. 4.4);

and to indicate how the existence of λ^* implies the corollary in the introduction (see Sect. 4.1.2).

4.1 *Some known facts from the general theory of nonuniformly hyperbolic systems*

The material in this subsection is not particular to the Hénon maps. We consider an arbitrary C^2 diffeomorphism $F: M \circlearrowleft$ of a finite dimensional Riemannian manifold preserving an ergodic Borel probability measure μ with compact support. It will be assumed throughout that (F, μ) has no zero Lyapunov exponents.

4.1.1 *Stable and unstable manifolds*

In [P1] Pesin proved the existence of stable and unstable manifolds in this nonuniform setting and studied their properties. (Pesin assumed that μ is equivalent to Lebesgue.) The case of arbitrary invariant measures is considered by Ruelle

[Ru2].) We recall here a couple of their results, giving precise statements only of what we will use and leaving out much more that is proved.

We write $T_xM = E^u(x) \oplus E^s(x)$ wherever it makes sense, and for $\delta > 0$, we let $B^u_\delta(x)$ and $B^s_\delta(x)$ denote the balls of radius δ about 0 in $E^u(x)$ and $E^s(x)$ respectively. Let $B_\delta(x) = B^u_\delta(x) \times B^s_\delta(x)$. It is sometimes convenient geometrically to introduce a new inner product $\langle \cdot , \cdot \rangle'_x$ on T_xM: under $\langle \cdot , \cdot \rangle'_x$, $E^u(x)$ and $E^s(x)$ are perpendicular; whereas restricted to $E^u(x)$ and $E^s(x)$, $\langle \cdot , \cdot \rangle'_x$ agrees with the given Riemannian metric. Let $\| \cdot \|'_x$ denote the corresponding norm. The following is true:

There exist Borel subset $\Gamma_1 \subset \Gamma_2 \subset \ldots$ of M with $\mu(\bigcup \Gamma_i) = 1$ and sequences of positive numbers δ_n, ε_n and θ_n, possibly $\downarrow 0$ as $n \uparrow \infty$, such that (1) and (2) below hold for every $x \in \Gamma_n$. (Think of the Γ_n's as uniformly hyperbolic sets that are not necessarily invariant, with the strength of hyperbolicity deteriorating as $n \to \infty$.)

(1) Let $\Gamma_n(x) = \{y \in \Gamma_n : d(x, y) < \varepsilon_n\}$. For $y \in \Gamma_n(x)$, let $W^u_x(y)$ denote the connected component of $(\exp_x^{-1} W^u(y)) \cap B_{\delta_n}(x)$ that contains $\exp_x^{-1} y$. Then for all $y \in \Gamma_n(x)$, $W^u_x(y)$ is the graph of a function $\varphi : B^u_{\delta_n}(x) \to B^s_{\delta_n}(x)$ with $\| D\varphi \|'_x \leqq \frac{1}{100}$. Moreover, as a C^1 embedded disk, $W^u_x(y)$ varies continuously with y. An analogous statement holds for $W^s_x(y)$.

(2) For $i = 1, 2$, let Σ_i be either $W^u_x(y)$ for some $y \in \Gamma_n(x)$ or a plane in $B_{\delta_n}(x)$ parallel to $E^u(x)$. Let $\Sigma'_1 = \Sigma_1 \cap \bigcup_{y \in \Gamma_n(x)} W^s_x(y)$, and let π be the map that takes $z \in \Sigma'_1$ to Σ_2 by sliding along $W^s_z(\cdot)$. Then for every Borel set $A \subset \Sigma'_1$.

$$\mathrm{Leb}(\pi A) \geqq \theta_n \mathrm{Leb}(A) .$$

Property (2) is the precise statement of what is called the "absolute continuity of the W^s-foliation". It is proved for "dissipative" systems in [PS].

4.1.2 Generic points for SBR measures

Let $F : (M, \mu) \circlearrowleft$ be as above. A point $x \in M$ is said to be *future-generic with respect to μ* or simply *μ-generic* if for every continuous function $\varphi : M \to \mathbb{R}$, $n^{-1} \sum_{i=0}^{n-1} \varphi \circ F^i(x) \to \int \varphi \, d\mu$ as $n \to \infty$. In particular, if μ is ergodic, then μ-a.e. x is μ-generic, and if x is μ-generic, then every $y \in W^s(x)$ is μ-generic as well.

The corollary stated in the introduction is an immediate consequence of the following general fact.

Proposition 3 *Let $F : (M, \mu) \circlearrowleft$ be as above. If μ is an ergodic* SBR *measure with no zero Lyapunov exponents, then there is a Borel subset $Y \subset M$ with positive Riemannian measure such that every $y \in Y$ is μ-generic.*

Proof. Since μ is an SBR measure, there is a piece of unstable manifold γ and a set $A \subset \gamma$ with $mA > 0$ such that every $x \in A$ is μ-generic. (As usual, m denotes the induced Riemannian measure on γ.) Using the absolute continuity of the W^s-foliation discussed in the last subsection, we see that

$$Y := \bigcup_{x \in A} W^s(x)$$

has the desired properties. □

4.1.3 *Ergodic properties of SBR measures*

The following is proved by Pesin [P2] when μ is equivalent to Riemannian volume and generalized by Ledrappier [L] to the situation where μ has absolutely continuous conditional measures on W^u:

Let μ be as above and (F, μ) be ergodic. Then there are pairwise disjoint Borel sets $A_1, \ldots, A_n \in M$, such that

(1) $F(A_i) = A_{i+1}$ for $i < n$, $F(A_n) = A_1$;

and

(2) $(F^n | A_i, \mu | A_i)$ is Bernoulli for all i.

4.2 *The support of* λ^*

Recall the sequence of choices leading to the selection of "good" parameters in [BC2]: First $\delta > 0$ is fixed. Then $a_0 < a_1 < 2$ are chosen with a_0 very near 2. Next, b is chosen sufficiently small depending on a_0 and a_1; and finally, for fixed b, "good" maps $T_{a,b}$ are selected by varying $a \in (a_0, a_1)$.

Consider first the 1-d situation. Assume that $\delta > 0$ is fixed, so that $\Delta^{\pm}$, the outermost intervals on the partition of $(-\delta, \delta)$ are determined. (See Subsect. 3.1 for definitions.) Let $\hat{x}_a$ be the unique fixed point of $f_a : [-1, 1] \circlearrowleft$. For $a = 2$ since f_2 is topologically conjugate to its piecewise linear model, $\bigcup_{n \geq 0} f_2^{-n} \hat{x}_2$ is dense in $[-1, 1]$. So $\exists N_0 \in \mathbb{Z}^+$, intervals $\tilde{\Delta}^{\pm} \subset \Delta^{\pm}$, and a neighborhood V of $\hat{x}_2$, such that $f_2^{N_0} \tilde{\Delta}^{\pm} \supset V$. Moreover, one can choose $N_0, \tilde{\Delta}^{\pm}$ and V such that for all $x \in \tilde{\Delta}^{\pm}, f_2^{N_0} x$ is "free". Chose a_0 such that for all $a \in (a_0, 2)$, this picture persists with the same $N_0, \tilde{\Delta}^{\pm}$ and V.

Returning now to our Hénon maps $T_{a,b}$, we assume b is sufficiently small that the fixed point $\hat{z}_{a,b}$ lies in $V \times \mathbb{R}$, and that if γ is a curve with small Hausdorff distance from either $\Delta^+ \times \{0\}$ or $\Delta^- \times \{0\}$, then $T_{a,b}^{N_0} \gamma$ contains a curve with small Hausdorff distance from $V \times \{0\}$. Moreover, it is clear from our derivative estimates is Sect. 2 that if, in addition, $\gamma \subset W$ and is a free segment, then $T^{N_0} \gamma$ is a $C^2(b)$ curve, which must then intersect $W^s_{\text{loc}}(\hat{z})$ with an angle $\geq \pi/4$ in at least one point.

From our construction of λ^* in Sect. 3, it follows that the support of λ^* contains a curve γ near $\Delta^+ \times \{0\}$ or $\Delta^- \times \{0\}$ that is the C^1-limit of free segments in W. This guarantees that γ intersects $W^s(\hat{z})$ transversally, which in turn implies that $W^u_{\text{loc}}(\hat{z}) \subset \operatorname{supp} \lambda^*$. Since $\Lambda = \bar{W}$, it follows that $\Lambda \subset \operatorname{supp} \lambda^*$. The reverse inclusion is obvious.

4.3 *Uniqueness of SBR measures*

Suppose that λ^* is not unique, so that there is another SBR measure μ on Λ. Without loss of generality we may assume that μ is ergodic. Fix $N_1 \in \mathbb{Z}^+$ with $\mu \Gamma_{N_1} > 0$. (See 4.1.1 for definition.) We claim that $\exists z_1 \in \Gamma_{N_1}$ such that in a neighborhood of z_1 we have the following picture: (For notational simplicity let us confuse $\Lambda \subset B_{\delta_{N_1}}(z_1)$ with $\exp_{z_1} \Lambda$ in the next few paragraphs.)

In $B_{\delta_{N_1}}(z_1)$, there is a "rectangle" two of whose sides, γ_1 and γ_2, are $W^u_{z_1}$-manifolds. Let us assume that $\gamma_1 = W^u_{z_1}(z_1)$. The other two "sides" of this "rectangle" are sets of the form $W^s_{z_1}(A_i)$, $i = 1, 2$, where A_i is a subset of γ_1 and $W^s_{z_1}(A_i) := \bigcup_{z\in A_i} W^s_{z_1}(z)$. The sets A_1 and A_2 are to have the following properties:

(1) $m(A_1), m(A_2) > 0$ $\quad(m = \text{Leb on } \gamma_1)$;
(2) every $z \in A_1 \cup A_2$ is *generic* with respect to μ.

Moreover, there is at least one point $w \in \Lambda$ in the "interior" of our "rectangle". (See Fig. 1.)

Let us assume this picture for now and complete the proof. We claim that there is a W^u-leaf γ with the following properties:

(3) m-a.e. $z \in \gamma$ is generic with respect to λ^*;
(4) γ connects the "outside" of our "rectangle" to the "inside".

To see that this claim is valid, recall the geometric properties of $\hat{\lambda}$ in Sect. 3 and the argument in Sect. 4.2 showing that $W^u_{\text{loc}}(\hat{z})$ is the C^1 limit of curves with property (3). Iterating forward, we see that every compact segment of W is the C^1 limit of curves with property (3). Now $w \in \Lambda = \bar{W}$ is "inside" our "rectangle", whereas we may assume that $\hat{z}$ is "outside". (3) and (4) should now be obvious.

Since γ clearly cannot intersect γ_1 or γ_2, it must intersect $W^s_{z_1}(A_1) \cup W^s_{z_2}(A_2)$. If we show that this intersection has positive m-measures in γ, then we will have proved that a positive m-measure set in γ is generic with respect to both λ^* and μ, forcing $\mu = \lambda^*$.

Two points need to be justified. First, the "rectangle" in our picture. Let $D = \{z \in \mathbb{R}^2 : z \text{ is generic with respect to } \mu\}$. Chose $z_1 \in \Gamma_{N_1}$ such that

(i) $\mu\{z \in \Gamma_{N_i} : d(z, z_1) < \varepsilon\} > 0$ for all $\varepsilon > 0$,
(ii) $m\{z \in W^u_{\text{loc}}(z_1) : z \in \Gamma_{N_1} \cap D$ and $d^u(z, z_1) < \varepsilon\} > 0$ for all $\varepsilon > 0$. (Here $d^u :=$ dist along $W^u(z_1)$.)

Let $\gamma_1 = W^u_{z_1}(z_1)$. Then A_1 and A_2 can be chosen as subsets of $\gamma_1 \cap \Gamma_{N_1} \cap D$. To complete our "rectangle" and to guarantee that some point of Λ lies inside, it suffices to argue that arbitrarily near γ_1, there are infinitely many $W^u_{z_1}$-curves. If this was not the case, then by (i) above we must have $\mu\gamma_1 > 0$. A standard argument in nonuniform hyperbolic theory then tells us that for some $n > 0$, $T^n\gamma \supset \gamma$ and $T^n|_{T^{-n}\gamma}$ is, in suitable coordinates, expanding. (See e.g. [K] for more details.) This implies that $\mu\{z_1\} > 0$, contradicting our assumption that μ is SBR.

The other point which perhaps needs some justification is our assertion that

$$m(\gamma \cap (W^s_{z_1}(A_1) \cup W^s_{z_1}(A_2))) > 0 .$$

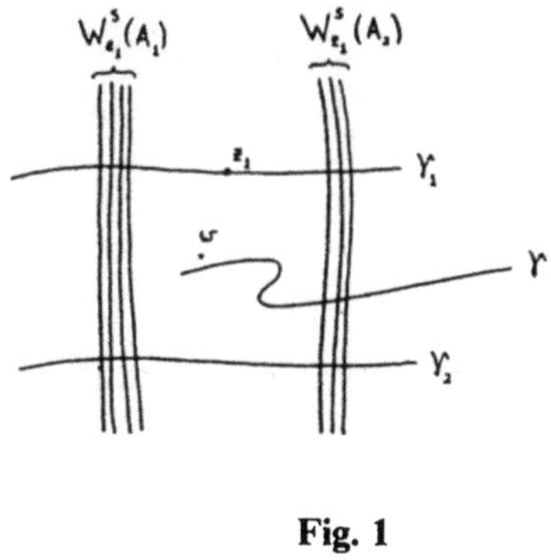

Fig. 1

We may assume that A_1 is a Cantor set, and is $= \bigcap_{n=1}^{\infty} E_n$ where each E_n is the disjoint union of a finite number of curve segments $\{E_{n,i}: 1 \leqq i \leqq i_n\}$ in γ_1. For each (n, i) let $y_{n,i}^l$ and $y_{n,i}^r$ be the left and right endpoint of $E_{n,i}$, and let $S_{n,i}$ be the strip in $B_{\delta_{N_1}}(z_1)$ bounded by $W_{z_1}^s(y_{n,i}^l)$, $W_{z_1}^s(y_{n,i}^r)$ and $\partial B_{\delta_{N_1}}(z_1)$. Then clearly $\bigcap_n \bigcup_i S_{n,i} = W_{z_1}^s(A_1)$. Let us assume that γ connects the two side of $S_{1,1}$ and let τ_1 be a subsegment of γ joining $W_{z_1}^s(y_{1,1}^l)$ to $W_{z_1}^s(y_{1,1}^r)$. For each (n, i) with $n > 1$, choose inductively $\tau_{n,i} \subset$ some $\tau_{n-1,j}$ such that $\tau_{n,i}$ joins $W_{z_1}^s(y_{n,i}^l)$ to $W_{z_1}^s(y_{n,i}^r)$. Then for every (n, i), we have

$$m(\tau_{n,i}) \geqq \text{min dist between } W_{z_1}^s(y_{n,i}^l) \text{ and } W_{z_1}^s(y_{n,i}^r)$$

$$\geqq c \cdot \text{dist}(L \cap W_{z_1}^s(y_{n,i}^l), L \cap W_{z_1}^s(y_{n,i}^r)),$$

where L is any line in $B_{\delta_{N_1}}(z_1)$ parallel to $E^u(z_1)$ and $c > 0$ is a constant depending only on the angle between $E^u(z_1)$ and $E^s(z_1)$. Using the absolute continuity of the W^s-foliation (see Sect. 4.1.1) we may then conclude that

$$m(\tau_{n,i}) \geqq c \cdot \theta_{N_1} \cdot m(A_1 \cap E_{n,i})$$

and hence

$$m(\gamma \cap W_{z_1}^s(A_1)) \geqq m\left(\bigcap_n \bigcup_i \tau_{n,i}\right) \geqq c \cdot \theta_{N_1} \cdot mA_1 .$$

4.4 *Bernoulliness of* (T, λ^*)

In light of Sect. 4.1.3 it suffices to show that (T^n, λ^*) is ergodic for all $n \geqq 1$. Let us fix $n_0 \in \mathbb{Z}^+$ and let μ_0 be one of the ergodic components of (T^{n_0}, λ^*) with absolutely continuous conditional measures on W^u-manifolds. From our construction of λ^* it follows that for some $k \in \mathbb{Z}^+$, $\mu_0(T^k X) > 0$, where X is the set in Sect. 3.4.3. Using again the fact that points in the same W^u-leaf belong in the same ergodic component of T^{n_0}, we see that $W_{\text{loc}}^u(\hat{z}) \subset \text{supp}\, \mu_0$, from which it follows that $\text{supp}\, \mu_0 = \Lambda$. Let $\bar{\mu}_0$ be μ_0 normalized. Repeating the argument in Sect. 4.3 with $(T^{n_0}, \bar{\mu}_0)$ in place of (T, λ^*), we see that $\bar{\mu}_0$ is the unique SBR measure for T^{n_0}. Thus (T^{n_0}, λ^*) has only one ergodic component and our proof is complete.

4.5 *Further properties of* (T, λ^*) *and an open problem*

We mention a couple of facts about (T, λ^*) that follow from general nonuniform hyperbolic theory. Let $\chi_1 > 0 > \chi_2$ be the Lyaponov exponents of (T, λ^*). We have the entropy formula

$$h_{\lambda^*}(T) = \chi_1 ,$$

where the quantity on the left is metric entropy. (See [P2, LS].) Using [Y1] and the entropy formula above, we obtain the following formula for the dimension of λ^*:

$$\text{HD}(\lambda^*) = h_{\lambda^*}(T) \cdot \left(\frac{1}{\chi_1} - \frac{1}{\chi_2}\right) = 1 - \frac{\chi_1}{\chi_2} = 1 + \frac{\chi_1}{\chi_1 - \log b} .$$

We finish by mentioning a problem the resolution of which would give a more complete geometric picture of these "good" Hénon maps. Let

$$B = \{z \in \mathbb{R}^2 : d(T^n, \Lambda) \to 0 \text{ as } n \to \infty\} .$$

The set B is called the *basin of attraction* of Λ and is known to contain an open neighborhood of Λ. Proposition 3 in Sect. 4.1 tells us that a positive Lebesgue measure subset of B consists of points that are λ^*-generic. That is to say, the statistics of these orbits are completely governed by the invariant measure λ^*. It would be nice to know if this property holds not just on a large set but almost everywhere in B.

References

[B] Bowen, R.: Equilibrium states and the ergodic theory of Anosov diffeomorphisms. (Lect. Notes Math., vol. 470) Berlin Heidelberg New York Springer: 1975

[BC1] Benedicks, M., Carleson, L.: On iterates of $x \mapsto 1 - ax^2$ on $(-1, 1)$. Ann. Math. **122**, 1–25 (1985)

[BC2] Benedicks, M., Carleson, L.: The dynamics of the Hénon map. Ann. Math. **133**, (1991) 73–169

[BM] Benedicks, M., Moeckel, R.: An attractor for the Hénon map. Zürich: ETH (Preprint)

[BY] Benedicks, M., Young, L.S.: Random perturbations and invariant measures for certain one-dimensional maps. Ergodic Theory Dyn. Syst. (to appear)

[L] Ledrappier, F.: Propriétés ergodiques des mesures de Sinai. Publ. Math., Inst. Hautes Etud. Sci. **59**, 163–188 (1984)

[LS] Ledrappier, F., Strelcyn, J.-M.: A proof of the estimation from below in Pesin's entropy formula. Ergodic Theory Dyn. Sys **2**, 203–219 (1982)

[LY] Ledrappier, F., Young, L.-S.: The metric entropy of diffemorphisms. Part I. Characterization of measures satisfying Pesin's formula. Part II. Relations between entropy, exponents and dimension. Ann. Math. **122**, 509–539, 540–574 (1985)

[MV] Mora, L., Viana, M.: Abundance of strange attractors. IMPA reprint (1991)

[K] Katok, A.: Lyapunov exponents, entropy and periodic orbits for diffeomorphisms. Publ. Math., Inst. Hautes Étud. Sci. **51**, 137–174 (1980)

[P1] Pesin, Ja.G.: Families of invariant manifolds corresponding to non-zero characteristic exponents. Math. USSR, Izv. **10**, 1261–1305 (1978)

[P2] Pesin, Ja.G.: Characteristic Lyaponov exponents and smooth ergodic theory. Russ. Math. Surv. **32.4**, 55–114 (1977)

[PS] Pugh, C., Shub, M.: Ergodic Attractors. Trans. Am. Math. Soc. **312**, 1–54 (1989)

[Ro] Rohlin, V.A.: On the fundamental ideas of measure theory. Transl., Am. Math. Soc. **10**, 1–52 (1962)

[Ru1] Ruelle, D.: A measure associated with Axiom A attractors. Am. J. Math. **98**, 619–654 (1976)

[Ru2] Ruelle, D.: Ergodic theory of differentiable dynamical systems. Publ. Math., Inst. Hautes Étud. Sci. **50**, 27–58 (1979)

[S1] Sinai, Ya. G.: Markov partitions and C-diffeomorphisms. Func. Anal. Appl. **2**, 64–89 (1968)

[S2] Sinai, Ya. G.: Gibbs measures in ergodic theory. Russ. Math. Surv. **27:4**, 21–69 (1972)

[Y1] Young, L.-S.: Dimension, entropy and Lyapunov exponents. Ergodic Theory Dyn. Syst. **2**, 163–188 (1982)

[Y2] Young, L.-S.: A Bowen-Ruelle measure for certain piecewise hyperbolic maps. Trans. Am. Math. Soc. **287**, 41–48 (1985)

SIAM J. NUMER. ANAL.
Vol. 36, No. 2, pp. 491–515

ON THE APPROXIMATION OF COMPLICATED DYNAMICAL BEHAVIOR*

MICHAEL DELLNITZ[†] AND OLIVER JUNGE[†]

Abstract. We present efficient techniques for the numerical approximation of complicated dynamical behavior. In particular, we develop numerical methods which allow us to approximate Sinai–Ruelle–Bowen (SRB)-measures as well as (almost) cyclic behavior of a dynamical system. The methods are based on an appropriate discretization of the Frobenius–Perron operator, and two essentially different mathematical concepts are used: our idea is to combine classical convergence results for finite dimensional approximations of compact operators with results from ergodic theory concerning the approximation of SRB-measures by invariant measures of stochastically perturbed systems. The efficiency of the methods is illustrated by several numerical examples.

Key words. computation of invariant measures, approximation of the Frobenius–Perron operator, computation of SRB-measures, almost invariant set, cyclic behavior

AMS subject classifications. 58F11, 65L60, 58F12

PII. S0036142996313002

1. Introduction.

The approximation of the behavior of a dynamical system is typically done by direct simulation. This method is particularly useful in the situation where a specific trajectory has to be approximated for a finite period of time. However, if one is interested in the long term behavior and if the underlying system exhibits complicated dynamics, then the information derived from a single trajectory is not always satisfying. Rather in this case it seems more appropriate to determine a statistical description of the dynamical behavior, and this information is encoded in an underlying (natural) invariant measure.

In this paper we describe a numerical method for the approximation of such invariant measures based on a discretization of the *Frobenius–Perron operator*. Using the fact that invariant measures are fixed points of this operator, we first approximate it by a Galerkin projection and then compute eigenvectors of the discretized operator corresponding to the eigenvalue 1. This allows us to identify regions in state space where trajectories are likely to be observed or, on the other hand, hardly observed. In addition to this information we show how to use other parts of the spectrum of the Frobenius–Perron operator to determine the dynamical behavior of the system. First we describe how to decompose an invariant set into components which are *cyclically permuted* by the dynamics. Second we develop techniques for the approximation of *almost invariant sets*, that is, regions in state space which are visited for a "long" period of time before the dynamical process leads to different areas. More generally the same techniques allow us to detect *almost cyclic behavior*, that is, to identify components of invariant sets which are "frequently" cyclically permuted by the dynamical process. Moreover, we can quantify the probability by which the cycle occurs depending on the absolute value of a corresponding eigenvalue of the Frobenius–Perron operator. Roughly speaking, we construct an approximation of the *essential dynamical behavior*,

*Received by the editors November 27, 1996; accepted for publication (in revised form) February 6, 1998; published electronically February 19, 1999. This research was partially supported by the Deutsche Forschungsgemeinschaft under grant De 448/5-2.

http://www.siam.org/journals/sinum/36-2/31300.html

[†]Mathematisches Institut, Universität Bayreuth, D-95440 Bayreuth (dellnitz@uni-bayreuth.de, http://www.uni-bayreuth.de/departments/math/~mdellnitz; junge@uni-bayreuth.de, http://www.uni-bayreuth.de/departments/math/~ojunge).

that is, the dynamics modulo complex (unpredictable) behavior which is due to the presence of chaos.

From a practical point of view the most important invariant measures are the so-called *SRB-measures*. The reason is that for these measures the spatial and temporal averages of observables are identical for a set of initial conditions which has positive Lebesgue measure. The introduction of the underlying concept goes back to Y. Sinai (see [27]), and the existence of these measures has been shown for Axiom A systems by D. Ruelle and R. Bowen [26, 3]. In this article we suggest a numerical method for the approximation of SRB-measures, and in this context their stochastic stability is particularly important: first we use this fact as an analytical tool in our main convergence result in section 4, and second it is of practical importance if we view the numerical approximation as a small random perturbation. Indeed, stochastic stability of SRB-measures is guaranteed for Axiom A systems [17, 18].

More precisely there are two essential mathematical ingredients which allow us to develop a numerical method for the approximation of SRB-measures. We use a result of Yu. Kifer on the convergence of invariant measures in stochastic perturbations of the underlying dynamical system to the SRB-measure (see [17]) and combine this with results on the convergence of eigenspaces of discretized compact operators (see, e.g., [25]). The same technique is used for the approximation of the subsets in state space which are (almost) cyclically permuted by the dynamical process. With respect to the approximation of SRB-measures a similar result has previously been obtained by F. Hunt (see [15]). However, our methods are quite different from the ones used in that work. In particular, the results stated here cover the important situation where the random perturbations have a probability distribution with local support. In fact, this is the relevant case having in mind that the round off error in the numerical approximation can be interpreted as such a local perturbation. Another approach for the computation of SRB-measures—avoiding the approximation of the Frobenius–Perron operator—has recently been suggested by Yu. Kifer [19].

As mentioned above, in addition to the approximation of SRB-measures the main development in this paper is a numerical method which allows us to identify (almost) cyclic behavior. To accomplish this we use a Galerkin method to discretize the Frobenius–Perron operator in such a way that the discretization has the same cyclic properties as the operator itself. More precisely, if the underlying dynamical system has a cycle of order r, then the rth roots of unity are eigenvalues of the Frobenius–Perron operator, and we will show that the corresponding eigenmeasures $\nu_0, \dots, \nu_{r-1}$ yield the desired information on the cyclic components: these components can be identified as supports of probability measures obtained by specific linear combinations of $\nu_0, \dots, \nu_{r-1}$. Our Galerkin approximation respects the cyclic behavior in the sense that the rth roots of unity are also eigenvalues of the discretized operator and that the corresponding eigenvectors converge to the eigenmeasures $\nu_0, \dots, \nu_{r-1}$ with increasing dimension of the approximating space. We will illustrate how to use these results to determine the subsets in state space which are almost cyclically permuted.

Finally, let us remark that the results on the approximation of the essential dynamical behavior obtained in this article can also be used to compute other statistical quantities such as the *entropy*, *dimensions* (depending on the particular invariant measure), or *Lyapunov exponents*. In fact, the efficient numerical use of invariant measures for the computation of Lyapunov exponents is currently under investigation.

An outline of the paper is as follows. In section 2 we begin with a brief review of the results on Markov processes which will be needed later on. The Frobenius–Perron operator is introduced in section 3. In that section we also describe the Galerkin projection that we use in the numerical approximation. In section 4 we use Kifer's result on small random perturbations of diffeomorphisms to prove convergence of the approximations to an SRB-measure in the hyperbolic case (Theorem 4.4). In the main section of this article, section 5, we show how to extract numerically the information on the (almost) cyclic components from the spectrum of the Frobenius–Perron operator. In particular, we present a method to identify regions in phase space where, on average, trajectories stay for a long period of time. Finally, in section 6 we illustrate by several examples the usefulness of our methods as tools in the numerical analysis of dynamical behavior.

2. Stochastic transition functions. For our main theoretical results we are using the concept of *small random perturbations* of dynamical systems. Since we assume that the typical reader is not familiar with this concept we begin by recalling some basic notions and results on Markov processes that will be needed later on. For a detailed introduction the reader is referred to [11].

Invariant measures. Our aim is to approximate the dynamical behavior of discrete dynamical systems of the form

$$x_{i+1} = f(x_i), \quad i = 0, 1, 2, \dots,$$

where $f : X \to X$ is a diffeomorphism on a compact subset $X \subset \mathbb{R}^n$. We denote by $\mathcal{B}$ the Borel σ-algebra on X and by m the Lebesgue measure on $\mathcal{B}$. Moreover, let $\mathcal{M}$ be the space of probability measures on $\mathcal{B}$. Recall that a measure $\mu \in \mathcal{M}$ is *invariant* if

$$\mu(B) = \mu(f^{-1}(B)) \quad \text{for all } B \in \mathcal{B}.$$

On the other hand, a set $A \in \mathcal{B}$ is *invariant* if

$$f(A) \subset A.$$

We now turn our attention to the more general stochastic framework.

DEFINITION 2.1. *A function $p : X \times \mathcal{B} \to [0,1]$ is a* stochastic transition function *if*

(i) $p(x, \cdot)$ *is a probability measure for every* $x \in X$,

(ii) $p(\cdot, A)$ *is Lebesgue-measurable for every* $A \in \mathcal{B}$.

Let δ_y denote the Dirac measure supported on the point $y \in X$. Then $p(x, A) = \delta_{h(x)}(A)$ is a stochastic transition function for every m-measurable function h. We will see below that the specific choice $h = f$ represents the deterministic situation in this more general setup.

We set $p^{(1)}(x, A) = p(x, A)$ and define recursively the *i-step stochastic transition function* $p^{(i)} : X \times \mathcal{B} \to \mathbb{R}$ by

$$p^{(i+1)}(x, A) = \int p^{(i)}(y, A)\, p(x, dy), \quad i = 1, 2, \dots,$$

where $p(x, dy)$ indicates that the integration is done with respect to the measure $p(x, \cdot)$. It is easy to see that $p^{(i)}$ is indeed a stochastic transition function. In particular, for the case where $p(x, A) = \delta_{f(x)}(A)$ we obtain for $i \geq 1$

$$p^{(i)}(x, A) = \delta_{f^i(x)}(A).$$

We now define the notion of an invariant measure in the stochastic setting.

DEFINITION 2.2. *Let p be a stochastic transition function. If $\mu \in \mathcal{M}$ satisfies*

$$\mu(A) = \int p(x, A)\, d\mu(x)$$

for all $A \in \mathcal{B}$, then μ is an invariant measure of p.

Remark 2.3.

(a) In the literature on Markov processes (e.g., [11]) an invariant measure is typically referred to as a *stationary absolute probability measure*. However, having the situation in mind that we consider stochastically perturbed dynamical systems we prefer the notion of an invariant measure (see also [17, 24]).

(b) If μ is an invariant measure of p, then it follows that

$$\mu(A) = \int p^{(i)}(x, A)\, d\mu(x)$$

for all $i = 1, 2, \ldots$.

The following example illustrates the previous remark that we recover the deterministic situation in the case where $p(x, \cdot) = \delta_{f(x)}$.

EXAMPLE 2.4. *Suppose that $p(x, \cdot) = \delta_{f(x)}$ and let μ be an invariant measure of p. Then we compute for $A \in \mathcal{B}$*

$$\mu(A) = \int p(x, A)\, d\mu(x) = \int \delta_{f(x)}(A)\, d\mu(x) = \int \chi_A(f(x))\, d\mu(x) = \mu(f^{-1}(A)),$$

where we denote by χ_A the characteristic function of A. Hence μ is an invariant measure for the diffeomorphism f.

DEFINITION 2.5. *A set $A \in \mathcal{B}$ is called a* consequent set *of x if $p^{(i)}(x, A) = 1$ for all $i \geq 1$. The set A is* invariant *if it is the consequent set of all of its points. Furthermore if $C \in \mathcal{B}$ is a set for which*

$$\lim_{i \to \infty} p^{(i)}(x, C) = 0 \quad \textit{for all } x \in X,$$

then C is called a transient set.

Considering our guiding example let $p(x, \cdot) = \delta_{f(x)}$ and let A be an invariant set for p. Then we have for $y \in A$

$$1 = p(y, A) = \delta_{f(y)}(A).$$

Hence $f(A) \subset A$ and A is an invariant set for the diffeomorphism f.

Absolutely continuous stochastic transition functions. Now we assume that for every $x \in X$ the probability measure $p(x, \cdot)$ is absolutely continuous with respect to the Lebesgue measure m. Hence we may write $p(x, \cdot)$ as

$$p(x, A) = \int_A k(x, y)\, dm(y) \quad \text{for all } A \in \mathcal{B},$$

with an appropriate *transition density function* $k : X \times X \to \mathbb{R}$. Obviously,

$$k(x, \cdot) \in L^1(X, m) \quad \text{and} \quad k(x, y) \geq 0.$$

In this case we also call the stochastic transition function p *absolutely continuous.* Note that

$$\int k(x,y)\,dm(y) = p(x,X) = 1 \quad \text{for all } x \in X.$$

We let $k^{(1)}(x,y) = k(x,y)$ and define the *i-step transition density function* as

$$k^{(i+1)}(x,y) = \int k(x,\xi)k^{(i)}(\xi,y)\,dm(\xi), \quad i = 1,2,\ldots.$$

With this definition we obtain for $A \in \mathcal{B}$

$$p^{(i)}(x,A) = \int_A k^{(i)}(x,y)\,dm(y),$$

that is, the i-step transition density function $k^{(i)}$ is the stochastic transition density function for $p^{(i)}$.

The following theorem provides a characterization of all invariant measures of a certain class of stochastic transition functions.

THEOREM 2.6. *Let p be an absolutely continuous stochastic transition function with density function k. Suppose that $k(x,y) \leq M$ for some positive constant $M > 0$ and all $x,y \in X$.*

Then X can be decomposed into finitely many disjoint invariant sets $E_1, E_2, \ldots, E_e$ and a transient set $F = X - \cup_{j=1}^{e} E_j$ such that for each E_j there is a unique probability measure $\pi_j \in \mathcal{M}$ with $\pi_j(E_j) = 1$ and

$$\lim_{N\to\infty} \frac{1}{N}\sum_{i=1}^{N} p^{(i)}(x,A) = \pi_j(A) \quad \text{for all } A \in \mathcal{B} \text{ and all } x \in E_j. \tag{2.1}$$

Furthermore the left-hand side in (2.1) *exists uniformly in x and defines for every fixed $x \in X$ an invariant measure. Finally, every invariant measure of p is a convex combination of the π_j's.*

A proof of this theorem can be found in [11].

Remark 2.7.

(a) The E_j's are called the *ergodic sets* of p.

(b) One can show that the invariant measures π_j are absolutely continuous with density functions $\kappa_j \in L^1$, that is, we have

$$\pi_j(A) = \int_A \kappa_j(y)\,dm(y), \quad j = 1,\ldots,e.$$

It follows that for every fixed $x \in X$ the limit in (2.1) is also absolutely continuous with a density function $\ell(x,\cdot) \in L^1$.

3. Approximation of the Frobenius–Perron operator. The main purpose of this section is to describe an appropriate Galerkin method for the approximation of the Frobenius–Perron operator. But first we introduce this operator and derive certain spectral properties.

The Frobenius–Perron operator.

DEFINITION 3.1. *Let p be a stochastic transition function. Then the* Frobenius–Perron operator *$P : \mathcal{M}_C \to \mathcal{M}_C$ is defined by*

$$P\mu(A) = \int p(x,A)\,d\mu(x),$$

where $\mathcal{M}_\mathbb{C}$ is the space of bounded complex valued measures on $\mathcal{B}$. If p is absolutely continuous with density function k, then we may define the Frobenius–Perron operator P on L^1 by

$$Pg(y) = \int k(x,y)g(x)\, dm(x) \quad \text{for all } g \in L^1.$$

Remark 3.2.

(a) By definition a measure $\mu \in \mathcal{M}$ is invariant if and only if it is a fixed point of P. In other words, invariant measures correspond to eigenmeasures of P for the eigenvalue 1.
Moreover, let $\lambda \in \mathbb{C}$ be an eigenvalue of P with corresponding eigenmeasure ν, that is, $P\nu = \lambda\nu$. Then in particular

$$\lambda\nu(X) = P\nu(X) = \int p(x,X)\, d\nu(x) = \nu(X)$$

since $p(x,X) = 1$ for all $x \in X$. It follows that $\nu(X) = 0$ if $\lambda \neq 1$.

(b) Observe that in the deterministic situation where $p(x,\cdot) = \delta_{f(x)}$ we obtain

$$P\mu(A) = \int p(x,A)\, d\mu(x) = \mu(f^{-1}(A))$$

(cf. Example 2.4). This is indeed the standard definition of the Frobenius–Perron operator in the deterministic setting (see, e.g., [21]).

(c) Note that in the case where p is absolutely continuous we have $P : L^1 \to L^1$ since for each $g \in L^1$

$$\begin{aligned}\int Pg(y)\, dm(y) &= \iint k(x,y)g(x)\, dm(x)\, dm(y)\\ &= \int g(x) \int k(x,y)\, dm(y)\, dm(x)\\ &= \int g(x)\, dm(x) < +\infty.\end{aligned}$$

Correspondingly, a nonnegative fixed point $g \in L^1$ of P with $\|g\|_1 = 1$ is the density of an invariant probability measure and, conversely, the density of every absolutely continuous invariant probability measure is a fixed point of P.

We are particularly interested in approximating cyclic dynamical behavior of the underlying dynamical system. In the stochastic setting this corresponds to the situation where there are disjoint compact subsets $X_j \subset X$, $j = 0, \dots, r-1$, such that

$$X = \bigcup_{j=0}^{r-1} X_j,$$

and for which the stochastic transition function p satisfies

$$p(x, X_{j+1 \bmod r}) = \begin{cases} 1 & \text{if } x \in X_j, \\ 0 & \text{otherwise.} \end{cases} \tag{3.1}$$

We now relate the cyclic dynamical behavior described by (3.1) to spectral properties of the corresponding Frobenius–Perron operator P.

PROPOSITION 3.3. *If the stochastic transition function p satisfies* (3.1), *then we have the following for the corresponding Frobenius–Perron operator* P:

(a) *The* r*th power* P^r *has an eigenvalue* 1 *of multiplicity at least* r. *Moreover, there are* r *corresponding invariant measures* $\mu_k \in \mathcal{M}$, $k = 0, 1, \dots, r-1$, *with support on* X_k, *that is,* $\operatorname{supp}(\mu_k) \subset X_k$. *These measures can be chosen to satisfy*

$$\mu_k = P^k \mu_0, \quad k = 0, 1, \dots, r-1.$$

(b) *The* r*th roots of unity* ω_r^k, $k = 0, 1, \dots, r-1$, *where* $\omega_r = e^{2\pi i/r}$, *are eigenvalues of* P.

Proof.

(a) Observing that for each $\mu \in \mathcal{M}_C$

$$P^j \mu(A) = \int p^{(j)}(x, A)\, d\mu(x),$$

where $p^{(j)}$ is the j-step stochastic transition function, the existence of the measures μ_k, $k = 0, 1, \dots, r-1$, follows from standard results on Markov processes (see, e.g., [11, Chap. V]). Moreover, these measures can be chosen so that

$$\mu_{k+1 \bmod r} = P\mu_k, \quad k = 0, 1, \dots, r-1.$$

Simply note that if μ_k is invariant for P^r, then

$$P^r(P\mu_k) = P(P^r\mu_k) = P\mu_k$$

and hence $P\mu_k$ is an invariant measure with support on X_{k+1}.

(b) Let μ be one of the probability measures which exist by part (a). We show that for $k \in \{0, 1, \dots, r-1\}$

$$\nu_k = \sum_{j=0}^{r-1} \omega_r^{-kj} P^j \mu \in \mathcal{M}_C \tag{3.2}$$

is an eigenmeasure of P for the eigenvalue ω_r^k. Indeed, using the fact that $P^r\mu = \mu$ we compute

$$\begin{aligned} P\nu_k &= P\mu + \omega_r^{-k} P^2\mu + \cdots + \omega_r^{-k(r-2)} P^{r-1}\mu + \omega_r^{-k(r-1)}\mu \\ &= \omega_r^k \left(\mu + \omega_r^{-k} P\mu + \cdots + \omega_r^{-k(r-1)} P^{r-1}\mu\right) \\ &= \omega_r^k \nu_k. \end{aligned}$$

Finally, $\nu_k \neq 0$ since $\nu_k(X_j) \neq 0$ for $j = 0, 1, \dots, r-1$. □

Approximation by a Galerkin method. We begin with the following observation which immediately follows from standard results on integral operators (see, e.g., [28, p. 277]).

LEMMA 3.4. *Suppose that the transition density function k satisfies*

$$\iint |k(x, y)|^2\, dm(x) dm(y) < \infty. \tag{3.3}$$

Then the Frobenius–Perron operator $P : L^2 \to L^2$ *is compact.*

From now on we consider the case where P is given by a dynamical process with a transition density function k satisfying the condition (3.3). The aim is to use a Galerkin method for the approximation of such a Frobenius–Perron operator together with its spectrum. More precisely, let V_d, $d \geq 1$, be a sequence of d-dimensional subspaces of L^2 and let $Q_d : L^2 \to V_d$ be a projection such that Q_d converges pointwise to the identity on L^2. If we define the approximating operators by $P_d = Q_d P$, then we have

$$\|P_d - P\|_2 \to 0 \quad \text{as } d \to \infty.$$

Denote by $\sigma(P)$ and $\rho(P)$ the spectrum and resolvent set of P, respectively, and by $R_z = (zI - P)^{-1}$, $z \in \rho(P)$, the resolvent operator. Let $\lambda \neq 0 \in \sigma(P)$ be a nonzero eigenvalue of P and let $\Gamma \subset \mathbb{C}$ be a circle in $\rho(P)$ with center λ such that no other point of $\sigma(P)$ is inside Γ. Then the operator defined by

$$E = E(\lambda) = \frac{1}{2\pi i} \int_\Gamma R_z(P)\, dz$$

is a projection onto the space of generalized eigenvectors associated with λ and P. The following theorem—which is a specific application of the main result of [25] on compact operators—allows us to approximate eigenvectors of P by eigenvectors of P_d.

THEOREM 3.5 (see [25]). *Let λ_d be an eigenvalue of P_d such that $\lambda_d \to \lambda$ for $d \to \infty$, and let g_d be a corresponding eigenvector of unit length. Then there is a vector $h_d \in R(E)$ and a constant $C > 0$ such that $(\lambda I - P)h_d = 0$ and*

$$\|h_d - g_d\|_2 \leq C \|(P - P_d)|_{R(E)}\|_2,$$

where $R(E)$ denotes the range of E.

Next we use Theorem 3.5 to approximate the eigenvalues of P which are lying on the unit circle. For this we construct a Galerkin projection which possesses the same cyclic behavior in the approximation. Suppose that (3.1) holds and let $\{\varphi_i^j\}$, $j = 0, 1, \ldots, r-1$, $i = 1, 2, \ldots, d_j$, be a basis of V_d with the following properties:

$$\text{(3.4)} \qquad \begin{aligned} &\text{(i)} \quad \operatorname{supp}(\varphi_i^j) \subset X_j \quad (j = 0, 1, \ldots, r-1, \quad i = 1, 2, \ldots, d_j), \\ &\text{(ii)} \quad \sum_{i=1}^{d_j} \varphi_i^j(x) = 1 \quad \text{for all } x \in X_j,\ j = 0, 1, \ldots, r-1. \end{aligned}$$

Remark 3.6.

(a) In section 5 we will see how to generate a basis satisfying (3.4). In that case, V_d consists of functions which are locally constant.

(b) Observe that by construction

$$\sum_{i,j} \varphi_i^j(x) = 1 \quad \text{for all } x \in X.$$

The Galerkin projection $Q_d g$ of $g \in L^2$ is defined by

$$(Q_d g, \varphi_i^j) = (g, \varphi_i^j) \quad \text{for all } i, j,$$

where $(\cdot, \cdot)$ is the usual inner product in L^2. The following result is a generalization of Lemma 8 in [10], where just the fixed point of P is considered. Recall that $\omega_r = e^{2\pi i/r}$.

PROPOSITION 3.7. *Suppose that the Galerkin projection satisfies* (3.4). *Then the approximating operators $P_d = Q_d P$ possess the eigenvalues ω_r^k, $k = 0, 1, \ldots, r-1$.*

Proof. Suppose that λ is an eigenvalue of P_d with corresponding eigenvector $\psi(x) = \sum_{i,j} \beta_i^j \varphi_i^j(x)$. Then $P_d\psi = \lambda\psi$ is equivalent to

$$\sum_{i_1,k_1} \beta_{i_1}^{k_1}(P\varphi_{i_1}^{k_1}, \varphi_{i_2}^{k_2}) = \lambda \sum_{i_1,k_1} \beta_{i_1}^{k_1}(\varphi_{i_1}^{k_1}, \varphi_{i_2}^{k_2}) \quad \text{for all } i_2, k_2.$$

Introducing the coefficient vector $\beta = (\beta_i^j)$ we may write this equation in matrix form as

$$M_1\beta - \lambda M_2\beta = 0, \tag{3.5}$$

where both M_1 and M_2 have nonnegative entries. Moreover, noting that

$$\int Q_d g\, dm = \int P_d g\, dm = \int g\, dm \quad \text{for every } g \in L^2,$$

and using the fact that $\sum_{i,j} \varphi_i^j(x) = 1$ we can proceed in the same way as in [10, Lemma 8], to see that $(1, 1, \ldots, 1)$ is a left eigenvector with eigenvalue 1 for the generalized eigenvalue problem (3.5). The fact that M_2 is invertible—since $\{\varphi_i^j\}$ is a basis of V_d—now implies that there is an eigenvector α with

$$M_2^{-1}M_1\alpha = \alpha.$$

We claim that $M_2^{-1}M_1$ has a cyclic structure so that $(M_2^{-1}M_1)^r$ is of block diagonal form where the blocks have the dimensions d_j, $j = 0, 1, \ldots, r-1$. Decomposing α with respect to this block structure we may proceed as in the proof of part (b) of Proposition 3.3 to show that ω_r^k, $k = 0, 1, \ldots, r-1$, are eigenvalues of (3.5) and hence of P_d as desired.

We now prove the claim. Since the matrix M_2 in (3.5) already has the desired block diagonal form (by (i) in (3.4)), it remains to show that the basis functions are cyclically permuted by P_d respecting the block structure of M_2. More precisely, we will show that $(P\varphi_{i_1}^{k_1}, \varphi_{i_2}^{k_2}) = 0$ if $k_2 \neq (k_1 + 1) \bmod r$.

By (3.1) we have

$$\int_{X_{j+1 \bmod r}} k(x,y)\, dm(y) = \begin{cases} 1 & \text{if } x \in X_j \\ 0 & \text{otherwise.} \end{cases}$$

It follows that

$$\int_{X_{k_2}} \int_{X_{k_1}} k(x,y)\, dm(x)dm(y) = \int_{X_{k_1}} \int_{X_{k_2}} k(x,y)\, dm(y)dm(x) = 0$$

if $k_2 \neq (k_1 + 1) \bmod r$, and therefore

$$(P\varphi_{i_1}^{k_1}, \varphi_{i_2}^{k_2}) = \int_{X_{k_2}} \int_{X_{k_1}} k(x,y)\varphi_{i_1}^{k_1}(x)\varphi_{i_2}^{k_2}(y)\, dm(x)dm(y) = 0$$

if $k_2 \neq (k_1 + 1) \bmod r$ as desired. □

Now we may combine Theorem 3.5 and Proposition 3.7 to obtain a convergence result for eigenvectors corresponding to eigenvalues of P of modulus one.

COROLLARY 3.8. *Suppose that P and its approximation P_d satisfy the hypotheses stated above. Then each simple eigenvalue $e^{2\pi ik/r}$ of P on the unit circle is an eigenvalue of P_d and there are corresponding eigenvectors g_d of P_d converging to an eigenfunction h of P. More precisely, there is a constant $C > 0$ such that for all $d \geq 1$*

$$\|h - g_d\|_2 \leq C\|P_d - P\|_2.$$

4. The computation of SRB-measures.

SRB-measures. Let us briefly recall the notion of an SRB-measure. In the existing literature several different definitions can be found which are all equivalent in the case where the underlying dynamical behavior is Axiom A. This is precisely the situation we will consider, and hence we can, without loss of generality, work with just one of them.

DEFINITION 4.1. *An ergodic measure μ is an* SRB-measure *if there exists a subset $U \subset X$ with $m(U) > 0$ and such that for each continuous function ψ*

$$\lim_{N\to\infty} \frac{1}{N} \sum_{j=0}^{N-1} \psi(f^j(x)) = \int \psi \, d\mu \tag{4.1}$$

for all $x \in U$.

Remark 4.2.

(a) Recall that (4.1) always holds for μ-a.e. $x \in X$ by the Birkhoff ergodic theorem. The crucial difference for an SRB-measure is that the temporal average equals the spatial average for a set of initial points $x \in X$ which has positive Lebesgue-measure. This is the reason why this measure is also referred to as the *natural* or the *physically relevant* invariant measure.

(b) The concept of SRB-measures in the context of Anosov systems had been introduced by Y. Sinai in the 1960s (e.g., [27]). Later the existence of SRB-measures was shown for Axiom A systems by R. Bowen and D. Ruelle (see [26, 3]). More recently, M. Benedicks and L.-S. Young have shown that the Hénon-map has an SRB-measure for a "large" set of parameter values [1]. However, it is still one of the major problems in ergodic theory to establish the existence of SRB-measures for a more general class of dynamical systems.

Small random perturbations. We specify concretely the stochastic transition function p underlying the numerical realization. Recall that the purpose is to approximate the Frobenius–Perron operator of a *deterministic* dynamical system represented by a diffeomorphism f. Hence the *stochastic* system that we consider should be a small perturbation of this original deterministic system.

For $\varepsilon > 0$ we set

$$k_\varepsilon(x,y) = \frac{1}{\varepsilon^n m(B)} \chi_B\left(\frac{1}{\varepsilon}(y-x)\right), \quad x, y \in X. \tag{4.2}$$

Here $B = B_0(1)$ denotes the open ball in $\mathbb{R}^n$ of radius one and χ_B is the characteristic function of B. Obviously $k_\varepsilon(f(x), y)$ is a transition density function and we may define a stochastic transition function p_ε by

$$p_\varepsilon(x, A) = \int_A k_\varepsilon(f(x), y) \, dm(y). \tag{4.3}$$

Remark 4.3. Note that $p_\varepsilon(x, \cdot) \to \delta_{f(x)}$ for $\varepsilon \to 0$ uniformly in x in a weak*-sense. Hence the Markov process defined by any initial probability measure μ and the transition function p_ε is a *small random perturbation* of the deterministic system f in the sense of Yu. Kifer [17].

Observe that we can apply the results from section 3 since

$$\iint |k_\varepsilon(f(x), y)|^2 \, dm(x) dm(y) \le \left(\frac{m(X)}{\varepsilon^n m(B)}\right)^2 < \infty,$$

and therefore the Frobenius–Perron operator $P_\varepsilon : L^2 \to L^2$ is compact (see Lemma 3.4).

Approximation of SRB-measures. We now combine Corollary 3.8 with a result of Yu. Kifer [17] to show that the approximations of the invariant measures converge to an SRB-measure with decreasing magnitude of the random perturbations.

Let us be more precise. Suppose that the diffeomorphism f possesses a hyperbolic attractor Λ with an SRB-measure μ_{SRB}, and let p_ε be a small random perturbation of f. Then, under certain hypotheses on p_ε, it is shown in [17] that the invariant measures of p_ε converge in a weak*-sense to μ_{SRB} as $\varepsilon \to 0$. On the other hand we can approximate the relevant eigenmeasures of P_ε by Corollary 3.8 and this leads to the desired result.

THEOREM 4.4. *Suppose that the diffeomorphism f has a hyperbolic attractor Λ, and that there exists an open set $U_\Lambda \supset \Lambda$ such that*

$$k_\varepsilon(x, y) = 0 \quad \text{if } x \in \overline{f(U_\Lambda)} \text{ and } y \notin U_\Lambda.$$

Then the transition function p_ε in (4.3) has a unique invariant measure π_ε with support on Λ, and the approximating measures

$$\mu_d^\varepsilon(A) = \int_A g_d^\varepsilon \, dm$$

converge in a weak-sense to the SRB-measure μ_{SRB} of f as $\varepsilon \to 0$ and $d \to \infty$,*

$$\lim_{\varepsilon \to 0} \lim_{d \to \infty} \mu_d^\varepsilon = \mu_{SRB}. \tag{4.4}$$

Proof. It is straightforward to check that the conditions of Theorem 1 in [17] are satisfied for the densities

$$q_x^\varepsilon(y) = k_\varepsilon(x, y),$$

provided $\varepsilon < 1$. Hence—denoting the unique invariant measure of the transition function p_ε with support in $\overline{U_\Lambda}$ by π_ε—this theorem implies that

$$\pi_\varepsilon \overset{\text{weak*}}{\longrightarrow} \mu_{SRB} \quad \text{for } \varepsilon \to 0. \tag{4.5}$$

By Remark 2.7 we know that π_ε is absolutely continuous, and we denote its density function by κ_ε. Then Corollary 3.8 guarantees that the fixed points g_d^ε of P_d^ε converge to κ_ε as $d \to \infty$. Therefore

$$\begin{aligned} \left| \int h \, d\mu_d^\varepsilon - \int h \, d\pi_\varepsilon \right| &= \left| \int h(g_d^\varepsilon - \kappa_\varepsilon) \, dm \right| \\ &\le \|h\|_2 \|g_d^\varepsilon - \kappa_\varepsilon\|_2 \to 0 \end{aligned}$$

as $d \to \infty$ for every $h \in L^2$ and, in particular,

$$\mu_d^\varepsilon \overset{\text{weak*}}{\longrightarrow} \pi_\varepsilon \quad \text{as } d \to \infty.$$

Combining this with (4.5) leads to (4.4), as desired. □

5. Extracting dynamical behavior. In the previous section we have seen that we can approximate the physically relevant invariant measure—the SRB-measure—of our original deterministic system by the computation of the invariant measure of a randomly perturbed system. However, dynamically the nonstationary behavior is also interesting, and we will now describe how to detect numerically components in state space which are (almost) cyclically permuted.

Extraction of cyclic behavior. Suppose that the stochastic transition function of the randomly perturbed dynamical system satisfies the cycle condition (3.1). Then the purpose is to identify the components X_j. By Proposition 3.7 we know that the approximating operator P_d^ε has the eigenvalues ω_r^k, $k = 0, 1, \ldots, r-1$, and we now show that the cyclic components can be approximated by certain linear combinations of the corresponding eigenvectors.

It is instructive to consider the simplest case where $r = 2$ before we turn our attention to the general situation. Suppose that we have two components X_0 and X_1 which are cyclically permuted by our process. Then the aim is to find approximations of eigenmeasures μ_0 and $\mu_1 = P_\varepsilon\mu_0$ of P_ε^2 with support on X_0 and X_1, respectively; see Proposition 3.3. By the same proposition we know that $\omega^0 = 1$ and $\omega^1 = -1$ are eigenvalues of P_ε. Let ν_0 and ν_1 be corresponding (real) eigenmeasures. Then, by (3.2), there are $\alpha_0, \alpha_1 \in \mathbb{R}$ such that

$$\nu_0 = \alpha_0(\mu_0 + P_\varepsilon\mu_0) \quad \text{and} \quad \nu_1 = \alpha_1(\mu_0 - P_\varepsilon\mu_0).$$

Rescaling ν_0 and ν_1 so that $\nu_0(X_0) = \nu_1(X_0) = 1$ we can compute μ_0 and μ_1 by

$$\mu_0 = \frac{1}{2}(\nu_0 + \nu_1) \quad \text{and} \quad \mu_1 = \frac{1}{2}(\nu_0 - \nu_1).$$

This process also shows how to construct eigenvectors of the Galerkin approximation for which linear combinations are appropriate approximations of the probability measures μ_0 and μ_1.

We now consider the general case. For $\ell = 0, 1, \ldots, r-1$ denote by $\mu_\ell = P_\varepsilon^\ell\mu_0$ the invariant measure of P_ε^r with support on X_ℓ (see Proposition 3.3).

LEMMA 5.1. *For $s \in \{0, 1, \ldots, r-1\}$ let*

$$\nu_k^s = \sum_{j=0}^{r-1} \omega_r^{-kj} P_\varepsilon^j \mu_s \tag{5.1}$$

be a specific choice for the eigenmeasures of P_ε corresponding to the eigenvalues ω_r^k, $k = 0, 1, \ldots, r-1$ (see (3.2)). Then

$$\frac{1}{r}\sum_{k=0}^{r-1} \omega_r^{\ell k}\nu_k^s = \mu_{\ell+s \bmod r}.$$

Proof. We compute

$$\frac{1}{r}\sum_{k=0}^{r-1} \omega_r^{\ell k}\nu_k^s = \sum_{j=0}^{r-1}\left(\frac{1}{r}\sum_{k=0}^{r-1}\omega_r^{(\ell-j)k}\right) P_\varepsilon^j\mu_s = P_\varepsilon^\ell\mu_s = \mu_{\ell+s \bmod r}.$$

Here we have used the identity

$$\frac{1}{r}\sum_{k=0}^{r-1}\omega_r^{(\ell-j)k} = \delta_{\ell j},$$

where $\delta_{\ell j}$ is the Kronecker symbol. □

The previous lemma indicates how to approximate the cyclic components of X: we have to find eigenvectors $v_0^s, \ldots, v_{r-1}^s$ of the matrix $M_2^{-1}M_1$ (see (3.5)) which are approximations of the eigenmeasures ν_k^s in (5.1) for an $s \in \{0, 1, \ldots, r-1\}$. Then we can compute

$$u_{\ell+s \bmod r} = \frac{1}{r}\sum_{k=0}^{r-1} \omega_r^{\ell k} v_k^s$$

for $\ell = 0, 1, \ldots, r-1$, and the positive components of u_j provide the desired information about the support of μ_j on X_j $(j = 0, 1, \ldots, r-1)$.

Hence it remains to describe how to construct eigenvectors $v_0^s, \ldots, v_{r-1}^s$ approximating the ν_k^s in (5.1) for an $s \in \{0, 1, \ldots, r-1\}$. We do this for the case where the eigenvalues ω_r^k are simple—the case of several coexisting cycles will be treated in the following section.

Suppose that we have a set of eigenmeasures ρ_k corresponding to the eigenvalues ω_r^k, $k = 0, 1, \ldots, r-1$. Since the eigenvalues are simple we know that for each $s \in \{0, 1, \ldots, r-1\}$ there is a constant $\alpha_k^s \in \mathbb{C}$ such that ρ_k can be written as

$$\rho_k = \alpha_k^s \nu_k^s.$$

Hence the task is to rescale ρ_k so that $\alpha_k^s = 1$ for all k. By (3.2) it is easy to see that for each $s \in \{0, 1, \ldots, r-1\}$

$$\rho_k(X_s) \neq 0.$$

We choose a particular s and rescale the ρ_k's by (complex) factors so that

$$\rho_k(X_s) = 1 \quad \text{for all } k = 0, 1, \ldots, r-1.$$

With this choice it follows that $\rho_k = \nu_k^s$.

In the realization of the approximation of the ν_k^s's we proceed with the eigenvectors of $M_2^{-1}M_1$ in an analogous way: by our choice of the Galerkin approximation we just need to find an index such that the corresponding components of the eigenvectors do not vanish. The scaling as described above can then be done in a similar way: find complex multiples of the eigenvectors so that they possess (real) positive components which add up to one for each vector. We will illustrate the method by examples in section 6.

Identification of several coexisting cycles or invariant sets. In applications it may occur that there exist several different cycles in the dynamical system under consideration. We now show how to identify these sets numerically. Replacing P_ε by an appropriate power P_ε^r if necessary we can, without loss of generality, restrict our attention to the case where there are different invariant sets.

Again we begin with the simplest case and assume that P_ε has two linearly independent invariant probability measures ν_1 and ν_2. Then, by Theorem 2.6, there are constants α_j^i $(i, j = 1, 2)$ with

$$\alpha_1^1\nu_1 + \alpha_2^1\nu_2 = \pi_1 \quad \text{and} \quad \alpha_1^2\nu_1 + \alpha_2^2\nu_2 = \pi_2,$$

where π_1 and π_2 are probability measures with $\pi_i(E_i) = 1$ for invariant sets E_1 and E_2. Let $B_i \subset E_i$ be subsets with $\nu_i(B_i) \neq 0$. Then the coefficients α_j^i can be found as the solutions of the equations

$$\alpha_1^1\nu_1(B_2) + \alpha_2^1\nu_2(B_2) = 0,$$
$$\alpha_1^2\nu_1(B_1) + \alpha_2^2\nu_2(B_1) = 0,$$

with the additional requirement that $\pi_i = \alpha_1^i \nu_1 + \alpha_2^i \nu_2$ are probability measures. Indeed, let β_j^i be constants such that

$$\alpha_1^1 \nu_1 + \alpha_2^1 \nu_2 = \beta_1^1 \pi_1 + \beta_2^1 \pi_2,$$
$$\alpha_1^2 \nu_1 + \alpha_2^2 \nu_2 = \beta_1^2 \pi_1 + \beta_2^2 \pi_2.$$

Then, in particular,

$$0 = \alpha_1^1 \nu_1(B_2) + \alpha_2^1 \nu_2(B_2) = \beta_1^1 \pi_1(B_2) + \beta_2^1 \pi_2(B_2) = \beta_2^1 \pi_2(B_2),$$
$$0 = \alpha_1^2 \nu_1(B_1) + \alpha_2^2 \nu_2(B_1) = \beta_1^2 \pi_1(B_1) + \beta_2^2 \pi_2(B_1) = \beta_1^2 \pi_1(B_1).$$

Since $\nu_i(B_i) \neq 0$, it follows that $\pi_i(B_i) \neq 0$ and therefore $\beta_2^1 = \beta_1^2 = 0$ as desired.

We now generalize this observation. Suppose that the eigenspace of the Frobenius–Perron operator corresponding to the eigenvalue 1 is e-dimensional. Then, by Theorem 2.6, there are e distinct invariant sets $E_1, E_2, \ldots, E_e$ and invariant measures $\pi_1, \pi_2, \ldots, \pi_e$ such that $\pi_j(E_j) = 1$.

LEMMA 5.2. *Let ν_j, $j = 1, \ldots, e$, be invariant measures spanning the e-dimensional eigenspace of the Frobenius–Perron operator corresponding to the eigenvalue 1. Let $B_i \subset E_{k_i}$ be subsets such that $\nu_i(B_i) \neq 0$, $i = 1, 2, \ldots, e$. We have that*

(a) *the matrix $M = (\nu_j(B_i))_{i,j=1,\ldots,e}$ has full rank if and only if*

$$\{k_1, k_2, \ldots, k_e\} = \{1, 2, \ldots, e\}; \tag{5.2}$$

(b) *if $\text{rank}(M) = e$, then the invariant measures π_j, $j = 1, \ldots, e$, are given by*

$$\pi_{k_\ell} = \sum_{j=1}^{e} \alpha_j^\ell \nu_j,$$

where $\alpha^\ell = (\alpha_j^\ell)$ is the (rescaled) null vector of $M_\ell = (\nu_j(B_i))_{i,j=1,\ldots,e,\ i\neq\ell}$.

Proof. Suppose that (5.2) holds. Let $\alpha \in \mathbb{R}^e$ be an element of the kernel of the matrix $M = (\nu_j(B_i))$, that is,

$$\sum_{j=1}^{e} \alpha_j \nu_j(B_i) = 0, \quad i = 1, 2, \ldots, e.$$

Using the fact that there are constants β_j, $j = 1, \ldots, e$, such that

$$\sum_{j=1}^{e} \alpha_j \nu_j = \sum_{j=1}^{e} \beta_j \pi_j,$$

we obtain

$$0 = \sum_{j=1}^{e} \beta_j \pi_j(B_i) = \beta_{k_i} \pi_{k_i}(B_i), \quad i = 1, 2, \ldots, e.$$

Since $0 \neq \nu_i(B_i) = \gamma_i \pi_{k_i}(B_i)$ for appropriate constants γ_i, it follows that $\beta_i = 0$ for $i = 1, 2, \ldots, e$ (here we have used (5.2)). Hence $\sum_{j=1}^{e} \alpha_j \nu_j = 0$, and since the ν_j's span the e-dimensional eigenspace of the Frobenius–Perron operator corresponding to the eigenvalue 1, we may conclude that $\alpha = 0$.

To complete the proof of part (a) it remains to show that $M = (\nu_j(B_i))$ is singular if (5.2) is not satisfied, that is, if there is a $k \in \{1, 2, \ldots, e\}$ which is not in $\{k_1, k_2, \ldots, k_e\}$. Writing π_k as

$$\pi_k = \sum_{j=1}^{e} \alpha_j \nu_j$$

and using $B_i \not\subset E_k$ we obtain for $i = 1, \ldots, e$

$$0 = \pi_k(B_i) = \sum_{j=1}^{e} \alpha_j \nu_j(B_i).$$

Hence $\alpha = (\alpha_j)$ is a nontrivial null vector of M. The statement in part (b) is an immediate consequence. □

For the numerical identification of the sets $E_1, \ldots, E_e$ we choose nonvanishing components of the e approximations $v^1, \ldots, v^e$ of eigenmeasures $\nu_1, \ldots, \nu_e$ in such a way that the matrix $M = (v_i^j)_{i,j=1,\ldots,e}$ is nonsingular. Then we identify the distinct invariant components as the support of the eigenmeasures approximated by the scaled null vectors of the matrices $M_\ell = (v_i^j)_{i,j=1,\ldots,e,\ i\neq\ell}$.

Extraction of almost cyclic behavior. We distinguish two different scenarios by which an *almost cyclic* behavior may occur in a dynamical system:

(i) The first scenario we have in mind is that cyclic components $X_0, X_1, \ldots, X_{r-1}$ merge while a control parameter is varied in the system. If this has happened, then the cyclic behavior can frequently be observed although it is strictly no longer present.

(ii) Second, it may happen that two *different* cycles merge while a control parameter is varied. For instance, if two invariant sets "collide," then immediately after the collision there are still two subsets in state space which are *almost invariant.*

In section 6 we will illustrate both scenarios by numerical examples.

We know that the rth roots of unity are eigenvalues of P_ε if there are r cyclic components. If these components merge, then these eigenvalues leave the unit circle. The main purpose of this section is to relate the modulus of these eigenvalues to the probability that the cyclic behavior is still observed. As in the previous subsection, we may just consider *almost invariant* sets by replacing P_ε by P_ε^r if necessary. In this case a cluster of eigenvalues moves away from one along the real line while several (precisely) invariant sets disappear.

DEFINITION 5.3. *A subset $A \subset X$ is δ-*almost invariant with respect to $\rho \in \mathcal{M}$ *if $\rho(A) \neq 0$ and*

$$\int_A p_\varepsilon(x, A)\, d\rho(x) = \delta\rho(A).$$

Remark 5.4.

(a) Using the definition of p_ε we compute for a subset $A \subset X$

$$p_\varepsilon(x, A) = \frac{m(A \cap B_{f(x)}(\varepsilon))}{m(B_0(\varepsilon))}.$$

Hence

$$\delta = \frac{1}{\rho(A)} \int_A \frac{m(A \cap B_{f(x)}(\varepsilon))}{m(B_0(\varepsilon))}\, d\rho(x).$$

(b) We have seen that $p_\varepsilon(x,\cdot) \to \delta_{f(x)}$ for $\varepsilon \to 0$. Thus, we obtain in the deterministic limit

$$\int_A p_0(x, A)\, d\rho(x) = \int_A \delta_{f(x)}(A)\, d\rho(x) = \rho(f^{-1}(A) \cap A).$$

Therefore in this case δ is the relative ρ-measure of the subset of points in A which are mapped into A.

According to the classification of the occurrence of almost cyclic behavior given at the beginning of this section we identify almost cyclic behavior in the numerical realization as follows:

(i) If cyclic components $X_0, X_1, \ldots, X_{r-1}$ merge while a control parameter is varied in the dynamical system, then we use the same linear combinations as in the unperturbed case to identify the components of the "almost-cycle." The overlap of the different components indicates the subset of points which no longer follow the cyclic behavior.

(ii) If two invariant sets merge, then we use the results obtained for the identification of coexisting invariant sets. However, observe that in the perturbed case we can no longer expect that an approximation of the matrix M in Lemma 5.2 will be singular although the chosen subsets do not properly represent the (almost) cyclic behavior. In the numerical realization we take this into account and use as a criterion the condition number of these matrices: in the construction of the matrix M we choose the components of the approximating vectors in such a way that the condition number of M is as small as possible.

From now on we assume that $\lambda \neq 1$ is an eigenvalue of P_ε with corresponding real valued eigenmeasure $\nu \in \mathcal{M}_{\mathbb{C}}$, that is,

$$P_\varepsilon \nu = \lambda \nu.$$

Recall that in this case $\nu(X) = 0$ (see Remark 3.2 (a)). The aim is to relate the value of this eigenvalue to the probability δ in Definition 5.3. We begin with the following elementary observation.

LEMMA 5.5. *Suppose that ν is scaled so that $|\nu| \in \mathcal{M}$, and let $A \subset X$ be a set with $\nu(A) = \frac{1}{2}$. Then $\nu = |\nu|$ on A.*

Proof. Obviously, $\nu(B) \leq |\nu|(B)$ for all measurable B. For contradiction let $B \subset A$ be a measurable set with $\nu(B) < |\nu|(B)$. It follows that there is a $C \subset A$ with $\nu(C) < 0$. Hence we have for $E = A - C$

$$\nu(E) > \nu(A) = \frac{1}{2}.$$

Using $\nu(X) = 0$ and $|\nu|(X) = 1$, this leads to a contradiction

$$1 = |\nu|(E) + |\nu|(X - E) \geq |\nu(E)| + |\nu(X - E)| > \frac{1}{2} + \frac{1}{2} = 1. \qquad \square$$

Remark 5.6. Observe that by the *Hahn decomposition* (see, e.g., [28]) the existence of a set A with $\nu(A) = \frac{1}{2}$ is guaranteed.

PROPOSITION 5.7. *Suppose that ν is scaled so that $|\nu| \in \mathcal{M}$, and let $A \subset X$ be a set with $\nu(A) = \frac{1}{2}$. Then*

$$\delta + \sigma = \lambda + 1, \tag{5.3}$$

if A is δ-almost invariant and $X - A$ is σ-almost invariant with respect to $|\nu|$.

Proof. By Lemma 5.5 we have

$$\int_A p_\varepsilon(x, A)\, d\nu(x) = \int_A p_\varepsilon(x, A)\, d|\nu|(x) = \delta\nu(A)$$

since $\nu(A) = \frac{1}{2}$. Similarly,

$$\begin{aligned}\int_{X-A} p_\varepsilon(x, A)\, d\nu(x) &= \int_{X-A} 1 - p_\varepsilon(x, X - A)\, d\nu(x)\\ &= \nu(X - A) - \sigma\nu(X - A)\\ &= (\sigma - 1)\nu(A),\end{aligned}$$

since $\nu(X - A) = -\nu(A)$. Finally, using the fact that ν is an eigenmeasure we compute

$$\begin{aligned}\lambda\nu(A) &= \int_A p_\varepsilon(x, A)\, d\nu(x) + \int_{X-A} p_\varepsilon(x, A)\, d\nu(x)\\ &= \delta\nu(A) + (\sigma - 1)\nu(A) = (\delta + \sigma - 1)\nu(A),\end{aligned}$$

yielding (5.3). □

Remark 5.8.

(a) Observe that in the case where λ is close to one we may assume that the probability measure $|\nu|$ is close to the invariant measure μ of the system. In this sense we have derived the desired relation between the eigenvalue λ and the probability that the system is still behaving in a cyclic way.

(b) In our numerical computations we will work with the unperturbed equations rather than introduce noise artificially. Thus, it would be important to know whether the eigenvalues of P_0 and P_ε are close to each other for small ε. First results concerning the stochastic stability of the spectrum of the Frobenius–Perron operator are obtained in [2].

In (5.3) both δ and σ occur, and in general there will be no relation between these constants. However, if the underlying system possesses an additional symmetry, then we can express one of them in terms of the other one.

To illustrate this fact let us consider the simplest case where we have a symmetry transformation κ in the problem with $\kappa^2 = id$. In that case

$$p_\varepsilon(x, B) = p_\varepsilon(\kappa x, \kappa B) \quad \text{for all measurable } B \subset X, \tag{5.4}$$

which implies that for any $\rho \in \mathcal{M}$

$$\int_B p_\varepsilon(x, B)\, d\rho = \int_{\kappa B} p(x, \kappa B)\, d\kappa^* \rho.$$

Hence we have the following corollary.

COROLLARY 5.9. *Suppose in addition to the assumptions in Proposition* 5.7 *that*

(i) *p_ε is symmetric, that is,* (5.4) *holds,*

(ii) *the set A satisfies $\kappa A = X - A$, and*

(iii) *the measure $|\nu|$ is κ-symmetric, that is, $\kappa^*|\nu| = |\nu|$.*

Then $X - A$ is δ-almost invariant with respect to $|\nu|$ if and only if A is δ-almost invariant, and in particular

$$\delta = \frac{\lambda + 1}{2}. \tag{5.5}$$

Numerical solution of the eigenvalue problems. Since in this article we were mainly interested in the description of how to extract numerically information on the dynamical behavior from the spectrum of P_ε, we just outline the algorithmic steps which are necessary for the numerical approximation and solution of the eigenvalue problem (3.5). For the details concerning the implementation the reader is asked to consult the references listed below.

All the algorithms described in the following are integrated into the software package

GAIO (Global Analysis of Invariant Objects),

which can be obtained online via http. (See the first page of this paper for the authors' homepages.)

(i) *Construction of a box-covering.* We begin with the construction of a box-covering of the interesting (randomly perturbed) dynamics in state space. This can be done either by a subdivision technique (see [12, 6, 5, 7]) or by a cell-mapping approach (see, e.g., [20, 14]). This way we obtain a collection of boxes B_k, $k = 1, 2, \ldots, N$, such that the part of state space containing the interesting dynamics is covered by their union.

Remark 5.10. To simplify the description we assume that all the boxes have the same volume. However, it turned out that a box-covering can be constructed in an even more efficient way if the size of the boxes is chosen in an adaptive way in each step of the subdivision procedure. This fact is explored in [9].

(ii) *Galerkin approximation.* The basis functions we have chosen are the characteristic functions of the B_k's,

$$\varphi_k = \chi_{B_k}.$$

We assume that the Hausdorff distance between X and the covering $\cup B_k$ is small enough so that the assumption (3.4) on the φ_k's is satisfied.

Remark 5.11. In practice we use the subdivision technique from [6] to construct a box-covering for which (3.4) holds. In fact, we simply neglect those boxes which have measure zero once a certain number of steps in the subdivision algorithm has been performed.

(iii) *Approximation and solution of the eigenvalue problem.* We approximate the coefficients of the matrices M_1 and M_2 in (3.5) by a numerical evaluation of the integrals which are involved. Observe that by our specific choice of the boxes and the φ_k's the matrix M_2 is a multiple of the identity. Hence the main numerical effort lies in the computation of the inner products to approximate the operator P_ε. This is done either by a Monte Carlo method or by an exhaustion technique as described in [13].
For the computation of eigenvalues and corresponding eigenvectors we use ARPACK which provides an iterative eigenvalue solver for sparse matrices (see [22]). Alternatively we use an approach based on *bordered matrices* for finding specific single eigenvectors. For the solution of the corresponding system of linear equations we use an iterative method taking into account the fact that the matrix M_1 is extremely sparse.

6. Examples. In this section we illustrate our numerical methods by three examples in which the transition from a true cyclic behavior to an almost cycling one becomes apparent. First we use the well-known Hénon map as an example to analyze a 2-cycle and an almost 2-cycle. The second example is a $\mathbb{Z}_3$-equivariant mapping in the complex plane which shows a cycling behavior of period six. Finally we investi-

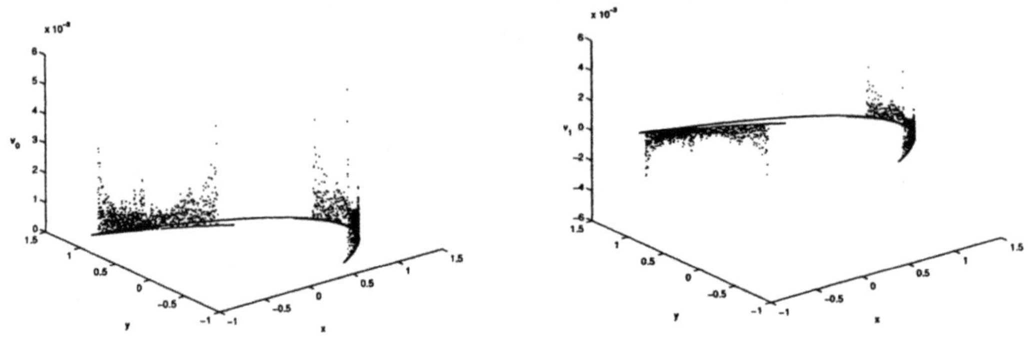

(a) v_0 (b) v_1

FIG. 6.1. *Eigenvectors of the approximation of the Frobenius–Perron operator for the Hénon map* ($a = 1.2$, $b = 0.2$).

gate numerically the *Chua circuit*. Since this is an ordinary differential equation, no nontrivial cycling is expected to be seen. However, in this case we will identify two sets in phase space which are almost invariant with respect to the flow.

All the computations were done without an artificial introduction of noise. Rather it turned out to be sufficient for our purposes to interpret the round off error as a small random perturbation.

A 2-cycle in the Hénon map. We consider a scaled version of the well-known *Hénon map*,

$$f(x, y) = (1 - ax^2 + y/5,\ 5bx),$$

where we fix $b = 0.2$ and vary a. For $a = 1.2$ this map possesses a 2-cycle, and we have used the approximation procedure described in section 5 to identify the two components X_0 and X_1. In Figure 6.1 we show the approximations v_0 and v_1 of the two eigenmeasures of the Frobenius–Perron operator corresponding to the eigenvalues $\lambda_0 = 1$ and $\lambda_1 = -1$.

By Lemma 5.1

$$u_0 = \frac{1}{2}(v_0 + v_1) \quad \text{and} \quad u_1 = \frac{1}{2}(v_0 - v_1)$$

are approximations of probability measures μ_0 and μ_1 which have support on X_0 and X_1, respectively. These are shown in Figure 6.2.

Remark 6.1. In the computation the box-covering was obtained by the continuation algorithm described in [5]. The boxes were of size $1/2^{10}$ in each coordinate direction and the continuation was restricted to the square $Q = [-2, 2]^2 \subset \mathbb{R}^2$. This way we have produced a covering of the closure of the one-dimensional unstable manifold of the hyperbolic fixed point in the first quadrant by 2525 boxes.

Next we set $a = 1.272$. For this parameter value the 2-cycle has disappeared, but in simulations the cycling behavior can still be observed for most iterates. Correspondingly we find that $\lambda_1 = -0.9944$ is an eigenvalue of the approximation of the Frobenius–Perron operator. Using the same notation as before we show in Figures 6.3

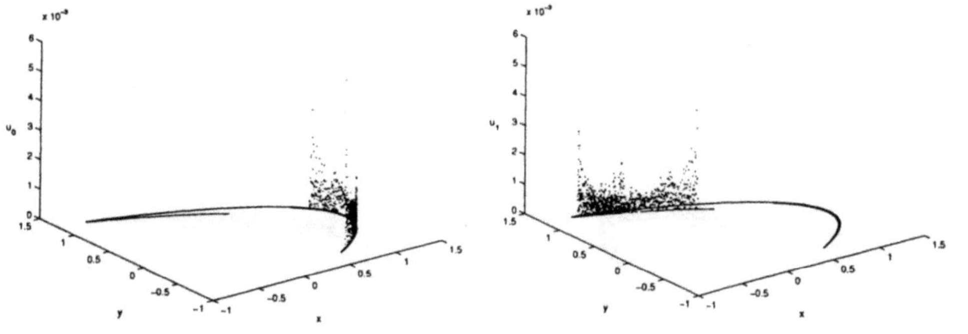

(a) u_0 (b) u_1

FIG. 6.2. *Approximations of probability measures with support on the two components of the 2-cycle* ($a = 1.2$, $b = 0.2$).

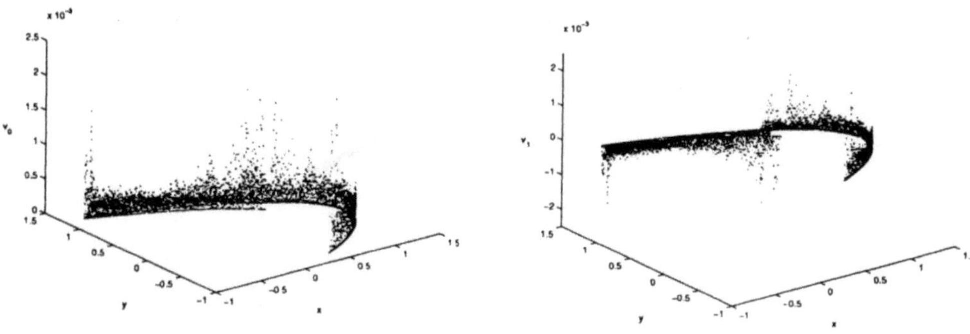

(a) v_0 (b) v_1

FIG. 6.3. *Eigenvectors of the approximation of the Frobenius–Perron operator for the Hénon map* ($a = 1.272$, $b = 0.2$).

and 6.4 the approximations of the eigenmeasures. In this case the box-covering has 3101 elements. Note that the supports of u_0 and u_1 have a nonempty intersection. This fact is illustrated in Figure 6.5 where we have marked by black circles all boxes on which $u_0 > 10$ and $u_1 > 10$.

A period six cycle. As the second example we slightly modify a mapping from [4] and consider the dynamical system $f : \mathbb{C} \to \mathbb{C}$,

$$f(z) = e^{-\frac{2\pi i}{3}} \left((|z|^2 + \alpha)z + \frac{1}{2}\bar{z}^2 \right),$$

for the parameter value $\alpha = -1.7$. For the computation of the box-covering we have used the subdivision algorithm described in [6]. Starting with the square $Q = [-1.5, 1.5]^2$ we have subdivided Q seven times by bisection in each coordinate direction

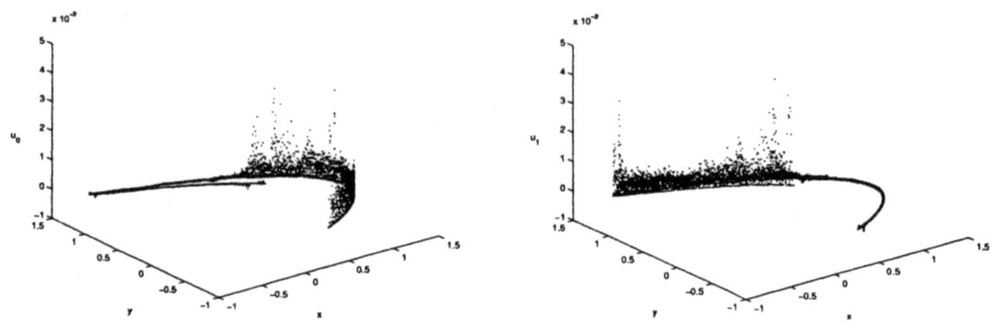

(a) u_0 (b) u_1

FIG. 6.4. *Approximations of probability measures which correspond to the two components of the almost 2-cycle* ($a = 1.272$, $b = 0.2$).

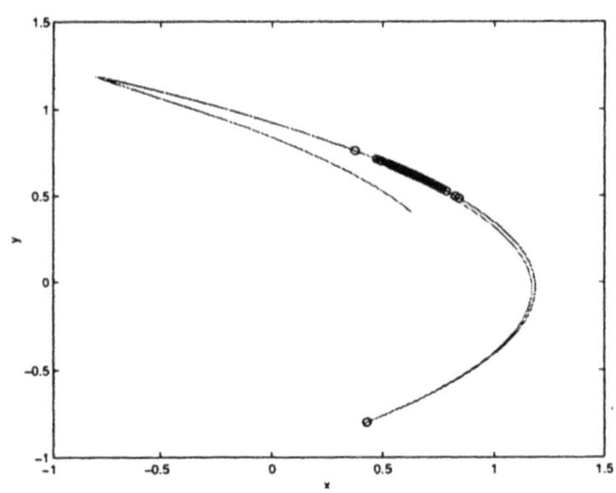

FIG. 6.5. *Approximation of a subset which is in the supports of both probability measures corresponding to the almost 2-cycle.*

which leads to a box-covering by 3606 boxes. In Figure 6.6 we show the approximation of the invariant measure, that is, the eigenvector v_0 corresponding to the eigenvalue $\lambda_0 = 1$ of the discretized Frobenius–Perron operator. In this case this operator additionally has the eigenvalues ω_6^k, $k = 1, \ldots, 5$, and hence we may use Lemma 5.1 to compute approximations $v_0, \ldots, v_5$ of the probability measures with support on the cyclic components $X_0, \ldots, X_5$. These supports are shown in Figure 6.7.

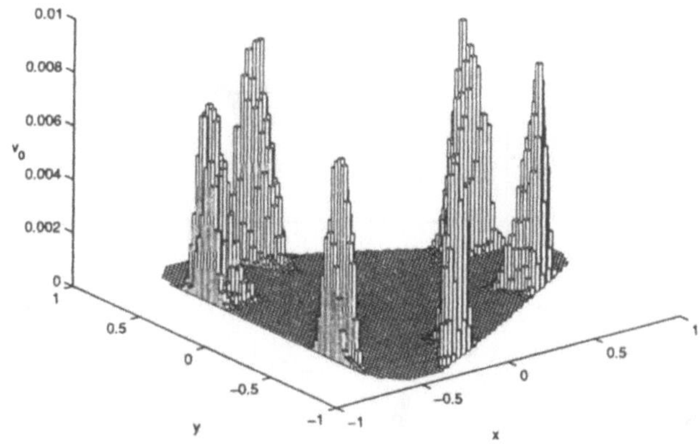

FIG. 6.6. *Approximation of the invariant measure for* $\alpha = -1.7$.

TABLE 1

j	λ_j	$\lvert\lambda_j\rvert$
0	1	1
1,5	$0.4918 \pm 0.8534i$	0.985
2,4	$-0.4880 \pm 0.8437i$	0.9747
3	-0.9709	0.9709

Now we vary the parameter and set $\alpha = -1.8$. For this value of α the strict cyclic behavior disappears and there is an almost 6-cycle. In Figure 6.8(a) we show the "essential" supports of the approximations $v_0, \ldots, v_5$ of the six almost cyclic components. More precisely we show all boxes B_ℓ for which $(v_i)_\ell > 0.1$, $i = 0, \ldots, 5$ (by $(v_i)_\ell$ we denote the ℓth component of the vector v_i). In Figure 6.8(b) we demonstrate that the intersection of these supports is nonempty. We have shown all boxes B_ℓ for which there are at least two indices $i, j \in \{0, \ldots, 5\}$, $i \neq j$, such that $(v_i)_\ell > 0.1$ and $(v_j)_\ell > 0.1$.

In Table 1 we list the corresponding eigenvalues together with their absolute values. We remark that for this parameter value the subdivision algorithm leads to a covering by 4364 boxes.

Two almost invariant sets in the Chua circuit. Finally we present a system of three first order ordinary differential equations in which two almost invariant sets can be identified numerically. The system that we are considering is the *Chua circuit*,

$$\begin{aligned}\dot{x} &= \alpha\left(y - m_0 x - \frac{1}{3} m_1 x^3\right),\\ \dot{y} &= x - y + z,\\ \dot{z} &= -\beta y,\end{aligned}$$

where we have chosen the parameter values $\alpha = 18$, $\beta = 33$, $m_0 = -0.2$, and $m_1 = 0.01$.

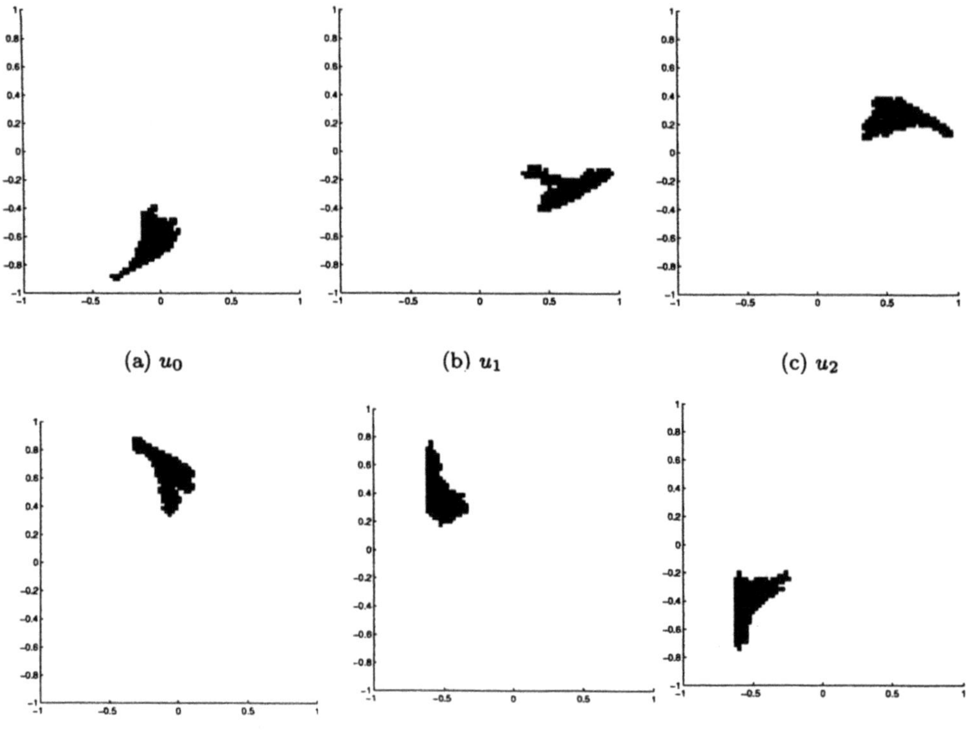

(a) u_0 (b) u_1 (c) u_2

(d) u_3 (e) u_4 (f) u_5

FIG. 6.7. *Approximation of the cyclic components* $X_0, \ldots, X_5$ *for* $\alpha = -1.7$.

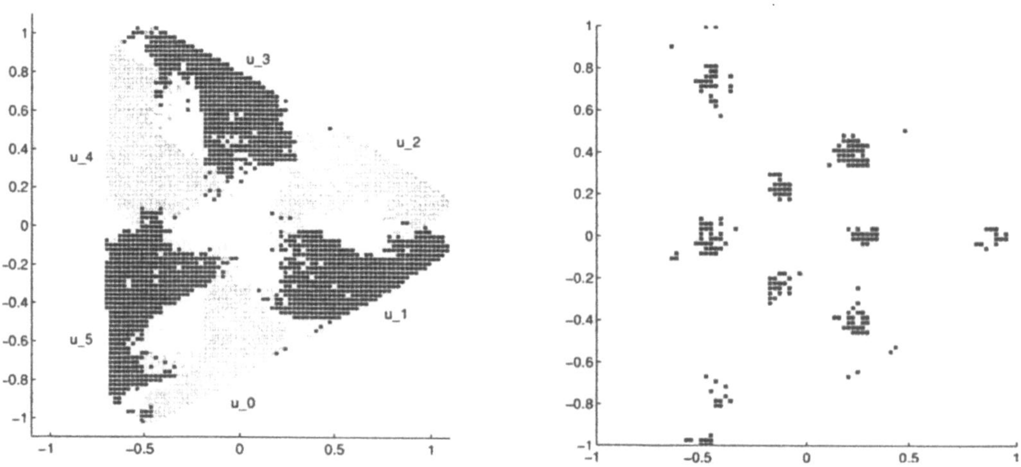

(a) "Essential" supports (b) Intersection of the supports

FIG. 6.8. *Approximation of the almost cyclic components for* $\alpha = -1.8$.

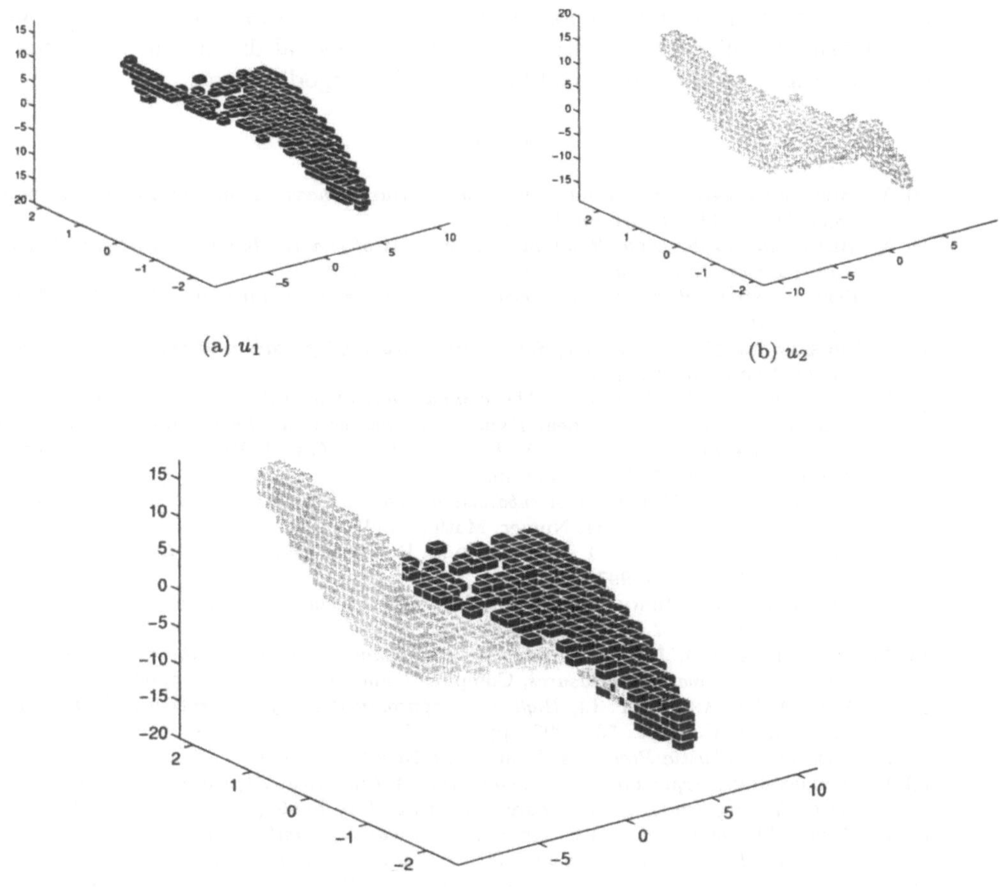

(a) u_1 (b) u_2

(c) u_3

FIG. 6.9. *Illustration of the existence of two almost invariant sets in the Chua circuit.* (a) *Boxes corresponding to components of the approximating densities with value bigger than* 10^{-4}; (b) *boxes corresponding to components of the approximating densities with value less than* -10^{-4}; (c) *superposition of the two almost invariant sets.*

A detailed discussion of the dynamical behavior of this system can be found in [16]; see also [23]. We consider the time-0.1 map and—using the continuation method described in [6]—cover the unstable manifold of the origin by 10372 boxes. In addition to the eigenvalue 1 the discretized Frobenius–Perron operator does also possess the eigenvalue $\lambda_1 = 0.9272$. We may conclude from this result that there are two almost invariant sets. Indeed, a numerical approximation of the corresponding regions in phase space leads to the result shown in Figure 6.9. A detailed numerical study of this particular example can be found in [8].

Acknowledgments. We are grateful to Bernold Fiedler, who inspired our work on the detection of cyclic behavior during a provocative dinner conversation, and we thank Gerhard Keller for his patience throughout several discussions concerning our questions related to the theoretical background in ergodic theory.

REFERENCES

[1] M. Benedicks and L.-S. Young, *Sinai-Bowen-Ruelle measures for certain Hénon maps*, Invent. Math., 112 (1993), pp. 541–576.

[2] M. Blank and G. Keller, *Random perturbations of chaotic dynamical systems: Stability of the spectrum*, Nonlinearity, 11 (1998), pp. 1351–1364.

[3] R. Bowen and D. Ruelle, *The ergodic theory of axiom a flows*, Invent. Math., 29 (1975), pp. 181–202.

[4] P. Chossat and M. Golubitsky, *Symmetry-increasing bifurcation of chaotic attractors*, Phys. D, 32 (1988), pp. 423–436.

[5] M. Dellnitz and A. Hohmann, *The computation of unstable manifolds using subdivision and continuation*, in Nonlinear Dynamical Systems and Chaos, Progr. Nonlinear Differential Equations Appl. 19, H. W. Broer, S. A. van Gils, I. Hoveijn, and F. Takens, eds., Birkhäuser, Basel, 1996, pp. 449–459.

[6] M. Dellnitz and A. Hohmann, *A subdivision algorithm for the computation of unstable manifolds and global attractors*, Numer. Math., 75 (1997), pp. 293–317.

[7] M. Dellnitz, A. Hohmann, O. Junge, and M. Rumpf, *Exploring invariant sets and invariant measures*, Chaos, 7 (1997), p. 221.

[8] M. Dellnitz and O. Junge, *Almost invariant sets in Chua's circuit*, Internat. J. Bifur. Chaos Appl. Sci. Engrg., 7 (1997), pp. 2475–2485.

[9] M. Dellnitz and O. Junge, *An adaptive subdivision technique for the approximation of attractors and invariant measures*, Comput. Visual. Sci., 1 (1998), pp. 63–68.

[10] J. Ding, Q. Du, and T. Y. Li, *High order approximation of the Frobenius-Perron operator*, Appl. Math. Comput., 53 (1993), pp. 151–171.

[11] J. L. Doob, *Stochastic Processes*, John Wiley, New York, 1960.

[12] M. Eidenschink, *Exploring Global Dynamics: A Numerical Algorithm Based on the Conley Index Theory*, Ph.D. thesis, Georgia Institute of Technology, Atlanta, GA, 1995.

[13] R. Guder, M. Dellnitz, and E. Kreuzer, *An adaptive method for the approximation of the generalized cell mapping*, Chaos Solitons Fractals, 8 (1997), pp. 525–534.

[14] H. Hsu, *Global analysis by cell mapping*, Internat. J. Bifur. Chaos Appl. Sci. Engrg., 2 (1992), pp. 727–771.

[15] F. Y. Hunt, *Approximating the invariant measures of randomly perturbed dissipative maps*, J. Math. Anal. Appl., 198 (1996), pp. 534–551.

[16] A. Khibnik, D. Roose, and L. O. Chua, *On periodic orbits and homoclinic bifurcations in Chua's circuit with a smooth nonlinearity*, Internat. J. Bifur. Chaos Appl. Sci. Engrg., 3 (1993), pp. 363–384.

[17] Yu. Kifer, *General random perturbations of hyperbolic and expanding transformations*, J. Anal. Math., 47 (1986), pp. 111–150.

[18] Yu. Kifer, *Random Perturbations of Dynamical Systems*, Birkhäuser, Basel, 1988.

[19] Yu. Kifer, *Computations in dynamical systems via random perturbations*, Discrete Continuous Dynam. Systems, 3 (1997), pp. 457–476.

[20] E. Kreuzer, *Numerische Untersuchung nichtlinearer dynamischer Systeme*, Springer-Verlag, New York, 1987.

[21] A. Lasota and M. C. Mackey, *Chaos, Fractals and Noise*, Springer-Verlag, New York, 1994.

[22] R. B. Lehoucq, *Analysis and Implementation of an Implicitly Restarted Arnoldi Iteration*, Rice University Technical Report TR95-13, Houston, TX, 1995.

[23] T. Matsumoto, L. O. Chua, and M. Komuro, *The double scroll*, IEEE Trans. Circuits Systems, 32 (1985), pp. 797–818.

[24] S. P. Meyn and R. L. Tweedie, *Markov Chains and Stochastic Stability*, Springer-Verlag, New York, 1993.

[25] J. E. Osborn, *Spectral approximation for compact operators*, Math. Comp., 29 (1975), pp. 712–725.

[26] D. Ruelle, *A measure associated with axiom a attractors*, Amer. J. Math., 98 (1976), pp. 619–654.

[27] Y. G. Sinai, *Gibbs measures in ergodic theory*, Russian Math. Surveys, 166 (1972), pp. 21–69.

[28] K. Yosida, *Functional Analysis*, Springer-Verlag, New York, 1980.

Commun. Math. Phys. 208, 605 – 622 (2000)

Communications in
Mathematical Physics

Absolutely Continuous Invariant Measures for Piecewise Real-Analytic Expanding Maps on the Plane

Masato Tsujii

Department of Mathematics, Hokkaido University, Sapporo 060-0810, Japan.
E-mail: tsujii@math.sci.hokudai.ac.jp

Received: 5 June 1998 / Accepted: 11 May 1999

Abstract: We prove the existence of absolutely continuous invariant measures for piecewise real-analytic expanding maps on bounded regions in the plane.

1. Introduction

Expanding properties of dynamical systems give rise to chaotic behavior of the orbits. On the other hand, they often lead to good ergodic properties such as the existence of absolutely continuous invariant measures. One typical example is the fork-lore theorem [12] that shows the existence of a smooth ergodic measure for every expanding C^2 self-map on a closed manifold. Hence, one interest in the study of chaotic dynamical systems is the relations between expanding properties and the ergodic properties they produce.

Lasota and Yorke showed, in their famous work [11], the existence of absolutely continuous invariant measures for piecewise C^2 expanding maps on intervals. They made use of the Perron–Frobenius operator and functions of bounded variation, and their idea has been used extensively in the study of one dimensional dynamical systems. This paper concerns the generalization of their result towards higher dimension. Though it is natural to expect similar results, it has turned out that things are not simple in higher dimension. In fact, at present, we do not know whether piecewise C^2 expanding maps on bounded regions in higher dimensional Euclidean space always have absolutely continuous invariant measures. The main difficulty in higher dimension is the fact that the partition of the domain into the regions where an iteration of the map is smooth can be very complicated.

Gerhard Keller treated piecewise C^2 expanding maps on bounded regions in the plane in his thesis [7,8] and gave some criterion for the existence of absolutely continuous invariant measure. The most effective result we have so far is that of Góra and Boyarski [4], which gives a lower bound for the minimum expansion rate that assures the existence of absolutely continuous invariant measures. Their result is valid for arbitrary dimension.

But their lower bound depends on the minimal angle on the boundaries of the regions in the partition associated to the map. See [1] for a modification of their result.

In this article, we consider the problem for piecewise real- analytic maps on bounded regions in the plane. (We will give the definition of piecewise real-analytic maps in the next section.) The real-analytic property somewhat relaxes the difficulty we mentioned above. In fact, we can prove the following theorem as the main result of this paper.

Theorem 1. *An absolutely continuous invariant finite measure exists for every piecewise real-analytic expanding map on a bounded region in the plane.*

This result improves a theorem of Keller in his thesis [7,8], which gives the same conclusion under one additional assumption that the map is piecewise conformal.

Actually we will prove the so-called Lasota–Yorke type inequality for some iterations of piecewise real-analytic expanding maps. It is known that we can derive many other properties of the maps from that kind of inequality. We will mention some of them in the appendix.

The author learned from Gerhard Keller that Jérôme Buzzi (C.N.R.S., Institut de Mathématiques de Luminy) obtained a similar result [2] when he was preparing the manuscript of this paper.

2. Piecewise Real-Analytic Map

We call a map $c : [a, b] \to \mathbb{R}^2$ a real-analytic curve if it is a restriction of a real-analytic map defined on a neighborhood of $[a, b]$ and satisfies $c'(t) \neq 0$ for $t \in [a, b]$. In what follows, we will assume

$$\|c'(t)\| \equiv 1 \quad \text{for } t \in [a, b] \tag{1}$$

for real-analytic curves, by real-analytic change of the variable t. Also we will denote the image of a real-analytic curve $c : [a, b] \to \mathbb{R}^2$ by the same symbol c, as an abuse of the symbol.

A continuous map $c : [a, b] \to \mathbb{R}^2$ is called a piecewise real-analytic curve if there is a sequence $a = \xi_0 < \xi_1 < \xi_2 < \cdots < \xi_n = b$ such that the restrictions $c|_{[\xi_i, \xi_{i+1}]}$, $0 \leq i < n$, are real-analytic curves.

Let D be a region on the plane $\mathbb{R}^2$ whose boundary consists of finite simple closed piecewise real-analytic curves. We consider a finite (quasi)-partition $\xi = \{D_i\}_{i=1}^k$ of the domain D such that

- $D_i \subset D$ is a region whose boundary is a finite union of simple closed piecewise real-analytic curves,
- $D_i \cap D_j = \emptyset$ if $i \neq j$, and
- $\cup_{i=1}^k \overline{D_i} = \overline{D}$, where $\overline{D}$ and $\overline{D_i}$ denote the closures of D and D_i respectively.

We call such a partition a real-analytic partition of D. We denote $E = \cup_{i=1}^k \partial D_i = \overline{D} - \cup_{i=1}^k D_i$.

For a real-analytic partition $\xi = \{D_i\}_{i=1}^k$, we can choose a finite set of real-analytic curves $\{\gamma_i : [a_i, b_i] \to E \subset \mathbb{R}^2\}_{i=1}^m$ in E with the following properties

- each γ_i is a simple curve, that is, has no multiple point,
- the boundary of each region D_j, $1 \leq j \leq k$, is a union of the image of some γ_i's, and

- the images of curves γ_i, $1 \le j \le k$, except their end points are mutually disjoint, that is,
$$\gamma_i(t) \neq \gamma_j(s) \quad \text{if } i \neq j, t \in (a_i, b_i) \text{ and } s \in (a_j, b_j).$$

We call these curves the dividing curves of the partition ξ. Remark that there are only finitely many points that belong to more than two dividing curves.

A map $f : D \to D$ is called a piecewise real-analytic map on D if there is a real-analytic partition $\xi = \{D_i\}_{i=1}^k$ of D as above such that each restriction $f|_{D_i}$ of f to D_i, $1 \le i \le k$, can be extended to a neighborhood of $\overline{D}_i$ as a real-analytic map. We will denote, by f_{D_i}, the real- analytic extension of $f|_{D_i}$ to a neighborhood of $\overline{D}_i$.

For a tangent vector v at $x \in D - E$, we define its expansion rate $\rho(v, f)$ by
$$\rho(v, f) = \frac{\|Df(v)\|}{\|v\|}.$$
The minimum expansion rate $\rho(f)$ of the map f is the infimum of the expansion rate over all non-zero vectors at all points in $D - E$. If $\rho(f) > 1$ for a piecewise real-analytic map f, the map f is called a piecewise real-analytic expanding map. We will fix a piecewise real-analytic expanding map f, the partitions $\xi = \{D_i\}_{i=1}^k$ and the dividing curves $\{\gamma_i : [a_i, b_i] \to \mathbb{R}^2\}_{i=1}^m$ throughout this paper.

An important consequence from the above definitions is the fact that the iterations f^n of a piecewise real-analytic map f are also piecewise real-analytic maps. We will use this fact repeatedly in the proof of Theorem 1.

Remark 2. Iterations of piecewise C^r maps with $r \le \infty$ are not necessarily piecewise C^r maps since the partition associated to them may have infinitely many connected components. We refer to [14] for examples of piecewise C^r expanding maps with $r < \infty$ that have singular ergodic properties.

3. Germs of Real-Analytic Curves

Let p be a point on the plane $\mathbb{R}^2$, and let $c_i : [0, \epsilon_i] \to \mathbb{R}^2$, $i = 1, 2$, be two real-analytic curves satisfying $c_i(0) = p$ and (1). We say that these two curves give the same germ at p if $c_1(t) = c_2(t)$ for $0 \le t < \min\{\epsilon_1, \epsilon_2\}$. This is an equivalence relation between real-analytic curves c satisfying $c(0) = p$ and (1). The equivalence classes are called germs of real-analytic curves at p.

We say that two open subsets U_1 and U_2 on the plane $\mathbb{R}^2$ give the same germ at p if there exists $\delta > 0$ such that $U_1 \cap B(p, \delta) = U_2 \cap B(p, \delta)$, where $B(p, \delta) = \{x \in \mathbb{R}^2; \|x - p\| < \delta\}$. This is also an equivalence relation. We call the equivalence classes germs of open subsets at p.

Let β_1 and β_2 be distinct germs of real-analytic curves at p, and let $b_i : [0, \epsilon_i] \to \mathbb{R}^2$, $i = 1, 2$ be simple real-analytic curves that represent β_i respectively. If $\delta > 0$ is sufficiently small, the open set $B(p, \delta)\backslash(b_1 \cup b_2)$ consists of two connected components. The germs of an open subset represented by the connected component of $B(p, \delta)\backslash(b_1 \cup b_2)$ that is located in the counterclockwise direction of the curve b_1 is called the region between β_1 and β_2. From this definition, the region U between β_1 and β_2 and that between β_2 and β_1 are complementary. The germs of real-analytic curves β_1 and β_2 are called the boundary curves of the region U.

Let $\mathrm{Angle}_1(\beta_1, \beta_2) \in [0, 2\pi]$ be the angle that is formed by the region between β_1 and β_2 at p. So we have $\mathrm{Angle}_1(\beta_1, \beta_2) = 2\pi - \mathrm{Angle}_1(\beta_2, \beta_1)$. If $\mathrm{Angle}_1(\beta_1, \beta_2) \neq 0$

we define $\mathrm{Ord}(\beta_1, \beta_2) = 1$. On the other hand, if $\mathrm{Angle}_1(\beta_1, \beta_2) = 0$, we define $\mathrm{Ord}(\beta_1, \beta_2)$ as the contact order of the two germs of real-analytic curves β_1 and β_2 at p, that is,

$$\begin{aligned}\mathrm{Ord}(\beta_1, \beta_2) &= \lim_{t\to+0} \frac{\log \min\{\|b_1(t) - b_2(s)\| \mid s \in [0, \epsilon_2)\}}{\log t}\\ &= \lim_{t\to+0} \frac{\log \min\{\|b_2(t) - b_1(s)\| \mid s \in [0, \epsilon_1)\}}{\log t}.\end{aligned}$$

When $\mathrm{Ord}(\beta_1, \beta_2) = d > 1$, we define $\mathrm{Angle}_d(\beta_1, \beta_2)$ by

$$\begin{aligned}\mathrm{Angle}_d(\beta_1, \beta_2) &= \lim_{t\to+0} \frac{\min\{\|b_1(t) - b_2(s)\| \mid s \in [0, \epsilon_2)\}}{t^d}\\ &= \lim_{t\to+0} \frac{\min\{\|b_2(t) - b_1(s)\| \mid s \in [0, \epsilon_1)\}}{t^d}.\end{aligned}$$

For the region U between β_1 and β_2, we define $\mathrm{Ord}(U) = \mathrm{Ord}(\beta_1, \beta_2)$ and $\mathrm{Angle}_d(U) = \mathrm{Angle}_d(\beta_1, \beta_2)$. We will need the following elementary lemmas.

Lemma 3. *Let $H : W \to \mathbb{R}^2$ be a real-analytic map defined on a neighborhood W of a point $p \in \mathbb{R}^2$. Assume that $\|DH(p)w\|/\|w\| \geq 1$ for all tangent vectors $w \neq 0$ at p. Let β_1 and β_2 be germs of real-analytic curves at p, and let v_i, $i = 1, 2$, be the unit tangent vectors of them at the point p respectively. Let U be the region between β_1 and β_2.*

(a) *If* $0 < \mathrm{Angle}_1(\beta_1, \beta_2) < 2\pi$, *we have*

$$|\det DH(p)| = \left|\frac{\sin(\mathrm{Angle}_1(H(U)))}{\sin(\mathrm{Angle}_1(U))}\right| \cdot \rho(v_1, H)\rho(v_2, H). \tag{2}$$

(b) *If* $\mathrm{Ord}(U) = d > 1$, *we have*

$$|\det DH(p)| = \frac{\mathrm{Angle}_d(H(U))}{\mathrm{Angle}_d(U)} \cdot \rho(v_1, H)^{d+1}. \tag{3}$$

Remark 4. In the claim (b), $v_1 = v_2$ and $\rho(v_1, H) = \rho(v_2, H)$.

Proof. In the case $\mathrm{Ord}(U) = 1$, the formula (2) says nothing but the fact that the absolute value of the determinant of H at p is the ratio between the area of the infinitesimal parallelogram at p spanned by the vectors v_1 and v_2 and that of its image under $DH(p)$. Let us consider the case $\mathrm{Ord}(U) = d > 1$. Since two curves b_i, $i = 1, 2$, representing the germs β_i are almost parallel in small neighborhoods of the point p, the minimum $\min\{\|b_1(t) - b_2(s)\| | s \in [0, \epsilon_2)\}$ is attained when $b_1(t) - b_2(s)$ is almost orthogonal to the vector $v_1 = v_2$ when t is small. Hence we can see, by elementary geometric argument,

$$\begin{aligned}\lim_{t\to+0} \frac{\min\{\|H \circ b_1(t) - H \circ b_2(s)\| \mid s \in [0, \epsilon_2)\}}{\min\{\|b_1(t) - b_2(s)\| \mid s \in [0, \epsilon_2)\}} &= \langle DH(v_1^{\perp}), DH(v_1)^{\perp}\rangle\\ &= \frac{|\det DH(p)|}{\rho(v_1, H)},\end{aligned}$$

where $v_1^{\perp}$ and $DH(v_1)^{\perp}$ are the unit vectors that are orthogonal to the vectors v_1 and $DH(v_1)$ respectively and $\langle\cdot,\cdot\rangle$ is the inner product in the ordinary sense. Taking into account the change of the variable t so that the map $H \circ b_1(t)$ satisfies the condition (1), we obtain the claim (b). □

Under the assumption of Lemma 3, we define $\rho(U, H)$ as the maximum of the expansion rate $\rho(v, H) = \|DH(v)\|/\|v\|$ over all tangent vectors v at p that is in between v_1 and v_2 or, in other words, that is contained in the closure of the angle formed by U at p.

Lemma 5. *Under the same assumption as in the last lemma, we have*

$$\frac{\mathrm{Angle}_{\mathrm{Ord}(U)}(H(U))}{\mathrm{Angle}_{\mathrm{Ord}(U)}(U)} \leq 2\pi \cdot \frac{|\det DH(p)|}{\rho(U,H)\rho_0}, \tag{4}$$

where $\rho_0 \geq 1$ is the minimum of the expansion rate $\rho(w, H)$ over all tangent vectors $w \neq 0$ at p.

Proof. If $\mathrm{Ord}(U) > 1$, (4) is obvious from the last lemma. Let us consider the case $\mathrm{Ord}(U) = 1$. We first prove, for $K = 2\pi$,

$$\frac{\mathrm{Angle}_1(H(U))}{\mathrm{Angle}_1(U)} \leq K \cdot \frac{|\det DH(p)|}{\rho(v_1,H)\rho_0}. \tag{5}$$

If $\mathrm{Angle}_1(U) > \pi/2$, we get (5) for $K = 4$ because the left-hand side is smaller than $2\pi/(\pi/2) = 4$, while the right-hand side except K is not smaller than 1. If $\mathrm{Angle}_1(U) \leq \pi/2$ and $\mathrm{Angle}_1(H(U)) \leq \pi/2$, we get (5) for $K = \pi/2$ from the last proposition because

$$2/\pi \leq \sin(x)/x \leq 1 \quad \text{for } 0 < x \leq \pi/2.$$

Finally, we consider the case when $\mathrm{Angle}_1(U) \leq \pi/2$ and $\mathrm{Angle}_1(H(U)) > \pi/2$. In this case, we can take a unit tangent vector v between v_1 and v_2 such that $DH(p)v$ is orthogonal to $DH(p)v_1$. Let γ be a germ of real-analytic curve passing through U that is tangent to the vector v at p, and let V be the region between β_1 and γ. Then we can apply the above argument to V and get (5) with U replaced by V for $K = \pi/2$. Since

$$\frac{\mathrm{Angle}_1(H(U))}{\mathrm{Angle}_1(U)} \leq \frac{2\pi}{\mathrm{Angle}_1(V)} = 4\frac{\mathrm{Angle}_1(H(V))}{\mathrm{Angle}_1(V)},$$

we get (5) for $K = 2\pi$. Therefore we have (5) for $K = 2\pi$ in any case.

Remark that we can replace v_1 by v_2 in (5) by symmetry. Now let us take a vector v at p that is in between v_1 and v_2 and satisfy $\rho(v, H) = \rho(U, f)$. If $v = v_1$ or $v = v_2$, the conclusion of the lemma is nothing but (5) or that with v_1 replaced by v_2. Otherwise, we consider a germ γ of a real-analytic curve that is tangent to v at p. The germ of curve γ divide U into two regions U_1 and U_2. Applying (5) to these regions, we get

$$\begin{aligned}\frac{\mathrm{Angle}_1(H(U))}{\mathrm{Angle}_1(U)} &\leq \max\left\{\frac{\mathrm{Angle}_1(H(U_1))}{\mathrm{Angle}_1(U_1)}, \frac{\mathrm{Angle}_1(H(U_2))}{\mathrm{Angle}_1(U_2)}\right\} \\ &\leq 2\pi \cdot \frac{|\det DH(p)|}{\rho(v,H)\rho_0}.\end{aligned}$$

The lemma is proved. □

4. Weighted Multiplicity

In this section we introduce what we call weighted multiplicity that count the multiplicity of the intersection of dividing curves $\{\gamma_i\}_{i=1}^m$ with appropriate weight. Let p be a point in $E = \cup_{i=1}^k \partial D_i$. Let $\gamma_i : [a_i, b_i] \to \mathbb{R}^2$ be a dividing curve. If the curve γ_i passes through the point p, it gives germs of real-analytic curves at p in the following manner: If $\gamma_i(t) = p$ for $t \in (a_i, b_i)$, the curve γ_i gives two germs of real-analytic curve at p represented by the curves $s \mapsto \gamma_i(t+s)$ and $s \mapsto \gamma_i(t-s)$. If $\gamma_i(a_i) = p$ (resp. $\gamma_i(b_i) = p$), the curve γ_i gives one germ represented by a curve $s \mapsto \gamma_i(a_i + s)$ (resp. $s \mapsto \gamma_i(b_i - s)$).

Let $\mathcal{B}(p) = \{\beta_i(p)\}_{i=1}^{m(p)}$ be the collection of the distinct germs of real-analytic curves given in such a way by all dividing curves. Remark that $m(p) = 2$ for all points $p \in E$ except for finite points. These germs of real-analytic curves are called the germs of curves at p given by the dividing curves. We always assume that the germs of curves $\beta_i(p)$, $i = 1, 2, \cdots, m(p)$, are arranged in counterclockwise order around the point p.

Let U_i, $1 \le i < m(p)$, be the region between $\beta_i(p)$ and $\beta_{i+1}(p)$, and let $U_{m(p)}$ be that between $\beta_{m(p)}(p)$ and $\beta_1(p)$. We denote the set of these regions by $\mathcal{U}(p) = \{U_i(p)\}_{i=1}^{m(p)}$. For $U \in \mathcal{U}(p)$, let f_U be the germ of a real-analytic map at p that is obtained as the real-analytic extension of the restriction of f to a representative of U.

We define the weight $W(U_i(p))$ of the region $U_i(p) \in \mathcal{U}(p)$ by

$$W(U_i(p)) = \frac{\|Df_{U_i(p)}(p)v_1\|/\|v_1\| + \|Df_{U_i(p)}(p)v_2\|/\|v_2\|}{|\det Df_{U_i(p)}(p)|}$$

for $1 \le i \le m(p)$, where v_1 and v_2 are the tangent vectors of the boundary curves of $U_i(p)$ at p. The weighted multiplicity $M(p, f)$ at a point $p \in E$ is defined by

$$M(p, f) = \sum_{i=1}^{m(p)} W(U_i(p)).$$

The weighted multiplicity $M(f)$ of a piecewise real-analytic expanding map f is the supremum of $M(f, p)$ over all points $p \in E$. Remark again that $M(f, p) \le 4\rho(f)^{-1}$ for all $p \in E$ except for finite points. Weighted multiplicity $M(f)$ is the quantity that we are most concerned with in the argument below.

5. Functions of Bounded Variation

We use the theory of bounded variation functions in higher dimensional space, which is developed in the book [3]. We recall some definitions and properties of functions of bounded variation from [3].

Let U be an open subset of the plane $\mathbb{R}^2$. Let $C^r(U, \mathbb{R}^2)$ be a set of bounded vector-valued C^r functions $g = (g_1, g_2) : U \to \mathbb{R}^2$ and let $C_0^r(U, \mathbb{R}^2)$ be the subset of $C^r(U, \mathbb{R}^2)$ that consists of functions with compact support. Similarly, let $\Omega^r(U)$ be the set of 1-forms $\Psi = \Psi_1 dx + \Psi_2 dy$ of class C^r on U and let Ω_0^r be the subset of $\Omega^r(U)$ that consists of 1-forms with compact support. We denote the d-dimensional Hausdorff measure by μ_d. We define the variation $\mathrm{Var}(\varphi, U)$ of the function $\varphi \in L^1(U)$ as the supremum of

$$\int_U \varphi(z) \mathrm{Div} g(z) d\mu_2(z) \tag{6}$$

over all $g = (g_1, g_2) \in C^1_0(U, \mathbb{R}^2)$ satisfying $\|g(z)\| \le 1$ for $z \in U$, where

$$\mathrm{Div} g(x, y) = \frac{\partial}{\partial x} g_1(x, y) + \frac{\partial}{\partial y} g_2(x, y).$$

A function $\varphi \in L^1(U)$ is said to be of bounded variation if $\mathrm{Var}(f, \mathbb{R}^2) < \infty$. We denote, by $\mathcal{BV}(U)$, the set of functions $\varphi \in L^1(U)$ of bounded variation. Sometimes it is convenient to write the variation $\mathrm{Var}(\varphi, U)$ as

$$\mathrm{Var}(\varphi, U) = \sup\left\{\int_U \varphi d\Psi \ ; \ \Psi \in \Omega^1_0(U) \text{ and } \|\Psi(z)\| \le 1 \text{ for } z \in U.\right\}, \quad (7)$$

where $\|\Psi(z)\| = \sqrt{\Psi_1^2(z) + \Psi_2^2(z)}$ for $\Psi = \Psi_1 dx + \Psi_2 dy$. We can obtain this formula from the correspondence between $g = (g_1, g_2) \in C^1_0(U, \mathbb{R}^2)$ and $\Psi = -g_2 dx + g_1 dy \in \Omega^1_0(U)$.

Remark that, for $\varphi \in \mathcal{BV}(U)$, the functional

$$\Phi_\varphi : g \in C^1_0(U, \mathbb{R}^2) \mapsto \int_U \varphi \mathrm{Div} g d\mu_2 \in \mathbb{R}$$

satisfies

$$|\Phi(g_1) - \Phi(g_2)| \le \left(\sup_{z \in U} \|g_1(z) - g_2(z)\|\right) \cdot \mathrm{Var}(\varphi, U).$$

Hence Φ can be extended uniquely as a continuous linear functional on $C^0_0(U, \mathbb{R}^2)$. We can consider this extension as a vector-valued Radon measure on U with total variation $\mathrm{Var}(\varphi, U)$. (See [10, Ch.6] for example.) We denote this vector-valued Radon measure by $D\varphi$. Let $\int_U \langle g, D\varphi\rangle$ be the integration of a vector-valued function $g \in C^0(U, \mathbb{R}^2)$ with respect to the vector-valued measure $D\varphi$. Let $|D\varphi|$ be the measure that is obtained as the total variation of $D\varphi$. Obviously we have $\int_U \langle g, D\varphi\rangle \le \int \|g\| \cdot |D\varphi|$ for $g \in C^0(U, \mathbb{R}^2)$, where $\|g\|$ denotes the function $\|g\|(x) = \|g(x)\|$ on U.

The bounded variation norm of the function $g \in \mathcal{BV}(U)$ is defined as

$$\|\varphi\|_{BV(U)} = \mathrm{Var}(\varphi, U) + \int_U |\varphi| d\mu_2.$$

This norm makes $\mathcal{BV}(U)$ a Banach space. See [3, Remark 1.12]. We make use of the following fact when we prove the existence of an absolutely continuous invariant measure in Sect. 6.

Proposition 6. *Let $U \subset \mathbb{R}^2$ be a bounded open set with C^1 boundary. Then sets of functions in $\mathcal{BV}(U)$ that are uniformly bounded in the bounded variation norm $\|\cdot\|_{BV(U)}$ are relatively compact in $L^1(U)$.*

Another important property of functions of bounded variation is that they give traces on the boundary. Let $U \subset \mathbb{R}^2$ be a bounded region whose boundary is a finite union of real-analytic simple closed curves. We denote by $L^1(\partial U)$ the set of functions that is integrable with respect to the one dimensional Hausdorff measure μ_1. We put $B(x, r) = \{y \in \mathbb{R}^2 \mid \|x - y\| < r\}$. Then we have

Proposition 7. *For $\varphi \in \mathcal{BV}(U)$, there is a unique function $\varphi^- \in L^1(\partial U)$ such that*

$$\lim_{r\to 0} \mu_2(B(x,r))^{-1} \int_{B(x,r)\cap U} |\varphi(z) - \varphi^-(x)| d\mu_2(z) = 0$$

for μ_1-almost all $x \in \partial U$. Moreover,

(a) *for $\zeta \in C_0^1(\mathbb{R}^2, \mathbb{R}^2)$, we have*

$$\int_{\partial U} \varphi^-(z)\langle\zeta(z), \nu(z)\rangle d\mu_1(z) = \int_U \varphi(z)\mathrm{Div}\zeta(z) d\mu_2(z) + \int_U \langle\zeta, D\varphi\rangle,$$

where $\nu(z)$ is the unit outer normal vector for the boundary ∂U at z,

(b) *if we define $\varphi(z) = 0$ for $z \notin U$, we have*

$$\mathrm{Var}(\varphi, \mathbb{R}^2) = \mathrm{Var}(\varphi, U) + \int_{\partial U} |\varphi^-(x)| d\mu_1(x).$$

The function $\varphi^- \in L^1(\partial U)$ in the above proposition is called the trace of φ on the boundary ∂U. We refer to Theorem 1.19 of [3] for Proposition 6, and Theorem 2.10 and Remark 2.14 of [3] for Proposition 7.

Remark 8. In Theorem 2.10 and Remark 2.14 of [3], the boundary of the region U is assumed to be Lipschitz. Hence Proposition 7 is not a direct consequence of that theorem when there are cusps on the boundary of U. But, with slight modification in the proof, we can derive Proposition 7, because the proposition is essentially a local one.

6. An Existence Theorem for Absolutely Continuous Invariant Measures

Let $f : D \to D$ be a piecewise real-analytic expanding map and let $\xi = \{D_i\}_{i=1}^k$ be the partition of the domain D associated to it as in Sect. 2. In this section we prove

Theorem 9. *If f satisfies*

(a) *$M(f) + \rho(f)^{-1} < 1$ and,*

(b) *the continuous extension of each restriction $f|_{D_i}$, $1 \le i \le k$, to the closure $\overline{D_i}$ is injective,*

then there exists an absolutely continuous invariant finite measure for f.

Theorem 9 above is a modification of the result of Góra and Boyarski in [4], and the essential part of the proof below is a repetition of the argument in [4]. We define the Perron–Frobenius operator $P_f : L^1(D) \to L^1(D)$ by

$$P_f(\varphi)(x) = \sum_{f(y)=x} \frac{\varphi(y)}{|\det Df(y)|},$$

where the sum is taken over all $y \in \cup_{i=1}^k D_i$ such that $f(y) = x$. Remark that if there exists a non-negative valued function $h \neq 0$ in $L^1(D)$ such that $P_f(\varphi) = \varphi$, the measure $h \cdot \mu_2$ is an absolutely continuous invariant finite measure for f. From the definition of Perron–Frobenius operator, we have

$$\int_D P_f g d\mu_2 = \int_D g d\mu_2 \tag{8}$$

for a non-negative valued function $g \in L^1(D)$. In what follows, we consider each element $\varphi \in L^1(D)$ as an element of $L^1(\mathbb{R}^2)$ by defining $\varphi(x) = 0$ for $x \notin D$. The following is the key in the proof of Theorem 9.

Proposition 10. *For any $\epsilon > 0$, there exists a constant $C > 0$ such that*

$$\begin{aligned}\mathrm{Var}(P_f\varphi, \mathbb{R}^2) &\leq \sum_{j=1}^{k} \mathrm{Var}(P_f(\varphi \cdot \chi_{D_j}), \mathbb{R}^2)\\ &\leq (M(f) + \rho^{-1}(f) + \epsilon)\mathrm{Var}(\varphi, \mathbb{R}^2) + C\|\varphi\|_{L^1} \qquad (9)\end{aligned}$$

for $\varphi \in \mathcal{BV}(D)$.

This kind of inequality appeared in the original work of Lasota and Yorke and can be seen in the papers of Keller [7,8] and Góra&Boyarski [4].

First, we prove that Proposition 10 implies Theorem 9. Take a small number $\epsilon > 0$ such that $M(f) + \rho(f)^{-1} + \epsilon < 1$. Then take the constant C in Proposition 10 for that ϵ. Let $\varphi \in \mathcal{BV}(D)$ be a non-negative valued function such that $\int_D \varphi d\mu_2 = 1$. From (9) and (8), we obtain

$$\mathrm{Var}(P_f^n\varphi, \mathbb{R}^2) \leq (1 - M(f) - \rho(f)^{-1} - \epsilon)^{-1}C + \mathrm{Var}(\varphi, \mathbb{R}^2)$$

by induction. Hence the set of functions $\{\zeta_n = n^{-1}\sum_{i=0}^{n-1} P_f^i\varphi\}_{n=1}^{\infty}$ are contained in $\mathcal{BV}(D)$ and uniformly bounded in the bounded variation norm $\|\cdot\|_{BV(\mathbb{R}^2)}$ on $\mathbb{R}^2$. Applying Theorem 6 to a bounded open subset $U \supset D$ with C^1 boundary, we can find a subsequence ζ_{n_k} that converges to a function $\varphi_\infty \in L^1(D)$ in the L^1 norm. Obviously φ_∞ is non-negative valued. We have $\int_D \varphi_\infty d\mu_2 = 1$ from (8). Moreover, φ_∞ is a fixed point of the Perron–Frobenius operator P_f because

$$\begin{aligned}\|P_f\varphi_\infty - \varphi_\infty\|_{L^1} &= \lim_{k\to\infty}\left\| \frac{1}{n_k}\sum_{i=0}^{n_k-1} P_f^{i+1}\varphi - \frac{1}{n_k}\sum_{i=0}^{n_k-1} P_f^i\varphi \right\|_{L^1}\\ &\leq \lim_{k\to\infty}\frac{1}{n_k}\|P_f^{n_k}\varphi - \varphi\|_{L^1} \leq \lim_{k\to\infty}\frac{2}{n_k} = 0.\end{aligned}$$

Therefore $\varphi_\infty \cdot \mu_2$ is an absolutely continuous invariant finite measure for f. Theorem 9 is proved.

Now let us go into the proof of Proposition 10. We first study the situation that the image of a dividing curve $\gamma_i : [a_i, b_i] \to \mathbb{R}^2$ is contained in the boundary of some region $D_j \in \xi$. From Proposition 7, a function $\varphi \in \mathcal{BV}(D)$ viewed as a function on D_j gives the trace φ_j^- on the curve γ_i. We consider one side of the tubular neighborhood of the curve γ_i,

$$\Gamma_{ij} : [a_i, b_i] \times [0, \delta] \to \mathbb{R}^2, \quad (t, s) \mapsto \gamma_i(t) + s \cdot \nu(t),$$

where $\nu(t)$ is the unit inner normal vector for the boundary ∂D_j at $\gamma_i(t)$ and $\delta > 0$ is a small constant that we will specify in the argument below. We first take $\delta > 0$ so small that Γ_{ij} is a diffeomorphism. We will denote the image of Γ_{ij} by the same symbol Γ_{ij}.

Let $V_i(x)$ be the unit tangent vector of the curve γ_i at $x \in \gamma_i$. We define a real-analytic function $h_{ij} : \gamma_i \to \mathbb{R}$ by

$$h_{ij}(x) = \frac{\rho(V_i(x), f_{D_j})}{|\det Df_{D_j}(x)|}.$$

Let $\pi : [a_i, b_i] \times [0, \delta] \to [a_i, b_i]$ be the projection. We define a function

$$\tilde{h}_{ij} = (0, h_{ij} \circ \Gamma_{ij} \circ \pi) : [a_i, b_i] \times [0, \delta] \to \mathbb{R}^2.$$

Put $\|\tilde{h}_{ij}\|(x) = \|\tilde{h}_{ij}(x)\|$. Then we have

Lemma 11. *If $\varphi \in \mathcal{BV}(D)$, the composition $\varphi \circ \Gamma_{ij}$ viewed as a function on the open rectangle $(a_i, b_i) \times (0, \delta)$ is of bounded variation. We have*

$$\int_{\gamma_i} h_{ij} \cdot \varphi_j^- d\mu_1 \leq \frac{1}{\delta} \int_{[a_i,b_i]\times[0,\delta]} \|\tilde{h}_{ij}\| \cdot |\varphi \circ \Gamma_{ij}| d\mu_2 + \int_{(a_i,b_i)\times(0,\delta)} \|\tilde{h}_{ij}\| \cdot |D(\varphi \circ \Gamma_{ij})|.$$

Proof. We can get the first claim easily from formula (7). For $y \in (0, \delta)$, let us consider the rectangle $R_y = [a_i, b_i] \times [0, y]$. The function $\varphi \circ \Gamma_{ij}$, viewed as a function on the interior of R_y, gives the trace G_y^- on the boundary ∂R_y of the rectangle. Remark that the restriction of G_y^- on the edge $[a_i, b_i] \times \{0\}$ does not depend on y and equals the function $\varphi_i^- \circ \Gamma_{ij}$ on $[a_i, b_i] \times \{0\}$ from the definition of the trace. Obviously, we have

$$\int_{[a_i,b_i]\times\{0\}} G_y^- \cdot h_{ij} \circ \Gamma_{i,j} \circ \pi d\mu_1 = \int_{\gamma_i} h_{ij} \cdot \varphi_j^- d\mu_1.$$

Let us define a function $B : (a_i, b_i) \times (0, \delta) \to \mathbb{R}$ by $B(x, y) = G_y^-(x, y)$. Then, from Lebesgue's theorem [15, Theorem 1.3.8], $B(x, y) = \varphi \circ \Gamma_{ij}(x, y)$ for almost every $(x, y) \in (a_i, b_i) \times (0, \delta)$. Applying Proposition 7 to $\varphi \circ \Gamma_{ij}(x, y)$ and $\tilde{h}_{ij}$ on R_y, we obtain

$$\left| \int_{[a_i,b_i]\times\{y\}} B \cdot h_{ij} \circ \Gamma_{ij} \circ \pi d\mu_1 - \int_{\gamma_i} h_{ij} \cdot \varphi_j^- d\mu_1 \right| \leq \int_{\mathrm{int} R_y} \|\tilde{h}_{ij}\| \cdot |D(\varphi \circ \Gamma_{ij})| \tag{10}$$

because $\operatorname{Div} \tilde{h}_{ij} \equiv 0$. Since $\|\tilde{h}_{ij}\| = h_{ij} \circ \Gamma_{ij} \circ \pi$, we get

$$\delta \left(\int_{\gamma_i} h_{ij} \cdot \varphi_j^- d\mu_1 - \int_{(a_i,b_i)\times(0,\delta)} \|\tilde{h}_{ij}\| \cdot |D(\varphi \circ \Gamma_{ij})| \right) \leq \int_{[a_i,b_i]\times[0,\delta]} \|\tilde{h}_{ij}\| \cdot |\varphi \circ \Gamma_{ij}| d\mu_2$$

from Fubini's theorem. This implies the lemma. □

Let us take a small number $\eta > 0$ such that $(1+\eta)^2 M(f) < M(f) + \epsilon$. Remark that, if $\delta > 0$ is sufficiently small, the map Γ_{ij} is almost an isometry on $[a_i, b_i] \times [0, \delta]$. Hence we can take $\delta > 0$ so small that

$$(1+\eta)^{-1}\|v\| < \|D\Gamma_{ij}(v)\| \leq (1+\eta)\|v\| \tag{11}$$

for all tangent vectors v at all points in $[a_i, b_i] \times [0, \delta]$. Let us define a function $\hat{h}_{ij} : \mathbb{R}^2 \to \mathbb{R}$ by

$$\hat{h}_{ij}(z) = \begin{cases} h_{ij} \circ \Gamma_{ij} \circ \pi \circ \Gamma_{ij}^{-1} & \text{for } x \in \Gamma_{ij}; \\ 0 & \text{for } x \notin \Gamma_{ij}. \end{cases}$$

Since $\hat{h}_{ij} \circ \Gamma_{ij} = \|\tilde{h}_{ij}\|$ for $x \in \Gamma_{ij}$, we obtain, from (11),

$$\int_{(a_i,b_i)\times(0,\delta)} \|\tilde{h}_{ij}\| \cdot |D(\varphi \circ \Gamma_{ij})| \leq (1+\eta) \int_{int\Gamma_{ij}} \hat{h}_{ij}|D\varphi|.$$

Therefore we get, from the last lemma,

Proposition 12. *We have*

$$\int_{\gamma_i} h_{ij} \cdot \varphi_j^- d\mu_1$$
$$\leq (1+\eta)^2 \left(\frac{1}{\delta} \int_{\Gamma_{ij}} \hat{h}_{ij}|\varphi| d\mu_2 + \int_{\mathbb{R}^2} \hat{h}_{ij}|D\varphi| \right).$$

Next we prove the following proposition.

Proposition 13. *Let U be one of the regions in the partition $\xi = \{D_j\}_{j=1}^k$. Let $V(x)$ be the unit tangent vector for the boundary ∂U at $x \in \partial U$. Let $\varphi \in \mathcal{BV}(U)$ be non-negative valued. Let φ^- be the trace of φ on the boundary ∂U. Then we have*

$$\mathrm{Var}(P_f\varphi, \mathbb{R}^2) \leq \rho(f)^{-1}\mathrm{Var}(\varphi, U) + C(f, U)\|\varphi\|_{L^1}$$
$$+ \int_{\partial U} \frac{\rho(V(x), f_U)}{|\det Df_U(x)|} \varphi^-(x) d\mu_1(x),$$

where $C(f, U)$ is a constant depending only on the restriction $f|_U$ of f (defined in (12) below).

Proof. We have

$$\int_{\mathbb{R}^2} P_f\varphi d\Psi = \int_{\mathbb{R}^2} \frac{\varphi}{|\det Df|} df^*\Psi$$
$$= \int_{\mathbb{R}^2} \varphi\, d\left(\frac{f^*\Psi}{|\det Df|}\right) - \int_{\mathbb{R}^2} \varphi\, d\left(\frac{1}{|\det Df|}\right) \wedge f^*\Psi$$

for $\Psi \in \Omega_0^1(f(U))$. Hence we get, from formula (7),

$$\mathrm{Var}(P_f\varphi, f(U)) \leq \rho(f)^{-1}\mathrm{Var}(\varphi, U) + C(f, U)\|\varphi\|_{L^1}$$

if we put $\sigma(f, U) = \sup\{\|Df(v)\|/\|v\| \mid 0 \neq v \in T_x\mathbb{R}^2, x \in U\}$ and

$$C(f, U) = \sigma(f, U) \sup\{\|D((\det Df)^{-1})(x)\| \mid x \in U\}. \tag{12}$$

On the other hand, we have

$$\int_{\partial f(U)} (P_f\varphi)^- d\mu_1 = \int_{\partial U} \frac{\rho(V(x), f_U)}{|\det Df_U(x)|}\varphi^-(x)d\mu_1(x).$$

From these and Proposition 7(b), we obtain the conclusion. □

Now we complete the proof of Proposition 10. From Proposition 13,

$$\begin{aligned}\mathrm{Var}(P_f\varphi, \mathbb{R}^2) &\leq \sum_{j=1}^{k} \mathrm{Var}(P_f(\varphi \cdot \chi_{D_j}), \mathbb{R}^2)\\ &\leq \sum_{j=1}^{k}\left(\frac{\mathrm{Var}(\varphi, D_j)}{\rho(f)} + C(f, D_j)\|\varphi\|_{L^1} + \sum_{i\sim j}\int_{\gamma_i} h_{ij}\varphi_j^- d\mu_1\right),\end{aligned}$$

where $\sum_{i\sim j}$ is the sum over i satisfying $\gamma_i \subset \partial D_j$. We have

$$\sum_{j=1}^{k} \mathrm{Var}(\varphi, D_j) \leq \mathrm{Var}(\varphi, \mathbb{R}^2).$$

Hence, in order to prove Proposition 10, we show

$$\sum_{j=1}^{k}\sum_{i\sim j}\int_{\gamma_i} h_{ij}\varphi_j^- d\mu_1 \leq (M(f)+\epsilon)\mathrm{Var}(\varphi, \mathbb{R}^2) + K\|\varphi\|_{L^1}$$

for some constant $K > 0$. But, from Proposition 12, it is sufficient to show

$$(1+\eta)^2\sum_{j=1}^{k}\sum_{i\sim j}\int_{\mathbb{R}^2} \hat{h}_{ij}|D\varphi| \leq (M(f)+\epsilon)\mathrm{Var}(\varphi, \mathbb{R}^2)$$

or, more simply,

$$(1+\eta)^2\sum_{j=1}^{k}\sum_{i\sim j}\hat{h}_{ij}(x) \leq M(f)+\epsilon \quad \text{for } x \in D. \tag{13}$$

Notice that $\hat{h}_{ij}(x) = h_{ij}(x)$ on the dividing curves γ_i. From the definition of weighted multiplicity $M(p, f)$, we have

$$\sum_{j=1}^{k}\sum_{i\sim j}\hat{h}_{ij}(x) \leq M(x, f)$$

for $x \in E$, if δ is sufficiently small. Let F be the set of points that is contained in more than two dividing curves. From the choice of η, we can take a small open neighborhood W of the finite set F in $\overline{D}$ such that the left-hand side of (13) is not larger than $(M(f)+\epsilon)$ on W. If δ is sufficiently small, the intersections of two distinct subsets in $\{\Gamma_{ij} \mid \gamma_i \subset D_j\}$ are contained in W. By continuity, we easily see that the left-hand side of (13) is smaller than $M(f)+\epsilon$ for $x \in \overline{D} - W$ if δ is sufficiently small. Therefore (13) holds for sufficiently small δ. Proposition 10 is proved. □

7. Estimates of the Weighted Multiplicity for the Iterations

In this last section, we complete the proof of Theorem 1 by considering the iterations of a piecewise real-analytic expanding map f. First, remark that a map has an absolutely continuous finite invariant measure if and only if an iteration of it does. Since f^n is a piecewise real-analytic map with minimum expansion rate $\rho(f^n) \geq \rho(f)^n$, we can assume

$$\rho(f) > 64\pi$$

without loss of generality. We prove the following theorem.

Theorem 14. $M(f^n) \to 0$.

Subdividing the partition ξ by real-analytic curves artificially, we can assume that f satisfies condition (b) in Theorem 9. It means that all iterations of f also satisfy that condition. Thus, from Theorem 14 above, iterations f^n of f satisfy assumptions (a) and (b) of Theorem 9 if n is sufficiently large. Therefore we can get Theorem 1 from Theorem 14.

Let us prepare some notations in order to prove Theorem 14. Let $\xi_n = \{D_i^{(n)}\}_{i=1}^{k(n)}$ be the real-analytic partition associated to the piecewise real-analytic map f^n. Let $E(n) = \overline{D} - \cup_{i=1}^{k(n)} D_i^{(n)} = \cup_{i=1}^{k(n)} \partial D_i^{(n)}$. For $p \in E(n)$, let $\{\beta_i(p,n)\}_{i=1}^{m(p,n)}$ be the germs of real-analytic curves at $p \in E(n)$ given by the dividing curves of the partition ξ_n. We assume that these germs of curves are arranged in counterclockwise order around p as before. Let $U_i(p,n)$ be the region between $\beta_i(p,n)$ and $\beta_{i+1}(p,n)$ for $1 \leq i \leq m(p,n)$. Let us denote $\mathcal{U}(p,n) = \{U_i(p,n)\}_{i=1}^{m(p,n)}$. For $U \in \mathcal{U}(p,n)$, let f_U^n be the germ of the real-analytic map at p obtained as a real-analytic continuation of the restriction of f^n to a representative of U.

Let Δ be the maximum of $\mathrm{Ord}(U_i(p,1))$ over all $1 \leq i \leq m(p,1)$ and all $p \in E(1)$. Let θ and Θ be the minimum and maximum of

$$\mathrm{Angle}_{\mathrm{Ord}(U_i(p,1))}(f(U_i(p,1)))$$

over all $1 \leq i \leq m(p,1)$ and all $p \in E(1)$ respectively. Let μ be the maximum of $m(p,1)$ over all $p \in E(1)$. Thus Δ, θ, Θ and μ depend only on the single map f.

Let us consider a point $p \in E(n)$. Each $V \in \mathcal{U}(p,n+1)$ is contained in some $U \in \mathcal{U}(p,n)$ as a germ. Remark that, if $V \subset U$ and $V \neq U$, the image $f_U^n(p)$ is contained in $E(1)$ and a dividing curve of the partition $\xi = \xi_1$ passing through $f_U^n(p)$ divides $f^n(U)$ into more than two regions.

We say that $V \in \mathcal{U}(p,n+1)$ is a kid of $U \in \mathcal{U}(p,n)$ if $V \subset U$. If V is a kid of U and if V and U have at least one germ of real-analytic curve as a boundary curve in common additionally, we say V is a daughter of U. Especially, if $V = U$, V is a daughter of U. Obviously, each $U \in \mathcal{U}(p,n)$ has at most two daughters. If $\mathrm{Ord}(V) > \mathrm{Ord}(U)$, we say that V is a small kid of U.

The reason why we distinguish daughters is the following. Let $V \in \mathcal{U}(n+1,p)$ be a kid of $U \in \mathcal{U}(n,p)$ and assume that $U \neq V$. If V is not a daughter, $f^n(V)$ should coincide with an element of $\mathcal{U}(f_U^n(p),1)$. So $\mathrm{Angle}_{\mathrm{Ord}(V)}(f^{n+1}(V)) \leq \Theta$. On the other hand, if V is a daughter and it is small, we can not expect such anestimate on $\mathrm{Angle}_{\mathrm{Ord}(V)}(f^{n+1}(V))$. For the same reason, we put the following definition. An element U of $\mathcal{U}(p,n)$ is called special if $\mathrm{Ord}(U) > \Delta$ or if there is a chain $U_i \in \mathcal{U}(p,n-\ell+i)$, $i = 0,1,2,\cdots,\ell$, of regions with length $\ell+1 \geq 2$ such that

- $U_\ell = U$,

- U_1 is a small kid and daughter of U_0, and
- U_{i+1} is a daughter of U_i for $1 \le i < \ell$.

In order to estimate $M(f^n)$, we introduce what we call the modified weight $\mathcal{W}(U)$ of $U \in \mathcal{U}(p, n)$ in the following manner. We fix a small number $0 < \eta < 1$ that will be specified later in the condition (16). We define the level $\ell(U)$ of $U \in \mathcal{U}(p, n)$, $p \in E(n)$, by

$$\ell(U) = \begin{cases} 2\min\{\mathrm{Ord}(U), \Delta + 1\} & \text{if } U \text{ is not special;} \\ 2\min\{\mathrm{Ord}(U), \Delta + 1\} - 1 & \text{if } U \text{ is special.} \end{cases}$$

Remark that we always have $\ell(V) \ge \ell(U)$ if V is a kid of U. If $U \in \mathcal{U}(p, n)$ is special, we define

$$\mathcal{W}(U) = \frac{\eta^{\ell(U)} \rho(U, f_U^n)}{|\det Df_U^n(p)|}.$$

(For the definition of $\rho(U, f_U^n)$, see Sect. 3.) On the other hand, if U is not special, we define

$$\mathcal{W}(U) = \frac{\eta^{\ell(U)} \rho(U, f_U^n)}{|\det Df_U^n(p)|} \left[\frac{\mathrm{Angle}_{\mathrm{Ord}(U)}(f^n(U))}{\theta} + 1 \right],$$

where $[\cdot]$ is Gauss' symbol. We put

$$\mathcal{M}(p, f^n) = \sum_{U \in \mathcal{U}(p,n)} \mathcal{W}(U, f^n) \quad \text{and} \quad \mathcal{M}(f^n) = \sup_{p \in E(n)} \mathcal{M}(p, f^n).$$

Clearly, we have $M(f^n) \le 2\eta^{-2\Delta-2}\mathcal{M}(f^n)$. In order to prove Theorem 14, it is enough to show the following proposition.

Proposition 15. *If we take $\eta > 0$ sufficiently small, we have*

$$\sum_{V:\text{a kid of } U} \mathcal{W}(V) \le \mathcal{W}(U)/2 \tag{14}$$

for all $U \in \mathcal{U}(p, n)$, $p \in E(n)$, $n \ge 1$.

In fact, if this is true, we have

$$\mathcal{M}(p, f^{n+1}) \le (1/2)\mathcal{M}(p, f^n) \le (1/2)\mathcal{M}(f^n)$$

for $p \in E(n)$. On the other hand, we have

$$\mathcal{M}(p, f^{n+1}) \le \rho(f)^{-n}\mathcal{M}(f^n(p), f) \le (1/2)^n \mathcal{M}(f)$$

for $p \in E(n+1) - E(n)$. These show $\mathcal{M}(f^n) \le (1/2)^n \mathcal{M}(f)$ inductively. Therefore $M(f^n) \le (1/2)^{n-1}\eta^{-2\Delta-2}\mathcal{M}(f) \to 0$ as $n \to \infty$.

Proof. We prove Proposition 15. Let us consider a region $U \in \mathcal{U}(p, n)$ and its kids. We assume that $\mathrm{Ord}U \le \Delta$ until the end of this proof where we treat the case $\mathrm{Ord}U > \Delta$. We classify the kids V of U into the following four classes:

1. V is a daughter of U, and V is a small kid of U,
2. V is not a daughter of U, and V is a small kid of U,
3. V is a daughter of U, and V is not a small kid of U, and
4. V is not a daughter of U, and V is not a small kid of U.

We estimate the sums of modified weights over the kids V in each class.

First we consider the kids in class 1. Kids V in this class is special. Hence we have

$$\mathcal{W}(V) \leq \frac{\rho(f^n(V), f_{f^n(U)})}{|\det Df_{f^n(U)}(f_U^n(p))|}\mathcal{W}(U) \leq \rho(f)^{-1}\mathcal{W}(U).$$

Since the number of kids in this class is at most 2, the sum of $\mathcal{W}(V)$ over the kids V in this class is not larger than $2\rho(f)^{-1}\mathcal{W}(U) < (1/8)\mathcal{W}(U)$.

We consider class 2. Note that $\ell(V) \geq \ell(U) + 1$ in this case. The number of the kids in this class is not larger than μ. For each kid V in this class, we have

$$\text{Angle}_{\text{Ord}(V)}(f^{n+1}(V)) \leq \Theta.$$

Thus it holds

$$\begin{aligned}\mathcal{W}(V) &\leq \eta(\Theta/\theta + 1)\frac{\rho(f^n(V), f_{f^n(V)})}{|\det Df_{f^n(V)}(f_V^n(p))|}\mathcal{W}(U)\\ &\leq \eta(\Theta/\theta + 1)\mathcal{W}(U).\end{aligned}$$

Hence the sum of $\mathcal{W}(V)$ over the kids V in this class is not larger than $\mu\eta(\Theta/\theta + 1)\mathcal{W}(U)$.

We consider case 3. Remark that U is special if and only if so is V. In the case U is special, we easily see that

$$\mathcal{W}(V) \leq \frac{\rho(f^n(V), f)}{|\det Df_U(p)|}\mathcal{W}(U) \leq \rho(f)^{-1}\mathcal{W}(U).$$

And the sum is not larger than $2\rho(f)^{-1}\mathcal{W}(U) < (1/8)\mathcal{W}(U)$. In case U is not special, we have

$$\begin{aligned}\mathcal{W}(V) &\leq \frac{[\text{Angle}_{\text{Ord}(V)}(f^{n+1}(V))/\theta] + 1}{[\text{Angle}_{\text{Ord}(U)}(f^n(U))/\theta] + 1}\frac{\rho(f^n(V), f_{f^n(V)})}{|\det Df_{f^n(V)}(f_V^n(p))|}\mathcal{W}(U)\\ &\leq \left\{1 + \frac{\text{Angle}_{\text{Ord}(V)}(f^{n+1}(V))}{\text{Angle}_{\text{Ord}(V)}(f^n(V))}\right\}\frac{\rho(f^n(V), f_{f^n(V)})}{|\det Df_{f^n(V)}(f_V^n(p))|}\mathcal{W}(U).\end{aligned}$$

Here we used the fact $\text{Angle}_{\text{Ord}(U)}(f^n(U)) \geq \text{Angle}_{\text{Ord}(V)}(f^n(V))$ and an inequality

$$\frac{[y+1]}{[x+1]} \leq \frac{y+1}{\max\{x, 1\}} \leq \frac{y}{x} + 1 \quad \text{for } x > 0 \text{ and } y > 0.$$

Using Lemma 5 in case $\frac{\text{Angle}(f^{n+1}(V))}{\text{Angle}(f^n(V))} > 1$, we get

$$\mathcal{W}(V) \leq 4\pi\rho(f)^{-1}\mathcal{W}(U).$$

Since the number of kids in this class is at most 2, the sum is not larger than $8\pi\rho(f)^{-1}\mathcal{W}(U) < (1/8)\mathcal{W}(U)$.

We see class 4. In this case, let $V_i, i = 1, 2, \cdots, \ell$, be the kids of this class and let $d = \text{Ord}(U)$. Remark that $V_i, i = 1, 2, \cdots, \ell$, are not special. In case U is special, we have $\ell(V_i) \geq \ell(U) + 1$ for all i. Hence, in this case, we can see that the sum $\sum_{i=1}^{\ell}\mathcal{W}(V_i)$

is not larger than $\mu\eta(\Theta/\theta+1)\mathcal{W}(U)$ by just the same argument as above for class 2. Let us consider the case that U is not special. Obviously we have

$$\sum_{i=1}^{\ell} \mathrm{Angle}_d(f^n(V_i)) \le \mathrm{Angle}_d(f^n(U)). \tag{15}$$

Since $\mathrm{Angle}_d(f^{n+1}(V_i)) \ge \theta$ for $i = 1, 2, \cdots, \ell$, we have

$$[\mathrm{Angle}_d(f^{n+1}(V_i))/\theta + 1] \le 2\mathrm{Angle}_d(f^{n+1}(V_i))/\theta.$$

By using Lemma 5, we obtain

$$\begin{aligned}
\frac{\mathcal{W}(V_i)}{\mathcal{W}(U)} &\le \frac{[\mathrm{Angle}_d(f^{n+1}(V_i))/\theta+1]}{[\mathrm{Angle}_d(f^n(U)/\theta+1]} \cdot \frac{\rho(f^n(V_i), f)}{|\det Df_{f^n(V_i)}(f^n_{V_i}(p))|} \\
&\le \frac{[\mathrm{Angle}_d(f^{n+1}(V_i))/\theta+1]}{[\mathrm{Angle}_d(f^n(U))/\theta+1]} \cdot \frac{2\pi\,\mathrm{Angle}_d(f^n(V_i))}{\rho(f)\mathrm{Angle}_d(f^{n+1}(V_i))} \\
&\le \frac{4\pi\,\mathrm{Angle}_d(f^n(V_i))}{\rho(f)\mathrm{Angle}_d(f^n(U))}.
\end{aligned}$$

Thus we have, from (15),

$$\sum_{i=1}^{\ell} \mathcal{W}(V_i) \le 4\pi\rho(f)^{-1}\mathcal{W}(U)$$

in this case.

Summing up all the above arguments for the four classes, we obtain (14) for the case $\mathrm{Ord}(U) \le \Delta$, if we take $\eta > 0$ so small that

$$\mu\eta(\Theta/\theta+1) < 1/8 \tag{16}$$

because the sums of modified weights over each of four classes are smaller than $(1/8)\mathcal{W}(U)$.

Finally, let us consider the case $\mathrm{Ord}(U) > \Delta$. In this case every kid of U should be a daughter. So the number of the kids is at most 2. Since U and its kid V are special, we have

$$\mathcal{W}(V) \le \frac{\rho(f^n(V), f)}{|\det Df_{f^n(V)}(f^n_V(p))|}\mathcal{W}(U) \le \rho(f)^{-1}\mathcal{W}(U).$$

Therefore the left-hand side of (14) is smaller than $2\rho(f)^{-1}\mathcal{W}(U) < (1/2)\mathcal{W}(U)$. We completed the proof of Proposition 15. □

Appendix: Other Ergodic Properties of Piecewise Real-Analytic Expanding Maps

Proposition 10 and Theorem 9 imply that, if n is sufficiently large, the iteration f^n satisfies the inequality (9) with f replaced by f^n and the coefficient $(M(f^n)+\rho^{-1}(f^n)+\epsilon)$ is smaller than 1. As is pointed out in the papers of Keller [7] and Góra & Boyarski [4], once we get such an inequality, we can derive many properties of Perron–Frobenius operator P_f and those of the absolutely continuous invariant measures for f. In fact, we can apply an ergodic theorem of Ionescu-Tulcea and Marinescu [6] and show that, if we put $\mathcal{BV}(U, \mathbb{C}) = \{\xi + \sqrt{-1}\eta \mid \xi, \eta \in \mathcal{BV}(U)\}$ and $\|\xi + \sqrt{-1}\eta\|_{\mathcal{BV}(U,\mathbb{C})} = \|\xi\|_{\mathcal{BV}(U)} + \|\eta\|_{\mathcal{BV}(U)}$,

- the operator $P_f : \mathcal{BV}(U) \to \mathcal{BV}(U)$ has finitely many eigenvalues $\lambda_1, \lambda_2, \cdots, \lambda_r$ of modulus 1, and the corresponding eigenspaces $E_i = \{\varphi \in \mathcal{BV}(U, \mathbb{C}) \mid P_f\varphi = \lambda_i\varphi\}$ are of finite dimension,
- the natural extension of P_f to $\mathcal{BV}(U, \mathbb{C})$ is written as

$$P_f = \sum_{i=1}^{r} \lambda_i \Psi_i + Q,$$

where Ψ_i are projections to E_i and
  - $\limsup_{n\to\infty} \sqrt[n]{\|Q^n\|_{\mathcal{BV}(U,\mathbb{C})}} < 1$,
  - $\Psi_i \circ \Psi_j = 0$ for $i \neq j$ and $\Psi_i \circ Q = Q \circ \Psi_i = 0$ for $1 \le i \le r$.

Also one of the eigenvalues λ_i must be 1. See [7] for details. Let us assume $\lambda_1 = 1$. Since P_f does not increase L^1-norm of functions and since $\mathcal{BV}(U)$ is dense in $L^1(U)$, we can see, by approximation argument,

- For any $\varphi \in L^1(U)$, the sequence $1/n \sum_{k=0}^{n-1} P_f^k(\varphi)$ converges in L^1 to an element in the eigenspace E_1 for $\lambda_1 = 1$.

Especially the density function of each absolutely continuous invariant measure is contained in the finite dimensional space E_1. So we have

- there exists only finitely many absolutely continuous ergodic measures $\mu_k, 1 \le k \le q$, and all other absolutely continuous invariant measures are convex combinations of them.

Furthermore, using the argument in [5, 13], we can derive the following ergodic properties of the measure μ_i's.

- For each μ_k, there exist a positive integer p and Borel measurable mutually disjoint subsets C_i, $1 \le i \le p$, such that $\mu_k(C_i) = 1/p$, $f(C_i) \subset C_{i+1}$ for $1 \le i \le p-1$, $f(C_p) \subset C_1$ and $f^p|_{C_i}$ are exact.

We refer to [4, Sect. 3], [9] and the references given there for further results.

References

1. Adl-Zarabi, K.: Absolutely continuous invariant measure for piecewise expanding C^2 transformations in $\mathbb{R}^n$ on domains with cusps on the boundaries. Ergod. Th.& Dynam. Sys. **16**, 1–18 (1996)
2. Buzzi, J.: A.C.I.M.'S for arbitrary expanding piecewise $\mathbb{R}$-analytic mappings of the plane. Ergod. Th.& Dynam. Sys. (to appear)
3. Giusti, E.: *Minimal Surfaces and Functions of Bounded Variation*. Monographs in Mathematics, Vol. **80**, Boston: Birkhauser, 1984
4. Góra, P., Boyarski, A.: Absolutely continuous invariant measures for piecewise expanding transformations in $\mathbb{R}^N$. Israel J. Math. **67**, 272–276 (1989)
5. Hofbauer, F. and Keller, G.: Ergodic properties of invariant measures for piecewise monotonic transformations. Math. Z. **180**, 119–140 (1982)
6. Ionescu-Tulcea, C., and Marinescu, G.: Theorie ergodique pour des classes d'operations non completement continues. Ann. Math.(2) **52**, 140–147 (1950)
7. Keller, G.: *Propriété ergodique des endomorphismes dilatants, C^2 par morceaux, des rǵions bornées du plan.* Thesis, Universite de Rennes, 1979
8. Keller, G.: Ergodicité et mesures invariantes pour les transformations dilatantes par morceaux d'une région bornée du plan. C.R.Acad. Sci. Paris **289** Serie A, 625–627 (1979)
9. Keller, G.: Generalized bounded variation and applications to piecewise monotonic transformations. Z. Wahr. verw. Geb. **69**, 461–478 (1985)

10. Lang, S.: *Real and Functional Analysis*. Graduate Text in Math. **142**, Berlin–Heidelberg–New York: Springer, 1993
11. Lasota, A., Yorke, J.: On the existence of invariant measure for piecewise monotonic transformations. Trans. A.M.S. **186**, 481–488 (1973)
12. Renyi,A.: Representation of real numbers and their ergodic properties. Acta. Math. Akad. Sc. Hungar. **8**, 477–493 (1957)
13. Rychlik, M.: Bounded variation and invariant measures. Studia math. **76**, 69–80 (1983)
14. Tsujii, M.: Piecewise expanding maps on the plane with singular ergodic properties. Ergod. Th.& Dynam. Sys. (to appear)
15. Ziemer, W.: *Weakly Differentiable Functions*. Graduate Text in Math. 120, Berlin–Heidelberg–New York: Springer, 1989

Communicated by Ya. G. Sinai

Invent. math. 140, 351–398 (2000)
Digital Object Identifier (DOI) 10.1007/s002220000057

Inventiones mathematicae

SRB measures for partially hyperbolic systems whose central direction is mostly expanding*

José F. Alves[1], **Christian Bonatti**[2], **Marcelo Viana**[3]

[1] Dep. Matemática Pura, Faculdade de Ciências do Porto, P-4099-002 Porto, Portugal (e-mail: jfalves@fc.up.pt)
[2] Lab. Topologie, CNRS – UMR 5584, Univ. Bourgogne, B.P. 47870, 21078 Dijon Cedex, France (e-mail: bonatti@satie.u-bourgogne.fr)
[3] IMPA, Estrada D. Castorina 110, Jardim Botânico, 22460-320 Rio de Janeiro RJ, Brazil (e-mail: viana@impa.br)

Oblatum 16-IV-1999 & 29-X-1999
Published online: 21 February 2000 –

Abstract. We construct Sinai-Ruelle-Bowen (SRB) measures supported on partially hyperbolic sets of diffeomorphisms – the tangent bundle splits into two invariant subbundles, one of which is uniformly contracting – under the assumption that the complementary subbundle is *non-uniformly* expanding. If the rate of expansion (Lyapunov exponents) is bounded away from zero, then there are only finitely many SRB measures. Our techniques extend to other situations, including certain maps with singularities or critical points, as well as diffeomorphisms having only a dominated splitting (and no uniformly hyperbolic subbundle).

1. Introduction

The following approach has been most effective in studying the dynamics of complicated systems: one tries to describe the average time spent by typical orbits in different regions of the phase space. According to the ergodic theorem of Birkhoff, such times are well defined for almost all point, with respect to any invariant probability measure. However, the notion of typical orbit is usually meant in the sense of volume (Lebesgue measure), which is not always captured by invariant measures. Indeed, it is a fundamental open problem to understand under which conditions the behaviour of typical points is well defined, from this statistical point of view.

* Work carried out at the Laboratoire de Topologie, CNRS-UMR 5584, Dijon, and IMPA, Rio de Janeiro. Partially supported by IMPA and PRONEX-Dynamical Systems, Brazil, Université de Bourgogne, France, and Praxis XXI-Física Matemática and Centro de Matemática da Universidade do Porto, Portugal.

This problem can be precisely formulated by means of the following notion, introduced by Sinai, Ruelle, and Bowen. Let us consider discrete time systems, namely maps $f : M \to M$ on a manifold M. Given an f-invariant Borelian probability μ in M, we call *basin* of μ the set $B(\mu)$ of the points $x \in M$ such that the averages of Dirac measures along the orbit of x converge to μ in the weak* sense:

$$\lim_{n\to+\infty} \frac{1}{n} \sum_{j=0}^{n-1} \varphi(f^j(x)) = \int \varphi \, d\mu \quad \text{for any continuous } \varphi : M \to \mathbb{R}. \qquad (1)$$

Then we say that μ is a *physical* or *Sinai-Ruelle-Bowen (SRB) measure* for f if the basin $B(\mu)$ has positive Lebesgue measure in M. Then, one would like to know whether, for most systems, the basins of all SRB measures cover a full Lebesgue measure subset of the whole manifold.

This question has an affirmative answer in the context of uniformly hyperbolic systems, after [Sin72,BR75,Bow75,Rue76]. A detailed picture is also available for maps of the circle or the interval, see [Jak81,Lyu]. However, in higher dimensions the problem is mostly open, outside the uniformly hyperbolic setting, despite substantial progress in the study of certain classes of maps and flows with some properties of non-uniform hyperbolicity, including the Lorenz-like attractors, see e.g. [Pes92,Sat92,Tuc99], and the Hénon-like attractors, see [BC91,BY92,BV].

In this work we deal with diffeomorphisms admitting *partially hyperbolic* invariant sets: the tangent bundle over the set has an invariant dominated splitting into two subbundles, one of which is uniformly hyperbolic (contracting or expanding). Precise definitions are given in the next subsection. This property yields a fair amount of geometric information, e.g. invariant foliations, so that it is natural to try to recover for these systems as much as possible of the geometric and ergodic properties of the hyperbolic ones.

Indeed, some knowledge of such properties is already available, through works of several authors. For the foundations concerning invariant foliations see [BP74,HPS77]. Gibbs u-states were constructed by [PS82], and used by [You90,Car93,BV99] to construct SRB measures for some types of partially hyperbolic systems. SRB measures and decay of correlations were also studied by [Alv97,Cas98,Dol]. Moreover, [Shu71,Mañ78,GPS94,Kan94,BD96,Via97,SW] constructed examples where partial hyperbolicity is used to get robustness (stability) of topological or ergodic properties, which is also a main topic in [BP74]. Conversely, [Mañ82,DPU99,BDP], showed that partial hyperbolicity (or, in high dimensions, existence of a dominated splitting) is in fact a necessary condition for robust topological transitivity.

The interest in this class of systems was further stressed by these recent results, that suggest that partial hyperbolicity (or, at least, existence of a dominated splitting) should be a crucial ingredient in a global theory of Dynamics. A program towards such a theory has been proposed a few years ago by [Pal99], the cornerstone of which is a conjecture containing the following statement: every system can be approximated by one having finitely many SRB measures, whose basins cover a full Lebesgue measure subset of the phase space.

In the present paper we obtain results of existence and finiteness of SRB measures for a large class of partially hyperbolic maps. Previous constructions depended on the existence of a uniformly expanding (strong-unstable) invariant subbundle. In fact, except for the situation of [Via97,Alv97], the SRB measures coincide with the Gibbs u-states (invariant measures absolutely continuous in the strong-unstable direction) constructed by [PS82]. As shown in [BV99], this happens whenever the central subbundle is (at least) non-uniformly contracting.

One main novelty here is that we do not assume existence of a strong-unstable direction. Using only a condition of non-uniform expansion in an invariant (centre-unstable) subbundle, we are able to construct invariant measures absolutely continuous in this centre-unstable direction. Actually, this does not even require the full strength of partial hyperbolicity, it suffices to have a dominated splitting. In some cases, these *Gibbs cu-states* are SRB measures: in particular, this is always the case if the complementary subbundle is uniformly contracting on the support of the measure.

1.1. *Partially hyperbolic diffeomorphisms*

Let $f : M \to M$ be a C^1 diffeomorphism on a manifold M. Here we say that a compact set $K \subset M$ is *partially hyperbolic* for f if it is positively invariant, i.e. $f(K) \subset K$, and there exists a continuous Df-invariant splitting $T_K M = E^{ss} \oplus E^{cu}$ of the tangent bundle restricted to K and a constant $\lambda < 1$ satisfying (for some choice of a Riemannian metric on M)

1. E^{ss} is uniformly contracting: $\|Df \mid E_x^{ss}\| \le \lambda$ for all $x \in K$;
2. E^{cu} is dominated by E^{ss}: $\|Df \mid E_x^{ss}\| \cdot \|Df^{-1} \mid E_{f(x)}^{cu}\| \le \lambda$ for all $x \in K$.

We call E^{ss} strong-stable subbundle and E^{cu} centre-unstable subbundle.

Theorem A. *Let $f : M \to M$ be a C^2 diffeomorphism having a partially hyperbolic set K. Assume that f is non-uniformly expanding along the centre-unstable direction, meaning that*

$$\limsup_{n\to+\infty} \frac{1}{n} \sum_{j=1}^{n} \log \left\|Df^{-1} \mid E_{f^j(x)}^{cu}\right\| < 0 \qquad (2)$$

for all x in a positive Lebesgue measure set $H \subset K$. Then f has some ergodic SRB measure with support contained in $\cap_{j=0}^{\infty} f^j(K)$. In fact, Lebesgue almost every point in H belongs in the basin of some such SRB measure.

As we shall explain below, for the proof it is enough to have condition (2) on a positive Lebesgue measure subset of some disk transverse to the strong-stable direction.

The SRB measures produced by this theorem have dim E^{cu} positive Lyapunov exponents and dim E^{ss} negative Lyapunov exponents. Moreover, they have absolutely continuous conditional measures along corresponding Pesin unstable manifolds (which are tangent to the centre-unstable subbundle E^{cu} at almost every point).

We ignore whether the conclusion of Theorem A remains true if the non-uniform expansion condition is replaced by

$$\limsup_{n\to+\infty} \frac{1}{n} \log \left\| Df^{-n} \mid E^{cu}_{f^n(x)} \right\| < 0 \tag{3}$$

(i.e. positivity of all the Lyapunov exponents in the centre-unstable direction). Clearly, the two formulations (2) and (3) are equivalent when the non-uniformly expanding direction is 1-dimensional, that is, when E^{cu} can be split as $E^c \oplus E^{uu}$ with E^c having dimension 1 and E^{uu} being uniformly expanding. Let us also note that for a set of points with full probability (full measure with respect to any invariant probability measure) condition (3) implies condition (2) for some f^k, $k \geq 1$. Of course, the theorem is not affected when f is replaced by any positive iterate of it.

As a consequence of the proof of the theorem we also get

Corollary B. *Under the assumptions of Theorem A, if the limit in (2) is bounded away from zero, then the set H is contained in the union of the basins of finitely many SRB measures, up to a zero Lebesgue measure subset.*

These results have the following curious consequence: in the setting of Theorem A the set of values of the limit in (2) over a full Lebesgue measure subset of H is discrete, with zero as the unique possible accumulation value.

The assumption of partial hyperbolicity in Theorem A can be somewhat relaxed, as we explain in Sect. 6. Assuming only the existence of a dominated splitting $E^{cs} \oplus E^{cu}$ with E^{cu} non-uniformly expanding, we prove that the diffeomorphism admits invariant probability measures with absolutely continuous conditional measures along the centre-unstable direction. This may be thought of as a non-uniform version of the results of [PS82] on existence of Gibbs u-states. We also show that in some cases where E^{cs} is non-uniformly contracting these invariant measures are SRB measures for the corresponding system.

1.2. Maps with singular sets

Similar methods allow us to construct SRB measures for certain maps with singularities and/or critical points. Apart from a condition of non-uniform expansion on a positive Lebesgue measure subset H, similar to (2), we also need the points of H not to spend most of the time too close to the singular set. This is properly expressed in (6) below. Before that, let us explain what we mean by singular set of a map.

Let M be a compact manifold, $\mathcal{S} \subset M$ a compact subset, and $f : M \setminus \mathcal{S} \to M$ a C^2 map on $M \setminus \mathcal{S}$. We assume that f behaves like a power of the distance to $\mathcal{S}$ close to the *singular set* $\mathcal{S}$, in the following sense: there exist constants $B > 1$ and $\beta > 0$ such that

(S1) $\dfrac{1}{B}\operatorname{dist}(x, \mathcal{S})^{\beta} \le \dfrac{\|Df(x)v\|}{\|v\|} \le B \operatorname{dist}(x, \mathcal{S})^{-\beta}$;

(S2) $|\log\|Df(x)^{-1}\| - \log\|Df(y)^{-1}\|| \le B\dfrac{\operatorname{dist}(x, y)}{\operatorname{dist}(x, \mathcal{S})^{\beta}}$;

(S3) $|\log|\det Df(x)| - \log|\det Df(y)|| \le B\dfrac{\operatorname{dist}(x, y)}{\operatorname{dist}(x, \mathcal{S})^{\beta}}$;

for every $v \in T_xM$ and $x, y \in M \setminus \mathcal{S}$ with $\operatorname{dist}(x, y) < \operatorname{dist}(x, \mathcal{S})/2$.

Given $\delta > 0$ and $x \in M \setminus \mathcal{S}$ we define the *δ-truncated distance* from x to $\mathcal{S}$

$$\operatorname{dist}_{\delta}(x, \mathcal{S}) = \begin{cases} 1 & \text{if } \operatorname{dist}(x, \mathcal{S}) \ge \delta \\ \operatorname{dist}(x, \mathcal{S}) & \text{otherwise.} \end{cases} \tag{4}$$

Note that this is not really a distance function: $\operatorname{dist}(x, y) + \operatorname{dist}_{\delta}(y, \mathcal{S})$ may be smaller than $\operatorname{dist}_{\delta}(x, \mathcal{S})$. We also denote $\mathcal{S}_{\infty} = \cup_{n=0}^{\infty} f^n(\mathcal{S})$.

Theorem C. *Assume that f satisfies (S1), (S2), (S3) and is non-uniformly expanding, in the sense that*

$$\limsup_{n\to+\infty} \frac{1}{n} \sum_{j=0}^{n-1} \log\|Df(f^j(x))^{-1}\| < 0 \tag{5}$$

for all x in a positive Lebesgue measure set $H \subset M \setminus \mathcal{S}_{\infty}$. Assume moreover that, given any $\varepsilon > 0$ there exists $\delta > 0$ such that for every $x \in H$

$$\limsup_{n\to+\infty} \frac{1}{n} \sum_{j=0}^{n-1} -\log \operatorname{dist}_{\delta}(f^j(x), \mathcal{S}) \le \varepsilon. \tag{6}$$

Then Lebesgue almost every point in H belongs in the basin of some ergodic absolutely continuous invariant measure.

As a by-product, corresponding to the case when the singular set is empty,

Corollary D. *Let $f : M \to M$ be a C^2 covering map (local diffeomorphism) on a compact manifold M, which is non-uniformly expanding: (5) holds for all x in a set $H \subset M$ with positive Lebesgue measure. Then Lebesgue almost every point $x \in H$ belongs in the basin of some ergodic absolutely continuous invariant measure.*

In the settings of these last two results, we also have an analogue of Corollary B: if the limit in (5) is bounded away from zero then H is covered by the basins of finitely many ergodic absolutely continuous invariant measures. See also the comments at the end of Sect. 5 concerning similar results for partially hyperbolic maps with singularities.

For unimodal maps of the interval with negative Schwarzian derivative, results in a similar spirit had been obtained by [Led81] and [Kel90]. In particular, [Kel90] contains a strong version of Corollary D for such maps.

This paper is organized as follows. To construct the SRB measures in Theorem A we fix any C^2 disk transverse to the strong-stable direction and intersecting H on a positive Lebesgue measure subset, and we consider the sequence of averages of forward iterates of Lebesgue measure restricted to such a disk. We prove, in Sect. 3, that a definite fraction of each average corresponds to a measure that is absolutely continuous with respect to Lebesgue measure along the iterates of the disk, with uniformly bounded densities. This uses distortion bounds that we obtain in Sect. 2, together with the key notion of *hyperbolic times*, first introduced in [Alv97]. The construction of the SRB measures is completed in Sect. 4, where we show that the absolute continuity property passes to the limit. In Sect. 4 we also obtain enough information about the basins of these measures to prove the finiteness statement in Corollary B.

In Sect. 5 we explain how condition (6) allows us to bypass the difficulty caused by the presence of singularities, and obtain Theorem C and Corollary D, as well as the results on partially hyperbolic maps with singularities that we mentioned before. Sect. 6 contains related results for systems with a dominated splitting $E^{cs} \oplus E^{cu}$ where neither of the factors needs have any uniform subbundle, that we also mentioned above. In the Appendix we give a few simple criteria allowing to check the assumptions of our results in specific situations.

2. Curvature and distortion

A first main step in the proof of Theorem A is to prove a bounded distortion property for iterates of f over disks whose tangent space is contained in a centre-unstable cone at each point. This section is devoted to the precise statement and proof of this property.

Remark 2.1. Throughout this section, as well as in Sect. 3 and Subsect. 4.1, we do not need the full strength of partial hyperbolicity. In Subsect. 2.1 we only use existence of a dominated splitting $E^{cs} \oplus E^{cu}$. Subsections 2.2 and 2.3, Sect. 3, and Subsect. 4.1 depend also on the condition of non-uniform expansion along the centre-unstable direction (2). Existence of a strong-stable subbundle E^{ss} is used for the first time in Subsect. 4.2.

We fix continuous extensions of the two subbundles E^{cs} and E^{cu} to some neighbourhood V_0 of K, that we denote $\tilde{E}^{cs}$ and $\tilde{E}^{cu}$. It should be noted that we do not require these extensions to be invariant under Df. Then, given $0 < a < 1$, we define the *centre-unstable cone field* $C_a^{cu} = \left(C_a^{cu}(x)\right)_{x \in V_0}$ *of width* a by

$$C_a^{cu}(x) = \left\{v_1 + v_2 \in \tilde{E}_x^{cs} \oplus \tilde{E}_x^{cu} \text{ such that } \|v_1\| \leq a\|v_2\|\right\}. \tag{7}$$

We define the *centre-stable cone field* $C_a^{cs} = \left(C_a^{cs}(x)\right)_{x \in V_0}$ *of width* a in a similar way, just reversing the roles of the subbundles in (7).

We fix $a > 0$ and V_0 small enough so that, up to slightly increasing $\lambda < 1$, the domination condition 2. in Subsect. 1.1 remains valid for any pair of vectors in the two cone fields:

$$\|Df(x)v^{cs}\| \cdot \|Df^{-1}(f(x))v^{cu}\| \leq \lambda \|v^{cs}\| \, \|v^{cu}\|$$

for every $v^{cs} \in C_a^{cs}(x)$, $v^{cu} \in C_a^{cu}(f(x))$, and any point $x \in V_0 \cap f^{-1}(V_0)$. Note that the centre-unstable cone field is positively invariant: $Df(x)C_a^{cu}(x) \subset C_a^{cu}(f(x))$ whenever x and $f(x)$ are in V_0. Indeed, the domination property together with the invariance of $E^{cu} = (\tilde{E}^{cu} \mid K)$ imply

$$Df(x)C_a^{cu}(x) \subset C_{\lambda a}^{cu}(f(x)) \subset C_a^{cu}(f(x)),$$

for every $x \in K$, and this extends to any $x \in V_0 \cap f^{-1}(V_0)$ by continuity.

Wherever we presume E^{cs} to be uniformly contracting (as already mentioned, this will happen not happen before Subsect. 4.2), we denote it by E^{ss} instead, and represent by $\tilde{E}^{ss}$ its extension to V_0. Moreover, in that case the cone field C_a^{cs} is denoted C_a^{ss}, and called *strong-stable*.

2.1. Hölder control of the tangent direction

We say that an embedded C^1 submanifold $N \subset V_0$ is *tangent to the centre-unstable cone field* C_a^{cu} if the tangent subspace to N at each point $x \in N$ is contained in the corresponding cone $C_a^{cu}(x)$. Then $f(N)$ is also tangent to the centre-unstable cone field, if it is contained in V_0.

The tangent bundle of N is said to be Hölder continuous if $x \mapsto T_xN$ defines a Hölder continuous section from N to the corresponding Grassman bundle of M. In this subsection we show that the tangent bundle of the iterates of a C^2 submanifold are Hölder continuous (as long as they do not leave V_0), with uniform Hölder constants.

The basic idea is contained in the following observation, which the reader may easily check. Let E_1, E_2 be two Euclidean spaces and L be a linear isomorphism on $E_1 \oplus E_2$ leaving both factors invariant. Assume that we have the domination property

$$\|L \mid E_1\| \cdot \|L^{-1} \mid E_2\| < 1.$$

Then there exist $C > 0$ and $0 < \zeta \leq 1$ such that if $\Gamma \subset E_1 \oplus E_2$ is the graph of a $C^{1+\zeta}$ map $\phi : E_1 \to E_2$, with Hölder constant C, then the same is true for $L(\Gamma)$. In fact, it suffices to take any ζ such that $\|L \mid E_1\| \cdot \|L^{-1} \mid E_2\|^{1+\zeta} < 1$.

In order to apply similar arguments in our situation, it is useful to express the notion of Hölder variation of the tangent bundle in local coordinates, as follows.

We choose $\delta_0 > 0$ small enough so that the inverse of the exponential map $\exp_x$ is defined on the δ_0 neighbourhood of every point x in V_0. From now on we identify this neighbourhood of x with the corresponding neighbourhood U_x of the origin in T_xN, through the local chart defined by $\exp_x^{-1}$. Accordingly, we identify x with $0 \in T_xN$. Reducing δ_0, if necessary, we may suppose that $\tilde{E}_x^{cs}$ is contained in the centre-stable cone $C_a^{cs}(y)$ of every $y \in U_x$. In particular, the intersection of $C_a^{cu}(y)$ with $\tilde{E}_x^{cs}$ reduces to the zero vector. Then, the tangent space to N at y is parallel to the graph of a unique linear map $A_x(y) : T_xN \to \tilde{E}_x^{cs}$. Given constants $C > 0$ and $0 < \zeta \leq 1$, we say that *the tangent bundle to N is (C, ζ)-Hölder* if

$$\|A_x(y)\| \leq C d_x(y)^\zeta \quad \text{for every } y \in N \cap U_x \text{ and } x \in V_0 . \tag{8}$$

Here, $d_x(y)$ denotes the distance from x to y along $N \cap U_x$, defined as the length of the shortest curve connecting x to y inside $N \cap U_x$.

Recall that we have chosen the neighbourhood V_0 and the cone width a sufficiently small so that the domination property remains valid for vectors in the cones $C_a^{cs}(z)$, $C_a^{cu}(z)$, and for any point z in V_0. Then, there exist $\lambda_1 \in (\lambda, 1)$ and $\zeta \in (0, 1]$ such that

$$\|Df(z)v^{cs}\| \cdot \|Df^{-1}(f(z))v^{cu}\|^{1+\zeta} \leq \lambda_1 < 1 \tag{9}$$

for every norm 1 vectors $v^{cs} \in C_a^{cs}(z)$ and $v^{cu} \in C_a^{cu}(z)$, at any $z \in V_0$. Then, up to reducing $\delta_0 > 0$ and slightly increasing $\lambda_1 < 1$, (9) remains true if we replace z by any $y \in U_x$, $x \in V_0$ (taking $\| \cdot \|$ to mean the Riemannian metric in the corresponding local chart).

We fix ζ and λ_1 as above in all that follows. Then, given a C^1 submanifold $N \subset V_0$, we denote

$$\kappa(N) = \inf\{C > 0 : \text{the tangent bundle of } N \text{ is } (C, \zeta)\text{-Hölder}\}. \tag{10}$$

Proposition 2.2. *There exist $\lambda_0 < 1$ and $C_0 > 0$ so that if $N \subset V_0 \cap f^{-1}(V_0)$ is any C^1 submanifold tangent to the centre-unstable cone field then*

$$\kappa(f(N)) \leq \lambda_0 \kappa(N) + C_0 .$$

Proof. Of course, we only need to consider the case when $\kappa(N)$ is finite, that is, the tangent bundle of N is (C, ζ)-Hölder for some $C > 0$. Let $x \in N$ be fixed. We use $(u, s) \in T_xN \oplus \tilde{E}_x^{cs}$ and $(u_1, s_1) \in T_{f(x)}f(N) \oplus \tilde{E}_{f(x)}^{cs}$, respectively, to represent the local coordinates in U_x and $U_{f(x)}$ introduced above. We write the expression of our map in these local coordinates as

$f(u,s) = (u_1(u,s), s_1(u,s))$. Observe that if $x \in K$ then the partial derivatives of u_1 and s_1 at the origin $0 \in T_xN$ are

$$\partial_u u_1(0) = Df|T_xN, \quad \partial_s u_1(0) = 0, \quad \partial_u s_1(0) = 0, \quad \partial_s s_1(0) = Df|\tilde{E}^{cs}_x.$$

This is because $E^{cs}_x = \tilde{E}^{cs}_x$ is mapped to $E^{cs}_{f(x)} = \tilde{E}^{cs}_{f(x)}$ under $Df(x)$ and, similarly, T_xN is mapped to $T_{f(x)}N$. Then, given any small $\varepsilon_0 > 0$ we have that

$$\|\partial_u u_1(y) - Df|T_xN\|, \ \|\partial_s u_1(y)\|, \ \|\partial_u s_1(y)\|, \ \|\partial_s s_1(y) - Df|\tilde{E}^{cs}_x\|, \tag{11}$$

are all less than ε_0 for every $x \in V_0$ and $y \in U_x$, as long as δ_0 and V_0 are small. Taking the cone width a also small, we get

$$\begin{aligned} \left\|Df|T_yN - Df|\tilde{E}^{cu}_x\right\| &\leq \varepsilon_0 \quad \text{and} \\ \left\|Df^{-1}|T_{f(y)}f(N) - Df^{-1}|\tilde{E}^{cu}_{f(x)}\right\| &\leq \varepsilon_0, \end{aligned} \tag{12}$$

for every $x \in V_0$ and $y \in U_x$. Since f is C^2, there is also some constant $K_2 > 0$ such that

$$\|\partial_s u_1(y)\| \leq K_2 d_x(y) \quad \text{and} \quad \|\partial_u s_1(y)\| \leq K_2 d_x(y). \tag{13}$$

For y_1 in $U_{f(x)}$, let $A_{f(x)}(y_1)$ be the linear map from $T_{f(x)}f(N)$ to $\tilde{E}^{cs}_{f(x)}$ whose graph is parallel to $T_{y_1}f(N)$. We are going to prove that, fixing ε_0 sufficiently small, then $A_{f(x)}(y_1)$ satisfies (8) for any $C > \lambda_0\kappa(N) + C_0$, with convenient λ_0 and C_0. Let us begin by noting that $\|A_{f(x)}(y_1)\|$ is bounded by some uniform constant $K_1 > 0$, since $f(N)$ is tangent to the centre-unstable cone field. We will choose the constant $C_0 \geq K_1/(\delta_0/\|Df^{-1}\|)^\zeta$, so that (8) is immediate when $d_{f(x)}(y_1) \geq \delta_0/\|Df^{-1}\|$:

$$\|A_{f(x)}(y_1)\| \leq K_1 \leq C_0(\delta_0/\|Df^{-1}\|)^\zeta \leq C_0 d_{f(x)}(y_1)^\zeta.$$

Here $\|Df^{-1}\|$ is the supremum of all $\|Df^{-1}(z)\|$ with $z \in U_w$, $w \in V_0$, where the norms are taken with respect to the Riemannian metrics in the local charts. This permits us to restrict to the case when $d_{f(x)}(y_1) < \delta_0/\|Df^{-1}\|$ in all that follows. Let Γ_1 be any curve on $f(N) \cap U_{f(x)}$ joining $f(x)$ to y_1 and whose length approximates $d_{f(x)}(y_1)$. Then $\Gamma = f^{-1}(\Gamma_1)$ is a curve in $N \cap U_x$ joining x to $y = f^{-1}(y_1)$, with length less than δ_0. In fact, cf. (12),

$$d_x(y) \leq \text{length}(\Gamma) \leq \left(\|Df^{-1}|\tilde{E}^{cu}_{f(x)}\| + \varepsilon_0\right)\text{length}(\Gamma_1).$$

This shows that $d_x(y) \leq (\|Df^{-1}|\tilde{E}^{cu}_{f(x)}\| + \varepsilon_0)d_{f(x)}(y_1)$.

Now we observe that

$$A_{f(x)}(y_1) = \left[\partial_u s_1(y) + \partial_s s_1(y) \cdot A_x(y)\right] \cdot \left[\partial_u u_1(y) + \partial_s u_1(y) \cdot A_x(y)\right]^{-1}.$$

On the one hand, by (11) and (13),

$$\begin{aligned}\|\partial_u s_1(y) + \partial_s s_1(y)\cdot A_x(y)\| &\le K_2 d_x(y) + \big(\|Df|\tilde{E}^{cs}_x\| + \varepsilon_0\big)\kappa(N) d_x(y)^\zeta \\ &\le \big(K_2 + \big(\|Df|\tilde{E}^{cs}_x\| + \varepsilon_0\big)\kappa(N)\big) d_x(y)^\zeta.\end{aligned}$$

On the other hand, $\|\partial_s u_1(y)\cdot A_x(y)\| \le \epsilon_0 K_1$, which can be made much smaller than $1/\|(\partial_u u_1(y)^{-1}\|$. As a consequence, recall (12) and (13),

$$\big\|\big[\partial_u u_1(y) + \partial_s s_1(y)\cdot A_x(y)\big]^{-1}\big\| \le \big\|Df^{-1}|\tilde{E}^{cu}_{f(x)}\big\| + \varepsilon_1,$$

where ε_1 can be made arbitrarily small by reducing ε_0. Putting these bounds together, we conclude that $\|A_{f(x)}(y_1)\|\, d_{f(x)}(y_1)^{-\zeta}$ is less than

$$\frac{\big(\|Df|\tilde{E}^{cs}_x\| + \varepsilon_0\big)\big(\|Df^{-1}|\tilde{E}^{cu}_{f(x)}\| + \varepsilon_1\big)}{\big(\|Df^{-1}|\tilde{E}^{cu}_{f(x)}\| + \varepsilon_0\big)^{-\zeta}}\kappa(N) + \frac{K_2\big(\|Df^{-1}|\tilde{E}^{cu}_{f(x)}\| + \varepsilon_1\big)}{\big(\|Df^{-1}|\tilde{E}^{cu}_{f(x)}\| + \varepsilon_0\big)^{-\zeta}}.$$

So, choosing δ_0, V_0, a sufficiently small, we can make ε_0, ε_1, sufficiently close to zero so that the factor multiplying $\kappa(N)$ is less than some $\lambda_0 \in (\lambda_1, 1)$; recall (9). Moreover, the second term in the expression above is bounded by some constant that depends only on f. We take C_0 larger than this constant. □

Remark 2.3. The proof remains valid if the diffeomorphism f is only of class $C^{1+\zeta}$: it suffices to replace (13) by $\|\partial_s u_1(y)\|$, $\|\partial_u s_1(y)\| \le K_2 d_x(y)^\zeta$, which is sufficient for the rest of the argument.

Corollary 2.4. *There exists $C_1 > 0$ such that, given any C^1 submanifold $N \subset V_0$ tangent to the centre-unstable cone field,*

a) there exists $n_0 \ge 1$ such that $\kappa(f^n(N)) \le C_1$ for every $n \ge n_0$ such that $f^k(N) \subset V_0$ for all $0 \le k \le n$;
b) if $\kappa(N) \le C_1$, then the same is true for every iterate $f^n(N)$, $n \ge 1$, such that $f^k(N) \subset V_0$ for all $0 \le k \le n$;
c) in particular, if N and n are as in b), then the functions

$$J_k : f^k(N) \ni x \mapsto \log\big|\det\big(Df \mid T_x f^k(N)\big)\big|, \quad 0 \le k \le n,$$

are (L_1, ζ)-Hölder continuous with $L_1 > 0$ depending only on C_1 and f.

Proof. It suffices to choose any $C_1 \ge C_0/(1-\lambda_0)$. □

Remark 2.5. Suppose that we have the following stronger form of domination

$$\|Df|E^{cs}_x\| \cdot \|Df^{-1}|E^{cu}_{f(x)}\|^i \le \lambda \quad \text{for } i = 1, 2,$$

and any $x \in K$. Then, assuming that f is a C^2 diffeomorphism, we may take $\zeta = 1$ in the previous arguments. In that case, $\kappa(N)$ yields a bound on the curvature tensor of N. So, Corollary 2.4 asserts that if N is C^2 then the curvature of all its iterates $f^n(N)$, $n \ge 1$, is bounded by some constant that depends only on the curvature of N.

2.2. Hyperbolic times and distortion bounds

The following notion will allow us to derive *uniform behaviour* (expansion, distortion) from the hypothesis of non-uniform expansion in (2).

Definition 2.6. *Given $\sigma < 1$, we say that n is a σ-*hyperbolic time *for a point $x \in K$ if*

$$\prod_{j=n-k+1}^{n} \|Df^{-1} \mid E^{cu}_{f^j(x)}\| \leq \sigma^k \qquad \text{for all } 1 \leq k \leq n.$$

In particular, if n is a σ-hyperbolic time for x then $Df^{-k} \mid E^{cu}_{f^n(x)}$ is a contraction for every $1 \leq k \leq n$:

$$\|Df^{-k}|E^{cu}_{f^n(x)}\| \leq \prod_{j=n-k+1}^{n} \|Df^{-1}|E^{cu}_{f^j(x)}\| \leq \sigma^k.$$

Moreover, if a is taken sufficiently small in the definition of our cone fields, and we choose $\delta_1 > 0$ also small (in particular, the δ_1-neighbourhood of K should be contained in V_0), then, by continuity,

$$\|Df^{-1}(f(y))v\| \leq \frac{1}{\sqrt{\sigma}}\|Df^{-1}|E^{cu}_{f(x)}\| \, \|v\| \tag{14}$$

whenever $x \in K$, $\text{dist}(x, y) \leq \delta_1$, and $v \in C^{cu}(y)$.

Let D be any C^1 disk contained in V_0 and tangent to the centre-unstable cone field. We use $\text{dist}_D(\cdot, \cdot)$ to denote distance between two points in the disk, measured along D. The distance from a point $x \in D$ to the boundary of D is $\text{dist}_D(x, \partial D) = \inf_{y \in \partial D} \text{dist}_D(x, y)$.

Lemma 2.7. *Given any C^1 disk $D \subset V_0$ tangent to the centre-unstable cone field, $x \in D \cap K$, and $n \geq 1$ a σ-hyperbolic time for x,*

$$\text{dist}_{f^{n-k}(D)}(f^{n-k}(y), f^{n-k}(x)) \leq \sigma^{k/2} \text{dist}_{f^n(D)}(f^n(y), f^n(x)).$$

for any point $y \in D$ with $\text{dist}(f^n(x), f^n(y)) \leq \delta_1$.

Proof. Let η_0 be a curve of minimal length in $f^n(D)$ connecting $f^n(x)$ to $f^n(y)$. For $1 \leq k \leq n$ write $\eta_k = f^{n-k}(\eta_0)$. We prove the lemma by induction. Let $1 \leq k \leq n$ and assume that

$$\text{length}(\eta_j) \leq \delta_1 \quad \text{for } 0 \leq j \leq k-1.$$

Denote by $\dot{\eta}_0(z)$ the tangent vector to the curve η_0 at the point z. Then, in view of the choice of δ_1 in (14) and the definition of σ-hyperbolic times,

$$\|Df^{-k}(z)\dot{\eta}_0(z)\| \leq \sigma^{-k/2} \|\dot{\eta}_0(z)\| \prod_{j=n-k+1}^{n} \|Df^{-1}|E^{cu}_{f^j(x)}\| \leq \sigma^{k/2}\|\dot{\eta}_0(z)\|.$$

As a consequence,

$$\operatorname{length}(\eta_k) \leq \sigma^{k/2} \operatorname{length}(\eta_0) = \sigma^{k/2} \operatorname{dist}_{f^{n-k}(D)}(f^{n-k}(y), f^{n-k}(x)) \leq \delta_1 .$$

This completes our induction, thus proving the lemma. □

Proposition 2.8. *There exists $C_2 > 1$ such that, given any C^1 disk D tangent to the centre-unstable cone field with $\kappa(D) \leq C_1$, and given any $x \in D \cap K$ and $n \geq 1$ a σ-hyperbolic time for x, then*

$$\frac{1}{C_2} \leq \frac{|\det Df^n \mid T_y D|}{|\det Df^n \mid T_x D|} \leq C_2$$

for every $y \in D$ such that $\operatorname{dist}(f^n(y), f^n(x)) \leq \delta_1$.

Proof. For $0 \leq i < n$ and $y \in D$, we denote $J_i(y) = |\det Df \mid T_{f^i(y)} f^i(D)|$. Then,

$$\log \frac{|\det Df^n \mid T_y D|}{|\det Df^n \mid T_x D|} = \sum_{i=0}^{n-1} \big(\log J_i(y) - \log J_i(x) \big)$$

By Corollary 2.4, $\log J_i$ is (L_1, ζ)-Hölder continuous, for some uniform $L_1 > 0$. Moreover, by Lemma 2.7, the sum of all $\operatorname{dist}_D(f^j(x), f^j(y))^\zeta$ over $0 \leq j \leq n$ is bounded by $\delta_1/(1 - \sigma^{\zeta/2})$. Now it suffices to take $C_2 = \exp(L_1 \delta_1/(1 - \sigma^{\zeta/2}))$. □

2.3. *Curvature at hyperbolic times*

It is possible to obtain control of the curvature at hyperbolic times, without having to assume the stronger form of domination in Remark 2.5. As before, we assume that f is a C^2 diffeomorphism with a partially hyperbolic set K. Let $\sigma < 1$ be fixed, and $\delta_1 > 0$ be chosen as in (14).

Proposition 2.9. *Let D be a C^2 disk tangent to the centre-unstable cone field, $x \in D \cap K$, and $n \geq 1$ be a σ-hyperbolic time for x. Then, the curvature of the δ_1-neighbourhood of $f^n(x)$ in $f^n(D)$ is bounded by a constant $K_0 > 0$ that depends only on f, σ, and the curvature of D. In fact, if n is sufficiently large then K_0 may be taken depending only on f and σ.*

The main idea in the proof of this proposition is to show that, up to conformal changes of the Riemannian metric, we may suppose that $Df \mid E^{cu}$ is uniformly expanding at every point $f^j(x)$, $0 \leq j < n$. As a consequence, the domination condition 2. in Subsect. 1.1 implies the condition in Remark 2.5 (with respect to the modified metrics). In doing this, it is important that all metric changes can be done by dilation, which is due to the hyperbolic time condition.

Lemma 2.10. *Let $n \geq 1$ and $a_1, \ldots, a_n, c_0$, be real numbers such that*

$$\sum_{j=k+1}^{n} a_j \geq (n-k)c_0 \quad \textit{for all } 0 \leq k < n. \tag{15}$$

Then, there exist $b_1, \ldots, b_n$ such that

1. $|b_j| \leq \sup_{1\leq j\leq n} |c_0 - a_j|$ *for all* $1 \leq j \leq n$;
2. $a_j + b_j \geq c_0$ *for all* $1 \leq j \leq n$;
3. $\sum_{j=1}^{k} b_j \geq 0$ *for* $1 \leq k < n$ *and* $\sum_{j=1}^{n} b_j = 0$.

Proof. Define b_j by recurrence, through

$$b_1 = \max\{0, c_0 - a_1\} \quad and \quad b_j = \max\{-\sum_{i=1}^{j-1} b_j, c_0 - a_j\}$$
$$\text{for } j = 2, \ldots, n.$$

The first condition in the statement is clear, in the case when $b_j = c_0 - a_j$. Otherwise, $b_j = -\sum_{i=1}^{j-1} b_i$ which, by construction, is always non-positive. So, in this second case we must have $0 \geq b_j \geq c_0 - a_j$, so that the bound in 1. remains valid. The second condition follows immediately from the construction, and the same is true for the first statement in 3. To obtain the last claim, we begin by proving the following fact, by induction on j:

$$\sum_{i=1}^{j} b_i \leq \sum_{i=j+1}^{n} (a_i - c_0) \quad \text{for } j = 1, \ldots, n-1. \tag{16}$$

In view of (15),

$$\sum_{i=2}^{n}(a_i - c_0) \geq 0 \quad \text{and} \quad \sum_{i=2}^{n}(a_i - c_0) \geq \sum_{i=1}^{n}(a_i - c_0) + (c_0 - a_1) \geq (c_0 - a_1).$$

This gives $\sum_{i=2}^{n}(a_i - c_0) \geq \max\{0, c_0 - a_1\} = b_1$, corresponding to case $j = 1$. Now we suppose that, by recurrence, $\sum_{i=1}^{j-1} b_i \leq \sum_{i=j}^{n}(a_i - c_0)$. Then, either $b_j = c_0 - a_j$ in which case, adding b_j to both sides of the previous inequality immediately gives the conclusion. Or else, $b_j = -\sum_{i=1}^{j-1} b_i$ and then

$$\sum_{i=1}^{j} b_i = 0 \leq \sum_{i=j+1}^{n} (a_i - c_0),$$

due to our assumption (15). This completes the proof of (16). In particular, taking $j = n-1$, we get that $-\sum_{i=1}^{n-1} b_i \geq c_0 - a_n$, and so $b_n = -\sum_{j=1}^{n-1} b_i$. □

Now we use this lemma to prove Proposition 2.9.

Proof. Let D_n be the neighbourhood of radius δ_1 around $f^n(x)$ in $f^n(D)$, and $D_j = f^{j-n}(D_n)$ for $0 \le j < n$. Take $c_0 = -\log\sigma$ and $a_j = -\log \|Df^{-1}|E^{cu}_{f^j(x)}\|$, $1 \le j \le n$, in Lemma 2.10. Let b_j, $1 \le j \le n$, be the corresponding sequence, and denote $t_j = \exp(\sum_{i=1}^{j} b_i)$ for $1 \le j \le n$, and $t_0 = 1$. Conclusion 3. in the lemma implies that

$$t_j \ge 1 \text{ for every } j \quad \text{and} \quad t_n = 1.$$

In all that follows $\|\cdot\|_j$ denotes the metric obtained by multiplying the initial Riemannian metric of M by t_j, $0 \le j \le n$. Accordingly, we denote $\|Df\|_{j-1,j}$ and $\|D^2f\|_{j-1,j}$ the norms of the derivatives of f from $(M, \|\cdot\|_{j-1})$ to $(M, \|\cdot\|_j)$. We use similar notations for the restrictions of Df, Df^{-1} to the subbundles $\tilde{E}^{cu}$ and $\tilde{E}^{cs}$. Observe that

$$\|Df\|_{j-1,j} = \exp b_j \|Df\| \quad \text{and} \quad \|D^2f\|_{j-1,j} = \frac{\exp b_j}{t_{j-1}} \|D^2 f\|.$$

Since the b_j are bounded, and the t_j are larger than 1, $\|Df\|_{j-1,j}$ and $\|D^2f\|_{j-1,j}$ are bounded by constants that depend only on f and σ. Note also that the domination property is not affected by this conformal change of metrics:

$$\left\|Df|\tilde{E}^{cs}_y\right\|_{j-1,j} \cdot \left\|Df^{-1}|\tilde{E}^{cu}_{f(y)}\right\|_{j,j-1} = \left\|Df|\tilde{E}^{cs}_y\right\| \cdot \left\|Df^{-1}|\tilde{E}^{cu}_{f(y)}\right\| \le \lambda, \tag{17}$$

at every point y where these subbundles are defined.

Conclusion 2. in the lemma now means that

$$\left\|Df^{-1}|E^{cu}_{f^j(x)}\right\|_{j,j-1} = \exp(-b_j - a_j) \le \sigma.$$

So, by (14) and Lemma 2.7, $\|Df^{-1}|\tilde{E}^{cu}_{f^j(y)}\|_{j,j-1} \le \sqrt{\sigma} < 1$ for every y in D_0 and $1 \le j \le n$. Together with (17), this gives

$$\left\|Df|\tilde{E}^{cs}_{f^{j-1}(y)}\right\|_{j-1,j} \cdot \left\|Df^{-1}|\tilde{E}^{cu}_{f^j(y)}\right\|^2_{j,j-1} \le \lambda\sqrt{\sigma} \le \lambda$$

for every $y \in D_0$ and $1 \le j \le n$. This means that a strong domination property as in Remark 2.5 is valid, with respect to the relevant modified metrics, at every point of $D_0 \cup f(D_0) \cup \cdots \cup f^{n-1}(D_0)$. Since we already checked that the first and second derivatives have uniformly bounded norms relative to these modified metrics, the arguments in Proposition 2.2 carry on completely to the present context to prove Proposition 2.9. □

Closing this section we observe that these arguments could also be used to give an alternative proof of the distortion bounds we obtained in the previous section.

3. Lebesgue measure at hyperbolic times

The following lemma, due to Pliss [Pli72], will permit us to prove that a point x satisfying assumption (2) has many (positive density at infinity) hyperbolic times.

Lemma 3.1. *Given $A \geq c_2 > c_1 > 0$, let $\theta_0 = (c_2 - c_1)/(A - c_1)$. Then, given any real numbers $a_1, \dots, a_N$ such that*

$$\sum_{j=1}^{N} a_j \geq c_2 N \qquad \text{and} \qquad a_j \leq A \text{ for every } 1 \leq j \leq N,$$

there are $l > \theta_0 N$ and $1 < n_1 < \cdots < n_l \leq N$ so that

$$\sum_{j=n+1}^{n_i} a_j \geq c_1(n_i - n) \quad \text{for every } 0 \leq n < n_i \text{ and } i = 1, \dots, l.$$

Proof. (cf. [Mañ87, Section 2]) Define $S(n) = \sum_{j=1}^{n}(a_j - c_1)$, for each $1 \leq n \leq N$, and also $S(0) = 0$. Then define $1 < n_1 < \cdots < n_l \leq N$ to be the maximal sequence such that $S(n_i) \geq S(n)$ for every $0 \leq n < n_i$ and $i = 1, \dots, l$. Note that l can not be zero, since $S(N) > S(0)$. Moreover, the definition means that

$$\sum_{j=n+1}^{n_i} a_j \geq c_1(n_i - n) \quad \text{for } 0 \leq n < n_i \text{ and } i = 1, \dots, l.$$

So, we only have to check that $l > \theta_0 N$. Observe that, by definition,

$$S(n_i - 1) < S(n_{i-1}) \quad \text{and so} \quad S(n_i) < S(n_{i-1}) + (A - c_1)$$

for every $1 < i \leq l$. Moreover, $S(n_1) \leq (A - c_1)$ and $S(n_l) \geq S(N) \geq N(c_2 - c_1)$. This gives,

$$N(c_2 - c_1) \leq S(n_l) = \sum_{i=2}^{l} \big(S(n_i) - S(n_{i-1})\big) + S(n_1) < l(A - c_1),$$

which completes the proof. □

Clearly, the set H in the statement of Theorem A may be taken positively invariant under f. Given any $\sigma < 1$, let $H(\sigma)$ be the set of points in H for which the limit in (2) is smaller than $3 \log \sigma$. Then $H(\sigma)$ is positively invariant and, since we are assuming that H has positive Lebesgue measure, $H(\sigma)$ must also have positive Lebesgue measure if σ is close enough to 1. Then, there exists some small C^2 disk D transverse to the centre-stable subbundle, and intersecting $H(\sigma)$ in a set with positive Lebesgue measure inside D. Up to replacing it by some small disk contained in $f^l(D)$ for some large enough $l \geq 1$, we may suppose that D is tangent to the centre-unstable cone field and $\kappa(D) \leq C_1$. Here $\kappa(\cdot)$ is the Hölder constant defined by (10), recall Corollary 2.4. We fix such a disk D, once and for all.

Corollary 3.2. *There is $\theta > 0$, depending only on σ and f, such that, given any $x \in D \cap H(\sigma)$ and any sufficiently large $N \geq 1$, there exist σ-hyperbolic times $1 \leq n_1 < \cdots < n_l \leq N$ for x, with $l \geq \theta N$.*

Proof. Since $x \in H(\sigma)$, we have

$$\sum_{j=1}^{N} \log \left\| Df^{-1} | E^{cu}_{f^j(x)} \right\| \leq 2N \log \sigma,$$

for all $N \geq 1$ sufficiently large. Now it suffices to take $c_1 = |\log \sigma|$, $c_2 = 2c_1$, $A = \sup \left| \log \|Df^{-1}|E^{cu}\| \right|$, and $a_j = -\log \|Df^{-1}|E^{cu}_{f^j(x)}\|$ in the previous lemma. □

Let $\delta_1 > 0$ be the small number introduced prior to Lemma 2.7. In particular, we requested that the δ_1-neighbourhood of the set K be contained in the domain V_0 of the invariant cone fields C^{cs} and C^{cu}. Reducing δ_1 if necessary, we may suppose that the subset $A = A(D, \sigma, \delta_1)$ of points $x \in D \cap H(\sigma)$ such that $\mathrm{dist}_D(x, \partial D) \geq \delta_1$ still has positive Lebesgue measure in D.

We consider the sequence

$$\mu_n = \frac{1}{n} \sum_{j=0}^{n-1} f^j_* \, \mathrm{Leb}_D \tag{18}$$

of averages of forward iterates of Lebesgue measure on D. A main idea is to decompose μ_n as a sum of two measures, to be denoted ν_n and η_n, such that ν_n is uniformly absolutely continuous on iterates of the disk D (uniformly bounded density with respect to Lebesgue measure) and has total mass uniformly bounded away from zero for all large n. This is done as follows.

Given integers $n \geq 1$, define the following subset of $D \cap H(\sigma)$

$$H_n = \{x \in A : n \text{ is a } \sigma\text{-hyperbolic time for } x\}.$$

It follows from Lemma 2.7 that if $x \in H_n$ then the distance from $f^n(x)$ to the boundary of $f^n(D)$ is larger than δ_1. For $\delta > 0$, we denote $\Delta_n(x, \delta)$ the δ-neighbourhood of $f^n(x)$ inside $f^n(D)$. Clearly, $(f^n_* \, \mathrm{Leb}_D) \mid \Delta_n(x, \delta_1)$ is absolutely continuous with respect to Lebesgue measure on $\Delta_n(x, \delta_1)$. Moreover, if $x \in H_n$ and one normalizes both measures, then Proposition 2.8 means that the density of the former with respect to the latter is uniformly bounded from below and above.

Proposition 3.3. *There exists a constant $\tau > 0$ such that for any n there exists a finite subset $\widehat{H}_n$ of H_n such that the balls of radius $\delta_1/4$ in $f^n(D)$ around the points $x \in f^n(\widehat{H}_n)$ are two-by-two disjoint, and their union Δ_n satisfies*

$$f^n_* \, \mathrm{Leb}_D(\Delta_n \cap H(\sigma)) \geq f^n_* \, \mathrm{Leb}_D(\Delta_n \cap f^n(H_n)) \geq \tau \, \mathrm{Leb}_D(H_n) \, .$$

This is obtained taking $N = f^n(D) \cap V_0$, $\omega = f^n_* \operatorname{Leb}_D$, $r = \delta_1$, $\Omega = f^n(H_n)$, and $\widehat{H}_n = I$ in the following abstract result.

Lemma 3.4. *There exist $\tau > 0$ and $r_0 > 0$ such that the following holds. Let $N \subset V_0$ be a C^1 embedded submanifold of M tangent to the centre-unstable cone field, and ω be a finite Borelian measure in N. Let $0 < r \le r_0$ and $\Omega \subset N$ be a measurable subset with compact closure, whose distance to the boundary of N is larger than $r > 0$.*

Then there exists a finite subset $I \subset \Omega$ such that the balls $\Delta(x, r/4)$ in N around the points of I are two-by-two disjoint, and their union Δ satisfies

$$\omega(\Delta \cap \Omega) \ge \tau\omega(\Omega) .$$

Proof. By the continuity of the centre-unstable cone field, we may fix $r_0 > 0$ small enough so that the connected component of the intersection of N with the ball of radius r_0 around each point $z \in N$ that contains z coincides with the graph of a map g_z from a neighbourhood of 0 in E^{cu}_z to E^{cs}_z (or, to be more precise, with the image of such a graph under the exponential map $\exp_z$). Moreover, as long as r_0 is small enough, then g_z is a Lipschitz continuous map, with Lipschitz constant depending only on the constant a in the definition of our cone fields. As a consequence, there exists a constant $R > 0$ such that, given $0 < r_1 < r_2 \le r_0$, any ball of radius r_2 in N can be covered by at most $(r_2/r_1)^d R$ balls of radius r_1, with $d = \dim E^{cu}$. We assume that $\omega(\Omega) > 0$, since otherwise there is nothing to prove.

Let $0 < r \le r_0$ and $z_1 \in N$ be such that $\omega(\Omega \cap \Delta(z_1, r))$ is larger than $\omega(\Omega \cap \Delta(z, r))/2$ for any other point $z \in N$. By the previous remarks, we may find a point $y_1 \in N$ such that the ball $\Delta(y_1, r/8)$ of radius $r/8$ around y_1 intersects $\Delta(z_1, r)$ and

$$\omega(\Delta(y_1, r/8) \cap \Omega) \ge \frac{1}{R8^d}\,\omega(\Omega \cap \Delta(z_1, r)).$$

In particular $\Delta(y_1, r/8)$ contains some point $x_1 \in \Omega$. We take it to be the first point in our set I. Observe that, due to the choice of z_1,

$$\begin{aligned}\omega(\Delta(x_1, r/4) \cap \Omega) &\ge \omega(\Delta(y_1, r/8) \cap \Omega)\\ &\ge \frac{1}{R8^d}\,\omega(\Omega \cap \Delta(z_1, r))\\ &\ge \frac{1}{2R8^d}\,\omega(\Omega \cap \Delta(x_1, r)).\end{aligned} \tag{19}$$

Now we consider $\Omega_1 = \Omega \setminus \Delta(x_1, r)$. Either this set has zero ω measure, in which case we stop, or we may apply the same construction as before to determine a second point $x_2 \in \Omega$. Observe that the balls of radius $r/4$ around x_1 and x_2 are disjoint, since x_2 belongs in Ω_1. Repeating this procedure, we find a sequence x_i, $i \ge 1$, of points in Ω whose balls of radius $r/4$ are two-by-two disjoint. By compactness, this sequence is necessarily finite.

Moreover, by construction, Ω is contained in the union of the balls of radius r around these x_i. So, using the bound in (19),

$$\omega(\Omega) \le 2R8^d \sum_i \omega(\Delta(x_i, r/4)) \cap \Omega) = 2R8^d \omega(\Delta \cap \Omega).$$

This means that we may take $\tau = 1/(2R8^d)$. □

In the sequel we shall denote $\mathcal{D}_n$ the family of balls of radius $\delta_1/4$ in $f^n(D)$ around the points $x \in f^n(\widehat{H}_n)$, that form Δ_n. Now we define

$$\nu_n = \frac{1}{n}\sum_{j=0}^{n-1}(f_*^j \operatorname{Leb}_D) \mid \Delta_j\,, \tag{20}$$

and $\eta_n = \mu_n - \nu_n$.

Proposition 3.5. *There is $\alpha > 0$ such that $\nu_n(H(\sigma)) \ge \alpha$ for all n large enough.*

Proof. Recall that we took H and $H(\sigma)$ positively invariant under f. By Proposition 3.3 we have that $\nu_n(H(\sigma))$ is bounded from below by the product of τ by $n^{-1}\sum_{i=0}^{n-1}\operatorname{Leb}_D(H_i)$. So, it suffices to prove that this last expression is larger than some positive constant, for n large.

For all $k > 0$, denote A_k the set of points $x \in A$ such that, for any $n \ge k$ the sum $\sum_{j=1}^{n}\log\|Df^{-1}|E^{cu}_{f^j(x)}\|$ is smaller than $2n\log\sigma$. As A is the increasing union of the A_n, there is $k_0 \ge 1$ such that the Lebesgue measure of A_k is nonzero for all $k \ge k_0$. Given any $k \ge k_0$, let ξ_n be the measure in $\{1, \ldots, n\}$ defined by $\xi_n(B) = \#B/n$, for each subset B. Then, using Fubini's theorem

$$\begin{aligned}\frac{1}{n}\sum_{i=0}^{n-1}\operatorname{Leb}_D(H_n) &= \int\left(\int \chi(x,i)\, d\operatorname{Leb}_D(x)\right) d\xi_n(i)\\ &= \int\left(\int \chi(x,i)\, d\xi_n(i)\right) d\operatorname{Leb}_D(x),\end{aligned}$$

where $\chi(x,i) = 1$ if $x \in H_i$ and $\chi(x,i) = 0$ otherwise. Now, Corollary 3.2 means that the integral with respect to $d\xi_n$ is larger than $\theta > 0$, as long as $k \ge k_0$. So, the expression on the right hand side is bounded from below by $\theta\operatorname{Leb}_D(D)$. □

Remark 3.6. We proved a slightly stronger fact, that will be useful in Sect. 4: $\nu_n\big(\bigcup_{i=0}^{n-1} f^i(D \cap H(\sigma))\big) \ge \alpha$ for every large n.

We consider some subsequence $(n_k)_k$ such that μ_{n_k} and ν_{n_k} converge to measures μ and ν, respectively. It is easy to see that μ is a probability and f-invariant. Moreover $\nu(K) \ge \limsup_k \nu_k(K) \ge \alpha > 0$. We shall prove, in the next section, that ν has a property of absolute continuity along certain (fairly large) disks contained in its support. Here are some preparatory comments.

Recall that each ν_n is supported on a finite union $\cup_{j=0}^{n-1}\Delta_j$ of disks whose size is bounded from below and from above. Then the support of ν is contained in the set

$$\Delta_\infty = \cap_{n=1}^\infty \text{ closure } \left(\cup_{j\geq n}\Delta_j\right)$$

of accumulation points of such Δ_j. Given $y \in \Delta_\infty$ then there exist $(j_i)_i \to \infty$, disks $D_i = \Delta_{j_i}(x_i, \delta_1/4) \subset \Delta_{j_i}$, and points $y_i \in D_i$ converging to y as $i \to \infty$. Up to considering subsequences, we may suppose that the centers x_i converge to some point x and, using the theorem of Ascoli-Arzela, the D_i converge to a disk $D(x)$ of size $\delta_1/4$ around x. Then y is in the closure $\bar{D}(x)$ of $D(x)$, and $\bar{D}(x) \subset \Delta_\infty$.

We shall denote $\mathcal{D}_\infty$ the family of disks $D(x)$ obtained in this way. Observe that these points x are in $\hat{H}_\infty = \cap_{n=1}^\infty \text{ closure } \left(\cup_{j\geq n} f^j(\hat{H}_j)\right)$. Since every $\hat{H}_j$ is contained in K, which is compact and positively invariant, $\hat{H}_\infty$ is a subset of $\cap_{n=1}^\infty f^n(K)$. According to the next lemma, $D(x)$ depends only on x and not on the various choices we made in the construction.

Lemma 3.7. *The subspace E_x^{cu} is uniformly expanding: $\|Df^{-k} \mid E_x^{cu}\| \leq \sigma^{k/2}$ for all $k \geq 1$. The disk $D(x)$ is contained in the corresponding strong-unstable manifold $W^{uu}(x)$, and so it is uniquely defined by x. Moreover, $D(x)$ is tangent to the centre-unstable subbundle at every point of $\cap_{n=1}^\infty f^n(K) \cap D(x)$.*

Proof. Let $j_i \to \infty$, $x_i \to x$, and $D_i \to D(x)$ be as in the construction of $D(x)$ above. Note that D_i is contained in the j_i-iterate of D, which was taken tangent to the centre-unstable cone field. So, the domination property implies that the angle between D_i and E^{cu} goes to zero as $i \to \infty$, uniformly on $\cap_{n=1}^\infty f^n(K)$. By Lemma 3.7, given any $k \geq 1$ then f^{-k} is a $\sigma^{k/2}$-contraction on D_i for every large i. Passing to the limit, we get that every f^k is a $\sigma^{k/2}$-contraction on $D(x)$, and $D(x)$ is tangent to the centre-unstable subbundle at every point in $\cap_{n=1}^\infty f^n(K) \cap D(x)$, including x.

In particular, we have shown that the subspace E_x^{cu} is indeed uniformly expanding for Df. The domination property means that any expansion Df may exhibit along the complementary direction E^{cs} is weaker than this. Then, see [Pes76], there exists a unique *strong-unstable* manifold $W_{loc}^{uu}(x)$ tangent to E^{cu} and which is contracted by the negative iterates of f: for every $y \in W^{uu}(x)$, $\text{dist}(f^{-k}(x), f^{-k}(y))$ decreases at least as $\|Df^{-k} \mid E_x^{cu}\| \leq \sigma^{k/2}$ when k gets large. To see that $D(x)$ is contained in $W^{uu}(x)$ is suffices to recall that it is contracted by every f^{-k}, and that all its negative iterates are tangent to centre-unstable cone field. □

4. Existence and finiteness of SRB measures

This section contains the proofs of Theorem A and Corollary B. First we show that the measure ν obtained in the previous section has an absolute continuity property on disks as in Lemma 3.7. Then, using the uniformly contracting bundle, we conclude that some ergodic component of the invariant measure μ is an SRB measure.

4.1. Absolute continuity

We write $u = \dim E^{cu}$ and $s = \dim E^{cs}$, and use B^u, B^s to represent the unit compact balls in the Euclidean space of dimension u, s, respectively. In what follows we call cylinder any diffeomorphic image of $B^u \times B^s$.

Proposition 4.1. *There exists a cylinder $\mathcal{C} \subset M$, and there exists a family $\mathcal{K}_\infty$ of disjoint disks contained in $\mathcal{C} \cap \Delta_\infty$ and which are graphs over B^u, such that*

1. *the union of all the disks in $\mathcal{K}_\infty$ has positive ν-measure; in fact, the intersection of K with this union also has positive ν-measure;*
2. *the restriction of ν to that union has absolutely continuous conditional measures along the disks in $\mathcal{K}_\infty$.*

The first step of the proof is to construct a covering of the support of ν by cylinders, one of which will be the $\mathcal{C}$ in the statement. Let us point out that these cylinders we shall obtain are not small: each one contains some ball with radius uniformly bounded away from zero, depending only on the diffeomorphism f.

As we have seen, given any $y \in \Delta_\infty$ there exists a point $x \in \hat{H}_\infty$ and a disk $D(x)$ of size $\delta_1/4$ around x such that $y \in \bar{D}(x) \subset \Delta_\infty$. For any such x and $r > 0$ small, let $C_r(x)$ be the tubular neighbourhood of $\bar{D}(x)$, defined as the union of the images under the exponential map at each point $z \in \bar{D}(x)$ of all vectors orthogonal to $\bar{D}(x)$ at z and with norm less or equal than r. We take r to be sufficiently small, so that $C_r(x)$ is a cylinder and it is endowed with a canonical projection $\pi : C_r(x) \to D(x)$. Slightly adjusting r if necessary, we may also suppose that the boundary of $C_r(x)$ has zero ν-measure.

The covering of the support of ν by cylinders that we mentioned above will be obtained decomposing each of these $C_r(x)$ into a sufficient number of domains with small diameter in the centre-unstable direction, as we now explain.

Recall that each set Δ_j, $j \geq 0$, consists of a finite union of disks of radius $\delta_1/4$ inside $f^j(D)$. For any small $\varepsilon > 0$, we denote $\Delta_{j,\varepsilon}$ the subset of Δ_j obtained by removing the ε-neighbourhood of the boundary from each one of these disks. Moreover, for $n \geq 1$, we denote $\nu_{n,\varepsilon}$ the restriction of ν_n to $\cup_{j=0}^{n-1} \Delta_{j,\varepsilon}$. Let $\alpha > 0$ be as in Proposition 3.5.

Lemma 4.2. *If $\varepsilon > 0$ is sufficiently small then $\nu_{n,\varepsilon}(K) \geq \alpha/2$ for every large n.*

Proof. This is a simple consequence of Proposition 3.5. Indeed, the proposition implies that the ν_n-measure of K is greater or equal than α, for all large n. On the other hand, if ε is small then the Lebesgue measure of the ε-neighbourhood of the boundary of each disk in Δ_j is a small fraction of the Lebesgue measure of that disk. Then, in view of the distortion bound given by Proposition 2.8, the same is true with $f_*^j \operatorname{Leb}_D$ in the place of Lebesgue measure. So, taking ε small enough, we are certain to have $(f_*^j \operatorname{Leb}_D)(\Delta_j \setminus \Delta_{j,\varepsilon}) \leq \alpha/2$ for every $j \geq 0$. Then, by the definitions of ν_n and $\nu_{n,\varepsilon}$,

$$\nu_n(K) - \nu_{n,\varepsilon}(K) \leq \frac{1}{n}\sum_{j=0}^{n-1}(f_*^j \operatorname{Leb}_D)(\Delta_j \setminus \Delta_{j,\varepsilon}) \leq \frac{\alpha}{2}.$$

This completes the proof. □

In the sequel, we fix $\varepsilon > 0$ as in the lemma. Let $x \in \hat{H}_\infty$ and $\pi : C_r(x) \to \bar{D}(x)$ be as above. We fix a covering of $\bar{D}(x)$ by finitely many domains $D_{x,l} \subset \bar{D}(x), l = 1, \ldots, N$, small enough so that the intersection of each $C_{x,l} = \pi^{-1}(D_{x,l})$ with any smooth disk γ tangent to the centre-unstable cone has diameter less than ε inside γ. We take the $D_{x,l}$ diffeomorphic to the compact ball B^u, so that every $C_{x,l}$ is a cylinder.

We say that a disk γ *crosses* $C_{x,l}$ if π maps $\gamma \cap C_{x,l}$ diffeomorphically onto $D_{x,l}$. For each $j \geq 0$, let $K_j(x, l)$ be the union of the intersections of $C_{x,l}$ with all the disks in $\mathcal{D}_j$ (the disks in the support Δ_j of $f_*^j \operatorname{Leb}_D$) that cross $C_{x,l}$. Similarly, let $K_\infty(x, l)$ be the union of the intersections of $C_{x,l}$ with all the disks $D(x)$ in $\mathcal{D}_\infty$ that cross $C_{x,l}$.

The next lemma asserts that, for at least one of the cylinders $C_{x,l}$, the part of the measure ν that is carried by the disks in $K_\infty(x, l)$ gives positive weight to the set K. Recall that $\nu = \lim \nu_{n_k}$ for some $(n_k)_k$.

Lemma 4.3. *There exist (x, l) and $\alpha_1 > 0$ such that $\nu\big(K \cap K_\infty(x, l)\big) \geq \alpha_1$, and $\nu_n\big(K \cap \cup_{j=0}^{n-1} K_j(x, l)\big) \geq \alpha_1$ for n in some subsequence of $(n_k)_k$.*

Proof. For each $n \geq 1$, let $\tilde{\nu}_{n,\varepsilon}$ be the restriction of $\nu_{n,\varepsilon}$ to K, i.e., the measure defined by $\tilde{\nu}_{n,\varepsilon}(E) = \nu_{n,\varepsilon}(K \cap E)$ for every measurable subset E of M. Up to replacing $(n_k)_k$ by some subsequence, we may suppose that $\tilde{\nu}_{n_k,\varepsilon}$ converges to some measure $\tilde{\nu}_\varepsilon$. On the one hand, Lemma 4.2 means that $\tilde{\nu}_{n,\varepsilon}(M) \geq \alpha/2$ for every large n, and so $\tilde{\nu}_\varepsilon(M) \geq \alpha/2$. On the other hand, the support of $\tilde{\nu}_\varepsilon$ is contained in $\cap_{n=1}^\infty$ closure $\big(\cup_{j \geq n} \Delta_{j,\varepsilon}\big)$, and this set is covered by the interiors of the cylinders $C_r(x)$. By compactness, the support of $\tilde{\nu}_\varepsilon$ is contained in the union of a finite number of these $C_r(x)$, and so it is also contained in the union of finitely many cylinders $C_{x,l}$.

As a consequence, there must be (x, l) such that $\tilde{\nu}_\varepsilon(C_{x,l}) > 0$. We are going to show that any such (x, l) satisfies the conclusion of the lemma, if $\alpha_1 < \tilde{\nu}_\varepsilon(C_{x,l})$.

Given any disk D_j in $\mathcal{D}_j$, $j \geq 1$, let $D_{j,\varepsilon}$ be the subset obtained by removing from D_j the ε-neighbourhood of the boundary. As a consequence of the way we chose these cylinders, we have that if $D_{j,\varepsilon}$ intersects $C_{x,l}$ then D_j must cross $C_{x,l}$. This implies that

$$\tilde{\nu}_{n,\varepsilon}\big(C_{x,l}\big) = \nu_{n,\varepsilon}\big(K \cap C_{x,l}\big) \leq \nu_n\big(K \cap \cup_{j=0}^{n-1} K_j(x, l)\big)$$

for every $n \geq 1$. Since the boundary of $C_{x,l}$ has zero measure for ν, and $\tilde{\nu}_\varepsilon \leq \nu$,

$$\lim_k \tilde{\nu}_{n_k,\varepsilon}\big(C_{x,l}\big) = \tilde{\nu}_\varepsilon(C_{x,l}) > \alpha_1 .$$

Combining this with the previous inequality, we get the second part of the lemma. To get the first part, we observe that the accumulation set of $\cup_{j=0}^{n-1} K_j(x, l)$, as $n \to \infty$, is contained in $K_\infty(x, l)$. So, since K is compact,

$$\limsup_k \nu_n\big(K \cap \cup_{j=0}^{n-1} K_j(x, l)\big) \leq \nu(K \cap K_\infty(x, l)).$$

Thus, $\nu(K \cap K_\infty(x, l)) \geq \tilde{\nu}_\varepsilon(C_{x,l}) > \alpha_1$ as we claimed. □

In what follows, we fix (x, l) as in the lemma. We take the cylinder $\mathcal{C}$ in Proposition 4.1 to be $C_{x,l}$, and we let $\mathcal{K}_\infty$ be the family of disks forming $K_\infty(x, l)$. To complete the proof of the proposition, we now show that the restriction of ν to $K_\infty(x, l)$ has absolutely continuous conditional measures along the disks in $\mathcal{K}_\infty$.

Lemma 4.4. *There exists $C_3 > 1$ and a family of conditional measures $(\nu_\gamma)_\gamma$ of $\nu \mid K_\infty(x, l)$ along the disks $\gamma \in \mathcal{K}_\infty$, such that ν_γ is absolutely continuous with respect to Lebesgue measure Leb_γ on γ, with $1/C_3 \,\mathrm{Leb}_\gamma(B) \leq \nu_\gamma(B) \leq C_3 \,\mathrm{Leb}_\gamma(B)$ for any Borel set $B \subset \gamma$.*

Proof. Let us introduce $\widehat{K}(x, l) = \cup_{0 \leq j \leq \infty} K_j(x, l) \times \{j\}$. In this space, we consider the sequence of (finite) measures $\hat{\nu}_n$ defined by

$$\hat{\nu}_n \left(B_0 \times \{0\} \cup \cdots \cup B_{n-1} \times \{n-1\}\right) = \frac{1}{n} \sum_{j=0}^{n-1} f_*^j \,\mathrm{Leb}_D(B_j),$$

and $\hat{\nu}_n(B) = 0$ whenever B is contained in $\cup_{n \leq j \leq \infty} K_j(x, l) \times \{j\}$. We also consider a sequence of partitions $\mathcal{P}_k$ in $\widehat{K}(x, l)$ constructed as follows. Fix an arbitrary point z in $D_{x,l}$, and let V be the inverse image $\pi^{-1}(z)$ under the canonical projection. Fix also a sequence $\mathcal{V}_k$, $k \geq 1$, of increasing partitions of V with diameter going to zero. Then, by definition, two points (x, m), $(y, n) \in \widehat{K}(x, l)$ are in a same atom of the partition $\mathcal{P}_k$ if

1. the disk in $\mathcal{D}_m$ containing x, and the disk in $\mathcal{D}_n$ containing y intersect a same element of $\mathcal{V}_k$;
2. either $m \geq k$ and $n \geq k$ or $m = n < k$.

It is clear from the construction that for any point $\xi \in K_j(x,l)$, $0 \le j \le \infty$,

$$\mathcal{P}_1(\xi) \supset \cdots \supset \mathcal{P}_k(\xi) \supset \cdots$$

and $\cap_{k=1}^{\infty} \mathcal{P}_k(\xi)$ coincides with the intersection of the cylinder $C_{x,l}$ with the disk in $\mathcal{D}_j$ that contains ξ. Let $\hat{\pi} : \widehat{K}(x,l) \to D_{x,l}$ be defined by $\hat{\pi}(x,j) = \pi(x)$. We claim that there exits $C_3 > 1$ such that, given any Borel subset B of $D_{x,l}$, $k \ge 1$, and $\xi \in \widehat{K}(x,l)$,

$$\frac{1}{C_3}\nu_n(\mathcal{P}_k(\xi))\,\mathrm{Leb}(B) \le \hat{\nu}_n\left(\hat{\pi}^{-1}(B) \cap \mathcal{P}_k(\xi)\right) \le C_3\nu_n(\mathcal{P}_k(\xi))\,\mathrm{Leb}(B)\,. \tag{21}$$

Indeed, by definition each atom $\mathcal{P}_k(\xi)$ is a union of sets $\gamma \times \{j\}$, where γ is the intersection of the cylinder with a disk in $\mathcal{D}_j$. Since the projection π maps γ diffeomorphically onto $D_{x,l}$,

$$\begin{aligned}\frac{1}{C_4}\frac{\mathrm{Leb}(B)}{\mathrm{Leb}(D_{x,l})} &\le \frac{\mathrm{Leb}(\hat{\pi}^{-1}(B)\cap(\gamma\times\{j\}))}{\mathrm{Leb}(\gamma\times\{j\})}\\ &= \frac{\mathrm{Leb}(\pi^{-1}(B)\cap\gamma)}{\mathrm{Leb}(\gamma)} \le C_4\frac{\mathrm{Leb}(B)}{\mathrm{Leb}(D_{x,l})},\end{aligned}$$

for some uniform constant C_4. By Proposition 2.8, the density of $f_*^j\,\mathrm{Leb}_D$ with respect to Lebesgue measure on each disk in Δ_j is bounded from below and from above. So, the previous inequality implies

$$\frac{1}{C_2^2C_4}\frac{\mathrm{Leb}(B)}{\mathrm{Leb}(D_{x,l})} \le \frac{(f_*^j\,\mathrm{Leb}_D)(\hat{\pi}^{-1}(B)\cap(\gamma\times\{j\}))}{(f_*^l\,\mathrm{Leb}_D)(\gamma\times\{j\})} \le C_2^2C_4\frac{\mathrm{Leb}(B)}{\mathrm{Leb}(D_{x,l})},$$

Since this holds for every γ, we get (21) with $C_3 = C_2^2C_4/\mathrm{Leb}(D_{x,l})$.

Clearly, any accumulation measure of the sequence $\hat{\nu}_n$ must be supported in $K_\infty(x,l) \times \{\infty\}$. We have chosen a sequence n_k such that ν_{n_k} converges to some measure ν, and it is easy to see that this is just the same as saying that $\hat{\nu}_k$ converges to the measure $\hat{\nu}_\infty$ defined by $\hat{\nu}_\infty(B \times \{\infty\}) = \nu(B)$, for any Borel set $B \subset C_{x,l}$. Then, by (21) and the theorem of Radon-Nikodym, the disintegration of $\hat{\nu}$ along the disks $\cap_{k=1}^{\infty}\mathcal{P}_k(\xi)$ is absolutely continuous with respect to Lebesgue measure on those disks, with densities almost everywhere bounded from above by C_3 and from below by $1/C_3$. Since $\hat{\nu}$ is naturally identified with ν, this gives the conclusion of the lemma. □

At this point we completed the proof of Proposition 4.1.

4.2. *Ergodicity and basin of attraction*

Let us introduce some notations that are useful for the proof of the next lemma. We denote by R the set of *regular points* z of f: this means that, given any continuous function $\varphi : M \to \mathbb{R}$, both forward and backward time averages

$$\lim_{n\to+\infty} \frac{1}{n} \sum_{j=0}^{n-1} \varphi(f^j(z)) \quad \text{and} \quad \lim_{n\to+\infty} \frac{1}{n} \sum_{j=0}^{n-1} \varphi(f^{-j}(z))$$

exist and they coincide. The ergodic theorem ensures that R has full measure, with respect to any invariant probability. We say that two points $z, w \in R$ are in a same *accessibility class*, see [BP74,PS89], if there exist $N \geq 1$ and points $z = z_0, z_1, \dots, z_{N-1}, z_N = w$ in R such that $f^{t_i}(z_i)$ is in the union $W^s(z_{i-1}) \cup W^u(z_{i-1})$ of the stable and the unstable sets of z_{i-1}, for some integer t_i and every $1 \leq i \leq N$. It follows from the definition, that accessibility classes are invariant sets. Since forward averages are constant on stable sets, and backward averages are constant on unstable sets, the restriction of any invariant probability measure to an accessibility class is an ergodic measure (possibly identically zero).

Lemma 4.5. *The invariant measure $\mu = \nu + \eta$ has some ergodic component μ_* whose Lyapunov exponents are all non-zero, and whose conditional measures along local unstable manifolds are absolutely continuous with respect to Lebesgue measure. Moreover, we may choose μ_* with support contained in $\cap_{j=0}^{\infty} f^j(K)$ and* $\mathrm{Leb}_D(B(\mu_*) \cap H) > 0$.

Proof. Since R has full μ-measure, it must also have full ν-measure. In particular, up to replacing $\mathcal{K}_\infty$ by a convenient sub-family of disks γ, whose union has full ν measure in K_∞, we may suppose that ν_γ almost every point in γ is regular, for every $\gamma \in \mathcal{K}_\infty$. In particular, Leb_γ almost every point in any disk γ is regular. Using the fact that the strong-stable foliation is absolutely continuous, cf. [BP74, Section 2], we conclude that all such regular points are in a same accessibility class. Moreover, this accessibility class $\mathcal{A}$ has positive ν-measure, and so also positive μ-measure, by Lemma 4.3.

We let μ_* be the normalized restriction of μ to $\mathcal{A}$: $\mu_*(B) = \mu(B \cap \mathcal{A})/\mu(\mathcal{A})$ for every Borel set B. Then μ_* is an invariant ergodic probability measure. It follows from Lemma 3.7 that the Lyapunov exponents of μ_* along the tangent space of the disks γ are positive. Of course, the Lyapunov exponents along the strong-stable direction are all negative. Since μ_* is ergodic, its conditional measures along local unstable manifolds in $\mathcal{K}_\infty$ are either almost everywhere singular or almost everywhere absolutely continuous (with respect to Lebesgue measure). This is a well known fact, whose proof can be sketched as follows.

Suppose there is $A \subset K_\infty(x, l)$ such that $m_\gamma(A \cap \gamma) = 0$ for all $\gamma \in \mathcal{K}_\infty$, and yet $\mu_*(A) > 0$ (hence $\mu_{*\gamma}(A \cap \gamma) > 0$ for many $\gamma \in \mathcal{K}_\infty$). Let

$B = \cup_{j=-\infty}^{+\infty} f^j(A)$. By ergodicity, $\mu_*(B) = 1$, and so $\mu_{*,\gamma}(B \cap \gamma) = 1$ for $\hat{\mu}_*$-almost all γ in $\mathcal{K}_\infty$. On the other hand, $m_\gamma(B \cap \gamma) = 0$ for every $\gamma \in \mathcal{K}_\infty$. This is because f is a diffeomorphism, unstable manifolds are an invariant family of submanifolds, and B is given by a countable union. So, in this case, $\mu_{*\gamma}$ is singular with respect to m_γ for $\hat{\mu}_*$-almost all $\gamma \in \mathcal{K}_\infty$. Now suppose that, on the contrary, every measurable set $A \subset K_\infty(x, l)$ satisfying $m_\gamma(A \cap \gamma) = 0$ for all $\gamma \in \mathcal{K}_\infty$ has zero μ_*-measure. Then, restricted to $K_\infty(x, l)$, μ_* is absolutely continuous with respect to the product measure $m_\gamma \times \hat{\mu}_*$. Consequently, in this second case, the conditional measures $\mu_{*,\gamma}$ are absolutely continuous with respect to Lebesgue measure m_γ for $\hat{\mu}_*$-almost every γ in $\mathcal{K}_\infty$.

In the setting we are dealing with, the singular case is easily excluded: the conditional measures of μ_* can be written as the sum of the conditional measures of the restrictions of η and ν to the accessibility class, and the latter are equivalent to Lebesgue measure at least on $K_\infty(x, l)$. So, the conditional measures of μ_* must be almost everywhere absolutely continuous.

Since $\nu(K \cap K_\infty(x, l))$ is positive, by Lemma 4.3, $\mu(K \cap \mathcal{A}) > 0$ and so $\mu_*(K) > 0$. As K is compact and positively invariant, the ergodicity of μ_* implies that $\mu_*(\cap_{n=0}^\infty f^n(K)) = 1$, and the support of μ_* is contained in $\cap_{j=0}^\infty f^j(K)$.

To prove the last statement in the lemma we need the following fact:

Claim: *There exists a disk D_∞ inside $K_\infty(x, l)$ such that Lebesgue almost all the points in D_∞ are in the basin of μ_*, and there exists a sequence D_k of disks in $K_{j_k}(x, l)$ accumulating on D_∞ and such that* $\mathrm{Leb}_{D_k}(D_k \cap f^{j_k}(H(\sigma) \cap D))$ *is uniformly bounded away from zero.*

We assume this for a while, and explain how to conclude the proof of the lemma from it. Since forward averages of continuous functions are constant on strong-stable leaves, the basin of μ_* contains the union of all strong-stable leaves through Lebesgue almost all points in D_∞. As this foliation is absolutely continuous, and the D_k accumulate on D_∞, such union intersects D_k in a subset whose relative Lebesgue measure inside D_k goes to 1 when k goes to infinity. In particular, $\mathrm{Leb}_{D_k}(D_k \cap f^{j_k}(H(\sigma) \cap D) \cap B(\mu_*))$ is positive for every large k. Of course, the basin is invariant by f, so we may conclude that $\mathrm{Leb}_D(H(\sigma) \cap B(\mu_*)) > 0$. □

All that is left to do is to prove the Claim above.

Proof. We use Remark 3.6:

$$\nu_n\big(\cup_{i=0}^{n-1} f^i(D \cap H(\sigma))\big) \geq \alpha.$$

It follows that if $\epsilon > 0$ is fixed sufficiently small then there exists a subset of disks in the support of ν_n with total ν_n-mass larger than $\alpha/2$ and such that a fraction larger than ϵ of any such disk corresponds to points coming from $D \cap H(\sigma)$. Then, by the same argument as at the end of the proof of Lemma 4.3, the union E of the disks in $\mathcal{D}_\infty$ that are accumulated by disks as above has ν-mass larger than $\alpha/2$. Then some of these disks must be such that Lebesgue almost all points in it are in the basin of μ_*. Indeed, since μ_* is ergodic, its basin has full μ_*-measure. Then, a full measure subset of E consists of disks where almost all points, with respect to the conditional measure of μ_* on the disk, are in $B(\mu_*)$. Since we know that the conditional measures of μ_* along the disks in $\mathcal{K}_\infty$ are bounded away from zero (because the same is true for ν, cf. Lemma 4.4), we conclude that Lebesgue almost all points in some disk in E is in $B(\mu_*)$. □

As a consequence, we also get that

$$\lim_{n\to+\infty} \frac{1}{n} \sum_{j=1}^{n-1} \log \left\| Df^{-1} | E^{cu}_{f^j(x)} \right\| = \int \log \left\| Df^{-1} | E^{cu}_y \right\| d\mu_*(y) < 2 \log \sigma \tag{22}$$

for μ_*-almost every $x \in M$ (and for every $x \in B(\mu_*)$ that remains in the neighbourhood V_0 where E^{cu} has a meaning). This is a simple consequence of the ergodicity of μ_*, and the fact that its basin intersects $H(\sigma)$. In particular, all the Lyapunov exponents of μ_* in the centre-unstable direction are larger than $-\log \sigma$.

Finally, we deduce the following result which completes the proof of Theorem A and also gives Corollary B.

Corollary 4.6. *For any $\sigma < 1$, a full Lebesgue measure subset of $H(\sigma)$ is contained in the union of finitely many SRB measures supported in $\cap_{j=0}^{\infty} f^j(K)$.*

Proof. First we observe that the set of points in $H(\sigma)$ which do not belong in the basin of some SRB measure as in the statement must have zero Lebesgue measure. Indeed, otherwise we could apply the previous arguments with this set in the place of $H(\sigma)$: we would get, cf. Lemma 4.5, an extra positive Lebesgue measure subset in the basin of some SRB measure, contradicting the definition.

The main point to obtain the finiteness statement is to note that the choice of δ_1 in the context of (14) depends only on σ. Using this remark, we can deduce that for any SRB measure μ_* as we constructed above there exists a disk $D(\mu_*)$ of fixed radius δ_1, tangent to the centre-unstable cone field and such that Lebesgue almost every point in $D(\mu_*)$ is in the basin of μ_*. For this, we recall that for the SRB measures we constructed above there exist disks D_0 containing some point $x \in H(\sigma) \cap B(\mu_*)$ and on which Lebesgue almost every point is in the basin $B(\mu_*)$. In view of (22), x has many σ-hyperbolic times n. We may take $D(\mu_*)$ to be the disk of radius δ_1 around $f^n(x)$ inside $f^n(D_0)$, for any such n sufficiently large.

Then, the union of all strong-stable leaves through the points in $D(\mu_*) \cap B(\mu_*)$ is contained in $B(\mu_*)$. Using the absolute continuity property of the strong-stable foliation, we may conclude that this union contains a subset of a neighbourhood of $D(\mu_*)$ with volume bounded away from zero by some constant that depends only on δ_1 and f. For this, we fix some neighbourhood of $D(\mu_*)$ with size uniformly bounded from below, as well as a smooth foliation of it by disks C^1 close to $D(\mu_*)$. For instance, $D(\mu_*)$ could be contained in one of the leaves of this foliation. Given any leaf D, the strong-stable manifolds through the points of $D(\mu_*)$ intersect D on a subdisk D' whose Lebesgue measure inside D is bounded away from zero, by a constant that depends only on the size δ_1 of D, and on the map f. This is just by continuity of the strong-stable foliation. Moreover, absolute continuity implies that Lebesgue almost every point of D' is in the strong-stable manifold of a point of $D(\mu_*) \cap B(\mu_*)$. Now, our claim that the Lebesgue measure of the union of these strong-unstable leaves is uniformly bounded away from zero follows from Fubini's theorem.

Of course, basins of different SRB measures are two-by-two disjoint. So the conclusion of the previous paragraph implies that there can only be finitely many such measures (even if we do not assume M to be compact), since small neighbourhoods of the compact set K have finite volume. □

5. Maps with singular or critical points

Here we explain how the previous arguments can be adapted to prove Theorem C and Corollary D. A main difference concerns the notion of hyperbolic times. A key point in the previous sections was that *if n is a hyperbolic time for a point x then there exists a neighbourhood of x, in the disk D, which is mapped onto a ball of fixed radius around $f^n(x)$, in $f^n(D)$, diffeomorphically and with uniformly bounded distortion*. This was a consequence of the contraction property in Definition 2.6. Now, in the presence of a singular set $\mathcal{S}$, in order to have a similar property we must also ensure that iterates $f^j(x)$ with $0 \le j < n$ are not too close to $\mathcal{S}$.

Let $B > 1$ and $\beta > 0$ be as in the hypotheses (S1), (S2), (S3). In what follows b is any fixed constant such that $0 < b < \min\{1/2, 1/(2\beta)\}$.

Definition 5.1. *Given $\sigma < 1$ and $\delta > 0$, we say that n is a (σ, δ)*-hyperbolic time *for a point $x \in M \setminus \mathcal{S}_\infty$ if, for all $1 \le k \le n$,*

$$\prod_{j=n-k}^{n-1} \|Df(f^j(x))^{-1}\| \le \sigma^k \quad \textit{and} \quad \operatorname{dist}_\delta(f^{n-k}(x), \mathcal{S}) \ge \sigma^{bk}.$$

Let us begin by proving that this notion does imply the key property above:

Lemma 5.2. *Given $\sigma < 1$ and $\delta > 0$, there exists $\delta_1 > 0$ such that if n is a (σ, δ)-hyperbolic time for a point $x \in M \setminus \mathcal{S}_\infty$, then there exists a neighbourhood V_x of x such that*

1. *f^n maps V_x diffeomorphically onto the ball of radius δ_1 around $f^n(x)$;*
2. *for every $1 \le k < n$ and $y, z \in V_x$,*

$$\operatorname{dist}(f^{n-k}(y), f^{n-k}(z)) \le \sigma^{k/2} \operatorname{dist}(f^n(y), f^n(z)).$$

Proof. We shall prove, by induction on $j \geq 1$, that if δ_1 is chosen small enough then there exists a well defined branch of f^{-j} on the ball of radius δ_1 around $f^n(x)$, mapping $f^n(x)$ to $f^{n-j}(x)$. In addition, this branch is a $\sigma^{j/2}$-contraction. The precise condition δ_1 should satisfy is given by the following statement:

Claim: *Fix $\delta_1 > 0$ so that $4\delta_1 < \delta$ and $4B\delta_1 < \delta^\beta |\log \sigma|$. Then,*

$$\|Df(y)^{-1}\| \leq \sigma^{-1/2} \|Df(f^{n-j}(x))^{-1}\| \tag{23}$$

for any $1 \leq j < n$ and any point y in the ball of radius $2\delta_1 \sigma^{j/2}$ around $f^{n-j}(x)$.

Proof. By hypothesis $\text{dist}_\delta(f^{n-j}(x), \mathcal{S}) \geq \sigma^j$. According to the definition of the truncated distance, this means that

$$\text{dist}(f^{n-j}(x), \mathcal{S}) = \text{dist}_\delta(f^{n-j}(x), \mathcal{S}) \geq \sigma^{bj}$$
$$\text{or else} \quad \text{dist}\,(f^{n-j}(x), \mathcal{S}) \geq \delta.$$

In either case, $\text{dist}(y, f^{n-j}(x)) < \text{dist}(f^{n-j}(x), \mathcal{S})/2$ because we chose $b < 1/2$ and $\delta_1 < \delta/4 < 1/4$. Therefore, we may use (S2) to conclude that

$$\log \frac{\|Df(y)^{-1}\|}{\|Df(f^{n-j}(x))^{-1}\|} \leq B \frac{\text{dist}(y, f^{n-j}(x))}{\text{dist}(f^{n-j}(x)), \mathcal{S})^\beta} \leq B \frac{2\delta_1 \sigma^{j/2}}{\min\{\sigma^{b\beta j}, \delta^\beta\}}.$$

Since δ and σ are smaller than 1, and we took $b\beta < 1/2$, the term on the right hand side is bounded by $2B\delta_1\delta^{-\beta}$. Moreover, our second condition on δ_1 means that this last expression is smaller than $\log \sigma^{-1/2}$. □

Starting the induction argument to prove Lemma 5.2, we note that for $j = 1$ the Claim gives

$$\|Df(y)^{-1}\| \leq \sigma^{-1/2} \|Df(f^{n-1}(x))^{-1}\| \leq \sigma^{1/2},$$

since n is a hyperbolic time for x. This means that f is a $\sigma^{-1/2}$-dilation in the ball of radius $2\delta_1\sigma^{1/2}$ around $f^{n-1}(x)$. As a consequence, there exists some neighbourhood $V(n-1)$ of $f^{n-1}(x)$ contained in that ball of radius $2\delta_1\sigma^{1/2}$, that is mapped diffeomorphically onto the ball of radius δ_1 around $f^n(x)$.

Now, given any $j > 1$, let us suppose that we have constructed a neighbourhood $V(n-j+1)$ of $f^{n-j+1}(x)$ such that the restriction of f^{j-1} to $V(n-j+1)$ is a diffeomorphism onto the ball of radius δ_1 around $f^n(x)$, with

$$\|Df(f^i(z))^{-1}\| \leq \sigma^{-1/2} \|Df(f^{n-j+i+1}(x))^{-1}\| \tag{24}$$

for all z in $V(n-j+1)$ and $0 \leq i < j$. Then, by the Claim and the hypothesis that n is a hyperbolic time for x,

$$\|Df^j(y)^{-1}\| \leq \prod_{i=0}^{j-1} \|Df(f^i(y))^{-1}\| \leq \prod_{i=0}^{j-1} \sigma^{-1/2} \|Df(f^{n-j+i}(x))^{-1}\| \leq \sigma^{j/2}$$

for any point y in the ball of radius $2\delta_1\sigma^{j/2}$ whose image $z = f(y)$ is in $V(n-j+1)$.

Now we can construct an inverse branch of f^j on the ball of radius δ_1 around $f^n(x)$, by lifting geodesics in the following way. Given a geodesic γ connecting $f^n(x)$ to a point in the boundary of the ball, there is a well defined lift of the restriction of γ to a small neighbourhood of $f^n(x)$, into a curve starting at $f^{n-j}(x)$. Moreover, as far as this curve does not leave the ball of radius $2\delta_1\sigma^{j/2}$, the derivative on it is a $\sigma^{-j/2}$-dilation. This means that the length of the lifted curve is less than $\delta_1\sigma^{j/2}$, and so the curve is actually contained in a smaller ball. This proves that the lift is well defined on the whole geodesic γ. Thus, we have a well defined branch of f^{-j} on the ball of radius δ_1 around $f^n(x)$ as we claimed. We call $V(n-j)$ the image of that inverse branch. By construction, $V(n-j)$ is contained in the $2\delta_1\sigma^{j/2}$-ball around $f^{n-j}(x)$ and its image under f coincides with $V(n-j+1)$. So, in view of the Claim, we also recovered the induction assumption (24) for points in $V(n-j)$ and times $0 \le i \le j$.

In this way, we construct neighbourhoods $V(n-j)$ of $f^{n-j}(x)$ as above, for all $1 \le j \le n$. The lemma follows taking $V_x = V(0)$. □

Corollary 5.3. *There exists $C_5 > 0$ such that for every $x \in M \setminus \mathcal{S}_\infty$, any n that is a (σ, δ)-hyperbolic time for x, and every $y, z \in V_x$*

$$\frac{1}{C_5} \le \frac{|\det Df^n(y)|}{|\det Df^n(z)|} \le C_5 .$$

Proof. By construction, for $0 \le k < n$, the distance from $f^k(x)$ to either $f^k(y)$ or $f^k(z)$ is less than $\delta_1\sigma^{(n-k)/2}$, which is much smaller than $\sigma^{b(n-k)} \le \operatorname{dist}(f^k(x), \mathcal{S})$. So, assumption (S3) implies

$$\log\frac{|\det Df^n(y)|}{|\det Df^n(z)|} = \sum_{k=0}^{n-1}\log\frac{|\det Df(f^k(y))|}{|\det Df(f^k(z))|} \le \sum_{k=0}^{n-1} 2B\frac{\delta_1\sigma^{(n-k)/2}}{\sigma^{b\beta(n-k)}}.$$

Now, it suffices to take $C_5 \ge \exp\left(\sum_{i=1}^{\infty} 2B\delta_1\sigma^{(1/2-b\beta)i}\right)$, recall that $b\beta < 1/2$. □

Let $\sigma < 1$ be fixed. The assumptions of Theorem C imply that, if σ is close enough to 1, then the set $H(\sigma)$ of points $x \in M \setminus \mathcal{S}_\infty$ for which the limit in (5) is less than $3\log\sigma$ has positive Lebesgue measure. The next lemma asserts that points in $H(\sigma)$ have many (σ, δ)-hyperbolic times.

Lemma 5.4. *There are $\theta > 0$ and $\delta > 0$, depending only on σ and on the map f, such that given any $x \in H(\sigma)$ and any sufficiently large $N \ge 1$ there exist (σ, δ)-hyperbolic times $1 \le n_1 < \cdots < n_l \le N$ for x, with $l \ge \theta N$.*

Proof. The strategy is to use Lemma 3.1 twice, first for the sequence given by $a_j = -\log\|Df(f^{j-1}(x))^{-1}\|$ (up to a cut off that makes it bounded from above), and then with $a_j = \log\operatorname{dist}_\delta(f^{j-1}(x), \mathcal{S})$ for a convenient $\delta > 0$.

We prove that there exist many times n_i for which the conclusion of Pliss' Lemma 3.1 holds, simultaneously, for both sequences. Then we check that any such n_i is a (σ, δ)-hyperbolic time for x.

Let $x \in H(\sigma)$. By definition of $H(\sigma)$, for every large N we have

$$\sum_{j=0}^{N-1} -\log \|Df(f^j(x))^{-1}\| \geq 2|\log \sigma| N .$$

Fix any $\rho > \beta$. Then (S1) implies that

$$\big|\log \|Df(x)^{-1}\|\big| \leq \rho\,|\log \operatorname{dist}(x, \mathcal{S})| \tag{25}$$

for every x in a neighbourhood V of $\mathcal{S}$. Fix $\varepsilon_1 > 0$ so that $\rho\varepsilon_1 \leq |\log \sigma|/2$, and let $r_1 > 0$ be small enough so that

$$\sum_{j=0}^{N-1} -\log \operatorname{dist}_{r_1}(f^j(x), \mathcal{S}) \leq \varepsilon_1 N . \tag{26}$$

Assumption (6) ensures that this is possible. Fix any $H_1 \geq \rho\,|\log r_1|$ large enough so that it is also an upper bound for $-\log \|Df^{-1}\|$ on the complement of V. Then let E be the subset of times $1 \leq j \leq N$ such that $-\log \|Df(f^{j-1}(x))^{-1}\| > H_1$, and define

$$a_j = \begin{cases} -\log \|Df(f^{j-1}(x))^{-1}\| & \text{if } j \notin E \\ 0 & \text{if } j \in E. \end{cases}$$

By construction, $a_j \leq H_1$ for $1 \leq j \leq N$. Note that if $j \in E$ then $f^{j-1}(x) \in V$. Moreover, for $j \in E$ we have $\operatorname{dist}(f^{j-1}(x), \mathcal{S}) < r_1$:

$$\rho\,|\log r_1| \leq H_1 < -\log \|Df(f^{j-1}(x))^{-1}\| < \rho\,|\log \operatorname{dist}(f^{j-1}(x), \mathcal{S})|.$$

In particular, $\operatorname{dist}_{r_1}(f^{j-1}(x), \mathcal{S}) = \operatorname{dist}(f^{j-1}(x), \mathcal{S}) < r_1$ for all $j \in E$. Therefore, by (25) and (26),

$$\sum_{j\in E} -\log \|Df(f^{j-1}(x))^{-1}\| \leq \rho \sum_{j\in E} |\log \operatorname{dist}(f^{j-1}(x), \mathcal{S})| \leq \rho\,\varepsilon_1 N.$$

We have chosen ε_1 in such a way that the last term is less than $|\log \sigma| N/2$. As a consequence,

$$\sum_{j=1}^{N} a_j = \sum_{j=1}^{N} -\log \|Df(f^{j-1}(x))^{-1}\| - \sum_{j\in E} -\log \|Df(f^{j-1}(x))^{-1}\| \geq \frac{3}{2}\,|\log \sigma| N .$$

Thus, we have checked that we may apply Lemma 3.1 to a_j, with $c_1 = |\log\sigma|$, $c_2 = 3|\log\sigma|/2$, and $A = H_1$. The lemma provides $\theta_1 > 0$ and $l_1 \geq \theta_1 N$ times $1 \leq p_1 < \cdots < p_{l_1} \leq N$ such that

$$\sum_{j=n+1}^{p_i} -\log\|Df(f^{j-1}(x))^{-1}\| \geq \sum_{j=n+1}^{p_i} a_j \geq (p_i - n)\,|\log\sigma| \qquad (27)$$

for every $0 \leq n < p_i$ and $1 \leq i \leq l_1$.

Now fix $\varepsilon_2 > 0$ small enough so that $\varepsilon_2/(b|\log\sigma|) < \theta_1$, and let $r_2 > 0$ be such that, cf. condition (6),

$$\sum_{j=0}^{N-1} \log \operatorname{dist}_{r_2}(f^j(x), \mathcal{S}) \geq -\varepsilon_2 N\,.$$

Let $c_1 = b\log\sigma$, $c_2 = -\varepsilon_2$, $A = 0$, and

$$\theta_2 = \frac{c_2 - c_1}{A - c_1} = 1 - \frac{\varepsilon_2}{b\,|\log\sigma|}\,.$$

Applying Lemma 3.1 to the sequence $a_j = \log \operatorname{dist}_{r_2}(f^{j-1}(x), \mathcal{S})$, we conclude that there are $l_2 \geq \theta_2 N$ times $1 \leq q_1 < \cdots < q_{l_2} \leq N$ such that

$$\sum_{j=n}^{q_i-1} \log \operatorname{dist}_{r_2}(f^j(x), \mathcal{S}) \geq b\log\sigma\,(q_i - n) \qquad (28)$$

for every $0 \leq n < q_i$ and $1 \leq i \leq l_2$.

Finally, our condition on ε_2 means that $\theta_1 + \theta_2 > 1$. Let $\theta = \theta_1 + \theta_2 - 1$. Then there exist $l = (l_1 + l_2 - N) \geq \theta N$ times $1 \leq n_1 < \cdots < n_l \leq N$ at which (27) and (28) occur simultaneously:

$$\sum_{j=n}^{n_i-1} \log\|Df(f^j(x))^{-1}\| \leq (n_i - n)\log\sigma$$

and

$$\sum_{j=n}^{n_i-1} \log \operatorname{dist}_{r_2}(f^j(x), \mathcal{S}) \geq b\log\sigma(n_i - n),$$

for every $0 \leq n < n_i$ and $1 \leq i \leq l$. Therefore, given $1 \leq i \leq l$ and $1 \leq k \leq n_i$,

$$\prod_{j=n_i-k+1}^{n_i} \|Df^{-1}(f^j(x))\| \leq \sigma^k \quad \text{and} \quad \operatorname{dist}_{r_2}(f^{n_i-k}(x), \mathcal{S}) \geq \sigma^{bk}.$$

In other words, all those n_i are (σ, δ)-hyperbolic times for x, for $\delta = r_2$. □

Now we prove Theorem C, the same argument gives Corollary D.

Proof. Let H be the positive Lebesgue measure set in the statement, and $H(\sigma)$ be as above.

Lemma 5.5. *Suppose σ is close enough to 1 so that $H(\sigma)$ has positive Lebesgue measure. Then f admits some invariant probability measure μ_0 absolutely continuous with respect to Lebesgue measure and giving positive weight to $H(\sigma)$.*

Proof. According to Lemma 5.4, there exists $\delta > 0$ depending only on σ, such that for any point x in $H(\sigma)$ there exist many (σ, δ)-hyperbolic times. We let μ_n be the averages of the positive iterates of Lebesgue measure on M, and ν_n be part of μ_n carried on disks of radius δ_1 around points $f^j(x)$ such that $1 \leq j \leq n$ is a (σ, δ)-hyperbolic time for x. More precisely, arguing as in Lemma 3.4 and Proposition 3.3, we may find for each $j \geq 1$ a finite set of points $x_1, \ldots, x_N$ admitting j as a (σ, δ)-hyperbolic time, such that

1. $V_{x_1}, \ldots, V_{x_N}$ are two-by-two disjoint;
2. the Lebesgue measure of $W_j = V_{x_1} \cup \ldots \cup V_{x_N}$ is larger than the Lebesgue measure of the set of points in $H(\sigma)$ having j as a (σ, δ)-hyperbolic time, up to a uniform multiplicative constant $\tau > 0$.

Then we take

$$\nu_n = \frac{1}{n} \sum_{j=0}^{n-1} f_*^j(\text{Leb} \,|\, W_j).$$

As before in Proposition 3.5, each of these ν_n has total mass bounded away from zero, in fact, $\nu_n(H(\sigma)) \geq \alpha$ for some uniform $\alpha > 0$. Moreover, as a consequence of the distortion Corollary 5.3, every $f_*^j(\text{Leb} \,|\, W_j)$ is absolutely continuous with respect to Lebesgue measure, with density uniformly bounded from above, and so the same is true for every ν_n.

Now take $n_k \to \infty$ such that both μ_{n_k} and ν_{n_k} converge to measures μ and ν, respectively, in the weak* sense. Then μ is an invariant probability measure, $\mu = \nu + \eta$ for some measure η, ν is absolutely continuous with respect to Lebesgue measure, and $\nu(H(\sigma)) \geq \alpha > 0$. Now, if $\eta = \eta_{ac} + \eta_s$ denotes the Lebesgue decomposition of η (as the sum of an absolutely continuous and a completely singular measure, with respect to Lebesgue measure), then $\mu_{ac} = \nu + \eta_{ac}$ gives the absolutely continuous component in the corresponding decomposition of μ. By uniqueness of the Lebesgue decomposition, and the fact that the push-forward under f preserves the class of absolutely continuous measures, we may conclude that μ_{ac} is an invariant measure. Clearly, $\mu_{ac}(H(\sigma)) \geq \nu(H(\sigma)) > 0$. □

Up to replacing μ_0 by its normalized restriction to the (positively invariant) set $H(\sigma)$ we may suppose that $\mu_0(H(\sigma)) = 1$. The next lemma will allow us to show that $H(\sigma)$ is covered by the basins of finitely many ergodic absolutely continuous invariant measures.

Lemma 5.6. *For any positively invariant set $G \subset H(\sigma)$ there exists some disk Δ with radius $\delta_1/4$ such that* $\mathrm{Leb}(\Delta \setminus G) = 0$.

Proof. It suffices to prove that there exist disks of radius $\delta_1/4$ where the relative measure of G is arbitrarily close to 1.

Let ϵ be some small number, G_c be a compact subset of G, and G_o be a neighbourhood of G_c such that both $G \setminus G_c$ and $G_o \setminus G_c$ have Lebesgue measure less than $\epsilon \,\mathrm{Leb}(G)$. By Lemma 5.4 and Fubini's theorem, there exist arbitrarily large values of $j \geq 1$ such that the Lebesgue measure of the subset G_j of points in G for which j is a (σ, δ)-hyperbolic time has Lebesgue measure is at least $\theta \,\mathrm{Leb}(G)$. So, as long as ϵ is fixed sufficiently small, the Lebesgue measure of $G_c \cap G_j$ is larger than $(\theta/2)\,\mathrm{Leb}(G)$. Assume that j is large enough so that for any point x in $G_c \cap G_j$, the neighbourhood V_x is contained in G_o. Here V_x is the neighbourhood of x constructed in Lemma 5.2: it is mapped diffeomorphically onto the ball of radius δ_1 around $f^j(x)$ by f^j. Let $W_x \subset V_x$ be the pre-image of the ball of radius $\delta_1/4$ under this diffeomorphism. Let $x_1, \ldots, x_N \in G_c \cap G_j$ be such that $W_{x_1}, \ldots, W_{x_N}$ cover the compact set $G_c \cap G_j$. Up to reordering, we may suppose that $W_{x_1}, \ldots, W_{x_n}$, some $n \leq N$, is a maximal sub-family whose elements are two-by-two disjoint. Notice that the $V_{x_1}, \ldots, V_{x_n}$ cover $G_c \cap G_j$, since their union contains every W_{x_i}, $1 \leq i \leq N$. Indeed, every W_{x_i} must intersect some W_{x_k} with $k \leq n$. Then its image under f^j intersects the ball of radius $\delta_1/4$ around $f^j(x_k)$ and so it is contained in the corresponding ball of radius δ_1. This means, precisely, that W_{x_i} is contained in V_{x_k}.

By the bounded distortion property, $\mathrm{Leb}(W_x)$ is larger than the product of $\mathrm{Leb}(V_x)$ by some uniform constant $\tau > 0$ (independent of x or j). So, the Lebesgue measure of $W_{x_1} \cup \cdots \cup W_{x_n}$ is larger than $\tau \,\mathrm{Leb}(G_c \cap G_j)$. If $\xi > 0$ is such that $\mathrm{Leb}(W_{x_i} \setminus (G_c \cap G_j)) \geq \xi \,\mathrm{Leb}(W_{x_i})$ for every $1 \leq i \leq n$, then

$$\mathrm{Leb}\left(W_{x_1} \cup \cdots \cup W_{x_n}) \setminus (G_c \cap G_j)\right) \geq \xi\tau \,\mathrm{Leb}(G_c \cap G_j) \geq \xi\tau\theta \,\mathrm{Leb}(G) .$$

On the other hand, since the W_{x_i} are contained in G_o and $G_c \cap G_j \subset G$, this measure must be smaller than $\epsilon \,\mathrm{Leb}(G)$. This means that by reducing ϵ (which we may, by increasing j), we can force ξ to be arbitrarily small. In other words, we may find j and W_{x_i} such that the relative Lebesgue measure of $W_{x_i} \cap G_c \cap G_j$ in W_{x_i} is arbitrarily close to 1. Then, by bounded distortion, the relative Lebesgue measure of $G \supset f^j(G_c \cap G_j)$ in the ball of radius $\delta_1/4$ around $f^j(x_i)$ is also arbitrarily close to 1. So the proof of the lemma is complete. □

Finally, we may conclude the proof of Theorem C and Corollary D.

Let μ_0 be any absolutely continuous invariant measure with $\mu_0(H(\sigma)) = 1$. If μ_0 is not ergodic then we may decompose $H(\sigma)$ into two disjoint invariant sets H_1, H_2 both with positive μ_0-measure. In particular, both H_1 and H_2 have positive Lebesgue measure. Let μ_1, μ_2 be the normalized restrictions of μ_0 to H_1, H_2, respectively. Clearly, they are also absolutely continuous invariant measures. If they are not ergodic, we continue decomposing them, in the same way as we did for μ_0. On the other hand, by Lemma 5.6, each one of the invariant sets we find in this decomposition has full Lebesgue measure in some disk with fixed radius. Since these disks must be disjoint, and the ambient manifold is compact, there can only be finitely many of them. So, the decomposition must stop after a finite number of steps, giving that μ_0 can be written $\mu_0 = \sum_{i=1}^{s} \mu_0(H_i)\mu_i$ where $H_1, \ldots, H_s$ is a partition of $H(\sigma)$ into invariant sets with positive measure and each $\mu_i(\cdot) = \mu_0(\cdot \cap H_i)/\mu_0(H_i)$ is an ergodic measure.

This completes the proof of Theorem C and Corollary D, and it also gives the finiteness result stated in Subsect. 1.1 right after the corollary. □

Having in mind important classes of maps with singularities, such as Poincaré return maps of singular (or generalized Lorenz) attractors, [ABS77,GW79,Rov93,MPP98,BPV97] we now propose a natural extension of the previous results for *partially hyperbolic maps with singularities*.

Let us a manifold M, a compact subset $\mathcal{S}$, and a C^2 diffeomorphism (onto its image) $f : M \setminus \mathcal{S} \to M$. We suppose f has a compact positively invariant subset K, in the sense that $f(K \setminus \mathcal{S}) \subset K$, such that the tangent bundle of M restricted to $K \setminus \mathcal{S}$ has a Df-invariant dominated splitting $T_{K\setminus\mathcal{S}}M = E^{ss} \oplus E^{cu}$ such that E^{ss} is uniformly contracting.

We assume that f behaves like a power of the distance to $\mathcal{S}$ *along the centre-unstable* direction: for every $x \in K \setminus \mathcal{S}$ and $v \in E^{cu}_x$

(R1) $\frac{1}{B}\operatorname{dist}(x, \mathcal{S})^{\beta} \le \frac{\|Df(x)v\|}{\|v\|} \le B\operatorname{dist}(x, \mathcal{S})^{-\beta}$;

(R2) $\|D(Df(x))\| \le B\operatorname{dist}^{cu}(x, \mathcal{S})^{-\beta}$ and
$\|D(Df(x)^{-1})\| \le B\operatorname{dist}^{cu}(x, \mathcal{S})^{-\beta}$

(R3) $|\log\|Df^{-1} \mid E^{cu}_{f(x)}\| - \log\|Df^{-1} \mid E^{cu}_{f(y)}\|| \le B\frac{\operatorname{dist}^{cu}(x,y)}{\operatorname{dist}^{cu}(x,\mathcal{S})^{\beta}}$, if x and y are in a same disk tangent to the centre-unstable cone field, and $\operatorname{dist}^{cu}(x, y) < \operatorname{dist}^{cu}(x, \mathcal{S})/2$.

Here dist^{cu} denotes the shortest distance measured along curves tangent to the centre-unstable cone field, and we also define the truncated version $\operatorname{dist}^{cu}_{\delta}$ of dist^{cu}, in the same way as in (4). Let $\mathcal{S}_{\infty} = \cup_{n=-\infty}^{+\infty} f^n(\mathcal{S})$.

Although we did not try to check all the details, it seems that the following statement can be obtained by combining the arguments in the proofs of Theorems A and C:

Let f be as above, and assume that it is non-uniformly expanding along the centre-unstable direction, in the sense that (2) holds for all x in a positive Lebesgue measure set $H \subset M\setminus\mathcal{S}_{\infty}$ Assume moreover that, given any $\varepsilon > 0$ there exists $\delta > 0$ such that for every $x \in H$

$$\limsup_{n\to+\infty} \frac{1}{n}\sum_{j=0}^{n-1} -\log\operatorname{dist}^{cu}_{\delta}(f^j(x), \mathcal{S}) \le \varepsilon.$$

Then Lebesgue almost every point in H is in the basin of some SRB measure.

The main technical point in giving a full proof, is to control the curvature of iterates of disks tangent to the centre-unstable cone field, similarly to what we did in Subsect. 2.1.

6. Diffeomorphisms with a dominated splitting

Let $f : M \to M$ be a $C^{1+\zeta}$ diffeomorphism on a manifold M. Here we suppose that f has a compact positively invariant set $K \subset M$ with a continuous invariant *dominated splitting* $T_K M = E^{cs} \oplus E^{cu}$: there exists a constant $\lambda < 1$ and some choice of a Riemannian metric on M such that

- $\|Df \mid E^{cs}_x\| \cdot \|Df^{-1} \mid E^{cu}_{f(x)}\| \leq \lambda$ for all $x \in K$.

We call E^{cs} centre-stable subbundle and E^{cu} centre-unstable subbundle.

As we did at the beginning of Sect. 2, we can extend the subbundles continuously to a neighbourhood V_0 of K, and then consider cone fields C^{cs}_a and C^{cu}_a with small width $a > 0$ around these extended subbundles. As before, we assume that f is non-uniformly expanding along the centre-unstable direction:

$$\limsup_{n\to+\infty} \frac{1}{n} \sum_{j=1}^{n} \log \left\|Df^{-1}|E^{cu}_{f^j(x)}\right\| < 0 \tag{29}$$

for every x in a positive Lebesgue measure subset H of $D \cap K$, where D is some C^2 disk tangent to the centre-unstable cone field. We fix any $\sigma < 1$ such that $H(\sigma)$, defined as in Sect. 3, has positive Lebesgue measure in D.

Everything we did in Sect. 2 through Subsect. 4.1 applies immediately in this context. So, cf. Proposition 4.1, there exist measures μ, ν, η, a cylinder $\mathcal{C}$ and a family $\mathcal{K}_\infty$ of disjoint disks crossing $\mathcal{C}$ such that

1. μ is an invariant probability measure and $\mu = \nu + \eta$;
2. the disks in $\mathcal{K}_\infty$ are accumulated by sub-disks of radius δ_1 in $f^n(D)$, around points $f^n(x)$ such that n is a σ-hyperbolic time for $x \in H(\sigma)$;
3. the union K_∞ of all the disks in $\mathcal{K}_\infty$ intersects K in a set with positive ν-measure;
4. the restriction of ν to K_∞ has absolutely continuous conditional measures along the disks in $\mathcal{K}_\infty$.

Let us recall a few well-known notions and facts that are useful for the proof of the next lemma. Given a point x, let us denote μ_x the probability measure given by the time average along the orbit of x:

$$\int \varphi \, d\mu_x = \lim_{n\to+\infty} \frac{1}{n} \sum_{j=0}^{n-1} \varphi(f^j(x)) \quad \text{for every continuous } \varphi : M \to \mathbb{R}. \tag{30}$$

According to the ergodic decomposition theorem, see [Mañ87, Section 2.6], μ_x is well-defined and ergodic for every x in a set $\Sigma \subset M$ that has full measure with respect to any invariant measure ξ. Moreover, $x \mapsto \int g\, d\mu_x$ is measurable and

$$\int g\, d\xi = \int \left(\int g\, d\mu_x \right) d\xi(x)$$

for every measurable bounded function $g : M \to \mathbb{R}$. In fact, for any such g the integral $\int g\, d\mu_x$ coincides almost everywhere with the time average of g over the orbit of x.

Let R be the set of regular points of f, as introduced in Subsect. 4.2: $z \in R$ if and only if the forward and backward time averages of each continuous function over the orbit of z exist and coincide. R has full measure for any f-invariant probability measure ξ, as a consequence of Birkhoff's ergodic theorem. Let us point out that μ_x is constant on the intersection of R with every disk γ of $\mathcal{K}_\infty$. This is because these disks are (exponentially) contracted by negative iterates, cf. property 2. above and Lemmas 2.7 and 3.7, and so points in a same $\gamma \in \mathcal{K}_\infty$ have the same backward average (hence points in $R \cap \gamma$ also have the same forward average) for each continuous function φ.

Lemma 6.1. *There exists $z \in K_\infty \cap K \cap \Sigma \cap R$ such that $\mu_z(K_\infty \cap K) > 0$ and μ_z has absolutely continuous conditional measures along the disks in $\mathcal{K}_\infty$. In particular, the support of μ_z is contained in $\cap_{j=0}^{\infty} f^j(K)$.*

Proof. Fix B to be some measurable subset of M such that

$$m_\gamma(B \cap \gamma) = 0 \quad \text{for every} \quad \gamma \in \mathcal{K}_\infty, \tag{31}$$

and $\mu(B)$ is maximal among all measurable sets with this property. For instance, $B = \cup_n B_n$ where the $B_n, n \geq 1$, are measurable sets with property (31) such that $\mu(B_n)$ converges to the largest value compatible with that property. Observe that $\nu(B) = 0$, because ν is absolutely continuous along the leaves of $\mathcal{K}_\infty$. Let $Z_\infty = K_\infty \cap K \cap \Sigma \cap R \setminus B$. Then,

$$\mu(Z_\infty) \geq \nu(Z_\infty) = \nu(K_\infty \cap K \cap \Sigma \cap R) = \nu(K_\infty \cap K) > 0.$$

Let $(\mu \mid Z_\infty)$ be the restriction of μ to Z_∞: by definition $(\mu \mid Z_\infty)(E) = \mu(E \cap Z_\infty)$ for any measurable set E in M.

Let A be any measurable subset of Z_∞ such that $m_\gamma(A \cap \gamma) = 0$ for every $\gamma \in \mathcal{K}_\infty$. Then $\mu(A)$ must be zero, since we took $\mu(B)$ maximal. This means that $(\mu \mid Z_\infty)$ is absolutely continuous with respect to the product $m_\gamma \times \hat{\mu}$, where $\hat{\mu}$ stands for the quotient measure induced by $(\mu \mid Z_\infty)$ on $\mathcal{K}_\infty$. As a consequence, the conditional measures $\tilde{\mu}_\gamma$ of $(\mu \mid Z_\infty)$ on the disks $\gamma \in \mathcal{K}_\infty$ are absolutely continuous with respect to Lebesgue measure

m_γ for $\hat{\mu}$-almost all $\gamma \in \mathcal{K}_\infty$. On the other hand, for any measurable set $A \subset Z_\infty$,

$$\mu(A) = \int \mu_x(A)\, d\mu(x) \tag{32}$$

where the integral is taken over M or, more precisely, over the full measure subset Σ. We want to express this in terms of an integral over Z_∞. As we mentioned before,

$$\mu_x(A) = \int \mathcal{X}_A\, d\mu_x = \lim_{n\to+\infty} \frac{1}{n} \sum_{j=0}^{n-1} \mathcal{X}_A(f^j(x))$$

almost everywhere, with respect to any invariant measure. So, up to disregarding a zero μ-measure set of points, $\mu_x(A)$ can be non-zero only if x has some iterate in $A \subset Z_\infty$. Let $k(z)$ denote the first backward return time of a point $z \in Z_\infty$, that is, the smallest positive integer such that $f^{-k(z)}(z) \in Z_\infty$. This is defined μ-almost everywhere, by Poincaré's recurrence theorem. Observe also that $\mu_z = \mu_{f^j(z)}$ for any z and any integer j. Thus, we can rewrite (32) as

$$\mu(A) = \int_{Z_\infty} k(z)\mu_z(A)\, d\mu(z)$$

for any measurable subset A of Z_∞. The next lemma can be inferred from §3 of Rokhlin [Rok62]. For the reader's convenience, we state it explicitly and prove it, after completing the proof of Lemma 6.1.

Lemma 6.2. *Let λ be a finite measure on a measure space Z, with $\lambda(Z) > 0$. Let $\mathcal{K}$ be a measurable partition of Z, and $(\lambda_z)_{z\in Z}$ be a family of finite measures on Z such that*

1. *the function $z \mapsto \lambda_z(A)$ is measurable, and it is constant on each element of $\mathcal{K}$, for any measurable set $A \subset Z$*
2. *$\{w : \lambda_z = \lambda_w\}$ is a measurable set with full λ_z-measure, for every $z \in Z$.*

Assume that $\lambda(A) = \int \ell(z)\lambda_z(A)\, d\lambda$ for some measurable function $\ell : Z \to \mathbb{R}_+$ and any measurable subset A of Z. Let $\{\tilde{\lambda}_\gamma, \gamma \in \mathcal{K}\}$, and $\{\tilde{\lambda}_{z,\gamma}, \gamma \in \mathcal{K}\}$, be disintegrations of λ and λ_z, respectively, into conditional probability measures along the elements of the partition $\mathcal{K}$. Then

$$\tilde{\lambda}_{z,\gamma} = \tilde{\lambda}_\gamma$$

for λ-almost every $z \in Z$ and $\hat{\lambda}_z$-almost every γ, where $\hat{\lambda}_z$ is the quotient measure induced by λ_z on $\mathcal{K}$.

We take $Z = Z_\infty$, $\lambda = (\mu \mid Z_\infty)$, $\mathcal{K} = \mathcal{K}_\infty$, $\lambda_z = (\mu_z \mid Z_\infty)$, and $\ell(z) = k(z)$, for each $z \in Z_\infty$. It is easy to check that the hypotheses of Lemma 6.1 are satisfied. The first part of assumption 1 is contained in the ergodic decomposition theorem, and the second part follows from (30), as we explained before. Let $\mathcal{D}$ be any countable dense subset of the space of continuous functions on M. Given $z, w \in Z_\infty$, then $\mu_z = \mu_w$ if and only for every $\varphi \in \mathcal{D}$ and every $p \geq 1$ there exists $q \geq 1$ such that

$$\left|\frac{1}{n}\sum_{j=0}^{n-1}\varphi(f^{-j}(z)) - \frac{1}{n}\sum_{j=0}^{n-1}\varphi(f^{-j}(w))\right| < \frac{1}{p} \qquad \text{for any } n \geq p.$$

This gives the measurability condition in assumption 2. In this case the last part of assumption 2 is just a restatement of the fact that $\lambda_z = \mu_z$ is ergodic.

Then, according to Lemma 6.2, the conditional probability measures $\tilde{\mu}_{z,\gamma}$ of $(\mu_z \mid Z_\infty)$ along the disks $\gamma \in \mathcal{K}_\infty$ coincide almost everywhere with the corresponding conditional measures $\tilde{\mu}_\gamma$ of $(\mu \mid Z_\infty)$. Recall that we had already shown that the latter are almost everywhere absolutely continuous with respect to Lebesgue measure m_γ. We also have that

$$\int_{Z_\infty} k(z)\mu_z(Z_\infty)\, d\mu = \mu(Z_\infty) > 0.$$

It follows that there exists a positive μ-measure subset of points $z \in Z_\infty$ such that $\mu_z(K_\infty \cap K) \geq \mu_z(Z_\infty) > 0$, and the restriction of μ_z to Z_∞ has conditional measures with respect to $\mathcal{K}$ that are μ_z-almost everywhere absolutely continuous with respect to Lebesgue measure on the corresponding disk $\gamma \in \mathcal{K}$. Thus, any such z satisfies the first two claims in the statement of the lemma.

Finally, since K is compact and positively invariant, ergodicity implies that the support of μ_z is contained in $\cap_{j=0}^{\infty} f^j(K)$. □

Now we prove Lemma 6.2:

Proof. The idea is quite simple. Let $\mathcal{E}$ be the partition of the set Z into equivalence classes for the equivalence relation $z \sim w \Leftrightarrow \lambda_z = \lambda_w$. Assumption 2 ensures that the elements of $\mathcal{E}$ are measurable sets, and assumption 1 implies that every $\gamma \in \mathcal{K}$ is contained in some element of $\mathcal{E}$. Given any $e \in \mathcal{E}$ we define $\lambda_e = \lambda_z$, where z is an arbitrary point in e. We show that, up to normalization, $\{\lambda_e, e \in \mathcal{E}\}$ is a disintegration of λ with respect to the partition $\mathcal{E}$. Now, $\{\tilde{\lambda}_{e,\gamma} = \tilde{\lambda}_{z,\gamma}, \gamma \in \mathcal{K}, \gamma \subset e\}$ is a disintegration of $\lambda_e = \lambda_z$, with respect to the partition induced by $\mathcal{K}$ on each $e \in \mathcal{E}$. It follows that $\{\tilde{\lambda}_{e,\gamma}, \gamma \in \mathcal{K}, \gamma \subset e, e \in \mathcal{E}\}$ is a disintegration of λ with respect to $\mathcal{K}$ (obtained by conditioning first to $\mathcal{E}$, then to $\mathcal{K}$). By (essential) uniqueness of the disintegration into conditional probability measures, we must have $\tilde{\lambda}_{z,\gamma} = \tilde{\lambda}_\gamma$ almost everywhere.

Now we give the detailed argument. Let $\pi_\gamma : Z \to \mathcal{K}$ and $\pi_e : \mathcal{K} \to \mathcal{E}$ be the canonical projections. We represent by $\mathcal{B}(\mathcal{E})$ the σ-algebra generated by $\mathcal{E}$. Let $g : Z \to \mathbb{R}_+$ be a conditional expectation of ℓ relative to $\mathcal{E}$, that is, a Radon-Nikodym derivative, with respect to the restriction of λ to $\mathcal{B}(\mathcal{E})$, of the measure defined by

$$\mathcal{B}(\mathcal{E}) \ni E \mapsto \int_E \ell \, d\lambda.$$

In other words, g is a $\mathcal{B}(\mathcal{E})$-measurable function satisfying

$$\int_E \ell \, d\lambda = \int_E g \, d\lambda \quad \text{for every } E \in \mathcal{B}(\mathcal{E}). \tag{33}$$

$\mathcal{B}(\mathcal{E})$-measurability implies that g is constant on elements of $\mathcal{E}$. Set $g(e) = g(z)$ for any $e \in \mathcal{E}$ and $z \in e$. Let us consider the set

$$\{h : Z \to \mathbb{R} \text{ such that } \int \ell h \, d\lambda = \int gh \, d\lambda\}.$$

By (33), every characteristic function of an element of $\mathcal{B}(\mathcal{E})$ is in this set. Using linearity of the integral and the dominated convergence theorem, we conclude that the set contains any bounded $\mathcal{B}(\mathcal{E})$-measurable function. In particular, it contains $h(z) = \lambda_z(A)$, for any measurable set $A \subset Z$. Therefore,

$$\lambda(A) = \int \ell(z)\lambda_z(A) \, d\lambda(z) = \int g(z)\lambda_z(A) \, d\lambda(z) = \int g(e)\lambda_e(A) \, de, \tag{34}$$

where $de = (\pi_e \circ \pi_\gamma)_* \lambda$ is the quotient measure induced by λ on $\mathcal{E}$. Assumption 2 implies that

$$g(e)\lambda_e(Z \setminus e) = 0 \quad \text{for any } e \in \mathcal{E}. \tag{35}$$

Then,

$$g(e)\lambda_e(Z) = g(e)\lambda_e(e) = 1 \quad \text{for } de\text{-almost all } e \in \mathcal{E}. \tag{36}$$

Indeed, let $\delta > 0$ and F_δ be the set of all $e \in \mathcal{E}$ for which $g(e)\lambda_e(Z) \geq 1 + \delta$. Denote $E_\delta = (\pi_e \circ \pi_\gamma)^{-1}(F_\delta)$. Then, using (34) and (35),

$$de(F_\delta) = \lambda(E_\delta) = \int g(e)\lambda_e(E_\delta) \, de = \int_{F_\delta} g(e)\lambda_e(Z) \, de \geq (1 + \delta) de(F_\delta),$$

which implies $de(F_\delta) = 0$. Analogously, the set of $e \in \mathcal{E}$ for which $g(e)\lambda_e(Z)$ is less than $1 - \delta$ has zero de-measure for any $\delta > 0$. This proves (36). In this way, we have shown that $\{g(e)\lambda_e, \, e \in \mathcal{E}\}$ is a disintegration of λ with respect to the partition $\mathcal{E}$.

Now, for each $e \in \mathcal{E}$, let $\hat{\lambda}_e = (\pi_\gamma \mid e)_* \lambda_e$ be the quotient measure of λ_e on $(\mathcal{K} \mid e)$. Moreover, let $\{\tilde{\lambda}_{e,\gamma},\ \gamma \in \mathcal{K},\ \gamma \subset e\}$ be a disintegration of λ_e with respect to $(\mathcal{K} \mid e)$. Of course, $\tilde{\lambda}_{e,\gamma} = \tilde{\lambda}_{z,e}$ for any $z \in e$. Then

$$\lambda_e(A) = \int \tilde{\lambda}_{e,\gamma}(A)\, d\hat{\lambda}_e(\gamma) \quad \text{and so}$$

$$g(e)\lambda_e(A) = \int \tilde{\lambda}_{e,\gamma}(A)\, d(g(e)\hat{\lambda}_e)(\gamma)$$

for any measurable set $A \subset Z$. Replacing this in (34), we find

$$\lambda(A) = \int \int \tilde{\lambda}_{e,\gamma}(A)\, d(g(e)\hat{\lambda}_e)(\gamma)\, de. \tag{37}$$

Denote $d\gamma = (\pi_\gamma)_* \lambda$, the quotient measure of λ on $\mathcal{K}$. Note that $de = (\pi_e)_* d\gamma$, that is, de coincides with the quotient measure of $d\gamma$ on $\mathcal{E}$. Moreover,

$$d\gamma(\Gamma) = \lambda\big(\pi_\gamma^{-1}(\Gamma)\big) = \int g(e)\lambda_e\big(\pi_\gamma^{-1}(\Gamma)\big)\, de = \int g(e)\hat{\lambda}_e(\Gamma)\, de$$

for every measurable set $\Gamma \subset \mathcal{K}$. This means that $\{g(e)\hat{\lambda}_e,\ e \in \mathcal{E}\}$, is a disintegration of $d\gamma$ with respect to the partition $\mathcal{E}$. Thus (37) gives

$$\lambda(A) = \int \tilde{\lambda}_{e,\gamma}(A)\, d\gamma,$$

and so $\{\tilde{\lambda}_{e,\gamma},\ \gamma \in \mathcal{K}\}$, is a disintegration of λ with respect to the partition $\mathcal{K}$. Since disintegrations into conditional probability measures, when they exist, are uniquely defined almost everywhere, it follows that $\tilde{\lambda}_{e,\gamma} = \tilde{\lambda}_\gamma$ for $d\gamma$-almost every $\gamma \in \mathcal{K}$. Equivalently, this holds for de-almost every $e \in \mathcal{E}$ and $d\hat{\lambda}_e$-almost every $\gamma \in (\mathcal{K} \mid e)$, which is just the same as the conclusion of the lemma. □

Let $z \in K_\infty \cap K$ be as in Lemma 6.1. Property 2. above implies that the measure $\mu_* = \mu_z$ has dim E^{cu} Lyapunov exponents larger than $-\log \sigma$. The domination condition implies that all the other exponents are less than $-\log \sigma + \log \lambda < -\log \sigma$. So, by Pesin theory [Pes76], μ_z-almost every point x has a local strong-unstable manifold which is an embedded disk whose backward orbits approach the backward of x at the exponential rate $\log \sigma$. Moreover, the disks $\gamma \in \mathcal{K}_\infty$ contain the local strong-unstable manifolds of points in their interior.

Combining these remarks with Lemma 6.1, we get

Theorem 6.3. *Let f be a C^2 diffeomorphism admitting a positively invariant compact set with a dominated splitting $E^{cs} \oplus E^{cu}$. Assume that f is non-uniformly expanding along the centre-unstable direction, cf. (29). Then f has some ergodic Gibbs cu-state μ_* supported in $\cap_{j=0}^{\infty} f^j(K)$: μ_* is an* *invariant probability measure whose* $\dim E^{cu}$ *larger Lyapunov exponents are positive and whose conditional measures along the corresponding local strong-unstable manifolds are almost everywhere absolutely continuous with respect to Lebesgue measure on these manifolds.*

If the remaining $\dim E^{cs}$ Lyapunov exponents of μ_* are all negative then μ_* is an SRB measure. This is a well known consequence of the absolute continuity property of μ_* and absolute continuity of the stable lamination [Pes76]: the union of the stable manifolds through the points whose time averages are given by μ_* is a positive Lebesgue measure set contained in the basin of μ_*.

Clearly, the centre-stable Lyapunov exponents are indeed negative whenever the subbundle E^{cs} is uniformly contracting, which was precisely our setting in Sects. 2 through 4. In general, if one assumes that E^{cs} is non-uniformly contracting

$$\limsup_{n\to+\infty} \frac{1}{n} \sum_{j=0}^{n-1} \log \left\| Df \,|\, E^{cs}_{f^j(x)} \right\| < 0 \tag{38}$$

on a positive Lebesgue measure subset of $H \subset D$, it is not clear whether this information can be passed to a limit Gibbs *cu*-state. There are however some cases where this can be done, and so our methods do yield SRB measures supported in K.

A sufficient condition is that there exist a positive Lebesgue measure set of points in H with many (positive density at infinity) *simultaneous* σ*-hyperbolic times* with respect to the two subbundles:

$$\prod_{j=n-k+1}^{n} \left\| Df^{-1} \,|\, E^{cu}_{f^j(x)} \right\| \le \sigma^k \quad \text{and} \quad \prod_{j=n-k}^{n-1} \left\| Df \,|\, E^{cs}_{f^j(x)} \right\| \le \sigma^k$$

for every $0 \le k \le n$, for some $\sigma < 1$.

Proposition 6.4. *In the setting of Theorem 6.3, suppose that every point H has many (positive density at infinity) simultaneous σ-hyperbolic times for some $\sigma < 1$. Then ergodic Gibbs cu-states can be constructed as in the theorem which are SRB measures, and whose basins cover a full Lebesgue measure subset of H.*

Proof. Let $\sigma < 1$ be fixed such that the subset $H(\sigma)$ of points with many simultaneous σ-hyperbolic times has positive Lebesgue measure. Cf. (14) if n is a σ-hyperbolic time (with respect to E^{cu}) then the tangent space at every point in the ball of radius δ_1 around $f^n(x)$ in $f^n(D)$ is uniformly contracted by the first n negative iterates of f. Up to reducing δ_1, we may also suppose that, whenever n is a simultaneous σ-hyperbolic time then the centre-stable subbundle is $\sigma^{j/2}$-expanded by these iterates f^{-j}, $1 \le j \le n$, at every point in that ball. We construct measures ν_n as before, except that we take into account only simultaneous hyperbolic times. As a consequence, E^{cs} is $\sigma^{j/2}$-expanded by all negative iterates f^{-j}, $j \geq 1$, at every point in the support of the limit measure ν. In particular, any ergodic Gibbs cu-state for which the support of ν has positive measure must have $\dim E^{cs}$ negative Lyapunov exponents, and so it is an SRB measure.

Moreover, as we have seen in the proof of the Claim in Lemma 4.5, a definite fraction of measures ν_n is carried by disks where points in the corresponding iterate of $H(\sigma)$ occupy a subset with relative Lebesgue measure bounded away from zero. Then it is easy to verify that the content of that Claim is valid for, at least, some Gibbs cu-state μ_* charging the support of ν: there exists a sequence D_k of disks in $f^{j_k}(D)$ in which $f^{j_k}(H(\sigma))$ has relative Lebesgue measure bounded away from zero, converging to some disk D_∞ in the support of μ_*, tangent to the centre-unstable direction and such that almost every point in D_∞ is in the basin of μ_*. By Pesin theory (absolute continuity of the stable lamination [Pes76]) the union of the stable manifolds of these points in $D_\infty \cap B(\mu_*)$ cuts D_k in a subset with relative Lebesgue measure going to 1 as D_k approaches D_∞. In particular, since these stable manifolds are contained in $B(\mu_*)$, the basin must contain a positive Lebesgue measure subset of $H(\sigma)$. This proves that Lebesgue almost every point in $H(\sigma)$ is in the basin of some SRB measure, for every $\sigma < 1$.

□

Finally, we describe a simple condition on the diffeomorphism f implying existence of many simultaneous hyperbolic times. This condition is satisfied by a non-empty C^1 open set of diffeomorphisms of the 4-torus admitting an invariant set with a dominated splitting (without uniformly hyperbolic subbundles), as will be shown in the Appendix.

Proposition 6.5. *Let f be a C^2 diffeomorphism admitting a dominated splitting $E^{cs} \oplus E^{cu}$ on some positively invariant compact set K. Let*

$$A^u = \sup_{f(K)} - \log \|Df^{-1} \mid E^{cu}\| \quad \textit{and} \quad A^s = \sup_{K} - \log \|Df \mid E^{cs}\|.$$

Suppose that there exist positive constants c^u and c^s such that

$$\frac{c^u}{A^u} + \frac{c^s}{A^s} > 1 \tag{39}$$

and

$$\limsup_{n\to+\infty} \frac{1}{n} \sum_{j=1}^{n} \log \left\|Df|E^{cu}_{f^j(x)}\right\| \leq -c^u,$$

$$\limsup_{n\to+\infty} \frac{1}{n} \sum_{j=0}^{n-1} \log \left\|Df|E^{cs}_{f^j(x)}\right\| \leq -c^s$$

for some point $x \in K$. Then there exists $\sigma < 1$ such that the simultaneous σ-hyperbolic times of x have positive density at infinity.

Proof. This is a direct consequence of the expression of the density bound θ in Lemma 3.1: given σ close to 1 then, for every large N, there exist $\theta^s N$ σ-hyperbolic times with respect to E^{cs} and $\theta^u N$ σ-hyperbolic times with respect to E^{cu} in the time interval $\{1, \dots, N\}$, with

$$\theta^u = \frac{c^u + \log\sigma}{A^u + \log\sigma} \quad \text{and} \quad \theta^s = \frac{c^s + \log\sigma}{A^s + \log\sigma}.$$

If σ is close enough to 1 then $\theta^u + \theta^s > 1$ and the simultaneous hyperbolic times have density $\theta^u + \theta^s - 1 > 0$. □

A. Appendix: Applications

Here we present a few simple conditions implying the assumptions of Theorem A, Corollary D, and Propositions 6.4 and 6.5. They allow us to exhibit some robust (C^1 open) classes of maps to which these results apply.

Lemma A.1. *Given a real number σ_1 and integers $p, q \geq 1$ with $\sigma_1 > q$, there exists $\varepsilon_0 > 0$ such that the following holds. Let M be a manifold with finite volume, $f : M \to M$ be a C^1 map, and $\{B_1, \dots, B_p, B_{p+1}, \dots, B_{p+q}\}$ be a covering of M by measurable sets, such that*

1. *$|\det Df(x)| \geq \sigma_1$ for every x in $B_{p+1} \cup \cdots B_{p+q}$;*
2. *$(f \mid B_i)$ is injective for all $1 \leq i \leq p+q$.*

Then the orbit of Lebesgue almost every point $x \in M$ spends a fraction ε_0 of the time in $B_1 \cup \cdots \cup B_p$: that is, $\#\{0 \leq j < n : f^j(x) \in B_1 \cup \cdots \cup B_p\} \geq \varepsilon_0 n$ for every large n.

Proof. Let n be fixed, for the time being. Given a sequence $\underline{i} = (i_0, i_1, \dots, i_{n-1})$ in $\{1, \dots, p+q\}$, we denote

$$[\underline{i}] = B_{i_0} \cap f^{-1}(B_{i_1}) \cap \cdots \cap f^{-n+1}(B_{i_{n-1}}).$$

Moreover, we define $g(\underline{i})$ to be the number of values of $0 \leq j \leq n-1$ for which $i_j \leq p$. We begin by noting that, given any $\varepsilon_0 > 0$, the total number of sequences $\underline{i}$ for which $g(\underline{i}) < \varepsilon_0 n$ is bounded by

$$\sum_{k<\varepsilon_0 n} \binom{n}{k} p^k q^{n-k} \leq \sum_{k\leq\varepsilon_0 n} \binom{n}{k} p^{\varepsilon_0 n} q^n .$$

A standard application of Stirling's formula (see e.g. [BV99, Section 6.3]) gives that the last expression is bounded by $e^{\gamma_0 n} p^{\varepsilon_0 n} q^n$, where γ_0 depends only on ε_0 and goes to zero when ε_0 goes to zero. On the other hand, as a consequence of assumptions 1 and 2, $\text{Leb}([\underline{i}]) \leq \text{Leb}(M)\, \sigma_1^{-(1-\varepsilon_0)n}$. Then the measure of the union I_n of all the sets $[\underline{i}]$ with $g(\underline{i}) < \varepsilon_0 n$ is less than

$$\text{Leb}(M)\sigma_1^{-(1-\varepsilon_0)n} e^{\gamma_0 n} p^{\varepsilon_0 n} q^n .$$

Since we supposed $q < \sigma_1$, we may fix ε_0 small so that $e^{\gamma_0} p^{\varepsilon_0} q < \sigma_1^{1-\varepsilon_0}$. This means that the Lebesgue measure of I_n goes to zero exponentially fast as $n \to \infty$. Thus, by the lemma of Borel-Cantelli, Lebesgue almost every point $x \in M$ belongs in only finitely many sets I_n. Clearly, any such point x satisfies the conclusion of the lemma. □

Proposition A.2. *Given real numbers $\sigma_1, \sigma_2 > 0$ and integers $p, q \geq 1$ such that $\sigma_1 > q \geq 1 > \sigma_2$, there exist $\delta_0 > 0$ and $c_0 > 0$ such that the following holds. Let M, $f : M \to M$, and $B_1, \dots, B_{p+q}$ be as in Lemma A.1, and assume that*

1. $\|Df(x)^{-1}\| \leq \sigma_2$ if $x \in B_i$, $1 \leq i \leq p$, and
2. $\|Df(x)^{-1}\| \leq 1 + \delta_0$ if $x \in B_i$, $p+1 \leq i \leq p+q$.

Then f is non-uniformly expanding: for Lebesgue almost every point $x \in M$

$$\limsup_{n\to+\infty} \frac{1}{n} \sum_{j=0}^{n-1} \log \|Df(f^j(x))^{-1}\| \leq -c_0 .$$

Proof. Let $\varepsilon_0 > 0$ be the constant given by Lemma A.1. Then, fix $\delta_0 > 0$ small enough so that $\sigma_2^{\varepsilon_0}(1+\delta_0) \leq e^{-c_0}$ for some $c_0 > 0$. Let x be any point satisfying the conclusion of the lemma. Then

$$\prod_{j=0}^{n-1} \|Df(f^j(x))^{-1}\| \leq \sigma_2^{\varepsilon_0 n}(1+\delta_0)^{(1-\varepsilon_0)n} \leq e^{-c_0 n}$$

for every large enough n, This means that x satisfies the conclusion of the proposition, so the proof is complete. □

Remark A.3. We also proved that, for f as in Proposition A.2, the Lebesgue measure of the set

$$\left\{x \in M : \|Df^j(x)^{-1}\| > e^{-c_0 j} \text{ for some } j \geq n\right\}$$

goes to zero exponentially fast when $n \to \infty$.

With the aid of this proposition we can exhibit an explicit construction of a C^1 open class of maps satisfying the hypotheses of Corollary D. Let M be any compact manifold supporting some uniformly expanding map f_0: there exists $\sigma_2 < 1$ such that

$$\|v\| < \sigma_2 \|Df_0(x)v\| \quad \text{for every } x \in M \text{ and } v \in T_x M.$$

For instance, M could be the d-dimensional torus T^d. Let $V \subset M$ be some small compact domain, so that the restriction of f_0 to V is injective.

Corollary A.4. *Let f_1 be any C^1 map coinciding with f_0 outside V, and such that f_1 is volume expanding everywhere, $|\det Df_1(x)| > 1$ for every $x \in M$, and f is not too contracting on V: $\|Df_1(x)^{-1}\| \leq 1 + \delta_0$ for every $x \in V$ and some small enough $\delta_0 > 0$. Then every map f in a C^1-neighbourhood $\mathcal{N}$ of f_1 is non-uniformly expanding.*

Proof. Taking the C^1-neighbourhood sufficiently small, we may assume that there exists $\sigma_1 > 1$ such that the Jacobian of every f in it is bounded from below by σ_1. Moreover, $\|Df(x)^{-1}\| \le \sigma_2$ for every x outside V. Let $B_1, \dots, B_p, B_{p+1} = V$ be any partition of M into domains such that f is injective on each B_j, $1 \le j \le p+1$. The claim follows from Proposition A.2, with $q = 1$. □

Remark A.5. Maps f_1 as in the statement can be obtained, e.g. through deformation of f_0 by isotopy inside V. In general, these maps are not expanding: deformation can be made in such way that f_1 have periodic saddles.

Using very similar ideas one can also construct robust classes of partially hyperbolic diffeomorphisms (or, more generally, diffeomorphisms with a dominated splitting) whose centre-unstable direction is non-uniformly expanding. We just sketch the main points.

This time we start with a linear Anosov diffeomorphism f_0 on the d-dimensional torus $M = T^d$, $d \ge 2$. We write $TM = E^u \oplus E^s$ the corresponding hyperbolic decomposition. Let V be a small closed domain in M, in the following sense: there exist unit open cubes K^0 and K^1 in $\mathbb{R}^d$ such that $V \subset \pi(K^0)$ and $f_0(V) \subset \pi(K^1)$, where $\pi : \mathbb{R}^d \to T^d$ is the canonical projection. Now, let f be a diffeomorphism on T^d such that

(a) f admits invariant cone fields C^{cu} and C^{cs}, with small width $\alpha > 0$ and containing, respectively, the unstable bundle E^u and the stable bundle E^s of the Anosov diffeomorphism f_0;
(b) there is $\sigma_1 > 1$ so that $|\det(Df \mid T_x\mathcal{D}^{cu})| > \sigma_1$ and $|\det(Df \mid T_x\mathcal{D}^{cs})| < \sigma_1^{-1}$ for any $x \in M$ and any disks $\mathcal{D}^{cu}$ and $\mathcal{D}^{cs}$ through x tangent, respectively, to the centre-unstable cone field C^{cu} and to centre-stable cone field C^{cs}.
(c) f is C^1-close to f_0 in the complement of V, so that there exists $\sigma_2 < 1$ satisfying

$$\|(Df \mid T_x\mathcal{D}^{cu})^{-1}\| < \sigma_2 \quad \text{and} \quad \|(Df \mid T_x\mathcal{D}^{cs})\| < \sigma_2$$

for $x \in (M \setminus V)$ and any disks $\mathcal{D}^{cu}$, $\mathcal{D}^{cs}$ tangent to C^{cu}, C^{cs}, respectively.
(d) there exists some small $\delta_0 > 0$ satisfying

$$\|(Df \mid T_x\mathcal{D}^{cu})^{-1}\| < (1+\delta_0) \quad \text{and} \quad \|(Df \mid T_x\mathcal{D}^{cs})\| < (1+\delta_0)$$

for any $x \in V$ and any disks $\mathcal{D}^{cu}$, $\mathcal{D}^{cs}$ tangent to C^{cu}, C^{cs}, respectively.

Closeness in (c) should be enough to ensure that $f(V)$ is also contained in the projection of a unit open cube.

For instance, if f_1 is a torus diffeomorphism satisfying (a), (b), (d), and coinciding with f_0 outside V, then any map f in a C^1 neighbourhood of f_1 satisfies all the previous conditions. The C^1 open classes of transitive non-Anosov diffeomorphisms presented in [BV99, Section 6], as well as other robust examples from [Mañ78], are constructed in this way and they fit in the present setting: both these diffeomorphisms and their inverses satisfy (a)–(d).

In what is left of this appendix, we argue that any f satisfying (a)–(d) is non-uniformly expanding along its centre-unstable direction. More precisely, condition (2) in Theorem A holds, with limit bounded away from zero, on a full Lebesgue set of points $x \in M$.

To explain this, let $B_1, \dots, B_p, B_{p+1} = V$ be any partition of T^d into small domains, in the same sense as before: there exist open unit cubes K_i^0 and K_i^1 in $\mathbb{R}^d$ such that

$$B_i \subset \pi(K_i^0) \quad \text{and} \quad f(B_i) \subset \pi(K_i^1). \tag{40}$$

Let $\mathcal{F}_0^u$ be the unstable foliation of f_0, and $\mathcal{F}_j = f^j(\mathcal{F}_0^u)$ for every $j \geq 0$. By (a), each $\mathcal{F}_j$ is a foliation of T^d tangent to the centre-unstable cone field C^{cu}. For any subset E of a leaf of $\mathcal{F}_j$, $j \geq 0$, we denote $\mathrm{Leb}_j(E)$ the Lebesgue measure of E inside that leaf.

Fix any small disk D_0 contained in a leaf of the foliation $\mathcal{F}_0$. Then, for any sequence $\underline{i} = (i_0, \dots, i_{n-1})$ in $\{1, \dots, p, p+1\}$, define

$$[\underline{i}] = \{x \in D_0 : f^j(x) \in B_{i_j} \text{ for } 0 \leq j < n\}.$$

Claim: *There exists* $C_0 > 0$ *depending only on* f *such that* $\mathrm{Leb}_0([\underline{i}]) \leq C_0 \sigma_1^{-n}$ *for every sequence* $\underline{i}$ *as above.*

Proof. Indeed, let $\tilde{\mathcal{F}}_j$ be the lift to $\mathbb{R}^d$ of $\mathcal{F}_j$, for $j \geq 0$. Using (40) one can easily conclude, by induction on j, that $f^j([\underline{i}])$ is contained in the image $\pi(K_{j-1}^1 \cap \tilde{F}_j)$ of the intersection of K_{j-1}^1 with some leaf $\tilde{F}_j$ of $\tilde{\mathcal{F}}_j$, for every $0 \leq j \leq n$. So, using (b) and the fact that $(\pi \mid K_{n-1}^1)$ is a diffeomorphism and an isometry onto its image,

$$\mathrm{Leb}_0([\underline{i}]) \leq \sigma_1^{-n} \mathrm{Leb}_n(f^n([\underline{i}])) \leq \sigma_1^{-n} \mathrm{Leb}_n(F_n \cap K_{n-1}^1). \tag{41}$$

Recall that we took f_0 linear, so that its unstable foliation $\mathcal{F}_0^u$ lifts to a foliation $\tilde{\mathcal{F}}_0^u$ of $\mathbb{R}^d$ by affine hyperplanes. The leaves of every $\tilde{\mathcal{F}}_n$ are C^1 submanifolds of $\mathbb{R}^d$ transverse to these hyperplanes, with angles uniformly bounded away from zero at every intersection point. Consequently, the intersection of a leaf of $\tilde{\mathcal{F}}_n$ with any unit cube in $\mathbb{R}^d$ has Lebesgue measure (inside the leaf) bounded by some uniform constant C_0. In particular, the last factor in (41) is bounded by C_0. □

Now, using the same arguments as in Lemma A.1, we may conclude that Leb_0-almost every point $x \in D_0$ spends a positive fraction ε_0 of the time outside the domain V. Then, using assumptions (c) and (d) above, there exists $c_0 > 0$ such that

$$\limsup_{n\to\infty} \frac{1}{n} \sum_{j=0}^{n-1} \log \left\| \left(Df \middle| E^{cu}_{f^j(x)}\right)^{-1} \right\| \leq -c_0$$

for Leb_0-almost every point $x \in D_0$. Since D_0 was an arbitrary disk inside a leaf of $\mathcal{F}^s_0$, and the latter is an absolutely continuous foliation, we conclude that f is non-uniformly expanding along E^{cu}, Lebesgue almost everywhere in $M = T^d$.

Remark A.6. These arguments also show that f is non-uniformly contracting along the centre-stable direction, if it satisfies (a)–(d): Lebesgue almost every $x \in M$ has

$$\limsup_{n\to\infty} \frac{1}{n} \sum_{j=0}^{n-1} \log \left\| Df \middle| E^{cs}_{f^j(x)} \right\| \leq -c_0 .$$

Finally, reducing δ_0 if necessary (this can be done without changing c_0), we can make $A^u = \sup -\log \|Df^{-1} \mid E^{cu}\|$ and $A^s = \sup \log \|Df \mid E^{cs}\|$ arbitrarily close to zero. In particular, $c_0/A^u + c_0/A^s > 1$, as in Proposition 6.5.

References

[ABS77] V. S. Afraimovich, V. V. Bykov, L. P. Shil'nikov. On the appearence and structure of the Lorenz attractor. Dokl. Acad. Sci. USSR, **234**:336–339, 1977

[Alv97] J. F. Alves. SRB measures for nonhyperbolic systems with multidimensional expansion. PhD thesis, IMPA, 1997. To appear Ann. Sci. École Norm. Sup.

[BC91] M. Benedicks, L. Carleson. The dynamics of the Hénon map. Annals of Math., **133**:73–169, 1991

[BD96] C. Bonatti, L. J. Díaz. Nonhyperbolic transitive diffeomorphisms. Annals of Math., **143**:357–396, 1996

[BDP] C. Bonatti, L. J. Díaz, E. Pujals. A C^1-generic dichotomy for diffeomorphisms: weak forms of hyperbolicity or infinitely many sinks or sources. Preprint, 1999

[Bow75] R. Bowen. Equilibrium states and the ergodic theory of Anosov diffeomorphisms, volume 470 of Lect. Notes in Math. Springer Verlag, 1975

[BP74] M. Brin, Ya. Pesin. Partially hyperbolic dynamical systems. Izv. Acad. Nauk. SSSR, **1**:177–212, 1974

[BPV97] C. Bonatti, A. Pumariño, M. Viana. Lorenz attractors with arbitrary expanding dimension. C. R. Acad. Sci. Paris, **325**, Série I:883–888, 1997

[BR75] R. Bowen, D. Ruelle. The ergodic theory of Axiom A flows. Invent. math., **29**:181–202, 1975

[BV] M. Benedicks, M. Viana. Solution of the basin problem for Hénon-like attractors. Preprint, 1999

[BV99] C. Bonatti, M. Viana. SRB measures for partially hyperbolic systems whose central direction is mostly contracting. Israel Journal Math., 1999. To appear

[BY92] M. Benedicks, L.-S. Young. Absolutely continuous invariant measures and random perturbations for certain one-dimensional maps. Ergod. Th. & Dynam. Sys., **12**:13–37, 1992

[Car93] M. Carvalho. Sinai-Ruelle-Bowen measures for n-dimensional derived from Anosov diffeomorphisms. Ergod. Th. & Dynam. Sys., **13**:21–44, 1993

[Cas98] A. A. Castro. Backward inducing and exponential decay of correlations for partially hyperbolic attractors with mostly contracting central direction. PhD thesis, IMPA, 1998

[Dol] D. Dolgopyat. On dynamics of mostly contracting diffeomorphisms. Preprint, 1998

[DPU99] L. J. Díaz, E. Pujals, R. Ures. Partial hyperbolicity and robust transitivity. Acta Math., 1999. To appear

[GPS94] M. Grayson, C. Pugh, M. Shub. Stably ergodic diffeomorphisms. Annals of Math., **140**:295–329, 1994

[GW79] J. Guckenheimer, R. F. Williams. Structural stability of Lorenz attractors. Publ. Math. IHES, **50**:59–72, 1979

[HPS77] M. Hirsch, C. Pugh, M. Shub. Invariant manifolds, volume 583 of Lect. Notes in Math. Springer Verlag, 1977

[Jak81] M. Jakobson. Absolutely continuous invariant measures for one-parameter families of one-dimensional maps. Comm. Math. Phys., **81**:39–88, 1981

[Kan94] I. Kan. Open sets of diffeomorphisms having two attractors, each with an everywhere dense basin. Bull. Amer. Math. Soc., **31**:68–74, 1994

[Kel90] G. Keller. Exponents, attractors and Hopf decompositions. Ergod. Th. & Dynam. Sys., **10**:717–744, 1990

[Led81] F. Ledrappier. Some properties of absolutely continuous invariant measures on an interval. Ergod. Th. & Dynam. Sys., **1**:77–93, 1981

[Lyu] M. Lyubich. Almost every real quadratic map is either regular or stochastic. Preprint, 1997

[Mañ78] R. Mañé. Contributions to the stability conjecture. Topology, **17**:383–396, 1978

[Mañ82] R. Mañé. An ergodic closing lemma. Annals of Math., **116**:503–540, 1982

[Mañ87] R. Mañé. Ergodic theory and differentiable dynamics. Springer Verlag, 1987

[MPP98] C. Morales, M. J. Pacifico, E. Pujals. On C^1 robust singular transitive sets for three-dimensional flows. C. R. Acad. Sci. Paris, **326**, Série I:81–86, 1998

[Pal99] J. Palis. A global view of Dynamics and a conjecture on the denseness of finitude of attractors. Astérisque, **261**:339–351, 1999

[Pes76] Ya. Pesin. Families of invariant manifolds corresponding to non-zero characteristic exponents. Math. USSR. Izv., **10**:1261–1302, 1976

[Pes92] Ya. Pesin. Dynamical systems with generalized hyperbolic attractors: hyperbolic, ergodic and topological properties. Ergod. Th. & Dynam. Sys., **12**:123–151, 1992

[Pli72] V. Pliss. On a conjecture due to Smale. Diff. Uravnenija, **8**:262–268, 1972

[PS82] Ya. Pesin, Ya. Sinai. Gibbs measures for partially hyperbolic attractors. Ergod. Th. & Dynam. Sys., **2**:417–438, 1982

[PS89] C. Pugh, M. Shub. Ergodic attractors. Trans. Amer. Math. Soc., **312**:1–54, 1989

[Rok62] V.A. Rokhlin. On the fundamental ideas of measure theory. A. M. S. Transl., **10**:1–52, 1962. Transl. from Mat. Sbornik **25** (1949), 107–150

[Rov93] A. Rovella. The dynamics of perturbations of the contracting Lorenz attractor. Bull. Braz. Math. Soc., **24**:233–259, 1993

[Rue76] D. Ruelle. A measure associated with Axiom A attractors. Amer. J. Math., **98**:619–654, 1976

[Sat92] E. A. Sataev. Invariant measures for hyperbolic maps with singularities. Russ. Math. Surveys, **471**:191–251, 1992

[Shu71] M. Shub. Topologically transitive diffeomorphisms on T^4. volume 206 of Lect. Notes in Math., page 39. Springer Verlag, 1971

[Sin72] Ya. Sinai. Gibbs measure in ergodic theory. Russian Math. Surveys, **27**:21–69, 1972

[SW] M. Shub, A. Wilkinson. Pathological foliations and removable zero exponents. Preprint, 1998

[Tuc99] W. Tucker. The lorenz attractor exists. C. R. Acad. Sci. Paris, **328**, Série I:1197–1202, 1999

[Via97] M. Viana. Multidimensional nonhyperbolic attractors. Publ. Math. IHES, **85**:69–96, 1997

[You90] L.-S. Young. Large deviations in dynamical systems. Trans. Amer. Math. Soc., **318**:525–543, 1990

ELSEVIER

Physica D 170 (2002) 50–71

PHYSICA D

www.elsevier.com/locate/physd

SLYRB measures: natural invariant measures for chaotic systems

Brian R. Hunt[a,b,*], Judy A. Kennedy[c], Tien-Yien Li[d], Helena E. Nusse[a,1]

[a] *Institute for Physical Science and Technology, University of Maryland, College Park, MD 20742, USA*
[b] *Department of Mathematics, University of Maryland, College Park, MD 20742, USA*
[c] *Department of Mathematics, University of Delaware, Newark, DE 19716, USA*
[d] *Department of Mathematics, Michigan State University, East Lansing, MI 48824, USA*

Received 13 August 2001; received in revised form 18 January 2002; accepted 18 March 2002
Communicated by E. Kostelich

In honor of James A. Yorke's 60th birthday

Abstract

In many applications it is useful to consider not only the set that constitutes an attractor but also (if it exists) the asymptotic distribution of a typical trajectory converging to the attractor. Indeed, in the physics literature such a distribution is often assumed to exist. When it exists, it is called a "natural invariant measure". The results by Lasota and Yorke, and by Sinai, Ruelle and Bowen represent two approaches both of which establish the existence of an invariant measure. The goal of this paper is to relate the "Lasota–Yorke measure" for chaotic attractors in one-dimensional maps and the "Sinai–Ruelle–Bowen measure" for chaotic attractors in higher-dimensional dynamical systems. We introduce the notion of "*SLYRB measure*". (We pronounce the term "SLYRB" as a single word "slurb".) The SRB concept of measure can be motivated by asking how a trajectory from a typical initial point is distributed asymptotically. Similarly the SLYRB concept of measure can be motivated by asking what the average distribution is for trajectories of a large collection of initial points in some region not necessarily restricted to a single basin. The latter is analogous to ask where all the rain drops from a rain storm go and the former asks about where a single rain drop goes, perhaps winding up distributed throughout a particular lake.

Keywords: Chaotic attractor; Natural invariant measure; SLYRB measure

1. Introduction

In this paper, we relate the topics of "chaotic attractors" and the corresponding "natural invariant measures". We start off with a simple yet challenging example. Consider the map $F: \mathbb{R}^2 \to \mathbb{R}^2$ defined by

$$F(x, y) = (x^2 - y^2 + ax + by,\ 2x + cx + dy), \tag{1}$$

where $a, b, c, d \in \mathbb{R}$. We select $a = 0.9$, $b = -0.6013$, $c = 2$ and $d = 0.5$. Pick any initial condition (x_0, y_0) and compute the orbit (trajectory) $\{(x_n, y_n)\}_{n=0}^{N}$ for some large N, where $(x_{n+1}, y_{n+1}) = F(x_n, y_n)$. Choosing the

* Corresponding author. Present address: Institute for Physical Science and Technology, University of Maryland, College Park, MD 20742, USA.
E-mail address: bhunt@ipst.umd.edu (B.R. Hunt).
[1] Permanent address: Department of Econometrics, University of Groningen, P.O. Box 800, NL-9700 AV Groningen, The Netherlands.

0167-2789/02/$ – see front matter
PII: S0167-2789(02)00445-1

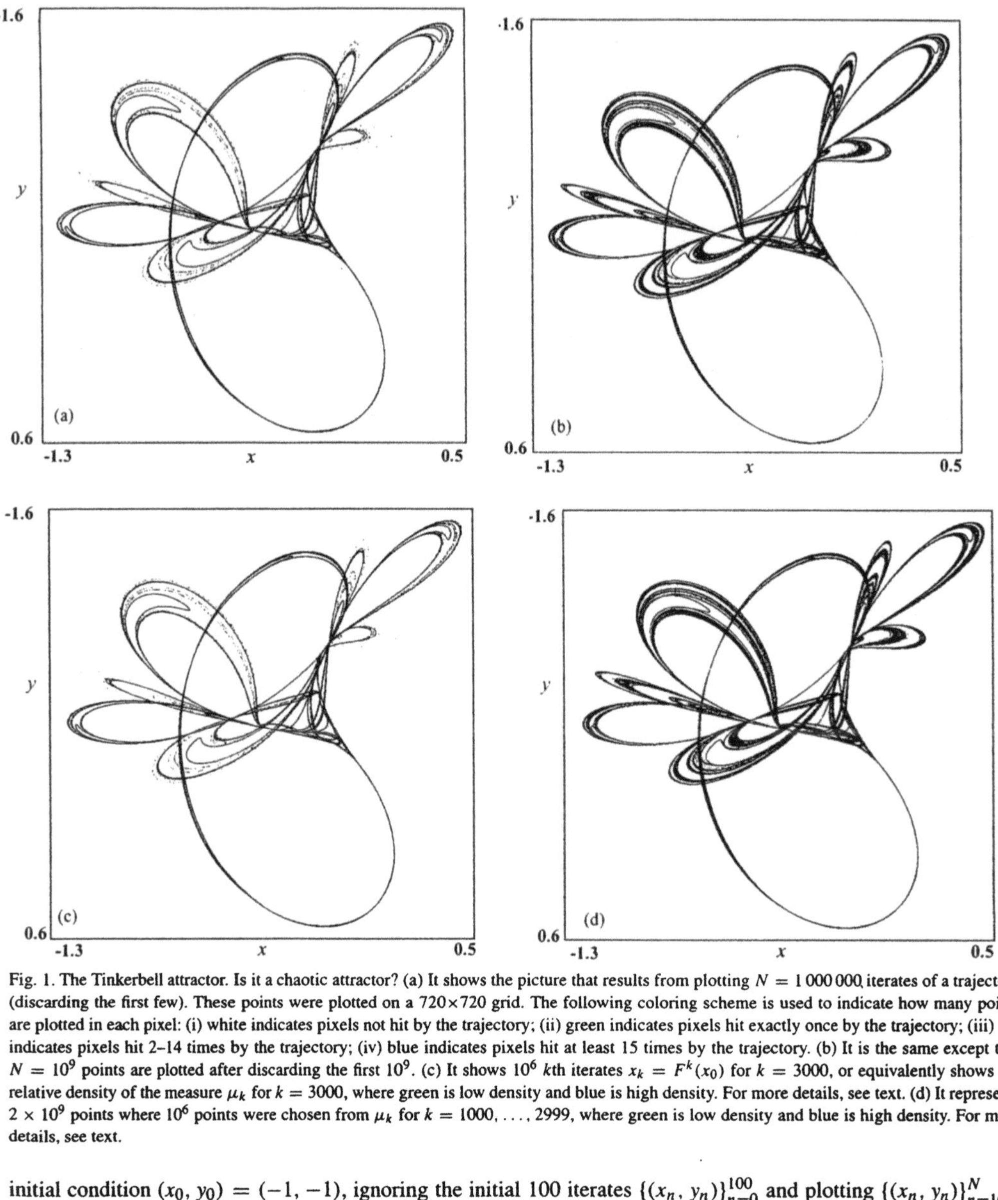

Fig. 1. The Tinkerbell attractor. Is it a chaotic attractor? (a) It shows the picture that results from plotting $N = 1\,000\,000$ iterates of a trajectory (discarding the first few). These points were plotted on a 720×720 grid. The following coloring scheme is used to indicate how many points are plotted in each pixel: (i) white indicates pixels not hit by the trajectory; (ii) green indicates pixels hit exactly once by the trajectory; (iii) red indicates pixels hit 2–14 times by the trajectory; (iv) blue indicates pixels hit at least 15 times by the trajectory. (b) It is the same except that $N = 10^9$ points are plotted after discarding the first 10^9. (c) It shows 10^6 kth iterates $x_k = F^k(x_0)$ for $k = 3000$, or equivalently shows the relative density of the measure μ_k for $k = 3000$, where green is low density and blue is high density. For more details, see text. (d) It represents 2×10^9 points where 10^6 points were chosen from μ_k for $k = 1000, \ldots, 2999$, where green is low density and blue is high density. For more details, see text.

initial condition $(x_0, y_0) = (-1, -1)$, ignoring the initial 100 iterates $\{(x_n, y_n)\}_{n=0}^{100}$ and plotting $\{(x_n, y_n)\}_{n=101}^{N}$, the resulting picture is shown in Fig. 1 with $N = 10^6$ in Fig. 1(a) and $N = 10^9$ in Fig. 1(b). The picture shows a compact set, some parts of which are heavily filled with dots and some parts of which are only sparsely filled. For every initial condition we tried in a certain region B, we got essentially the same picture. Presumably it is an example of a "Milnor attractor" for the map F. For the purpose of this paper, any continuous map $G : \mathbb{R}^m \to \mathbb{R}^m$

we call a compact set A a *Milnor attractor* for G if and only if (1) $G(A) = A$, and (2) for every open neighborhood U of A there exists a set $V \subset U$ of positive Lebesgue measure such that every initial condition $p_0 \in V$ satisfies $\text{dist}(p_n, A) \to 0$ as $n \to \infty$, where $p_n = G^n(p_0)$ is the nth iterate of p_0. We call a Milnor attractor A a *transitive Milnor attractor* if and only if A contains a trajectory that is dense in A. (In the literature there exists several (nonequivalent) definitions of the notion "attractor". For a discussion of this topic see, for example, the article by Milnor [1].) Assuming the picture in Fig. 1 represents a Milnor attractor Y^* (i.e. there is a positive probability that the trajectory of a randomly chosen point $p_0 = (x_0, y_0) \in B$ will be attracted to Y^*, so $\text{dist}(F^n(p_0), Y^*) \to 0$ as $n \to \infty$), the attractor Y^* is the Tinkerbell attractor [2], and is an example of a so-called "chaotic attractor".

In many applications it is useful to consider not only the set that constitutes an attractor but also (if it exists) the asymptotic distribution of a typical trajectory converging to the attractor. Indeed, in the physics literature such a distribution is often assumed to exist. When it exists, it is called a "natural invariant measure" [3]. Thus for the Tinkerbell attractor, we can ask: Is it true, as numerical study suggests, that for almost all (in the sense of Lebesgue measure) $p_0 \in B$ the orbit $\{p_n\}_{n=0}^{\infty}$ has a well-defined distribution and that this distribution is independent of the choice of the initial condition? We challenge the reader to come up with a right answer. Perhaps we should challenge future generations. This is a difficult question and currently we are unable to answer it. Instead, we will discuss some aspects of natural invariant measures. We want to concentrate on results by Lasota and Yorke, and by Sinai and by Ruelle and Bowen, which establish the existence of an invariant measure. *The goal is to relate the "Lasota–Yorke measure" for chaotic attractors in one-dimensional maps and the "Sinai–Ruelle–Bowen measure" for chaotic attractors in higher-dimensional dynamical systems. In Section 5 we introduce the notion of "SLYRB measure".* (We pronounce the term "SLYRB" as a single word "slurb".) We say that a system is *SLYRB chaotic* if it admits a chaotic attractor with a SLYRB. (The ordering of the letters S-LY-RB is based on the dates of three pivotal papers of Sinai, Lasota and Yorke, and Bowen and Ruelle; see Sections 3–5 for more details.) The organization of the paper is as follows. In Section 2, we discuss briefly the two approaches and concentrate on the results by Lasota and Yorke and by Sinai and by Ruelle and Bowen which establish the existence of an invariant measure. We also present a few classical examples of maps having an invariant measure. Section 3 is devoted to the Lasota and Yorke approach to invariant measures illustrated by a simple example. In Section 4, we first discuss briefly some of the important results on invariant measures due to Bowen, Ruelle and Sinai in uniformly hyperbolic systems. Thereafter, we give a partial review on the existence of invariant measures for one-dimensional quadratic maps. Finally, we give a flavor of results on invariant measures for certain two-dimensional nonhyperbolic systems having chaotic nonhyperbolic attractors, namely the well-known Lozi maps and Hénon maps. In Section 5 we give our conclusions and introduce the notion of SLYRB measure.

2. Invariant measures

Much of this paper is devoted to a discussion of existing theorems on invariant measures. The results representing the two approaches we want to concentrate on are the results by Lasota and Yorke (Theorem LaY in Section 3), and by Sinai and by Ruelle and Bowen (Theorem SRB in Section 4), which establish the existence of an invariant measure. In this section, we discuss briefly the two approaches and also review a few classical examples of systems having an invariant measure. We start off with some definitions.

Let X be a space with Lebesgue measure m. Throughout this paper m will denote Lebesgue measure. A measure μ (on X) is said to have *density* f if $\mu(E) = \int_E f(x)\,dx$ for all measurable sets $E \subset X$, where $f : X \to [0, \infty)$ is a measurable function. A measure μ is called *absolutely continuous* if it has a density. In this paper we assume that all densities satisfy $\int_X f(x)\,dx < \infty$. (Ref. [4] requires $\int_X f(x)\,dx = 1$.) For a density f, write μ_f for the measure that has f as its density. A measure μ is a *probability measure* if $\mu(X) = 1$. Let $M : X \to X$ be a piecewise smooth map. We write $M^{-1}(E) = \{x \in X : M(x) \in E\}$ for the pre-image of E under M. A measure μ is said to be an *invariant measure* for M if $\mu(M^{-1}(E)) = \mu(E)$ for every measurable set $E \subset X$.

2.1. Invariant measures derived from Lebesgue measure: two approaches

We now briefly discuss the two approaches. Let $M : X \to X$ be a piecewise smooth map. One approach follows ideas of Sinai, who since the beginning of the 1970s, systematically developed a Markov partition method for dynamical systems. In 1972 [5], he proved the existence of invariant measures for certain systems having a Markov partition. Bowen and Ruelle [6–9] extended these techniques to Axiom A diffeomorphisms and flows in papers that were published during the period 1973–1976. The resulting measure is often called an SRB measure. To describe such an invariant measure, let $U \subset X$ be a set of positive Lebesgue measure that is mapped into itself and let $A \subset U$ be a Milnor attractor such that for every $x \in U$, $M^n(x) \to A$ as $n \to \infty$. They consider the average value along a trajectory $\{M^i(x)\}_{i=0}^{\infty}$ of any continuous bounded function φ, i.e. $\langle\varphi\rangle_x = \lim_{n\to\infty}(1/n)\sum_{i=0}^{n-1}\varphi(M^i(x))$. If for some probability measure μ, $\langle\varphi\rangle_x$ exists and equals $\int\varphi\,d\mu$ for all φ, then μ represents the asymptotic distribution of the trajectory of x. They are interested in cases in which Lebesgue almost all initial conditions $x \in U$ yield the same measure. Thus when an SRB measure μ exists, it is described by $\lim_{n\to\infty}(1/n)\sum_{i=0}^{n-1}\varphi(M^i(x)) = \int\varphi\,d\mu$ for Lebesgue almost every x in U and for every continuous bounded function φ on U (see Section 4). Hence, the significance of such an SRB measure is that it describes the asymptotic distribution of the trajectory of initial points in a set of positive Lebesgue measure in the phase space.

An alternative, but closely related, approach (based on the Perron–Frobenius techniques) was used, for example, by Lasota and Yorke [10]. Here the "evolution" of a measure μ_0 under M is studied. To explain that term, choose an initial point x_0 at random using a probability measure μ_0 and examine the trajectory defined by $x_{n+1} = M(x_n)$, $n = 0, 1, 2, \ldots$. Then x_1 has a probability distribution μ_1, where $\mu_1(E) = \mu_0(M^{-1}(E))$ for every $E \subset X$. The measure μ_n for the distribution of x_n can be defined iteratively, since the probability that $x_{n+1} \in E$ equals the probability that $x_n \in M^{-1}(E)$. Hence

$$\mu_{n+1}(E) = \mu_n(M^{-1}(E)) \quad \text{for } n = 0, 1, 2, \ldots. \tag{2}$$

In this approach an invariant measure is described as the limit of $(1/n)\sum_{i=0}^{n-1}\mu_i$ (see Section 3).

The Frobenius–Perron operator $\mathcal{P}$ for M (also denoted by $\mathcal{P}_M$) is used to describe how the corresponding densities map, i.e. if μ_0 and $\mu_1 = \mu_0(M^{-1}(\cdot))$ are as above, and if they have densities f_0 and f_1, respectively, then

$$f_1 = \mathcal{P}f_0. \tag{3}$$

When the Jacobian matrix $DM(x)$ of M at x exists and $\det DM(x) \neq 0$ for almost all x (w.r.t. the Lebesgue measure), we can equivalently define $\mathcal{P}\ (= \mathcal{P}_M)$ by

$$\mathcal{P}_M(f)(x) = \sum_{y\in M^{-1}(x)} \frac{f(y)}{|\det DM(y)|}. \tag{4}$$

The Frobenius–Perron operator is very useful for studying the evolution of densities under iteration of M. If an initial point x is chosen using a probability distribution with density f, then $\mathcal{P}_M f$ is the density for $M(x)$. A good reference is, for example, Lasota and Mackey [4].

We now illustrate the two approaches. Let F be the Tinkerbell map defined in Eq. (1). Let the measure μ_0 denote a uniform distribution on the rectangle shown in red in Fig. 2. The rectangle is approximately 400 pixels×400 pixels. In Fig. 2 ($k = 0$) we randomly chose 10^6 points x_0 from this distribution and plotted them. The coloring scheme is as follows. A pixel is colored white if it is hit by none of these 10^6 points, green if it is hit by exactly

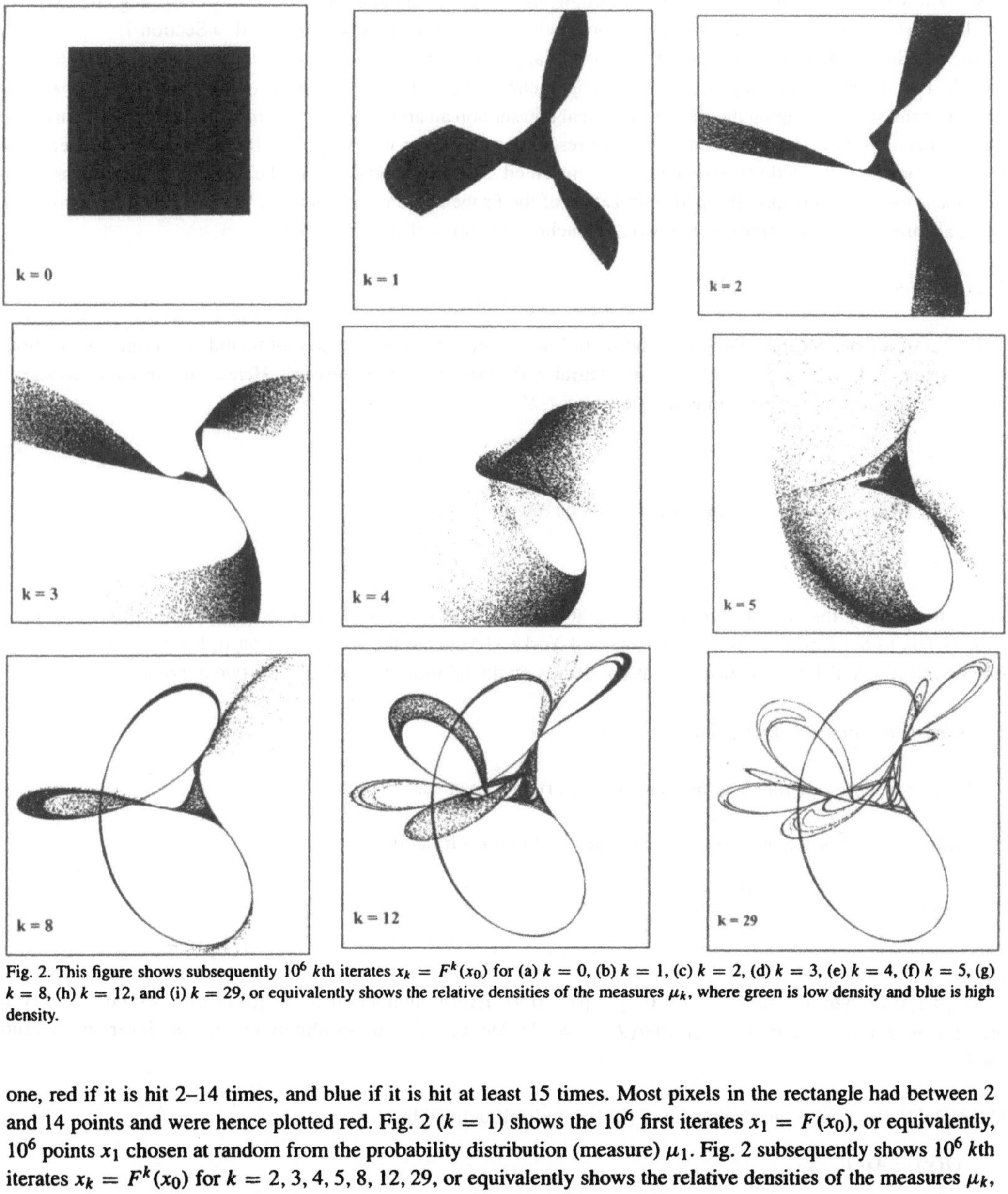

Fig. 2. This figure shows subsequently 10^6 kth iterates $x_k = F^k(x_0)$ for (a) $k = 0$, (b) $k = 1$, (c) $k = 2$, (d) $k = 3$, (e) $k = 4$, (f) $k = 5$, (g) $k = 8$, (h) $k = 12$, and (i) $k = 29$, or equivalently shows the relative densities of the measures μ_k, where green is low density and blue is high density.

one, red if it is hit 2–14 times, and blue if it is hit at least 15 times. Most pixels in the rectangle had between 2 and 14 points and were hence plotted red. Fig. 2 ($k = 1$) shows the 10^6 first iterates $x_1 = F(x_0)$, or equivalently, 10^6 points x_1 chosen at random from the probability distribution (measure) μ_1. Fig. 2 subsequently shows 10^6 kth iterates $x_k = F^k(x_0)$ for $k = 2, 3, 4, 5, 8, 12, 29$, or equivalently shows the relative densities of the measures μ_k, where green is low density and blue is high density. Fig. 1(c) shows this distribution for μ_k where $k = 3000$ and Fig. 1(d) represents 2×10^9 points where 10^6 points were chosen from μ_k for $k = 1000, \ldots, 2999$. Note that there is still a low density region in Fig. 1(d). Although Fig. 1(b) is generated by a single trajectory, Fig. 1(b) and (d) displays certain similarities. This may be of some help in answering the question posed in Section 1.

The way the two approaches describe the measures appear quite different. However, if the system has one attractor, then the two measures resulting from the two approaches are equal. The theoretical construction of a global SRB measure consists of averaging the iterates of a local measure on an arbitrary unstable manifold (see the construction for the Lozi map in Section 4). This construction resembles the Lasota and Yorke's pushing forward of the Lebesgue measure construction. Dellnitz and Junge [11] presented efficient techniques for the numerical approximation of SRB measures based on appropriate discretization of the Frobenius–Perron operator. We show how one may think of the apparent difference between the two approaches as being a change of variables.

2.2. A standard change of variables

The set of measures can be viewed as the set of linear operators on the space $\mathcal{C}$ of bounded continuous functions on the space X. Every linear operator is an integral with respect to some measure. Hence, the measure associated with $\mathcal{P}_M f$ for the map M is the operator (letting $g \in \mathcal{C}$)

$$g \mapsto \int_X (\mathcal{P}_M f)(x) g(x)\, dx. \tag{5}$$

Assuming M is piecewise invertible, Eq. (5) can be rewritten as

$$g \mapsto \int_X f(y) g(M(y))\, dy \tag{6}$$

as can be seen by the standard change of variables $x = M(y)$. We will see that the representation in Eq. (5) corresponds to the approach used by Lasota and Yorke [10] while the representation in Eq. (6) was utilized by Bowen and Ruelle [7]. See Section 5 for more details on the relationship between the two approaches.

2.3. Classical examples of invariant measures

We first give three examples of maps having invariant measures.

Example 1. Let T be a tent map of the interval [0, 1] into itself defined by

$$T(x) = \begin{cases} 2x & \text{if } 0 \le x \le 0.5, \\ 2(1-x) & \text{if } 0.5 < x \le 1. \end{cases}$$

For every subinterval $J \subset [0, 1]$ we have that $T^{-1}(J)$ is a union of two intervals J_1 and J_2. For example, if $J = [0.2, 0.4]$ then $T^{-1}(J) = [0.1, 0.2] \cup [0.8, 0.9]$. Every subinterval $J \subset [0, 1]$ satisfies $m(T^{-1}(J)) = m(J_1) + m(J_2) = (1/2)m(J) + (1/2)m(J) = m(J)$. Hence, m is an absolutely continuous invariant measure for T.

Example 2. Let Q be a quadratic map of [0, 1] into itself defined by

$$Q(x) = 4x(1-x).$$

In an abstract being a preliminary report, Ulam and Von Neumann [12] wrote in 1947 "... starting with almost every x_1 (in the sense of Lebesgue measure) and iterating the function $f(x) = 4x(1-x)$ one obtains a sequence of numbers on (0, 1) with a computable algebraic distribution". Although one often encounters statements suggesting that an invariant measure for Q was first constructed by Ulam and Von Neumann [12], in fact that the algebraic expression is not stated in the published text. There is a well-known simple change of variables by which one can reduce the study of the map Q to that of T in Example 1. The earliest description of this change of variables seems to be in Rechard [13], who reported in 1956 that Ulam has observed that the map Q is conjugate to T under the homeomorphism φ: $[0, 1] \to [0, 1]$ defined by $\varphi(x) = 2/\pi \arcsin \sqrt{x}$, i.e. $\varphi \circ Q \circ \varphi^{-1} = T$. (In later papers, Ulam refers to [12,13] for this conjugation φ.) To study the evolution of densities under Q, Rechard considers the Frobenius–Perron operator $\mathcal{P}_Q$ from $L^1([0, 1], m)$ to itself defined by

$$\mathcal{P}_Q f(x) = \frac{1}{4\sqrt{1-x}}\left[f\left(\frac{1}{2}(1-\sqrt{1-x})\right) + f\left(\frac{1}{2}(1+\sqrt{1-x})\right)\right].$$

Defining $\rho : (0, 1) \to \mathbb{R}$ by $\rho(x) = \varphi'(x) = 1/\pi\sqrt{x(1-x)}$ and observing that $\rho(x) = \rho(1-x)$, one verifies

$$\int_0^{4x(1-x)} \rho(t)\,\mathrm{d}t = \varphi \circ Q(x) = T \circ \varphi(x) = \begin{cases} 2\varphi(x) = \int_0^x \rho(t)\,\mathrm{d}t + \int_{1-x}^1 \rho(t)\,\mathrm{d}t & (0 \le x \le \frac{1}{2}), \\ 2[1-\varphi(x)] = \int_x^1 \rho(t)\,\mathrm{d}t + \int_1^{1-x} \rho(t)\,\mathrm{d}t & (\frac{1}{2} \le x \le 1). \end{cases}$$

This implies that $\mathcal{P}_Q\rho = \rho$. Thus the measure m^* defined by $m^*(A) = \int_A (\mathrm{d}x/\pi\sqrt{x(1-x)})$ is an absolutely continuous invariant measure for the map Q.

Example 3. In 1969, Krzyzewski and Szlenk [14] proved the existence of a unique absolutely continuous invariant probability measure for C^2-expanding maps on compact, connected differentiable manifolds. The following is a special case of their result. Let F be a differentiable map of the two-dimensional torus into itself defined by $F(x, y) = (ax + by + \varepsilon f(x, y), cx + dy + \varepsilon g(x, y))(\bmod 1 \text{ in } x \text{ and } y)$, $0 \le x, y < 1$, where (a) $a, b, c, d \in \mathbb{Z}$; (b) the eigenvalues of

$$\begin{bmatrix} a & b \\ c & d \end{bmatrix}$$

are real and their moduli are greater than 1; (c) the maps $f, g : \mathbb{R}^2 \to \mathbb{R}$ are of class C^2 and each of them is periodic with period 1 with respect to each variable; (d) ε is a positive number. If ε is sufficiently small, then the map F is expanding and so it admits an absolutely continuous invariant probability measure.

3. Lasota–Yorke measures for chaotic attractors

We first concentrate on the following example. Let S: $\mathbb{R} \to \mathbb{R}$ be the "skew tent" map defined by

$$S(x) = \begin{cases} ax + 1 & \text{for } x \le 0, \\ bx + 1 & \text{for } x > 0. \end{cases} \tag{7}$$

Yorke reports privately that the goal of [10] initially was only to handle maps like tent maps given by Eq. (7) where $1 < a < 2$ and $b = -a$, but their techniques allowed them to prove the result for more general scalar maps. The theorem of Krzyzewski and Szlenk had considered only maps for which there is in fact a C^2 invariant density. That is not the case for most tent maps. It can be shown that the absolutely continuous invariant densities for the tent maps are discontinuous at the points that are the image of $x = 0$ under τ^k for $k \ge 1$ and continuous elsewhere. For most choices of such parameter values a, the trajectory of 0 is dense in some interval, and so the support of the density f^* has a dense set of discontinuities of f^*. The set of discontinuities of f^* has Lebesgue measure zero and therefore f^* is continuous Lebesgue almost everywhere. Note that if $a = \sqrt{2}$ and $b = -a$, then the third iterate of the turning point $x = 0.5$ equals the fixed point p given by $p = 1/(1-b)$, and if $1 < a < \sqrt{2}$ and $b = -a$, the support is always disconnected. In the latter case, the attractor lies in two intervals, one lying wholly to the left of the fixed point p and the other to the right of p.

We will present the interesting case of the skew tent map (7) with values $a = 0.5$, $b = -4.4$, because there the support of f^* is disconnected. The points $p_1 = -(1+b+b^2)/(ab^2-1) < 0$, $p_2 = -(1+a+ab)/(ab^2-1) > 0$, and $p_3 = -(1+b+ab)/(ab^2-1) > 0$ constitute a period-3 orbit. Straightforward computation yields

$$S(p_1) = -a(1+b+b^2)/(ab^2-1)+1 = -(1+a+ab)/(ab^2-1) = p_2,$$
$$S(p_2) = -b(1+a+ab)/(ab^2-1)+1 = -(1+b+ab)/(ab^2-1) = p_3,$$
$$S(p_3) = -b(1+b+ab)/(ab^2-1)+1 = p_1.$$

Hence, the map S displays Li–Yorke chaos [15]. In particular for every $n \in \mathbb{N}$, the map S has a period-n orbit and so there are infinitely many periodic points. Plotting any trajectory (or orbit) $\{x_n\}_{n=0}^{\infty}$, it seems that it fills in three intervals. The third iterate of S, S^3, is defined by

$$S^3(x) = \begin{cases} a[a(ax+1)+1]+1 & \text{if } x \le c_1 = -6, \\ b[a(ax+1)+1]+1 & \text{if } c_1 < x \le c_2 = -2, \\ b[b(ax+1)+1]+1 & \text{if } c_2 < x \le c_3 = -2-2b^{-1}, \\ a[b(ax+1)+1]+1 & \text{if } c_3 < x \le c_4 = 0, \\ a[b(bx+1)+1]+1 & \text{if } c_4 < x \le c_5 = -b^{-1}-b^{-2}, \\ b[b(ax+1)+1]+1 & \text{if } c_5 < x \le c_6 = -b^{-1}, \\ b[a(bx+1)+1]+1 & \text{if } c_6 < x \le c_7 = -3b^{-1}, \\ a[a(bx+1)+1]+1 & \text{if } c_7 < x \end{cases} \qquad (8)$$

for $a = 0.5$, $b = -4.4$. Its graph is shown in Fig. 3. It is easy to verify that for every $x \in \mathbb{R}$, there exists an $n \in \mathbb{N}$ such that $S^n(x) \in [-4, 1]$ and S maps the interval $[-4, 1]$ into itself. From now on, we consider the map S restricted to the interval $I = [-4, 1]$. One can easily verify that for every $x \in I \setminus \{c_n\}_{n=1}^{7}$, $|(S^3)'(x)| \ge |a^2 b| > 1$.

Before we are able to state the fact that S has a "chaotic attractor", we need some basic notions. The trajectory (or orbit) of a point $x_0 \in I$ under the map S is the set $\{S^n(x_0)\}_{n=0}^{\infty}$, where S^n denotes the nth iterate of S. We define the Lyapunov number $L(x_0)$ of the trajectory of x_0 to be $L(x_0) = \lim_{n\to\infty}|(S^n)'(x_0)|^{1/n}$, where we assume that the limit exists. The Lyapunov exponent $\lambda(x_0)$ of the trajectory of x_0 is the logarithm of the Lyapunov number and so $\lambda(x_0) = \log L(x_0)$. Hence, $\lambda(x_0) = \lim_{n\to\infty}(1/n)\sum_{i=1}^{n}\log|S'(S^{i-1}(x_0))|$. Observe that if $\lambda(x_0) > 0$ then $|(S^n)'(x_0)|$ grows exponentially as $n \to \infty$. The trajectory of a point x is called a chaotic trajectory if (1) the trajectory of x is bounded and x is not asymptotically periodic, and (2) the trajectory of x has a positive Lyapunov exponent. Therefore, a chaotic trajectory consists of infinitely many different points. We call a compact set A a *chaotic attractor* (for S) if and only if (a) A is a Milnor attractor, and (b) A contains a chaotic trajectory that is dense in A. The following property easily follows from known results, but we provide a proof that is in the spirit of this paper.

Property 1. The map S has a chaotic attractor.

Proof. For $1 \le i \le 3$, let K_i be the interval of which one end point is the period-3 point p_i and the other point is q_i defined by $S^3(q_i) = p_i$ as shown in Fig. 3. We write $G = S^3$. One easily verifies $G(c_2) \in (q_1, p_1)$, $G(c_4) \in (q_2, p_2)$, and $G(c_7) \in (p_3, q_3)$. Therefore, G maps each of the intervals K_i into itself. Define $D = \{x \in I : G(x) \notin \cup_{i=1}^{3}\mathrm{Int}\, K_i\}$, i.e. D is the collection of points in I that are not mapped into the interior of K_i $(1 \le i \le 3)$ when the

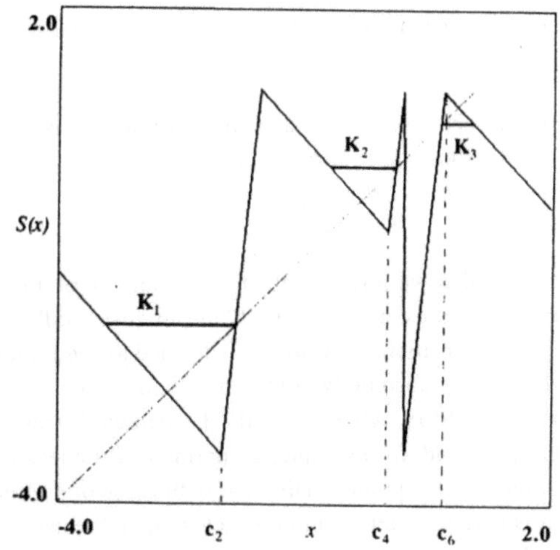

Fig. 3. The third iterate S^3 of the map S, where $S : [-4, 2] \to [-4, 2]$ is defined by $S(x) = 0.5x + 1$ if $x \le 0$ and $S(x) = -4.4x + 1$ if $x > 0$. Each interval K_i is mapped into itself when S^3 is applied ($1 \le i \le 3$).

map G is applied. Note that $c_i \notin D$ ($1 \le i \le 7$). We have that G is expanding on D, i.e. $|G'(x)| \ge |a^2b| > 1$ for all $x \in D$. The collection of points that stay in D when the map G is applied repeatedly, has Lebesgue measure zero, see [16]. (We note that there are quite a number of results in the literature which establish this fact too.) We now conclude that the trajectory of Lebesgue almost every point will enter the union of the three intervals K_i ($1 \le i \le 3$). We write $J_1 = [G(c_2), G^2(c_2)] \subset K_1$, $J_2 = [G(c_4), G^2(c_4)] \subset K_2$, and $J_3 = [G(c_7), G^2(c_7)] \subset K_3$. These compact intervals satisfy $G(J_i) = J_i$ ($1 \le i \le 3$). One easily verifies that for every $x \in \text{Int}\, K_i$, there exists an $N_x \in \mathbb{N}$ such that $G^n(x) \in J_i$ for all $n \ge N_x$ ($1 \le i \le 3$). We write $A = J_1 \cup J_2 \cup J_3$. We have $S(A) = A$ and the collection of points whose trajectories do not converge to A has Lebesgue measure zero. Therefore, A is a Milnor attractor for S.

In order that A be a chaotic attractor, it remains to be shown that there is a chaotic trajectory that is dense in A. We now concentrate on J_2. The two fixed points of G in K_2 are p_2 and $p_2^* = (ab + a + 1)/(1 - a^2b) = -1/3$. Hence, the map G has one fixed point in J_2, namely $p_2^* = -1/3$. A simple computation yields $G(c_4) = G(0) = -0.7$, $G^2(0) = 0.07$ and $G^3(0) = -0.0224$. Applying the property "the map G has a dense trajectory in $J_2 \Leftrightarrow p_2^* < G^3(c_4)$" (see, for example, [16]), we now get that the map G has a dense trajectory in J_2. Therefore, the map S has a dense trajectory in A. We now have that A is a transitive Milnor attractor for S. Let $\{x_n\}_{n=0}^{\infty} \subset A$ be a trajectory that is dense in A such that for every $n \in \mathbb{N}$, $x_n \ne 0$. Since $|S'(x_n)| \ge |a^2b| > 1$, the Lyapunov exponent $\lambda(x_0)$ of this trajectory is positive. Therefore, A has a chaotic trajectory that is dense in A. We conclude that A is a chaotic attractor. □

Property 2. The map S has a Lasota–Yorke invariant measure.

("Lasota–Yorke measure" is the invariant measure in Theorem LaY stated below.)

Proof. Consider the Frobenius–Perron operator $\mathcal{P}_S$ from $L^1([-4, 1], m)$ to itself defined by $\mathcal{P}_S f(x) = \sum_{y \in S^{-1}(x)} (f(y)/|S'(y)|)$. Let $f \in L^1([-4, 1], m)$ be a density function, so $f \ge 0$ and $\int_{-4}^{1} f(s)\, ds = 1$. Applying the theorem due to Lasota and Yorke (Theorem LaY below) yields $(1/n) \sum_{i=0}^{n} \mathcal{P}_S^i f \to f^* \in L^1([-4, 1], m)$

as $n \to \infty$, and the measure $d\mu = f^* \, dm$ is invariant under S, i.e. $\mu^*(S^{-1}(E)) = \mu(E)$ for each measurable set E. □

Conclusion. The map S has a chaotic attractor and admits a natural invariant measure, the "Lasota–Yorke invariant measure". The support of this measure is the chaotic attractor A.

3.1. The method used by Lasota and Yorke

The construction of Lasota and Yorke that we just described is an example of a method that is now commonly used for constructing natural invariant measures, namely that of "pushing forward" Lebesgue measure and Cesaro averaging. In the general case, the natural invariant measure μ_* is formed as the limit (in an appropriate topology) as $n \to \infty$ of $(1/n)\sum_{i=0}^{n-1}\mu_i$, where μ_0 is absolutely continuous with respect to Lebesgue measure and μ_i is the measure with density function $f_i = \mathcal{P}^i f_0$, where $\mathcal{P}$ is the Frobenius–Perron operator and f_0 the density function for μ_0. The initial measure μ_0 could, for example, be normalized Lebesgue measure on some compact set, representing a uniform distribution of initial points on this set. In the case of a Milnor attractor, $(1/n)\sum_{i=0}^{n-1}\mu_i$ may converge to a measure μ_* that is not absolutely continuous even though each of the μ_i is. Lasota and Yorke investigated maps g on an interval I for which $(1/n)\sum_{i=0}^{n-1} f_i$ converges (in the L^1-topology) to a function f_*, which is the density function for an absolutely continuous invariant measure μ_*. Recall that the Frobenius–Perron operator $\mathcal{P}_g\colon L^1(I,m) \to L^1(I,m)$ describes the evolution of these densities under g, namely $f_{n+1} = \mathcal{P}_g f_n$ where $\mathcal{P}_g f(x) = \sum_{y\in g^{-1}(x)}(f(y)/|g'(y)|)$. If $\lim_{n\to\infty}(1/n)\sum_{i=0}^{n-1}\mathcal{P}_g^i f = f_*$, then a formal calculation yields:

$$\mathcal{P}_g\left(\lim_{n\to\infty}\frac{1}{n}\sum_{i=0}^{n-1}\mathcal{P}_g^i f\right) = \lim_{n\to\infty}\frac{1}{n}\sum_{i=1}^{n}\mathcal{P}_g^i f,$$

$$\mathcal{P}_g f_* - f_* = \mathcal{P}_g\left(\lim_{n\to\infty}\frac{1}{n}\sum_{i=0}^{n-1}\mathcal{P}_g^i f\right) - \lim_{n\to\infty}\frac{1}{n}\sum_{i=0}^{n-1}\mathcal{P}_g^i f = \lim_{n\to\infty}\frac{1}{n}[\mathcal{P}_g^n f - f].$$

Since $[\mathcal{P}_g^n f - f]$ is the difference of two densities, its L^1-norm is at most 2, and hence $\int_X |\mathcal{P}_g f_* - f_*|\, dm = 0$. Therefore, $\mathcal{P}_g f_* = f_*$ m-almost everywhere.

3.2. The Lasota–Yorke results

We now want to present the main result due to Lasota and Yorke [10]. Let I be a compact interval $[a, b]$. Let $\tau\colon I \to I$ be a measurable nonsingular transformation, i.e. $m(\tau^{-1}(E)) = 0$ whenever $m(E) = 0$ for any measurable set E. The map $\tau\colon I \to I$ is called a piecewise C^2-map if there exist finitely many points c_i $(0 \le i \le N)$ such that $c_{i-1} < c_i$ and the restriction τ_i of τ to the open interval (c_{i-1}, c_i) is a C^2-map that can be extended to the compact interval $[c_{i-1}, c_i]$ as a C^2-map, where $1 \le i \le N$ and $I = [c_0, c_N]$. For any piecewise C^2-map $\tau\colon I \to I$, consider the Frobenius–Perron operator $\mathcal{P}_\tau$ from $L^1(I, m)$ to itself. Note that

$$\mathcal{P}_\tau f(x) = \sum_{y\in\tau^{-1}(x)}\frac{f(y)}{|\tau'(y)|} = \frac{d}{dx}\int_{\tau^{-1}([a,x])} f(s)\, ds.$$

Recall that $f \in L^1(I, m)$ is a density (function) $\Leftrightarrow$ $f \ge 0$ and $\int_a^b f(s)\, ds = 1$. Here is the first result due to Lasota and Yorke [10] published in 1973.

Theorem 1. *Let $\tau\colon I \to I$ be a piecewise C^2-map such that*

$$\inf\{|\tau'(x)| : x \in I \setminus \{c_i\}_{i=0}^N\} > 1.$$

Then, for every density function $f \in L^1(I, m)$, *the sequence* $(1/n)\sum_{i=0}^{n} \mathcal{P}_\tau^i f$ *converges in the* L^1*-norm to a function* $f^* \in L^1(I, m)$ *as* $n \to \infty$. *The limit function* f^* *has the following properties:*

(a) f^* *is a density function;*
(b) $\mathcal{P}_\tau f^* = f^*$ *and so* f^* *is a fixed point of the Frobenius–Perron operator* $\mathcal{P}_\tau$ *and consequently the measure* m_f^* *defined by* $\mathrm{d}m_f^* = f^*\,\mathrm{d}m$ *is invariant under* τ;
(c) *the function* f^* *is of bounded variation.*

Lasota and Yorke prove this theorem by establishing convergence for functions f of bounded variation and using the fact that these functions are dense in $L^1(I, m)$. Their original idea of considering the space of functions of bounded variation and the properties of the Frobenius–Perron operator on this space has been used many times since. We now combine a number of results in [10] into the following theorem. Note that Theorem 1 is a special case hereof.

Theorem LaY (1973). *Let* $\tau: I \to I$ *be a piecewise* C^2*-map such that for some* $K \in \mathbb{N}$, $\inf\{|(\tau^K)'(x)| : x \in I$ *at which* $(\tau^K)'(x)$ *is defined* $\} > 1$. *Then, for every density function* $f \in L^1(I, m)$, *the sequence* $(1/n)\sum_{i=0}^{n} \mathcal{P}_\tau^i f$ *converges in the* L^1*-norm to a function* $f^* \in L^1(I, m)$ *as* $n \to \infty$. *The limit function* f^* *has the following properties:*

(a) f^* *is a density function;*
(b) $\mathcal{P}_\tau f^* = f^*$ *and so* f^* *is a fixed point of the Frobenius–Perron operator* $\mathcal{P}_\tau$ *and consequently the measure* m_f^* *defined by* $\mathrm{d}m_f^* = f^*\,\mathrm{d}m$ *is invariant under* τ;
(c) *the function* f^* *is of bounded variation.*

We refer to the measure m_f^* in Theorem LaY as a *Lasota–Yorke measure*. This measure is an absolutely continuous invariant measure. It is not difficult to show that for such a map τ, each chaotic attractor has a unique absolutely continuous invariant measure and each measure m_f^* is a linear combination of these [17].

3.3. Application to tent maps

An example to which this theorem can be applied is the one-parameter family of tent maps $T_\alpha : [0, 1] \to [0, 1]$ defined by $T_\alpha(x) = \alpha x$ if $0 \le x \le 0.5$ and $T_\alpha(x) = \alpha(1 - x)$ if $0.5 < x \le 1$, where $1 < \alpha \le 2$. The above theorem guarantees that the tent map has an absolutely continuous invariant measure for every $1 < \alpha \le 2$.

Note. Lasota and Yorke also gave the following example. Consider the map $R : [0, 1] \to [0, 1]$ defined by $R(x) = x/(1 - x)$ if $0 \le x \le 0.5$ and $R(x) = 2x - 1$ if $0.5 < x \le 1$. Although $R'(0) = 1$, $R'(x) > 1$ for all $x \in (0, 1) \setminus \{0.5\}$, yet this map does not admit a (bounded) absolutely continuous invariant measure.

4. Sinai–Ruelle–Bowen invariant measures for chaotic attractors

In this section, we first discuss briefly some of the important results on invariant measures due to Bowen, Ruelle and Sinai in uniformly hyperbolic systems. Thereafter, we give a partial review on the existence of invariant measures for one-dimensional quadratic maps. Finally, we give a flavor of results on invariant measures for certain two-dimensional nonhyperbolic systems having chaotic nonhyperbolic attractors, namely the well-known Lozi maps and Hénon maps.

4.1. Sinai–Ruelle–Bowen measures for hyperbolic systems

Let $F : \mathbb{R}^n \to \mathbb{R}^n$ be a C^1-map. If F is a linear map then F is called *hyperbolic* if none of its eigenvalues lie on the unit circle. If F is nonlinear, then $p \in \mathbb{R}^n$ is a *hyperbolic fixed point* for F if and only if (1) $F(p) = p$ and (2) $DF(p)$ is a hyperbolic linear map. If $p \in \mathbb{R}^n$ is a hyperbolic fixed point for F, then one of the following possibilities occurs: (1) p is a fixed point *attractor* (all eigenvalues of $DF(p)$ are inside the unit circle), or (2) p is a fixed point *repellor* (all eigenvalues of $DF(p)$ are outside the unit circle), or (3) p is a fixed point *saddle* (at least one eigenvalue of $DF(p)$ is inside and at least one eigenvalue of $DF(p)$ is outside the unit circle). Similarly, a hyperbolic periodic point (of period m) is a hyperbolic fixed point for F^m. The notion of a "hyperbolic invariant set" is a generalization of the notion of hyperbolic fixed point. From now on, let F be a C^2-diffeomorphism (i.e. F is an invertible map and both F and F^{-1} are C^2-functions) and let $A \subset \mathbb{R}^n$ be a compact F-invariant set and so $F(A) = A$ and $F^{-1}(A) = A$. The map F is called *uniformly hyperbolic* on A if there exists a continuous splitting of the tangent bundle over A into a direct sum of two DF-invariant subbundles, written $TA = E^u \oplus E^s$, so that the following hold: there exist constants $0 \leq \lambda < 1$ and $C > 0$ such that for every $x \in A$ and $n \in \mathbb{N}$, if $v \in E^u(x)$ then $\|DF_x^{-n} v\| \leq C\lambda^n \|v\|$ and $v \in E^s(x)$ implies that $\|DF_x^n v\| \leq C\lambda^n \|v\|$. The set A is called a *uniformly hyperbolic attractor* or an *Axiom A attractor* for $F \Leftrightarrow$ (1) F is uniformly hyperbolic on A; (2) there exists a compact neighborhood U of A such that $F(U) \subset U$ and $A = \cap_{n=0}^{\infty} F^n(U)$; (3) F is topologically transitive on A, i.e. there exists a point in A whose orbit is dense in A. In a more general setting, F is defined on a Riemannian manifold M. The results in the following theorem were first proved for Anosov diffeomorphisms by Sinai, see, e.g. [5]. (Anosov diffeomorphisms are maps that are hyperbolic on the entire manifold M.) They were later generalized to Axiom A attractors by Ruelle and Bowen (see, e.g. [6–8]). Let m denote the Lebesgue measure.

Theorem SRB (1975). *Let $F : \mathbb{R}^m \to \mathbb{R}^m$ be a C^2-diffeomorphism. Let $U \subset \mathbb{R}^m$ be a compact region having nonempty interior such that $F(U) \subset \mathrm{Int}(U)$. Let $A \subset U$ be an Axiom A attractor such that for every $x \in U$, $F^n(x) \to A$ as $n \to \infty$.*

Then there exists a unique F-invariant Borel probability measure μ such that (1) $\operatorname{supp} \mu = A$ *and* (2) *for m-almost every x in U and for every continuous function φ on U, one has* $\lim_{n\to\infty}(1/n)\sum_{i=0}^{n-1}\varphi(F^i(x)) = \int \varphi \, \mathrm{d}\mu$.

4.2. The name "SRB"

We remark that conclusion (2) in Theorem SRB is similar to the conclusion of the Birkhoff ergodic theorem, but the crucial difference is that it holds for Lebesgue almost every x in U, as opposed to μ-almost every x. This means that a Lebesgue typical initial condition chosen near the attractor A generates a trajectory that is asymptotically distributed according to μ, i.e. μ is a natural invariant measure. An invariant measure with this property has also been called a Bowen–Ruelle measure by Young [18] and a Bowen–Ruelle–Sinai (or BRS) measure by Tsujii [19]. On the other hand, Benedicks and Young [20] define a Sinai–Bowen–Ruelle (or SBR) measure as an invariant measure with a positive Lyapunov exponent and absolutely continuous conditional measures along unstable manifolds (see [5] or [20] for an explanation of the latter condition). This property also holds under the hypotheses of Theorem SRB. Above we have given just a few examples of how the names of Bowen, Ruelle, and Sinai have been applied in various orders to invariant measures with one or both of the properties we described. In the recent literature, SRB measure seems to be used more often than BRS measure or SBR measure. We now quote from Bowen [6]: "The notes as a whole constitute a version of Sinai's program [5] for applying statistical mechanics to diffeomorphisms. It was Ruelle who carried Sinai's work on Anosov diffeomorphisms over to Axiom A attractors [9] and brought in the formalism of equilibrium state [8]". Based on this statement, one could argue that SRB is a good choice for the order of the three symbols B, R and S. Another source for a definition of SRB measure is Pesin's book [21].

Note. If the attractor A in the theorem is a periodic orbit, then the SRB measure is the Dirac measure equally distributed on this orbit.

4.3. Natural invariant measures for one-dimensional quadratic maps

We now review some results for the family of quadratic maps $f_a : [0, 1] \to [0, 1]$ defined by $f_a(x) = ax(1-x)$, where $1 \leq a \leq 4$. (The proofs of these results are beyond the scope of this paper.) Note that f_a is not uniformly expanding for any a. In fact, near the critical point $c = 0.5$, f_a transforms any bounded density in one step to a density with inverse square-root singularity. We discussed the case of $a = 4$ in Section 2. One of the first general results on the existence of absolutely continuous invariant measures for quadratic maps is the following one due to Ruelle [22]. Let $3 < a \leq 4$ for which there exists $n \in \mathbb{N}$ such that $f_a^n(c) = p_0$, where p_0 is an expanding periodic point (i.e. there is an $m \geq 1$ such that $f_a^m(p_0) = p_0$ and $|(f_a^m)'(p_0)| > 1$). Then f_a admits an absolutely continuous invariant measure. In 1981, Misiurewicz [23] published a result for a class of maps including the members of the quadratic family. Let $3 < a \leq 4$ such that every iterate $f_a^n(c)$ $(n \geq 1)$ is bounded away from c (for example, if the forward orbit of c is trapped in an expanding invariant set). Then f_a admits an absolutely continuous invariant measure. Concerning the quadratic family, Misiurewicz's result can be thought of as a generalization of Ruelle's result. Related to these results is the following result due to Collet and Eckmann [24]. Let $3 < a \leq 4$ such that the orbit of $f_a(c)$ has a positive Lyapunov exponent. Then f_a admits an absolutely continuous invariant measure. An important question is: What can be said about the collection of parameters a $(3 < a \leq 4)$ for which f_a admits an absolutely continuous invariant measure? A very important result dealing with this issue is the famous theorem by Jakobson [25]. The set in the parameter space given by $\{a \in [3, 4]$: f_a admits an absolutely continuous invariant measure$\}$ has positive Lebesgue measure. For an alternative proof, see, for example [26]. The argument used in the proof of Jakobson's result is that the critical orbit does not approach the critical point too closely too quickly, and the collection of those parameters for which this approach is sufficiently controlled has positive Lebesgue measure. Therefore, these "Jakobson parameter values" include the "Misiurewicz parameter values".

In describing the distribution of the orbit $\{f_a^n(x)\}_{n=0}^{\infty}$, Tsujii [19] uses the Dirac measures δ_{x_i} concentrated at the trajectory points $x_i = f_a^i(x)$ and he examines the sequence of probability measures $\mu_n(x) = (1/n)\sum_{i=0}^{n-1}\delta_{x_i}$, $n \geq 1$. If this sequence converges to a probability measure μ as $n \to \infty$, Tsujii calls μ the *asymptotic distribution* for the orbit of x. Here the convergence is in the sense of weak topology, i.e. $\int \varphi\, d\mu_n = (1/n)\sum_{i=0}^{n-1}\varphi(f_a^i(x)) \to \int \varphi\, d\mu$ as $n \to \infty$ for every continuous function φ on the interval $[0, 1]$. Hence, the statistical properties of the orbit are given by the asymptotic distribution μ, if it exists. Tsujii calls a probability measure μ on the interval the *Bowen–Ruelle–Sinai measure* for f_a if the asymptotic distribution of the orbit exists and equals μ for almost every point in the interval with respect to the Lebesgue measure. Hence, f_a has a unique BRS measure μ if and only if $\lim_{n\to\infty}(1/n)\sum_{i=0}^{n-1}\delta_{x_i} = \mu$ for Lebesgue almost all points $x_0 \in [0, 1]$, where x_i denotes the ith iterate of x_0, so $x_i = f_a^i(x_0)$ $(i \geq 0)$. Tsujii studies how the BRS measure depends on the parameter a in the family $\{f_a\}$ and considers two cases: (1) f_a has an attracting hyperbolic periodic orbit, and (2) f_a admits an absolutely continuous invariant probability measure. In case (1) the orbit is unique and the invariant probability measure on it is the BRS measure for f_a. In case (2) the measure is unique and it is the BRS measure for f_a. Tsujii explored the continuity of BRS measures in the quadratic family. If μ_a denotes the BRS measure for f_a when it exists, then it depends continuously on a when f_a has an attracting hyperbolic periodic orbit, while μ_a does not depend continuously on a when f_a admits an absolutely continuous invariant measure. In fact, the dependence of the BRS measure μ_a on the parameter a is quite irregular when

$a \to 4$. In other recent papers on one-dimensional maps, this BRS measure is called SBR measure or SRB measure.

Another breakthrough in the analysis of the one-parameter family of quadratic maps is the following result due to Graczyk and Swiatek [27,28] and Lyubich [29]. The collection of parameters $a \in [1, 4]$ for which f_a has an attracting periodic orbit is dense in $[1, 4]$. The Lebesgue measure of this set is less than 3, since the collection of parameter values for which f_a has an absolute continuous invariant measure is positive. However, for almost every $a \in [1, 4]$, the map f_a has either an attracting periodic orbit or has an absolutely continuous invariant measure [30].

4.4. *Chaotic trajectories and chaotic attractors*

Let $F: \mathbb{R}^2 \to \mathbb{R}^2$ be a diffeomorphism and let $R \subset \mathbb{R}^2$ be a compact region such that F maps R into itself. Let $A \subset R$ be a compact F-invariant set which is a Milnor attractor such that for every $z \in R$, $F^n(z) \to A$ as $n \to \infty$. Recall that the collection of all $z \in \mathbb{R}^2$ whose orbits converge to A is called the *basin of attraction* of A. Notice R is contained in the basin of attraction. The trajectory $\{F^n(z_0)\}_{n=0}^{\infty}$ of any initial point z_0 has (if the relevant limits exist) two Lyapunov exponents $\lambda_1(z_0)$ and $\lambda_2(z_0)$ describing the average stretching or contracting behavior of the derivative DF along the trajectory. The trajectory of a point x is called a *chaotic trajectory* if (1) the trajectory of x is bounded and x is not asymptotic to either a fixed point or a periodic orbit, and (2) the trajectory of x has a positive Lyapunov exponent. Similar to a chaotic attractor for the skew tent map, we call a compact set C a *chaotic attractor* (for F) if and only if (a) C is a Milnor attractor, and (b) C contains a chaotic trajectory that is dense in C. Hence, the Milnor attractor A is a chaotic attractor (for F) if it contains a chaotic trajectory that is dense in A. If the trajectory of some initial condition $z_0 \in R$ is a chaotic trajectory, may we conclude that there is a chaotic attractor?

4.5. *Hénon attractors and invariant measures*

To illustrate the difficulty of determining if there is a chaotic attractor, we consider the Hénon map $H_{a,b} : \mathbb{R}^2 \to \mathbb{R}^2$ defined by

$$H_{a,b}(x, y) = (1 - ax^2 + by, x). \tag{9}$$

For every $a, b \in \mathbb{R}$ with $-1 < b < 1$, the map $H_{a,b}$ is a C^2-diffeomorphism. For $a = 1.4$, $b = 0.3$, Hénon [31] gave the coordinates of the four corner points of a quadrilateral having the property that it is mapped into itself when the map $H_{a,b}$ is applied once. Based on this information, we consider the quadrilateral Q defined by the points $P_1 = (-1.862, 1.96)$, $P_2 = (1.848, 0.6267)$, $P_3 = (1.743, -0.6533)$ and $P_4 = (-1.484, -2.3333)$. The superposition of the quadrilateral $Q = P_1P_2P_3P_4$ and its first iterate are shown in Fig. 4(a). The intersection of all forward iterates $H^* = \cap_{n=0}^{\infty} H_{a,b}^n(Q)$ of the quadrilateral region Q is a compact set of Lebesgue measure zero, since $|\det DH_{a,b}(z)| = 0.3 < 1$ for all $z \in \mathbb{R}^2$. Hence, for all $z \in Q$, $H_{a,b}^n(z) \to H^*$ as $n \to \infty$, so H^* is a Milnor attractor. In the literature, H^* is known as the *Hénon attractor*.

Choosing initial point $z_0 = (1, 1)$, ignoring the initial 100 iterates and plotting the next 10^6 iterates, the resulting picture is shown in Fig. 4(b). The Hénon attractor is believed to be a chaotic attractor because (according to numerical calculations) it contains a trajectory which is dense and which has a positive Lyapunov exponent. However, there exists no proof of such a fact. One has to prove that there exists a chaotic trajectory that is dense in H^*. Numerical simulations give the strong impression this is true. One may ask the following question. Does there exists a natural $H_{a,b}$-invariant Borel probability measure μ such that (1) the support of the measure μ is H^* and (2) for m-almost every x in Q and every continuous function φ on Q, $(1/n)\sum_{i=0}^{n-1}\varphi(H_{a,b}^i(x)) \to \int \varphi \, d\mu$? If there exists such a measure, then for m-almost all $z_0 \in Q$ the orbit $\{z_n\}_{n=0}^{\infty}$ has a well-defined distribution and that this distribution is independent of the choice of the initial condition. Unfortunately, the answer is still unknown.

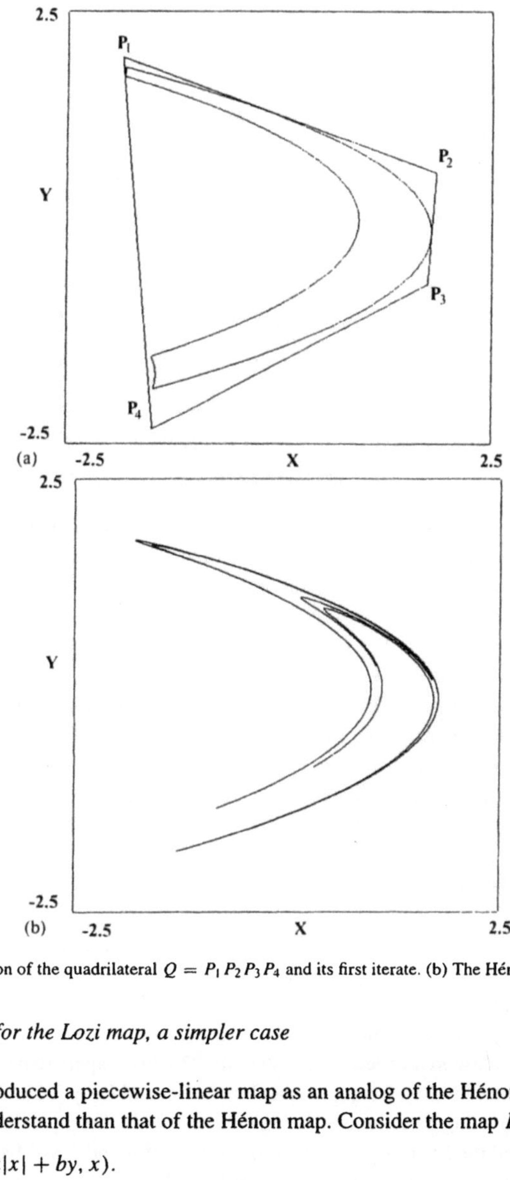

Fig. 4. (a) The superposition of the quadrilateral $Q = P_1P_2P_3P_4$ and its first iterate. (b) The Hénon attractor. Is it a chaotic attractor?

4.6. A chaotic attractor for the Lozi map, a simpler case

In 1978 Lozi [32] introduced a piecewise-linear map as an analog of the Hénon map. The dynamics of the Lozi map may be easier to understand than that of the Hénon map. Consider the map $L_{a,b} : \mathbb{R}^2 \to \mathbb{R}^2$ defined by

$$L_{a,b}(x, y) = (1 - a|x| + by, x). \tag{10}$$

For every $a, b \in \mathbb{R}$ with $-1 \le b \le 1$, the map $L_{a,b}$ is a continuous, piecewise C^2-diffeomorphism (it is a C^2-diffeomorphism on $\mathbb{R}^2 \setminus \{(x, y) \in \mathbb{R}^2 : x = 0\}$). For $a = 1.7$, $b = 0.5$, numerical simulations suggested the existence of a strange attractor (we will discuss this term later, but you may think of a chaotic attractor). Choosing initial point $z_0 = (1, 1)$, ignoring the initial 100 iterates and plotting the next 10^6 iterates, the resulting picture is

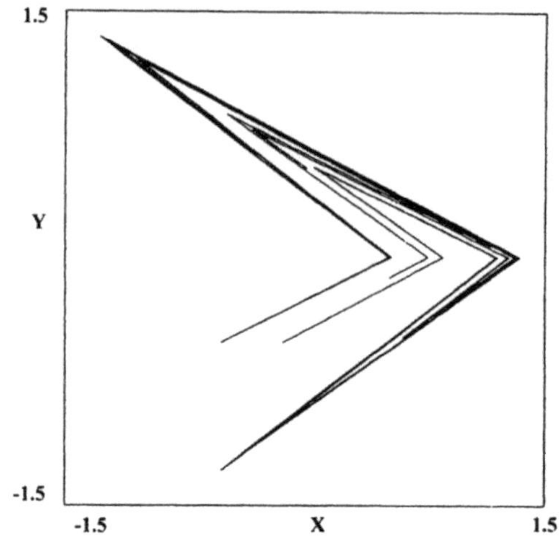

Fig. 5. The Lozi attractor.

shown in Fig. 5. If $0 < b < 1$, $a > 0$ and $a + b > 1$, the map $L_{a,b}$ is a homeomorphism. For an open set of parameter values (which does not include $a = 1.7$, $b = 0.5$), Misiurewicz [33] showed there is a trapping region, i.e. he constructed a compact region Q such that Q is mapped into the interior of Q when the map $L_{a,b}$ is applied once. The intersection of all forward iterates $L^* = \cap_{n=0}^{\infty} L_{a,b}^n(Q)$ of the region Q is a compact set of Lebesgue measure zero. Hence, L^* is a Milnor attractor. Let P be the fixed point of $L_{a,b}$ that has two positive coordinates. Let $W^u(P)$ be the unstable "manifold" of P, i.e. $W^u(P)$ is the collection of points whose orbits converge to P as the map is iterated backwards. Misiurewicz [33] proved that L^* is a strange attractor by showing that the map had a hyperbolic structure and that L^* equals the closure of the unstable "manifold" $W^u(P)$. (Since the map is not differentiable at $z = (0, y)$, its hyperbolic structure must be understood only as the existence of a hyperbolic splitting at those points at which it may exist.) It appears that the Lozi map is an intermediate stage between the Axiom A systems and more complicated systems like the Hénon map.

4.7. *An invariant measure for the Lozi map*

Bowen and Ruelle [7] assumed the maps were smooth in the expanding directions. The measure they described however was typically not smooth because the map had the property of having contracting directions. In contrast, Lasota and Yorke [10] studied maps which were expanding. Their measure was typically not smooth because the maps had the property of being only piecewise smooth, and the successive iterates of the non-C^2 points (points at which the map is not C^2) could be dense in a collection of intervals. The first papers to deal with maps having both properties were the papers by Rychlik [34,35], Young [18] and by Collet and Levy [36]. They dealt with maps that had the properties of having a contracting direction and being non-smooth in the expanding direction.

Collet and Levy [36] investigated the Lozi maps for the same values of a and b as Misiurewicz but assumed in addition that b is small. Recall that $L_{a,b}$ is a homeomorphism and that $W^u(P) = \{z \in \mathbb{R}^2 : \lim_{n\to\infty} L_{a,b}^{-n}(z) = P\}$ is the unstable "manifold" of P. The maximal smooth component $W_{\text{loc}}^u(P)$ of $W^u(P)$ containing P is called the *local unstable manifold* of P. Collet and Levy [36] construct an invariant measure as follows. They start off with the local unstable manifold of P. Let $B_0 = W_{\text{loc}}^u(P)$. From Misiurewicz results one knows that L^* is the closure of $\cup_{n=0}^{\infty} L_{a,b}^n(B_0) = W^u(P)$. Therefore, it is natural to try to obtain an invariant measure by iterating the Lebesgue measure supported by B_0. They define a sequence $\{\mu'_n\}_{n\in\mathbb{Z}}$ of probability measures on $\mathcal{M}_{L^*}$ by $\mu'_n(A) = \ell(L_{a,b}^{-n}(A) \cap B_0)/\ell(B_0)$ for $A \in \mathcal{M}_{L^*}$, where $\mathcal{M}_{L^*}$ denotes the Borel sub σ-algebra of $\mathbb{R}^2$ corresponding to L^* and ℓ is the one-dimensional Lebesgue measure. Note that for every n, μ'_n has support in $L_{a,b}^n(B_0)$. The sequence $\{\mu_n\}_{n\in\mathrm{N}}$ defined by $\mu_n = (1/n)\sum_{i=0}^{n-1}\mu'_i$ is a sequence of probability measures on $\mathcal{M}_{L^*}$. Since L^* is compact, one can extract a subsequence which converges in the weak star topology to an invariant Borel probability measure μ. They show that μ is unique, that it is absolutely continuous with respect to the Lebesgue measure in the unstable direction, and that it is in fact a Bowen–Ruelle measure (i.e. a measure generated by the orbits of almost all points). This approach resembles the Lasota and Yorke's pushing forward of the Lebesgue measure construction.

The existence of invariant measures for the Lozi maps for parameter values studied by Misiurewicz, was shown independently by Rychlik [34,35] by a method independent of Young [18] and Collet and Levy [36]. The results were reported in [34]. The paper [34] contains only sketches of the arguments and the full proofs can be found in Rychlik's dissertation [35], which is available on the author's website. The paper [34] and the dissertation [35] present a hybrid approach to the "Lasota–Yorke" measures and the "Sinai–Ruelle–Bowen" measures. In fact, they contain an abstract theorem which implies both the existence of Lasota–Yorke measures for $C^{1+\varepsilon}$ piecewise expanding maps of an interval and Sinai–Ruelle–Bowen measures for Lozi mappings.

4.8. Generalized Lozi maps and their invariant measures

Young [18] introduced a much more general class of piecewise C^2-diffeomorphisms on the region $R = [0, 1] \times [0, 1]$ which are continuous one-to-one maps and which map the region R into itself. These maps are hyperbolic piecewise smooth maps. She based them on the geometric properties of Lozi maps and calls them "generalized Lozi maps". Recall that the above result of Lasota and Yorke states that piecewise expanding maps of the interval have absolutely continuous invariant measures (Lasota–Yorke invariant measure). Young proves an analogous statement for her generalized Lozi maps. A part of the proof follows a combination of the methods used by Sinai, and by Lasota and Yorke. Young's result is that the generalized Lozi maps admit certain invariant measures which imply that the maps have "Bowen–Ruelle measures", i.e. measures generated by the orbits of almost all initial points. More precisely, she defines a Borel probability measure μ on R to be a *Bowen–Ruelle measure* if there is a set $U \subset R$ of positive Lebesgue measure such that for every continuous function φ: $R \to R$, $(1/n)\sum_{i=0}^{n-1}\varphi(f^i(x)) \to \int\varphi\,\mathrm{d}\mu$ for Lebesgue-almost every $x \in U$.

Note. Jakobson and Newhouse [37] also studied invariant measures for hyperbolic piecewise smooth mappings of a rectangle.

4.9. Invariant measures for the Hénon map

We now return to the Hénon map $H_{a,b}$. For the parameter values $a = 1.4, b = 0.3$, numerical simulations suggest that for almost every initial condition in a certain region, the orbit has a well-defined distribution as $n \to \infty$, and this distribution is independent of the choice of initial condition. However, in this case as well as for other parameter values for which there seems to be a strange attractor, the situation is very delicate. The question is whether there exists an invariant measure that reflects the statistical behavior of Lebesgue-almost every point in a much larger set. But why should such an invariant measure exist? On the other hand, if (one can show that) such a measure exists, it is of great significance.

In the proofs of Theorem SRB, Markov partitions were used to connect the dynamics of F to certain one-dimensional lattice systems in statistical mechanics, and μ was realized as a Gibbs state or equilibrium state. For an introduction to the applications in nonequilibrium statistical mechanics of chaotic dynamics, see [38]. However, Markov partitions are not available for systems more general than Axiom A, like the Hénon map. To see why it is so difficult to prove the existence of a natural invariant measure for the Hénon map, consider the following. Let P be the fixed point of $H_{a,b}$ that has two positive coordinates. Let $W^u(P)$ be the unstable manifold of P. Recall that $W^u(P)$ is the collection of points whose orbits converge to P as the map is iterated backwards. Similar to the construction of an invariant measure by Collet and Levy [36] for the Lozi map, suppose we start with the local unstable manifold $W^u_{\mathrm{loc}}(P) = B_0$ of the fixed point P and push forward Lebesgue measure on B_0. The definition of a local unstable manifold says that if $x \in B_0$ then $H^{-n}_{a,b}(x) \to P$ as $n \to \infty$, but generally, one has no idea what happens to points on B_0 under forward iteration of the map. For example, it is easy to imagine that parts of $W^u(P) = \cup_{n=0}^{\infty} H^n_{a,b}(B_0)$ belong to basins of attraction of periodic orbits, so the orbits of certain points of B_0 converge to a periodic orbit attractor when the map $H_{a,b}$ is applied repeatedly. Hence, the question of existence of natural invariant (SRB) measures is very delicate. (The Lozi map question is simpler since it has a uniformly expanding direction.) It is in some sense analogous to the problem of existence of absolutely continuous invariant measures for nonuniformly expanding maps (Jakobson theorem) only that it is more complicated because one has to consider not only rates of expansion but their directions as well.

Despite these difficulties, there have been some major breakthroughs. Benedicks and Carleson [39] developed elaborate machinery for analyzing the dynamics of $H_{a,b}$ for a positive measure set of parameters near $a = 2$, $b = 0$. Subsequently, Benedicks and Young [20] constructed SRB measures for these attractors. Here we state their important results.

Theorem 2 ([39]). *Let $H_{a,b}$ be the Hénon map. Then there is a rectangle $\Delta = (a_0, a_1) \times (0, b_1)$ in parameter space such that for all $(a, b) \in \Delta$, $H_{a,b}$ has an attractor A with a dense orbit.*

Theorem 3 ([20]). *Let $H_{a,b}$ be the Hénon map. Then for all sufficiently small $b > 0$, there exists a positive Lebesgue measure set $\Delta_b \subset \mathbb{R}^2$ such that for all $a \in \Delta_b \subset \Delta$, $H_{a,b}$ admits an SRB measure λ^*. Morevoer, λ^* is unique and it is supported on the entire attractor.*

Corollary ([20]). *For $b > 0$ small and $a \in \Delta_b$, the map $H = H_{a,b}$ has the following property: Let U be a neighborhood of the attractor. Then there exists a set $\tilde{U} \subset U$ with positive Lebesgue measure such that for all continuous functions $\varphi: U \to \mathbb{R}$, $(1/n) \sum_{i=0}^{n-1} \varphi \circ H^i(x) \to \int \varphi \, \mathrm{d}\lambda^*$ for every $x \in \tilde{U}$.*

We quote from [20]: "The property expressed in this corollary is often assumed to be true in physics; it is also taken for granted in numerical experiments".

Note. More recently, Young [40] and Alves et al. [41] investigated diffeomorphisms which are not hyperbolic, but partially hyperbolic. These papers contain important results that are beyond the scope of this brief overview. Under quite general assumptions, Young [40] proves that correlations for hyperbolic systems with singularities decay exponentially. If Lyapunov exponents are bounded away from zero, Alves et al.[41] prove that there are only finitely many natural invariant (Sinai–Ruelle–Bowen) measures for such maps and Lebesgue almost every point is in one of the corresponding basins.

5. SLYRB chaos and SLYRB measures

5.1. Strange attractors

In the 1980s a lot of attention was spent on attractors in dynamical systems. Conferences devoted to this single topic were held. The notion of "strange attractor" was originally introduced by Ruelle and Takens [42]. It meant an attractor which is not a fixed point attractor, a periodic orbit or a smooth manifold without boundary such as a circle or torus. Nowadays the terms "chaotic attractor" and "strange attractor" are quite common. Generally, a *chaotic attractor* is a Milnor attractor on which the dynamics is chaotic, i.e. the system has at least one positive Lyapunov exponent. Frequently, the two notions are used as synonyms. Indeed, a chaotic attractor is always a strange attractor in the sense of Ruelle and Takens. However, an example of a strange attractor which is not a chaotic attractor is the so-called Feigenbaum attractor which occurs in the one-parameter family of quadratic maps $f_\alpha(x) = \alpha x(1-x)$. (The *Feigenbaum attractor* is a Milnor attractor which is a Cantor set and the Lyapunov exponent of Lebesgue almost every trajectory is zero.) On the other hand, one encounters in the literature the following. A *strange attractor* is a Milnor attractor that has a Cantor-like structure. Of course, under this definition a chaotic attractor need not be strange. The chaotic attractor consisting of three intervals in the example of skew tent map above demonstrates this. There is also a large literature on "strange non-chaotic attractors" in systems that have terms that are quasi-periodic in time (see, e.g. [43]). We concentrate on chaotic attractors.

We recall that the Lasota–Yorke measures are absolutely continuous invariant measures, while the Sinai–Ruelle–Bowen (or SRB) measures are not. Hence, from the point of view of measure theory applied to dynamics, those invariant measures are considered to be different. However, from the point of view of dynamics using measure theory, the existence of a chaotic attractor implies in certain systems the existence of the special invariant measures as discussed in the previous sections. We propose the following: a dynamical system displays "SLYRB chaos" whenever there is a chaotic attractor which admits a certain natural invariant measure (e.g. a Lasota–Yorke measure or a Sinai–Ruelle–Bowen measure). The notion SRB varies somewhat from paper to paper. In a recent paper, Young [40] limits the definition of SRB measures to the context of no zero Lyapunov exponents and states a definition of Sinai–Ruelle–Bowen measure. She notes that this definition is a generalization of the ideas in Sinai [5] and first appeared in a paper by Ledrappier [44].

5.2. *SLYRB (we pronounce it "slurb" as a single word)*

After comparing the construction of SRB measures and its properties to the construction and properties of Lasota–Yorke measure, we suggest that the natural invariant measure supported on any chaotic attractor might be called a "SLYRB measure". (The ordering of the letters S-LY-RB correspond to the dates of the papers 1972–1973–1975.) We propose the following. A measure μ is an *SLYRB measure* if and only if μ is the weighted sum $\mu = \sum_{i\geq 1} \alpha_i \mu_i$ of invariant measures μ_i such that each μ_i is the asymptotic distribution of a trajectory for a set of initial points of positive measure, where $a_i \geq 0$.

The SRB concept of measure can be motivated by asking how a trajectory from a typical initial point is distributed asymptotically. Similarly the SLYRB concept of measure can be motivated by asking what the average distribution is for trajectories of a large collection of initial points in some region not necessarily restricted to a single basin. (The coefficient a_i is the fraction of initial conditions in the ith basin B_i.) The latter is analogous to asking where all the rain drops from a rain storm go and the former asks about where a single rain drop goes, perhaps winding up distributed throughout a particular lake. Consider a system that has exactly one attractor A and almost all initial points chosen near the attractor A generate trajectories that are asymptotically distributed according to a single measure μ. Then μ would be both an SRB measure (by some standards) and a SLYRB measure. Also, if the SLYRB measure μ is unique, then the system has exactly one attractor A, and a Lebesgue typical initial point chosen near the attractor A generates a trajectory that is asymptotically distributed according to μ. Recall that an invariant measure with this latter property for certain two-dimensional systems has been called a Bowen–Ruelle measure by Young [18] and for certain one-dimensional systems a Bowen–Ruelle–Sinai (or BRS) measure by Tsujii [19]. Therefore, the Bowen–Ruelle measure and Tsujii's BRS measure are special cases of SLYRB measures. Furthermore, the SBR measure defined by Benedicks and Young [20] being an invariant measure with a positive Lyapunov exponent and absolutely continuous conditional measures along unstable manifolds [20] is also an example of a SLYRB measure. It is not hard to show that the Lasota–Yorke measure is an example of a SLYRB measure too.

To see that a Lasota–Yorke measure must also be a SLYRB measure, recall that according to their hypotheses the map τ^K is piecewise C^2 and expanding wherever its derivative is defined. Each attractor for τ^K must contain at least one point where τ^K is not expanding; since there are only finitely many such points, τ^K has only finitely many attractors. Thus τ itself has only finitely many attractors, and Lebesgue almost every initial condition is in the basin of one of the attractors, see also [17]. Applying Theorem LaY to an initial density function whose support is in the basin of a particular attractor, we conclude that each attractor must support an absolutely continuous invariant measure μ_i. By the Birkhoff ergodic theorem, μ_i-almost every initial condition, and hence Lebesgue almost every initial condition within the attractor, generates a trajectory that is asymptotically distributed according to μ_i. We can then make the same statement about Lebesgue almost every initial condition in the basin of the attractor because, since τ is piecewise expanding, Lebesgue almost every point in the basin has a neighborhood that maps diffeomorphically onto an interval in the attractor under a finite number of iterations of τ. Finally, an arbitrary initial density function f can be decomposed into a sum of functions f_i, each of which is supported in the basin of a single attractor; the Lasota–Yorke measure associated to f by Theorem LaY is then a linear combination of the measures μ_i with coefficients determined by the integrals of the f_i (see also [17]).

5.3. The relationship of the LaY approach and the SRB approach

Assume that the measures exist and processes converge for the system in question. If an initial point x is in a basin, iterating it forward will in certain circumstances yield a trajectory distributed according to some measure μ_x according to the SRB approach and almost all x in that basin will yield the same measure. The Lasota–Yorke approach is to choose a density f that need not be restricted to one basin. Pushing it forward and averaging yields a measure m^*_f that depends on f. (Theorem LaY guarantees that the sequence $(1/n)\sum_{i=0}^{n}\mathcal{P}^i f$ converges in the L^1-norm to a function $f^* \in L^1(I, m)$ as $n \to \infty$ and the measure m^*_f is defined by $\mathrm{d}m^*_f = f^*\,\mathrm{d}m$.)

We now want to relate the Lasota–Yorke and SRB measures. Set aside the differences in the settings of the Lasota–Yorke and SRB measures and compare the similarities. Theorem LaY deals with functions that can have several (disjoint) chaotic attractors, while Theorem SRB deals with maps that have one attractor. How could one change the SRB theorem to handle that more general kind of situation when the location of the attractors and their basins are not known? Depending on the precise hypotheses used, there might still be an invariant measure μ_x for almost every x but it would depend on x. Therefore one could choose a probability density function f and define a measure which is an average of μ_x weighted by f, i.e. $m_f = \int f(x)\mu_x\,\mathrm{d}x$. Then the (generalized) SRB theorem would say that for every integrable f and bounded continuous φ,

$$\lim_{n\to\infty}\int f(x)\frac{1}{n}\sum_{i=0}^{n-1}\varphi(F^i(x))\,\mathrm{d}x = \int \varphi\,\mathrm{d}m_f \tag{11a}$$

for some measure m_f (where m_f depends linearly on f). Eq. (11a) is equivalent to

$$\lim_{n\to\infty}\frac{1}{n}\sum_{i=0}^{n-1}\int f(x)\varphi(F^i(x))\,\mathrm{d}x = \int \varphi\,\mathrm{d}m_f. \tag{11b}$$

Since F is one-to-one, the Frobenius–Perron operator for F is simply $(\mathcal{P}f)(x) = f(F^{-1}(x))/|\det DF(x)|$. Using the change of variables $y = F^{-i}(x)$ of Eq. (6), we have that $\int f(x)\varphi(F^i(x))\,\mathrm{d}x = \int(\mathcal{P}^i f)(y)\varphi(y)\,\mathrm{d}y$. Hence,

Eq. (11b) can be rewritten as

$$\lim_{n\to\infty}\frac{1}{n}\sum_{i=0}^{n-1}\int \mathcal{P}^i f(y)\varphi(y)\,\mathrm{d}y = \int \varphi\,\mathrm{d}m_f. \tag{12a}$$

Eq. (12a) is equivalent to

$$\lim_{n\to\infty}\int \frac{1}{n}\sum_{i=0}^{n-1}\mathcal{P}^i f(y)\varphi(y)\,\mathrm{d}y = \int \varphi\,\mathrm{d}m_f, \tag{12b}$$

which is the formula used by [10] in one dimension.

Finally, we propose: a dynamical system displays *SLYRB chaos* whenever there exists a chaotic attractor that supports a SLYRB measure.

All the computer assisted pictures were made by using the program Dynamics [2].

Acknowledgements

This work was supported in part by the National Science Foundation (Grant No. 0104087), and by the W.M. Keck Foundation.

References

[1] J. Milnor, On the concept of attractor, Commun. Math. Phys. 99 (1985) 177–195;
J. Milnor, Comments on the concept of attractor: corrections and remarks, Commun. Math. Phys. 102 (1985) 517–519.
[2] H.E. Nusse, J.A. Yorke, Dynamics: Numerical Explorations, 2nd Edition, Accompanied by Software Program Dynamics 2, An Interactive Program for IBM Compatible PCs and Unix Computers (the Unix version of the program is by B.R. Hunt, E.J. Kostelich), Applied Mathematical Sciences, Vol. 101, Springer, New York, 1998.
[3] D.J. Farmer, E. Ott, J.A. Yorke, The dimension of chaotic attractors, Physica D 7 (1983) 153–180.
[4] A. Lasota, M. Mackey, Chaos, Fractals and Noise, Stochastic Aspects of Dynamics, 2nd Edition, Applied Mathematical Sciences, Vol. 97, Springer, New York, 1994.
[5] Ya.G. Sinai, Gibbs measures in ergodic theory, Russ. Math. Surv. 27 (1972) 21–69.
[6] R. Bowen, Equilibrium states and the ergodic theory of Anosov diffeomorphisms, Springer Lecture Notes in Mathematics, Vol. 470, Springer, New York, 1975.
[7] R. Bowen, D. Ruelle, The ergodic theory of Axiom A flows, Invent. Math. 29 (1975) 181–202.
[8] D. Ruelle, Statistical mechanics on a compact set with Z^ν action satisfying expansiveness and specification, Trans. Am. Math. Soc. 187 (1973) 237–251.
[9] D. Ruelle, A measure associated with Axiom A attractors, Am. J. Math. 98 (1976) 619–654.
[10] A. Lasota, J.A. Yorke, On the existence of invariant measures for piecewise monotonic transformations, Trans. Am. Math. Soc. 186 (1973) 481–488.
[11] M. Dellnitz, O. Junge, On the approximation of complicated dynamical behavior, SIAM J. Numer. Anal. 36 (1999) 491–515.
[12] S.M. Ulam, J. von Neumann, On combination of stochastic and deterministic processes, Abstract 403, Bull. Am. Math. Soc. 53 (1947) 1120.
[13] O.W. Rechard, Invariant measures for many-one transformations, Duke Math. J. 23 (1956) 477–488.
[14] K. Krzyzewski, W. Szlenk, On invariant measures for expanding differentiable mappings, Studia Math. 33 (1969) 83–92.
[15] T.-Y. Li, J.A. Yorke, Period three implies chaos, Am. Math. Monthly 82 (1975) 985–992.
[16] J. Guckenheimer, Sensitive dependence to initial conditions for one-dimensional maps, Commun. Math. Phys. 70 (1979) 133–160.
[17] T.-Y. Li, J.A. Yorke, Ergodic transformations from an interval into itself, Trans. Am. Math. Soc. 235 (1978) 183–192.
[18] L.-S. Young, Bowen–Ruelle measures for certain piecewise hyperbolic maps, Trans. Am. Math. Soc. 287 (1985) 41–48.
[19] M. Tsujii, On continuity of Bowen–Ruelle–Sinai measures in families of one-dimensional maps, Commun. Math. Phys. 177 (1996) 1–11.
[20] M. Benedicks, L.-S. Young, Sinai–Bowen–Ruelle measures for certain Hénon maps, Invent. Math. 112 (1993) 541–576.
[21] Y.B. Pesin, Dimension Theory in Dynamical Systems, Contemporary Views and Applications, University of Chicago Press, Chicago, 1997.
[22] D. Ruelle, Applications conservant une measure absolument continue par raport a dx sur [0,1], Commun. Math. Phys. 55 (1977) 47–51.

[23] M. Misiurewicz, Absolutely continuous measures for certain maps of the interval, IHES Publ. Math. 53 (1981) 17–51.
[24] P. Collet, J.-P. Eckmann, Positive Lyapunov exponents and absolutely continuity, Ergod. Th. Dynam. Syst. 3 (1983) 13–46.
[25] M. Jakobson, Absolutely continuous invariant measures for one-parameter families of one-dimensional maps, Commun. Math. Phys. 81 (1981) 39–88.
[26] M.R. Rychlik, Another proof of Jakobson's theorem and related results, Ergod. Th. Dynam. Syst. 8 (1988) 93–109.
[27] J. Graczyk, G. Swiatek, Generic hyperbolicity in the logistic family, Ann. Math. 146 (1997) 1–52.
[28] J. Graczyk, G. Swiatek, The Real Fatou Conjecture, Annals of Mathematical Studies, Vol. 144, Princeton University Press, Princeton, NJ, 1998.
[29] M. Lyubich, Dynamics of quadratic polynomials, Parts I and II, Acta Math. 178 (1997) 185–297.
[30] M. Lyubich, Almost every real quadratic map is either regular or stochastic, Preprint IMS at Stony Brook, No. 1997/1998, Ann. Math, in press. Available at http://www.math.sunysb.edu/~mlyubich/.
[31] M. Hénon, A two-dimensional mapping with a strange attractor, Commun. Math. Phys. 50 (1976) 69–77.
[32] R. Lozi, Un attracteur étrange (?) du type attracteur de Hénon, J. Phys. (Paris) 39 (Coll. C5) (1978) 9–10.
[33] M. Misiurewicz, Strange attractors for Lozi mappings, in: R.H.G. Helleman (Ed.), Nonlinear Dynamics, Annals of New York Academy of Sciences, Vol. 357, New York Academy of Sciences, New York, 1980, pp. 348–358.
[34] M. Rychlik, Measures invariantes et principe variationel pour les applications de Lozi, CR Acad. Sci. (Paris) 296 (1983) 19–22.
[35] M. Rychlik, Invariant measures and variational principle for Lozi applications, Ph.D. Dissertation, University of California, Berkeley, 1983. http://alamos.math.arizona.edu/~rychlik/publications.html.
[36] P. Collet, Y. Levy, Ergodic properties of the Lozi mappings, Commun. Math. Phys. 93 (1984) 461–481.
[37] M.V. Jakobson, S.E. Newhouse, A two-dimensional version of the folklore theorem, Am. Math. Soc. Trans., Ser. 2 171 (1996) 89–105; M.V. Jakobson, S.E. Newhouse, Asymptotic measures for hyperbolic piecewise smooth mappings of a rectangle, Astérisque 261 (2000) 103–159.
[38] J.R. Dorfman, An introduction to chaos in nonequilibrium statistical mechanics, Cambridge Lecture Notes in Physics, Vol. 14, Cambridge University Press, Cambridge, 1999.
[39] M. Benedicks, L. Carleson, The dynamics of the Hénon map, Ann. Math. 133 (1991) 73–169.
[40] L.-S. Young, Statistical properties of dynamical systems with some hyperbolicity, Ann. Math. 147 (2) (1998) 585–650.
[41] J.F. Alves, C. Bonatti, M. Viana, SRB measures for partially hyperbolic systems whose central direction is mostly expanding, Invent. Math. 140 (2000) 351–398.
[42] D. Ruelle, F. Takens, On the nature of turbulence, Commun. Math. Phys. 20 (1971) 167–192.
[43] E. Ott, Chaos in Dynamical Systems, Cambridge University Press, Cambridge, 1993.
[44] F. Ledrappier, Propriétés ergodiques des measures de Sinai, IHES Publ. Math. 59 (1984) 163–188.

Credits

We would like to extend our thanks to the following Authors and Publishers for granting permission to reprint the following articles:

E.N. Lorenz, Deterministic nonperiodic flow, J. Atm. Sc. 20 (1963), 130–141. Reprinted with permission from American Meteorological Society.

K. Krzyzewski and W. Szlenk On invariant measures for expanding differentiable mappings, Studia Math. 33 (1969), 83–92. Reprinted with the permission of the Institute of Mathematics, Polish Academy of Sciences.

A. Lasota and J.A. Yorke, On the existence of invariant measures for piecewise monotonic transformations Trans. Amer. Math. Soc. 186 (1973), 481–488. Reprinted with permission from the American Mathematical Society.

R. Bowen and D. Ruelle, The ergodic theory of Axiom A flows, Invent. Math. 29 (1975), 181–202. Reprinted with permission from Springer-Verlag.

T.-Y. Li and J.A. Yorke, Period three implies chaos. Amer. Math. Monthly 82 (1975), 985–992. Reprinted with permission from Mathematical Association of America.

R.M. May, Simple mathematical models with very complicated dynamics, Nature 261 (1976), 459–467. Reprinted with permission from Macmillan Publishers Limited.

M. Henon, A two-dimensional mapping with a strange attractor, Commun. Math. Phys. 50 (1976), 69–77. Reprinted with permission from Springer-Verlag.

E. Ott, Strange attractors and chaotic motions of dynamical systems, Rev. Modern Phys. 53 (1981), 655–671. Reprinted with permission from American Physical Society.

F. Hofbauer and G. Keller, Ergodic properties of invariant measures for piecewise monotonic transformations Math. Z. 180 (1982), 119–140. Reprinted with permission from Springer-Verlag.

D. J. Farmer, E. Ott and J.A. Yorke, The dimension of chaotic attractors, Physica D 7 (1983), 153–180. Reprinted with permission from Elsevier Science.

P. Grassberger and I. Procaccia, Measuring the strangeness of strange attractors, Physica D 9 (1983), 189–208. Reprinted with permission from Elsevier Science.

M. Rychlik, Invariant measures and Variational Principle for Lozi Applications, Ph.D. dissertation, University of California at Berkeley (1983). Unpublished, reprinted with permission from the author.

P. Collet and Y. Levy, Ergodic properties of the Lozi mappings, Commun. Math. Phys. 93 (1984), 461–481. Reprinted with permission from Springer-Verlag.

J. Milnor, On the concept of attractor, Commun. Math. Phys. 99 (1985), 177–195; Comments "On the concept of attractor": corrections and remarks, Commun. Math. Phys. 102 (1985), 517–519. Reprinted with permission from Springer-Verlag.

L.-S. Young, Bowen-Ruelle measures for certain piecewise hyperbolic maps, Trans. Amer. Math. Soc. 287 (1985), 41–48. Reprinted with permission from American Mathematical Society.

J.-P. Eckmann and D. Ruelle Ergodic theory of chaos and strange attractors, Rev. Modern Phys. 57 (1985), 617–656. Reprinted from permission from American Physical Society.

M.R. Rychlik, Another proof of Jakobson's theorem and related results, Ergodic Theory Dynamical Systems 8 (1988), 93–109. Reprinted with permission from Cambrige University Press.

C. Grebogi, E. Ott, and J.A. Yorke, Unstable periodic orbits and the dimensions of multifractal chaotic attractors, Phys. Rev. A 37 (1988), 1711–1724. Reprinted with permission from American Physical Society.

P. Gora and A. Boyarsky, Absolutely continuous invariant measures for piecewise expanding C^2 transformation in R^N, Israel J. Math. 67 (1989), 272–286. Reprinted with permission from The Hebrew University, Magness Press.

M. Benedicks and L.-S. Young, Sinai-Bowen-Ruelle measures for certain Hwnon maps, Invent. Math. 112 (1993), 541–576. Reprinted with permission from Springer-Verlag.

M. Dellnitz and O. Junge, On the approximation of complicated dynamical behavior, SIAM J. Numer. Anal. 36 (1999), 491–515. Reprinted with permission from SIAM.

M. Tsujii, Absolutely continuous invariant measures for piecewise real-analytic expanding maps on the plane, Comm. Math. Phys. 208 (2000), 605–622. Reprinted with permission from Springer-Verlag.

J.F. Alves, C. Bonatti and M. Viana, SRB measures for partially hyperbolic systems whose central direction is mostly expanding, Invent. Math. 140 (2000), 351–398. Reprinted with permission from Springer-Verlag.

B.R. Hunt, J.A. Kennedy, T.-Y. Li and H.E. Nusse, SLYRB measures: natural invariant measures for chaotic systems, Physica D 170 (2002), 50–71. Reprinted with permission from Elsevier Science.

MIX
Papier aus verantwortungsvollen Quellen
Paper from responsible sources
FSC® C105338

If you have any concerns about our products,
you can contact us on
ProductSafety@springernature.com

In case Publisher is established outside the EU,
the EU authorized representative is:
Springer Nature Customer Service Center GmbH
Europaplatz 3, 69115 Heidelberg, Germany

Printed by Libri Plureos GmbH
in Hamburg, Germany